Methods in Cell Biology

VOLUME 75

Cytometry, 4th Edition: New Developments

Series Editors

Leslie Wilson

Department of Molecular, Cellular, and Developmental Biology
University of California
Santa Barbara, California

Paul Matsudaira

Whitehead Institute for Biomedical Research
Department of Biology
Division of Biological Engineering
Massachusetts Institute of Technology
Cambridge, Massachusetts

Methods in Cell Biology

Prepared under the auspices of the American Society for Cell Biology

VOLUME 75

Cytometry, 4th Edition: New Developments

Edited by

Zbigniew Darzynkiewicz

Brander Cancer Research Institute
New York Medical College
Hawthorne, New York

Mario Roederer

Vaccine Research Center
National Institute of Allergy and Infectious Diseases
National Institutes of Health
Bethesda, Maryland

Hans Tanke

Department of Molecular Cell Biology
Leiden University Medical Center
Leiden, The Netherlands

ELSEVIER
ACADEMIC
PRESS

AMSTERDAM • BOSTON • HEIDELBERG • LONDON
NEW YORK • OXFORD • PARIS • SAN DIEGO
SAN FRANCISCO • SINGAPORE • SYDNEY • TOKYO

Elsevier Academic Press
525 B Street, Suite 1900, San Diego, California 92101-4495, USA
84 Theobald's Road, London WC1X 8RR, UK

This book is printed on acid-free paper. 

For all information on all Academic Press publications
visit our Web site at www.books.elsevier.com

ISBN: 0-12-480283-4

PRINTED IN THE UNITED STATES OF AMERICA
04 05 06 07 08 9 8 7 6 5 4 3 2 1

CONTENTS

PART III Immunology/T-Cell Responses

PART V Other Topics in Immunology

PART VI Cytogenetics

CONTRIBUTORS

Numbers in parentheses indicate the pages on which authors' contributions begin.

John D. Altman (433), Emory University, School of Medicine, Department of Microbiology and Immunology, and The Emory Vaccine Center, Atlanta, Georgia 30329

David Ambrozak (221), Vaccine Research Center, National Institute of Allergy and Infectious Diseases, National Institutes of Health, Bethesda, Maryland 20895

Gabriela M. Baerlocher (719), Terry Fox Laboratory, British Columbia Cancer Agency, Vancouver, British Columbia, V5Z 1L3, Canada and Department of Hematology and Department of Clinical Research, University Hospital Bern, 3010 Bern, Switzerland

Karsten Bahlmann (23), Massachusetts Institute of Technology, Department of Mechanical Engineering, Division of Botanical Engineering, Cambridge, Massachusetts 02139, and TissueVision, Inc., Somerville, Massachusetts 02143

Nicole Baumgarth (3, 643), Center for Comparative Medicine, University of California, Davis, California 95616

Annette Becker (375), Erft-Verband, 50126 Bergheim, Germany

Michael R. Betts (497), Immunology Laboratory, Vaccine Research Center, National Institute of Allergy and Infectious Diseases, National Institutes of Health, Bethesda, Maryland 20892

Marty Bigos (3), The J. David Gladstone Institute of Virology and Immunology, San Francisco, California 94103

Jason M. Brenchley (481), Human Immunology Section, Vaccine Research Center, National Institute of Allergy and Infectious Disease, National Institutes of Health, Bethesda, Maryland 20874

Martijn H. Breuning (751), Center for Human and Clinical Genetics, Leiden University Medical Center, 2333 AL Leiden, Nederland

Marcel P. Bruchez (171), Quantum Dot Corporation, Hayward, California 94545

Nigel P. Carter (769), The Wellcome Trust Sanger Institute, Wellcome Trust Genome Campus, Hinxton, Cambridge CB10 1SA, United Kingdom

Georgy Cherepnev (453), Institute for Medical Immunology, Charite Universitätsmedizin Berlin, Campus Charité Mitte, Berlin 10117, Germany

Zbigniew Darzynkiewicz (241, 307), Braner Cancer Research Institute, New York Medical College, Hawthorne, New York 10532

Stephen C. De Rosa (577), University of Washington School of Medicine, Department of Laboratory Medicine, Seattle, Washington 98109

Johan T. den Dunnen (751), Center for Human and Clinical Genetics, Leiden University Medical Center, 2333 AL Leiden, Nederland

Jurek W. Dobrucki (41), Laboratory of Confocal Microscopy and Image Analysis, Department of Biophysics, Faculty of Biotechnology, Jagiellonian University, 30-387 Kraków, Poland

Daniel C. Douek (481), Human Immunology Section, Vaccine Research Center, National Institute of Allergy and Infectious Disease, National Institutes of Health, Bethesda, Maryland 20874

Heike Fiegler (769), The Wellcome Trust Sanger Institute, Wellcome Trust Genome Campus, Hinxton, Cambridge CB10 1SA, United Kingdom

Jonathan Flint (799), The Wellcome Trust Centre for Human Genetics, Churchill Hospital, Headington, Oxford, Oxfordshire OX3 7BN, United Kingdom

David W. Galbraith (153), University of Arizona, Department of Plant Sciences and Institute for Biomedical Science and Biotechnology, Tucson, Arizona 85721

Wojciech Gorczyca (595, 665), Director of Hematopathology/Oncology Services, Genzyme Genetics/IMPATH, New York, New York 10019

Ignacy Gryczynski (73), Center for Fluorescence Spectroscopy, University of Maryland at Baltimore, Department of Biochemistry and Molecular Biology, Baltimore, Maryland 21201

Zygmunt Gryczynski (73), Center for Fluorescence Spectroscopy, University of Maryland at Baltimore, Department of Biochemistry and Molecular Biology, Baltimore, Maryland 21201

Mats Gullberg (787), Department of Genetics and Pathology, University of Uppsala, S-751-85 Uppsala, Sweden

Melvin Henriksen (185), CompuCyte Corporation, Cambridge, Massachusetts 02139

Elena Holden (185), CompuCyte Corporation, Cambridge, Massachusetts 02139

Xuan Huang (307), Brander Cancer Research Institute, New York Medical College, Hawthorne, New York 10532

Jonas Jarvius (787), Department of Genetics and Pathology, University of Uppsala, S-751-85 Uppsala, Sweden

Malin Jarvius (787), Department of Genetics and Pathology, University of Uppsala, S-751-85 Uppsala, Sweden

Louis Kamentsky (185), CompuCyte Corporation, Cambridge, Massachusetts 02139

Kathryn L. Kellar (409), Biotechnology Core Facility Branch, Scientific Resources Program, National Center for Infectious Diseases, Centers for Disease Control and Prevention, Atlanta, Georgia 30333

Florian Kern (453), Institute for Medical Immunology, Charite Universitätsmedizin Berlin, Campus Charité Mitte, Berlin 10117, Germany

Ki Hean Kim (23), Massachusetts Institute of Technology, Department of Mechanical Engineering, Division of Biological Engineering, Cambridge, Massachusetts 02139

Malcolm A. King (307), Brander Cancer Research Institute, New York Medical College, Hawthorne, New York 10532, and Department of Clinical Immunology, Royal North Shore Hospital, St. Leonards NSW 2065, Australia

Mariam Klouche (679), Bremer Zentrum für Laboratoriumsmedizin GmbH, D-28205 Bremen, Germany

Samantha J. L. Knight (799), The Wellcome Trust Centre for Human Genetics, Churchill Hospital, Headington, Oxford, Oxfordshire OX3 7BN, United Kingdom

Richard A. Koup (221, 497), Immunology Laboratory, Vaccine Research Center, National Institute of Allergy and Infectious Diseases, National Institutes of Health, Bethesda, Maryland 20892

Joseph R. Lakowicz (73), Center for Fluorescence Spectroscopy, University of Maryland at Baltimore, Department of Biochemistry and Molecular Biology, Baltimore, Maryland 21201

Ulf Landegren (787), Department of Genetics and Pathology, University of Uppsala, S-751-85 Uppsala, Sweden

Peter M. Lansdorp (719), Terry Fox Laboratory, British Columbia Cancer Agency, Vancouver, British Columbia, V5Z 1L3, Canada and Department of Medicine, University of British Columbia, Vancouver, British Columbia, V5Z 4E3, Canada

Chatarina Larsson (787), Department of Genetics and Pathology, University of Uppsala, S-751-85 Uppsala, Sweden

Peter P. Lee (513), Department of Medicine, Division of Hematology, Stanford University School of Medicine, Stanford, California 94305

Pui Lee (343), Division of Nephrology, Hypertension, and Transplantation, Department of Medicine, University of Florida, Gainesville, Florida 32610

Karin Loré (623), Vaccine Research Center, National Institutes of Health, Bethesda, Maryland 20892–3022

Ed Luther (185), CompuCyte Corporation, Cambridge, Massachusetts 02139

Holden T. Maecker (535), Becton Dickinson Biosciences, San Jose, California 95131

Vernon C. Maino (535), Becton Dickinson Biosciences, San Jose, California 95131

Joanna Malicka (73), Center for Fluorescence Spectroscopy, University of Maryland at Baltimore, Department of Biochemistry and Molecular Biology, Baltimore, Maryland 21201

Maria del Carmen Martinez-Ballesta (375), University of Leipzig, Institute of Botany, Department of Plant Physiology, D-04103 Leipzig, Germany

John R. Mascola (709), Vaccine Research Center, National Institute of Allergy and Infectious Diseases, National Institutes of Health, Bethesda, Maryland 20892

János Matkó (105), Department of Immunology, Eötvös Loránd University, Budapest H-1117, Hungary

Evgenia Matveeva (73), Center for Fluorescence Spectroscopy, University of Maryland at Baltimore, Department of Biochemistry and Molecular Biology, Baltimore, Maryland 21201

Armin Meister (375), Institute of Plant Genetics and Crop Plant Research, D-06466 Gatersleben, Germany

Mats Nilsson (787), Department of Genetics and Pathology, University of Uppsala, S-751-85 Uppsala, Sweden

Kazimierz Nowaczyk (73), Center for Fluorescence Spectroscopy, University of Maryland at Baltimore, Department of Biochemistry and Molecular Biology, Baltimore, Maryland 21201

Masaki Okafuji (307), Brander Cancer Research Institute, New York Medical College, Hawthorne, New York 10532, and Department of Oral and Maxillofacial Surgery, Yamaguchi University, School of Medicine, Ube 755-8505, Japan

Peggy L. Olive (355), Department of Medical Biophysics, British Columbia Cancer Research Centre, Vancouver, British Columbia, Canada V5Z 1L3

Kerry G. Oliver (409), Radix BioSolutions, Ltd., Georgetown, Texas 78626

David R. Parks (241), Department of Genetics, Stanford University, Stanford, California 94305

Stephen P. Perfetto (221), Vaccine Research Center, National Institute of Allergy and Infectious Diseases, National Institutes of Health, Bethesda, Maryland 20892

Louis J. Picker (535), Vaccine and Gene Therapy Institute and the Oregon National Primate Research Center at the Oregon Health & Science University, West Campus, Beaverton, Oregon 97006

Timothy Ragan (23), Massachusetts Institute of Technology, Department of Mechanical Engineering, Division of Biological Engineering, Cambridge, Massachusetts 02139, and TissueVision, Inc., Somerville, Massachusetts 02143

Mario Roederer (221, 241), Vaccine Research Center, National Institute of Allergy and Infectious Diseases, National Institutes of Health, Bethesda, Maryland 20892

Gregor Rothe (679), Bremer Zentrum für Laboratoriumsmedizin GmbH, D-28205 Bremen, Germany

Ola Söderberg (787), Department of Genetics and Pathology, University of Uppsala, S-751-85 Uppsala, Sweden

Ingrid Schmid (221), David Geffen School of Medicine, University of California, Los Angeles, Department of Hematology/Oncology, Los Angeles, California 90095

Mark S. Segal (343), Division of Nephrology, Hypertension, and Transplantation, Department of Medicine, University of Florida, Gainesville, Florida 32610

Peter T. C. So (23), Massachusetts Institute of Technology, Department of Mechanical Engineering, Division of Biological Engineering, Cambridge, Massachusetts 02139

János Szöllősi (105), Department of Biophysics & Cell Biology, and Cell Biophysics Research Group of the Hungarian Academy of Sciences, University of Debrecen, Debrecen H-4012, Hungary

Jörg Toepel (375), University of Leipzig, Institute of Botany, Department of Plant Physiology, D-04103 Leipzig, Germany

Frank Traganos (257), Brander Cancer Research Institute, New York Medical College, Hawthorne, New York 10532

György Vereb (105), Department of Biophysics & Cell Biology, and Cell Biophysics Research Group of the Hungarian Academy of Sciences, University of Debrecen, Debrecen H-4012, Hungary

Hans-Dieter Volk (453), Institute for Medical Immunology, Charite Universitätsmedizin Berlin, Campus Charité Mitte, Berlin 10117, Germany

Joshua M. Walker (535), Vaccine and Gene Therapy Institute and the Oregon National Primate Research Center at the Oregon Health & Science University, West Campus, Beaverton, Oregon 97006

Stefan J. White (751), Center for Human and Clinical Genetics, Leiden University Medical Center, 2333 AL Leiden, Nederland

Christian Wilhelm (375), University of Leipzig, Institute of Botany, Department of Plant Physiology, D-04103 Leipzig, Germany

Brent Wood (559), Department of Laboratory Medicine, University of Washington, Seattle, Washington 98195

Xingyong Wu (171), Quantum Dot Corporation, Hayward, California 94545

PREFACE

The four editions amounting to six volumes of the Methods in Cell Biology (MCB) devoted to cytometry published within the span of 15 years (volumes 33, 41, 42, 63, 64 and 75) provide evidence of the explosive growth of cytometric methods and of the multitude of their applications during the past two decades. These volumes altogether contain 213 chapters that describe progress in the development of instrumentation as well as a variety of cytometric methods and their applications. Although the term "edition" is suggestive of the updating of earlier volumes, in fact, well over 90% chapters of each successive edition of CYTOMETRY are devoted to entirely new topics. These volumes of MCB, thus, represent the most inclusive collection of articles on different cytometric methods and the associated instrumentation. Whereas the first two editions were focused on flow cytometry the recent editions include the cytometric methods that probe individual cells not necessarily in flow.

There is an abundance of the methodology books presenting particular methods in a form of technical protocols (e.g. Current Protocols by Wiley-Liss, "Practical Approach" books by Oxford Press, Method series by Humana Press, *etc*). Protocols are also often included with the commercially available reagent kits. The latter are generally cryptic, and because of the proprietary nature of some reagents, do not inform about chemistry of the components or mechanistic principles of the kit. While the protocols provide the guidance to reproduce a particular assay, their "cook-book" format is restrictive and does not allow one to explain in detail principles of the methodology, discuss its limitations and possible pitfalls. Likewise the discussion on optimal choice of the assay for a particular task and/or cell system, or review of the method applications, is limited. Yet such knowledge is of importance for rational use of the methodology and for extraction of maximal relevant information from the experiment.

The chapters in CYTOMETRY MCB volumes, including this 4th edition, fill up this niche by providing more comprehensive, and often complementary, description of particular methods, compared to the protocol-format series. The authors were invited to present not only technical protocols but also to discuss the aspects of the methodology that cannot be included in the typical protocols, explain theoretical foundations of the described methods, their applicability in experimental laboratory and clinical setting, common traps and pitfalls, problems with data interpretation, comparison with alternative assays, *etc*. Some chapters review applications of cytometry and complementary methodologies to particular biological problems or clinical tasks.

The 35 chapters presented in CYTOMETRY 4th Edition cover a wide range of diverse topics. Several chapters describe progress in technology of fluorescence measurement. The novel phenomenon of the surface-plasmon coupled emission (SPCE) presented in one of these chapters, in combination with nanophotonic technology, is expected to open new era for biophysical and biochemical applications of fluorescence. Another chapter introduces quantum dot technology, the new approach that most likely will revolutionize the fluorochrome-labeling of cells and molecules for a variety of applications. The chapter on cytometry of fluorescence resonance energy transfer (FRET) describes the theoretical foundations- and uncovers further analytical possibilities- of this methodology. The chapter on fluorescent proteins is an exhaustive review of accomplishments and possibilities offered in this rapidly expanding field. The critical assessment of quantitative analytical capabilities of confocal microscopy, optimization of emission optics and further progress in development of laser scanning cytometry instrumentation are the topics of other chapters focused on methodology of fluorescence measurement.

Another group of chapters describe different cytometric methods and cytometry applications in studies of cell death, particularly by mode of apoptosis, mechanism of antitumor drug action, and DNA damage. The rapidly growing application of cytometry in phytoplankton is also assessed in great detail. The chapters on biohazard sorting and data analysis guidelines will be of interest to a variety of researchers in different fields.

Immunophenotyping represents the most common clinical application of cytometry. It is not surprising, therefore, that over a third of all chapters are devoted to this topic. There have been two major areas of technological advances in immune monitoring since the last 3rd edition: high-end multicolor immunophenotyping and single-cell functional analysis. Several chapters review the additional complexities arising from 6+ color immunophenotyping—and cover the enormous increase in information content provided by such analysis. Several important functional analyses (proliferation, cytokine responses, cytolytic potential) have also been adapted to flow cytometric analysis—the combination of these functions with multicolor phenotyping provides an unparalleled view of the immune function of antigen-specific cells. This combination of function and phenotype is becoming particularly relevant to the burgeoning field of vaccination aimed at inducing T cell responses, where the search for correlates of immune protection are relying on surrogate markers such as antigen-specific T cell function.

The field of cytogenetics and molecular genetics is represented by several chapters. Presented are the methods for telomere length measurement and genomic array technology as well as multiplex amplifiable probe hybridization (MAPH) and multiplex ligation–dependent probe amplification (MLPA), the latter to probe the copy number changes (deletions, multiplications) in genomic DNA. A very exhaustive is the chapter describing the use of subtelomeric probes in studies of mental retardation.

In tradition with the earlier CYTOMETRY editions, the chapters were prepared by the colleagues who either developed the described methods, contributed to their modification, or found new applications and have extensive experience in their use. The list of authors, thus, is a continuation of "Who's Who" directory in the field of cytometry. We are thankful to all contributing authors for the time they devoted to share their knowledge and experience.

Zbigniew Darzynkiewicz
Mario Roederer
Hans J. Tanke

PREFACE TO THE THIRD EDITION

This is the third edition of cytometry volumes in the *Methods in Cell Biology* series. The first, single-volume edition (*Flow Cytometry, Methods in Cell Biology*, Volume 33, 1990) appeared a decade ago. The continuing rapid growth of this methodology prompted us to prepare the second, two-volume edition (*Flow Cytometry, Methods in Cell Biology*, Volumes 41 and 42, 1994), which introduced a variety of new methods developed since the publication of the first edition. The growth and applications of this methodology have continued at an accelerating pace. This progress and the demand for the first two editions, which have become the "bible" for researchers who utilize the presented methods in a variety of fields of biology and medicine, prompted us to prepare the third edition.

This two-volume set differs from the earlier editions in several respects. The title is changed to *Cytometry* to indicate its wider scope. Several chapters describe methods and instrumentation that are not particular to cell analysis in "flow." Also changed are the scope and specifics of many chapters. Specifically, with the appearance of similar series of books on methods by other publishers (e.g., *Current Protocols in Cytometry* by Wiley-Liss or the "Practical Approach" books by Oxford Press), there was no point in duplicating them by focusing on presentation of individual methods in a cookbook form only. The authors, therefore, were requested to prepare their chapters in a form that presented not only technical protocols but also different aspects of the methodology that cannot be included in the protocols format. Thus, theoretical foundations of the described methods, their applicability in experimental laboratory and clinical settings, traps and pitfalls common to particular methods, problems with data interpretation, comparison with alternative assays, etc., are all presented in greater detail in many chapters. Furthermore, some chapters review applications of cytometry and complementary methodologies to particular biological problems or clinical tasks.

With few exceptions, nearly all 56 chapters in the present edition are novel, describing methods that were not included in the earlier editions. The present edition thus complements rather than merely updates the earlier edition. Because most of the methods described in Volumes 41 and 42 have not changed much since publication and are still in wide use, the combination of the earlier two volumes and these two new volumes becomes the most comprehensive collection of all methods in cytometry ever published.

The chapters presented in *Cytometry* cover a wide range of topics. The first several chapters are introductory. They describe principles of flow cytometry, laser scanning cytometry (LSC), confocal microscopy, and general approaches in cell measurement and data acquisition. Newcomers to the field of cytometry may find these chapters particularly useful, as they provide the foundation needed to understand specific methods and more complex data analysis. The next chapters address the issue of cell preparation for analysis by cytometry, quality assurance, and standardization. Of special interest may be the chapters focused on strategies for cell permeabilization and fixation to detect intracellular components, quantitation of the immunocytochemical staining reactions, and standardization in cytometry in general. Unfortunately, these important issues are neglected in many studies utilizing cytometric methods.

Analysis of cell proliferation is the subject of several other chapters. All the methods presented in these chapters are used extensively in experimental and clinical research. The methods include measurements of mitotic activity by flow cytometry, assays of cell kinetics by analysis of BrdU incorporation, analysis of the history of cell proliferation of the progeny cells from geometric dilution of the probe integrated into parent cells, applications of Ki-67 and PCNA antibodies as proliferation markers, and analysis of the lymphocyte activation antigens. Further chapters are devoted to methods of analysis of cell death, primarily by apoptosis. They include probing of mitochondria, activation of caspases, and analysis of apoptosis in the plant kingdom. A review of common problems, difficulties, and pitfalls encountered in analysis of apoptosis, with the key information of how to avoid them, also is provided.

Another group of methods is focused on analysis of cell-to-cell interactions and interactions of cells with the extracellular matrix. This is an exciting cross-disciplinary area that is revealing new directions for cytometry. This section includes cell migration assays, cytometric analysis of cell contact and adhesion, and measurement of extracellular matrix degradation. Also included in this group is a review of extracellular matrix substrates for culturing cells and analysis of cell interactions in three dimensions.

The field of cytogenetics and molecular genetics is represented by several chapters that present methods for sorting plant chromosomes, quantitative DNA fiber mapping, primed *in situ* (PRINS) methodology, individual chromosome telomere length analysis, and approaches to detecting products of chromosomal translocation in individual interphase cells. Functional cell assays, such as probing mitochondria with new markers of the electrochemical transmembrane potential, and measurement of RNA synthesis by immunocytochemical detection of the incorporated BrU, as well as new approaches to monitoring erythropoiesis or proliferation and differentiation of human progenitor cells, are all the subjects of additional chapters.

A large group of chapters is devoted to applications of cytometry in experimental oncology. Presented here are the methods to study interactions between antitumor drugs and intracellular targets, monitoring cellular resistance to chemotherapy in solid tumors and in spheroids related to their three-dimensional architecture, and analyzing DNA damage in individual cells caused by ionizing radiation. The methods specifically designed for studying effects of hyperthermia on tumor cells are also presented.

The most numerous chapters are those on applications of cytometry in the clinic. Indisputably, immunophenotyping is the most common application of cytometry in the clinical setting, and three chapters are devoted to this subject. A very exhaustive chapter on multiparametric analysis of human leukocytes describes the approaches to identifying the cells in different hematological malignancies. Adaptation of laser scanning cytometry to achieve similar tasks is the topic of another chapter on this subject. The third chapter addresses the specific issue of utility of a CD45 gate for identification of malignant cells in acute leukemias. Two insightful and exhaustive reviews, one that critically assesses clinical impact of analysis of different proliferation markers in human solid tumors and another that covers applications of flow cytometry and complementary methodologies in detection of minimal residual disease in leukemias, will be of great value for oncologists. There are also two reviews on applications of laser scanning cytometry to analysis of human tumors, one presented from the perspective of the pathologist and another from the surgeon's perspective. Applications of flow cytometry to monitoring HIV-infected patients is also a subject of thorough review. The last chapter in this group (Chapter 53) may be of particular interest to all researchers, regardless of the discipline. This chapter presents a unique approach to multiparameter data analysis in the clinic that often reveals unexpected correlations with high impact on disease prognosis.

The last group of chapters describes the methods and applications of cytometry in studies of microorganisms. They include flow cytometric analysis of microorganisms and monitoring of bacterial susceptibility to antibiotics. Applications of these assays are expected to rapidly expand and become routine tools in microbiology with wide application in the field of infectious diseases as well as in monitoring environmental contaminations.

As in the earlier editions, the chapters were prepared by colleagues who developed the described methods, contributed to their modification, or found new applications and have extensive experience in their use. The list of authors, as before, represents a "Who's Who" directory in the field of cytometry. On behalf of the readers, we express our gratitude to all contributing authors for the time they devoted to sharing their knowledge and experience.

Zbigniew Darzynkiewicz
Harry A. Crissman
J. Paul Robinson

PREFACE TO THE SECOND EDITION

The first edition of this book appeared four years ago (*Methods in Cell Biology*, Vol. 33, *Flow Cytometry*, Z. Darzynkiewicz and H. A. Crissman, Eds., Academic Press, 1990). This was the first attempt to compile a wide variety of flow cytometric methods in the form of a manual designed to describe both the practical aspects and the theoretical foundations of the most widely used methods, as well as to introduce the reader to their basic applications. The book was an instant publishing success. It received laudatory reviews and has become widely used by researchers from various disciplines of biology and medicine. Judging by this success, there was a strong need for this type of publication. Indeed, flow cytometry has now become an indispensable tool for researchers working in the fields of virology, bacteriology, pharmacology, plant biology, biotechnology, toxicology, and environmental sciences. Most applications, however, are in the medical sciences, in particular immunology and oncology. It is now difficult to find a single issue of any biomedical journal without an article in which flow cytometry has been used as a principal methodology. This book on methods in flow cytometry is therefore addressed to a wide, multidisciplinary audience.

Flow cytometry continues to rapidly expand. Extensive progress in the development of new probes and methods, as well as new applications, has occurred during the past few years. Many of the old techniques have been modified, improved, and often adapted to new applications. Numerous new methods have been introduced and applied in a variety of fields. This dramatic progress in the methodology, which occurred recently, and the positive reception of the first edition, which became outdated so rapidly, were the stimuli that led us to undertake the task of preparing a second edition.

The second edition is double the size of the first one, consisting of two volumes. It has a combined total of 71 chapters, well over half of them new, describing techniques that had not been presented previously. Several different methods and strategies for analysis of the same cell component or function are often presented and compared in a single chapter. Also included in these volumes are selected chapters from the first edition. Their choice was based on the continuing popularity of the methods; chapters describing less frequently used techniques were removed. All these chapters are updated, many are extensively modified, and new applications are presented.

From the wide spectrum of chapters presented in these volumes it is difficult to choose those methods that should be highlighted because of their novelty, possible high demand, or wide applicabilities. Certainly those methods that offer new tools for molecular biology belong in this category; they are presented in chapters on fluorescence *in situ* hybridization (FISH), primed *in situ* labeling (PRINS), mRNA species detection, and molecular phenotyping. Detection of intracellular viruses and viral proteins and analysis of bacteria, yeasts, and plant cells are broadly described in greater detail than before in separate chapters. The chapter on cell viability presents and compares ten different methods for identifying dead cells and discriminating between apoptosis and necrosis, including a new method of DNA gel electrophoresis designed for the detection of degraded DNA in apoptotic cells. The chapter describing analysis of enzyme kinetics by flow cytometry is very complete. The subject of magnetic cell sorting is also described in great detail.

Numerous chapters that focus on the analysis of cell proliferation also should be underscored. The subjects of these chapters include univariate DNA content analysis (using a variety of techniques and fluorochromes applicable to cell cultures, fresh clinical samples, or paraffin blocks), the deconvolution of DNA content frequency histograms, multivariate (DNA vs protein or DNA vs RNA content) analysis, simple and complex assays of cell cycle kinetics utilizing BrdUrd and IdUdr incorporation, and studies of the cell cycle based on the expression of several proliferation-associated antigens, including the G_1- and G_2-cyclin proteins. Approaches to discriminating between cells having the same DNA content but at different positions in the cell cycle (e.g., noncycling G_0 vs cycling G_1, G_2 vs M, and G_2 of lower DNA ploidy vs G_1 of higher ploidy) are also presented.

Many of the methods described in these volumes will be used extensively in the fields of toxicology and pharmacology. Among these are the techniques designed for analysis of somatic mutants, formation of micronuclei, DNA repair replication, and cumulative DNA damage in sperm cells (DNA *in situ* denaturability). The latter is applicable as a biological dosimetry assay. A plethora of methods for analysis of different cell functions (functional assays) will also find application in toxicology and pharmacology.

The largest number of chapters is devoted to methods having clinical applications, either in medical research or in routine practice. Chapters dealing with lymphocyte phenotyping, reticulocyte and platelet analysis, analysis and sorting of hemopoietic stem cells, various aspects of drug resistance, DNA ploidy, and cell cycle measurements in tumors are very exhaustive. Diagnosis and disease progression assays in HIV-infected patients, as well as sorting of biohazardous specimens, new topics of current importance in the clinic, are also represented in this book.

Individual chapters are written by the researchers who developed the described methods, contributed to their modification, or found new applications and have extensive experience in their use. Thus, the authors represent a "Who's Who" directory in the field of flow cytometry. This ensures that the essential details of each methodology are included and that readers may easily learn these techniques by following the authors' protocols. We express our gratitude to all contributing authors for sharing their knowledge and experience.

The chapters are designed to be of practical value for anyone who intends to use them as a methods handbook. Yet, the theoretical bases of most of the techniques are presented in detail sufficient for teaching the principle underlying the described methodology. This may be of help to those researchers who want to modify the techniques, or to extend their applicability to other cell systems. Understanding the principles of the method is also essential for data evaluation and for recognition of artifacts. A separate section of most chapters is devoted to the applicability of the described method to different biological systems. Another section of most chapters covers the critical points of the procedure, possible pitfalls, and experience of the author(s) with different instruments. Appropriate controls, standards, instrument adjustments, and calibrations are the subjects of still another section of each chapter. Typical results, frequently illustrating different cell types, are presented and discussed in yet another section. The Materials and Methods section of each chapter is exhaustive, providing a detailed, step-by-step description of the procedure in a protocol or cookbooklike format. Such exhaustive treatment of the methodology is unique; there is no other publication on the subject of similar scope.

We hope that the second edition of *Flow Cytometry* will be even more successful than the first. The explosive growth of this methodology guarantees that soon there will be the need to compile new procedures for a third edition.

Zbigniew Darzynkiewicz
Harry A. Crissman
J. Paul Robinson

PREFACE TO THE FIRST EDITION

Progress in cell biology has been closely associated with the development of quantitative analytical methods applicable to individual cells or cell organelles. Three distinctive phases characterize this development. The first started with the introduction of microspectrophotometry, microfluorometry, and microinterferometry. These methods provided a means to quantitate various cell constituents such as DNA, RNA, or protein. Their application initiated the modern era in cell biology, based on quantitative—rather than qualitative, visual—cell analysis. The second phase began with the birth of autoradiography. Applications of autoradiography were widespread and this technology greatly contributed to better understanding of many functions of the cell. Especially rewarding were studies on cell reproduction; data obtained with the use of autoradiography were essential in establishing the concept of the cell cycle and generated a plethora of information about the proliferation of both normal and tumor cells.

The introduction of flow cytometry initiated the third phase of progress in methods development. The history of flow cytometry is short, with most advances occurring over the past 15 years. Flow cytometry (and associated with it electronic cell sorting) offers several advantages over the two earlier methodologies. The first is the rapidity of the measurements. Several hundred, or even thousands, of cells can be measured per second, with high accuracy and reproducibility. Thus, large numbers of cells from a given population can be analyzed and rare cells or subpopulations detected. A multitude of probes have been developed that make it possible to measure a variety of cell constituents. Because different constituents can be measured simultaneously and the data are recorded by the computer in list mode fashion, subsequent bi- or multivariate analysis can provide information about quantitative relationships among constituents either in particular cells or between cell subpopulations. Still another advantage of flow cytometry stems from the capability for selective physical sorting of individual cells, cell nuclei, or chromosomes, based on differences in the variables measured. Because some of the staining methods preserve cell viability and/or cell membrane integrity, the reproductive and immunogenic capacity of the sorted cells can be investigated. Sorting of individual chromosomes has already provided the basis for development of chromosomal DNA libraries, which are now indispensable in molecular biology and cytogenetics.

Flow cytometry is a new methodology and is still under intense development, improvement, and continuing change. Most flow cytometers are quite complex and not yet user friendly. Some instruments fit particular applications better than others, and many proposed analytical applications have not been extensively tested on different cell types. Several methods are not yet routine and a certain degree of artistry and creativity is often required in adapting them to new biological material, to new applications, or even to different instrument designs. The methods published earlier often undergo modifications or improvements. New probes are frequently introduced.

This volume represents the first attempt to compile and present selected flow cytometric methods in the form of a manual designed to be of help to anyone interested in their practical applications. Methods having a wide immediate or potential application were selected, and the chapters are written by the authors who pioneered their development or who modified earlier techniques and have extensive experience in their application. This ensures that the essential details are included and that readers may easily master these techniques in their laboratories by following the described procedures.

The selection of chapters also reflects the peculiarity of the early phase of method development referred to previously. The most popular applications of flow cytometry are in the fields of immunology and DNA content–cell cycle analysis. While the immunological applications are now quite routine, many laboratories still face problems with the DNA measurements, as is evident from the poor quality of the raw data (DNA frequency histograms) presented in many publications. We hope that the descriptions of several DNA methods in this volume, some of them individually tailored to specific dyes, flow cytometers, and material (e.g., fixed or unfixed cells or isolated cell nuclei from solid tumors), may help readers to select those methods that would be optimal for their laboratory setting and material. Of great importance is the standardization of the data, which is stressed in all chapters and is a subject of a separate chapter.

Some applications of flow cytometry included in this volume are not yet widely recognized but are of potential importance and are expected to become widespread in the near future. Among these are methods that deal with fluorescent labeling of plasma membrane for cell tracking, flow microsphere immunoassay, the cell cycle of bacteria, the analysis and sorting of plant cells, and flow cytometric exploration of organisms living in oceans, rivers, and lakes.

Individual chapters are designed to provide the maximum practical information needed to reproduce the methods described. The theoretical bases of the methods are briefly presented in the introduction of most chapters. A separate section of each chapter is devoted to applicability of the described method to different biological systems, and when possible, references are provided to articles that review the applications. Also discussed under separate subheads are the critical points of procedure, including the experience of the authors with different instruments, and the appropriate controls and standards. Typical results, often illustrating different cell types, are presented and discussed in the "Results" section. The "Materials and Methods" section of each chapter is the most extensive, giving a detailed description of the method in a cookbook format.

Flow cytometry and electronic sorting have already made a significant impact on research in various fields of cell and molecular biology and medicine. We hope that this volume will be of help to the many researchers who need flow cytometry in their studies, stimulate applications of this methodology to new areas, and promote progress in many disciplines of science.

Zbigniew Darzynkiewicz
Harry A. Crissman

PART I

Instrumentation/Fluorochromes

CHAPTER 1

Optimization of Emission Optics for Multicolor Flow Cytometry

Nicole Baumgarth* and Marty Bigos[†]

*Center for Comparative Medicine, University of California
Davis, California 95616

[†]The J. David Gladstone Institute of Virology and Immunology
San Francisco, California 94103

I. Introduction

In the mid to late 1990s a large expansion in the simultaneous measurement capabilities of flow cytometers was achieved through hardware design and fluorochrome development that now allows the simultaneous measurement of up to 12

METHODS IN CELL BIOLOGY, VOL. 75

0091-679X/04 $35.00

colors and two scatter parameters (Anderson *et al.*, 1998; Bigos *et al.*, 1999; De Rosa *et al.*, 2001; Roederer *et al.*, 1997). Before that time, the maximum number of reported simultaneous measurements by flow cytometry had been six colors. This development had a particular impact on immunophenotyping (i.e., the identification of cells and cell subsets based on their [surface] receptor expression profiles). Identification of even smaller cell subsets and different functional cell subpopulations became feasible. There is a growing interest in this technology for a number of different applications. This means either upgrading and extending existing flow cytometers or purchasing newer flow cytometers that are designed to facilitate multicolor immunophenotyping. Whatever the choice may be, optimization of optical paths, though important even for two- to three-color analyses, becomes crucial for multicolor flow cytometry. Nobody should rely on the commercial flow cytometers, including those sold for this purpose, to be optimized for individual uses. Every user group uses different reagent and dye combinations or analyzes different materials that make it important to evaluate each flow cytometer for quality of emission signal collection.

Some of the newer dyes are inherently "less bright" than standard fluorescein thiocyanate (FITC), phycoerythrin (PE), and allophycocyanin (APC) dyes used for three- to four-color analyses. Thus, optimizing light collection for those reagents might mean the difference between detecting a signal and not detecting a signal. Equally important, simultaneous use of multiple colors increases the amount of spectral overlaps between measurement parameters, thereby degrading the quality of the measurements achieved with each dye. The more light and the better the signal that one can collect, the better the compensated data will look. In this chapter, we aim at providing a guide for the optimization of light paths, that is, the optical setup of flow cytometers. To provide a context for the approaches we discuss at the end of this chapter, we first briefly review the available fluorochromes and the hardware properties of the individual components of flow cytometers that affect light signal generation and detection. It is important to note, however, that even the best-designed flow cytometers cannot compensate for mistakes made in the choice of dye and reagent combinations for staining. We refer you elsewhere (Baumgarth and Roederer, 2000) for a review on reagent choices and staining design.

II. Fluorochromes

A. Spectral Characteristics

Few fluorochromes are available for use with flow cytometers, presenting a major hurdle for the number of simultaneous measurements that can be undertaken. Two important characteristics determine the usefulness of fluorochromes for flow cytometry: their ability to be excited by light of a certain wavelength (*excitation spectrum*) and their ability to emit light of a longer wavelength after excitation (*emission spectrum*). Figure 1 summarizes some commonly used fluorochromes and their excitation and emission spectra.

In flow cytometry, lasers are commonly used as light sources to excite fluorochromes that are coupled to antibodies or other proteins. The light emitted from the laser-excited fluorochromes is then passed to a detector (photomultiplier tubes [PMTs]) where it is converted into electronic signals, which are processed to yield the data points for the measured objects. Engineering complexities and cost limit the number of lasers to usually one to four per machine. Hence, to simultaneously measure a large number of fluorochromes, multiple dyes have to be excited by the same laser (i.e., they have to have similar excitation spectra). In contrast, to be measured accurately, each fluorochrome should emit light at a wavelength distinct from that of any other fluorochrome, so they have to have distinct emission spectra.

In reality, most fluorochromes have emission spectra that overlap at least partially with those of other fluorochromes. These partial overlaps, or "spillovers," are corrected in flow cytometry in two ways: first, by blocking out unwanted light before data collection by installing filters that selectively pass light of a narrow band of light onto a detector (*bandpass filters*), and second, through a process of measurement correction called *compensation* (Roederer, 2001). Compensation mathematically removes spillover signals emitted from fluorochromes other than those to be measured. Compensation is applied to flow cytometric measurements before ("hardware compensation") or after data acquisition ("software compensation").

B. Chemical Characteristics

Apart from spectral characteristics, the chemical properties of fluorochromes are also important to consider. One can distinguish small organic molecules (FITC, Texas red [TR], cascade blue, cascade yellow, Alexa dyes) and large organic molecules (PE, APC), which are directly excited and emit measurable light, from so-called *tandem dyes*. The latter consist of two molecules that are covalently linked to achieve close proximity. Of these two, the donor is excited by the laser light and transfers its energy to the other linked molecule (acceptor) instead of fluorescing. The transferred energy from the donor excites the acceptor, resulting in fluorescence, which is the signal that can be collected. This process is termed *resonance energy transfer* and is reviewed in detail in chapter 5.

The advantage of tandem dyes is that they use a limited number of donors, mostly PE and APC. Thus, they have the same excitation spectra, but because of the differences in the molecules linked to them, they obtain distinct emission spectra. Examples are PE-cyanate-5 (Cy5) and PE-cyanate-7 (Cy7). In both cases PE donor molecules are linked to an acceptor, either Cy5 or Cy7. The donor PE molecule is excited by a 488-nm laser light, and energy transfer to Cy5 and Cy7 results in emission spectra that are distinct from each other and distinct from PE itself (Fig. 1). However, the energy transfer process is not 100% efficient, causing

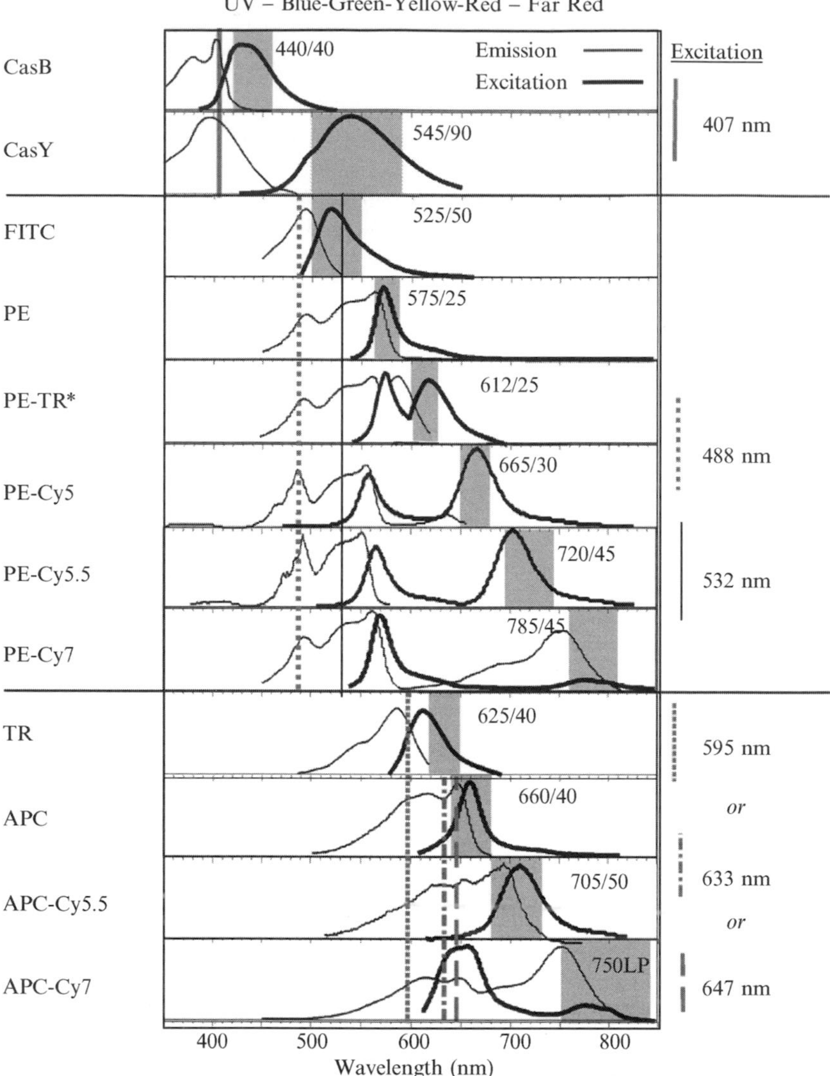

Fig. 1 Spectral characteristics of fluorochromes used for multicolor flow cytometry. Shown are the excitation (*thin lines*) and emission spectra (*thick lines*) of various fluorochromes. Dashed lines represent a number of excitation wavelengths often used on flow cytometers. Gray areas and numbers show the areas of light collection and the specificities of bandpass filters that can be used for simultaneous signal collection. *Note that PE-TR (ECD) and TR cannot be used simultaneously because of strong spectral overlaps. APC, allophycocyanin; CasB, cascade blue; CasY, cascade yellow; Cy, cyanate; FITC, fluorescein thiocyanate; PE, phycoerythrin; TR, Texas red.

some light emission by the donor itself. Therefore, emission spectra of tandem dyes always contain a certain component of donor emission (in the example PE; Fig. 1) that must also be compensated. Optimization of tandem-dye conjugations aims at maximizing signals from the tandem dye while minimizing emission from the donor itself. It is important to realize that batch-to-batch and source-to-source variations between tandem-dye conjugations exist, because it is difficult to keep the donor/acceptor ratio exactly the same. Thus, each tandem-dye lot might have slightly different emission profiles. Therefore, compensation controls for these colors need to precisely match the batch or reagent used in the experiment itself to achieve correct results.

C. Relative Brightness

Assuming adequate spectral characteristics, the relative brightness of a fluorochrome is in many respects its most important property that determines its usefulness for flow cytometry (Baumgarth and Roederer, 2000). It is important to understand that the term *brightness* is used to denote the amount of signal emitted from stained cells or particles compared to the amount of signal emitted from the unstained sample. More accurate is *signal to noise*, but this engineering nomenclature has not found acceptance in the biological research community. Brightness of a reagent depends on a number of characteristics, including the absolute amount of photons emitted and collected from the fluorochrome and the amount of "background" or "autofluorescence" that is emitted at the collected wavelength. For example, red-emitting APC conjugates usually yield very bright staining, despite the fact that the absolute number of photons collected is relatively low. The brightness is due mainly to the very low autofluorescence signals at the collected wavelengths, hence creating a large differential between stained and unstained samples.

As discussed in more detail later in this chapter, knowing the basic characteristics of the dyes that need to be evaluated for multicolor flow cytometry is essential to design optimal collection optics. Newer dyes such as the increasing number of available Alexa dyes (Molecular Probes, Eugene, Oregon) have similar but not always identical properties to previously measured fluorochromes. Thus, changes in the hardware might be necessary when switching from one dye to another. For example, the fluorochromes Cy5-PE and peridinin chlorophyll protein (PerCP) are usually measured on the same detector of the flow cytometers, although peak emission of PerCP is 10 nm longer than that of Cy5-PE. Thus, depending on the light path of a particular machine, optimization for each fluorochrome might enhance the signals received from these dyes. Multicolor flow cytometry requires that light collection for each dye is optimized so that staining combinations can be designed that allow one to take full advantage of the large amount of information and increased sensitivity that come with this application. If the characteristics of the various dyes are not taken into consideration, the results will likely be disappointing at best.

III. Hardware Components Affecting Light Collection

Lasers as the primary light source for the excitation of fluorochromes, emission collection lenses, dichroic mirrors, bandpass filters, and PMTs are the main components of the optical layout of a flow cytometer. Following is a brief outline of their properties that need to be considered when aiming to optimize light collection at a flow cytometer. Although the user cannot easily change all these components, knowledge of their functions, their characteristics, and their limitations aid in making correct design decisions for optimization purposes.

A. Lasers

Lasers are commonly used on flow cytometers as light sources for fluorochrome excitation. Most frequently, a three-laser setup is used for multicolor flow cytometry. This includes a blue laser emitting at approximately 490 nm, a red laser emitting between 595- and 647-nm laser light, and a third one tuned to give either 350-nm or low 400-nm emission. Figure 1 shows which dyes are excited by which laser source. A recent alteration to this setup is the use of a fourth laser exciting at 532 nm. This laser excites PE and its derivatives better than a 488-nm laser (Table I). The 488-nm laser, however, is still necessary if FITC is to be used.

Some complexities in choosing which red-emitting laser to use concern cost and ease of use, whereas others concern the limited number and types of dyes that can be excited. The high-power 647-nm line of a krypton laser is well matched to APC and its tandem conjugates, but it also directly excites the Cy5 and Cy5.5

Table I
Effects of Laser Power and Wavelength on Signal Intensities

	100 mW 488 nm	100 mW 532 nm
FITC[a]	3.6-fold[b]	N/A
PE	1.1	1.7-fold[c]
TR-PE	1.4	4.5
Cy5.5-PE	1.9	1.2
Cy5.5-PerCP	1.1	0.5
PerCP	0.6	0.4
Cy7-PE	2.7	2.6

Note: Cy5.5, cyanate 5.5; FITC, fluorescein thiocyanate; PE, phycoerythrin; PerCP, TR, Texas red.
[a]Human peripheral blood mononuclear cells stained with CD4 or CD8 anti-human antibodies conjugated to the indicated fluorochromes were measured on a modified LSR-II (Becton Dickinson).
[b]Shown are fold-changes in staining brightness (see Fig. 4 for details) for each fluorochrome compared to measurements obtained at a laser power of 20 mW.
[c]Shown are fold-changes in staining brightness obtained with a 532-nm laser (Compass 315M, Coherent Inc, Mountain View, California) compared to results obtained with a 100-mW 488-nm laser (Saphire 499-200, Coherent Inc).

components of the PE tandem conjugates. A dye laser tuned to 595 nm does not have this problem, but the cost and complexity of using this type of laser are high. An advantage is that the 595-nm dye laser directly excites TR, a relatively bright dye with little spectral overlaps except into APC. On the other hand, PE-TR reagents cannot be used on machines with that laser. The HeNe and diode lasers are a good choice for excitation in the low to mid 630-nm range, but their power output is only 10% or less compared to the other choices. Because higher powered lasers enhance the number of photons emitted by the fluorochromes, the quality of results obtained are better compared to those obtained with lower powered lasers (see later discussion). For cytometers with highly efficient collection optics, such as bench-top analysis machines, a low-powered solid-state laser can work well enough. For those flow cytometers with less efficient optics, like jet-in-air sorters, the higher powered lasers are preferable (Table I).

Several fluorochromes are excited by violet light, including cascade blue, Alexa 405, Alexa 430 (Anderson *et al.*, 1998), Pacific blue, and Am-cyan. In addition, there are several efforts currently underway to excite some of the red-emitting QuantumDot reagents (QuantumDot, Hayward, California) using violet light. Violet excitation can work well with the Hoechst dyes to measure DNA content and the stem cell–enriched "side populations." In addition, violet excitation is useful for molecular biology applications, for example, excitation of cyan fluorescent protein and the CCF2 substrate for measuring β-lactamase expression (Zlokarnik *et al.*, 1998). Violet-enhanced krypton lasers can provide up to 200 mW of 407/413-nm light and work well with the applications described, but they are relatively expensive. Inexpensive 405-nm diode lasers are available but provide only 15–20 mW of power. These are adequate for immunophenotyping applications in cytometers with efficient collection optics but do not work well with the Hoechst dyes. Thus, the best choice for violet excitation is a violet-enhanced krypton laser.

Ultraviolet (UV) excitation, especially around 350 nm, has no substitute for certain applications. The most critical one is Indo-1 for measuring calcium flux. The argon and krypton lasers provide excitation in the 350- to 360-nm range and are ideal for Indo-1. Moreover, they give excellent results with the Hoechst dyes and other DNA probes, making them the lasers of choice for these applications. Solid-state UV lasers are just becoming available. Further work is needed to determine whether they can substitute for the high-powered water-cooled gas ion lasers.

B. Collection Optics

Microscope-style objectives, usually with a magnification of 10×, collect and focus emitted light for processing by the rest of the optical system. Although the user has little choice in this matter, the care of the objective is important, especially with jet-in-air sorters, with which it is easy for liquid from the jet to splash onto the lens. This can seriously degrade the light collection, especially in the red and far-red region, reducing the signal by half or even more. Frequent cleaning with lens paper and methanol can prevent such signal degradation.

C. Filters

Light emitted from fluorescent particles that are excited by the laser light of flow cytometers must be appropriately collected onto the PMT. This is accomplished by a set of reflecting dichroic mirrors and bandpass filters. The mirrors are designed to reflect light of certain wavelengths and to pass light of differing wavelengths. They are constructed by depositing multiple layers of thin films on a high-quality (HQ) glass substrate, providing high transmission of the desired light and blocking the undesired light by destructive interference in the thin film layers. One distinguishes "long-pass" from "short-pass" mirrors (i.e., mirrors that let the longer wavelengths pass and reflect the shorter wavelengths and the other way around). The specific wavelength at which the mirror changes from passing to reflecting characterizes each dichroic mirror. Thus, a 530 long-pass filter passes light at 530 nm or more and reflects light of wavelengths less than 530 nm.

Dichroics split the fluorescence emissions into different optical paths on flow cytometers (see Fig. 3). To measure the appropriate wavelengths for only one fluorochrome at each PMT, additional filters are used to block unwanted light. These filters are usually *bandpass filters*, which block all but a small band of light. Bandpass filters are designated by the peak light that they pass and the widths around that peak light collection. A 530/50 bandpass filter, for example, collects laser light of 530 nm plus wavelengths up to 25 nm below and above 530 nm. Hence, it is designed to pass light between 505 and 555 nm. Bandpass filters are constructed in the same manner as dichroics; in essence they are both a long-pass and a short-pass filter put together. The number of layers used in the bandpass determines how quickly the filter changes from passing light to blocking light and the amount of blocking (optical density) provided.

It is almost impossible to design the optical components of a flow cytometer to ensure that there is no scattered stray light from the lasers. Bandpass filters must ensure that this stray light does not reach the detector. This is accomplished by specifying that the bandpass has an optical density (O.D.) of 6 (O.D. is logarithmic. An O.D. of 6 indicates attenuation of 10^{-6}) at the nearby laser line. For this, the bandpass must be some distance away from the laser line. What that distance is depends on the wavelength being considered and the skill of the filter manufacturer in depositing multiple layers onto glass. In reality, this will require about 10–15 nm spacing away from the laser line. For example, an HQ 525/50 bandpass (Chroma Technology Corp, Rockingham, Vermont) will block the 488-nm laser line and is an optimal filter for measuring FITC, but a broader filter centered on 525 nm will start to pass the laser light. It should be noted that the bandpass filters need to block laser light from all lasers on the instrument. Even small amounts of scattered light could easily raise the background of the desired signal to unacceptable levels.

D. Photomultiplier Tubes

Light signals in flow cytometers are converted to electronic pulses mainly by the use of PMTs. These are low-noise and high-amplification (the amplification being determined by the voltage it operates at) devices. The light-converting element in the PMT (the cathode) can be made from different materials. Some are more sensitive in the red part of the spectrum than others. To optimally measure red-emitting fluorochromes, it is imperative to use "red-sensitive" PMTs and all flow cytometer manufacturers now do so. What is not as well known is that these are more efficient across the entire visible spectrum. Thus, using these throughout the flow cytometer, except in the scatter channels where sensitivity is not an issue, will result in better overall measurements.

IV. Maximizing Signal Intensities

In most cases, the optical bench of a flow cytometer cannot be changed, so this aspect of optimization can rarely be conducted. However, if a bench layout can be designed, it should be kept in mind that each time the emitted light passes a dichroic mirror, one has to estimate about 10% of light loss. In contrast, there is almost no loss to the reflected light. Hence, each optical layout design must aim at minimizing the number of dichroics emitted light has to pass before it is collected at a PMT. The other consideration to keep in mind is that long-pass dichroics in general have a sharper transition from reflection to transmission than short-pass filters. Thus, using long-pass instead of short-pass dichroics may allow better light collection.

V. Maximizing Signal Collection Quality

The remaining design issues concern the choice of dichroic mirrors and bandpass filters to ensure that each PMT receives maximal amounts of the signal to be measured and minimal amounts of other light. An optimally designed dichroic and bandpass filter allows most of the emission peak through to the PMT, thereby facilitating maximal amount of light collection. To choose appropriate bandpass filters, we should know the exact emissions spectra of the dyes to be measured. Useful sources for emission spectra include the Molecular Probes web site (www.probes.com) and the Becton Dickinson spectra viewing applet (www.bdbiosciences.com/spectra). Because bandpass filters are generally easily exchanged, it is desirable to have bandpass filters that are optimized for each fluorochrome used. It might be useful to have two bandpass filters for the same PMT, if more than one fluorochrome is measured in different experiments. For example, PE-Cy5 and PerCP are usually measured on the same PMT, although their spectral characteristics are distinct and using them with dyes that have a further red spectrum would require different bandpass filters for optimal signal detection.

Unfortunately, the choice of bandpass filters that allows maximal light collection for each of the fluorochromes cannot be the only design criterion. In addition to choosing bandpass filters on their ability to let desired light through, they must also block unwanted light. Significant amounts of unwanted light come from two sources: emission from other fluorochromes and scattered laser light. With regard to the laser light, probably the most problematic aspects are those concerning the red laser choices. To name but one example, the choice of a dye laser tuned to give an emission of 595 nm has the great advantage of exciting TR/Alexa 594, a dye with little spectral overlaps in any dyes other than APC, making it a good color to add to a multicolor staining combination. However, having to block a 595-nm light source means that one has to sacrifice some of the PE signal because it has a significant signal at that wavelength (see Fig. 1). As discussed earlier, one needs to stay about 10 nm away from the laser light wavelengths to completely block any scattered laser light. Thus, the PE collection in a setup containing a 595-nm emitting laser cannot be done with a filter that collects light above 585–590 nm. In this case, the trade-off might be well worth it, because PE is one of the "brightest" dyes available, and thus, a small reduction in its measured signal might not affect its usefulness. This example illustrates the point, however, that by just combining multiple excitation sources a seemingly unrelated measurement can be negatively affected.

Another inherent light source is the autofluorescence emitted by the cell/molecule to be analyzed in flow cytometry. For the purposes of hardware design, one must consider that cellular autofluorescence is triggered particularly by laser light in the UV and violet. Cellular autofluorescence is also particularly strong in the violet and blue range but is very low in the red and far-red channels. Considerations of the autofluorescence levels are important when determining the precise characteristics of the bandpass filters. Specifically how broad should one choose the filters? Is it advantageous to design wide filters that enable the collection of maximal amounts of light, or should narrow filters be chosen that collect only the peak emission, thereby reducing collection of other unwanted light signals?

There seems to be more than one correct answer to this question, and it might well depend on exactly the parameter one tries to optimize. Obviously, the goal is to achieve maximum signals from the positive events and minimum "background" staining from the negative events. When collecting a narrow band around the peak emission, one achieves the largest differential of stain versus background. However, the question is whether this narrow band of peak emission alone is of sufficient intensity to segregate negative from positive events or whether more signal needs to be collected. Because of the relatively high levels of autofluorescence in some channels, better measurements might not be achieved by increasing the width of their bandpass. Although the positive signals might be stronger, any differential expression between negative and positive staining could become inconsequential because of the large increase in autofluorescence that is also measured at that PMT. On the other hand, collecting more light using a broader filter will aid in producing better compensated data, because the broadening of the compensated measurements (see later discussion) will be reduced.

More complex sources of unwanted light come from the emission signals of other simultaneously measured fluorochromes, so-called *spillovers*. Those signals might be caused by overlaps in emission spectra between two dyes excited by the same laser (Fig. 1). Alternatively, they might be due to a second excitation peak or a broad excitation peak of a dye. An example of the latter is PE-Cy5. This tandem dye is excited via the PE component by 488-nm laser light, which transfers energy to the Cy5 component. Any Cy5 components can also be excited directly by a red laser with emission spectra similar to those of APC. Hence, significant signals from PE-Cy5 will appear in the APC channel. Proper design of bandpass filters and dichroics can avoid, or at least minimize, spectral overlap problems resulting from excitation with the same laser. Overlaps resulting from "double" excitation are difficult to address in hardware design.

A mathematical process termed *compensation* must correct all spectral overlaps that cannot or are not addressed by the design of the optical paths. Although compensation allows signal correction, it cannot correct the actual differences in signals measured because of the additional spillover signals from other dyes. One of the negative, noncorrectable consequences of these additional signals is a broadening of the spillover channels. Figure 2A shows uncompensated data of human peripheral blood lymphocytes stained with anti-CD8 Cy7-APC. There was significant spillover of this reagent into the APC measurement channel because of the emission of the APC component of this tandem dye. With the filters and gains used to make this measurement, the spillover into APC was 12% (the ratio of Cy7-APC/APC). When compensated, the $CD8^+$ cells had a significantly higher background in APC than the $CD8^-$ cells (Fig. 2B). Importantly, this was not because compensation was not applied correctly, as shown in Fig. 2C. Here, the compensated data are displayed on a rescaled plot that combines both a linear and a logarithmic display. This allows those data to be viewed better than those that are close to the bottom measurement range (Tung *et al.*, 2004). One can see that the data are correctly compensated, because the mean APC signals of the $CD8^-$ and $CD8^+$ populations are equal. However, compensation introduced a larger spread of the APC signal among the $CD8^+$ compared to the $CD8^-$ cells. Although the spread of the $CD8^-$ signal in APC is approximately 1 decade, the spread of the $CD8^+$ signal is almost 2 decades. This broadening in effect reduces the usefulness of a given channel, because it makes it difficult or impossible to distinguish dull-positive from negative staining. The red and far-red dyes are more affected by this compensation effect, because the coefficient of variation (i.e., the measurement spread) is inversely proportional to the square root of the amount of light (photons) for that measurement (Steen, 1992). And as discussed earlier, only relatively low numbers of photons are collected for the red dyes. Whenever possible, spectral overlap problems should be addressed first in

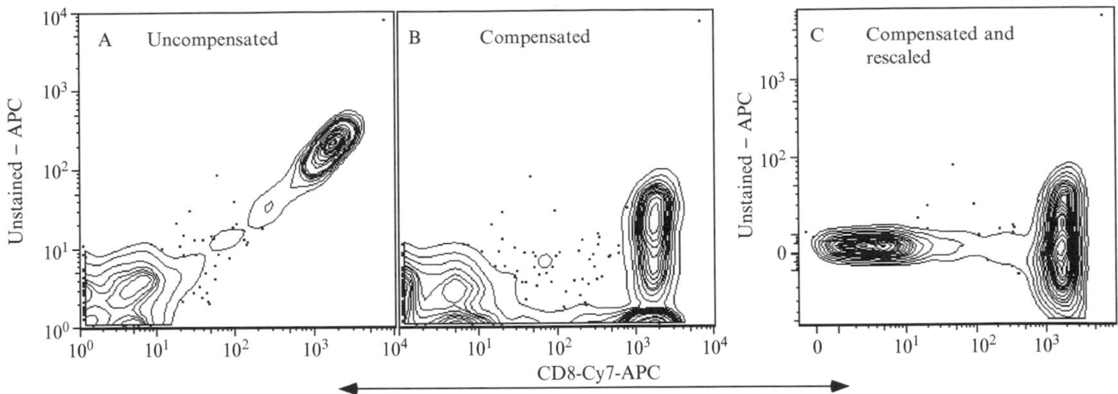

Fig. 2 Effects of compensation on background measurement. Human peripheral blood lymphocytes were stained with an anti-CD8 (RPA-T8) Cy7-APC reagent and analyzed on a FACSDiva (Becton Dickinson). Data were analyzed and compensation was applied using FlowJo (Treestar Inc) software. (A) A 5% contour plot of uncompensated data. (B) A 5% contour plot of compensated data. (C) A 5% contour plot of compensated data plotted with logarithmic and linear y-axis scaling to show the measurement spread of the CD8$^+$ cells that occurs when correctly applying compensation.

hardware, because it will increase the quality of the true measured signals. Remaining issues then need to be addressed in the reagent choices for each multicolor staining combination.

VI. Emission Collection Optimization

Figure 3 provides an example in which an optical path was reconfigured resulting in increased signals from all of the fluorochromes measured within this path. The figure shows the original layout of the 488-nm excited emission collection optics and the optimized emission paths on a jet-in-air multicolor flow cytometer (MoFlo, Cytomation, Fort Collins, Colorado). The changes that were undertaken to optimize light collection were as follows: (1) Change of a dichroic that splits light between FITC/PE and CY5-PE/Cy5.5-PE/Cy7-PE (from a 605SP to a 630SP) effectively enhanced the PE signal 1.5 times. The increased signal was likely due to the use of better quality dichroics rather than their declared specificities. (2) Moving the Cy5-PE detector to allow the use of long-pass rather than short-pass dichroics. This improved Cy7-PE signals despite that one more component is now in its path. (3) Allowing maximal signal detection of Cy5-PE by better matching the specificity of the dichroic with the bandpass filter. (4) Changing the light path and installing HQ filters also significantly increased Cy5.5-PE signals. The results of the comparison are summarized in Table II and are explained in more detail later in this chapter.

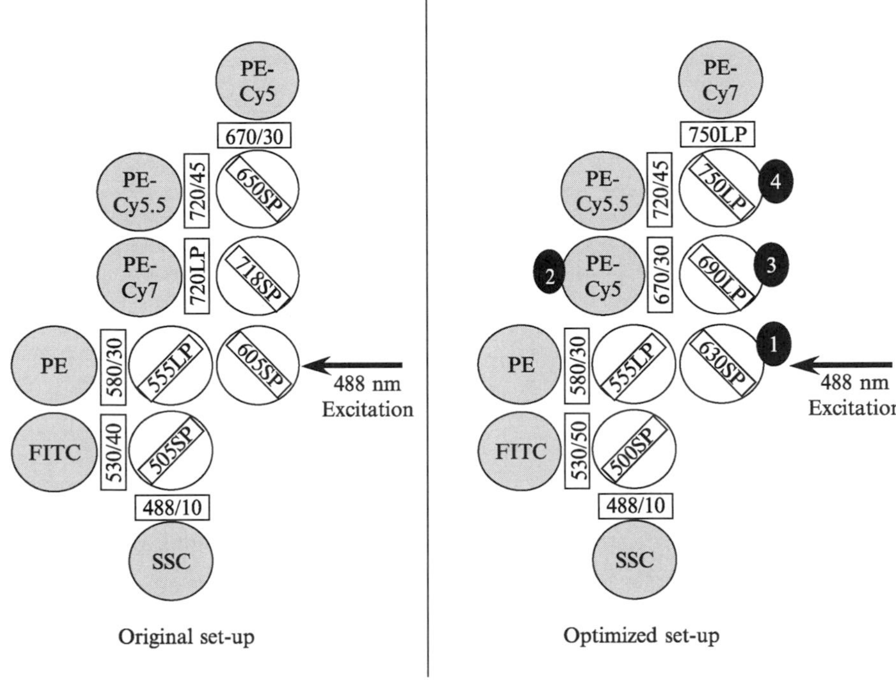

Fig. 3 Optimization of emission optics. Shown is one arm of an optical bench of a MoFlo Cytometer (Cytomation, Ft. Collins, Colorado) that collects emissions from fluorescent signals excited by a 488-nm Enterprise (Coherent) laser line providing 200 mW of laser power. Other lasers on this machine are a Spectrum laser (Coherent) tuned to give 647 nm with a power of 200 mW and a violet-enhanced Krypton laser tuned to 407-nm laser light and 150-mW power. Left panel shows the original set-up as installed on purchase. The right panel shows the optimized set-up. Gray circles represent photomultiplier tubes. Fluorochromes for which light collection was optimized are indicated. White circles represent dichroics and squares indicate bandpass filters. Optical characteristics indicated. LP, long-pass dichroics; SP, short-pass dichroics. Results of comparative measurements using these two optical paths are summarized in Table I. Significant changes between the layouts are indicated by numbers: (1) Change of dichroic. (2) Move of Cy5-PE and Cy7-PE detectors. (3) Change in light pass from SP to LP. (4) Optimization of dichroic for Cy5.5-PE light collection.

It is difficult to measure and predict all sources of stray light that might interfere with the light collection. Therefore, identification of the best bandpass filters might involve measuring light collection with a set of different filters (see later discussion). In the example shown in Fig. 3, a wider bandpass filter in front of the PE PMT should have provided increased signals. When this was attempted using a 580/45 filter, the signal was greatly decreased because of large increases in "background" (data not shown). The source of this increase in background light was not immediately apparent and thus could not have been predicted. Nonetheless, studying the emission spectra of the dyes to be measured and knowing the

Table II
Comparison of Optical Emission Paths

	Original MFI pos	Original MFI neg	Ratio pos/neg	Optimum MFI pos	Optimum MFI neg	Ratio pos/neg	% Original signal
FITC	321[a]	7.2[b]	44	334	7.1	47	106
PE	104	2.6	41	411	6.5	63	155
Cy5-PE	34	2.4	14	150	3.2	47	336
Cy5.5-PE	26	2.5	11	54	3.0	18	169
Cy7-PE	35	2.5	14	54	3.1	18	127

Note: Comparison of fluorescent signals collected on a MoFlo (Cytomation) using the original and optimized emission optics shown in Fig. 3. Excitation was done with a 488-nm laser line of an Enterprise laser (Coherent) at 200 mW. Cy5.5, cyanate 5.5; Cy7, cyanate 7; MFI, mean fluorescent intensity; FITC, fluorescein thiocyanate; PE, phycoerythrin.
[a]Mean fluorescent intensity of murine spleen cells positively identified with a direct conjugate to CD45R (FITC, PE, Cy5-PE), CD21 (Cy5.5PE), or CD4 (Cy7PE).
[b]MFI of cells not identified (unstained) with the conjugate.

wavelengths of the lasers that are used to excite these dyes provides the information needed to design a good optical layout that might just need minor adjustments.

VII. Testing Emission Optics

Having summarized the main fluorochrome and hardware issues that need to be considered when designing the optical bench of flow cytometers, in this last part we describe how to determine whether any given alteration in the optical layout improves the measurements, or for new measurements, whether they create HQ data.

Any data acquisition is best performed on a machine that has been appropriately standardized. This is crucial when evaluating the performance of the instrument. Basic set-up conditions should be consistent between experiments so that the only variable is the actual specimen to be analyzed. This means adjusting laser power and PMT voltages so a given signal always results in measurements of the same signal intensity. Practically, this means adjusting the PMT voltages while acquiring data from a set of beads with known signal intensities to achieve the same signals from run to run. Using cells rather than beads for these adjustments should be avoided, because of their inherently greater variability. PMT voltages and laser power should be recorded. They can be a useful diagnostic tool for data evaluation and for monitoring the performance of the machine. There is little reason to adjust laser power rather than PMT voltages for day-to-day signal calibration. However, the optimal laser power should be established for each of the dyes to be analyzed at least once. Obviously, this is possible only on machines with adjustable laser powers.

A. PMT Voltage Optimization

By convention, the PMT voltages are adjusted so most of the unstained or negative events fall within the first decade of a 4-decade logarithmic plot. This is best done with a set of unstained cells of the same type that are run in the experiments. Different cell populations differ in their levels of autofluorescence, which could significantly affect optimal PMT voltage settings. One should avoid choosing PMT settings that result in most of the negative events sitting on the left-hand axis. This might not always be possible, however, particularly for the far-red dyes, because there is often very little light collected for the negative events. Accurate compensation requires an accurate measurement of the mean fluorescence intensities of the negative events. If most events are on the axis, this measurement cannot be done correctly.

B. Laser Power Optimization

The general idea is simple: Single-color–stained cells are measured at different laser powers and the "brightness" is recorded and compared to choose the optimal power. Both the background signal levels and the background signal widths need to be accounted for. For most phenotyping applications, one is interested in measuring signal above the brightest of the background (unstained) signal. There are various ways of doing this; one is shown in Fig. 4. The first step is to determine the widths of the background (unstained) peak for each dye at the lowest laser power to be tested. This should be done by determining the distance between the 50th and the 95th percentile of the negative peak "W" (this would be well off the measurement axis). Then one should determine the distance "D" between the median of the stained and the unstained cells. By dividing D by W, one obtains a good measure of the differential between the negative and positive population. These measurements and calculations should then be repeated with increasing laser powers.

An example of optimal laser power determination is shown in Table I. With a few exceptions, the higher the excitation power, the brighter the fluorochrome measurement. In this example, the exception is PerCP, which is known to easily "bleach" at higher laser powers. However, at a certain level, increased laser power increases both the background and the stained signal equally, thus not further increasing the brightness. For the fluorochromes commonly used in flow cytometry, this tends to occur between 400 and 800 mW (data not shown).

C. Testing Emission Optics

Having standardized the machine, a run using the standardization beads should be recorded. Then a single-color–stained sample for each of the fluorochromes to be tested should be run and data recorded. The purpose is to measure the mean fluorescence intensities of negative and positive events for each PMT. Therefore, the samples should be stained so a good proportion of both negative and positive

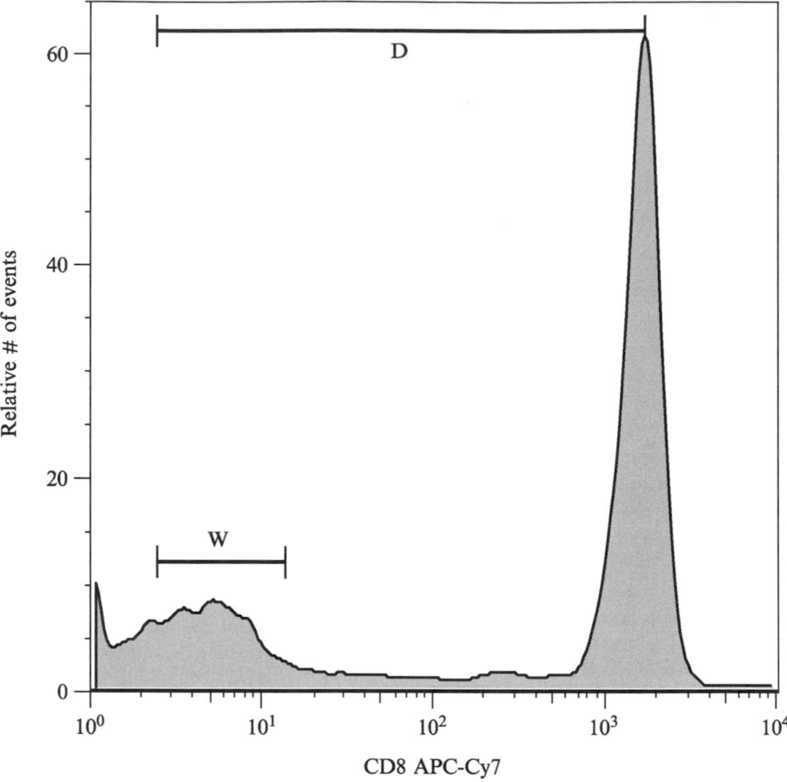

Fig. 4 Determination of the "brightness (signal/noise ratio) of a signal. Indicated on the histogram plot is the distance, *W*, between the 95th and the 50th percentiles of human peripheral blood mononuclear cells not stained for CD8. *D* is the distance between the median fluorescence intensities of the $CD8^+$ and the unstained cells. D/W can be calculated as a measure of the "brightness" of the (CD8 APC-Cy7) signal.

events can be recorded. An unstained sample should also be run, because it helps to determine background fluorescence intensities (excluding background staining) in each channel. Antibody-capture beads might also be used instead of cells for such an analysis, particularly if tandem-dye conjugates are used or when the cell population to be tested is very small. These commercially available beads capture the conjugate onto the bead surface usually via anti–light-chain antibodies and thus are usable with a large number of different reagents.

Any changes in dichroics and filters should then be made and the identical samples should be rerun and recorded without changing any other machine settings. In the case of immunophenotyping with a "new" dye, it is advisable to choose an antibody conjugated to that dye that is known to strongly identify a substantial fraction of cells or to use antibody-capture beads. If the new set-up results in events being outside the measurable range (appearing on the right- or left-hand axis of a data plot), readjust the PMT voltages, record the new settings, and perform a third run under those changed conditions.

D. Data Analysis

The first step in data analysis requires determination of the mean fluorescent intensities (MFIs) of the positive and the negative cell populations for each single-stained sample. Inconsistencies in the gating strategy must be avoided because they can greatly affect data interpretation. In some software packages, like FlowJo (Treestar Inc), fluorescence intensities on a certain percentile of events can be retrieved. For example, in a stain in which 30% of cells are negative and 70% are positive, one would determine fluorescence intensities of the 15th and 65th percentile of events to obtain the median fluorescence value of the bottom and top population. Any rerun of that sample would then be analyzed the same way. Increases in fluorescent signals from the positive events are useful only if the signal from the negative events (background) does not also increase significantly. Hence, the ratio between the fluorescent signals from the positive and the negative population is a better indicator of performance than simply measuring the fluorescent signal of the positive population. Alternatively, as discussed earlier, the differential of the widths of the negative and the positive signals might be determined. Increases in the median fluorescent ratios or larger D/W ratios would indicate a better separation between positive and negative events, hence, better performance, whereas a decrease in the ratios would indicate reduced measurement performance.

Table II provides a summary for the comparison of the two optical paths shown in Fig. 3. The data were generated using individual single-stained samples. In this case, mouse spleen cells were stained with an antibody conjugate recognizing T- or B-cell markers (see legend to Table II). In each sample, a significant proportion of cells were positive and negative for that conjugate. The MFIs of the negative and positive cells for each of the stains were recorded. The differential (ratio) of the positive and negative signals was calculated (MFI positive/MFI negative). The larger the ratio, the better the ability to distinguish positive from negative events. The same samples were run under the original and the altered (optimized) conditions and ratios were established for each fluorochrome and each run. A proper comparison of the performance of the optical path for each fluorochrome was then possible by comparing the calculated ratios. In the example given, the relative minor changes introduced produced signals up to 3.6 times (Cy5-PE) as good as those measured before optimization.

If one uses as analysis criterion solely the position of the positive stained cells, the results might be incorrectly interpreted. Figure 5 shows a comparison of two set-ups in which the bandpass filter for cascade blue detection on a BD FACSAria (Becton Dickinson, San Jose, California) was changed from a standard 440/40 bandpass filter to an HQ 440/50 filter (Chroma, Fort Collins, Colorado).

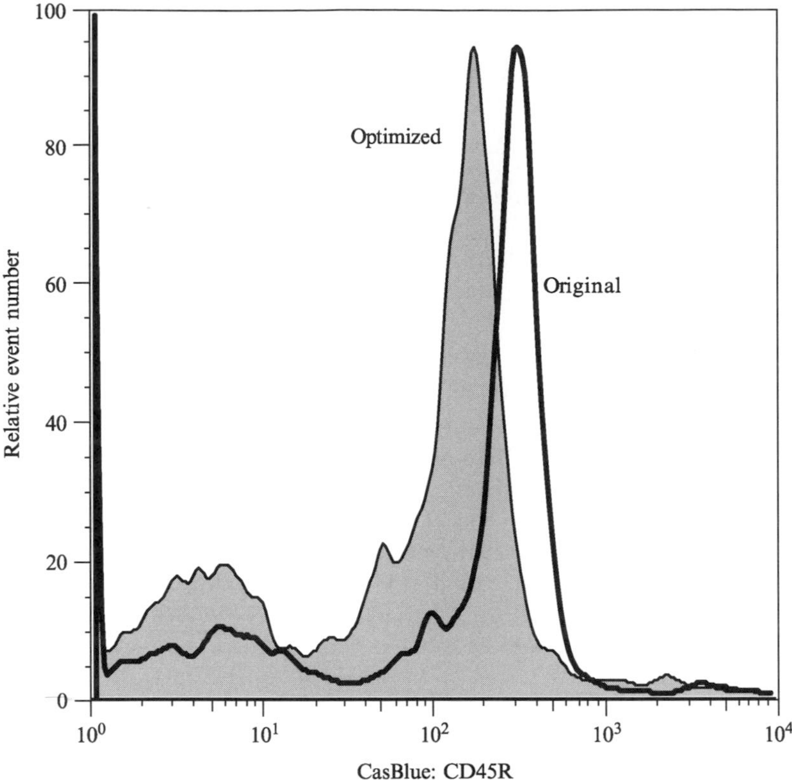

Fig. 5 Optimization of emission optics might reduce background staining rather than increase fluorescent intensities of positive events. Shown are the histogram profiles of mouse splenic lymphocytes stained with a monoclonal antibody (RA3-6B2) to mouse CD45R (B220) conjugated to cascade blue and analyzed on a FACSAria (Beckton Dickinson). In the original setup (gray histogram) a 440/40 bandpass filter was used, in the optimized setup (thick line, open histogram) a high-quality 440/50 bandpass filter. Change of the bandpass filter resulted in a reduction of the mean fluorescent intensity (MFI) of the positive population from 313 to 148. At the same time, however, the MFI of the negative population decreased from 3.3 to 1.0. This resulted in a large increase of events on the left-hand axis (*arrow*) and creating an increase in the ratio of positive to negative stained cells from 101 to 147, thus a 46% better separation of positive to negative events.

The positive peak in the "original setting" is shifted about twofold to the right of that run under the "optimized" setting. However, the mean fluorescent signals of the negative populations under the latter condition are threefold lower compared to the original run. Thus, the optimized setting generates a nearly 1.5-fold better separation of positive and negative events (147-fold difference vs 101-fold difference). Because the analyses were done under identical PMT voltage settings, many of the negative cells under the altered conditions were recorded on the left-hand axis, indicating lower background staining. To visualize the negative populations better, one could increase the PMT voltage to bring these cells on scale.

In fact, any large increases or decreases in signal detection (for either positive or negative signals) likely require PMT voltage adjustments. Similar or increased signal intensities that are achieved with reduced PMT voltages indicate increased light collection. Although this is usually desirable, it can become a problem if too much background light is collected that reduces the differential between negative and positive events. Increases in backgrounds on single-stained samples are caused usually by stray laser light or by increased collection of autofluorescence. The former can be easily tested by switching off, one-by-one, the lasers that are not exciting the dye of interest (one might have to change the laser that measures forward light scatter). Nonspecific shifts for certain PMTs can easily be observed with the unstained samples.

To determine how much of a certain positive signal "spills over" and is measured on the other PMTs, their signal should be recorded also on all other measurement channels. Thus, for each stained cell population, one determines the mean fluorescence intensity for each of the channels. Any large changes in fluorescent signal intensities on channels that are not designed to collect those signals indicate increased "spillover" and thus are a requirement for increased compensation.

Final analysis of any multicolor setup will have to be done with a multicolor-stained sample or a mix of single-colored beads. It is crucial to determine how compensation will affect each signal and how the combination of multiple colors affects the background measurement on each channel. This is affected as much by the design of the staining combination as by the optical paths of the flow cytometers. A properly set-up optical bench that is optimized for the particular set of fluorochromes to be analyzed is an important starting point for obtaining high quality data that can reveal novel information.

Acknowledgments

N. B. thanks Paul Millman of Chroma Technologies for his generous help with providing filters and dichroics for optimizing a MoFlo flow cytometer, and David R. Parks and Dick Stovel, Stanford University Flow Cytometry Development Group, for their help and guidance. M. B. acknowledges Larry Duckett of BD Biosciences for configuring an LSR-II with blue and green lasers.

References

Anderson, M. T., Baumgarth, N., Haugland, R. P., Gerstein, R. M., Tjioe, T., and Herzenberg, L. A. (1998). Pairs of violet-light-excited fluorochromes for flow cytometric analysis. *Cytometry* **33,** 435–444.

Baumgarth, N., and Roederer, M. (2000). A practical approach to multicolor flow cytometry for immunophenotyping. *J. Immunol. Methods* **243,** 77–97.

Bigos, M., Baumgarth, N., Jager, G. C., Herman, O. C., Nozaki, T., Stovel, R. T., Parks, D. R., and Herzenberg, L. A. (1999). Nine color eleven parameter immunophenotyping using three laser flow cytometry. *Cytometry* **36,** 36–45.

De Rosa, S. C., Herzenberg, L. A., and Roederer, M. (2001). 11-color, 13-parameter flow cytometry: Identification of human naive T cells by phenotype, function, and T-cell receptor diversity. *Nat. Med.* **7,** 245–248.

Roederer, M. (2001). Spectral compensation for flow cytometry: Visualization artifacts, limitations, and caveats. *Cytometry* **45,** 194–205.

Roederer, M., De Rosa, S., Gerstein, R., Anderson, M., Bigos, M., Stovel, R., Nozaki, T., Parks, D., and Herzenberg, L. (1997). 8 color, 10-parameter flow cytometry to elucidate complex leukocyte heterogeneity. *Cytometry* **29,** 328–339.

Steen, H. B. (1992). Noise, sensitivity, and resolution of flow cytometers. *Cytometry* **13,** 822–830.

Tung, J. W., Parks, D. R., Moore, W. A., and Herzenberg, L. A. (2004). New approaches to fluorescence compensation and visualization of FACS data. *Clin. Immunol.* **110,** 277–283.

Zlokarnik, G., Negulescu, P. A., Knapp, T. E., Mere, L., Burres, N., Feng, L., Whitney, M., Roemer, K., and Tsien, R. Y. (1998). Quantitation of transcription and clonal selection of single living cells with beta-lactamase as reporter. *Science* **279,** 84–88.

CHAPTER 2

Two-Photon Tissue Cytometry

Timothy Ragan,*,† Ki Hean Kim,* Karsten Bahlmann,*,† and Peter T. C. So*

*Massachusetts Institute of Technology
Department of Mechanical Engineering
Division of Biological Engineering
Cambridge, Massachusetts 02139

†TissueVision, Inc.
Somerville, Massachusetts 02143

I. Introduction

Although the native environment of cells is within the three-dimensional (3D) environment of their host tissue, standard preparation for both flow and image cytometry involves the dissociation of cells from their tissue matrix. Flow cytometry (FCM) is conducted with cells in suspension, whereas image cytometry studies are typically conducted in two-dimensional (2D) culture. It is well known that the biochemical, mechanical, and intracellular inputs present in a tissue can all dramatically affect the behavior of a cell. Once the cell is removed from this

METHODS IN CELL BIOLOGY, VOL. 75

0091-679X/04 $35.00

tissue environment, these inputs are lost, along with much of the native cell morphology and any metainformation about its place in the 3D architecture of the tissue. The differences in the cellular behavior in their native tissue setting and in an artificial environment are relevant in the interpretation of clinical and basic research results. In pharmaceutical development, these differences are also partially responsible for the high failure rate of drug candidates in tests at the tissue level inside organisms. For these reasons, there is a tremendous need for the development of cytometric tools that can effectively characterize cells in their native tissue state.

Two-photon microscopy (TPM) is a high-resolution fluorescence microscopy technique that provides 3D images of cells and tissues. Its strengths include inherent 3D sectioning, low background, high penetration depth, and low phototoxicity. It is particularly well suited for imaging tissues. By incorporating high-speed image capabilities, it provides images of large populations of cells within their native tissue environment using TPM and extends standard 2D image cytometry into 3D. We have termed this technique *two-photon tissue cytometry*.

In this chapter, we give a brief overview of some of the early cytometry techniques such as flow and image cytometry. In addition, we discuss the underlying principles and instrumentation of two-photon tissue cytometry and describe some emerging applications of this technology.

II. Technical Development of Two-Photon Tissue Cytometry

A. Cytometry

Cytometry is defined as the quantification of the physical and biochemical features of cells. The characteristics of cells most commonly measured are optical such as fluorescent and scattering properties, which relate back to the morphological, biochemical, and gene expression states of the cells. For the purposes of this chapter, we focus on cytometry techniques that are capable of examining large populations of cells in order to generate quantitative statistics of the population.

1. Flow Cytometry

One of the earliest cytometry techniques that enabled researchers to quantitatively and statistically assay large populations of cells was FCM (Givan, 2001; Kamentsky and Melamed, 1967; Melamed, 2001; Van Dilla *et al.*, 1969). It has proven itself to be a valuable tool in both the research and the clinical setting.

In FCM, a suspension of cells is prepared and then flowed through the focus of a laser beam or other excitation source in a fine stream so individual cells sequentially pass through the focus. Both scattering and fluorescence parameters can be collected from the interaction of the cell within the excitation source (Rieseberg *et al.*, 2001). The instrument has an excellent time response and can examine millions of cells in an experimental run. Furthermore, because the cells are examined individually, it is possible to investigate rare cells that would otherwise remain obscured in the larger population. Cellular populations as large as 1×10^8 have been examined at a rate on the order of kilohertz (Bajaj *et al.*, 2000). The major strengths of FCM can be summarized as follows: (1) a high-throughput rate that can characterize large populations on a cell-by-cell basis, (2) a high specificity that allows frequencies as low as one cell in 10^7 to be identified, and (3) with the addition of an in-line cell sorter, cell populations can be physically separated based on their morphological and spectroscopic signatures.

Despite these capabilities, FCM does have some disadvantages: First, it is generally not possible to monitor time-dependent changes of a given cell because once a particular cell has been analyzed, it is no longer possible to uniquely identify the same cell again for further analysis. This prevents relabeling cells and correlating new measurements with previous measurements on a cell-by-cell basis. Second, FCM has no spatial resolution, which excludes the direct measurement of any spatial features, and morphological parameters have to be inferred from measurements such as scattering coefficients. Finally, the isolation procedure required to prepare a suspension of cells suitable for analysis in FCM is often laborious and can introduce artifacts into the study.

2. Image Cytometry

Motivated by some of these limitations, the technique of image cytometry was developed (Darzynkiewicz *et al.*, 1999; Kamentsky and Kamentsky, 1991; Kamentsky *et al.*, 1997). It shares some important characteristics with FCM, like the ability to investigate large cell populations and acquire multiparameter data on individual cells. In image cytometry, a population of cells is imaged on a microscope coverslip or other type of solid support rather than analyzed in suspension. It differs from standard microscopy studies in that a much larger region is imaged at lower resolution in order to be able to gather data on a large statistically relevant population of cells. For example, a laser scanning cytometer described by Kamentsky (2001) used a 25× objective and had a spot size for the excitation volume of 2.5 μm full width half maximum in the radial direction and extent of 171 μm in the axial direction. A 0.5-μm step size was employed for the raster scan and an area of 4.25 mm^2 was imaged per minute (Kamentsky, 2001).

Image cytometry offers several advantages over FCM. First, because it is an imaging technique, it acquires morphological information about the cells in the population. This not only provides valuable information about the biological state of the cell, but it also makes it possible to spatially localize and associate biochemical signals with specific cellular compartments such as the nucleus or the cytoplasm and to exclude cellular and matrix debris from the analysis. An additional advantage of image cytometry is the ability to relocate cells of interest. This makes it feasible to do time-series studies or to relabel the cells and acquire additional parameters on the cell. A disadvantage of image cytometry is that it is generally slower than FCM. However, the performance of imaging cytometers is increasing and the gap in processing speed between image cytometry and FCM is decreasing. For example, a laser scanning cytometer has previously demonstrated the capability to detect fetal nucleated red blood cells with frequencies as low as 1 in 20 million cells at a frequency of almost a 1-kHz rate (Bajaj *et al.*, 2000).

An additional advantage of image cytometry is the capability to study cells in intact tissues (Gorczyca *et al.*, 1997; Hendricks, 2001). As previously mentioned, this has very important research and clinical implications because image cytometers can be used as an important quantitative aid to standard histopathology.

3. Tissue-Based Cytometry

Investigating the behavior of cells in intact tissues offers several advantages. First, because it is not necessary to extract the cells from the tissue by mechanical or enzymatic means, the risk of introducing artifacts into the analysis is minimized. Second, cellular morphology is drastically different between cells inside native tissues and the same cells grown on 2D tissue culture. Third, the removal of cells from a tissue also necessarily eliminates many of the mechanical and biochemical signal inputs that are known to dramatically affect cell behavior and gene expression profile. Fourth, cell–cell interactions inside an intact tissue can also be an important factor in cell function that is impossible to study once the cells have been removed from the tissue. Finally, imaging intact tissues preserves the 3D architecture of the tissue sample and provides a great deal of spatial and contextual information that is otherwise lost.

All of these factors will affect the behavior and state of the cell and, depending on the question being asked, could easily have important biological or clinical ramifications. In many cases, the gene expression profile of the cell will have been changed from its native state in the tissue. Many cell-based studies may, thus, be potentially predicated on a fundamentally faulty data set. For example, Bennett *et al.* (1996) and Piston *et al.* (1999) used TPM to monitor nicotinamide adenine dinucleotide phosphate (reduced form) (NADPH) level activity in β cells in intact pancreatic islets. Previous research on β cells dispersed onto coverslips had revealed significant β-cell heterogeneity in response to differing glucose levels as indicated by the NADPH activity within the cells. These 2D culture experiments led to metabolic models of glucose metabolism based on stepwise recruitment of individual β cells. However, when Bennett *et al.* (1996) imaged β cells within intact islets, they found a far more homogeneous glucose response, which suggests the model based on the results in 2D culture may not be relevant to the physiological insulin response in the pancreas of an organism. This has obvious implications for the pharmaceutical design of agents to treat diabetes. Another example is in the field of liver research where researchers have been unable to construct a liver model based on hepatocytes grown in 2D culture. To approximate the expected behavior of hepatocytes, researchers have had to engineer 3D liver cell constructs (Powers *et al.*, 2002).

Another important area in which the *in situ* tissue properties of cells are important is in field of cancer research. Cancer is a disease that has a very strong spatial component to its etiology (Liotta and Kohn, 2001). Cancer cells can invade the stroma of the surrounding tissue and recruit nonmalignant cells to differentiate and support the growing tumor. The arrangement of healthy tissue boundaries becomes pathogenic as the expression profiles of the surrounding cells are altered by cell signaling from the malignant cells (Gohongi *et al.*, 1999; Knezevic *et al.*, 2001). The realization of the importance of the *in situ* properties of cancer, in terms of both the expression profiles of the individual cells and their spatial arrangement with respect to one another, has led to the development of such techniques such as laser capture microdissection (Bonner *et al.*, 1997; EmmertBuck *et al.*, 1996). Laser capture microdissection is a tissue microdissection that physically removes cells from the tissue on a cell-by-cell basis depending on the cell's morphological or spectroscopic properties or on its location within the bulk tissue. In this way, we can directly assay the gene expression profile of cells from spatially distinct portions of a tissue.

Other important research areas in which it is vital to study cells within the context of the tissue in which the process is occurring include wound healing and embryology. In summary, traditional cell culture–based studies provide a wealth of information, but for many biological processes, they provide only a partial picture of the phenomena under study.

Despite the obvious advantages of studying cells within tissue, most cytometry studies have been limited to cells in suspension are 2D culture. This is mainly due to technical limitations in the traditional instrumentation, which cannot easily characterize large populations of cells within the 3D environment of macroscopic thick tissues. Most high-resolution imaging techniques are not well suited for tissue imaging. Standard wide-field microscopy suffers from very poor resolution and contrast when attempting to image more than several microns from the surface of the tissue. We can use deconvolution techniques, but these are generally computationally prohibitive and work well only for thin specimens. Confocal laser scanning microscopy is a technique that has shown some success in imaging intact tissue. However, in practice the axial imaging depth is less than 40 microns for many tissue types. In addition, it has high phototoxicity and photobleaching of out-of-focus planes, which further limits its usefulness. Another imaging technique that has proven far more suitable for studying tissues is TPM.

B. Two-Photon Excitation Microscopy

TPM is a proven optical microscopy technique for imaging in highly scattering media and has a depth penetration of more than 200 μm. TPM yields 3D-resolved microscopic images in deep tissues with minimal photodamage (So *et al.*, 2000). In addition to providing tissue morphological information, two-photon spectroscopy based on autofluorescence allows the monitoring of tissue metabolism and the identification of biochemical composition (Masters *et al.*, 1997). Two-photon imaging and spectroscopy provide tissue structural and functional information. Denk *et al.* (1990) introduced two-photon excitation microscopy in which fluorophores can be excited by the simultaneous absorption of two photons, each having half the energy needed for the excitation transition. Because the two-photon excitation probability is significantly less than the one-photon probability, two-photon excitation occurs only at appreciable rates in regions of high temporal and spatial photon concentration. The high spatial concentration of photons can be achieved by focusing the laser beam with a high numerical aperture objective to a diffraction-limited spot. The high temporal concentration of photons is made possible by the availability of high-peak power mode–locked lasers.

In general, TPM allows 3D biological structures to be imaged with a resolution comparable to that found in confocal microscopy but with a number of significant advantages: (1) Conventional confocal techniques obtain 3D resolution by using a detection pinhole to reject out of focal plane fluorescence. In contrast, two-photon excitation achieves a similar effect by limiting the excitation region to a submicron volume at the focal point. This capability of limiting the region of excitation instead of the region of detection is critical. Because out-of-plane chromophores are not excited, photobleaching and photodamage of biological specimens is restricted to the focal point. (2) Two-photon excitation wavelengths are typically red shifted to about twice the corresponding one-photon excitation wavelength. The significantly lower absorption and scattering coefficients at these longer wavelengths ensure deeper tissue penetration. (3) The wide separation between the excitation and emission spectra ensures that the excitation light and the Raman scattering can be rejected without filtering any of the fluorescence photons. This sensitivity enhancement improves the detection signal/background ratio.

Depth discrimination is the most important feature of TPM. In TPM more than 80% of the total fluorescence intensity comes from a 1-μm thick region about the focal point for objectives with a numerical aperture of 1.25. Thus, 3D images can be constructed as in confocal microscopy, but without confocal pinholes that cause light loss in highly scattering specimens. This depth discrimination effect of the two-photon excitation arises from the quadratic dependence of the two-photon fluorescence intensity upon the excitation photon flux, which decreases rapidly away from the focal plane.

1. High-Speed Two-Photon Microscopy

The incorporation of high-speed imaging capacity is critical for imaging large areas in a reasonable amount of time. Conventional two-photon systems have image acquisition rates on the order of a few seconds, which is far too slow for imaging macroscopic samples. The basic challenge in the engineering of this system is to achieve high-speed imaging (>10–30 Hz) while maintaining sufficient detection sensitivity to visualize weakly autofluorescent tissue structures. It is instructive to survey the three categories of video-rate two-photon systems developed. The first category of video-rate two-photon imaging systems is based on the line-scanning approach. Image acquisition time is reduced by covering the image plane with a line instead of a point (Brakenhoff *et al.*, 1996; Guild and Webb, 1995). The line focus is typically achieved by using a cylindrical element in the excitation beam path. The resulting fluorescent line image is acquired with a spatially resolved detector such as a CCD camera. The main drawback associated with line scanning is the inevitable degradation of the image point spread function, especially in the axial direction.

A second method is based on the raster scanning of a single diffraction-limited spot using either a high-speed resonance scanner (Fan *et al.*, 1999) or a rotating polygonal mirror (Kim *et al.*, 1999). In this case, a large single point detector, such as a photomultiplier tube (PMT) or an avalanche photodiode, can be used. The spatial information is encoded by the timing design of the raster scan pattern. By replacing the CCD imager with a single pixel detector, the image resolution may be improved by removing the dependence on the emission point spread function. This is particularly important in turbid specimens, in which the scattered fluorescence signal is not confined to a single pixel of the CCD imager and may degrade the image resolution.

A third approach, termed *multiphoton multifocal microscopy* (MMM) (Bewersdorf *et al.*, 1998; Brakenhoff *et al.*, 1996), is analogous to the Nipkow disk-based confocal systems. This elegant approach is based on using a custom-fabricated multiple lenslet array in place of the scan lens that focuses the incident laser into multiple foci at the field aperture plane. The lenslets in the array are arranged in a pattern similar to the traditional Nipkow disk design. Upon the rotation of the lenslet array, the projection of the lenses will uniformly cover the field aperture plane. A CCD camera registers the spatial distribution of the resulting fluorescent spots and integrates them into a coherent image. The ability to excite multiple sample regions simultaneously reduces total data acquisition time. This technique still suffers from some resolution degradation as compared with the single-point method but is a significant improvement over line scanning. It also has the added advantage of being extremely robust.

2. High-Speed Two-Photon Microscopy Based on a Polygonal Scanner

We have previously designed a video-rate system based on raster scanning of a single diffraction-limited focal spot using a high-speed polygonal mirror (Fig. 1) (Kim *et al.*, 1999). A femtosecond titanium-sapphire is used to induce two-photon fluorescence. The laser beam is rapidly raster scanned across the sample plane by two scanners. A fast rotating polygonal mirror accomplishes high-speed line scanning (x-axis) and a slower galvanometer-driven scanner with a 500-Hz bandwidth correspondingly deflects the line-scanning beam along the sample's y-axis. The spinning disk of the polygonal mirror is composed of 50 aluminum-coated facets ($2\,mm^2$), arranged contiguously around the perimeter of the disk. In the

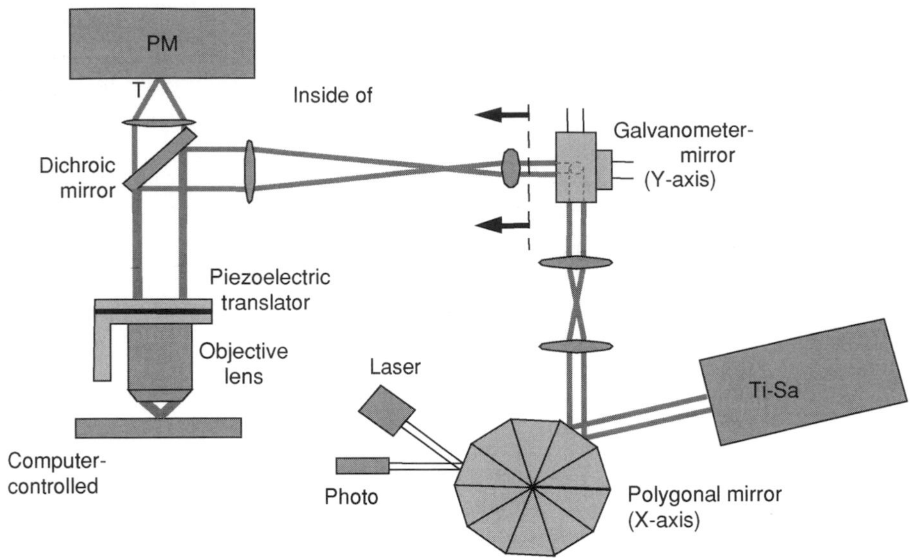

Fig. 1 A schematic of a three-dimensional two-photon image cytometer.

fastest mode, the corresponding scanning speed is 40 μs/line allowing the acquisition of approximately one hundred 256- × 256-pixel images per second. The image acquisition rate is 100 times faster than conventional scanning systems. The laser beam enters the epiluminescence light path of an upright microscope (Axioscope, Zeiss, Thornwood, New York). The beam is reflected by the dichroic mirror toward the objective and is focused in the specimen. The induced fluorescence signal is collected by the same objective and passes through the dichroic mirror. An additional barrier filter removes residual scattered light and a PMT operating in analog mode records the signal fluorescence. To perform 3D volume scans, the objective is mounted on a computer-controlled piezoelectric objective translator with an approximate bandwidth of 1 kHz (P-721.00, Physik Instrumente, Waldbronn, Germany). By translating the objective axially, z-stacks of x,y-plane images are obtained.

To demonstrate the acquisition of deep tissue images using a video-rate two-photon microscope based on the polygonal mirror approach, we have studied the dermal structures in *ex vivo* human skin. The skin was previously frozen. We studied the collagen/elastin fiber structures in the dermal layer. One hundred images were taken at depths between 80 and 120 μm below the skin surface. The frame acquisition time was 90 ms and the whole stack was imaged with a data acquisition time of 9 seconds. The collagen/elastin fibers can be clearly observed. Representative images of the collagen/elastin fiber structures are shown in Fig. 2.

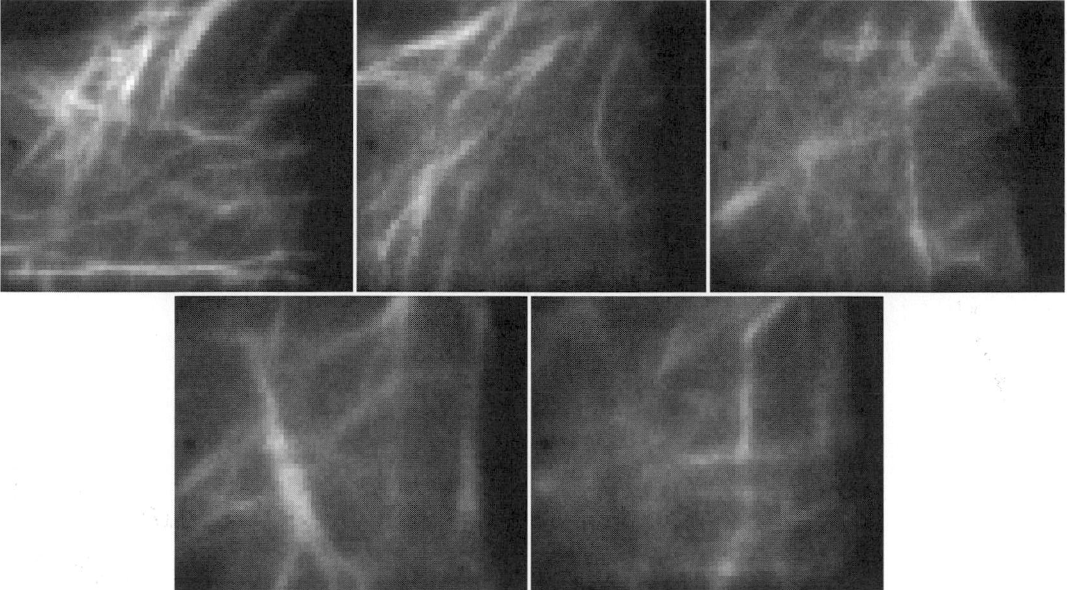

Fig. 2 Collagen/elastin fibers in the dermis of frozen human skin acquired at successive depths using two-photon video-rate microscopy. The image contrast is based on autofluorescence. The five panels (from left to right) are images acquired at successive depths of 80, 100, 120, 140, 160 μm below the skin surface. The size of each image is $80 \times 100\ \mu m^2$.

3. High-Speed Two-Photon Microscopy Based on Multifoci Excitation

An alternative strategy to increase the speed of TPM is to scan multiple foci across the sample. In this design, an array of foci is generated at the field aperture plane. The emission from each focal point is again collimated by the excitation tube lens and is directed to the objective through a dichroic mirror. The collimated rays are designed to overlap at the back aperture of the objective and to overfill it. The objective projects the individual focal points into the specimen. The scanning of the foci can be achieved either by rotating or translating a lenslet array or by using galvanometric scanners in the light path. The emission light is collected by the objective in an epi-illumination mode, is deflected by the dichroic mirror, and is focused by the emission tube lens. A CCD camera or PMT array can be placed at the image plane to record the image (Fig. 3).

4. Two-Photon Microtomy

Although the TPM penetration depth of more than 200 μm is superior to other high-resolution optical microscopy techniques, it is still insufficient to address a number of biological questions. For example, tumor metastasis has been studied

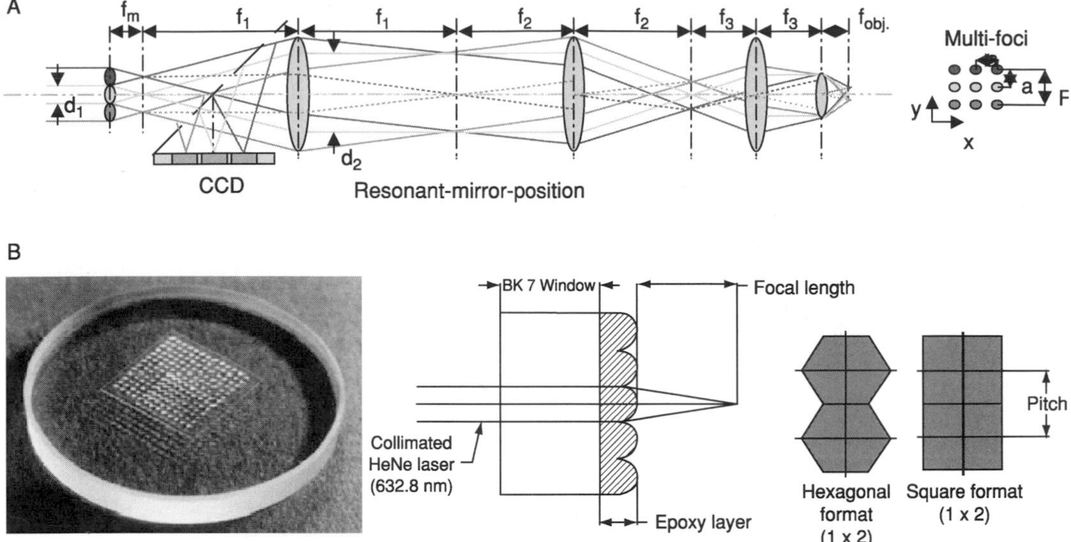

Fig. 3 (A) Schematic of a multiphoton multifocal microscopy system. (B) A microlens array is used to produce the multiple foci that will be scanned across the sample. (See Color Insert.)

in animal models and has been shown to occur over great distances (Hoffman, 2002). Additionally, there is growing interest in understanding how the information contained in the genome is deployed throughout the 3D environment of an organism (Streicher and Muller, 2001). To effectively understand how the gene expression varies over the organ requires imaging techniques that can image throughout the entire organ and relate the morphological information to the underlying expression profile. Our research in this area has addressed this limitation by incorporating an automated microtome into a high-throughput multiphoton microscope. By imaging serial sections 200–300 microns into a fixed tissue block and then sectioning the top layer and imaging the newly exposed section, it is possible to image through an entire fixed sample and obtain high-resolution 3D images of macroscopic volumes (Fig. 4).

III. Experimental Demonstration of Two-Photon Tissue Cytometry

A. Rare Cell Detection in Three Dimension

To test the basic instrument sensitivity to quantitatively characterize cell populations in a 3D environment, we prepared collagen matrix tissue samples seeded with two populations of mouse 3T3 NIH fibroblasts. One of the populations was labeled with the cytoplasm marker CellTracker Green, and the nuclei label

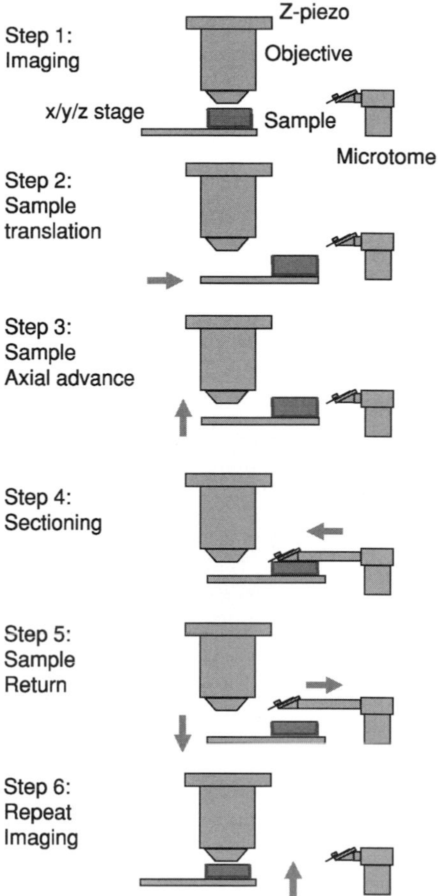

Fig. 4 Schematic demonstrating the acquisition of axially extended volumes.

Hocchst 33342. The other population was labeled with Hoechst 33342 only. Cell mixtures with varying relative concentrations ranging from 1:1 up to 1:10,000 green cells per total cells were prepared. These cell preparations were then seeded into collagen matrixes.

Macroscopic areas of the tissue matrix on the order of millimeters were imaged down to a depth of 80 microns. This was accomplished by imaging individual blocks with dimensions of 200 × 200 microns (at 256 × 256 pixels) in the radial direction and 80 microns (40 layers with a 2-micron separation between planes) in the axial direction. An overlap of approximately 10 microns between image stacks was kept, which allowed a mosaic of the images into a single large volumetric image. The emission light was separated into two channels using a dichroic mirror at 500 nm.

Figure 5A is a ratiometric image of the two-channel data 20 μm into the surface of a 7- $\times$ 5.5-mm area of a collagen matrix, which was seeded with one green cell per 1000 cells. Figure 5B shows an enlargement of a region within the larger image containing a single cell labeled with CellTracker green. Figure 5D is a 3D visualization of the cell within the collagen matrix.

Five samples were prepared at concentrations of 1:1, 1:10, 1:100, 1:1000, and 1:10,000 green cells per total cells. The images were analyzed by computer using image analysis algorithms written in Igor Pro (Wavemetrics, Lake Oswego, Oregon) and C++. Briefly, the images were first separated into their two component channels. A median filter was applied to the blue channel containing the cell

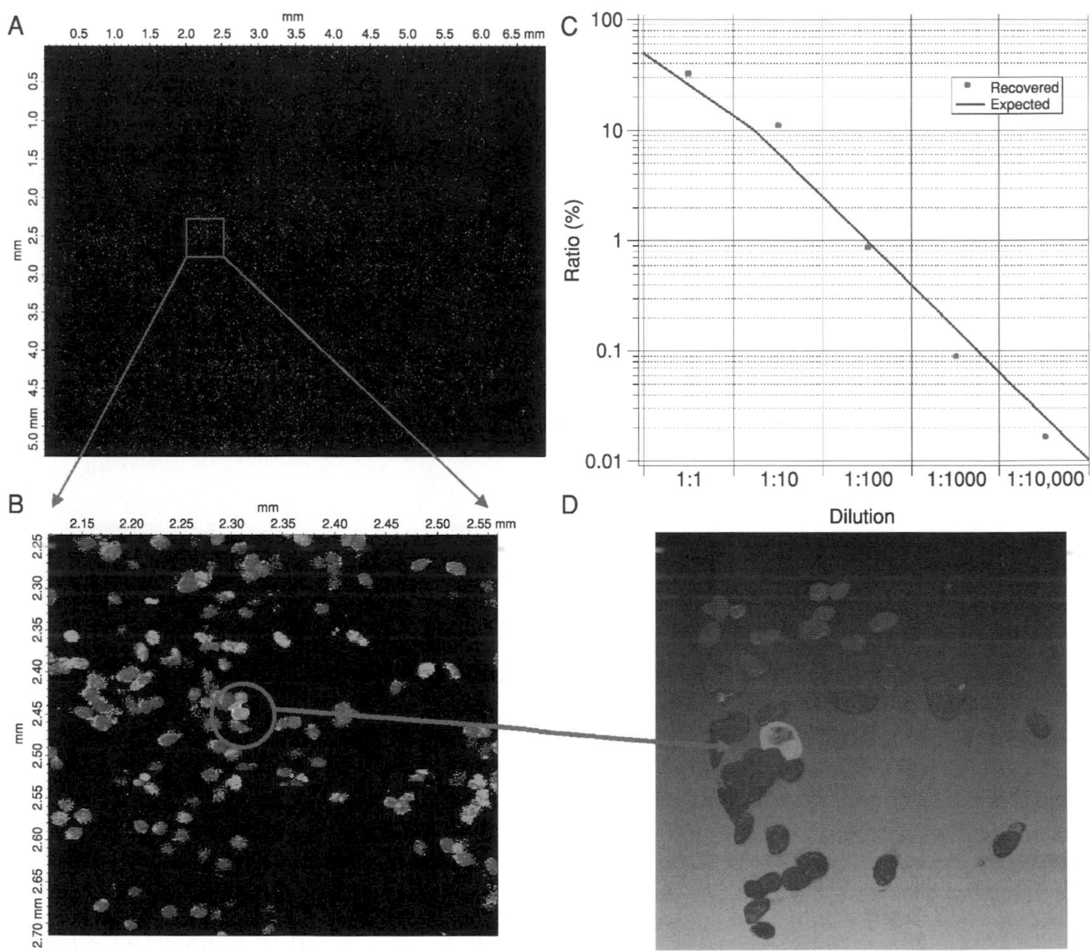

Fig. 5 Rare cell detection of a labeled cell within a three-dimensional (3D) collagen matrix. (A) Created by stitching together many smaller images to create a single larger image. (D) A 3D plot of the labeled cell using the 3D visualization program, Amira, from TGS, Inc. (See Color Insert.)

nuclei, and then the image was thresholded using an adaptive thresholding algorithm based on the method of Otsu (1979). A distance transform was then applied to the image to calculate the distance between each feature pixel and the nearest background pixel. This transformed image was again thresholded to identify markers for the nuclei. Finally, a labeling algorithm was applied to the image to count the number of disconnected components in the image and record their volumes. To identify the green cells, a median filter was applied to the second channel of the original image, and a ratiometric image was produced by "ratioing" the two median filtered channels. The green cells could then be identified and thresholded on the basis of their spectroscopic signature. A labeling algorithm was then used on the thresholded image to count the number of green cells in the image. Figure 5C is a plot of the recovered ratios of the samples that were imaged.

B. Quantification of Tissue Cytometric States

To demonstrate the ability to image macroscopic tissue volumes with the two-photon microtome system, we imaged a portion of a mouse brain that had been transfected to coexpress green fluorescent protein (GFP) along with actin throughout to highlight the brain tissue structure (Fig. 6). The brain had been fixed with 4% paraformaldehyde and embedded in paraffin. The fixed sample was mounted on the microscope and an imaging depth of 90 μm and a sectioning depth of 60 μm were chosen. The radial resolution was 0.78 μm and the optical

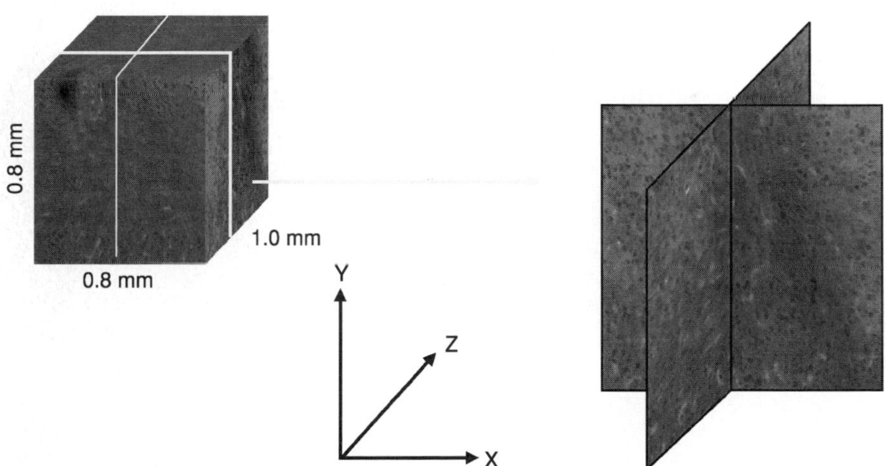

Fig. 6 An approximately 1-mm^3 chunk of brain tissue from a mouse that had been engineered to coexpress green fluorescent protein along with actin in the brain. The image was created by stitching together separate sections that were produced by alternating optical and mechanical sectioning. (See Color Insert.)

z-sections were chosen to be 3 μm apart. A finer resolution can be chosen and a smaller overlap between successive physical sections if desired. The total imaging time including the mechanical sectioning was 45 minutes. Assuming rough cell dimensions on the order of a 20-μm spherical diameter and relatively close cell packing, the tissue volume contains roughly 1×10^5 individual cells. The procedure is completely automated once the sample has been mounted on the microscope, and after data acquisition, we can view and analyze the entire 3D image on the computer.

In Fig. 7A, we show a portion of a mouse heart tissue that has been fixed in 4% paraformaldehyde and embedded in paraffin. The heart cell nuclei have been labeled with Hoechst 33342 with intravital labeling. We were able to image the tissue morphological features by taking advantage of intrinsic autofluorescence of the fixed and embedded tissue. Abnormal growth of cardiomyocytes in heart tissue underlies different aspects of heart pathology in conditions such as hypertension. It would be useful in such cases to be able to accurately quantify the morphological changes of cardiomyocytes. Unfortunately traditional 2D histopathology is ill suited to make such measurements because cardiomyocytes have dimensions on the order of 100 μm, and a single 2D slice taken at a random orientation provides insufficient information to accurately measure the volume of the cell. We and our collaborators have developed an approach to label extracellular matrix of heart tissue using a Texas red maleimide conjugate. This type of labeling is useful for demarcating the boundaries of individual cardiomyocytes to facilitate accurate cell volume measurements on a cell-by-cell basis (Fig. 7B). This 3D histopathology approach avoids artifacts associated with traditional 2D

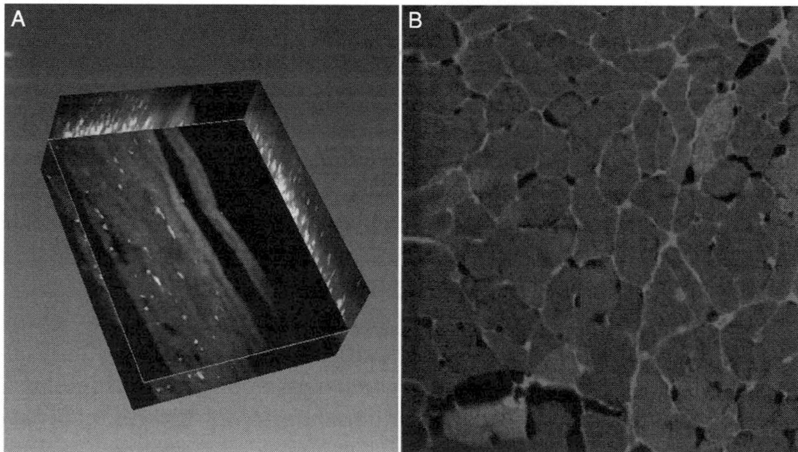

Fig. 7 (A) Mouse heart tissue that has been intravitally labeled with Hoechst 33342 to label the nuclei. (B) Mouse heart tissue that has been intravitally labeled with a Texas red–maleimide conjugate, which labels the proteoglycans in the extracellular matrix throughout the mouse. (See Color Insert.)

histopathology by allowing the cell boundaries to be directly measured in 3D over hundreds of microns of tissue or even throughout the entire heart if desired. This approach can greatly facilitate the quantification at the tissue level of *in situ* models of heart disease.

IV. Future Outlook

High speed two-photon tissue cytometry is well positioned to extend the capabilities of traditional 3D tissue cytometry and histopathology. The ability to quickly assay macroscopic tissue volumes at submicron resolution opens a number of previously closed research avenues. Investigators can now begin to quantitatively characterize the 3D morphology of a statistically relevant population of cells within a tissue while recording the native morphology and tissue ultrastructure. The data we have presented are based mainly on single-point scanning modalities. However, multifoci techniques will dramatically decrease the imaging time, making it possible to image an entire organ at submicron resolution on the order of hours.

The addition of multispectral capabilities will allow the profiling of 3D gene expression patterns throughout the tissue on multiple gene products simultaneously. This is particularly important in the context of the genomics revolution where there is an increasing push to understand how the linear DNA code is deployed in the 3D environment of an organ. Furthermore, there is a growing consensus that the one-gene one-disease model is insufficient to explain the etiology of a number of diseases. Imaging the spatial distribution of a number of gene products simultaneously may provide some important clues to the underlying basis of a disease that may have gone unnoticed using more traditional approaches. This type of analysis can be extremely useful in areas such as pharmaceutical development where one of the primary challenges is to validate potential drug targets at the tissue level where they ultimately must function. A major area of development in this field is to improve 3D data analysis algorithms for analyzing tissue structure to make it possible to formulate various classes of assays addressing different problems in areas such as gene expression, cancer etiology, drug delivery, heart disease, tissue engineering, and drug development.

V. Conclusion

We have described a new cytometry technique based on the combination of TPM and high-speed imaging techniques. This new technology will be a useful improvement on traditional image cytometry techniques for investigating the 3D morphology and biochemical states of cells within their native tissue environment up to the entire organ level and allow researchers to explore many questions that were previously impossible or cost prohibitive to address.

Acknowledgments

We would like to acknowledge our collaborators Dr. Rich Lee, Mr. Jeremy Sylvan, and Dr. Hayden Huang at Brigham Women's Hospital in the heart tissue data study; our collaborators Dr. Elly Nedivi and Mr. Wei-Chung Allen Lee for the brain tissue study; and our collaborator Dr. Bevin Engelward for the cell mixture study. Peter So and Timothy Ragan acknowledge National Institutes of Health (NIH) support: R21/R33 CA84740-01, PO1 HC 64858-01A1, and R33 CA091354-01A1. Tim Ragan and Karsten Bahlmann further acknowledge NIH SBIR funding CA97670.

References

Bajaj, S., Welsh, J. B., Leif, R. C., and Price, J. H. (2000). Ultra-rare-event detection performance of a custom scanning cytometer on a model preparation of fetal nRBCs. *Cytometry* **39,** 285–294.

Bennett, B. D., Jetton, T. L., Ying, G., Magnuson, M. A., and Piston, D. W. (1996). Quantitative subcellular imaging of glucose metabolism within intact pancreatic islets. *J. Biol. Chem.* **271,** 3647–3651.

Bewersdorf, J., Pick, R., and Hell, S. W. (1998). Mulitfocal multiphoton microscopy. *Opt. Lett.* **23,** 655–657.

Bonner, R. F., EmmertBuck, M., *et al.* (1997). Cell sampling—laser capture microdissection: Molecular analysis of tissue. *Science* **278,** 1481–1483.

Brakenhoff, G. J., Squier, J., *et al.* (1996). Real-time two-photon confocal microscopy using a femtosecond, amplified Ti:sapphire system. *J. Microsc.* **181**(Pt 3), 253–259.

Darzynkiewicz, Z., Bedner, E., Li, X., Gorczyca, W., and Melamed, M. R. (1999). Laser-scanning cytometry: A new instrumentation with many applications. *Exp. Cell Res.* **249,** 1–12.

Denk, W., Strickler, J. H., and Webb, W. W. (1990). Two-photon laser scanning fluorescence microscopy. *Science* **248,** 73–76.

EmmertBuck, M. R., Bonner, R. F., *et al.* (1996). Laser capture microdissection. *Science* **274,** 998–1001.

Fan, G. Y., Fujisaki, H., Miyawaki, A., Tsay, R. K., Tsien, R. Y., and Ellisman, M. H. (1999). Video-rate scanning two-photon excitation fluorescence microscopy and ratio imaging with chameleons. *Biophys. J.* **76,** 2412–2420.

Givan, A. L. (2001). Principles of flow cytometry: An overview. *Methods Cell Biol.* **63,** 19–50.

Gohongi, T., Fukumura, D., Boucher, Y., Yun, C. O., Soff, G. A., Compton, C., Todoroki, T., and Jain, R. K. (1999). Tumor-host interactions in the gallbladder suppress distal angiogenesis and tumor growth: Involvement of transforming growth factor beta1. *Nat. Med.* **5,** 1203–1208.

Gorczyca, W., Darzynkiewicz, Z., and Melamed, M. R. (1997). Laser scanning cytometry in pathology of solid tumors—A review. *Acta. Cytol.* **41**(1), 98–108.

Guild, J. B., and Webb, W. W. (1995). Line scanning microscopy with two-photon fluorescence excitation. *Biophys. J.* **68,** 290a.

Hendricks, J. B. (2001). Quantitative histology by laser scanning cytometry. *J. Histotechnol.* **24,** 59–62.

Hoffman, R. M. (2002). Green fluorescent protein imaging of tumour growth, metastasis, and angiogenesis in mouse models. *Lancet Oncol.* **3,** 546–556.

Kamentsky, L. A. (2001). Laser scanning cytometry. *Methods Cell Biol.* **63,** 51–87.

Kamentsky, L. A., Burger, D. E., Gershman, R. J., Kamentsky, L. D., and Luther, E. (1997). Slide-based laser scanning cytometry. *Acta. Cytol.* **41,** 123–143.

Kamentsky, L. A., and Kamentsky, L. D. (1991). Microscope-based multiparameter laser scanning cytometer yielding data comparable to flow cytometry data. *Cytometry* **12,** 381–387.

Kamentsky, L. A., and Melamed, M. R. (1967). Spectrophotometric cell sorter. *Science* **156,** 1364–1365.

Kim, K. H., Buehler, C., and So, P. T. C. (1999). High-speed, two-photon scanning microscope. *Appl. Opt.* **38,** 6004–6009.

Knezevic, V., Leethanakul, C., *et al.* (2001). Proteomic profiling of the cancer microenvironment by antibody arrays. *Proteomics* **1,** 1271–1278.

Liotta, L. A., and Kohn, E. C. (2001). The microenvironment of the tumour-host interface. *Nature* **411,** 375–379.

Masters, B. R., So, P. T., and Gratton, E. (1997). Multiphoton excitation fluorescence microscopy and spectroscopy of *in vivo* human skin. *Biophys. J.* **72,** 2405–2412.

Melamed, M. R. (2001). A brief history of flow cytometry and sorting. *Methods Cell Biol.* **63,** 3–17.

Otsu, N. (1979). Threshold selection method from gray-level histograms. *IEEE Trans. Syst. Man Cybernetics* **9,** 62–66.

Piston, D. W., Knobel, S. M., Postic, C., Shelton, K. D., and Magnuson, M. A. (1999). Adenovirus-mediated knockout of a conditional glucokinase gene in isolated pancreatic islets reveals an essential role for proximal metabolic coupling events in glucose-stimulated insulin secretion. *J. Biol. Chem.* **274,** 1000–1004.

Powers, M. J., Janigian, D. M., *et al.* (2002). Functional behavior of primary rat liver cells in a three-dimensional perfused microarray bioreactor. *Tissue Engineering* **8,** 499–513.

Rieseberg, M., Kasper, C., Reardon, K. F., and Scheper, T. (2001). Flow cytometry in biotechnology. *Appl. Microbiol. Biotechnol.* **56,** 350–360.

So, P. T., Dong, C. Y., Masters, B. R., and Berland, K. M. (2000). Two-photon excitation fluorescence microscopy. *Annu. Rev. Biomed. Eng.* **2,** 399–429.

Streicher, J., and Muller, G. B. (2001). 3D modelling of gene expression patterns. *Trends Biotechnol.* **19,** 145–148.

Van Dilla, M. A., Trujillo, T. T., Mullaney, P. F., and Coulter, J. R. (1969). Cell microfluorometry: A method for rapid fluorescence measurement. *Science* **163,** 1213–1214.

CHAPTER 3

Confocal Microscopy: Quantitative Analytical Capabilities

Jurek W. Dobrucki

Laboratory of Confocal Microscopy and Image Analysis
Department of Biophysics
Faculty of Biotechnology
Jagiellonian University
30-387 Kraków, Poland

METHODS IN CELL BIOLOGY, VOL. 75

0091-679X/04 $35.00

I. Introduction

An optical microscope may seem just a very good magnifying glass, while a confocal microscope may be thought of as simply a very good microscope. Yet contemporary optical microscopes are no longer just tools to see a magnified view of a small world that cannot be seen by the naked eye. They are sophisticated analytical devices, capable of measuring a number of physical parameters simultaneously within a living cell. A name fully describing the measuring capabilities of a contemporary optical microscope could be so long that it is impractical—for instance, a "microbiospectromultifluorimeter"—or an even more bizarre neologism. Although we obviously still use the original name, we should be amazed at the sophistication of the contemporary descendant of instruments constructed in the seventeenth century by Antony van Leeuvenhoek and Robert Hooke. The spectrum of physical parameters that can be measured and the biological parameters that can be subsequently calculated from them has expanded enormously in the last 2 decades.

The basic parameters used to characterize a microscope, such as resolution or magnification, may be misleading for a layperson. Maximum magnification is not a parameter per se describing microscope capabilities. That parameter is the maximum achievable resolution, i.e, capability to visually resolve two very small neighboring entities in the field of view. Another critical parameter characterizing a contemporary fluorescence microscope is sensitivity to light and spatial precision (Piston, 1998). Spatial resolution is limited by diffraction to a few hundred nanometers, yet in a microscope equipped with a sensitive light detector, single molecules can be visualized and structures many times smaller than the diffraction limit can be studied. Paradoxically, a confocal microscope does not dramatically improve resolution in the horizontal plane beyond what is achieved by a standard optical microscope. It offers the capability to optically section a specimen into slices approximately 700 nm thick and thus minimizes blurring from structures located at a distance from the plane of focus. The elimination of blur increases image contrast and consequently makes possible the full utilization of the spatial resolution offered by the optical components of the microscope.

In the simplest case, a fluorescence confocal microscope is used to record fluorescence intensity as a function of position in two-dimensional (2D) or three-dimensional (3D) space. If we assume that a sample is a fixed preparation that contains some structure stained with a nonbleaching fluorescent label emitting in a narrow spectral band, that there is no autofluorescence, that the optical filters are selected optimally, and that the instrument and the sample are not changing over time, the task of recording an image should be relatively simple. In biological research, the sample is never that simple and several parameters may need to be recorded. These parameters could be the intensities of several fluorescence signals in 3D space versus time, fluorescence lifetimes, intensity and wavelength of scattered light, and so on. From these physical parameters, an experimenter can estimate biologically relevant values. They include a local concentration of a probe, which can then be translated into the concentration of a target molecule (e.g., DNA or protein), a distance in space, properties of the molecular environment like membrane fluidity, diffusion of small molecules, and fluorescent proteins, bound and unbound pool of an active protein in a living cell, and many more.

The limitations to measurements in a confocal microscope are complex, especially for the study of live cells. An understanding of these limitations is a *condicio sine qua non* for correct interpretation of data. In a real experiment, the sample is not stable during image acquisition; the signal decays because of photobleaching, live cells move and suffer phototoxic damage, fluorescent probes may undergo saturation, dimerization, photoconversion, or detachment from the target, intensity of excitation light fluctuates, and the focal plane drifts. These factors impose limits on what is achievable in practice by confocal microscopy.

In this chapter, I discuss capabilities and limitations pertinent to measuring various parameters that either standard or sophisticated specially adapted confocal instruments can measure. I emphasize applications in live cell studies and the limitations imposed by live biological samples. The subject is vast, so this text does not cover the full field; the choice of and emphasis on various aspects of quantitative confocal studies is arbitrary.

II. Cartesian Coordinates, Distance, Size, Shape, and Volume in Confocal Microscopy

A. Representation of Small Objects in 2D and 3D Space

Let us consider a very small (e.g., 3 nm in diameter), strongly fluorescent sphere that is placed in a dark 3D space (Fig. 1A). An observer who looks at this sphere through a standard optical microscope will see a circle of light, bright in the center and darker at the periphery. There may be a faint ring detectable around this central bright area (Fig. 1B).

If an observer were to calibrate the instrument, he would find that the central bright region has a diameter corresponding to a few hundred nanometers—2 orders of magnitude larger than the faithful image of a 3-nm object ought to be (Fig. 1B). Larger objects with a diameter of 10–50 nm would still be represented

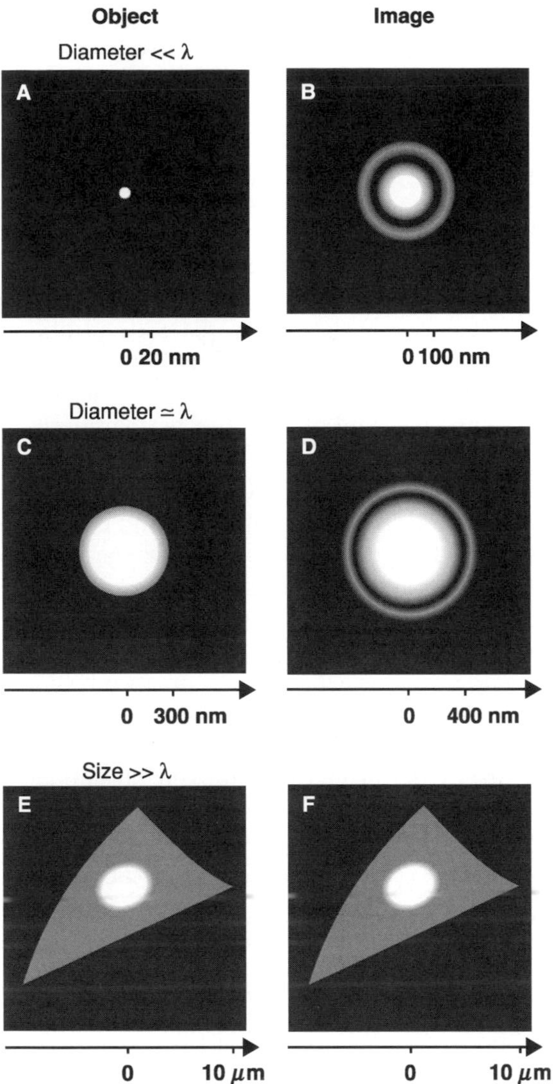

Fig. 1 A schematic representation of small objects (left) and their images (right) generated by a standard fluorescence microscope. An image of a bright 3-nm diameter sphere (A) is a large circle ~250 nm in diameter, accompanied by interference fringes (B). When a sphere with a diameter comparable to the wavelength of light (a few hundred nanometers) (C) is imaged, the diameter of the image reflects the real size reasonably well (D), that is, in correct proportion to the field of view; interference effects appear at the edges. When animal cells *in vitro* (size 10–30 μm) (E) are imaged, their size and shape are reflected correctly and interference effects are less noticeable (F). Note that the object (A) and the corresponding image (B) are not drawn to scale. If proportions between a 3-nm sphere and the corresponding image were to be maintained, the image (B) would have to be larger than the page on which it is printed. The intensity of interference fringes is exaggerated in diagrams (B) and (D).

by bright circles overestimating the size of the original. If the studied object were larger still, a few hundred or thousand nanometers across, such as an erythrocyte or a cell nucleus, the size and shape would be represented reasonably faithfully (Fig. 1C–F and Fig. 2).

If an observer using a wide-field fluorescence microscope collected a series of images and reconstructed a 3D image of the 3-nm sphere, he would see a complicated shape extending for a few hundred nanometers along a line parallel to the optical axis of the microscope (Keller, 1995). The length of this figure would be several times greater than the diameter of the circle of light seen in a horizontal plane (i.e., the plane of the microscope stage). The size and shape of this figure would in no way resemble that of the sphere (Fig. 3A, B).

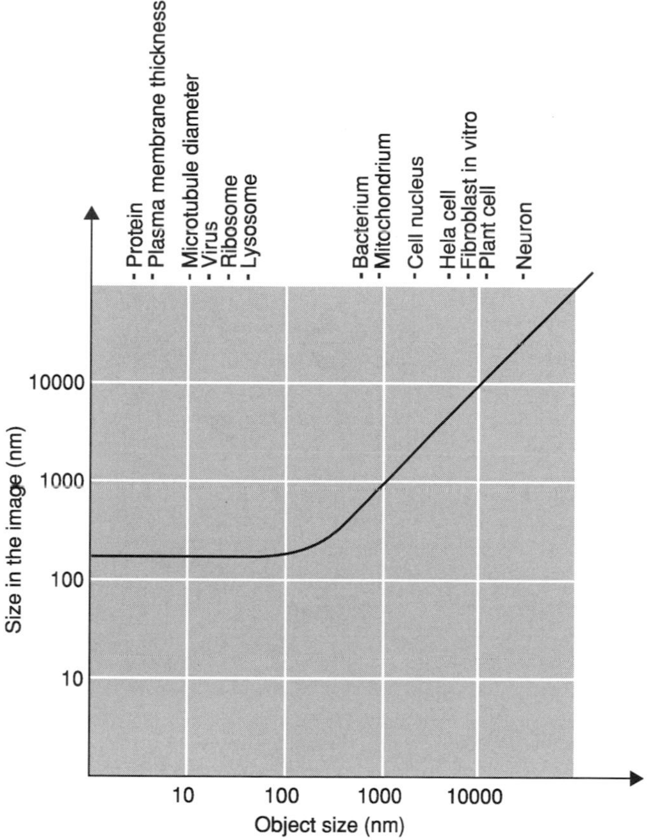

Fig. 2 Relationship between the true size of the object and the size of this object in a fluorescence microscope image. The smaller the object, the greater, relatively, the distortion. The relative contribution of interference fringes is also greater in images of small objects. Cell components like protein molecules (if detectable at all) or the thickness of the plasma membrane, for instance, will appear too large in relation to bigger objects like the nucleus and to the whole cell in the same image.

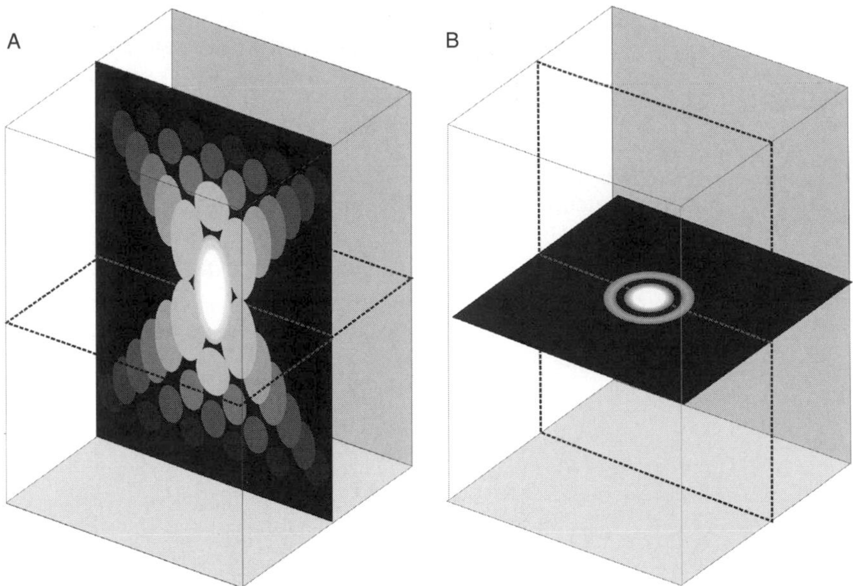

Fig. 3 Schematics of an image of a light-emitting point in a wide-field fluorescence microscope. The image is a three-dimensional (3D) figure resulting from interference of diffracted light. Vertical (A) and horizontal (B) cross sections are shown. (In practice, registering a reasonable 3D image of a subresolution bright sphere is technically quite difficult for reasons described in subsequent sections.)

If a 3D image of this sphere were created using a confocal microscope (with a pinhole set at 1 Airy disk), the image would resemble an ellipsoid for which the short and long axes would extend to no less than approximately 250 and 600 nm, respectively (Fig. 4).

Misrepresentation of size and shape for such small structures imaged by a standard optical microscope arises from interference of diffracted light. A point-scanning confocal microscope can significantly reduce this misrepresentation (although it cannot eliminate it). In other words, a confocal microscope, even if it were equipped with a very highly sensitive light detector, would be capable of detecting the presence of small well-spaced fluorescent objects within a specimen but would still misrepresent their size and shape. This effect is rarely observed, because the sensitivity of standard confocal microscopes is too low to detect single fluorescent molecules under typical imaging conditions. A reasonably faithful representation of dimensions is achieved only for detectable structures greater than the wavelength of visible light (400–700 nm), when the high NA (where $NA = n \cdot \sin\alpha$; see section F) objective is used and the optical elements are optimally aligned (Pawley, 1995).

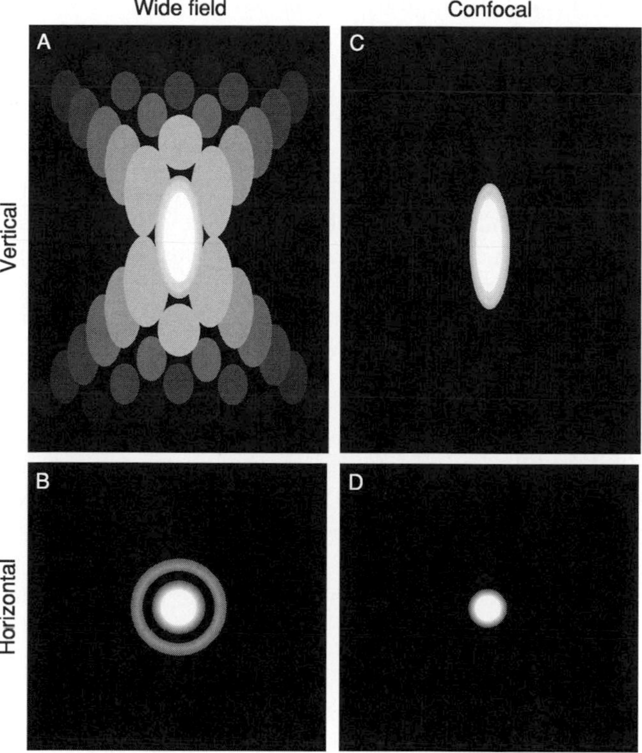

Fig. 4 Schematics of a three-dimensional image of a light-emitting point in a wide-field (A, B) and in a confocal (C, D) fluorescence microscope. Vertical (A, C) and horizontal (B, D) cross sections are shown. A confocal aperture eliminates most image elements derived from first-order and higher-order interference of diffracted rays.

B. Shape of Objects and Image Resolution

The reasoning described in the previous section leads to the following conclusions: (1) a fluorescence confocal microscope inflates the size and shape of objects smaller than the wavelength of light, (2) this inflation is greater along the optical axis of the microscope than in the focus (horizontal) plane, and (3) the capability to distinguish two small bright objects located close to each other in 3D space is limited. We refer to this limitation as the *horizontal* and *axial resolution.* Two small objects separated by a distance many times greater than their size may be imaged as a single cloud of light; they will not be resolved.

C. The "Height" of Objects and Surface Profiles

If an oil-immersion objective lens is moved closer and closer to a coverslip, the point that is imaged moves deeper into the specimen. Let us assume that a sample is a thick water-containing biological specimen. In this case, the distance traveled by the imaged point is different from the distance traveled by the lens. In other words, the movement of the objective cannot be directly used as a way to measure the "height" of the structures in the specimen. The imaged structures will appear too short in relation to the horizontal dimensions of the object (Fig. 5). Moreover, spherical aberration will seriously degrade image quality (Fig. 5C).

This discrepancy occurs because of the way light rays travel through the water/glass and immersion oil before they form an image. Because of the difference between the refractive indices of glass and water, the light rays are refracted as shown in Fig. 5B. A similar phenomenon will occur if a dry lens (i.e., a lens designed to be used with a layer of air separating it from the coverslip) is used for imaging a sample mounted in a high-refractive index medium. In this case, the height of objects in a 3D image will be overestimated.

D. Volume of 3D Objects

The volume of a 3D confocal image can be measured by counting the voxels within a selected region of interest. A correction is required to offset the misrepresentation of height in the imaged sample. Many commercial software packages used in confocal microscopy include volume measurements as a standard feature.

E. Surface Area of 3D Objects

Measurements of the surface area of 3D objects within a confocal image require fitting a surface to the selected object and calculating its area. This operation can be performed by some specialist software packages (ImagePro+, Voxblast, VoxelView). Other approaches to this problem are based on stereology techniques (Cox, 1999; Underwood, 1970).

One can argue that surface measurements on objects in 3D confocal images may result in a gross underestimate of the real value, because of the limited spatial resolution of a microscope. The surface area of a glass-attached cell can be used as an example. Electron microscopy (EM) micrographs reveal a highly uneven surface for such cells, with microvilli, protrusions, and invaginations, features generally not resolved by optical microscopy. Thus, a measurement of "cell surface" by optical microscopy can provide only an estimate of a "cell contour area" rather than the true surface area of a plasma membrane.

F. Factors Affecting Resolution, Distance, and Shape in Confocal Images

The Abbé formula describes the best resolution that can be achieved in the plane of a periodic object with an objective of high NA, using light of wavelength λ (Inoué, 1995; Oldfield, 1994):

$$d = a \cdot \lambda [n \cdot \sin\alpha]^{-1} \quad (1)$$

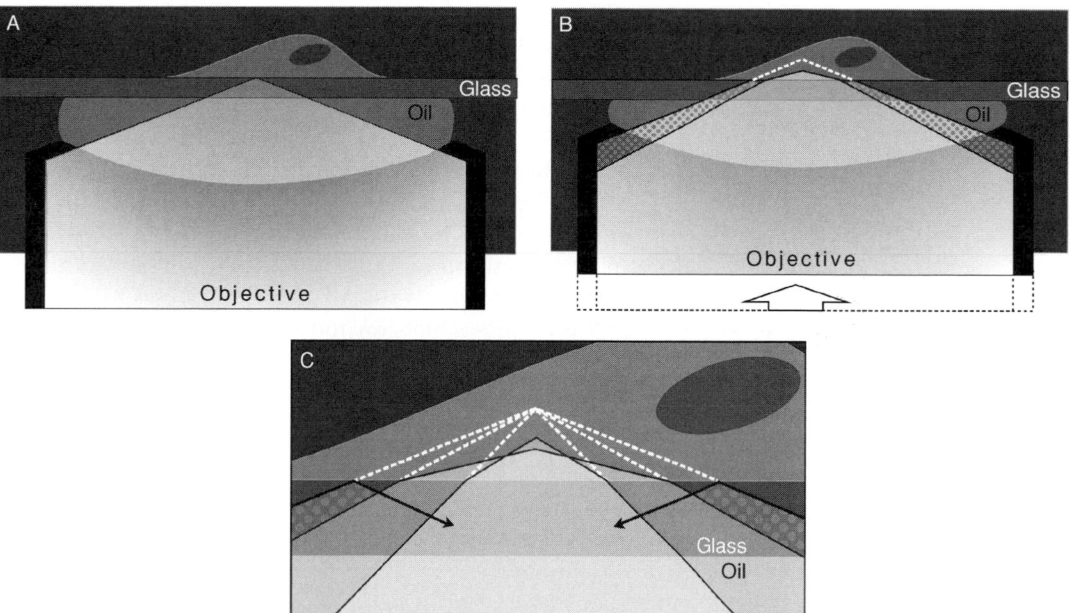

Fig. 5 A schematic drawing showing how light rays travel between the objective lens and a living cell attached to a coverslip as the imaged point is brought deeper into the sample. The light rays traveling from the lens through the immersion oil into the coverslip do not change direction (A), although some refraction occurs at the interface between glass and immersion oil if their refraction indices do not match perfectly. When the lens is brought closer to the coverslip (B), the light rays are refracted at the interface between glass and water or glass and cell (solid vs dotted line). The oil immersion lens is designed to work properly only just on the far side of a coverslip of a specified thickness. Note that the outermost rays are reflected from the glass–water interface and the solid angle of the cone of light decreases, resulting in a decrease of the effective numerical aperture (see C for a magnified view of the reflected and refracted rays). It is worth pointing out that a similar phenomenon of light refraction and "cell-mediated focusing" occurs when excitation and emission rays travel between regions differing in refractive index in living cells, for instance, between cytoplasm and cell nucleus or lipid droplets.

where

d = the smallest distance between two points that can be visually resolved (resolution)

a = a constant, dependent on the adopted criterion of resolution (a = 0.5 in classic Abbé theory)

λ = the wavelength of light that forms the image

n = index of refraction of the medium between the specimen and the objective lens

α = half the solid angle between the outermost rays entering the lens

However, under conditions typically encountered, resolution is worse than predicted by Eq. 1 (i.e., worse than "diffraction-limited"). A number of other factors that are not explicit in this equation influence the achievable resolution and the capability to faithfully image the distance and shape in 3D space. Horizontal resolution in a confocal microscope depends on the wavelength of both excitation and emission light. Vertical resolution is proportional to $(NA)^{-2}$ and is typically three times worse at NA = 1.3. A poor fluorescence signal can severely limit image resolution. The number of photons must exceed the background many times to obtain reasonable contrast. Several measurements of the number of photons emitted in a pixel must be averaged to minimize the influence of Poisson noise on contrast and eventually on resolution. Spherical aberration is also an important factor influencing the ability to resolve small image features. Optimization of a lens type, coverslip thickness, and immersion and embedding media is critical for minimizing the effects of spherical aberration. Above all, the optimal alignment of optical components of a confocal microscope is the prerequisite for obtaining the highest resolution for an image (Cox, 1999).

When collecting a digital image, one must ensure that sampling frequency is optimized to fully use the optical resolution of the microscope. The Nyquist theorem is interpreted to mean that more than two measurements must be made per optically resolved distance in the image (Inoué and Spring, 1997), although often higher sampling rates are required. However, it may not be possible to satisfy this criterion (i.e., to collect the required number of frames), because of photobleaching of the fluorescent probe and phototoxic effects exerted on living cells. Thus, ironically, one of the principal factors limiting image resolution is not the optical system and diffraction of light, but the photophysics of the fluorescent probe.

When the sample under study is a fixed or a live cell submerged in physiological saline or culture medium, resolution and signal level deteriorate as distance from the coverslip increases. If an oil-immersion lens is used, the effect is dramatic (Fig. 6).

Imaging live cells in optimal growth conditions requires maintaining the sample at physiological temperature. To achieve that, either the specimen is maintained at 37 °C, while other components of the microscope are kept at room temperature, or the whole microscope is enclosed in a housing and maintained at 37 °C. The first solution creates temperature gradients between specimen and objective lens. In both cases the optical and metal components expand as temperature increases and the lens (and the immersion oil if this is a standard one) is used at a temperature higher than that for which it was designed. Periodic changes and gradients of temperature may adversely affect performance of the optical system, and image resolution and fidelity may be compromised.

Live cells contain regions differing in refractive index. Subcellular structures become part of an optical imaging system. It follows that the effect shown in Fig. 5 will occur as light rays travel through the imaged cell. Imaging of a flat surface through a cell nucleus yields a concave shape, not a flat shape, a striking example of misrepresentation arising from the optical inhomogeneity of the cell itself (Pawley, 2002).

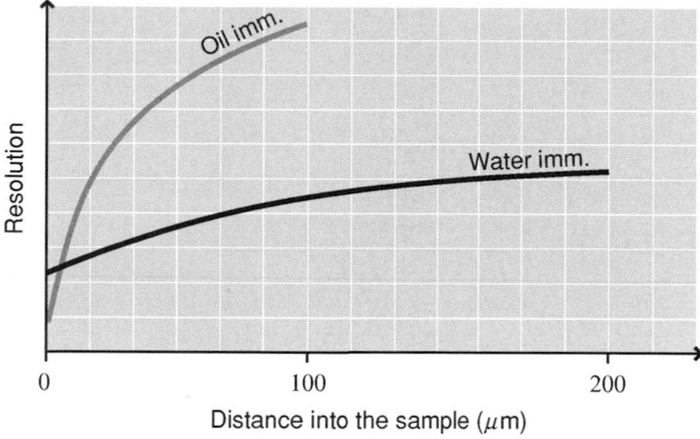

Fig. 6 A schematic representation of loss of resolution occurring when imaging deep within water-containing specimen, using an oil-immersion (NA = 1.4) or a water-immersion (NA = 1.2) objective lens.

In summary, the size and shape of small structures and the distance between them can be faithfully represented only by a typical, optimally aligned point-scanning confocal microscope for a range of sizes exceeding the wavelength of visible light. This statement should not be considered a dogma operating under all conditions, however. New microscopy approaches exist that are capable of going beyond this limit (Hell, 2003).

G. Measuring Distances Smaller than "Diffraction-Limited Resolution"

Let us assume that two 3-nm spheres emit light of different wavelengths and are separated by only 50 nm (Fig. 7). Their images will overlap almost entirely in 3D space but can be recorded separately because of the different colors. The diameters of the ellipsoid images will measure a few hundred nanometers. It is reasonable to expect that the centers of both images would still be located 50 nm from each other in a multicolor image, assuming that a correction for register shift and chromatic aberration has been applied. Thus, the distance between the geometric centers of both images (or more precisely, the bary centers of the 3D images) is still equal to the distance that separates the subresolution spheres.

This principle has been applied in standard confocal microscopy (Bornfleth *et al.*, 1997) and is called *spectral precision distance microscopy.* It was developed further using a spatially modulated illumination approach (Albrecht *et al.*, 2002) and applied to measure small distances between structures labeled with different fluorescent probes. Although this technique does not provide point images of point objects, the distance between them, which is significantly less than the wavelength of light and as small as a few nanometers, can still be measured accurately.

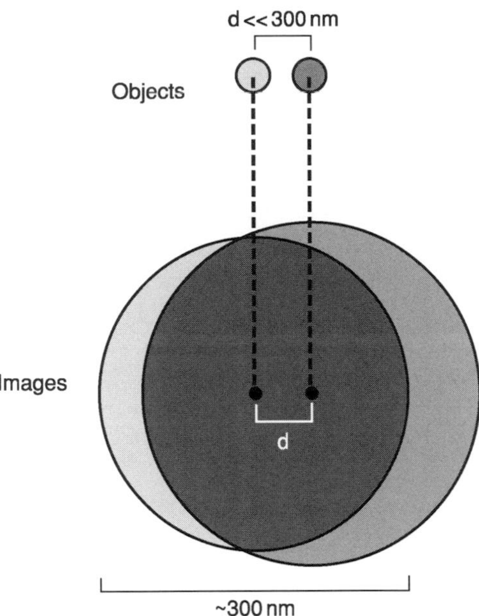

Fig. 7 Measurements of small distances in confocal microscopy ("spectral precision distance microscopy"). Two small structures, separated by a distance *d* smaller than the wavelength of light, emit fluorescence at different wavelengths. The images overlap almost entirely in space. The positions of the bary centers of both images are calculated and the distance *d* between them is measured.

H. Imaging with Resolution Beyond the Diffraction Limit

Images with resolution far beyond that of a standard or a confocal microscope can be created using new microscopy techniques: 4Pi and stimulated emission depletion (STED).

In a conventional epifluorescence microscope, excitation light is delivered and fluorescence collected by the same lens. The resolution of images created in this system is described by the classic Abbé formula (Eq. 1, earlier in this chapter). It follows that resolution is increased as the solid angle over which data are collected is increased. Up to a limit, resolution can be improved by using a lens of higher NA and a matching immersion oil and by mounting a sample in a high RI mounting medium. Even a high-NA lens collects only a fraction of the incident light over a limited solid angle. In 4Pi microscopy, a second lens is placed on the far side of the specimen and used for collecting more of fluorescence. Both lenses are focused at the same focal point. Both light beams are made to travel equal distances and they interfere at the focus to produce an image. In this arrangement, fluorescence is gathered over a larger solid angle than in a standard point-scanning confocal microscope. The axial resolution of a 4Pi image is greater than that of an image created with one lens (Schrader *et al.*, 1998). Several related techniques that incorporate this principle exist (Gustafsson, 1999; Lindek *et al.*, 1995). An application of 4Pi microscopy that incorporates two-photon excitation to studies of microtubules in fixed fibroblasts has been demonstrated (Nagorni and Hell, 1998; Schrader *et al.*, 1998). A study of live cells by 4Pi microscopy has also been reported (Bahlman *et al.*, 2001). Wide applicability of 4Pi methods in cell biology and especially in live cell studies remains to be explored.

Stimulated emission depletion (STED) microscopy has been shown to generate images of fixed cells with a resolution of 50 nm (Dyba *et al.*, 2003) and of living cells with approximately 100-nm resolution, in both axial and lateral directions (Egner *et al.*, 2002). In this technique, fluorescence in the sample is excited by a focused beam of visible light emitted by a pulsed laser. These pulses are followed by stimulated emission-depletion pulses of near-infrared radiation. The STED beam is shaped so that it is intense around the center and dark in the center. The fluorophores excited at the periphery of the spot by the excitation pulse are quenched by the STED pulses and fluorescence in these regions is stopped. In effect, fluorescence can be detected only from the central area of the original excitation spot. Images of live yeast and bacterial cells have demonstrated resolution approximately three times greater (in all directions) than those produced with a standard confocal microscope. The intensity of visible light (532-nm, 10-ps pulses) used in live cell experiments was 10 μW, and near-infrared (793-nm, 107-ps pulses) 5.2 mW per lens (Dyba *et al.*, 20023). Because the mechanisms of photobleaching and phototoxicity of fluorescent dyes used in microscopy are poorly understood, it is difficult to predict whether such light intensities will permit long-term studies of live intact mammalian cells stained with fluorescent probes. Because improved resolution implies more voxels of smaller size, this increased sampling frequency may also cause more photodynamic damage. Again, wide applicability is yet to be demonstrated, but if successful, the method may become a dramatic step forward in the development of new imaging techniques in biology (Hell, 2003).

III. Measurements of Fluorescence Intensity in Confocal Microscopy

Reading fluorescence intensity anywhere within the 2D digital confocal image is in principle straightforward. Thus, relative local concentrations of fluorescently labeled species within the same field of view or relative changes of concentration at the same point over time can be measured. Similar comparisons between samples are much more difficult, though. Measurements of absolute concentrations are virtually impossible for the reasons described in the following paragraphs.

A. Relative Fluorescence Intensity in Space and Time

Let us consider a simple idealized experiment. DNA in fixed cells is stained with a bright fluorescent label that binds to DNA in a stoichiometric manner. The goal of the experiment is to compare the local concentrations of DNA at two selected points within one nucleus. To achieve this goal, we need to measure the integrated intensity of fluorescence, or more precisely, the number of photons emitted by DNA-bound fluorescent probe in two selected areas of the image at a given time. The fundamental limitation to the accuracy of such determinations arises from the quantomechanical nature of fluorescence. A series of measurements of a nonbleaching stable fluorophore in the same pixel will yield a distribution of results described by Poisson statistics, that is, $N \pm N^{1/2}$. Thus, the error of such measurements will be relatively large in comparison with errors encountered in measurements of many macroscopic physical parameters. In practice, in confocal microscopy only a few (up to 20) photons are usually collected per bright pixel in one scan (Pawley, 2000). Moreover, the detector has a limited quantum efficiency and the electronics may have its own fluctuations. These factors are additional sources of error in measurements of fluorescence intensity.

As a consequence, accurate measurements of fluorescence intensities in space or time require a large number of replicate measurements and the averaging and deconvolution of the data. Unfortunately, acquiring and averaging high numbers of images is hampered by several factors, including the following:

1. Cell movement
2. Photobleaching of fluorescent probes in fixed and live cells (Bernas *et al.*, 2004; Song *et al.*, 1995)
3. Phototoxic effects exerted on live cells (Tsien and Waggoner, 1995; Dobrucki, 2001)
4. Thermal drift of the microscope stage (i.e., a specimen) in relation to the objective lens (Adler and Pagakis, 2003)
5. Fluctuations of the intensity of excitation light reaching the specimen (Zucker and Price, 2001)
6. Vibrations of the instrument

Other factors responsible for limited sensitivity and accuracy of measurements of fluorescence intensity in single-color and multicolor confocal images include the following:

1. Cellular autofluorescence
2. Detector noise and dark current
3. Detector nonlinearity
4. Curvature of the field of view
5. Inner filtering effect (Lakowicz, 1999) and cell-mediated focusing (Fig. 5)
6. Self-quenching, fluorescence resonance energy transfer (FRET), and saturation of fluorescent probes (Lakowicz, 1999)
7. Collisional quenching of fluorescence through collision with other molecules in the sample (e.g., oxygen and formaldehyde)

8. Other environmental factors (e.g., pH, which affects the intensity of fluorescence of fluorescein and derivatives)
9. Chromatic aberration and register shift (Fricker and White, 1992)

Depending on the type of specimen, fluorescent label, and instrument settings, the influence of these factors on measurements of fluorescence intensity may vary from negligible to dramatic. A review of the list given previously shows some factors that influence the measured fluorescence intensity, which are very difficult to control (e.g., stage drift), whereas other factors are simply beyond the control of the experimenter (e.g., cell-mediated focusing, inner filtering, quenching by intracellular quenchers, including oxygen, and pH in intracellular compartments of living cells).

B. Calibration of Fluorescence Intensity

When fluorescence intensities in all pixels are measured as accurately as possible, the experimenter might want to calibrate the intensity scale. It would be wrong to think, however, that a solution of the same fluorophore or a suspension of beads with known numbers of molecules of a fluorophore could be used to generate a calibration curve specific for the lens and type of specimen used. One has to consider that the quantum efficiency of the fluorophore bound to a target within a cell may be different than in a simple solution. In our example, it would be necessary to use a solution of propidium bound to DNA, because the fluorescence of an unbound dye is many-fold weaker (more precisely, the quantum efficiency is lower). Also, the measured intensity of fluorescence depends, to some degree, on the distance from the center of the field of view (i.e., in the horizontal plane), especially if optical alignment is imperfect. It obviously depends strongly on vertical position, that is, on the location relative to the center of the confocal plane. Bleaching properties of a fluorophore bound to a cell target are likely to be different than in solution or on a bead (Bernas *et al.*, 2004; Song *et al.*, 1995). Because these and other factors listed earlier cannot be controlled satisfactorily, calibration and measurements of absolute concentrations of fluorophores in confocal microscopy appear impossible, with one exception. A technique called *ratio imaging* allows one to minimize the influence of local variation in probe concentration to permit absolute measurement of pH and calcium cations concentration (Tsien and Waggoner, 1995).

C. Fluorescence Intensity Ratio Measurements

Let us consider a fluorescent probe whose emission spectrum changes when it interacts with a soluble cell constituent X (Fig. 8). The intensity of fluorescence at λ_1 (I_1) is roughly independent of the concentration of X, but the intensity at λ_2 (I_2) decreases with increasing concentration of X (Fig. 8A, B). A ratio of I_2/I_1 is thus independent of probe concentration and increases with concentration of

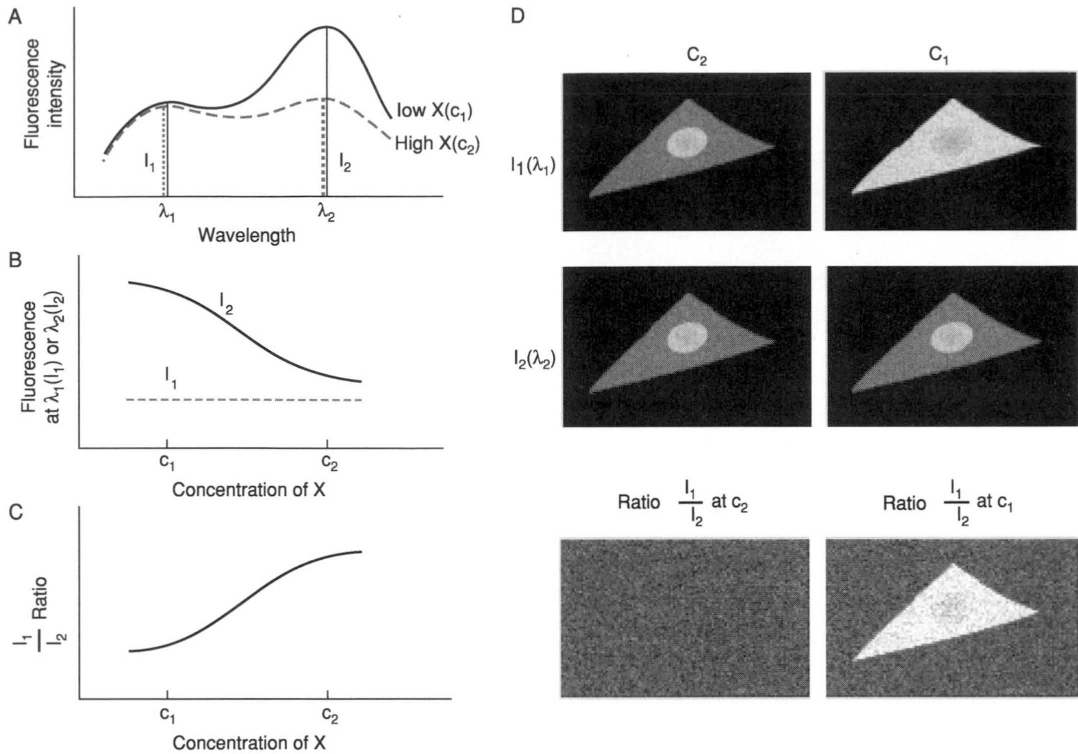

Fig. 8 The principle of ratio measurements. (A) The emission spectrum of a fluorophore is a function of the concentration of X. (B) At the wavelength λ_1 the emission intensity I_1 is independent of X, whereas at λ_2 the emission intensity I_2 decreases with concentration of X. (C) The ratio of emissions at I_1 and I_2 is a measure of the concentration of X. (D) Images collected at λ_1 and λ_2, at two different concentrations of X (c_1, c_2), and a "ratio image" in which each pixel depicts a ratio between values of I_2 over I_1, reveals a concentration of X in the cell.

X (Fig. 8C). Fluorescent signals I_1 and I_2 can be used to build separate images. The ratio I_2/I_1 at any point in space is, thus, a measure of the local concentration of X and is independent of the local concentration of the fluorescent probe itself (as long as neither signal approaches zero). Ratio imaging is, in principle, free of many limitations and problems listed in the previous section. Most importantly, the unavoidable loss of fluorescent probe due to photobleaching or rapid changes in subcellular distribution ought not to influence determinations of X.

Practical difficulties that need to be solved to implement this approach include the following:

1. Maintaining both detection channels in register
2. Accounting for different photobleaching rates and subcellular distributions of the bound and unbound forms (Opitz *et al.*, 1994)

3. Proper subtraction of background signals and autofluorescence

4. Avoiding artifacts caused by phototoxicity and light-induced reactions (Knight *et al.*, 2003)

Ratio measurements are widely used for measurements of intracellular calcium and pH. Technical aspects are treated in Bright *et al.* (1989), Opitz *et al.* (1994), and Opas and Dziak (1999).

IV. Time Resolution in Confocal Microscopy

It takes approximately 1 second to collect a 512-× 512-pixel, 8-bit confocal image of a selected area of a biological specimen in a typical point-scanning confocal microscope. Because the intensity of fluorescence is usually quite low, one needs to average or sum up several frames to collect an image with a reasonable signal/noise ratio. In practice, the time required to collect one image at maximum resolution may be 5–30 seconds. If one collects a 3D image of a cell nucleus 10 μm in diameter, the number of confocal sections should be approximately 30–50. Thus, the time needed to collect such an image may range from 3 to 20 minutes or more. Viewed from this perspective, a typical confocal microscope may seem to be unsuitable for studies of processes occurring in times measured in minutes or even tens of minutes. This is not the case, however, because phenomena occurring on a shorter time scale can be studied in small regions of space. In this case, the instrument can measure fluorescence intensities along the same selected line within a cell or follow changes in a very small rectangular area by repeatedly scanning only this region. When one uses this regimen of data collection, changes occurring within milliseconds can be detected.

Confocal microscopes capable of recording images at a greater rate have been constructed. For example, 2D images can be collected at video rate or even faster. Although constructing such a fast scanning instrument is more complicated and costly, the real obstacle is something else. In most biological specimens, the number of photons collected from a diffraction-limited voxel is too small to record a signal from a 512-×-512 raster in less than 1 second. Thus, fast scanning can yield usable images only if the intensity of the fluorescence is very high.

V. Confocal Techniques Based on Measurements of Fluorescence Intensity vs Time

Measurements of fluorescence intensity within the spatial and temporal limits outlined in the preceding sections can be employed to solve several classes of biological problems. Accumulation or loss of proteins in selected areas of a cell can be investigated. Examples are studies of accumulation of transcription/repair factor in ultraviolet-damaged foci in cell nuclei (Mone *et al.*, 2001), exchange of histones in heterokaryons (Kimura and Cook, 2001), or accumulation and subcellular distribution of the fluorescent antitumor antibiotic daunomycin in living cells (Dobrucki *et al.*, 2003, unpublished observations).

A. FRAP

Mobility of fluorescent species in living cells can be measured using a technique originally called fluorescence photobleaching recovery (FPR) (Axelrod *et al.*, 1976), and currently known as fluorescence recovery after photobleaching (FRAP) (Reits and Neefjes, 2001) or fluorescence redistribution after photobleaching (Houtsmuller and Vermuelen, 2001).

FRAP was first used some 30 years ago for measurements of mobility of fluorescently labeled proteins and lipids in plasma membranes of living cells (Axelrod *et al.*, 1976). Currently, this approach is used primarily for studies of the dynamics of chimeric fluorescent proteins.

The FRAP technique is based on the following principle (Fig. 9). At the start of the experiment, the imaged cell contains a population of fluorescently tagged molecules (Fig. 9A, D). The tag can be a fluorescent compound covalently linked to a lipid or protein. Recently, the most popular tags are fluorescent proteins fused with a protein of interest. A small area of the cell is illuminated with light of intensity sufficiently high to rapidly bleach the fluorescent tags (Fig. 9B, E). The bleached region may be a strip, a small rectangle (Misteli *et al.*, 2000), or an area as large as half the nucleus (Lever *et al.*, 2000). In effect, two subpopulations among the tagged molecules are created: the fluorescent and the nonfluorescent. Subsequently a series of images of the cell are recorded and changes (i.e., recovery) of fluorescence intensity are measured within a bleached spot (Fig. 9C, F, G). These images reveal the mobility of molecules still carrying an unbleached fluorescent tag. If all molecules of the protein under study are permanently attached to stable subcellular structures, the bleached area does not fill with new fluorescently tagged protein molecules (Fig. 9A–C). If, however, some or all molecules of the protein studied are unbound and are, thus, free to exchange and diffuse, the bleached area fills with fluorescence over time (Fig. 9D–F). The rate at which this process occurs is a function of the diffusion constant and the rate(s) of exchange between different binding sites of this protein.

FRAP can measure association and dissociation constants, distribution of a protein between mobile and immobilized pools, and the effective diffusion coefficient of the molecule under study (Carrero *et al.*, 2003; Lippincott-Schwartz *et al.*, 2001). The actual diffusion constant, as defined in chemistry, cannot be determined (Reits and Neefjes, 2001). Originally, the mathematical model describing FRAP was designed for lateral diffusion in membranes. Because FRAP is now used for studies of fluorescent proteins diffusing between and within 3D intracellular compartments, appropriate mathematical models have been proposed (Carrero *et al.*, 2003).

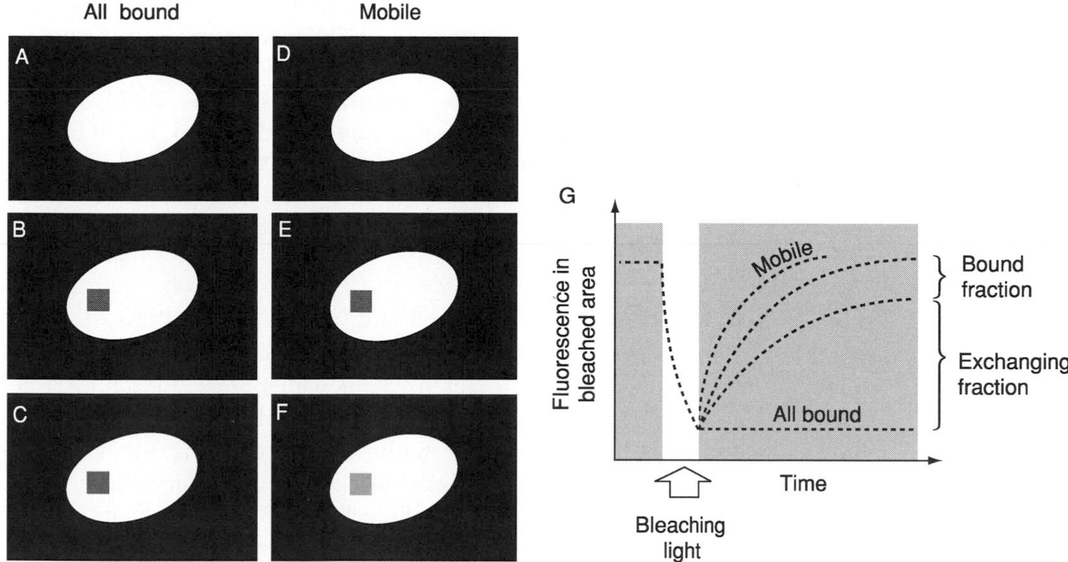

Fig. 9 Principle of the fluorescence recovery after photobleaching (FRAP) technique as used in studies of mobility of fluorescently tagged proteins. A small area of the cell is bleached. If the tagged molecules are immobile, the bleached region does not fill with fluorescence (A–C; G curve marked "All bound"). If a fraction or all tagged molecules are mobile, the fluorescence in the bleached region recovers (D–F; G curves marked "Mobile" depicting various rates of recovery of fluorescence and an incomplete recovery when a fraction of molecules does not exchange with the mobile fraction).

The relevance of FRAP studies hinges on several assumptions. It is assumed that the native untagged and the modified tagged molecules do not differ significantly with respect to their biological role and biophysical properties, including diffusion rates, binding, distribution within subcellular compartments, capacity to catalyze reactions, etc. It is also assumed that the process of photobleaching does not significantly alter the structure and function of the studied cell. In the case of green fluorescent protein (GFP)–tagged proteins, the first group of assumptions can be verified by investigating the function of a tagged protein in cells where the native gene is inactive. Surprisingly, fusing a 238-amino acid fluorescent protein with a native protein has been shown not to cause measurable impairment to the functions of many studied proteins. However, examples of adverse effects and artifactual subcellular redistribution of fluorescently tagged proteins have also been reported (Carrero *et al.*, 2003). Potential phototoxic effects may depend strongly on the fluorescent tag used. In the case of low-molecular-weight fluorescent tags, the formation of oxygen radicals and photo-oxidized forms of fluorophores is to be expected and may lead to disruption of physiological functions and sublethal or lethal damage to cells. However, there is some evidence to suggest that, given the right linker segment and binding site the phototoxicity of GFP is considerably less pronounced than that of other popular low-molecular-weight fluorescent compounds. Mechanisms of photobleaching of fluorescent proteins may be different from that of low-molecular-weight fluorophores and probably do not involve oxygen (Bernas *et al.*, 2004). Phototoxic effects of fluorescent proteins that do occur are not understood. In FRET, phototoxicity may arise not only from illumination of fluorescent fusion proteins, but also from simultaneous photoreactions of mitochondrial flavins, other native fluorescent constituents of living cells, and riboflavin in culture medium.

B. FLIP

In a fluorescence loss in photobleaching (FLIP) approach, a large region of a cell is continuously illuminated with intense light, which causes bleaching (Fig. 10).

Measurements of fluorescence intensity are repeatedly collected from the area that is not bleached (Fig. 10E). If the fluorescent molecule is not bound and is highly mobile, the concentration of this molecule falls rapidly in both the bleached and the unbleached part (Fig. 10F) of the cell. If all fluorescent molecules are permanently bound and cannot exchange, the fluorescence in the bleached part is lost and does not recover, whereas the unbleached part remains fluorescent. If a fraction of molecules is bound and does not exchange, while the remaining molecules are highly mobile, continuous illumination of part of a cell will bleach out the mobile fraction and reveal the bound molecules in the unbleached region (Fig. 10D). FLIP can measure diffusion (continuity) between compartments (i.e., mobile and immobile pools) and can reveal subcellular localization of the immobile pool of a studied protein.

The FRAP method measures influx of a tagged protein into a previously "bleached-out" area of the cell. It is difficult to prove that illumination, which led to bleaching of a probe, had no effect on cell structure and function in the illuminated volume and its immediate vicinity. FLIP focuses on the area that was not previously illuminated and that suffered less or no photodynamic damage. Thus, FLIP has the potential advantage of avoiding artifacts arising from interaction between intense light and a fluorescent tag.

C. FRET

Fluorescence resonance energy transfer (FRET) is the process whereby an excited molecule (called a *donor*) transfers energy to a chromophore on an adjacent molecule (called an *acceptor*) through dipole–dipole interactions. Thus, transfer of energy in FRET does not involve emission of a photon. If the emission spectrum of a donor and the absorbance spectrum of an acceptor overlap significantly (Fig. 11), FRET can occur between the two molecules if their dipoles are aligned and they are within a distance of about 1–10 nm. Nonradiative transfer of energy from a donor to an acceptor is manifested by a decrease of the intensity of donor fluorescence, the shortening of its fluorescence lifetime, and an increase in

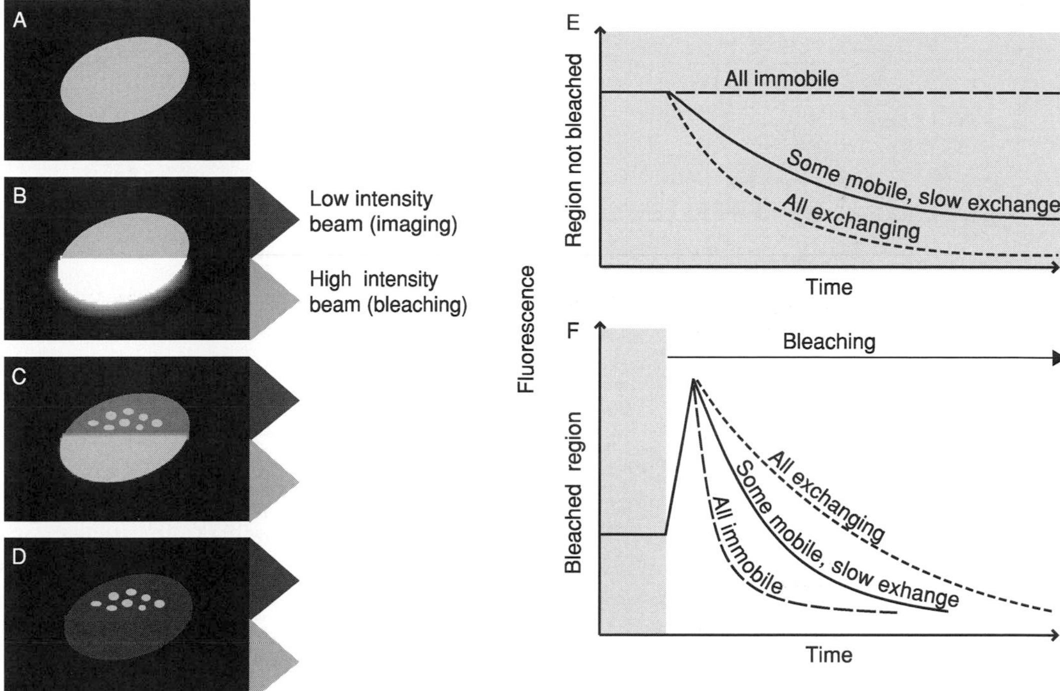

Fig. 10 The principle of the fluorescence loss in photobleaching (FLIP) method. Fluorescent molecules are distributed throughout a cell, depicted by the gray oval (A). The laser beam is scanning the field of view in a raster fashion. Intensity of the scanning laser beam is low when scanning the upper part of the cell and high when scanning the lower part (B). Fluorescence excited by the low-intensity beam is used to build a standard image of the upper part of the cell. The high-intensity beam causes photobleaching of fluorescent molecules in the lower region of the cell. The PMT black level and gain are optimized for imaging the upper part of the cell. At this PMT gain, the lower part is overexposed, the PMT is saturated, and this part of the image initially appears very bright (B). As bleaching progresses, the mobile fraction of fluorescent molecules is depleted in the whole cell and the intensity of fluorescence of both the upper and lower parts decreases (C). The bound molecules in the unbleached upper part remain in place and the location of these molecules is revealed (C, D). The rate of loss of fluorescence in the upper part (E) of the cell is a function of binding constants and diffusion rates of the tagged molecules. Fluorescence intensity in the lower, bleached part of the cell is also shown (F).

the intensity of acceptor fluorescence. The rate of energy transfer in FRET decreases with the sixth power of the distance between the donor and acceptor. Thus, measurements of FRET can be used to assess changes of distance between interacting fluorescent molecules. Hence, the technique acquired the name *spectroscopic ruler*. FRET can be observed in fixed and live cells and has been applied in studies of protein–protein interactions, membrane potentials, and calcium metabolism (Herman *et al.*, 2001; Pollock and Heim, 1999).

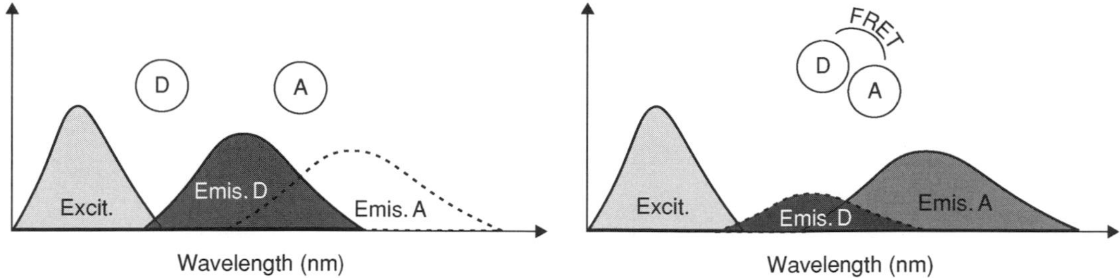

Fig. 11 Fluorescence resonance energy transfer: schematics of excitation and emission spectra for a donor (D) and an acceptor (A).

In principle, FRET can be observed by exciting a donor and collecting the emission of the acceptor. In practice, detecting the genuine FRET signal requires careful optimization of experimental conditions and refined data collection strategies. In microscopy, FRET is detected by measuring changes in the relative intensities of emission at the wavelength selected for the donor and acceptor. The ratio between the two emission intensities is a function of the average distance between donor and acceptor. Photobleaching must be carefully controlled because loss of fluorescent donor molecules may affect this ratio. Self-quenching of a fluorescent label present at a high local concentration may also introduce artifacts.

A pair of fluorescent proteins, cyan fluorescent protein (CFP) and yellow fluorescent protein (YFP), provides a good example of problems and solutions associated with detection of FRET in confocal microscopy (Karpova *et al.*, 2003) (Fig. 12).

If CFP is excited using the 458-nm line from a low-power argon laser, CFP will also be excited. Because the emission spectrum of CFP extends far into the YFP emission range, it is not possible to excite the donor exclusively using a popular argon laser, nor is it possible to collect only the emission from the acceptor. Even with optimal filter selection, some direct excitation of the acceptor will occur, and when acceptor fluorescence is collected, it will contain a significant contribution from donor emission. As a result, FRET cannot be observed directly by looking at the acceptor emission. Bleed-through of CFP emission into the YFP detection channel has to be corrected and the effects of direct excitation of YFP during excitation of CFP must be accounted for. An approach applied successfully in studies of FRET between CFP and YFP relies on acceptor photobleaching. In this technique, the acceptor is photobleached, and as a result, the intensity of donor fluorescence increases, as shown schematically in Fig. 12C. The advantage of this technique is that bleed-through is eliminated by bleaching.

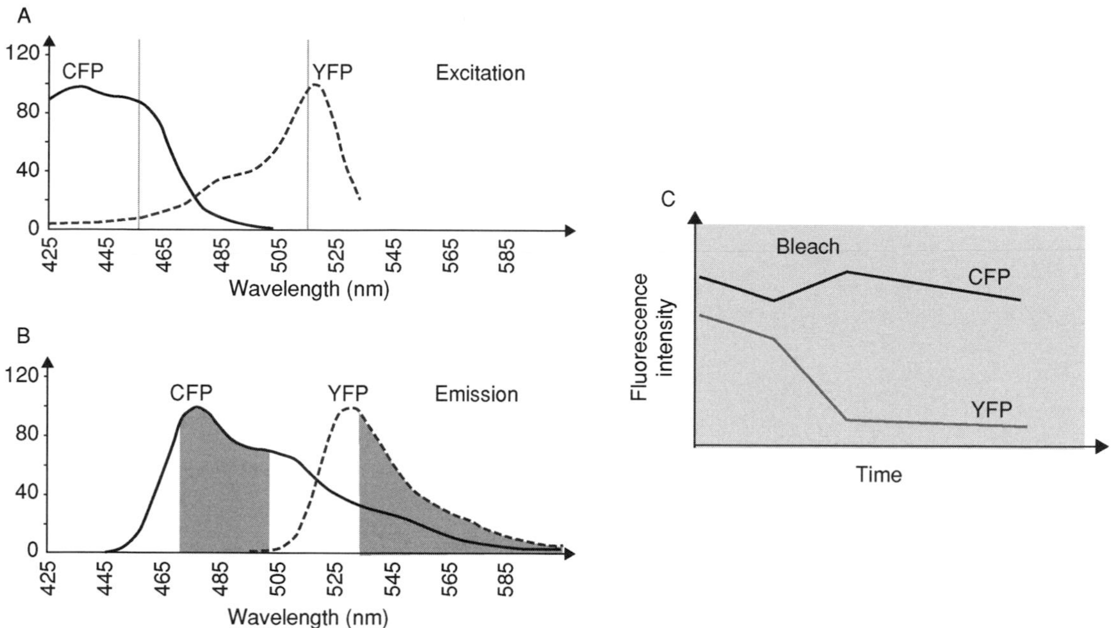

Fig. 12 Fluorescence resonance energy transfer of CFP to YFP. Excitation spectra (A), emission spectra (B), and a diagram showing changes of intensity of CFP before and after bleaching YFP (C). (Karpova *et al.*, 2003.)

D. FLIM

In a typical confocal microscopy experiment, a sample is illuminated with excitation light of a constant intensity. If a pulsed laser and a detector capable of high-frequency modulation or fast gating is used (Lakowicz, 1999) (Fig. 13), the decay of fluorescence can be studied. Fluorescence lifetimes are measured in all voxels of a selected sample volume. The time required for fluorescence to decay is sensitive to environmental conditions but is independent of the fluorescent probe concentration and the length of the excitation light path. Because the same fluorescent probe yields different lifetimes in different subcellular regions characterized by different molecular environments, fluorescence lifetime imaging microscopy (FLIM) adds a new parameter in multidimensional confocal microscopy (Bastiaens and Squire, 1999; Hanley *et al.*, 2001). It has been applied to measurements of a number of parameters in living cells, including intracellular pH, calcium, and oxygen concentrations (Gerritsen *et al.*, 1997). Because of poor signal/noise ratios, relevant studies of FLIM in live cells require careful optimization of conditions to avoid phototoxicity (Dobrucki, 2001).

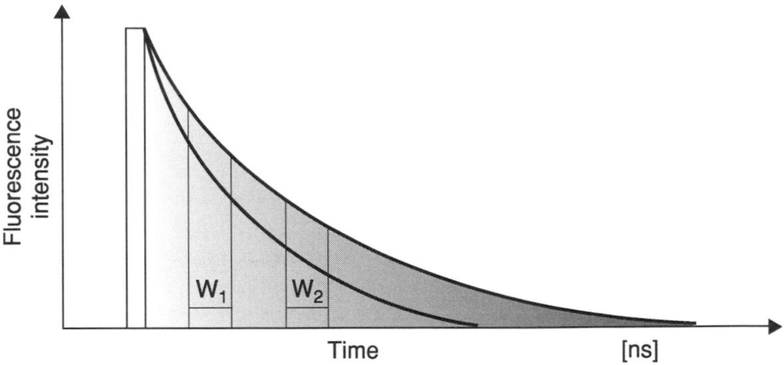

Fig. 13 Principle of fast-gating measurements of fluorescence lifetimes. A pulse of light (*vertical white bar*) excites fluorescence, which decays within a few nanoseconds. The time of decay depends on the type of molecule and on interactions with other molecules in the immediate vicinity (Lakowicz, 1999). Two measurements of intensity are collected at time intervals W_1 and W_2 and a fitting procedure is applied to calculate the half-life of fluorescence.

E. FCS

Fluorescence correlation spectroscopy (FCS) is based on measurements of fluctuations in the number of photons collected as fluorescently labeled molecules diffuse in and out of the confocal volume (Lippincott-Schwartz *et al.*, 2001; Schwille, 2001). These fluctuations reflect changes in the average number of fluorescent molecules in the volume. FCS can measure concentrations and diffusion constants in intracellular compartments, as well as free and bound fractions of several diffusing fluorescent species simultaneously.

VI. Spectrally Resolved Confocal Microscopy

A. Channel Bleed-Through Correction

Two fluorescent probes can be used to label two types of molecules and two distinct subcellular structures in a cell simultaneously. Their fluorescent signals of different wavelengths can be separated by optical filters and detected by separate light detectors. The images then are displayed as separate color panels or are overlaid. This way, the relative positions of two subcellular structures can be analyzed and the distribution of two molecules of interest between these structures can be determined. The task of multicolor imaging is relatively easy if the emission spectra of both fluorescent probes are well separated. It is common, however, for the emission (and excitation) spectra to exhibit significant overlap (Fig. 14A). In this case, photons emitted by fluorescent probe G associated with a structure 1 reach not only detector (PMT) 1, but also detector 2 (Fig. 14C).

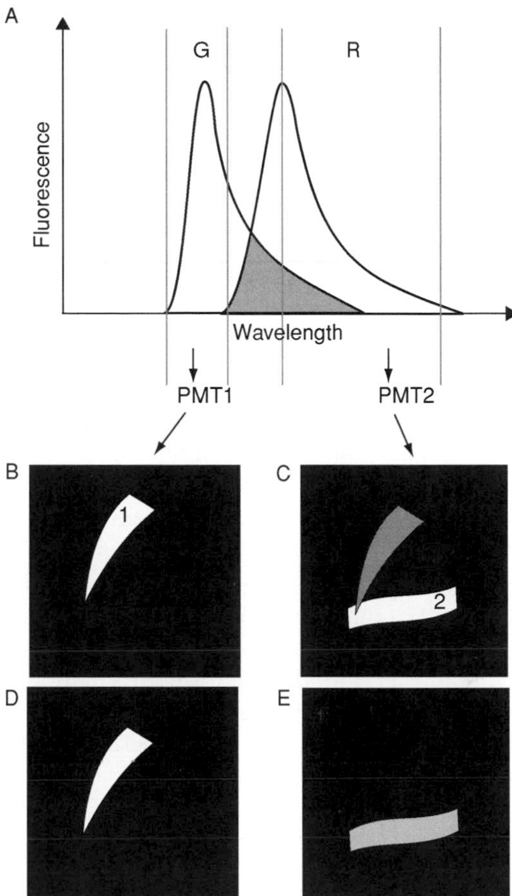

Fig. 14 Bleed-through in two-color confocal microscopy. Subcellular structure 1 is stained with fluorescent probe G, and structure 2 with fluorescent probe R. Probe G emits mostly in the green but has some emission in the red; probe R has only emission in the red (A). Fluorescence of probes G and R is excited simultaneously. The resulting green image (B) correctly represents structure 1 stained with G; the red image (C) shows—incorrectly—that structures 1 and 2 are stained with R. When 15% of the value of signal G is subtracted from signal R in every pixel, the image in channel 2 depicts only structure 2 (E).

This small proportion of photons will be displayed within a multicolor image, erroneously indicating the presence of fluorescent probe R in structures that were stained only with G (Fig. 14C). This phenomenon is known as *bleed-through* or *crosstalk.* The standard way to circumvent this problem is to excite each fluorophore separately and collect emission using optimally selected filters. If excitation wavelengths appropriate for sequential excitation are not available, mathematical normalization for bleed-through can be performed. If the spectral shape of G and

R are both constant in all pixels of the image, and if the light detector has a linear response, the relative contribution of the "unwanted" G signal appearing in channel 2 should be the same in all pixels of the image. Thus, the bleed-through can be corrected by subtracting a known contribution (arising from spectral overlap) of signal G from signal R. This operation can be performed live during data collection by subtracting a predetermined percentage of signal G recorded in channel 1 from signal R recorded in channel 2. This approach is used for correcting small crosstalk contributions in one direction, that is, when signal G bleeds into detector R, but there is no crosstalk of R into G. This way, the spatial distribution of both fluorescent probes and their binding to structures 1 and 2 can be determined.

Unfortunately, the assumption of uniform shape of the spectrum throughout the image is not always justified. It is known that the shape of the spectrum of many probes changes in response to microenvironmental conditions. Moreover, other factors influence the intensity of spectra and the linearity of detection, including the contribution from autofluorescence, which varies from pixel to pixel, different photobleaching rates in regions of different dye concentrations (Bernas *et al.*, 2004), different singlet-state saturation properties of fluorescent probes, chromatic aberration and register shift between channels, and nonideal alignment of optical components. Thus, one can argue that using one flat value for bleed-through correction in all pixels of the image yields a good approximation but cannot give totally accurate information about spatial distribution of probes with overlapping spectra (Brelje *et al.*, 1993).

B. Colocalization Analysis

Assuming that a reasonably accurate bleed-through correction has been achieved, the multicolor confocal image can be used to assess the proportion of regions in the sample in which two types of molecules occur in proximity and to identify the regions stained exclusively by one of the probes. The parameter describing the proportion of image pixels containing signals of both fluorescent probes is called *colocalization.* Typically, colocalization is quantified using the Pearson's correlation coefficient. Software packages supplied with confocal microscopes include colocalization modules for calculating colocalization coefficients. Calculations of colocalization coefficients are sensitive to noise and selected signal thresholds and may be difficult to interpret. It has been pointed out that a more refined definition and display of colocalization is required. Several methods of calculating colocalization have been proposed (Adler *et al.*, 2003; Manders *et al.*, 1993).

The practical meaning of *colocalization* deserves some discussion. Colocalization between two fluorescent labels may be understood as evidence that the labeled structures are in the same position in 3D space. In fact, however, colocalization indicates only the presence of both structures in the smallest distinguishable volume element (see earlier discussion)—a voxel, or a group of voxels. Thus, two fluorescently labeled molecules may be imaged in one voxel, even when the distance separating them is two orders of magnitude greater than their size. For example, colocalization of two labels in the whole nucleus can be interpreted correctly as proof of the simultaneous presence of both labeled molecules within this nucleus. However, when a labeled protein colocalizes with the nuclear membrane, it may actually reside within the membrane itself or in the reticulum just outside the nucleus. When a label colocalizes with the plasma membrane, it is impossible to determine whether it is submerged within the membrane or is located tens of nanometers from one or another side. These questions, relevant to structure and function, cannot be resolved (optically and logically) exclusively on the basis of colocalization coefficients derived from a confocal multicolor image.

C. Multispectral Imaging and Unmixing

Let us consider an experiment in which subcellular structures are labeled with several, not just two, probes emitting fluorescence of different colors. The goal of the experiment is to determine the binding of these probes to various subcellular structures, the spatial distribution of the probes to identify the position of their targets, and the local concentration of target molecules. The individual spectra of these fluorophores are shown schematically in Fig. 15.

The local concentrations of these fluorophores are expected to be very different, reflecting different concentrations of their targets. Thus, the intensity of light emitted at various wavelength bands will vary from pixel to pixel. Consequently, the shape of a convoluted emission spectrum will differ from pixel to pixel as well. Because the overlap of the spectra is extensive, it is not possible to design a set of interference emission filters capable of separating the light bands derived from each of the fluorophores. With this degree of spectral overlap and number of fluorophores, any filter is bound to pass photons originating from more than one fluorophore. Under such conditions, separation of signals originating from different fluorophores can be achieved only by splitting fluorescence into several optimally selected emission bands (Neher and Neher, 2004), registering images in each band, and performing subsequent mathematical analysis.

Detection of discrete spectral bands is achieved by placing either a dispersive element such as a grating or prism or a selection of narrow-band filters in the path of the emitted light. Fluorescence is split into several spectral bands that are simultaneously detected by a multidetector array or by photomultipliers. This way, a series of images is collected, each corresponding to one selected spectral band (this series is often called a *lambda stack*). Using just a large number of narrow spectral windows may discard photons in favor of spectral separation and result in very weak signals reaching the detector, thereby defeating the purpose of the whole exercise. Thus, the optimal selection of spectral detection bands is crucial (Neher and Neher, 2004).

The spectral data acquired in the way described above can now be deconvoluted or "unmixed." The relative contribution of each fluorophore to each detection

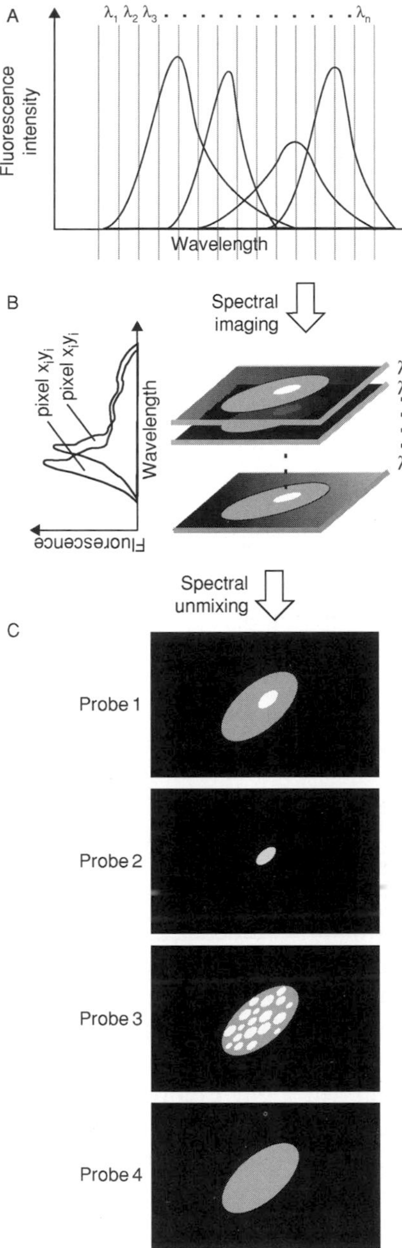

Fig. 15 Spectrally resolved confocal microscopy. (A) Overlapping spectra of four different fluorophores. (B) Spectral imaging: a set of images is collected through narrow spectral windows (a lambda stack); a schematic representation of different spectra emitted in different pixels of the image is shown. (C) Unmixing: spectral components are deconvoluted, bleed-through (crosstalk) between all probes is corrected, and images based on fluorescence signals originating from each probe alone are displayed.

channel must be measured in the absence of other fluorescing species during calibration experiments. The number of detection channels must be equal to or greater than the number of fluorophores (Neher and Neher, 2004). Eventually, the contribution from each of the overlapping fluorophores to the emission detected in each individual pixel is calculated. Thus, images based on emission of each probe alone can be constructed (Fig. 15C).

The principal advantage of spectral imaging and unmixing is the ability to image several fluorophores with overlapping emission bands using one excitation wavelength and a suboptimal set of filters. This technique can work well only when signal level is sufficiently high to preserve a good signal/noise ratio in each spectral band. Thus, separation of weak signals from autofluorescence may pose a problem. Over time, the different rates of photobleaching of different fluorophores distort spectral information. The long time required to collect the lambda stack and overcome low signal/noise ratios may result in phototoxicity when live cells are studied.

VII. Conclusion

A standard fluorescence confocal microscope can faithfully represent size, distance, and shape, position in 3D space, surface profiles, and volume of structures exceeding the wavelength of visible light; the smaller structures may be detected and their position can be established accurately, but their size and shape cannot be accurately conveyed. Measurements of relative fluorescence intensities require that numerous instrumental and experimental factors be carefully controlled; calculating absolute concentrations of fluorescent probes and their cellular targets on the basis of their measured fluorescence intensities is not possible, with the exception of ratio measurements. Signals of multiple fluorescent labels with overlapping emission spectra can be deconvoluted and analyzed independently using spectrally resolved confocal imaging. Dynamic processes and molecular interactions in living cells, including protein diffusion, mobility, exchange, and binding, can be studied in subcellular compartments of living cells, using FRAP, FLIP, and FRET techniques in standard confocal microscopes, or using FLIM and FCS in specially adapted advanced instruments. The spatial resolution of new optical imaging techniques such as 4Pi and STED goes beyond the limit of diffraction and reaches nanometer scale. These approaches may open new areas of biological confocal imaging.

Acknowledgments

I gratefully acknowledge financial support of The Wellcome Trust (London), the Polish State Committee for Scientific Research (Warsaw), and the Foundation for Polish-German Cooperation (Warsaw). I thank Drs. J. Pawley, G. van den Engh, J. P. Robinson, and B. Rajwa for helpful discussions and critical comments on the manuscript; I am indebted to Ms. G. Lawler for language corrections and Ms. J. Deszcz for the artwork.

References

Adler, J., Parmryd, I., McKay, I., and Pagakis, S. N. (2003). "On the quantification and representation of colocalisation." Abstract presented at: Conference Focus on Microscopy, Genoa.

Adler, J., and Pagakis, S. N. (2003). Reducing image distortions due to temperature-related microscope stage drift. *J. Microsc.* **210,** 131–137.

Albrecht, B., Failla, A. V., Schweitzer, A., and Cremer, C. (2002). Spatially modulated illumination microscopy allows axial distance resolution in the nanometer range. *Appl. Opt.* **41,** 80–87.

Axelrod, D., Koppel, D. E., Schlessinger, J., Elson, E., and Webb, W. W. (1976). Mobility measurement by analysis of fluorescence photobleaching recovery kinetics. *Biophys. J.* **16,** 1055–1069.

Bahlman, K., Jakobs, S., and Hell, S. W. (2001). 4Pi-confocal microscopy of live cells. *Ultramicroscopy* **87,** 155–164.

Bastiaens, P. I., and Squire, A. (1999). Fluorescence lifetime imaging microscopy: Spatial resolution of biochemical processes in the cell. *Trends Cell Biol.* **9,** 48–52.

Bernas, T., Zarebski, M., Cook, P. R., and Dobrucki, J. W. (2004). Minimizing photobleaching during confocal microscopy of fluorescent probes bound to chromatin: Role of anoxia and photon flux. *J. Microsc.* **215**(3), 281–296.

Bornfleth, H., Sätzler, K., Eils, R., and Cremer, C. (1997). High precision distance measurements and volume-conserving segmentation of objects near and below the resolution limit in three-dimensional confocal fluorescence microscopy. *J. Microsc.* **189,** 118–136.

Bright, G. R., Fischer, G. W., Rogowska, J., and Taylor, D. L. (1989). Fluorescence ratio imaging microscopy. *Methods Cell Biol.* **30,** 157–192.

Brelje, T. C., Wessendorf, M. W., and Sorenson, R. L. (1993). Multicolor laser scanning confocal immunofluorescence microscopy: Practical application and limitations. *Methods Cell Biol.* **38,** 97–181.

Carrero, G., McDonald, D., Crawford, E., de Vries, G., and Hendzel, M. J. (2003). Using FRAP and mathematical modeling to determine the *in vivo* kinetics of nuclear proteins. *Methods* **29,** 14–28.

Cox, G. C. (1999). Measurements in the confocal microscope. *In* "Confocal Microscopy Methods and Protocols" (S. W. Paddock, ed.). Humana Press, Totowa, NJ.

Dobrucki, J. W. (2001). Interaction of oxygen-sensitive luminescent probes $Ru(phen)_3^{2+}$ and $Ru(bipy)_3^{2+}$ with animal and plant cells *in vitro.* Mechanism of phototoxicity and conditions for non-invasive oxygen measurements. *J. Photochem. Photobiol. B.* **65,** 136–144.

Dyba, M., Jakobs, S., and Hell, S. W. (2003). Immunofluorescence stimulated emission depletion microscopy. *Nat. Biotechnol.* **21,** 1303–1304.

Egner, A., Jakobs, S., and Hell, S. W. (2002). Fast 100-nm resolution three-dimensional microscope reveals structural plasticity of mitochondria in live yeast. *Proc. Natl. Acad. Sci. USA* **99,** 3370–3375.

Fricker, M. D., and White, N. S. (1992). Wavelength considerations in confocal microscopy of botanical specimens. *J. Microsc.* **166,** 29–42.

Gerritsen, H. C., Sanders, R., Draaijer, A., Inee, C., and Levine, Y. K. (1997). Fluorescence lifetime imaging of oxygen in living cells. *J. Fluoresc.* **7,** 11–15.

Gustafsson, M. G. (1999). Extended resolution fluorescence microscopy. *Curr. Opin. Struct. Biol.* **9,** 627–634.

Hanley, Q. S., Subramaniam, V., Arndt-Jovin, D. J., and Jovin, T. M. (2001). Fluorescence lifetime imaging: Multi-point calibration, minimum resolvable differences, and artifact suppression. *Cytometry* **43,** 248–260.

Hell, S. W. (2003). Toward fluorescence nanoscopy. *Nat. Biotechnol.* **21,** 1347–1355.

Herman, B., Gordon, G., Mahajan, N., and Centonze, V., Measurement of fluorescence resonance energy transfer in the optical microscope. *In* "Methods in Cellular Imaging" (A. Periasamy, ed.). Oxford University Press, Oxford.

Houtsmuller, A. B., and Vermeulen, W. (2001). Macromolecular dynamics in living cell nuclei revealed by fluorescence redistribution after photobleaching. *Histochem. Cell Biol.* **115,** 13–21.

Inoué, S., and Spring, K. R. (1997). Video microscopy: The fundamentals, 2nd ed. Plenum Press, New York.

Inoué, S., (1995). Foundations of confocal scanned imaging in light microscopy. *In* "Handbook of Biological Confocal Microscopy" (J. B. Pawley, ed.), 2nd ed. Plenum Press, New York.

Karpova, T. S., Baumann, C. T., He, L., Wu, X., Grammer, A., Lipsky, P., Hager, G. L., and McNally, J. G. (2003). Fluorescence resonance energy transfer from cyan to yellow fluorescent protein detected by acceptor photobleaching using confocal microscopy and a single laser. *J. Microsc.* **209,** 56–70.

Keller, E. H. (1995). Objective lenses for confocal microscopy. *In* "Handbook of Biological Confocal Microscopy" (J. B. Pawley, ed.), 2nd ed. Plenum Press, New York.

Kimura, H., and Cook, P. R. (2001). Kinetics of core histones in living human cells: Little exchange of H3 and H4 and some rapid exchange of H2B. *J. Cell. Biol.* **153,** 1341–1353.

Knight, M. M., Roberts, S. R., Lee, D. A., and Bader, D. L. (2003). Live cell imaging using confocal microscopy induces intracellular calcium transients and cell death. *Am. J. Physiol. Cell Physiol.* **284,** C1083–C1089.

Lakowicz, J. R. (1999). "Principles of fluorescence spectroscopy," 2nd ed. Kluwer Academic/Plenum Publishers, New York.

Lever, M. A., Th'ng, J. P., Sun, X., and Hendzel, M. J. (2000). Rapid exchange of histone H1.1 on chromatin in living human cells. *Nature* **408,** 873–876.

Lindek, S., Stelzer, E. H., and Hell, S. W. (1995). Two new high-resolution confocal fluorescence microscopies (4Pi, Theta) with one and two-photon excitation. *In* "Handbook of Biological Confocal Microscopy" (J. B. Pawley, ed.), 2nd ed. Plenum Press, New York.

Lippincott-Schwartz, J., Snapp, E., and Kenworthy, A. (2001). Studying protein dynamics in living cells. *Nat. Rev. Mol. Cell Biol.* **2,** 444–456.

Manders, E. M., Verbeek, F. J., and Aten, J. A. (1993). Measurement of co-localisation of objects in dual-colour confocal images. *J. Microsc.* **169,** 375–382.

Misteli, T., Gunjan, A., Hock, R., Bustin, M., and Brown, D. T. (2000). Dynamic binding of histone H1 to chromatin in living cells. *Nature* **408,** 877–881.

Mone, M. J., Volker, M., Nikaido, O., Mullenders, L. H., van Zeeland, A. A., Verschure, P. J., Manders, E. M., and van Driel, R. (2001). Local UV-induced DNA damage in cell nuclei results in local transcription inhibition. *EMBO Rep.* **2,** 1013–1017.

Nagorni, M., and Hell, S.W. (1998). 4Pi-confocal microscopy provides three-dimensional images of the microtubule network with 100- to 150-nm resolution. *J. Struct. Biol.* **123,** 236–247.

Neher, R., and Neher, E. (2004). Optimizing imaging parameters for the separation of multiple labels in a fluorescence image. *J. Microsc.* **213,** 46–62.

Oldfield, R. (1994). Light microscopy. An illustrated guide. Wolfe Publishing, London, England.

Opas, M., and Dziak, E. (1999). Intracellular pH and Ca measurement. *In* "Confocal Microscopy Methods and Protocols" (S. W. Paddock, ed.). Humana Press, Totowa, NJ .

Opitz, N., Merten, E., and Acker, H. (1994). Evidence for redistribution-associated intracellular pK shifts of the pH-sensitive fluoroprobe carboxy-SNARF-1. *Pflugers Arch.* **427,** 332–342.

Piston, D. W. (1998). Choosing objective lenses: The importance of numerical aperture and magnification in digital optical microscopy. *Biol. Bull.* **195,** 1–4.

Pawley, J. B. (1995). Fundamental limits in confocal microscopy. *In* "Handbook of Biological Confocal Microscopy" (J. B. Pawley, ed.), 2nd ed. Plenum Press, New York.

Pawley, J. B. (2000). The 39 steps: A cautionary tale of quantitative 3-D fluorescence microscopy. *BioTechniques* **5,** 884–886.

Pawley, J. B. (2002). Limitations on optical sectioning in live-cell confocal microscopy. *Scanning* **24,** 241–246.

Pollock, B. A., and Heim, R. (1999). Using GFP in FRET-based applications. *Trends Cell Biol.* **9,** 57–60.

Reits, E. A., and Neefjes, J. J. (2001). From fixed to FRAP: Measuring protein mobility and activity in living cells. *Nat. Cell Biol.* **3,** E145–E147.

Schrader, M., Bahlmann, K., Giese, G., and Hell, S. W. (1998). 4Pi-confocal imaging in fixed biological specimens. *Biophys. J.* **75,** 1659–1668.

Song, L., Hennink, E. J., Young, I. T., and Tanke, H. J. (1995). Photobleaching kinetics of fluorescein in quantitative fluorescence microscopy. *Biophys. J.* **68,** 2588–2600.

Schwille, P. (2001). Fluorescence correlation spectroscopy and its potential for intracellular applications. *Cell Biochem. Biophys.* **34,** 383–408.

Tsien, R. Y., and Waggoner, A. (1995). Fluorophores for confocal microscopy. *In* "Handbook of Biological Confocal Microscopy" (J. B. Pawley, ed.), 2nd ed. Plenum Press, New York.

Underwood, E. E. (1970). Quantitative Stereology. Addison-Wesley, Reading, MA.

Zucker, R. M., and Price, O. (2001). Evaluation of confocal microscopy system performance. *Cytometry* **44,** 273–294.

CHAPTER 4

Surface-Plasmon–Coupled Emission: New Technology for Studying Molecular Processes

Zygmunt Gryczynski, Ignacy Gryczynski, Evgenia Matveeva, Joanna Malicka, Kazimierz Nowaczyk, and Joseph R. Lakowicz

Center for Fluorescence Spectroscopy
University of Maryland at Baltimore
Department of Biochemistry and Molecular Biology
Baltimore, Maryland 21201

I. Introduction

The use of fluorescence technology in biology, biotechnology, and medicine has been rapidly increasing over the last decade. Fluorescence has become the foundation of numerous analyses in sensing, medical diagnostics, biotechnology, and

METHODS IN CELL BIOLOGY, VOL. 75

0091-679X/04 $35.00

gene expression. As examples, DNA sequencing by fluorescence was first reported in 1987 (Prober *et al.*, 1987; Smith *et al.*, 1986), resulting in completion of the sequence of the human genome by 2001, just 14 years later (The Human Genome Feb. 15 and Feb. 16, 2001). Fluorescence detection has replaced radioactivity in most medical testing (Ichinose *et al.*, 1987; Kricka *et al.*, 1997). Fluorescence methods have been extended to high-throughput studies of gene expression using the so-called *gene chips*, which can contain more than 30,000 individual DNA sequences (Brown and Botstein, 1999; Lipschutz *et al.*, 1999), almost equal to the number of human genes. These applications rely on the free-space spectral properties of the fluorophores, which is the emission into a transparent nonabsorbing medium (solutions). In free space, the excited fluorophores emit nearly isotropically in all directions with the radiative decay rate (Γ) given by the Strickler and Berg relation (Lakowicz, 1994). Under such conditions, it is practically impossible to collect more than a small percentage of the emitted photons, unless one uses expensive microscope objectives and optics.

In spite of high sensitivity of fluorescence detectors, there is always a need for increased sensitivity to detect fewer target molecules. Several factors limit the sensitivity of fluorescence methods. The most important are photodestruction of the fluorophores, extent of background fluorescence, and efficiency of collection of emitted photons. We have reported different approaches to modify the emissive properties of fluorophores to increase detection sensitivity. In several papers, we describe the large effects on the emission properties of fluorophores in proximity of conducting metallic particles (silver) (Lakowicz *et al.*, 2001, 2002, 2003a). In particular, the observed effects included enhanced excitation, increased quantum yield, and photostability accompanied by the dramatic decrease of fluorescence lifetime. These results indicate a substantial increase of radiative decay rate (Γ), indicating real possibilities for effective control of fluorophore radiative processes (Lakowicz, 2001).

A valuable approach to improve sensitivity is to increase the fraction of the total emission collected by the detector. In this chapter, we describe a new opportunity for fluorescence detection that results from the interaction of excited fluorophores with the nearby thin continuous metallic surface. Close proximity of fluorophore to metal–dielectric interface results in high-efficiency emission coupling through a thin metal film with a strongly directional emission, which we call *surface-plasmon–coupled emission* (SPCE). This remarkable phenomenon is a consequence of the near-field interaction of fluorophores with nearby semi-transparent planar noble metal surfaces and/or metal-coated periodic structures (gratings) that provides a new pathway for creation of far-field radiation. The resulting SPCE is highly polarized with precisely defined direction either back into the substrate or into the sample media, depending on the details of the optical system. This new approach can provide 50% light collection efficiency and intrinsic wavelength resolution with the use of very simple optics. These properties can result in a wide range of simple, inexpensive, and robust devices with generic usefulness in biology and medicine. We stress that the directional SPCE is *not* due to reflections, but to coupling of the oscillating dipoles of the excited fluorophores with surface plasmons on the metallic surfaces, which in turn radiate into the substrate or sample.

In this chapter, we describe interaction of fluorophores with metallic surfaces that results in SPCE, which significantly increases detection sensitivity. This new approach allows direction of a large fraction of the emission toward the detector by coupling the emission to surface plasmons on the metallic surfaces. The coupling angle and efficiency of SPCE strongly depend on the interface conditions. This new approach will be very useful for detecting biomolecular binding on the surfaces, interaction, and conformational changes of biomolecular systems and will provide a novel platform for biomedical assay development.

II. Overview of Surface Plasmon Resonance

The effects of metallic surfaces on fluorescence have been described in the optical physics literature (Amos and Barnes, 1997; Barnes, 1998; Drexhage, 1974; Haroche and Kleppner, 1989; Haroche and Raimond, 1993; Hinds, 1991; Worthing and Barnes, 1999). However, we believe SPCE is more closely related to surface-plasmon resonance (SPR). The phenomenon of SPR is complex and the underlying principles are new to many individuals with experience in biochemistry and/or fluorescence spectroscopy (Raether, 1977). SPCE can be understood from the theory and practice of SPR, which is used to measure biomolecule binding to surfaces, as in the Biacore apparatus (www.biacore.com). To avoid confusion, we refer to this application of SPR as surface-plasmon resonance analysis (SPRA), which is widely used as a generic approach to measurement of biomolecule interactions on surfaces (Cooper, 2002; Frutos and Corn, 1998; Liedberg and Lundstrom, 1993; Melendez *et al.*, 1996; Salamon and Tollin, 1996; Wegner *et al.*, 2002). A schematic description of SPRA is shown in Fig. 1. The measurement is based on the interaction of light with thin metal films on a glass substrate. The film is typically made of gold 40–50 nm thick. The chosen thickness is determined by the optical constants of the metal and substrate. The gold analysis surface, distal from the glass prism, contains capture biomolecules, which have affinity for the analyte of interest. The capture biomolecules are typically covalently bound to dextran, which is covalently bound to the gold surface. The analysis substrate is optically coupled to a hemispherical prism by an index matching fluid. Light impinges on the gold film through the prism, which is called the *Kretschmann configuration*. The instrument measures the reflectivity of the gold film at various angles of incidence (θ), with the same angle used for observation (θ). Other configurations can be used, such as a triangular prism or more complex optical geometry and a position-sensitive detector. In any event, the measurement is the same: reflectivity of the gold surface versus an angle of incidence.

The usefulness of SPRA is due to the strong dependence of the gold film reflectivity (in particular the angle for minimum reflectivity) on the refractive index of the solution (medium) immediately above the 47-nm gold film, a gold

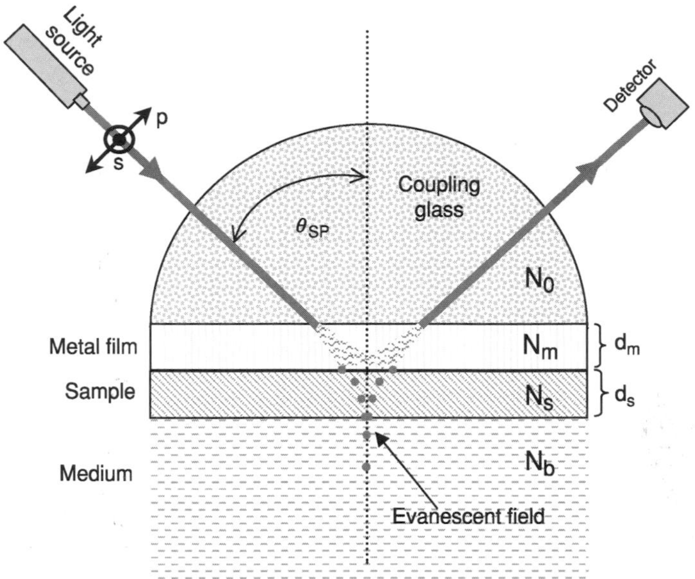

Fig. 1 Typical configuration for surface-plasmon resonance analysis. (See Color Insert.)

thickness thought to be ideal for SPRA with visible wavelengths (Frey *et al.*, 1995; Frutos and Corn, 1998; Jordan *et al.*, 1994). Binding of macromolecules above the gold film causes small changes in the refractive index, which result in changes in reflectivity. Figure 2 shows typical SPRA data, a plot of reflectivity versus the angle of incidence. The reflectivity minimum at given angle (θ_{SP}) is called the *SPR*. The SPR angles change as the gold surface is coated with 11-mercaptoundecanoic acid (MU), then biotinylated poly-lysine (PL), and finally avidin. The changes in the SPR angle are due to changes in the refractive index near the gold surface due to the adsorbed layers. The angular changes seen in Fig. 2 are much larger than observed in a typical SPRA experiment.

It is important to realize why the minimum reflectivity angle θ_{SP} is sensitive to the refractive index in the aqueous medium when the light is reflected by the gold film. This sensitivity is due to an evanescent field, which penetrates approximately 200 nm into the solution (Fig. 1). The evanescent field appears whenever there is resonance between the incident beam and the gold surface and is not present when there is no plasmon resonance, that is, at angles where the reflectivity is high. The evanescent field above the metal due to SPR is much more intense than the field due to TIR, which is not a resonance phenomenon. Interaction of the evanescent field with adsorbed biomolecules above the gold surface shifts θ_{SP} because of the change in the effective refractive index, sensed by the evanescent field.

An important characteristic of the attenuated reflectivity angles in SPR is that they are strongly dependent on wavelength. Figure 3 shows the reflectivity curves

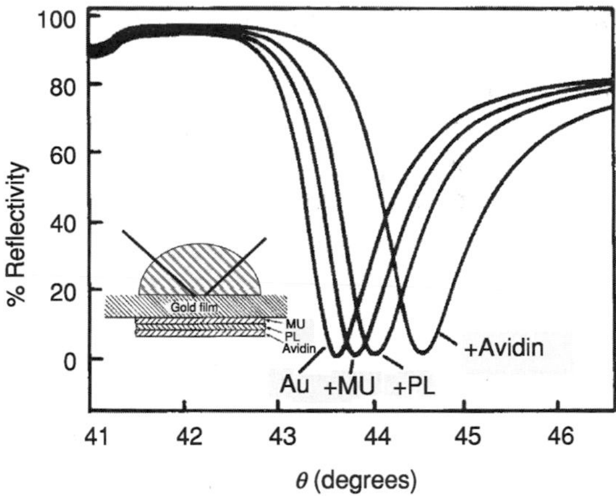

Fig. 2 Surface-plasmon resonance reflectivity curves for a 47-nm gold film on BK-7 glass. Illumination at 633 nm. The gold film was progressively coated with 11-mercaptoundecanoic acid, followed by biotinylated poly-lysine, then avidin. [Adopted from Frutos and Corn (1998) and Lakowicz (2004).]

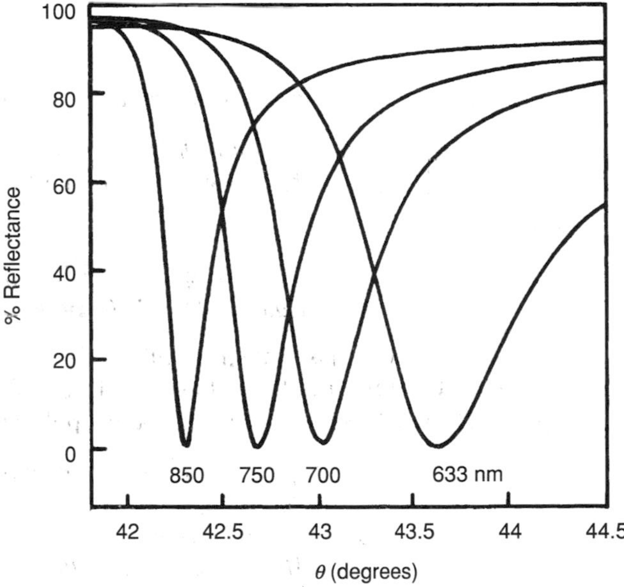

Fig. 3 Calculated wavelength-dependent reflectivity for a 47-nm thick gold film. [Adopted from Natan and Lyon (2002) and Lakowicz (2004).]

of a gold film for several wavelengths (Natan and Lyon, 2002). The surface-plasmon angle increases as the wavelength decreases. The dependence on wavelength can be understood in terms of the refractive indexes (dielectric constants) of the metals and dielectrics, which depend on the wavelength (frequency) of incident light and the wavelength of the surface plasmon. This dependence of θ_{SP} on wavelength results in intrinsic spectral resolution when observing plasmon-coupled emission.

A. Physical Basis of SPR

There are several ways of understanding the physical basis for the appearance of attenuated reflectivities in a conventional SPR experiment. We adopt the approach that has been used in earlier descriptions of the SPR phenomena by Salamon and Tollin (1996) and Salamon *et al.* (1997a) and is a straightforward consequence of the application of electromagnetic field theory to thin-film systems (MacLeod, 2001). We consider a simple case of two thin layers between the prism and emerging mediums as shown in Fig. 1. In this case, the metallic thin film is characterized by a complex refractive index, N_m, which includes the refractive index n_m and the extinction coefficient k, ($N_m = n_m - ik$). Propagation of a plane, monochromatic, linearly polarized, and homogeneous electromagnetic field through a multilayer thin-film system can be described by the set of Maxwell's equations with proper boundary conditions (Macleod, 2001). For our system of two layers as shown in Fig. 1, the optical admittance, Y, is given by

$$Y = \frac{C}{B}, \tag{1}$$

where B and C are normalized electric and magnetic field, respectively, given by

$$\begin{bmatrix} B \\ C \end{bmatrix} = \begin{bmatrix} \cos\theta_m & i(\sin\theta_m)/y_m \\ iy_m\sin\theta_m & \cos\theta_m \end{bmatrix} \begin{bmatrix} \cos\theta_s & i(\sin\theta_s)/y_s \\ iy_s\sin\theta_s & \cos\theta_s \end{bmatrix} \begin{bmatrix} 1 \\ y_b \end{bmatrix}, \tag{2}$$

where

$$\begin{aligned} \theta_m &= 2\pi N_m d_m \cos\nu_m/\lambda = 2\pi(n_m - ik_m)d_m\cos\nu_m/\lambda \\ \theta_s &= 2\pi N_s d_s \cos\nu_s/\lambda = 2\pi(n_s - ik_s)d_s\cos\nu_s/\lambda \end{aligned} \tag{3}$$

corresponds to the phase thickness of metallic layer (m) and sample (dielectric) layer (s) at the appropriate angle of incidence (ν) and the wavelength (λ), and respective substrate admittances for p and s polarizations are given by

$$\begin{aligned} y_m &= (n_m - ik_m)/\cos\nu_m \quad \textit{for p-polarization} \\ y_s &= (n_s - ik_s)/\cos\nu_s \\ y_b &= (n_b - ik_b)/\cos\nu_b \\ y_m &= (n_m - ik_m)\cos\nu_m \quad \textit{for s-polarization} \\ y_s &= (n_s - ik_s)\cos\nu_s \\ y_b &= (n_b - ik_b)\cos\nu_b \end{aligned} \tag{4}$$

The indexes *m, s,* and *b* relate to metal film layer, sample layer, and emergent medium, respectively. The reflectance of the four-layer system can be calculated from the following relationship involving the optical admittance:

$$R = \left(\frac{y_0 - Y}{y_0 + Y}\right)\left(\frac{y_0 - Y}{y_0 + Y}\right)^* = \frac{(y_0 - Y)^2}{(y_0 + Y)^2} \tag{5}$$

where y_0 is the admittance of the incident medium (glass) and astrisk (∗) denotes complex conjugate.

Equation 5 describes the reflectance for a plane-wave incident on layers of thin films as a function of the incident angle. It is useful to consider reflectivity of our system, which consists of a glass prism coated with the metal film coated with the variable thickness dielectric layer (sample) (Fig. 1). The simplest case where the thickness of the coating layer is small or zero corresponds to the conventional conditions of SPR experiment. In this case, coupling of the incident light to the surface wave on the metal layer requires the optical admittances of the adjoining medium to be positive imaginary and of magnitude very close to that of the extinction coefficient *k* of the metal. In practice, this means that for visible light only metals with a small value of the refractive index *n* and large value of *k*, such as silver and gold will generate surface wave.

Figure 4 shows calculated reflectivity curves at 500-nm wavelengths for our four-layer model system shown on Fig. 1 (all parameters for the simulations are specified in the figure legend). Detailed analysis of optical admittance shows that for thin dielectric layer (below 100 nm) evanescent wave beyond the critical angle has positive imaginary value of admittance for *p* polarization and negative imaginary for *s* polarization. In this simple case, the surface-plasmon–coupled resonance (attenuated reflection) occurs only for *p* polarization for the very narrow range of angles (Fig. 4, first two panels), and for *s* polarization, there is no corresponding resonance. However, by increasing the thickness of the dielectric coating layer, one can transform the admittance of the emergent medium so that the admittance presented to the metal is positive imaginary also for *s* polarization. This has been achieved in the thin-film wave guides in optical couplers or surface-plasmon–waveguide resonators (SPWRs) (Salomon and Tollin, 1996; Salomon *et al.*, 1997a). When the thicknesses of adjoining medium become comparable to the light wavelength (in this medium), the presented to metal admittance may have a positive imaginary value also for the *s* polarization. Coupling of the *s*-polarized incident light to the surface wave results similarly to the *p*-polarized light in sharp dip in reflectivity curve. Figure 4 shows the reflectivity curves for *p* and *s* polarizations for the layer thicknesses from 10 to 500 nm. It is interesting to note that the thick dielectric layer allows multiple resonance for *p*- and s-polarized surface plasmons appearing in the alternating fashion.

Already this brief examination of the distribution of electric field amplitudes through the entire layer system shows that the admittance matching layer (or system of layers) is a very important component of the resonant system. For thick

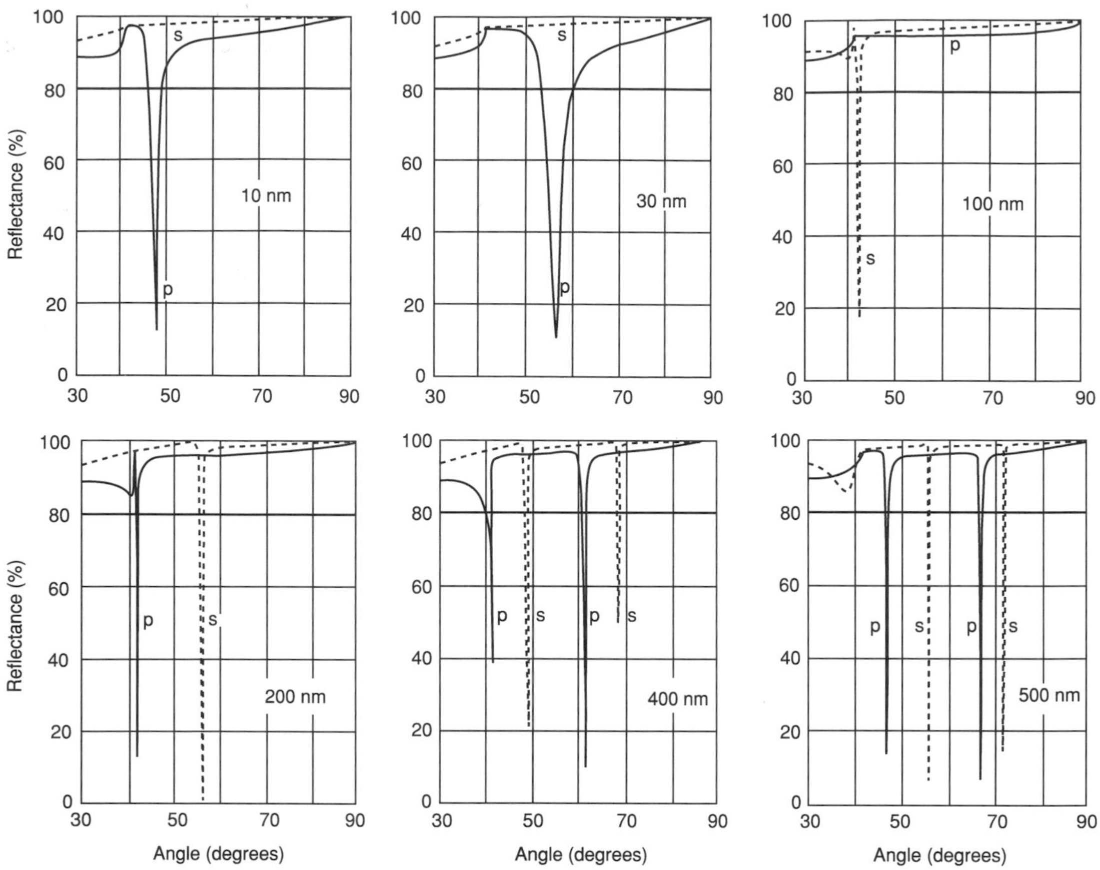

Fig. 4 Calculated reflectivity curves for *p* polarization (*solid line*) and *s* polarization (*dashed line*) for different thicknesses of coating layers. Assumed parameters for simulations were wavelength 500 nm; PVA refractive index 1.5; silver refractive index 0.05–2.87i.

dielectric, both *p* and *s* polarizations present positive imaginary admittance, so multiple plasmon resonance may occur (Fig. 4, bottom).

III. Surface-Plasmon–Coupled Emission

We now describe how the phenomenon of SPR can be applied to fluorescence detection. If an incident beam couples with the evanescent field of a surface plasmon, it seems reasonable to expect that an excited fluorophore will couple to the surface plasmon throughout an outgoing (emissive) beam. Such a possibility is shown in Fig. 5. This reasoning suggests that proximity of a fluorophore to a

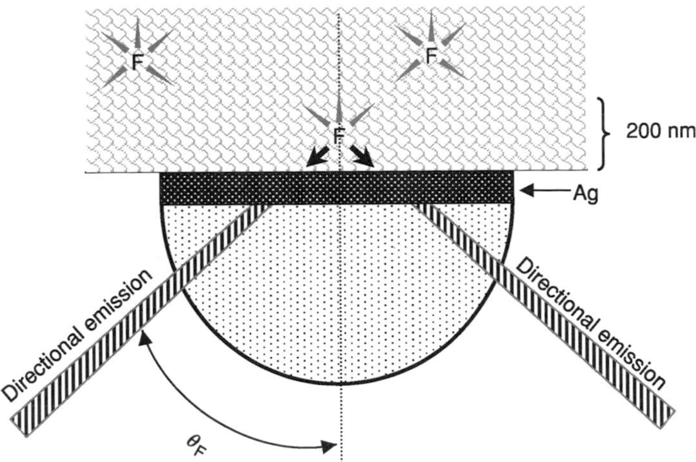

Fig. 5 Concept of surface-plasmon–coupled emission. F is a fluorophore.

metallic film should result in the emission becoming directional with a sharply defined cone angle (θ_F), as shown on Fig. 5. These angles would be different from θ_{SP} because the wavelengths are different. For example, suppose a fluorophore was excited at 633 nm and observed at 700 nm. From Fig. 3 we can see that if the fluorescence is excited at the reflectivity minimum $\theta_{SP} = 43.6$ degrees, the emission (700 nm) would couple back into the prism with $\theta_F = 43$ degrees. This difference may not sound large but average change in θ_{SP} in a conventional SPRA measurement is much less than 0.1 degree.

The dependence of θ_F on wavelength suggests that different fluorophores will display directional emission at plasmon angles determined by the emission maxima. If SPCE displays the same characteristics as SPR, then fluorophores near the metal film will emit into the prism at angles defined by emission wavelength and by the optical properties of the layer system at that wavelength. This coupling should occur for fluorophores within the same region where the evanescent waves exist, out to about 100–500 nm from the metal surface. Based on the wavelength dependence seen in Fig. 3, we expect the longer wavelength Stokes' shifted emission to occur at smaller angles from the z-axis (Fig. 6, top). Based on experiments (Gryczynski *et al.*, 2004a; Lakowicz *et al.*, 2003b), we expect θ_F to be independent of the distance from the metal surface, except to the extent due to the thickness and dielectric constant above the metal film.

Because the emission process is completely independent from the excitation, coupling of the emission to the surface plasmon is also independent from the mode (method) of excitation. In practice, an excited fluorophore near the metal film would not know how it was excited, that is, the emission should be the same whether the fluorophore is excited by the evanescent SPR field (Fig. 6, top) or from a light source not coupled to the surface plasmon (Fig. 6, bottom) at all. The fact

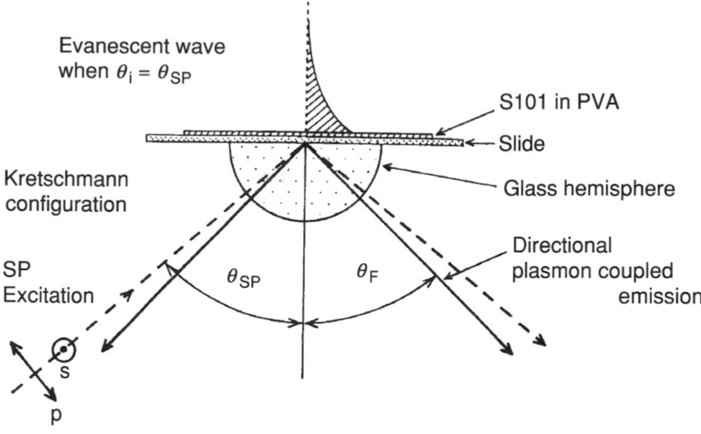

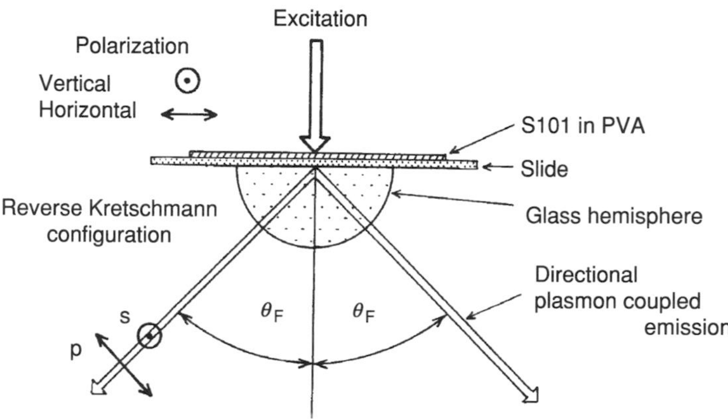

Fig. 6 Surface-plasmon–coupled emission with excitation by the evanescent wave (Kretschmann configuration) and from the side opposite the prism (reverse-Kretschmann configuration).

that SPCE can occur without the surface plasmon excitation may greatly simplify the applications and devices based on this phenomenon.

Consider excitation from the sample side opposite the prism (Fig. 6, bottom). This excitation configuration is called the *reverse-Kretschmann configuration*. Because the fluorophore emits without remembering its source of excitation, there is no plane of incidence for the plasmon-coupled emission. The SPCE will be the same for all azimuthal angles (θ_A) around the z-axis. Suppose the prism is a hemisphere rather than a hemicylinder (Fig. 7). The emission seen along the z-axis will appear as a cone with an equal distribution of emission at all azimuthal angles θ_A. Additionally, different emission wavelengths will appear at different cone angles (θ_F), then a multicolor sample will display a rainbow-like pattern (Fig. 7).

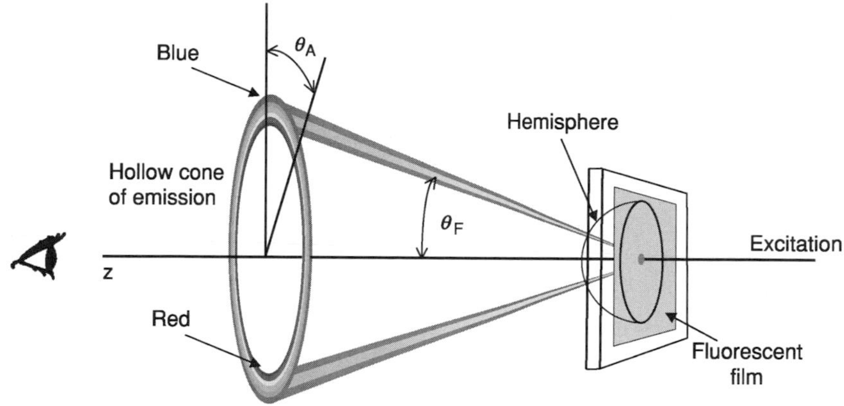

Fig. 7 Cone of surface-plasmon–coupled emission for a fluorophore near a metallic film, θ_A is the azimuthal angle. (See Color Insert.)

Plasmon-coupled emission displays interesting polarization properties. Conventional SPR occurs only for the *p*-polarized component of the incident light because the projection of this component on the metal plane depends on the incident angle. Similarly, only the *p*-polarized component of the emission will couple with the surface plasmon. Hence, the polarization of the cone will always point away from the normal z-axis (Fig. 8, left). SPCE is expected to be *p* polarized at all angles around the cone, independently of the mode of excitation. For thicker layers of adjoining medium (wave-guide conditions), dipoles with the *s* orientation will also couple into the plasmon with a specific matching angle. In this case, the coupled emission will be *s* polarized at all angles around the cone and polarization will always be a tangent to the circle formed by the cone (Fig. 8, right).

IV. Experimental Studies of Surface-Plasmon–Coupled Emission

To measure SPCE we constructed an apparatus that allows angle-dependent observations (Fig. 9). This apparatus allowed excitation by the Kretschmann or the reverse-Kretschmann mode (Fig. 6) (Gryczynski *et al.*, 2004a; Raether, 1988). The Kretschmann mode refers to excitation through the prism (hemicylinder) at the plasmon angle θ_{SP} to obtain an evanescent field. The reverse-Kretschmann mode refers to the direct excitation of the sample from the air side, without excitation of a surface plasmon in the metal. The angular distribution of the emission was obtained using fiber optics on the rotating stage.

We present two cases for our experimental results: one, a thin dielectric layer (plasmon mode) corresponding to conventional SPR conditions where only *p* polarization couples to surface plasmon, and second, a thick dielectric layer

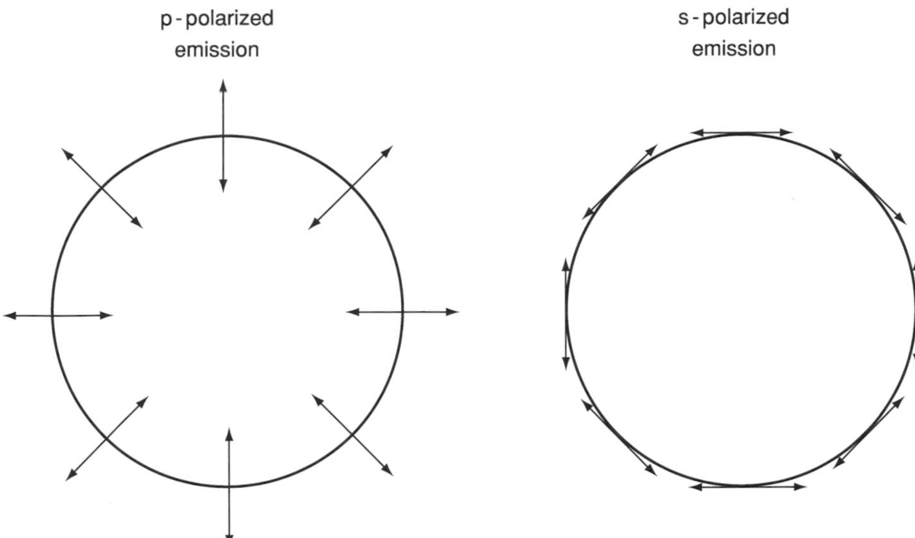

Fig. 8 Polarization of the emission cone for p-polarized (left) and s-polarized (right) surface-plasmon–coupled emission.

(wave-guide mode) corresponding to SPWR conditions (Salomon and Tollin, 1996; Salomon *et al.*, 1997b) when both *p* and *s* polarizations will couple to surface plasmon.

A. Plasmon Mode: Thin Layer

A thin dielectric layer (thickness much smaller than a wavelength) over a metallic surface has a pronounced effect on the SPR response. To test the effect of sample layer thickness on SPCE, we used different thicknesses of polyvinyl alcohol (PVA)–containing fluorophores (sulforhodamine 101-S101) (Gryczynski *et al.*, 2004a). The PVA solution containing S101 was spin-coated to a thickness of 15 and 30 nm on a glass microscope slide that was silver coated (50 nm) by vapor deposition by the EMF Corporation (Ithaca, New York). The sample was illuminated normal to the mirror surface with an argon ion laser at 514 nm. Under these conditions, there is no excitation of surface plasmon. The emission was detected through a laser-notch filter to eliminate scattered light and a long-pass filter to select the emission (Fig. 10).

Angular distributions of the emission of sulforhodamine 101 (S101) in PVA spin coated on the silver glass plate to a final thickness of about 30 and 15 nm are shown in Fig. 11 for the respective PVA thicknesses. The emission is sharply directed back into the hemicylinder and is distributed equally on both sides of the normal axis with the sharp angular distribution. By integration of the emission

Fig. 9 Rotation stage and sample holder for directional excitation and emission measurements.

observed at all angles, we estimated that 30–50% of the emission coupled back into the hemicylinder for the 30- and 15-nm coatings. This result indicates that simple optics based on surface-coupled emission can collect emission as efficiently as larger expensive optics with free-space emission. The angles are about 3 degrees different for the 30-nm (50-degree) and 15-nm (47-deg) PVA coatings. This difference is due to the difference in thickness of the PVA above the silver film. The evanescent wave extends about 200 nm above the silver film, so it samples both PVA and air. These angles are roughly equal to the plasmon angle calculated for 600-nm emission with the refractive index of silver $N_m = 0.06 - 3.75i$ (Fig. 11, right). This dependence on effective refractive index is the effect usually measured on SPRA, but with smaller angular differences.

We measured emission spectra to confirm that the observed signal was due to S101 and not scattered light. The emission spectra for the SPCE and the free-space emission are the same, showing that the signal was due to S101 (Fig. 10). Because of its sharp angular distribution, SPCE is more than 10-fold more intense than

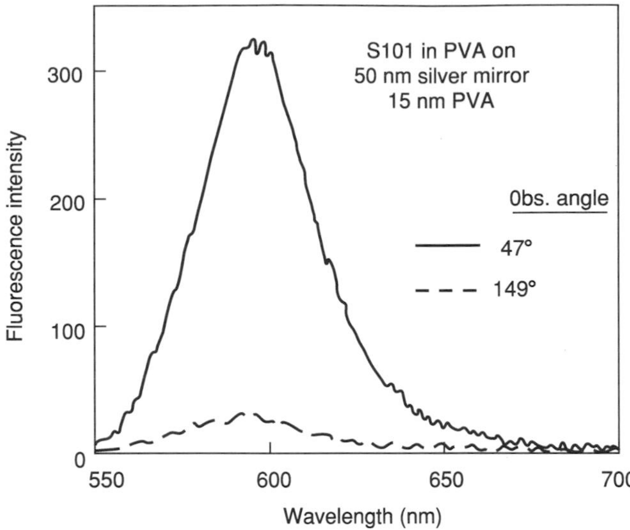

Fig. 10 Emission spectra of plasmon-coupled emission and free-space emission of sulforhodamine 101 (S101) in PVA film.

the free-space emission. We found the same angular emission distribution for the Kretschmann and reverse-Kretschmann configurations of the experiment. These observations support our claim that the surface-plasmon coupling occurs in the excited state and does not depend on the evanescent wave for excitation. By observation through a polarizer, we found that the emission was displaying *p* polarization at any place around the cone of the emission and was independent of the polarization of the excitation around the azimuthal angle (not shown). While collecting these spectra we recognized a favorable property of SPCE with reverse-Kretschmann excitation. The incident light can be polarized perpendicular to the SPCE so that observation through a polarizer will effectively reject scattered light.

1. Background Rejection with SPCE

We anticipate that a valuable attribute of SPCE will be background suppression. Surface-plasmon coupling allows selectively collect emission from molecules near the metal, and the assay chemistry can be localized on this surface. To test background suppresion, we used S101 in PVA directly on the metal film and a solution of pyridine 2 (Py2) in ethanol adjacent to the PVA. The solubility of PVA in ethanol is low and the ethanol did not dissolve PVA or wash out the S101 during these experiments. Py2 emits at a much longer wavelength ($\sim$700 nm) than S101 (610 nm). We measured the spectra of the directional emission and the free-space emission (Fig. 12). We found the free-space emission dominated by

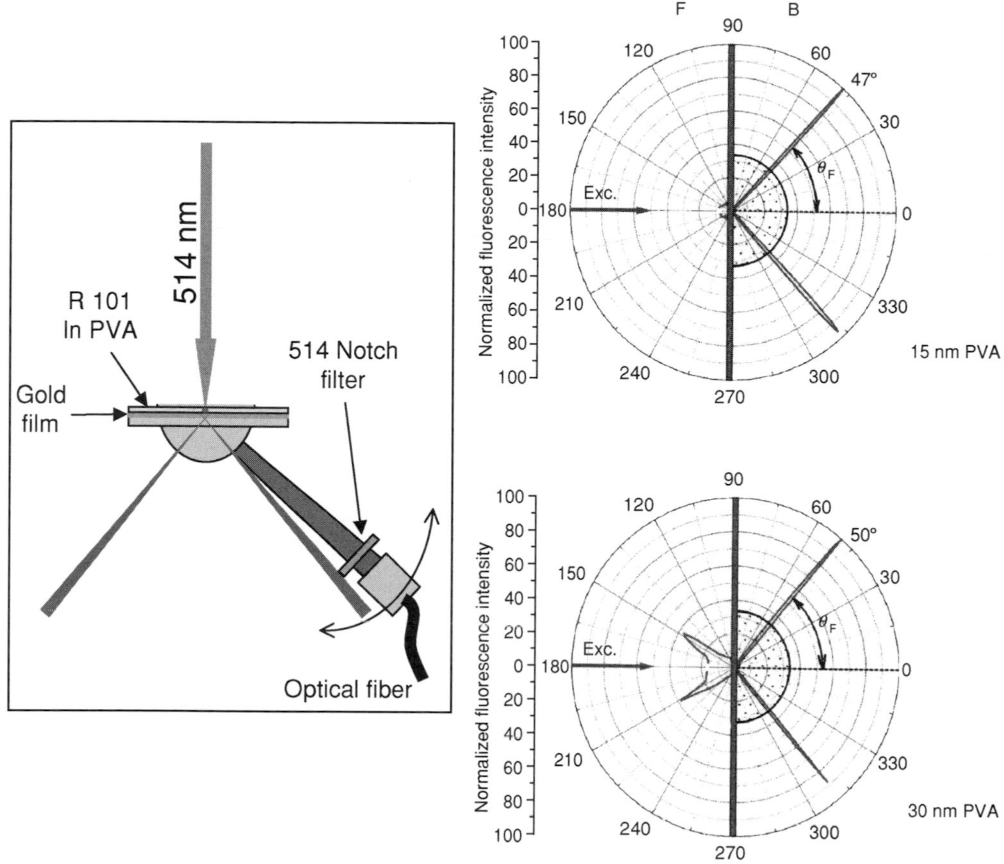

Fig. 11 (Left) Experimental configuration. (Right) Measured angular distribution of the emission from S101 in PVA films.

the mock Py2 impurity, and the directional emission was mostly due to S101, which was near the silver film. This result shows that a simple silver coating can be used for efficient signal collection and suppression of background from fluorophores not close to the metal surfaces. Because excitation and emission occur through the prism, SPCE could be observed from optically dense samples such as whole blood or samples like plasma with high levels of autofluorescence.

2. Intrinsic Wavelength Resolution

We tested the wavelength resolution of SPCE by measuring the emission spectra of multiple fluorophores at different observation angles. We spin coated a mixture of rhodamine 123 (R123), S101, and pyridine 2 (Py2) in PVA on the silver film. The concentrations were adjusted to yield approximately similar intensities for

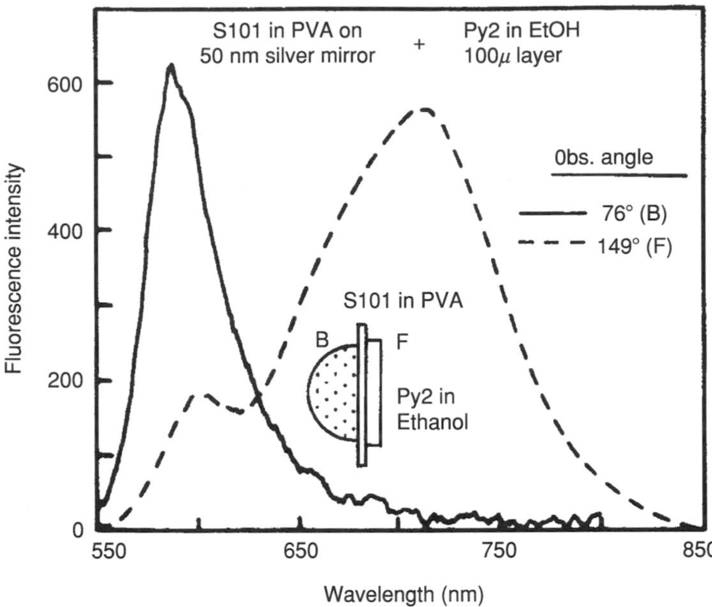

Fig. 12 Free-space (*dashed line*) and plasmon-coupled emission (*solid line*) spectra of S101 with the rhodamine 6G (R6G) background.

each fluorophore using 514-nm excitation. The free-space emission showed contributions from all three fluorophores (Fig. 13, top panel). For the surface-plasmon-coupled emission, the spectra changed dramatically as the observation angle was changed from 52 to 42 degrees (Fig. 13), with the peak shifting from 540 nm (R123) to 590 nm (S101) and to 660 nm (Py2). This effect is so dramatic that one can see the change in color by looking around the hemicylinder axis. Photographs taken at different observation angles are also shown in Fig. 13, but the photos under-represent the visual color differences. Figure 14 shows the image of a fluorescence ring for the system shown in Fig. 13 that made with the hemisphere prism. It is striking how excellent resolution can be achieved with a simple digital camera and the intrinsic wavelength dispersion of SPCE.

B. Wave-Guide Mode: Thick Layer

Theoretical consideration shows that for a thick dielectric layer we can expect multiple surface plasmon resonances that correspond to different angles and polarizations. In fact, when measuring SPCE with the thicker sample (~600 nm), we observed increased emission at four angles (Fig. 15) (Gryczynski *et al.*, 2004b). Figure 16 shows calculated reflectivity curves for 47- and 595 nm thick PVA. For an assumed thickness of 595 nm, we calculated four-reflectivity minima, with alternating *s* and *p* polarizations. The angles and polarizations

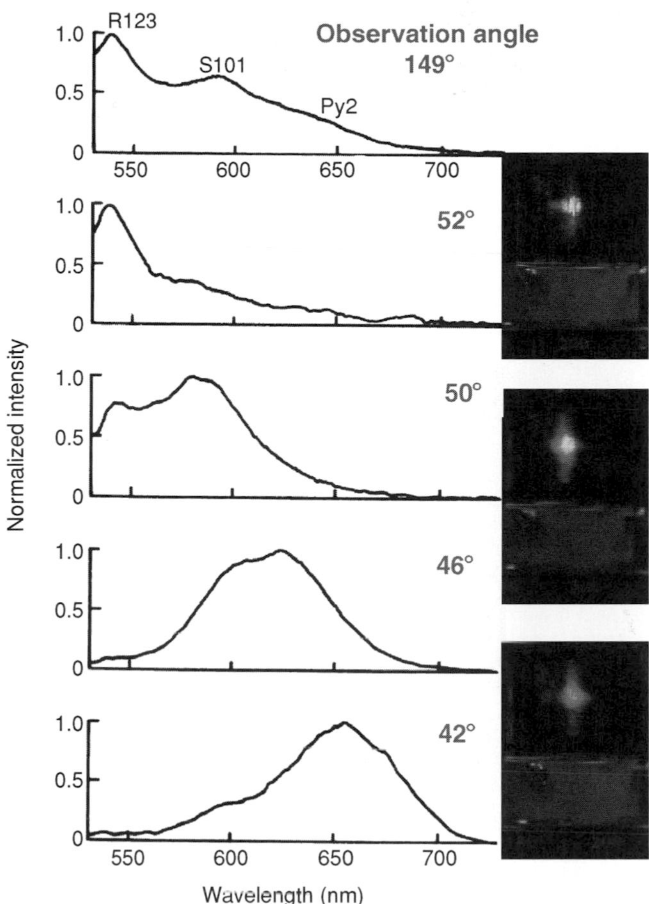

Fig. 13 Wavelength selection using the observation angle. The PVA film contained rhodamine 123 (R123), S101, and pyridine 2 (Py2). Photography of the directional emission seen from the different angle (right panel). (See Color Insert.)

shown in Fig. 15 precisely match those predicted by simulation in Fig. 16. The alternative *s* and *p* polarizations can be seen from the photographed cone of emission (Fig. 17, bottom). Only the two most intense rings are visible in the photograph. The outer ring is *p* polarized and the inner ring *s* polarized, as can be seen from photographs taken through a polarizer (Fig. 17, bottom). We conclude that the wave-guide modes present in the thick PVA samples are the origin of the multiple rings and polarizations of the coupled emission.

SPCE in optical wave guides can have numerous practical applications. The SPCE angles, polarizations, and intensities are very sensitive to the sample thickness. Hence, any process that alters the sample thickness or affects refractive index will change the angles. The effective sample thickness could be changed by

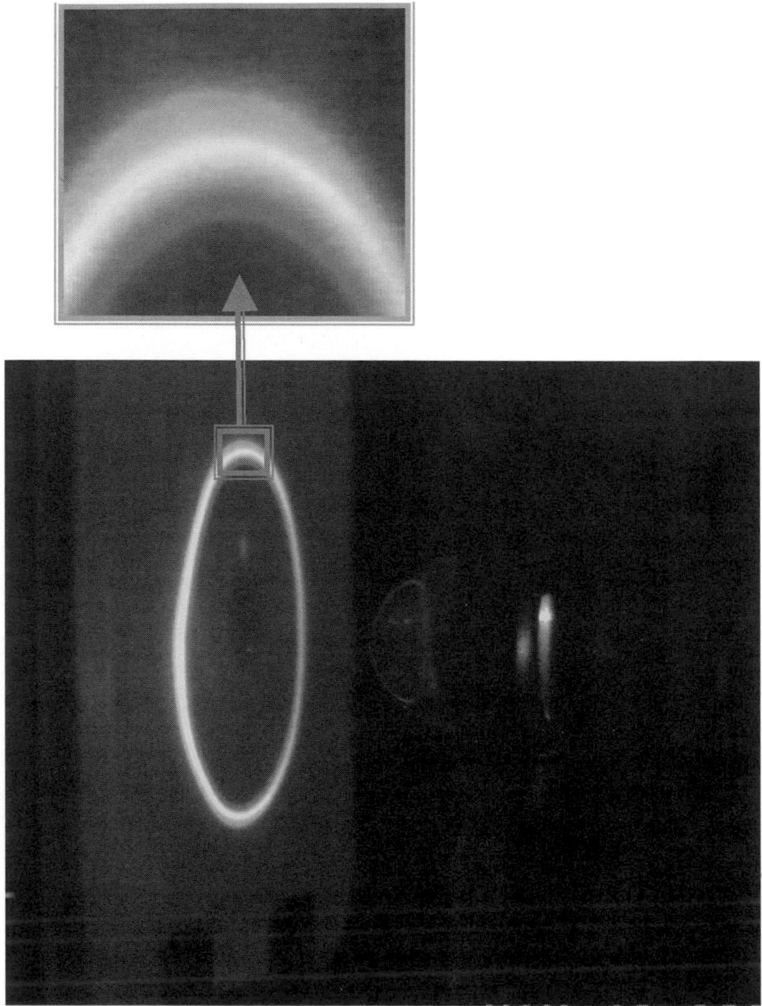

Fig. 14 Photography of surface-plasmon–coupled emission cone for multifluorophore sample. (See Color Insert.)

vapor binding to a polymer, by biomolecule binding to a bioaffinity surface, or by changes in ionic strength, to name a few.

V. Practical Applications

Fluorescence immunoassays are extensively used in medical diagnostics (Gosling, 1980; Hemmila, 1991; Van Dyke *et al.*, 1990; Vo-Dinh *et al.*, 1993). As a practical application, we described a novel approach to immunoassays using

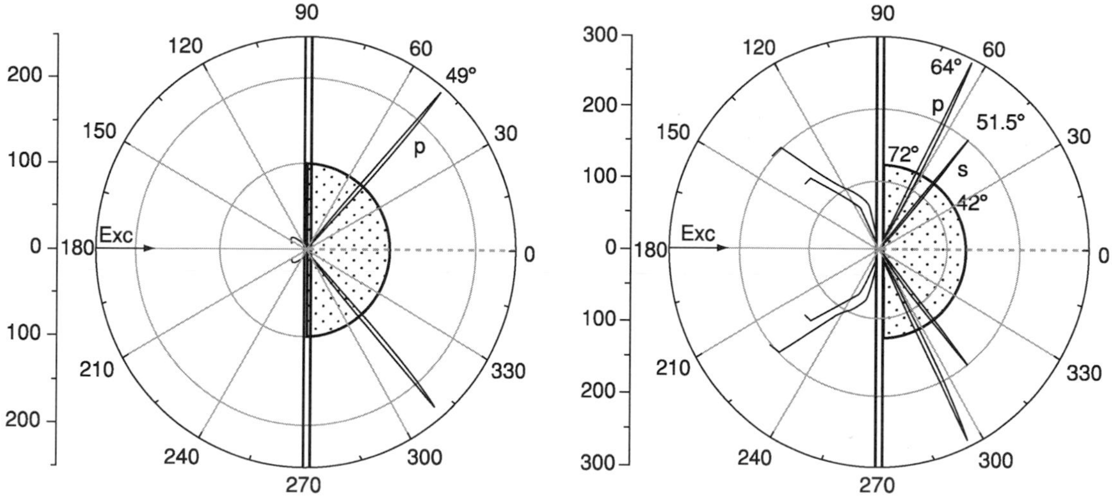

Fig. 15 Angular distribution of the emission from nile blue (NB) in PVA films. (Left) Thin 44-nm PVA layer and (right) thicker sample layer ~600 nm.

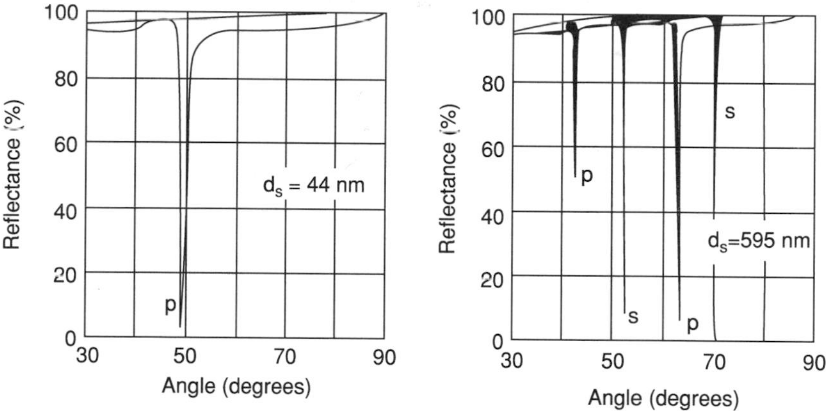

Fig. 16 Theoretical reflection calculated for thin PVA coating (44 nm) and thick PVA coating (595 nm). The calculations were performed for a four-phase system: glass ($n_p = 1.514$), 50 nm silver ($N_m = 0.06 - 3.75i$), 44 or 595 nm PVA ($n_s = 1.50$), and air ($n_o = 1.0$) using TF Calc. Software.

SPCE. Many immunoassays are designed on transparent substrate surfaces with the layer of antibodies bound to that surface. Such designs are compatible with SPCE technology. Fluorescence is mostly isotropic in space, and sensitivity is limited in part by the light collection efficiency. By the use of SPCE we can convert a significant part of emission into a conelike directional beam in a glass substrate and highly increase efficiency of signal collection. For our experiments, we used

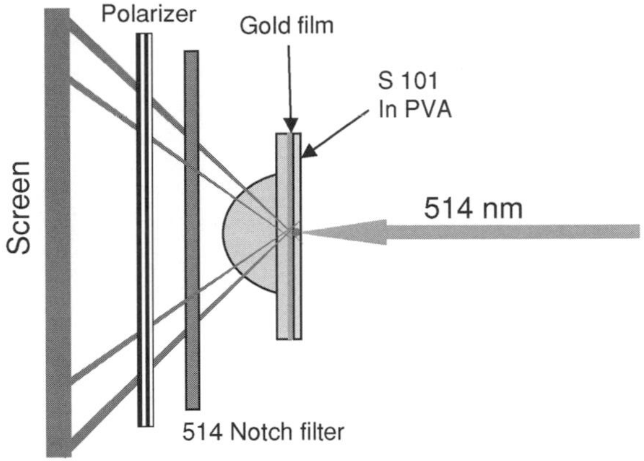

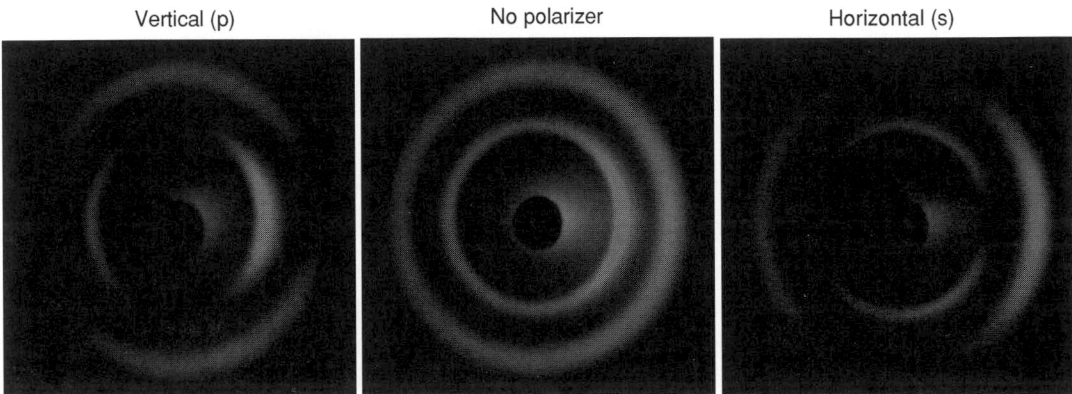

Fig. 17 Emission rings of surface-plasmon–coupled emission for the thick sample. (Top). Experimental configuration. (Bottom) *p* (left) or *s* (right) emission polarizer, (center) no emission polarizer. Only two most intense rings are seen at 51.5 degrees and 64 degrees. (See Color Insert.)

a model affinity assay using labeled anti-rabbit immunoglobulin G (IgG) antibodies against rabbit IgG bound to a 50-nm thick silver film. A schematic of the immunoassay principles is shown in Fig. 18.

A. End-Point Binding Experiment

Dye-labeled conjugate rhodamine red-X-anti-rabbit IgG was added to the slide coated with rabbit IgG and incubated at 37 °C in a humid chamber for 1–2 hours and washed (Matveeva *et al.*, 2004a). Then, a rubber ring (7-mm diameter and 9-mm height) was placed on the metallic side of the slide and covered with a second glass slide (Fig. 19). About 1.5 ml of the Na-phosphate buffer (50 mM, pH-7.4) was added inside the rubber ring chamber using a needle, and fluorescence

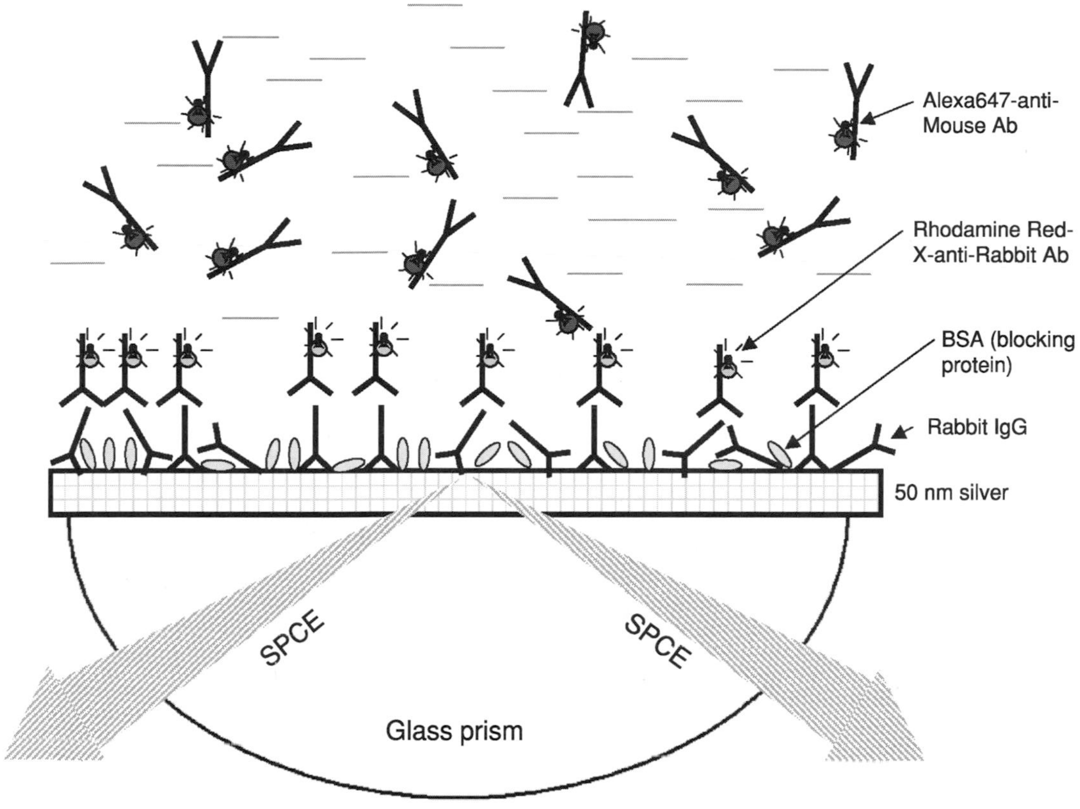

Fig. 18 Immunoassay scheme.

measurements were performed at two optical configurations (Kretschmann and reverse Kretschmann). The angular distribution of the emission is presented in Fig. 20.

The emission spectrum of the SPCE was characteristic of the rhodamine probe (Fig. 21) and was not corrupted by scattered light at the excitation wavelength. An important characteristic of SPCE is the complete polarization in the *p* direction, meaning the electric vector is oriented parallel to the plane of incidence. Figure 21 shows the emission spectra collected through an emission polarizer oriented *p* or *s*. The orientation of the excitation polarizer did not affect the relative intensities. The *p*-polarized intensity is 20-fold more intense, resulting in a very high polarization of about 0.9. This large value is a result of the fact that only *p*-polarized light may couple with the surface substrate. This effect is much more efficient than photoselection in an isotropic media. Also, this value is independent of the orientation of the excitation polarization. This *p* polarization proves that the

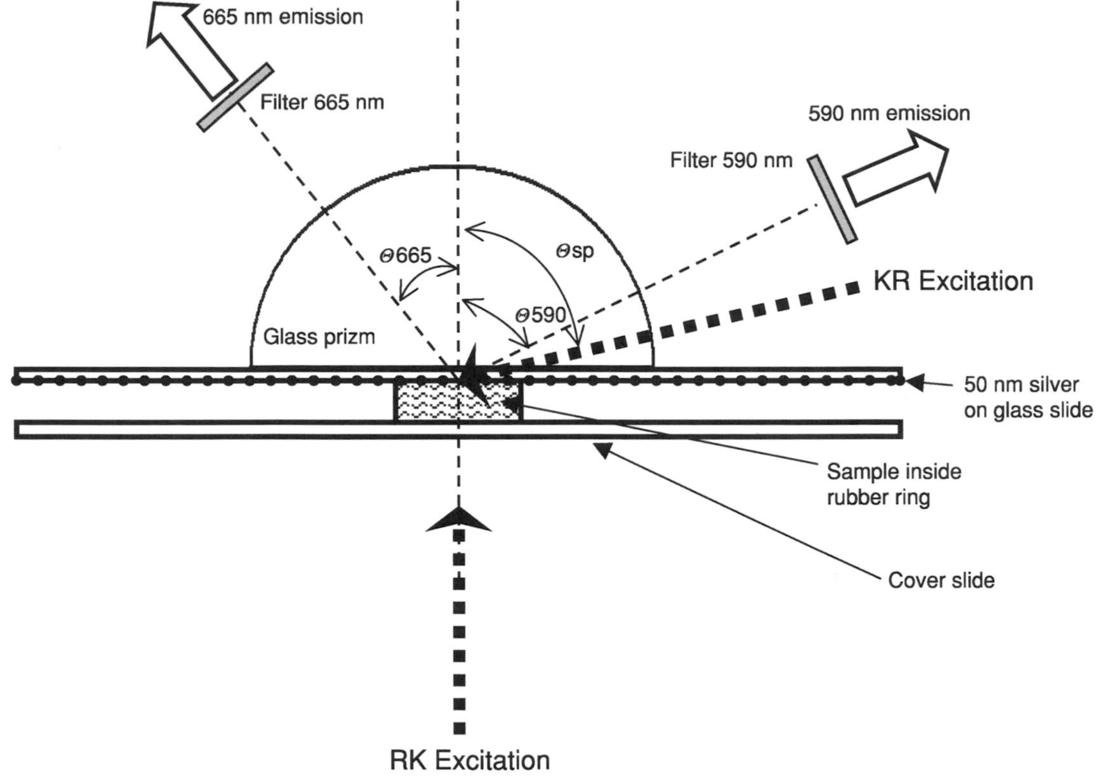

Fig. 19 Experimental scheme for measurement of surface-plasmon–coupled emission using Kretschmann (KR) and reverse-Kretschmann (RK) configurations. Directions Θ_{590} and Θ_{665} correspond to SPCE collected at 590 nm and 665 nm, respectively.

emission is due to surface plasmon, which under these conditions cannot emit *s*-polarized light.

B. Kinetic Binding Experiment

We also tested the use of SPCE to measure the binding kinetics of the rhodamine-labeled antibodies to the surface-bound antigen. The kinetics trace measured and experimental system are shown in Fig. 22. The bottom panel shows the emission intensity changes after adding the labeled antibody. The emission climbs rapidly and reaches a limiting value. It is important to recognize that this 10-fold change in intensity is not the result of a change in the rhodamine quantum yield upon binding. We measured the effect of binding the rhodamine-labeled antibody to the antigen while both were free in solution and found the intensity decrease due to binding was about 25%. This indicates the intensity change is due to

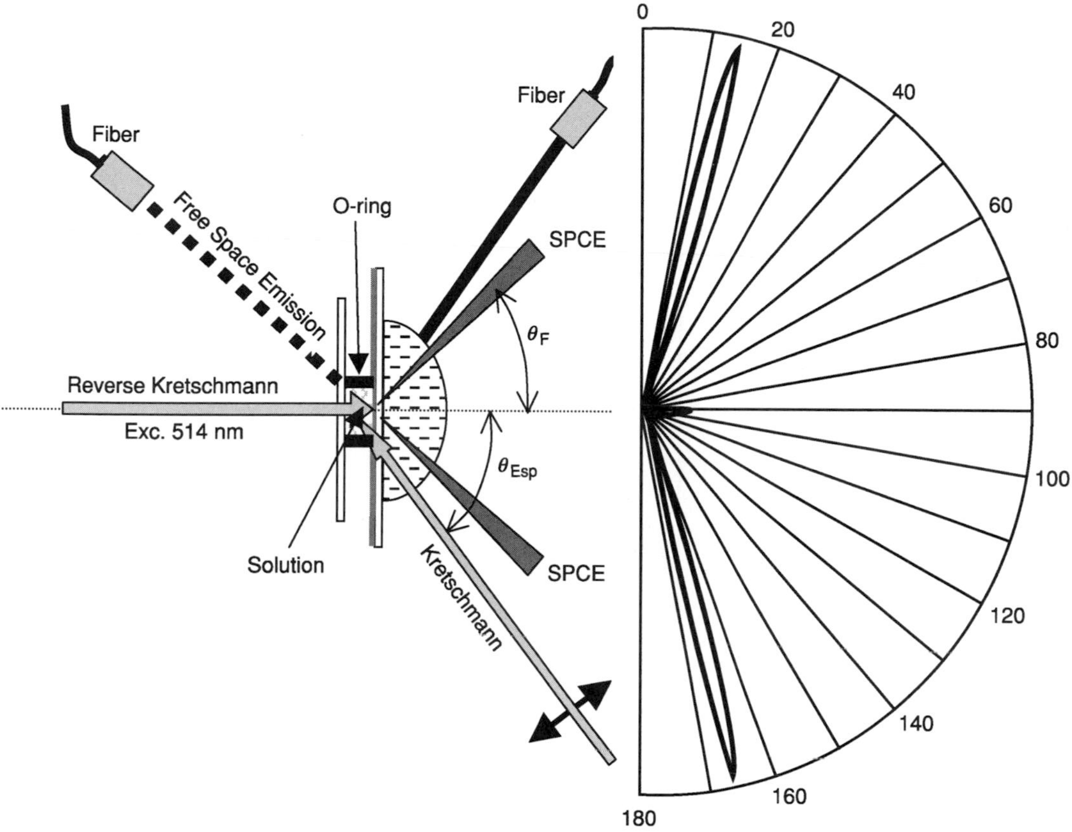

Fig. 20 Angular distribution of the 590-nm fluorescence emission of rhodamine red-X–labeled anti-rabbit antibodies bound to the rabbit IgG immobilized on the 50-nm silver mirror surface. Left shows the experimental configuration.

localization of the probe in the evanescent field near the silver. This proves SPCE is a generic method for detection of surface localization by a change in intensity, which does not require any change in the fluorophore quantum yield upon binding to the surface.

C. Background Rejection Experiment

To test the effect of a background, we used two solutions added inside the rubber ring cell in Fig. 22 (top): (1) a highly absorbing bovine hemoglobin (HbBv) solution as nonfluorescent background or (2) a highly fluorescent solution of

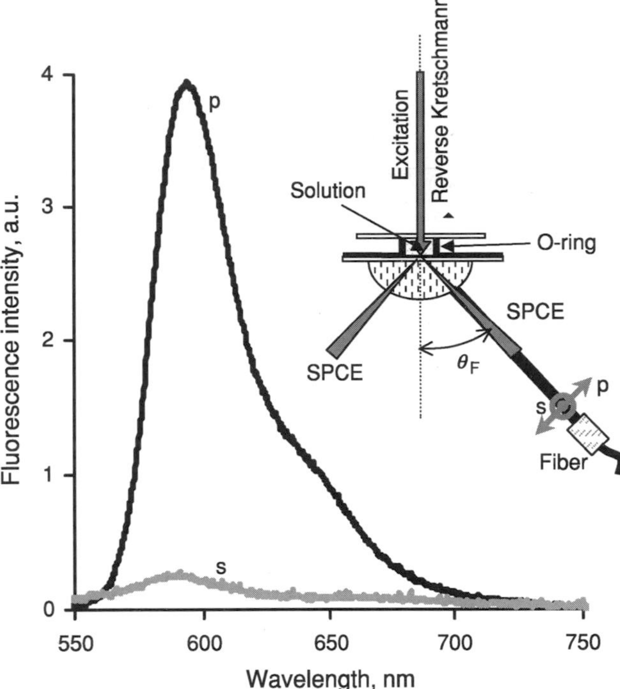

Fig. 21 Fluorescence spectra of the rhodamine red-X–labeled anti-rabbit antibodies bound to the immobilized rabbit IgG observed at 77 degrees in reverse-Kretschmann surface-plasmon–coupled emission configuration (see Fig. 20) seen through the polarizer; *p* and *s* refer to the orientation of the emission polarizer.

non-binding Alexa fluor-647–antimouse IgG conjugate. The HbBv solution was used undiluted and Alexa fluor-647–anti-mouse IgG conjugate was used at 500-fold dilution.

1. Highly Absorbing Background

In medical testing, it is often desirable to perform homogenous assays without separation steps, sometimes in whole blood. We reasoned that SPCE should be detectable in optically dense media because this signal arises from the sample within 200 nm of the surface. To mimic whole blood, we added 17% HbBv solution, which had optical densities of 5 and 3 at 514 and 590 nm, respectively. In a 1.0-mm thick sample, these optical densities would attenuate the fluorescence signal almost a million fold. Using SPCE the signal was attenuated less than three-fold (Fig. 23). These results show the potential of using SPCE in optically dense samples.

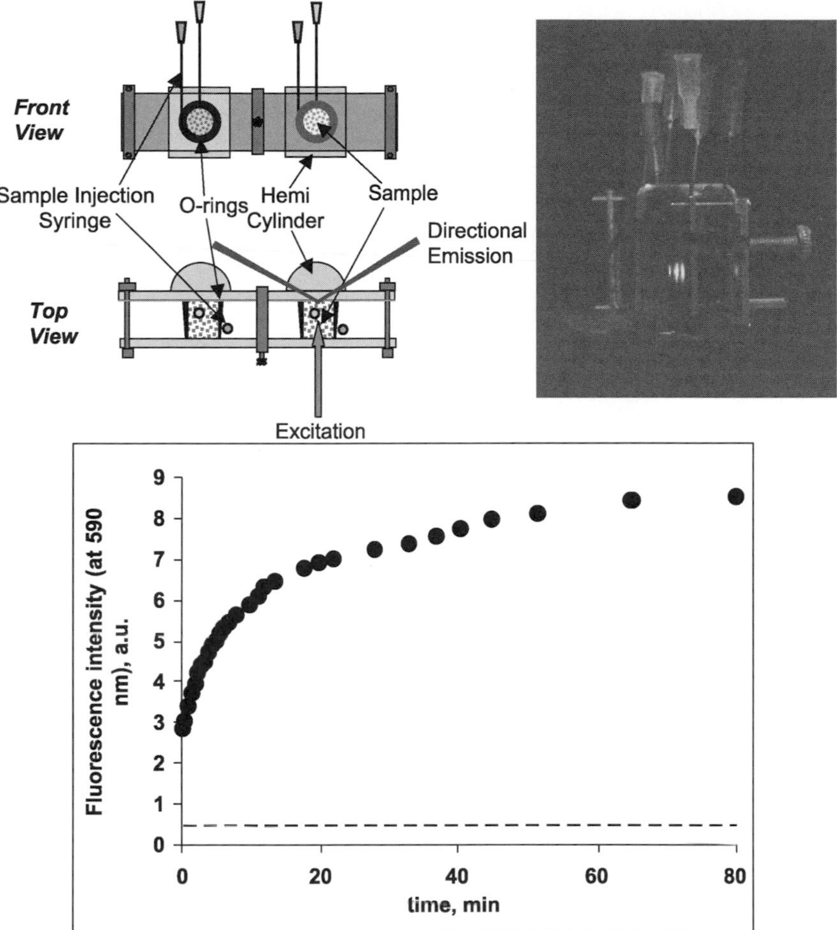

Fig. 22 (Top) Experimental setup (schematic and photography) for kinetic measurements of free-space and surface-plasmon–coupled emission with reverse-Kretschmann (*RK*) and Kretschmann (*KR*) configurations. (Bottom) Binding kinetics of the rhodamine red-X–labeled anti-rabbit antibodies bound to rabbit IgG immobilized on a 50-nm silver mirror surface observed with KR/SPCE configuration (top). (See Color Insert.)

2. Highly Fluorescent Background

We tested several optical configurations to determine the relative intensities and extent of background rejection using SPCE. These three configurations are shown in the top panels in Fig. 24. This sample consisted of the surface saturated with rhodamine-labeled antibody. We then added Alexa-647–labeled antibody (not binding to the surface) to mimic autofluorescence from the sample. The 0.03-μM

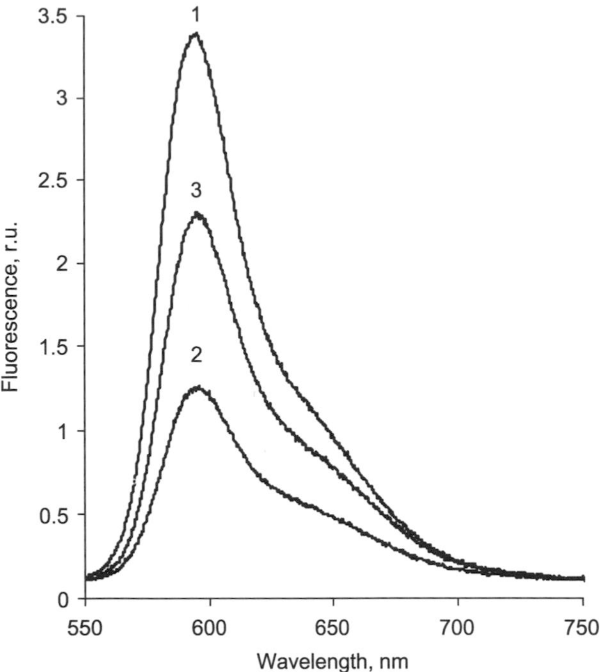

Fig. 23 Fluorescence spectra (SPCE) of the rhodamine red-X–labeled anti-rabbit antibodies bound to the rabbit IgG immobilized on a 50-nm silver mirror surface in absence (1) of highly absorbing background; (2) after substituting buffer with highly absorbing background (bovine hemoglobin) observed with the reverse-Kretschmann/SPCE configuration; (3) same as (2) with 10-fold diluted hemoglobin.

concentration of this antibody (0.13 μM of Alexa dye) resulted in a dominant free-space fluorescence signal from the sample. First, the sample was excited using the RK configuration, and the free-space emission was observed from the same water side of the sample. Compared to subsequent measurements, the intensity of the desired rhodamine antibody below the signals was weak. The free-space emission was dominated by the emission from Alexa-647 at 665 nm, with only weak rhodamine emission at 590 nm. We then measured the emission spectrum of the SPCE signal (middle panel) while still using RK illumination. The emission spectrum was dramatically changed from a 10-to-1 excess of the unwanted background to a 1-to-5-excess of the desired signal. Hence, the use of SPCE resulted in selective detection of the rhodamine-labeled antibody near the silver film.

We then changed the mode of excitation to the KR configuration (Fig. 24, right panel). In this case, the sample was illuminated at θ_{SP}, creating an evanescent field in the sample. The overall intensity was increased about 10-fold while further suppressing the unwanted emission from Alexa-647. The increased intensity and

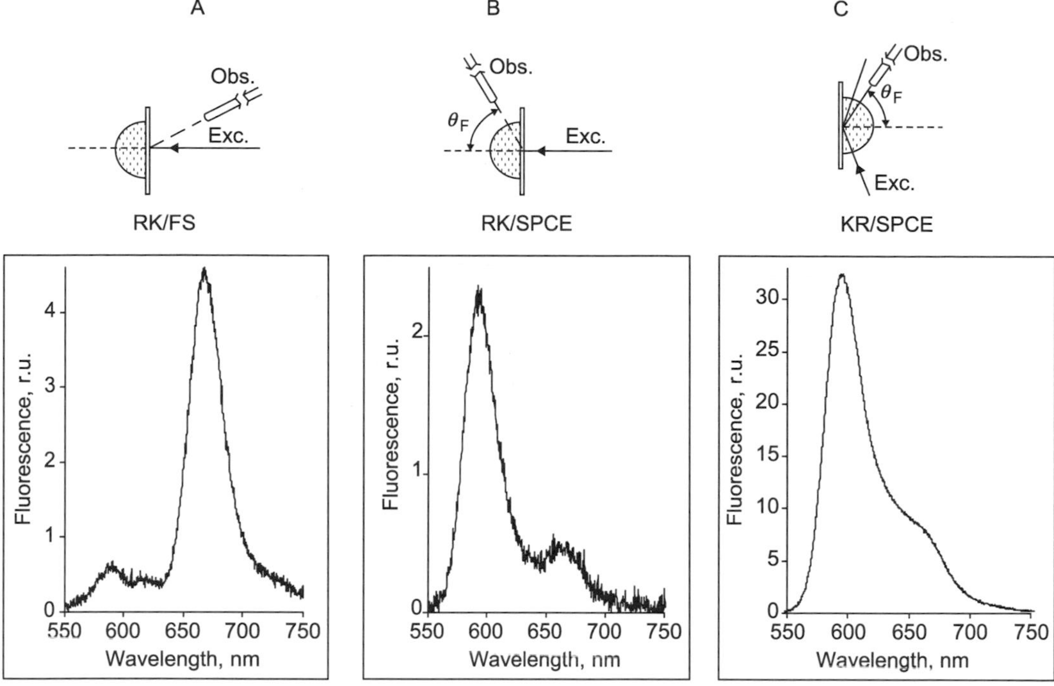

Fig. 24 Effect of the fluorescing background (Alexa fluor-647) on the fluorescence of the rhodamine red-X–labeled anti-rabbit antibodies bound to the immobilized on the surface antigen (rabbit IgG): (A) Reverse Kretschmann (RK) configuration, front face (sample side) observation; (B) RK configuration, SPCE; (C) KR configuration, SPCE.

decreased background are the result of localized excitation by the resonance-enhanced field near the metal. In this case, the emission was due almost entirely to the rhodamine, with just a minor contribution from the Alexa-labeled protein.

D. Multi-Wavelength Immunoassay

As a final example of novel SPCE application to immunoassay, we describe multi-wavelength immunoassays (Matveeva *et al.*, 2004b). Coupling of excited fluorophores with a nearby thin metal film, in our case silver, results in strongly directional emission into the underlying glass substrate. The experimental configuration used for two-wavelength SPCE is shown in Fig. 19. The protein-coated silver surface can be illuminated at the surface-plasmon angle through the glass prism (the Kretschmann configuration) or from the sample side (reverse-Kretschmann configuration). The angle at which the emitted radiation propagates through the prism depends on the surface-plasmon angle for the relevant wavelength. These angles depend on emission wavelength, allowing measurement of

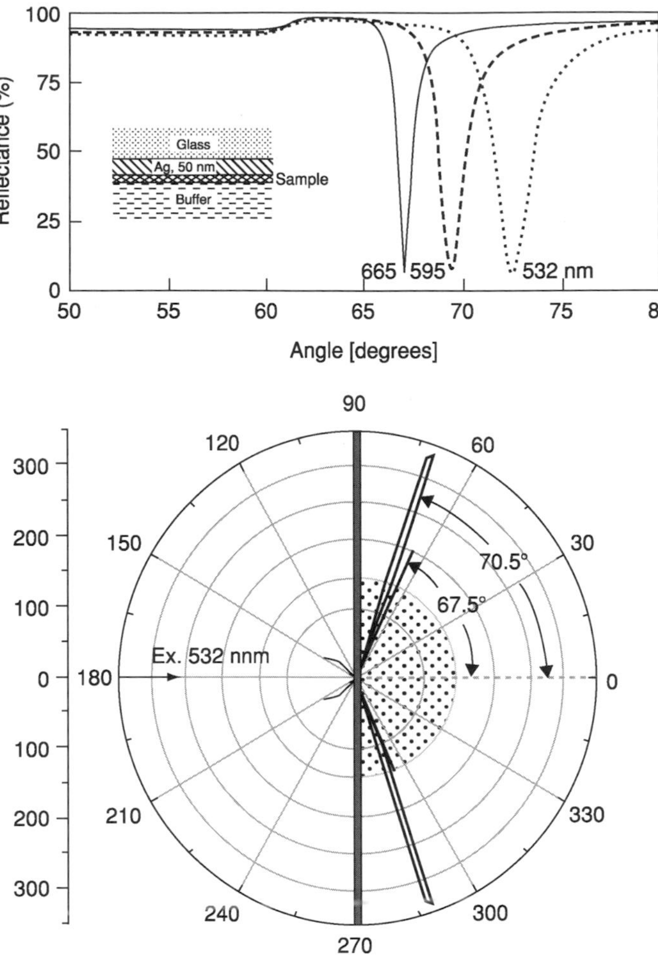

Fig. 25 (Top) Calculated reflectivity of a 50-nm silver film on BK7 glass ($n_p = 1.52$). The sample (protein layers) was assumed to be 15-nm thick ($n_s = 1.50$). The buffer thickness was taken as infinite with $n_w = 1.33$. For silver phase, we used $N_m^{532} = 0.055 - 2.87i$, $N_m^{595} = 0.06 - 3.75i$ and $N_m^{665} = 0.07 - 4.2i$. (Bottom). Angle-dependent emission from surface-bound rhodamine red-X antibody and Alexa antibody measured at 595 and 665 nm. The sample was excited at 532 nm using the reverse Kretschmann configuration.

multiple analytes using multiple emission wavelengths. We demonstrated this possibility using antibodies labeled with either rhodamine red-X or Alexa fluor-647. These antibodies were directed against an antigen protein bound to the silver surface. The emission from each labeled antibody occurred at a different angle on the glass prism, allowing independent measurement of surface binding of each antibody. This method of SPCE immunoassay can be readily extended to four or more wavelengths.

We calculated the reflectivity expected for the 50-nm thick silver films and our optical configuration using Eqn. 5. The reflectivity curves can be also calculated using either software available on the Internet (Nelson *et al.*, 1999, http://corninfo.chem.wisc.edu) or commercial software (TF Calc.), which we found to yield equivalent results. The reflectivity minima were found at 72.5 degrees for 532 nm, and at 69 and 67 degrees for 595 and 665 nm, respectively. Hence, we expected to obtain excitation of surface plasmons with a 532-nm incident angle of 72.5 degrees, and to observe the rhodamine red-X-antibody and alexa-antibody emission at 69 degrees and 67 degrees, respectively.

To demonstrate the wavelength resolution for the emission, we excited the sample through the aqueous phase (reverse Kretschmann). Observed angle-dependent emission signal is shown in Fig. 25. The emission from each labeled antibody was strongly directional in the prism at the surface-plasmon angle for the emission wavelength. Similar results were obtained for the Kretschmann configuration for excitation. This result demonstrates that SPCE is due to an interaction of the excited fluorophores with the metal surface and does not depend on creation of surface plasmon by the incident light.

The angle-dependent intensity in Fig. 25 was collected through an emission filter to isolate the emission from each labeled antibody. However, these measurements did not resolve the emission spectra of each antibody. Figure 26 shows emission spectra collected using observation angles of 71, 69.5, and 68 degrees.

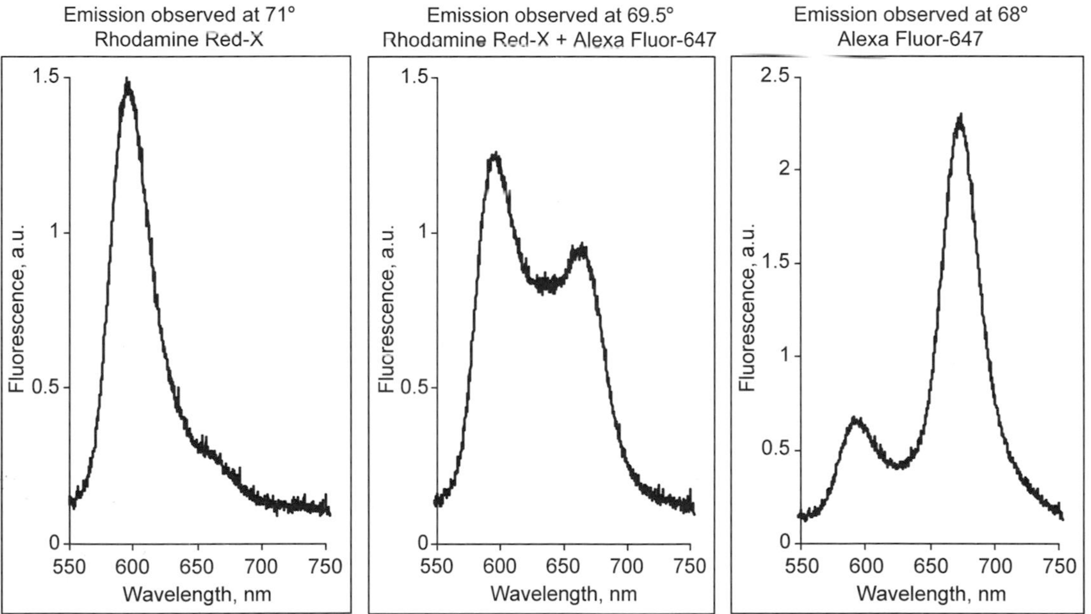

Fig. 26 Emission spectra from a surface containing rhodamine red-X antibody and Alexa antibody measured at three observation angles, using the reverse-Kretschmann configuration.

At 71 degrees, the emission is almost completely due to rhodamine red-X antibody with an emission maximum of 595 nm. At 68 degrees, the emission is due mostly to Alexa antibody at 665 nm, with a residual component from rhodamine red-X antibody at 595 nm. At the intermediate angle of 69.5 degrees, the emission from both labeled antibodies is seen. These emission spectra show that the desired emission wavelength can be selected by adjustment of the observation angle.

VI. Conclusion

The SPCE offers new opportunities for high-sensitivity fluorescence detection. Its intrinsic simplicity for very efficient fluorescence signal collection opens novel approaches for a simpler design of fluorescence-based detection devices. In particular, SPCE may find application for detecting biomolecular binding on the surface (analogous to conventional SPR), medical assay development, and DNA hybridization. SPCE can be easily implemented to high-throughput screening and fluorescence microscopy to facilitate single molecule detection.

Acknowledgments

This work was supported by the National Institute of Biomedical Imaging and Bioengineering, EB-00682, EB-00981, and EB-1690 the National Center for Research Resource, RR-08119 and Philip Morris USA, Inc.

References

Amos, R. M., and Barnes, W. L. (1997). Modification of the spontaneous emission rate of Eu^{3+} ions close to a thin metal mirror. *Phys. Rev. B* **55,** 7249–7254.

Barnes, W. L. (1998). Topical review. Fluorescence near interfaces: The role of photonic mode density. *J. Modern Optics* **45,** 661–699.

Brown, P. O., and Botstein, D. (1999). Exploring the new world of the genome with DNA microarrays. *Nat. Gent.* **21**(Suppl.), 33–37.

Cooper, M. A. (2002). Optical biosensors in drug discovery. *Nat. Rev.* **1,** 515–528.

Drexhage, K. H. (1974). Interaction of light with monomolecular dye lasers. *In* "Progress in optics" (E. Wolfe, ed.). North Holland, Amsterdam.

Frey, B. L., Jordan, C. E., Kornguth, S., and Corn, R. M. (1995). Control of the specific adsorption of proteins onto gold surfaces with poly(l-ysine) monolayers. *Anal. Chem.* **67,** 4452–4457.

Frutos, A. G., and Corn, R. M. (1998). SPR of utrathin organic films. *Analytical Chem.* 449A–455A.

Gosling, J. P. (1980). A decade of development in immunoassay methodology. *Clin. Chem.* **36,** 14808–14827.

Gryczynski, I., Malicka, J., Gryczynski, Z., and Lakowicz, J. R. (2004a). Radiative decay engineering 4. Experimental studies of surface plasmon coupled directional emission. *Anal. Biochem.* **324,** 170–182.

Gryczynski, I., Malicka, J., Nowaczyk, K., Gryczynski, Z., and Lakowicz, J. R. (2004b). Waveguide effects on surface plasmon-coupled emission. *J. Biomed. Optics* (submitted).

Haroche, S., and Raimond, J. M. (1993). Cavity quantum electrodynamics. *Sci. Am.* 54–62.

Haroche, S., and Kleppner, D. (1989). Cavity quantum electrodynamics. *Phys. Today* 24–30.

Hemmila, I. A. (1991). Applications of Fluorescence in Immunoassays. John Wiley & Sons, New York.

Hinds, E. A. (1991). Cavity quantum electrodynamics. *Adv. At. Mol. Opt. Phys.* **28,** 237–289.

Ichinose, N., Schwedt, G., Schnepel, F. M., and Adachi, K. (1987). *Fluorometric Analysis in Biomedical Chemistry*. John Wiley & Sons, Inc., New York.

Jordan, C. E., Frey, B. L., Kornguth, S., and Corn, R. M. (1994). Characterization of poly-L-lysine adsorption onto alkanethiol-modified gold surfaces with polarization-modulation fourier transform infrared spectroscopy and surface plasmon resonance measurements. *Langmuir* **10,** 3642–3648.

Kricka, L. J., Skogerboe, K. J., Hage, D. A., Schoeff, L., Wang, J., Sokol, L. J., Chan, D. W., Ward, K. M., and Davis, K. A. (1997). Clinical chemistry. *Anal. Chem.* **69,** 165R–229R.

Lakowicz, J. R. (2001). Radiative decay engineering. Biophysical and biomedical applications. *Anal. Biochem.* **298,** 1–24.

Lakowicz, J. R., Shen, Y., D'Auria, S., Malicka, J., Gryczynski, Z., and Gryczynski, I. (2002). Radiative decay engineering 2: Effects of silver island films on fluorescence intensity, lifetimes and resonance energy transfer. *Anal. Biochem.* **301,** 261–277.

Lakowicz, J. R., Malicka, J., D'Auria, S., and Gryczynski, I. (2003a). Release of the self-quenching of fluorescence near silver metallic surfaces. *Anal. Biochem.* **320,** 13–20.

Lakowicz, J. R., Shen, B., Gryczynski, Z., D'Auria, S., and Gryczynski, I. (2001). Intrinsic fluorescence from DNA can be enhanced by metallic particles. *Biochem. Biophys. Res. Commun.* **286,** 875–879.

Lakowicz, J. R. (ed.) (1994). *Topics in fluorescence spectroscopy: Probe design and chemical sensing*. Plenum Press, New York.

Lakowicz, J. R. (2004). Radiative decay engineering 3. Surface plasmon coupled directional emission. *Anal. Biochem.* **324,** 153–169.

Lakowicz, J. R., Malicka, J., Gryczynski, I., and Gryczynski, Z. (2003b). Directional surface plasmon-coupled emission: New method for high sensitivity detection. *BBRC* **307,** 435–439.

Liedberg, B., and Lundstrom, I. (1993). Principles of biosensing with an extended coupling matrix and surface plasmon resonance. *Sensors and Actuators B* **11,** 63–72.

Lipschutz, R. J., Fodor, S. P. A., Gineras, T. R., and Lockhart, D. J. (1999). High density synthetic oligonucleotide arrays. *Nat. Genet. Suppl.* **1,** 20–24.

MacLeod, H. L. (2001). Thin Film Optical Filters, 3rd. ed., Institute of Physics Publishing, Bristol, UK.

Matveeva, E., Gryczynski, Z., Gryczynski, I., Lakowicz, J. R. (2004a). Immunoassays based on directional surface plasmon coupled emission. *J. Immunol. Methods* **286,** 133–140.

Matveeva, E., Malicka, J., Gryczynski, I., Gryczynski, Z., Lakowicz, J. R. (2004b). Multi-wavelength immunoassays using surface plasmon coupled emission. Biochem. *Biophys. Res. Comm.* **313,** 721–726.

Melendez, J., Carr, R., Bartholomew, D. U., Kukanskis, K., Elkind, J., Yee, S., Furlong, C., and Woodbury, R. (1996). A commercial solution for surface plasmon sensing. *Sensors and Actuators B* 35–36, 212–216.

Natan, M. J., and Lyon, L. A. (2002). Surface plasmon resonance biosensing with colloidal au amplification. *In* "Metal nanoparticles: Synthesis, characterization, and applications" (D. L. Feldheim and C. A. Foss, eds.). Marcel Dekker, New York.

Nelson, B. P., Frutos, A. G., Brockman, J. M., and Corn, R. M. (1999). Near-infrared surface plasmon resonance measurements of ultrathin films. 1. Angle shift and SPR imaging experiments. *Anal. Chem.* **71,** 3928–3934.

Prober, J. M., Trainor, G. L., Dam, R. J., Hobbs, F. W., Robertson, C. W., Zagursky, R. J., Cocuzza, A. J., Jensen, M. A., and Baumeister, K. (1987). A system for rapid DNA sequencing with fluorescent chain-terminating dideoxynucleotides. *Science* **238,** 336–343.

Raether, H. (1977). Surface plasma oscillations and their applications. *In* "Physics of thin films, advances in research and development" (G. Hass, M. H. Francombe, and R. W. Hoffman, eds.), Vol. 9. Academic Press, New York.

Raether, H. (1988). *Surface plasmons on smooth and rough surfaces and on gratings*. Springer-Verlag, New York.

Salamon, Z., and Tollin, G. (1996). Surface plasmon resonance studies of complex formation between cytochrome c and bovine cytochrome c oxidase incorporated into a supported planer lipid bilayer. 1.

Binding of cytochrome c to cardiolipin/phosphatidylcholine membranes in the absence of oxidase. *Biophys. J.* **71,** 848–857.

Salamon, Z., MacLeod, H. A., and Tollin, G. (1997a). Surface plasmon resonance spectroscopy as a tool for investigating the biochemical and biophysical properties of membrane protein systems. I: Theoretical principles. *Biochim. et. Biophys. Acta* **1331,** 117–129.

Salamon, Z., MacLeod, H. A., and Tollin, G. (1997b). Coupled plasmon-waveguide resonators: A new spectroscopic tool for probing proteolipid film structure and properties. *Biophys. J.* **73,** 2791–2797.

Smith, L. M., Sanders, J. Z., Kaiser, R. J., Hughes, P., Dodd, C., Connell, C. R., Heiner, C., Kent, S. B. H., and Hood, L. E. (1986). Fluorescence detection in automated DNA sequence analysis. *Nature* **321,** 674–679.

TF Calc., Software Spectra, Inc., Portland, Oregon.

The Human Genome. Science February 16, 2001, 1177–1351.

The Human Genome. Nature February 15, 2001, 813–958.

Van Dyke, K., and Van Dyke R. (eds.) (1990). *Luminescence immunoassay and molecular applications.* CRC Press, Boca Raton, Florida.

Vo-Dinh, T., Sepaniak, M. J., Griffin, G. D., and Alarie, J. P. (1993). Immunosensors: Principles and applications. *Immunomethods* **3,** 85–92.

Wegner, G. J., Lee, H. J., and Corn, R. M. (2002). Characterization and optimization of peptide arrays for the study of epitope-antibody interactions using surface plasmon resonance imaging. *Anal. Chem.* **74,** 5161–5168.

Worthing, P. T., and Barnes, W. L. (1999). Spontaneous emission within metalclad microcavities. *J. Opt. A Pure Appl. Opt.* **1,** 501–506.

CHAPTER 5

Cytometry of Fluorescence Resonance Energy Transfer

György Vereb,* János Matkó,† and János Szöllősi*

*Department of Biophysics & Cell Biology, and Cell Biophysics Research Group
of the Hungarian Academy of Sciences
University of Debrecen
Debrecen H-4012, Hungary

†Department of Immunology
Eötvös Loránd University
Budapest H-1117, Hungary

I. Introduction

With the onset of modern proteomics, hundreds of pairs of cellular proteins that are capable of interacting with each other *in vitro* have been identified. However, the extent to which this possibility of interaction reflects their behavior in living cells is not clear. Nonetheless, it is abundantly clear that these protein–protein

METHODS IN CELL BIOLOGY, VOL. 75

0091-679X/04 $35.00

interactions are crucial in both maintaining the stable "resting" state of cells and driving activation processes to eventually give cellular responses to external stimuli. Hence, it becomes more and more important to be able to detect such interactions *in situ* inside or on the surface of cells. Fluorescence techniques are widely used to quantify molecular parameters of various biochemical and biological processes *in vivo* because of their inherent sensitivity, specificity, and temporal resolution. The combination of fluorescence spectroscopy with flow and image cytometry provided a solid basis of rapid and continuous improvements of these technologies. A major asset in studying molecular-level interactions was the application of fluorescence resonance energy transfer (FRET) to cellular systems. FRET is a special phenomenon in fluorescence spectroscopy during which energy is transferred from an excited donor molecule to an acceptor molecule under favorable spectral, proximity, and orientational conditions. Applying donor- and acceptor-labeled antibodies, lipids, and various types of fluorescent proteins (such as green fluorescent protein [GFP] analogs), the FRET technique can be used to determine intermolecular and intramolecular distances of cell surface components in biological membranes and molecular associations within live cells. For reviews see Szöllősi *et al.* (1998, 2002), Selvin (2000), Jares-Erijman and Jovin (2003), Scholes (2003), Sekar and Periasamy (2003), and Szöllősi and Alexander (2003). With the help of FRET, we can measure molecular dimensions and determine them in functioning live cells, providing information that would be impossible to obtain with other classic approaches such as with electron microscopic methods.

Excellent reviews are available on the applicability of FRET to biological systems, as well as descriptions and comparison of various approaches (Bastiaens and Squire, 1999; Berney and Danuser, 2003; Clegg, 1995, 2002; Gordon *et al.*, 1998; Matkó and Edidin, 1997; Szöllősi *et al.*, 1998; Vereb *et al.*, 2003).

On the following pages, we describe three flow cytometric FRET (FCET) and three image cytometric FRET approaches in technical details (Bastiaens *et al.*, 1996; Nagy *et al.*, 1998a,b, 2002; Sebestyen *et al.*, 2002; Szöllősi *et al.*, 1984; Tron *et al.*, 1984; Vereb *et al.*, 1997). We have selected these approaches because either we have played seminal roles in their elaboration or we have used these techniques extensively in biological systems. Also, the methods detailed have proved to be rather straightforward and robust in answering biological questions about molecular associations *in situ* in/on the cell. After a brief introduction to the theory of FRET, we explain details and relevant protocols of these FRET techniques. This is followed by several demonstrative FRET applications used by us or others. Specific advantages and limitations of FRET approaches are also discussed, together with perspectives and advances that may provide new approaches for the detection of FRET in the future. You will also be referred to most relevant papers and reviews—both classic and very recent—on FRET methodologies.

II. Theory of FRET

In FRET, an excited fluorescent dye, called a *donor,* donates energy to an acceptor dye, a phenomenon first described *correctly* by Förster (1946), as a dipole–dipole resonance energy transfer mechanism. For the process to occur, a set of conditions have to be fulfilled:

- The emission spectrum of the donor has to overlap with the excitation spectrum of the acceptor. The larger the overlap, the higher the rate of FRET is.
- The emission dipole vector of the donor and the absorption dipole vector of the acceptor need to be close to parallel. The rate of FRET decreases as the angle between the two vectors increases. In biological situations where molecules are free to move (rotate), we generally assume that dynamic averaging takes place; that is, during its excited lifetime, the donor assumes many possible steric positions, among them those that can yield an effective transfer of energy.
- The distance between the donor and acceptor should be between 1 and 10 nm.

This latter phenomenon is the basis of the popularity of FRET in biology: The distance over which FRET occurs is small enough to characterize the proximity of possibly interacting molecules; under special circumstances, it even provides quantitative data on exact distances and information on the spatial orientation of molecules or their domains. Hence, the very apropos term from Stryer and Haugland (1967), who equated FRET with a "spectroscopic ruler."

Figure 1 illustrates some major aspects and qualities of the FRET process. The usual term for characterizing the efficiency of FRET is E, which is the ratio of excited state molecules relaxing by FRET to the total number of excited molecules. The energy transfer rate is dependent on the negative sixth power of the distance R between the donor and the acceptor, resulting in a sharply dropping curve when plotting E against R, centered around R_0, the distance where $E = 0.5$, that is, where there is a 50% chance that the energy of the excited donor will be transferred to the acceptor. As separation between the donor and acceptor increases, E decreases, and at $R = 2R_0$, E is already getting negligible. Conversely, as the distance reaches values below 1 nm, strong ground–state interactions or transfer by exchange interactions becomes dominant at the expense of FRET (Dexter, 1953).

When looking at the Jablonski scheme of FRET (Fig. 1), it is easy to see that its occurrence has profound consequences on the fluorescence properties of both the donor and the acceptor. An additional de-excitation process is introduced in the donor, with rate constant k_t, in addition to fluorescent (k_f) and nonfluorescent relaxations (k_{nf}). This decreases the fluorescence lifetime and the quantum

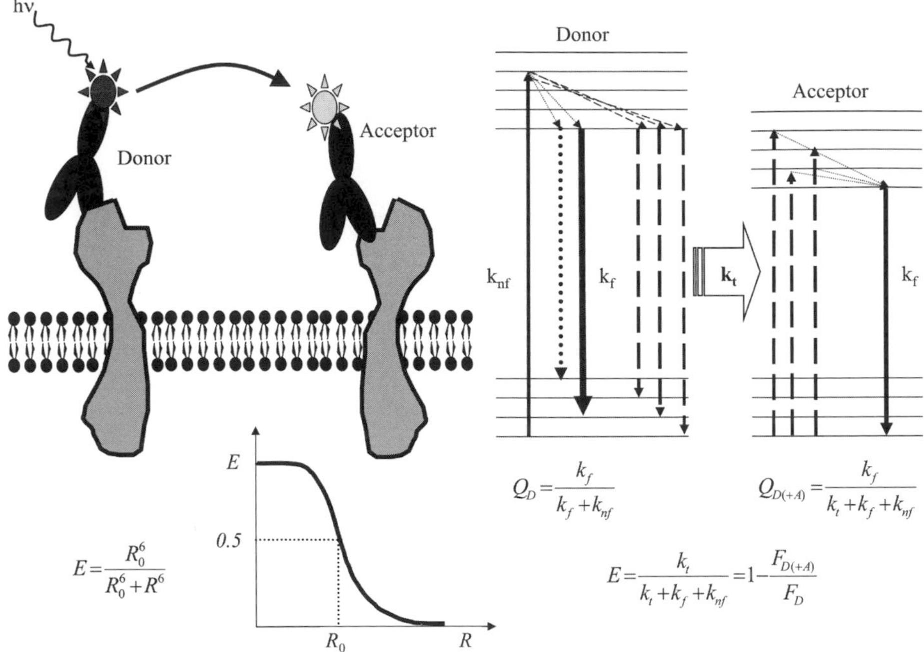

Fig. 1 The fluorescence resonance energy transfer (FRET) primer. FRET occurs when two suitable dye molecules that serve as donor and acceptor in the process come into sufficiently close molecular proximity of each other, the donor emission spectrum overlaps with the acceptor absorption spectrum and the spatial orientation of the respective emission, and absorption dipoles is favorable. The efficiency of FRET, E, is the function of the inverse sixth power of donor–acceptor separation, and thus, FRET occurs only over short R distances in the range of 1–10 nm, distributed around R_0, which marks 50% efficiency for a given donor–acceptor pair. The requirement for molecular proximity makes the phenomenon ideal for detecting association or aggregation of various macromolecules even in functioning cells. The Jablonski diagram for FRET shows that this process is an additional de-excitation pathway (rate constant = k_t) for the excited donor that otherwise can relax by radiative (k_f) and nonfluorescent (k_{nf}) processes. Henceforth, the Q_D quantum efficiency of the donor decreases in the presence of FRET ($Q_{D(+A)}$). It ensues that measuring donor fluorescence, a parameter proportional to Q, in the presence and absence of acceptor is one way to assess FRET efficiency.

efficiency of the donor ($Q_{D(+A)} < Q_D$), rendering it less fluorescent. The decrease in donor fluorescence (often termed *donor quenching*) can be one of the most easily measured spectroscopic characteristic that indicates the occurrence of FRET. Additionally, because the acceptor is excited as a result of FRET, acceptors that are fluorescent will emit photons (proportional to their quantum efficiency) also when FRET occurs. This is called *sensitized emission* and can also be a sensitive measure of FRET.

III. Measuring FRET

The various approaches that have been used to quantitate FRET can be categorized based on the spectrofluorometric parameter detected and whether the donor or the acceptor is investigated:

- Measurements based on intensity changes upon FRET
  - Measuring donor quenching based on donor fluorescence with and without the acceptor
    - Based on samples labeled with and without acceptor (Turcatti *et al.*, 1996)
    - Based on photobleaching the acceptor (Bastiaens *et al.*, 1996; Vereb *et al.*, 1997)
    - Based on reversibly switching on and off the acceptor (Giordano *et al.*, 2002)
    - Measuring sensitized acceptor emission (Suzuki, 2000)
  - Measuring both donor quenching and sensitized acceptor emission
  - Measuring integrated emission in spectral bands through filters (Nagy *et al.*, 1998b; Tron *et al.*, 1984)
  - Measuring and fitting emission spectra (Clegg *et al.*, 1992)
- Measurements of decreased donor lifetime
  - Direct measurement of donor lifetime with and without the acceptor (Bastiaens *et al.*, 1996; Clegg *et al.*, 2003)
  - Estimation of donor lifetime based on photobleaching rate of the donor (Gadella and Jovin, 1995; Jovin and Arndt-Jovin, 1989; Vereb *et al.*, 1997)
  - Estimation of donor lifetime based on fluorescence anisotropy (Matkó *et al.*, 1993)
- Measurements of increased acceptor photobleaching rate (Mekler, 1994)
- Measurements of decreased fluorescence anisotropy of identical fluorophores (Clayton *et al.*, 2002; Lidke *et al.*, 2003).

These techniques vary greatly in their applicability, sensitivity, accuracy, and robustness as was demonstrated in a review by Berney and Danuser (2003). The authors systematically compared eight methods for calculating FRET efficiency and seven methods for calculating FRET indices. Using Monte Carlo simulation, representative sets of FRET efficiencies and indices were validated in different experimental settings. It is highly recommended to consult the results of this paper when choosing the right FRET approach for a special experimental system (Berney and Danuser, 2003).

FRET can be measured both in microscopic imaging and in flow cytometry. FCET carries the advantage of examining large cell populations in a short time and provides a FRET efficiency value averaged over the population or on a cell-by-cell basis, but averaged over each cell. The microscopic approaches have the ability to provide subcellular detail and the possibility to correlate FRET values with other biological information gained from fluorescent labeling, on a pixel-by-pixel basis. However, the sample size is restricted, because data acquisition and processing are rather time consuming. Also, biological variation may influence the composition of the cell population selected for imaging, which may or may not be advantageous, depending on our understanding of selection criteria.

In a review about FRET imaging, Jares-Erijman and Jovin (2003) classified 22 approaches to quantifying FRET in a systematic way. Six of these techniques have not been tested in a biological system yet. The characterization is based on the fact that in exploiting FRET in image cytometry, two fundamental challenges should be faced. First, the formalism must be appropriate for quantifying FRET under conditions of arbitrary, generally unknown, intramolecular and/or intermolecular stoichiometries, distributions and microenvironments of donor and acceptor. Second, to provide temporal resolution, continuous methods of observation of FRET are needed in most studies of live cells. The techniques fall in two major groups: group I includes 18 methods and is based on donor quenching and/or acceptor sensitization; group II, having only four methods, is based on measuring emission anisotropy of either the donor or the acceptor. Of the 22, eight techniques can be applied in flow cytometry as well (Jares-Erijman and Jovin, 2003).

A. Flow Cytometric FRET

1. Donor-Quenching FRET

If one quickly and easily needs to estimate the efficiency of FRET between two epitopes, the easiest way is to label one of them with a donor and the other with an acceptor fluorophore, bound to monoclonal antibodies (mAbs) or their Fab fragments. Suitable donor–acceptor pairs are mostly selected based on the availability of conjugated antibodies and excitation lines in the flow cytometer. The classic dye pair used for FRET is fluorescein and tetramethylrhodamine (TR). Other spectrally similar dyes such as Alexa fluor 488 or Cy2 can be used in place of fluorescein, and TR can be replaced by Cy3 or Alexa fluor 546. Suitable pairs of visible fluorescent proteins (like cyan FP and yellow FP) fused to the target proteins are also applicable, but their usage is mostly advised after the cellular system has already been explored using specific antibodies.

When applied for the measurement of donor quenching, only the fluorescence of fluorescein (the donor) is measured, and thus, the 488 Ar laser line most commonly available in flow cytometers is adequate. We need to measure the *background-corrected* fluorescence from a donor-only–labeled sample (F_D) and from a sample double labeled with donor and acceptor ($F_{D(+A)}$). FRET efficiency is calculated as

$$E = 1 - \frac{F_{D(+A)}}{F_D} \tag{1}$$

Since F_D and $F_{D(+A)}$ are measured on distinct samples, only the histogram means from the two populations can be considered here. This is one of the main disadvantages of the method, so it is only suggested as a quick and rough estimate of whether FRET and thus molecular proximity occurs. However, it is a quite useful approach when signals are low compared to background/autofluorescence.

There is always one additional control to make, and that is to check for competition of the antibody carrying the acceptor with donor labeling. This should be done with the unlabeled antibody against the "acceptor" epitope, and any decrease of donor fluorescence caused by adding the unlabeled antibody should be attributed to competition rather than donor quenching caused by FRET. Needless to say, competition between labeling antibodies is likely also a sign of molecular proximity, albeit not as readily quantitated as FRET.

2. Classic FCET Based on Both Donor Quenching and Acceptor Sensitization

Although the measurement of FRET-induced donor quenching does not need either dual-laser instruments or complicated evaluation, quenching cannot be used for cell-by-cell data analysis of FRET efficiency. In the classic approach introduced for the cell-by-cell measurement of FRET (Damjanovich *et al.*, 1983; Szöllősi *et al.*, 1984; Tron *et al.*, 1984) either a dual-laser flow cytometer with Argon ion lasers tuned to 488 and 514 nm or a single laser device modified with a prism to split these two lines during multiline operation can be used. The method, later adapted to digital microscopy as well (Nagy *et al.*, 1998b), is based on the following model.

In a sample labeled with both the donor and the acceptor, the intensity measured when exciting the donor (say, at 488 nm) and detecting donor emission (e.g., with a bandpass filter centered around 540 nm) can come from the donor (I_D), but also from the acceptor if that is excited at 488 nm and its emission spectrum overlaps with the detection bandpass of the donor. In the ideal situation, which is the case for the fluorescein–TR dye pair, the acceptor TR is not detected in the donor excitation/emission regimen. (As a point of caution, Cy3 as an acceptor is less red shifted than TR, so using it as an acceptor in place of TR requires either different filters or a correction for crosstalk in the model of calculation.)

If FRET is present, the unquenched donor intensity I_D is diminished to the $1 - E$ fraction of its original value.

$$I_1(ex : D, em : D) = I_D(1 - E) \tag{2}$$

The emission measured in the acceptor emission channel (here, >590 nm) has two possible sources of excitation. We can generate fluorescence not only by exciting at 514 nm (see I_3 in Eq. 9), but also by exciting at 488. In this latter case, we have two sources of emission. The donor will emit light above 590 nm, and this is proportional to the quantity of donor dyes, I_D, its quenching by FRET $(1 - E)$ and a factor S_1 that accounts for the ratio of signals generated by the same amount of donor molecules in the acceptor and the donor emission channels:

$$S_1 = \frac{I_2(donor)}{I_1(donor)} \tag{3}$$

$$I_2(ex : D, em : A) = I_D(1 - E)S_1 + I_A S_2 + I_D E\alpha \tag{4}$$

The second two components of this I_2 emission originate from the acceptor. If the acceptor were to emit in the acceptor channel after excited by its proper (here, 514 nm) excitation wavelength (*ex: A, em: A*), the intensity measured in the absence of FRET would be I_A. Exciting the acceptor at the donor wavelength (*ex : D*) is also possible, but it will be less efficient, the factor of proportionality being S_2:

$$S_2 = \frac{I_2(acceptor)}{I_3(acceptor)} \tag{5}$$

This factor can be obtained from measuring the I_2 and I_3 (see later discussion) intensities for samples that are labeled with acceptor only.

The emission of acceptor has yet another source in addition to excitation at *ex:* D = 488 nm. It gains energy from the donor via FRET, resulting in an emission component $I_D E\alpha$. The signal arising from sensitized emission of the acceptor is proportional to the number of excitation quanta transferred from the donor to the acceptor and is therefore proportional to $I_D E$. The α factor defines the detection sensitivity of fluorescence from an excited acceptor molecule with respect to the sensitivity to detect fluorescence from an excited donor molecule and can be determined from the ratio of the following signals: the I_2 signal arising from N pieces of excited acceptor molecules and the I_1 signal arising from the same number of excited donor molecules. It depends on the fluorescence quantum yields of the used dyes (Q_D for the donor and Q_A for the acceptor), and the overall detection efficiencies in the donor (η_D) and the acceptor (η_A) detection channels for photons with wavelength distribution of the donor and the acceptor emission, respectively:

$$\alpha = \frac{Q_A \eta_A}{Q_D \eta_D} \tag{6}$$

Experimentally α can be determined by comparing the I_1 signal of a donor-only–labeled sample to the I_2 signal of a sample labeled with acceptor only:

$$\alpha = \frac{I_{2A}}{I_{1D}} \cdot \frac{B_D}{B_A} \cdot \frac{L_D}{L_A} \cdot \frac{\varepsilon_D}{\varepsilon_A} \tag{7}$$

where B denotes the mean number of receptors per cell labeled by the corresponding antibody, L stands for the mean number of dye molecules attached to an antibody molecule, and ε is the molar extinction coefficient of the dyes at 488 nm; subscripts A and D refer to acceptor and donor, respectively. The ratios of B, L, and ε correct for the different number of photons absorbed by the acceptor- and donor-only–labeled samples. The α factor has to be determined in each experiment, because it depends on the instrument setup and the dye pair. In the equation for α, there is a contribution from the sample labeled only with acceptor excited at the donor wavelength. Usually, this fluorescence intensity is rather small, thus giving the main error source in the calculations. To decrease this error, α should be determined using a protein that is abundant on our cells and recalculated for the actual antibodies used in the experiment.

The fluorescence quantum yields of the dyes may depend on the type of antibody they are attached to and even on the labeling ratio L, thereby affecting the value of α. The α factor determined for a given donor–acceptor antibody pair can be used for other antibody pairs labeled with the same dyes, provided its value is corrected for the differences in the quantum yields:

$$\alpha_2 = \alpha_1 \frac{Q_{A,2}}{Q_{A,1}} \frac{Q_{D,1}}{Q_{D,2}} \tag{8}$$

where subscript 1 refers to the antibody pair for which α has been determined previously, and subscript 2 refers to the new antibody pair.

With this said, we finally progress to analyzing the signal in the acceptor channel detected after exciting specifically the acceptor:

$$I_3(ex : A, em : A) = I_D(1 - E)S_3 + I_A + I_D E \alpha S_x \tag{9}$$

The first component is the contribution of quenched donor emission, $I_D(1-E)$, to this channel, which is weighted by the factor S_3, obtained on donor-only–labeled samples similarly to S_1.

$$S_3 = \frac{I_3(donor)}{I_1(donor)} \tag{10}$$

The second component is the acceptor emission without sensitization, I_A itself, and the third component is the sensitized acceptor emission. This is the same as in Eq. 4 for I_2, except an S_x factor has to be introduced to correct for exciting the donor at the acceptor's optimum wavelength rather than at its own (as for I_2):

$$S_x = \frac{I_3(donor)}{I_2(donor)} \tag{11}$$

Incidentally, S_x can be expressed from S_3 and S_1 that were already introduced, so we can reformulate Eq. 9 as

$$I_3(ex : A, em : A) = I_D(1 - E)S_3 + I_A + I_D E\alpha \frac{S_3}{S_1} \tag{12}$$

As already hinted at, the factors S_1, S_2, S_3, and α can be determined from measuring the intensities I_1, I_2, and I_3 for donor-only and acceptor-only–labeled samples. Now it only remains to measure the intensities I_1, I_2, and I_3 on samples labeled with both donor and acceptor (then, of course, subtract the appropriate fluorescence values measured on unlabeled control cells), and calculate, from the list-mode data file, an "A" value for each cell:

$$A = \frac{1}{\alpha}\left[\frac{(I_2 - S_2 I_3)}{\left(1 - \frac{S_3 S_2}{S_1}\right)I_1} - S_1\right] \tag{13}$$

From Eqs. 2, 4, and 12, it follows that the FRET efficiency E is related intimately to A:

$$E = \frac{A}{1 + A} \tag{14}$$

And hence FRET efficiency on a cell-by-cell basis can be calculated.

3. FCET with Cell-by-Cell Autofluorescence Correction

Both the accuracy and the reproducibility of FRET measurements are compromised if the ligands for the fluorescently labeled probes are expressed at low levels. In such cases, the contribution of autofluorescence may be significant relative to the specific signal. To improve the applicability of FRET in low signal-to-noise systems, one strategy is to apply a red-shifted donor–acceptor pair such as Cy3 and Cy5, or Alexa fluor 546 and Alexa fluor 633. This approach is particularly useful because cellular autofluorescence is higher in the blue and green spectral regions than in the red edge of the visible spectrum (Loken *et al.*, 1987). Furthermore, a cell-by-cell correction for autofluorescence can be achieved using yet another fluorescence channel in addition to those specific for the donor and the acceptor mentioned in the previous section (Sebestyen *et al.*, 2002).

The mathematical model can be developed based on that described for FCET earlier, with the following modifications. Because Cy3 and Cy5 are used as the donor-acceptor pair, the 488 nm line of an argon ion laser and the 635-nm line of a red diode laser can efficiently excite the donor and acceptor dyes, respectively. The emission maximum of Cy3 (565 nm) falls in the range of the *FL2* detection channel (e.g., through an 585 nm bandpass filter), whereas the maximum of Cy5 emission (667 nm) is in the range of the *FL3* (670-nm long-pass filter) and *FL4* (661-nm bandpass filter) channels. The *FL1* channel, which does not overlap with the emission spectrum of Cy5 and has very little overlap with the spectrum of

Cy3, can be used for detecting autofluorescence. Thus, *FL*2, *FL*3, and *FL*4 will replace I_1, I_2, and I_3 of the previous section, and for each of these intensities, an additional contributing component, that from autofluorescence, is also considered. *FL*1 is introduced as the channel specific for autofluorescence, and the contribution of autofluorescence to each of the other channels is estimated from weight factors B_2, B_3, and B_4, similar to the S_1, S_2, and S_3 factors already used. Clearly, the contribution of fluorescent labels to the autofluorescence channel also needs to be considered and consequentially further S factors—S_5 and S_6—need to be introduced. Furthermore, the acceptor Cy5 may contribute to the donor-specific *FL*2 signal depending on the optical setup, so its contribution has to be taken into account using an S_4 factor. The equation system that evolves is as follows:

$$FL1(488, 530) = AF + I_D(1 - E) \cdot S_5 + I_A \cdot S_6 + I_D \cdot E \cdot \alpha \cdot \frac{S_6}{S_2} \tag{15}$$

$$FL2(488, 585) = AF \cdot B_2 + I_D(1 - E) + I_A \cdot S_4 + I_D \cdot E \cdot \alpha \cdot \frac{S_4}{S_2} \tag{16}$$

$$FL3(488, > 670) = AF \cdot B_3 + I_D(1 - E) \cdot S_1 + I_A \cdot S_2 + I_D \cdot E \cdot \alpha \tag{17}$$

$$FL4(635, 661) = AF \cdot B_4 + I_D(1 - E) \cdot S_3 + I_A + I_D \cdot E \cdot \alpha \cdot \frac{S_3}{S_1} \tag{18}$$

The numbers in brackets refer to the excitation and detection wavelengths. The autofluorescence (native to *FL*1), the unquenched donor fluorescence (native to *FL*2), and the directly excited acceptor fluorescence (native to *FL*4) are denoted by *AF*, I_D, and I_A, S_1, S_3, and S_5 are determined using samples labeled only with Cy3:

$$S_1 = \frac{FL3}{FL2} \tag{19}$$

$$S_3 = \frac{FL4}{FL2} \tag{20}$$

$$S_5 = \frac{FL1}{FL2} \tag{21}$$

S_2, S_4, and S_6 are determined on cells labeled only with Cy5:

$$S_2 = \frac{FL3}{FL4} \tag{22}$$

$$S_4 = \frac{FL2}{FL4} \tag{23}$$

$$S_6 = \frac{FL1}{FL4} \tag{24}$$

B_2, B_3, and B_4 are determined on unlabeled cells:

$$B_2 = \frac{FL2}{FL1} \tag{25}$$

$$B_3 = \frac{FL3}{FL1} \tag{26}$$

$$B_4 = \frac{FL4}{FL1} \tag{27}$$

When using the filters and dichroic mirrors available in the BD FACSCalibur, S_3, S_4, and S_6 are zero, which simplifies Eqs. 15–18. Solving the simplified equations yields the following expression for the parameter $A = E/(1-E)$:

$$\frac{FL1 \cdot (B_2S_1 + B_4S_2 - B_3) + FL2 \cdot (B_3S_5 - S_1 - B_4S_2S_5) + FL3 \cdot (1 - B_2S_5) + FL4 \cdot S_2 \cdot (B_2S_5 - 1)}{\alpha(FL2 - FL1 \cdot B_2)} \tag{28}$$

Using this formula, we can easily calculate the value of E for each cell either in a spreadsheet of imported list-mode data or in a specific program where gating and determination of the S, B, and α factors is considerably easier. An example showing the improvement resulting from this approach in the case of relatively low receptor expression is depicted in Fig. 2.

4. Sample Experiment for FCET Measurement

Two molecular epitopes on the surface of cells are chosen for testing their nanometer-scale proximity. They are labeled with Cy3 and Cy5 conjugated dyes. Samples are made where both, either, or none of the epitopes are labeled. The samples are run in a flow cytometer and the appropriate emission intensities are collected in list-mode files. Then the samples are analyzed using a custom-made program (available to the public from my web site at www.biophys.dote.hu/research.htm on the menu "instruments, software," and submenu "FLEX") to assess whether the labeled epitopes are in each other's molecular proximity. The analysis yields a cell-by-cell distribution of FRET efficiency values. Such analysis can also be performed in a spreadsheet table if one prefers.

In general, the following samples are necessary:

1. Unlabeled cells
2. Cells with one epitope labeled using a Cy3 conjugated antibody (or alternatives like Alexa fluor 546)
3. Cells with the same epitope labeled using a Cy5 conjugated antibody (or alternatives like Alexa fluor 633)
4. Cells with the other epitope labeled using a Cy3 (or alternative) conjugated antibody

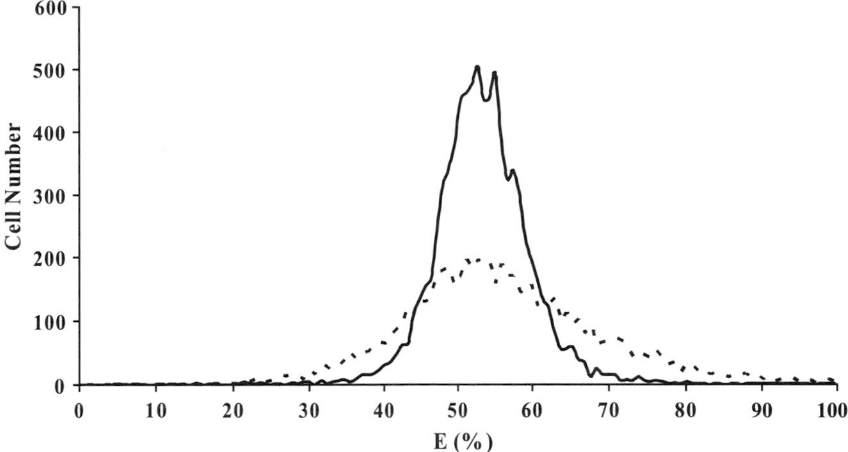

Fig. 2 Improving fluorescence resonance energy transfer (FRET) analysis using cell-by-cell autofluorescence correction. Distribution histograms of FRET efficiency are produced from flow cytometric measurements on breast cancer cells expressing relatively low numbers (~40,000/cell) of ErbB2 molecules. Two epitopes on ErbB2 were tagged with Alexa fluor 546 and Alexa fluor 647 conjugated monoclonal Fab fragments. Consequently, FRET distribution histograms peak at about 50%, indicating efficient intramolecular energy transfer. Cell-by-cell autofluorescence correction (*solid line*) vastly improves the dispersion of data as compared to analysis with subtraction of an average autofluorescence value (*dashed line*), but the mean remains unchanged. This approach is especially useful when signals from the specific label are low relative to autofluorescence.

5. Cells with the other epitope labeled using a Cy5 (or alternative) conjugated antibody
6. Cells double labeled with the pair of antibodies Cy3 on epitope 1 and Cy5 on epitope 2
7. Cells double labeled with the pair of antibodies Cy5 on epitope 1 and Cy3 on epitope 2.

This set allows for measuring FRET in both directions, which is useful if expression levels of the two epitopes are not comparable. If a positive control with known molecular interactions is known, it is advisable to have a set of samples of the same cell type similar to those listed earlier. For example, if we want to know about the molecular interaction of major histocompatibility complex (MHC) class I and class II on a lymphocytic cell, we need the following:

1. Unlabeled cells
2. Cells with MHC class I labeled using a Cy3 conjugated antibody
3. Cells with MHC class I labeled using a Cy5 conjugated antibody
4. Cells with MHC class II labeled using a Cy3 conjugated antibody
5. Cells with MHC class II labeled using a Cy5 conjugated antibody

6. Cells double labeled with anti-MHC-I–Cy3 and anti-MHC-II–Cy5 antibodies
7. Cells double labeled with anti-MHC-I–Cy5 and anti-MHC-II–Cy3 antibodies.

As for the positive control, we know that MHC-I binds the β_2-microglobulin (β2m), so we make further samples:

8. Cells with β2m labeled using a Cy3 conjugated antibody (this replaces MHC-II as the donor, when MHC-I is the acceptor, and the reason to make β2m the donor is that surely all β2m is bound to an MHC-I, but not all MHC-I may bind β2m)
9. Cells double labeled with anti–β2m-Cy3 and anti-MHC-I–Cy5.

Additional information needed for the calculations is as follows:

- The dye/protein molar ratio of all antibodies used
- The molar absorption coefficients of all dyes used
- The quantum efficiencies of the dyes used, if any of the labels give intensity less than three to five times the background.

a. Materials

Fluorochromes

- Cyanine dyes Cy3 and Cy5 or Alexa fluor dyes Alexa 546 and Alexa 633 as donor and acceptor, respectively

Cells

- Lymphocytic cells in suspension.
- Adherent cells should be detached (trypsinized, or treated with collagenase) before washing and labeling.

Solutions

- PBS (for washing)
- PBS + 0.1% BSA (for labeling)
- PBS + 1% formaldehyde (for fixation)
- Antibodies W6/32 against MHC-I, L368 against the β2m, L243 against MHC-II

b. Sample Preparation

Harvesting Cells

- 75-ml flasks of adherent cells are trypsinized and medium with FCS is added to stop the trypsin.

• Cells are left to recover 20 minutes in the flask. It has been determined that after gentle trypsinization, most cell surface proteins are either unchanged or totally recovered within 20–30 minutes.

• One needs to determine labeling intensity as a function of time after trypsinization for the particular proteins examined.

Labeling

- Wash cells with ice-cold PBS and centrifuge suspension
- Repeat washing
- Add 1 million cells per sample tube and store on ice
- Label cells with (usually) 5–50 μg/ml final concentration of antibodies (should be above saturating concentration that was determined previously) in 50 μg total volume of PBS–0.1% BSA mixture for 15–30 minutes on ice
- Wash cells twice with ice-cold PBS and centrifuge suspension
- Fix cells with PBS + 1% formaldehyde in 500–1000 μl
- Store samples in refrigerator or cold room until measurement

Before measurement, resuspend the cells with gentle shaking, and if upon examination in the microscope clumps are detected, run the suspension through a fine sieve.

Always examine labeled cells dropped on a microscopic slide in the fluorescence microscope to verify proper cellular position (e.g., membrane) of the label.

c. *Instrument Settings*

- Use sample 1 (background) as a negative control
- Set FSC and SSC in linear mode to see your population on the scatter plot (*SSC/FSC* dot plot)
- Set FL1, FL2, FL3, and FL4 in logarithmic mode
- Set FL1, FL2, FL3, and FL4 voltages so that mean fluorescence intensities are about 10^1
- Run sample 2 (donor) and adjust FL1, FL2, and FL3 voltages so that mean fluorescence intensities are in the 10^2–10^4 range
- Run sample 4 (acceptor) and adjust FL3 and FL4 voltages so that mean fluorescence intensities are in the 10^2–10^4 range
- Save instrument settings
- Draw the following plots:

- *FSC/SSC*
- *FL*1, *FL*2, *FL*3, and *FL*4 for intensity check
- *FL*2/*FL*3, *FL*2/*FL*4, and *FL*3/*FL*4 for correlation check

- Run sample 1 and draw a gate on the *FSC/SSC* dot plot around viable cells

- Format plots so that only R1-gated events will appear
- Define statistics window to show the mean fluorescence intensities of all histograms from the R1-gated events
- Set the machine to acquire 20,000 events
- Run samples from 1 to 9.

d. Analysis

During data analysis, it is advised to follow a general scheme. First, one needs to determine the mean background intensities and the autofluorescence correction factors. Then calculate the α factor and spectral overspill parameters (S factors) from the acceptor- and donor-labeled samples. With these parameters in hand, now the energy transfer efficiency can be determined on a cell-by-cell basis.

B. Image Cytometric FRET

When measuring FRET in a flow cytometer, one should always keep in mind that FRET efficiency is calculated from fluorescence intensities that are averaged over each cell. Thus, the FCET technique inherently excludes the possibility of gaining information about the subcellular distribution of molecular proximities. Thus, the idea of measuring FRET in microscopic imaging has been introduced quite early in the joint history of cytometry and FRET (Jovin and Arndt-Jovin, 1989). All possible approaches to imaging FRET—both those already implemented and those only in the proposal stage—are reviewed in Jares-Erijman and Jovin (2003). We concentrate on three approaches that can be implemented using rather conventional imaging equipment and that have been applied extensively to answer biological questions. The common feature in these techniques is that they exploit the change of fluorescence intensity caused by the occurrence of FRET. A fourth approach that has successfully been used in cell biology is fluorescence lifetime imaging (Bastiaens and Jovin, 1996; Bastiaens and Squire, 1999; Gadella *et al.*, 1994; Haj *et al.*, 2002; Van Munster and Gadella, 2004). For an excellent overview, see Clegg *et al.* (2003).

As for intensity-based approaches, the first two measure donor fluorescence, whereas the third is similar in principle to the ratiometric technique described for FCET in section III.A.3. When we observe fluorescently stained microscopic samples, one labeled with donor and the other with both donor and acceptor, the samples differ in two basic features (Fig. 3). First, those labeled with both donor and acceptor will show (on average) less intense donor fluorescence because of quenching by the acceptor. Second, excited state reactions resulting in irreversible photobleaching of the donor—a phenomenon usually most annoying to microscopists—will occur at different rates. This latter is related to the decreased excited state lifetime of the donor in the presence of the acceptor, which confers some degree of "protection" from photobleaching to the donor, thereby lengthening the

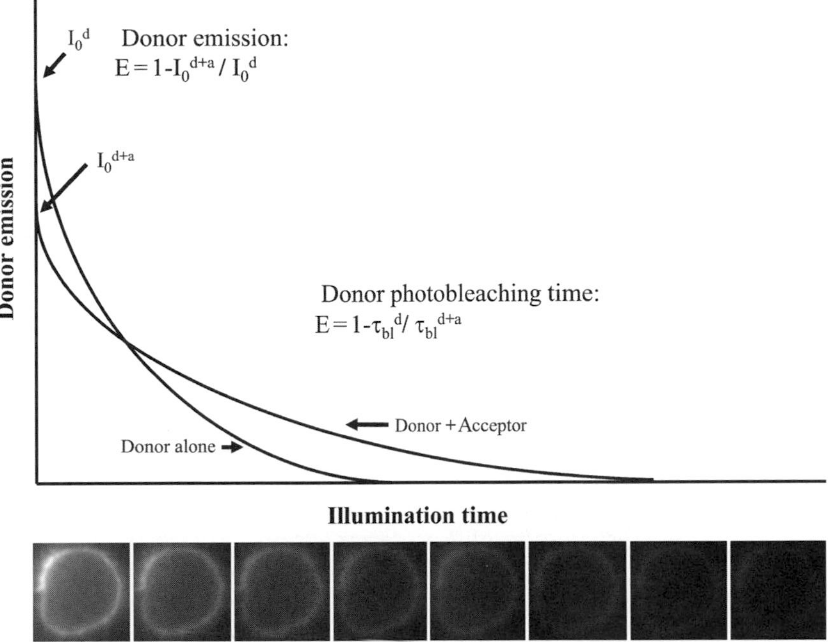

Fig. 3 Fluorescence resonance energy transfer (FRET) measured from donor parameters. Comparing donor fluorescence of donor with donor- and acceptor-labeled samples in the microscope yields two basic differences, each exploitable for the determination of FRET efficiency. On a single observation, lower fluorescence intensity in the donor- and acceptor-labeled sample because of "donor quenching" will be apparent. During continuous or repeated observations, photobleaching of the dye can occur (see also image sequence on lower inset—MHC-I labeled on human lymphocyte), the rate of which is inversely proportional to the excited state lifetime of the donor. Both these phenomena can be exploited to assess the change of donor lifetime upon the occurrence of the acceptor and thus the efficiency of FRET.

time constant of photobleaching. Both of these features can be exploited for the measurement of FRET efficiency, and because excited state reactions are central to both, these techniques are often called *photobleaching FRET* (pbFRET).

1. Donor Photobleaching FRET

As depicted in Fig. 3, a fluorescently labeled cellular sample is subjected to photobleaching under continuous observation in the fluorescence microscope. The rate of photobleaching depends on the concentration of free radicals, the illumination flux, and the excited state lifetime. Keeping the former two constant, one is able to deduce the relation between the photobleaching time constant τ_{bl}, which is inversely proportional to the excited state lifetime τ, and the E efficiency of FRET (Bastiaens and Jovin, 1998; Jovin and Arndt-Jovin, 1989):

$$E = 1 - \frac{\tau^{D+A}}{\tau^{D}} = 1 - \frac{\tau_{bl}^{D}}{\tau_{bl}^{D+A}} \tag{29}$$

Here the D and $D + A$ upper indices refer to donor-only– and donor-plus-acceptor–labeled samples, respectively. It ensues that for this measurement approach, we need to prepare two samples, one with the epitope in question labeled with donor and the other with both donor and acceptor. Both samples are observed and imaged in the microscope using donor-specific optical filters, taking a sequence of images until the donor is totally bleached and hence its intensity reaches background. Intensity usually decays according to an exponential function to which a pixel-by-pixel fit can be made. Because the control (donor only) bleaching time constants originate from a separate sample, one needs to average these time constants and use the average as a reference τ_{bl}^{D} value for the calculation of a pixel-by-pixel FRET efficiency according to Eq. 29. These E values can then be evaluated with respect to subcellular localization, or they can be pooled to form distribution histograms.

The method carries inherent advantages and disadvantages. Although subcellular distribution of E is derived from the measurement, this distribution should be judged carefully, because pixel-by-pixel variations of bleaching times can also result from variations of the local molecular environment, oxygenation, and previous bleaching of neighboring cells or even pixels. For this reason, the method offers more reliable results in full-field microscopy, whereas confocal laser scanning microscopes (LSMs) tend to fare worse in implementing this approach. Nonetheless, donor pbFRET is relatively simple to implement and is rather sensitive; FRET efficiencies of 2–4% can be reasonably well measured if labeling is good and images are free of noise. Adherent cells are the best targets for investigations with this approach, similar to other image cytometric FRET measurements. However, suspension cells and cell lines can also be measured after making the cells adhere to a substrate, either by sedimentation onto poly-L-lysine or collagen-coated coverslips or by using a cytocentrifuge.

Some disadvantages and pitfalls should also be considered when choosing and implementing this method. Primarily, the measurements are not self-controlled in the conventional sense, so care should be taken to execute bleaching sequences alternately between the donor-only– and the donor-plus-acceptor–labeled samples—even more so because fluctuations of temperature, illumination light intensity, and oxygenation greatly influence the photobleaching rate. A dye that is easily photobleached should be chosen as the donor, for example, fluorescein. This will hopefully minimize movement artifacts that hinder the proper exponential fitting. Should such artifacts persist, the microscope needs to be checked for mechanical stability, and cells for proper adherence. Sequential images can be corrected for registration, but in the case of a long sequence (and about 30 images are necessary for a good fitting), this may be cumbersome even if using a simple fast Fourier transform (FFT)–based algorithm.

Special consideration ought to be given to the actual kinetics of photobleaching. This mostly depends on the number and nature of various excited state reactions the donor can undergo. Fluorescein, for example, shows a rather complex behavior in this respect (Song *et al.*, 1995), so although it offers the advantage of fast bleaching, a multiexponential fitting may be necessary to obtain the proper bleaching time constants. In practice, quite often a double exponential works very well. Here, to be able to easily compare pixels or cells, an amplitude weighted average can be calculated from the two bleaching time constants (Nagy *et al.*, 1998a; Szabó *et al.*, 1995). Another consideration is the initial bleaching that occurs during the adjustment of the microscope, localization of the spot to measured, and focusing. If the bleaching is not monoexponential, an overestimation of the time constant (and underestimation of E) may result. Furthermore, it is most necessary to choose a highly photostable acceptor, because photobleaching of the acceptor after excitation via FRET will primarily destroy the nearest acceptors and eliminate FRET between the tightest donor–acceptor pairs, leading to an underestimation of E. A sign that hints at such a possibility is the relative stability or unexpected increase of donor fluorescence in the initial phases of the bleaching curve.

Comparison between FCET measurements and the donor photobleaching image cytometric approach revealed that consistently higher transfer values are obtained with the pbFRET method. This overestimation was independent of the pixel size: Pixel sizes as large as a cell gave similar results to those obtained with smaller ones, reinforcing the view that energy transfer values are independent of fluorescence intensity in the samples. Some of this discrepancy can be attributed to the different weighting of energy transfer values in the pbFRET and the ratiometric energy transfer methods. Using Monte Carlo simulation, Nagy *et al.* (1998b) demonstrated that this overestimation in pbFRET is proportional to the heterogeneity in the pixel-by-pixel FRET efficiency values. Therefore, discrepancies between FRET efficiency values obtained with pbFRET and ratiometric approaches should be interpreted with caution (Nagy *et al.*, 1998b).

2. Acceptor Photobleaching FRET

Some of the pitfalls of donor pbFRET can be addressed by using the acceptor pbFRET technique (Bastiaens and Jovin, 1998; Bastiaens *et al.*, 1996, 1997; Vereb *et al.*, 1997). Most importantly, the acceptor bleaching is a reasonably simple technique that offers the unique advantage of self-controlled measurements on a pixel-by-pixel basis. The idea is to label a sample with both donor and acceptor, measure the fluorescence of the donor, and after having used an acceptor-specific excitation wavelength targeted at a given region of interest (ROI), measure it again to assess the increase in donor fluorescence. This is easily done in a (confocal) LSM. The energy transfer efficiency E can be calculated as

$$E_{(i,j)} = 1 - \frac{F^{I}_{D(i,j)}}{\gamma \cdot F^{II}_{D(i,j)}}, \tag{30}$$

where $F^{I}_{D(i,j)}$ and $F^{II}_{D(i,j)}$ are the background-subtracted donor fluorescence values of the $(i,j)^{\text{th}}$ pixel in the entire image before (I) and after (II) photobleaching the acceptor. In the denominator, γ is a correction factor that takes into consideration the photobleaching of the donor during the whole protocol. It can be calculated as

$$\gamma = \left\langle F^{ref,I}_{D(i,j)} \right\rangle / \left\langle F^{ref,II}_{D(i,j)} \right\rangle, \tag{31}$$

where $\langle F^{ref,I}_{D(i,j)} \rangle$ and $\langle F^{ref,II}_{D(i,j)} \rangle$ are the mean, background-subtracted donor intensities in pixels (i,j) above threshold of a reference sample labeled with donor only, before (I) and after (II) running an identical acceptor photobleaching protocol. This constant must be determined for each experimental setting.

In addition to determining this constant, the following controls and measures of caution need to be implemented. (1) During the image acquisition protocol, we must confirm that the acceptor was totally bleached. (2) Using an acceptor-only–labeled specimen, the whole protocol should be run, and no change in the donor channel should be seen. (3) It is recommended to leave the acceptor intact in part of the image and generate a FRET efficiency histogram from that area alone to see whether after applying the appropriate correction (γ), the histogram mean is zero. (4) If the photobleached acceptor can still absorb in the spectral range of donor emission, "dark transfer" can occur, which does not manifest in sensitized acceptor fluorescence, but nonetheless quenches the donor. (5) It follows from the points above that the ideal dye pair for acceptor pbFRET is made of an extremely photostable donor with high absorption and quantum efficiency, and an acceptor that is easy to bleach, does not accept FRET after bleaching, and does not spectrally interfere with the donor after bleaching. With the proper choice of filter, the indocarbocyanine dyes Cy3 and Cy5 reasonably well match these criteria. (6) Donor images have to be checked (and, if necessary, corrected) for proper spatial registration before implementing pixel-by-pixel calculations. Failure to do so will usually cause edge effects that grossly distort FRET distribution histograms. (7) Depending on the noise in our raw images, it may help to do a Gaussian filtering possibly using a filter size (usually 3×3) that is still less than the optical resolving power of our microscope. (8) From the prebleach donor and acceptor images an acceptor/donor ration image should also be formed and its correlation with the FRET image checked. It is expected that higher acceptor/donor ratios can yield higher E values without decreasing donor–acceptor separation. This is especially important to note in the case of measuring FRET between identical epitopes, in which equal donor/acceptor ratios are expected throughout the sample.

The acceptor photobleaching technique can be easily implemented using an LSM, and the image processing that yields a 2D map of FRET efficiency is not too complicated. However, in addition to the several controls that one has to perform, the somewhat low sensitivity of the method is also a disadvantage. In our experience, E values higher than 5% can be determined safely with a well-set LSM system and proteins that are expressed at more than about 20,000 per cell. The great advantage of self-controlled measurements can be further exploited by combining this method with the donor bleaching method. Briefly, the ROI is divided into two halves: In one half the acceptor is bleached and a map of FRET is calculated. Then in the whole ROI the donor bleaching kinetics is measured, and a map of FRET is calculated in the region with the acceptor still present, based on the average bleaching time constant measured in the area already void of acceptor. The histogram and subcellular distribution of E in the two half-ROIs can then be compared. Those interested should consult Bastiaens and Jovin (1998) for details.

3. Intensity-Based Ratiometric FRET

Both donor and acceptor photobleaching approaches carry the inherent drawback of irreversibly destroying fluorophores in the sample. It follows that measurements cannot be repeated at different times in the same sample, which may well be a problem if the time course of interactions is to be followed in a self-controlled manner. Strategies have already been suggested to reversibly switch on or off the donor or the acceptor (Jares-Erijman and Jovin, 2003). Lifetime imaging of the donor in the constant presence of the acceptor, using a sample without acceptor as a reference has been successful to follow in time intracellular protein interactions (Haj *et al.*, 2002; Legg *et al.*, 2002). Among the relatively less complicated intensity-based measurement modalities, the microscopic intensity–based ratiometric FRET (MI-FRET) offers similar advantages. This approach has been elaborated to exploit acceptor sensitization for estimating FRET efficiency on a pixel-by-pixel basis without losing temporal resolution (Nagy *et al.*, 1998b). The method originally used fluorescein as donor and rhodamine as acceptor, but it can be applied to other donor and acceptor pairs (e.g., Cy3 and Cy5) with slight modification. To achieve quantitative FRET measurements, the autofluorescence of cells needs to be assessed on a pixel-by-pixel basis. Because the autofluorescence of cells has a wide excitation and emission spectrum and is fairly stable, it is possible to define ratios (B_1, B_2, and B_3) of fluorescence intensities recorded with the fluorescein (I_1), energy transfer (I_2), and rhodamine (I_3) filter sets to the fluorescence intensity (I_{Bg}) called *background intensity* in this terminology, recorded with ultraviolet (UV) excitation at (360-nm) and 450-nm emission:

$$B_x = \frac{I_x}{I_{Bg}}, \quad x = 1, 2, 3. \tag{32}$$

These autofluorescence correction ratios are determined with unlabeled cells. Calculation of B_x values is not carried out on a pixel-by-pixel basis, but the pixel values are summed in the images, especially because the fluorescence intensities in single pixels are sometimes very low. With the B factors in hand, the following compensation is carried out on a pixel-by-pixel basis:

$$I_x(\textit{without cellular autofluorescence}) = I_x(\textit{with cellular autofluorescence}) - B_x I_{Bg}, \tag{33}$$

where I_x stands for I_1, I_2, and I_3 (fluorescein, energy transfer, and rhodamine intensities, respectively), I_{Bg} is the UV-excited background, and B_x is the ratio described earlier. This is valid because our fluorescent dyes have negligible absorption in the relevant UV range. This correction eliminates the contribution of autofluorescence from the I_1, I_2, and I_3 intensities.

In microscopy, intensity-based ratiometric determination of energy transfer usually relies on the measurement of sensitized acceptor emission, by exciting the donor (fluorescein) and measuring the fluorescence emission from the acceptor (rhodamine, I_2 in the equations below). This, however, is complicated by the fact that I_2 has a background component ($B_2 I_{Bg}$) and a contribution from the direct emission of the donor ($I_F(1-E)S_1$) and the acceptor ($I_R S_2$) even with the best-designed filter sets. To overcome this problem, four independent images of the same field should be measured. The relationship between the background image [I_{Bg} (360,450)], fluorescein image (I_1), rhodamine image (I_3), energy transfer image (I_2), and the FRET efficiency (E) are presented in the following equations:

$$I_1(490, 535) = B_1 I_{Bg} + I_F(1 - E) + I_R S_4 \tag{34}$$

$$I_2(490 > 590) = B_2 I_{Bg} + I_F(1 - E)S_1 + I_R S_2 + I_F E\alpha \tag{35}$$

$$I_3(546 > 590) = B_3 I_{Bg} + I_F\,(1 - E)S_3 + I_R \tag{36}$$

In the above equations, I_F is the unquenched fluorescein emission, and I_R is the direct rhodamine emission. The numbers in parentheses refer to the wavelengths of excitation and emission. The $S_1 - S_4$ factors characterize the spectral overlap between different channels. S_1 and S_3 are determined using samples labeled with fluorescein only according to the following equations (Szöllősi *et al.*, 1984; Tron *et al.*, 1984):

$$S_1 = \frac{I_2}{I_1}, \quad S_3 = \frac{I_3}{I_1} \tag{37, 38}$$

Fortunately, using standard fluorescein and rhodamine filter and dichroic mirror setup, S_3 is usually zero, making Eqs. 34–36 simpler. In spite of this, we retained this designation to make our equations compatible with those of the flow cytometric approach (see section III.A.2). S_2 and S_4 are determined on samples labeled with rhodamine only according to the following equations:

$$S_2 = \frac{I_2}{I_3}, \quad S_4 = \frac{I_1}{I_3} \tag{39, 40}$$

The fluorescence intensities in Eqs 37–40 represent summed fluorescence intensities from the same part of the images. For example, to calculate S_1, some bright cells are selected, and the intensity of these cells is summed in the energy transfer (I_2) and the fluorescein (I_1) image. The proportionality α factor is the ratio of the fluorescence intensity of a given number of excited rhodamine molecules measured in the I_2 channel to the fluorescence intensity of the same number of excited fluorescein molecules detected in the I_1 channel. This is constant for a given experimental setup and a particular pair of donor and acceptor conjugated ligands and must be measured for every defined case (Nagy *et al.*, 1998b).

Although photobleaching is useful in pbFRET, it presents a problem in the case of intensity-based ratiometric measurements. When the energy transfer values are calculated using the intensity-based approach, correction for photobleaching should be taken into account. Because the bleaching rate of rhodamine is very low with usual full-field illumination intensities, this correction needs to be applied only for the fluorescein and the energy transfer images. Fluorescein and energy transfer images are sequentially recorded. When calculating the energy transfer in an image, relevant fluorescein intensity is computed as the average of the previous and the next fluorescein image. This linear approximation of bleaching is reasonable only for short exposure times. From these images, the FRET efficiency can be calculated on a pixel-by-pixel basis using two different approaches.

First, FRET can be calculated in a similar way to the flow cytometric approach, which has already been discussed in section III.A.2. In this case, $I_R S_2$ is subtracted from the energy transfer image, and the remaining $I_F E\alpha$ is divided by $\alpha(I_1 - B_1 I_{Bg} - I_R S_4) = I_F(1 - E)\alpha$. This gives $E/(1-E)$, which is termed β in Eq. 41:

$$\frac{I_2 - B_2 I_{Bg} - (I_1 - B_1 I_{Bg})S_1 + (I_3 - B_3 I_{Bg})(S_1 S_4 - S_2)}{\alpha(I_1 \quad B_1 I_{Bg} - (I_3 - B_3 I_{Bg})S_4)} = \beta \tag{41}$$

.

In the denominator, we have the I_1 image corrected for background and rhodamine contribution multiplied by α. In the numerator, we have the I_2 image corrected for background and fluorescein contribution ($I_2 - B_2 I_{Bg} - [I_1 - B_1 I_{Bg}]S_1$) and the background-corrected I_3 image multiplied by ($S_1 S_4 - S_2$).

Second, E can also be calculated according to the classic equation, which is modified slightly so that it can be applied for microscopes not using single-wavelength excitation, but bandpass filters:

$$\frac{F_{AD}}{F_A} = 1 + \frac{A_D c_D}{A_A c_A} E, \tag{42}$$

where F_A is the direct acceptor fluorescence (in our case $I_R S_2$) measured at the same excitation wavelength as the sensitized emission, A_D and A_A are the integrated absorptions of the donor and the acceptor in the fluorescein excitation range, respectively, and c_D and c_A are the concentrations of the donor and the acceptor, respectively. Solving this equation gives

$$\left(\frac{I_2 - B_2 I_{Bg} - (I_1 - B_1 I_{Bg} - (I_3 - B_3 I_{Bg})S_4)S_1}{(I_3 - B_3 I_{Bg})S_2} - 1\right)\frac{A_A c_A}{A_D c_D} = E. \quad (43)$$

To be able to calculate E according to Eq. 43, we need to calculate the absorption ratio (A_A/A_D) and the pixel-by-pixel acceptor/donor ratio (c_A/c_D). Determination of these parameters is not trivial, and a detailed experimental approach is given by Nagy *et al.* (1998b).

The intensity-based ratiometric FRET approach uses accurate background correction and subtraction of spectral overlaps from the energy transfer channel. This measurement results in images similar to those obtained by the donor photobleaching method. Important advantages of the ratiometric method are that it requires much less image-storing capacity and it is less sensitive to environmental factors than photobleaching. Photophysics of bleaching is far from being understood and sometimes pbFRET results give anomalously high values, especially when cell surface proteins are very dense. One drawback of the ratiometric microscopic technique is that for accurate determination of FRET, it requires higher fluorescence intensity than pbFRET does, because sensitized emission is only a small fraction of the I_2 intensity in the FRET image. The method is also very sensitive to precise determination of the S factors and the pixel-by-pixel donor/acceptor ratio, which also demands high levels of fluorescence intensity. A further drawback of the technique is that the absorption ratio is instrument dependent, and when lamps are used for excitation instead of lasers, this absorption ratio may be altered as the spectral characteristics of the lamp change with time (Nagy *et al.*, 1998b).

4. Sample Experiment for Image Cytometric FRET Measurement

a. *Materials*

1. Prepare 1% formaldehyde solution in PBS. This solution should be made fresh.

2. Prepare fluorescently conjugated mAbs according to the method described in Szöllösi *et al.* (1989), Nagy *et al.* (1998b), and Sebestyen *et al.* (2002). The concentration of the stock solution should be in the range of 0.5–1.0 mg/ml. The dye/protein labeling ratio for donor- or acceptor-labeled antibody should preferentially be between 2 and 4. Store stock solutions in the presence of 0.1% NaN_3 at 4 °C.

3. Before fluorescent labeling of cells, remove antibody aggregates by centrifuging the diluted fluorescently conjugated antibody solution (100 μg/ml) at 100,000 g for 30 minutes.

b. Sample Preparation

1. Deposit cells on microscope slide by growing them on the slide or by attaching suspended cells to polylysine-coated slide as described earlier (Damjanovich *et al.*, 1995).

2. Label cells with fluorescently tagged mAbs at saturating concentration (1–2 μM depending on the type of antibody) for 30 minutes on ice in the dark. The incubation volume should be 80 μl to cover the cells at a 15×15 mm^2 area. At least four samples should be prepared: unlabeled cells, cells labeled with donor only, cells labeled with acceptor only, and cells labeled with both donor and acceptor (FRET sample). For the double-labeled sample, donor and acceptor conjugated antibodies should be mixed in the needed concentration ratio before addition to the cells.

3. The excess mAb should be removed by washing cells in PBS twice.

4. Cells should be fixed in 1% formaldehyde solution in PBS for at least 1 hour.

5. Mount cells under a coverslip by adding a drop of a solution containing 90% glycerol in PBS and seal the preparation by melted paraffin. Alternatively, a commercial antifade mounting medium (e.g., Vectashield, Vector laboratories, Inc., Burlingame, California) can be used. When intensity-based and pbFRET are compared on the same set of samples, avoid using antifade solution.

c. Fluorescence Measurements and Analysis

1. Choose the appropriate excitation, emission, and dichroic filters according to your donor and acceptor dye.

2.A. For donor pbFRET, take a donor-only and a donor-plus-acceptor–labeled sample. Take a sequence of images of the donor in both samples until donor is bleached to background. Finally take an image of the acceptor in the double-labeled sample.

2.B. For acceptor pbFRET, use a double-labeled sample and a donor-only–labeled sample. (For the sake of simplicity, we assume that for the Cy3–Cy5 pair, all aforementioned controls are done and negative.) Run the following protocol on both samples: take donor image, take acceptor image, bleach acceptor in consecutive scans, take acceptor image to confirm it is totally bleached, and take donor image again.

2.C. For ratiometric FRET, collect images (I_1, I_2, and I_3 and I_{Bg}) for all samples (donor, acceptor, and double labeled). For autofluorescence correction, one extra set of images is to be taken from nonlabeled cells. For each sample, take the autofluorescence (I_{Bg}) image first, then the acceptor (I_3), the donor (I_1), and the FRET (I_2) images. If there is photobleaching, take the FRET and the donor images, in this order, again.

3. Correct all the images for camera dark current by subtracting images taken with closed shutters.

4. Because of the filter changes (especially for dichroic filters), check images for image registration. Correct images for any pixel shift.

5. Correct images for the background caused by the fluorescence of the optical elements of the microscope, using cell-free areas of the images.

5.A. For donor pbFRET, do a pixel-by-pixel monoexponential or biexponential fit for both donor-only and donor-plus-acceptor sample sequences. Calculate average bleaching time constant from the donor-only image and determine E for every pixel in the donor-plus-acceptor sample using Eq. 29.

5.B. For acceptor pbFRET, determine the γ correction factor according to Eq. 31 from images of the donor-only sample. Then calculate E for every pixel using γ and applying Eq. 30 to images of the double-labeled sample.

5.C. For ratiometric FRET, analyze unlabeled cells first, determining the spectral overlap factors of autofluorescence. Then analyze donor-only and acceptor-only samples. Using all four images, correct the images for autofluorescence first. Then determine the spectral overlap factors for the donor spectrum from the donor-only sample and for the acceptor spectrum from the acceptor-only sample. Consecutively, analyze all four images of the double-labeled samples, and correct for autofluorescence, then for donor emission and acceptor emission. After all the corrections, calculate the FRET efficiency on a pixel-by-pixel basis. If there is photobleaching during image recording, use the average of donor and FRET images that were taken in alternated sequence.

C. Limitations of FRET

Although FRET can provide very useful information about molecular proximity and associations, it has its own limitations. The most serious drawback of FRET is that it has restricted capacity in determining absolute distances because FRET efficiency depends not only on the actual distance between the donor and acceptor, but also on their relative orientation (κ^2). It is still quite good at determining relative distances, namely, whether the two labels are getting closer or farther on a certain stimulus/effect. Even when measuring relative distances, care must be taken to ensure that the orientation factor (κ^2) does not change between the two systems to be compared. If the fluorescent dye is attached to an antibody or Fab fragment via a carbon linker having 6–12 carbon atoms, the linker often allows relatively free rotation of the dye, which minimizes uncertainty of κ^2.

Another problem is that FRET has a very sharp distance dependence, making it difficult to measure relatively long distances because the signal gets very weak. At the same time, energy transfer tends to occur on an all or none basis; if the donor and acceptor are within $1.63 \times R_0$ distance, energy transfer is detectable, and if they are farther apart, energy is transferred with very little efficiency. Because of this sharp decrease, the absence of FRET is not direct proof of the absence of molecular proximity between the epitopes investigated. Also, absence of FRET can be caused by sterical hindrance even for neighboring molecules or protein domains. On the other hand, presence of FRET to any appreciable extent above the experimental error of measurement is strong evidence of molecular interactions.

Indirect immunofluorescent labeling strategies may be applied to FRET measurements if suitable fluorophore-conjugated mAbs are not available or as an approach to enhancing the specific fluorescence signal. In such cases, special attention should be paid to the fact that the size of the antibody complexes used affects the measured FRET efficiency values. Application of a larger antibody complex causes a decrease in FRET efficiency as a result of the geometry of the antibody complexes, because when antibody or F(ab') complexes become larger, the actual distance between the donor and acceptor fluorophores increases (Sebestyen *et al.*, 2002). This explains the decrease of FRET efficiency when fluorescent secondary $F(ab')_2$ fragments are used on both the donor and the acceptor side as opposed to using direct-labeled primary antibodies. It also explains a further decrease in FRET efficiency when intact fluorescent secondary antibodies are used. Such findings underline the notion that FRET values cannot be compared directly to each other if they are obtained using different labeling strategies (Sebestyen *et al.*, 2002).

To increase the signal, we can use phycoerythrin (PE)- or allophycocyanine (APC)-labeled antibodies, because PE and APC have exceptional brightness. However, the size of these molecules is comparable to or even greater than whole antibodies. Because of steric limitations, the measurable FRET efficiency values can be low, at the border of detection limit (Batard *et al.*, 2002). Nonetheless, it should be noted that even these low FRET efficiency values might have a biological meaning, because the accuracy of measurements is greatly improved by the high level of specific signals. Appropriate positive and negative controls can help only make the decision of whether given molecules are associated or not on the basis of the measured FRET efficiency values (Batard *et al.*, 2002).

When studying cells labeled with donor and acceptor conjugated mAbs, we must perform averaging at different levels. The first averaging follows from the random conjugation of the fluorescent label. An additional averaging is brought about by the actual distribution of separation distances between the epitopes labeled with mAbs. This multiple averaging, an inevitable consequence of the nonuniform stoichiometry, explains why the goals of FRET measurements are so uniquely different in the case of purified molecular systems on the one hand, and in the case of *in situ*–labeled membrane or cytoplasmic molecules on the other hand. In the former case, FRET efficiency values can be converted into absolute distances, whereas in the latter, relative distances and their changes are investigated.

Calculation of distance relationships from energy transfer efficiencies is easy in the case of a single-donor/single-acceptor system if the localization and relative orientation of the fluorophores are known. If the FRET measurements are performed on the cell surface or inside the cell, many molecules might not be labeled at all; many could be single without a FRET pair, and others may be in smaller groups of hetero-oligomers, creating higher rates of FRET than expected from stand-alone pairs. A large number of epitopes binding the acceptor increases E simply by increasing the rate of transfer rather than actually meaning that the two epitopes investigated are closer to each other. If, in these cases, reversing the labeling still results in large E values, the proximity can be considered verified. However, if reversing the ratio of donor/acceptor to more than 1 makes FRET disappear, chances are that we have previously seen random colocalizations because of the high number of acceptors. In addition, if cell membrane components are investigated, a 2D restriction applies to the labeled molecules. Analytical solutions for randomly distributed donor and acceptor molecules and numerical solutions for nonrandom distribution have been elaborated by different groups (Dewey and Hammes, 1980; Snyder and Freire, 1982; Wolber and Hudson, 1979). To differentiate between random and nonrandom distributions, energy transfer efficiencies have to be determined at different acceptor concentrations. However, the donor/acceptor ratios should be in the range of 0.1–10.0 (Berney and Danuser, 2003) if we want to obtain reliable FRET measurements. Outside this range, noise and data irreproducibility propagate unfavorably, rendering accurate FRET efficiency calculations impossible. Berney and Danuser (2003) also suggest that to obtain stable FRET measurements, energy transfer has to be observed in the FRET channel, that is, by excitation of the donor and a measurement of the acceptor emission. Methods estimating FRET from the donor signal in the presence and absence of acceptor are less robust. Donor–acceptor dye pairs should be chosen with the maximal spectral overlap, although this will increase the crosstalk between the detection channels. However, spectral overlaps can be corrected on the basis of samples labeled with donor or acceptor only (Berney and Danuser, 2003).

IV. Applications

A. Mapping Receptor Oligomerization

FRET technology has two major advantages: It can provide spatial information about molecular proximities/interactions in a distance range not covered by conventional optical microscopies (i.e., the range below the ~250-nm diffraction limit of resolution); additionally FRET data have a unique specificity for the two labels (donor and acceptor dyes) sitting on specified sites of two labeled macromolecules. This technique, thus, provides a special tool for *in situ,* semiquantitative mapping of oligomerization, or clustering of cell membrane molecules (proteins or lipids) in live cells, without any significant distortion of cellular integrity often experienced, for example, in electron microscopy because of the special sample preparation procedures. Since the introduction and adaptation of the FRET methodology to cellular studies in the early 1980s, the method gradually became very popular in cell biology, immunology, and neurobiology.

In the following section, we focus on several selected biological questions concerning homotropic and heterotropic associations of MHC molecules as antigen-presenting entities, colocalization of antigen receptors (e.g., T-cell receptor [TCR]) and co-receptors (e.g., CD4 and CD8) on lymphoid cells or functional homotropic and heterotropic associations of several other cell surface receptors in distinct cell types first assessed and characterized by FRET. We pay special attention to specific technical details of various experimental strategies and the lessons and problems that have arisen during these investigations.

1. Clustering of Cell Surface MHC Molecules

The very first targets of cellular FRET studies were, among others, the class I and class II MHC molecules, because they are expressed on all nucleated human cells (MHC-I) or their expression is linked to and regulated by pathological processes, such as viral/bacterial infection (MHC-II). In addition, in FRET studies these molecules also proved favorable from the technical point of view, as MHC-I expression on distinct cell types is usually relatively high ($>10^5$ copies/cell) and the MHC-II level is also significantly upregulated by a number of pathogenic stimuli. Moreover, a large panel of mAbs reactive to different epitopes on their extracellular domains are also available. Thus, MHC glycoproteins served as ideal molecules for adaptation and optimization of FRET measurements to the live cell's surface. Early FRET studies (mostly FCET analysis) on the lateral membrane organization of MHC-I and MHC-II molecules (Bene *et al.*, 1994; Chakrabarti *et al.*, 1992; Matkó *et al.*, 1994, 1995; Mátyus *et al.*, 1995; Szöllősi *et al.*, 1989, 1996) in different cells (B and T lymphocytes, lymphomas, hybridomas, fibroblasts, etc. of mouse and human origin), in accordance with findings by other methods—lateral diffusion (Edidin *et al.*, 1991), single particle tracking (Smith *et al.*, 1999), and immunobiochemical analysis (Triantafilou *et al.*, 2000)—unequivocally demonstrated their inherent ability to form dimers and/or higher-order oligomers at the cell surface. In addition, FRET experiments revealed several heterotropic interactions of MHC-I molecules with various other cell surface receptors of diverse function, such as the insulin receptor (Liegler *et al.*, 1991), the interleukin-2 (IL-2) receptor (Matkó *et al.*, 2002; Szöllősi *et al.*, 1987; Vereb *et al.*, 2000), the epidermal growth factor receptor (Schreiber *et al.*, 1984), or the transferrin receptor (Mátyus *et al.*, 1995). These heterotropic MHC-I interactions revealed originally by FRET were confirmed later by other methods or functional assays (e.g., Ramalingam *et al.*, 1997).

Donor pbFRET and FCET were both essentially useful in exploring further functional aspects of MHC-I clustering in antigen-presenting cells. In works by Chakrabarti *et al.* (1992) and Bodnár *et al.* (2003), regulation of cell surface MHC-I clusters by exogenous β2m (light chain) was revealed on B lymphoblast antigen-presenting cells, in relation to its effect on the functional activity of cytotoxic effector T lymphocytes. In the experiments, *homo-FRET* between intact MHC-I molecules, as well as *hetero-FRET* between β2m-free MHC-I heavy chains (FHC), and intact MHC-I were both analyzed and found sensitive to the level/binding of exogenous β2m. (Bodnár *et al.*, 2003). The high hetero-FRET efficiency, in accordance with the high degree of colocalization of FHC and intact heterodimer (α chain + β2m light chain) shown by high resolution scanning near-field optical microscopy (SNOM), suggested that the β2m-free heavy chains (FHC) are likely nucleating/initiating the large MHC-I clusters, because their recombination by exogenous β2m resulted in a remarkable β2m-concentration–dependent decrease in the efficiency of both homo-FRET and hetero-FRET, simultaneously with a decreased antigen presentation function.

Several important technical lessons could also be drawn from these basic FRET studies on MHC organization: (1) The FRET efficiency (E) measured for a given donor–acceptor dye pair coupled to antibodies (either whole immunoglobulin G [IgG] or Fab fragment) against the light (β2m) and heavy (α) chains of MHC-I, respectively, may serve as a *cell type–* and *cell cycle–independent internal standard for FRET measurements with comparative goals* (e.g., studies on different cells derived from experimental animals and measurements at different days or if the same cells/cell type is analyzed in different laboratories). This standard can be applied to most of the nucleated cells, because they all express MHC-I at their surface. (2) The efficiency of FRET between the heavy and light chains of MHC-I may be, however, asymmetrical in some cells: E was slightly lower when the heavy chain of intact MHC-I molecules was tagged by donor label and the β2m light chain by acceptor than in the case of reversed tagging (e.g., on B and T lymphoma cells with high FHC expression). The reason for this is likely the aforementioned mixed clustering of FHC and intact heterodimeric forms of MHC-I molecules, resulting also in an unequal number of accessible heavy and light chain epitopes on intact MHC-I. Therefore, we propose to use FRET from the β2m light chain (donor) to the α heavy chain (acceptor) for *internal standard.* (3) These data also demonstrate that using the same donor–acceptor combination on the same cells in different days, FRET efficiency was highly reproducible. Using the same conditions, Bodnár *et al.* (2003) experienced similarly good agreement between the results of FCET and microscopic pbFRET measurements (4) The efficiency of FRET for a given fixed pair of antibody-tagged epitopes of MHC on a particular cell was, however, very sensitive to changes in the donor or acceptor dyes or in the tagging carrier (whole antibody or Fab fragment). Changing either the donor or the acceptor dye may result in an altered spectral overlap (R_0) or an altered dynamic orientational averaging (κ^2), both critically influencing the measurable efficiency of FRET, besides the actual distance of the donor and acceptor dipolar dyes (R). Some recent analogs of the classic fluorescein and rhodamine dyes (e.g., succinylated derivatives SFX or TAMRA-X, respectively, Molecular Probes, Eugene, Oregon) were developed with spacer arms, which favors their full dynamic orientational averaging when bound to antibodies and simultaneously reduces their aspecific binding (for review of these questions, see Matkó and Edidin [1997]).

A novel and intriguing aspect of FRET studies with MHC molecules was reported recently by Gáspár *et al.* (2001), who tried to fit *in situ* FRET data (on the MHC-I homotropic association and the relative spatial arrangement of membrane-bound MHC-I epitopes to the surface of the plasma membrane) to the x-ray crystallographic model of the given MHC-I molecule (human leukocyte antigen A2 [HLA-A2]), as well as other counteracting molecules, such as CD8 and TCR. This approach was based on the analysis of FRET efficiency (by FCET) between distinct structurally identified epitopes of MHC-I molecules relative to each other and between a given MHC-I epitope relative to the surface of the cell membrane labeled by fluorochrome-conjugated lipids (e.g., Bodipy-PC), using a modified Stern–Volmer analysis (Yguerabide, 1994).The structural models for the applied antibody and its Fab fragment were available and helped to define their docking site on MHC heavy chain. Fitting x-ray diffraction-derived structural models of macromolecules to each other while having informative *in situ* FRET data on their molecular interactions and positions relative to the plane of the membrane in hand proved to be a very powerful approach to generate models, approaching the physiological conditions for supramolecular clusters of MHC at cellular contacts, such as immunological synapses (Bromley *et al.*, 2001). Among the several thousands of computer-simulated possibilities for different molecular geometries/orientations of MHC-I molecules in homo-oligomers, only one to two matched the *in situ* FRET data measured on live cells (Gáspár *et al.*, 2001). Thus, besides giving a realistic model for supramolecular activation clusters at the contact site of antigen-presenting cells and cytotoxic T lymphocytes (CTLs), this approach provides a strategic template for studies on modeling the 3D structure of supramolecular receptor complexes appearing at the surface (or at a contact surface) of live cells, with a goal of designing drugs with effects on these molecular interactions.

It should be noted here that reliable models are available for analysis of FRET data obtained for *transverse (vertical) distance distributions* (Yguerabide, 1994), as well as for analysis of *lateral homotransfer (oligomerization)* in the case of membrane-bound proteins (Runnels and Scarlata, 1995).

As another special application of cellular FRET, *intercellular FRET* experiments should also be mentioned here. In this work, Bacso *et al.* (1996) labeled different molecular species (e.g., MHC-I and intercellular adhesion molecule-1 [ICAM-1] on antigen-presenting cells, CD8, and leukocyte function–associated antigen-1 [LFA-1] on T cells) with proper donor and acceptor dyes on antigen-presenting cells and T cells, respectively, and then measured intercellular FRET between these labeled molecules after initiating conjugation (synapse formation) of these cells. These FRET measurements had to be implemented inevitably with microscopic FRET, because formation of cell–cell contacts also had to be detected simultaneously. These intercellular pbFRET experiments (kinetic measurement of donor photobleaching) contributed important quantitative data to our view of immunological synapses. Namely, in the cytotoxic T-cell synapses the molecule pairs in point contact (e.g., CD8/MHC-I) displayed detectable nanometer scale FRET efficiency ($E = 17$–19%), while the longer adhesion molecules (e.g., ICAM-1/LFA-1 pair) forming focal contacts and spanning larger distance range (≥ 25 nm) did not ($E < 1.5\%$). These FRET results were later confirmed by several pieces of evidence from confocal and electron microscopic analysis of immunological synapses (Monks *et al.*, 1998; Wulfing *et al.*, 1998). It is noteworthy that the FRET efficiency histogram for the CD8/MHC-I molecule pair obtained from this pbFRET analysis was broad and heterogenous (E = 0–30%), suggesting the existence of molecular associations with varying tightness at the contact zone of antigen-presenting cells–T-cell conjugates.

Finally, we note that a "shorter spectroscopic ruler" alternative of FRET, the long-range electron transfer (LRET) technique (detecting electron transfer between fluorescent donors and nonfluorescent electron acceptor spin label radicals), was also successfully applied in characterizing cell surface MHC associations, in good quantitative agreement with FRET data on the same cells [for details and review, see Matkó *et al.* (1992, 1995) and Matkó and Edidin (1997)].

2. Antigen Receptor–Co-receptor Interactions in T-Cell Immunology

Early FCET (Mittler *et al.*, 1989) and line-scanning microscopic pbFRET (Szabó *et al.*, 1995) studies convincingly demonstrated that the antigen receptor of T cells (i.e., TCR) physically associates with the CD4 co-receptor molecules on helper T cells upon ligation of TCR. That this association is dynamic was nicely confirmed by measuring FRET between fluoresceinated anti-CD4 and rhodaminated anti-CD3 at physiological (37 °C) and low (4 °C) temperatures. The efficiency of FRET at 37 °C was closely 10%, whereas at 4 °C it decreased to less than 1% (Mittler *et al.*, 1989), suggesting that this association requires a dynamic redistribution of interacting molecules in the membrane. Using cells with mutant CD4 (lacking the cytoplasmic domain), the absence of FRET nicely demonstrated that this domain of CD4 is essential for its interaction with TCR. Further FRET analysis of this question revealed that cytoplasmic CD4 motifs that bind $p56^{lck}$ are critical in binding CD4 to TCR (Collins *et al.*, 1992).

These studies turned our attention to a critical point of FRET analysis. Care should be taken in the choice of the antibody used to tag our receptor. Here, for example, the whole anti-CD3 IgG molecule (145-2c11) induced molecular redistribution and microaggregation (involving TCR, CD4, and $p56^{lck}$) in the plasma membrane of T cells. Therefore, to avoid such artifacts, for labeling of TCR, its rhodaminated $F(ab')_2$ fragment should be used. Hence, we propose to check all antibodies against multichain immune-recognition receptors (MIRRs) for such an activating/stimulating effect before their use as a simple fluorescent tag for FRET.

For further analyzing the role of CD4 co-receptors in antigen-specific T-cell activation, Zal *et al.* (2002) introduced a digital imaging deconvolution microscopic FRET approach. They applied a special labeling technique to attach donor and acceptor dyes to the TCR–CD3 complex and CD4 molecules. The CD3ζ chain was fused with a green fluorescent protein (GFP) mutant, ECFP (donor, Clontech) through a six amino acid linker, and the CD4 molecules were fused at their intracellular terminus, through a five amino acid linker, with another GFP mutant, EYFP (acceptor, Clontech). The cyan and yellow variants of GFP provide a good spectral overlap for FRET measurements, although their emissions also show some overlap (crosstalk). The double-labeled T cells were analyzed for FRET in the absence or presence of antigen-presenting cells in thermostated chambers, using an inverted fluorescence microscope equipped with a sensitive cooled CCD camera and computer-controlled motorized stage using an interactive control software driving also the 3D deconvolution (z-axis optical slicing) program. The images of the double-labeled cells were recorded in three optical channels of the microscope and then compensated for crosstalk between the channels, and finally the donor-normalized FRET information was derived based on earlier procedures described for FCET or FRET-microscopy (Gordon *et al.*, 1998; Tron *et al.*, 1984). Generating a 30-minute time lapse of the corrected FRET signal, it was shown that TCR and CD4 rapidly accumulated into the contact area of the antigen-presenting cell and T cell (synapse) and were localized in close proximity (exhibiting high FRET), strictly depending on the nature of the antigen peptide. Only agonist peptide ligands promoted the TCR–CD4 interaction, whereas the antagonist ligands inhibited it [for more details, see Zal *et al.* (2002)]. Several technical lessons emerge from this study: (1) FRET analysis should preferably be made on non-deconvoluted images, although they are of lower z-resolution, because of the possible nonlinearity of the deconvolution procedure; (2) cell areas where the donor (cyan fluorescent protein [CFP]) fluorescence level is less than 10 times noise level, as well as areas where the acceptor to donor fluorescence ratio is beyond the 1:1 to 3:1 range, are recommended for exclusion from the analysis; and (3) using such imaging FRET analysis, the corrected E_{app}—FRET signal—is not a real FRET efficiency, but a number proportional to FRET efficiency.

3. Oligomerization of Diverse Membrane Proteins

The different variations of FRET measurements (pbFRET, FCET, MI-FRET, fluorescence lifetime imaging [FLIM], bioluminescence resonance energy transfer [BRET], described later, etc.) have all been extensively applied in studies of oligomerization, heterotropic interactions, and membrane compartmentation of various receptors, adhesion, accessory, or signal molecules in the plasma membrane of a number of human or murine cell types. Just to mention several examples, the cell surface distribution of FcεRI (IgE receptor) and its relationship to MAFA (a potential regulatory molecule) was nicely analyzed in detail on mast cells, using the donor pbFRET approach: The IgE receptor was found partially associated with MAFA, depending on FcεRI oligomerization/cross-linking by antigen (allergen), forming this way a platform for negative regulation of mast cell activation by MAFA (Jurgens *et al.*, 1996). *FRET* and *BRET* were both successfully applied for a number of G-protein–coupled hormone receptors (GPCR) to reveal their functionally critical dimerization (Eidne *et al.*, 2002; Milligan *et al.*, 2003). As an interesting observation, the human thyrotropin receptor (thyroid-stimulating hormone receptor [TSHR]) was found to be naturally inhibited by its higher oligomeric surface distribution, whereas it became active when these forms dissociated to dimers or monomers, upon binding their ligand (thyroid-stimulating hormone [TSH]). These conclusions were drawn from FRET measurements between the GFP–TSHR fusion construct (donor) and a TSHR–Myc tag construct + anti–Myc-Cy3 (acceptor), from microscopic intensity-based ratiometric FRET analysis, where the FRET efficiency signal fell to less than 1% in ligated cells from about a 20% value in unstimulated cells (Latif *et al.*, 2002).

As new research areas taking an interest in the application of FRET, we must mention *neurobiology* and *intracellular organelle research*. This might be due to both the increasingly wide scale of microscopic FRET techniques and the technical development of microscopes. For example, subunit assembly and its regulation by nicotine of the $\alpha_4\beta_2$-*nicotinic acetylcholine receptor* in neurons was successfully characterized by MI-FRET measurements, using CFP and yellow FP (YFP) constructs of the individual subunits after transfection to midbrain neurons (Nashmi *et al.*, 2003). A novel intracellular version of FRET—detected by FLIM between labeled CD44 (hyaluronan receptor) and ezrin (cytoskeletal linker protein)—revealed a new PKC-regulated mechanism by which CD44-ezrin coupling and the directional cell motility is controlled (Legg *et al.*, 2002). FLIM detection of FRET between GFP-fusion constructs of epidermal growth factor (EGF) receptor or platelet-derived growth factor (PDGF) receptor (transmembrane receptor tyrosine kinases [RTKs]) and Cy3 antibody–labeled protein tyrosine phosphatase 1B (PTP1B) clearly demonstrated that the activated RTKs meet the PTP1B after their internalization, mainly on the surface of the endoplasmic reticulum, suggesting that the scenes of their activation and inactivation are spatially and temporally isolated within the cells (Haj *et al.*, 2002).

B. Defining Plasma Membrane Microdomains (Lipid Rafts)

The plasma membranes of most mammalian cells exhibit small microdomains (called *lipid rafts*) that are enriched in glycosphingolipids (GSLs), sphingomyelin (SM), and cholesterol (CHOL), while the polyunsaturated phospholipids are very rare in these domains. Lipid rafts are also rich in glycosyl-phosphatidylinositol (GPI)–anchored proteins. These domains were first defined chemically as detergent-resistant, light buoyant density membrane fractions and then were characterized as diffusion barriers or co-patches of fluorescently labeled lipids and proteins (Brown and London, 1998; Edidin, 2001; Jacobson and Dietrich, 1999; Matkó and Szöllősi, 2002; Simons and Ikonen, 1997; Subczynski and Kusumi, 2003; Vereb *et al.*, 2003).

FRET was also introduced into the studies directed to define existence, composition, and size (a highly controversial question) of lipid rafts. The starting strategy was to investigate homo-FRET between labeled GPI-anchored proteins—supposed to be enriched in raft microdomains—as a function of their cell surface density. First, applying an indirect approach, Varma and Mayor (1998) detected fluorescence *anisotropy* of GPI-anchored folate receptors (FR-GPI) in CHO cells with microscopy, using a fluorescent folate analog (N$^{\alpha}$-pteroyl-N$^{\varepsilon}$-(4′-fluorescein-thiocarbamoyl)-L-lysine [PLF]) for labeling. For comparison, chimeric transmembrane forms of FR (FR-TRM) were also investigated. Their working hypothesis was that if the FR-GPI was randomly distributed at the surface of CHO cells, the measured fluorescence anisotropy should depend on the actual surface density of FR (controlled by transfection-directed expression), due to concentration depolarization (homo-FRET) effects above a critical molecular proximity (within the 10-nm Förster-distance). In contrast, if the FR-GPI molecules are compartmented/concentrated in lipid microdomains, no density-dependent change is expected in the anisotropy, because only the number of microdomains increases with increased expression levels, but the mutual proximity of FR molecules does not. Interestingly, they detected density-dependent anisotropy in the case of the FR-TRM form, whereas the physiological FR-GPI form showed a constant anisotropy over a wide range of expression levels, consistent with the microdomain-localization of FR-GPI. This was further confirmed by the observed density dependence of FR-GPI anisotropy in cells, where the membrane was depleted of cholesterol (stabilizer of rafts) before anisotropy measurements. The size of an individual domain was estimated from these measurements as being smaller than 70 nm (containing <50 GPI-anchored molecules) (Varma and Mayor, 1998).

Conflicting with these data, Kenworthy and Edidin (1998) measured homo-FRET between GPI-anchored 5′-nucleotidase proteins, using Cy3 (donor)- and Cy5 (acceptor)-conjugated antibodies, in MDCK cells, by the acceptor-photobleaching method. A relatively strong correlation was found between the homo-FRET efficiency and the cell surface density of 5′-nucleotidase. These data suggest that either the GPI-anchored 5′-nucleotidase is not constitutively compartmented in raft microdomains or their microdomains are not stable, that is, have a short lifetime. Further addressing this question, Kenworthy *et al.* (2000) analyzed homo-FRET between fluorescent cholera toxin B–labeled GSL components, GM_1 gangliosides, and their hetero-FRET with GPI-anchored proteins on three cell lines together with the dependence of FRET on cell surface densities of acceptor-labeled lipid or protein species. Surprisingly, in most cell types even the GM_1 ganglioside (considered a major lipid constituent of raft microdomains) showed density-dependent homo-FRET, indicative of its weak, hardly detectable clustering. This might mean that either raft microdomains occupy only a very small fraction of the cell surface area on the investigated cell types or the domains of the investigated GPI proteins are only transient (short lifetime) assemblies. This might be, however, highly cell specific, because the amount of cell surface raft constituents (e.g., GSL) may vary with cell type or differentiation, at least in lymphoid cells (e.g., Tuosto *et al.*, 2001). Nevertheless, most microdomains that we can see in the fluorescence microscope are likely composed of aggregates of small elementary raft microdomains, either upon biological signaling of the cell or just upon modulation of raft size by the procedure and agents used to label raft marker components.

Another often used FRET strategy to investigate raft (microdomain) compositions on live cells is to measure hetero-FRET by microscopy (or by FCET) between a selected fluorescently labeled membrane protein (receptor) and raft markers proteins (e.g., GPI-anchored proteins) or GM_1 gangliosides. Such an approach was used to show that a large fraction of CD25 (α chain of the IL-2 receptor) is constitutively raft associated in lymphoma/leukemia T cells, using CD48 as a raft marker. In these FRET experiments, *Fab* fragments were used to label the two proteins, to minimize possible modulation of raft size by cross-linking with bivalent antibodies. Although on these T cells CD25 was expressed at a higher level (twofold) than CD48, the FRET efficiency was similar in both directions (donor on CD25 or on CD48). This, together with the pronounced decrease of FRET efficiency upon depleting membrane cholesterol, suggests that the two proteins are colocalized within common microdomains (Matkó and Szöllösi, 2002; Matkó *et al.*, 2002; Vereb *et al.*, 2000).

C. Receptor Tyrosine Kinases

The type I family of transmembrane RTKs comprises four members: EGFRs (EBFR or ErbB1), ErbB2 (HER2 or Neu), ErbB3, and ErbB4 (Citri *et al.*, 2003; Nagy *et al.*, 1999b; Vereb *et al.*, 2002; Yarden and Sliwkowski, 2001). Within a given tissue, these receptors are rarely expressed alone but are found in various combinations. Members of the family form homoassociations and heteroassociations at the cell surface. Their ligands belong to three groups: EGF-like ligands bind only to ErbB1, heregulin-like (or neuregulin-like) ligands bind only to ErbB3 and ErbB4, and EGF- and neuregulin-like ligands bind ErbB1 and ErbB4. ErbB3 shares growth factor–binding specificity with ErbB4 but lacks intrinsic kinase activity. ErbB2 is an orphan receptor; no soluble physiological ligand specific to ErbB2 has been detected. Despite this, ErbB2 participates actively in ErbB receptor combinations, and receptor complexes including the ErbB2–ErbB3 dimer that appears to be more potent than any other combination (Citri *et al.*, 2003; Nagy *et al.*, 1999b; Sliwkowski *et al.*, 1999; Vereb *et al.*, 2002; Yarden and Sliwkowski, 2001). Thus, both ligand and receptor expression can vary by tissue and will, in part, determine the specificity and the potency of cellular signals.

Molecular scale physical associations among ErbB family members have been studied by classic biochemical (Sliwkowski *et al.*, 1994; Tzahar *et al.*, 1996), molecular biological, and biophysical methods (Gadella and Jovin, 1995; Nagy *et al.*, 1998a,b, 2002). When isolated from cells, members of the ErbB family self-associate (homoassociate) and associate with other family members (heteroassociate) (Tzahar *et al.*, 1996). However, experiments on isolated proteins are inherently unable to detect interactions in cellular environments *in vivo* and *in situ* and cannot detect heterogeneity within or among cells. FRET measurements detected dimerization of ErbB1 receptors in fixed (Gadella and Jovin, 1995; Gadella *et al.*, 1994) and living cells (Gadella *et al.*, 1994). FRET was also applied to monitor the association pattern of ErbB2 in breast tumor cells (Nagy *et al.*,

1998a,b, 2002). First, we applied FCET measurements (Szöllősi *et al.*, 1984) to reveal cell-to-cell heterogeneity within a cell population. Classic FCET (section III.A.2) and autofluorescence-corrected FCET (section III.A.3) gave practically the same results, because the expression level of ErbB2 molecules is so high on these cell lines (SKBR-3, BT474, MDA453) that autofluorescence is practically negligible compared to the signal.

We estimated the association state of ErbB2 and assessed how it was affected by EGF treatment in breast tumor cell lines by measuring FRET between fluorescent mAbs or Fab fragments. There was considerable homoassociation of ErbB2 and heteroassociation of ErbB2 with EGFR in quiescent breast tumor cells. ErbB2 homoassociation was enhanced by EGF treatment in SKBR-3 cells and in the BT474 subline BT474M1 with high tumorigenic potential, whereas the original BT474 line was resistant to this effect. These differences correlated well with EGFR expression. Because we calculated these measurements in flow cytometry, we obtained one single FRET efficiency value for each cell analyzed. To reveal heterogeneity in the homoassociation pattern of ErbB2 within a single cell, we had to use one of the microscopic FRET approaches. We have used donor pbFRET microscopy to visualize FRET efficiency within single cells with spatial resolution limited only by diffraction in the optical microscope (Bastiaens *et al.*, 1996; Nagy *et al.*, 1998b; Vereb *et al.*, 1997). This allows detailed analysis of the spatial heterogeneity of molecular interactions. At first, we applied the donor pbFRET in wide-field microscopy and revealed extensive pixel-by-pixel heterogeneity in ErbB2 homoassociation (Nagy *et al.*, 1998a). In our measurements, ErbB2 homoassociation was also heterogeneous in unstimulated breast tumor cells; and membrane domains with ErbB2 homoassociation had mean diameters of less than 1 μm (Nagy *et al.*, 1998a, 1999a). It was not clear whether the domain size was imposed by the optical resolution limit of wide-field microscopy in the X-Y plane or whether it originated from the actual size of ErbB2 aggregates. To refine the size estimate of domains containing ErbB2 molecules, we turned to SNOM and later combined FRET measurements with confocal microscopy.

To improve spatial resolution, first we studied the cell surface distribution of ErbB2 RTKs using SNOM (Kirsch *et al.*, 1996; Monson *et al.*, 1995; Vereb *et al.*, 1997). This technique is not limited by diffraction optics and can readily image objects in the 0.1- to 1.0-μm range, including submicrometer lipid and protein clusters in the plasma membrane (Edidin, 1997). ErbB2 was concentrated in irregular membrane patches, with a mean diameter of approximately 500 nm, containing up to 1000 ErbB2 molecules in nonactivated SKBR-3 and MDA453 human breast tumor cells. The mean cluster diameter increased to 600–900 nm when SKBR-3 cells were treated with EGF, heregulin, or a partially agonistic anti-ErbB2 antibody. The increase in cluster size was inhibited by an EGFR-specific tyrosine kinase inhibitor. Because the domain size was well within the resolution limit of the confocal microscopy (250–300 nm in the X-Y plane), we were able to confirm the SNOM results with a confocal LSM (CLSM) (Nagy *et al.*, 1999a).

We then implemented acceptor pbFRET in a CLSM equipped with three lasers, to study the correlation between the association of ErbB2 with ErbB2 and ErbB3 and the local density of these RTKs. From fluorescence intensities generated in two optical channels, we were able to determine the FRET efficiency values on a pixel-by-pixel basis, while the third laser beam provided the signal for measuring the expression level of ErbB3. The homoassociation of ErbB2 correlated positively with the local concentration of ErbB2 but negatively with ErbB3 local density. This negative correlation suggests that ErbB2–ErbB3 heterodimers compete with ErbB2 homodimers, and therefore, a high number of ErbB3 molecules can disassemble ErbB2 homodimers (Nagy *et al.*, 1999a).

We also investigated the colocalization of lipid rafts (section IV.B) and ErbB2 clusters using similar approaches. The size of lipid rafts was identified with either fluoresceinated or Cy5-labeled subunit B of cholera toxin (CTX-B), which binds to the GSL GM_1 ganglioside. Once again, we used the signal from one laser beam to monitor the lipid rafts and signals from the other two laser beams to reveal the homoassociation pattern of ErbB2. In this case, we applied both the donor and the acceptor pbFRET, and they gave very similar results. Our observations suggest that like ErbB1, ErbB2 is localized mostly in lipid rafts. However, there is a negative correlation between ErbB2 homoassociation and the local density of the lipid raft marker CTX-B. This environment could alter the association properties of ErbB2, similar to our findings regarding other membrane receptors (Vereb *et al.*, 2000). Because stimulating ErbB2 increases the size of ErbB2 clusters (Nagy *et al.*, 1999a) and lipid rafts (Nagy *et al.*, 2002), the amount of ErbB2 concentrated in rafts is very likely related to the function of the protein. Localization of ErbB2 in lipid rafts is dynamic, because it can be dislodged from rafts by cholera toxin–induced raft cross-linking (Nagy *et al.*, 2002). The association properties and biological activity of ErbB2 expelled from rafts differ from those inside rafts. For example, 4D5 (parent murine version of trastuzumab)-mediated internalization of ErbB2 is blocked in cholera toxin–pretreated cells, whereas its antiproliferative effect is not. These results emphasize that alterations in the local environment of ErbB2 strongly influence its association properties, which are reflected in its biological activity and in its behavior as a target for therapy (Nagy *et al.*, 2002).

V. Perspectives

Advances in developing new fluorescent probes, instrumentation, and methodologies have greatly improved the applicability of FRET to the systematic exploration of the localization, translocation, and association of signaling proteins in living or intact cells.

The newly developed fluorescent probes provide high sensitivity and great versatility while minimally perturbing the cell under investigation. Genetically encoded reporter constructs that are derived from GFPs and red fluorescent proteins are leading a revolution in the real-time visualization and tracking of various cellular events (Aoki *et al.*, 2004; Scarlata and Dowal, 2004; Zhang *et al.*, 2002). Some advances include the continued development of "passive" markers for the measurement of biomolecule expression and localization in live cells and "active" indicators for monitoring more complex cellular processes such as small molecule messenger dynamics, enzyme activation, and protein–protein interactions (Lippincott-Schwartz and Patterson, 2003; Lippincott-Schwartz *et al.*, 2003). A review published by Subramaniam *et al.* (2003) provides an excellent summary of the photophysical properties of GFPs and red fluorescent proteins and their application in quantitative microscopy. A mutant protein, which is capable of unique irreversible photoconversion from the nonfluorescent to a stable bright-red fluorescent form, has also been described: This "kindling fluorescent protein" can be used for precise *in vivo* photolabeling to track the movements of cells, organelles, and proteins and as an acceptor in future FRET experiments (Chudakov *et al.*, 2003a,b).

An interesting and promising modification of GFPs is when diverse antibody-derived binding loops are inserted into four of the exposed loops at one end of GFP creating "fluorobodies" that combine the advantages of antibodies (high affinity and specificity) with those of GFP (intrinsic fluorescence, high stability, expression, and solubility). These "fluorobodies" can be used effectively, among others, in flow and image cytometry and show affinities as high as antibodies. These new probes will probably gain widespread application in various fields of cytomics and proteomics (Zeytun *et al.*, 2003).

One of the most exotic versions of FRET techniques is the photochromic FRET (pcFRET), when diheteroarylethenes are used as acceptors (Giordano *et al.*, 2002; Song *et al.*, 2002). In pcFRET, the fluorescent emission of the donor is modulated by cyclical transformations of a photochromic acceptor. Light induces a reversible change in the structure and, concomitantly, in the absorption properties of the acceptor. The corresponding variation in the overlap integral (and, thus, critical transfer distance R_0) between the two states provides the means for reversibly switching the process of FRET on and off, allowing direct and repeated evaluation of the relative changes in the donor fluorescence quantum yield. Diheteroarylethenes demonstrate excellent stability in aqueous media, an absence of thermal back reactions, and negligible fatigue. Photochromic FRET is applicable for monitoring the time course of changes in FRET efficiencies, that is, the temporal dynamics of protein translocation and association processes (Giordano *et al.*, 2002).

The various kinds of image cytometric FRET (i.e., donor pbFRET, acceptor pbFRET, and intensity-based ratiometric FRET) can be applied either in wide-field microscopy or in laser-scanning confocal microscopy (Bastiaens and Jovin, 1998; Nagy *et al.*, 1998a,b, 2002; Vereb *et al.*, 1997). The major limitation of the wide-field microscopy technique is that fluorescence light below and above the focal plane contributes to the useful signal, thereby degrading the quality of the image. Confocal microscopy can overcome this limitation because of its capability to reject signals from outside the focal plane (Elangovan *et al.*, 2003; Nagy *et al.*, 2002). This capability provides a significant improvement in vertical resolution and allows for the use of serial optical sectioning of the living cells. The disadvantage of confocal microscopy is potentially serious photobleaching and photodamage of light-sensitive fluorophores, which can be detrimental during optical sectioning. Two-photon FRET microscopy can eliminate this problem; because two-photon excitation occurs only in the focal volume, the detected fluorescence signal is exclusively in-focus light. In addition, because two-photon excitation uses longer wavelength light, it is less damaging to living cells, thereby limiting problems associated with fluorophore photobleaching and photodamages, as well as intrinsic autofluorescence of cellular components (Elangovan *et al.*, 2003). The advantages of two-photon excitation can be further improved when it is combined with FLIM. Advantages of FLIM approaches are that lifetime measurements are less sensitive to changes in probe concentration and photobleaching. In addition, FLIM can discriminate fluorescence coming from different dyes that have similar absorption and emission characteristics but show a difference in fluorescence lifetime. In two-photon FRET FLIM, the reduction in donor lifetime in the presence of acceptor reveals the dimerization of protein molecules and determines more precisely the distance between the donor and the acceptor. This methodology allows for studying the dynamic behavior of protein–protein interactions in living cells and tissues (Chen and Periasamy, 2004).

A new development in FRET modalities is when FRET measurements are combined in spectral imaging. Spectral imaging and linear unmixing extends the possibilities to discriminate distinct fluorophores with highly overlapping emission spectra and, thus, the possibilities of multicolor imaging. Spectral imaging also offers advantages for fast multicolor time-lapse microscopy and FRET measurements in living samples (Zimmermann *et al.*, 2003).

The sensitivity of FRET can be extended to the single-molecule level by detecting FRET between a single-donor fluorophore and a single acceptor fluorophore. Single-pair FRET (spFRET) provides a unique means of observing conformational fluctuations and interactions of molecules at a single-molecule resolution even in the native, *in vivo* context of a living cell (Ha, 2001; Heyduk, 2002).

Another new FRET modality is the homotransfer or energy migration FRET (emFRET). This approach exploits fluorescence polarization measurements in flow cytometry, in wide-field LSMs, and in cLSMs, or in the form of anisotropy fluorescence lifetime imaging microscopy (rFLIM). These methods permit the assessment of rotational motion, association, and proximity of cellular proteins *in vivo*. They are particularly applicable to probes generated by fusions of visible fluorescence proteins and are capable of monitoring homoassociations of various signaling proteins (Clayton *et al.*, 2002; Lidke *et al.*, 2003; Tramier *et al.*, 2003).

The sensitivity of the FRET can be greatly enhanced when bioluminescence is used as a source of excited donor. In BRET, one protein is fused to Renilla luciferase and the other protein to a mutant of GFP (Xu *et al.*, 1999). The luciferase can be activated by addition of its substrate, and if the proteins in question interact, resonance energy transfer occurs between the excited luciferase and the mutant GFP. BRET can be detected by monitoring the fluorescence signal emitted by the mutant GFP. By choosing the proper luciferase/GFP mutant combinations, BRET can be used to measure protein–protein interactions *in vitro* and *in vivo*. BRET is perfectly suited for cell-based proteomics applications, including receptor research and mapping signal transduction pathways (Devost and Zingg, 2003; Issad *et al.*, 2003; Xu *et al.*, 1999).

SNOM provides many interesting imaging possibilities, which can be further increased by combining it with FRET. The simplest combination is when instead of the classic microscope, SNOM is used to detect FRET between dye molecules bound to cell surfaces (Kirsch *et al.*, 1999). In the other approach, the acceptor dye of a FRET pair is attached to the tip of a near-field fiberoptic probe. Light exiting the SNOM probe, which is not absorbed by the acceptor dye, excites the donor dye introduced into a sample. As the tip approaches the sample containing the donor dye, energy transfer from the excited donor to the tip-bound acceptor produces a red-shifted fluorescence. By monitoring this red-shifted acceptor emission, one can observe a dramatic reduction in the sample volume probed with an uncoated SNOM tip (Vickery and Dunn, 1999). Local fluorescence probes based on CdSe semiconductor nanocrystals have already been tested as FRET-SNOM sources with the prospect of single-molecule resolution (Shubeita *et al.*, 2003). It should be noted that in the last two examples the FRET phenomenon is used only for enhancing the resolution of the SNOM technique.

VI. Conclusions

A new and critically important strategic approach of today's cell biology is what we usually call "nanobiotechnology," that is, the efforts to resolve spatial details of living cells at nanometer resolution and to follow the kinetics of biochemical reactions or molecular interactions at a nanosecond (or even shorter) time scale. As we can see from this chapter, the dipole resonance energy transfer mechanism described by Förster in 1946 became by now successfully adopted to cellular systems and with the past few years' enormous technical developments in laser and microscope technologies, as well as in ultrafast electronics, it became one of the most powerful approaches in detecting interactions between different molecular constituents of cells. Classic modalities of FRET measurements (FCET, pbFRET, MI-FRET) allowed us to assess many intriguing and important questions about the molecular architecture of the plasma membrane of cells and about its dynamic heterogeneity (microdomain structure). Advances in the development of fluorescent probes with special design for FRET (the natural fluorescent proteins, GFP family, or their "fluorobody" constructs with the advantage of high affinity and specificity of antibody binding, different tandem fluorophores, etc.) and the novel innovative microscopic FRET modalities (confocal, two-photon microscopy, time-resolved FLIM-FRET, etc.) allow us now to explore interactions of signal molecules in the cytoplasm or at intracellular organelles. These, together with some approaches presently evaluated in biological applications (e.g., FRET combined with spectral imaging, to allow the detection of multiple interactions at the same time or FRET combined with SNOM to offer spatial resolution exceeding the diffraction limit), will certainly revolutionize nanobiology research.

Acknowledgments

We have been supported by the following research grants: OTKA T37831, T43061, T034393, TS40773, TS044711, ETT 532/2003, 524/2003, EU FP5 QLG1-CT-2000-01260, and Békésy Fellowship to G.V.

References

Aoki, K., Nakamura, T., and Matsuda, M. (2004). Spatio-temporal regulation of Rac1 and Cdc42 activity during nerve growth factor–induced neurite outgrowth in PC12 cells. *J. Biol. Chem.* **279,** 713–719.

Bacso, Z., Bene, L., Bodnár, A., Matkó, J., and Damjanovich, S. (1996). A photobleaching energy transfer analysis of CD8/MHC-I and LFA-1/ICAM-1 interactions in CTL-target cell conjugates. *Immunol. Lett.* **54,** 151–156.

Bastiaens, P. I. H., and Jovin, T. M. (1996). Microspectroscopic imaging tracks the intracellular processing of a signal transduction protein: Fluorescent-labeled protein kinase C beta I. *Proc. Natl. Acad. Sci. USA* **93,** 8407–8412.

Bastiaens, P. I. H., Majoul, I. V., Verveer, P. J., Söling, H. D., and Jovin, T. M. (1996). Imaging the intracellular trafficking and state of the AB_5 quaternary structure of cholera toxin. *EMBO J.* **15,** 4246–4253.

Bastiaens, P. I. H., Majoul, I. V., Verveer, P. J., Söling, H.-D., and Jovin, T. M. (1997). Imaging the intracellular trafficking and state of the AB_5 quaternary structure of cholera toxin. *EMBO J.* **15,** 4246–4253.

Bastiaens, P. I. H., and Jovin, T. M. (1998). Fluorescence resonance energy transfer microscopy, *In* "Cell biology: A laboratory handbook" (J. E. Celis, ed.), 2nd ed., vol 3, Academic Press, New York.

Bastiaens, P. I. H., and Squire, A. (1999). Fluorescence lifetime imaging microscopy: Spatial resolution of biochemical processes in the cell. *Trends Cell Biol.* **9,** 48–52.

Batard, P., Szöllosi, J., Luescher, I., Cerottini, J. C., MacDonald, R., and Romero, P. (2002). Use of phycoerythrin and allophycocyanin for fluorescence resonance energy transfer analyzed by flow cytometry: Advantages and limitations. *Cytometry* **48,** 97–105.

Bene, L., Balázs, M., Matkó, J., Most, J., Dierich, M. P., Szöllosi, J., and Damjanovich, S. (1994). Lateral organization of the ICAM-1 molecule at the surface of human lymphoblasts: A possible model for its co-distribution with the IL-2 receptor, class I and class II HLA molecules. *Eur. J. Immunol.* **24,** 2115–2123.

Berney, C., and Danuser, G. (2003). FRET or no FRET: A quantitative comparison. *Biophys. J.* **84,** 3992–4010.

Bodnár, A., Bacso, Z., Jenei, A., Jovin, T. M., Edidin, M., Damjanovich, S., and Matkó, J. (2003). Class I HLA oligomerization at the surface of B cells is controlled by exogenous beta(2)-microglobulin: Implications in activation of cytotoxic T lymphocytes. *Int. Immunol.* **15,** 331–339.

Bromley, S. K., Burack, W. R., Johnson, K. G., Somersalo, K., Sims, T. N., Sumen, C., Davis, M. M., Shaw, A. S., Allen, P. M., and Dustin, M. L. (2001). The immunological synapse. *Annu. Rev. Immunol.* **19,** 375–396.

Brown, D. A., and London, E. (1998). Structure and origin of ordered lipid domains in biological membranes. *J. Membr. Biol.* **164,** 103–114.

Chakrabarti, A., Matkó, J., Rahman, N. A., Barisas, B. G., and Edidin, M. (1992). Self-association of class I major histocompatibility complex molecules in liposome and cell surface membranes. *Biochemistry* **31,** 7182–7189.

Chen, Y., and Periasamy, A. (2004). Characterization of two-photon excitation fluorescence lifetime imaging microscopy for protein localization. *Microsc. Res. Technol.* **63,** 72–80.

Chudakov, D. M., Belousov, V. V., Zaraisky, A. G., Novoselov, V. V., Staroverov, D. B., Zorov, D. B., Lukyanov, S., and Lukyanov, K. A. (2003a). Kindling fluorescent proteins for precise *in vivo* photolabeling. *Nat. Biotechnol.* **21,** 191–194.

Chudakov, D. M., Feofanov, A. V., Mudrik, N. N., Lukyanov, S., and Lukyanov, K. A. (2003b). Chromophore environment provides clue to "kindling fluorescent protein" riddle. *J. Biol. Chem.* **278,** 7215–7219.

Citri, A., Skaria, K. B., and Yarden, Y. (2003). The deaf and the dumb: The biology of ErbB-2 and ErbB-3. *Exp. Cell Res.* **284,** 54–65.

Clayton, A. H., Hanley, Q. S., Arndt-Jovin, D. J., Subramaniam, V., and Jovin, T. M. (2002). Dynamic fluorescence anisotropy imaging microscopy in the frequency domain (rFLIM). *Biophys. J.* **83,** 1631–1649.

Clegg, R. M., Murchie, A. I., Zechel, A., Carlberg, C., Diekmann, S., and Lilley, D. M. (1992). Fluorescence resonance energy transfer analysis of the structure of the four-way DNA junction. *Biochemistry* **31,** 4846–4856.

Clegg, R. M. (1995). Fluorescence resonance energy transfer. *Curr. Opin. Biotechnol.* **6,** 103–110.

Clegg, R. M. (2002). FRET tells us about proximities, distances, orientations and dynamic properties. *J. Biotechnol.* **82,** 177–179.

Clegg, R. M., Holub, O., and Gohlke, C. (2003). Fluorescence lifetime-resolved imaging: Measuring lifetimes in an image. *Methods Enzymol.* **360,** 509–542.

Collins, T. L., Uniyal, S., Shin, J., Strominger, J. L., Mittler, R. S., and Burakoff, S. J. (1992). p56lck association with CD4 is required for the interaction between CD4 and the TCR/CD3 complex and for optimal antigen stimulation. *J. Immunol.* **148,** 2159–2162.

Damjanovich, S., Tron, L., Szöllősi, J., Zidovetzki, R., Vaz, W. L., Regateiro, F., Arndt-Jovin, D. J., and Jovin, T. M. (1983). Distribution and mobility of murine histocompatibility H-2Kk antigen in the cytoplasmic membrane. *Proc. Natl. Acad. Sci. USA* **80,** 5985–5989.

Damjanovich, S., Vereb, G., Schaper, A., Jenei, A., Matkó, J., Starink, J. P., Fox, G. Q., Arndt-Jovin, D. J., and Jovin, T. M. (1995). Structural hierarchy in the clustering of HLA class I molecules in the plasma membrane of human lymphoblastoid cells. *Proc. Natl. Acad. Sci. USA* **92,** 1122–1126.

Devost, D., and Zingg, H. H. (2003). Identification of dimeric and oligomeric complexes of the human oxytocin receptor by co-immunoprecipitation and bioluminescence resonance energy transfer. *J. Mol. Endocrinol.* **31,** 461–471.

Dewey, T. G., and Hammes, G. G. (1980). Calculation on fluorescence resonance energy transfer on surfaces. *Biophys. J.* **32,** 1023–1035.

Dexter, D. L. (1953). A theory of sensitized luminescence in solids. *J. Chem. Phys.* **21,** 836–850.

Edidin, M., Kuo, S. C., and Sheetz, M. P. (1991). Lateral movements of membrane glycoproteins restricted by dynamic cytoplasmic barriers. *Science* **254,** 1379–1382.

Edidin, M. (1997). Lipid microdomains in cell surface membranes. *Curr. Opin. Struct. Biol.* **7,** 528–532.

Edidin, M. (2001). Shrinking patches and slippery rafts: Scales of domains in the plasma membrane. *Trends Cell Biol.* **11,** 492–496.

Eidne, K. A., Kroeger, K. M., and Hanyaloglu, A. C. (2002). Applications of novel resonance energy transfer techniques to study dynamic hormone receptor interactions in living cells. *Trends Endocrinol. Metab.* **13,** 415–421.

Elangovan, M., Wallrabe, H., Chen, Y., Day, R. N., Barroso, M., and Periasamy, A. (2003). Characterization of one- and two-photon excitation fluorescence resonance energy transfer microscopy. *Methods* **29,** 58–73.

Förster, T. (1946). Energiewanderung und Fluoreszenz. *Naturwissenschaften* **6,** 166–175.

Gadella, T. W., Jr., Clegg, R. M., and Jovin, T. M. (1994). Fluorescence lifetime imaging microscopy: pixel-by-pixel analysis of phase-modulation data. *Bioimaging* **2,** 139–159.

Gadella, T. W., Jr., and Jovin, T. M. (1995). Oligomerization of epidermal growth factor receptors on A431 cells studied by time-resolved fluorescence imaging microscopy. A stereochemical model for tyrosine kinase receptor activation. *J. Cell Biol.* **129,** 1543–1558.

Gáspár, R., Jr., Bagossi, P., Bene, L., Matkó, J., Szöllősi, J., Tozser, J., Fesus, L., Waldmann, T. A., and Damjanovich, S. (2001). Clustering of class I HLA oligomers with CD8 and TCR: Three-dimensional models based on fluorescence resonance energy transfer and crystallographic data. *J. Immunol.* **166,** 5078–5086.

Giordano, L., Jovin, T. M., Irie, M., and Jares-Erijman, E. A. (2002). Diheteroarylethenes as thermally stable photoswitchable acceptors in photochromic fluorescence resonance energy transfer (pcFRET). *J. Am. Chem. Soc.* **124,** 7481–7489.

Gordon, G. W., Berry, G., Liang, X. H., Levine, B., and Herman, B. (1998). Quantitative fluorescence resonance energy transfer measurements using fluorescence microscopy. *Biophys. J.* **74,** 2702–2713.

Ha, T. (2001). Single-molecule fluorescence resonance energy transfer. *Methods* **25,** 78–86.

Haj, F. G., Verveer, P. J., Squire, A., Neel, B. G., and Bastiaens, P. I. (2002). Imaging sites of receptor dephosphorylation by PTP1B on the surface of the endoplasmic reticulum. *Science* **295,** 1708–1711.

Heyduk, T. (2002). Measuring protein conformational changes by FRET/LRET. *Curr. Opin. Biotechnol.* **13,** 292–296.

Issad, T., Boute, N., Boubekeur, S., Lacasa, D., and Pernet, K. (2003). Looking for an insulin pill? Use the BRET methodology! *Diabetes Metab.* **29,** 111–117.

Jacobson, K., and Dietrich, C. (1999). Looking at lipid rafts? *Trends Cell Biol.* **9,** 87–91.

Jares-Erijman, E. A., and Jovin, T. M. (2003). FRET imaging. *Nat. Biotechnol.* **21,** 1387–1395.

Jovin, T. M., and Arndt-Jovin, D. J. (1989). FRET microscopy: Digital imaging of fluorescence resonance energy transfer. Application in cell biology. *In* "Cell structure and function by microspectrofluorimetry" (E. Kohen and J. G. Hirschberg, eds.). Academic Press, San Diego.

Jurgens, L., Arndt-Jovin, D., Pecht, I., and Jovin, T. M. (1996). Proximity relationships between the type I receptor for Fc epsilon (Fc epsilon RI) and the mast cell function-associated antigen (MAFA) studied by donor photobleaching fluorescence resonance energy transfer microscopy. *Eur. J. Immunol.* **26,** 84–91.

Kenworthy, A. K., and Edidin, M. (1998). Distribution of a glycosylphosphatidylinositol-anchored protein at the apical surface of MDCK cells examined at a resolution of <100 Å using imaging fluorescence resonance energy transfer. *J. Cell Biol.* **142,** 69–84.

Kenworthy, A. K., Petranova, N., and Edidin, M. (2000). High-resolution FRET microscopy of cholera toxin B-subunit and GPI-anchored proteins in cell plasma membranes. *Mol. Biol. Cell* **11,** 1645–1655.

Kirsch, A., Meyer, C., and Jovin, T. M. (1996). Integration of optical techniques in scanning probe microscopes: The scanning near-field optical microscope (SNOM). *In* "Proceedings of NATO Advanced Research Workshop: Analytical use of fluorescent probes in oncology, Miami, Fl. Oct. 14–18 1995" (E. Kohen and J. G. Kirschberg, eds.). Plenum Press, New York.

Kirsch, A. K., Subramaniam, V., Jenei, A., and Jovin, T. M. (1999). Fluorescence resonance energy transfer detected by scanning near-field optical microscopy. *J. Microsc.* **194,** 448–454.

Latif, R., Graves, P., and Davies, T. F. (2002). Ligand-dependent inhibition of oligomerization at the human thyrotropin receptor. *J. Biol. Chem.* **277,** 45059–45067.

Legg, J. W., Lewis, C. A., Parsons, M., Ng, T., and Isacke, C. M. (2002). A novel PKC-regulated mechanism controls CD44 ezrin association and directional cell motility. *Nat. Cell Biol.* **4,** 399–407.

Lidke, D. S., Nagy, P., Barisas, B. G., Heintzmann, R., Post, J. N., Lidke, K. A., Clayton, A. H., Arndt-Jovin, D. J., and Jovin, T. M. (2003). Imaging molecular interactions in cells by dynamic and static fluorescence anisotropy (rFLIM and emFRET). *Biochem. Soc. Trans.* **31,** 1020–1027.

Liegler, T., Szöllősi, J., Hyun, W., and Goodenow, R. S. (1991). Proximity measurements between H-2 antigens and the insulin receptor by fluorescence energy transfer: Evidence that a close association does not influence insulin binding. *Proc. Natl. Acad. Sci. USA* **88,** 6755–6759.

Lippincott-Schwartz, J., Altan-Bonnet, N., and Patterson, G. H. (2003). Photobleaching and photoactivation: Following protein dynamics in living cells. *Nat. Cell Biol.* Suppl. S7–S14.

Lippincott-Schwartz, J., and Patterson, G. H. (2003). Development and use of fluorescent protein markers in living cells. *Science* **300,** 87–91.

Loken, M. R., Keij, J. F., and Kelley, K. A. (1987). Comparison of helium-neon and dye lasers for the excitation of allophycocyanin. *Cytometry* **8,** 96–100.

Matkó, J., Ohki, K., and Edidin, M. (1992). Luminescence quenching by nitroxide spin labels in aqueous solution: Studies on the mechanism of quenching. *Biochemistry* **31,** 703–711.

Matkó, J., Jenei, A., Mátyus, L., Ameloot, M., and Damjanovich, S. (1993). Mapping of cell surface protein-patterns by combined fluorescence anisotropy and energy transfer measurements. *J. Photochem. Photobiol. B.* **19,** 69–73.

Matkó, J., Bushkin, Y., Wei, T., and Edidin, M. (1994). Clustering of class I HLA molecules on the surfaces of activated and transformed human cells. *J. Immunol.* **152,** 3353–3360.

Matkó, J., Jenei, A., Wei, T., and Edidin, M. (1995). Luminescence quenching by long range electron transfer: A probe of protein clustering and conformation at the cell surface. *Cytometry* **19,** 191–200.

Matkó, J., and Edidin, M. (1997). Energy transfer methods for detecting molecular clusters on cell surfaces. *Methods Enzymol.* **278,** 444–462.

Matkó, J., Bodnár, A., Vereb, G., Bene, L., Vamosi, G., Szentesi, G., Szöllősi, J., Gáspár, R., Horejsi, V., Waldmann, T. A., and Damjanovich, S. (2002). GPI-microdomains (membrane rafts) and signaling of the multi-chain interleukin-2 receptor in human lymphoma/leukemia T cell lines. *Eur. J. Biochem.* **269,** 1199–1208.

Matkó, J., and Szöllősi, J. (2002). Landing of immune receptors and signal proteins on lipid rafts: A safe way to be spatio-temporally coordinated? *Immunol. Lett.* **82,** 3–15.

Mátyus, L., Bene, L., Heiligen, H., Rausch, J., and Damjanovich, S. (1995). Distinct association of transferrin receptor with HLA class I molecules on HUT-102B and JY cells. *Immunol. Lett.* **44,** 203–208.

Mekler, V. M. (1994). A photochemical technique to enhance sensitivity of detection of fluorescence resonance energy transfer. *Photochem. Photobiol.* **59,** 615–620.

Milligan, G., Ramsay, D., Pascal, G., and Carrillo, J. J. (2003). GPCR dimerisation. *Life Sci.* **74,** 181–188.

Mittler, R. S., Goldman, S. J., Spitalny, G. L., and Burakoff, S. J. (1989). T-cell receptor–CD4 physical association in a murine T-cell hybridoma: Induction by antigen receptor ligation. *Proc. Natl. Acad. Sci. USA* **86,** 8531–8535.

Monks, C. R., Freiberg, B. A., Kupfer, H., Sciaky, N., and Kupfer, A. (1998). Three-dimensional segregation of supramolecular activation clusters in T cells. *Nature* **395,** 82–86.

Monson, E., Merritt, G., Smith, S., Langmore, J. P., and Kopelman, R. (1995). Implementation of an NSOM system for fluorescence microscopy. *Ultramicroscopy* **57,** 257–262.

Nagy, P., Bene, L., Balázs, M., Hyun, W. C., Lockett, S. J., Chiang, N. Y., Waldman, F., Feuerstein, B. G., Damjanovich, S., and Szöllősi, J. (1998a). EGF-induced redistribution of erbB2 on breast tumor cells: Flow and image cytometric energy transfer measurements. *Cytometry* **32,** 120–131.

Nagy, P., Vamosi, G., Bodnár, A., Lockett, S. J., and Szöllősi, J. (1998b). Intensity-based energy transfer measurements in digital imaging microscopy. *Eur. Biophys. J.* **27,** 377–389.

Nagy, P., Jenei, A., Kirsch, A. K., Szöllősi, J., Damjanovich, S., and Jovin, T. M. (1999a). Activation-dependent clustering of the erbB2 receptor tyrosine kinase detected by scanning near-field optical microscopy. *J. Cell Sci.* **112,** 1733–1741.

Nagy, P., Jenei, A., Damjanovich, S., Jovin, T. M., and Szolosi, J. (1999b). Complexity of signal transduction mediated by ErbB2: Clues to the potential of receptor-targeted cancer therapy. *Pathol. Oncol. Res.* **5,** 255–271.

Nagy, P., Vereb, G., Sebestyen, Z., Horvath, G., Lockett, S. J., Damjanovich, S., Park, J. W., Jovin, T. M., and Szöllősi, J. (2002). Lipid rafts and the local density of ErbB proteins influence the biological role of homo- and heteroassociations of ErbB2. *J. Cell Sci.* **115,** 4251–4262.

Nashmi, R., Dickinson, M. E., McKinney, S., Jareb, M., Labarca, C., Fraser, S. E., and Lester, H. A. (2003). Assembly of alpha4beta2 nicotinic acetylcholine receptors assessed with functional fluorescently labeled subunits: Effects of localization, trafficking, and nicotine-induced upregulation in clonal mammalian cells and in cultured midbrain neurons. *J. Neurosci.* **23,** 11554–11567.

Ramalingam, T. S., Chakrabarti, A., and Edidin, M. (1997). Interaction of class I human leukocyte antigen (HLA-I) molecules with insulin receptors and its effect on the insulin-signaling cascade. *Mol. Biol. Cell* **8,** 2463–2474.

Runnels, L. W., and Scarlata, S. F. (1995). Theory and application of fluorescence homotransfer to melittin oligomerization. *Biophys. J.* **69,** 1569–1583.

Scarlata, S., and Dowal, L. (2004). The use of green fluorescent proteins to view association between phospholipase C beta and G protein subunits in cells. *Methods Mol. Biol.* **237,** 223–232.

Scholes, G. D. (2003). Long-range resonance energy transfer in molecular systems. *Annu. Rev. Phys. Chem.* **54,** 57–87.

Schreiber, A. B., Schlessinger, J., and Edidin, M. (1984). Interaction between major histocompatibility complex antigens and epidermal growth factor receptors on human cells. *J. Cell Biol.* **98,** 725–731.

Sebestyen, Z., Nagy, P., Horvath, G., Vamosi, G., Debets, R., Gratama, J. W., Alexander, D. R., and Szöllősi, J. (2002). Long wavelength fluorophores and cell-by-cell correction for autofluorescence significantly improves the accuracy of flow cytometric energy transfer measurements on a dual-laser benchtop flow cytometer. *Cytometry* **48,** 124–135.

Sekar, R. B., and Periasamy, A. (2003). Fluorescence resonance energy transfer (FRET) microscopy imaging of live cell protein localizations. *J. Cell Biol.* **160,** 629–633.

Selvin, P. R. (2000). The renaissance of fluorescence resonance energy transfer. *Nat. Struct. Biol.* **7,** 730–734.

Shubeita, G. T., Sekatskii, S. K., Dietler, G., Potapova, I., Mews, A., and Basche, T. (2003). Scanning near-field optical microscopy using semiconductor nanocrystals as a local fluorescence and fluorescence resonance energy transfer source. *J. Microsc.* **210,** 274–278.

Simons, K., and Ikonen, E. (1997). Functional rafts in cell membranes. *Nature* **387,** 569–572.

Sliwkowski, M. X., Schaefer, G., Akita, R. W., Lofgren, J. A., Fitzpatrick, V. D., Nuijens, A., Fendly, B. M., Cerione, R. A., Vandlen, R. L., and Carraway, K. L.3rd (1994). Coexpression of erbB2 and erbB3 proteins reconstitutes a high affinity receptor for heregulin. *J. Biol. Chem.* **269,** 14661–14665.

Sliwkowski, M. X., Lofgren, J. A., Lewis, G. D., Hotaling, T. E., Fendly, B. M., and Fox, J. A. (1999). Nonclinical studies addressing the mechanism of action of trastuzumab (Herceptin). *Semin. Oncol.* **26,** 60–70.

Smith, P. R., Morrison, I. E., Wilson, K. M., Fernandez, N., and Cherry, R. J. (1999). Anomalous diffusion of major histocompatibility complex class I molecules on HeLa cells determined by single particle tracking. *Biophys. J.* **76,** 3331–3344.

Snyder, B., and Freire, E. (1982). Fluorescence energy transfer in two dimensions. A numeric solution for random and nonrandom distributions. *Biophys. J.* **40,** 137–148.

Song, L., Hennink, E. J., Young, I. T., and Tanke, H. J. (1995). Photobleaching kinetics of fluorescein in quantitative fluorescence microscopy. *Biophys. J.* **68,** 2588–2600.

Song, L., Jares-Erijman, E. A., and Jovin, T. M. (2002). A photochromic acceptor as a reversible light-driven switch in fluorescence resonance energy transfer (FRET). *J. Photochem. Photobiol. A* **150,** 177–185.

Stryer, L., and Haugland, R. P. (1967). Energy transfer: A spectroscopic ruler. *Proc. Natl. Acad. Sci. USA* **58,** 719–726.

Subczynski, W. K., and Kusumi, A. (2003). Dynamics of raft molecules in the cell and artificial membranes: Approaches by pulse EPR spin labeling and single molecule optical microscopy. *Biochim. Biophys. Acta* **1610,** 231–243.

Subramaniam, V., Hanley, Q. S., Clayton, A. H., and Jovin, T. M. (2003). Photophysics of green and red fluorescent proteins: Implications for quantitative microscopy. *Methods Enzymol.* **360,** 178–201.

Suzuki, Y. (2000). Detection of the swings of the lever arm of a myosin motor by fluorescence resonance energy transfer of green and blue fluorescent proteins. *Methods* **22,** 355–363.

Szabó, G., Jr., Weaver, J. L., Pine, P. S., Rao, P. E., and Aszalos, A. (1995). Cross-linking of CD4 in a TCR/CD3-juxtaposed inhibitory state: A pFRET study. *Biophys. J.* **68,** 1170–1176.

Szöllősi, J., Tron, L., Damjanovich, S., Helliwell, S. H., Arndt-Jovin, D., and Jovin, T. M. (1984). Fluorescence energy transfer measurements on cell surfaces: A critical comparison of steady-state fluorimetric and flow cytometric methods. *Cytometry* **5,** 210–216.

Szöllősi, J., Damjanovich, S., Goldman, C. K., Fulwyler, M. J., Aszalos, A. A., Goldstein, G., Rao, P., Talle, M. A., and Waldmann, T. A. (1987). Flow cytometric resonance energy transfer measurements support the association of a 95-kDa peptide termed T27 with the 55-kDa Tac peptide. *Proc. Natl. Acad. Sci. USA* **84,** 7246–7250.

Szöllősi, J., Damjanovich, S., Balázs, M., Nagy, P., Tron, L., Fulwyler, M. J., and Brodsky, F. M. (1989). Physical association between MHC class I and class II molecules detected on the cell surface by flow cytometric energy transfer. *J. Immunol.* **143,** 208–213.

Szöllősi, J., Horejsi, V., Bene, L., Angelisova, P., and Damjanovich, S. (1996). Supramolecular complexes of MHC class I, MHC class II, CD20, and tetraspan molecules (CD53, CD81, and CD82) at the surface of a B cell line JY. *J. Immunol.* **157,** 2939–2946.

Szöllősi, J., Damjanovich, S., and Mátyus, L. (1998). Application of fluorescence resonance energy transfer in the clinical laboratory: Routine and research. *Cytometry* **34,** 159–179.

Szöllősi, J., Nagy, P., Sebestyen, Z., Damjanovich, S., Park, J. W., and Mátyus, L. (2002). Applications of fluorescence resonance energy transfer for mapping biological membranes. *J. Biotechnol.* **82,** 251–266.

Szöllősi, J., and Alexander, D. R. (2003). The application of fluorescence resonance energy transfer to the investigation of phosphatases. *Methods Enzymol.* **366,** 203–224.

Tramier, M., Piolot, T., Gautier, I., Mignotte, V., Coppey, J., Kemnitz, K., Durieux, C., and Coppey-Moisan, M. (2003). Homo-FRET versus hetero-FRET to probe homodimers in living cells. *Methods Enzymol.* **360,** 580–597.

Triantafilou, K., Triantafilou, M., Wilson, K. M., and Fernandez, N. (2000). Human major histocompatibility molecules have the intrinsic ability to form homotypic associations. *Hum. Immunol.* **61,** 585–598.

Tron, L., Szöllősi, J., Damjanovich, S., Helliwell, S. H., Arndt-Jovin, D. J., and Jovin, T. M. (1984). Flow cytometric measurement of fluorescence resonance energy transfer on cell surfaces. Quantitative evaluation of the transfer efficiency on a cell-by-cell basis. *Biophys. J.* **45,** 939–946.

Tuosto, L., Parolini, I., Schroder, S., Sargiacomo, M., Lanzavecchia, A., and Viola, A. (2001). Organization of plasma membrane functional rafts upon T cell activation. *Eur. J. Immunol.* **31,** 345–349.

Turcatti, G., Nemeth, K., Edgerton, M. D., Meseth, U., Talabot, F., Peitsch, M., Knowles, J., Vogel, H., and Chollet, A. (1996). Probing the structure and function of the tachykinin neurokinin-2 receptor through biosynthetic incorporation of fluorescent amino acids at specific sites. *J. Biol. Chem.* **271,** 19991–19998.

Tzahar, E., Waterman, H., Chen, X., Levkowitz, G., Karunagaran, D., Lavi, S., Ratzkin, B. J., and Yarden, Y. (1996). A hierarchical network of interreceptor interactions determines signal transduction by Neu differentiation factor/neuregulin and epidermal growth factor. *Mol. Cell Biol.* **16,** 5276–5287.

Van Munster, E. B., and Gadella, T. W. (2004). phiFLIM: A new method to avoid aliasing in frequency-domain fluorescence lifetime imaging microscopy. *J. Microsc.* **213,** 29–38.

Varma, R., and Mayor, S. (1998). GPI-anchored proteins are organized in submicron domains at the cell surface. *Nature* **394,** 798–801.

Vereb, G., Meyer, C. K., and Jovin, T. M. (1997). Novel microscope-based approaches for the investigation of protein–protein interactions in signal transduction. *In* Interacting protein domains, their role in signal and energy transduction. NATO ASI series (L. M. G. Heilmeyer, Jr., ed.). vol H102, Springer–Verlag, New York.

Vereb, G., Matkó, J., Vamosi, G., Ibrahim, S. M., Magyar, E., Varga, S., Szöllősi, J., Jenei, A., Gáspár, R., Jr., Waldmann, T. A., and Damjanovich, S. (2000). Cholesterol-dependent clustering of IL-2Ralpha and its colocalization with HLA and CD48 on T lymphoma cells suggest their functional association with lipid rafts. *Proc. Natl. Acad. Sci. USA* **97,** 6013–6018.

Vereb, G., Nagy, P., Park, J. W., and Szöllősi, J. (2002). Signaling revealed by mapping molecular interactions: Implications for ErbB-targeted cancer immunotherapies. *Clin. Appl. Immunol. Rev.* **2,** 169–186.

Vereb, G., Szöllősi, J., Matkó, J., Nagy, P., Farkas, T., Vigh, L., Mátyus, L., Waldmann, T. A., and Damjanovich, S. (2003). Dynamic, yet structured: The cell membrane three decades after the Singer-Nicolson model. *Proc. Natl. Acad. Sci. USA* **100,** 8053–8058.

Vickery, S. A., and Dunn, R. C. (1999). Scanning near-field fluorescence resonance energy transfer microscopy. *Biophys. J.* **76,** 1812–1818.

Wolber, P. K., and Hudson, B. S. (1979). An analytic solution to the Förster energy transfer problem in two dimensions. *Biophys. J.* **28,** 197–210.

Wulfing, C., Sjaastad, M. D., and Davis, M. M. (1998). Visualizing the dynamics of T cell activation: Intracellular adhesion molecule 1 migrates rapidly to the T cell/B cell interface and acts to sustain calcium levels. *Proc. Natl. Acad. Sci. USA* **95,** 6302–6307.

Xu, Y., Piston, D. W., and Johnson, C. H. (1999). A bioluminescence resonance energy transfer (BRET) system: Application to interacting circadian clock proteins. *Proc. Natl. Acad. Sci. USA* **96,** 151–156.

Yarden, Y., and Sliwkowski, M. X. (2001). Untangling the ErbB signalling network. *Nat. Rev. Mol. Cell Biol.* **2,** 127–137.

Yguerabide, J. (1994). Theory for establishing proximity relations in biological membranes by excitation energy transfer measurements. *Biophys. J.* **66,** 683–693.

Zal, T., Zal, M. A., and Gascoigne, N. R. (2002). Inhibition of T cell receptor–coreceptor interactions by antagonist ligands visualized by live FRET imaging of the T-hybridoma immunological synapse. *Immunity* **16,** 521–534.

Zeytun, A., Jeromin, A., Scalettar, B. A., Waldo, G. S., and Bradbury, A. R. (2003). Fluorobodies combine GFP fluorescence with the binding characteristics of antibodies. *Nat. Biotechnol.* **21,** 1473–1479.

Zhang, J., Campbell, R. E., Ting, A. Y., and Tsien, R. Y. (2002). Creating new fluorescent probes for cell biology. *Nat. Rev. Mol. Cell Biol.* **3,** 906–918.

Zimmermann, T., Rietdorf, J., and Pepperkok, R. (2003). Spectral imaging and its applications in live cell microscopy. *FEBS Lett.* **546,** 87–92.

CHAPTER 6

The Rainbow of Fluorescent Proteins

David W. Galbraith

University of Arizona
Department of Plant Sciences and Institute for Biomedical Science and Biotechnology
Tucson, Arizona 85721

I. The Fluorescent Proteins

Fluorescent proteins (FPs) can be defined as the class of proteins that exhibit fluorescence as a consequence of fluorophore formation by an intramolecular reaction of the amino acid side chains contained within these proteins. This class comprises approximately 30 members, all of which have been isolated from aquatic organisms. Specifically excluded from this class are all proteins whose fluorescence requires the presence of nonprotein prosthetic groups. Classifying

0091-679X/04 $35.00

FPs in this manner focuses attention on their unique ability to produce fluorescence after expression of the polypeptide structure without requiring additional components, including specific chaperones. Consequently, FPs comprise uniquely flexible transgenic markers capable of producing fluorescence within an essentially limitless range of species and cell types, in both prokarya and eukarya (Chalfie *et al.*, 1994; Miyawaki *et al.*, 2003; Tsien, 1998). A subset of FP proteins, termed *chromoproteins* (Labas *et al.*, 2002), display intense absorption but without subsequent fluorescence emission, but variants of these can be isolated that behave as FPs. The evolutionary relationship between chromoproteins and FPs is further discussed in section I.F.

A. Green Fluorescent Protein

The prototype FP is the green FP (GFP) of the jellyfish *Aequorea victoria*, which was first described 40 years ago (Johnson *et al.*, 1962; Shimomura *et al.*, 1962). At least four species names exist for *Aequorea* (*Aequorea victoria*, *Aequorea aequorea*, *Aequorea coerulescens*, and *Aequorea forskalea*). This taxonomy is based on morphological features that are both plastic and developmentally regulated (for a complete discussion, see Mills, 1999). Clarification of species relationships awaits application of molecular methods.

GFP is a small (~26-kd) protein comprising 238 amino acids, organized as eleven β sheets forming a capped barrel (Ormö *et al.*, 1996; Yang *et al.*, 1996) that encloses the fluorophore and serves to exclude solvent molecules. This structure also confers unusual stability on the protein, which remains fluorescent up to 65 °C, at a pH of eleven, and in 1% sodium dodecylsulfate, or in 6M guanidinium hydrochloride, and which is largely resistant to proteases. The fluorophore, a *p*-hydroxy benzylidene imidazolinone, is formed by an intramolecular rearrangement and oxidation involving ser65, tyr66, and gly67 (Tsien, 1998). Within the jellyfish, GFP serves to absorb broad-spectrum blue light produced by aequorin, a Ca^{2+}-dependent photoluminescent protein.

B. Modifications to Green Fluorescent Protein

Early work established the feasibility of altering the optical properties of GFP through modification of the protein primary sequence (Heim *et al.*, 1994; Tsien, 1998). Early modifications resulted in changes to shape of the bimodal absorbance spectrum of wild-type GFP to a unimodal peak center either in the ultraviolet (UV) or near 488 nm. Other amino acid changes, single and multiple, were found to alter the wavelengths of maximal absorption and emission of the fluorochrome. The naming conventions of mutant GFPs are confusing (Galbraith *et al.*, 1998). Some convergence has occurred though, with the following names for commonly encountered mutant GFP forms being generally recognized: cyan FP (CFP; absorbance maximum 439 nm and emission maximum 476 nm), blue

FP (BFP; absorption maximum 384 nm and emission maximum 448 nm), and yellow FP (YFP; absorption maximum 512 nm and emission maximum 529 nm). The spectral maxima given here should be regarded as approximations, because slightly different versions of these three classes of mutants have been produced via different combinations of amino acid substitutions, and these differ in the positions of their absorption and emission maxima (Tsien, 1998).

A number of desirable physicochemical and biological changes to the properties of FPs have also been achieved through alterations to the coding sequence of GFP. These include improving the translatability of the messenger RNA (mRNA) through codon optimization, eliminating cryptic introns (Haseloff *et al.*, 1997), increasing the rate of folding of the translated products, decreasing misfolding and aggregation, and modifying observed sensitivities to inorganic anions and to changes in pH (Tsien, 1998; Zhang *et al.*, 2002). All of these modifications should be considered within the context of the organism and cell type of interest, as well as of the subcellular location within which the FP is to be expressed. Thus, for example, targeting to certain locations may involve encountering regions of particularly low pH levels (e.g., the plant cell surface and the lysosomal or vacuolar lumen), which can decrease FP fluorescence.

Many modifications work reasonably well across species and even kingdoms. For example, GFP codon usage optimized for mammalian expression is also compatible with expression in plants and serves to eliminate a cryptic intron, which would otherwise be spliced (Haseloff *et al.*, 1997).

Further modifications to the FP primary structure include topogenic alterations consequent to addition of targeting sequences. FPs can also be adapted to act as specific sensors of cellular physiology. These modifications, which do not involve changes to the core primary structure of the FP, are described in section I.E.

C. Nonhydrozoan Fluorescent Proteins

A large number of additional sources of FPs have been found within the *Cnidaria*, particularly the anthozoa (Matz *et al.*, 1999), and a comprehensive phylogeny has been constructed (Labas *et al.*, 2002; Shagin *et al.*, 2003) (see also section I.F). Some have GFP-like spectra, for example, that of *Renilla reniformis*. However, a particular advantage offered by the FPs of other anthozoans relates to the fact that none of the spectral variants of *Aequorea* GFP have emission maxima longer than about 529 nm (Miyawaki *et al.*, 2003). In contrast, a number of anthozoan FPs have longer emission maxima, including the commercially available DsRed, AsRed, HcRed, and Kindling Red (see section I.G). Fluorescence emission in the red offers some advantages with animal cells, because they generally lack autofluorescence at this wavelength. The presence of chlorophyll, which produces intense red fluorescence, reduces the utility of these FPs in photosynthetic organisms.

D. Photokinetic and Photodynamic Fluorescent Proteins

Other intriguing features of FPs include the observation of changes in excitation and emission spectra over time (Labas *et al.*, 2002; Terskikh *et al.*, 2000) and after actinic irradiation (Ando *et al.*, 2002; Chudakov *et al.*, 2003; Zhang *et al.*, 2002). This behavior is particularly useful for defining promoter activities and charting subcellular dynamics of labeled proteins, because most FPs exhibit long half-lives and are, therefore, unsuited for measurement of rapid changes in protein levels or locations.

E. Modifying Fluorescent Proteins to Access Specific Cellular Locations and Specific Physiological Processes, and to Act as Specific Ligands

FPs in general readily accommodate additions to their N- and C-terminals, and linker peptide sequences are not required. The close physical proximity of the N- and C-terminus permits circular permutation of GFP (Baird *et al.*, 1999), allowing addition of sequences at equatorial locations relative to the barrel structure. A large variety of topogenic motifs have been successfully employed with FPs, with most work being done with GFP, and these allow FP targeting to most, perhaps all, subcellular compartments of eukaryotes (Lippincott-Schwartz and Patterson, 2003; Tsien, 1998). Measurement of dynamic changes in FP levels and intracellular locations is facilitated by use of photokinetic and photodynamic FPs. FP stability can also be modified by fusion to proteins of short half-life or to peptide sequences directing proteolytic degradation (Li *et al.*, 1998). A potential complication to the use of FPs is that most naturally exist as oligomers (Zhang *et al.*, 2002). Mutations have been identified that reduce the extent of intermolecular interactions, thereby decreasing or eliminating oligomer formation (Campbell *et al.*, 2002; Zhang *et al.*, 2002).

FP pairs having overlapping emission and excitation spectra are amenable to fluorescence resonance energy transfer (FRET). This has formed the basis for the family of calcium sensors termed *cameleons* pioneered by the group of Roger Tsien in San Diego (Miyawaki *et al.*, 1997, 1999) and by others (Romoser *et al.*, 1997). Cameleons couple calcium-specific modulation of calmodulin interactions to alterations in the spatial relationship of a pair of FPs synthesized within the context of a single polypeptide chain. This results in ratiometric alterations in the emission spectra of the composite molecule as a function of calcium concentration. Cameleons have been successfully employed to monitor calcium transients in nonmammalian systems, including plants (Allen *et al.*, 1999, 2000, 2001). An alternative to FRET-based measurements of calcium ions involves insertion of calmodulin in an equatorial position into the backbone of YFP, resulting in a non-ratiometric calcium indicator (camgaroo-1), which increases eightfold in intensity on saturation with calcium (Baird *et al.*, 1999). An improved version of this reporter has decreased pH sensitivity and chloride interference and increased photostability and levels of productive expression at 37 °C (Griesbeck *et al.*,

2001). Camgaroos have been successfully employed in nonmammalian systems (Yu *et al.*, 2003a).

The intramolecular FRET strategy inherent to cameleons emerged from one that was initially devised for sensing protease action (Heim and Tsien, 1996) and that now has been extended for measurement of the concentrations and activities of many signaling molecules, including cyclic guanosine monophosphate (cGMP), Ras, Rap1, Ran, and protein kinases and phosphatases, as well as for measurement of membrane potential. Attaching topogenic motifs to these reporter constructions further permits localized determination of changes in the concentrations of the sensed molecules and ions. Intermolecular FRET measurements using FP pairs have also been employed for measurement of cyclic adenosine monophosphate (cAMP) levels and for detection of protein–protein interactions. For a comprehensive review of intermolecular and intramolecular FP FRET, see Zhang *et al.* (2002). Reports are also emerging of the measurement of protein–protein interactions using reconstitution of complementary fragments of GFP and YFP (Ghosh *et al.*, 2000; Hu *et al.*, 2002).

Developments of particular relevance to flow cytometry is the modification of GFP to attach specific binding sites, thereby permitting its direct use as a fluorescent ligand. This has been done through fusing of single-chain Fv proteins (Hink *et al.*, 2000) or through direct attachment of loops derived from the complementarity determining region 3 of the antibody heavy chain, followed by selection using phage display (Zeytun *et al.*, 2003).

F. Evolutionary Aspects of Fluorescent Proteins

With the identification of increasing numbers of GFP-like proteins within a variety of species, comparative sequence analyses can be used to examine evolutionary relationships and their ecological relevance (Labas *et al.*, 2002; Shagin *et al.*, 2003). Analyses of this type using all available GFP-like proteins, including the nonfluorescent chromoproteins of non-bioluminescent anthozoans, GFP-like proteins from anthozoan medusae and planktonic *Copepoda*, and colorless extracellular proteins of Bilaterians containing GFP-like protein-binding domains (Shagin *et al.*, 2003), suggest that GFP-like proteins comprise a large superfamily whose molecular function primarily revolves around pigmentation. The biological relevance of pigmentation and its evolutionary significance remain unclear. The superfamily can be divided into two lineages: one containing all FPs and the other leading to the colorless extracellular proteins. The phylogenetic analysis is consistent with evolution of FPs before the separation of *Cnidaria* and *Bilatera*. Within the *Cnidaria*, an extensive series of gene duplications then led to the appearance of a large and diverse set of FP paralogs, which underwent extensive evolutionary change during separation of Anthozoan and Hydrozoan subclasses. This gave rise, from the ancestral GFP, to the cyans, reds, and yellows, with at least 15 color diversification events being recognized (Shagin *et al.*, 2003). It is relevant for protein engineering to consider that different organisms have evolved GFP-like proteins of the same color, but that are based on different structural principles, because these different structures might provide suitable starting points for production of FPs having desired biotechnological characteristics (Shagin *et al.*, 2003).

G. Commercial Availability of Fluorescent Proteins

Three companies are sources of most of the anthozoan and hydrozoan FPs. Table I provides the optical properties of these FPs.

1. *Becton Dickinson/Clontech* (www.clontech.com): A large number of vectors encoding various FPs are available from this company (www.bdbiosciences.com/clontech/gfp/index.shtml for details). These include *Aequorea* GFP and the *Aequorea* BFP, CFP, and YFP variants, and a number of anthozoan FPs. DsRed1

Table I
Optical Properties of Commercially Available Fluorescent Proteins

Name	Excitation maximum (nm)	Emission maximum (nm)	Quantum yield	Extinction coefficient ($M^{-1}cm^{-1}$)	Comments	Reference
CopGFP	482	502	0.60	70,000		www.evrogen.com
PhyYFP	525	537	0.40	130,000		
Kindling Red	580	600	0.07	59,000	Activated by green illumination	
HcRed tandem	590	637	0.04	160,000		
EBFP	383	445	0.25	31,000		Patterson *et al.* (2001)
ECFP	434	477	0.40	26,000		
EGFP	489	508	0.60	55,000		
EYFP	514	527	0.61	84,000		
dsRed	558	583	0.68	72,500		
AmCyan1	458	489	0.75	39,000		www.bdbiosciences.com/clontech/products/literature/pdf/brochures/LivingColors.pdf; Clontech technical services
ZsGreen1	493	505	0.91	43,000		
ZsYellow1	529	539	0.65	20,000		
DsRed-Express	557	579	0.90	19,000		
DsRed2	563	582	NA	NA		
AsRed2	576	592	0.21	61,000		
HcRed1	588	618	0.03	20,000		
ECFP	439	476	0.40	26,000		
EGFP	484	510	0.60	55,000		
EYFP	512	529	0.60	84,000		
hrGFP	498	509	0.8	270,000		Ward and Cormier (1979)

and DsRed2 encode the red fluorescent protein (RFP) from the IndoPacific sea anemone relative, *Discosoma* species. A variant of this, pTimer, exhibits the previously described time-dependent changes in fluorescent emission from green to red (Terskikh *et al.*, 2000). AmCyan encodes a variant of the wild-type *Anemonia majano* (Aiptasia anemone) CFP, AsRed encodes a variant of the *Anemonia sulcata* RFP, ZsGreen encodes a variant of the wild-type *Zoanthus* species GFP, ZsYellow encodes a variant of the wild-type *Zoanthus* species YFP. Finally, HcRed was generated by mutagenesis of a nonfluorescent chromoprotein from *Heteractis crispa*. All variants are available in codon-optimized form for expression in mammalian systems, and a variety of different constructions are available for purposes such as targeting the proteins to different subcellular locations, providing bifunctional reporters, retroviral expression, and so on. The company is also a source of various anti-FP monoclonal and polyclonal antibodies.

2. *Stratagene* (www.stratagene.com): Stratagene provides plasmids based around the coding sequence of the *Renilla reniformis* GFP. It has been suggested that expression of Renilla GFP in mammalian cells may be less toxic than that of *Aequorea* GFP (www.stratagene.com; however, see Kirsch *et al.*, 2003).

3. *Evrogen* (www.evrogen.com): Evrogen provides plasmids encoding HcRed and a variety of other novel cnidarian FPs. Cop-green is a GFP-like FP from the copepod *Pntelina plumata* (phylum Arthropoda, subphylum Crustacea, class Maxillopoda, subclass Copepoda, order Calanoida, family Pontellidae). Phi yellow is a GFP-like FP from the hydromedusa *Phialidium* species (class Hydrozoa, order Hydroida, suborder Leptomeduzae, family Campanulariidae). Kindling Red is derived from a mutant of the chromoprotein isolated from *Anemonia sulcata* (Chudakov *et al.*, 2003a,b; Lukyanov *et al.*, 2000). Ase green is a fluorescent mutant of a naturally nonfluorescent GFP-like protein from *A. coerulescens* (Gurskaya *et al.*, 2003).

II. Flow Analysis Using Fluorescent Proteins

In this chapter, the discussion is only of expression of FPs within multicellular eukaryotes. Reviews have been published by Galbraith *et al.* (1998, 1999) and Hawley *et al.* (2001a).

A. Flow Analysis of Fluorescent Protein–Expressing Cells

A prerequisite to flow analysis is the ability to express sufficient quantities of FPs within the cell type of interest for the cells to be detectable using standard flow instrumentation. Various methods exist for transgenic expression in eukaryotic organisms. These can be roughly divided according to whether they confer transient or permanent expression, and further specification of the latter includes whether this involves homologous gene replacement. Expression of FPs evidently should ideally not perturb normal cellular functions or those functions and pathways under specific investigation. Reports concerning the effects of FP expression are contradictory; on the one hand, many of these describe the production of viable transgenic organisms expressing FP levels that are readily detected even by the eye, some of which have penetrated the nonscientific marketplace (www.ekac.org/gfpbunny.html; www.glofish.com/). Therefore, FP expression clearly is not grossly deleterious to living organisms (and obviously not within their source organisms). On the other hand, phenotypic effects of FP expression are sometimes noted (Gong *et al.*, 2003; Kirsch *et al.*, 2003; Liu *et al.*, 1999; Torbett, 2002). The availability of genome-wide methods for analysis of gene expression, such as microarrays, should permit detection of subtle alterations in gene expression that are FP specific. It is recommended that any studies involving microarrays include adequate replication to allow determination of statistical significance. Regarding potential variables, excessive overexpression of any protein (FP or otherwise, and especially one that has the potential to oligomerize or precipitate) can be toxic to the organism of interest, and the transgenic effects of FPs may also be a function of the ambient level of illumination of the transgenic organisms, a parameter not routinely (at least for animals) reported.

Shortly following the first reports of the expression of GFP in eukaryotic cells, methods were devised and described for flow cytometric detection of GFP in plants and animals (Anderson *et al.*, 1996; Galbraith *et al.*, 1995; Lybarger *et al.*, 1996; Ropp *et al.*, 1996; Sheen *et al.*, 1995). As with all flow cytometric procedures, single-cell suspensions must be employed. In the latter case, this involves production of wall-less cells (protoplasts) via enzymatic digestion (Galbraith *et al.*, 1999).

Most flow cytometers are equipped with single lasers producing excitation light at 488 nm, which is sufficient for detecting expression of wild-type GFP within transfected protoplasts (Galbraith *et al.*, 1995; Sheen *et al.*, 1995). Introducing mutations into the GFP coding sequence to eliminate either of its two peaks of absorbance allows selective fluorescence excitation at 409 nm (GFP containing S202F, T203I, and V163A replacements within its coding sequence) or at 488 nm (GFP containing S65T and V163A replacements) (Anderson *et al.*, 1998). The former is particularly suited for illumination using the 407-nm line produced by the krypton-ion laser or by the newer low-power violet laser diodes. Lybarger *et al.* (1996) described conditions for flow analysis of transfected 3T3 NIH cells expressing GFP, using laser excitation at 488 nm. Ropp *et al.* (1996) reported analysis of GFP- and YFP-expressing 293 cells using excitation at 407 nm. Beavis and Kalejta (1999) extended this work through evaluation of different single-laser wavelengths for discrimination of mammalian cells expressing GFP, YFP, and CFP. They found that excitation at 458 nm was suitable for any combination of these FPs, with electronic compensation adequately eliminating spectral overlap. Excitation of YFP at either 407 or 457 nm is certainly nonoptimal, based on the absorbance spectrum, and thus may be appropriate only for cells expressing high levels of this protein. Zhu *et al.* (1999) described the use of two-line excitation from an argon-ion laser (360 and 488 nm) for two-color (BFP/YFP) and three-color (BFP/GFP/YFP) flow analysis of transfected mammalian cells.

A disadvantage of excitation at 457 nm is that it requires an expensive tunable argon-ion laser. In contrast, argon-ion lasers producing single-wavelength (488-nm) light are relatively cheap. Stull *et al.* (2000) illustrated use of 488-nm illumination for the combined immunophenotypic discrimination of transfected human CD34$^+$ mononuclear cells expressing GFP and YFP. Hawley *et al.* (2001b) have described conditions for the analysis of three FPs using 488-nm excitation. In this work, discrimination was possible between fluorescent signals of DsRed, YFP, and GFP produced by stably transformed mammalian cell lines. Multilaser analysis of FP expression will be facilitated by the availability of solid-state lasers, which provide cost-conscious excitation at appropriate wavelengths. Telford *et al.* (2003) have demonstrated the suitability of a violet laser diode for flow analysis of cells expressing CFP. Table II provides a summary of this information.

Given the increasing availability of routine methods for flow analysis of FP expression of animal cells, numerous reports are now emerging concerning their use in specialized assays. Selected examples cover all aspects of biology, for example, including analysis of steroid-regulated gene expression (Necela *et al.*, 2003), of B-cell differentiation (Guglielmi *et al.*, 2003), cancer metastasis (Zhang *et al.*, 2003), RNA interference (Mousses *et al.*, 2003; Nagy *et al.*, 2003), mechanisms of viral replication (Voronin and Pak, 2003), parental imprinting (Preis *et al.*, 2003), manipulation of Cre-Lox recombination (Van den Plas *et al.*, 2003),

Table II
Laser Wavelengths Employed for Analysis of Fluorescent Proteins within Mammalian Cells

Laser excitation wavelengths (nm)	Fluorescent proteins analyzed	Comments	Reference
488, 407	GFP		Anderson *et al.*, 1996
488, 407	GFP, YFP		Ropp *et al.*, 1996
488	GFP		Lybarger *et al.*, 1996
360, 488	BFP, GFP		Yang *et al.*, 1998
488	GFP, YFP		Lybarger *et al.*, 1998
360, 488	BFP, GFP, YFP		Zhu *et al.*, 1999
458, 488, 514	GFP, YFP, CFP		Beavis and Kalejta, 1999
488	GFP, YFP		Stull *et al.*, 2000
458, 488, 568	YFP, GFP, CFP, DsRed		Hawley *et al.*, 2001b
408	CFP	Violet laser diode	Telford *et al.*, 2003
413, 514	CFP, YFP	FRET	Chan *et al.*, 2001
458	CFP, YFP	FRET	He *et al.*, 2003

and measurement of the extent of syncytium formation of placental trophoblast cells (Kudo *et al.*, 2003).

Progress in higher plants has included the quantitative analysis of abscisic acid–inducible gene expression (Hagenbeek and Rock, 2001), extension of multiparametric and clustering analyses (Bohanec *et al.*, 2002), and the global characterization of cell-type–specific gene expression (Birnbaum *et al.*, 2003) (see later discussion). Beyond mammals and plants, FP expression in other cell types and its detection by flow cytometry has been reported for fish (Kobayashi *et al.*, 2004; Molina *et al.*, 2002; Takeuchi *et al.*, 2002), and *Drosophila* (Banks *et al.*, 2003).

For certain routine flow cytometric procedures (e.g., analysis of the cell cycle in mammalian systems), the cells are fixed before analysis. This creates problems for concurrent FP analysis because it has been reported that the fluorescence of most FPs is abolished by precipitative fixatives containing organic solvents (Tsien, 1998; Ward *et al.*, 1980). Kalejta *et al.* (1997, 1999) have described the use of membrane-associated GFP fusions for simultaneous analysis of transfected cells and the cell cycle in ethanol-fixed cells. Chu *et al.* (1999) described the use of formaldehyde fixation before ethanol fixation as a means of retaining cytoplasmic GFP fluorescence within transfected leukemic cells. It is evidently difficult to reconcile these reports with those concerning the structural instability of GFP in the presence of ethanol, although some FPs may be resistant to its denaturing effects (Yu *et al.*, 2003b).

Overall, flow cytometric analysis of FP-expressing cells is clearly now largely routine. As such, the general rules for detection of cell-associated fluorescence apply: Laser-excitation wavelengths should be matched to the excitation spectra of the FPs of interest, dichroic and barrier filters should be selected (custom-made if necessary) to optimally discriminate between the different emission FP spectra, and consideration should be given to the relative proportions of the different fluorescent proteins that are likely to be produced and their comparative brightness, to roughly balance the signals being produced and analyzed during flow cytometry.

B. Flow Analysis of FRET

Chan *et al.* (2001) established the feasibility of employing flow cytometric FRET analysis to characterize interactions between tumor necrosis factor (TNF) family members and between receptor and downstream signaling proteins. This involved fusing the interacting proteins to CFP and YFP and employing two-laser illumination at 413 and 514 nm. He *et al.* (2003) have extended this work by developing a flow cytometric method to detect FRET between CFP and YFP using the 458-nm excitation from a single tunable argon-ion laser.

C. Flow Analysis of Fluorescent Protein–Accumulating Organelles

FP-based flow analysis can also be applied to subcellular organelles, although in general reports of flow cytometric measurements of organelles are uncommon relative to those of cells. We have employed nuclear targeting of FPs to highlight plant nuclei within protoplasts and transgenic plants of tobacco (Grebenok *et al.*, 1997a,b) and have successfully used flow cytometry for the analysis of GFP accumulation within these nuclei in subcellular homogenates. In this work, we targeted the nucleoplasm using a nuclear localization signal, which required increasing the size of the GFP-fusion protein beyond the exclusion limit of the nuclear pores by further fusing it to the β-glucuronidase coding sequence. When applied to *Arabidopsis thaliana*, which has a nuclear genome approximately one-thirtieth times that of tobacco, a technical difficulty is that this composite molecule leaks from the nuclei in homogenates and this impedes flow cytometric analysis. FP fusions to proteins that are structural components of the nucleus avoid this problem (Zhang and Galbraith, 2003, unpublished observation).

Sirk *et al.* (2003) have described a novel assay for measuring the import and turnover of nuclear-encoded mitochondrial proteins in living mammalian cells. This involved production of transgenic cells within which the expression of a mitochondrially targeted form of GFP was placed under the regulation of an inducible promoter. This GFP fusion has the additional feature of being non-fluorescent before import into the mitochondria because of the presence of a signal peptide. This means that flow analysis of the extent of cellular GFP fluorescence emission, with PI gating to eliminate the contribution of dead cells, provides a means of quantifying mitochondrial accumulation of this protein within living cells.

III. Flow Sorting Using Fluorescent Proteins

Given the ability to successfully discriminate FP-labeled cells or subcellular organelles using flow cytometry, we can then use flow sorting to selectively purify these components for further study. Sorting of mammalian cells according to the presence of detectable fluorescent signals is evidently routine, and signals from FP expression create no exceptions. Previous reviews include Galbraith *et al.* (1998, 1999) and more generally Boeck (2001).

Examples of applications of GFP sorting in mammalian systems can be found in Van Tendeloo *et al.* (2000), who described the use of GFP expression as an alternative to co-selection methods for production of stably transfected hematopoietic cell lines. De Angelis *et al.* (2003) employed flow sorting of GFP-labeled cells for isolation of myogenic progenitor cells. Similar approaches have been used for isolation of neural precursors (Aubert *et al.*, 2003) and pluripotential progenitor cells (Nunes *et al.*, 2003), hepatic progenitor cells (Fujikawa *et al.*, 2003), hematopoietic stem cells (Ma *et al.* 2002), and primary aortic endothelial cells (Magid *et al.*, 2003). Zharkikh *et al.* (2002) have reported the use of flow sorting of GFP-labeled cells for collection of different kidney cell types, and Hirai *et al.* (2003) for purification of hemogenic and non-hemogenic endothelial cells. Other applications include the generation of cell lines containing inducible promoter systems (Lai *et al.*, 2003).

In higher plants, Birnbaum *et al.* (2003) have described the use of flow cytometry and sorting for the global characterization of cell-type–specific gene expression in Arabidopsis roots. This involved use of transgenic plants in which cell-type–specific expression of GFP within the root was under the control of specific promoter or enhancer sequences. Protoplasts prepared from the different transgenic plants were sorted according to GFP fluorescence. RNA was then prepared from the sorted protoplasts and employed for hybridization to Affymetrix whole-genome microarrays. Expression data were also obtained from root segments corresponding to different distances from the root tip. Combining the results allowed creation of a road map of global gene expression of the entire root. Within this road map, it was possible to assign much of the changes in gene expression to eight characteristic local expression domains. Identification of the presence of specific genes within these domains allows hypothesis generation with respect to root development, specifically concerning the involvement of phytohormones and the roles of transcription factor networks.

Reports of flow sorting based on FP expression in other organisms have appeared. Takeuchi *et al.* (2002) and Kobayashi *et al.* (2004) have described the development of methods for isolation of primordial germ cells from rainbow trout. Finally, Makridou *et al.* (2003) described the use of sorting based on GFP and RFP expression for the production of stably transformed cell lines in *Drosophila*.

IV. Conclusion

Is there a pot of gold at the end of the rainbow of fluorescent colors? Most certainly for the investigator; as scientists, we have only just begun to employ FP expression as a means of defining and exploring the properties of living cells and their subcellular components. The tools of flow cytometry and fluorescence-activated sorting provide unique ways to investigate the contributions of individuals within populations. Further work is needed to evaluate the greatly expanded numbers of FPs for their suitability in transient and transgenic expression. Continued expansion of the numbers of different FPs is, of course, anticipated, and these in turn will require evaluation. The emergence of novel flow-based assays for different cellular functions and components will continue, and improved sensitivity of end-point measurements should permit lowering the total numbers of cells or subcellular organelles required to be purified by flow sorting. Finally, further integration of these techniques is anticipated with established and emerging high-throughput methods, including genomics, proteomics, and metabolomics.

Acknowledgments

Some of the work described in this chapter has been supported by grants to D. G. from the National Science Foundation, primarily from the Plant Genome Program, and from the U.S. Department of Agriculture.

References

Allen, G. J., Kwak, J. M., Chu, S. P., Llopis, J., Tsien, R. Y., Harper, J. F., and Schroeder, J. I. (1999). Cameleon calcium indicator reports cytoplasmic calcium dynamics in Arabidopsis guard cells. *Plant J.* **19,** 735–747.

Allen, G. J., Chu, S. P., Schumacher, K., Shimazaki, C. T., Vafeados, D., Kemper, A., Hawke, S. D., Tallman, G., Tsien, R. Y., Harper, J. F., Chory, J., and Schroeder, J. I. (2000). Alteration of stimulus-specific guard cell calcium oscillations and stomatal closing in *Arabidopsis det3* mutant. *Science* **289,** 2338–2342.

Allen, G. J., Chu, S. P., Harrington, C. L., Schumacher, K., Hoffman, T., Tang, Y. Y., Grill, E., and Schroeder, J. I. (2001). A defined range of guard cell calcium oscillation parameters encodes stomatal movements. *Nature* **411,** 1053–1057.

Anderson, M. T., Baumgarth, N., Haugland, R. P., Gerstein, R. M., Tjioe, T., Herzenberg, L. A., and Herzenberg, L. A. (1998). Pairs of violet-light–excited fluorochromes for flow cytometric analysis. *Cytometry* **33,** 435–444.

Anderson, M. T., Tjioe, I. M., Lorincz, M. C., Parks, D. R., Herzenberg, L. A., Nolan, G. P., and Herzenberg, L. E. (1996). Simultaneous fluorescence-activated cell sorter analysis of two distinct transcriptional elements within a single cell using engineered green fluorescent proteins. *Proc. Natl. Acad. Sci. USA* **93,** 8508–8511.

Ando, R., Hama, H., Yamamoto-Hino, M., Mizuno, H., and Miyawaki, A. (2002). An optical marker based on the UV-induced green-to-red photoconversion of a fluorescent protein. *Proc. Natl. Acad. Sci. USA* **99,** 12651–12656.

Aubert, J., Stavridis, M. P., Tweedie, S., O'Reilly, M., Vierlinger, K., Li, M., Ghazal, P., Pratt, T., Mason, J. O., Roy, D., and Smith, A. (2003). Screening for mammalian neural genes via fluorescence-activated cell sorter purification of neural precursors from Sox1-gfp knock-in mice. *Proc. Natl. Acad. Sci. USA* **100**(Suppl. 1), 11836–11841.

Baird, G. S., Zacharias, D. A., and Tsien, R. Y. (1999). Circular permutation and receptor insertion within green fluorescent proteins. *Proc. Natl. Acad. Sci. USA* **96,** 11241–11246.

Banks, D. J., Hua, G., and Adang, M. (2003). Cloning of a *Heliothis virescens* 110 kDa aminopeptidase N and expression in *Drosophila* S2 cells. *Insect Biochem. Mol. Biol.* **33,** 499–508.

Beavis, A. J., and Kalejta, R. F. (1999). Simultaneous analysis of the cyan, yellow and green fluorescent proteins by flow cytometry using single-laser excitation at 458 nm. *Cytometry* **37,** 68–73.

Birnbaum, K., Shasha, D. E., Wang, J. Y., Jung, J. W., Lambert, G. M., Galbraith, D. W., and Benfey, P. N. (2003). A gene expression map of the Arabidopsis root. *Science* **302,** 1956–1960.

Boeck, G. (2001). Current status of flow cytometry in cell and molecular biology. *Int. Rev. Cytol.* **204,** 239–298.

Bohanec, B., Luthar, Z., and Rudolf, K. (2002). A protocol for quantitative analysis of green fluorescent protein–transformed plants, using multiparameter flow cytometry with cluster analysis. *Acta Biol. Cracov. Ser. Bot.* **44,** 145–153.

Chalfie, M., Tu, Y., Euskirchen, G., Ward, W. W., and Prasher, D. C. (1994). Green-fluorescent protein as a marker for gene expression. *Science* **263,** 802–805.

Campbell, R. E., Tour, O., Palmer, A. E., Steinbach, P. A., Baird, G. S., Zacharias, D. A., and Tsien, R. Y. (2002). A monomeric red fluorescent protein. *Proc. Natl. Acad. Sci. USA* **99,** 7877–7882.

Chan, F. K., Siegel, R. M., Zacharias, D., Swofford, R., Holmes, K. L., Tsien, R. Y., and Lenardo, M. J. (2001). Fluorescence resonance energy transfer analysis of cell surface receptor interactions and signaling using spectral variants of the green fluorescent protein. *Cytometry* **44,** 361–368.

Chu, Y.-W., Wang, R., Schmid, I., and Sakamoto, K. M. (1999). Analysis with flow cytometry of green fluorescent protein expression in leukemic cells. *Cytometry* **36,** 333–339.

Chudakov, D. M., Feofanov, A. V., Mudrik, N. N., Lukyanov, S., and Lukyanov, K. A. (2003a). Chromophore environment provides clue to "kindling fluorescent protein" riddle. *J. Biol. Chem.* **278,** 7215–7219.

Chudakov, D. M., Belousov, V. V., Zaraisky, A. G., Novoselov, V. V., Staroverov, D. B., Zorov, D. B., Lukyanov, S., and Lukyanov, K. A. (2003b). Kindling fluorescent proteins—a novel tool for precise *in vivo* photolabeling. *Nature Biotech.* **21,** 191–194.

De Angelis, M. G. C., Balconi, G., Bernasconi, S., Zanetta, L., Boratto, R., Galli, D., Dejana, E., and Cossu, G. (2003). Skeletal myogenic progenitors in the endothelium of lung and yolk sac. *Exptl. Cell Res.* **290,** 207–216.

Fujikawa, T., Hirose, T., Fujii, H., Oe, S., Yasuchika, K., Azuma, H., and Yamaoka, Y. (2003). Purification of adult hepatic progenitor cells using green fluorescent protein (GFP)–transgenic mice and fluorescence-activated cell sorting. *J. Hepatol.* **39,** 162–170.

Galbraith, D. W., Harkins, K. R., and Knapp, S. (1991). Systemic endopolyploidy in *Arabidopsis thaliana. Plant Physiol.* **96,** 985–989.

Galbraith, D. W., Grebenok, R. J., Lambert, G. M., and Sheen, J. (1995). Flow cytometric analysis of transgene expression in higher plants: Green Fluorescent Protein. *Methods Cell Biol.* **50,** 3–12.

Galbraith, D. W., Anderson, M. T., and Herzenberg, L. A. (1998). Flow cytometric analysis and sorting of cells based on GFP accumulation. *Methods Cell Biol.* **58,** 315–341.

Galbraith, D. W., Herzenberg, L. A., and Anderson, M. (1999). Flow cytometric analysis of transgene expression in higher plants: Green fluorescent protein. *Methods Enzymol.* **320,** 296–315.

Ghosh, I. Hamilton, A. D., and Regan, L. (2000). Antiparallel leucine zipper–directed protein reassembly: Application to the green fluorescent protein. *J. Am. Chem. Soc.* **122,** 5658–5659.

Gong, Z., Wan, H., Tay, T. L., Wang, H., Chen, M., and Tan, Y. (2003). Development of transgenic fish for ornamental and bioreactor by strong expression of fluorescent proteins in the skeletal muscle. *Biochem. Biophys. Res. Commun.* **308,** 58–63.

Grebenok, R. J., Pierson, E. A., Lambert, G. M., Gong, F.-C., Afonso, C. L., Haldeman-Cahill, R., Carrington, J. C., and Galbraith, D. W. (1997a). Green-fluorescent protein fusions for efficient characterization of nuclear localization signals. *Plant J.* **11,** 573–586.

Grebenok, R. J., Lambert, G. M., and Galbraith, D. W. (1997b). Characterization of the targeted nuclear accumulation of GFP within the cells of transgenic plants. *Plant J.* **12,** 685–696.

Griesbeck, O., Baird, G. S., Campbell, R. E., Zacharias, D. A., and Tsien, R. Y. (2001). Reducing the environmental sensitivity of yellow fluorescent protein—mechanism and applications. *J. Biol. Chem.* **276,** 29188–29194.

Guglielmi, L., Le Bert, M., Comte, I., Dessain, M. L., Drouet, M., Ayer-Le Lievre, C., Cogne, M., and Denizot, Y. (2003). Combination of 3′ and 5′ IgH regulatory elements mimics the B-specific endogenous expression pattern of IgH genes from pro-B cells to mature B cells in a transgenic mouse model. *Biochim. Biophys. Acta Mol. Cell Res.* **1642,** 181–190.

Gurskaya, N. G., Fradkov, A. F., Pounkova, N. I., Staroverov, D. B., Bulina, M. E, Yanushevich, Y. G., Labas, Y. A., and Lukyanov, S. and Lukyanov, K. A. (2003). A colorless GFP homologue from the non-fluorescent hydromedusa *Aequorea coerulescens* and its fluorescent mutants. *Biochem. J.* **373,** 403–408.

Hagenbeek, D., and Rock, C. D. (2001). Quantitative analysis by flow cytometry of abscisic acid-inducible gene expression in transiently transformed rice protoplasts. *Cytometry* **45,** 170–179.

Haraguchi, T., Koujin, T., and Hiraoka, Y. (2000). Application of GFP: Time-lapse multi-wavelength fluorescence imaging of living mammalian cells. *Acta. Histochem. Cytochem.* **33,** 169–175.

Haseloff, J., Siemering, K. R., Prasher, D. C., and Hodge, S. (1997). Removal of a cryptic intron and subcellular localization of green fluorescent protein are required to mark transgenic *Arabidopsis* plants brightly. *Proc. Natl. Acad. Sci. USA* **94,** 2122–2127.

Hawley, T. S., Telford, W. G., and Hawley, R. G. (2001a). "Rainbow" reporters for multispectral marking and lineage analysis of hematopoietic stem cells. *Stem Cells* **19,** 18–124.

Hawley, T. S., Telford, W. G., Ramezani, A., and Hawley, R. G. (2001b). Four-color flow cytometric detection of retrovirally expressed red, yellow, green, and cyan fluorescent proteins. *Biotechniques* **30,** 1028–1034.

He, L. S., Bradrick, T. D., Karpova, T. S., Wu, X. L., Fox, M. H., Fischer, R., McNally, J. G., Knutson, J. R., Grammer, A. C., and Lipsky, P. E. (2003). Flow cytometric measurement of fluorescence (Förster) resonance energy transfer from cyan fluorescent protein to yellow fluorescent protein using single-laser excitation at 458 nm. *Cytometry* **53A,** 39–54.

Heim, R., Prasher, D. C., and Tsien, R. Y. (1994). Wavelength mutations and posttranslational autoxidation of Green Fluorescent Protein. *Proc. Natl. Acad. Sci. USA* **91,** 12501–12504.

Heim, R., and Tsien, R. Y. (1996). Engineering green fluorescent protein for improved brightness, longer wavelengths and fluorescence resonance energy transfer. *Curr. Biol.* **6,** 178–182.

Hink, M. A., Griep, R. A., Borst, J. W., van Hoek, A., Eppink, M. H. M., Schots, A., and Visser, A. J. W. G. (2000). Structural dynamics of green fluorescent protein alone and fused with a single chain Fv protein. *J. Biol. Chem.* **275,** 17556–17560.

Hirai, H., Ogawa, M., Suzuki, N., Yamamoto, M., Breier, G., Mazda, O., Imanishi, J., and Nishikawa, S. (2003). Hemogenic and nonhemogenic endothelium can be distinguished by the activity of fetal liver kinase (Flk)-1 promoter/enhancer during mouse embryogenesis. *Blood* **101,** 886–893.

Hu, C.-D., Chinenov, Y., and Kerppola, T. K. (2002). Visualization of interactions among bZIP and Rel family proteins in living using bimolecular fluorescence complementation. *Mol. Cell* **9,** 789–798.

Johnson, F. H., Gershman, L. C., Waters, J. R., Reynolds, G. T., Saiga, Y., and Shimomura, O. (1962). Quantum efficiency of cypridina luminescence, with a note on that of *Aequorea*. *J. Cell Comp. Physiol.* **60,** 85–104.

Kalejta, R. F., Shenk, T., and Beavis, A. J. (1997). Use of a membrane-localized green fluorescent protein allows simultaneous identification of transfected cells and cell cycle analysis by flow cytometry. *Cytometry* **29,** 286–291.

Kalejta, R. F., Brideau, A. D., Banfield, B. W., and Beavis, A. J. (1999). An integral membrane green fluorescent protein marker, Us9-GFP, is quantitatively retained in cells during propidium iodide–based cell cycle analysis by flow cytometry. *Exptl. Cell Res.* **248,** 322–328.

Kirsch, P., Hafner, M., Zentgraf, H., and Schilling, L. (2003). Time course of fluorescence intensity and protein expression in HeLa cells stably transfected with hrGFP. *Mol. Cells* **15,** 341–348.

Kobayashi, T., Yoshizaki, G., Takeuchi, Y., and Takeuchi, T. (2004). Isolation of highly pure and viable primordial germ cells from rainbow trout by GFP-dependent flow cytometry. *Mol. Reprod. Dev.* **67,** 91–100.

Kudo, Y., Boyd, C. A. R., Kimura, H., Cook, P. R., Redman, C. W. G., and Sargent, I. L. (2003). Quantifying the syncytialisation of human placental trophoblast BeWo cells grown *in vitro*. *Biochim. Biophys. Acta. Mol. Cell Res.* **1640,** 25–31.

Labas, Y. A., Gurskaya, N. G., Yanushevich, Y. G., Fradkov, A. F., Lukyanov, K. A., Lukyanov, S. A., and Matz, M. V. (2002). Diversity and evolution of the green fluorescent protein family. *Proc. Natl. Acad. Sci. USA* **99,** 4256–4261.

Lai, J. F., Juang, S. H., Hung, Y. M., Cheng, H. Y., Cheng, T. L., Mostov, K. E., and Jou, T. S. (2003). An ecdysone and tetracycline dual regulatory expression system for studies on Rac1 small GTPase-mediated signaling. *Am. J. Physiol. Cell Physiol.* **285,** C711–C719.

Li, X., Zhao, X., Fang, Y., Jiang, X., Duong, T., Huang, C.-C., and Kain, S. R. (1998). Generation of destabilized enhanced green fluorescent protein as a transcription reporter. *J. Biol. Chem.* **273,** 34970–34975.

Lippincott-Schwartz, J., and Patterson, G. H. (2003). Development and use of Fluorescence Protein markers in living cells. *Science* **300,** 87–91.

Liu, H. S., Jan, M. S., Chou, C. K., Chen, P. H., and Ke, N. J. (1999). Is Green Fluorescent Protein toxic to the living cell? *Biochem. Biophys. Res. Commun.* **260,** 712–717.

Lukyanov, K. A., Fradkov, A. F., Gurskaya, N. G., Matz, M. V., Labas, Y. A., Savitsky, A. P., Markelov, M. L., Zaraisky, A. G., Zhao, X., Fang, Y., Tan, W., and Lukyanov, S. A. (2000). Natural animal coloration can be determined by a non-fluorescent GFP homolog. *J. Biol. Chem.* **275,** 25879–25882.

Lybarger, L., Dempsey, D., Franek, K. J., and Chervenak, R. (1996). Rapid generation and flow cytometric analysis of stable GFP-expressing cells. *Cytometry* **25,** 211–220.

Ma, X. Q., Robin, C., Ottersbach, K., and Dzierzak, E. (2002). The Ly-6A (Sca-1) GFP transgene is expressed in all adult mouse hematopoietic stem cells. *Stem Cells* **20,** 514–521.

Magid, R., Martinson, D., Hwang, J., Jo, H., and Galis, Z. S. (2003). Optimization of isolation and functional characterization of primary murine aortic endothelial cells. *Endothel. NY* **10,** 103–109.

Makridou, P., Burnett, C., Landy, T., and Howard, K. (2003). Hygromycin B–selected cell lines from GAL4-regulated pUAST constructs. *Genesis* **36,** 83–87.

Matz, M. V., Fradkov, A. F., Labas, Y. A., Savitsky, A. P., Zaraisky, A. G., Markelov, M. L., and Lukyanov, S. A. (1999). Fluorescent proteins from nonbioluminescent Anthozoa species. *Nat. Biotech.* **17,** 969–973.

Mills, C. E. (1999–present). Bioluminescence of *Aequorea*, a hydromedusa. Available at: http://faculty.washington.edu/cemills/Aequorea.html.

Miyawaki, A., Llopis, J., Heim, R., McCaffery, J. M., Adams, J. A., Ikura, M., and Tsien, R. Y. (1997). Fluorescent indicators for Ca^{2+} based on green fluorescent protein and calmodulin. *Nature* **388,** 882–887.

Miyawaki, A., Griesbeck, O., Heim, R., and Tsien, R. Y. (1999). Dynamic and quantitative Ca^{2+} measurements using improved cameleons. *Proc. Natl. Acad. Sci. USA* **96,** 2135–2140.

Miyawaki, A., Sawano, A., and Kogure, T. (2003). Lighting up cells: Labeling proteins with fluorophores. *Nat. Cell Biol.* **5,** S1–S7.

Molina, A., Carpeaux, R., Martial, J. A., and Muller, M. (2002). A transformed fish cell line expressing a green fluorescent protein-luciferase fusion gene responding to cellular stress. *Toxicol. In Vitro* **16,** 201–207.

Mousses, S., Caplen, N. J., Cornelison, R., Weaver, D., Basik, M., Hautaniemi, S., Elkahloun, A. G., Lotufo, R. A., Choudary, A., Dougherty, E. R., Suh, E., and Kallioniemi, O. (2003). RNAi microarray analysis in cultured mammalian cells. *Genome Res.* **13,** 2341–2347.

Nagy, P., Arndt-Jovin, D. J., and Jovin, T. M. (2003). Small interfering RNAs suppress the expression of endogenous and GFP-fused epidermal growth factor receptor (erbB1) and induce apoptosis in erbB1-overexpressing cells. *Exptl. Cell Res.* **285,** 39–49.

Necela, B. M., and Cidlowski, J. A. (2003). Development of a flow cytometric assay to study glucocorticoid receptor–mediated gene activation in living cells. *Steroids* **68,** 341–350.

Nunes, M. C., Roy, N. S., Keyoung, H. M., Goodman, R. R., McKhann, G., Jiang, L., Kang, J., Nedergaard, M., and Goldman, S. A. (2003). Identification and isolation of multipotential neural progenitor cells from the subcortical white matter of the adult human brain. *Nat. Med.* **9,** 439–447.

Ormö, M., Cubitt, A. B., Kallio, K., Gross, L. A., Tsien, R. Y., and Remington, S. J. (1996). Crystal structure of the *Aequorea victoria* Green Fluorescent Protein. *Science* **273,** 1392–1395.

Patterson, G., Day, R. N., and Piston, D. (2001). Fluorescent protein spectra. *J. Cell Sci.* **114,** 837–838.

Preis, J. I., Downes, M., Oates, N. A., Rasko, J. E. J., and Whitelaw, E. (2003). Sensitive flow cytometric analysis reveals a novel type of parent-of-origin effect in the mouse genome. *Curr. Biol.* **13,** 955–959.

Romoser, V. A., Hinkle, P. M., and Persechini, A. (1997). Detection in living cells of Ca^{2+}-dependent changes in the fluorescence emission of an indicator composed of two Green Fluorescent Protein variants linked by a calmodulin-binding sequence: A new class of fluorescent indicators. *J. Biol. Chem.* **272,** 13270–13274.

Ropp, J. D., Donahue, C. J., Wolfgang-Kimball, D., Hooley, J. J., Chin, J. Y. W., Cuthbertson, R. A., and Bauer, K. D. (1996). Aequorea green fluorescent protein: Simultaneous analysis of wild-type and blue-fluorescing mutant by flow cytometry. *Cytometry* **24,** 284–288.

Shagin, D. A., Barsova, E. V., Yanushevich, Y. G., Fradkov, A. F., Lukyanov, K. A., Labas, Y. A., Ugalde, J. A., Meyer, A., Nunes, J. M., Widder, E. A., Lukyanov, S. A., and Matz, M. V. (2004). GFP-like proteins as ubiquitous metazoan superfamily: Evolution of functional features and structural complexity. *Mol. Biol. Evol.* **21,** 841–850.

Sheen, J., Hwang, S., Niwa, Y., Kobayashi, H., and Galbraith, D. W. (1995). Green Fluorescent Protein as a new vital marker in plant cells. *Plant J.* **8,** 777–784.

Shimomura, O., Johnson, F. H., and Saiga, Y. (1962). Extraction, purification and properties of Aequorin, a bioluminescent protein from luminous hydromedusan. *Aequorea. J. Cell Comp. Physiol.* **59,** 223–239.

Sirk, D. P., Zhu, Z. P., Wadia, J. S., and Mills, L. R. (2003). Flow cytometry and GFP: A novel assay for measuring the import and turnover of nuclear-encoded mitochondrial proteins in live PC12 cells. *Cytometry* **56A,** 15–22.

Stull, R. A., Hyun, W. C., and Pallavicini, M. G. (2000). Simultaneous flow cytometric analysis of enhanced green and yellow fluorescent proteins and cell surface antigens in doubly transduced immature hematopoietic cell populations. *Cytometry* **40,** 126–134.

Takeuchi, Y., Yoshizaki, G., Kobayashi, T., and Takeuchi, T. (2002). Mass isolation of primordial germ cells from transgenic rainbow trout carrying the green fluorescent protein gene driven by the vasa gene promoter. *Biol. Reprod.* **67,** 1087–1092.

Telford, W. G., Hawley, T. S., and Hawley, R. G. (2003). Analysis of violet-excited fluorochromes by flow cytometry using a violet laser diode. *Cytometry* **54A,** 48–55.

Terskikh, A., Fradkov, A., Ermakova, G., Zaraisky, A., Tan, P., Kajava, A. V., Zhao, X., Lukyanov, S., Matz, M., Kim, S., Weissman, I., and Siebert, P. (2000). "Fluorescent timer": Protein that changes color with time. *Science* **290,** 1585–1588.

Torbett, B. E. (2002). Reporter genes: Too much of a good thing? *J. Gene Med.* **4,** 478–479.

Tsien, R. Y. (1998). The Green Fluorescent Protein. *Annu. Rev. Biochem.* **67,** 509–544.

Van Tendeloo, V. F. I., Ponsaerts, P., Van Broeckhoven, C., Berneman, Z. N., and Van Bockstaele, D. R. (2000). Efficient generation of stably electrotransfected human hematopoietic cell lines without drug selection by consecutive FACsorting. *Cytometry* **41,** 31–35.

Van den Plas, D., Ponsaerts, P., Van Tendeloo, V., Van Bockstaele, D. R., Berneman, Z. N., and Merregaert, J. (2003). Efficient removal of LoxP-flanked genes by electroporation of Cre-recombinase mRNA. *Biochem. Biophys. Res. Commun.* **305,** 10–15.

Voronin, Y. A., and Pathak, V. K. (2003). Frequent dual initiation of reverse transcription in murine leukemia virus–based vectors containing two primer-binding sites. *Virology* **312,** 281–294.

Ward, W. W., and Cormier, M. J. (1979). An energy transfer protein in coelenterate bioluminescence: Characterization of the *Renilla* green-fluorescent protein. *J. Biol. Chem.* **254,** 781–788.

Ward, W. W., Cody, C. W., Hart, R. C., and Cormier, M. J. (1980). Spectrophotometric identity of the energy-transfer chromophores in Renilla and Aequorea green-fluorescent proteins. *Photochem. Photobiol.* **31,** 611–615.

Yang, F., Moss, L. G., and Phillips, G. N., Jr. (1996). The molecular structure of Green Fluorescent Protein. *Nat. Biotechnol.* **14,** 1246–1251.

Yu, D. H., Baird, G. S., Tsien, R. Y., and Davis, R. L. (2003a). Detection of calcium transients in Drosophila mushroom body neurons with camgaroo reporters. *J. Neurosci.* **23,** 64–72.

Yu, Y. A., Oberg, K., Wang, G., and Szalay, A. A. (2003b). Visualization of molecular and cellular events with green fluorescent protein in developing embryos: A review. *Luminescence* **18,** 1–18.

Zharkikh, L., Zhu, X. H., Stricklett, P. K., Kohan, D. E., Chipman, G., Breton, S., Brown, D., and Nelson, R. D. (2002). Renal principal cell-specific expression of green fluorescent protein in transgenic mice. *Am. J. Physiol. Renal Physiol.* **283,** F1351–F1364.

Zeytun, A., Jeromin, A., Scalettar, B. A., Waldo, G. S., and Bradbury, A. R. M. (2003). Fluorobodies combine GFP fluorescence with the binding characteristics of antibodies. *Nat. Biotechnol.* **21,** 1473–1479.

Zhang, J. H., Tang, J., Wang, J., Ma, W. L., Zheng, W. L., Yoneda, T., and Chen, J. (2003). Over-expression of bone sialoprotein enhances bone metastasis of human breast cancer cells in a mouse model. *Intl. J. Oncol.* **23,** 1043–1048.

Zhu, J., Musco, M. L., and Grace, M. J. (1999). Three-color flow cytometry analysis of tricistronic expression of eBFDP, eGFP, and eYFP using EMCV-IRES linkages. *Cytometry* **37,** 51–59.

CHAPTER 7

Labeling Cellular Targets with Semiconductor Quantum Dot Conjugates

Xingyong Wu and Marcel P. Bruchez
Quantum Dot Corporation,
Hayward, California 94545

I. Introduction

Fluorescence is a well-established tool for examining biomolecular expression in cells using various detection platforms including light microscopes, flow and scanning cytometers, and microplate readers. Traditionally, detection ligands such as antibodies or oligonucleotides have been conjugated to fluorochromic organic dye molecules such as fluorescein, rhodamine, or cyanine (Cy) and Alexa dyes. As a scientific tool, fluorescence methods have provided the bulk of our understanding of modern cell biology. Despite the huge successes, current fluorochromes have limitations and require elaborate equipment and expertise. Another fluorochrome technology has been developed based on the extraordinary optical properties of nanometer (10^{-9} meter) scale semiconductor nanocrystalline particles, called *quantum dots* (QDs) (Alivisatos, 1996).

0091-679X/04 $35.00

As fluorescence labels, QDs have several key advantages over traditional organic fluorochromes. First, they absorb light with efficiencies that are orders of magnitude beyond organic dye molecules. Second, if properly engineered, they have an extremely high efficiency of converting excitation energy into emitted photons, that is, quantum yields approaching 100%. These two features make QDs much brighter than any organic fluorochromes. Third, a unique and beneficial feature of QDs is that they can be efficiently excited by any wavelength bluer than their emission. Furthermore, their emission spectra can be engineered by changing the size and/or the composition of the dots. Hence, their emission spectra can be designed or "tuned" during synthesis and typically are narrow and symmetrical. Thus, by exciting at a single blue/ultraviolet (UV) wavelength, all colors of QDs in a sample will be excited and will emit at a color related to its size and composition. This extraordinary feature enables multicolor labeling of cellular targets with minimal spectral overlap and with greatly simplified imaging systems. Finally, the fluorescence of properly engineered QDs is extremely stable so continuous long-term imaging or repeated measurements can be performed without loss of signal. Commercially available Qdot Conjugates have allowed several research groups to exploit these benefits in various immunofluorescence applications (Dahan *et al.*, 2003; Lidke *et al.*, 2004; Ness *et al.*, 2003; Tokumasu and Dvorak, 2003). QDs are rapidly becoming the fluorochrome of choice for even the most routine of applications.

With this success, many researchers are seeking to use QDs in place of existing fluorochromes and have quickly realized that they behave differently in terms of biological preparation methods and imaging methods. In this chapter, we discuss some key considerations for effective use of QDs to label cellular targets for immunofluorescence in live and fixed cells.

II. Selection of QDs and Their Conjugates

A single QD is a nanometer scale crystal of semiconductor material, typically cadmium selenide (although other materials can be used), which is coated with a ZnS semiconductor outer shell (Dabbousi, 1997; Watson *et al.*, 2003). The nascent core-shell QD is very hydrophobic and must be coated with a layer of water-miscible material for use in biological applications. Many coatings have been described in the literature with limited success in preserving the particle integrity and performance. Advances through rigorous engineering have described a polymer layer (Watson *et al.*, 2003; Wu *et al.*, 2003) that affords water solubility, allows the dots to be conjugated to biological affinity molecules, and controls the nonspecific binding. This coating ensures the high stability, brightness, and overall performance of QDs of various colors in biological applications.

Because the QDs are nanoscale engineered particles, there are some critical distinctions between these materials and organic dye molecules. Because they are the same molecules, organic fluorochromes (e.g., fluorescein) from different manufactures have essentially the same chemical and optical properties, such as stability, excitation, and emission spectra. In contrast, QDs from different sources, even though they are made from the same materials and may have the same color (emission peak), can be substantially different because of differences in core-shell synthesis and surface-coating chemistry. High-quality QDs should be stable in aqueous solutions and have high extinction coefficients, high quantum yield, and narrow emission spectra (expressed as full width at half maximum [FWHM]). Qdot Conjugates are the only commercially available QDs conjugated to biological affinity molecules. They remain stable and monodispersed in aqueous buffers for months, display high fluorescence brilliance, and have narrow emission spectra (FWHM is consistently as low as 27 nm for some products).

Qdot Conjugates are available with various affinity molecules, including primary antibodies, secondary antibodies, streptavidin, biotin, and protein A. There are six colors of Qdot QDs available with emission peaks at 525, 565, 585, 605, 655, and 705 nm. Qdot QDs are named after their emission peaks. For example, the green color conjugates are named Qdot 525 because their maximum emission is centered at 525 nm. This naming method has an advantage of providing users the exact emission peaks of different dots, which is important for users when selecting the appropriate color of dots for their instruments and applications. However, it needs to be pointed out that this naming system is different from that used for Alexa organic dyes. Alexa dyes are named after their excitation peaks rather than emission maximums. For instance, Alexa 488 is a green light emitting dye with excitation maximum at 488 nm but with an emission peak at around 520 nm. Because each color of organic dye usually can only be effectively excited by light in a specific narrow range, the information about excitation maximum is important for using organic dyes. In contrast, QDs are very flexible for excitation. Any wavelength from the UV end of the spectrum to the emission peak of a specific color of dots can be used to excite these dots.

III. Labeling of Fixed Cells for Fluorescence Microscopy

A. Sample Preparation Methods for Using QDs

One major challenge in detecting intracellular antigens or oligonucleotide sequences with fluorescent probes is the ability of the probes to penetrate the tissue space to the site of the biomolecule of interest. In the case of immunofluorescence, access of a labeling antibody, a large molecule, to the target inside a cell requires permeabilizing the cell membrane without disrupting the cell architecture and antigen distribution (Melan, 1999). This challenge is even more acute when one uses QD conjugates to label intracellular targets. Starting with a core/shell ranging from 3 to 10 nm followed by a water-miscible layer and conjugated molecular surface, the final QD can be approximately 10 nm larger than its initial core (Watson *et al.*, 2003). Monodispersed materials are substantially smaller than aggregated materials, so preparations of dots that have significant aggregation may produce inferior labeling results. In contrast, organic fluorochrome conjugates are smaller. Hence, the fixation and permeabilization for dye conjugates and QD conjugates may be different in some applications.

There are two basic types of chemical fixatives for cells: coagulants and cross-linking reagents (Melan, 1999). Coagulants such as methanol, acetone, or higher-molecular-weight polyalcohols, dehydrate proteins, and macromolecules in cells lead to precipitation or condensation on cellular architectures. They also make the cell permeable to various molecules by removing membrane lipids. Cross-linking fixatives such as formaldehyde form bridges between protein molecules (Melan, 1999). Cross-linking reagents usually preserve cell structures better than coagulants but at the same time may chemically destroy binding sites or cause steric blockage of antibody access to epitopes. Most cross-linking fixatives require some form of permeabilization with detergents or other chemicals for labeling intracellular targets.

For immunolabeling with QD conjugates, the fixation and permeabilization conditions varied with the cellular location of the antigen. For detecting surface antigens, fixative selection was more flexible because the epitopes are exposed on the cell surface. The Her2 receptor antigen on the surface of SK-BR-3 cells was labeled equally well with both methanol and formaldehyde fixation. For tubulin epitopes in microtubules and mitochondrial proteins in the cytoplasm, fixation with cold methanol at $-20\,^{\circ}\mathrm{C}$ or with 4% formaldehyde containing 0.2% Triton X-100 provided satisfactory results. However, for some nuclear antigens, fixation with cold acetone at $-20\,^{\circ}\mathrm{C}$ resulted in better labeling. Hence, as with organic fluorochrome conjugates, the selection of the best fixation and permeabilization conditions for labeling biomolecules in cells and tissues with QD conjugates will continue with each new antigen to be labeled. Such an example is shown in Fig. 1.

B. Mounting Media

A common problem in performing immunofluorescence detection with conventional fluorochromes is photobleaching. When exposed to excitation light, all fluorescent dyes will fade at some rate depending on the dye and its environment. Photobleaching is irreversible and appears to be associated with the presence of molecular oxygen (Murphy, 2001). The degree of photobleaching depends on both the intensity and the duration of illumination. To deal with this limitation of traditional fluorochromes, antifade chemicals that inhibit the generation and diffusion of reactive oxygen species have been added to mounting media (Ono, 2001).

In contrast, high-quality QDs exhibit exceptional photostability (Bruchez *et al.*, 1998; Chan and Nie, 1998; Wu *et al.*, 2003). Continuous illumination of Qdot QDs with intense UV light for hours results in little or no change in emission intensity (Xingyong Wu, 2003, unpublished data). Therefore, fixed cells or tissue sections can be mounted in 50–90% glycerol in PBS without any antifade. To avoid the inconvenience of sealing the edges of the coverslip with nail polish or other sealing

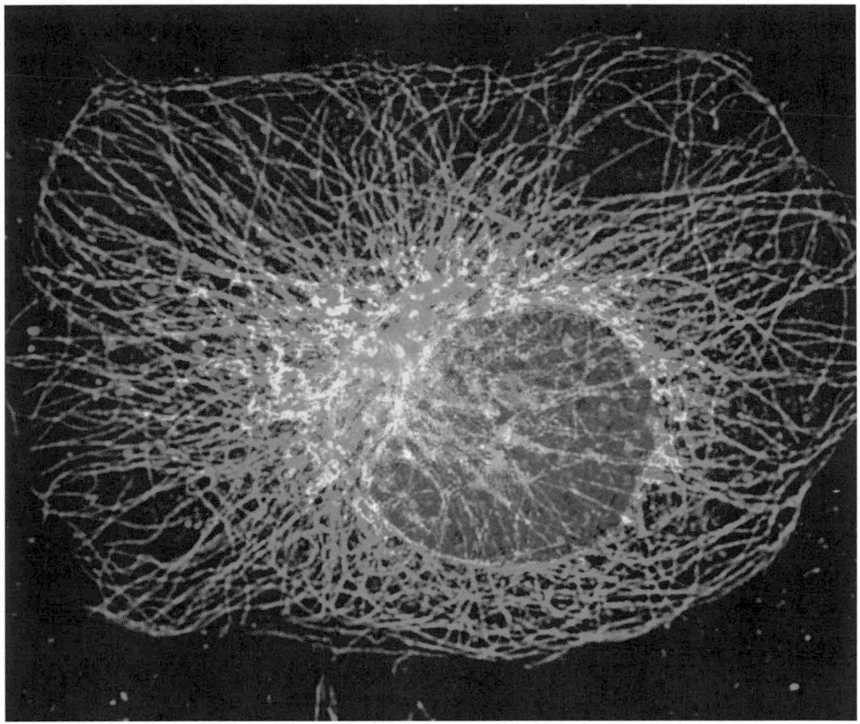

Fig. 1 High-quality staining with Qdot Conjugates at all levels of cellular structure: simultaneous detection of three cellular targets in HeLa cells fixed with 70% methanol/30% acetone using Qdot Conjugates. Microtubules (green), mitochondria (red), and nucleosomes (pseudo-colored blue) were labeled with Qdot 525 Streptavidin Conjugate, Qdot 605 anti-human conjugate, and Qdot 655 anti-mouse conjugate, respectively, after the cells were incubated with appropriate primary and secondary antibodies. (See Color Insert.)

reagents, one can use self-sealing polyvinyl alcohol-based mounting media, such as the Polyvinyl Alcohol Mounting Medium containing 1-4-diazabicyclo [2,2,2]-octane (DABCO) from Sigma-Aldrich (St. Louis, Missouri). This mounting medium is also suitable for using QDs in combination with organic dyes because it provides antifade protection. All traditional mounting media are specifically formulated for organic fluorochromes and some of them may not be compatible with QDs. Users need to test their specific mounting medium with QDs before attempting to archive or mount samples stained with dyes and QDs.

C. Detection with a Fluorescence Microscope

QDs have a broad excitation range and have very strong absorption coefficient in the blue and UV portions of the spectrum. A wide range of light sources such as mercury, xenon, or mercury–xenon lamp combinations or blue UV lasers are suitable for QDs. Illuminating QDs with blue or UV light not only greatly increases their fluorescence intensity but also results in a large Stokes shift, the distance between the peaks of excitation and emission. This large Stokes shift improves detection sensitivity by removing filter bleed-through and eliminating background artifacts caused by scatter. Two major fluorescence filter manufacturers are providing filters specifically for detection of Qdot Conjugates. For excitation, Omega Optical (Brattleboro, Vermont) has two excitation filters: a 100-nm broad bandpass filter (415BW100) and a 45-nm narrow bandpass filter (425DF45). Chroma Technology (Brattleboro, Vermont) provides 460SPUV, a short-pass filter reflecting all light shorter than 460 nm to the QD-labeled sample. With a mercury lamp, both the 415BW100 and the 460SPUV provide very strong excitation for high-intensity emission signals. These two filters may be more suitable for detecting low-abundance antigens or for use with low magnification objectives (10× and 20×). With high-magnification oil objectives, these two filters may induce sample heating and photo-oxidation, causing potential damage to the sample. This damage can be minimized by using either neutral density filters to reduce the intensity of the light from the light source or the narrower bandpass excitation 425DF45 filter. This is especially important for observing living cells.

The narrow and symmetrical emission spectra of QDs (Bruchez *et al.*, 1998; Watson *et al.*, 2003) make them ideal fluorochromes for multicolor fluorescent labeling. Conversely, conventional fluorochromes have broad emission spectra with long red tails and, thus, require sophisticated image acquisition equipment and image analysis software for spectral unmixing. Multicolor detection with QDs is much easier. An example of this is shown in Fig. 2. Cultured SK-BR-3 breast cancer cells expressing high levels of the Her2/neu receptor were first incubated with a mouse anti-Her2 antibody followed by a biotinylated goat anti-mouse immunoglobulin G (IgG), then aliquots of cells were labeled separately with Qdot Streptavidin Conjugates with emissions at 525, 565, 605, and 655 nm. These cells were washed and pooled into a single tube, and the cell mixture was examined under an epifluorescence microscope. When observed with a 425DF45 excitation filter and a long-pass emission filter (510ALP), all cells were clearly visible with four distinguishable colors (Fig. 2A). When the cells in the same field were examined with individual 20-nm bandpass emission filters centered at the emission peak of each color of QDs, only the cells labeled with corresponding QDs were visible (Fig. 2B–E). Because the emission spectra of the QDs are narrow, 20-nm bandpass filters did not compromise the signal intensity. Furthermore, because the QDs exhibit symmetrical emission spectra and lack a long emission tail typical of conventional organic dyes, there was no obvious crosstalk between the four detection channels, although the emission peaks of any two adjacent colors of QDs are separated only by a 40- or 50-nm spectral space. Thus, multicolor detection and discrimination can be dramatically simplified using Qdot conjugate detection. The optical properties of these materials have been designed to eliminate the need for spectral unmixing for five or more colors using simple bandpass emission filtering.

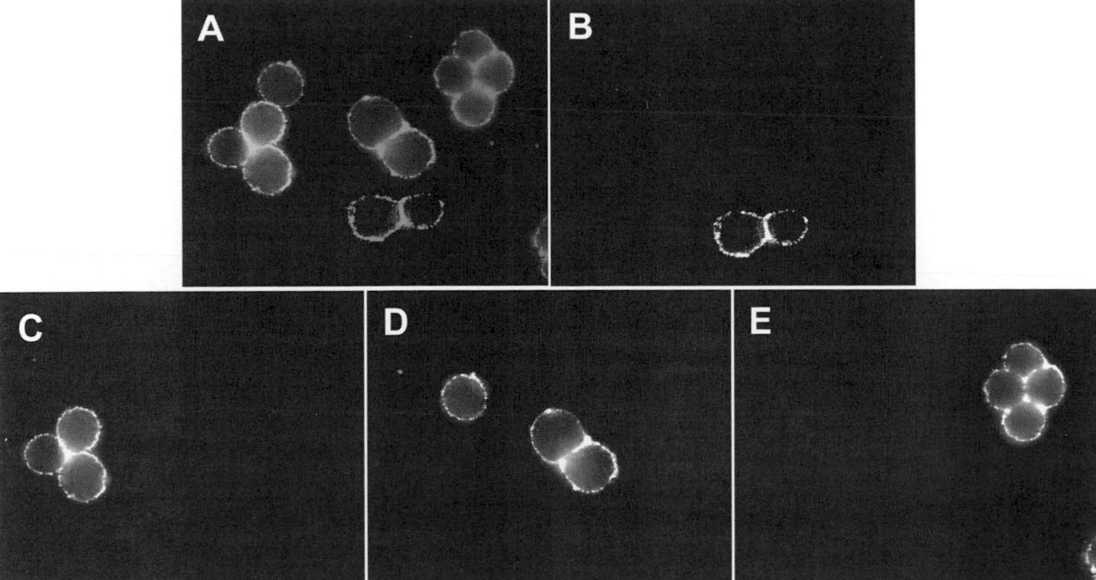

Fig. 2 Qdot Conjugates provide effective color discrimination with simple bandpass emission filters: detection of Her2 on the surface of SK-BR-3 cells with Qdot Conjugates. The cells were sequentially incubated with a mouse anti-Her2 antibody and biotinylated goat anti-mouse immunoglobulin G and the targets were labeled with 525-, 565-, 605-, and 655-nm Qdot Streptavidin Conjugates separately. Cells labeled with different colors of QDs were then mixed and examined with an epifluorescent microscope. When the mixed cells were observed with a long-pass emission filter, all cells in the field of view were visible with four distinguishable colors (A: Cells were pseudo-colored from gray scale images to mimic the real colors viewed under the microscope). When the cells in the same field of view were examined with individual 20-nm bandpass filters centered at 525, 565, 605, and 655 nm (B–E), only the cells labeled with quantum dots having an emission peak in the detection range of the filter were visible. (See Color Insert.)

D. Imaging QDs with a Laser Confocal Microscope

Laser scanning confocal microscopy (LSCM) is an important tool for providing optical sections and three-dimensional (3D) images of cellular structures and protein distributions. LSCM systems are typically equipped with multiple lasers that are tuned to the excitation maximums of conventional fluorochromes to produce the best images possible from these weak emitters. Because all colors of QD have a very broad absorbance profile from the UV region of the spectrum to its emission peak, it is possible to excite all of them with a single laser light in the blue or UV region. To demonstrate this, we sequentially incubated acetone-fixed human epithelial cells with a monoclonal mouse anti-nucleosome antibody and biotinylated goat anti-mouse IgG and labeled the nucleosomes in the nuclei of six groups of cells separately using six colors of Qdot Streptavidin Conjugates with emission maximums at 525, 565, 585, 605, 655, and 705 nm. We used the 488-nm line from an argon laser, the most common laser for LSCM systems to excite these labeled cells and collect the fluorescent signals for each color of conjugate. As shown in Fig. 3, all colors of QD were effectively excited with the 488-nm laser line. These results indicate that one can use a single laser for an LSCM system with multicolor detection capacity when using QDs as fluorescent labels. It simplifies an LSCM system and makes LSCM systems more affordable for multicolor detection. This also avoids potential issues of laser alignment or excitation chromatic effects encountered in a multilaser LSCM system. Several manufacturers have introduced LSCM systems with 405-nm excitation lasers that should be ideal for imaging QD-labeled tissue.

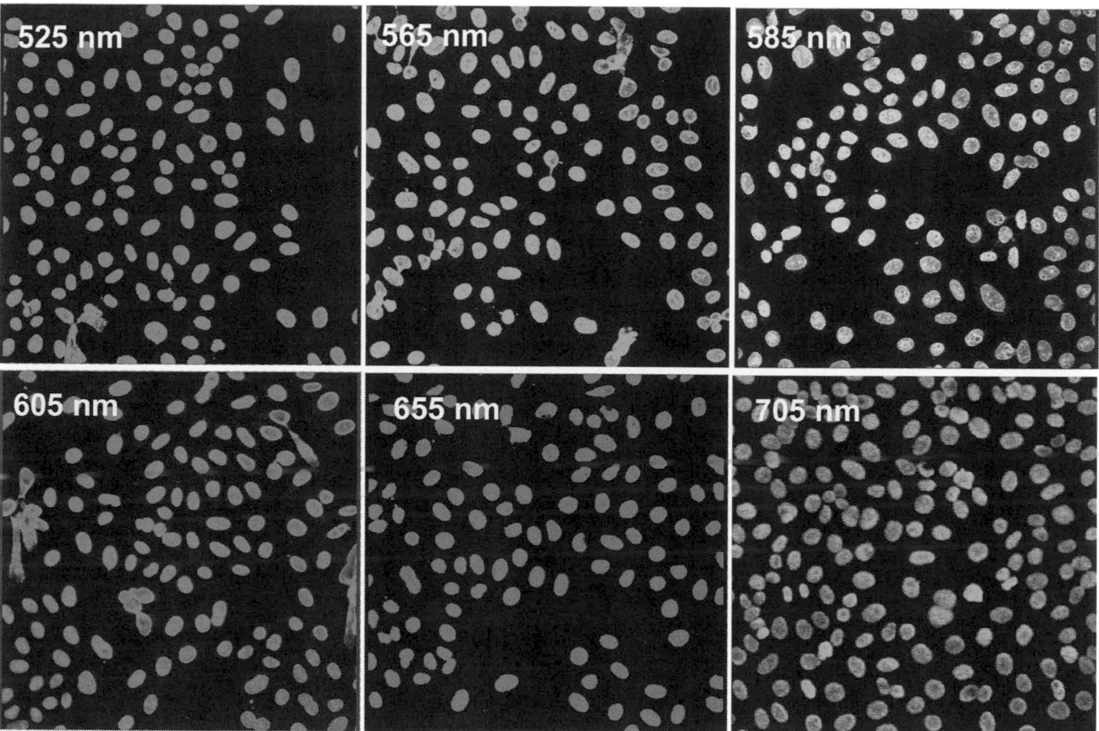

Fig. 3 Qdot Conjugates use single-laser excitation source for multicolor confocal microscopy: Nucleosomes in fixed human epithelial cells were labeled separately with Qdot 525, 565, 585, 605, 655, and 705 Streptavidin Conjugates. The labeled cells were examined with a Leica TCS SP 2 confocal microscope. Because quantum dots (QDs) have broad absorption profiles, all colors of the QDs were effectively excited with the 488-nm line from an argon laser. Color in each panel was adjusted to mimic the colors seen through a long-pass filter. A gray-scale image was used for Qdot 705 because this color is not visible to human eyes. (See Color Insert.)

IV. Measurements on Living Cells Using QD Conjugates

Imaging and measurement of biological responses of living cells in real time has posed a special challenge for optical and fluorescence imaging techniques. In these systems, the events of interest may take place on time scales ranging from milliseconds to hours, and the process that an investigator is hoping to see may occur hours after applying the initial stimulus. Constitutively expressed protein labels, such as the visible fluorescent proteins (Cubitt *et al.*, 1995), have provided enabling tools that allow study of synthesis and trafficking of distinctly engineered proteins within the cells, although these materials do not allow long-term tracking of a single protein. In addition, these tools can be applied only to proteins that are synthesized within a cell and cannot be introduced efficiently from outside the cell. The unique optical properties of the QDs make them particularly well suited for investigations in living cells.

The absorbance and quantum yield of the QDs combine to produce a material with a very high brightness in imaging applications. This can be harnessed in several important ways to image live cells. One persistent problem with dynamic microscopy of living cells is phototoxicity, where the exposure of the cells to high intensities often required for sensitive fluorescence detection results in a biological response or even in cell death (Knight *et al.*, 2003). For applications in which typical excitation conditions could be used to image fluorescence dyes, substantially lower intensities could be used to image Qdot Conjugates (typically 20–100 times lower intensities because of the increased brightness of the Qdot Conjugates). This allows the researcher to do substantially less photodamage to the cells while collecting data over substantially longer times. If, on the other hand, the researcher desires higher video-rate data to measure fast biological processes, then keeping the high intensity allows much faster microscopy to be performed with no loss of sensitivity. These properties have allowed researchers to image single Qdot Conjugates with time resolutions of less than 0.1 seconds for many minutes in a single experiment (Dahan *et al.*, 2003; Lidke *et al.*, 2004).

The photostability of the Qdot Conjugates has contributed another important property to the use of these materials in dynamic microscopy. High-quality QDs have essentially perfect photostability, with measured photostability time scales reported from minutes to hours, but in all cases with essentially no or very little loss of fluorescence intensity over the time scale measured (Bruchez *et al.*, 1998; Wu *et al.*, 2003). This allows detailed tracking of the dynamics of proteins in real time. Techniques such as fluorescence recovery after photobleaching (FRAP) and photosensitizable fluorescent proteins (Ando *et al.*, 2002; Lippincott-Schwartz *et al.*, 2003; Tsien, 1998) have been used to study large-scale phenomena such as protein migration and trafficking. These techniques are instrumentally intensive and represent a set of tools available only to the most sophisticated imaging users. They can also be used only to look for large-scale changes, because tracking a single "photodestroyed" GFP is impossible, whereas tracking photosensitized

GFP molecules can be performed only for a short time because of the intrinsic photostability limits of fluorescent proteins. Using Qdot Conjugates, one can track a single protein from the moment it begins interaction with the cell to its ultimate fate within the cell (Dahan *et al.*, 2003; Lidke *et al.*, 2004). Harnessing the brightness and photostability of these materials and the potential for multiplexing remain a significant opportunity in live cell imaging.

The chemical and energy-transfer properties of QD materials are in the early phases of exploration (Clapp *et al.*, 2004; Medintz *et al.*, 2003, 2004) but pose a significant opportunity for development as imaging tools for cytometry. Studies indicate that it should be possible to prepare materials that can be switched in the presence of an analyte using fluorescence resonance energy transfer properties, in which the QDs serve as an energy-transfer donor. These could be multiplexed quite easily, and the readout would consist of a single excitation that would excite all of the QDs, and where appropriate, these would transfer their energy to the acceptor, resulting in sensitized fluorescence. This would provide a very powerful set of biochemical tools, and if it were possible to deliver these materials directly into the cytoplasm, would provide an exciting opportunity for real-time cell physiology measurements. In addition, preparing the reverse materials (i.e., materials in which a cleavage event would release a signal that was previously transferred or quenched) would provide an extremely bright, selectively sensitized probe that could be used for both physiology and trafficking, as described earlier.

The QD materials are truly nanoscale materials, on the order of large proteins such as immunoglobulins. Even small proteins that are not directly targeted to receptors cannot easily be loaded directly into cells without accompanying endocytosis. Some methods have been proposed and successfully demonstrated (Tkachenko *et al.*, 2003) that can deliver colloidal materials through the endocytic machinery and allow escape and direct delivery into the cytoplasmic milieu. It is yet undemonstrated whether these techniques can be applied to QD probes, although the common size scale suggests that this would be a promising approach. The diffusion dynamics of QD-labeled receptors did not prove markedly different than the diffusion dynamics of receptors measured independently, although these studies have all applied to extracellular QD labeling. It remains to be seen whether the "ball-and-chain" effect will play a role in determining the diffusion rate of QD-conjugated biological molecules when the conjugates have to be pulled through the relatively viscous environment of the cytoplasm.

A growing number of publications demonstrate the power of Qdot Conjugates for live cell imaging. In studies of the glycine receptor (GlyR) on living neurons, Dahan *et al.* (2003) demonstrated that the Qdot Conjugates provided incredibly high signal/noise ratios at a single molecule level, and that they could be tracked for many minutes with no degradation in the quality of the signal. These properties allowed a single GlyR to be tracked while diffusing the length of a neuron, sampling space within and without the synapse. Long-term imaging allowed careful determination that GlyR demonstrates markedly different diffusion when it is in the synapse, compared to when it is outside the synapse. These statistics would have been very difficult to measure without the optical properties of the QDs enabling the continuous observation for many minutes (Dahan *et al.*, 2003). Similar studies using epidermal growth factor (EGF)–labeled Qdot particles in combination with visible fluorescent proteins to track the dynamics of the erbB family of receptors revealed a new mechanism of transport along the filopodia of EGF-bound erbB2 receptors (Lidke *et al.*, 2004). In all of these studies, the use of QDs to track receptors required very low levels of Qdot Conjugates, as opposed to those used to label immunofluorescence targets at saturation. Typically, the Qdot Conjugates are used at concentrations less than 1 nM, as compared to 10–20 nM concentrations in staining of fixed antigens.

Dynamic tracking of cell populations in chemotaxis and development has been demonstrated by two independent groups, where the QDs are delivered into the cells, and then tracked as an optical probe of cell movement and cell lineage. Using different colors of QDs to tag distinct cell populations of *D. discoideum,* one starved and one healthy, and then mixing them in culture, Jaiswal *et al.* (2003) showed the selective chemoattraction of the starved cells, a process driven by chemotaxis to released cyclic adenosine monophosphate. In these experiments, the starved cells were seen to migrate selectively by observing motion of the QD fluorescence of one color, whereas the nonstarved cells remained static in culture, signaled by the lack of motion in the alternative color. In an elegant set of experiments, Dubertret *et al.* (2002) prepared phospholipid-coated QDs and introduced these dots into selected cells in the blastocyst phase of development of *Xenopus*. They were able to track the fluorescence of these cells through to the developed tadpole. These studies demonstrated that the QDs were stable to physiological conditions, were well retained by the cells and passed to subsequent generations, were tolerated as particles, and were nontoxic to the highly controlled and sensitive process of development.

The use of QD materials to study processes in living cells is an area ripe for exploration. Many of the properties of the QDs have not been fully exploited and present exciting opportunities for developing an understanding of dynamic cellular processes as they occur. The coming years will witness the use of these materials to understand a wide variety of biological processes. An increased understanding of the optical and chemical properties of these QDs will pave the way for an even more sophisticated understanding of the underlying network of signaling pathways that determines the interactions of a cell with its environment.

V. Conclusion

After years of intensive research and development, QDs and their conjugates are available for various uses in biological investigation. No longer simply a research subject for chemists and physicists, these materials have become powerful tools for researchers in life and biomedical sciences. Although QDs have potential advantages over conventional fluorochromes in all immunofluorescence detection areas, microscope-based imaging is the first application area in which scientists have quickly adopted this technology. Compared with other fluorescence detection instruments such as flow cytometers and microplate readers, the filter configuration of a microscope system can be much more easily modified by users to effectively detect QDs. The optical and biological properties of the QDs have enticed manufacturers of fluorescence detection instruments, including flow cytometers, to consider this new generation of fluorochromes in their next-generation instrument designs. They are now modifying the filter and illumination configurations of their systems to allow more flexible and effective detection of QDs. These bright QDs will find their way into a much broader array of immunofluorescence applications.

References

Alivisatos, A. P. (1996). Perspectives on the physical chemistry of semiconductor nanocrystals. *J. Phys. Chem.* **100,** 13226–13239.

Ando, R., Hama, H., Yamamoto-Hino, M., Mizuno, H., and Miyawaki, A. (2002). An optical marker based on the UV-induced green-to-red photoconversion of a fluorescent protein. *Proc. Natl. Acad. Sci. USA* **99,** 12651–12656.

Bruchez, M. Jr., Moronne, M., Gin, P., Weiss, S., and Alivisatos, A. P. (1998). Semiconductor nanocrystals as fluorescent biological labels. *Science* **281,** 2013–2016.

Chan, W. C., and Nie, S. (1998). Quantum dot bioconjugates for ultrasensitive nonisotopic detection. *Science* **281,** 2016–2018.

Clapp, A. R., Medintz, I. L., Mauro, J. M., Fisher, B. R., Bawendi, M. G., and Mattoussi, H. (2004). Fluorescence resonance energy transfer between quantum dot donors and dye-labeled protein acceptors. *J. Am. Chem. Soc.* **126,** 301–310.

Cubitt, A. B., Heim, R., Adams, S. R., Boyd, A. E., Gross, L. A., and Tsien, R. Y. (1995). Understanding, improving and using green fluorescent proteins. *Trends Biochem. Sci.* **20,** 448–455.

Dabbousi, B. O., Rodriguez-Viejo, J., Mikulec, F. V., Heine, J. R., Mattoussi, H., Ober, R., Jensen, K. F., and Bawendi, M. G. (1997). (CdSe)ZnS core-shell quantum dots: Synthesis and characterization of a size series of highly luminescent nanocrystallites. *J. Phys. Chem. B* **101,** 9463–9475.

Dahan, M., Levi, S., Luccardini, C., Rostaing, P., Riveau, B., and Triller, A. (2003). Diffusion dynamics of glycine receptors revealed by single-quantum dot tracking. *Science* **302,** 442–445.

Dubertret, B., Skourides, P., Norris, D. J., Noireaux, V., Brivanlou, A. H., and Libchaber, A. (2002). *In Vivo* imaging of quantum dots encapsulated in phospholipid micelles. *Science* **298,** 1759–1762.

Jaiswal, J.K., Mattoussi, H., Mauro, J. M., and Simon, S. M. (2003). Long-term multiple color imaging of live cells using quantum dot bioconjugates. *Nat. Biotechnol.* **21,** 47–51.

Knight, M. M., Roberts, S. R., Lee, D. A., and Bader, D. L. (2003). Live cell imaging using confocal microscopy induces intracellular calcium transients and cell death. *Am. J. Physiol. Cell Physiol.* **284,** C1083–C1089.

Lidke, D. S., Nagy, P., Heintzmann, R., Arndt-Jovin, D. J, Post, J. N., Grecco, H. E., Jares-Erijman, E. A, and Jovin, T. M. (2004). Quantum dot ligands provide new insights into erbB/HER receptor–mediated signal transduction. *Nat. Biotechnol.* **22,** 198–203.

Lippincott-Schwartz, J., Altan-Bonnet, N., and Patterson, G. H. (2003). Photobleaching and photoactivation: Following protein dynamics in living cells. *Nat. Cell Biol.* Suppl, S7–S14.

Medintz, I. L., Clapp, A. R., Mattoussi, H., Goldman, E. R., Fisher, B., and Mauro, J. M. (2003). Self-assembled nanoscale biosensors based on quantum dot FRET donors. *Nat. Mater.* **2,** 630–638.

Medintz, I. L., Trammell, S. A., Mattoussi, H., and Mauro, J. M. (2004). Reversible modulation of quantum dot photoluminescence using a protein-bound photochromic fluorescence resonance energy transfer acceptor. *J. Am. Chem. Soc.* **126,** 30–31.

Melan, M. A. (1999). Overview of cell fixatives and cell membrane permeants. *In* Method in molecular biology, (Javois, L. C. ed.). Vol. 115, pp. 44–55. Human Press, Totowa, NJ.

Murphy, D. B. (2001). Fluorescence microscopy. *In* "Fundamentals of light microscopy and electronic imaging," (D. B. Murphy, ed.), pp. 177–203. Wiley-Liss, New York.

Ness, J. M., Akhtar, R. S., Latham, C. B., and Roth, K. A. (2003). Combined tyramide signal amplification and quantum dots for sensitive and photostable immunofluorescence detection. *J. Histochem. Cytochem.* **51,** 981–987.

Ono, M., Murakami, T., Kudo, A., Isshiki, M., Sawada, H., and Segawa, A. (2001). Quantitative comparison of anti-fading mounting media for confocal laser scanning microscopy. *J. Histochem. Cytochem.* **49,** 305–311.

Tkachenko, A. G., Xie, H., Coleman, D., Glomm, W., Ryan, J., Anderson, M. F., Franzen, S., and Feldheim, D. L. (2003). Multifunctional gold nanoparticle-peptide complexes for nuclear targeting. *J. Am. Chem. Soc.* **125,** 4700–4701.

Tokumasu, F., and Dvorak, J. (2003). Development and application of quantum dots for immunocytochemistry of human erythrocytes. *J. Microsc.* **211**(Pt 3), 256–261.

Tsien, R. Y. (1998). The green fluorescent protein. *Annu. Rev. Biochem.* **67,** 509–544.

Watson, A., Wu, X., and Bruchez, M. P. (2003). Lighting up cells with quantum dots. *Biotechniques* **34,** 296–300, 302–303.

Wu, X., Liu, H., Liu, J., Haley, K. N., Treadway, J. A., Larson, J. P., Ge, N., Peale, F., and Bruchez, M. P. (2003). Immunofluorescent labeling of cancer marker Her2 and other cellular targets with semiconductor quantum dots. *Nat. Biotechnol.* **21,** 41–46.

CHAPTER 8

Next-Generation Laser Scanning Cytometry

Ed Luther, Louis Kamentsky, Melvin Henriksen, and Elena Holden

CompuCyte Corporation
Cambridge, Massachusetts 02139

METHODS IN CELL BIOLOGY, VOL. 75

0091-679X/04 $35.00

I. Introduction

In the past, flow cytometry (FC) has been at the forefront of quantitative cytometric analysis. Recent experimental needs in the life sciences demand a combination of quantitative cytometry and imaging cytometry. This demand has been fulfilled by the development of laser scanning cytometry (LSC).

The most important features of the next generation of laser scanning cytometers described in this chapter are as follows:

1. Cellular features are extracted from large populations at variable speeds.
2. Morphological and biochemical profiles of cells are obtained on specimens placed on carriers of any type.
3. LSC platforms are highly amenable to the use of fluorescent functional probes on live cells.
4. Assays are being developed to evaluate even subtle experimental effects of extrinsic agents on cell populations.
5. Unbiased, rule-based selection of "typical" exemplar cells can be employed.
6. Analysis of multiple specimen features is facilitated.

These features provide laser scanning cytometers with great potential for use in drug development, biomarker discovery, and clinical pathology.

II. LSC-Hybrid Technology

Three technologies can quantify cell constituents using fluorescence: fluorescence image analysis (FIA), FC, and LSC. Each can rapidly assay multiple cellular constituents by automatically measuring at multiple wavelengths the fluorescence of cells treated with one or more fluorescent dyes. Two of these, FC and LSC, can measure both light scatter and fluorescence, which in FC and LSC result from interaction of the cells with a laser beam.

A. LSC vs Flow Cytometry

During the past 3 decades, FC and cell sorting have become common tools in various disciplines of biology, medicine, and biotechnology. However, because cells are measured while suspended in a stream of liquid and are often discarded after the measurement, the analytical capability of FC is limited for many applications for the following reasons:

1. Events requiring time resolution such as enzyme kinetics, drug uptake, or efflux cannot be analyzed on individual cells.
2. Morphology of the measured cell may be assessed only after sorting, a cumbersome process that is not always available.

3. Subcellular localization of the fluorochrome cannot be quantified.

4. Once measured, the cell cannot be reanalyzed with another probe(s).

5. Analysis of solid tissue requires cell or nucleus isolation, leading to loss of information on tissue architecture.

6. Small samples such as those from fine-needle aspirates or spinal fluid are seldom analyzed by FC because sample processing requires a large number of cells per sample.

7. The measured sample is lost and cannot be stored for archival preservation.

The microscope-based laser scanning cytometers manufactured since the mid-1990s by CompuCyte Corporation (Cambridge, Massachusetts) offer many of the advantages of FC, without the limitations listed above. The analytical capabilities of LSC are comparable to those of FC and extend the use of cytometry in many applications (Darzynkiewicz *et al.*, 2000; Kamentsky *et al.*, 1991, 2001; Roman *et al.*, 2002.).

Fundamentally, the advantages of LSC result from the ability to provide the quantification capabilities of FC to specimens on a solid substrate. This adds the following capabilities to traditional FC analysis:

1. Rescanning of specimens at will, thus viewing them over time and with the opportunity for intermediate manipulation.

2. Visualization of specimens, thus showing relationships between the quantitative assessment and the visualized morphology.

3. Additional morphological quantification with analytical features.

4. Incorporation of image-processing techniques into the analysis.

5. Recording of images of analyzed samples for future analysis and archival purposes.

B. LSC vs Imaging

In FIA, the cells are uniformly illuminated, preferably by a mercury or xenon arc epi-illuminator. Fluorescence is imaged at high resolution and low depth of field to a sensitive CCD camera. A wavelength bandpass filter is used to isolate the fluorescence at specific bandwidths, and this filter can be mechanically changed to measure cell fluorescence at multiple wavelengths.

The lasers of both FC and LSC provide an intense concentration of monochromatic excitation light at the optical plane of the cell to allow better separation of fluorescence emission from excitation. As a result, cell constituents can generally be detected with higher measurement sensitivity than in FIA. Additionally, LSC can quantify laser light scatter and absorption along with the fluorescence emissions and employ laser light scatter in a unique bright-field visualization mode.

C. LSC vs Confocal Microscopy

LSC technology uses similar components as laser-based confocal microscopes, but there is a significant difference in the optical characteristics and the applications for which the two technologies are used.

Confocal microscopy uses sharply focused laser beams to illuminate as small an area of the sample as possible at any given time. In addition, apertures are used to limit the observation to light emitting only from near the focal plane. This optical sectioning allows excellent spatial resolution but is not very applicable to constituent quantification. Confocal technology is usually used for detailed analysis of small numbers of cells.

In LSC technology, the beam is as collimated as possible to allow a depth of field of typically 20–30 μm. This allows quantification of all of the fluorescent light emitted from the entire cell depth at each spatial location. This design permits scanning relatively large sample areas without the need for refocusing the instrument, making the instrument suited for automated experiments where large numbers of cells are examined. LSC technology transforms the microscope from a qualitative to a quantitative tool for cell biology.

III. Description of the LSC Platform

A. Instruments

CompuCyte Corporation manufactures a product line of laser scanning cytometers consisting of the original LSC® Laser Scanning Cytometer and two newer, next-generation systems, the *iCyte™ Automated Imaging Cytometer* (Fig. 1), and the *iCys™ Research Imaging Cytometer* (Fig. 2).

1. LSC Laser Scanning Cytometer

CompuCyte's original LSC Laser Scanning Cytometer is designed around an Olympus BX 51 microscope. The digitized fluorescence (and scatter) data are assembled into a two-dimensional (2D) computer memory array from the dual motion of the scanning laser and the stage progression. Many of the microscope's functions are left intact, giving LSC technology visualization capabilities that are not possible in FC, including the ability to associate images with any of the events in the population data.

The data structure for the LSC consists of three types of files. A protocol file (.pro) contains the information regarding the instrument settings for a particular experiment. Included is a list of the lasers and photomultiplier tubes selected, their relevant settings, what segmentation strategies are used, and what type of information is extracted for the segmented events.

The display file (.dpr) contains the parameters related to the display of the data on the computer screen. The main data file is the list mode data file, which follows

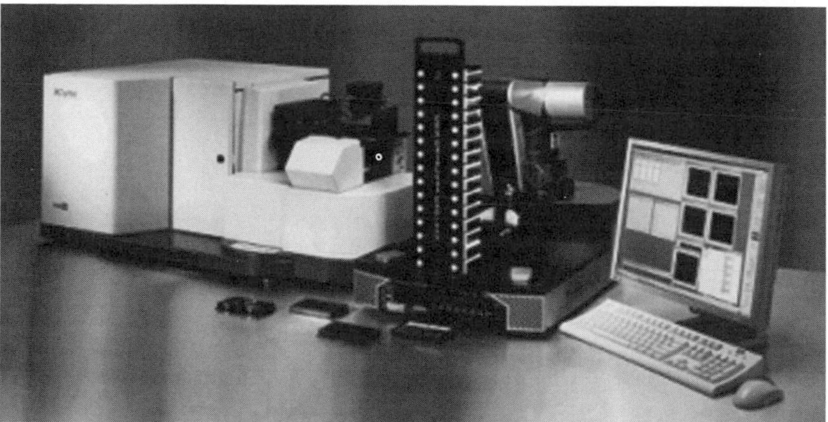

Fig. 1 The iCyte automated imaging cytometer. The electronics console on the left houses the laser light sources, photomultiplier tubes, and electronic circuitry. An inverted microscope base serves as the mount for the automated stage. There are no microscope viewing optics, but instead, laser scan imaging provides images of cells. This instrument is shown with the robotic arm for walkaway analysis of up to 45 carriers.

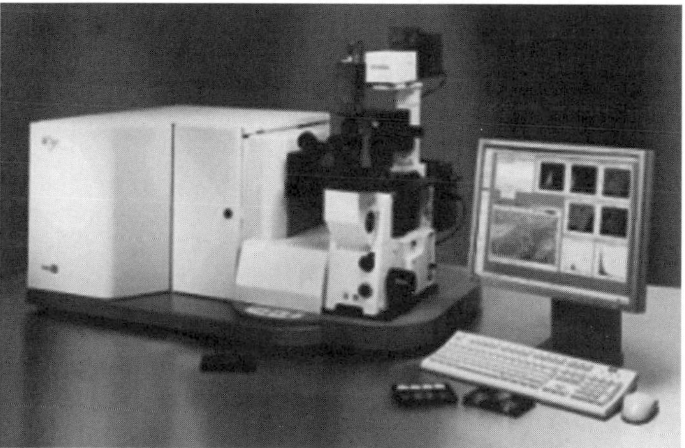

Fig. 2 The iCys research imaging cytometer. This instrument retains the scanning and imaging capabilities of the iCyte but also has the microscope viewing optics and illuminators for interactive analysis, viewing, and micromanipulation of samples.

the FC standard for data reporting (.fcs) and contains a list of features for each event.

Galleries of cell images can be obtained for user-selected regions in the data analysis displays either by using a video camera and frame grabber board or by rescanning and obtaining laser-scan images. These images can be exported into a report or captured using the computer clipboard.

2. *iCyte Automated Imaging Cytometer*

The *iCyte Automated Imaging Cytometer* is based on the fundamental design of the LSC but incorporates several key enhancements, which make it suitable for applications requiring higher throughput and automation.

The instrument employs an inverted format and has stage adapters that allow the analysis of samples in microtiter plates, chamber slides, and Petri dishes and on microscope slides. A utility allows the user to "define" additional carriers that fit within the microtiter plate footprint.

An autofocus system allows in-focus scanning of these large-footprint carriers without operator intervention. Because of the improved nature of the laser-scan imaging compared to the original LSC, microscope visualization, and imaging camera systems are not needed in this system for the intended automated applications.

An optional robotic arm can be used for large-scale experiments so up to 45 carriers can be automatically loaded onto the instrument for batch analysis. These carriers are typically microtiter plates, but any carrier with the same footprint as a plate can be used.

The automated nature of the analysis paradigms required extensive modifications to the data structure compared to the original LSC analysis, as shown in Fig. 3.

The changes are as follows:

- The .pro and .dpr file equivalents of LSC analysis are combined into a single entity called the *Workspace file* that contains all instrument settings. The list-mode data file is now called an *Event file* (.evt). Because the instrument is designed for assays and experiments consisting of multiple samples—for example, microtiter plates, multiple microscope slides, or tissue microarrays (TMAs)—a new data element, *Well Features,* was developed for reporting the population demographics of individual wells. For example, Well Features may be the mean area of the cells, the total number of cells in the area scanned, the mean fluorescence of all the cells in a well, the number of cells falling into a particular region, and so on. Well Features are saved in a delimited text file and may be viewed in the iCyte application software, the iBrowser Data Analysis software, or in spreadsheet software such as Microsoft Excel.
- Data storage options include the ability to store all of the scan images for all of the detectors employed in an analysis in a raw data file, thus enabling the reanalysis of data under different analysis conditions. The raw data files can be saved as either 16-bit resolution images or, to conserve data storage space, at 8-bit resolution as .jpg files. The conversion of the 16-bit resolution images to 8 bit is not equivalent to doing original 8-bit digitations.

Data files

	LSC		ICyte/iCys			
Control files	Prototocol file	Display file	Workspace file	Run file		
Image files			Raw data images	Mosaic images	Selected scans	Image galleries
Processed data	Report		Histograms	2D Histograms	Report	
Population data	Event file		Event file	Well features file		

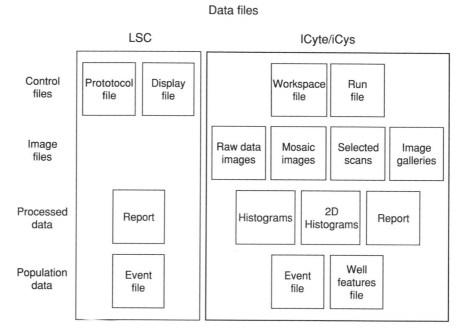

Fig. 3 Laser scanning technology data structures. The laser scanning cytometry (LSC) data file structure is multilevel, with control files that contain the instrument and computer display settings, image files of varying size and complexity, processed data files, and population data files. In both the LSC and the iSeries instruments, users can select which data elements are stored for analysis.

• Mosaic images—low-resolution images of the entire scan area—can be created by applying a transfer function to the individual images and then stitching them together.

• Selected scan fields can be saved, essentially as a subset of the raw data file images. Individual channels can be selected for storage and a single representative set of images can be obtained for each well. In addition, virtual channels (combinations of the image data from one or more channels) can be saved as scan images.

• Galleries of cell images from selected regions can also be stored. Unlike LSC analysis, the regions are predefined and the images are inserted into the galleries in real time as the scanning is done.

• Histogram frequency distribution files are also stored, as well as two-parameter frequency distribution files and 2D feature expression maps.

3. *iCys Research Imaging Cytometer*

The *iCys Research Imaging Cytometer*, shares with the *iCyte* an expanded carrier portfolio, autofocus, and enhanced data acquisition and analysis software. Developed for interactive and research applications, this instrument has full microscope capabilities; in this case the inverted-format Olympus IX71 microscope, including epi-fluorescence, bright-field illumination, and CCD camera image capture are used. The combination of advances in the analysis software with the microscope features provides the iCys with a unique mode of analysis. Once the initial scanning is complete, and following any desired reanalysis of the scanned data, the microscope stage may be automatically moved to the location of any event of interest. The event can then be viewed using the microscope optics while simultaneously viewing the laser scan images. The user may further use any number of microscope accessories available, including optomechanical devices for micromanipulation and cell capture.

B. Multiple Operational Modes: Interactive and Walkaway

As briefly discussed, the iCys and iCyte systems provide for either interactive (iCys) or walkaway (iCyte) analysis. The iCyte, with its emphasis on quantification of laser-scanned imagery and robot-aided batch analysis, can be the workhorse that processes significant numbers of specimen carriers per *day* or *night*.

While still retaining the depth of data and the autofocus capability for large-format specimen carriers, the iCys allows for visible light microscopy to complement the fluorescence quantification and imagery. This configuration results in an interactive research approach.

C. Obtaining the Images

The scanning and data acquisition process is similar in all of the CompuCyte cytometers. As shown in the block diagram of the *iCyte* optical system (Fig. 4), the beams from a 488-nm argon ion laser, a 633 nm HeNe laser and a 405-nm solid-state laser are first combined at a dichroic mirror and then steered to a second dichroic mirror, which reflects the laser wavelengths and transmits other wavelengths. Each laser beam is shuttered under computer control. A single beam or two combined beams are steered to a computer-controlled scanning mirror that is oscillating at a nominal rate of 350 Hz, to create a line scan at the specimen. After passing through a scan lens, the beam enters the side port of an Olympus IX71 inverted microscope and is imaged by a tube lens and the microscope's objective lens to the focal plane at the specimen. A multiplexing mode, with subsequent scans of the same area of the slide using alternate combinations of laser beams, can be enabled when multiple lasers are employed.

The line scan at the focal plane has a 10-μ diameter $1/e^2$ width and 768 μ extent using a 10× objective, a 5-μ diameter width, and 384-μ extent using a 20× objective, or a 2.5-μ diameter width and 192-μ extent using a 40× objective.

The specimen carrier is mounted in a holder on a computer-controlled, stepper motor–driven stage equipped with home position sensors. By initially homing the stage, the computer can maintain a record of the position of the slide throughout the scan. Because the scan mirror is also computer controlled, the stage and scan

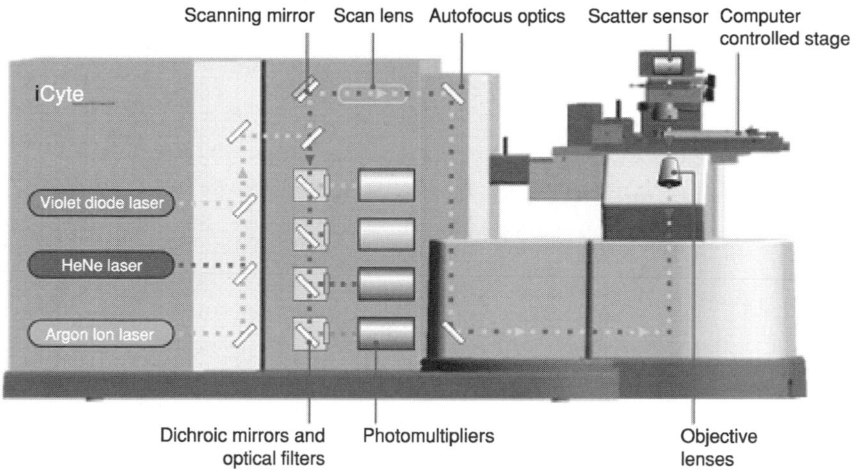

Fig. 4 Block diagram of the iCyte. Three lasers are used in a multiplexed fashion to illuminate the specimen from underneath. Fluorescent light is returned along the same pathway and is diverted and directed by dichroic mirrors and filters to four photomultiplier tubes measuring distinct spectral bandwidths.

mirror positions can be combined to record the coordinates of the object with respect to the laser beam. The stage is moved in adjustable steps, nominally of 0.5-μ per scan line in a direction perpendicular to the line scan. Fluorescent energy emitted by each cell is collected by the objective lens and steered through the scan lens to the scanning mirror. The collimated fluorescence emission passes through a series of dichroic mirrors and optical interference filters to up to four photomultipliers, each detecting a specific fluorescence wavelength range. As the laser beam intersects a cell, scattered light is collected by a lens and directed to an assembly containing a solid-state photo sensor.

The four sensor signals are simultaneously digitized at 625,000 Hz. By adjusting both the scan beam velocity and the scan length, sampling is made at 0.5-μ spatial intervals (for a 20× objective lens) along the scan. The sensor signals are digitized by an analog-to-digital converter to 14-bit digital pixel values.

A series of 1000 steps is assembled into a *scan field* for visualization and analysis. It should be noted that in the earlier versions of the LSC, scan fields were not intended to be visual elements but served the purpose of intermediate elements in the analytical data quantification. Technological developments, including high-precision mechanical stages, beam-shaping optics to reduce the effective beam size, and proprietary laser scatter bright-field imaging, changed that. The scan fields serve as digital microscopy images, as seen in Fig. 5, where a green fluorescence image of Chinese hamster ovary (CHO) cells labeled with fluorescein isothiocyanate (FITC) nonspecifically binding to the protein is shown

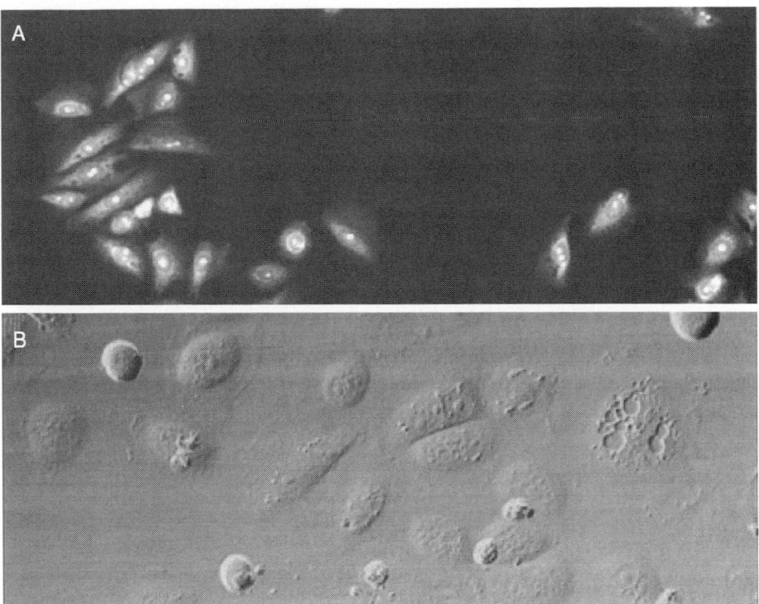

Fig. 5 Laser scan memory arrays of green fluorescence and laser light scatter. Although these have the appearance of digital images, they are technically memory arrays of photomultiplier or photodiode outputs, digitized to correspond to 0.25 by 0.5 spatial resolution. The array dimensions are 768 by 1000 pixels.

in the upper image, and a laser scatter image of colchicine-treated CHO cells is shown below.

Virtual channels may be defined using mathematical or logical operators on one or more real channels to obtain a processed resultant image. The operators available are as follows:

- Add—Calculates the sum of the two constituent channels
- Subtract—Calculates the difference between the two constituent channels
- Multiply—Multiplies the pixel values from a single channel by a specified operand
- Max—For each pixel, retains the greater of the two pixel values from each constituent channel
- Min—For each pixel, retains the lesser of the two pixel values from each constituent channel
- Invert—Calculates the inverted pixel values; thus, a pixel value that is saturated (white) is displayed as black
- ShiftPeak—Normalizes the pixel values of an image to a specified value

Virtual channels are treated like any other channel; they may be used to generate contours or even as constituent channels in the definition of other virtual channels (*nested virtual channels*). They have utility in far-ranging applications from correcting for autofluorescence in tissue sections to performing spectral overlap compensation in immunostaining experiments. Virtual channels defined using the Add or Max functions can be used as segmentation channels in experiments where the cell types do not share any features in common, such as in live/dead cell assays.

CompuColor images are a special type of virtual channel, where the outputs from numerous detectors are color-coded and merged into a single color image.

D. Segmentation and Feature Extraction

To process events to obtain cell features, the scan data are first segmented to associate specific pixels with each event. Data from one selected channel, either real or virtual (the contour channel), are used to draw a contour around all of the adjacent pixels above a preset threshold value set by the user to define an event. These threshold contours can be visualized on the scan data display as geometric shapes surrounding the data corresponding to each event.

In Fig. 6A, the first inner contour around the cell image is this threshold contour and is shown in red. For each event, this contour developed from the contour channel data is used to calculate features from all other channels.

For robust analysis, it is advisable to have the segmentation contour well separated from the background noise level; thus, this contour probably will not capture all of the segmented event's fluorescence. For this reason, a second contour, a fixed number of pixels outside the threshold contour, is established to include all cell pixels in the total fluorescence computation. The pixels for each sensor within this *integration* contour, shown in green in Fig. 6A, arc summed to compute the cell's *total scatter* or *total fluorescence* feature, the estimate of the amount of cell constituent resulting from the scatter or fluorescent dye measured by the sensor. The number of pixels in the *integration* contour line is used as the event's *perimeter* feature. The values of each sensor's maximum pixel within the contour are used as MaxPixel (*maximum pixel)* features.

The instrument was designed to provide accurate determinations of cell-constituent values. To this end, the software provides two necessary computations on the summed pixel values. The first corrects for differences in excitation intensity and emission-detection efficiency as a function of scan mirror position. The laser beam is off the optical axis at all but central pixel values of the scan motion. This, combined with movement of the scan beam over optical surfaces, creates irregularities in light excitation and collection. For this reason, the efficiency of excitation and emission detection is established by scanning uniform calibration particles. The total fluorescence of the particles is measured as a function of scan position and used to generate a permanently stored table to automatically correct each cell's total fluorescence based on its position along the scan line.

To accurately determine cell-constituent and maximum pixel values in the presence of variable background fluorescence or scatter, it is necessary to establish a background value for each sensor for each event. This "dynamic background"

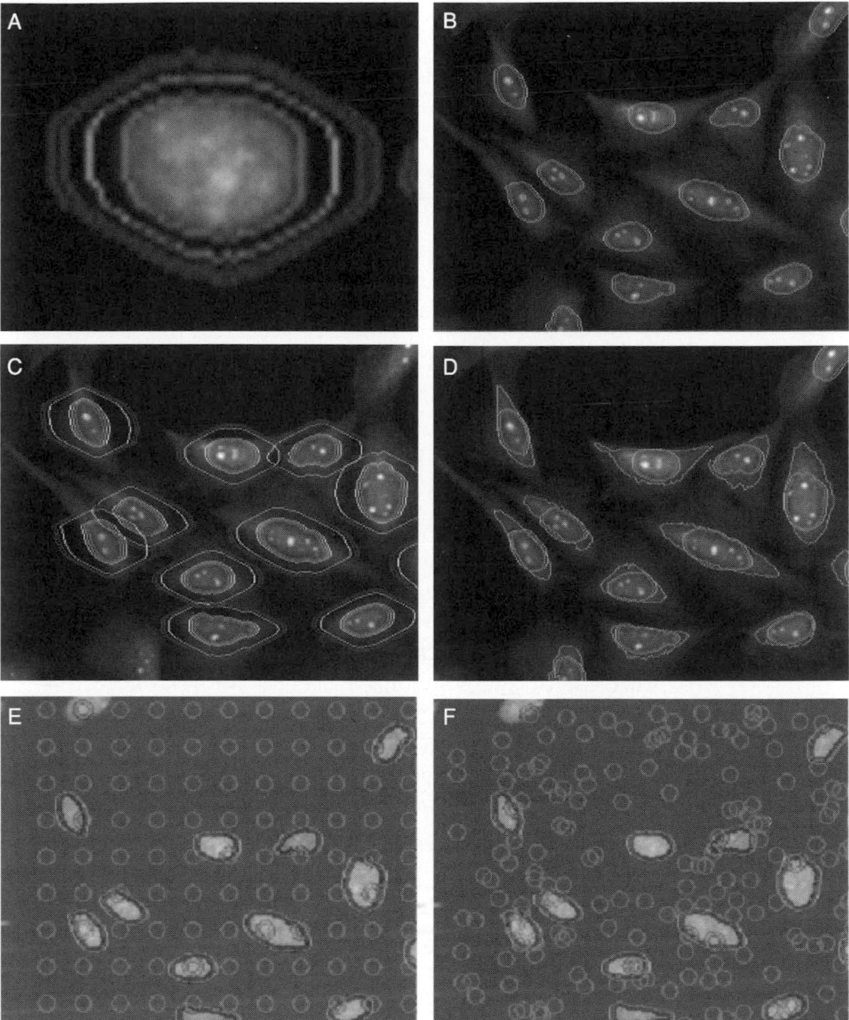

Fig. 6 Segmentation strategies. (A) The basic segmentation strategy employed by laser scanning cytometry technology. All of the adjacent pixels above a preset threshold level are used to define an event, with a red contour drawn around a nucleus. Other contours are used to indicate the area of integration and the areas where the local background is measured. (B, C) Intranuclear and extranuclear event quantification are shown. (D) shows secondary segmentation on an independent parameter. (E, F) Variations of random sampling, with the sampling points either in a regular lattice or randomly placed. (See Color Insert.)

determination is made using third and fourth contours constructed some number of pixels outside the integration contour. These are shown in blue in Fig. 6A. A function of all of the pixel values lying between these contours is used to determine and correct for each event's local background.

The software can generate additional sets of contours within each data contour, using channels other than those used to establish the data contour within a segmented event (Fig. 6B). This allows the analysis of specimens, primarily those with probe spots from fluorescence *in situ* hybridization (FISH) preparations, which contain localized constituents that can be independently stained, and their fluorescence differentiated from the fluorescence of a counterstain. Once the probe spot contours are determined, the data within these contours are summed to determine the *spot total fluorescence* feature. The number of pixels within each contour is used as its *spot area* feature. The maximum pixel value within the probe spot contour is used as *spot maximum pixel* feature.

A second method for compartmentalizing the data to estimate cytoplasmic constituents is also available (Fig. 6C). An algorithm is included in the software to sum pixel values between two contours of user-determined spacing set in position outside the data contour. These are called *peripheral contours*. The sum of these pixel values determines the *peripheral contour integral* and *peripheral MaxPixel* features and can be used to estimate constituents outside the nucleus of each cell if contouring was made using nuclear fluorescence.

The iSeries software includes the ability to separately segment on different channels and relate the independently segmented events to each other. Figure 6D shows an example of primary segmentation on the cell nuclei (red) and secondary segmentation on the cellular cytoplasm. Segmentation of cellular events is supplemented by a noncellular analysis approach based on random sampling methods. Phantom contours were originally developed for measuring instrument background in quantitative analysis and consist of creating circular contours of a user-determined size and number per field. Features are calculated for each of these phantom contours as if they were cells and are treated like segmented data in the analysis. The phantom contours can be generated in either a lattice (Fig. 6E) or a random (Fig. 6F) pattern. They have been found to have utility in mapping tissue constituents and in cellular analysis.

E. Data Analysis

Both the iCyte and the iCys system generates and stores multilevel types of data.

1. Event Data

Event Data may be displayed in the iCyte and iCys in various ways: scattergrams, histograms, expression maps (two parameter frequency distribution plots), or in the statistics table. Gating regions may be drawn on either of these graphs and populations isolated by these gates may then be inherited by other

Table I
Event Features from iCyte and iCys

Feature	Meaning
Area	Area in square microns of each event
Area (Sub*)	For primary events if there are subcontours, the total area, in square microns, of all subevents within the primary event
Background	Background fluorescence or signal nearby each event
Circularity	How close the event is to a perfect circle
Count (Sub*)	For primary events if there are subcontours, a count of the number of subevents within the primary event
Event ID	A unique identifier for each event
Integral	Integrated fluorescence per event for each saved channel
Integral (Sub*)	For primary events if there are subcontours, the integrated fluorescence per subevent for each saved channel
MaxPixel	Brightest pixel value per event for each saved channel
Parent ID	Subcontours only: the WellID of the primary event to which this subcontour is related
Perimeter	The number of pixels that comprise the perimeter of the event
Peripheral Integral	Peripheral contours only: the integral of the fluorescence in the area defined by the peripheral contour
Peripheral Max	Peripheral contours only: the MaxPixel in the area defined by the peripheral contour
Scan	Pixel number, from 0 to 768, indicating the y-position of the scanning mirror where the event was detected
Time	Time, in seconds, at which event was acquired
Well #	Well number (or scan area number) of event
X position	X position in the well or slide for each event
Y position	Y position in the well or slide for each event

scattergrams and histograms. Any of the event features (or ratios of two event features) may be displayed in these graphs (see Table I).

Statistics of event features or their subpopulations may be displayed in a statistics table. Detailed event feature information about a particular event may be displayed by selecting the event (or set of events) from a graph (see Table I).

2. Well Features

Well Features are summary statistics derived from the event features on a per-well or per-tissue-array element basis. Examples of the Well Features derived from a drug titration experiment are shown in Fig. 7A and B, and with the corresponding histograms from the experiment shown in Fig. 7C.

The changes in the cell number due to a loss of proliferation and the increase in the mean DNA content are key to interpreting the results of the experiment. The following Well Features may be generated for each of the channels employed in an analysis:

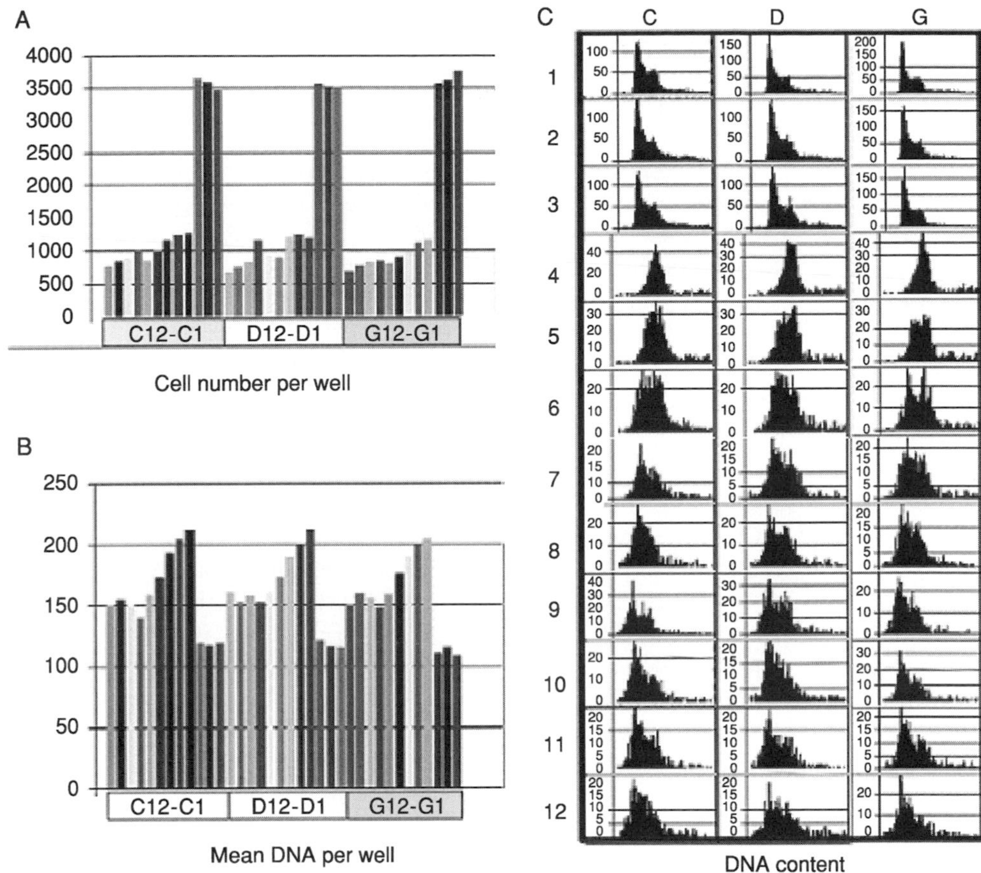

Fig. 7 Well features. Well features are summary statistical data for each of the wells or tissue elements in an analysis. (A) DNA content histograms from CHO cells in a camptothecin titration experiment. The cell counts for the analysis are plotted as Well Features in (B) and as the mean DNA content in (C).

- Mean
- Count
- Standard deviation
- Coefficient of variation
- Minimum
- Maximum
- Sum
- Full width half maximum

These can be calculated on the entire population of events in a well or on any gated subpopulation.

Well Feature values may be shown for an entire specimen carrier in terms of color-coded expression levels in the applications program carrier maps and in the iBrowser data management program.

The iCyte and iCys systems have the option to store all of the image data as a raw data image file for each of the selected hardware channels during scanning. These images may be reanalyzed and the user can modify any of the analysis parameters that are not directly related to data acquisition (such as the PMT or scatter sensor settings). As a result, once data from a specimen have been acquired at optimum settings, the analysis itself (selecting threshold levels, gallery images, etc.) may be optimized without rescanning the actual specimen, by simply reanalyzing the raw data files.

3. Repetitive Scanning

The iCyte and iCys can be set to perform *repetitive scans* or *rescans* of a given scan area, either for observing specimens at times separated by significant intervals or for manipulating the specimen (such as restaining or incubating) between scans.

The LSC provides the ability to merge event files (list-mode files) from multiple assays, based on the location of the events on the slide, to associate additional measurements with previous measurements. This foundational approach to combining data from multiple assay runs has been modified in the iCyte and iCys systems. Rather than merging the data after the event data have already been generated, iCyte and iCys merge the *image* data. In this way, the hardware channels from two separate runs may be treated as though they were from a single run. Contouring may be done on a channel from either run (or in the case of subcontours or the iNovator, both runs). Virtual Channels may be defined using constituents from either run.

F. Data Management Software and Reporting

As discussed earlier, the user has the ability to define and save numerous types of data files, both numerical and image. Each type of data element may have several entities, and a complete set of the data elements is obtained for all the samples in an analysis, such as wells in microtiter plate analysis or elements in a TMA. In practice, a very large amount of data can be produced in a single experiment, resulting in gigabytes of data spread throughout thousands of files. Although these data files are readable by many third-party programs and they are located in a logical hierarchy of directories, there is a need for a convenient method to access and review the data.

A *run* file serves as a "master index" for all the data associated with a given run. This *run* file may be opened using CompuCyte's *iBrowser*™ *data analysis software*.

The iBrowser allows the user to display the run data and perform additional analyses.

1. Displaying Summary Data

The iBrowser software displays summary data for a specific run, including annotations for that run, the run name and ID number, and a table of the Well Features for all wells or scan areas for that run. Figure 8 shows the portion the run file for an experiment in the starting page of the iBrowser analysis software. The experimental annotations are displayed, as well as the actual Well Feature data.

2. Displaying Carrier Data

Carrier data are data relevant to the experiment as a whole. A series of tabs for the various data elements are provided. Within each tab, the data element is laid out to approximate the physical arrangement of the wells or scan areas for the carrier used for that run. The tabs (and data elements) available are as follows:

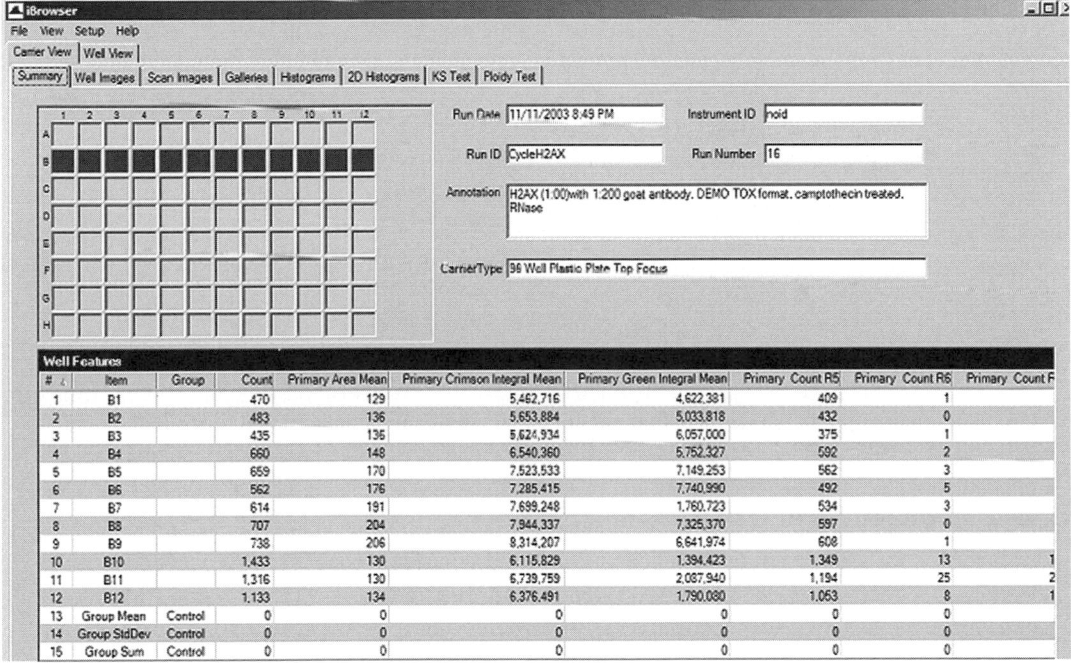

#	Item	Group	Count	Primary Area Mean	Primary Crimson Integral Mean	Primary Green Integral Mean	Primary Count R5	Primary Count R6	Primary Count F
1	B1		470	129	5,462,716	4,622,381	409	1	
2	B2		483	136	5,653,884	5,033,818	432	0	
3	B3		435	136	5,624,934	6,057,000	375	1	
4	B4		660	148	6,540,360	5,752,327	592	2	
5	B5		659	170	7,523,533	7,149,253	562	3	
6	B6		562	176	7,285,415	7,740,990	492	5	
7	B7		614	191	7,699,248	1,760,723	534	3	
8	B8		707	204	7,944,337	7,325,370	597	0	
9	B9		738	206	8,314,207	6,641,974	608	1	
10	B10		1,433	130	6,115,829	1,394,423	1,349	13	1
11	B11		1,316	130	6,739,759	2,087,940	1,194	25	2
12	B12		1,133	134	6,376,491	1,790,080	1,063	8	1
13	Group Mean	Control	0	0	0	0	0	0	
14	Group StdDev	Control	0	0	0	0	0	0	
15	Group Sum	Control	0	0	0	0	0	0	

Fig. 8 iBrowser Data Management Software start screen. An iCyte analysis can generate thousands of data and image files in the course of characterizing the test samples. The iBrowser enables the user to access the files, view, and analyze the data and generate reports.

- Well Images
- Scan-Field Images
- Galleries
- Histograms
- Two-Dimensional Histograms
- Statistical Analysis

Statistical tests include a modified form of the Kolmogorov-Smirnov test, where the user defines a group of control wells. The test will then compare each of the histograms in the data set to the control value and generate a new D-value histogram.

A ploidy test determines the percentage of cells in G_1, S, and the G_2/M phases of the cell cycle using a set of control samples to determine the DNA values used to partition the various cell cycle phases.

Data generated on multiple carriers within a single run (using the iCyte robot plate handler) may be displayed one of two ways. First, the data may be displayed on a carrier-by-carrier basis, the user selecting which carrier's data to display for each data tab. Alternatively, all the wells or scan areas may be displayed together, allowing comparisons between wells of different carriers.

3. Displaying Well Data

Any of the stored Well Features can be graphed as Well Feature histograms or as 2D scattergrams. Various elements such as scan-field images, histograms, and KS test plots may be displayed on a single page for a given well. An interactive carrier map allows the user to choose the well for which to display data.

A group of wells or a scan area may be defined so specified statistics may be applied to data from the entire group. Groups may be substituted for individual wells in the well-data display so graphical data of an *average of the group members* (as opposed to a single well) are displayed.

Users may select from any of the stored data elements and generate a report format that can be applied to any of the wells or tissue elements or to groups of them. The reports can be printed or stored as files.

G. iNovator Application Development Module

The *iNovator* application development module adds significantly to the capabilities of the *iCyte* and *iCys* systems. With the iNovator, the user can (1) employ imaging tools to the segmentation and data analysis process, (2) control this process with visually oriented macros, and (3) perform multiscale scanning and analysis.

The iNovator workspace (Fig. 9) provides multiple modules that control the various functions of data acquisition, image processing, segmentation, and event generation. By inserting these modules into the macro workspace and

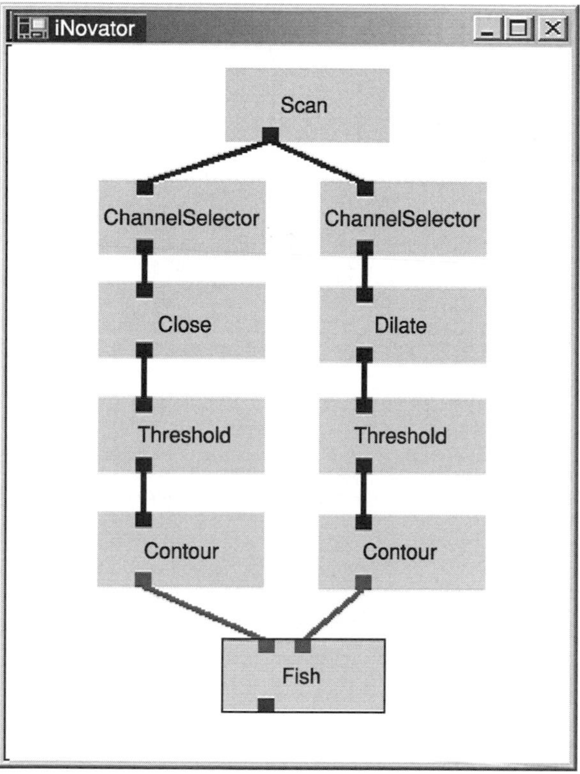

Fig. 9 The iNovator Applications Development System allows instrument scanning, image processing enhancement, event segmentation, and feature calculation protocols to be defined using a modular graphic interface.

connecting them, analysis paths may be constructed to control the scanning and data acquisition process.

IV. Selected Applications

A number of applications have been developed for the new iCys and iCyte platforms. They highlight the changes in the capabilities of the new generation of instrumentation.

A. Cell Cycle

Cell cycle analysis was one of the earliest applications developed on the LSC platform and continues to be one of the most important ones. As in FC, the total amount of DNA per cell can be precisely and stoichiometrically determined to

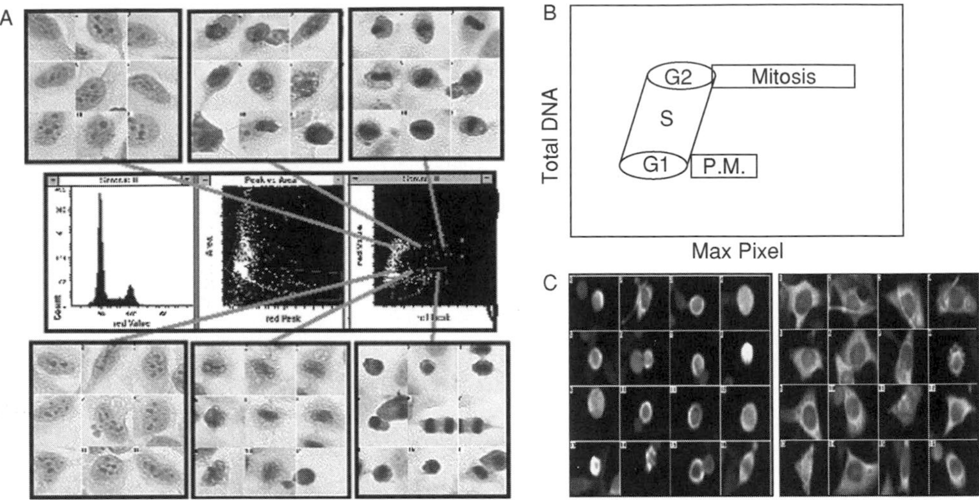

Fig. 10 Laser scanning cytometry technology allows quantitative analysis of the cell cycle along with morphometric feature analysis and the ability to produce images of cells from any of the events in the "scattergrams" (A). (B) shows how the full cell cycle is resolved using the MaxPixel and the total DNA features. (C) Automated detection of the translocation of cyclin B1 from the cytoplasm to the nucleus is also shown.

obtain cell cycle distributions. In addition, one of the morphometric features obtained for segmented nuclei, the MaxPixel, is directly correlated with the condensation of the chromatin in nuclei and can be used to differentiate interphase cells from mitotic cells (Luther *et al.*, 1996) (Fig. 10A). In plots of either the nuclear area or the DNA content versus the chromatin condensation, extensions of dots with increasing MaxPixel can be seen extending to the right from the G_1 and G_2 populations. Morphological examination of cells relocated from within regions set to correspond to increasing MaxPixel content confirms that they resolve distinct phases of the cell cycle, as seen in the hematoxylin-eosin–stained cell images in the galleries. Figure 10B illustrates how the entire cell cycle is resolved using a single detector, leaving other detectors available to measure additional related markers.

B. Automated Cell Cycle

The iCyte and iCys cytometers often use microtiter plates, enabling higher throughput and large sample numbers in a cell cycle experiment. Data from a toxicological study are shown in Fig. 11, as a portion of the data from an analysis of a plate of adherent CHO cells stained with propidium iodide and treated with RNAse. The three left-hand columns are untreated cell controls, and the other five columns are data from wells that were treated with increasing amounts of colchicine.

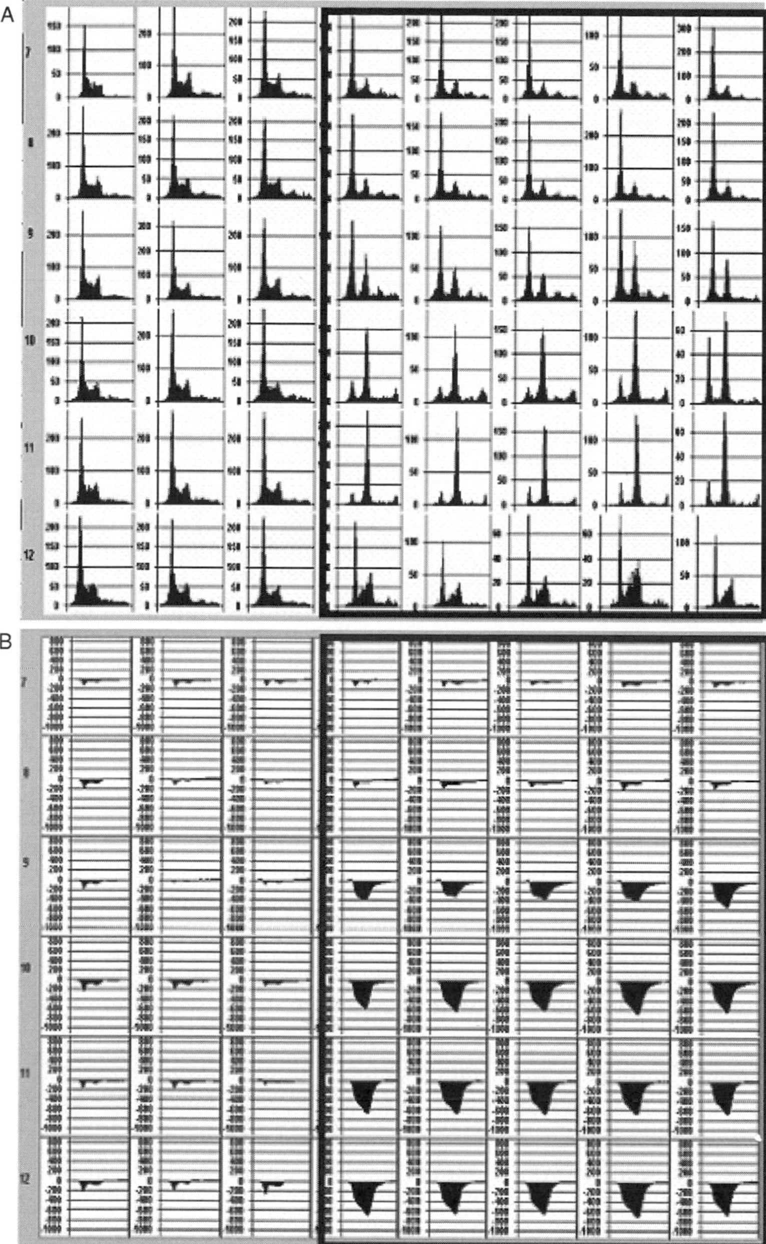

Fig. 11 Cell cycle Kolmogorov-Smirnov test. (A) DNA staining histograms from an analysis where cells were treated with colchicine, stained and then analyzed. The drug-treated wells (in the box) show a blockage of the cell cycle in G_2. (B) KS D-value histograms are shown for the same data set. The amount of deviation from the centerline is indicative of the amount of difference between the sample and the defined control value.

In Fig. 11A, the raw DNA histograms are shown, with good precision as evidenced by the consistent location of the G_1 peak within the various histograms. Blockages of the cells in the G_2 phase are seen in the histograms within the box. In Fig. 11B, the same data set is shown as Kolmogorov-Smirnov D-value histograms. The effects of the drug are clearly displayed as the "shark-fin" distributions at the higher drug dosages. These analyses also produce numerical values that can be exported to spreadsheet programs.

C. Cell Cycle Combined with Multiple Fluorescent Markers

1. Multiple Color Cell Cycle

Using a single detector to obtain the complete cell cycle information leaves the other detectors free to evaluate additional cell cycle and apoptosis-related markers. Cyclin B1 is a controlling molecule at the S/G_2 checkpoint and originates in the cytoplasm of cells and migrates to the nucleus as the cells enter mitosis. The peripheral contouring technique was developed to automatically quantify this movement, in a process that has come to be known as *compartmentalization*. In Fig. 10C, images of cyclin B1–positive cells are shown in galleries of images relocated after automated compartmentalization of the FITC-labeled cyclin B1 fluorescence (Kakino *et al.*, 1996; Kamentsky *et al.*, 1997). Cyclins D and E have also been correlated with the cell cycle in multicolor studies (Juan and Darzynkiewicz, 1998).

2. Multicolor Function-Related Markers

Many antibodies have been developed that measure the functional state of molecules. Examples of phosphorylation-state antibodies include antibodies to histone H3, a cell cycle–dependent state that occurs near metaphase in mitosis, as well as histone γH2AX, which is phosphorylated around sites of double-stranded DNA breakage. Other activation antibodies recognize epitopes that are exposed only after molecules are activated, such as antibodies that are specific for the cleaved form of caspase-3. Combined with cell cycle information and other methods of detecting apoptosis such as the TUNEL assay, events leading to apoptosis can be dissected and the effects of various drugs can be determined. Studies from Huang *et al.* (2003) and Dmitrieva *et al.* (2004) demonstrate the time sequence for the transition from DNA damage to apoptosis in a leukemia-derived cell line as a response to DNA damaging topoisomerases, where low numbers of drug-induced strand breaks appear before caspase-3 activation and the induction of apoptosis-related strand breaks. Other studies have used LSC to examine the signal transduction cascades that result in DNA damage reported by the phosphorylation of H2AX (Huang *et al.*, 2004). Figure 12 shows the iCyte analysis screen from similar studies, but as a dose response to camptothecin in the adherent CHO cell line. A sample scan image is shown, and in the scattergrams there is a clear demarcation between H2AX-positive and H2AX-negative cells.

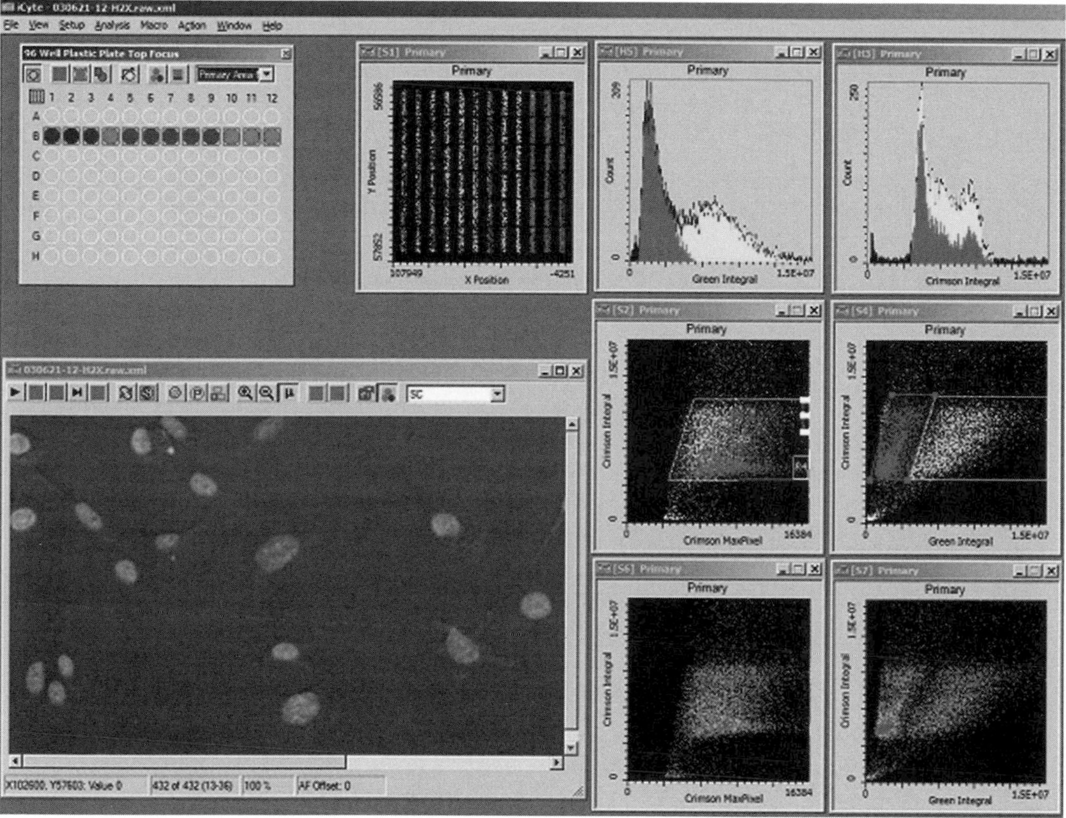

Fig. 12 Analysis screen from an iCyte analysis. Many standard flow cytometry population data analysis tools are available, such as histograms, scattergrams, and density expression maps, and color gating. These are combined with images and image processing techniques to provide high content analysis. (See Color Insert.)

Several methods, including scattergrams and plate expression maps, exist within the application for color-coding the expression of constituents. Further analysis can be done in the iBrowser software, as in Fig. 13, where the DNA content and H2AX expression are shown as histograms, along with their KS D-value histograms.

In this model system, there are dramatic changes in the DNA profiles as a function of drug dosage. At low levels of camptothecin, the cells are blocked in the S phase of the cell cycle. At higher dosages, there appears to be a loss of the S-phase cells. Monitoring the cell numbers, caspase-3 immunostaining, and nuclear morphology in time-course studies show the difference in cell numbers caused by the lack of proliferation, as opposed to the loss of cells from the system to

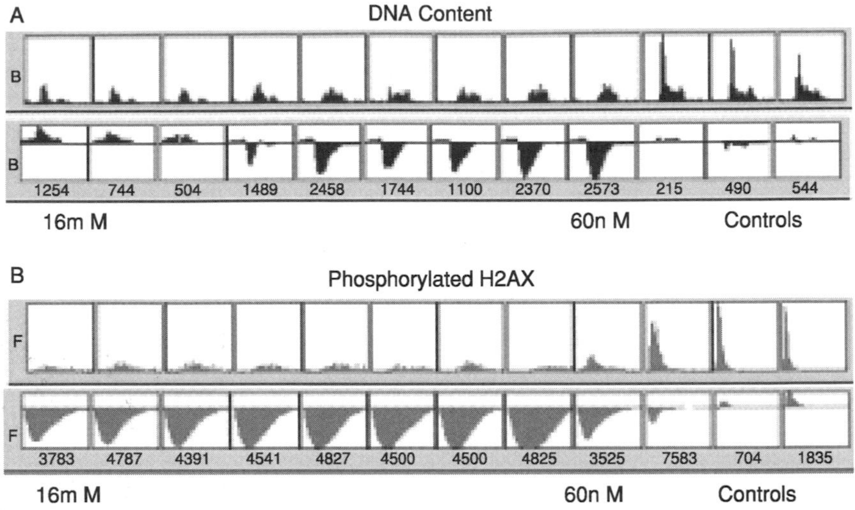

Fig. 13 Kolmogorov-Smirnov analysis. (A) KS analysis of cells that were treated with a range of dosages of camptothecin. Differential effects of the drug can be seen, with the greatest differences being in the lower dosage levels. (B) Green fluorescence data from the same wells labeled with antisera to phosphorylated histone H2AX.

apoptosis, suggesting differential pathways for cell death at different dosage levels of the camptothecin.

3. Subtle DNA Damage: Clastogens

The evaluation of clastogenic effects of drugs is usually done by some form of the micronucleus assay, where broken or detached chromosomes are separated from the spindle apparatus. After cells undergo mitosis, the fragments become trapped in the cytoplasm of cells, where a nuclear membrane forms around them and they become micronuclei. Traditionally, micronuclei are enumerated by manual visualization to assay the clastogenicity of compounds, but methods have been developed for performing automated analysis of micronuclei using the LSC (Smolewski *et al.*, 2001; Styles *et al.*, 2001). The iNovator software module has been employed to automate the cell-line micronucleus assay. Figure 14 shows the segmentation steps in the process, where primary segmentation identifies cells based on protein staining of the cytoplasm with FITC. Secondary segmentation segments nuclei on images that are binary masks of the DNA staining. Watershed algorithms are used to subsegment the nuclei and separate the nuclei of binucleate cells (a result of the cytochalasin B that is commonly employed in the assay) and the micronuclei. In the analysis procedure, quantitative DNA content measurements are used to denote eligible parent cells, and for each of the parent cells, the

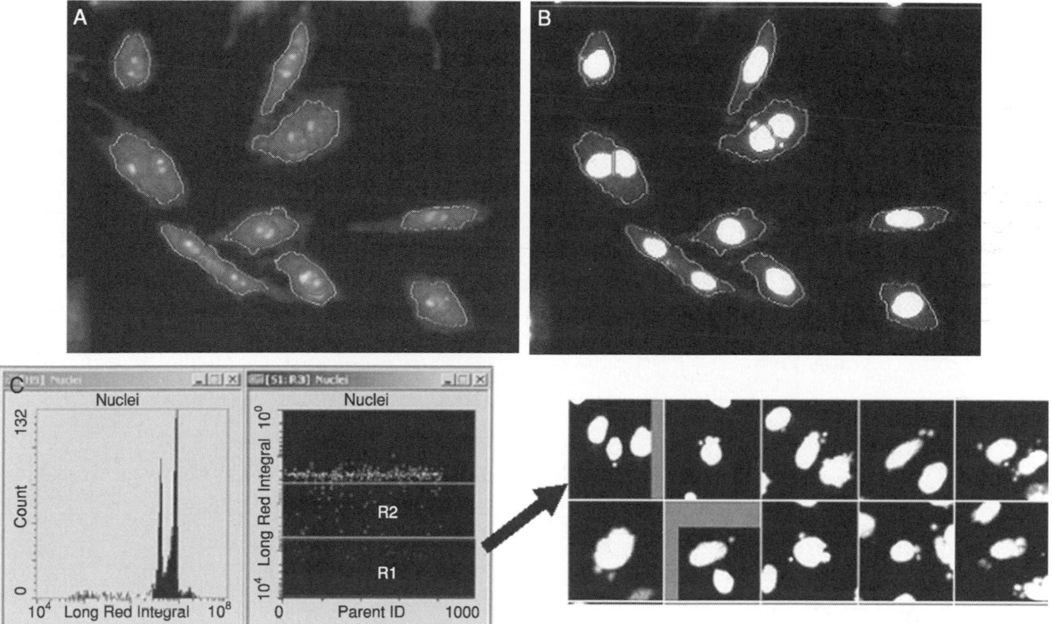

Fig. 14 Micronucleus quantification. Complex segmentation and image processing techniques are used to quantify the amount of micronuclei in test samples. Cells are segmented based on cytoplasmic staining, and nuclei are subsegmented based on DNA staining. Watershed algorithms are used to differentiate individual nuclei and micronuclei.

number and DNA content of the nuclei are determined. Damaged, or sub-G_1, nuclei can be classified as distinct entities from micronuclei, and counts of the number of cells with micronuclei are obtained. Images of the micronuclei are obtained for assay validation purposes.

Results from a known clastogen (etoposide) titration are shown in Fig. 15. For each well, the percentages of micronuclei and damaged nuclei are calculated, as well as the mean DNA content per cell and the mean nuclear area. The data show differential dosage responses to the drug, with micronuclei formation more prevalent at low dosages and other indicators of cytotoxicity appearing at higher dosages.

D. Live Cell Analysis

The analysis of live cells is greatly facilitated by the inverted format of the iCyte and iCys. Cells can be analyzed in carriers (microtiter plates, chamber slides, Petri dishes, and even tissue culture flasks) that allow them to remain immersed in suitable amounts of culture fluid.

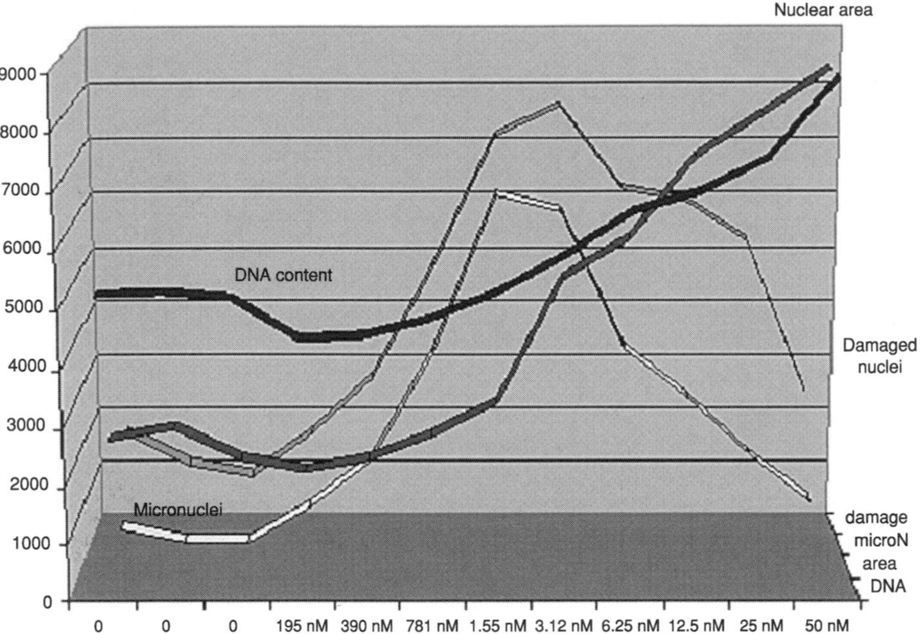

Fig. 15 Micronucleus count combined with other *Well Features*. In this titration study of the drug etoposide, micronuclei count is combined with the count of damaged nuclei, DNA content and the nuclear area to assay the differential effects of the drug at different concentrations. Micronuclei formation, an indicator of clastogenic activity, occurs at the lower dosages. At higher levels, the drug effects are toxicological.

Live cell studies are typically performed as homogeneous assays, with the staining solutions left in the supporting media and numerous dyes combined. A three-color system is illustrated in Fig. 16A. DNA content was measured with Hoechst 33342, a dye traditionally excited by ultraviolet lasers but has been found to be adequately excited by the 405-nm violet diode laser. The dye gives cell cycle information, and necrotic and apoptotic cells exhibit hyperstaining. FITC-labeled Annexin V (Serologicals, Georgia) was used to detect membrane inversion and phosphatidyl serine exposure, a midrange apoptotic event. Finally, propidium iodide (Molecular Probes, Eugene, Oregon) was used to assay membrane permeability, a late apoptotic or necrotic event.

Numerous dyes can be used to measure mitochondrial membrane potential to make this a four-color system. In Fig. 16B, Mitoshift reagent (Trevigen, Gaithersburg, Maryland) was added to the dye combination, excited with the blue laser, and orange fluorescence was quantified. In viable cells the mitochondria are perinuclear, but immediately after insult with valinomycin, they aggregate into

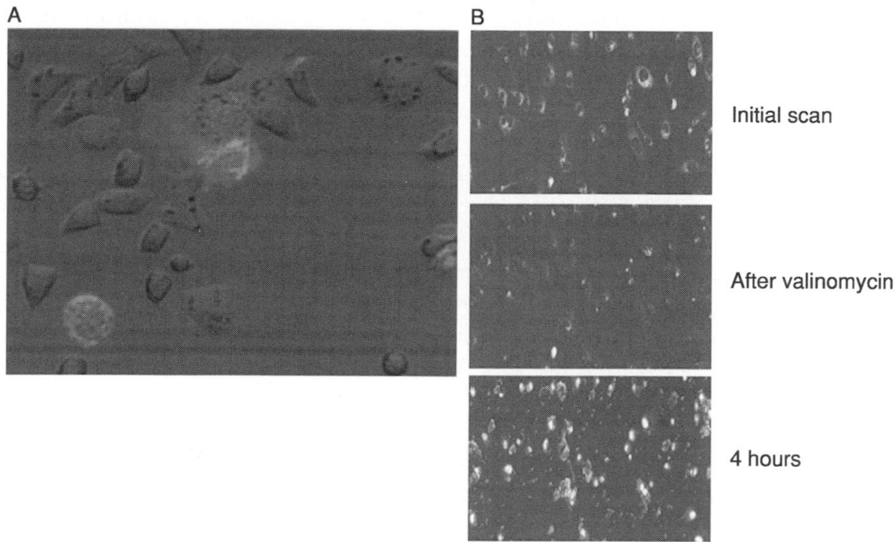

Fig. 16 Live cell analysis. Live cells can be analyzed in a multilabel homogeneous system. (A) Cell labeled with Hoechst 33342 for DNA content, Annexin V for apoptosis-induced membrane inversion, and propidium iodide for membrane permeability. (B) The same dye cocktail with a mitochondrial marker dye added. (See Color Insert.)

clusters. After 4 hours, the mitochondrial staining is abrogated, and Annexin V and propidium iodide staining is evident.

Many other features of live cells can be measured. A powerful technique was developed where cells are analyzed first when they are in the live state, then removed from the instrument, fixed, restained with a new set of reagents, and reanalyzed. The resultant data files are "merged," where the location of the cells on the carrier is used as an identifier, and the live and fixed analysis features are available for each cell (Li *et al.*, 1999).

E. CFP/YFP FRET

The cloning of the green fluorescent protein (GFP) gene led to a revolutionary new approach in cellular analysis which allows cells to be engineered via molecular biology techniques to produce their own fluorescent labels. A report used a GFP fusion with transmembrane receptors whose expression level was analyzed during the cell cycle using the LSC. In brief, the GFP can be inserted at will within a cell's DNA, and if the particular segment DNA is actively transcribed, the cell will synthesize GFP as a reporter molecule. The subsequent development of spectrally distinct variants, such as cyan fluorescent protein (CFP) and yellow fluorescent protein (YFP), has led to even more sophisticated techniques relying on FRET. If two fluorescent molecules have overlapping emission and excitation spectra and the molecules are very close spatially, excitation of the donor molecule will lead to the emission of fluorescent light by the acceptor molecule at its emission wavelength.

An illustrative model of the power of the technique involves the key apoptosis effector molecule caspase-3. In the unactivated form, the caspase-3 molecule exists as a dimer. Activation by any one of multiple signaling pathways leads to cleavage of the dimer. The cleaved fragments activate other molecules to commit to the apoptosis pathway. A construct was developed that inserts a CFP molecule on one side of the DEVD cleavage site, CP3, and a YFP molecule on the other side of the site (Barteneva *et al.*, 2004). In intact caspase-3, the CFP and YFP are close enough that FRET occurs. In cleaved molecules, the proximity requirement for FRET is not met, and there is no FRET.

Figure 17A shows that the amount of FRET can be quantified on the iCys as the slope of the events in scattergrams plotting violet-excited versus blue-excited yellow fluorescence. Scattergrams for YFP-only transfected cells, untreated CP3-transfected cells, drug-treated CP3 cells, and CP3-transfected cells that were exposed to the caspase inhibitor z-VAD-FMK before drug treatment are shown. The analysis criterion is the slope of the points in the scattergrams, and the results shown are consistent with the expected biological results. Untreated samples established the baseline for the amount of FRET, the drug-treated samples showed a loss of FRET, and the inhibitor-treated samples maintained the amount of FRET.

The preceding evaluation of FRET relied on only two of the variables available in FRET analysis. The amount of violet-excited blue fluorescence is a measure of the amount of CFP in the cells. The ability of LSC technology to define ratios of parameters allows the transformation of multidimensional data into two dimensions, with a corresponding increase in the amount of separation between the intact and cleaved populations in scattergrams (Fig. 17B). Several methods are available for quantifying the amount of FRET as a dose response in titration studies, including plotting the number of cells in defined "FRET" and "no-FRET" regions. The result is a sensitive method for quantifying drug dosage responses (Fig. 17C).

F. Automated Tissue Analysis and Tissue Micro Arrays

LSC technology allows analysis of samples with preserved tissue architecture, such as touch preparations (tissue imprints), TMAs, and tissue sections. Often the nucleus is used to segment or identify events that serve as the basis for quantifying the elements of the tissue. In many types of tissue, especially in tissues with well-separated nuclei such as epidermal and liver tissues, this paradigm works well.

Other types of tissue present problems in segmentation. Many tumors are characterized by large irregular nuclei with very little cytoplasm separating them from each other. These "molded" nuclei create difficulties in segmentation.

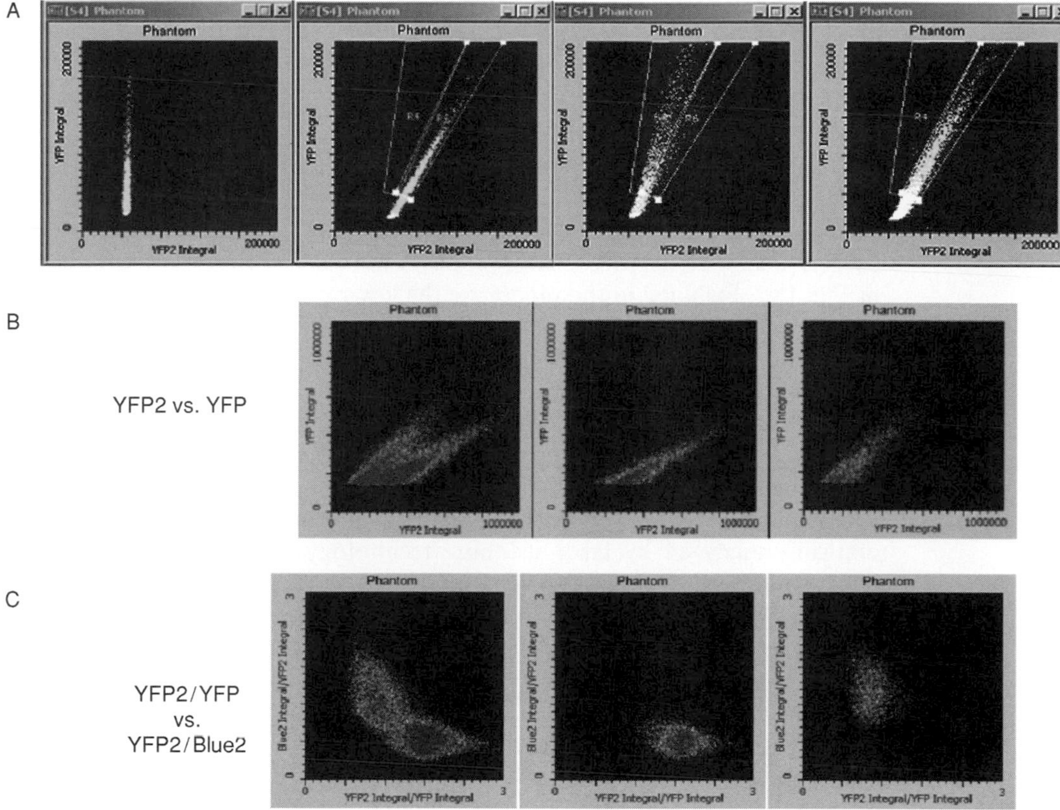

Fig. 17 Fluorescence resonance energy transfer (FRET)–based caspase-3 analysis. A biomarker construct using cyan fluorescent protein to yellow fluorescent protein FRET to monitor the activation of caspase-3 is employed as an early marker of apoptosis induction. In (A) The amount of FRET is empirically determined, and its loss is quantified as the caspases are cleaved. (B) Employing ratios of parameters increases the separation between live and apoptotic cells in scattergrams.

Lymphoid tissues such as spleen and tonsil are also difficult to segment because the lymphocytes that comprise them are densely packed, with little cytoplasm to separate them. In these types of samples, phantom segmentation is often used to map the constituents.

Automation of the tissue analysis process is often desirable to increase throughput and introduce objectivity. One application, using the iCyte and the automated robotic arm, is the analysis of a large number of liver tissue sections in pharmaceutical toxicological studies. Up to 45 holders containing four slides each may be prepared and stored in the robotic magazine. The holders are loaded onto the instrument, and an initial low-resolution scout scan is done to identify the pieces of tissue on the slides. The low-resolution scan employs the same 40× objective lens used for the high-resolution analysis to maintain the instrument sensitivity, but the stage steps of the instrument are increased dramatically, typically 40-fold, to decrease the required scanning time. The low-resolution scans are aspect-ratio corrected and the pieces of tissue are automatically segmented and regions for high-resolution scans are generated within them. This strategy was adopted to avoid edge effects in the staining of the tissues as shown in Fig. 18A.

After the high-resolution analysis is completed, the experimental results are made available to the iBrowser program. A typical element report screen is shown in Fig. 18B. Included in the report are the low-resolution scan image of the slide, mosaic images for all of the scanned areas, representative high-resolution scans, and histogram data, both raw and corrected for autofluorescence. A final expression map of the cellular data is also shown. One of the major benefits of this system is that the entire analysis is done in a walkaway fashion. Quantitative data regarding the protein of interest are generated automatically, and documentation is obtained for each step of the analysis procedure.

TMAs are becoming essential tools in proteomic research and can be automatically analyzed by laser scanning technology. As in the previous section, low-resolution scans are performed to identify the location of the array elements (Fig. 19A), typically using a composite virtual signal that is the result of the summation of all of the fluorescence channels being employed.

Tissue elements are then actively identified and high-resolution analysis of the found tissue plugs is done (Fig. 19B). Multiple features are quantified for each of the arrays by using multiple color analysis, and serial sections of array blocks can be analyzed to further increase the depth of features. The resultant features can be displayed graphically (Fig. 19C) or quantitatively or, more likely, exported to external databases where the data are combined with the known clinical and pathological information about the tissue elements. Multiple resolution image data are also obtained for all of the tissue elements. In Fig. 19D, scan images are shown for HER-2/neu–stained breast tissue plugs developed with either chromatic or fluorescent dyes. The chromatic dyes are visualized and to a certain extent quantified, by laser light absorption contrasted to the fluorescence of the counterstain to give a hybrid image.

The examples shown demonstrate the utility of automated tissue analysis in nonclinical diagnostic settings. The development of new analysis techniques and the availability of many new molecular markers make it possible to obtain objective quantitative data along with image data for assay verification and documentation.

G. Immunophenotyping

Immunophenotyping has become one of the mainstream applications for flow cytometric analysis. There are certain areas where LSC technology is the preferred method, for example, small size samples that do not contain enough material to be eligible for a flow cytometric analysis. These include samples

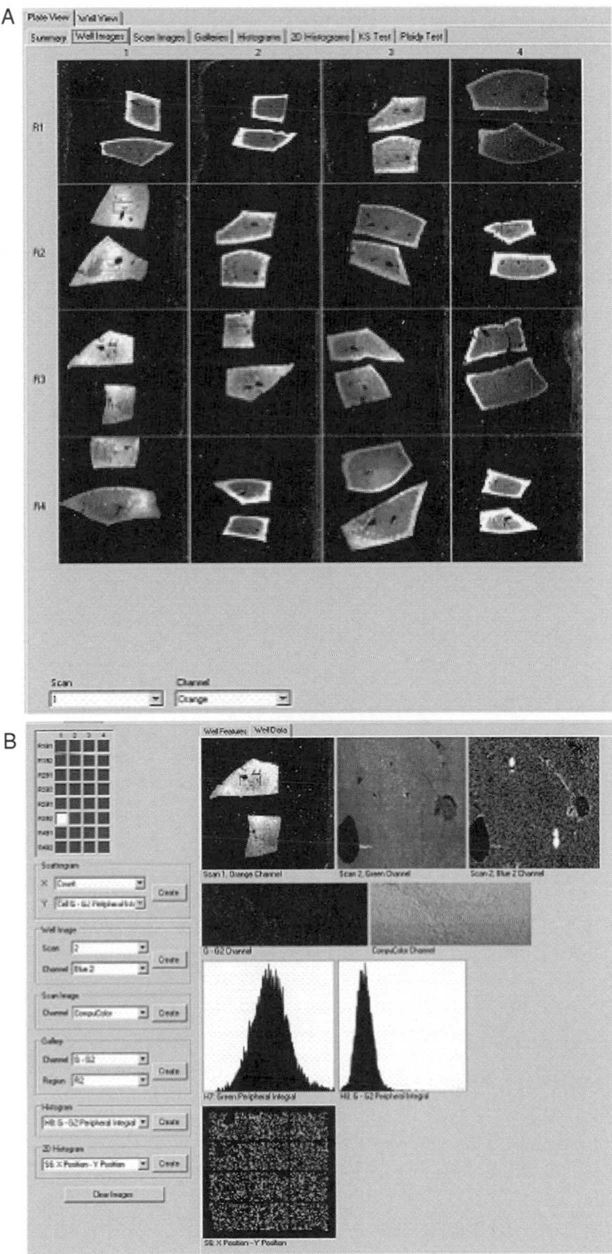

Fig. 18 (A) Automated tissue analysis. A fully automated system has been designed to automatically load trays of microscope slides, do a high-speed low-resolution scan to find the tissue elements and assign an area for a high-resolution scan, and then perform the high-resolution analysis. (B) A report for one of the tissue elements.

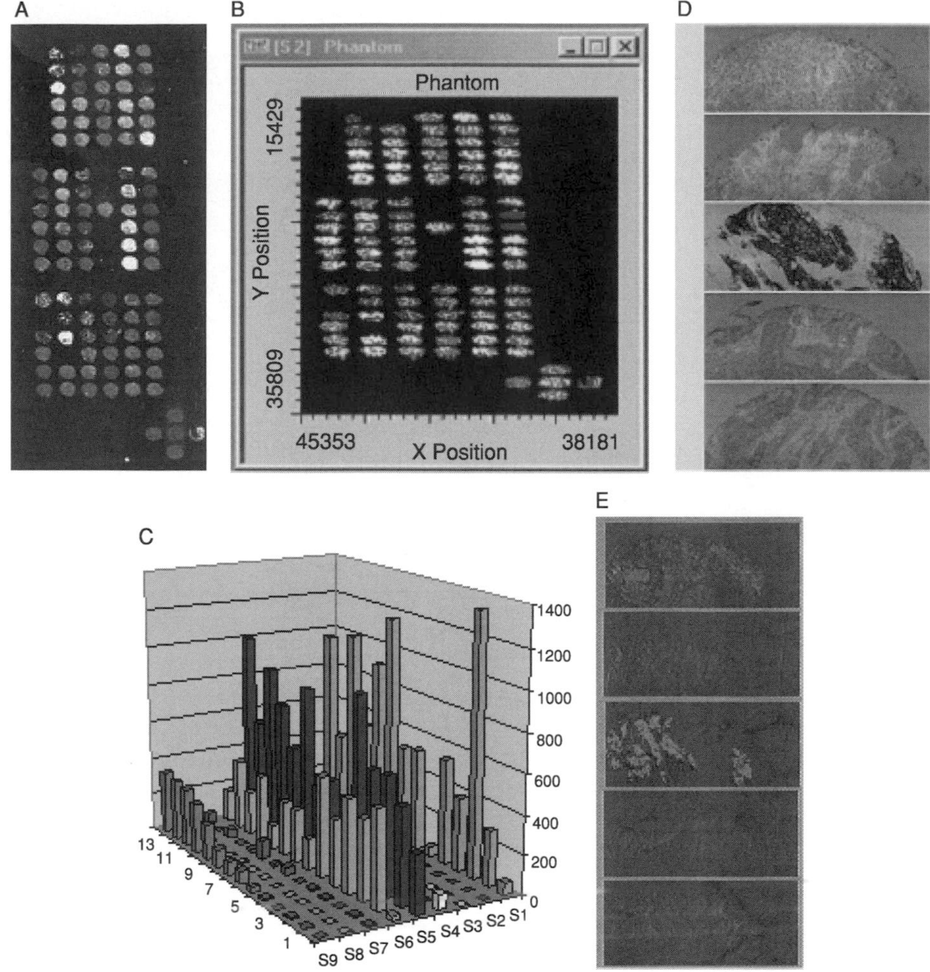

Fig. 19 Tissue microarray analysis (TMA). Tissue microarrays are also analyzed by first finding the elements with a high-speed scan (A), and then scanning the individual elements at high resolution, in this case using phantom segmentation. Quantitative, high content data are generated, as well as image data for each of the TMA elements. (See Color Insert.)

from hematopathological biopsies and fine needle aspirates. LSC methods have been developed where the cells are adhered to chambers on a microscope slide, stained and washed in place, and then analyzed in a single automated analysis (Clatch *et al.*, 1997, 2001). The same principles are being transferred to the adherence of cells in microtiter wells, staining, and analysis *in situ* with the iCyte and iCys instruments.

V. Utility of Solid-Phase Imaging Cytometry in Life Science Research and Drug Discovery

Current laser scanning technology is a powerful hybrid, incorporating the strengths of the related technologies of FC, digital fluorescence microscopy, bright-field microscopy, and image analysis into a single platform. These combined attributes give it a degree of versatility and flexibility that is not found in any single alternative technology.

In life sciences research, LSC technology combines quantitative analytical data with imaging capabilities, as well as the ability to use the microscope to visualize and manipulate events of interest. It can be used as a stand-alone platform but complements other technologies such as FC and confocal microscopy in core laboratory facilities.

LSC technology also plays a useful role in drug discovery. The concept of high content cellular screening and analysis has been presented to the pharmaceutical industry by a number of instrument manufacturers as a method to speed the drug discovery process. LSC technology provides significant utility in the early stages of the drug discovery process, including target identification, lead optimization, and predictive/investigative toxicology.

References

Barteneva, N., Kim, M. Luo, K., Chang, D., SumenCarman, C., Ketman, K., Brochu, M., Jr., Telford, W., and Luther, E. (2004). Cytometry analysis of living cells expressing FRET (fluorescence resonance energy transfer)–caspase sensors (in preparation).

Clatch, R. J. (2001). Immunophenotyping of hematological malignancies by laser scanning cytometry. *In* "Methods in Cell Biology" (Z. Darzynkiewicz, J. Robinson, and H. Crissman, eds.), Vol. 64. Academic Press, San Diego.

Clatch, R. J., and Walloch, J. L. (1997). Multiparameter immunophenotypic analysis of fine needle aspiration biopsies and other hematologic specimens by laser scanning cytometry. *Acta Cytol.* **41,** 109–122.

Darzynkiewicz, Z., Bedner, E., Li, X., Gorczya, W., and Melamed, M. (2000). Laser scanning cytometry—a new instrumentation with many applications. *Exp. Cell Res.* **249,** 1–12.

Dmitrieva, N. I., Cai, Q., and Burg, M. (2004). Cells adapted to high NaCl have many DNA breaks and impaired DNA repair both in cell culture and *in vivo*. *Proc. Natl. Acad. Sci. USA* **101,** 2317–2322.

Huang, X., Traganos, F., and Darzynkiewicz, Z. (2003). DNA damage induced by DNA topoisomerase I- and topoisomerase II inhibitors detected by histone H2AX phosphorylation in relation to the cell cycle phase and apoptosis. *Cell Cycle* **2,** 614–619.

Huang, X., Okafuji, M., Traganos, F., Luther, E., Holden, E., and Darzyienkiewicz, Z. (2004). Assessment of histone H2AX phosphorylation induced by DNA topoisomerase I and II inhibitors topotecan and mitoxantrone and by DNA cross-linking agent cisplatin. *Cytometry* (in press).

Juan, G., and Darzynkiewicz, Z. (1998). Detection of cyclins in individual cells by flow and laser scanning cytometry. *Methods Mol. Biol.* **91,** 67–75.

Kakino, S., Sasaki, K., Kurose, A., and Ito, H. (1996). Intracellular localization of cyclin B1 during the cell cycle in glioma cells. *Cytometry* **24,** 49–54.

Kamentsky, L. A., and Kamentsky, L. D. (1991). Microscope-based multiparameter laser scanning cytometer yielding data comparable to flow cytometry data. *Cytometry* **12,** 381–387.

Kamentsky, L. A., Burger, D. E., Gershman, R. J., Kamentsky, L. D., and Luther, E. (1997). Slide-based laser scanning cytometry. *Acta Cytol.* **41,** 123–143.

Kamentsky, L. A. (2001). Laser scanning cytometry. *In* "Methods in Cell Biology" (Z. Darzynkiewicz, J. Robinson, and H. Crissman, eds.), Vol. 63. Academic Press, San Diego.

Li, X., and Darzynkiewicz, Z. (1999). The Schrodinger's cat quandary in cell biology: Integration of live cell functional assays with measurements of fixed cells in analysis of apoptosis. *Exp. Cell Res.* **249,** 404–412.

Luther, E., and Kamentsky, L. A. (1996). Resolution of mitotic cells using laser scanning cytometry. *Cytometry* **23,** 272–278.

Roman, D., Greiner, B., Ibrahim, M., Pralet, D., and Germann, P. G. (2002). Laser technologies in toxicopathology. *Toxicol. Pathol.* **30,** 11–14.

Smolewski, P., Ruan, Q., Vellon, L., and Darzynkiewicz, Z. (2001). Micronuclei assay by laser scanning cytometry. *Cytometry* **45,** 19–26.

Styles, J. A., Clark, H., Festing, M. F., and Rew, D. A. (2001). Automation of mouse micronucleus genotoxicity assay by laser scanning cytometry. *Cytometry* **44,** 153–155.

PART II

General Techniques

CHAPTER 9

Biohazard Sorting

Ingrid Schmid,[*] Mario Roederer,[†] Richard A. Koup,[†] David Ambrozak,[†] and Stephen P. Perfetto[†]

[*]David Geffen School of Medicine, University of California Los Angeles
Department of Hematology/Oncology
Los Angeles, California 90095

[†]Immunology Laboratory
Vaccine Research Center
National Institute of Allergy and Infectious Diseases
National Institute of Health
Bethesda, Maryland 20892

I. Introduction

Flow sorters are instruments capable of separating cell populations based on their physical properties or by exploiting differences in cell surface receptors, intracellular structures, or molecular expression. When these instruments are used for sorting unfixed samples known to harbor pathogens, the term *biohazard sorting* has been created to encompass aspects of this process pertinent to the protection of sorter operators, others involved in these experiments, and the environment. Although *biohazard sorting* is mostly associated with the processing of samples containing infectious agents (Giorgi, 1994), it also applies to cell sorting of unfixed

METHODS IN CELL BIOLOGY, VOL. 75

0091-679X/04 $35.00

human cell preparations or unfixed cells from other species that may carry pathogenic organisms known to infect humans (Schmid *et al.*, 1997b, 2003). Samples that have been treated with fixatives to inactivate infectious agents (Shapiro, 2003) are considered non-biohazardous. However, a careful review of the effectiveness of a given fixation procedure against a certain pathogen is essential before classifying a sort as non-hazardous, and it is always advisable to use prudent safety practices even when sorting fixed cells (Schmid, 2000; Schmid *et al.*, 1997b, 2003).

II. Critical Aspects of the Procedure

A. Sample Handling

1. General Considerations

In the United States, all laboratories processing human blood must follow universal precautions as outlined in the Federal Code regulation: Occupational Exposure to Bloodborne Pathogens (National Committee for Clinical Laboratory Standards, 1997; United States Federal Code Regulation, 1991) and additional local and institutional regulations. Other countries have developed their own regulatory standards or adopted aspects of regulations for working with biological agents as mandated in the United States. Levels of containment (known as *biological safety level* [BSL]) combine safety equipment, laboratory practices, and facility design. For specimens carrying infectious agents, safety levels are assessed according to the criteria set forth in the Centers for Disease Control and Prevention (CDC) publication *Biosafety in microbiological and biomedical laboratories*, 4th ed., available online at www.cdc.gov/od/ohs. As a general rule, pathogens encountered in a typical cell-sorting laboratory are transmitted by the parenteral route and require BSL-2 containment, for example, human immunodeficiency viruses (HIV-1,-2), hepatitis viruses (B,C,D [delta]) (Schmid *et al.*, 2003). Relevant details for the preparation of infectious samples containing HIV for flow cytometric analysis such as shipping and receiving of specimens, local sample transport, staining, and disposal have been described (Schmid *et al.*, 1999). Although cell sorting of HIV-infected specimens can be performed in a BSL-2 facility, because of the generation of aerosols during cell sorting, BSL-3 practices and personal protective equipment are mandatory for sorting HIV-positive samples to protect operators from potential exposure (Ferbas *et al.*, 1995; Giorgi, 1994; Perfetto *et al.*, 2003). Complete BSL-3 containment is required for the sorting of infectious agents causing serious or potentially lethal disease that are transmitted by the inhalation route (CDC, 1999).

2. Sample Preparation and Sort Acquisition

For proper operation of any cell sorter, samples must be prepared as single-cell suspensions. Aggregates of cells that are present in the sample to be sorted tend to accumulate at the interface of sample fluid and sheath fluid and can partially or completely clog sorting nozzles. Any interruption of a biohazard sort enhances the risk of operator exposure to pathogens contained in the sort sample due to potential splashes and escape of sort aerosols during the manipulations required to resume sorting. Consequently, it is imperative that formation of cell aggregates during sample preparation be minimized and clumped cells be prevented from interfering with normal instrument operation. The composition of the culture medium used for sample resuspension can influence clumping. For instance, murine cells aggregate in the presence of EDTA and, therefore, should be sorted only in EDTA-free media. Passing samples through narrow-gauge syringes, sample vortexing before sorting, and sample agitation during sorting can help to disperse cell aggregates, but mixing should not be too intense or overused because it may cause cells to break apart. Furthermore, highly concentrated cell suspensions have an increased tendency to clump, so cells should be diluted to the lowest possible density for the sort speed used. Frozen cell samples that are thawed for sorting frequently contain dead cells, which may release DNA into the media. DNA then binds to the surface of live cells, and after centrifugation, these samples form solid aggregates leading to nozzle clogging problems. In these situations, adding 20 μg/ml of DNAse for 10 minutes at 37 °C can help. Furthermore, spinning samples at 300 g for 5–10 minutes is sufficient to pellet cells. Higher centrifugation speeds can damage cells and compact cells so densly that they are difficult to break up into singlets. Sort samples are often chilled to preserve cellular structures and prevent capping of antibodies bound to cell surface receptors. However, the cold can aggravate clumping, so sorting at an intermediate temperature such as 15 °C may be preferable over sorting at 4 °C. Options to remove cell aggregates include filtration through nylon mesh filters, such as using meshes with different pore sizes available from Small Parts, Inc. (Hialeah, Florida), 12 × 75-mm tubes with 100 μm cell strainer caps (Becton Dickinson, Falcon), or individual 70-μm cell strainers (Becton Dickinson, Falcon). Filtering samples immediately before the sort gives cell less time to reassociate. For large cell numbers, it is advisable to distribute cell aliquots into separate tubes and filter each sample individually before placing it onto the sorter. If feasible, an in-line filter, such as from BD Biosciences (San Jose, California), Cytek Developments (Fremont, California), or made in the laboratory by heating the end of a clipped-off pipet tip and fusing it with nylon mesh, put on the uptake port can prevent cell aggregates from reaching the nozzle tip. Selection of a sort tip with the appropriate nozzle size for the cell size to be sorted is another essential factor. Smaller nozzle sizes provide optimal signal resolution and easy sort setup; however, to avoid sort interruptions resulting from clogged fluid lines, it is recommended that the nozzle orifice be at least four times bigger than the cell diameter (Stovel, 1977).

B. Instrumentation

1. Background

The standard jet-in-air technology used for flow sorting involves a liquid stream that contains the cells to be sorted exiting through a nozzle vibrating at high frequency. The stream is broken into individual droplets passing by high-voltage plates that electrostatically charge the droplets carrying the cells that were preselected by the operator and deflect them into the designated receptacles. Thus, jet-in-air sorters produce droplets during normal operation that can be aerosolized. The size of the sort droplets depends on the instrument operating pressure, the size of the nozzle orifice, and the vibration frequency (Ibrahim and van den Engh, 2003). Newer high-speed sorters operate with higher instrument pressures and generate large numbers of small droplets. The droplet size of an aerosol determines its movement and velocity in air and the speed of its settlement out of air due to gravitational forces (Andersen, 1958). The aerosol droplet size is also directly related to the biohazard potential because it defines particle deposition during inhalation. When inhaled, droplets larger than 5 μm remain in the upper respiratory tract while smaller ones penetrate into the lung of the exposed individual (Andersen, 1958; Sattar and Ijaz, 1987). Secondary aerosols of various droplet sizes are formed when the undeflected center stream and the side streams splash into their receptacles. During failure modes of the sorter, such as a partial nozzle block, the stream exiting the nozzle can strike a hard surface and the production of aerosols increases substantially. Again, larger amounts of secondary aerosols are produced during high-speed sorting because of the higher operating pressure compared to regular-speed low-pressure sorting. When aerosols escape from the instrument into the environment, the infectious agents they contain could be harmful if inhaled (Andersen, 1958; Musher, 2003; Schmid *et al.*, 1997a; Vecchio *et al.*, 2003). Thus, for operator safety, it is important to prevent aerosol escape through instrument design or attachment of optional safety devices and verify the efficiency of aerosol containment using appropriate testing. Further critical aspects of biohazard sorting are the placement of physical barriers between operator and hazard (personal protective equipment) and mandatory operator training in the operating procedures established in the laboratory (see later discussion). Instruments that sort cells by a fluidic switching mechanism use an enclosed fluid system, such as FACSort (BD Biosciences), and do not produce aerosols, but cannot achieve the sort speeds needed for many research applications.

2. Instrument Design Features

Since the early 1980s when Merrill (1981) published the first paper describing the generation of aerosols by flow sorters and a method to assess their production and escape during the application of various aerosol control measures instrument manufacturers have modified cell sorter design keeping operator safety in mind. Instrument features that reduce the generation of aerosols, remove aerosols from the operator area, and place primary barriers between the operator and the potential hazard are important because they greatly reduce the risk of exposure of instrument operators and others who are present during the sort. All modern sorters have a vacuum-evacuated catch tube for the undeflected center stream and an enclosed sort compartment. Many have stream-view cameras to facilitate operator observation of sort-stream stability without the need to come close to the area of the instrument that poses the greatest hazard. Systems such as the Accudrop System (BD Biosciences), which illuminates the center stream and the deflected streams near the sort collection vials with a low-powered laser beam, allow the operator to monitor an increase in aerosol production due to a shift in stream positions and fanning. In addition, some newer model jet-in-air sorters have completely enclosed sample ports and/or auxiliary vacuum pumps that remove aerosols from the sort area as optional attachments. Useful custom modifications for improved aerosol containment on any instrument are the sealing of any openings in the sort chamber door and the sort collection area and the installation of an air evacuation system where aerosols are generated. A removable containment hood that is vented by a high-efficiency particulate air filtration (HEPA) filter/fan unit and covers the sort area and the sample introduction port (Cytek Development, Fremont, California) has become commercially available to improve containment on FACStar, FACSVantage, and FACSDiVa (BD Biosciences) cell sorters. Aerosol production is most intense during failure modes of operation; thus, sorters that shut off the sample stream when the nozzle is partially clogged offer an additional safety margin. DakoCytomation (Fort Collins, Colorado) provides a class I biosafety cabinet attachment for their MoFlo high-speed cell sorter. Although standard cell sorters with water-cooled lasers are too large to fit into biosafety cabinets, the newest benchtop sorter, FACSAria from BD Biosciences, equipped with solid-state lasers, is small enough to be completely enclosed into a walk-in clean air and biocontainment biological safety enclosure (BioPROtect II, The Baker Co., Sanford, Maine).

3. Testing the Efficiency of Aerosol Containment during Cell Sorting

a. Background

For biosafety reasons, it is important to verify that the aerosol control measures on the sorter are effective in complete prevention of aerosol escape into the environment. The first protocol described in this chapter is based on the standard gravitational force method, which uses lytic T4 bacteriophage and petri dishes with T4-susceptible *Escherichia coli* lawns (Ferbas *et al.*, 1995; Giorgi, 1994; Merrill, 1981; Schmid *et al.*, 1997a,b). A highly concentrated suspension of T4 bacteriophage is run through the instrument to tag aerosol droplets. *E. coli* lawn-containing petri dishes are placed in, on, and near the flow sorter where aerosols are formed and could potentially escape. Aerosols are detected by plaque formation in the *E. coli* lawn, resulting from T4 bacteriophage landing on the petri dishes and lysing *E. coli*. This technique measures aerosols near the aerosol source and detects droplets that rapidly settle from air, which in general constitute most of the aerosols produced during sorting. Submicrometer particles (i.e., droplet nuclei) that may contain inorganic, organic material, or infectious agents from dehydrated small (less than 5-μm) droplets can also be generated and may stay suspended in air for prolonged periods of time (Sattar and Ijaz, 1987). These droplet nuclei need to be measured with active air sampling methods (Andersen, 1958). As described in the second protocol, an Andersen Air Sampler can be used to direct room air onto petri dishes containing T4-susceptible

E. coli lawns. Tagging aerosol droplets with bacteriophages is an established technique that provided the phage titer is sufficiently high, ensures that all of the droplets generated during sorting contain T4. Because it has been established that a single phage is sufficient to generate one plaque (Merrill, 1981), the assay provides high sensitivity and the assessment of containment results by counting plaques is straightforward (Fig. 1). However, the method requires intermediate knowledge of microbiological techniques, relies on the performance of biological materials, and even when all the materials have been pre-prepared, results take at least several hours.

A novel nonbiological method that uses melamine copolymer resin particles for rapid visualization of aerosol production and escape has been developed (Oberyszyn, 2002; Oberyszyn and Robertson, 2001). A suspension of highly fluorescent Glo Germ resin particles (Glo Germ, Inc., Moab, Utah) with an approximate size range between 1 and 10 μm that simulates a biological sample is sorted on the instrument to be tested. Aerosol containment is measured by placing microscope slides around the instrument where aerosols are produced and can potentially escape and examining the slides for the presence of Glo Germ particles under a fluorescent microscope. Perfetto *et al.* (2003) have increased the sensitivity and reproducibility of the original Glo Germ method by using a viable microbial particle sampler that draws room air onto a microscope slide placed into the device. Microscopic readout of the containment test results is facilitated by concentrating the collected Glo Germ particles onto the areas on the slide located underneath the intake ports of the particle sampler. This technique described in the third protocol in this chapter is suitable to be performed immediately before starting a biohazardous sort but for this requires ready access to a fluorescent microscope. Glo Germ particles are highly fluorescent and, therefore, are easily

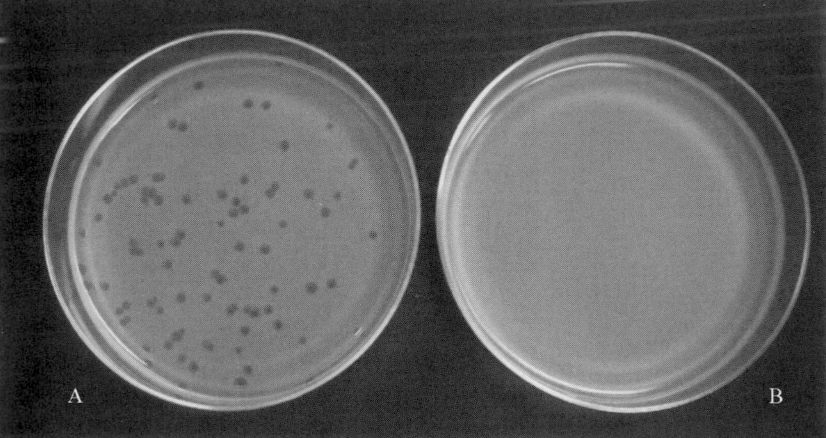

Fig. 1 Plaque formation by T4 bacteriophages on a confluent *Escherichia coli* lawn. (A) Plaques appear where T4 phages have lysed *E. coli*; (B) Confluent *E. coli* lawn on a control petri dish.

detected (Fig. 2). However, meticulous cleaning and handling of the air sampler and microscope slides are essential to avoid false positives, and diligent scanning of the entire slide is needed to reliably detect escape of single particles.

b. Testing the Efficiency of Aerosol Containment During Cell Sorting Using T4 Bacteriophage Using Gravitational Deposition of Droplets

Testing is performed in regular sort mode and in failure mode to simulate conditions such as a partially blocked sort nozzle or air in the system, which result in a considerable increase of aerosol production. Aerosol containment testing should be repeated every 1–3 months and whenever the sorter was modified.

Materials and Methods.

1. *E. coli*, ATCC# 11303 (*E. coli*) (The American Type Culture Collection [ATCC], Rockville, Maryland)
2. Bacteriophage T4 (T4 phage), ATCC# 11303-B4 (ATCC)
   Prepare growth media by autoclaving the mixtures at 121°C for 15 minutes.
3. Bottom agar:
   10 g minimal agar Davis (Difco, Becton Dickinson, Franklin Lakes, New Jersey)
   13 g bacto tryptone (Difco)
   8 g sodium chloride

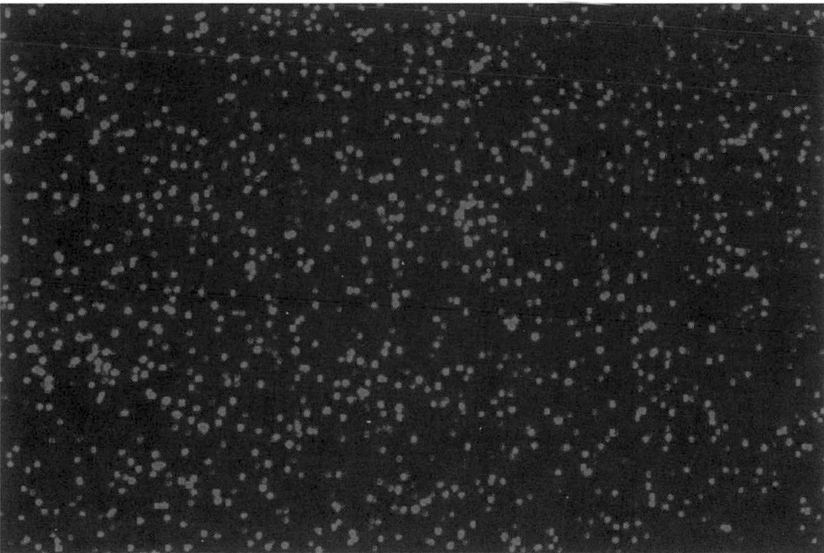

Fig. 2 Visualization of Glo Germ particles by fluorescence microscopy using the setup and magnification as described in the text.

2 g sodium citrate
1.3 g glucose
distilled H_2O to 1 liter

4. Nutrient broth:
   8 g bacto nutrient broth (Difco)
   5 g sodium chloride
   1 g glucose
   distilled H_2O to 1 liter
5. Soft agar:
   6.5 g minimal agar Davis (Difco)
   13 g of bacto tryptone (Difco)
   8 g sodium chloride
   2 g sodium citrate
   3 g glucose
   distilled H_2O to 1 liter
6. Chloroform
7. Sheath fluid for the flow sorter, which is compatible with maintaining T4 phage viability, such as phosphate buffered saline (PBS), Hank's balanced salt solution (HBSS). Note that sheath fluids containing detergents and/or preservatives are incompatible with T4 phage viability.

i. Preparation for Aerosol Containment Testing. All reagents, supplies, and equipment must be sterile and proper sterile technique must be used. Although the preparation of the materials needed for the testing takes several days, the actual aerosol containment test can be completed in 1 day.

Preparation of T4 Bacteriophage Stock.

1. Add 0.3–0.4 ml of nutrient broth to one vial of T4-susceptible *E. coli* and mix well. Unused *E. coli* suspension can be frozen at −20 °C for at least 6 months.

2. Initiate an overnight culture in a culture flask by inoculating approximately 100 ml of nutrient broth with 10–20 μl of reconstituted *E. coli*. Incubate the flask overnight at 37 °C in a warm room on an orbital shaker at 150 rpm or a shaking incubator at the equivalent temperature and speed.

3. Next morning, take out 2.5 ml of the *E. coli* culture and transfer it to a flask containing 50 ml of nutrient broth. For multiple cultures, set up several 50-ml flasks and incubate them at 37 °C in a warm room on an orbital shaker.

4. After 1 hour, start to monitor *E. coli* growth by measuring absorbance of the culture at 600 nm in a spectrophotometer as compared to blank nutrient broth. When the culture reaches an optical density of 0.5, it has entered log-phase growth. For optimal propagation of T4 phage, phage needs to be added to the *E. coli* culture

at this time. Rehydrate one vial of lyophilized T4 bacteriophage with 0.5 ml of nutrient broth and mix well. Add 0.1 ml of the T4 phage suspension to each culture flask. The remaining T4 phage suspension can be stored at 4 °C for at least 6 months.

5. Incubate the culture on the orbital shaker at 37 °C for 6–8 hours until the culture starts to look more transparent.

6. Add 12 drops of chloroform to each flask, shake vigorously, transfer each culture to a 50 ml polypropylene tube, and spin at 2000 g for 20 minutes.

7. Take off the supernatant and transfer it to screw-cap glass vials, keeping the airspace above the liquid to a minimum. Vials containing T4-phage stock can be stored indefinitely at 4 °C in the dark, but it is advisable to retitrate the stock before each aerosol containment test.

Titration of T4 Bacteriophage Stock.

1. Prepare petri dishes with bottom agar by first heating the agar to 50 °C. Using a sterile pipet put approximately 20 ml of liquified agar in each dish and let them cool with the lid ajar to avoid the accumulation of moisture. Close the lid and store dishes until needed at 4 °C. For long-term storage, wrap bottom agar dishes in plastic to prevent dehydration.

2. Initiate a new overnight *E. coli* culture by repeating step 2 from the previous section.

3. Next morning, place 0.9 ml of nutrient broth into nine 12- × 75-mm tubes. Add 0.1 ml of the T4-phage suspension to the first tube, mix well, and make a serial dilution of the stock.

4. Heat approximately 100 ml of soft agar to 50 °C. Do not exceed this temperature, because *E. coli* is heat sensitive. Add 2 ml of the *E. coli* broth culture, mix well, and transfer 4 ml of the mixture to a 15-ml culture tube. The number of tubes will depend on the number of dilutions of the T4-phage stock to be plated. Initially, it may be advisable to plate all nine dilutions of the T4-phage stock. As a general rule, when the T4-phage stock suspension has been grown to the expected titer (10^{11}–10^{12} plaque-forming units [PFU] per ml), it is sufficient to plate only the three lowest concentrations of the stock as the higher ones will produce too many plaques for accurate counting.

5. Quickly add 0.1 ml of the T4-phage dilutions to the mixture, mix gently, and pour the entire content of each of the 15-ml culture tubes onto petri dishes containing bottom agar.

6. Incubate petri dishes at 37 °C for plaque formation. Plaques can be seen within 4–5 hours, but dishes can be incubated overnight, if this is more convenient.

7. Select a petri dish with an intermediate number of plaques, for example, 10–100 for the calculation of PFU of T4 phage per ml. PFU/ml are calculated by multiplying the number of plaques with 100 times the inverse of the dilution

factor. Note that the dilution factor is increased by 10^{-2} because only 0.1 ml of the T4-phage stock suspension is used and only 0.1 ml of the diluted stock is plated.

ii. Testing the Efficiency of Aerosol Containment on a Flow Sorter.
Generation of an E. coli Log Growth Culture and Preparation of E. coli Lawns.

1. Initiate an overnight culture in a culture flask by inoculating approximately 100 ml of nutrient broth with 10–20 μl of reconstituted *E. coli*. Incubate the flask overnight at 37 °C in a warm room on an orbital shaker at 150 rpm or a shaking incubator at the equivalent temperature and speed.
2. Heat approximately 100 ml of soft agar to 40–50 °C. Add approximately 2 ml of the *E. coli* broth culture to the soft agar and mix by swirling gently. Add 4 ml of the mixture to petri dishes with bottom agar and incubate them for approximately 2 hours at 37 °C until dishes appear slightly opaque indicating the formation of confluent *E. coli* lawns. The number of petri dishes to be prepared depends on how many are to be used in the aerosol containment test on the sorter (see later discussion).

Preparation of the Instrument.

1. Set up the sorter in a fast flow rate by measuring the consumption of nutrient broth using an electronic balance. Note that the flow rate varies according to the type of instrument, the differential pressure, the sort nozzle size, and the instrument pressure used. For repeated experiments, set up the sorter at a similar flow rate.

Determine the Expected Throughput of T4 Phage.

1. Calculate the expected throughput of viable T4-phage in PFU/min based on the concentration of the T4 stock suspension determined by serial dilution and the measured instrument flow rate.

Measure the Actual Throughput of Viable T4 Phage through the Sorter.

1. Equip the sorter with T4-phage–compatible sheath fluid, such as PBS and HBSS, and place a culture tube containing T4-phage stock suspension onto the instrument.
2. Weigh an empty culture tube. While the phage suspension is running through the sorter, collect the center stream exiting the sort nozzle for exactly 1 minute Re-weigh to determine the amount of liquid collected.
3. Mix well and immediately transfer 0.1 ml into 0.9 ml of nutrient broth. This step will minimize T4-phage activity loss through its exposure to sheath fluid.
4. Perform a titration of the T4-phage suspension by serial dilution as described in the section "Titration of T4 Bacteriophage Stock." Note that there will be a decrease in titer due to the dilution of the T4-phage suspension with sheath fluid and some loss of activity due to the stress of shearing forces. A titer more than 1×10^7 of PFU/ml is required for a successful aerosol containment test.

Testing the Instrument.
Perform aerosol containment testing in mock regular sort mode:

1. Distribute petri dishes with confluent *E. coli* lawns with the lids still in place onto the instrument where aerosols are generated and where they could potentially escape. Label each petri dish to identify its location during the test. Initially, petri dishes may be placed at various locations within the sorter or around the room. When aerosol containment has been established, for a standard retest of a cell sorter, two petri dishes may be placed within the sort chamber, two on top of the instrument, and two near the work area of the operator.

2. Place sufficient T4-phage stock suspension onto the sorter to perform testing of the instrument for 1–2 hours. Previously published protocols have used various lengths of time for testing the instruments in mock regular sorting mode. The time chosen should take into account the settlement rate of the aerosol produced in the sorter and average sort times used in the laboratory.

3. Set up a mock sort by generating side streams and placing sort collection vials into the sort receptacles. Close the sort chamber and remove the lids from all the petri dishes.

4. After the designated time, stop the sort, place the lids onto the petri dishes and incubate them at 37 °C for plaque formation. Plaques can be observed already after 4–5 hours of incubation, but dishes can be left overnight, if this is more convenient. Dishes that were placed outside the sort chamber should not show any plaques, if aerosol containment on the sorter was complete. If aerosol containment was incomplete, modify the sorter until satisfactory results are achieved during retesting.

Perform aerosol containment testing in failure mode:

1. Repeat steps 1 and 2 from aerosol testing in regular sort mode.

2. Set up the sorter in failure mode by either directing the center stream toward the waste catcher or generating fanning of the streams. Close the sort chamber door and remove the lids from all the petri dishes.

3. After 15 minutes, stop the sort, place the lids onto the petri dishes, and incubate them at 37 °C for plaque formation. Dishes that were placed outside the sort chamber should not show any plaques if aerosol containment on the sorter was complete. If aerosol containment was incomplete, modify the sorter until satisfactory results are achieved during retesting.

b. Testing the Efficiency of Aerosol Containment During Cell Sorting Using T4 Bacteriophage by Using an Air Sampler

1. Set up the air sampler (e.g., Andersen single-stage air sampler, model N6IACFM, Grayseby-Andersen, Smyrna, Georgia) according to the manufacturer's instruction. The air sampler should be equipped with a two-port manifold to allow for simultaneous testing of room air in close vicinity of the sorter and further away.

2. Prepare six petri dishes containing bottom agar with confluent *E. coli* lawn as described above in paragraphs 1 and 2 under "Generation of an *E. coli* Log Growth Culture and Preparation of *E. coli* Lawns" and label them carefully to indicate their positions and test conditions. Place one petri dish into the sampling stage and sample the room air for 10 minutes before starting containment testing on the sorter. This petri dish will serve as negative control.

3. Take two new uncovered petri dishes with *E. coli* lawns. Place one into one sampling stage close to the sorting chamber, the other approximately 1 m away from the instrument. Start the aerosol containment testing in mock regular sort mode by running T4 phage through the sorter as described above and sample the room air for 10 minutes. Sampling times longer than 10 minutes can be used; however, depending on the prevalence of spores in the room air, petri dishes may overgrow with airborne contaminants.

4. Remove the petri dishes from the air sampler and replace their lids.

5. Take two new uncovered petri dishes with *E. coli* lawns. Place one into one sampling stage close to the sorting chamber, the other approximately 1 m away from the instrument. Start the aerosol containment testing in mock failure mode by running T4 phage through the sorter as described above and sample the room air for 10 minutes.

6. Take one more uncovered petri dish containing a confluent *E. coli* lawn into one sampling stage close to the sorting chamber, repeat the mock failure sort mode with the sort chamber door open, and sample the air for 10 minutes. This petri dish will serve as a positive control.

7. Collect all petri dishes and incubate at 37 °C for plaque formation.

8. Testing aerosol containment using an active air sampling method can be combined with the settle plate method or, if desired, be performed separately.

c. *Testing the Efficiency of Aerosol Containment During Cell Sorting using Glo Germ Particles and a Microbial Particle Sampler*

Testing is performed before each biohazardous sort. The instrument is placed in failure mode to simulate conditions such as a partially blocked sort nozzle or air in the system, which result in a considerable increase of aerosol production. Aerosol containment results are recorded as pass/fail before each sort.

Materials and Methods.

1. AeroTech 6TM viable microbial particle sampler, Cat. No. 6TM (AeroTech Laboratories, Inc., Phoenix, Arizona (www.aerotechlabs.com)

2. Vacuum source, for example, in-house or a vacuum pump capable of drawing 45 L/min of air

3. Matheson flow meter: Vacuum Meter, Cat. No. 5083R60, Thomas Scientific, Swedesboro, New Jersey

4. Glo Germ Particles (GLO GERM, Inc., Cat. No. GGP (www.glogerm.com)

a. Glo Germ stock particles (5-μm melamine copolymer resin beads in a 5-ml volume of ethanol to yield a stock concentration ranging from 200 to 400 $\times 10^6$ particles per ml) were washed two times in 100% ethanol by centrifugation at 900 $\times$ g for 10 minutes. If Glo Germ is in powder form resuspend in 5 ml of 100% ethanol before proceeding.

b. Resuspend in 100 ml of wash media (10% FCS + 1% TWEEN 20 + 1 mg/ml sodium azide in PBS).

c. Store particles in an opaque glass container at 4 °C for up to 1 year.

d. Particles are filtered through a 100-μm filter to create a stock suspension. This suspension is then diluted with PBS (~1:20) before sorting to be able to achieve a high acquisition rate (e.g., 20 000 particles/sec).

Measurement of Containment.

1. Add a clean slide to a clean petri dish without a lid and place both into the AeroTech 6TM viable microbial particle sampler. Close the top lid of the sampler and carefully secure clasps on each side. The particle sampler is kept inside a laminar flow biosafety cabinet until containment testing begins. Note: Slides must be very clean and can be cleaned with 70% ethanol before use. Care must be taken not to touch their surface. The microbial particle sampler needs to be cleaned meticulously before each containment test by sonication for 5 minutes in a water bath followed by a wash with a mild detergent and distilled water.

2. Place a clean microscope slide inside the sort collection chamber. This slide will be used as the positive control.

3. Adjust the vacuum to the AeroTech 6TM viable microbial particle sampler to 45 L/minutes (as measured by the flow meter).

4. Place a freshly made suspension of Glo Germ particles by adding 7–10 drops of concentrated suspension to 3 ml of PBS onto the instrument and adjust the flow rate to a high setting, 20 000 events per second. Note: Glo Germ particles are small, so they require a higher forward scatter voltage setting compared to human leukocytes.

5. With the sort chamber door shut and deflection plates on, adjust the center stream to glance off of the waste catcher to create as much aerosol as possible. This situation is considered the "failure mode" of operation.

6. Turn on all auxiliary vacuum sources as applicable and close all containment barriers according to the manufacturer guidelines.

7. Place the AeroTech 6TM viable microbial particle sampler at three locations for 10 minutes each location. Recommended locations are directly in front of the sort chamber door, on top of the sort chamber, and two feet away from the front of the sorter, usually on the operator's chair.

8. Stop the sort and remove the microscope slide from the AeroTech6TM particle sampler and the positive control slide from inside the sort chamber. Note: After every sort, wait before opening the sort chamber until the aerosol has

cleared from the sort collection chamber. The wait time required will vary and depend on the rate of air exchange within the chamber (see later discussion under "Standard Operating Procedures"). Do not allow the sample tube to back-drip to avoid contamination of the outside area with Glo Germ particles. To remove excess Glo Germ particles from the sample tubing, place 70% ethanol on the sample station and run it through the sample tubing for 5 minutes while under sort containment.

9. Examine the entire slide for the presence of particles using a fluorescent microscope with a fluorescein-exciter filter (450/490 nm) scanning with a 10 × objective and a fluorescein emission filter (520–640 nm).

10. All tolerances must be achieved before proceeding with a viable, potentially infectious sort.

11. Tolerance of particles outside the sort collection chamber: zero tolerance, no particles on entire slide. Any positive result must be investigated and resolved and the sorter must be retested before proceeding with sorting potentially infectious samples.

12. Tolerance of particles inside the sort collection chamber (positive control): Greater than 50 per 10× objective field, however, results may vary with slide location.

Notes and Comments. As additional control for the AeroTech 6TM viable microbial particle sampler, it is recommended to collect particles in failure mode with the sort chamber door open for a 10-minute period. This slide should be positive for particles and indicates that the collection system is working correctly.

As an alternative to employing a fluorescein-exciter filter, it is possible to use a DAPI-exciter filter (e.g., 405 nm), which will produce very bright orange GloGerm fluorescence for an easy microscopic readout.

C. Laboratory Facilities

Appropriate sort facility design contributes to the protection of operators and provides a barrier to protect persons outside the room from the potential release of aerosols as a result of cell sorting potentially infectious samples. Because of the possible escape of aerosols generated during cell sorting, it is highly recommended that all biohazardous cell sorting be performed in a BSL-3 laboratory facility (CDC, 1999). BSL-3 facilities are only accessible through two self-closing doors connected through a passageway and contain washable walls, ceilings, and floors, and a ducted HEPA-filtered ventilation system. However, few older institutions have such rooms available and renovation costs are often prohibitive. When planning a new or remodeling an existing sort laboratory, facility design should take into account the BSL of future experiments and incorporate higher levels of containment, whenever possible. In the absence of a BSL-3 sort room, sorting of pathogenic samples requiring BSL-2 containment can be performed in a separate lockable room with negative air pressure equipped with easy-to-clean work surfaces and floors and a sink for hand washing (Giorgi, 1994; Schmid *et al.*, 2003). Access to the room must be limited and warning signs must be posted when experiments are in progress to prevent entry of persons not wearing the required protective clothing (see later discussion). Alternatively, some newer sorters are small enough to fit inside a Biocontainment Biological Safety Enclosure (The Baker Co.) and thus allow for BSL-3 containment without the cost of building a BSL-3 facility.

D. Standard Operating Procedures

All laboratories performing cell sorting on unfixed samples containing known or unknown human pathogens need to establish standard written operating procedures (SOP) to be followed diligently by all personnel involved in these experiments. Whenever there is a change in the biohazard potential of the sort samples, laboratory directors need to reassess the SOP to reflect these changes in practices. Various aspects to be addressed in detail are instrument operation, personal protection, and operator training.

1. Instrument Operation, Troubleshooting, Decontamination, and Maintenance

Instrument procedures for biohazardous sorting will include standard startup, alignment, calibration, sterilization, and shut down as established in the facility for nonhazardous sorting. In addition, it must detail the specific steps to be followed for sorting of potentially infectious specimens. Practices may involve turning on the auxiliary vacuum source at the proper setting to ensure sufficient negative airflow. The setting needs to be high enough for the removal of aerosols from the sort chamber, yet low enough to not disrupt the sort streams. Depending on the instrument, configuration and SOP setup for biohazardous sorting may include attaching the containment hood to the sorter and/or performing aerosol containment testing either before the start of each sort (Perfetto *et al.*, 2003) or at regular intervals as outlined in previous publications (Giorgi, 1994; Schmid *et al.*, 1997a,b). The waste tank needs to be filled with concentrated bleach (5.25% sodium hypochlorite solution) in sufficient quantity to achieve a final concentration of 10% bleach when the tank is full. However, bleach corrodes metal tanks, so substitution with a less corrosive disinfectant such as povidone iodine (10% final concentration) can prolong tank life. Steps to be taken during an instrument malfunction such as a nozzle blockage must be clearly defined. For example, the wait time required before opening the sort door after a clog has occurred will depend on the speed of aerosol clearance. This time can be checked with bottled smoke (Lab Safety Supply, Inc., Janesville, Wisconsin). Clearance time can also be measured by collecting Glo Germ particles (as outlined earlier) with the sort chamber door open after the sort is stopped and the sample is no longer pressurized.

Rigorous cleaning after each sort, for example, by running an alkaline solution (0.1 N sodium hydroxide) followed by an enzymatic solution, such as Coulter Clenz (Beckman-Coulter, Miami, Florida), and distilled water, clears the fluid lines and sort nozzles from residual cells and cellular debris. In addition, after sorting cells of epithelial origin a hyaluronidase solution (e.g., made by dissolving 8 μg of hyaluronidase [Sigma-Aldrich, Cat. # H2126] in 25 ml of distilled water [Steve Merlin, personal communication]) can be useful for elimination of mucin-producing cells. Furthermore, utilization of a strong detergent solution, such as Contrad 70 (Decon Laboratories, Inc., Bryn Mawr, Pennsylvania), can provide effective removal of organic material.

These cleaning practices improve the effectiveness of subsequent disinfection of contaminated fluid lines and sorter parts, which is an important measure to eliminate the potential for the spread of infectious agents. Per regulations outlined in the Bloodborne Pathogen Standard (U.S. Federal Code Regulation, 1991) appropriate disinfectants to be used for decontamination of equipment or work surfaces exposed to blood or other potentially infectious materials include diluted bleach, Environmental Protection Agency (EPA)–registered tuberculocides, EPA-registered sterilants or products registered to be effective against HIV or HBV as listed online at www.epa.gov/oppad001/chemregindex.html. Commonly used laboratory disinfectants (Rutala, 1996; Vesley and Lauer, 1995) and their properties are listed in Table I. Alcohols are not considered high-level disinfectants, because they are not able to inactivate bacterial spores and penetrate protein-rich materials, and isopropanol cannot kill hydrophilic viruses. In addition to disinfection of fluid lines, it is critical to decontaminate the sample uptake port in sorters with sample line back-drip. Before designing a specific sorter decontamination protocol, it is important to check with the manufacturer of a given model sorter to clarify that all instrument components that will come in contact with the disinfectant solution can tolerate exposure. Running distilled water through the sample line for at least 10 minutes after sorter disinfection is essential, because residual disinfectant solution may compromise cell viability in subsequent sort experiments. To this end, it is also recommended that sterile sort media be run in sufficient quantity through the sample tubing each time before starting to sort the sample of interest.

Maintaining the flow sorter in excellent working condition is critical for processing biohazardous samples because presumably the greatest operator hazard comes from sudden instrument malfunctions during a sort such as a broken fluid line, a stuck valve, a blocked waste line, or a damaged or clogged HEPA filter for air evacuation. Thus, a regular preventive maintenance schedule must be followed, either as part of an instrument service contract offered by manufacturers or performed by personnel of the sort laboratory. Nozzle tips must be cleaned frequently by sonication and should be replaced whenever they show wear and tear to minimize instrument trouble.

2. Personal Protection

Personal protection refers to the placement of physical primary barriers between flow sorter operators and biohazard as a means to protect the operator from exposure in the event of a breakdown of mechanical barriers (engineering

Table I
Properties of Common Disinfectants for Use in Cell Sorter Decontamination

Compound	Chlorine compounds	Ethyl alcohol	Isopropyl alcohol	Hydrogen peroxide	Iodophors
			Practical Requirements		
Use dilution	1/10-1/100 dilution of 0.71 M sodium hypochlorite, ~50–500 ppm[a]	70–85%[b]	70–85%[b]	3–6%	5–10%, 800–1600 ppm[a]
Contact time (min)[c]	5–30	10–30	10–30	10–30	10–30[d]
Inactivation by organic material[e]	Yes	No	No	Yes	Yes
Stability[f]	No[g]	Yes	Yes	Yes	Yes
Corrosive	Yes	No	No	Yes[h]	Yes
			Inactivation Profile		
Vegetative bacteria	Yes	Yes	Yes	Yes	Yes
Bacterial spores	Yes	No	No	Yes	Yes
Tubercle bacilli	Yes	Yes	Yes	Yes	Yes
Enveloped (lipophilic) viruses	Yes	Yes	Yes	Yes	Yes
Nonenveloped (hydrophilic) viruses	Yes	Yes[i]	No	Yes	Yes
Fungi	Yes	Yes	Yes	Yes	Yes
Protozoal parasites	Yes	No	No	No	No

Note: From Vesley and Lauer, 1995; Rutala, 1996; www.ianr.unl.edu/animaldisease/g1410.htm.

[a]Available halogen.

[b]Activity drops sharply when diluted below 50%.

[c]Optimal exposure time depends on the amount of contamination and type of contaminant.

[d]Iodophors may require prolonged contact time for inactivation of bacterial spores, tubercle bacilli, and certain fungi.

[e]Before application of the disinfectant cleaning with an enzymatic or lipophilic detergent necessary.

[f]Shelf life of solutions more than one week if protected from heat, light, and air.

[g]If stored at ambient room temperature concentrated bleach solutions retain their sodium hypochlorite content of 5.25% for 3 months after production. Stability of diluted bleach solutions is highly variable depending on the purity of the water used for dilution and the protection from light and air; for optimal disinfectant activity use freshly diluted solutions within 24 hours.

[h]Mildly corrosive.

[i]Variable results depending on the virus.

devices). Extent and types of personal protective equipment used differ among laboratories that routinely sort samples from individuals infected with HIV (Ferbas *et al.*, 1995; Giorgi, 1994; Perfetto *et al.*, 2003). Nevertheless, during biohazardous sorting, at a minimum operator's safety equipment must conform with BSL-3 recommendations (CDC, 1999) and consist of a disposable wrap-around laboratory coat, gloves, safety glasses with side shields, and a respiratory mask appropriate for aerosol protection, such as N95 NIOSH-approved particulate respirators (e.g., 2300N95 Moldex respirators [Zee Medical Service, Inc., Irvine, California]). A plastic shield may be worn over the respirator mask to provide an additional safety margin. Perfetto *et al.* (2003) described the use of a complete Depuy Bio-Hazard Respiratory System (DePuy Chesapeake Surgical, Ltd., Sterling, Virginia), which consists of a body suit, a helmet, and its battery-powered respiratory system with electrostatic filter media, for performing biohazardous sorting. Any personal protective equipment must always be removed whenever the operator leaves the sort room or the adjacent anteroom.

Immunization against infectious agents should be offered to personnel, if available. Prophylactic HBV vaccination is highly recommended. Postexposure prophylaxis should be available in any laboratory involved in sorting of HIV-positive specimens (Mikulich and Schriger, 2002; Schriger and Mikulich, 2002; Wang *et al.*, 2000) and should always follow the latest recommendations from CDC available online at www.cdc.gov/mmwr. Drawing a baseline serum sample from personnel before work is started, monitoring the health status of individuals, in particular compromised immunity, and periodically evaluating serum samples may be an appropriate risk assessment for laboratories that routinely sort samples containing infectious agents.

3. Operator Training and Experience

Only operators with considerable experience of frequent sorting of nonhazardous samples on the type of instrument to be used for hazardous sorting should start performing separations of samples known to carry human pathogens (Giorgi, 1994; Schmid *et al.*, 2003). The time needed for trainees to become proficient in sorter operations varies considerably, but on average 6 months to 1 year is common. Ideally, operators should also be experienced in handling samples infected with the pathogen present in the samples to be sorted (Evans *et al.*, 1990). Mandatory operator training in all the relevant safety aspects of biohazardous sorting, including the procedures involved in testing the efficiency of aerosol containment on the cell sorter is essential for reducing hazard risks to sort personnel and others involved in these experiments. Training records need to be kept on file, and retraining must be performed whenever the SOP changes because of a new sorter or cell sorting of samples with altered biohazard potential.

III. Applications and Future Directions

Flow sorting of unfixed samples is frequently used in infectious disease studies to separate leukocyte subsets on the basis of differential cell surface expressions. Sorted subpopulations can then be examined for their response to the pathogen of interest, for cellular mechanisms of its pathogenesis, or to identify or characterize cells infected with the pathogenetic organism (Giorgi, 1994). Recent applications include investigations of gene expression patterns of cells that either carry a pathogen or have been transfected with vectors that contain genetic sequences of an infectious organism using fluorescent reporter molecules to select the cells to be isolated by sorting (Herzenberg *et al.*, 2002). Emerging applications involve preparative cell sorting of clinical samples for therapeutic interventions (Leemhuis and Adams, 2000; Lopez, 2002). Clinical cell sorting not only requires protection of instrument operators from known or unknown pathogens contained in the samples but sorting has to be performed using Good Manufacturing Practices under clean room conditions to prevent contamination of the sorted product to be re-infused into the patient (Ibrahim and van den Engh, 2003; Keane-Moore *et al.*, 2002).

Acknowledgments

This work was supported by National Institutes of Health awards CA-16042 and AI-28697.

References

Andersen, A. A. (1958). New sampler for the collection, sizing, and enumeration of viable airborne particles. *J. Bacteriol.* **76,** 471–484.

Centers for Disease Control and Prevention. (1999). Biosafety in microbiological and biomedical laboratories. Government Printing Office, Washington, U.S.

Evans, M. R., Henderson, D. K., and Bennett, J. E. (1990). Potential for laboratory exposures to biohazardous agents found in blood. *Am. J. Public Health* **80,** 423–427.

Ferbas J., Chadwick K. R., Logar A., Patterson A. E., Gilpin R. W., Margolick J. B. (1995). Assessment of aerosol containment on the ELITE flow cytometer. *Cytometry* 22: 45–47.

Giorgi, J. V. (1994). Cell sorting of biohazardous specimens for assay of immune function. *Methods Cell Biol.* **42,** 359–369.

Herzenberg, L. A., Parks, D., Sahaf, B., Perez, O., Roederer, M., and Herzenberg, L. A. (2002). The history and future of the fluorescence activated cell sorter and flow cytometry: A view from Stanford. *Clin. Chem.* **48,** 1819–1827.

Ibrahim, S. F., and van den Engh, G. (2003). High-speed cell sorting: Fundamentals and recent advances. *Curr. Opin. Biotechnol.* **14,** 5–12.

Keane-Moore, M., Coder, D., and Marti, G. (2002). Public meeting and workshop on safety issues pertaining to the clinical application of flow cytometry to human-derived cells. *Cytotherapy* **4,** 89–90.

Leemhuis, T., and Adams, D. (2000). Applications of high-speed sorting for CD34+ hematopoietic stem cells. *In* "Emerging tools for single-cell analysis" (G. Durack and J. P. Robinson, eds.), pp. 73–93. Wiley-Liss, New York.

Lopez, P. A. (2002). Basic aspects of high-speed sorting for clinical applications. *Cytotherapy* **4,** 87–88.

Merrill, J. T. (1981). Evaluation of selected aerosol-control measures on flow sorters. *Cytometry* **1,** 342–345.

Mikulich, V. J., and Schriger, D. L. (2002). Abridged version of the updated US Public Health Service guidelines for the management of occupational exposures to hepatitis B virus, hepatitis C virus, and human immunodeficiency virus and recommendations for postexposure prophylaxis. *Ann. Emerg. Med.* **39,** 321–328.

Musher, D. M. (2003). How contagious are common respiratory tract infections? *N. Engl. J. Med.* **348,** 1256–1266.

National Committee for Clinical Laboratory Standards. (1997). Protection of laboratory workers from infectious disease transmitted by blood, body fluids, and tissue. Document M29-A.

Oberyszyn, A. S. (2002). Method for visualizing aerosol contamination in flow sorters. *In* "Current protocols in cytometry" (J. P. Robinson, Z. Darzynkiewicz, P. N. Dean *et al.*, eds.) Unit 3.5.1–3.5.7, John Wiley & Sons, New York.

Oberyszyn, A. S., and Robertson, F. M. (2001). Novel rapid method for visualization of extent and location of aerosol contamination during high-speed sorting of potentially biohazardous samples. *Cytometry* **43,** 217–222.

Perfetto, S. P., Ambrozak, D. R., Koup, R. A., and Roederer, M. (2003). Measuring containment of viable infectious cell sorting in high-velocity cell sorters. *Cytometry* **52A,** 122–130.

Rutala, W. A. (1996). APIC guidelines for infection control practice. *Am. J. Infect. Control* **24,** 313–342.

Sattar, S. A., and Ijaz, M. K. (1987). Spread of viral infections by aerosols. *CRC Crit. Rev. Environ. Control* **17,** 89–131.

Schmid, I. (2000). Biosafety in the flow cytometry laboratory. *In* "In living color, protocols in flow cytometry and cell sorting" (R. A. Diamond and S. DeMaggio, eds.), pp. 655–665. Springer, New York.

Schmid, I., Hultin, L. E., and Ferbas, J. (1997a). Testing the efficiency of aerosol containment during cell sorting. *In* "Current protocols in cytometry" (J. P. Robinson, Z. Darzynkiewicz, P. N. Dean *et al.*, eds.) Unit 3.3, John Wiley & Sons, New York.

Schmid, I., Kunkl, A., and Nicholson, J. K. (1999). Biosafety considerations for flow cytometric analysis of human immunodeficiency virus-infected samples. *Cytometry* **38,** 195–200.

Schmid, I., Merlin, S., and Perfetto, S. P. (2003). Biosafety concerns for shared flow cytometry core facilities. *Cytometry* **56A,** 113–119.

Schmid, I., Nicholson, J. K. A., Giorgi, J. V., *et al.* (1997b). Biosafety guidelines for sorting of unfixed cells. *Cytometry* **28,** 99–117.

Schriger, D. L., and Mikulich, V. J. (2002). The management of occupational exposures to blood and body fluids: Revised guidelines and new methods of implementation. *Ann. Emerg. Med.* **39,** 319–321.

Shapiro, H. M. (2003). Parameters and probes: fixation: Why and how. *In* "Practical Flow Cytometry" 4th ed., pp. 302–306. John Wiley & Sons, New York.

Stovel, R. T. (1977). The influence of particles on jet breakoff. *J. Histochem. Cytochem.* **25,** 813–820.

United States Federal Code Regulation. (1991). Occupational exposure to bloodborne pathogens. 29 CFR PART. 1910.1030.

Vecchio, D., Sasco, A. J., and Cann, C. I. (2003). Occupational risk in health care and research. *Am. J. Ind. Med.* **43,** 369–397.

Vesley, D., and Lauer, J. L. (1995). Decontamination, sterilization, disinfection, and antisepsis. *In* "Laboratory safety" (D. O. Fleming, J. H. Richardson, J. J. Tulis, and D. Vesley, eds.), 2nd ed, pp. 219–237. American Society for Microbiology Press, Washington, D.C.

Wang, S. A., Panlilio, A. L., Doi, P. A., White, A. D., Stek, M. Jr., and Saah, A. (2000). Experience of healthcare workers taking postexposure prophylaxis after occupational HIV exposures: Findings of the HIV postexposure prophylaxis registry. *Infect. Control Hosp. Epidemiol.* **21,** 780–785.

CHAPTER 10

Guidelines for the Presentation of Flow Cytometric Data

Mario Roederer,* Zbigniew Darzynkiewicz,† and David R. Parks‡

*Vaccine Research Center
National Institute of Allergy and Infectious Diseases
National Institutes of Health
Bethesda, Maryland 20892

†Braner Cancer Research Institute
New York Medical College
Hawthorne, New York 10532

‡Department of Genetics
Stanford University
Stanford, California 94305

I. Introduction

In this chapter, we outline some of the concepts underlying the presentation of flow cytometric data in a consistent and informative manner. These guidelines are primarily aimed at preparation of data for publication in print but are applicable to the presentation of data in slides (e.g., during seminars). We cover

METHODS IN CELL BIOLOGY, VOL. 75

0091-679X/04 $35.00

some fundamental aspects of data presentation and then discuss issues particularly relevant to the presentation of data arising from flow cytometric experiments. We limit this discussion to the presentation of "raw" cytometric data; we will not cover more general issues of presenting derived data like charts or line plots, even though they are derived from flow cytometric analyses.

A number of excellent books and papers discuss appropriate and inappropriate graphical data presentation guidelines. One that is highly recommended is Edward Tufte's *The Visual Display of Quantitative Information* (2001), a highly educational and entertaining treatise on data graphing. Many of the guidelines presented in this chapter derive from suggestions that Tufte makes.

Although we touch on topics like control samples, details of good experimental design are outside the scope of this chapter. Understanding which controls are necessary to interpret data (and are, therefore, important to publish) is critical to successful publication, but our assumption is that such decisions have already been made by the author. These guidelines are to help decide *how* to present data, not *which* data to present. Most importantly, these guidelines should be used to decide what additional information needs to be present in publications to allow fully informed interpretation of the data and thus of the experiments and conclusions that are drawn.

Many of these guidelines, when implemented, may lengthen chapters. It should be noted that many journals now have mechanisms for supplying supporting data as electronic documents that can be accessed by readers over the Internet. This mechanism should be employed to provide substantiating graphs (and statistics) that will assist readers in making a full interpretation of the data.

The guidelines presented here are heavily weighted toward the presentation of immunophenotyping data. This kind of data presentation is the most common in cytometry. However, the topics covered here, to a large part, apply to other kinds of cytometric data presentation including cell cycle analysis, analysis of time-resolved events, and so on. Because these analyses have somewhat different complexities associated with their interpretation, they are the subject of a separate discourse. The topic of presentation of the data specifically related to cell cycle analysis is addressed in chapter 11 of this volume.

Finally, we note that this represents only a beginning. The Data Presentation Standards Committee of the International Society for Analytical Cytology will strive to update, revise, and expand these guidelines and to encompass topics of a more specialized nature (e.g., cell cycle, time-resolved kinetics analysis, cell proliferation analysis, and compensation). The goal of this committee is to develop a strict set of publication guidelines that journal publishers can use to ensure consistent and accurate presentation of flow cytometric data.

Throughout this discussion, specific guidelines are set as italicized paragraphs. A summary of all guidelines (and page references of their discussion) is given in Table I.

Table I
Summary of Guidelines

Topic	Guideline	Page
General	• Quantitative information must be presented using standard methods (numbers or charts), not by inference from univariate or bivariate distribution plots.	244
	• The identity and version of the software used to generate graphs in a publication must be given.	244
Graphical Displays	• The number of events displayed in any graph must be indicated.	244
	• To convey quantitative representation of subsets from graphical displays, a calculated frequency of gated events must be displayed. The graph itself cannot convey such information.	245
	• The choice of smoothing and specific display type is up to the author. Choose whichever graph and display options most readily convey the information needed to interpret the experiments, but be consistent across all graphs within an analysis.	249
	• The type of scaling on plot axes must be obvious. Numerical values for axis ticks are rarely relevant and can be eliminated except when necessary to clarify the scaling. The number of logs spanning the whole scale should be evident.	251
	• A percentage value shown on a graph that indicates a proportion of gated cells refers to the fraction of the total cells shown in that graph unless otherwise indicated on the graph.	252
Univariate Displays	• Unless explicitly noted, the abscissa (y-axis) for a one-dimensional histogram is assumed to be linear starting at zero. In most cases, numerical axis values are irrelevant and should not be included with zero-based linear axes.	245
Gating	• Whenever gated analyses are performed, an illustration of the gating process should be shown. A back-gating display is highly recommended.	255
	• Unless otherwise explicitly stated, gating is assumed to have been performed subjectively.	255
	• The use of control samples to set gates should be shown; the algorithm to place gates should be explicitly defined if it was not subjective.	256
Statistical Analyses	• When reporting intensity measurements, it is necessary to explicitly define the statistic (e.g., mean, median, a particular percentile) that is applied. All statistics should be applied to the "scaled" intensity measurement, and not to "channel" numbers.	252
	• In general, statistical significance of results should be reported using standard experimental procedures (replicate analysis).	253

II. General Principles of Graphical Presentation

It is important to understand the role of graphical presentation in scientific chapters. Graphs are *not* drawn to allow users to make quantitative estimates. We should never rely on a reader to look at graphs and, for example, conclude that a particular phenotype of cells is present at a greater (or lower) frequency in one sample than in another. All quantitative information should be derived from statistical analyses that are presented in addition to the graphics—in the form of numerical annotations on the graphics or in the accompanying text (e.g., in the figure legend, results section text, or in the footnote).

Quantitative information must be presented using standard methods (numbers or charts), not by inference from univariate or bivariate distribution plots.

The most important role of graphs is to show *patterns*. The human mind is particularly adept at pattern recognition; thus, use graphs to illustrate variations in patterns (if this is relevant). Do not show 10 nearly identical graphs on 10 samples simply to illustrate the expression patterns. Rather, show a single sample and then present the results of a statistical test to convey the variation.

A. Types of Graphs

There are many types of graphs, but only a few are common and, therefore, readily interpretable by readers. Every software package can generate these graphs; however, each software has unique algorithms for generating the specific display. Hence, the same data set can appear subtly (or not so subtly) different when graphed by different software packages. This in itself is not a concern; however, all graphics of a particular type that are presented in a publication must be generated using the same algorithms (i.e., the same software), to ensure no reader confusion with altered displays. In addition, it is critical to note in a publication which software package and which version of it was used to generate the graphs.

The identity and version of the software used to generate graphs must be given in the publication.

No matter which type of graphics is chosen to illustrate data, the number of events that the software uses to construct the graph can affect how the graphics will appear. This is particularly true for dot plots. Therefore, the number of events displayed in a graph must be evident to the reader. It is not necessary that each graph the exact number of events; however, the reader must be able to estimate the number of events based on a report of, say, the total number of events collected and the fraction of events that fall into the displayed subset. The main purpose is to give the reader an order-of-magnitude estimate on the number of events to help understand the approximate distribution of underlying subsets that the graphic shows. Furthermore, when comparing distributions between different sets of data, one must indicate whether a substantially different number of events (e.g., higher or lower by 50%) is displayed in each graphic. More on this topic can be found in the discussion on smoothing, later in this chapter.

The number of events displayed in any graph must be indicated.

Univariate graphs (one-dimensional histograms) have less utility than generally accepted. The primary reason for showing a histogram is to illustrate that an expression distribution is, or is not, uniform, thereby justifying a particular statistic that is used to quantify the expression in that distribution. A secondary reason is that histograms lend themselves to overlays quite well, allowing readers to quickly grasp the differences in the pattern of expression between samples.

It is important to understand, however, that there are serious limitations to the quantitative information that can be derived from such graphs. The first important point is that the abscissa scale is *relative*. In other words, the actual numbers on the abscissa have little meaning. The height of a histogram depends on three things: the number of events displayed, the breadth of a peak (i.e., CV), and the resolution of the histogram itself. In most software packages, this resolution is the same for all displays, so it can generally be ignored. However, when one compares histograms from different software packages, the same data could have different y-axis values because of differing histogram resolution. This will not affect statistics in any way and does not change the shape of the graph. The different scale values can be ignored.

But the convolution of events and CV mean that you could collect five times as many events in one sample as another, and the overlayed histograms would have the same peak height because the second sample's peak was much broader than the first one. Therefore, do not rely on a histogram's peak height to convey quantitative information (i.e., number of cells in the peak). A corollary of this is that the scale numbers on a histogram abscissa are, therefore, not valuable.

To convey quantitative representation of subsets from graphical displays, a calculated frequency of gated events must be displayed. The graph itself cannot convey such information.

Note that some software packages allow the abscissa to be a nonlinear (e.g., logarithmic) scale, to allow for the expansion of very uncommon populations. The presence of a nonlinear scaling must be made obvious by axis tick placement or by adding numerical values to the abscissa.

Unless explicitly noted, the abscissa (y-axis) for a one-dimensional histogram is assumed to be linear starting at zero. In most cases, numerical axis values are irrelevant and should not be included with zero-based linear axes.

Bivariate graphs come in many flavors. One of the most commonly used is a simple dot plot, where each cell gets a single black dot on the graph. This graph has the advantage that most readers are familiar with it and can readily interpret it. However, a serious disadvantage of dot plots is that they do not accurately convey relative density of subsets when too many events are displayed; the problem is that a second event appearing in the same place as a previous event is hidden (Fig. 1). Therefore, in general, dot plots should be drawn with a maximum of about 10,000 dots (events); most software has an option to limit the number of events drawn. (Note that limiting the number of events drawn can cause low-frequency populations to "disappear" because their events are no longer visible.) The number of dots plotted, thus, is dictated by a compromise between demonstrating as many cell subpopulations as possible and limiting the dots overlap in dense areas of the scatterplot. Remember to note how many events are displayed!

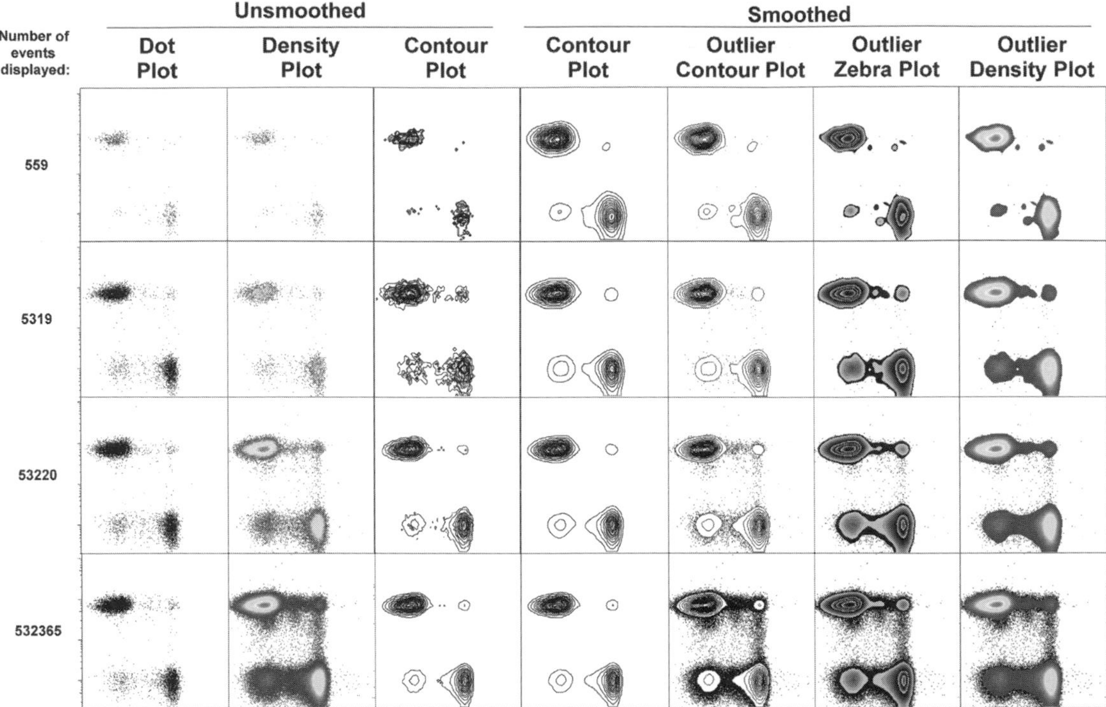

Fig. 1 Different bivariate displays of the same data. Four data sets comprising 500–500,000 events are shown in seven displays. See the discussion for advantages and disadvantages of various display methods. Note that the color scheme is not particularly relevant except to convey information to the reader; hence, the "Zebra plot" (which simply alternates black to gray transitions) is as interpretable as the density plot that employs a color scale corresponding to event density. Illustrated here is the power of advanced density estimation (smoothing); note that the smoothed graphs for 500 events look remarkably similar to the unsmoothed graphs where 1000 times as many events were analyzed. This illustrates that the smoothing algorithm employed here does not hamper the interpretation of the data distributions; in fact, it may significantly aid in visualization when low event numbers are present or when plots representing different numbers of events are to be compared. (See Color Insert.)

The disadvantage of dot plots can be overcome using density plots. Density plots are dot plots that have colored dots that convey information about the number of events that are at the same location. Density plots are readily interpreted by readers who are used to dot plots and should be used whenever possible. Historically, the limitation on publishing density plots has been printing in black and white. Virtually all journals can publish color figures now, albeit some at substantial cost. Although gray scale can be used in place of color, it is less desirable; as color publishing costs come down, the preference should be to use color density plots in place of dot plots whenever possible. Most journals do offer free color figures on the electronically accessed papers.

Note that there are different ways in which colors can be assigned to the dots; different software packages will use different algorithms (and different color mapping). Thus, the same data set can appear somewhat differently in various software representations. The specific algorithm is unimportant; what is important is solely whether the graphical representation conveys the point that the author is trying to make (i.e., the distribution of expression); any color scheme that conveys this information is acceptable.

A third option for bivariate graphs are contour plots (Fig. 1). Contour plots have the desirable feature that they maintain features across a wide dynamic range of displayed events, making them particularly useful for comparing distributions from different samples. Contour plots can also reveal subtle expression patterns that the human brain does not pick out from dot plots or density plots (such as clusters of relatively less-frequent subsets or correlated expression patterns). Historically, a criticism of contour plots was that the selection of the levels at which to draw each contour line was user controlled; therefore, different analyses within the same presentation could be adjusted to look more or less alike simply by choice of levels. However, all software packages now automatically choose contour levels in an algorithmic way, so this bias no longer occurs. Different software may use unique methods to choose the levels, leading to somewhat different displays of the same data. However, as long as the same algorithm is used for all displays in a presentation, the precise algorithm for choosing the contour levels becomes unimportant.

A downside of contour plots is that events outside the last contour do not enter the display. Hence, rare populations are invisible; changes in such subsets then do not become apparent. To solve this problem, many software packages allow a combination of contour and dot plots, where any events outside the last contour plot are drawn as dots (Fig. 1). This "outlier" graph combines the advantages of contour plots (uniform density estimation over a wide dynamic range of events) with that of dot plots (revealing rare populations) and avoids the disadvantages of both methods.

Overlaying bivariate displays is often difficult. The order in which overlayed subsets are drawn significantly affects the display (i.e., one subset's events can be hidden by the other subset). Density plots cannot be overlayed, because color must be used to convey subset identity. Contour plots and dot plots can be overlayed with some success, but care must be taken that information is not hidden. In general, bivariate overlays are most useful when showing analyses such as "back-gating"—showing where a subpopulation distributes among the various combinations of parameters that were used to gate that subpopulation from the main event collection (Fig. 2).

Should displays be smoothed? Smoothing of data is a topic that can be highly controversial. Many advocate that the only fair presentation is that of the "raw" data (although what constitutes raw data is itself controversial). Historically, smoothing of data displays was disparaged because the software algorithms allowed user-controlled extent of smoothing (and resmoothing). This meant that

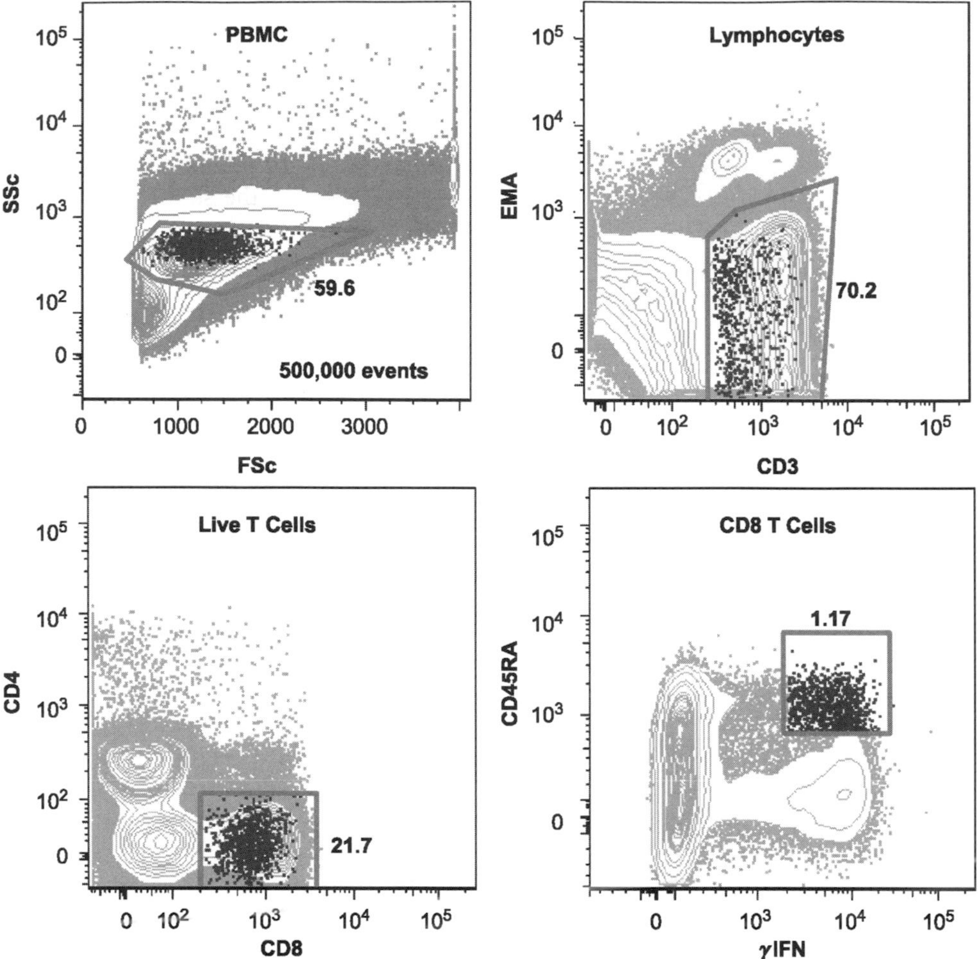

Fig. 2 Back-gating analysis. This data set is the same as that shown in Fig. 4. Instead of the typical graphical representation of the gating schema, a "back-gating" display is shown. Here, the final gated population is overlayed on each gating step as blue dots on the outlier contour plot. This allows readers to easily interpret the relative expression of the dependent markers (i.e., CD3, CD4, CD8, and EMA). (See Color Insert.)

different displays within the same analysis could be smoothed to different extents, causing significantly different interpretation of the data.

At present, however, most software for flow cytometry offers more sophisticated density estimation procedures rather than generic smoothing. Such methods selectively use data from the vicinity of each point to improve the estimated density at that point (Moore and Kautz, 1986). Points with high counts use only close data for estimation, whereas in data-sparse areas a wider span of local data values are used. This process retains sharp features that are statistically justified while suppressing random noise. As shown in Fig. 1, such algorithms can be remarkably accurate, in that the smoothed displays generated from a few hundred events are quite similar to the unsmoothed displays generated from half a million events. For displays of immunofluorescence and light scatter data, density estimation is generally quite successful and is particularly helpful when data sets to be compared contain different numbers of events.

There is no reason to mandate either smoothed or unsmoothed displays. However, it is important that all displays in an analysis are created identically—using the same smoothing setting. Authors should try to avoid displaying statistical variations as though they were actual features of the distribution. In particular, unsmoothed displays for contour plots should be avoided because the contour details tend to reflect statistical "noise," leading to complex contours that are difficult to interpret.

The choice of smoothing and specific display type is up to the author. Choose whichever graph and display options most readily convey the information needed to interpret the experiments, but be consistent across all graphs within an analysis.

There are other kinds of data display options for flow cytometric data that are far less commonly used; these are beyond the scope of this chapter. In general, however, authors should consider the likelihood that using uncommon types of displays may make data interpretation more difficult for readers.

B. Scaling Options for Display

Nearly all immunophenotyping data have been shown on a logarithmically scaled axis. The primary reason for this is that the dynamic range of the signal is so great; typically, brightly stained cells can be more than 1000 times as bright as unstained cells. To visualize both populations simultaneously, one must employ a scale that compresses the high-intensity measurements. A number of properties of the logarithmic scaling have made it the best choice.

Nonetheless, logarithmic scaling can have undesirable effects on visualization of data, particularly after fluorescence compensation. The details of the reasons for this effect are discussed extensively elsewhere (Roederer, 2001). Basically, however, the problem derives from the observation that at very low fluorescence values, the distribution of events becomes linear normal, rather than log normal, which is typical for positive immunofluorescence distributions. In particular, net fluorescence distributions for compensated cell populations often extend below zero on the dye signal axis. (There is nothing improper about measurement values below zero. Compensation involves the subtraction of correction terms, which have associated errors, so when the error in the result is greater than the mean level of fluorescence, some events inevitably end up below zero.) A log scale is particularly poor for representing low-mean, linear-normal distributions, and negative values, of course, cannot be represented on a log scale (Fig. 3).

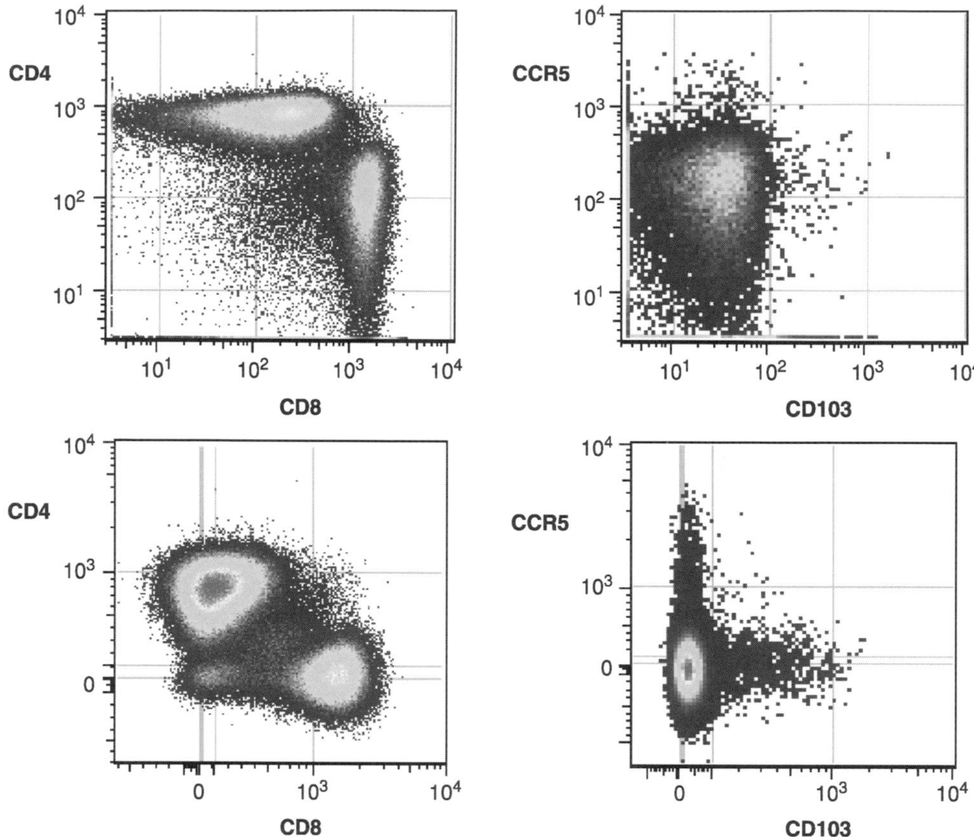

Fig. 3 Logarithmic and Logicle scaled displays of two data sets. In the logarithmic CD4/CD8 plot, note the pileup of events on the axes. All of the events appearing below or left of the zero lines in the corresponding Logicle plot are contained in the axis pileup in the logarithmic plot. Also, note the distinct $CD4^-/CD8^-$ population near {0,0} in the Logicle display that is not apparent in the logarithmic version. In the CCR5/CD103 logarithmic plot, it looks like there is a set of $CD103^+$ events on the CCR5 baseline and another at about 200 units in the CCR5 dimension. The Logicle plot reveals these events to be a single group centered near zero in CCR5. (See Color Insert.)

To solve this, Parks and Moore (in preparation) have devised a new scaling algorithm they term "Logicle" scaling. It is derived from the hyperbolic sine function and combines near-logarithmic compression for high-intensity values with near-linear scaling for low and negative values. As shown in Fig. 3, the Logicle scaling imparts a significantly more interpretable rendering of the data than the standard logarithmic scaling. Note that the Logicle scaling in itself does not affect statistics, gating, or other analyses; it is simply a tool for visualizing data. In some cases, use of Logicle displays should lead to more accurate selection of gating boundaries and improve statistical results by retaining very low and negative data values that would otherwise be truncated by the software.

An important feature of the Logicle scale is the presence of zero on the axis. This is not possible on a standard logarithmic axis, and its position is important. Hence, at least this tick must be clearly identified on a Logicle axis. However, the numerical values of the other markers are not critical and could be eliminated if desired. As for a logarithmic axis, the placement of the tick marks conveys the relative scaling quite well.

The type of scaling on plot axes must be obvious. Numerical values for axis ticks are rarely relevant and can be eliminated except when necessary to clarify the scaling. The number of logs spanning the whole scale should be evident.

III. Statistics

There are two main purposes of statistical analysis in our work. One is to reduce a complex data set to a manageable number of summary values—for example, the average fluorescence intensity of a stained population. The second, which may use these summary values, is to test for statistical significance.

Flow cytometric measurements are invariably rich with quantitative information. In common experiments, half a dozen measurements are made on each of tens of thousands of cells. These cells are rarely homogeneous; thus, after identifying subsets of cells, we are left with a multitude of possible descriptors for each type of cell. Therefore, inevitably, all analyses include significant data-reduction steps in which the key features of the sample are summarized by one or a few numerical values. Although this practice is common and necessary, we should never forget that there may be considerably more information available on our data sets and should strive to ensure that our statistical reduction is appropriate for the sample.

For example, a statistic such as average fluorescence intensity does not convey the *distribution* of the expression; is it very uniform, is it broad, is it bimodal? Of course, this is where a graphical representation of the data to accompany the statistics is important: The graph conveys distribution information not retained by the selected statistics and can support the selection of a particular statistic as an appropriate representation of the sample.

Flow cytometric analyses always center on the measurement of intensity from a cell (fluorescence or scattered light). When we report the *frequency* of a cell population, what we mean is the proportion of cells that express some combination of ranges of intensities. For example, CD4 T cells may be defined as $CD4^{+}$ (more than background CD4 fluorescence), $CD3^{+}$, $CD8^{-}$ (no more than background CD8 fluorescence). Because all three measurements are important to defining the CD4 population, all three should be graphically illustrated to demonstrate that the gating was performed appropriately.

The simplest statistic reported is the proportion (percentage) of cells that falls within a certain combination of gates. However, even such a simple value is considerably more complex than it first appears. Consider the gating scheme shown in Fig. 4, to identify the expression of interferon-γ (IFN-γ) in stimulated CD8 T cells. The gating scheme is shown to illustrate how CD8 T cells were identified, and finally, how the cytokine expression was defined (as positive or negative). In the end, 11.2% of CD45RA$^-$ CD8 T cells were identified as positive for the cytokine. However, there are additional ways to interpret these events: the CD8$^+$CD45RA$^-$IFNγ^+ events are 82% of all CD8 T cells that secreted IFN-γ, 2.4% of all T cells, 1.7% of lymphocytes, and 1.0% of collected events. Which of these percentages should be reported is dependent on the particular interpretation that the presenter wants to make. However, if the percentage reported is not the fraction of events gated on that particular graph, then it must be made immediately obvious.

A percentage value shown on a graph that indicates a proportion of gated cells refers to the fraction of the total cells shown in that graph unless otherwise indicated on the graph.

Outside of statistics related to representation of a subset of cells, there is a wealth of information regarding the distribution of expression by cells (e.g., of a fluorescent marker). Often, the simplest analysis is to calculate how much of the marker is expressed on a per-cell basis. This reduction of the fluorescence measurement to a single value is an attempt to convey the central tendency of a population (e.g., the mean, median, or mode). Another common statistic is a measure of the population uniformity (such as the CV or robust CV). (Note: A "robust CV" is a measure of population uniformity that is more reliable on logarithmically scaled data; discussion of its use can be found in Parks and Bigos [1997].)

Nearly all software packages calculate such statistics on gated populations of cells. It is important to understand the differences between the statistics, particularly when applied to data that cover a wide dynamic range (i.e., that which is typically displayed on a logarithmic scale). It is not the intent here to cover the appropriateness of any given statistic for a distribution of data; however, it is wise to remember that with widely dispersed data, the median statistic is often a better estimate of the central tendency of a population than the mean. Statistical analyses on flow cytometric data are extensively discussed elsewhere (Parks and Bigos, 1997).

There has been some confusion (arising from older software packages) about whether it is appropriate to perform statistics on scale values (that are proportional to intensity) or "channel" values (which increase linearly across the scale, irrespective of whether the scaling is linear or logarithmic, and typically range from 0 to 256 or from 0 to 1024). All statistics reported should be on the scale values, *not* on channel values.

When reporting intensity measurements, it is necessary to explicitly define the statistic (e.g., mean, median, a particular percentile) that is applied. All statistics should be applied to the "scaled" intensity measurement and not to "channel" numbers.

A common question regards the statistical significance associated with the measured percentage of a subset. For example, consider an analysis such as that shown in Fig. 1 done on two subjects: one subject had 11.2% cells in the $CD8^{+}CD45RA^{-}IFN\text{-}\gamma^{+}$ gate with 500,000 events collected. For the second subject, for whom only 10,000 events were collected, this value is 8%. Are these different with statistical significance?

To answer this, we must know the precision with which each of these percentages was calculated; the precision is calculated based on the Poisson statistical distribution, from the total number of events in the *gated population,* and is represented by the square root of this number divided by the number of events. For the first subject, the number of cells counted was 500,000 × 59.6% (lymphocytes) × 70.2% (T cells) × 21.7% (CD8 T) × 11.2% (cytokine) = 5084. Thus, the precision on the measured percentage is the square root of 5084 divided by 5084, or 0.014. The measured value is thus 11.2 ± (11.2 × 0.014) or 11.2 ± 0.16. For the second subject, we can carry through the same calculation; assuming similar fractions of cells in the lymphocyte and T-cell gates, we find that the measurement is 8.0% ± 0.9%. Based solely on the measurement, there is a high degree of confidence that the two values are different.

Numerical analyses of this sort are useful in deciding how much data must be acquired to ensure that purely statistical accuracy will be adequate for the intended analysis. For actual results analysis, however, biological variation is typically much greater than this. In other words, it is highly unlikely, were the first sample tested many times, that the error on this fraction would be only 1.5% (±0.16%). Invariably, experimental and biological variation is much greater than that afforded by flow cytometric measurements; hence, replicate analyses are the best way to estimate the significance of differences between samples. This is true for subset percentages and fluorescence intensity measurements.

In general, statistical significance of results should be reported using standard experimental procedures (replicate analysis).

IV. Subset Analysis

Invariably, analysis and presentation of data is performed on a subset of the events that were collected. Initial subsetting is performed to eliminate debris, dead cells, and/or undesired subpopulations. Additional subsetting may be carried out to delineate subgroups within the population of interest. Because of the critical influence of this process on the interpretation of the data, an example of the gating approach should always be shown. Typically, this is done in a fashion such as that shown in Fig. 4, but a more informative (and useful) approach is a "back-gating" illustration such as that shown in Fig. 2. Here, it is immediately apparent not only how each subset was further divided, but also where within each defining gate the final population of interest resides. This is a powerful illustration that the gated subset is not an artefact of improperly positioned gates. It can also serve to

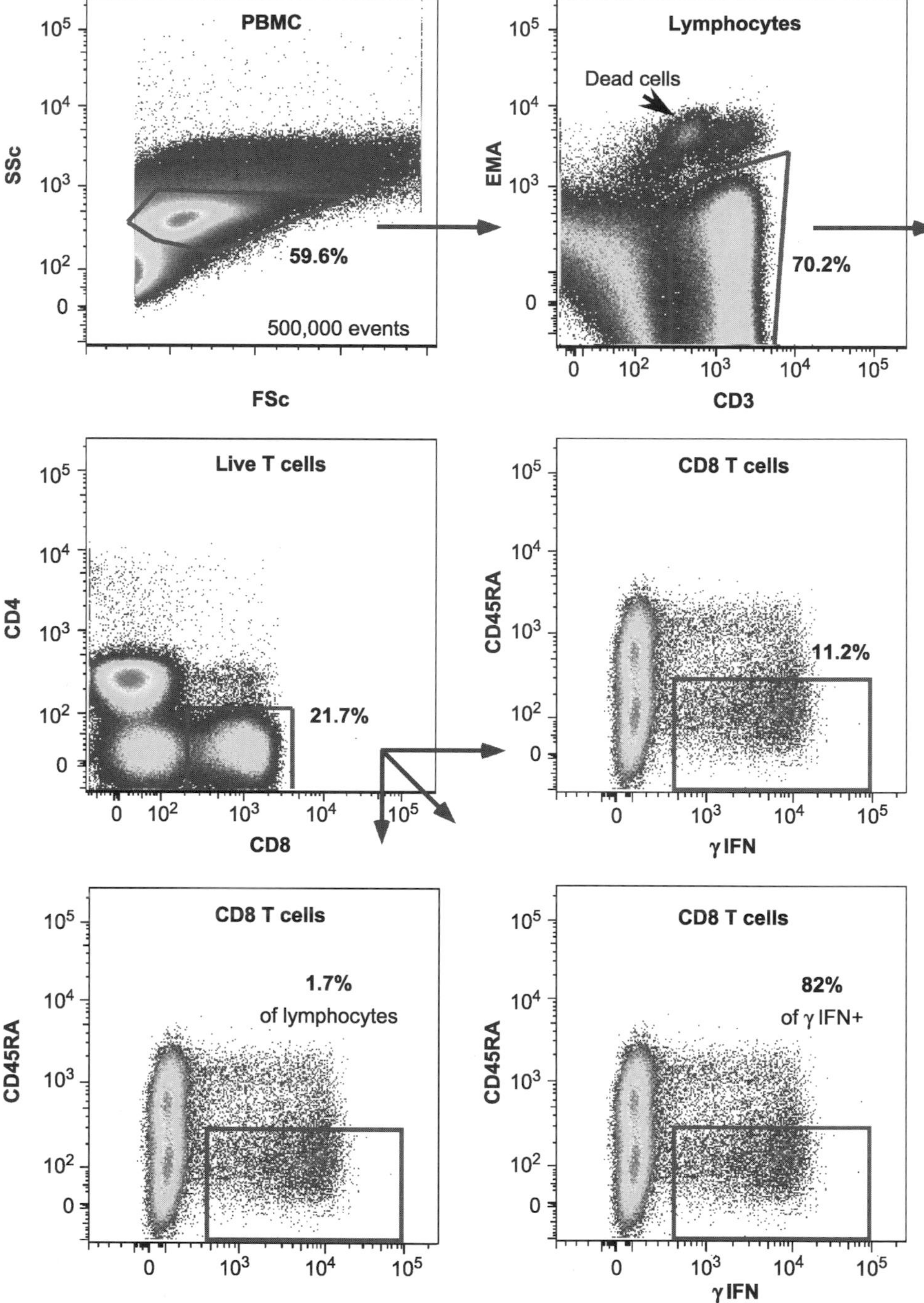
PBMC
SSc
FSc
59.6%
500,000 events
Lymphocytes
Dead cells
EMA
CD3
70.2%
Live T cells
CD4
CD8
21.7%
CD8 T cells
CD45RA
γ IFN
11.2%
CD8 T cells
1.7%
of lymphocytes
CD8 T cells
82%
of γ IFN+

illustrate more information about the underlying biology—in this example, that the gated cells, representing a stimulated population, have somewhat lower expression of CD3 and CD8 than the major CD3 or CD8 T-cell population.

Whenever gated analyses are performed, an illustration of the gating process should be shown. A back-gating display is highly recommended.

Because gating is nearly universal in flow cytometric analysis, it is necessary to justify the gate positions. Much analysis is done visually; the scientist draws a gate around a cluster of events for further characterization. There is nothing inherently wrong with this approach; the human mind is often much better at picking out clusters of events than automated approaches (and certainly often better than blindly relying on control staining samples). Nonetheless, knowledge of this process is essential to the interpretation of the chapter.

Unless otherwise explicitly stated, gating is assumed to have been performed subjectively.

If an objective method is used to position gates (e.g., based on control samples or based on a "fluorescence minus one" [FMO] control [Kantor and Roederer, 1997; Roederer, 2001]), then this process must be explained. In addition, an example of the gates applied to the control sample should be shown. Simply stating that such controls were used to apply the gates is insufficient, because the question remains how those controls were used. Were the gates positioned so 99% of the events in the control sample were below/within the gate or 95% of the events? The percentage of cells in the control sample that is defining the gate, which is then used for comparison of the experimental sample, should always be

---

Fig. 4 Reporting subset statistics in graphical presentations. In this experiment, human peripheral blood mononuclear cells were stained with reagents to detect dead cells (EMA), as well as subsets of stimulated T cells (CD3, CD4, CD8, CD45RA, and interferon-γ [IFN-γ]). A progressive gating scheme was used to identify CD8 T cells that do not express CD45RA and do express IFN-γ, with a lymphocyte gate **(top left)**, a live T-cell gate **(top right)**, a CD8 gate **(middle left)**, and a final subset gate **(middle right)**. In each graphic, the fraction of cells in the gate is shown as a percentage; this percentage is a fraction of the *displayed events in that graph*. Hence, in the top right, 70.2% of lymphocytes fall in the shown gate; this is 70.2% of 59.6% or 42.2% of the entire sample. The symbol "%" was appended to this fraction to explicitly designate what the number means. Arrows between a gating region in one panel and the entire next panel clarify the gating scheme. The bottom two graphs the same data as the final gated graph, but here the emphasis was to show the representation of the CD45RA^{-}IFN-γ^{+} cells as a fraction of lymphocytes or of CD8^{+}IFN-γ^{+} T cells. Hence, the percentage annotation is shown with a clarification as to what the denominator is. For these graphs, the axis values are labeled to illustrate the use of the Logicle scaling (see discussion); it would be acceptable to put a 0 under the zero tick mark; the placement of the logarithmically scaled tick above this is sufficient to convey the type of scaling used, and the values such as 10^2, 10^3, etc. could be dropped. Similarly, the numbers on the FSc (forward scatter) axis are unnecessary; the axis ticks placement conveys the linear scaling. Each graphic has been labeled with an annotation (e.g., "Lymphocytes") to illustrate that what is shown is a gated subset of the original collection; the gating scheme with percentages allows readers to estimate the number of events shown in each graph after the first (where it is explicitly shown). (See Color Insert.)

stated. In many cases, a control sample is analyzed and then the gate is still applied visually (subjectively); this fusion of objective and subjective methodology is acceptable but should be stated.

The use of control samples to set gates should be shown; the algorithm to place gates should be explicitly defined if it was not subjective.

V. Conclusion

In this chapter, we discussed a number of guidelines; following these guidelines is a beginning step toward ensuring that presentations of flow cytometric data adhere to a consistent and understandable format. These guidelines are only a skeletal framework; future efforts by the Committee on Data Presentation Standards will provide additional (and updated) suggestions.

References

Kantor, A., and Roederer, M. (1997). FACS analysis of lymphocytes. *In* "Handbook of experimental immunology," (L. A. Herzenberg, D. M. Weir, and C. Blackwell, eds.). 5th ed, Vol **2**, Blackwell Science, Cambridge.

Moore, W., and Kautz, R. (1986). Data analysis in flow cytometry. *In* "Handbook of experimental immunology," (D. M. Weir, L. A. Herzenberg, and C. Blackwell, eds.). 5th ed, Vol **2**, Blackwell Science, Cambridge.

Parks, D. R., and Bigos, M. (1997). Collection, display, analysis of flow cytometry data. *In* "Handbook of experimental immunology," 5th ed, Vol **2**, Blackwell Science, Cambridge.

Roederer, M. (2001). Spectral compensation for flow cytometry: visualization artifacts, limitations, and caveats. *Cytometry* **45,** 194–205.

Tufte, E. (2001). The visual display of quantitative information. Graphics Press, Cheshire.

CHAPTER 11

Mechanism of Antitumor Drug Action Assessed by Cytometry

Frank Traganos

Brander Cancer Research Institute
New York Medical College
Hawthorne, New York 10532

METHODS IN CELL BIOLOGY, VOL. 75

0091-679X/04 $35.00

I. Introduction

The confluence of a maturing technology, a dramatic increase in the understanding of the mechanisms underlying control of mammalian cell growth, and the ease with which it is now possible to obtain probes to virtually any cellular marker have allowed investigators to use increasingly sophisticated techniques to investigate the mechanism of action of potential antitumor agents. The sheer complexity of the control mechanisms operating in cells that affect cellular decisions to proliferate, differentiate, or undergo apoptosis is daunting. However, given the explosion in knowledge concerning the genes associated with these processes, it has become ever more important to use this information to develop rational and coherent strategies for investigating how exogenous agents perturb the various pathways controlling proliferation in both normal and tumor cells.

A host of approaches and techniques have been used to examine the effects of drugs on cell growth, but there exists no consensus on a variety of issues. Which is the appropriate model system? Which and how many cell lines should be used to test the activity of an agent. What techniques should be used? What should be measured at a minimum? How should the data be interpreted? Are the results obtained generally applicable or unique to a specific set of circumstances? Rather than provide a laundry list of techniques and their details, this chapter attempts to illustrate at least some of the variables that need to be considered when studying the mechanisms of drug activity. Because of the long history and extensive literature on this subject, the choice of topics and approaches is, of necessity, narrow and admittedly biased by my own experience. This chapter, however, attempts to provide some of the underlying principles that are important when studying drug activity in model systems and point out some of the problems and pitfalls that may arise when deciding which approaches to use or how to interpret experimental results. The structure of this chapter will likely deviate from the form of others in this volume in that potential problems and pitfalls are discussed within each topic rather than at the end.

II. Choice of a Model System

A. Issues to Consider when Choosing a Cell System

One of the first decisions one makes is to choose a cell system to test the activity of a prospective antitumor agent, a choice often influenced by the interest of the investigator or the nature of the agent. Generally, most studies use one or several immortal tumor cell lines. However, limiting the analysis to one or a few cell lines carries the possibility that the results obtained may be uniquely a function of that choice and not generally applicable to all cell types, or worse than any other cell type. One approach to overcome this theoretical bias was the development of the "NCI-60" cell lines, a group of 60 human tumor cell lines promulgated in 1988 as a panel for use in disease-oriented drug screening (Shoemaker *et al.*, 1988). The cell lines include diverse cell lineages (lung, renal, colorectal, ovarian, breast, prostate, central nervous system, melanoma, and hematological malignancies) and have been used to test more than 100,000 compounds since 1990 (Roschke *et al.*, 2003). However, this screening tool was generated more for identifying agents with "activity," rather than analyzing in any detail the mechanism by which an agent effects tumor cell growth. Thus, detailed studies of drug mechanism of action tend to focus on a few cell lines at best.

B. Normal Cells

Interestingly, many studies in this area fail to include normal cells in the test. After all, it is not difficult to find agents toxic to cells. However, the point of antitumor drug testing is to identify drugs that may provide a therapeutic advantage, that is, a difference in toxicity for tumor versus normal cells. Thus, an approach that fails to include a normal cell counterpart to the tumor cells being used begs the question of whether the agent is generally toxic or specifically toxic to tumor cells.

Although the decision to test a normal cell counterpart to the tumor cells being studied is important, even that decision must be carefully examined. Generally, it is preferable to use normal human cells with a finite life span (i.e., a limited number of generations in culture) available from various sources. Even when using such cells, one must try to keep reproducibility within a narrow range of doublings because these cell lines tend to change growth characteristics and even karyotype as they reach their maximum life span (passage).

C. Tumor Cell Lines

There are many more considerations when choosing a tumor cell line for use in drug studies. The first order of business is to ensure that the cell line you think you are using is, in fact, the cell line you have. Older investigators may well remember the studies of Walter Nelson-Rees in the mid-1970s who identified scores of cell lines as misidentified; they were all HeLa cells (Nelson-Rees and Flandermeyer, 1976). This led to the identification of hundreds of scientific papers that were based on fraudulent cell lines resulting in some $10 million worth of research being discredited (O'Brien, 2001). The assumption that this happened a long time ago and is no longer relevant is a mistake. In 1999 it was reported that 18% of 252 new cell cultures deposited at a German cell line repository were contaminated by other cell lines, most notably HeLa, T-24, SK-HEP-1, U-937, and HT-29 (MacLeod *et al.*, 1999). Again, these misidentified cell lines gave rise to numerous studies being performed on inappropriate cell systems. Therefore, care should be exercised when obtaining cell lines; in most cases they should be obtained directly from a cell repository that has used molecular markers to rule out misidentification.

Even after establishing the origin of a cell line, one has an additional consideration to make that may have a bearing on interpretation of experimental results. Cell lines derived from solid tumor are often assumed to be clonal, that is, to represent a single genotype. However, reports have documented the heterogeneous nature of such cell lines (O'Brien, 2001). The NCI-60 cell lines were all observed to have a wide variety of ploidy levels, numerical changes, and structural rearrangements arising principally from a striking lability at centromeric regions among epithelial tumor cell lines (Roschke *et al.*, 2003). For instance, ploidy heterogeneity was so substantial that in 29 of 58 cell lines 1–10% of the populations had ploidy difference, whereas in 23 cell lines, the differences in ploidy existed in more than 10% of the cells (Roschke *et al.*, 2003). These same kinds of ranges were observed for numerical (chromosome copy number) and structural (chromosome rearrangement) heterogeneity among the cell lines. Taken together, this indicates that any one cell line presents a complex target for any potential drug, especially because rearrangements and losses or additions of entire chromosomes were common in a substantial minority of cells in almost all the cell lines in this screen. Variations in gene dosage, mutation, and inappropriate activation or inactivation of important genes involved in control of cell proliferation or apoptosis are likely to affect the response of a cell line to a particular agent, especially if the genes in question involve the target or a molecule in a pathway targeted by such a compound.

Another consequence of chromosomal heterogeneity is the propensity of cell lines to undergo cultural "drift." Thus, genotypic variations occurring in culture are liable to provide a small subset of cells with a proliferative advantage. Occasionally, when cells are put under stress such as alterations in the growth conditions, selection for cells whose mutation has provided them with a growth advantage (e.g., shorter doubling time and ability to grow in the presence of lower serum concentrations or oxygen tension) occurs. This can lead to the evolution of a new cell line with a different modal genotype that may no longer respond to challenge with an agent the same way as cells of the initial genotype did. This shift can be subtle and occur over a relatively short period (e.g., a month). One way to avoid such problems is to freeze aliquots of the cell line soon after obtaining it and then at periodic intervals discard the current batch and defrost and start a fresh culture from the original aliquot.

D. Suspension vs Adherent Cell Cultures

Typically, hematopoietic cell lines grow in suspension culture, and epithelial cells grow attached to their substrate (adherent cultures). Often the choice of which cell system to use is dictated by the interest of the investigator; those interested in treating leukemias, for instance, would likely work with suspension cultures, whereas those interested in prostate cancer research are forced to work with adherent cell cultures. The compound itself may also determine the kind of cell system to be used; hormone agonists or antagonists (e.g., tamoxifen) would normally be tested in solid tumor models, which tend to be adherent cell cultures, whereas potential differentiating agents (e.g., retinoic acid) would most often be tested in leukemia cells growing in suspension culture, although there are adherent cell systems that can undergo differentiation in culture (e.g., preadipocytes). Both types of systems have their unique attributes.

Suspension cultures are easily processed for cytometric examination because they already exist in cell suspension obviating the need for the use of enzymes to detach the cells from the growth surface. Because it is easy to remove small aliquots from a single culture multiple times, it is easier to do cell and viability counts at multiple time points and to assay changes in cell kinetics without the variations associated with setting up multiple cultures, as would be required for adherent cells. However, suspension cultures are more difficult to synchronize (unless you use an elutriator).

Adherent cell cultures require physical or enzymatic treatment to remove cells from the surface upon which they grow, so it is impractical to attempt to assay a single culture multiple times. As a result, you must set up multiple aliquots if it will be necessary to monitor the effects after various lengths of exposure to a compound. The very nature of setting up multiple cultures, plates, or wells and having to aliquot compounds to individual cultures increases the potential for variability in the results due, for instance, to pipetting errors. Cell counting is also more difficult with suspension cultures, often requiring estimates to be made of the percentage of surface area occupied by the culture at various points. However, adherent cultures are generally more amenable to complex treatment protocols in which compounds are added or removed at specific times; the media can be poured off, the cells rinsed with fresh media, and either fresh media or media containing an additional compound or compounds added. Caution should be exercised when using such a strategy, however, because as cells begin to undergo cell death by apoptosis they lose their ability to remain attached to the substrate and may inadvertently be discarded when replacing media. Mitotic cells are also loosely attached to the substrate, allowing one to mechanically synchronize some adherent cultures by mitotic detachment, which provides a population of cells that are identical in "time since mitosis" (Darzynkiewicz *et al.*, 1982).

E. Cell Concentration Effects

In either type of culture, reproducibility of conditions is important when studying drug effects. Generally, the most reproducible conditions for studying cell cycle effects of antitumor agents are in cultures undergoing asynchronous exponential growth. All cell cultures have an ideal concentration at which their doubling time is at a minimum. Concentrations of cells considerably lower than the ideal will often be associated with a delay in growth, whereas higher than ideal cell concentrations lead to slow down or cessation of cell growth.

All suspension cultures have a cell concentration that they will grow to but not exceed. Cells stop growing at high cell densities because they either exhaust the nutrients in the media or have become too crowded. This is termed *plateau phase*. Plateau phase growth is not useful for examining changes in cell proliferation or redistribution of cells within the cell cycle after addition of antitumor agents because the cells have stopped cycling. Typically, depending on the nature of the culture (i.e., normal or tumor cells), plateau cultures tend to accumulate in the G_1 phase. Adherent cultures are also affected by cell concentration. Thus, when adherent cultures cover the surface of their vessel, they tend to slow or cease growing either because they quickly exhaust the growth factors in the media or because cell–cell contact provides a growth inhibitory signal often referred to as *contact inhibition of growth*.

In both suspension and adherent cultures, allowing cell to grow to high cell density before dilution down to the cell concentration appropriate for exponential growth may result in a partially synchronized culture because in both conditions, high cell density leads to fewer cells in S and generally higher percentages of G_1 cells. This partial synchronization may persist through the first or several cell cycles, with the extent and duration of this synchronization increasing with the length of time spent in plateau phase.

Also important is that most immortal cell lines can be thought of as having a growth fraction of 100% under ideal growth conditions. Therefore, it is not unreasonable to assume that cells growing exponentially and asynchronously will all be potential targets for the action of drugs, especially those that target proliferating cells. *In vivo*, however, conditions tend not to favor a growth fraction that high (Mendelsohn, 1960) because by the time most tumors become apparent, they will have grown sufficiently so that in the absence of inducing new blood vessel formation, they will contain hypoxic or anoxic cells that proliferate slowly or not at all and are significantly more resistant to most chemotherapeutic agents and radiation (Thomlinson and Gray, 1955). Some fraction of these cells, if not killed, appear capable of reentering the cell cycle.

In reality, the *in vitro* system that most closely replicates that of solid tumors is the solid tumor spheroid, that is, tumor cells that grow in three dimensions in culture much as they grow *in vivo* (Durand and Olive, 2001). Except for the presence of a vascular network, spheroids recreate the kind of conditions likely to be operating in small tumors where diffusion of oxygen and nutrients controls proliferation. Tumor spheroids are heterogeneous in terms of their growth rate, which diminishes the closer one gets to the center of the spheroid. As such, cells grow with different kinetics as a function of their distance from oxygen and nutrients, providing for a complex target for chemotherapy, but one that is probably more realistic than attempting to draw a conclusion of drug efficacy tested on isolated tumor cells growing under ideal conditions.

In addition to spheroids, there are alternative three-dimensional (3D) culture systems (matricis) that are similar in their ability to provide information concerning "the interplay of architecture, environment, and intercellular contacts as modulators of antitumor therapies" (Durand and Olive, 2001).

III. Drug Dose and Length of Exposure

Information derived from cell growth/viability curves is always informative, often necessary, and occasionally indispensable in interpreting cell kinetic data. Interpretation of single-parameter DNA distributions of cultures treated with compounds that may effect cell proliferation cannot be understood in the absence of cell growth curves and should be used in determining the range of dose to be used in the analysis of drug mechanism of action.

A. Selecting a Dose Range

There are several considerations when choosing the range of doses to test for activity. Exposing cells to a compound and doing growth and viability curves should provide information on the minimum concentration necessary to effect proliferation, the dose that prevents growth, and doses that induce some degree of cell death (Fig. 1). Changes to the slope of the growth curve that still result in positive growth (Fig. 1, curves 2 and 3) are often associated with changes in the transit of cells through the cell cycle but could also be the result of modest cell death or even cell differentiation (i.e., entrance of cells into a quiescent state). Higher doses that result in a plateau in (Fig. 1, curve 4) or a negative slope to the cell growth curve often involve a combination of effects that involve both a block in cell transit through the cycle and apoptosis. Still higher doses lead to increased

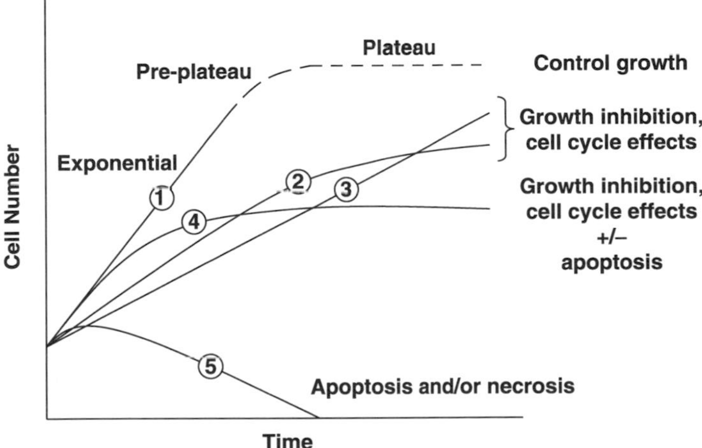

Fig. 1 Cell growth curves. Curve 1 is typical of an asynchronous, exponentially growing culture (*straight line portion*). As cells approach high cell density, they begin to slow their rate of growth and enter a pre-plateau phase followed by complete cessation of growth, or plateau phase. Curves 2 and 3 result from a drug-induced slowdown in cell cycle progression generally observed at lower drug concentrations. Intermediate drug concentrations lead to both a slowdown of growth and possibly a small amount of cell death, as illustrated in curve 4. When drug treatment causes substantial cell death by apoptosis or necrosis, the growth curve inverts as illustrated by curve 5.

levels of apoptosis or, in extreme cases, cell necrosis (Fig. 1, curve 5). Determining the mechanism of drug action is almost always best accomplished at concentrations that effect cell cycle traverse but are accompanied by minimal or no cell death. When significant levels of apoptosis occur, cell cycle analysis is dependent on knowing the extent of cell loss. Because apoptosis dose not generally occur synchronously and is of finite duration, dead cells have a tendency to disappear from the analysis, particularly at longer treatment times. Therefore, any attempts at analyzing drug effects at concentrations likely to produce more than minimal cell death require that the duration of exposure be short enough to ensure that all cells undergoing cell death are still present and can be accounted for.

B. Pulse vs Continuous Exposure

Most studies tend to expose cells to agents of interest for extended lengths of time, such as continuous exposure. However, the half-life of many clinically useful chemotherapeutic agents is actually quite short, sometimes less than 1 hour (e.g., adriamycin) (Greene *et al.*, 1983). As a result, there are several approaches for studying potential antitumor agents.

Pulse exposures mean exposing the cells to a compound for a time considerably less than the doubling time of the culture, removing the compound by washing the cells, and returning the cells to culture in drug-free media. Some investigators believe that this approach more closely approximates the circumstances *in vivo*. Typically, exposure times of 1–2 hours are used to mimic the half-lives of some chemotherapeutic agents. If the target for a particular compound is limited to a specific cell cycle phase, then the expected response will be generally limited to the percentage of cells that happened to be in that cell cycle phase and, therefore, vulnerable. For instance, if the target of a compound is the mitotic spindle, because cells spend a relatively small amount of time in mitosis (perhaps 1–4% of the cell's total cycling time), then for a short pulse, it is unlikely that more than that proportion of cells will be effected. The longer the target is present or the longer the exposure time, the greater the percentage of cells that are effected.

Not all cells that are targets for cell cycle phase–specific agents are necessarily affected equally during pulse exposure. Assume for a moment that the target for a particular agent is the S phase and that one of the cell systems used in the study has an S-phase duration of 12 hours. If the pulse of drug exposure is 2 hours, then cells that have been in the S phase from 0 to 10 hours will all receive the maximum dose of the drug. However, those cells leaving the S phase over the course of the 2-hour pulse will get less than the total dose, with those closest to G_2 receiving the least. The same is true of cells entering the S phase. Cells entering 1 hour after the start of the pulse would be vulnerable to the drug for only 1 hour compared to those that were already in the S phase and received the full 2-hour pulse of drug. If being in contact with the drug for only seconds is sufficient to affect cell survival or block cell proliferation, then in this example, the window of cell vulnerability would be 14 hours (the 12 hours of S phase plus the 2 hours in which cells entered the S phase).

If, however, the cells needed to spend the entire 2 hours in the presence of the drug before an effect was observed, then the window of vulnerability would be reduced to 10 hours (the cells leaving S phase in the first 2 hours would not receive the full complement of drug nor would the cells entering the S phase during the pulse).

Obviously, *continuous exposure* differs considerably from pulse exposure, especially when dealing with cell cycle phase–specific agents. If the length of exposure to the agents is greater than the duration of the cycle time of the cells under study, then in a simple model, all cells will have passed through the vulnerable cell cycle phase. This fact, however, does not make data interpretation any less complex. In this instance, although all cells should have passed through the window of vulnerability, some have done it early in the treatment, but others will have just gone into that cell cycle phase. Going back to the example used in the pulse experiment, if we assume that the culture with a 12-hour long S phase has a cell cycle duration of 24 hours and we treated the cells for 24 hours, then the cells in the S phase at the beginning of the experiment have been exposed to the drug for 24 hours, whereas there are cells that were in the G_2 phase when the drug was added that had to traverse G_2, divide, and pass through the G_1 phase (about 12 hours in this example) before they would begin to be affected by the drug. Thus, the latter population would be exposed for about 12 hours or half that of the cells already in the S phase at the beginning of the treatment.

IV. Single-Parameter DNA Histogram Analysis

Single-parameter DNA histograms are the oldest and perhaps most widely used approaches to studying drug effects on cell cycle kinetics and one of the earliest applications of flow cytometry (Rabinovitch, 1995). The frequency distribution of a population of cells stained for DNA content has a distinct shape if the population is growing exponentially. Figure 2A provides an example of a DNA frequency distribution typical for an asynchronously growing exponential population of cells: The initial "peak" at the lowest fluorescence values represents postmitotic G_1 (and under some circumstances, quiescent G_0) cells that have divided but have not yet initiated DNA synthesis; a second "peak" at approximately twice the fluorescence of the first that contains cells that have doubled their DNA content and have yet to divide (G_2) or are in the process of dividing (e.g., undergoing mitosis; M); and a distribution between the two peaks that contains cells with increasing levels of DNA, representing DNA synthesizing or S-phase cells. The position of the two peaks can be described in terms of DNA index (DI) with the G_1 peak having a value of 1.0 and the G_2M peak a value near 2.0; large variations from this ratio can indicate problems with either staining stoichiometry or photomultiplier linearity. The "widths" of the G_1 and G_2M peaks represent the coefficient of variation (CV) of the measurement ($CV = 100 \times$ standard deviation/mean of the peak). Ideally, because all cells in G_1 have the same DNA content, the fluorescence measurement of these cells should appear as a line-1 channel wide on the histogram. To the extent that the measurement of the fluorescence bound to

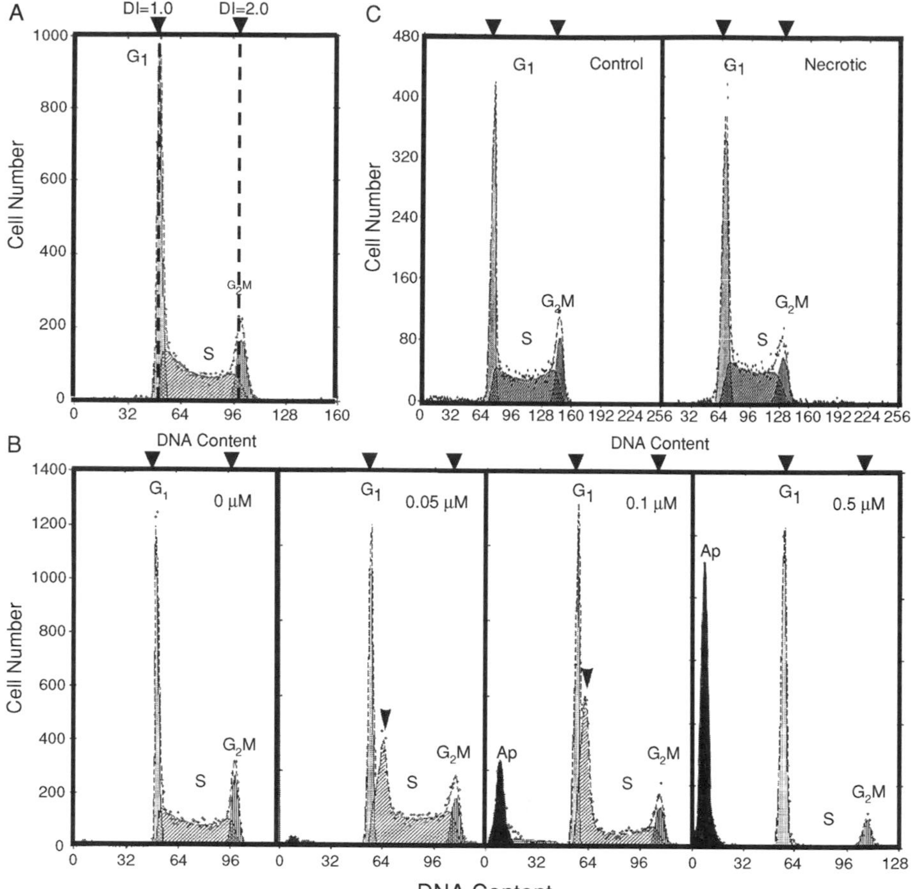

Fig. 2 DNA frequency distributions of exponentially growing control and drug-treated cells. (A) Cell cycle distribution of an untreated, exponentially growing culture illustration typical DNA frequency distribution with a G_1 peak at a DNA index (DI) of 1.0, a G_2M peak at a DI of 2, and S phase with intermediate DNA content. (B) Treatment of cells with increasing concentrations of drug for a standard period. Control, untreated cells are illustrated in the first panel. Low concentrations of an S-phase–active agent cause an accumulation of cells in early S phase (*arrow*). A higher drug concentration shows increased cell accumulation in early S phase, with fewer cells present in the late stages of S and the appearance of a sub-G_1 peak, indicating cells undergoing apoptosis. Higher drug concentrations result in a complete loss of S-phase cells that reappear as a sub-G_1 peak of apoptotic cells. (C) Comparison of histograms of control, untreated cells and cells exposed to toxic levels of a drug, leading to necrotic cell death. Note that the distributions are quite similar and would be difficult to distinguish in the absence of information on cell growth. All histograms in this figure are scaled to peak height.

the DNA in the cell (see later discussion) is variable and instrumentation (e.g., the optical alignment, stability of the sample stream, the shape and intensity of the exciting light beam, the choice of filters, and the quality of the photomultipliers) is less than ideal, the signal is more variable than the target (DNA) being measured (Hoffman, 2001). However, the better the DNA measurement (the lower the CV), the easier it is to analyze (deconvolute) the distribution and determine its component parts, G_0/G_1, S, and G_2M phase cells (Dressler and Seamer, 1994).

It should always be remembered, however, that DNA histograms represent snapshots in time and do not, in and of themselves, provide any information about rates of cell cycle progression. One histogram, taken on its own, even if it illustrates a dramatic redistribution of cells within the cell cycle does not provide unequivocal information about how that redistribution occurred. Histograms become increasingly informative though when they are paired with accurate cell and viability counts and are repeated during the course of the experiment. Obviously, the closer together in time samples are studied, the more informative they become.

A. Choice of DNA Stain

Although many fluorescent dyes are available for use in staining DNA in cells, the most useful are presented in Table I [for a review of DNA stains, see Crissman and Hirons (1994)]. However, there are two DNA dyes among this group that are used the most often: (1) the minor groove binding, ultraviolet (UV)-excited, blue-fluorescing stain 4′,6-diamidino-2-phenylindol (DAPI) and (2) the DNA intercalating, blue-excited, red-fluorescing stain propidium iodide (PI). Each of these dyes has advantages and disadvantages.

1. DAPI is a highly specific DNA stain that preferentially binds to AT regions of the DNA molecule and was first suggested for use in flow cytometry by Stohr *et al.* (1977) and Gohde *et al.* (1978). Although the dye requires cells to be fixed

Table I
Fluorescent Stains Useful for DNA Frequency Histogram Analysis by Cytometry

Cell attribute	Cell cycle compartments detected	Fluorochrome
DNA content, fixed cells	G_0/G_1, S, G_2M	Ethidium bromide (EB) Propidium iodide (PI)[a] Mithramycin Chromomycin A_3 DAPI[a] Thiazole orange (TOTO-1)
DNA content, live cells	G_0/G_1, S, G_2M	DRAQ 5 Hoechst 33342 Hoechst 33258
DNA and RNA content, fixed cells	G_0, G_1, S, G_2M	Acridine orange (AO)
Chromatin condensation, fixed cells	G_0, G_1, S, G_2M	Acridine orange (AO)

[a]DNA dyes most often used in bivariate and multiparameter analysis in which one of the components is DNA.

and an instrument with a UV light source, it does provide highly accurate measurement of DNA content (although remember it will be affected by the proportion of AT base pairs in the nucleus) (Otto, 1994). The dye can be especially useful in multiparameter measurements made with multiple light source instruments because the emission when bound to DNA (around 460 nm) is generally far removed from the emission of dyes that can be excited with a second, blue light source. Because DAPI does not bind to double-stranded RNA, it is not necessary to treat cells with RNase or isolate nuclei to obtain excellent DNA distributions.

2. PI intercalates between base pairs of double-stranded nucleic acids, which means that for PI to be used as a DNA stain, one must remove double-stranded RNA from cells. This has generally been accomplished either by (1) enzymatic digestion of double-stranded RNA with RNase (Crissman and Steinkamp, 1973) or by (2) isolation of nuclei by detergents and enzymes (Krishan, 1975; Vindelov, 1977). Obviously, the latter technique precludes the use of additional probes to stain cytoplasmic constituents or cell surface markers. Nevertheless, PI is a versatile and rather forgiving dye (i.e., it is fairly stoichiometric over a rather wide concentration range) and was first used in flow cytometry by Dittrich and Gohde (1969).

Both of these dyes are reasonably stoichiometric for DNA when used correctly, although neither can provide absolute DNA content values. DNA in cells interacts with chromatin proteins (e.g., histones), which, in effect, shield DNA from access by even small dye molecules such as DAPI or PI. PI is especially vulnerable to this shielding because it not only needs access to the DNA helix but it also requires the helix to unwind slightly for the dye to intercalate between the DNA base pairs; histone proteins stabilize the helix, making unwinding more difficult. As a result, PI binds to only about half (50%) of the total possible binding sites (Darzynkiewicz *et al.*, 1984); removal of chromatin proteins with 0.1N HCl increases the PI red DNA fluorescence by slightly more than 100%. Alternatively, similar treatment of the same cells led to only a 45% increase in DAPI fluorescence, indicating that DAPI is less sensitive to the degree of chromatin condensation caused by the binding of various chromatin proteins to DNA. Although the effect of dye accessibility to DNA in model cell systems is generally not a consideration (unless one is studying the effect of an inducer of differentiation), it should be clear that the fluorescence of a cell obtained with either dye does not provide a quantitative measurement of the amount of DNA present. Rather, both dyes can be used as a tool to follow changes in the cell cycle distribution of cells based on relative changes in fluorescence intensity.

B. Examples of Simple Drug Effects

Several examples are provided to illustrate some of the issues important to understanding how to interpret single-parameter DNA frequency histograms. As seen in Fig. 2A, histograms describing asynchronous exponential growth have a specific shape. In this instance, the cells are derived from an adult human T-cell (lymphocytic) leukemia. The cell line is hypertetraploid (~95 chromosomes), with most chromosomes present as two sets of two, each with several unique marker chromosomes (American Type Culture Collection). As noted in the figure, it is not possible to tell the ploidy of the cell line in the absence of a control, which might consist either of normal human mononuclear cells, of fluorescent beads, or of nucleated red blood cells (RBCs) from species like chicken or trout (Horan *et al.*, 1990). The use of mononuclear cells from a healthy volunteer would identify the position of the diploid cell precisely, whereas nucleated RBCs have less DNA than diploid human cells; fluorescent beads also tend to be chosen with mean fluorescence values lower than that of diploid human cells so as not to interfere with the analysis of the DNA histogram. However, when using nondiploid control particles or cells, a ratio must be established between the position of the peak for the control and that of normal diploid human cells so the apparent DNA content of unknown samples can then be determined. If it is important to establish the relative DNA content of cells, in those instances where cells have a peridiploid appearance (e.g., a peak slightly above or below the channel in which diploid cells appear), it is useful to admix some control cells in the sample to see whether it is possible to resolve two separate peaks. Of course, the ability to resolve the peaks of cells with nearly identical DNA contents will be a function of the CV of the stained cells (Rabinovitch, 1994).

For the purposes of examining the effects of a compound on cell growth, however, ploidy level is rarely important. Rather, it is the change in the distribution of the cells relative to the control exponentially growing culture. As an example, these same MOLT-4 cells were treated with varying concentrations of a drug for an identical length of time (3 hours). The drug in this instance was a topoisomerase-I inhibitor that was expected to have its most dramatic effects on S-phase cells. As can be seen in Fig. 2B, the DNA distribution of untreated control cells is more or less identical to that of a typical exponentially growing culture. At low concentrations of the compound (0.05 μM), the cell cycle was perturbed, as can be seen by the "peak" of cells in early S phase (arrow). This suggests that at least some cells fail to proceed as normal through the cycle and in turn would be expected to produce the kind of growth curves labeled "2" or "3" in Fig. 1. Doubling the drug concentration increased the effects of the "block" in early S phase (note the appearance of fewer cells to the right of the block in mid to late S phase when compared to the S-phase population of cells exposed to the lower drug concentration). In addition, these culture conditions have begun to give rise to some cells dying by apoptosis, as exemplified by the peak to the left of G_1 (sub-G_1 or hypodiploid peak) in the histogram. Cells growing under these conditions tend to have growth curves like that labeled "4" in Fig. 1. Finally, at a five times higher drug concentration, all cells in the S phase underwent apoptosis (about 50%), leaving the DNA distribution with only a G_1 and G_2M peak and a substantial sub-G_1 apoptotic peak.

Before discussing the impact of cell death on DNA histograms, it is necessary to note that histograms can be scaled in a variety of ways. As is the case in the example described above, many programs normalize histograms to "peak height," meaning the highest peak in any histogram will be chosen and everything else scaled to it. This is as opposed to comparing histograms that have been normalized to "area under the curve," meaning that the cumulative area under the histograms are all normalized to the same value. Under the latter conditions, changes in peak heights for a particular cell cycle phase are easier to detect and are more obvious. When histograms are scaled to "peak height," one must compare the cell number on the Y-axis before interpreting changes in the relative distributions of multiple histograms.

Identification of cells undergoing apoptosis in single-parameter DNA frequency distributions is dependent on various circumstances. Although the topic of apoptosis is covered in detail elsewhere in this volume, it is necessary to briefly visit some of the issues that effect interpretation of flow cytometric data as it relates specifically to analysis of drug effects. During the process of apoptosis, a caspase-specific DNase is activated, which in most (Arends *et al.*, 1990) but not all cases (Collins *et al.*, 1992) causes the degradation of internucleosomal stretches of DNA. These small stretches of DNA can be removed from the cell, lowering the apparent DNA stainability of those cells (Gong *et al.*, 1994a). Internucleosomal DNA degradation also gives rise to the so-called "ladder" during gel electrophoresis that represents single and multiples of the nucleosome—a core of two each of the four major histones with a stretch of 180 base pairs of DNA wrapped around it. The extent of the decrease in fluorescence for the population undergoing apoptosis is a function of the degree of degradation of the DNA between nucleosomes that gives rise to the small molecular weight fragments. Cell preparation and staining techniques that simply permeabilize the cell membrane with detergents or hypotonic buffers tend to liberate not only small-molecular-weight internucleosomal DNA but also cause the fragmentation of apoptotic nuclei into many apoptotic bodies and the release even of chromosomes from mitotic cells. This gives rise to stained particles with an apparent DNA content of $<0.1\%$ to several percentages of that of a G_1 cell that are often mistakenly counted as "apoptotic cells." Clearly, because a single human chromosome contains, on average, about 2% of the DNA of a normal G_1 cell, particles in the range of 0.1–1% of the stainability of a G_1 cells are likely to represent small apoptotic bodies, small chromosomes, fragmented chromosomes, or debris (clumps of stain). A single apoptotic cell can give rise to many particles falling within this range; that is, when investigators consider those particles to represent the apoptotic cells, they are dramatically overestimating the apoptotic index (i.e., the percentage of cells undergoing apoptosis at the moment of analysis). Therefore, beware of any DNA histograms plotted on a logarithmic scale. Unless the population of interest is not dividing but is undergoing polyploidization, there is no reason to plot DNA histograms on a logarithmic scale.

Occasionally, drug treatment gives rise to cell cycle distributions that appear not to have changed dramatically in shape when compared to the control distribution. One of several reasons can be the use of too high a drug concentration giving rise to necrotic cells, as depicted in Fig. 2C. In this instance, cells were purposely treated with a concentration of drug significantly higher than necessary to induce apoptosis. As a result, the cells stained much like normal untreated cells, and the cell cycle distribution failed to change, but a viability count of the culture would have resulted in a curve similar to the one labeled "5" in Fig. 1.

A second reason DNA frequency distributions might not change during treatment is if the agent freezes the cells in the cell cycle, completely preventing cell proliferation. Though rare, this type of effect was observed with the bis-intercalating antitumor agent ditercalinium (Traganos *et al.*, 1987). Thus, at concentrations that gave rise to a plateau in cell number (e.g., like curve 4 in Fig. 1), no apparent change occurred in the DNA distributions. To prove that cells were blocked in all cell cycle phases, Wilson *et al.* (1975) performed a stathmokinetic experiment, in which cells were treated with a drug (e.g., vinblastine) that collects cells in mitosis by virtue of its ability to prevent polymerization of the tubulin that forms the mitotic spindle. Although untreated cells accumulated in a time-dependent manner in mitosis (G_2M phase on a single-parameter DNA frequency distribution), no such accumulation was observed in ditercalinium-treated cultures. In fact, there was no apparent change in any of the cell cycle compartments, suggesting that the drug at concentrations sufficient to inhibit cell growth blocked cells from traversing all cell cycle phases.

Typically, cell cycle effects are more often easy to interpret, especially in the absence of significant cell death in the culture. Figure 3 represents an experiment in which MOLT-4 cells were exposed to the isoflavone genistein, an inhibitor of both tyrosine kinases and topoisomerase-II, among other targets (Traganos *et al.*, 1992). At subcytotoxic concentrations, genistein clearly slowed cell traverse of the S phase, resulting in an accumulation in the mid to late S phase. As is typical of

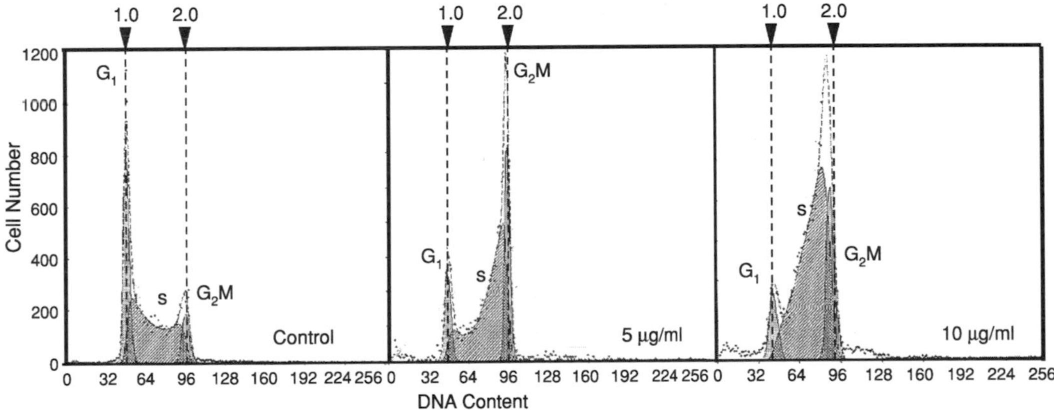

Fig. 3 Effect of an S-phase–active agent on DNA frequency distribution. (Left) The frequency distribution of control cells is illustrated. (Middle) Cells exposed to the drug begin to accumulate in the late S and G_2M phases. Treatment with this drug concentration slows progression through the S phase, resulting in an accumulation of cells in the G_2 phase. (Right) A higher concentration of the drug results in a more permanent S-phase block, with cells slowing down in early S and accumulating in late S phase; fewer cells were able to escape the block to accumulate in G_2M compared to the middle panel. Histograms are scaled to peak height.

these kinds of drug, the higher the dose, the slower the cells traverse the S phase and therefore the more prominent the accumulation in the S phase on DNA frequency distributions. Thus, treatment with 5 μg/ml slowed cell traverse of S, but at the time of analysis (i.e., 8 hours in this instance), cells appeared to have accumulated in the G_2M phase. When the drug concentration was doubled, the accumulation in the S phase appeared more prominent and lasted longer, giving rise to an increased proportion of S phase cells; the large accumulation in the later stages of the S phase suggests that traversal of the S phase was slowed dramatically with new cells entering the S phase from G_1 but few cells exiting S into G_2 phase.

Virtually all tumor cell lines contain a small and amazingly stable population of polyploid cells, as noted above for MOLT-4 cells, suggesting that cells in both ploidy levels have identical doubling times. Normal mammalian cells have a cell cycle checkpoint in place that requires that cells divide before entering a new round of DNA synthesis (Hartwell and Weinert, 1989) except under exceptional circumstances as, for instance, during megakaryocyte differentiation. Cell lines such as MOLT-4 occasionally develop a propensity for undergoing polyploidization when treated with certain agents. Such was the case when these cells were exposed to staurosporine, a potent inhibitor of protein kinases (Traganos *et al.*, 1994). Evidence had accumulated that this compound normally interfered with the G_1-to-S phase transition in normal cells (Bruno *et al.*, 1992; Crissman *et al.*, 1991), although higher concentrations block the entrance to mitosis of normal and tumor cells (Crissman *et al.*, 1991; Traganos *et al.*, 1994). An example of the result of treating MOLT-4 cells with this protein kinase inhibitor is shown in Fig. 4 (left, top, and bottom). What point is considered a high concentration (50 μg/ml), staurosporine caused MOLT-4 cells to cycle at a higher ploidy level. Interestingly, staurosporine had little in the way of cycle phase specificity because the treated cells had a more or less "normal" shape to their cell cycle distribution, which was confirmed by kinetic studies that demonstrated that the rate of cell entrance to G_2 phase in each ploidy was identical (Traganos *et al.*, 1994). Growth under these circumstances gave rise to a curve similar to the one labeled "4" in Fig. 1. As it turns out, the human monocytic leukemia cell line HL-60 also entered a higher ploidy level when treated with staurosporine under identical conditions, although the DNA distribution differed when compared to the control (Fig. 4, right, top, and bottom). In HL-60 cultures, some cells continued to divide at the lower ploidy level, as witnessed by the small peak at that locus, many cells remained with a G_2M phase DNA content (as would be typical following treatment with staurosporine), although a considerable number entered S and eventually the G_2M phase of the higher ploidy.

C. Interpreting DNA Histograms in the Face of Cell Death

Although, as noted above, DNA histograms should not be interpreted in the absence of cell growth information, there are instances when it is more difficult than others, especially in the absence of cell counts. Focusing on the far right

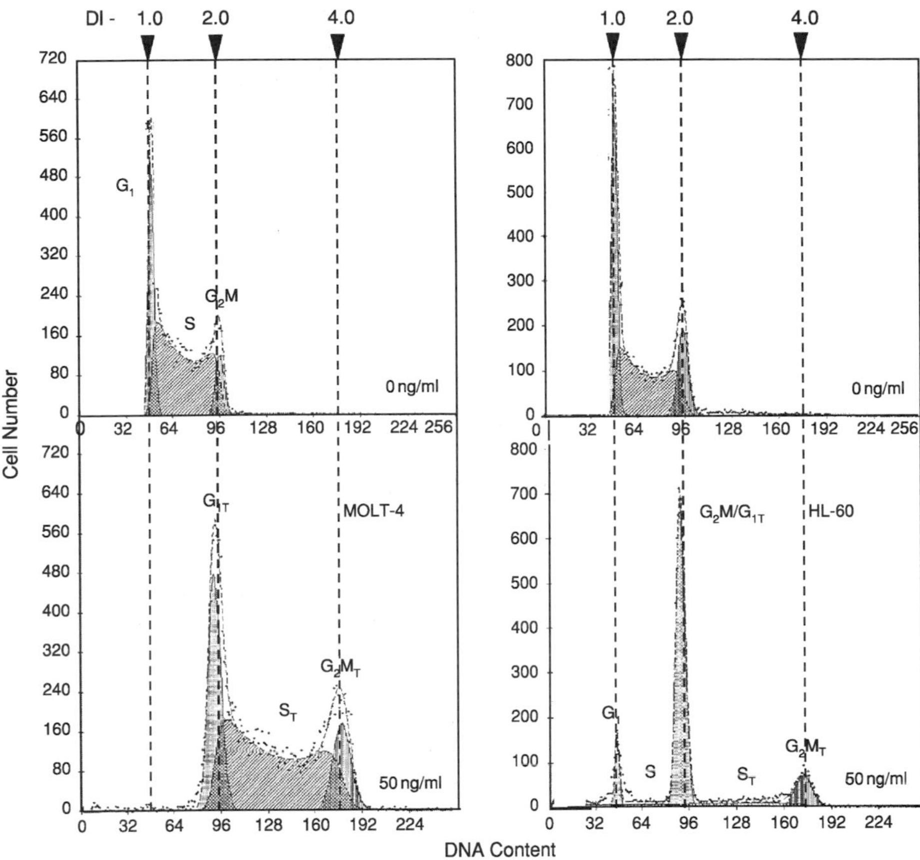

Fig. 4 Drug-induced growth of cells at a higher ploidy. Drug treatment led to growth of MOLT-4 cells (left, top and bottom panels) at a higher ploidy such the "G_1" phase of the treated cells appears with a DNA index (DI) of 2, indicating they now have a "tetraploid" level of DNA (G_{1T}) relative to untreated cells. Note that there was little or no change in the relative distributions of cells within the tetraploid cell cycle relative to control untreated cells. HL-60 cells (right, top and bottom panels) treated with the same drug also undergo a shift to higher ploidy but, as opposed to MOLT-4 cells, also demonstrate some proliferation at the lower ploidy (note treated cells in G_1), as well as an accumulation of cells in the G_2M/G_{1T} peak representing both cells in G_2 or M of the lower ploidy and G_1 of the higher ploidy. DIs are indicated on the figure at the top and with dashed lines. Histograms are scaled to peak height.

histogram in Fig. 2B, if one ignored the fact that there were apoptotic cells present in the analysis (or lost the apoptotic population or chose the wrong cell preparation method), it would be possible to interpret the remaining histogram as demonstrating cells were blocked in G_1 for an extended length of time, allowing for nearly all S-phase cells behind the block to exit S, pass through G_2M, and reenter G_1 phase. However, as we know, the absence of cells in S is due to the fact that they were killed by the drug, underwent apoptosis, and are now in the sub-G_1 peak to the left of the histogram. Obviously, these different scenarios would give rise to two very different models, leading to two very different mechanisms of action for the compound being studied. The only way the correct answer can be obtained is if the experiment is designed correctly, the drug concentrations are chosen based on cell growth/viability curves, and the data points are located fairly close together to help the researcher understand how the perturbed distribution evolved.

Although sophisticated computer programs initially developed in 1993 for Verity Software (Toposham, Maine) by Bruce Bagwell and for Phoenix Flow Systems (San Diego, California) by Peter Rabinovitch allow for deconvolution of complex DNA histograms including those containing multiple cell populations at different ploidies, apoptotic cells, and/or various kinds of debris, there still remain problematic assumptions inherent in analyzing single-parameter DNA frequency histograms.

V. Multiparameter Approaches

When discussing cell cycle analysis, multiparameter approaches always infer that although two or more parameters are measured for each cell, the cellular characteristic that is *always* measured is DNA content. Bivariate analysis of DNA content *versus* RNA content (Darzynkiewicz *et al.*, 1976) or protein content (Crissman and Steinkamp, 1973) as well as DNA content *versus* proliferation-associated proteins such as Ki-67 (Gerdes *et al.*, 1983) or PCNA (Celis *et al.*, 1984) have been described in detail elsewhere, in particular in earlier volumes of *Methods in Cell Biology* (volumes 41 and 63) and have been summarized in a historical review on cytometry of the cell cycle (Darzynkiewicz *et al.*, 2004). As a result, those approaches will not be addressed here. However, one of the most important bivariate approaches to the analysis of cell proliferation has been the use of halogenated pyrimidines incorporated into DNA to detect S-phase cells. Although this field has also been widely reviewed, there are some points concerning this approach worth reinforcing.

A. Measuring DNA Synthesis Using Bromodeoxyuridine

Historically, the study of cell cycle kinetics entailed the use of radioactively labeled thymidine to detect cells synthesizing DNA. Using autoradiography, radioactive thymidine incorporation allowed for the measurement of the labeling index (LI) and the fraction or percentage of labeled mitoses (PLM). The LI could be used to determine the fraction of cells synthesizing DNA (depending on S-phase duration), although the changing PLM curve provided information on the progression of labeled cells through the cycle (Terry and White, 2001). These tedious techniques have now been supplanted by the use of fluorescence-labeled halogenated pyrimidines as a replacement for radioactive thymidine. The most notable and widely used is bromodeoxyuridine (BrdUrd), although chlorodeoxyuridine and iododeoxyuridine have also been used either singly or in combinations. Detection of incorporation of these fluorescence-labeled analogs by flow cytometry allows for thousands or tens of thousands of cells to be analyzed in detail, providing statistically relevant data.

In virtually all cases, incorporation of halogenated pyrimidines is detected by tagging them with fluorescence-labeled monoclonal antibodies. Initially, however, techniques for detecting BrdUrd incorporation into DNA involved the fact that certain DNA stains (e.g., Hoechst 33358) were quenched in its presence (Latt, 1974). The decrease in fluorescence of the quenched dye could be combined with a DNA-specific dye whose fluorescence emission was unaffected by the presence of BrdUrd. Analysis of the binding of both dyes provided for a more precise quantitation of the S-phase population in the cell cycle distribution (Bohmer and Ellwart, 1981). This approach allowed for measurement of a number of parameters and was especially useful because more than one cell cycle could be analyzed in the same measurement (Kubbies *et al.*, 1985).

As noted, detection of BrdUrd incorporation into DNA has been based on immunocytochemical techniques. In this approach, partial DNA denaturation by strong acid (e.g., 2 M HCl) or heat is required to make the incorporated halogenated pyrimidine accessible to the monoclonal antibody probe, which can be either directly or indirectly labeled with fluorescein. The non-denatured DNA can then be stained, typically with an intercalating fluorochrome such as PI. The combination of PI red fluorescence and fluorescein green fluorescence provides a bivariate plot of cellular DNA content *versus* incorporated BrdUrd (Dolbeare *et al.*, 1983).

However, because immunocytochemical detection of incorporated BrdUrd requires the rather harsh conditions of low pH or high temperature, these treatments often affected cell morphology, resulted in significant cell loss during staining, and damaged many epitopes that could be used to detect other cellular constituents. To avoid either nuclei isolation or the harsh conditions required for DNA denaturation, techniques have evolved that use restriction enzymes to provide access for the antibody to the incorporated halogenated pyrimidine (Dolbeare and Gray, 1988). However, all these techniques present a trade-off between the maximum amount of incorporated BrdUrd that can be detected and the accurate measurement of total DNA content. Thus, the more successful one is in denaturing or digesting DNA to uncover incorporated BrdUrd, the less intact DNA will be left to bind the DNA stain, which is almost always an intercalator like PI that requires DNA to be in a double-stranded conformation. More accurate DNA content determinations requiring intact double-stranded DNA mean that BrdUrd incorporation cannot be optimally detected.

Another approach does not require uncovering the site of BrdUrd incorporation by either denaturation or digestion, thereby avoiding the compromise inherent in techniques that require such treatments. In this approach, BrdUrd is incorporated into DNA, as in the previous techniques. However, although typically one is cautioned against allowing the cells to be exposed to light after BrdUrd incorporation (Terry and White, 2001), UV light is used in this technique to induce DNA breaks by photolysis at the sites of BrdUrd incorporation (Li *et al.*, 1994). Next, the enzyme terminal deoxynucleotidyl transferase is used to add many (approximately 20) fluorochrome-labeled deoxynucleotides (typically BrdUrd) to each DNA break site. This means that the label is not on the antibody attempting to bind the single molecule of BrdUrd incorporated into the DNA, but that the fluorochrome is attached to one of many molecules being added at the site of incorporation, thereby dramatically magnifying the signal without degrading the DNA by denaturation or digestion in a way that would affect binding of the intercalating dye, PI. This approach is known as *strand break induction by photolysis* (SBIP) (Li *et al.*, 1994), an example of which is shown in Fig. 5. As can be seen in the figure, there is an approximately 10 times increase in fluorescence intensity of the cells incorporating BrdUrd compared to control cells grown in the absence of the label.

All these approaches have the potential for providing detailed information concerning cell cycle kinetics and are capable of being used in experiments designed to determine the cell cycle phase specificity of potential antitumor agents. It should also be noted that even in the absence of treatment, the distributions contain a wealth of information. The relative levels of BrdUrd incorporation into S-phase cells results in the distinct upside-down *U* shape of the distribution. However, the actual shape may vary depending on whether DNA synthesis is constant throughout S phase or it is faster in the beginning or end of S. In addition, the height of the S phase above the control also provides information concerning the rate of DNA synthesis and therefore the rate of cell traversal through S phase. As the S-phase lengthens (e.g., as the result of serum deprivation, growth to high cell density, and drug treatment), the height of the S-phase population will decrease, indicating a lower rate of BrdUrd incorporation per unit of time. In fact, in some instances, especially when studying cells *in vivo*, it is not unusual to observe cells with an S-phase DNA content that have incorporated little or no BrdUrd, suggesting they either are not cycling or are cycling so slow that they are not detectable under the experimental conditions being used. The width of the distribution of precursor incorporation provides information concerning the heterogeneity of the population. The more variable the measure of precursor incorporation, the more variable the rate of transit of cells through the cycle; incorporation seen as fluorescence intensity per DNA value provides information on the rate of synthesis per unit time.

B. Use of Cell Cycle Control Proteins to Assess Drug Action

Cyclins, cyclin-dependent kinases (CDKs), cyclin-dependent kinase inhibitors (CKIs), and proteins of the retinoblastoma protein (pRb) pathway are all involved in the molecular machinery of cell cycle progression and control cell transitions from one cell cycle phase to the next (Nasmyth, 2001; Nurse, 2000; Pines, 1999). Some of these proteins such as the cyclins are expressed discontinuously during

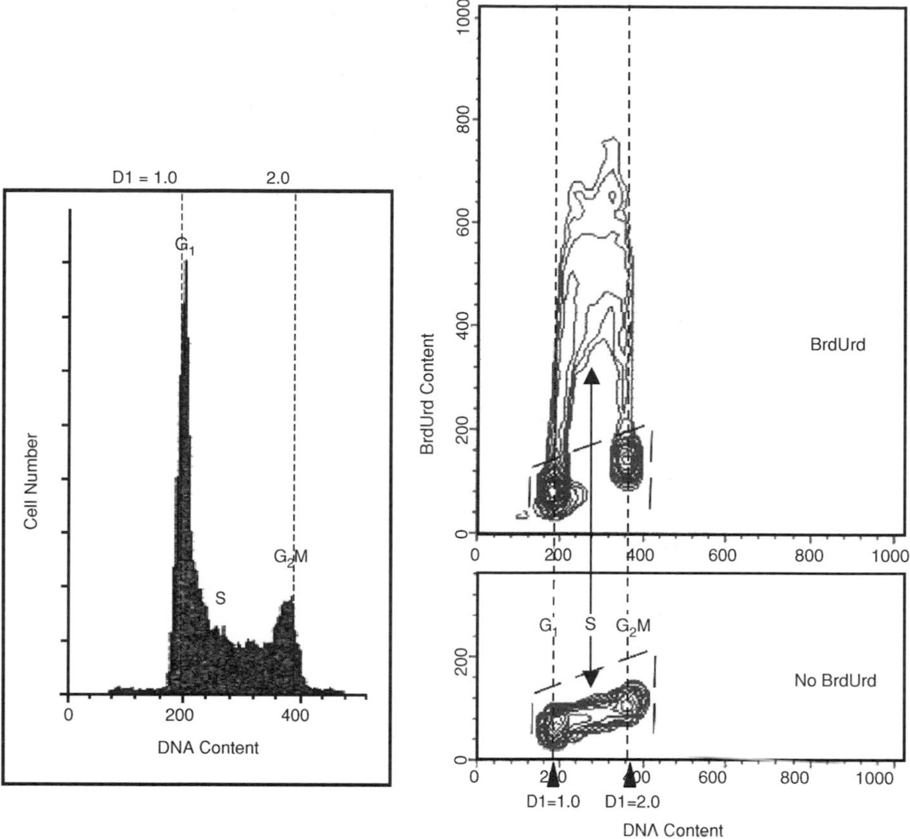

Fig. 5 Bivariate analysis of DNA content versus incorporation of bromodeoxyuridine (BrdUrd) into DNA of S-phase cells. (Left) The DNA frequency distribution of exponentially growing (MOLT-4) cells. (Right) The bivariate distribution in which S-phase cells have incorporated the thymidine analog BrdUrd into DNA, which was detected with a fluorescent probe (see text for details). Beneath is the distributions of the same cells grown in the absence of BrdUrd. Using this control, a compartment can be described that indicates the extent of unlabeled cell BrdUrd-associated fluorescence. This aids in the determination of the S-phase fraction in the top of the panel. Note that early and late S-phase cells have DNA fluorescence values that overlap with those of cells in the G_1 and G_2 population, respectively. The height and variability of fluorescence measuring BrdUrd incorporation represents the relative rate of traversal of the S phase and variability in the rate for cells in the S phase in the culture. The shape of the curve can vary if the relative rates of S-phase traversal differ for cells early or in the middle or late stages of the S phase. Although BrdUrd incorporation is plotted on a linear scale in this example, it is often plotted on a log scale.

the cell cycle and can be induced or suppressed by various environmental signals. Therefore, the analysis of bivariate distributions of DNA *versus* cell cycle proteins, either in terms of their abundance or modification, can provide insight into the mechanisms involved in cell cycle regulation and the perturbation of this regulation by various compounds including antitumor agents (Darzynkiewicz *et al.*, 1996, 2003; Frisa *et al.*, 2000).

Multiparameter cytometric approaches allow one to directly correlate, in individual cells, the expression of cyclins, CKIs, or proteins of the pRb pathway with the cell's position in the cycle or with DNA replication. This is typically accomplished through immunocytochemical labeling of the proteins, combined with measurement of either cellular DNA content or DNA replication. Because many of these proteins are expressed in specific cell cycle compartments, which can be "gated" according to DNA content, it is possible to estimate the relative abundance (or modification) of each protein as a function of cell cycle phase, based on the intensity of fluorescence. Analysis by multiparameter cytometry also allows one to (1) assess the intercellular variability in expression of these cell cycle control molecules, (2) detect cell subpopulations sharing similar or different features, and (3) determine whether there are thresholds in expression necessary for cells to enter the next cell cycle compartment. Another important advantage of this approach is that expression of these important cell cycle control molecules can be studied in asynchronously growing cells and do not require cell synchronization, which can often lead to unbalanced growth and misleading levels of protein expression (Gong *et al.*, 1995; Johnson *et al.*, 1993).

The discontinuous expression of the cyclin proteins is illustrated in the diagram at the top of Fig. 6. For the most part, the D family of cyclins is expressed early in the G_1 phase, binds to specific CDKs (CDK4 and CDK6), and when activated by the appropriate kinase, phosphorylates pRb, the retinoblastoma gene product. Typically, the expression of cyclin D remains high during G_1 because the complex also acts as a "sink," binding CKIs that would normally inactivate the next cyclin complex in the cycle: cyclin E/CDK2. Cyclin E expression peaks later in G_1/early in S phase and then begins to decline until it reaches background levels at the end of the S phase. Cyclin A expression begins rising at the beginning of the S phase (it also binds CDK2) and peaks in G_2 (where it switches to another CDK, CDK1). Cyclin A tends to be destroyed near the border between G_2 and mitosis. Finally, cyclin B family members begin increasing late in the S phase and peak in G_2 and the early stages of mitosis but must undergo proteolysis for the cell to undergo the transition from metaphase to anaphase.

1. *Normal expression of "G_1" cyclin proteins:* Although cyclin D expression is known to be involved in transducing growth signals from the cell surface to activate cell entrance into the cell cycle, cyclin E activation is directly associated with the transition of cells from G_1 to S phase (Nurse, 2000). The bivariate distribution of cyclin D in normal cells (fibroblasts) is shown in Fig. 6. According to expectations, cyclin D expression decreases to background levels as cells enter the S phase in this cell system. The bivariate flow cytometric pattern obtained for cells normally expressing cyclin E demonstrates that cyclin E must increase above a threshold (arrow) in G_1 before the cell can enter the S phase. As the cell continues through the S phase, cyclin E expression decreases continuously. One

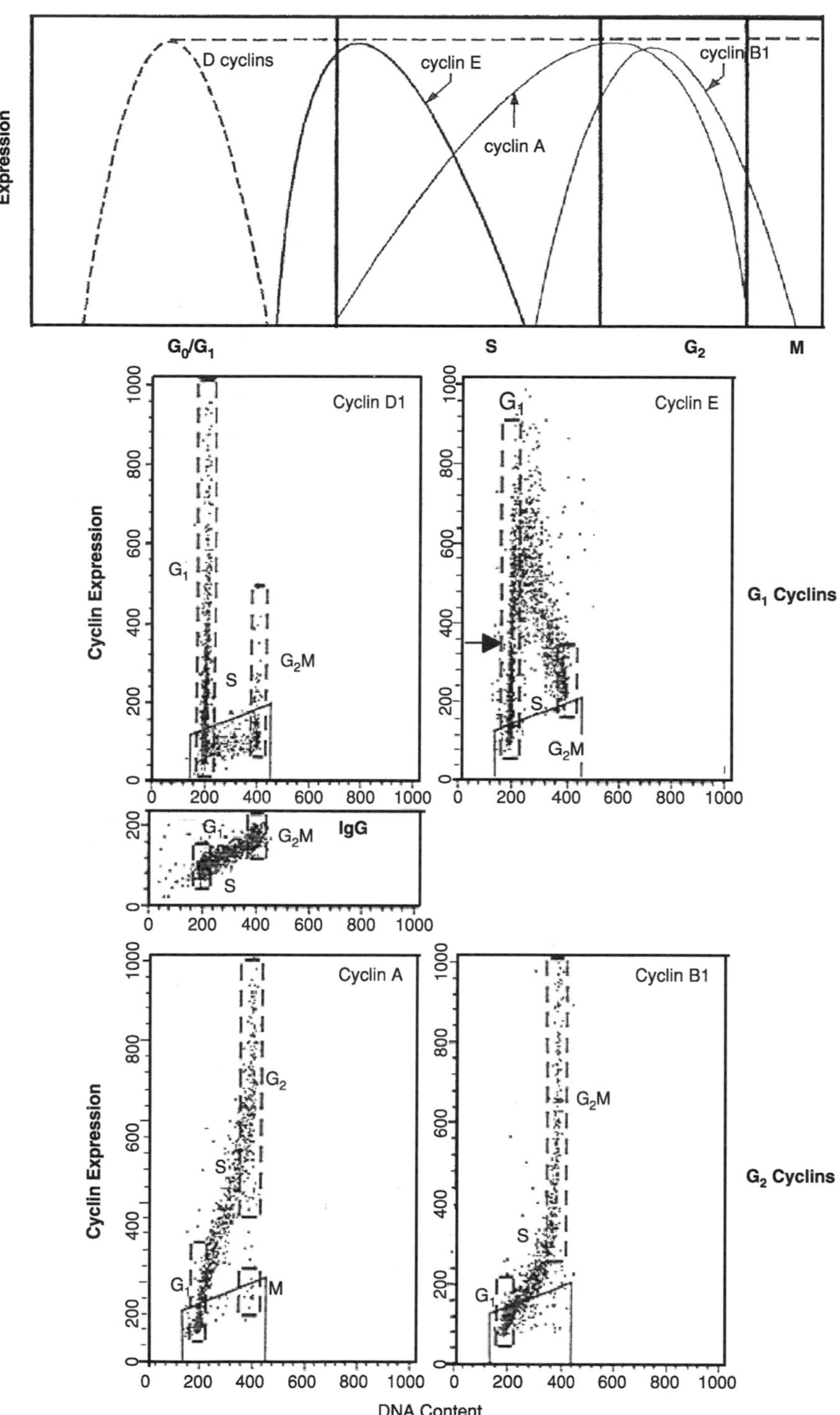

Fig. 6 (*Continued*)

way of visualizing the specificity of expression is by subtracting the control (nonspecifically bound antibody) from the cyclin E–stained sample leaving "cyclin E–specific expression" (Fig. 7A). After subtraction, there is no "threshold" required to be attained before cells enter the S phase. This would indicate that some cells upon division express little or no cyclin E (cyclin E^-) and only when expression increases over some threshold are (cyclin E^+) cells able to enter the S phase.

2. *Normal expression of "G_2" cyclins:* Cyclin A expression begins early in S phase and increases continuously as cells progress through S phase and into G_2 (Fig. 6). Mitotic cells are cyclin A negative. Cyclin B1 activates CDK1 and is responsible for controlling the transition of cells from G_2 to mitosis (along with cyclin A when it is bound to CDK1). Typically, cyclin B/CDK1 complexes reside in the cytoplasm during G_2 and are shuttled to the nucleus only when the signal occurs for entrance into mitosis. As with cyclin E, cyclin B1's expression occurs predominantly late in the cycle, peaking in G_2 phase and in dividing cells up to metaphase (Fig. 7B), at which point it is broken down providing the signal for completion of cell division (Nasmyth, 2001; Nurse, 2000; Pines, 1999).

3. *"Unscheduled" cyclin expression in tumor cells:* Many tumor cells do not express cyclins as do normal cells (Gong *et al.*, 1994; Juan *et al.*, 1996). Generally, this means that cyclins are expressed in inappropriate cell cycle phases. Figure 8 illustrates cyclin B1 expression in two human leukemic cell lines. In MOLT-4 cells, cyclin B1 expression appears normal (compare to normal cell expression in Fig. 7), whereas HL-60 expresses cyclin B1 in G_1 and early S phase. By subtracting the "normal" cyclin B1 fluorescence of the MOLT-4 cells from the HL-60 cells, it becomes clear where in the cell cycle there is "inappropriate" expression of cyclin B1. Because cyclin complexes provide one level of control over cell cycle checkpoints that govern the transition of cells from one phase to the next, inappropriate levels of scheduled expression or unscheduled expression in a particular cell cycle phase provides insight into the defects operating in that particular cell system.

Fig. 6 Cell cycle phase–specific expression of cyclin proteins. The top of the figure illustrates the expected expression of the different cyclins as a function of cell cycle phase. The "G_1 cyclins" D and E are expressed either exclusively and early in G_1 phase (cyclin D) or late in G_1 and in early S, after which expression falls to background levels as the cells traverse S phase (cyclin E). The "G_2 cyclins" A and B1 are expressed either early in S phase increasing until they peak in G_2 and disappear during mitosis (cyclin A) or later in S also peaking in G_2 and remaining until metaphase, at which point they are degraded (cyclin B1). The four panels beneath illustrate bivariate dot plots of each pair of G_1 and G_2 cyclins. As described above, the expression of each follows expectations in these normal cells (normal human fibroblasts for cyclin D and mitogen-stimulated human peripheral blood lymphocytes for the remaining cyclins). Nonspecific staining with the antibody was determined by using an irrelevant immunoglobulin G (IgG) (see panel beneath the cyclin D1 dot plot). The control nonspecific binding was used to identify the locus of cells with little or no cyclin binding in each of the other panels. The G_1 and G_2M populations are indicated by dashed rectangles.

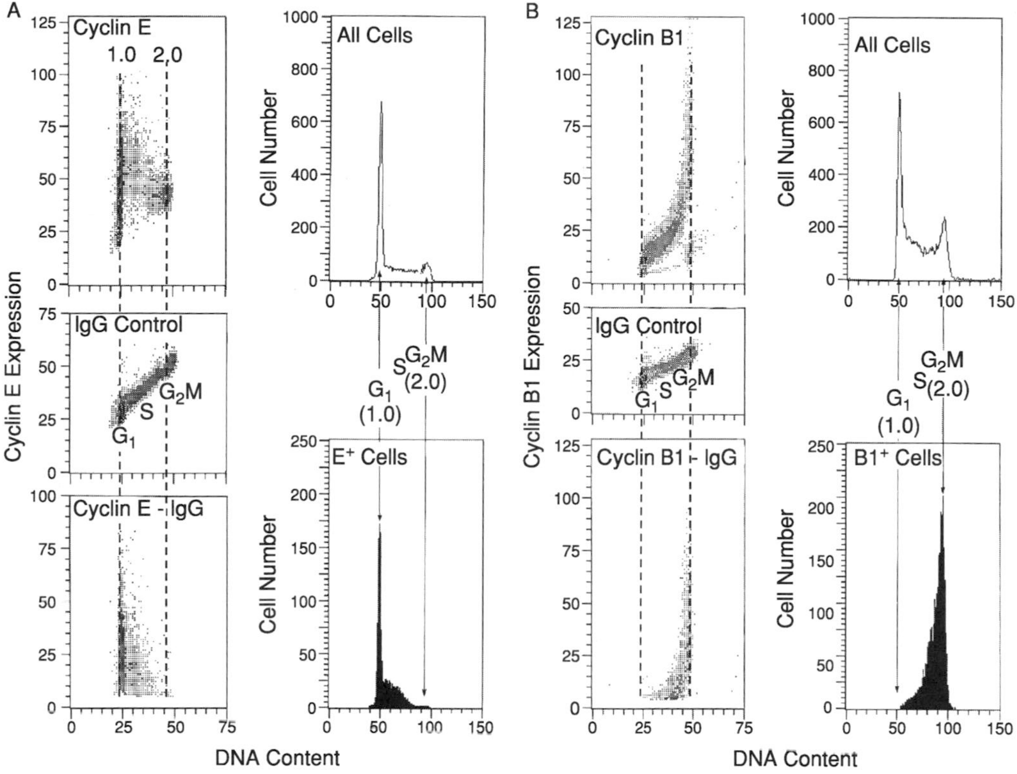

Fig. 7 Determination of cyclin E and B1 expression in normal cells by subtraction of control background fluorescence. To illustrate the cell cycle phase–specific expression of a G_1 and G_2 cyclin, cells were stained for DNA content either cyclin E (A) or cyclin B1 (B) or with an irrelevant antibody (immunoglobulin G [IgG]). By use of an Overton subtraction, it is possible to observe the cyclin expression of cells positive for each. As is obvious from the lower dot plot in (A), cyclin E expression is high in G_1 and early S and decreases to background as cells traverse S, whereas cyclin B1 is absent in early S-phase cells but increases dramatically in late S and G_2M phase. The DNA frequency distributions of the total populations (top) and the populations after subtraction of cells with little or no cyclin staining (bottom) fulfill expectations. All histograms are normalized to maximum peak height in this figure.

4. *Use of cyclin expression to characterize drug action:* Because cyclins are involved in complexes whose activities are required to traverse the various cell cycle checkpoints, the levels of cyclin protein expression can be useful in determining the activity of some cell cycle phase–specific drugs (Darzynkiewicz *et al.*, 1994; Gong *et al.*, 1994b,c,d). Although each of the cyclins can be used for this purpose, two cyclins, E and B1, representing a G_1 and a G_2 cyclin, respectively, have been chosen to provide examples of the principle.

Figure 9 illustrates the typical expression of the G_1 cyclin, cyclin E, in a cell system with normal cyclin expression. The box represented by the dashed line

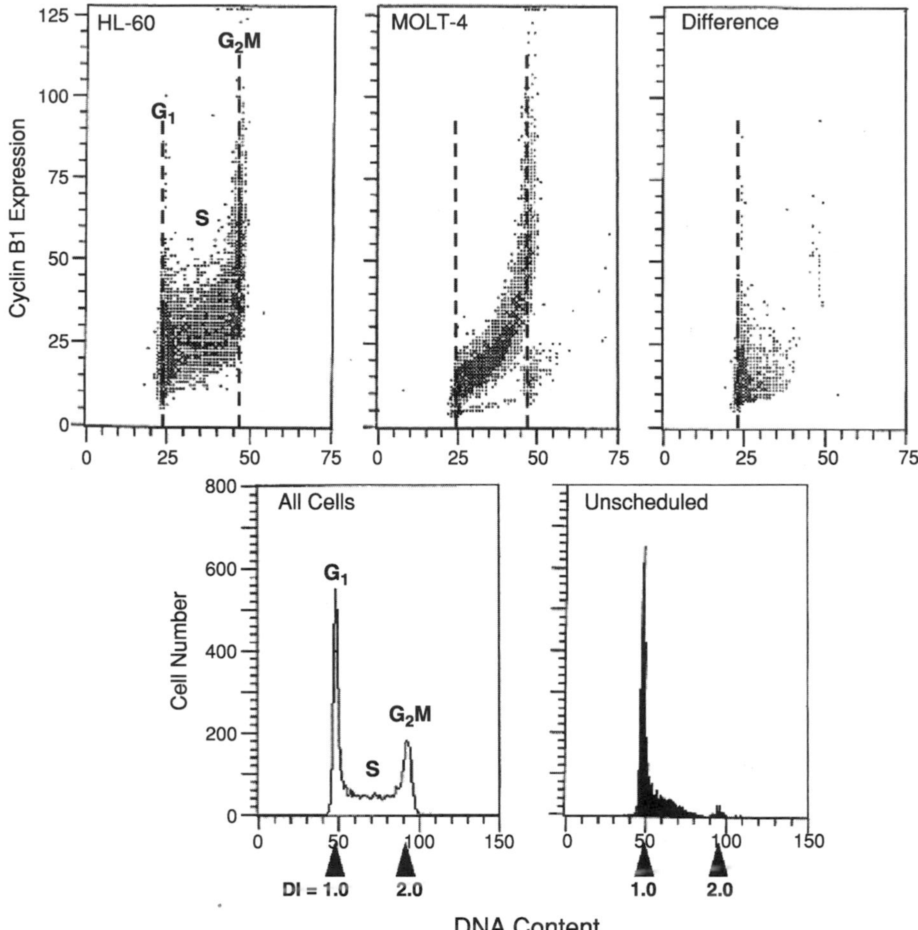

Fig. 8 An example of "unscheduled" cyclin B1 expression. MOLT-4 cells (center dot plot) express a normal pattern of cyclin B1 equivalent to that observed for normal human cells (see Fig. 6). Another human leukemic cell line HL-60 has a very different expression pattern for this cyclin (left). By subtracting the normal (MOLT-4) from the unusual (HL-60) cyclin distribution, one gets a difference plot. The DNA frequency distributions for the total (HL-60) and difference (HL-60 minus MOLT-4) are displayed beneath the dot plots. The difference plots illustrate that HL-60 appears to express the G_2 cyclin, cyclin B1, in G_1 and very early S-phase cells. The dashed lines in the dot plots indicate where DNA indexes (DIs) equivalent to 1 and 2 for these cell lines are located. The histograms are normalized to peak height.

illustrates the variation in cyclin E expression of G_1 cells from the threshold level required for cells to enter S phase (lower Y-axis boundary) to the maximum level attained by a G_1/early S phase cell (upper Y-axis boundary). The cells were treated with two compounds: (1) mimosine, a plant amino acid that arrests cells late in the G_1 phase (Watson *et al.*, 1992) perhaps by interacting with a protein necessary for initiation of DNA synthesis (Mosca *et al.*, 1992), and (2) sodium butyrate, an

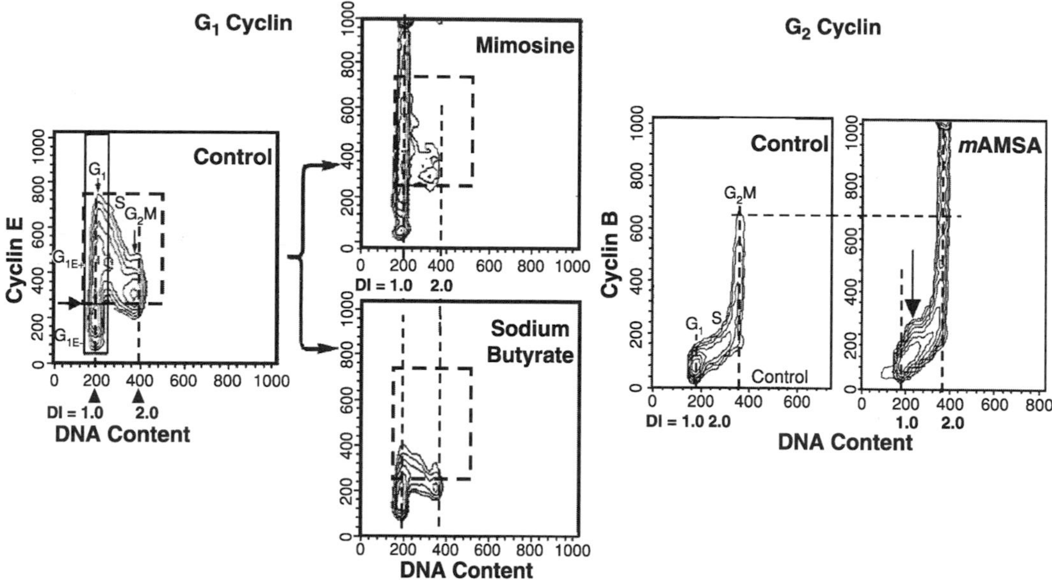

Fig. 9 Effect of drug treatment on cyclin expression. The G_1 cyclin, cyclin E, expression in normal cells is displayed on the left. The minimum amount of cyclin E expression required for cells to enter S phase and the maximum amount of expression for cells in the cycle form the lower and upper limits of box (*dashed lines*) describing cyclin E expression. Cells exposed to mimosine block cells in G_1 and express higher levels of cyclin E, suggesting the block is after the cyclin E restriction point. Sodium butyrate also blocks cells in G_1, but because cyclin E levels are lower in cells treated with this agent, its point of action must be before the cyclin E restriction point. The G_2 cyclin, cyclin B1, is expressed predominantly in late S, G_2, and M. Exposure to amsacrine (*m*AMSA) affects cell progression through S phase, resulting in cell accumulation in G_2 phase. Because cyclin B1 expression increased in the treated cells over the maximum values attained in control cultures (*dashed line*), the cells treated with *m*AMSA accumulated after the cyclin B1 restriction point.

inhibitor of histone deacetylase that arrests cells early in the G_1 phase presumably as a result of inner core histone hyperacetylation and/or histone H1 hypophosphorylation (D'Anna *et al.*, 1980). If one assumes that the onset of cyclin E synthesis in G_1 represents the cyclin E restriction point, then mimosine arrests cells in G_1 after the cyclin E restriction point because cyclin E continues to be synthesized and the protein reaches levels well above those normally attained by G_1 phase cells (above the box in Fig. 9). Because the cell cannot enter S phase, the onset of cyclin E degradation is delayed or inhibited. Alternatively, sodium butyrate's activity appears to occur before the cyclin E restriction point, as indicated by the fact that there was no increase but a decrease in the level of cycle E expression (below the box in Fig. 9).

The topoisomerase-II inhibitor amsacrine (*m*-AMSA) acts on cells in the S phase to slow their transit (note the bump in the early S phase indicated by the arrow) but also leads to accumulation of cells in the late G_2 phase after the cyclin B1 restriction point. As a result, cells blocked in G_2 continue to accumulate cyclin B1 (above the dashed line in Fig. 9) because cyclin B1 degradation begins only when cells transition from metaphase to anaphase. These same kinds of analyses can, of course, be done for the other major cyclins providing greater resolution of a drug's activity, for example, a compound might act after the cyclin D but before the cyclin E restriction point, resulting in increased accumulation of cyclin D but a decrease in cyclin E expression of cells accumulated in the G_1 phase. However, these kind of analyses can be carried out only with either normal cells or in tumor cell systems that have "scheduled" cyclin expression. "Unscheduled" expression of cyclins makes it difficult or impossible to interpret changes in expression as a result of drug action.

5. *Use of cyclin B1 to analyze cells growing at multiple ploidies:* As shown in Fig. 4, occasionally, compounds such as staurosporine induce polyploidy in some cell systems. The DNA frequency histograms created by such treatment occasionally give rise to a peak that contains G_2 and mitotic cells of the lower ploidy level and G_1 cells of the higher ploidy (G_{1Tet}), as in Fig. 10. Even the most sophisticated DNA histogram analysis programs cannot deconstruct such a peak correctly into its component parts. However, because cyclin B1 is positive for G_2 and mitotic cells (at least through metaphase) and negative for G_1 phase cells, it is possible to identify the G_1 cells of the higher ploidy among the cells in the G_2M/G_{1Tet} peak as a consequence of their reduced cyclin B1 expression (Fig. 10).

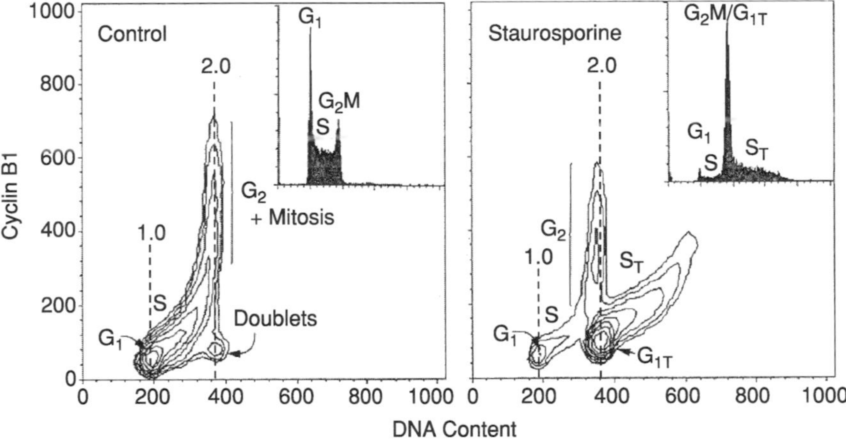

Fig. 10 Use of cyclin B1 to detect G_1 cells of a higher ploidy among a population of G_2M cells with the same DNA content. MOLT-4 cells exposed to staurosporine enter a higher ploidy level. Because cyclin B1 is normally expressed in G_2 cells and cells in mitosis up to metaphase but not in G_1 cells, it is possible to discriminate between the G_2M phase cells of the lower ploidy and the tetraploid G_1 (G_{1T}) cells of the higher ploidy.

As with any intracellular probe, binding specificity is important with antibodies to the various cyclins and should be checked and the antibody titer determined (Jacobberger, 2001). However, it is also important to recognize that cyclins are active only when in the nucleus. There are certain circumstances when cyclin/CDK complexes are shuttled intact into the cytoplasm to be sequestered and are shuttled back into the nucleus only when required. Cyclin B1/CDK1 is an example in which the complex formed in G_2 phase is inactivated by phosphorylation of threonine 14 and tyrosine 15 in the adenosine triphosphate (ATP) binding site and sequestered in the cytoplasm. When required (at the end of G_2 phase), the complex is activated by the action of the phosphatase cdc25c and returned to the nucleus, where its activity is necessary for cells to enter mitosis. Obviously, flow cytometric analysis of the fluorescence bound to cyclin B1 does not normally allow one to discriminate between its nuclear or cytoplasmic expression, although such analysis is possible on laser scanning cytometers that can discriminate fluorescence signals in the cytoplasm from those in the nucleus (Darzynkiewicz *et al.*, 1999; Kamentsky and Kamentsky, 1991).

At a more fundamental level, whether antibodies used in these analyses bind equally well to free or bound (complexed) cyclin is not known. Assuming these antibodies bind to cyclin complexed to their CDK, does the fact that whether the complex is activated (phosphorylated) or not affect binding? Even knowing that the epitope for the antibody is not, because of its location, likely to be unavailable to the probe does not ensure that subtle changes associated with complex formation does not change the affinity of the antibody for that epitope. For instance, binding of the cyclin to the CDK occurs only at the highly conserved cyclin box distant from both the substrate and the ATP binding sites. However, the conformation of both those binding sites is altered upon cyclin binding, meaning that a conformational change in the CDK is transmitted over some distance. Cyclin/CDK complexes can also have CKIs bound to them such as p16 (cyclin D/CDK4 or 6), p21, or p27 (all cyclin/CDK complexes). Although it is known that binding of CKIs alters the conformation of CDKs, it is not known whether their binding affects cyclin conformation and, if it did, antibody binding. Therefore, care should be taken when interpreting data obtained based on immunocytochemical staining of intracellular targets.

6. *Effect of cell synchronization on cell growth:* One of the clear advantages to the use of cytometric methods in studying cell cycle kinetics is that the analysis does not require cell synchronization. Because DNA content is invariably measured, it is a simple matter to "gate" along the DNA distribution to determine the relative expression of any molecule for virtually any of the cell cycle phases including mitosis (see later discussion). This contrasts with the necessity to attempt cell synchronization to identify different levels of molecular expression by biochemical means; in the absence of synchronization, bulk measurements such as obtained from Western blots provide an average for all cells and would miss minor subpopulations of cells expressing exceedingly high or low levels of a protein.

To avoid the averaging effect of bulk measurements, typically "double-thymidine" blocks are performed. This approach is expected to pile cells up at the G_1–S border. After synchronization, the cells are considered to be in late G_1, and once released, depending on the amount of elapsed time, cells can be collected predominantly in S phase, G_2, or mitosis. However, inhibition of DNA replication, although it arrests cells in the cycle, does not inhibit DNA transcription or translation to the same extent resulting in growth imbalance (Cohen and Studzinski, 1968; Johnson *et al.*, 1993). Although the DNA synthesis and cell growth cycles typically run in parallel, they are not inextricably linked. Unbalanced growth results from the uncoupling of the DNA synthesis cycle (inhibited by thymidine) from the cell growth cycle (little affected by thymidine). Using thymidine to arrest cells in G_1 phase by inhibiting DNA polymerase activity, MOLT-4 cells could be seen to accumulate significantly higher levels of cyclin proteins than would ever be observed in exponentially growing cells and to express some cyclins at levels inappropriate for their position in the cell cycle based on DNA content (Gong *et al.*, 1995).

Figure 11 illustrates the expression of cyclin E and B1, a G_1 and a G_2 cyclin, respectively, after synchronization by a double-thymidine block. Because the block occurs after the cyclin E restriction point, cyclin E expression continues to increase, although the cells remain blocked at the border to the S phase. Immediately before release cyclin E levels had dwarfed those typically obtained for this culture. Interestingly, 4 hours after release, some cells had already traversed the S phase and were entering the G_2 phase. In addition, the rate of cyclin E degradation sped up so that as cells reached the G_2 phase, their cyclin E expression had returned to levels typical of unsynchronized cells. Cyclin B1, not normally expressed in G_1 cells, had already reached a level indicative of cells late in the cycle (G_2 or M phase cells) before release. The uncoupling of the cell growth cycle from the DNA synthesis cycle had, in effect, resulted in cells expressing levels of cyclin B1 protein as if they were in the cell cycle phase they would have been in were it not for the thymidine block. Even though the cells before release had sufficient cyclin B1 expression equivalent to unsynchronized G_2 cells, they began synthesizing even more cyclin B1 after release and also sped through S phase with some cells arriving in G_2 phase within 4 hours.

7. *CKIs and drug activity:* Checkpoint control is not only exercised through the synthesis or degradation of cyclins and the activating/inhibiting phosphorylation of the CDK complexes. CDK inactivation also occurs through the binding of CKIs to the complexes (Harper and Elledge, 1996). There exists two families of CKIs, the INK family consisting principally of p15, p16, and p18 and the CIP/KIP family composed of p21, p27, and p57. The former group confines their inhibitor activity to complexes containing cyclin D, whereas the CIP/KIP family members bind to all cyclin–CDK complexes. Expression of these different inhibitors occur often during the normal course of the cell cycle or during cell activation, differentiation, or quiescence/senescence.

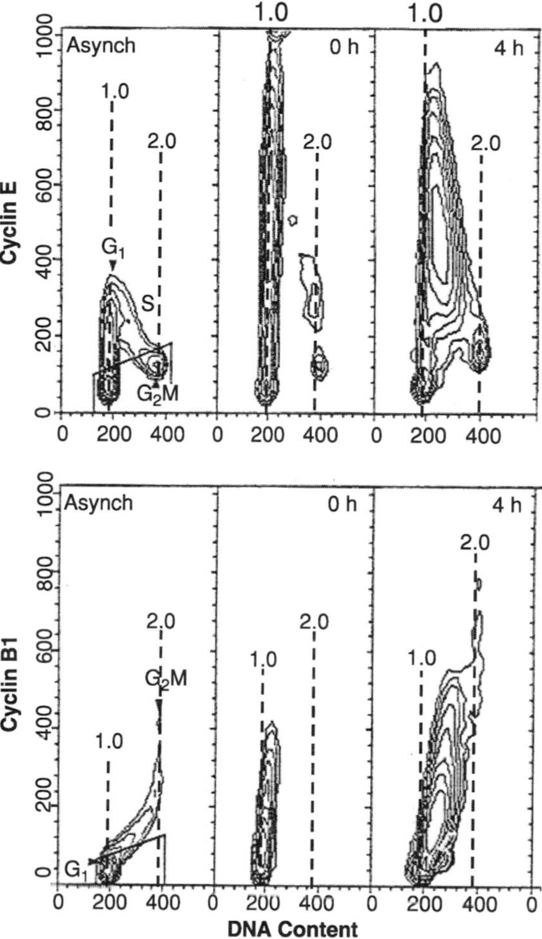

Fig. 11 The effect of chemically induced cell synchronization on cyclin expression. A double-thymidine block arrests cells in late G_1 phase at the entrance to S phase. Such treatment, however, induces unbalanced growth, resulting in the overexpression of some proteins including cyclins. Cyclin E expression in synchronized cells was much higher than for cells in asynchronous exponential growth. Release from the block resulted in an increased rate of transit through S phase and an increased rate of cyclin E degradation back to normal levels as the cells reached G_2 phase. Cyclin B1 in synchronized cultures was also present at very high levels in a cell cycle phase at which it is not expected to be present (G_1). Release of the cells from the block led to increased cyclin B1 synthesis even though the cells already contained more cyclin B1 than normal.

An example of the interplay between cyclins and CKI expression is shown in Fig. 12, which plots the expression of cyclin D3, cyclin E, p16, and p21 during mitogen stimulation of human lymphocytes into the cycle. Nonstimulated lymphocytes are quiescent (G_0) cells, some of which can be induced to proliferate

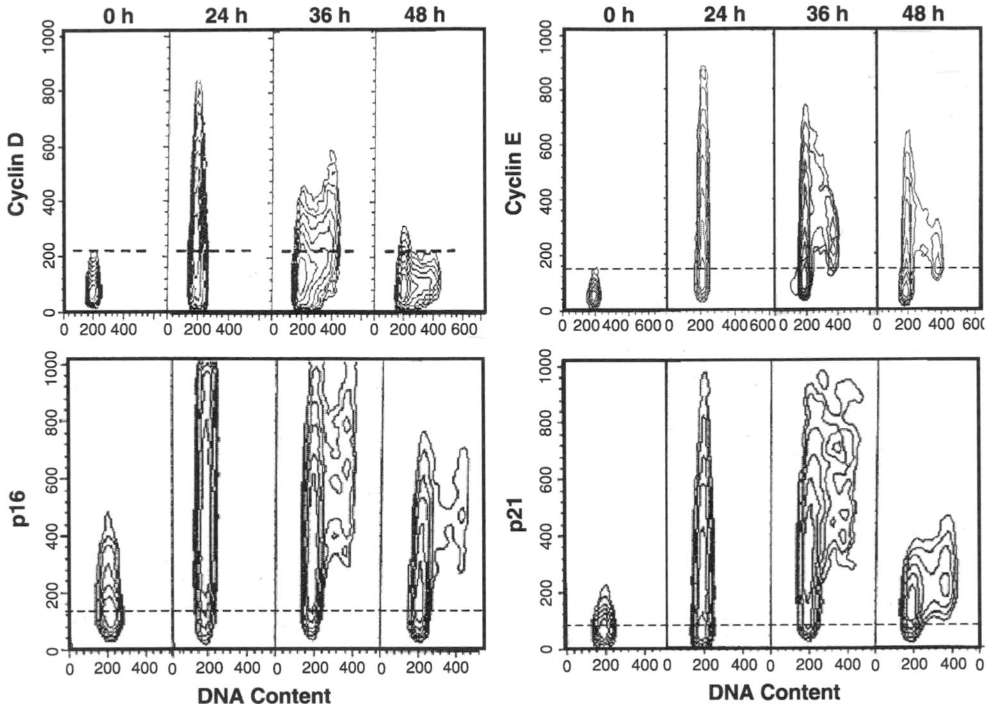

Fig. 12 Expression of the G_1 cyclins D and E and the cyclin-dependent kinase inhibitors (CKIs) p16 and p21 during the early stages of lymphocyte stimulation in culture. During mitogen stimulation of lymphocytes, cyclin D expression increases early (24 hours) then decreases as cells transit from G_0 to G_1. Cyclin E expression increases slightly later than cyclin D and remains high as cells proliferate. At the same time, the CKI p16 specific for cyclin D–CDK4 complex inhibition increases as cells reach G_1 and remains high, inactivating the cyclin D complexes. The p21 expression also increases and remains high until cells are actively cycling (48 hours).

by addition of various mitogens including phytohemagglutinin (PHA). Stimulation triggers a rather early (4 hours) induction of cyclin D3 that peaks at 12–24 hours (Fig. 12). The timing of cyclin D3 activation coincides with phosphorylation (inactivation) of its target, the retinoblastoma gene product (pRb). Inactivation of pRb results in, among other things, synthesis of cyclin E and enzymes required for DNA synthesis. As a result, by 24 hours, stimulated lymphocytes begin entering G_1 phase. Cyclin E expression reaches a maximum around 24 hours, but although cyclin D expression begins to decline as cells enter S phase at around 36 hours, cyclin E expression remains high and forms a pattern typical of exponentially growing cultures (Fig. 6).

Typically, lymphocyte stimulation is also associated with the upregulation of two CKIs, p16 and p21, members from each of the two families of inhibitors. The level of p16 increases dramatically at 24 hours, a time when the first lymphocytes are beginning to undergo their transition from G_0 to G_1 and at a time when cyclin D complexes need to be inactivated. The activity of p21 is more complex. Its induction early on reflects the requirement of this inhibitor for the formation of multifunctional complexes containing p21 in a 1:1 ratio with CDK2 and PCNA that are involved in regulation of DNA replication and repair (Ando *et al.*, 1993). Binding of a second p21 molecule to cyclin E–CDK 2 complexes would inactivate them. However, at the same time that cyclin D complexes are inactivated, they begin to play a role as a sink to which p21 can be bound, thereby lowering the availability of p21 and resulting in the activation of cyclin E–CDK2 complexes, consistent with the cells entering the S phase. As cyclin D–CDK4 complexes begin to degrade, there is no longer a necessity to retain high levels of p16, which begins to decrease at 48 hours, as does p21.

The same caveats described for staining cyclins in cells holds for CKIs. Thus, it is important to check the binding specificity and the stoichiometry of staining and to remember that the intracellular location of the molecule determines its activity. In addition, when studying a system like stimulated lymphocytes, it is important to recognize that literally half the cells present will not respond to the mitogen and therefore will be negative for the various probes throughout the experiment. For a more detailed discussion of the cytometric analysis of cell cycle regulatory proteins, see Darzynkiewicz *et al.* (2003).

C. Stathmokinetic Approach to Analyzing the Mechanism of Drug Action

1. Background

The stathmokinetic or metaphase arrest technique as proposed by Puck and Steffen (1963) provides an estimate of the rate of entry of cells to mitosis. The ability of flow cytometry to provide rapid, unbiased, and accurate measurements of large cell populations provided a distinct advantage compared to the cumbersome and time-consuming counting of mitotic figures by light or fluorescence microscopy. When Barfod and Barfod (1980) compared the rate of cell entry to mitosis as estimated by visual counts with the accumulation of cells in the $G_2 + M$ peak in DNA content frequency distributions, they found a good correlation for the rate of cell "production" generated by the two techniques in several cell lines. Dosik *et al.* (1981) also applied flow cytometric measurements to study stathmokinesis and provided in-depth mathematical analysis of the results. However, single-parameter measurements in stathmokinetic experiments provide for identification of fewer cell cycle compartments than can be obtained from multiparameter flow cytometric techniques.

Multivariate cell analysis makes it possible to detect numerous discrete compartments of the cell cycle. The combination of stathmokinesis and multivariate analysis offers the possibility of studying, in detail, not only the rate of cell entry to mitosis (instead of G_2M) but also the rates of cell traversal through other portions of the cell cycle related to those compartments. As a result, this approach has successfully been applied to investigate drug effects on the cell cycle. (Darzynkiewicz *et al.*, 1981; Traganos *et al.*, 1980a).

Detecting the cell birth rate or the rate of entry of cells into mitosis requires the use of agents that arrest cells at metaphase. Agents such as colchicine, Colcemid, or the periwinkle alkaloids vincristine or vinblastine affect polymerization of the mitotic spindle of the dividing cell, preventing the transition from the metaphase to the anaphase. The slope of the plot representing the accumulation of cells arrested in mitosis versus time of treatment with the stathmokinetic agent provides an estimate of the rate of cell entry to mitosis. Given the assumptions that the growth fraction of the cells is 100% (not unreasonable for cells in culture but rarely true *in vivo*) and that the cells are growing exponentially, the cell cycle time can be estimated from the slope of the mitosis accumulation curve (Wright and Appleton, 1980).

2. Multiparametric Analysis of Drug Action

The initial multiparametric approach used the metachromatic properties of the fluorochrome acridine orange (AO), which provided information about both the DNA content of a cell and the structure (degree of condensation) of the chromatin (Darzynkiewicz *et al.*, 1976). AO binds to double-stranded nucleic acids by intercalation between DNA base pairs and fluoresces green, whereas when binding to single-stranded nucleic acids, the dye molecules interact and the emission shifts to red luminescence when excited by blue light (Darzynkiewicz *et al.*, 1977). Thus, when RNA is enzymatically removed, AO can differentially stain single-versus double-stranded DNA. This is useful because the extent of chromatin condensation affects the sensitivity of DNA in cells to denature under conditions of increased temperature or decreased pH. It turned out that the condensed chromatin of metaphase cells compared to G_2 cells with the same DNA content differed in their staining with AO after partial denaturation, which allowed for discrimination of mitotic cells by flow cytometry (Darzynkiewicz *et al.*, 1977). At the same time, quiescent cells or cells having just divided also have a different degree of chromatin condensation that allowed for discrimination of subpopulations termed G_0 or G_{1A}, respectively (Darzynkiewicz *et al.*, 1979). An example of an exponentially growing cell line stained in this manner is presented in Fig. 13. As can be seen in the figure, mitotic cells are easily separated from interphase cells on the axis labeled α_t, which represents the degree of denaturation of the chromatin. This value is calculated by dividing the fluorescence of AO bound to single-stranded (denatured) DNA by the total luminescence—that is, that bound to double- and single-stranded DNA. Obviously, because DNA was either single- or double-stranded, the addition of the two values (green+red or total luminescence) provided a measure of DNA content on the other axis. With this approach, interphase cells could be divided into G_1, S, and G_2 phases based on total luminescence. An additional subpopulation was identified in G_1 phase by virtue of its increased denaturability (i.e., increased chromatin condensation). These cells

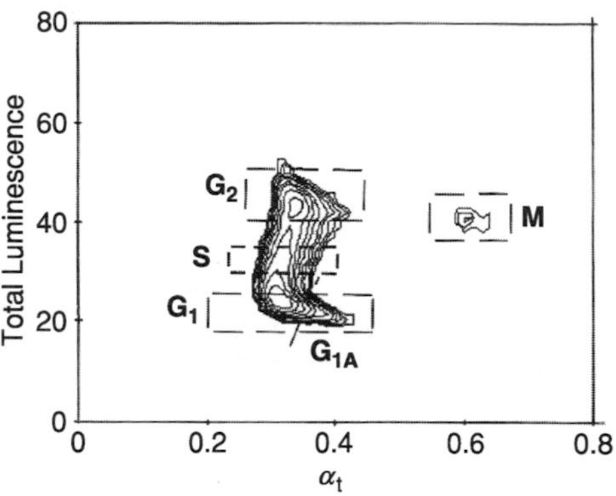

Fig. 13 Use of acridine orange to stain for DNA content versus the degree of chromatin condensation to identify cells in G_1, G_{1A}, S, G_2, and M. The bivariate distribution of cells stained with AO after partial denaturation demonstrates that increased chromatin condensation, as occurs in mitosis, can be identified. Condensed chromatin is more easily denatured so a larger proportion of AO fluoresces red compared to interphase cells, which contain more native double-stranded DNA that fluoresces green. α_t represents the ratio of red to red+green fluorescence (proportional to the level of denaturation), whereas total luminescence represents the combination of red+green (DNA content). Note that a portion of the G_1 population also has higher α_t values and represents immediately postmitotic G_{1A} cells.

were identified as immediately postmitotic cells and were defined as the G_{1A} compartment. The boxes drawn in the figure are comparable to those used for analysis of drug effects; the S phase can be divided into several windows in early, mid, and late S phase to better define the effects of drugs suspected of interfering with cell progression through this compartment.

Although this approach proved a very powerful tool to study drug effects on the cell cycle (Darzynkiewicz *et al.*, 1986; Traganos and Kimmel, 1990), the staining technique, because it is sensitive to dye concentration, proved technically difficult to measure on certain flow cytometers and cell sorters. The details of the technique and the caveats relating to stathmokinetic experiments in general and the particular use of AO are in a review by Darzynkiewicz *et al.* (1986).

3. New Approaches to Analyzing Stathmokinetic Experiments

The additional compartments identified by increased chromatin condensation (mitotic and G_{1A} and G_0 cells) can now be easily identified by other less complex and time-consuming staining techniques.

One approach uses the cell cycle specificity of cyclin proteins to provide information in addition to DNA content. As noted earlier, cyclin A expression increases throughout S phase until it peaks in G_2 phase and then is degraded as cells enter mitosis. The fact that essentially all cells in mitosis (with the possible exception of very early prophase) are cyclin A negative allows one to easily identify that population on a plot of DNA content versus cyclin A expression

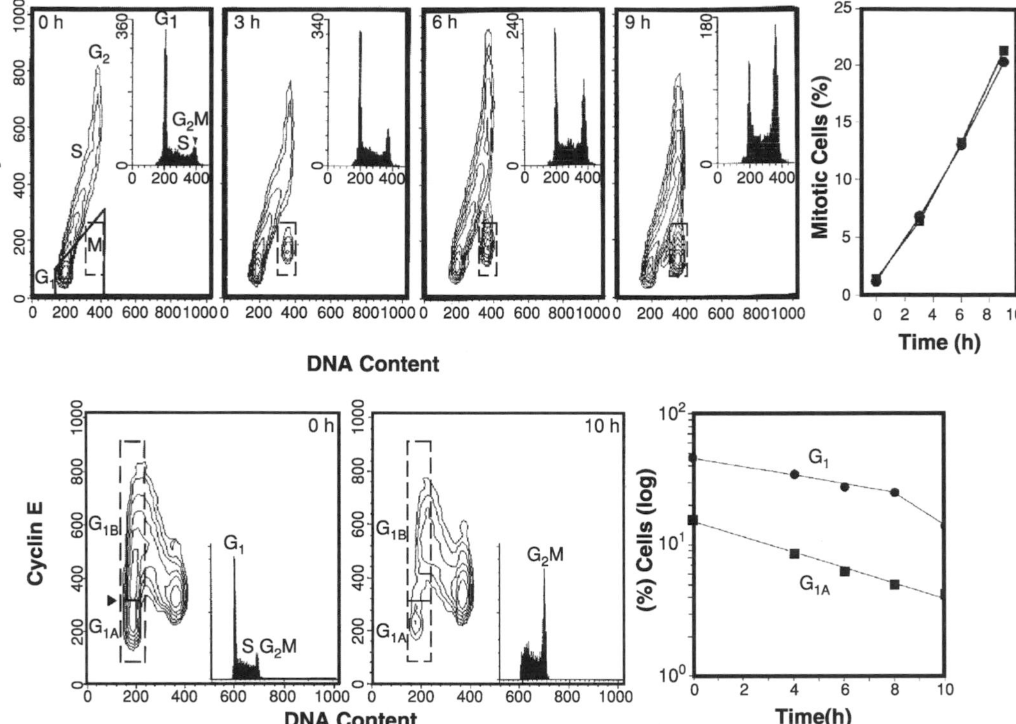

Fig. 14 Use of cyclins to identify M and G_{1A} cells during stathmokinetic experiments. Cyclin A is essentially not expressed in mitotic cells, although it is highly expressed in G_2 cells. Therefore, cultures exposed to a stathmokinetic agent contain increasing numbers of mitotic cells that can be identified as cells with a G_2 amount of DNA but no cyclin A expression. Cyclin E expression can differentiate between cells in G_{1A}, which do not express cyclin E and G_{1B} cells that express sufficient cyclin E to enter S phase. Upon exposure to a stathmokinetic agent, G_1 phase begins to empty. The kinetics of cell exit from G_1 ($G_{1A} + G_{1B}$), and the exit from the G_{1A} compartment alone can be determined during stathmokinesis, as demonstrated.

(Fig. 14, top). The percentage of cells determined to be in mitosis by this technique agreed almost exactly with the percentage of cells from the same cultures observed to be in mitosis by microscopy.

The same type of approach can be used to subdivide the G_1 population into the G_{1A} and G_{1B} compartments using bivariate plots of DNA content versus cyclin E expression, as in Fig. 14 (bottom). Cells that have yet to express a threshold level of cyclin E sufficient to enter S phase (arrow demarcating G_{1A} from G_{1B} cells) can be considered G_{1A} cells. Upon addition of a stathmokinetic agent, it is a simple matter of following the emptying of both the total G_1 and the G_{1A} and G_{1B} compartments.

Other flow cytometric approaches have also successfully differentiated mitotic cells from the remaining interphase cells. Thus, it is possible to identify mitotic cells by virtue of the fact that their condensed chromatin scatters light differently (Epstein *et al.*, 1988), or that they emit increased DNA fluorescence per pixel (Gorczyca *et al.*, 1996; Luther and Kamentsky, 1996), bind antibodies to p105 (Sramkoski *et al.*, 1999) or express high levels of phosphorylated histone H3 (Juan *et al.*, 1998b).

4. Plotting Results and Interpreting the Plots

Although discussion of these topics has been covered in detail elsewhere (Darzynkiewicz *et al.*, 1981, 1986; Traganos *et al.*, 1980), it is worth describing in general terms how one would interpret the kinds of data derived from stathmokinetic experiments. Figure 15 contains curves similar to those one would expect to get from experiments on exponentially growing cell cultures. The solid lines represent the curves one might obtain from an asynchronous exponentially growing culture after adding a stathmokinetic agent.

In Fig. 15A, one sees the kinetics of emptying of the G_1 compartment as a result of the fact that cells cannot reenter G_1 in the presence of an agent that blocks cells in mitosis. The curve representing the exit of cells from G_1 has two slopes. On a semilogarithmic plot, the first portion of the curve has a concave shape and the second part is a straight exponentially declining line. The exponentially declining slope of the curve provides an estimate of the half-time of cell residence in the indeterminate portion of G_1: The probabilistic model of the cell cycle (Smith and

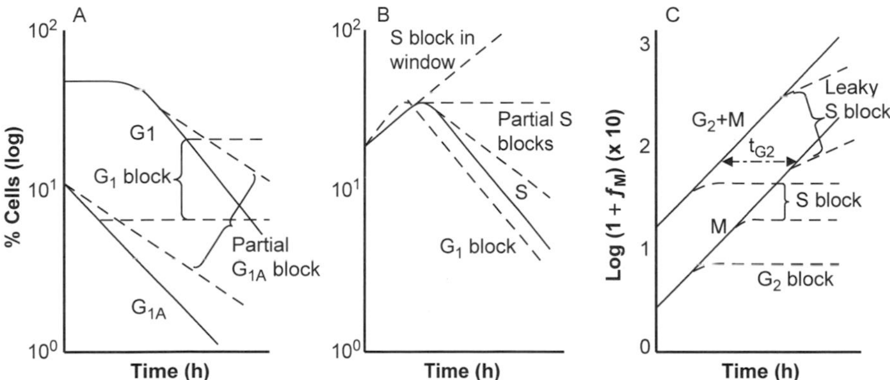

Fig. 15 Plots of data obtained from stathmokinetic experiments illustrate exit of cells from G_{1A} and G_1, transit through S, and accumulation in M or G_2M in the absence and presence of drugs. A schematic illustration of the curves of cell exit from G_1 and G_{1A} (A), transit through a window in S phase (B), and accumulation in M or G_2M (C) are illustrated based on a typical stathmokinetic experiment (*solid lines*). Deviations from these curves (*dashed lines*) are shown for various examples of drug interactions (described in the text).

Martin, 1973) predicts that there are two compartments in G_1 referred to as the *indeterminate* and *deterministic* compartments. The indeterminate compartment represents a portion of G_1 in which cells exit stochastically and, thus, have a variable transit time through this compartment. The concave portion of the curve represents the portion of G_1 through which cells pass with a finite and linear (deterministic) rate. The duration of the deterministic portion of G_1 (referred to as G_{1B}) can be determined by identifying where on the early portion of the curve the first deflection from the upward projection of the straight line portion of the curve occurs. The distance in time from the start of the experiment to this point is G_{1B}. The half-time of cell residence in G_{1A} can be established from the slope of the G_{1A} exit curve. Note that the slopes of the G_{1A} exit curve and the straight line portion of the G_1 exit curve tend to be parallel, indicating they represent the same kinetic event (Darzynkiewicz *et al.*, 1986).

The number of cells progressing through S phase may be calculated in "windows" selected in S phase, for instance, in early, mid, and/or late S phase. At first, the number of cells measured in a window (no matter where it is located) will increase if the cells are in asynchronous exponential growth (Fig. 15B). This is a reflection of the log–age distribution of the cells. The second phase is characterized by an exponentially declining slope, similar to the G_{1A} exit curve. The deflection point between the increase and decline in cells within an S-phase window occurs in relation to where in the S phase the window is selected and the rate of cell traversal of S phase.

When treated with a stathmokinetic agent, exponentially growing cells will give rise to a straight-line exponential increase in the accumulation of cells in mitosis (Fig. 15C). The slope of that line reflects the log–age distribution of the exponentially growing population (Puck and Steffen, 1963). Occasionally, a lag of 30 minutes to 1 hour can be observed at the beginning of the experiment before the increase in mitotic accumulation occurs, which may be the result of either a transient arrest of cells in G_2 or the time required for the stathmokinetic agent to interact with the mitotic spindle in sufficient quantity to block the metaphase to anaphase transition. Generally, the data can be plotted as the log of 1 plus the fraction of cells in mitosis (f_M). The curve should be a straight line on such a plot; a deviation from the straight line (in the absence of addition of a second agent) can indicate that the block is leaky, cells are not growing asynchronously, or there may be selective cell death in some part of the cycle. Cell entrance into G_2M should also be exponential. The slope of the M or G_2M accumulation curves provide an estimate of the duration of the cell cycle and the distance in time between them is equivalent to the length of G_2 phase. The scale of $\log(1 + f_M)$ or $\log(1 + f_{G2M})$ covers the range of 0 to 0.301 for the full cell cycle.

Drug-induced changes in cell cycle progression can easily be recognized and quantified from stathmokinetic experiments (Fig. 15). Comparison between the kinetic rates of the control, untreated culture and the drug-treated culture forms the basis for the analysis. By using a multiparameter approach to detail the kinetics of multiple points in the cycle simultaneously, one can detect even minor perturbations and secondary effects. The discussion later describes the character of the data and illustrates the potential of the method through hypothetical examples. The examples as described in the figure are taken from real examples of perturbations induced by various classes of drugs (Darzynkiewicz *et al.*, 1981; Traganos *et al.*, 1980a,b, 1981a,b,c, 1987a,b, 1992, 1993, 1994).

Figure 15A illustrates the effects of drugs on cell exit from G_1 (and G_{1A}). In cultures treated with G_1-arresting agents, cells may stop exiting G_{1A} and G_1 shortly after addition. Arrest of cells in G_{1B}, the deterministic portion of G_1, would be manifest as a change in the slope of the G_1 but not the G_{1A} exit curve. Conversely, a drug that affects cells only in G_{1A} would suppress the exit of cells from G_{1A} (change the slope of that curve) and would change the slope of the exponential portion of the G_1 exit curve. A full arrest of cells anywhere in G_1 would appear as a flattening of the exit curve, whereas a slowdown in traversal of G_1 would simply appear as an increase in the slope of the exit curve.

A family of curves can be generated to illustrate transit of cells through multiple windows in S phase, although only one such window is illustrated in Fig. 15B. Agents perturbing progression of cells through G_1 or S phase will produce a change in the ascending and descending slopes of the curve or in the position of the deflection point. Drugs that slow or block progression through S could change the curve, as illustrated by the dashed lines. If a slowdown or block in S occurred earlier in S phase than the "window" chosen for analysis, then the ascending portion of the curve would be less steep and the deflection point would occur later in time. If an agent blocked cells in G_1 but did not affect transit through S, the deflection of the curve would occur earlier than in control cultures. Drugs that have no effect on the exit of cells from G_1 but block cells within the window chosen to analyze S phase traversal would result in a continuously increasing ascending portion of the curve, similar to the curves for M and G_2M in Fig. 15C.

Cell arrest in G_2 would result in a change in the slope of the M accumulation curve at a time equivalent to the point in G_2 before mitosis that the block occurred, although the curve for G_2M accumulation would remain largely unchanged. A partial block in G_2 would result in a change in slope of cell accumulation in M while a slowdown in cell entrance into G_2 (S-phase effect) would cause either a flattening or a change in slope of both the M and G_2M accumulation curves at a time equivalent to the duration of G_2 and the fraction of S phase before G_2 that the block occurred.

5. Pitfalls

Successful acquisition and interpretation of stathmokinetic experiments as described above depend on a variety of factors, some of which are noted below:

1. In a normal analysis, it is assumed that the growth fraction of the population is 100% (i.e., all the cells are cycling).
2. The cells should be growing asynchronously and exponentially.

3. An appropriate stathmokinetic agent should be chosen because not all cell lines respond identically to all mitotic inhibitors.

4. It is also important to titrate the stathmokinetic agents so a sufficient amount is used to completely block cells in mitosis but not so much of the inhibitor is added that it affects cell progression through other cell cycle compartments.

5. Each cell system has a maximum length of time that the experiment can be run before cells blocked in mitosis begin dying, generally by apoptosis. As a result, generally cell systems with shorter doubling times provide more information because a larger portion of their cell cycle can be sampled before cell death occurs in the mitotic population. The loss of cells in mitosis would result in a change in slope of the control M accumulation curve and would complicate any analysis of the effects of an additional agent.

6. It is also important to consider that an agent being tested might act additively or synergistically with the stathmokinetic agent (e.g., as might be observed with the mitotic inhibitor Taxol).

7. When the experiments are being performed on cells growing in suspension, it is probably better to do the experiment in one large flask than in many small flasks, minimizing culture-to-culture variability in cell growth and drug concentrations.

Technical problems that arise typically result in disturbances in the control curves that are generally obvious unless it is the first time a particular cell line is being used.

D. Other Multiparameter Approaches

Various other approaches that incorporate measurements of important cell cycle control molecules can prove useful in studying the effects of drugs on cell cycle progression.

1. Phosphorylation of the retinoblastoma gene product, pRb. As noted earlier, pRb is an important molecule, regulating cell transition from G_1 to S phase. It protects the cell against unwarranted proliferation by binding to and inhibiting the E2F family of transcriptional activators. Thus, the unphosphorylated form of the molecule is active, whereas phosphorylation by cyclin–CDK complexes inactivates pRb, releasing the restriction against entrance to the S phase. There exist probes that bind to pRb (total pRb) regardless of its level of phosphorylation (Juan *et al.*, 1998a). Alternatively, a second antibody to pRb senses the state of phosphorylation and only binds to the unphosphorylated (active) form of the molecule. Using both probes, one can determine the relative state of activation of pRb by measuring the fraction of total pRb that is unphosphorylated. This type of measurement has been used to detect, for instance, how various compounds derived from natural sources affected growth of tumor cells in culture (Albino *et al.*, 2000; Juan *et al.*, 1998a).

2. DNA damage–p53. The importance of the p53 protein is suggested by its title "guardian of the genome." The p53 protein is a tumor suppressor activated in response to cellular stresses such as low oxygen levels, heat shock, and especially DNA damage. In response to such damage, p53 can initiate cell cycle arrest, senescence, and/or apoptosis. In as much as many chemotherapeutic agents act directly or indirectly to damage DNA, p53, and its downstream target p21 may be useful in characterizing drug activity not only indirectly by their effect on the cell cycle but by directly monitoring perturbations in the cellular levels of these proteins.

Wild-type p53 is present in small amounts but is stabilized and appears to increase in content when triggered by stress. Under such conditions, one would detect increasing fluorescence levels when using fluorescent-labeled antibodies to p53. However, the choice of antibody is important because many tumors contain mutated p53, which is almost always present at higher than normal levels. Therefore, it is important to use an antibody that can discriminate between wild-type and mutated p53. In addition, it is also important to be able to detect where in the cell p53 is present; p53 is active in the nucleus but can be inactivated by shuttling the molecule to the cytoplasm.

Topoisomerase inhibitors such as camptothecin cause DNA damage forming cleavable complexes (the drug covalently links to the 3′-OH ends of DNA nicks). Collisions between DNA polymerase during S phase or transcription-driven RNA polymerase at any time in interphase convert the complexes into unrepairable secondary lesions that are lethal to the cell (Liu *et al.*, 1996). Cells treated with agents such as camptothecin increase the levels of p53 in the nucleus in parallel with the transient arrest in G_1 phase caused by p53's upregulation of p21 expression. Using a laser scanning cytometer to monitor nuclear versus cytoplasmic protein expression, Deptala *et al.* (1999) observed that the increase in nuclear p53 expression can be dramatic (>20-fold) but occurs throughout the cell cycle, whereas p21 expression (>60-fold increase), as expected, remains confined to G_1 cells. Although the mechanism of action of camptothecin is well known, if similar effects were observed after treatment with an agent whose mechanism of action was unknown, it would point toward one of the cell stress pathways leading to DNA damage as a potential target.

Another interesting probe in the same family is the oncoprotein MDM2 (now known as HDM2 in humans) that is known to inactivate some of the functions of p53, shuttling it to the cytoplasm and resulting in its destabilization (Deb, 2003). Because this protein is often overexpressed in tumors and can cause a G_1 arrest in normal cells, it represents another interesting protein to monitor when attempting to assess the effect of a drug on cell cycle progression.

3. DNA damage—phosphorylation of histone H2AX. In addition to the downstream effects resulting from DNA damage, it is now possible to detect primary DNA double-strand breaks induced by a variety of clastogens. As previously noted, pairs of the four major histones make up nucleosomal cores around which

DNA is wrapped. A variant of one of these core histones, histone H2AX is phosphorylated (on serine 139) by ATM kinase in response to damage that induces double-stranded breaks in DNA (Rogakou *et al.*, 1998). Probes (antibodies) are now available that specifically detect phosphorylation of H2AX at Ser-139 and, therefore, can be used to detect DNA damage before either the upregulation of p53 or the initiation of apoptosis.

MacPhail *et al.* (2003) have demonstrated that the intensity of fluorescence of the antibody bound to the phosphorylated form of histone H2AX was linearly proportional to the dose of ionizing radiation used to produce the damage. DNA topoisomerases are well-known inducers of double-stranded breaks, causing apoptosis in 3–4 hours in leukemic cells such as MOLT-4 when present at a sufficient concentration (Fig. 2). The probe to phosphorylated histone H2AX detected DNA damage caused by lower concentrations of inhibitor after 1–2 hours of treatment (well before the first cells undergo apoptosis) was also proportional to dose (Huang *et al.*, 2003).

VI. Data Presentation

It should be self-evident that data presentation plays an important role in the transmittal of information in publications and presentations. It is, therefore, incumbent on authors and presenters to attempt to communicate their data in a straightforward and unambiguous fashion. Although it is not the purpose of this chapter and it is not possible to provide an all-inclusive discussion of data presentation, there are several standards that should be maintained when presenting data of this nature. A more extensive discussion of the points below and many additional areas of data interpretation and presentation can be found in Shapiro's *Practical Flow Cytometry* (2003).

A. Single-Parameter Data

Single-parameter DNA frequency histograms should meet the following standards:

1. *Axis labeling:* Axes should be appropriately labeled. The "DNA content" or X-axis should be labeled as such or, at a minimum, labeled with the dye used to obtain the DNA distribution. The terms *F11, F12*, etc. are meaningless and should *never* be used in either single-parameter or multiparameter plots. The Y-axis should be labeled "number of cells" or "cells/channel."

2. *Channel numbers:* The DNA content axis should be labeled with channel numbers and should *never* be truncated.

3. *Linear versus log scale:* DNA distributions should rarely, if ever, be presented on a log scale, especially when attempting to illustrate a sub-G_1 peak. Log scale

plots might be acceptable under unusual circumstances such as illustrating cell growth at multiple ploidies as during megakaryocyte differentiation.

4. *Coefficient of variation:* Generally, DNA frequency histograms with CVs in excess of 8 should not be used to determine cell cycle phase distributions because even with automatic deconvolution programs, the errors in cell cycle phase allocation tend to be large (Shankey *et al.*, 1993).

5. *Cell number:* It has been suggested (DNA Cytometry Consensus Conference Guidelines; Hedley *et al.*, 1993) that histograms consist of a minimum of 10,000 cells to increase the accuracy of analysis. Certainly, histograms containing too few cells (e.g., 1000) present a problem in accurately deconvoluting DNA frequency histograms. It should be noted that these numbers refer to the cells in the cycle. When the number of dead cells, aggregates, and/or debris is substantial, it is necessary to increase the total number of "events" acquired.

6. *Curve normalization:* When presenting multiple histograms, the author should note whether the histograms are normalized to "highest peak" or "area under the curve."

7. *DI:* The DI should be identified for the G_1 and G_2M (and higher ploidies if present) populations so the stoichiometry of staining for DNA can be appreciated.

8. *Diploid DNA position:* If an attempt is being made to determine the absolute ploidy of a sample, then either a known control (beads, etc.) or normal diploid cells should be added to the staining mixture or an indication of where that population would appear on the histogram provided.

B. Multiparameter (Bivariate) Data

Plots of DNA and a second or multiple other parameters should follow the rules noted above for single-parameter data with the addition of the following:

1. *Identification of subpopulations:* Clear indication of the limits used for the detection of a subpopulation should be provided on the plot. If the position of the subpopulation changes from plot to plot, the authors should provide an explanation for the lack of reproducibility.

2. *Criteria for selection of subpopulations:* An explanation of the criteria used to select a subpopulation should be provided in the text and/or legend.

VII. Summary and Future Directions

Multiparameter cytometric approaches to the study of the cell cycle can provide critical insights into the study of drug mechanism of action. With the discoveries of new cell cycle control molecules provided by basic research, new molecular probes are constantly arising that allow more detailed descriptions of the cell cycle, its control mechanisms, and the perturbations of those mechanisms associated with cancer. It is hopefully evident from this discussion that the cell cycle rather than become simpler to understand and study is becoming more complex the more we learn. To the extent that this complexity continues to evolve, it will become more important that those wishing to use cytometric techniques to study the cell cycle recognize the subtlety and interrelationships of these mechanisms.

The strength of multiparameter cytometry lies in its ability not only to measure the presence and amounts of important cellular attributes but also to detect the homogeneity or lack of it within cell populations. The strength of molecular biology is in elucidating molecules involved in cellular processes and evaluating their role in various pathways. Genomics and proteomics will help establish which genes and their products play a role in important cellular mechanisms that may become dysfunctional in tumor cells. However, differences in the numbers of messages for a protein do not necessarily translate into changes in expression. Even the presence or absence of proteins is not sufficient to understand their effect on cellular processes because as is evident from the study of the proteins involved in cell cycle regulation, their activation or inactivation is often a function of whether and in what ways they are modified (e.g., phosphorylated, acetylated, methylated, or ubiquitinated). In addition, even the presence of an appropriately modified protein does not necessarily fully explain its role in a particular pathway. It is becoming increasingly clear that in addition to knowing the level of expression of a protein and whether it is appropriately modified, it is perhaps more important to know where in the cell that protein is present. As we have seen, protein sequestration can occur as a temporary measure to keep a complex from acting until the precise moment it is needed, as with cyclin B1–CDK1 complexes required for the G_2 to M transition. However, protein sequestration can also occur as a result of an oncogenic event leading to inactivation of anti-oncogenes. Such is the case when HMD2 sequesters p53 in the cytoplasm, preventing it from detecting genetic damage and either blocking the cell cycle and marshaling the molecules necessary for repair or signaling to the cell that the damage is so extensive that it should initiate cell suicide. The status of p53 expression would not be evident from a genomic analysis of such cells because it is less a question of increased synthesis than of increased stabilization of the protein. It would also not be evident simply by looking at the level of protein expression in the cell unless it was with the understanding that other proteins act not only to modulate its activity (e.g., phosphorylation inhibits p53 activity) but also to remove it from the site of activity, the nucleus (e.g., HMD2). Notably, p53 is only one of an increasing number of important cell cycle control molecules in which modification and sequestration play an important role in determining the cell's fate. Thus, cytometry, or more appropriately, cytomics, has an important role to play in the study of how molecular mechanisms play out at the level of the cell, which is, after all, the basis of life and the site at which tumor formation begins.

References

Albino, A. P., Juan, G., Traganos, F., Reinhart, L., Connolly, J., Rose, D. P., and Darzynkiewicz, Z. (2000). Cell cycle arrest and apoptosis of melanoma cells by docosahexaenoic acid: Association with decreased pRb phosphorylation. *Cancer Res.* **60,** 4139–4145.

Ando, K., Ajchenbaum-Cymbalista, F., and Griffin, J. D. (1993). Regulation of G_1/S transition by cyclins D2 and D3 in hematopoietic cells. *Proc. Natl. Acad. Sci. USA* **90,** 9571–9575.

Arends, M. J., Morris, R. G., and Wyllie, A. H. (1990). Apoptosis. The role of the endonuclease. *Am. J. Pathol.* **136,** 593–608.

Barfod, I. H., and Barfod, N. M. (1980). Cell-production rates estimated by the use of vincristine sulphate and flow cytometry. I. An *in vitro* study using murine tumor cell lines. *Cell Tissue Kinet.* **13,** 1–8.

Bohmer R. M., and Ellwart J., (1981). Cell cycle analysis by combining 5-bromodeoxyuridine/33258 Hoechst technique with DNA-specific ethidium bromide staining. *Cytometry* **2,** 31–34.

Bruno, S., Ardelt, B., Skierski, J. S., Traganos, F., and Darzynkiewicz, Z. (1992). Different effects of staurosporine, an inhibitor of protein kinases, on the cell cycle and chromatin structure of normal and leukemic lymphocytes. *Cell Prolif.* **25,** 31–40.

Celis, J. E., Bravo, R., Larsen, P. M., and Fey, S. J. (1984). Cyclin: A nuclear protein whose level correlates directly with the proliferative state of normal as well as transformed cells. *Leuk. Res.* **8,** 143–157.

Cohen, L. S., and Studzinski, G. P. (1968). Correlation between cell enlargement and nucleic acid and protein content of HeLa cells in unbalanced growth produced by inhibitors of DNA synthesis. *Cell Physiol.* **69,** 331–339.

Collins, R. J., Harmon, B. V., Gobe, G. C., and Kerr, J. F. (1992). Internucleosomal DNA cleavage should not be the sole criteria for identifying apoptosis. *Int. J. Radiat. Biol.* **61,** 451–453.

Crissman, H. A., Gadbois, D. M., Tobey, R. A., and Bradbury, E. M. (1991). Transformed mammalian cells are deficient in kinase-mediated control of progression through the G_1 phase of the cycle. *Proc. Natl. Acad. Sci. USA* **88,** 7580–7584.

Crissman, H. A., and Hirons, G. T. (1994). Staining of DNA in live and fixed cells. *In* "Methods in cell biology" (Z. Darzynkiewicz, J. P. Robinson, and H. A. Crissman, eds.), Vol. 41. Academic Press, San Diego.

Crissman, H. A., and Steinkamp, J. A. (1973). Rapid, simultaneous measurement of DNA, protein, and cell volume in single cells from large mammalian cell populations. *J. Cell Biol.* **59,** 766–771.

D'Anna, J. A., Tobey, R. A., and Gurle, L. R. (1980). Concentration-dependent effects of sodium butyrate in Chinese hamster cells: cell-cycle progression, inner-histone acetylation, histone H1 dephosphorylation, and induction of an H1-like protein. *Biochemistry* **19,** 2656–2671.

Darzynkiewicz, Z., Traganos, F., Arlin, Z. A., Sharpless, T., and Melamed, M. R. (1976). Cytofluorimetric studies on conformation of nucleic acids *in situ*. II. Denaturation of deoxyribonucleic acid. *J. Histochem. Cytochem.* **24,** 49–56.

Darzynkiewicz, Z., Traganos, F., Sharpless, T., and Melamed, M. R. (1977). Recognition of cells in mitosis by flow cytofluorimetry. *J. Histochem. Cytochem.* **25,** 875–880.

Darzynkiewicz, Z., Traganos, F., Andreeff, M., Sharpless, T., and Melamed, M. R. (1979). Different sensitivity of chromatin to acid denaturation in quiescent and cycling cells as revealed by flow cytometry. *J. Histochem. Cytochem.* **27,** 478–485.

Darzynkiewicz, Z., Traganos, F., Xue, S., Staiano-Coico, L., and Melamed, M. R. (1981). Rapid analysis of drug effects on the cell cycle. *Cytometry* **1,** 279–286.

Darzynkiewicz, Z., Crissman, H., Traganos, F., and Steinkamp, J. (1982). Cell heterogeneity during the cell cycle. *J. Cell Physiol.* **113,** 465–474.

Darzynkiewicz, Z., Traganos, F., and Kimmel, M. (1986). Assay of cell cycle kinetics by multiparameter flow cytometry using the principle of stathmokinesis. *In* "Techniques in cell cycle kinetics," (J. E. Gray and Z. Darzynkiewicz, eds.). Humana Press, New York.

Darzynkiewicz, Z., Gong, J., and Traganos, F. (1994). Analysis of DNA content and cyclin protein expression in studies of DNA ploidy, growth fraction, lymphocyte stimulation and the cell cycle. *In* "Methods in cell biology" (A. Darzynkiewicz, J. P. Robinson, and H. A. Crissman, eds.), Vol. 41. Academic Press, San Diego.

Darzynkiewicz, Z., Gong, J., Juan, G., Ardelt, B., and Traganos, F. (1996). Cytometry of cyclin proteins. *Cytometry* **25,** 1–13.

Darzynkiewicz, Z., Traganos, F., Kapuscinski, J., StaianoCoico, L., and Melamed, M. R. (1984). Accessibility of DNA *in situ* to various flurochromes: Relationship to chromatin changes during erythroid differentiation of Friend leukemia cells. *Cytometry* **4,** 355–363.

Darzynkiewicz, Z., Bedner, E., Li, X., Gorczyca, W., and Melamed, M. R. (1999). Laser scanning cytometry. A new instrumentation with many application. *Exp. Cell Res.* **249,** 1–12.

Darzynkiewicz, Z., Juan, G., and Traganos, F. (2003). Cytometry of cell cycle regulatory proteins. *In* "Progress in cell cycle research," (L. Meijer, A. Jezequel, and M. Roberge, eds.). Landes Bioscience, Georgetown, Texas.

Darzynkiewicz, Z., Crissman, H., and Jacobberger, J. W. (2004). Cytometry of the cell cycle: Cycling through history. *Cytometry* **58A,** 21–32.

Deb, S. P. (2003). Cell cycle regulatory functions of the human oncoprotein MDM2. *Mol. Cancer Res.* **1,** 1009–1016.

Deptala, A., Li, X., Bedner, E., Cheng, W., Traganos, F., and Darzynkiewicz, Z. (1999). Differences in induction of p53, $p21^{WAF1}$ and apoptosis in relation to cell cycle phase of MCF-7 cells treated with camptothecin. *Int. J. Oncol.* **15,** 861–871.

Dittrich, W., and Gohde, W. (1969). Impulse fluorometry with single cells in suspension. *Z. Naturforsch.* **24B,** 360–361.

Dolbeare, F., Gratzner, H., Pallavicini, M. G., and Gray, J. W. (1983). Flow cytometric measurement of total DNA content and incorporated bromodeoxyuridine. *Proc. Natl. Acad. Sci. USA* **80,** 5573–5577.

Dolbeare, F., and Gray, J. W. (1988). Use of restriction endonucleases and exonuclease III to expose halogenated pyrimidines for immunochemical staining. *Cytometry* **9,** 631–635.

Dosik, G. M., Barlogie, B., White, R. A., Gohde, W., and Drewinko, B. (1981). A rapid automated stathmokinetic method for determination of *in vitro* cell cycle transit times.

Dressler, L. G., and Seamer, L. C. (1994). Controls, standards, and histogram interpretation in DNA flow cytometry. *In* "Methods in cell biology" (A. Darzynkiewicz, J. P. Robinson, and H. A. Crissman, eds.), Vol. 41. Academic Press, San Diego.

Durand, R. E., and Olive, P. L. (2001). Resistance of tumor cells to chemo- and radiotherapy modulated by the three-dimensional architecture of solid tumors and spheroids. *In* "Methods in cell biology" (A. Darzynkiewicz, H. A. Crissman, and J. P. Robinson, eds.), Vol. 64. Academic Press, San Diego.

Epstein, R. J., Watson, J. V., and Smith, P. J. (1988). Subpopulation analysis of drug-induced cell-cycle delay in human tumor cells using 90 degrees light scatter. *Cytometry* **9,** 349–358.

Frisa, P. S., Lanford, R. E., and Jacobberger, J. W. (2000). Molecular quantification of cell cycle–related gene expression at the protein level. *Cytometry* **39,** 79–89.

Gerdes, J., Schwab, U., Lemke, H., and Stein, H. (1983). Production of a mouse monoclonal antibody reactive with a human nuclear antigen associated with cell proliferation. *Int. J. Cancer.* **31,** 13–20.

Gohde, W., Schumann, Z., and Zante, J. (1978). The use of DAPI in pulse cytophotometry. *In* "Pulse cytophotometry" (D. Lutz, ed.). European Press, Ghent, Belgium.

Gong, J., Traganos, F., and Darzynkiewicz, Z. (1994a). A selective procedure for DNA extraction from apoptotic cells applicable for gel electrophoresis and flow cytometry. *Anal. Biochem.* **218,** 314–319.

Gong, J., Traganos, F., and Darzynkiewicz, Z. (1994b). Use of cyclin E restriction point to map cell arrest in G_1 induced by *n*-butyrate, cycloheximide, staurosporine, lovastatin, momosine and quercetin. *Int. J. Oncol.* **4,** 803–808.

Gong, J., Traganos, F., and Darzynkiewicz, Z. (1994c). Staurosporine blocks cell progression through G_1 between cyclin D and cyclin E restriction points. *Cancer Res.* **54,** 3136–3139.

Gong, J., Li, X., Traganos, F., and Darzynkiewicz, Z. (1994d). Expression of G_1 and G_2 cyclins measured in individual cells by multiparameter flow cytometry; a new tool in the analysis of the cell cycle. *Cell Prolif.* **27,** 357–371.

Gong, J., Traganos, F., and Darzynkiewicz, Z. (1995). Growth imbalance and altered expression of cyclins B1, A. E and D3 in MOLT-4 cells synchronized in the cell cycle by inhibitors of DNA replication. *Cell Growth Differ.* **6,** 1485–1493.

Gorczyca, W., Melamed, M. R., and Darzynkiewicz, Z. (1996). Laser scanning cytometer (LSC). analysis of fraction of labelled mitoses (FLM). *Cell Prolif.* **29,** 539–547.

Greene, R. F., Collins, J. M., Jenkins, J. F., Speyer, J. L., and Myers, C. E. (1983). Plasma pharmokinetics of adriamycin and adriamycinol: Implications for the design of *in vitro* experiments and treatment protocols. *Cancer Res.* **43,** 3417–3421.

Harper, J. W., and Elledge, S. J. (1996). Cdk inhibitors in development and cancer. *Curr. Opin. Genet. Dev.* **6,** 56–64.

Hartwell, L., and Weinert, T. (1989). Checkpoints: Controls that ensure the order of cell cycle events. *Science* **246,** 629–634.

Hedley, D. W., Shankey, T. V., and Wheeless, L. L. (1993). DNA cytometry consensus conference. *Cytometry* **14,** 471.

Horan, P. K., Muirhead, K. A., and Slezak, S. E. (1990). Standards and controls in flow cytometry. *In* "Flow cytometry and Sorting" (M. R. Melamed, T. Lindmo, and M. L. Mendelsohn, eds.), 2nd ed. Wiley-Liss, New York.

Hoffman, R. A. (2001). Standardization and quantitation in flow cytometry. *In* "Methods in cell biology," (Z. Darzynkiewicz, H. A. Crissman, and J. P. Robinson, eds.), Vol. 64. Academic Press, San Diego.

Huang, X., Traganos, F., and Darzynkiewicz, Z. (2003). DNA damage induced by DNA topoisomerase- and topoisomerase II-inhibitors detected by histone H2AX phosphorylation in relation to the cell cycle phase and apoptosis. *Cell Cycle* **2,** 614–619.

Jacobberger, J. W. (2001). Standardization, quality assurance. *In* "Methods in cell biology" (Z. Darzynkiewicz, H. A. Crissman, and J. P. Robinson, eds.), Vol. 64. Academic Press, San Diego.

Johnson, R. T., Downes, C. S., and Meyn, R. R. (1993). The synchronization of mammalian cells. *In* "The cell cycle. A practical approach," (P. Fantes and R. Brooks, eds.). Oxford University Press, Oxford, U. K.

Juan, G., Gong, J., Traganos, F., and Darzynkiewicz, Z. (1996). Unscheduled expression of cyclins D1 and D3 in human tumor cell lines. *Cell Prolif.* **29,** 259–266.

Juan, G., Ardelt, B., Mikulski, S. M., Shogen, K., Ardelt, W., Mittelman, A., and Darzynkiewicz, Z. (1998a). G_1 arrest of U937 cells by Onconase is associated with suppression of cyclin D3 expression, induction of $p16^{INK4A}$, $p21^{WAF1/CIP1}$ and $p27^{KIP}$ and decreased pRb phosphorylation. *Leukemia* **12,** 1241–1248.

Juan, G., Traganos, F., James, W. M., Ray, J. M., Roberge, M., Suave, D. M., Anderson, H., and Darzynkiewicz, Z. (1998b). Histone H3 phosphorylation and expression of cyclins A and B1 measured in individual cells during their progression through G_2 and mitosis. *Cytometry* **32,** 71–77.

Kamentsky, L. A., and Kamenstky, L. D. (1991). Microscope-based multiparameter laser scanning cytometer yielding data comparable to flow cytometry data. *Cytometry* **12,** 381–387.

Krishan, A. (1975). A rapid flow cytofluorometric analysis of mammalian cell cycle by propidium iodide staining. *J. Cell Biol.* **66,** 188–193.

Kubbies, M., Schindler, D., Hoehn, H., and Rabinovitch, P. S. (1985). Cell cycle kinetics by BrdU-Hoechst cytometry: An alternative to the differential metaphase labelling technique. *Cell Tissue Kinet.* **18,** 551–562.

Latt, S. A. (1974). Detection of DNA synthesis in interphase nuclei by fluorescence microscopy. *J. Cell Biol.* **62,** 546–550.

Li X., Traganos F., Melamed M. R., Darzynkiewicz Z. (1994). Detection of 5-bromo-2-deoxyuridine incorporated into DNA by labeling strand breaks induced by photolysis (SBIP). *Int. J. Oncol.* **4,** 1157–1161.

Liu, L. F., Duann, P., Lin, C.-T, D'Arpa, P., and Wu, J. (1996). Mechanism of action of camptothecin. *Ann. NY Acad. Sci.* **803,** 44–49.

Luther, E., and Kamentsky, L. A. (1996). Resolution of mitotic cells using laser scanning cytometry. *Cytometry* **23,** 272–278.

MacLeod, R. A. F., Dirks, W. G., Matsuo, Y., Kaufmann, M., Milch, H., and Drexler, H. G. (1999). Widespread intraspecies cross-contamination of human tumor cell lines arising at source. *Int. J. Cancer* **83,** 555–563.

MacPhail, S. H., Banath, J. P., Yu, Y., Chu, E., and Olive, P. L. (2003). Cell cycle-dependent expression of phosphorylated histone H2AX: reduced expression in unirradiated but not X-irradiated G_1 phase cells. *Radiat. Res.* **159,** 759–767.

Mendelsohn, M. L. (1960). The growth fraction: A new concept applied to tumors. *Science* **132,** 1497.

Mosca, P. J., Dijwell, P. A., and Hamlin, J. L. (1992). The plant amino acid mimosine may inhibit initiation at origins of replication in Chinese hamster cells. *Mol. Cell Biol.* **12,** 4375–4383.

Nasmyth, K. (2001). A prize for proliferation. *Cell* **107,** 689–701.

Nelson-Rees, W. A., and Flandermeyer, R. R. (1976). HeLa cultures defined. *Science* **191,** 96–98.

Nurse, P. (2000). A long twentieth century of the cell cycle. *Cell* **100,** 71–78.

O'Brien, S. J. (2001). Cell culture forensics. *Proc. Natl. Acad. Sci. USA* **98,** 7656–7658.

Otto, F. J. (1994). High-resolution analysis of nuclear DNA employing the fluorochrome DAPI. *In* "Methods in cell biology" (Z. Darzynkiewicz, H. A. Crissman, and J. P. Robinson, eds.), Vol. 41. Academic press, San Diego.

Pines, J. (1999). Four dimensional control of the cell cycle. *Nat. Cell Biol.* **1,** E73–E79.

Puck, T. T., and Steffen, J. (1963). Life cycle analysis of mammalian cells. I. A method for localizing metabolic events within the life cycle and its application to the action of Colcemid and sublethal doses of X-irradiation. *Biophys. J.* **3,** 379–397.

Rabinovitch, P. S. (1994). DNA content histogram and cell-cycle analysis. *In* "Methods in cell biology" (A. Darzynkiewicz, J. P. Robinson, and H. A. Crissman, eds.), Vol. 41. Academic Press, San Diego.

Rabinovitch, P. S. (1995). Analysis of flow cytometric DNA histograms. *In* "Cell growth and apoptosis" (G. P. Studzinski, ed.). Oxford University Press, New York.

Rogakou, E. P., Pilch, D. R., Orr, A. H., Ivanova, V. S., and Bonner, W. M. (1998). DNA double-strand breaks induce histone H2AX phosphorylation on serine 139. *J. Biol. Chem.* **273,** 5858–5868.

Roschke, A. V., Tonon, G., Gelhaus, K. S., McTyre, N., Bussey, K. J., Lababidi, S., Scudiero, D. A., Weinstein, J. N., and Kirsch, I. R. (2003). Karyotypic complexity of the NCI-60 drug screening panel. *Cancer Res.* **63,** 8634–8647.

Shankey, T. V., Rabinovitch, P. S., Bagwell, B., Dauer, K. D., Duque, R. E., Hedley, D. W., Mayall, B. H., and Wheeless, L. (1993). Guidelines for implementation of clinical DNA cytometry. *Cytometry* **14,** 472–477.

Shapiro, H. M. (2003). *Practical flow cytometry*, 4th ed. Wiley-Liss, New York.

Shoemaker, R. H., Monks, A., Alley, M. C., Scudiero, D. A., Fine, D. L., McLemore, T. L., Abbott, B. J., Paull, K. D., Mayo, J. G., and Boyd, M. R. (1988). Development of human tumor cell line panels for use in disease-oriented drug screening. *Prog. Clin. Biol. Res.* **276,** 265–286.

Smith, J. A., and Martin, L. (1973). Do cells cycle? *Proc. Natl. Acad. Sci. USA* **70,** 1263–1267.

Sramkoski, R. M., Wormsley, S. W., Bolton, W. E., Crumpler, D. C., and Jacobberger, J. W. (1999). Simultaneous detection of cyclin B1, p105, and DNA content provides complete cell cycle phase fraction analysis of cells that endoreduplicate. *Cytometry* **35,** 274–283.

Stohr, M., Schumann, J., Goerttler, K., and Vogt-Schaden, M. (1977). Extended application of flow microfluorometry by means of dual laser excitation. *Histochemistry* **51,** 310–315.

Terry, N. H., and White, R. A. (2001). Cell cycle kinetics estimated by analysis of bromodeoxyuridine incorporation. *In* "Methods in cell biology" (Z. Darzynkiewicz, H. A. Crissman, and J. P. Robinson, eds.), Vol. 41. Academic press, San Diego.

Thomlinson, R. H., and Gray, L. H. (1955). The histological structure of some human lung cancers and the possible implications for radiotherapy. *Br. J. Cancer* **9,** 539–549.

Traganos, F., and Kimmel, M. (1990). The stathmokinetic experiment: A single- and multiparameter flow cytometric analysis. *In* "Methods in cell biology" (Z. Darzynkiewicz and H. A. Crissman, eds.), Vol. 33. Academic Press, New York.

Traganos, F., Darzynkiewicz, Z., Staiano-Coico, L., Evenson, D., and Melamed, M. R. (1980a). Rapid, automatic analysis of the cell cycle point of action of cytostatic drugs. *In* "Flow cytometry IV" (O. D. Laerum, T. Lindmo, and E. Thorud, eds.). Universitetsforlaget, Bergen, Norway.

Traganos, F., Evenson, D. P., Staiano-Coico, L., Darzynkiewicz, Z., and Melamed, M. R. (1980b). Action of dihydroxyanthraquinone on cell cycle progression and survival of a variety of cultured mammalian cells. *Cancer Res.* **40,** 671–681.

Traganos, F., Darzynkiewicz, Z., and Melamed, M. R. (1981a). Effects of the L isomer (+)-1,2-bis(3,5-dioxopiperazine-1-yl) propane on cell survival and cell cycle progression of cultured mammalian cells. *Cancer Res.* **41,** 4566–4576.

Traganos, F., Staiano-Coico, L., Darzynkiewicz, Z., and Melamed, M. R. (1981b). Effects of dihydro-5-azacitidine on cell survival and cell cycle expression of cultured mammalian cells. *Cancer Res.* **41,** 780–789.

Traganos, F., Staiano-Coico, L., Darzynkiewicz, Z., and Melamed, M. R. (1981c). Effects of aclacinomycin on cell survival and cell cycle progression of cultured mammalian cells. *Cancer Res.* **41,** 2728–2737.

Traganos, F., Bueti, C., Darzynkiewicz, Z., and Melamed, M. R. (1987a). Effects of a new amsacrine derivative, *N*-5-dimethyl-9-(2-methoxy-4-methylsulfonylamino) phenylamino-4-acridinecarboxamide, on cultured mammalian cells. *Cancer Res.* **47,** 424–432.

Traganos, F., Bueti, C., Melamed, M. R., and Darzynkiewicz, Z. (1987b). Cytokinetic effects of the bifunctional antitumor intercalator ditercalinium on Friend erythroleukemia cells. *Leukemia* **1,** 411–416.

Traganos, F., Ardelt, B., Halko, N., Bruno, S., and Darzynkiewicz, Z. (1992). Effects of genistein on the growth and cell cycle progression of normal human lymphocytes and human leukemic MOLT-4 and HL-60 cells. *Cancer Res.* **52,** 6200–6208.

Traganos, F., Bueti, C., Melamed, M. R., and Darzynkiewicz, Z. (1987). Cytokinetic effects of the bifunctional antitumor intercalator ditercalinium on friend erythroleukemia cells. *Leukemia* **5,** 411–416.

Traganos, F., Knutti-Hotz, J., Hotz, M., Gorczyca, W., Ardelt, B., and Darzynkiewicz, Z. (1993). The protein kinase C inhibitor H7 blocks normal human lymphocyte stimulation and induces apoptosis of both normal lymphocytes and leukemic MOLT-4 cells. *Int. J. Oncol.* **2,** 47–59.

Traganos, F., Gong, J., Ardelt, B., and Darzynkiewicz, Z. (1994). Effect of staurosporine on MOLT-4 cell progression through G_2 and on cytokinesis. *J. Cell Physiol.* **158,** 535–544.

Vindelov, L. L. (1977). Flow microfluorometric analysis of nuclear DNA in cells from solid tumors and cell suspensions. A new method for rapid isolation and straining of nuclei. *Virchows Arch. B.* **24,** 227–242.

Watson, P. A., Hanauske-Abel, H. H., Flint, A., and Lalande, M. (1992). Mimosine reversibly arrests cell cycle progression at the G_1–S phase border. *Cytometry* **12,** 242–246.

Wilson, L., Creswell, K. M., and Chin, D. (1975). The mechanism of action of vinblastine. Binding of [acetyl-3H]vinblastine to embryonic chick brain tubulin and tubulin from sea urchin sperm tail outer doublet microtubules. *Biochemistry* **14,** 5586–5592.

Wright, N. A., and Appleton, D. R. (1980). The metaphase arrest technique. A critical review. *Cell Tissue Kinet.* **13,** 643–663.

CHAPTER 12

Cytometric Methods to Detect Apoptosis

Zbigniew Darzynkiewicz,* Xuan Huang,* Masaki Okafuji,*,† and Malcolm A. King*,‡

*Brander Cancer Research Institute
New York Medical College
Hawthorne, New York 10532

†Department of Oral and Maxillofacial Surgery
Yamaguchi University
School of Medicine
Ube 755-8505, Japan

‡Department of Clinical Immunology
Royal North Shore Hospital
St. Leonards NSW 2065, Australia

METHODS IN CELL BIOLOGY, VOL. 75

0091-679X/04 $35.00

I. Introduction

During the past decade the mechanism of cell death, particularly by mode of apoptosis, has become the subject of intense interest among researchers in different fields of biology and medicine. The interest covers a wide range of cell activities that predispose cells to apoptosis, regulate the initial still-reversible steps of the apoptotic process, trigger, carry out, and complete the cell execution, as well as engage the postmortem cell disposal mechanisms. The term *cell necrobiology* (biology of cell death) was introduced to collectively define all these cell activities (Darzynkiewicz *et al.*, 1997). The interest in necrobiology stems from the realization of active cell participation in its demise and of a multitude of complex regulatory mechanisms. Because the cell propensity to undergo apoptosis is the key element in many diseases, knowledge of these mechanisms offers the opportunity to intervene in the process of cell death and thus new therapeutic possibilities.

Applications of cytometry in necrobiology (Darzynkiewicz *et al.*, 1992, 1997, 2001; Ormerod, 1998; van Engeland *et al.*, 1998; Vermes *et al.*, 2000) have two goals. One goal is to elucidate molecular mechanisms associated with cell death. Along with this, goal cytometry is used to measure expression of the cell constituents involved in the regulation and/or execution of apoptosis or cell necrosis. The most frequently investigated are proapoptotic and antiapoptotic members of the Bcl-2 protein family, caspases, caspase inhibitors (e.g., IAPs), their antagonists (Smac/DIABLO), the protooncogenes (e.g., c-*myc* or *ras*), or tumor suppressor genes (e.g., p53 or pRB). Cytometric methods have been also developed to study changes in functional attributes of the cell that predispose to or accompany cell death, such as mitochondrial metabolism, oxidative stress, intracellular pH, or ionized calcium. The major advantage of cytometry in these applications is that by virtue of multiparameter analysis, it allows one to study correlation between the measured cell attributes. For example, when cellular DNA content (i.e., the parameter that reports the cell cycle position or DNA ploidy) is one of the measured attributes, an expression of other measured attributes can then be directly related to the cell cycle position (or cell ploidy) without the need for cell synchronization. Furthermore, the change in expression of particular cell constituents, or coexpression of different events, if strongly correlated with each other within the same cell, may yield clues regarding a possible cause–effect relationship between the detected events. Because individual cells rather than bulk of cell extracts are measured by cytometry, intercellular variability is assessed, cell subpopulations can be identified, and rare cells detected.

The second goal in applications of cytometry in necrobiology is to assay viability of individual cells in populations, including identification and quantification of dead cells and discrimination between apoptotic versus necrotic mode of death. Dead cell recognition is generally based on the change in cell morphology and/or on the presence of a biochemical or molecular marker characteristic for apoptosis, necrosis, or both. Morphological changes and biochemical and molecular events that occur during apoptosis are described in the first part of this chapter. Most of these changes are characteristic of the apoptotic mode of cell death and therefore serve as markers to identify and quantify apoptotic cells by cytometry. Because cytometry is often used to discriminate between the apoptotic versus the alternative necrotic ("accidental") mode of cell death, also discussed are major criteria for distinction of these diverse cell death mechanisms.

By combining advantages of flow and image cytometry, laser-scanning cytometry (LSC) (Darzynkiewicz *et al.*, 1999; Kamentsky, 2001) offers unique analytical capabilities in analysis of cell death. Major applications of LSC in studies of apoptosis and necrosis have been reviewed (Bedner *et al.*, 1999). These applications are also pertinent to the new generation of the LSC instrumentation, iCyt, which is described in chapter 8 of this volume. The major part of this chapter is devoted to the description of individual methods that have already found wide application either in investigations of mechanisms associated with cell death or for identification and quantification of dead cells. Limitations of these methods, their optimal choice for a particular task/cell system, and the pitfalls in data collection and interpretation are also described in this chapter.

Cytometric methods designed to probe cell death are often presented as collections of the specific protocols (Pozarowski *et al.*, 2003a). Although such protocols are quite useful to reproduce a particular assay, their "cookbook" format is restrictive and does not allow one to provide details about principles of the methodology, its limitations, possible pitfalls, or discussion of optimal choice of the assay for a particular task/cell system. Yet such knowledge is often important for rational use of the methodology and for extraction of maximal relevant information from the experiment.

Various kits are commercially available to identify dead/apoptotic cells. Reagents in the kits are prepackaged, in an easy-to-use form, and the procedures are described in step-by-step detail, making the kits simple to use. The protocols accompanying the kits, however, are generally cryptic, and because of the proprietary nature of some reagents, they often do not adequately inform about chemistry of the components or mechanistic principles of the kit. This chapter, therefore, complements the protocol-format literature by providing additional information about particular methods, comparing the methods, and discussing their advantages and limitations. Some details and steps of individual methods are discussed to underscore their critical role and possibility of pitfalls rather than provide their step-by-step detailed description. This chapter updates earlier reviews on application of cytometry in analysis of cell death (Darzynkiewicz *et al.*, 1992, 1994, 1997, 2001).

II. Physical and Molecular Features of Cells Dying by Apoptosis or Necrosis

A. Changes in Cell Morphology

Table I lists major morphological and molecular changes that occur during apoptosis. Many of these changes provide markers to identify and distinguish cells dying by either apoptosis or necrosis, adapted to microscopy, cytometry, or imaging techniques. Apoptosis and cell necrosis have been initially defined based on characteristic changes in cell morphology (Kerr *et al.*, 1972). Despite development of numerous biochemical, molecular, and cytometric

Table I
Changes in Cell Morphology and Molecular Events during Apoptosis

A. Morphological changes

Cell shrinkage/diminished size
Change in cell shape (spherical)
Nuclear chromatin condensation
  Loss of visually recognizable nuclear structure (framework)
  DNA hyperchromicity
Segregation of nucleoli
Dissolution of nuclear envelope
Nuclear fragmentation
Condensation of cytoplasm
Loss of cell surface structures (pseudopodia, microvilli)
Formation and detachment of apoptotic bodies (membrane "blebbing")
Detachment of cells in cultures
Phagocytosis of the apoptotic cell remains

B. Major biochemical and molecular events of apoptosis

Increased ratio of proapoptotic versus antiapoptotic members of the Bcl-2 family
- Mitochondrial changes
- Decrease/collapse of the transmembrane potential
- Leakage of cytochrome c and other mitochondrial constituents
- Oxidative stress (formation of ROI)

Intracellular Ca^{2+} rise
Cell dehydration
Loss of asymmetry in plasma membrane phospholipids/exposure of phosphatidylserine
Cascade activation of caspases
- Proteolysis of the "death substrates" (e.g., PARP)

Activation of serine protease(s)
Degradation of F-actin and proteins other than "death substrates"
Loss of DNA double-helix stability (susceptibility to denaturation)
- Appearance of single-stranded DNA sections

Endonucleolytic DNA degradation
- 50–300 kb fragments (DNA)
- Internucleosomal DNA cleavage

Activation of "tissue" transglutaminase

markers, the morphological changes are still considered the "gold standard" to define the mode of cell death (Majno and Joris, 1995).

The early event of apoptosis is cell dehydration. Water loss leads to condensation of the cytoplasm, followed by a change in cell size and often shape. The cells shrink, and when originally round, they become elongated or acquire an irregular shape. Another early event is segregation of the nucleolus. Nucleolar segregation defines separation and condensation of the granular and dense fibrillar nucleolar components, the structures composed of RNA, from the central fibrillar nucleolar component that contains DNA (Raipert *et al.*, 1999). This event initiates separation of RNA from DNA and packaging these nucleic acids into separate apoptotic bodies (Halicka *et al.*, 2000).

Another change is condensation of nuclear chromatin, considered one of the most characteristic features ("hallmarks") of apoptosis. The condensation starts at the nuclear periphery and the areas of condensed chromatin acquire a concave shape often resembling a crescent moon or a sickle. The condensed chromatin has a highly uniform, "smooth" appearance resulting from the loss of the texture typically seen in the interphase nucleus. DNA in the condensed (pyknotic) chromatin exhibits hyperchromasia, staining strongly with fluorescent or light-absorbing dyes (Fig. 1A). If mitotic cells undergo apoptosis (e.g., cells long arrested in mitosis by mitotic poisons), the structural features of individual chromosomes are lost, the chromosomes assume spherical shape (some may fuse with each other), and they appear as structureless chromatin droplets.

During progression of apoptosis of interphase cells, the nuclear envelope disintegrates, lamin proteins are degraded, and the nucleus undergoes fragmentation (karyorrhexis). Nuclear fragments stain strongly and uniformly with DNA dyes, are heterogeneous in size, and are scattered throughout the cytoplasm. With constituents of the cytoplasm (including intact organelles), the nuclear fragments are packaged and enveloped by fragments of the plasma membrane. These structures, called *apoptotic bodies*, subsequently detach from the surface of the dying cell by "budding" (Fig. 1B). Loss of cell surface structures such as pseudopodia or microtubules is also a characteristic feature of apoptosis (Endersen *et al.*, 1995). When the cells grow attached to flasks in tissue culture, they detach during apoptosis and float in the medium. When apoptosis occurs *in vivo*, apoptotic bodies are phagocytized by the neighboring cells. Their phagocytosis is facilitated by the exposure to phosphatidylserine on the plasma membrane surface, the mechanism that attracts the macrophages (Fadok *et al.*, 1992). Apoptotic cell death neither triggers an inflammatory reaction in the tissue nor induces scar formation (Blagosklonny, 2000; Danial and Korsmeyer, 2004; Kerr *et al.*, 1972; Majno and Joris, 1995; Searle *et al.*, 1982; Zorning *et al.*, 2001).

B. Biochemical and Molecular Events of Apoptosis

The induction and progression of apoptosis occurs along either the extrinsic or the intrinsic pathway. The extrinsic pathway is activated by one of the proteins of the tumor necrosis factor (TNF) superfamily (e.g., TNF-α, Fas, or Apo1

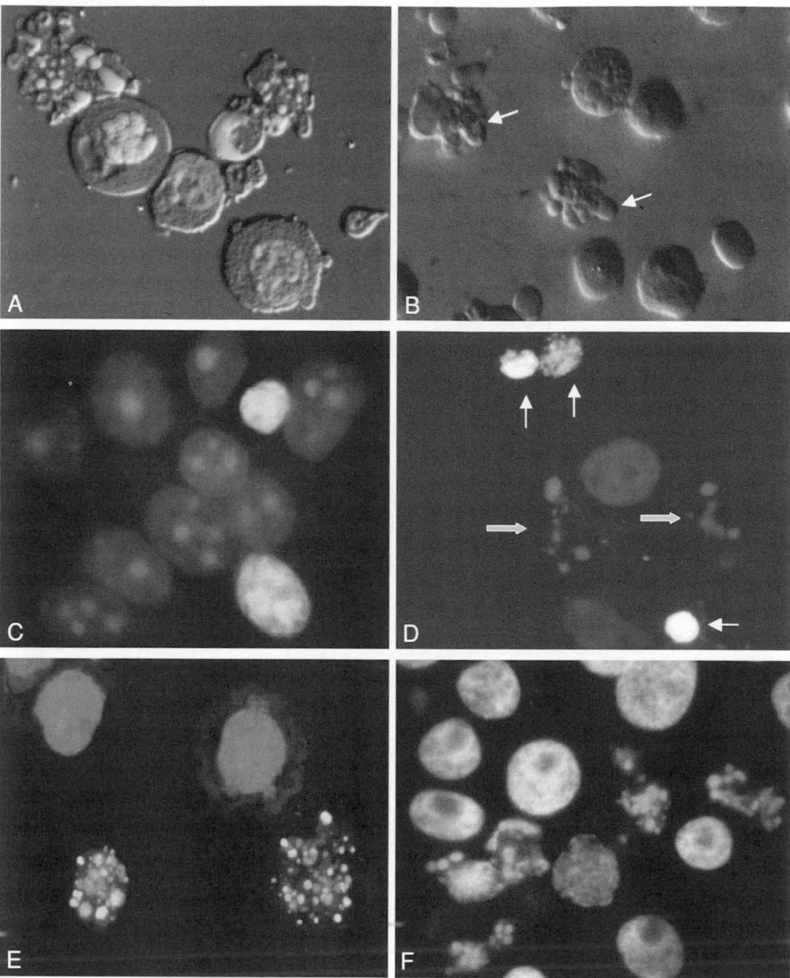

Fig. 1 Characteristic features of apoptotic cells. (A) Apoptosis of HL-60 cells was induced by treatment with DNA topoisomerase-I inhibitor camptothecin (CPT) under conditions that S-phase cells were affected preferentially (Gorczyca *et al.*, 1992). The cells were deposited on a microscope slide by cytocentrifugation and were fixed, and their DNA was stained with DAPI. The cells were viewed under combined illumination of incident ultraviolet light and interference contrast. Note condensation (DNA/DAPI hyperchromicity) and loss of structural framework of the chromatin of apoptotic cells and of apoptotic bodies, along with extensive nuclear fragmentation, compared with relatively unchanged chromatin of two interphase cells. DNA/DAPI hyperchromicity is also apparent in chromatin of the mitotic cell. (B) HL-60 cells treated with CPT were viewed under interference contrast microscope. Arrows point to the characteristic "budding" of plasma membrane of the cells undergoing apoptosis. (C) MCF-7 cells were treated with CPT to induce apoptosis and were then fixed, and the cleavage product of PARP (p89) was detected immunocytochemically using FITC-conjugated antibody p89 PARP. DNA was counterstained with 7-amino actinomycin D (7-AAD). Note preferential localization of p89 PARP in perinucleolar areas of the two apoptotic cells. (D) Apoptosis

ligand) upon its binding to the death receptor on cell surface (Sheikh and Fornace, 2000). Ligand binding causes trimerization of the receptor, which then recruits the "dead domain" (DD) containing respective adapter protein molecules (e.g., such as FADD) at the intracellular tail of the receptor, on which the initiator caspase-8 or caspase-10 becomes activated (Budihardjo *et al.*, 1999; Earnshaw *et al.*, 1999; Leist and Jaattela, 2001). The initiator caspase, in turn, activates the "executioner" (effector) caspase-3, caspase-6, and caspase-7, triggering proteolytic degradation of the "death" substrates.

The intrinsic mitochondrial pathway becomes activated in response to deprivation of growth and survival factors such as cytokines or upon internal stress to the cell such as DNA damage. One of the early changes is upregulation of the tumor suppressor p53, which among its many capabilities is able to activate transcription of the genes coding for the proapoptotic members of the Bcl-2 protein family such as Bax. A change in the ratio of the proapoptotic members (Bax, Bik, Bad, and Bcl-x_S) over the antiapoptotic members such as Bcl-2, Bcl-x_L, and Mcl-1 appears to be the critical event triggering the intrinsic pathway (Reed, 1995). Some of these proteins are localized on the outer mitochondrial membrane, nuclear envelope, and endoplasmic reticulum. Upon induction of apoptosis, they undergo translocation to mitochondria and dimerize with one another; the dimerization is modulated by their phosphorylation (Gajewski and Thompson, 1996). The exact mechanism by which these interactions lead to a release of cytochrome *c* from mitochondria is still not fully explained. The structure of Bcl-x_L suggests that this protein may form pores in biological membranes and that interactions between the antagonistic members of the Bcl-2 family may regulate the pore's permeability (Muchmore *et al.*, 1996). The loss of mitochondrial transmembrane potential ($\Delta\Phi_m$) observed early during apoptosis (Cossarizza *et al.*, 1994), appearance of reactive oxidative intermediates (ROIs) within the cell (Hedley and McCulloch, 1996), the release of cytochrome *c* (Kluck *et al.*, 1997), apoptotic protease-activating factor (Apaf-1)

---

of HL-60 cells was induced by CPT and during the final hour of culturing fluoresceinated cadaverine (F-CDV) was present in the medium. The cells were cytocentrifuged, fixed, and their DNA stained with PI in the presence of RNase (Grabarek *et al.*, 2002). Note three strongly F-CDV–labeled apoptotic cells (*thin arrows*). However, other apoptotic cells (with fragmented nuclei) are not labeled with F-CDV (*block arrows*). The cells undergoing apoptosis without TGase 2 activation are larger and become more flattened upon cytocentrifugation than the F-CDV–labeled cells. (E) HL-60 cells induced to apoptosis by CPT were fixed and their DNA strand breaks labeled with BrdUTP in the reaction catalyzed by terminal deoxynucleotidyltransferase (TUNEL) (Li and Darzynkiewicz, 1995). The incorporated BrdU was detected immunocytochemically by FITC-tagged antibody, DNA was stained with PI in the presence of RNase. The colocalization of PI (double-stranded DNA) and FITC (DNA strand breaks) in apoptotic cells with fragmented nuclei manifests in yellow fluorescence. (F) HL-60 cells were induced to apoptosis by CPT, and were then fixed and treated with acid to partially denature DNA and stained with acridine orange (AO) to differentially stain double-stranded versus single-stranded (denatured) DNA by fluorescing green versus red, respectively (Hotz *et al.*, 1992, 1994). Reflecting more extensive DNA denaturation, the chromatin of cells with fragmented nuclei had distinctly more intense red and less intense green fluorescence. (See Color Insert.)

(Zou *et al.*, 1999), and Ca^{2+} from mitochondria (McConkey *et al.*, 1998) are all presumed to be a consequence of the opening of pores in the mitochondrial membrane (Zamzani *et al.*, 1996, 1998). Release of cytochrome *c* from mitochondria is considered the critical event that triggers the irreversible steps of apoptosis. Namely, upon release from mitochondria, cytochrome *c* binds Apaf-1 and in the presence of adenosine triphosphate (ATP)/dATP, this complex, through the caspase-recruitment domain (CARD), recruits the initiator caspase-9 and all these components assemble to form an apoptosome on which procaspase-9 undergoes activation (Chinnaiyan *et al.*, 1997; Zou *et al.*, 1999). Caspase-9 triggers activation of the effector caspase-3, caspase-6, and caspase-7.

The endoplasmic reticulum (ER) stress-induced apoptosis represents another type of intrinsic pathway (Kaufmann, 1999; Rao *et al.*, 2004). Likewise in mitochondria, the proapoptotic and antiapoptotic proteins of the Bcl-2 family are also localized in the ER, where they modulate ER propensity to respond to the stress by induction of apoptosis. This pathway becomes activated in response to perturbed homeostasis of Ca^{2+} combined with accumulation of misfolded proteins in the ER. Unlike the mitochondrial pathway, however, the ER pathway is not mediated by cytochrome *c* and Apaf-1 (Rao *et al.*, 2004). In mice, the initiator is caspase-12; whether caspase-12 has the same function in human cells is unclear.

Activation of yet another intrinsic pathway takes place in anoikis. Anoikis defines apoptosis triggered by loss of interactions of the cell with the intercellular matrix, which leads to interruption of the integrin-mediated signals (Frisch and Screanton, 2001). In the absence of such interactions, Bax is rapidly recruited to mitochondria, followed by release of cytochrome *c* and caspase activation (Wang *et al.*, 2003).

A cross-talk between the intrinsic (mitochondrial) and extrinsic pathways of apoptosis can take place. Namely, caspase-8 activated along the extrinsic pathway cleaves the Bid protein and the cleavage product truncated Bid ([t]Bid) has the capability to engage the mitochondrial pathway (Esposti, 2002). The mitochondrial pathway, thus, is mobilized subsequently to the extrinsic pathway and serves as an amplification loop accelerating the process of apoptosis.

The cascade-like activation of caspases and serine proteases leads to proteolytic degradation of selected proteins, defined as "death substrates." Poly(ADP)ribose polymerase (PARP), lamin, actin, U1 small nuclear ribonucleoprotein (U1 snrp), rho-DGI, SREBP, and DNA-dependent protein kinase have been recognized among these substrates (Enari *et al.*, 1996; Lazebnik *et al.*, 1994). Involvement of serine proteases was observed in the specific degradation of two nuclear proteins, lamin B and NuMA (Weaver *et al.*, 1993).

The activation of an endonuclease that cleaves DNA is another characteristic event of apoptosis (Arends *et al.*, 1990). One of these endonucleases is caspase-activated DNase (CAD) (Nagata, 2000). Initially, DNA is cleaved at the sites of attachment of chromatin loops to the nuclear matrix, which results in the appearance of discrete 300- to 50-kb size fragments that can be detected by pulsed-field gel electrophoresis (Oberhammer *et al.*, 1993). Subsequently, DNA is cleaved between nucleosomes. The products are then discontinuous DNA fragments representing mononucleosomal-plus oligonucleosomal-sized DNA sections, which generate a characteristic "ladder" pattern during agarose gel electrophoresis. Because a part of DNA in apoptotic cells is fragmented, after cell permeabilization by detergents or prefixation in precipitating fixatives such as ethanol, this fraction of low-molecular-weight DNA can be extracted (Gong *et al.*, 1994). The fraction of nonfragmented, high-molecular DNA remains in the cell, and such apoptotic cells may be identified by cytometry as the cells with a deficit in their DNA content ("sub-G_1 cells" on the DNA content frequency histograms). It should be pointed out, however, that in some cell types, DNA cleavage does not proceed to internucleosomal-sized sections but stops at 300 to 50 kb size DNA fragments (Catchpoole and Stewart, 1993; Cohen *et al.*, 1992; Collins *et al.*, 1992).

Structural integrity and most of the plasma membrane transport function is preserved during the early phase of apoptosis. However, the permeability to certain fluorochromes, such as 7-AAD, Hoechst 33342, or Hoechst 33258, is increased (Ormerod *et al.*, 1993; Schmid *et al.*, 1992). The most characteristic change, however, is the loss of asymmetry of the phospholipids on the plasma membrane, leading to exposure of phosphatidylserine on its outer surface (Fadok *et al.*, 1992; Koopman *et al.*, 1994; van Engeland *et al.*, 1998). This change occurs early during apoptosis regardless of whether apoptosis is induced along the intrinsic or extrinsic pathway. Exposure of phosphatidylserine on the outer leaflet of the plasma membrane preconditions the remains of the apoptotic cell (apoptotic bodies) to become a target for phagocytizing cells. Loss of pseudopodia or microvilli is paralleled by degradation of F actin (Endersen *et al.*, 1995).

Still another event of apoptosis is activation of "tissue" transglutaminase (TGase 2). Its activation leads to cross-linking of cytoplasmic proteins (Piacentini *et al.*, 1995), resulting in a change in the physical properties of the cell, making cells stiffer and resistant to deformation under pressure. Protein cross-linking during apoptosis prevents their leakage in soluble form, which may protect from autoimmune reactions when apoptosis occurs *in vivo*. Elimination of TGase 2 (in knockout mice) impairs phagocytosis of apoptotic cell remnants (Szondy *et al.*, 2003).

C. "Accidental" Cell Death; Primary Cell Necrosis

"Accidental cell death" defines the mode of cell death that often is induced with overdose of radiation or in response to extreme stress. It also occurs when mitochondrial metabolism (oxidative phosphorylation) is inhibited and production of ATP prevented concurrently with subjecting the cells to the stress that otherwise induces apoptosis (Kim *et al.*, 2003). Other names are *cell necrosis, primary cell necrosis*, or *oncosis* (*oncos*, swelling in Greek) (Majno and Joris, 1995). Its characteristic feature is distortion of the mitochondrial membrane, mitochondrial and nuclear swelling, and cytoplasm vacuolization. The early loss of plasma membrane integrity is followed by leakage of cytosol and cytoplasm constituents, autolytic processes, and dissolution of remnants of chromatin (karyolysis). A patchy pattern of chromatin condensation may be apparent. Unlike apoptosis, necrotic cell death in the tissue attracts inflammatory cells and if extensive may lead to scar formation.

The primary cell necrosis should not be confused with the "necrotic phase of apoptosis," or "secondary necrosis," the terms used to define the late phase of apoptosis, the stage when integrity of the plasma membrane is compromised. The damaged plasma membrane that is unable to exclude cationic dyes such as trypan blue or propidium iodide (PI), thus, is the common lesion of both the primary cell necrosis and the late stage of apoptosis.

III. Different Time-Windows for Detection of Apoptosis: Possible Source of Error in Measurement of Incidence of Apoptosis

Apoptosis is a kinetic event of a variable induction and a variable execution period. The time-window between the induction and the onset of apoptosis varies depending of the cell type, for example, by as much as between 2 and more than 24 hours for HL-60 versus MCF-7 cells, respectively, treated with the same DNA-damaging drug (Del Bino *et al.*, 1999). This induction-to-execution time interval also varies depending on the inducer. For example, in HL-60 cells, the time window between the exposure to the death ligand (TNF-α) versus DNA-damaging drug and the onset of caspase activation was shown to be 30 and 90 minutes, respectively (Li and Darzynkiewicz, 2000). Furthermore, the duration of the apoptotic process, from the onset of the event to total cell disintegration, varies depending on the cell type and on whether apoptosis occurs *in vivo* or *in vitro*. In live tissue under conditions of cell homeostasis, when the cell death rate is balanced by proliferation rate, the mitotic index may exceed the apoptotic index (AI). This indicates that length of apoptosis cannot be greater than that of mitosis (i.e., about 1 hour). *In vitro*, however, apoptotic cells may remain for extended periods, enduring for many hours before their disintegration. Little is known about the duration of cell necrosis, although *in vivo* necrotic cells appear to be rapidly phagocytized.

Identification of apoptotic cells generally relies on a particular characteristic or a specific marker. The time interval when the cell undergoing apoptosis presents such a feature varies, depending on the feature. Some events are quite transient and others last longer. Caspase activation detected by cleavage of PARP, for example, occurs early, and in HL-60 cells induced to apoptosis by DNA damage, precedes DNA fragmentation by about 20 minutes (Li and Darzynkiewicz, 2000). Caspase activation (detected by binding of fluorochrome-labeled caspase inhibitors [FLICA]) precedes externalization of phosphatidylserine (Pozarowski *et al.*, 2003c). Decrease (loss) of the mitochondrial potential appears to be initially a transient event, followed by permanent collapse of the potential later during apoptosis (Li *et al.*, 2000).

One of the most common applications of cytometry in the analysis of apoptosis is to compare incidence of apoptosis in different samples, for example, after cell treatment with different drugs or different cell types treated with the same drug. This task is complicated by the variability of (1) the induction-to-execution interval, (2) the overall duration of apoptosis, and (3) the length of the time window of the manifestation of a particular apoptotic feature (marker). Namely, the snapshot measurement of the AI in different samples at the same time may fail to target them at the comparable time-windows of apoptosis, so the observed difference in the AI may not reflect the real difference in incidence of apoptosis in these samples. Attempts have been made, therefore, to obtain the cumulative AI, by measuring the rate (kinetics) of cell entrance to apoptosis and preventing disintegration of apoptotic cells (Smolewski *et al.*, 2002). The alternative solution is to count the absolute number of cells in the culture during the treatment (to account for cell loss) while estimating the AI.

IV. Light-Scattering Properties of Cells Dying by Apoptosis

Intensity of the forward light scatter signal measured by flow cytometry is correlated with cell size. On the other hand, the side (90-degree angle) scatter reports the cell refractive and reflective properties. Because of cell shrinkage, a decrease in intensity of forward light scatter is detected at the early stage of apoptosis (Ormerod *et al.*, 1995; Swat *et al.*, 1981). Early apoptotic cells, however, show either unchanged or even increased intensity of the side scatter signal, reflecting perhaps chromatin and cytoplasm condensation and nuclear fragmentation. When apoptosis is advanced and the cells become progressively smaller, the intensity of side scatter also decreases. Late apoptotic cells, therefore, are characterized by the markedly diminished ability to scatter light in both the forward and the right angle direction (Fig. 2).

It should be stressed that the change in light scatter is not a reliable marker of apoptosis or necrosis. Mechanically broken cells, isolated nuclei, cell debris, and individual apoptotic bodies may also display diminished light scatter properties. Further more, apoptosis can progress with or without activation of TGase 2 (Grabarek *et al.*, 2002), which has an effect on light-scattering properties of the cell. Namely, activation of TGase 2 results in protein cross-linking and extensive condensation within the cell, which is reflected by a transient increase in intensity of the side scatter signal and only a moderate decrease in forward scatter signal. Conversely, in the absence of TGase 2 activation, the decrease in both the forward and the side scatter signal is seen early and is very profound (Huang *et al.*, 2004). To provide more definite identification of apoptotic or necrotic cells, the analysis of light scatter has to be combined with other more specific markers of these death modes.

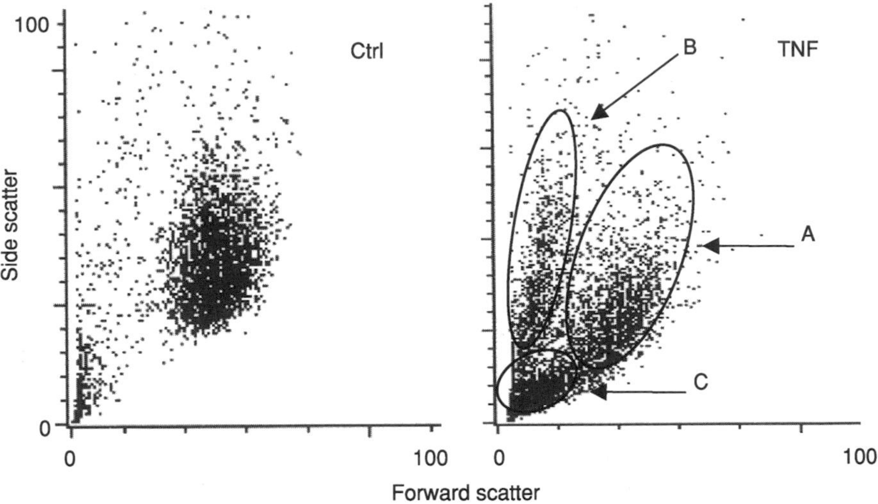

Fig. 2 Changes in light-scattering properties of cells during apoptosis. HL-60 cells were untreated (left panel) or treated with tumor necrosis factor-a (TNF-α) and cycloheximide (CHX) (right panel), as described (Grabarek *et al.*, 2002). Cells of the population marked A from the treated culture have similar light-scattering properties as the untreated cells. Early apoptotic cells (B) have diminished forward scatter, and their side scatter is less changed. The late apoptotic cells and apoptotic bodies (C) have greatly diminished forward and side scatter.

V. Mitochondrial Transmembrane Potential ($\Delta\Psi_m$)

Mitochondria play a critical role in the induction of apoptosis by releasing cytochrome *c* and apoptosis-inducing factor (AIF) (Liu *et al.*, 1996; Yang *et al.*, 1997). Although cytochrome *c* is essential for caspase-9 activation, AIF is also involved in activation of apoptosis-associated endonuclease (Susin *et al.*, 1997). Still another protein Smac/DIABLO that inactivates the inhibitors of caspases (IAPs), thereby promoting apoptosis, is released from mitochondria as well (Deng *et al.*, 2002). Dissipation (collapse) of mitochondrial transmembrane potential $\Delta\Psi_m$), called the *permeability transition* (PT), also occurs early during apoptosis (Cossarizza *et al.*, 1994; Kroemer, 1998; Zamzani *et al.*, 1996, 1998). However, there is conflicting evidence whether the latter event is mechanistically a prerequisite for the release of cytochrome *c* or AIF and activation of caspases (Finucane *et al.*, 1999; Li *et al.*, 2000; Scorrano *et al.*, 1999).

The lipophilic cationic fluorochromes rhodamine 123 (Rh123) or carboxycyanine dyes such as 3,3′-dihexiloxa-dicarbocyanine ($DiOC_6[3]$) can serve as cell-viability markers and as probes of ΔY_m (Darzynkiewicz *et al.*, 1981, 1982; Johnson *et al.*, 1980). They accumulate in mitochondria of live cells and the extent of their uptake, as measured by intensity of cellular fluorescence, reflects ΔY_m. A combination of Rh123 and PI was introduced as a viability assay that discriminates between live cells that stain with Rh123 but exclude PI versus early apoptotic cells that lost ability to accumulate Rh123 versus late apoptotic/necrotic cells that stain with PI only (Darzynkiewicz *et al.*, 1982, 1994). The specificity of Rh123 and $DiOC_6(3)$ as ΔY_m probes is higher at their lowest concentrations ($<0.5\ \mu M$ or $<0.1\ \mu g/ml$).

Other probes of ΔY_m are the J-aggregate–forming lipophilic cationic fluorochromes 5,5′,6,6′-tetrachloro-1,1′,3,3′-tetraethylbenzimidazolcarbocyanine iodide (JC-1) and 3,3′-dimethyl-α-nafthoxacarboxycyanine iodide (JC-9). Their uptake by charged mitochondria is characterized by orange/red fluorescence (aggregated form), while the loss of ΔY_m manifests by the shift to green fluorescence (Fig. 3), reflecting dissociation of the J-aggregates and transition to monomeric form (Cossarizza *et al.*, 2001).

A series of MitoTracker dyes (chloromethyltetramethylrosamine analogs) of different color were introduced by Molecular Probes as new mitochondrial probes (Haugland, 2003). One of them is MitoTracker Red CMXRos, a probe sensitive to ΔY_m (Poot *et al.*, 1997). Unlike the probes described earlier, the cells labeled with MitoTracker Red CMXRos retain the label after fixation with formaldehyde. It is possible, therefore to combine analysis of ΔY_m with the measurement of another cell attribute that requires cell fixation, such as DNA fragmentation detected by the TUNEL assay and immunocytochemical detection of intracellular protein.

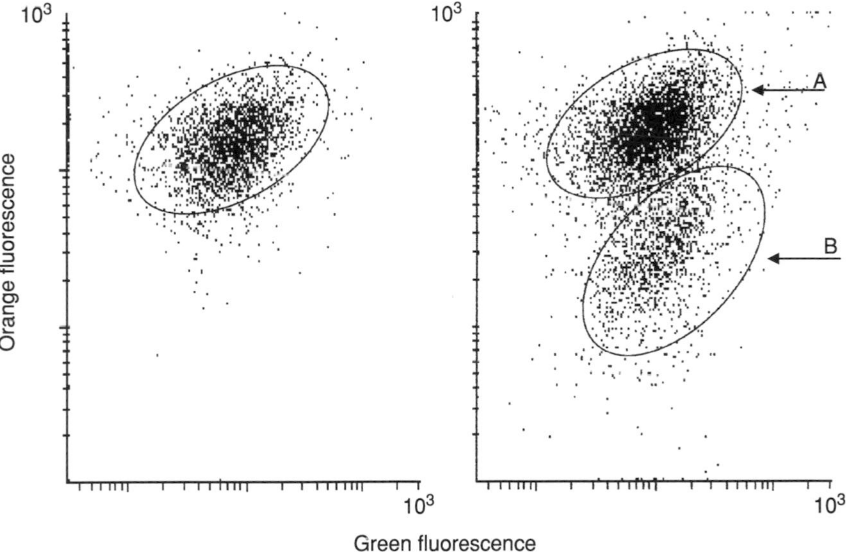

Fig. 3 Collapse of mitochondrial transmembrane potential (ΔY_m) detected by cell staining with the J-aggregate dye JC-1. HL-60 cells untreated (left) or treated (right) with camptothecin for 3 hours were stained with JC-1 and their orange and green fluorescence measured by flow cytometry (Pozarowski *et al.*, 2003a). Cells with collapsed ΔY_m have decreased intensity of orange fluorescence (B).

Still another useful probe of ΔY_m, similar to MitoTracker Red CMXRos, is tetramethylrhodamine methyl ester perchlorate (TMRM). The application of MitoTracker Red CMXRos combined with the marker of caspase activation, the FLICA (Pozarowski *et al.*, 2003c), is illustrated in Fig. 4.

Some mitochondrial probes such as 10-nonyl acridine orange, MitoFluor Green, and MitoTracker Green were reported to be markers of mitochondrial mass and not of ΔY_m (Poot *et al.*, 1997; Ratinaud *et al.*, 1988). It was proposed, therefore, to measure both ΔY_m and mitochondrial mass with a combination of ΔY_m-sensitive and ΔY_m-nonsensitive probes (Petit *et al.*, 1995). Further observations, however, indicated that 10-nonyl acridine orange, MitoFluor Green, and MitoTracker Green were sensitive to changes in ΔY_m, and therefore, either alone or in combination with ΔY_m-sensitive probes, cannot be used as markers of mitochondrial mass (Keiji *et al.*, 2000).

Measurement of ΔY_m, similar to other functional markers, is sensitive to changes in cell environment. The samples to be compared have to be incubated and measured under identical temperature, pH, time elapsed between the onset of incubation and fluorescence measurement, and other potential variables.

Another point to be considered in measuring ΔY_m is that the intracellular distribution of any cationic mitochondrial probe, in conformity with the Nernst equation, reflects the differences in the transmembrane potential across both the plasma membrane (i.e., between exterior vs interior of the cell) and the outer mitochondrial membrane (Shapiro, 2003). Thus, in addition to mitochondria the probes also accumulate in the cytosol. Their accumulation in the cytosol is

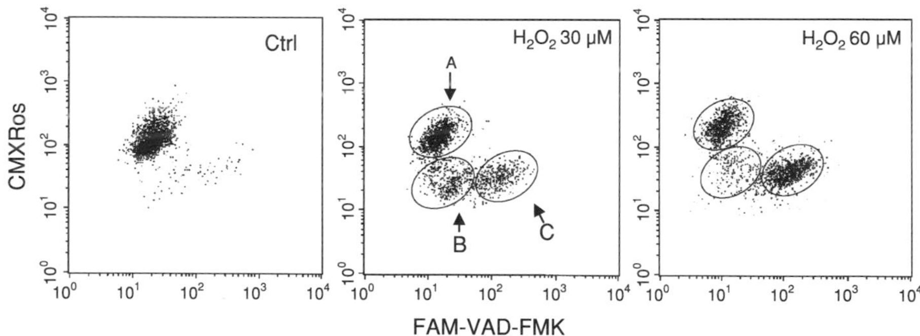

Fig. 4 Concurrent analysis of collapse of ΔY_m and caspase activation during apoptosis. Apoptosis of Jurkat cells was induced by oxidative stress (growth in the presence of 30- or 60-μM H_2O_2; control, untreated cells). The cells were then supravitally exposed to FAM-VAD-FMK and Mito Tracker Red CMXRos and then rinsed and their green and red fluorescence measured by flow cytometry (Pozarowski *et al.*, 2003c). Two subpopulations of apoptotic cells can be detected. Subpopulation B represents the cells that lost mitochondrial potential but did not activate caspases, whereas cells in subpopulation C are characterized by collapsed ΔY_m and caspase activation (FLICA binding). At 60-μM H_2O_2, caspase activation was accelerated as reflected by the increased proportion of cells in (C) compared to (B).

facilitated by a passive transport across the plasma membrane. The latter is probe-concentration dependent. To decrease the passive transport, therefore, one has to use the probes at their lowest possible concentration, that is, the minimal concentration that is still adequate to detect the cells with charged mitochondria, but may require relatively high settings of the photodetector sensitivities (photomultiplier voltage) and higher laser power output. The probes' specificity to ΔY_m can be assessed by pre-incubation of cells for 20–30 minutes with 10–50 μm protonophore carbonylcyanide m-chlorophenylhydrazone (CCCP), the agent that collapses the mitochondrial potential. Thus, treated cells may be used as a negative control.

VI. Activation of Caspases

A. Caspases: The "Killer Enzymes"

Activation of *c*ysteine-*a*spartic acid–*s*pecific prote*ases* (caspases) is one of the hallmarks of apoptosis (Alnemri *et al.*, 1996; Earnshaw *et al.*, 1999; Fadeel *et al.*, 2000; Kaufmann *et al.*, 1993; Shi, 2002). There are 14 members of the caspase family in mammals and many of them play an essential role during apoptosis. Although individual caspases have specific functions, some degree of overlapping specificity and redundancy among them is apparent. They remain in the cell in the form of inactive zymogen (procaspase) and become activated upon transcatalytic cleavage followed by assembly of the products of the cleavage. Specifically, the two cleaved procaspase molecules assemble to form a single enzymatically active hetero-tetramer molecule that has two active enzymatic sites on its opposite ends. Individual caspases recognize a four-amino acid sequence on their substrate proteins; the carboxyl end of aspartic acid within this sequence is the target for cleavage. Activation of caspases is considered the point of no return in the cell death process and several methods were developed to detect this event.

B. Fluorogenic and "Death" Caspase Substrates

One approach uses fluorogenic substrates of caspases. The peptide substrates are colorless or not fluorescent, but upon caspase-induced cleavage, they generate colored or fluorescing products (Gorman *et al.*, 1999; Hug *et al.*, 1999; Lee *et al.*, 2003; Liu *et al.*, 1999; Telford *et al.*, 2002). Many kits designed to measure activity of caspases using fluorometric or colorimetric assays are commercially available. Some kits detect activation of multiple caspases, and others are based on the substrates specific for either caspase-1, caspase-3, or caspase-8.

Another approach to study caspase activation that also is based on the analysis of the cleavage product relies on immunocytochemical detection of the cleaved PARP, one of the characteristic "death substrates" of caspases. PARP is a nuclear enzyme involved in DNA repair that is activated in response to DNA damage (de Murcia, 1994). Early in apoptosis, PARP is cleaved by caspases, primarily by caspase-3 (Alnemri *et al.*, 1996; Kaufmann *et al.*, 1993; Lazebnik *et al.*, 1994). The specific cleavage products are 89- and 24-kd PARP fragments. For years, their detection after electrophoresis of PARP was considered a hallmark of apoptotic mode of cell death. The antibody that recognizes the 89-kd product of PARP cleavage became commercially available and has been adapted to label apoptotic cells for detection by cytometry (Li and Darzynkiewicz, 2000) (Fig. 1C). As is evident in Fig. 5, the multiparameter analysis of the cells differentially stained for PARP p89 and DNA makes it possible not only to identify and score apoptotic cell populations but also to correlate apoptosis with the cell cycle phase or DNA ploidy.

C. Immunocytochemical Detection of Activated Caspases Epitope

Still another approach to detect caspase activation is based on immunocytochemical detection of the epitope of activated enzymes (enzymatically active hetero-tetramer). Antibodies specific to several activated caspases have been

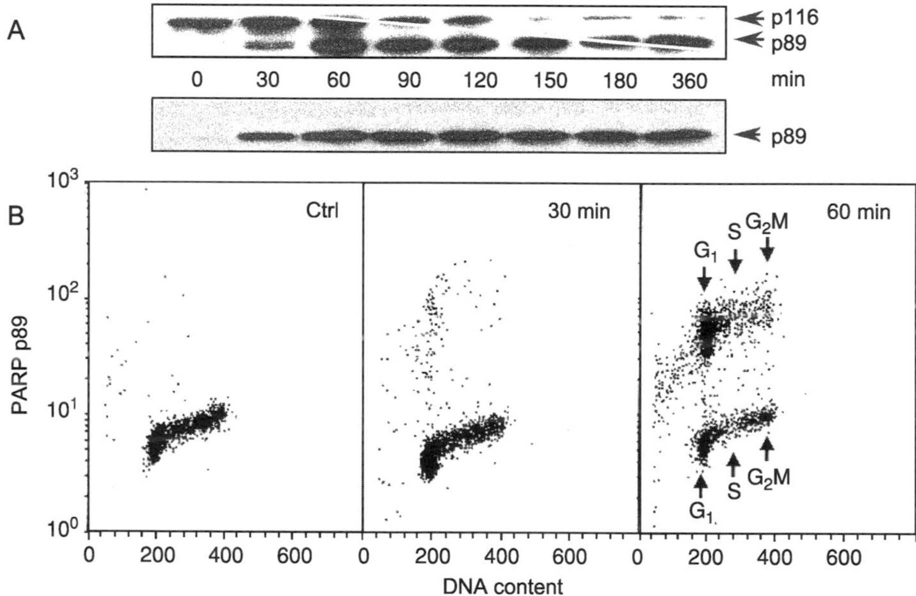

Fig. 5 Detection of PARP cleavage combined with DNA content (cell cycle) analysis. To induce apoptosis, HL-60 cells were treated with tumor necrosis factor-α (TNF-α) in the presence of CHX for 30–360 minutes (Li and Darzynkiewicz, 2000). (A) Immunoblots of the treated cells stained with PARP plus PARP p89—(upper gel) or PARP p89 only (lower gel)—antibody. (B) Bivariate distributions of PARP p89 versus DNA content (stained with PI) of the untreated (Ctrl) and treated for 30- and 60-minutes cells. Note the appearance of the first PARP p89—positive cells already after 30 minutes of treatment, coinciding in time with the detection of PARP cleavage on gels. There is no evidence of cell cycle phase specificity of apoptosis induced by TNF-α.

developed (Sallman *et al.*, 1997) and are now commercially available. Thus far most these antibodies have been used for immunoblotting. However, the caspase-3 antibody was tested as an immunocytochemical probe and was found to be useful to detect activation of this caspase by cytometry (Pozarowski *et al.*, 2003). Similar to immunocytochemical detection of PARP p89 (Fig. 5), the multiparameter analysis of caspase-3 activation and DNA content can be used to reveal the cell cycle phase specificity of its activation (Fig. 6).

D. Fluorochrome-Labeled Inhibitors of Caspases

Activation of caspases can also be revealed with the use of FLICAs (Bedner *et al.*, 2000; Pozarowski *et al.*, 2003c; Smolewski *et al.*, 2001). FLICAs are designed as affinity ligands to the active enzyme centers of individual caspases. Each FLICA has three functionally distinct domains: (1) the fluorochrome (fluorescein; FAM or sulforhodamine, SR) domain, (2) the caspase-recognition domain comprising a four amino acid peptide, and (3) the covalent binding moiety, which consists of either chloromethylketone (CMK) or fluoromethylketone (FMK), which irreversibly binds to cysteine of the respective caspase forming thiomethyl ketone (Thornberry *et al.*, 1997; Van Noorden, 2001). FLICAs are permeant to live cells and relatively nontoxic. The recognition peptide moiety of FLICAs is expected to provide specificity between a ligand and a particular caspase. Several FLICA kits are commercially available, including FAM (or SR) VAD-FMK, which contains the valyl alanyl aspartic acid residue sequence. This three amino

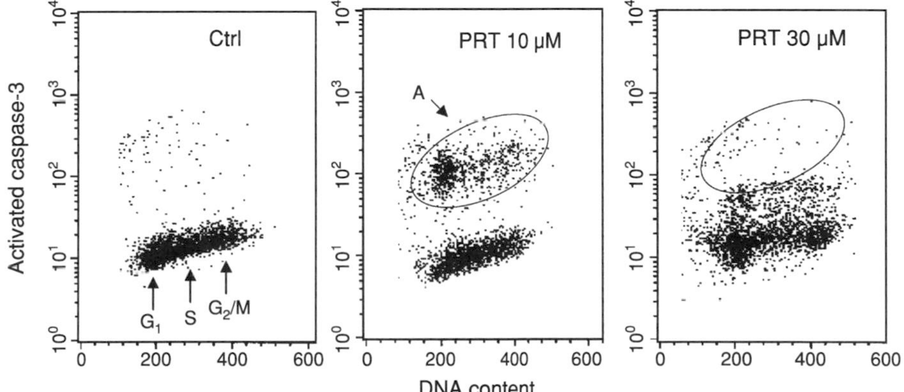

Fig. 6 Immunocytochemical detection of caspase-3 activation. To induce apoptosis, HL-60 cells were treated with 10 μM parthenolide (PRT), the inhibitor of nuclear factor-κB, while to induce necrosis, with 30 μM of PRT (Pozarowski *et al.*, 2003b). Caspase activation was detected using antibody that reacts with the epitope of the activated (cleaved) caspase-3. Subpopulation of cells with activated caspase-3 is apparent at 10-μM PRT (A). Necrotic cells in the culture treated with 30-μM PRT have much lower level of caspase-3 activation. No cell cycle phase specificity in caspase-3 activation is apparent.

acid target sequence allows the inhibitor to irreversibly bind to activated caspase-1, caspase-3, caspase-4, caspase-5, caspase-7, caspase-8, and caspase-9, making it a pan-caspase marker. Other inhibitors, such as those that contain DVAD, DEVD, VEID, YVAD, LETD, LEHD, or AEVD peptide residues, preferentially bind to activated caspase-2, caspase-3, caspase-6, caspase-1, caspase-8, caspase-9, or caspase-10, respectively.

Exposure of live cells to FLICAs results in the uptake of these reagents, followed by their covalent binding to activated caspases within the cells that undergo apoptosis. Unbound FLICAs are removed from the nonapoptotic cells that lack activated caspases by rinsing the cells with wash buffer. Concurrent probing of the plasma membrane permeability with PI allows one to distinguish at least two consecutive stages of apoptosis (Fig. 7). FLICA may also be used concurrently with a probe of mitochondrial potential, such as Mito Tracker Red CMXRos (Fig. 3).

Because FLICAs bind covalently, they remain in the cell after fixation (with formaldehyde) and cell permeabilization. The FLICA assay, thus, can be combined with the analysis of the cell attributes that can be measured only after fixation, such as DNA content (cell cycle phase) and DNA fragmentation (TUNEL assay).

Caution should be exercised, however, in interpreting the mechanism and specificity of FLICA binding. It was shown that certain facets of their binding suggest interactions with the intracellular targets other than caspases, which become accessible in the course of caspase activations (Pozarowski *et al.*, 2003c).

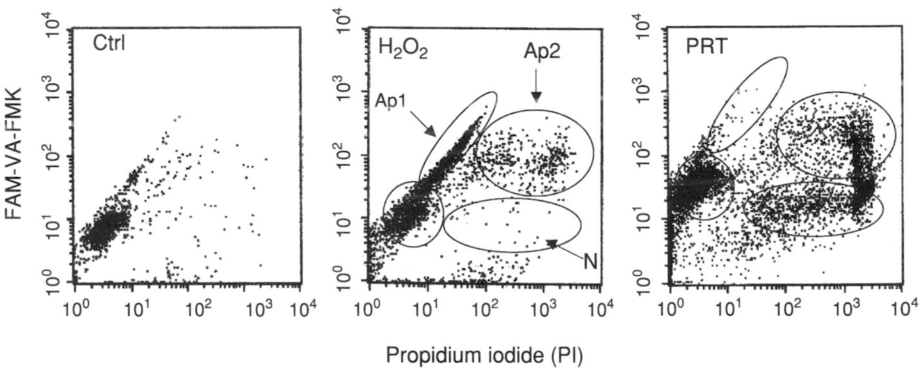

Fig. 7 Detection of caspase activation by affinity labeling enzyme active center with FAM-VAD-FMK combined with supravital staining with PI. Jurkat cells were either untreated (Ctrl) to induce apoptosis treated with H_2O_2 or to induce both apoptosis and necrosis or treated with parthenolide (PRT) as described (Pozarowski *et al.*, 2000b). The cells were then exposed in the culture to FAM-VAD-FMK and PI, and their green and red fluorescence measured by flow cytometry. Early apoptotic cells (Ap1) bind FAM-VAD-FMK but exclude PI. Late apoptotic cells (Ap2) are both FAM-VAD-FMK and PI positive (Ap2). Necrotic (N) and very late apoptotic cells stain with PI and are FAM-VAD-FMK negative.

E. Green Fluorescent Protein Fluorescence Resonance Energy Transfer Caspase Activation Assay

Activation of caspases can also be assayed using the tandem molecules of green fluorescent protein (GFP) and blue fluorescent protein (BFP), or cyan (CFP) and yellow fluorescent protein (YFP), covalently linked by a short peptide that is the target of a particular caspase (Jones *et al.*, 2000; Luo *et al.*, 2003; Tawa *et al.*, 2001; Xu *et al.*, 1998). The fluorescence resonance energy transfer (FRET) that occurs between the pairs of these fluorescent proteins when they are linked by the peptide is abruptly terminated upon cleavage of the peptide linker. Caspase activation, thus, manifests by loss of FRET. The major advantage of this methodology is that the marker of caspase activation is intrinsic, operating in the live cell. The method, thus, is simple and rapid, because there is no need for cell fixation or application of any external fluorochromes. The panel of different cell lines stably transfected to constitutively express the tandem FRET-fluorescing proteins, thus, would represent the optimal setting for drug screening, by monitoring caspase activation/apoptosis in the treated cells.

VII. Annexin V Binding

Phospholipids are asymmetrically distributed between inner and outer leaflets of the plasma membrane; phosphatidylcholine and sphingomyelin are exposed on the external leaflet, and phosphatidylserine is located on the inner surface of the lipid bilayer. This asymmetry is broken during apoptosis and phosphatidylserine becomes exposed on the outside (Fadok *et al.*, 1992; Koopman *et al.*, 1994; van Engeland *et al.*, 1998). Because the anticoagulant protein annexin V binds with high affinity to phosphatidylserine, fluorochrome-conjugated annexin V has found an application as a marker of apoptotic cells, in particular for their detection by flow cytometry (van Engeland *et al.*, 1998). The cells become reactive with annexin V before the loss of the plasma membrane's ability to exclude cationic dyes such as PI.

Live nonapoptotic cells, when stained with fluorescein isothiocyanate (FITC)–annexin V and PI, have minimal green (fluorescein–annexin V) fluorescence and minimal red (PI) fluorescence (Fig. 8). At early stages of apoptosis, cells stain green but still exclude PI and therefore continue to have no significant red fluorescence (Ap1). At late stages of apoptosis, the cells have intense green and red fluorescence (Ap2). It should be noted that isolated nuclei, cells with severely damaged membranes, and very late apoptotic cells stain rapidly and strongly with PI and may not bind annexin V. It should also be mentioned that even intact and live cells take up PI on prolonged incubation. Therefore, fluorescence measurement should be performed rather shortly after addition of this dye.

Interpretation of the results of this assay may be complicated by the presence of nonapoptotic cells with a damaged plasma membrane. Such cells may have

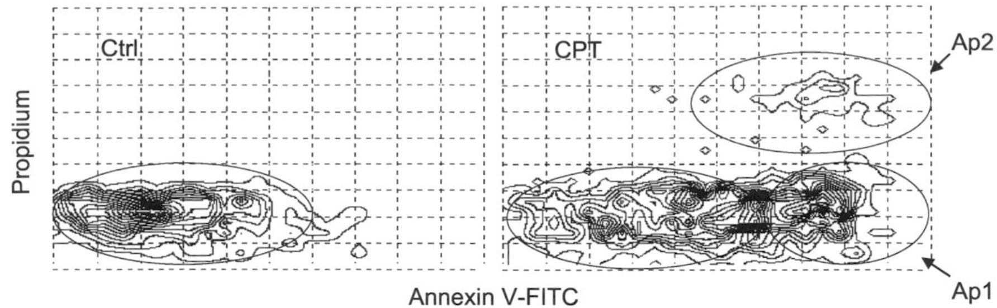

Fig. 8 Detection of early and late stages of apoptosis by concurrent staining with annexin V-fluorescein isothiocyanate (FITC) and PI. Untreated (Ctrl) and camptothecin-treated (CTP) HL-60 cells were exposed to annexin V–FITC conjugate and PI and their green and red fluorescence was measured by laser-scanning cytometry (LSC). At the early stage of apoptosis (Ap1), the cells bind annexin V–FITC but exclude PI. At the late stage of apoptosis (Ap2; also necrotic cells), they bind annexin V–FITC and stain with PI.

phosphatidylserine exposed on plasma membrane and, therefore, like apoptotic cells, bind annexin V. Mechanical disaggregation of tissues to isolate individual cells, extensive use of proteolytic enzymes to disrupt cell aggregates, to remove adherent cells from cultures, or to isolate cells from tissue, mechanical removal of the cells from tissue culture flasks (e.g., by "rubber policeman"), and cell electroporation all affect the binding of annexin V. Such treatments, therefore, may introduce experimental bias when using this assay. It should also be noted that otherwise live and healthy macrophages/monocytes, upon ingestion of apoptotic bodies or remains of apoptotic cells, also become annexin V positive and thus may be mistakenly identified as apoptotic cells (Marguet *et al.*, 1999).

Figure 9 illustrates multiparameter analysis of FLICA versus annexin V binding. The data distinctly show that upon induction of apoptosis, activation of caspases detected by FLICA binding precedes the loss of plasma membrane phospholipids asymmetry (Pozarowski *et al.*, 2003c). The time window of apoptosis detected by FLICA binding is much wider than by the annexin V–binding probe. These data also suggest that activation of caspases is a prerequisite for externalization of phosphatidylserine because essentially no FLICA-negative cells that bind annexin V are apparent.

VIII. DNA Fragmentation

A. Detection of Cells with Fractional ("Sub-G_1") DNA Content

Endonucleases activated during apoptosis cleave sections of internucleosomal DNA and cause extensive DNA fragmentation (Arends *et al.*, 1990; Kerr *et al.*, 1972; Nagata, 2000). The fragmented low-molecular-weight DNA is then extracted from the cells during the process of cell staining in aqueous solutions. Such extraction occurs when the cells are treated with detergent and/or hypotonic

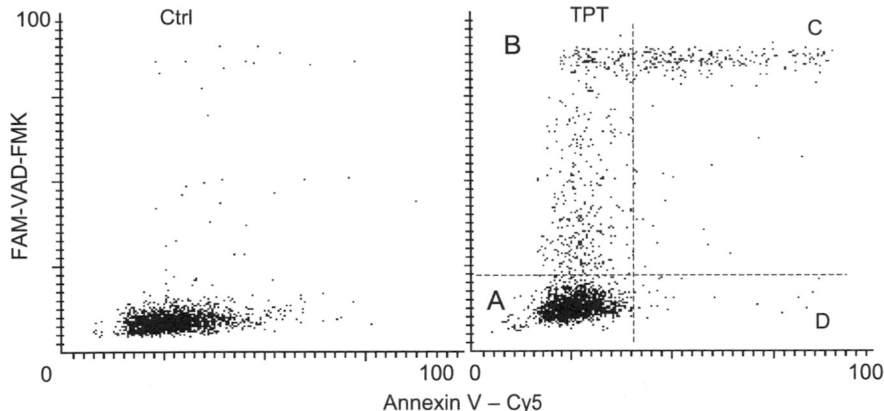

Fig. 9 Concurrent detection of caspase activation (fluorochrome-labeled inhibitors of caspase [FLICA] binding) and plasma membrane phosphatidylserine externalization (annexin V binding). To induce apoptosis, HL cells were treated with DNA topoisomerase-I inhibitor topotecan (TPT) as described (Pozarowski *et al.*, 2003c). The cells were then stained with annexin V–Cy5 and FAM-VAD-FMK and their far red and green fluorescence was measured by laser-scanning cytometry (LSC). More than 95% of the untreated cells (Ctrl) had annexin V–Cy5 and FAM-VAD-FMK fluorescence intensities below the thresholds (*dashed lines*). Note that essentially all cells that responded to TPT by increase in annexin V binding were FLICA positive. However, a large fraction of FLICA-positive cells did not bind annexin V. The data indicate that in the course of apoptosis, the time window for FLICA binding is wider than for annexin V binding and precedes the latter.

solution instead of fixation, to make them permeable to fluorochrome, or after fixation in precipitating fixatives such as ethanol. Fixation with cross-linking fixatives such as formaldehyde, on the other hand, results in the retention of low-molecular-weight DNA in the cell (which becomes cross-linked to protein) and therefore is incompatible with the "sub-G_1" assay. As a result of DNA extraction, apoptotic cells end up deficient in DNA content, and when stained with a DNA-specific fluorochrome, they can be recognized by cytometry as cells having fractional DNA content. On the DNA content frequency histograms, they often form a characteristic "sub-G_1" peak (Gong *et al.*, 1994; Nicoletti *et al.*, 1991; Umansky *et al.*, 1981). It should be noted that the shedding of apoptotic bodies containing fragments of nuclear chromatin may additionally contribute to the loss of DNA from apoptotic cells.

Optimally, the "sub-G_1 peak" representing apoptotic cells should be separated from the G_1 peak of the nonapoptotic cell population, with little or no overlapping between these two. However, the degree of extraction of low-molecular-weight DNA and consequently the content of DNA remaining in apoptotic cells for cytometric analysis vary markedly depending on the extent of DNA degradation (duration of apoptosis), the number of cell washings, and pH and molarity of the washing and staining buffers. Thus, the separation of "sub-G_1" is not always adequate. It has been noted that extraction of DNA from apoptotic cells can be enhanced by rinsing the cells in a high-molarity phosphate–citrate buffer before staining (Gong *et al.*, 1994). On the other hand, when DNA degradation does not proceed to internucleosomal regions but stops after generating 50–300 kb fragments (Oberhammer *et al.*, 1993), little DNA can be extracted and this method fails to detect such apoptotic cells. It also should be noted that if G_2, M, or even late S-phase cells undergo apoptosis, the loss of DNA from these cells may not be adequate to have them located at the "sub-G_1" peak because they may end up with DNA content equivalent of that of G_1 or early S-phase cells and therefore be indistinguishable from the latter.

It is a common practice to use detergents or hypotonic solutions instead of fixation in DNA staining protocols (Nicoletti *et al.*, 1991). This causes lysis of plasma membrane and nuclear isolation. This approach is simple and yields excellent resolution for DNA content analysis of nonapoptotic cells. However, it introduces significant bias when used to quantify apoptotic cells. Namely, nuclei of apoptotic cells are often fragmented, and upon cell lysis, a multiplicity of chromatin fragments are released from a single cell. Lysis of mitotic cells releases individual chromosomes or chromosome aggregates. In micronucleation (e.g., after cell irradiation or treatment with clastogens), the micronuclei are released as well. Each nuclear fragment, chromosome, or micronucleus is then recorded by flow cytometer as an individual object with a sub-G_1 DNA content. Such objects are then erroneously classified as individual apoptotic cells. This bias is particularly evident when the logarithmic scale is used to display DNA content on the histograms, which allows one to record the events with as little DNA content as 1% or even 0.1% of that of G_1 cells. These events certainly cannot be individual apoptotic nuclei, and their percentage grossly exceeds the percentage of apoptotic cells in the sample.

B. Detection of DNA Strand Breaks ("TUNEL" Assay)

During apoptosis, DNA fragmentation generates a multitude of DNA strand breaks (DSBs) in the nucleus (Arends *et al.*, 1990; Oberhammer *et al.*, 1993). The 3′-OH termini of the breaks may be marked by attaching a fluorochrome to them. This is generally done directly or indirectly (e.g., *via* biotin or digoxigenin) by using fluorochrome-labeled triphosphodeoxynucleotides in a reaction catalyzed preferably by exogenous terminal deoxynucleotidyltransferase (TdT) (Gorczyca *et al.*, 1992, 1993b; Li and Darzynkiewicz, 1995; Li *et al.*, 1996). The reaction is commonly known as *TUNEL* from "*T*DT-mediated d*U*TP-biotin *n*ick-*e*nd *l*abeling" (Gavrieli *et al.*, 1992). This acronym is a misnomer because the double-stranded (ds) breaks are labeled rather than the single-stranded (ss) *nicks*. Of all the deoxynucleotides, BrdUTP appears to be the most advantageous to label DSBs, in terms of high sensitivity, low cost, and simplicity of the assay (Li and Darzynkiewicz, 1995) (Fig. 10). BrdU attached to DSBs (as poly-BrdU) is detected with an FITC-conjugated anti-BrdU antibody, the very same antibody that is used to detect BrdU incorporated during DNA replication (Fig. 1E). Poly-BrdU at the DSBs,

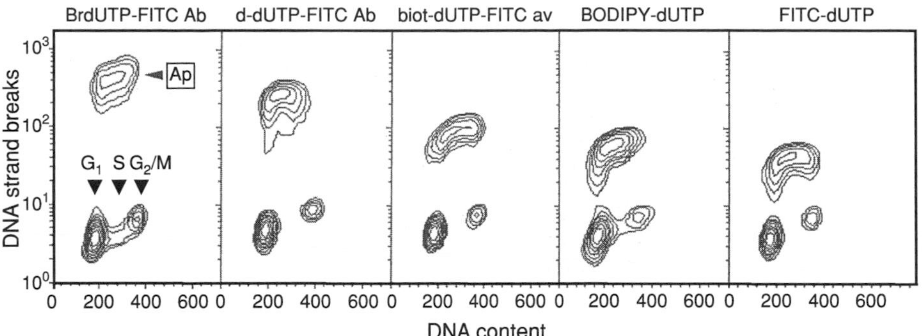

Fig. 10 Detection of DNA strand breaks ("TUNEL" assay) using different deoxynucleotides. Apoptosis of HL-60 cells was induced by camptothecin, which selectively targets S-phase cells (Li and Darzynkiewicz, 1995). In the reaction catalyzed by terminal deoxynucleotidyltransferase, the DNA strand breaks were labeled (from right to left): directly with fluorescein isothiocyanate (FITC)– and BODIPY-conjugated dUTP or indirectly with biotinylated (biot) dUTP detected by FITC–avidin, digoxigenin-conjugated dUTP detected by digoxigenin–FITC antibody, and with BrdUTP, detected by FITC–BrdU antibody. The highest resolution provides labeling with BrdUTP.

however, is accessible to the antibody without a need for DNA denaturation, which otherwise is required to detect the precursor incorporated during DNA replication.

The detection of DSBs by this assay requires prefixation of cells with a cross-linking agent such as formaldehyde. Unlike ethanol, formaldehyde prevents the extraction of small DNA fragments. Labeling DSBs in this procedure, which uses fluorescein-conjugated anti-BrdU antibody, can be combined with staining of DNA with the fluorochrome of another color (PI, red fluorescence). Cytometry of cells that are differentially stained for DNA strand breaks and for DNA allows one to distinguish apoptotic from nonapoptotic cell subpopulations and reveal the cell cycle distribution in these subpopulations, respectively (Gorczyca *et al.*, 1992, 1993b) (Fig. 10). However, late apoptotic cells may have diminished DNA content because of prior shedding of apoptotic bodies (which may contain nuclear fragments) or because there is such extensive DNA fragmentation that small DNA fragments cannot be retained in the cell even after fixation with formaldehyde. Such late apoptotic cells, thus, may have sub-G_1 DNA content. Several types of kits are commercially available that use either directly fluorochrome-tagged triphosphodeoxynucleotides or BrdUTP and BrdU antibody.

IX. Susceptibility of DNA to Denaturation

One of characteristic features of apoptosis associated with condensation of chromatin is increased susceptibility of DNA *in situ* to denaturation, which can be induced by heat, acid, or chemical denaturants such as formamide (Darzynkiewicz,

1994; Hotz *et al.*, 1992, 1994). DNA denaturation ("melting" of double helix) can be detected taking advantage of the metachromatic properties of acridine orange (AO). Namely, under proper equilibrium conditions, AO can differentially stain ds versus ss nuclei acids: When AO intercalates into dsDNA, it emits green fluorescence, but when it binds to ssDNA, it fluoresces in red wavelength (Darzynkiewicz, 1994). In this assay, the cells are briefly prefixed, treated with RNase to remove RNA (which otherwise also stains with AO), treated with heat or acid to denature DNA, and stained with AO at low pH (to prevent DNA renaturation). The propensity of DNA to denature in condensed chromatin of apoptotic cells is much greater that in interphase chromatin of nonapoptotic cells (Fig. 11). Hence, under the DNA-denaturing conditions, apoptotic cells have a much larger fraction of ssDNA that stains with AO by emitting red fluorescence and a smaller fraction of dsDNA that stains green than the nonapoptotic cells in the interphase (Fig. 11). Interestingly, apoptotic-like changes measured by both DNA denaturability and DNA fragmentation occur during defective spermatogenesis that leads to abnormal infertile spermatozoa (Gorczyca *et al.*, 1993c). DNA denaturation can also be detected immunocytochemically with an antibody reactive with ssDNA (Frankfurt, 1999).

The increased susceptibility of DNA to denaturation during apoptosis is also apparent if no internucleosomal DNA fragmentation occurs, for example, when

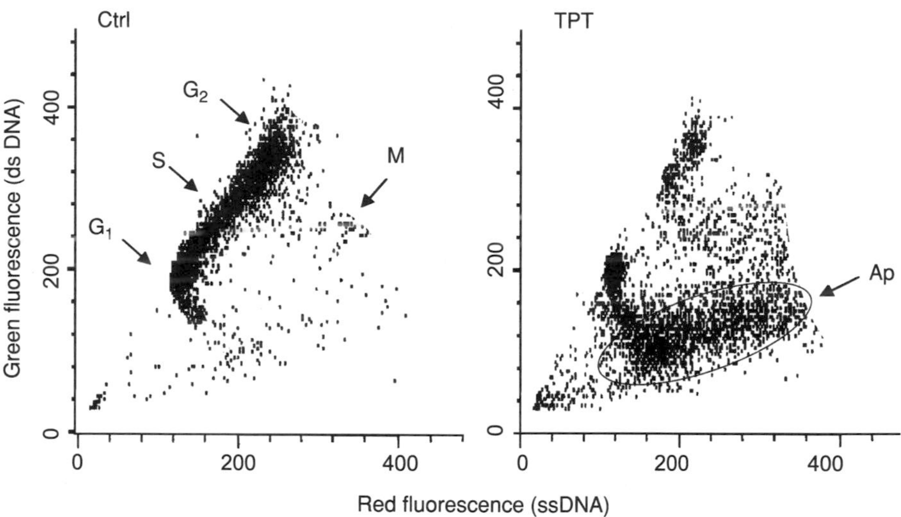

Fig. 11 Increased susceptibility of DNA in apoptotic cells to denaturation. Untreated (Ctrl) and topotecan (TPT)-treated cells were fixed and their DNA was partially denatured by acid. The cells were then stained with acridine orange (OA) to differentially stain double-stranded (ds) and single-stranded (ss) DNA to fluoresce green and red, respectively (Darzynkiewicz, 1994). In control culture, only M-phase and few apoptotic cells (spontaneous apoptosis) have increased red and decreased green fluorescence compared to most interphase cells. After treatment with TPT, most S-phase cells underwent apoptosis (Ap); they are characterized by increased red and decreased green fluorescence.

DNA fragmentation is prevented by inhibition of serine proteases (Bruno *et al.*, 1992; Hara *et al.*, 1996). The methods based on DNA denaturation, therefore, can be used when the methods that detect DNA fragmentation fail. It should be warned, however, that because DNA susceptibility to denaturation is a characteristic feature of condensed chromatin, the methods that rely on DNA denaturation cannot distinguish between apoptotic and mitotic cells because the latter also have highly condensed chromatin (Dobrucki and Darzynkiewicz, 2001). This limitation is of particular importance when one is studying cell death induced by mitotic poisons, that is, when mitotic cells are undergoing apoptosis.

X. Activation of "Tissue" Transglutaminase

TGases transfer acyl groups from (α-carboxamide moiety of protein glutamines to the ε-amino groups on the same or another protein molecule. These Ca^+-activated enzymes, thus, cause intramolecular and/or intermolecular protein cross-linking, thereby altering their physicochemical structure and mechanical properties (Aeschlimann and Paulsson, 1994). The polyamines spermidine and spermine, as well as their precursor putrescine, are natural substrates for TGases. Extensive protein cross-linking takes place during keratinization and during apoptosis. The ubiquitous TGase 2 tissue transglutaminase (tTGase) was identified as the enzyme that is activated during apoptosis (Fesus *et al.*, 1987; Melino and Piacentini, 1998). Because protein cross-linking lowers its solubility, it is presumed that activation of TGase 2 during apoptosis prevents release of soluble and immunogenic proteins from dying cells and thereby decreases a possibility of induction of autoimmune reaction. Furthermore, protein packaging into apoptotic bodies may be facilitated when they remain in solid state rather than in solution. The additional role of TGase 2, namely as one of the "executioner enzymes" during apoptosis, has been considered, but no clear evidence has been provided. Although earlier studies on TGase 2-null mice reported no major differences in terms of the ability of cells to undergo apoptosis, a recent study noted defects in the clearing of apoptotic bodies and the remains of apoptotic cells by macrophages in these mice (Szondy *et al.*, 2003).

It has been noticed that when live nonapoptotic cells were subjected to treatment with solutions of nonionic detergents, the lysis of their plasma membrane and release of the content of cytoplasm was complete, resulting in a suspension of isolated nuclei. In contrast, apoptotic cells resisted the detergent treatment; their cytoplasmic protein remained insoluble and attached to the nucleus in the form of a shell-like envelope (Grabarek *et al.*, 2002). Based on this observation, a simple method was developed that is based on the propensity of cross-linked protein in apoptotic cells to withstand treatment with detergents. After cell staining with DAPI and the protein fluorochrome sulforhodamine, 101 apoptotic cells were distinguished from the nuclei isolated from nonapoptotic cells by abundance of protein (red fluorescence) (Grabarek *et al.*, 2002). In addition, bivariate gating analysis of cellular DNA and protein content made it possible to reveal the cell cycle distribution separately, for the population of cells with cross-linked protein (activated TGase 2) and for the population of cells that did not show protein cross-linking (Grabarek *et al.*, 2002).

Another approach to detect activation of TGase 2 uses the fluorescein-tagged cadaverine, which upon TGase 2 activation becomes attached to the respective protein substrates within the cell (Lajemi *et al.*, 1998). This assay was adapted to flow cytometry and combined with concurrent analysis of cellular DNA content (Grabarek *et al.*, 2002). Like the detergent-based assay, this method is simple and offers good distinction of cells with activated TGase 2.

Interestingly, apoptosis can progress with or without TGase 2 activation even under identical conditions of induction of apoptosis. In fact, upon exposure to DNA topoisomerase inhibitors or hyperthermia, TGase 2 activation was detected in a fraction of apoptotic cells, whereas other apoptotic cells in the same cultures did not demonstrate the activation (Grabarek *et al.*, 2002).

XI. Measuring Incidence of Apoptosis: Which Method to Choose?

Which method is optimal depends on the cell type, the inducer of apoptosis, the desired information, and technical restrictions. For example, the need for sample transportation or prolonged storage before the cell staining and measurement requires cell fixation and, therefore, excludes the use of "supravital" methods such as the assays of plasma membrane integrity (exclusion of PI), membrane phospholipids asymmetry (annexin V binding), or mitochondrial potential. The need to obtain information on the cell cycle phase specificity of apoptosis eliminates the methods that do not allow one to concurrently measure cellular DNA content. Technical restrictions of the cytometer, such as a single laser and only few fluorescence detectors, limit the choice of fluorochromes and possibilities of multiparameter analysis.

Positive identification of apoptotic cells is not always straightforward. It was proposed to define apoptosis as a *caspase-mediated cell death* (Blagosklonny, 2000). Given the aforementioned, caspase activation would be the most specific marker of apoptosis (Shi, 2002). As discussed earlier in this chapter, there are numerous methods to detect activation of caspases, either directly (e.g., by the FLICA methodology, immunocytochemical detection of the activated caspase epitope, FRET, or fluorogenic substrate assays) or indirectly (e.g., by detection of PARP cleavage product p89). However, there are examples of cell death that very much resemble apoptosis, such as death mediated by AIF, and yet there is no evidence of caspase activation (Joza *et al.*, 2001). In such instances, other markers have to be used. Extensive DNA fragmentation also is considered a specific marker of apoptosis. The number of DNA strand breaks in apoptotic cells is so large that intensity of their labeling in the TUNEL reaction ensures their positive identification and discriminates them from the cells that underwent primary necrosis (Gorczyca *et al.*, 1992).

The internucleosomal DNA fragmentation also induces extensive phosphorylation of histone H2AX. The immunocytochemical detection of phosphorylated H2AX, especially when combined with the detection of caspase activation, can serve as a specific marker of apoptotic cells as well (Huang *et al.*, 2004). However, in the instances of apoptosis when internucleosomal DNA degradation does not occur (Catchpoole and Stewart, 1993; Cohen *et al.*, 1992; Collins *et al.*, 1992; Knapp *et al.*, 1999; Ormerod *et al.*, 1994), the intensity of cell labeling in TUNEL assay may be inadequate to identify apoptotic cells.

Annexin V binding is still another marker considered to be specific to apoptosis. However, as mentioned earlier, the use of the annexin V binding assay may generate false-positive data when the plasma membrane is damaged during cell preparation or storage, leading to the loss of asymmetry in distribution of phosphatidylserine across the membrane of otherwise live nonapoptotic cells. On the other hand, cells may undergo rapid DNA, nuclear, and cellular disintegration before exposing phosphatidylserine (King *et al.*, 2000). Furthermore, macrophages and other cells engulfing apoptotic bodies may also be positive in the annexin V assay (Marguet *et al.*, 1999).

Apoptosis can be recognized with greater confidence when the cells are subjected to several assays probing different apoptotic attributes (Hotz *et al.*, 1994). For example, the assay of plasma membrane integrity (exclusion of PI) and annexin V binding, combined with analysis of PARP cleavage or DNA fragmentation, may provide a more definitive assessment of the mode of cell death than can be determined when each of these methods is used alone. FLICA assay of caspase activation combined with the detection of mitochondrial potential by MitoTracker Red CMRos (Pozarowski *et al.*, 2003) (Fig. 4) is a simple procedure that reveals two quite early events of apoptosis and provides an early time window for the detection of apoptotic process. However, in light of the findings that the collapse of mitochondrial potential is not always a prerequisite for release of cytochrome *c*, AIF, and initiation of other apoptotic events (Finucane *et al.*, 1999; Li *et al.*, 2000; Scorrano *et al.*, 1999), one should also be cautious in interpreting the data of the assays using mitochondrial probes.

There are many other difficulties and potential pitfalls in analysis of apoptosis. It is quite difficult, for example, to distinguish late apoptosis (so-called "necrotic phase" of apoptosis) from primary necrosis by cytometry. In both cases the integrity of the plasma membrane is lost and the cells cannot exclude cationic dyes such as PI or trypan blue. As discussed earlier, often apoptosis is atypical, lacking a particular specific feature such as DNA fragmentation, release of cytochrome *c*, or activation of a given caspase. The method that identifies apoptotic cells based on such a feature cannot be used in these instances.

The estimation of incidence of apoptosis is also biased by selective loss of apoptotic cells during sample preparation and staining. Cell harvesting by trypsinization, mechanical or enzymatic cell disaggregation from tissues, and centrifugations each may lead to preferential loss of apoptotic cells. Because of cell dehydration (cytoplasm and chromatin condensation), the density of apoptotic cells is markedly increased and the volume decreased; this change should be taken under consideration, for example, when isolating cells by density (Ficoll-hypaque, Percoll) gradient centrifugations, or elutriation. All these potential problems, difficulties, and pitfalls, as well as the means to avoid them, are discussed at length elsewhere (Darzynkiewicz *et al.*, 2001).

The loss of cell surface antigens during apoptosis (Philippe *et al.*, 1997; Potter *et al.*, 1999) creates a particular problem in the studies aimed to identify the lineage of apoptotic cells by their immunophenotyping (Schmid *et al.*, 1992). The loss often occurs at early stages of apoptosis and, as in the loss of the intracellular antigens (Gorczyca *et al.*, 1993a), appears to be selective depending on the antigen and the inducer of apoptosis. Therefore, regardless of the apoptotic marker being used, attempts to identify lineage of apoptotic cells by immunophenotyping may be a subject of significant error.

Apoptosis was originally defined as a specific mode of cell death based on characteristic morphological changes (Kerr *et al.*, 1972). These changes are still considered the "gold standard" for identification of apoptosis. Therefore, although particular markers are used to assess incidence of apoptosis by flow cytometry, the mode of cell death should always be confirmed by inspection of cells by light or electron microscopy. The ambiguity regarding the mechanism of cell death often is resolved by analysis of cell morphology.

The characteristic morphological features of apoptotic cells are listed in Table I. The most specific (and generic to apoptosis) of these changes is chromatin condensation. The chromatin of apoptotic cells is "smooth" (structureless), lacking the structural framework that otherwise characterizes cell nucleus. Because of the condensation, chromatin shows strong hyperchromicity with any of the DNA-specific dyes. It should be noted, however, that after fixation with precipitating fixatives such as ethanol, the fragmented DNA may be extracted during the staining procedure and then hyperchromicity is not apparent. Prefixation with formaldehyde, the fixative that cross-links DNA and protein, prevents the extraction.

In addition to chromatin condensation, however, other changes are less generic to apoptosis and may not always be apparent. For example, although nuclear fragmentation is commonly observed during apoptosis of hematopoietic lineage cells, it may not occur during apoptosis of some epithelial- or fibroblast-lineage cells. Inhibitors of serine proteases prevent nuclear fragmentation but do not preclude chromatin condensation. Likewise, cell shrinkage, at least early during apoptosis, is not a universal marker of the apoptotic mode of cell death. It appears that cell shrinkage accompanies TGase 2 activation and its lack may indicate that apoptosis occurs in the absence of TGase 2 activation.

Cytospinning of live cells followed by their fixation and staining on slides generally provides a good preparation to assess apoptosis. The cells are then flat and their morphology is easy to appraise. In contrast, when the cells are initially fixed and stained in suspension, and then transferred to slides and analyzed under the microscope, their morphology is obscured by the unfavorable geometry; being spherical and thick, they require confocal microscopy to reveal details such as early signs of apoptotic chromatin condensation.

Differential staining of cellular DNA and protein of cells on slides with DAPI and sulforhodamine 101, respectively, is simple and rapid and provides good morphological resolution of apoptosis and necrosis (Darzynkiewicz *et al.*, 1997). Also attractive is a combination of 7-amino actinomycin D and FITC. Staining with AO allows one to probe susceptibility of DNA to denaturation, which reports chromatin condensation, one of the hallmarks of apoptosis (Hotz *et al.*, 1994).

Regardless of which cytometric method is used to identify apoptosis, it is advisable to confirm the mode of cell death by inspecting cells by light or electron microscopy. By having many attributes of both flow cytometry and image analysis, the LSC (Darzynkiewicz *et al.*, 1999; Kamentsky, 2001) appears to be an optimal instrument for analysis of apoptosis (Bedner *et al.*, 1999).

Acknowledgments

Supported by NCI grant CA RO1 28 704. M. A. K was supported by Ramsay Health Care Study Fellowship.

References

Alnemri, E. S., Livingston, D. I., Nicholson, D. W., Salvesen, G., Thornberry, N. A., Wong, W. W., and Yuan, J. (1996). Human ICE/CED-4 protease nomenclature. *Cell* **87,** 171–173.

Arends, M. J., Morris, R. G., and Wyllie, A. H. (1990). Apoptosis: The role of endonuclease. *Am. J. Pathol.* **136,** 593–608.

Aeschlimann, D., and Paulsson, M. (1994). Transglutaminases: Protein crosslinking enzymes in tissues and body fluids. *Tromb. Haemost.* **71,** 402–415.

Bedner, E., Li, X., Gorczyca, W., Melamed, M. R., and Darzynkiewicz, Z. (1999). Analysis of apoptosis by laser scanning cytometry. *Cytometry* **35,** 181–195.

Bedner, E., Smolewski, P., Amstad, P., and Darzynkiewicz, Z. (2000). Activation of caspases measured in situ by binding of fluorochrome-labeled inhibitors of caspases (FLICA): Correlation with DNA fragmentation. *Exp. Cell Res.* **259,** 308–313.

Blagosklonny, M. V. (2000). Cell death beyond apoptosis. *Leukemia* **14,** 1502–1508.

Bruno, S., Del Bino, G., Lassota, P., Giaretti, W., and Darzynkiewicz, Z. (1992). Inhibitors of proteases prevent endonucleolysis accompanying apoptotic death of HL-60 leukemic cells and normal thymocytes. *Leukemia* **6,** 1113–1120.

Budihardjo, I., Oliver, H., Lutter, M., and Luo, X. (1999). Biochemical pathways of caspase activation during apoptosis. *Annu. Rev. Cell Dev. Biol.* **15,** 269–290.

Catchpoole, D. R., and Stewart, B. W. (1993). Etoposide-induced cytotoxicity in two human T-cell leukemic lines. Delayed loss of membrane permeability rather than DNA fragmentation as an indicator of programmed cell death. *Cancer Res.* **53,** 4287–4296.

Chinnaiyan, A. M., O'Rourke, K., Lane, B. R., and Dixit, V. M. (1997). Interaction of CED-4 with CED-3 and CED-9: A molecular framework for cell death. *Science* **175,** 1122–1126.

Cohen, G. M., Su, X.-M., Snowden, R. T., Dinsdale, D., and Skilleter, D. N. (1992). Key morphological features of apoptosis may occur in the absence of internucleosomal DNA fragmentation. *Biochem. J.* **286,** 331–334.

Collins, R. J., Harmon, B. V., Gobe, G. C., and Kerr, J. F. R. (1992). Internucleosomal DNA cleavage should not be the sole criterion for identifying apoptosis. *Int. J. Radiat. Biol.* **61,** 451–453.

Cossarizza, A., Kalashnikova, G., Grassilli, E., Chiappelli, F., Salvioli, S., Capri, M., Barbieri, D., Troiano, L., Monti, D., and Franceschi, C. (1994). Mitochondrial modifications during rat thymocyte apoptosis: A study at a single cell level. *Exp. Cell Res.* **214,** 323–330.

Cossarizza, A., and Salvioli, S. (2001). Analysis of mitochondria during cell death. *Meth. Cell Biol.* **63,** 467–486.

Danial, N. N., and Korsmeyer, S. J. (2004). Cell death: Critical control points. *Cell* **116,** 205–219.

Darzynkiewicz, Z. (1994). Acid-induced denaturation of DNA *in situ* as a probe of chromatin structure. *Methods Cell Biol.* **41,** 527–542.

Darzynkiewicz, Z., Bedner, E., and Traganos, D. (2001). Difficulties and pitfalls in analysis of apoptosis. *Methods Cell Biol.* **63,** 527–546.

Darzynkiewicz, Z., Bedner, E., Li, X., Gorczyca, W., and Melamed, M. R. (1999). Laser scanning cytometry. A new instrumentation with many applications. *Exp. Cell Res.* **249,** 1–12.

Darzynkiewicz, Z., Bruno, S., Del Bino, G., Gorczyca, W., Hotz, M. A., Lassota, P., and Traganos, F. (1992). Features of apoptotic cells measured by flow cytometry. *Cytometry* **13,** 795–808.

Darzynkiewicz, Z., Juan, G., Li, X., Murakami, T., and Traganos, F. (1997). Cytometry in cell necrobiology: Analysis of apoptosis and accidental cell death (necrosis). *Cytometry* **27,** 1–20.

Darzynkiewicz, Z., Li, X., and Gong, J. (1994). Assays of cell viability. Discrimination of cells dying by apoptosis. *Methods Cell Biol.* **41,** 16–39.

Darzynkiewicz, Z., Staiano-Coico, L., and Melamed, M. R. (1981). Increased mitochondrial uptake of rhodamine 123 during lymphocyte stimulation. *Proc. Natl. Acad. Sci. USA* **78,** 2383–2387.

Darzynkiewicz, Z., Traganos, F., Staiano-Coico, L., Kapuscinski, J., and Melamed, M. R. (1982). Interactions of rhodamine 123 with living cells studied by flow cytometry. *Cancer Res.* **42,** 799–806.

Del Bino, G., Darzynkiewicz, Z., Degraef, C., Mosselmans, R., and Galand, P. (1999). Comparison of methods based on annexin V binding, DNA content or TUNEL for evaluating cell death in HL-60 and adherent MCF-7 cells. *Cell Prolif.* **32,** 25–37.

de Murcia, G., and Menissier-de Murcia, J. M. (1994). Poly(ADP-ribose) polymerase: A molecular nick sensor. *Trends Biochem. Sci.* **19,** 72–176.

Deng, Y., Lin, Y., and Wu, X. (2002). TRAIL-induced apoptosis requires Bax-dependent mitochondrial release of Smac/DIABLO. *Genes Dev.* **16,** 33–45.

Dobrucki, J., and Darzynkiewicz, Z. (2001). Chromatin condensation and sensitivity of DNA *in situ* to denaturation during cell cycle and apoptosis. A confocal microscopy study. *Micron* **32,** 645–652.

Earnshaw, W. C., Martins, L. M., and Kaufmann, S. H. (1999). Mammalian caspases: Structure, activation, substrates, and functions during apoptosis. *Annu. Rev. Biochem.* **68,** 383–424.

Enari, M., Talanian, R. V., Wong, W. W., and Nagata, S. (1996). Sequential activation of ICE-like and CPP32-like proteases during Fas-mediated apoptosis. *Nature* **380,** 723–726.

Endersen, P. C., Prytz, P. S., and Aarbakke, J. (1995). A new flow cytometric method for discrimination of apoptotic cells and detection of their cell cycle specificity through staining F-actin and DNA. *Cytometry* **20,** 162–171.

Esposti, M. D. (2002). The roles of Bid. *Apoptosis* **7,** 433–440.

Fadok, V. A., Voelker, D. R., Campbell, P. A., Cohen, J. J., Bratton, D. L., and Henson, P. M. (1992). Exposure of phosphatidylserine on the surface of apoptotic lymphocytes triggers specific recognition and removal by macrophages. *J. Immunol.* **148,** 22–29.

Fadeel, B., Orrenius, S., and Zhivotovsky, B. (2000). The most unkindest cut of all: On the multiple roles of mammalian caspases. *Leukemia* **14,** 1514–1525.

Fesus, L., Thomazy, V., and Falus, A. (1987). Induction and activation of tissue transglutaminase during programmed cell death. *FEBS Lett.* **224,** 104–108.

Finucane, D. M., Waterhouse, N. J., Amaranto-Mendes, G. P., Cotter, T. G., and Green, D. R. (1999). Collapse of the inner mitochondrial transmembrane potential is not required for apoptosis of HL-60 cells. *Exp. Cell Res.* **251,** 166–174.

Frankfurt, O. S. (1999). Immunoassay for single-stranded DNA in apoptotic cells. *Methods Mol. Biol.* **113,** 621–631.

Frisch, S. M., and Screaton, R. A. (2001). Anoikis mechanisms. *Curr. Opin. Cell Biol.* **13,** 555–562.

Gajewski, T. F., and Thompson, C. B. (1996). Apoptosis meets signal transduction: Elimination of BAD influence. *Cell* **87,** 589–592.

Gavrieli, Y., Sherman, Y., and Ben-Sasson, A. (1992). Identification of programmed cell death *in situ via* specific labeling of nuclear DNA fragmentation. *J. Cell Biol.* **119,** 493–501.

Gong, J., Traganos, F., and Darzynkiewicz, Z. (1994). A selective procedure for DNA extraction from apoptotic cells applicable for gel electrophoresis and flow cytometry. *Anal. Biochem.* **218,** 314–319.

Gorczyca, W., Ardelt, B., and Darzynkiewicz, Z. (1993a). Different rates of degradation of the proliferation-associated nuclear and nucleolar antigens during apoptosis of HL-60 cells induced by DNA topoisomerase inhibitors. *Int. J. Oncol.* **3,** 627–634.

Gorczyca, W., Bigman, K., Mittelman, A., Ahmed, T., Gong, J., Melamed, M. R., and Darzynkiewicz, Z. (1993b). Induction of DNA strand breaks associated with apoptosis during treatment of leukemias. *Leukemia* **7,** 659–670.

Gorczyca, W., Bruno, S., Darzynkiewicz, R., Gong, J., and Darzynkiewicz, Z. (1992). DNA strand breaks occurring during apoptosis: Their early *in situ* detection by the terminal deoxynucleotidyl transferase and nick translation assays and prevention by serine protease inhibitors. *Int. J. Oncol.* **1,** 639–648.

Gorczyca, W., Traganos, F., Jesionowska, H., and Darzynkiewicz, Z. (1993c). Presence of DNA strand breaks and increased sensitivity of DNA *in situ* to denaturation in abnormal human sperm. Analogy to apoptosis of somatic cells. *Exp. Cell Res.* **207,** 202–205.

Gorman, A. M., Hirt, U. A., Zhivotovsky, B., Orrenius, S., and Ceccatelli, S. (1999). Application of a fluorometric assay to detect caspase activity in thymus tissue undergoing apoptosis *in vivo*. *J. Immunol. Methods* **226,** 43–48.

Grabarek, J., Ardelt, B., Kunicki, J., and Darzynkiewicz, Z. (2002). Detection of *in situ* activation of transglutaminase during apoptosis: Correlation with the cell cycle phase by multiparameter flow- and laser scanning-cytometry. *Cytometry* **49,** 83–89.

Halicka, H. D., Bedner, E., and Darzynkiewicz, Z. (2000). Segregation of RNA and separate packaging of DNA and RNA in apoptotic bodies during apoptosis. *Exp. Cell Res.* **260,** 248–256.

Hara, S., Halicka, H. D., Bruno, S., Gong, J., Traganos, F., and Darzynkiewicz, Z. (1996). Effect of protease inhibitors on early events of apoptosis. *Exp. Cell Res.* **223,** 372–384.

Haugland, R. P. (2003). Handbook of fluorescent probes and research products, 9th ed, Eugene, Oregon, Molecular Probes.

Hedley, D. W., and McCulloch, E. A. (1996). Generation of oxygen intermediates after treatment of blasts of acute myeloblastic leukemia with cytosine arabinoside: Role of bcl-2. *Leukemia* **10,** 1143–1149.

Hotz, M. A., Gong, J., Traganos, F., and Darzynkiewicz, Z. (1994). Flow cytometric detection of apoptosis: Comparison of the assays of *in situ* DNA degradation and chromatin changes. *Cytometry* **15,** 237–244.

Hotz, M. A., Traganos, F., and Darzynkiewicz, Z. (1992). Changes in nuclear chromatin related to apoptosis or necrosis induced by the DNA topoisomerase II inhibitor fostriecin in MOLT-4 and HL-60 cells are revealed by altered DNA sensitivity to denaturation. *Exp. Cell Res.* **201,** 184–191.

Huang, X., Okafuji, M., Traganos, F., and Darzynkiewicz, Z. (2004). Cytometric assessment of antitumor-drug induced DNA damage versus apoptosis-associated DNA fragmentation detected by histone H2AX phosphorylation. *Cytometry* **58A,** 99–110.

Hug, H., Los, M., Hirt, W., and Debatin, K. M. (1999). Rhodamine 110-linked amino acids and peptides as substrates to measure caspase activity upon apoptosis induction in intact cells. *Biochemistry* **38,** 13906–13911.

Johnson, L. V., Walsh, M. L., and Chen, L. B. (1980). Localization of mitochondria in living cells with rhodamine 123. *Proc. Natl. Acad. Sci. USA* **77,** 990–994.

Jones, J., Heim, R., Hare, E., Stack, J., and Pollok, B. A. (2000). Development and application of a GFP-FRET intracellular caspase assay for drug screening. *J. Biolmol. Screen.* **5,** 307–318.

Joza, N., Susin, S. A., Gaugas, E., Stanford, W. L., Cho, S. K., Li, C. Y. I., Sasaki, T., Elia, A. J., Cheng, H.-Y. M., Ravagnan, L., Ferri, K. F., Zamzani, N., Wakeham, A., Hakem, R., Yoshida, H., Kong, Y.-Y., Mak, T. W., Zuniga-Pflucker, J. C., Kroemer, G., and Penninger, J. M. (2001). Essential role of the mitochondrial apoptosis-inducing factor in programmed cell death. *Nature* **410,** 549–554.

Kamentsky, L. A. (2001). Laser scanning cytometer. *Methods Cell Biol.* **63,** 51–87.

Kaufmann, S. H., Desnoyers, S., Ottaviano, Y., Davidson, N. E., and Poirier, G. G. (1993). Specific proteolytic cleavage of poly(ADP-ribose) polymerase: An early marker of chemotherapy-induced apoptosis. *Cancer Res.* **53,** 3976–3985.

Kaufman, R. J. (1999). Stress signaling from the lumen of endoplasmic reticulum: Coordination of gene transcriptional and translational controls. *Genes Dev.* **13,** 1211–1233.

Keiji, J. F., Bell-Prince, C., and Steinkamp, J. A. (2000). Staining of mitochondrial membranes with 10-nonyl acridine orange MitFluor Green, and Mito Tracker Green is affected by mitochondrial membrane potential altering drugs. *Cytometry* **39,** 203–210.

Kerr, J. F. R., Wyllie, A. H., and Curie, A. R. (1972). Apoptosis: A basic biological phenomenon with wide-ranging implications in tissue kinetics. *Br. J. Cancer* **26,** 239–257.

Kim, J. S., He, L., and Lemasters, J. J. (2003). Mitochondrial permeability transition: A common pathway to necrosis and apoptosis. *Biochem. Biophys. Res. Commun.* **304,** 463–470.

King, M. A., Radicchi-Mastroianni, M. A., and Wells, J. V. (2000). There is substantial nuclear and cellular disintegration before detectable phosphatidylserine exposure during the camptothecin-induced apoptosis of HL-60 cells. *Cytometry* **40,** 10–18.

Kluck, R. M., Bossy-Wetzel, E., Green, D. R., and Newmeyer, D. D. (1997). The release of cytochrome c from mitochondria: A primary site for Bcl-2 regulation of apoptosis. *Science* **275,** 1132–1136.

Koopman, G., Reutelingsperger, C. P. M., Kuijten, G. A. M., Keehnen, R. M. J., Pals, S. T., and van Oers, M. H. J. (1994). Annexin V for flow cytometric detection of phosphatidylserine expression of B cells undergoing apoptosis. *Blood* **84,** 1415–1420.

Knapp, P. E., Bartlett, W. P., Williams, L. A., Yamada, M., Ikenaka, K., and Skoff, R. P. (1999). Programmed cell death without DNA fragmentation in the jimpy mouse: Secreted factors can enhance survival. *Cell Death Differ.* **6,** 136–145.

Kroemer, G. (1998). The mitochondrion as an integrator/coordinator of cell death pathways. *Cell Death Diff.* **5,** 547–548.

Lazebnik, Y. A., Kaufmann, S. H., Desnoyers, S., Poirier, G. G., and Earnshaw, W. C. (1994). Cleavage of poly(ADP-ribose) polymerase by proteinase with properties like ICE. *Nature* **371,** 346–347.

Lajemi, M., Demignot, S., and Adolphe, M. (1998). Detection and characterization, using fluoresceincadaverine, of amine acceptor substrates accessible to active transglutaminase expressed by rabbit articular chondrocytes. *Histochem. J.* **30,** 499–508.

Lee, B. W., Johnson, G. L., Hed, S. A., Darzynkiewicz, Z., Talhouk, J. W., and Mehrotra, S. (2003). DEVDase detection in intact apoptotic cells using the cell permeant fluorogenic substrate $(z\text{-DEVD})_2$–cresyl violet. *Biotechniques* **35,** 1080–1085.

Leist, M., and Jaattela, M. (2001). Four deaths and a funeral: From caspases to alternative mechanisms. *Nat. Rev. Mol. Cell. Biol.* **2,** 589–598.

Li, X., and Darzynkiewicz, Z. (1995). Labeling DNA strand breaks with BdrUTP. Detection of apoptosis and cell proliferation. *Cell Prolif.* **28,** 571–579.

Li, X., and Darzynkiewicz, Z. (2000). Cleavage of poly(ADP-ribose) polymerase measured *in situ* in individual cells: Relationship to DNA fragmentation and cell cycle position during apoptosis. *Exp. Cell Res.* **255,** 125–132.

Li, X., Du, L., and Darzynkiewicz, Z. (2000). Caspases are activated during apoptosis independent of dissipation of mitochondrial electrochemical potential. *Exp. Cell Res.* **257,** 290–297.

Li, X., Melamed, M. R., and Darzynkiewicz, Z. (1996). Detection of apoptosis and DNA replication by differential labeling of DNA strand breaks with fluorochromes of different color. *Exp. Cell Res.* **222,** 28–37.

Li, X., Traganos, F., Melamed, M. R., and Darzynkiewicz, Z. (1995). Single-step procedure for labeling DNA strand breaks with fluorescein- or BODIPY-conjugated deoxynucleotides: Detection of apoptosis and bromodeoxyuridine incorporation. *Cytometry* **20,** 172–180.

Liu, J., Bhalgat, M., Zhang, C., Diwu, Z., Hoyland, B., and Klaubert, D. H. (1999). Fluorescent molecular probes V: A sensitive caspase-3 substrate for fluorometric assays. *Bioor. Med. Chem. Lett.* **9,** 3231–3236.

Liu, X., Kim, C. N., Yang, J., Jemmerson, R., and Wang, X. (1996). Induction of apoptotic program in cell-free extracts: Requirements for dATP and cytochrome *c*. *Cell* **86,** 147–157.

Luo, K. Q., Yu, V. C., Pu, Y., and Chang, D. C. (2003). Measuring dynamics of caspase-8 activation in a single living HeLa cells during THF-α–induced apoptosis. *Biochem. Biophys. Res. Commun.* **304,** 217–222.

Majno, G., and Joris, I. (1995). Apoptosis, oncosis, and necrosis. An overview of cell death. *Am. J. Pathol.* **146,** 3–16.

Marguet, D., Luciani, M.-F., Moynault, A., Williamson, P., and Chimini, G. (1999). Engulfment of apoptotic cells involves the redistribution of membrane phosphatidylserine on phagocyte and pray. *Nat. Cell Biol.* **1,** 454–456.

McConkey, D. J., Nicotera, P., Hartzell, P., Bellomo, G., Wyllie, A. H., and Orrenius, S. (1998). Glucocorticoids activate a suicide process in thymocytes through an elevation of cytosolic Ca^{2+} concentration. *Arch. Biochem. Biophys.* **269,** 365–370.

Melino, E., and Piacentini, M. (1998). "Tissue" transglutaminase in cell death: A downstream or multifunctional upstream effector? *FEBS Lett.* **430,** 59–63.

Muchmore, S. W., Sattlet, M., Liang, H., Meadows, R. P., Harlan, J. E., Yoon, H. S., Nettelsheim, D., Chang, B. S., Thompson, C. B., Wong, S. L., Ng, S. L., and Fesik, S. W. (1996). X-ray and NMR structure of human Bcl-x_L, an inhibitor of programmed cell death. *Nature* **381,** 335–341.

Nagata, S. (2000). Apoptotic DNA fragmentation. *Exp. Cell Res.* **256,** 12–18.

Nicoletti, I., Migliorati, G., Pagliacci, M. C., Grignani, F., and Riccardi, C. (1991). A rapid and simple method for measuring thymocyte apoptosis by propidium iodide staining and flow cytometry. *J. Immunol. Methods* **139,** 271–280.

Oberhammer, F., Wilson, J. M., Dive, C., Morris, I. D., Hickman, J. A., Wakeling, A. E., Walker, P. R., and Sikorska, M. (1993). Apoptotic death in epithelial cells: Cleavage of DNA to 300 and/or 50 kb fragments prior to or in the absence of internucleosomal fragmentation. *EMBO J.* **12,** 3679–3684.

Ormerod, M. G. (1998). The study of apoptotic cells by flow cytometry. *Leukemia* **12,** 1013–1025.

Ormerod, M. G., Cheetham, F. P. M., and Sun, X.-M. (1995). Discrimination of apoptotic thymocytes by forward light scatter. *Cytometry* **21,** 300–304.

Ormerod, M. G., O'Neill, C. F., Robertson, D., and Harrap, K. R. (1994). Cisplatin induced apoptosis in a human ovarian carcinoma cell line without a concomitant internucleosomal degradation of DNA. *Exp. Cell Res.* **211,** 231–237.

Ormerod, M. G., Sun, X.-M., Snowden, R. T., Davies, R., Fearhead, H., and Cohen, G. M. (1993). Increased membrane permeability of apoptotic thymocytes: A flow cytometric study. *Cytometry* **14,** 595–602.

Petit, J. M., Ratinaud, M. H., Cordelli, E., Spano, M., and Julien, R. (1995). Mouse testis cell sorting according to DNA and mitochondrial changes during spermatogenesis. *Cytometry* **19,** 304–312.

Philippe, J., Louagie, H., Thierens, H., Vral, A., Cornelissen, M., and De Ridder, L. (1997). Quantification of apoptosis in lymphocyte subsets and effect of apoptosis on apparent expression of membrane antigens. *Cytometry* **29,** 242–249.

Piacentini, M., Fesus, I., Farrace, M. G., Ghibelli, L., Piredda, L., and Melino, G. (1995). The expression of "tissue" transglutaminase in two human cancer cell lines is related to the programmed cell death (apoptosis) *Eur. J. Cell Biol.* **54,** 246–254.

Poot, M., Gibson, L. L., and Singer, V. L. (1997). Detection of apoptosis in live cells by Mito Tracker Red CMXRos and SYTO dyes flow cytometry. *Cytometry* **27,** 358–364.

Potter, A., Kim, C., Golladon, K. A., and Rabinovitch, P. S. (1999). Apoptotic human lymphocytes have diminished CD4 and CD8 receptor expression. *Cell. Immunol.* **193,** 36–47.

Pozarowski, P., Grabarek, J., and Darzynkiewicz, Z. (2003a). Flow cytometry of apoptosis. *Curr. Protocols Cytom.* **7,** 1–23.

Pozarowski, P., Halicka, D., and Darzynkiewicz, Z. (2003b). Cell cycle effects and caspase-dependent and independent death of HL-60 and Jurkat cells treated with the inhibitor of NF-κB parthenolide. *Cell Cycle* **2,** 377–383.

Pozarowski, P., Huang, X., Halicka, D. H., Lee, B., Johnson, G., and Darzynkiewicz, Z. (2003c). Interactions of fluorochrome-labeled caspase inhibitors with apoptotic cells. A caution in data interpretation. *Cytometry* **55A,** 50–60.

Rao, R. V., Poksay, K. S., Castro-Obregon, S., Schilling, B., Row, R. H., del Rio, G., Gibson, B. W., Ellerby, H. M., and Bredesen, D. E. (2004). Molecular components of a cell death pathway activated by endoplasmic reticulum stress. *J. Biol. Chem.* **279,** 177–178.

Raipert, S., Bennion, G., Hickman, J., and Allen, T. D. (1999). Nucleolar segregation during apoptosis of hematopoietic stem cell line FDCP-Mix. *Cell Death Differ.* **6,** 334–341.

Ratinaud, M. H., Leprat, P., and Julien, R. (1988). *In situ* cytometric analysis of nonyl acridine orange-stained mitochondria from splenocytes. *Cytometry* **9,** 206–212.

Reed, J. C. (1995). Regulation of apoptosis by bcl-2 family of proteins and its role in cancer and chemoresistance. *Curr. Opin. Oncol.* **7,** 541–546.

Sallman, F. R., Bourassa, S., Saint-Cyr, J., and Poirier, G. G. (1997). Characterization of antibodies specific for the caspase cleavage site on poly(ADP-ribose) polymerase: Specific detection of apoptotic fragments and mapping of the necrotic fragments of poly(ADP-ribose) polymerase. *Biochem. Cell Biol.* **75,** 451–458.

Schmid, I., Krall, W. J., Uittenbogaart, C. H., Braun, J., and Giorgi, J. V. (1992). Dead cell discrimination with 7-amino-actinomycin D in combination with dual color immunofluorescence in single laser flow cytometry. *Cytometry* **13,** 204–208.

Searle, J., Kerr, J. F. R., and Bishop, C. J. (1982). Necrosis and apoptosis: Distinct modes of cell death with fundamentally different significance. *Pathol. Ann.* **17,** 229–259.

Scorrano, L., Petronilli, V., Colonna, R., Di Lisa, F., and Bernard, P. (1999). Cloromethyltetramethylrosamine (Mitotracker Orange) induces the mitochondrial permeability transition and inhibits respiratory complex I. Implications for the mechanism of cytochrome *c* release. *J. Biol. Chem.* **274,** 24657–24663.

Shapiro, H. (2003). Practical flow cytometry. 4th ed, Wiley-Liss, New York.

Sheikh, M. S., and Fornace, A. J., Jr. (2000). Death and decoy receptors and p53-mediated apoptosis. *Leukemia* **14,** 1509–1513.

Shi, Y. (2002). Mechanisms of caspases activation and inhibition during apoptosis. *Mol. Cell* **9,** 459–470.

Smolewski, P., Bedner, E., Du, L., Hsieh, T.-C., Wu, J. M., Phelps, D. J., and Darzynkiewicz, Z. (2001). Detection of caspase activation by fluorochrome-labeled inhibitors: Multiparameter analysis by laser scanning cytometry. *Cytometry* **44,** 73–82.

Smolewski, P., Grabarek, J., Lee, B. W., Johnson, G. L., and Darzynkiewicz, Z. (2002). Kinetics of HL-60 cell entry to apoptosis during treatment with TNF-α or camptothecin assayed by stathmo-apoptosis method. *Cytometry* **47,** 143–149.

Susin, S. A., Zamzani, N., Larochette, N., Dallaporta, B., Marzo, I., Brenner, C., Hirsch, T., Petit, P. X., Geuskens, M., and Kroemer, G. (1997). A cytofluorometric assay of nuclear apoptosis induced in a cell-free system: Application to ceramide-induced apoptosis. *Exp. Cell Res.* **236,** 397–403.

Swat, W., Ignatowicz, L., and Kisielow, P. (1981). Detection of apoptosis of immature $CD4^{+}8^{+}$ thymocytes by flow cytometry. *J. Immunol. Methods* **137,** 79–87.

Szondy, Z., Sarang, Z., Molnar, P., Nemeth, T., Piacentini, M., Mastroberardino, P. G., Falasca, L., Aeschlimann, D., Kovacs, J., Kiss, J., Szegezdi, E., Lakos, G., Rajnavolgyi, E., Birckbichler, P. J., Melino, G., and Fesus, L. (2003). Transglutaminase 2−/− mice reveal a phagocytosis-associated crosstalk between macrophages and apoptotic cells. *Proc. Natl. Acad. Sci. USA* **100,** 7812–7817.

Tawa, P., Tam, J., Cassady, R., Nicholson, D. W., and Xanthoudakis, S. (2001). Quantitative analysis of fluorescent caspase substrated cleavage in intact cells and identification of novel inhibitors of apoptosis. *Cell Death Differ.* **8,** 30–37.

Telford, W. G., Komoriya, A., and Packard, B. Z. (2002). Detection of localized caspase activity in early apoptotic cells by laser scanning cytometry. *Cytometry* **47,** 81–88.

Thornberry, N. A., Rano, T. A., Peterson, E. P., Rasper, D. M., Timkey, T., Garcia-Calvo, M., Houtzager, V. M., Nordstrom, P. A., Roy, S., Vaillancourt, J. P., Chapman, K. T., and Nicholson, D. W. (1997). A combinatorial approach defines specificities of members of the caspase family and granzyme B. Functional relationships established for key mediators of apoptosis. *J. Biol. Chem.* **272,** 17907–17911.

Umansky, S. R., Korol', B. R., and Nelipovich, P. A. (1981). *In vivo* DNA degradation in the thymocytes of gamma-irradiated or hydrocortisone-treated rats. *Biochim. Biophys. Acta* **655,** 281–290.

van Engeland, M., Nieland, L. J. W., Ramaekers, F. C. S., Schutte, B., and Reutelingsperger, P. M. (1998). Annexin V–affinity assay: A review on an apoptosis detection system based on phosphatidylserine exposure. *Cytometry* **31,** 1–9.

Van Noorden, C. J. F. (2001). The history of Z-VAD-FMK, a tool for understanding the significance of caspase inhibition. *Acta Histochem.* **103,** 241–251.

Vermes, I., Haanen, C., and Reutelingsperger, C. (2000). Flow cytometry of apoptotic cell death. *J. Immunol. Methods* **243,** 167–190.

Wang, P., Valentijn, A. J., Gilmore, A. P., and Streuli, C. H. (2003). Early events in the anoikis program in the absence of caspase activation. *J. Biol. Chem.* **278,** 19917–19927.

Weaver, V. M., Lach, B., Walker, P. R., and Sikorska, M. (1993). Role of proteolysis in apoptosis: Involvement of serine proteases in internucleosomal DNA fragmentation in immature lymphocytes. *Biochem. Cell Biol.* **71,** 488–500.

Xu, X., Gerard, A. L., Huang, B. C., Anderson, D. C., Payan, D. G., and Luo, Y. (1998). Detection of programmed cell death using fluorescence energy transfer. *Nucleic Acids Res.* **26,** 2034–2035.

Yang, J., Liu, X., Bhalla, K., Ibrado, A. M., Cai, J., Peng, T. I., Jones, D. P., and Wang, X. (1997). Prevention of apoptosis by Bcl-2: Release of cytochrome *c* from mitochondria blocked. *Science* **275,** 1129–1132.

Zamzani, N., Brenner, C., Marzo, I., Susin, S. A., and Kroemer, G. (1998). Subcellular and submitochondrial mode of action of Bcl-2-like oncoproteins. *Oncogene* **16,** 2265–2282.

Zamzani, N., Susin, S. A., Marchetti, P., Hirsch, T., Gomez-Monterey, L., Castedo, M., and Kroemer, G. (1996). Mitochondrial control of nuclear apoptosis. *J. Exp. Med.* **183,** 1533–1544.

Zorning, M., Hueber, A.-D., Baum, W., and Evan, G. (2001). Apoptosis regulators and their role in tumorigenesis. *Biochim. Biophys. Acta* **1551,** F1–F37.

Zou, H., Li, Y., Liu, X., and Wang, X. (1999). An APAF-1, cytochrome *c* multimeric complex is a functional apoptosome that activates procaspase-9. *J. Biol. Chem.* **274,** 11549–11556.

CHAPTER 13

Real-Time Analysis of Apoptosis *In Vivo*

Pui Lee and Mark S. Segal

Division of Nephrology, Hypertension, and Transplantation
Department of Medicine
University of Florida
Gainesville, Florida 32610

I. Introduction

Apoptosis, or programmed cell death, is an important regulator of physiological processes including growth, development, immunity, and homeostasis. In contrast to necrosis, in which injured cells release their contents into the environment leading to inflammation, apoptosis is characterized by the organized dismantling of cellular components without affecting adjacent cells (Raff, 1998; Vaux and Strasser, 1996). Mechanisms controlling programmed cell death are tightly regulated, involving an intricate balance between antiapoptotic and proapoptotic signals. Defects in these pathways can result in either overwhelming apoptosis, as manifested by neuronal loss in patients with Parkinson's disease, or diminished apoptosis, as in certain cancers (Ethell and Buhler, 2003; Younes and Aggarwall, 2003). Methods to visualize cells undergoing apoptosis in real time would greatly

0091-679X/04 $35.00

enhance our understanding and study of this process (Stennicke and Salvesen, 2000).

Various cytometric markers to detect apoptosis have been developed in the past decade. Morphologically, programmed cell death can be recognized early by binding of annexin V to externalized phosphatidyl serine (PS), a membrane lipid normally found almost exclusively in the inner leaflet of the plasma membrane (Zhang *et al.*, 1998a). Downstream events in apoptosis result in cleavage of chromosomal DNA into 180–200 bp fragments that can be detected by TUNEL (TdT-mediated deoxyuridine triphosphate nick-end labeling) (Allen *et al.*, 1997). Enzymatically, apoptosis is often assessed by the activity of a family of cysteine proteases known as *caspases*. Caspase activity is most commonly measured with fluorometric and colorimetric substrates (Zhang, 1998b). Immunoblots and enzyme-linked immunosorbent assays have also been successfully used to assess caspase activation (Allen *et al.*, 1997). Despite being the most widely used protocols today, these methods share a common drawback: They allow evaluation only of *in vivo* apoptotic activity at a given time. With few exceptions, tracing apoptosis in the same cell population over time is problematic because the aforementioned methods require lysis or fixation of target cells (Lee *et al.*, 2002).

Several markers capable of measuring *in vivo* apoptosis in real time have been engineered. Though based on the underlying principles of the traditional methods, these markers allow for the detection of apoptosis in the same cells for hours and even days, without the need for fixation or lysis. These real-time apoptosis markers can be classified into membrane and caspase probes. Membrane probes such as radiolabeled annexin V and superparamagnetic iron oxide (SPIO) labeling of the C_2 domain of synaptotagmin detect alteration of membrane lipids (Blankenberg *et al.*, 1999b; Zhao *et al.*, 2001). Caspase probes such as the ubiquitin-based marker, fluorescence resonance energy transfer (FRET) substrates, and xanthene dye fluorescent substrates all detect caspase activity (Harpur *et al.*, 2001; Lee *et al.*, 2002; Packard *et al.*, 1996).

II. *In Vivo* Membrane Markers

Externalization of PS on the plasma membrane early in apoptosis has been well documented in human and murine cells. Normally confined on the inner layer, PS is actively translocated to the outer layer during programmed cell death (Fig. 1A). This feature of apoptosis occurs before membrane bleb formation, nuclear condensation, or DNA fragmentation. The externalized PS can be recognized by annexin V, a 36-kd human protein with high specificity for PS and little affinity for other major membrane lipids (Allen *et al.*, 1997; Zhang *et al.*, 1998a). When conjugated to fluorescent molecules such as fluorescein isothiocyanate (FITC), annexin V has been effectively used to study apoptosis *in vitro* using fluorescent microscopy or fluorescent-activated cell sort (FACS) (Allen *et al.*, 1997; Zhang *et al.*, 1998a).

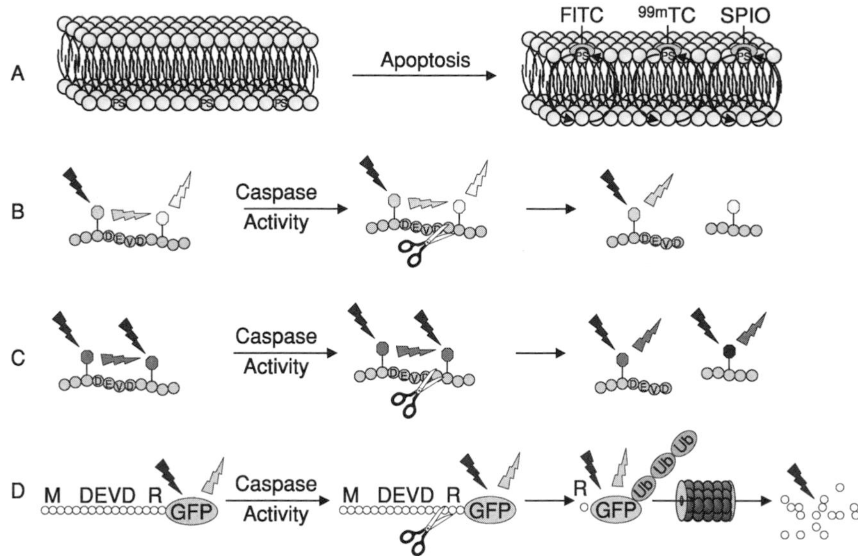

Fig. 1 Concepts behind real-time markers. (A) Annexin V or C_2 domain of recombinant synaptotagmin I. Phosphatidyl serine (PS) is normally found on the inner bilayer of the membrane. During apoptosis, the PS is flipped to the outer bilayer where it is accessible to annexin V or the C_2 domain of recombinant synaptotagmin I (blue donut) that is bound to one of several detector markers: fluorescein isothiocyanate (FITC) (detectable by fluorescence), technetium-99m (^{99m}Tc) (detectable by radionuclide imaging), or in the case of C_2, superparamagnetic iron oxide (SPIO) (detectable by magnetic resonance imaging). (B) Fluorescence resonance energy transfer (FRET) fluorophore peptide substrate. A donor and acceptor fluorophore are attached through a small peptide containing a caspase cleavage site, illustrated is the cleavage site of caspase-3, DEVD. When the FRET probe is excited with the optimal wavelength for the donor (cyan), acceptor fluorescence is emitted at the expense of donor fluorescence. When caspase activity is present, the FRET probe is cleaved, resulting in separation of donor and acceptor fluorophores. With this separation, there is increased emission of the donor fluorescence and decreased emission of acceptor fluorescence. (C) Rhodamine dimers. Two identical fluorophores are attached to a small peptide on either side of a caspase cleavage site. In this configuration, the identical fluorophores in the peptide substrate will quench each other. When caspase activity is present, the peptide is cleaved, separating the two fluorophores and leading to an increase in fluorescence. (D) Green fluorescent protein (GFP) substrate. The marker produced has a methionine as its N-terminal amino acid, leading to a GFP with a long half-life. When caspase activity is present, it cleaves after the DEVD site, producing a protein with an arginine as its N-terminal amino acid. The N-terminal arginine leads to rapid degradation of the GFP by the proteasome and decreased fluorescence. (See Color Insert.)

A. Radionuclide Markers

The functional properties of annexin V have also made it an attractive candidate in the development of a real-time, *in vivo* marker of apoptosis. Several groups have successfully demonstrated that radiolabeled annexin V can be used to detect apoptosis in living organisms by radionuclide imaging. One of the radioactive tags studied extensively for this purpose is technetium-99m (^{99m}Tc) (Blankenberg *et al.*,

1999b; Narula *et al.*, 2001; Ohtsuki *et al.*, 1999; Vriens *et al.*, 1998). In fact, ^{99m}Tc–annexin V is perhaps the most effective and most applicable real-time marker available (Fig. 1A).

Several means to label recombinant annexin V with ^{99m}Tc have been described. One method involves incubation of annexin V with a technetium diamide dimercaptide N_2S_2 chelate (TFP). The bifunctional TFP binds a lysine residue on annexin V on one moiety and ^{99m}Tc on the other (Kown *et al.*, 2001). Commercial kits to label annexin V using this method are available (Kasina *et al.*, 1991). An alternative protocol is to first make a derivative of annexin V with the bifunctional molecule hydrazinonicotinamide (HYNIC), followed by coupling with ^{99m}Tc (Blankenberg *et al.*, 1998). The latest and perhaps most efficient method to radiolabel annexin V uses succinic dihydrazide in the presence of tricine, nicotinic acid, PDTA, and stannous chloride (Subbarayan *et al.*, 2003). Although the last method reported the best yield, all three protocols produced stable protein conjugates without affecting the bioactivity of annexin V.

The *in vivo* applications of ^{99m}Tc–annexin V have been studied extensively in many animal models of apoptosis. Measuring uptake of the labeled marker via external radionuclear imaging, Blankenberg *et al.* (1998) tested ^{99m}Tc–annexin V in three mouse models: anti-Fas antibody–induced fulminant hepatic apoptosis, acute rejection of heterotopic PVG cardiac allografts, and cyclophosphamide treatment of transplanted 38C13 murine B-cell lymphomas. Data from all three models show clear and specific localization to regions of programmed cell death, where the uptake of ^{99m}Tc–annexin V was twofold to sixfold above control. Immunohistochemical staining for annexin V further confirmed binding of the radiolabeled marker. Results from these and other animal models, such as experimental rheumatoid arthritis and dexamethasone-induced thymic apoptosis, overwhelmingly showed ^{99m}Tc–annexin V to be a safe and effective way to measure *in vivo* programmed cell death (Brauer, 2003; Ohtsuki *et al.*, 1999; Post *et al.*, 2002).

In humans, ^{99m}Tc–annexin V imaging has been tested as a noninvasive tool to detect apoptosis associated with transplant rejection and certain types of cancer. A significant correlation between focal uptakes of ^{99m}Tc–annexin V and the severity of cardiac allograft rejection, as evaluated by clinical grading or immunohistochemical staining of biopsy specimens, has been described (Kown *et al.*, 2001; Narula *et al.*, 2001). In a study of patients with head and neck cancer, quantitative ^{99m}Tc–annexin V uptake showed good correlation with the number of apoptotic cells detected by TUNEL in biopsy samples (van de Wiele *et al.*, 2003). Moreover, increased tumor uptake of ^{99m}Tc–annexin V after chemotherapy has been associated with better clinical response and improved prognosis (Green and Steinmetz, 2002). Although the introduction of a radiolabeled exogenous protein can be considered invasive, the half-life of the protein conjugate is short and no adverse events have been reported from these studies (Blankenberg *et al.*, 1999a; Green and Steinmetz, 2002; Kemerink *et al.*, 2003). Taken together, the ability of ^{99m}Tc–annexin V to detect apoptosis can be applied not only as a diagnostic tool but also as a potential early prognostic indicator for the effectiveness of cancer treatments.

There are several limitations to the application of ^{99m}Tc–annexin V. Similar to annexin V used *in vitro*, ^{99m}Tc–annexin V is not specific for apoptotic cells because necrotic cells will also bind this marker (Zhang *et al.*, 1998a). Thus, a positive ^{99m}Tc–annexin V signal can indicate apoptotic cells, as a result of chemotherapy, or naturally occurring necrotic cells within a growing tumor. The accumulation of ^{99m}Tc–annexin V in organs such as the liver and kidneys may also interfere with accurate assessment of apoptosis occurring in those organs (Kemerink *et al.*, 2003). Finally, the relatively low resolution of radionuclear imaging of the marker limits *in vivo* applications of ^{99m}Tc–annexin V (Zhao *et al.*, 2001).

B. Magnetic Resonance Markers

Another frontier in the search for a real-time *in vivo* marker of apoptosis involves magnetic resonance technologies including magnetic resonance spectroscopy (MRS) and magnetic resonance imaging (MRI). MRI reflects the special distribution of proton nuclei within a tissue, whereas MRS contrasts the composition of magnetically active nuclei such as ^{1}H and ^{31}P in different chemical groups. Both methods have been shown to recognize signature biochemical changes related to apoptosis. The progress in developing MRS- and MRI-based programmed cell death markers has been extensively reviewed (Brauer, 2003). Despite their success in *in vitro* studies, few magnetic resonance–based markers have proceeded to *in vivo* testing because of either low sensitivity and poor spatial resolution or the inability to detect apoptosis in the early stages (Zhao *et al.*, 2001).

Zhao *et al.* (2001) described a noninvasive marker capable of detecting apoptosis *in vivo* at a high resolution using MRI. This marker, C_2–SPIO, is constructed by labeling the C_2 domain of recombinant synaptotagmin I with SPIO nanoparticles, an effective contrast agent for T2-weighted MRI. Similar to annexin V, the C_2 domain has high affinity for PS. The labeling protocol involves activation SPIO particles with periodate oxidation, followed by conjugation with recombinant rat C_2 domain. The conjugation reaction is stabilized using $NaBH_4$ reduction and the conjugated product is ultrafiltrated before analysis. Using the same principle, annexin V has been conjugated with SPIO (Schellenberger *et al.*, 2002).

C_2–SPIO was studied *in vivo* using mice inoculated with EL4 lymphoma cells and treated with a combination of cyclophosphamide and etoposide (Zhao *et al.*, 2001). Apoptosis of tumor cells after chemotherapy was evaluated by T2-weighted MRI. Results showed progressive and persistent changes in signal intensity in well-defined areas within the tumors. These areas corresponded to the distribution of apoptotic cells identified by the TUNEL method after tumor excision. More importantly, the resolution achieved with this method is 10-fold to 30-fold that obtained with ^{99m}Tc–annexin V imaging.

Because both SPIO-conjugate markers and ^{99m}Tc–annexin V function by binding to PS, they share the disadvantage of recognizing necrotic cells. SPIO conjugates also have a tendency to accumulate in the macrophages of the liver and spleen, making the assessment of apoptosis within these organs difficult (Brauer, 2003). The safety and ability of the SPIO conjugate to detect apoptosis in larger animals and humans have not been studied.

III. A New Generation of Caspase Markers

Programmed cell death is often assessed enzymatically by measuring the activity of caspases, a family of cysteine proteases central to many apoptotic pathways. Once activated from their precursors, caspases propagate and amplify the signal cascade, ultimately resulting in the dismantling of specific cellular targets (i.e., structural proteins and DNA repair enzymes) and further activation of destructive enzymes (DNases and other caspases) (Salvesen and Dixit, 1997; Stennicke *et al.*, 1998).

The most common way to measure caspase activity employs substrates made by conjugating a quencher compound via a specific caspase cleavage site to a fluorophore (i.e., 7-amino-4-trifluoromethyl coumarin [AFC]) or chromophore (*p*-nitroanilide [pNA]) (Zhang, 1998b). Caspase cleavage of the substrates frees the fluorophore or chromophore from the quencher molecule and caspase activity is reflected by the increase in fluorescence or light emission, respectively. For example, the activity of caspase-3 can be measured with the blue light–emitting substrate DEVD–AFC ($\lambda_{max} = 400$ nm). After proteolytic removal of DEVD by caspase-3, the remaining AFC molecule emits a yellow-green fluorescence ($\lambda_{max} = 505$ nm). Commercial kits for caspase detection using similar markers are widely available. However, the requirement of lysing a large number of cells limits assessment of caspase activity to a single time and renders the study of individual cells impossible. To solve these problems, three fluorescence-based markers capable of detecting real-time caspase activity in intact cells have been described.

A. Fluorescence Resonance Energy Transfer Caspase Probes

The phenomenon of FRET is often applied to measure protein interactions at a molecular level using a pair of fluorophores. Under this principle, a net transfer of energy takes place when the emission spectrum of one fluorophore (the donor) significantly overlaps the excitation spectrum of another (the acceptor), given that the two molecules are within 100 angstroms. FRET results in a net loss in the energy of emitted fluorescence from the donor and a net gain in the emission energy of the acceptor. Two enhanced green fluorescent protein (EGFP)–derived mutants, cyan fluorescent protein (CFP) and yellow fluorescent protein (YFP), are the donor–acceptor fluorophore pairs commonly used to construct caspase probes (Harpur *et al.*, 2001; Karpova *et al.*, 2003; Xu *et al.*, 1998).

In the intact probe, CFP (excitation 436 ± 10 nm, emission 480 ± 20 nm) and YFP (excitation 500 ± 20 nm, emission 535 ± 30 nm) are connected by a short linker peptide containing a caspase cleavage sequence (Karpova *et al.*, 2003). The close proximity of the fluorophores leads to quenching of CFP fluorescence and a simultaneous increase in YFP fluorescence emission, as predicted by the FRET phenomenon. With introduction of active caspases and cleavage of the linker region, the energy transfer diminishes as CFP and YFP separate in a rapid and efficient process (Fig. 1B.). The increase in CFP/YFP emission ratio (toward 1:1) due to FRET signal disruption serves as an indicator of caspase activity. The structure of the intact probe does not seem to interfere with caspase activity because similar kinetics are seen with endogenous caspase substrates such as poly-adenosine diphosphate–ribose polymerase (PARP) (Rehm *et al.*, 2002).

FRET caspase probes are constructed by designing plasmids encoding CFP and YFP with an appropriate intervening linker sequence. Transfection of the purified plasmids and expression of recombinant probes have been most efficient in HeLa and Cos-7 cells (Brophy *et al.*, 2002; Rehm *et al.*, 2002). Analysis of real-time caspase activity in transfected cells in response to different apoptotic stimuli such as staurosporin, tumor necrosis factor-α, and etoposide can then be performed in live intact cells. The change in fluorescence in cell populations can be measured by fluorometry, whereas confocal microscopy or flow cytometry can be used to study individual cells. By further selecting for stably transfected cells, apoptotic activity in the same cells can potentially be traced over a long period. A number of groups have successfully incorporated this method into their research of apoptosis and caspases (Brophy *et al.*, 2002; Luo *et al.*, 2001, 2003; Mahajan *et al.*, 1998; Takemoto *et al.*, 2003).

The main drawback of FRET caspase probes is their reliance on transfection, which restricts their application to *in vitro* studies. Also, the phenomenon of donor quenching has not been studied extensively and FRET efficiency may be difficult to calculate. Moreover, the efficiency of FRET probes also depends on the specific caspase being evaluated. For example, caspase-6 and caspase-2 are incapable of cleaving their respective FRET substrates (Rehm *et al.*, 2002).

B. Xanthene Dye–Based Caspase Probe

Aside from FRET substrates, understanding of fluorophore interactions has led to the development of another real-time caspase marker. This marker exploits the spectral characteristics of xanthene dye dimerization described by the exciton theory. When brought in close proximity, xanthene dyes from H-type dimers characterized by a blue shift in the absorption spectrum and a loss of fluorescence (Packard *et al.*, 1996, 1997).

Using this principle, Packard *et al.* (1996) have constructed protease probes by linking two identical fluorophores (i.e., 5′,6′-carboxytetramethylrhodamine) with a peptide containing a protease recognition sequence. The peptide linker can be synthesized using manual N-α-9-fluorenylmethyloxycarbonyl (Fmoc) chemistry then derivitized with the fluorophores in an organic solvent. Upon proteolytic cleavage of the recognition sequence, the H-type dimer formed by the fluorophores is disrupted, leading to the appearance of fluorescence (Fig. 1C). Thus, activity of the protease is reflected by the increased fluorescence and significant absorption changes.

A real-time marker for caspase-3 has been constructed using this system and commercialized by OncoImmunin, Incorporated under the trade name of PhiPhiLux (short for fluorescent-fluorescent light) (OncoImmunin, 2003). Available with green or red fluorescence, this marker incorporates the caspase-3 cleavage sequence DEVD in the bifluorophore-derivitized peptide. Cell membrane permeability of PhiPhiLux is advantageous because it allows the study of apoptosis in intact cells without the need for transfection. Analysis with flow cytometry or fluorescent microscopy can be performed after brief incubation of target cells with the marker. *In vitro* application of PhiPhiLux has been described in a number of cell types including JURKAT, T lymphocyte, oocytes, and thymocytes. Some data have also suggested the superiority in detection sensitivity of this method compared to annexin V. Similar markers for other members of the caspase family will be available soon.

Applied directly as membrane-permeable substrates, xanthene dye–based markers eliminate the need for transfection. However, their limited stability is a major drawback because tracing caspase activity in the same cells beyond a few hours is not feasible.

C. Ubiquitin-Based Marker

Functioning independent of fluorophore interactions, the latest real-time caspase marker exploits properties of ubiquitination and cellular protein degradation. With ubiquitination, the cellular enzyme ubiquitin hydrolase cleaves co-translationally after ubiquitin without regard to the subsequent amino acid, which then becomes the N-terminus of the remaining protein (Dantuma *et al.*, 2000a,b). The degradation rate of a protein is dependent on its N-terminal amino acid, as described by the N-end rule. For example, proteins with N-terminal methionine (M) are degraded via the proteosomal pathway at a rate at least 100-fold slower than proteins with an N-terminal arginine (R) (Bachmair *et al.*, 1986; Varshavsky, 1996).

Applying the principles of ubiquitination and the N-end rule, Lee *et al.* (2002) described a method that detects caspase activity by measuring the increase in EGFP stability. Upon transfection and expression of the marker plasmid, ubiquitin hydrolase co-translationally cleaves after the ubiquitin, leaving R-LETD-M-EGFP as the remaining protein. As predicted by the N-end rule, the N-terminal arginine directs this protein to rapid proteosomal degradation. However, in the presence of active caspases, the caspase recognition sequence LETD is cleaved and the remaining EGFP acquires stability with the N-terminal methionine. The increase in stability and EGFP fluorescence serves as the basis for the measurement of caspase activity (Fig. 1D). Alternatively, by simply switching the locations of the methionine and arginine residues adjacent to LETD, caspase activity can be assessed by a decrease in fluorescence.

Via fluorometry and flow cytometry, this marker accurately depicts an increase in caspase activity in cultured (human embryonic kidney) 293T cells and Cos-7 cells after the introduction of active caspase-8 (Lee *et al.*, 2002). Cleavage of the marker can be confirmed by immunoblotting for EGFP. Endogenous activity was also detected in nonapoptotic cells, perhaps as the result of baseline proteolytic activity of procaspases. The inducible nature of the marker plasmid allows real-time assessment of apoptosis in the same cells over a long period. Moreover, substitution of EGFP to a fluorescent protein with different spectral properties and replacement of LETD with the cleavage sequence of other caspases provides the potential for real-time colocalization of the different players involved in apoptosis.

In addition to the requirement for transfection, a major disadvantage of this marker is its complex kinetics involving simultaneous expression by the plasmid, cleavage by caspases, and degradation via the proteasome. In addition, the utilization of the proteosomal pathways may be problematic because these pathways have been implicated in many steps of apoptosis.

IV. Conclusion

Apoptosis plays an important role in many physiological processes. Defects in apoptotic pathways can result in various pathological conditions ranging from neuronal-loss Parkinson's disease to diminished apoptosis in certain cancers. The wide implications of program cell death have increased the demand for accurate methods to assess this cellular process. Although many protocols are available to detect apoptosis morphologically or enzymatically, most of them allow evaluation only of *in vivo* apoptotic activity at a given time, and lysis or fixation of a target cell is almost always required. Several markers capable of detecting apoptosis in real time have been engineered based on the underlying principles of traditional methods. These markers are effective in assessing *in vitro* programmed cell death in the same cells over time. More importantly, a few of these markers have shown great promise in *in vivo* mouse and human studies. Although each of the current markers has specific drawbacks, they all have paved the way in the development of real-time markers of this critical physiological process.

References

Allen, R. T., Hunter, W. J., 3rd, and Agrawal, D. K. (1997). Morphological and biochemical characterization and analysis of apoptosis. *J. Pharmacol. Toxicol. Methods* **37,** 215–228.

Bachmair, A., Finley, D., and Varshavsky, A. (1986). *In vivo* half-life of a protein is a function of its amino-terminal residue. *Science* **234,** 179–186.

Blankenberg, F. G., Katsikis, P. D., Tait, J. F., Davis, R. E., Naumovski, L., Ohtsuki, K., Kopiwoda, S., Abrams, M. J., Darkes, M., Robbins, R. C., Maecker, H. T., and Strauss, H. W. (1998). *In vivo* detection and imaging of phosphatidylserine expression during programmed cell death. *Proc. Natl. Acad. Sci. USA* **95,** 6349–6354.

Blankenberg, F., Ohtsuki, K., and Strauss, H. W. (1999a). Dying a thousand deaths. Radionuclide imaging of apoptosis. *Q. J. Nucl. Med.* **43,** 170–176.

Blankenberg, F. G., Katsikis, P. D., Tait, J. F., Davis, R. E., Naumovski, L., Ohtsuki, K., Kopiwoda, S., Abrams, M. J., and Strauss, H. W. (1999b). Imaging of apoptosis (programmed cell death) with ^{99m}Tc annexin V. *J. Nucl. Med.* **40,** 184–191.

Brauer, M. (2003). *In vivo* monitoring of apoptosis. *Prog. Neuropsychopharmacol. Biol. Psychiatry* **27,** 323–331.

Brophy, V. A., Tavare, J. M., and Rivett, A. J. (2002). Treatment of COS-7 cells with proteasome inhibitors or gamma-interferon reduces the increase in caspase 3 activity associated with staurosporine-induced apoptosis. *Arch. Biochem. Biophys.* **397,** 199–205.

Dantuma, N. P., Heessen, S., Lindsten, K., Jellne, M., and Masucci, M. G. (2000a). Inhibition of proteasomal degradation by the gly-Ala repeat of Epstein–Barr virus is influenced by the length of the repeat and the strength of the degradation signal. *Proc. Natl. Acad. Sci. USA* **97,** 8381–8385.

Dantuma, N. P., Lindsten, K., Glas, R., Jellne, M., and Masucci, M. G. (2000b). Short-lived green fluorescent proteins for quantifying ubiquitin/proteasome-dependent proteolysis in living cells. *Nat. Biotechnol.* **18,** 538–543.

Ethell, D. W., and Buhler, L. A. (2003). Fas ligand–mediated apoptosis in degenerative disorders of the brain. *J. Clin. Immunol.* **23,** 363–370.

Green, A. M., and Steinmetz, N. D. (2002). Monitoring apoptosis in real time. *Cancer J.* **8,** 82–92.

Harpur, A. G., Wouters, F. S., and Bastiaens, P. I. (2001). Imaging FRET between spectrally similar GFP molecules in single cells. *Nat. Biotechnol.* **19,** 167–169.

Karpova, T. S., Baumann, C. T., He, L., Wu, X., Grammer, A., Lipsky, P., Hager, G. L., and McNally, J. G. (2003). Fluorescence resonance energy transfer from cyan to yellow fluorescent protein detected by acceptor photobleaching using confocal microscopy and a single laser. *J. Microsc.* **209,** 56–70.

Kasina, S., Rao, T. N., Srinivasan, A., Sanderson, J. A., Fitzner, J. N., Reno, J. M., Beaumier, P. L., and Fritzberg, A. R. (1991). Development and biologic evaluation of a kit for preformed chelate technetium-99m radiolabeling of an antibody Fab fragment using a diamide dimercaptide chelating agent. *J. Nucl. Med.* **32,** 1445–1451.

Kemerink, G. J., Liu, X., Kieffer, D., Ceyssens, S., Mortelmans, L., Verbruggen, A. M., Steinmetz, N. D., Vanderheyden, J. L., Green, A. M., and Verbeke, K. (2003). Safety, biodistribution, and dosimetry of ^{99m}Tc-HYNIC-annexin V, a novel human recombinant annexin V for human application. *J. Nucl. Med.* **44,** 947–952.

Kown, M. H., Strauss, H. W., Blankenberg, F. G., Berry, G. J., Stafford-Cecil, S., Tait, J. F., Goris, M. L., and Robbins, R. C. (2001). *In vivo* imaging of acute cardiac rejection in human patients using (99m)technetium labeled annexin V. *Am. J. Transplant.* **1,** 270–277.

Lee, P., Beem, E., and Segal, M. S. (2002). Marker for real-time analysis of caspase activity in intact cells. *Biotechniques* **33,** 1284–1287, 1289–1291.

Luo, K. Q., Yu, V. C., Pu, Y., and Chang, D. C. (2001). Application of the fluorescence resonance energy transfer method for studying the dynamics of caspase-3 activation during UV-induced apoptosis in living HeLa cells. *Biochem. Biophys. Res. Commun.* **283,** 1054–1060.

Luo, K. Q., Yu, V. C., Pu, Y., and Chang, D. C. (2003). Measuring dynamics of caspase-8 activation in a single living HeLa cell during TNFalpha-induced apoptosis. *Biochem. Biophys. Res. Commun.* **304,** 217–222.

Mahajan, N. P., Linder, K., Berry, G., Gordon, G. W., Heim, R., and Herman, B. (1998). Bcl-2 and Bax interactions in mitochondria probed with green fluorescent protein and fluorescence resonance energy transfer. *Nat. Biotechnol.* **16,** 547–552.

Narula, J., Acio, E. R., Narula, N., Samuels, L. E., Fyfe, B., Wood, D., Fitzpatrick, J. M., Raghunath, P. N., Tomaszewski, J. E., Kelly, C., Steinmetz, N., Green, A., Tait, J. F., Leppo, J., Blankenberg, F. G., Jain, D., and Strauss, H. W. (2001). Annexin-V imaging for noninvasive detection of cardiac allograft rejection. *Nat. Med.* **7,** 1347–1352.

Ohtsuki, K., Akashi, K., Aoka, Y., Blankenberg, F. G., Kopiwoda, S., Tait, J. F., and Strauss, H. W. (1999). Technetium-99m HYNIC-annexin V: A potential radiopharmaceutical for the *in-vivo* detection of apoptosis. *Eur. J. Nucl. Med.* **26,** 1251–1258.

OncoImmunin, I. (2003). PhiPhiLux product information. Available at: www.phiphilux.com.

Packard, B. Z., Toptygin, D. D., Komoriya, A., and Brand, L. (1996). Profluorescent protease substrates: Intramolecular dimers described by the exciton model. *Proc. Natl. Acad. Sci. USA* **93,** 11640–11645.

Packard, B. Z., Toptygin, D. D., Komoriya, A., and Brand, L. (1997). Design of profluorescent protease substrates guided by exciton theory. *Methods Enzymol.* **278,** 15–23.

Post, A. M., Katsikis, P. D., Tait, J. F., Geaghan, S. M., Strauss, H. W., and Blankenberg, F. G. (2002). Imaging cell death with radiolabeled annexin V in an experimental model of rheumatoid arthritis. *J. Nucl. Med.* **43,** 1359–1365.

Raff, M. (1998). Cell suicide for beginners. *Nature* **396,** 119–122.

Rehm, M., Dussmann, H., Janicke, R. U., Tavare, J. M., Kogel, D., and Prehn, J. H. (2002). Single-cell fluorescence resonance energy transfer analysis demonstrates that caspase activation during apoptosis is a rapid process. Role of caspase-3. *J. Biol. Chem.* **277,** 24506–24514.

Salvesen, G. S., and Dixit, V. M. (1997). Caspases: Intracellular signaling by proteolysis. *Cell* **91,** 443–446.

Schellenberger, E. A., Hogemann, D., Josephson, L., and Weissleder, R. (2002). Annexin V-CLIO: A nanoparticle for detecting apoptosis by MRI. *Acad. Radiol.* **9**(Suppl. 2), S310–S311.

Stennicke, H. R., Jurgensmeier, J. M., Shin, H., Deveraux, Q., Wolf, B. B., Yang, X., Zhou, Q., Ellerby, H. M., Ellerby, L. M., Bredesen, D., Green, D. R., Reed, J. C., Froelich, C. J., and Salvesen, G. S. (1998). Pro-caspase-3 is a major physiologic target of caspase-8. *J. Biol. Chem.* **273,** 27084–27090.

Stennicke, H. R., and Salvesen, G. S. (2000). Caspase assays. *Methods Enzymol.* **322,** 91–100.

Subbarayan, M., Hafeli, U. O., Feyes, D. K., Unnithan, J., Emancipator, S. N., and Mukhtar, H. (2003). A simplified method for preparation of ^{99m}Tc-annexin V and its biologic evaluation for *in vivo* imaging of apoptosis after photodynamic therapy. *J. Nucl. Med.* **44,** 650–656.

Takemoto, K., Nagai, T., Miyawaki, A., and Miura, M. (2003). Spatio-temporal activation of caspase revealed by indicator that is insensitive to environmental effects. *J. Cell Biol.* **160,** 235–243.

van de Wiele, C., Lahorte, C., Vermeersch, H., Loose, D., Mervillie, K., Steinmetz, N. D., Vanderheyden, J. L., Cuvelier, C. A., Slegers, G., and Dierck, R. A. (2003). Quantitative tumor apoptosis imaging using technetium-99m-HYNIC annexin V single photon emission computed tomography. *J. Clin. Oncol.* **21,** 3483–3487.

Varshavsky, A. (1996). The N-end rule: Functions, mysteries, uses. *Proc. Natl. Acad. Sci. USA* **93,** 12142–12149.

Vaux, D. L., and Strasser, A. (1996). The molecular biology of apoptosis. *Proc. Natl. Acad. Sci. USA* **93,** 2239–2244.

Vriens, P. W., Blankenberg, F. G., Stoot, J. H., Ohtsuki, K., Berry, G. J., Tait, J. F., Strauss, H. W., and Robbins, R. C. (1998). The use of technetium Tc 99m annexin V for *in vivo* imaging of apoptosis during cardiac allograft rejection. *J. Thorac. Cardiovasc. Surg.* **116,** 844–853.

Xu, X., Gerard, A. L., Huang, B. C., Anderson, D. C., Payan, D. G., and Luo, Y. (1998). Detection of programmed cell death using fluorescence energy transfer. *Nucl. Acids Res.* **26,** 2034–2035.

Younes, A., and Aggarwall, B. B. (2003). Clinical implications of the tumor necrosis factor family in benign and malignant hematologic disorders. *Cancer* **98,** 458–467.

Zhang, G., Gurtu, V., Ma, J., and Kain, S. R. (1998a). Sensitive detection of apoptosis using enhanced color variants of annexin conjugates. *In* "Apoptosis detection and assay methods" (J. L. C. Zhu, ed.). Eaton Publishing, Palo Alto, Calif.

Zhang, G. Gurtu, V., Spencer, C., Ma, J., and Kain, S. R. (1998b). Detection of caspase activity associated with apoptosis using fluorimetric and colorimetric methods. *In* "Apoptosis detection and assay methods" (J. L. C. Zhu, ed.). Eaton Publishing, Palo Alto, Calif.

Zhao, M., Beauregard, D. A., Loizou, L., Davletov, B., and Brindle, K. M. (2001). Non-invasive detection of apoptosis using magnetic resonance imaging and a targeted contrast agent. *Nat. Med.* **7,** 1241–1244.

CHAPTER 14

Detection of DNA Damage in Individual Cells by Analysis of Histone H2AX Phosphorylation

Peggy L. Olive

Department of Medical Biophysics
British Columbia Cancer Research Centre
Vancouver, British Columbia, Canada V5Z 1L3

I. Introduction

Histone proteins are essential components of a dynamic nucleosomal structure that undergoes various chemical modifications associated with chromatin packaging, transcription, replication, and repair. Rogakou *et al.* (1998, 1999) reported that rapid phosphorylation of a specific serine at the C-terminal end of a minor

METHODS IN CELL BIOLOGY, VOL. 75

0091-679X/04 $35.00

nucleosomal histone protein, H2AX, occurred only at sites surrounding DNA double-stranded breaks. This very early event preceded the actions of most repair enzymes involved in homologous recombination and nonhomologous end-joining of these breaks. Subsequently, serine-139–phosphorylated H2AX, designated γH2AX, was found to co-localize with certain repair enzymes at sites of DNA damage (Celeste *et al.*, 2002, 2003; Paull *et al.*, 2000). Mice lacking H2AX were radiation sensitive, immune deficient, and growth retarded (Celeste *et al.*, 2002). Cells from these mice were also radiosensitive and chromosomally unstable (Bassing *et al.*, 2002). In mice deficient in the p53 tumor suppressor, loss of even a single H2AX allele decreased latency of development of lymphomas and other tumor types (Celeste *et al.*, 2003).

Mechanistically, members of the PI3 kinase family, primarily ATM and ATR, have been implicated in phosphorylation of H2AX (Burma *et al.*, 2001; Paull *et al.*, 2000; Redon *et al.*, 2002). Immediately after irradiation, microscopically visible spots, or "foci," begin to form, and within an hour, each focus contains thousands of γH2AX molecules covering about 2 Mb of DNA flanking the break. What is perhaps most exciting is that an individual DNA double-stranded break can now be visualized using antibody staining and fluorescence microscopy (Rothkamm and Lobrich, 2003; Sedelnikova *et al.*, 2002). In comparison, previous methods used to detect DNA double-stranded breaks, such as neutral filter elution, pulsed field gel electrophoresis, or neutral comet assay, had a detection limit of about 40–100 breaks/cell.

A model proposed by Redon *et al.* (2002) suggested that PI3 kinases like ATM are attracted to the DNA double-stranded break and phosphorylate H2AX molecules as they progress away from the break. The change in chromatin structure appears to activate ATM that can then phosphorylate H2AX (Bakkenist and Kastan, 2003). Then γH2AX or perhaps ATM itself (Andegeko *et al.*, 2001) attracts DNA repair enzymes, and once the break is rejoined, the kinase dissociated and phosphatases removed the phosphate group(s) of H2AX. Loss of γH2AX has been coupled to the rejoining of double-stranded breaks and the actions of protein phosphatase 1 (Nazarov *et al.*, 2003; Siino *et al.*, 2002). A modification to this model suggested that γH2AX may alter the efficiency or accuracy of repair, perhaps through interactions with intermediary proteins in the repair process and/or through chromatin structure modifications (Downs and Jackson, 2003). Greater sensitivity to killing by ionizing radiation has been associated with a slower rate of loss of γH2AX and a higher proportion of cells retaining γH2AX foci (MacPhail *et al.*, 2003a; Rothkamm and Lobrich, 2003; Taneja *et al.*, 2004).

II. Considerations in the Use of γH2AX as a Measure of Double-Stranded Breaks

The number of γH2AX foci formed in a cell appears to be directly related to the number of DNA double-stranded breaks (Sedelnikova *et al.*, 2002). Linear dose–response relationships can be generated by counting individual foci or by measuring γH2AX intensity using flow cytometry (MacPhail *et al.*, 2003a; Rothkamm and Lobrich, 2003). However, the level of the substrate H2AX is variable between different cell types and is dependent on DNA content (MacPhail *et al.*, 2003b; Rogakou *et al.*, 1998). In addition, the amount of γH2AX that is formed is dependent on the kinase and phosphatase activity of the cell (Paull *et al.*, 2000). Therefore, it is important to appreciate that the signal one sees using *flow cytometry* cannot be directly converted to the number of double-stranded breaks per cell. Moreover, the background level of γH2AX is cell-type–dependent and cell-phase–dependent; levels of γH2AX in G_o/G_1 phase cells are about three times lower than cells in S and G_2 phases (MacPhail *et al.*, 2003b). Conversely, damaged mitotic cells and some apoptotic cells express very high levels of this molecule (Rogakou *et al.*, 1999, 2000). Although γH2AX foci are described as covering 2 Mb surrounding a break, chromatin conformation has also been shown to influence foci sizes. Mitotic cells or cells treated with agents that alter chromatin structure (e.g., 0.5 M NaCl) develop much larger foci after irradiation (Reitsema *et al.*, 2004).

The sensitivity for detecting DNA double-stranded breaks using γH2AX is greatest using microscopic analysis of individual foci in intact nuclei. In theory, one double-stranded break can be detected in a single cell or even fractions of breaks within populations of damaged cells (Rothkamm and Lobrich, 2003). However, because S-phase cells can also demonstrate foci at sites of "background" damage, the greatest sensitivity for detecting radiation- or drug-induced damage is achieved using non-proliferating cells. Some transformed cell types, such as M059J human glioma cells, also show high background levels of foci, which reduces sensitivity for detecting induced double-stranded breaks (Paull *et al.*, 2000).

Table I compares γH2AX expression with a single-cell gel electrophoresis method, the comet assay—another method that can detect DNA damage in individual cells (Olive, 2002). Although the comet assay is more versatile because it can be modified to detect many more types of DNA damage, it is relatively insensitive for detecting double-stranded breaks (Olive *et al.*, 1991). It is, however, a more direct measure of double-stranded breaks that is not influenced by kinase and phosphatase activity or by variations in the amount of substrate available for phosphorylation, so damage can be directly compared between different cell types. The comet assay is, therefore, useful in conjunction with γH2AX detection.

Phosphorylation of γH2AX has been identified by binding of antibodies to phosphorylated serine-139 at the C-terminus of histone H2AX. Both monoclonal and polyclonal antibodies are commercially available, and detection is possible via immunoblotting, immunocytochemistry, immunohistochemistry, or flow cytometry. Immunoblotting has confirmed the specificity of the antibody for serine-139 histone H2AX, and it has been used to show that accessibility of chromatin to the antibody is not the cause for the cell cycle variation in γH2AX detected by flow cytometry (MacPhail *et al.*, 2003a,b). Immunocytochemistry is the most sensitive method for detecting DNA damage because individual foci can be recognized and counted. Information on foci size, number, and intensity is also possible with

Table I
Comparison between γH2AX Antibody Binding and Comet Assay for Detection of DNA Damage

Endpoint	γH2AX assay	Comet assay
Types of DNA damage detected	DNA double-stranded breaks, blocks to replication leading to double-stranded breaks, apoptosis (early)	Base damage, single- and double-stranded breaks, protein and interstrand crosslinks, apoptosis (all stages)
Sensitivity for detecting double-stranded breaks	1–5 per cell	40–100 per cell
Single cell detection	Yes	Yes
Independent of factors other than DNA damage or cell cycle position	No	Yes
Methods of detection	Flow and image cytometry immunoblotting (single cells and tissue sections)	Image cytometry (single cells only)
Influence of cell cycle position on double-stranded breaks		
Background	Three times lower in G_1 phase	Three times lower in S phase
Double stranded break induced	No effect	Three times lower in S phase
Ease of detection	Simple (using flow cytometry)	Time consuming (using image analysis)

appropriate software and good-quality confocal (or deconvolution) images (MacPhail *et al.*, 2003a). However, the accuracy of measuring discrete foci in whole nuclei is compromised for doses higher than 2 Gy (50 or so breaks per diploid cell). Examining individual optical sections can extend this limit higher, but with loss of accurate information on the total number of foci per cell.

Flow cytometry is useful for bivariate analysis of cell cycle responses, for ease of measurement of the kinetics of development and loss of γH2AX after DNA damage, and for rapid analysis of heterogeneity in response. For both flow and image analysis, samples can be counterstained with DNA binding dyes such as DAPI. For γH2AX detection, primary antibody staining is followed by staining with a secondary antibody that is typically fluorescein isothiocyanate (FITC) conjugated. The intensity of γH2AX staining increases linearly with staining up to about 10–20 Gy (250–500 breaks/diploid cell) (MacPhail *et al.*, 2003a,b). Using flow cytometry, saturation begins to occur after the damage exceeds about 500 breaks because there is a limit to the amount of H2AX that can be phosphorylated and neighboring 2 Mb foci will begin to overlap.

Maximum development of γH2AX foci size after exposure to ionizing irradiation requires about 30–60 minutes after treatment. After this time, there is a progressive loss of γH2AX foci as damage is repaired. The rate of loss is cell-type– and species-dependent, with typical half-times of 2–3 hours for radiation-resistant rodent cells but 6 hours or more for radiosensitive human tumor cells. Interestingly, the foci that remain hours after irradiation often appear larger. Rates of development and loss are similar for treatment with drugs that cause DNA double-stranded breaks with the caveat that if long drug treatments are used, foci numbers will likely represent the steady-state level of γH2AX formation and loss. Drugs like cisplatin, MNNG, and nitrogen mustard do not cause direct double-stranded breaks but produce increasing levels of γH2AX as a function of time after exposure (Olive and Banáth, 2004, unpublished observations). Expression of γH2AX after treatment with these drugs could be a result of γH2AX foci formed at stalled and/or damaged replication forks (Limoli *et al.*, 2002) intermediates in the repair of DNA interstrand crosslinks, base damage that stimulates γH2AX formation, or early changes associated with cell death (Bradley *et al.*, 1978).

There are some unresolved issues about the nature of the damage that stimulates γH2AX focus formation and exactly how these foci are able to promote DNA repair. However, the ability to accurately predict cell killing based only on γH2AX expression measured 1 hour after treatment with various drugs is remarkable (Banath and Olive, 2003). In that study, V79 fibroblasts were exposed for 30 minutes to six drugs known to produce double-stranded breaks and then allowed for form foci for 1 hour. The percentage of cells that showed γH2AX antibody binding levels greater than the untreated sample was found to be correlated with the percentage of cells that were killed by the treatment.

III. Methods of Analysis

A. Reagents

1. Tris buffered saline (TBS), pH 7.4.
2. TST permeabilizing buffer for flow analysis: TBS containing 4% fetal bovine serum and 0.1% triton X-100.
3. TTN permeabilizing buffer for slides: TBS containing 1% bovine serum albumin and 0.2% tween-20.
4. Primary antibody: Phosphoserine-139 H2AX monoclonal antibody (Upstate Biotechnology). Biotin and FITC-conjugated antibodies are also available.
5. Secondary antibody: Alexa 488 goat antimouse IgG (H + L)F(ab′)2 fragment conjugate or Alexa 594 goat antimouse immunoglobulin G (IgG) (H + L) F(ab′)2 fragment conjugate (Molecular Probes).
6. Stock solution of 4′ 6-diamidino-2-phenylindole dihydrochloride hydrate (DAPI; Sigma).
7. Fluorigard antifade solution (Bio-Rad).

B. Cell Preparation Procedures for Flow Cytometry Analysis of γH2AX

1. After treatment and incubation for repair, centrifuge 2–5 × 10^5 single cells at 600–800*g* and resuspend the cell pellet in 0.3-ml phosphate buffered saline in a 4.5-ml tube. Then add 0.7 ml 99.9% ethanol to the tube while vortexing. Fixed samples can be maintained at −20°C for at least 2 weeks before analysis.

2. On the day of analysis, add 1 ml of cold TBS to the cells in ethanol. Centrifuge cells at 800*g* and then resuspend the pellet in 1 ml of cold TST for 10 minutes to permeabilize and rehydrate the cells.

3. Centrifuge samples and resuspended the pellet in 200 μl mouse monoclonal anti-γH2AX antibody diluted 1:500 in TST. Place tubes on a shaker platform and incubate, covered, for 2 hours at room temperature.

4. Wash cells by centrifuging and resuspending the pellet in TST. Repeat.

5. Resuspend pellet in 200 μl of secondary antibody (e.g., Alexa 488–conjugated antimouse IgG, diluted 1:200 in TST), and return to the shaker platform, covered, for incubation for 1 hour at room temperature. During shaking, take care that cells remain in solution and do not form a meniscus above the liquid.

6. Wash cells by centrifuging and resuspending pellet in TBS with 4% fetal bovine serum.

7. Centrifuge and resuspend pellet in TBS containing 1 μg/ml of DAPI.

8. Analyze samples on a dual-laser flow cytometer using ultraviolet (UV) excitation for DAPI and 488-nm excitation for γH2AX. Perform gating on forward scatter and time of flight to eliminate signals from debris and doublets.

C. Cell Preparation Procedures for Image Cytometry Analysis of γH2AX

1. Grow cells on washed and sterile glass slides by first pipetting 10-μl cells (10^4 cells in serum-free medium) onto a localized region of the slide and allowing cells to attach for 30 minutes before adding complete medium to cover the coverslips. Cells are usually incubated overnight to maximize cell spreading and reduce nuclear thickness.

2. After treatment and allowing time for foci to develop, rinse coverslips in TBS. Then place slides for 15–30 minutes in freshly prepared 2% paraformaldehyde in TBS. Frozen sections of tissues are prepared after embedding tissue in OCT embedding compound. After allowing cut sections to dry for 1 minute, place slides in 2% paraformaldehyde for 15–30 minutes. After fixation, the following steps can be performed for cells and tumor sections attached to slides.

3. Rinse slides in TBS and permeabilize cells by submersing in −20 °C methanol for 1 minute.

4. Rinse slides in TBS and submerse in TTN for 20 minutes.

5. Deposit 25 μl anti-γH2AX diluted 1:500 in TTN on 1-cm^2 parafilm squares. Invert these squares over the cell or tissue region on the coverslips/slides. Place in a humidified chamber for 2 hours at room temperature or 4 °C overnight.

6. Rinse coverslips/slides with TBS and incubate with 25 μl fluorescent secondary antibody diluted 1:200 in TTN for 1 hour at room temperature in the dark.

7. Rinse coverslips in TBS several times for 5 minutes each time in the dark and immerse in 0.05 μg/ml of DAPI for 5 minutes followed by a 5-minute rinse in TBS. A final dip in fixative can be used.

8. Drain coverslips, add 10 μl of antifade mounting medium next to cells before inverting coverslip over the slide. Seal coverslip.

9. Slides are viewed using a fluorescent microscope with motorized focus capacity that allows deconvolution of about 50-nm slices through the cells. Our cell images were acquired using a 100× Neofluor objective and a Q-Imaging 1350 EX digital camera. Images are analyzed using Northern Eclipse 6.0 software (Empix Imaging). A macro was designed to identify foci positions and associated intensity and size after application of a thresholding algorithm and a mask to remove residual background staining (MacPhail *et al.*, 2003a). Alternatively, simply counting foci by eye is quite acceptable using coded slides.

D. Special Considerations

Using flow cytometry, γH2AX intensity can be expressed relative to the intensity of the control untreated population. These samples should be collected, stained, and analyzed at the same time as the test cells. When using asynchronous cells, radiation and drugs can block cells in various phases of the cell cycle. Because the control population will typically display a DNA histogram and γH2AX intensity consistent with asynchronous, unperturbed cells, simple comparisons of γH2AX intensity in treated populations with the mean fluorescence of controls may not be valid. In this situation, it may be necessary to limit analysis to cells in a specific phase of the cell cycle (e.g., based on DAPI staining intensity).

Foci in irradiated mitotic cells typically stain about five times more intensely than foci of cells in other phases of the cell cycle, and these cells can appear as a separate population above the G_2 phase cells. Apoptotic cells in which the membrane is "intact" also stain about 5–10 times more intensely with γH2AX antibodies than normal cells. In both cases, foci size appears to be increased. Both mitotic and apoptotic cells can be distinguished from the remaining cells within the population based on differential γH2AX staining and, if appropriate, omitted from analysis.

Some studies may be more easily interpreted if synchronized cells are used for analysis. Twofold differences in foci number will occur through the cell cycle simply as a (probabilistic) result of the increase in DNA at risk for damage. The

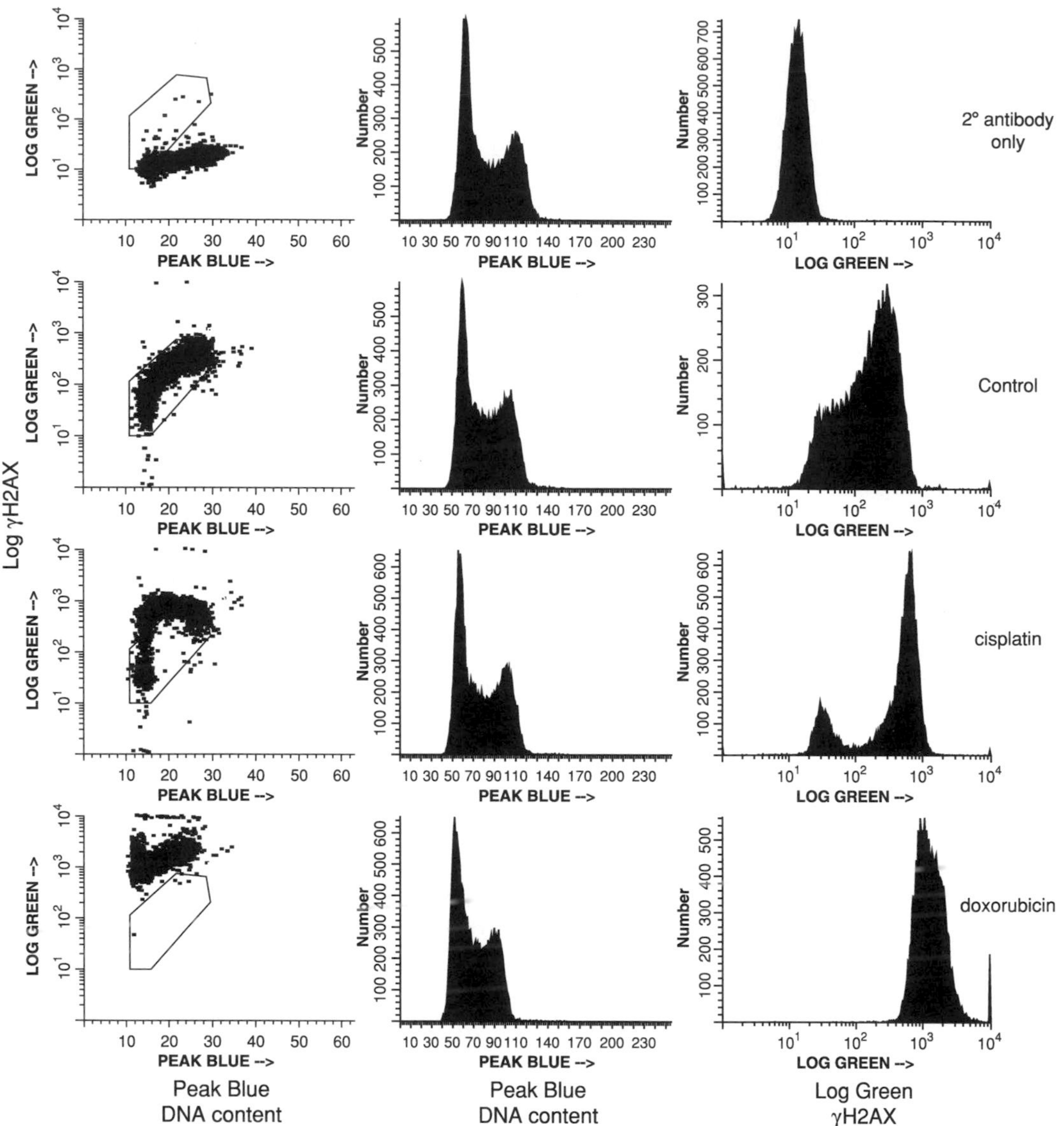

Fig. 1 Bivariate analysis of γH2AX and DNA content in hamster V79 cells. The control sample shows the typical increased expression of γH2AX in S/G_2 phase cells relative to G_1 phase cells. One hour after 2 hours of exposure to cisplatin, S-phase cells show increased expression of γH2AX. One hour after a 30-minute exposure to doxorubicin, all cells show increased expression of γH2AX.

situation is even more complex as cells enter S phase, because the initiation of DNA synthesis can result in a twofold to threefold higher background levels of γH2AX. Even with bivariate analysis of DAPI-stained cells, it is not possible to distinguish G_1 from early S-phase cells based on DNA content alone.

Colocalization of γH2AX with repair complexes may be a useful way to determine the kinetics of repair enzyme recruitment to sites of DNA damage. However, it is important to consider whether foci of γH2AX that colocalize with foci composed of various repair factors (e.g., BRCA1, Mre11, Rad51) might be an indication that repair of those particular breaks has not been successful. In this case, a higher percentage of colocalized foci could indicate poorer repair. The probability of (false) colocalization will increase as γH2AX foci size and number increase, so statistical approaches will be required. Application of optical sectioning methods will reduce the possibility that two foci in different planes can appear colocalized.

IV. Typical Results

A. Cell Cycle Variation in γH2AX Expression Detected Using Flow Cytometry

Bivariate analysis of γH2AX intensity versus DNA content indicates that untreated S- and G_2–M-phase cells have a higher level of expression of γH2AX than cells in G_1 phase (Fig. 1). Visually, the individual foci that are present in untreated S-phase cells typically appear much smaller than those in irradiated cells (Fig. 2). After drug or radiation treatment, two general patterns of γH2AX staining have been observed. After exposure to of asynchronously growing cultures to x-rays, bleomycin, doxorubicin, or etoposide, γH2AX expression is largely independent of cell cycle position. For these drugs, the relative increase in γH2AX intensity can be used as an indication of cell killing. Alternatively, one can determine the percentage of cells that show more γH2AX than the untreated controls (i.e., cells are outside a gate placed around the control population). The percentage of cells that fall into this gate has been shown to correlate with

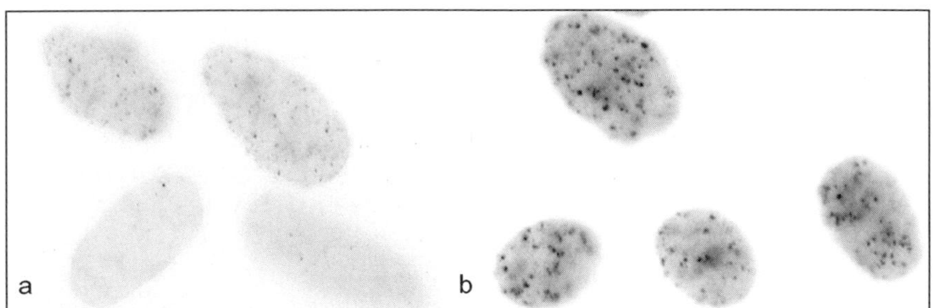

Fig. 2 Appearance of γH2AX foci in 4′,6-diamidino-2-phenylindole dihydrochloride hydrate (DAPI)–stained nuclei of human lung fibroblasts (HFL-1 cells) exposed to 0 Gy (A) or 1 hour after 2 Gy (B) S-phase cells in unirradiated samples show many smaller foci (two upper cells in panel [A]).

the (clonogenic) fraction of cells that survive treatment (Banáth and Olive, 2003). However, this approach needs to be established independently for each drug and is likely to depend on both treatment time and recovery time. For example, cells treated with cisplatin, tirapazamine, 4-nitroquinoline-*N*-oxide, or high-dose hydrogen peroxide show damage that is greatest in S-phase cells (but not excluded from G_1 cells). S-phase–dependent damage is likely to be a result of γH2AX produced at stalled replication forks, and foci development may be initiated as treated cells enter S phase. This makes timing an important consideration for determining whether cells have responded to some drugs by forming foci. In this case, the percentage of cells that retain foci measured after all cells have had an opportunity to enter S phase may be more informative for cell response than the relative increase in γH2AX measured soon after exposure.

B. Rate of Loss of γH2AX and Residual γH2AX as Measures of Cell Response to Radiation

The rate of loss of γH2AX after exposure to x-rays has been shown to correlate with radiation sensitivity for a series of human tumor cell lines—generally, tumor cells that are more radiation sensitive show a slower loss of γH2AX than radiation resistant cells (MacPhail *et al.*, 2003a). Figure 3A shows the kinetics of development and loss of γH2AX in SiHa cervical carcinoma cells as a function of time after exposure to various radiation doses. In this cell line as in most others, γH2AX loss rate after irradiation is dose independent. Using flow cytometry, background levels of γH2AX through the cell cycle and the radiation-induced G_2 block can complicate measurements of loss rate after doses below 2 Gy. Fig. 3B indicates that the maximum γH2AX signal has been reduced to 50% by about 5 hours after irradiation of this cell line, independent of dose. By 24 hours, those cells that retain foci can be counted and compared to the percentage of cells that survive irradiation (Fig. 3C). The percentage of cells that lack foci may be useful in providing a rough estimate of sensitivity to ionizing radiation. The appearance of selected SiHa cells examined 24 hours after exposure to 6 Gy is shown in Fig. 4. Note that foci can be quite large and variable in number, and occasionally they appear in micronuclei.

C. Double-Stranded Breaks in Human Lymphocytes

The excellent sensitivity of γH2AX for detecting DNA double-stranded breaks is illustrated with human lymphocytes exposed to low doses of ionizing radiation (Fig. 5). Damage by 5 cGy can be detected using flow cytometry (e.g., about one double-stranded break per cell). Blood was collected by venipuncture directly into gradient tubes (Becton Dickinson CPT tubes) for lymphocyte separation by centrifugation. After irradiation, 1 hour for focus formation, fixation, and antibody staining, samples were analyzed for γH2AX. To obtain the dose–response relationship, the mean fluorescence intensity of the irradiated samples was divided by the mean fluorescence of the control cells (note that the sensitivity for detecting double-stranded breaks is greatly increased in this case because of the absence of proliferating cells). In comparison, the limit of detection for double-stranded breaks in proliferating, asynchronous hamster V79 cells is about 20 cGy (MacPhail *et al.*, 2003a). A dose–response curve for white blood cells analyzed using the neutral comet assay is shown for comparison (Banath *et al.*, 1998); this is the only other method available for the measurement of initial DNA double-stranded breaks in individual cells and is almost two orders of magnitude less sensitive.

D. Detection of Apoptosis and Influence of Loss of Membrane Integrity

As previously mentioned, apoptotic cells have been found to stain intensely with γH2AX antibodies, which is not surprising because apoptosis results in extensive DNA fragmentation. Interestingly, TK6 cells that stained positive for antibodies

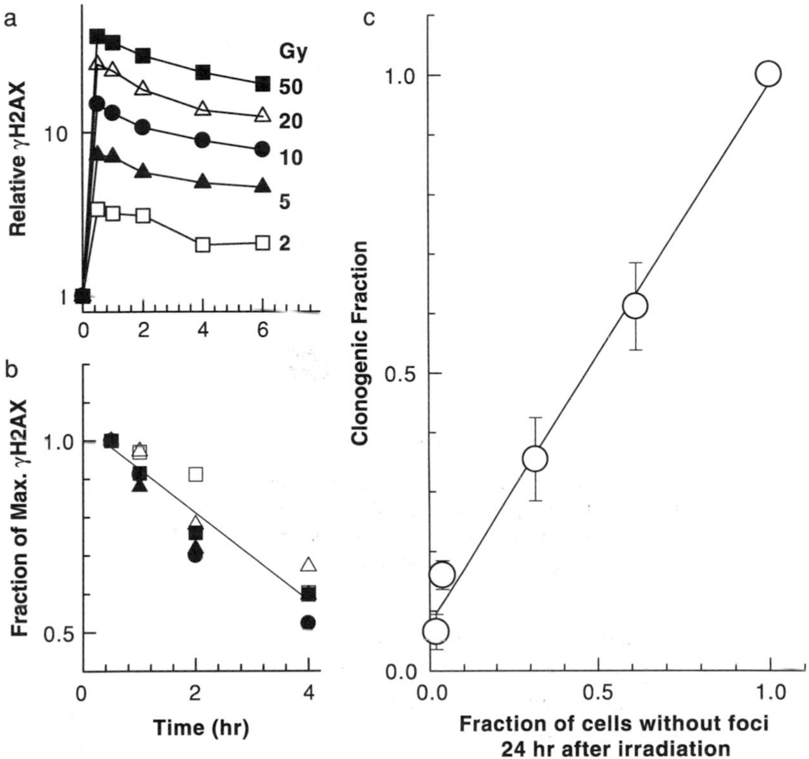

Fig. 3 Kinetics of development and loss of γH2AX measured by flow cytometry in exponentially growing SiHa cervical carcinoma cells examined as a function of time after exposure to 0–50 Gy. (A) Relative γH2AX intensity was calculated by dividing the mean fluorescence of the treated sample by the mean fluorescence of the control sample. (B) γH2AX intensity was divided by the maximum intensity of γH2AX achieved for each dose; the half-time for loss of γH2AX is independent of radiation dose. (C) The percentage of SiHa cells that lack foci 24 hours after irradiation was compared to the percentage of cells that survived to form colonies.

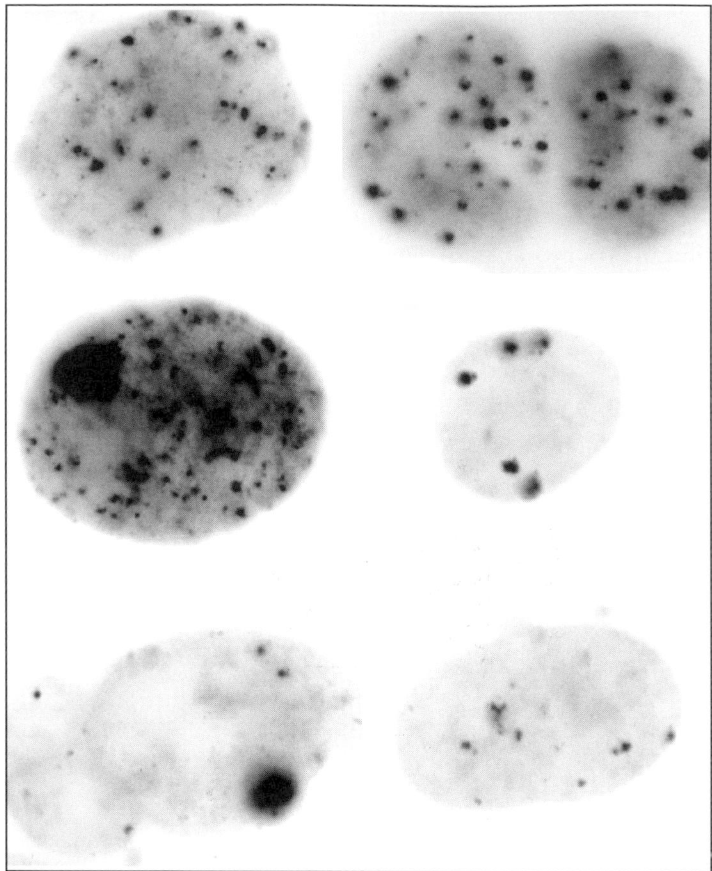

Fig. 4 Appearance of representative SiHa 4′,6-diamidino-2-phenylindole dihydrochloride hydrate (DAPI)–stained nuclei 24 hours after exposure to 6-Gy x-rays (only 10% of cells will remain clonogenic). The majority of nuclei exhibit γH2AX foci that are variable in both number and size. The percentage of cells that lacked foci correlated with the percentage that survived to form colonies (Fig. 4C).

against activated caspase-3 showed high levels of γH2AX when examined 4 hours after irradiation but near-normal levels 24 hours later despite that most of the cells had then undergone apoptosis (MacPhail *et al.*, 2003b). One explanation is that there is masking of the γH2AX epitope recognized by the antibody as a result of chromatin condensation (Rogakou *et al.*, 2003). It is also possible that at later times, apoptotic cells may no longer stain intensely because γH2AX is lost at the time when membrane integrity is lost. To examine this second possibility, asynchronous TK6 human lymphoblasts were exposed to 6-Gy x-rays and fixed 12 hours later. By this time, 24% of the (apoptotic) cells had become permeable to propidium iodide (PI), an indication that membrane integrity had been lost in those cells. Cells were sorted into two populations: those that were negative for PI

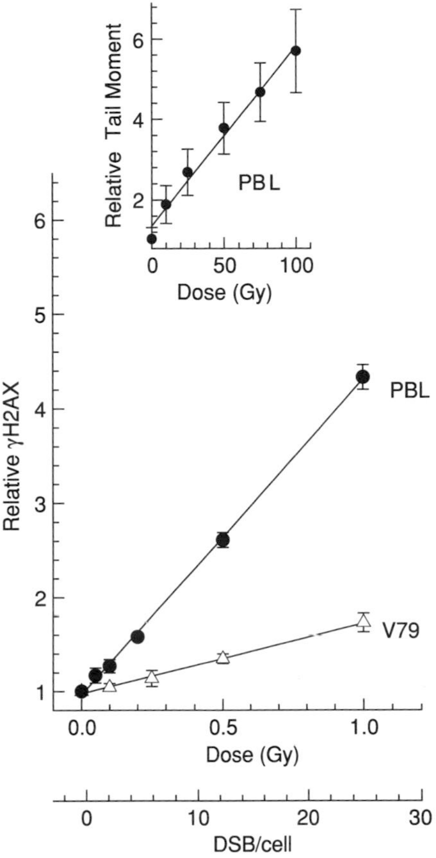

Fig. 5 Sensitivity for detecting double-stranded breaks in human peripheral blood lymphocytes (PBLs) exposed to x-rays. The response of PBL was measured 1 hour after exposure to x-rays. The decrease in sensitivity of hamster V79 cells (MacPhail *et al.*, 2003a) is largely a result of the higher background expression of γH2AX in S-phase cells, which reduces sensitivity for detecting radiation-induced damage. Shown for comparison purposes, (inset) the response of PBL in the neutral comet assay measured immediately after irradiation (Banáth *et al.*, 1998); from a comparison of slopes, the comet is 75 times less sensitive for detecting double-stranded breaks. Error bars are standard deviations for three experiments.

staining and those that were positive. Cells that were negative for PI showed only 12% apoptotic cells as measured using the neutral comet assay, and this value agreed well with the population of cells (15%) that expressed high levels of γH2AX (Fig. 6). However, PI-permeable cells that were 65% apoptotic/necrotic based on the comet assay showed only 12% of cells with high levels of γH2AX. Therefore, the presence of high levels of γH2AX in a cell can be an indication of apoptosis, but caution must be exercised if membrane integrity is lost. Lack of γH2AX expression in a cell that shows other evidence of apoptosis (e.g., positive for

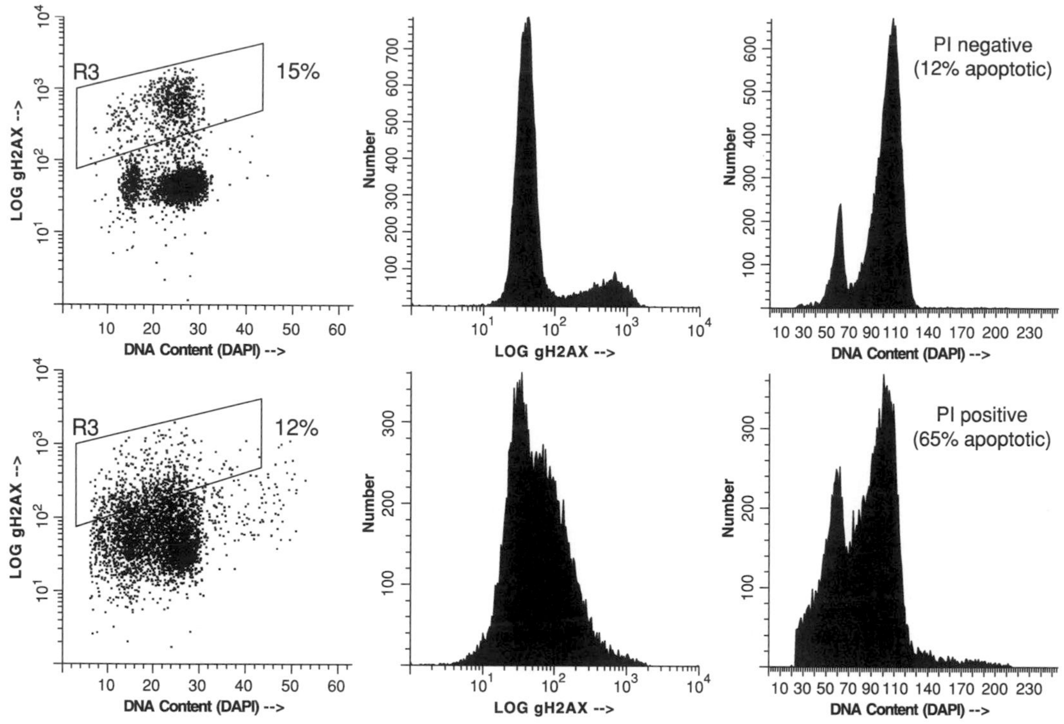

Fig. 6 Expression of γH2AX in apoptotic cells that retain membrane integrity. TK6 cells were exposed to 6 Gy and 12 hours later, and cells were incubated with the membrane impermeant dye, propidium iodide (PI). PI-positive and PI-negative cells were sorted, and sorted cells were then analyzed for the percentage of apoptotic cells using the comet assay (Olive *et al.*, 1993) or fixed and stained for γH2AX expression.

activated caspase-3 or positive in the comet assay) can be an indication that apoptosis has progressed to necrosis. This result does not rule out the possibility that γH2AX is still present but no longer readily detectable once cells have lost membrane integrity.

E. Analysis of DNA Damage in Cells from Tissues and Tumors

Both flow and image cytometry can be used to examine double-stranded breaks produced in cells from normal tissues and solid tumors exposed to DNA double-stranded breaking agents. Flow cytometry was used to analyze γH2AX in spleen cells and SCCVII tumor cells removed from C3H mice 1 hour after exposure to 60 mg of etoposide per kilogram of body weight (Fig. 7). This treatment killed about 30% of the tumor cells when measured using a clonogenic assay. The percentage of cells that showed γH2AX levels greater than the control level was consistent with this value. Interestingly, spleen cells removed from the same

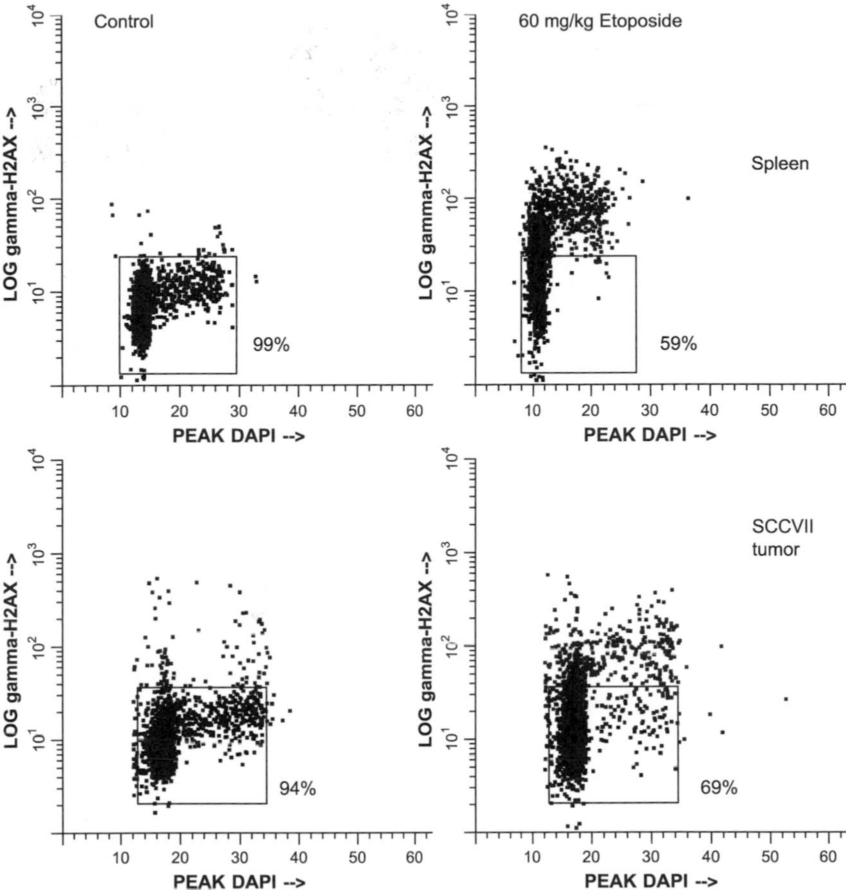

Fig. 7 Flow cytometry analysis of γH2AX in SCCVII tumor cells and spleen cells from C3H mice. Tissues were removed and fixed 1 hour after drug administration. Etoposide creates DNA double-stranded breaks in proliferating cells that contain its target enzyme topoisomerase II. The percentage of SiHa tumor cells that survived this treatment was 71% when measured using a clonogenic assay. This value correlates well with the percentage of cells that show control levels of γH2AX (69%). Note that spleen cells responded similarly to tumor cells from the same mouse.

treated mouse showed about the same enhancement of γH2AX staining relative to the control spleen cells, suggesting that the therapeutic ratio, at least by this criterion, may not be particularly favorable for proliferating normal tissues.

Fluorescent images of γH2AX tumor sections can also be prepared from tumors or normal tissues exposed to DNA damaging agents. In Fig. 8, C3H mice were exposed to x-rays, and 1 hour later, the DNA binding dye Hoechst 33342 was administered intravenously to stain nuclei of cells surrounding blood vessels. Frozen sections prepared from kidneys were fixed in paraformaldehyde and

Fig. 8 Identification of γH2AX in frozen sections of kidneys from C3H mice exposed to ionizing radiation. Images were obtained using a 20× neofluor objective (individual foci are not visible at this magnification). Blue is Hoechst 33342 nuclear staining; red is γH2AX antibody staining. (See Color Insert.)

stained for γH2AX. Virtually all nuclei show γH2AX antibody binding that is radiation dose dependent.

V. Possible Applications

Applications of this method might include but are not limited to the following:

1. Sensitive detection of DNA double-stranded breaks by ionizing radiation (applications in radiation biodosimetry)
2. Prediction of tumor cell killing by ionizing radiation (radiosensitivity) based on γH2AX loss rate or residual damage after treatment (MacPhail *et al.*, 2003a; Taneja *et al.*, 2004)
3. Prediction of tumor cell killing by drugs (chemosensitivity) based on level of γH2AX expression after treatment (Banath and Olive, 2003)
4. Identification of sites of double-stranded breaks for studies on the kinetics of recruitment of DNA repair enzymes to these sites (Fernandez-Capetillo *et al.*, 2003; Paull *et al.*, 2000)
5. Detection of cells in early phases of apoptosis (Huang *et al.*, 2003; MacPhail *et al.*, 2003b; Rogakou *et al.*, 2000)
6. Examination of the physical distribution of DNA damage within tumor sections using γH2AX immunohistochemistry (e.g., location of damaged cells relative to necrosis, blood vessels, and stroma)
7. Comparison between the amount of DNA damage produced in tumors in relation to normal tissues (therapeutic ratio)
8. Use as a tool for examining radiation bystander effects or adaptive responses that may involve DNA repair pathways and damage signaling (Ballarini *et al.*, 2002)
9. Identification of radiation-resistant hypoxic tumor cells (in combination with damage by ionizing radiation) (Olive and Banáth, 2004)

10. Use as a tool for examining chromosomal break points or telomere abnormalities (Takai *et al.*, 2003)

11. Characterization of genomic instability (Bassing *et al.*, 2002; Ivanov *et al.*, 2003)

12. Tool in studies of V(D)J recombination and meiotic recombination (Hamer *et al.*, 2003)

13. Use in examining the importance of higher-order chromatin organization in DNA damage and repair

VI. Conclusion

The formation of γH2AX foci at sites of double-stranded breaks is an important event in the recognition and signaling of the presence of the break. Exactly how these foci promote repair is not yet known, although both chromatin changes and repair enzyme recruitment are probably involved. Although there are unresolved questions concerning the nature of the physical damage that provokes γH2AX focus formation, this event offers an opportunity to measure, with exceptional sensitivity, a DNA lesion underlying chromosome damage and cell killing that is produced, directly or indirectly, by various agents. Colocalization studies with γH2AX provide an ideal way to visualize repair enzyme activity at sites of unrepaired breaks.

Acknowledgments

This work was supported by grants from the National Cancer Institute of Canada with funds provided by the Canadian Cancer Society. The expert collaboration of Dr. Judit P. Banáth and Susan H. MacPhail is gratefully acknowledged.

References

Andegeko, Y., Moyal, L., Mittelman, L., Tsarfaty, I., Shiloh, Y., and Rotman, G. (2001). Nuclear retention of ATM at sites of DNA double strand breaks. *J. Biol. Chem.* **276,** 38224–38230.

Bakkenist, C. J., and Kastan, M. B. (2003). DNA damage activates ATM through intermolecular autophosphorylation and dimer dissociation. *Nature* **421,** 499–506.

Ballarini, F., Biaggi, M., Ottolenghi, A., and Sapora, O. (2002). Cellular communication and bystander effects: A critical review for modelling low-dose radiation action. *Mutat. Res.* **501,** 1–12.

Banáth, J. P., Fushiki, M., and Olive, P. L. (1998). Rejoining of DNA single- and double-strand breaks in human white blood cells exposed to ionizing radiation. *Int. J. Radiat. Biol.* **73,** 649–660.

Banáth, J. P., and Olive, P. L. (2003). Expression of phosphorylated histone H2AX as a surrogate of cell killing by drugs that create DNA double-strand breaks. *Cancer Res.* **63,** 4347–4350.

Bassing, C. H., Chua, K. F., Sekiguchi, J., Suh, H., Whitlow, S. R., Fleming, J. C., Monroe, B. C., Ciccone, D. N., Yan, C., Vlasakova, K., Livingston, D. M., Ferguson, D. O., Scully, R., and Alt, F. W. (2002). Increased ionizing radiation sensitivity and genomic instability in the absence of histone H2AX. *Proc. Natl. Acad. Sci. USA* **99,** 8173–8178.

Bradley, M. O., Erickson, L. C., and Kohn, K. W. (1978). Non-enzymatic DNA strand breaks induced in mammalian cells by fluorescent light. *Biochim. Biophys. Acta* **520,** 11–20.

Burma, S., Chen, B. P., Murphy, M., Kurimasa, A., and Chen, D. J. (2001). ATM phosphorylates histone H2AX in response to DNA double-strand breaks. *J. Biol. Chem.* **276,** 42462–42467.

Celeste, A., Difilippantonio, S., Difilippantonio, M. J., Fernandez-Capetillo, O., Pilch, D. R., Sedelnikova, O. A., Eckhaus, M., Ried, T., Bonner, W. M., and Nussenzweig, A. (2003). H2AX haploinsufficiency modifies genomic stability and tumor susceptibility. *Cell* **114,** 371–383.

Celeste, A., Fernandez-Capetillo, O., Kruhlak, M. J., Pilch, D. R., Staudt, D. W., Lee, A., Bonner, R. F., Bonner, W. M., and Nussenzweig, A. (2003). Histone H2AX phosphorylation is dispensable for the initial recognition of DNA breaks. *Nat. Cell Biol.* **5,** 675–679.

Celeste, A., Petersen, S., Romanienko, P. J., Fernandez-Capetillo, O., Chen, H. T., Sedelnikova, O. A., Reina-San-Martin, B., Coppola, V., Meffre, E., Difilippantonio, M. J., Redon, C., Pilch, D. R., Olaru, A., Eckhaus, M., Camerini-Otero, R. D., Tessarollo, L., Livak, F., Manova, K., Bonner, W. M., Nussenzweig, M. C., and Nussenzweig, A. (2002). Genomic instability in mice lacking histone H2AX. *Science* **296,** 922–927.

Downs, J. A., and Jackson, S. P. (2003). Cancer: Protective packaging for DNA. *Nature* **424,** 732–734.

Fernandez-Capetillo, O., Celeste, A., and Nussenzweig, A. (2003). Focusing on foci: H2AX and the recruitment of DNA-damage response factors. *Cell Cycle* **2,** 426–427.

Hamer, G., Roepers-Gajadien, H.L., van Duyn-Goedhart, A., Gademan, I. S., Kal, H. B., van Buul, P. P., and de Rooij, D. G. (2003). DNA double-strand breaks and gamma-H2AX signaling in the testis. *Biol. Reprod.* **68,** 628–634.

Huang, X., Traganos, F., and Darzynkiewicz, Z. (2003). DNA damage induced by DNA topoisomerase I and topoisomerase II inhibitors detected by histone H2AX phosphorylation in relation to the cell cycle phase and apoptosis. *Cell Cycle* **2,** 614–619.

Ivanov, A., Cragg, M. S., Erenpreisa, J., Emzinsh, D., Lukman, H., and Illidge, T. M. (2003). Endopolyploid cells produced after severe genotoxic damage have the potential to repair DNA double strand breaks. *J. Cell Sci.* **116,** 4095–4106.

Limoli, C. L., Giedzinski, E., Bonner, W. M., and Cleaver, J. E. (2002). UV-induced replication arrest in the xeroderma pigmentosum variant leads to DNA double-strand breaks, gamma-H2AX formation, and Mre11 relocalization. *Proc. Natl. Acad. Sci. USA* **99,** 233–238.

MacPhail, S. H., Banath, J. P., Yu, T. Y., Chu, E. H., Lambur, H., and Olive, P. L. (2003a). Expression of phosphorylated histone H2AX in cultured cell lines following exposure to X-rays. *Int. J. Radiat. Biol.* **79,** 351–358.

MacPhail, S. H., Banath, J. P., Yu, T. Y., Chu, E., and Olive, P. L. (2003b). Cell cycle-dependent expression of phosphorylated histone H2AX: Reduced expression in unirradiated but not x-irradiated G(1)-phase cells. *Radiat. Res.* **159,** 759–767.

Nazarov, I. B., Smirnova, A. N., Krutilina, R. I., Svetlova, M. P., Solovjeva, L. V., Nikiforov, A. A., Oei, S. L., Zalenskaya, I. A., Yau, P. M., Bradbury, E. M., and Tomilin, N. V. (2003). Dephosphorylation of histone gamma-H2AX during repair of DNA double-strand breaks in mammalian cells and its inhibition by calyculin A. *Radiat. Res.* **160,** 309–317.

Olive, P. L., and Banáth, J. P. (2004). Phosphorylation of histone H2AX as a measure of radiosensitivity. *Int. J. Radiat. Oncol. Biol. Phys.* **58,** 331–335.

Olive, P. L. (2002). The comet assay. An overview of techniques. *Methods Mol. Biol.* **203,** 179–194.

Olive, P. L., Frazer, G., and Banáth, J. P. (1993). Radiation-induced apoptosis measured in TK6 human B lymphoblast cells using the comet assay. *Radiat. Res.* **136,** 130–136.

Olive, P. L., Wlodek, D., and Banáth, J. P. (1991). DNA double-strand breaks measured in individual cells subjected to gel electrophoresis. *Cancer Res.* **51,** 4671–4676.

Paull, T. T., Rogakou, E. P., Yamazaki, V., Kirchgessner, C. U., Gellert, M., and Bonner, W. M. (2000). A critical role for histone H2AX in recruitment of repair factors to nuclear foci after DNA damage. *Curr. Biol.* **10,** 886–895.

Redon, C., Pilch, D., Rogakou, E., Sedelnikova, O., Newrock, K., and Bonner, W. (2002). Histone H2A variants H2AX and H2AZ. *Curr. Opin. Genet. Dev.* **12,** 162–169.

Reitsema, T., Banath, J. P., MacPhail, S. H., and Olive, P. L. (2004). Chromatin condensation by hypertonic saline and expression of phosphorylated histone H2AX. *Radiat. Res.* **161,** 402–408.

Rogakou, E. P., Pilch, D. R., Orr, A. H., Ivanova, V. S., and Bonner, W. M. (1998). DNA double-stranded breaks induce histone H2AX phosphorylation on serine 139. *J. Biol. Chem.* **273,** 5858–5868.

Rogakou, E. P., Boon, C., Redon, C., and Bonner, W. M. (1999). Megabase chromatin domains involved in DNA double-strand breaks *in vivo*. *J. Cell Biol.* **146,** 905–916.

Rogakou, E. P., Nieves-Neira, W., Boon, C., Pommier, Y., and Bonner, W. M. (2000). Initiation of DNA fragmentation during apoptosis induces phosphorylation of H2AX histone at serine 139. *J. Biol. Chem.* **275,** 9390–9395.

Rogakou, E., Kannaar, R., and Konstantinopoulou, V. (2003). The gamma-H2AX masking factor in apoptosis is caspase dependent. *Cancer Res.* **44**(abstr).

Rothkamm, K., and Lobrich, M. (2003). Evidence for a lack of DNA double-strand break repair in human cells exposed to very low x-ray doses. *Proc. Natl. Acad. Sci. USA* **100,** 5057–5062.

Sedelnikova, O. A., Rogakou, E. P., Panyutin, I. G., and Bonner, W. M. (2002). Quantitative detection of (125)IdU-induced DNA double-strand breaks with gamma-H2AX antibody. *Radiat. Res.* **158,** 486–492.

Siino, J. S., Nazarov, I. B., Svetlova, M. P., Solovjeva, L. V., Adamson, R. H., Zalenskaya, I. A., Yau, P. M., Bradbury, E. M., and Tomilin, N. V. (2002). Photobleaching of GFP-labeled H2AX in chromatin: H2AX has low diffusional mobility in the nucleus. *Biochem. Biophys. Res. Commun.* **297,** 1318–1323.

Takai, H., Smogorzewska, A., and de Lange, T. (2003). DNA damage foci at dysfunctional telomeres. *Curr. Biol.* **13,** 1549–1556.

Taneja, N., Davis, M., Choy, J. S., Beckett, M. A., Singh, R., Kron, S. J., and Weichselbaum, R. R. (2004). Histone H2AX phosphorylation as a predictor of radiosensitivity and target for radiotherapy. *J. Biol. Chem.* **279,** 2273–2280.

CHAPTER 15

Cytometry of Freshwater Phytoplankton

Jörg Toepel,* Christian Wilhelm,* Armin Meister,† Annette Becker,‡ and Maria del Carmen Martinez-Ballesta*

*University of Leipzig
Institute of Botany
Department of Plant Physiology
D-04103 Leipzig, Germany

†Institute of Plant Genetics and Crop Plant Research
D-06466 Gatersleben, Germany

‡Erft-Verband
50126 Bergheim, Germany

METHODS IN CELL BIOLOGY, VOL. 75

0091-679X/04 $35.00

I. Introduction

A. Demands on Flow Cytometry

Flow cytometry (FCM) has become an increasingly important tool for phytoplankton analysis. FCM has been applied in taxonomic studies of marine and freshwater environments and in physiological analysis of phytoplankton. The main objective of marine research is the understanding of the structure and development of phytoplankton populations in the oceans, because phytoplankton primary production is one of the main carbon dioxide sinks on a global scale (Dyrssen, 2001; Falkowski *et al.*, 2000; Geider *et al.*, 2001). In contrast to marine ecosystems, freshwater is often characterized by a high phytoplankton concentration, which, depending on the season, is often dominated by organisms such as diatoms, chlorophytes, or cyanobacteria. Freshwater ecosystems are too small to have a global impact. However, water quality is essential for the health of billions of humans, and for this reason the analysis of freshwater environments is mainly focused on the risks of eutrophication (Smith, 2003). In addition to cell counting of phytoplankton samples, chlorophyll *a* (Chl *a*) is the basic and most common biological parameter of eutrophication analysis in freshwater ecosystems. However, the knowledge of total Chl *a* does not provide conclusions regarding the community structure of phytoplankton. The feasibility of a forecast, especially of toxic blooms caused by several algae species, is not possible via Chl *a* determination. Algal blooms are often caused by dominance of cyanobacteria, which can produce toxins that are extremely harmful to humans and animals (Dawson, 1998; Fleming *et al.*, 2002; Landsberg, 2002). The determination of toxic cyanobacteria in the early stages of bloom development offers the possibility to implement preventive measurements in water management.

In summary, the goals of FCM are the determination and characterization of phytoplankton abundance, composition, and physiological features. More specifically, in freshwater ecosystems used for drinking-water supply, FCM offers the capability of assessing the risk of toxic algae. In this chapter, we review the state of the art of FCM in freshwater analysis to determine phytoplankton structure and biomass, Chl *a* content per cell, the physiological state, and *in situ* hybridization (ISH) techniques for the detection of specific algal species.

B. Freshwater Analysis

Freshwater analysis focuses on the characterization of lakes, rivers, and other surface water bodies in respect to eutrophication. The measured parameters to obtain and characterize eutrophication are separated into three categories:

physical, chemical, and biological. To describe the development of the phytoplankton population, we commonly use the biological parameters Chl *a* content and biovolume. The enumeration of freshwater phytoplankton samples and the calculation of biovolume are routinely performed via microscopic analysis. Chl *a* is estimated spectrophotometrically via extraction in organic solvent or by means of calibrated *in situ* fluorescence monitors. Chl *a* determination is also done via high-pressure liquid chromatography (HPLC) analysis (Mackey *et al.*, 1996, 1998; Wilhelm *et al.*, 1995). The advantage of this technique is its ability to provide information about the population structure in the water sample based on the ratio between a taxon-specific marker pigment (xanthophyll) and Chl *a*. Because the results of microscopic enumeration and HPLC analysis refer to different parameters, they are not really equivalent. Chl *a* relates to photosynthesis, whereas the biovolume is much better correlated with total carbon. Therefore, the changes in phytoplankton composition caused by water mixing, and hence, time and spatial patchiness of phytoplankton distribution, could be monitored at best determining both the biovolume and the Chl *a*. Changes in the percentage contribution of the major algal taxa relative to Chl *a* can be measured online by fluorometric methods. Hence, several instruments used to estimate phytoplankton structure (e.g., using fluorometric methods) were directly tested in the natural environment. These instruments were developed by Beutler *et al.* (2002) or Leboulanger *et al.* (2002) and have been successfully tested in comparison to HPLC analysis (Wiltshire *et al.*, 1998). However, quantitative information concerning the taxonomic distribution of the biomass is difficult to obtain by these techniques. Therefore, alternative methods are highly desirable. For this purpose, FCM methods were developed to estimate all parameters in one step on the basis of fluorescence and scatter signals (Becker *et al.*, 2002).

C. Flow Cytometric Applications for Freshwater Analysis

Phytoplankton analysis via FCM focuses mainly on the determination of phytoplankton abundance. The production of different scatter and fluorescence signals after excitation with monochromatic light of different wavelengths provides the basis for the discrimination of phytoplankton cells. In most cases this light is produced by a laser, depending on the instrument in the spectral range between ultraviolet (UV) and red light.

Yentsch *et al.* (1983) were the first to use FCM to estimate phytoplankton composition in the sea. Studies have tried to extend the phytoplankton analysis in marine ecosystems to additional parameters. Blanchot *et al.* (1997, 2001) estimated phytoplankton composition and cell sizes on the basis of the forward scatter, and hence, carbon content per cell, via an empirical calculation. Mackey *et al.* (2002) estimated the structure of picoplankton in the Pacific Ocean via FCM in comparison to HPLC analysis using the CHEMTAX algorithm for quantification of the major algal taxa. To test the applicability of FCM, several cruises were performed to determine phytoplankton structure in different oceans and to study the impact of nutrient addition (Cavender-Bares *et al.*, 1999; Garrison *et al.*, 2000; Lebaron *et al.*, 2001; Li, 2002; Olson *et al.*, 2000; Sosik *et al.*, 2002;). Based on these studies it is still a matter of debate whether nutrient enrichment, especially iron fertilization in high nutrient low chlorophyll (HNLC) areas, is a realistic application for the stimulation of phytoplankton growth to increase the carbon sink in the oceans (Buesseler *et al.*, 2003; Dalton, 2002).

In freshwater analysis Becker *et al.* (2002) tested FCM to measure phytoplankton composition on the basis of biovolume and Chl *a*. The advantage of this method is the possibility to quickly evaluate water samples with a high sample throughput and, therefore, to replace the time-consuming cell counting. In combination with HPLC analysis and fluorometry (see earlier), FCM can be used to predict the biological development of freshwater ecosystems (e.g., by estimation of species-specific growth rates).

Given the risks associated with toxin-producing algae, cyanobacteria are of special interest. The detection and determination of such toxin-producing algae is very often erroneous because of the small cell size which requires microscopic analysis. Such critical species can be identified in a mixed-water sample much better via FCM if a set of optical properties (scatter and fluorescence values) (Becker *et al.*, 2002; Carr *et al.*, 1996) is processed in neuronal network systems (Steinberg *et al.*, 1998; Wilkins *et al.*, 1999). However, the high variability of phytoplankton cell sizes, ranging from a few to 200 microns, limits this method in many cases. Thus, ISH techniques for phytoplankton analysis have been introduced (Biegala *et al.*, 2003; Not *et al.*, 2002; Simon *et al.*, 1995, 2000) (see section II.E) to detect specific phytoplankton species in a mixed population. It is shown that recent progress allows the combination of FCM and ISH.

II. Background

A. Pigments

Both the marine and the freshwater phytoplankton community can contain representatives of at least 14 phyla. Figure 1 shows the absorption spectra of 11 representatives of major algal taxa together with the pigmentation pattern. The specific absorption spectra for the major taxa result from Chl *a*, which occurs ubiquitously in all algal cells and so-called *accessory pigments*. The latter can be either chlorophylls, carotenoids, or phycobilins, some of which are taxon specific. The accessory carotenoids transfer the absorbed light efficiently to Chl *a*, with the consequence that fluorescence is mainly emitted by Chl *a* (Büchel *et al.*, 1993). Hence, phytoplankton particles can be discriminated from all other particulate matter in a given water sample by the Chl *a* fluorescence emission at far red light. The comparison of absorption and fluorescence emission spectra shows that emission spectra, in contrast to the absorption spectra, are not in all cases sufficient to differentiate between phytoplankton phyla, because only energy-transferring pigments contribute to these

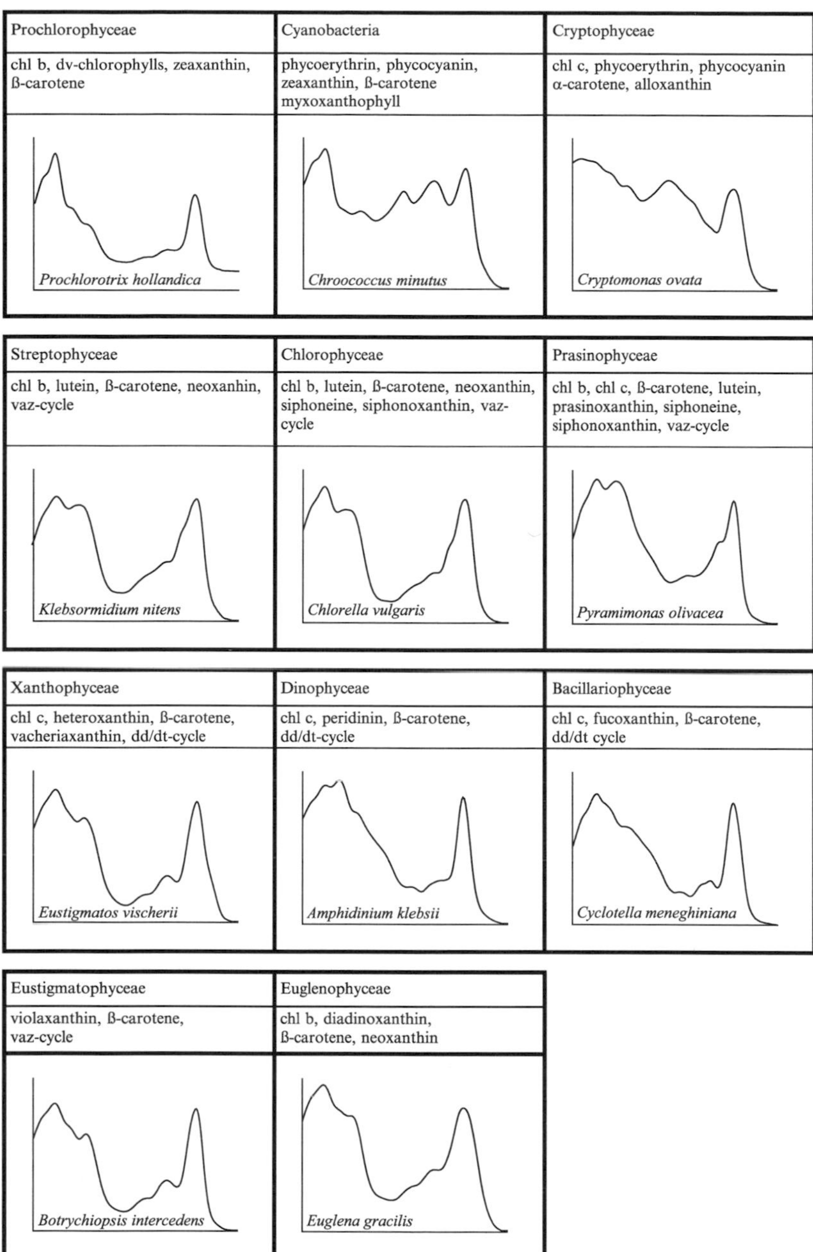

Fig. 1 Absorption spectra (between 400 and 700 nm) and pigmentation, inclusive xanthophyll cycle pigments for the major phytoplankton families (dv-chlorophyll, divinyl chlorophyll; vaz-cycle, violaxanthin, antheraxynthin zeaxanthin cycle; dd/dt cycle, diadinoxanthin/diatoxanthin cycle).

spectra. In contrast, absorption spectra yield information of Chl *a* and all accessory pigments (Hipkins and Baker, 1986).

The latter either can act as light-harvesting pigments, which funnel the absorbed excitons very efficiently to Chl *a* (e.g., fucoxanthin, peridinin, violaxanthin, or phycobiliproteins), or are involved in photoprotection (e.g., zeaxanthin, diatoxanthin, and β-carotene). The carotenoids transfer the absorbed energy with different efficiency to Chl *a*. In addition, the Chl *a* emission originates mainly from photosystem II, which has a highly variable quantum yield because of different physiological quenching processes. Therefore, the absorption spectra are less sensitive to changes in the physiological state than the excitation spectra.

In conclusion, if FCM is used to determine the autofluorescence of Chl *a* induced by simultaneous multiwavelength excitation, the excitation but not the absorption spectrum is decisive for the reproducibility and reliability of the signals. On the one hand, this fact is a strong disadvantage if quantitative estimates should be extracted from the phytoplankton community (e.g., if the amount of chlorophyll per taxon has to be estimated). On the other hand, the variability of the excitation spectrum in a given taxon can be used to extract physiological information from this subpopulation on the fly.

In Fig. 2, the excitation spectra of six typical freshwater taxa are presented. *Anacystis nidulans, Synechococcus rubescens*, and *Cryptomonas ovata* contain phycobilins with the typical absorption/fluorescence in the "green gap." Phycobilins also transfer their collected light energy with high efficiency to Chl *a*, but this efficiency is noticeably less than 100% (80–90% according to Hoffmann [1975], 90% according to Nobel [1974]), so a clear emission signal from these accessory pigments can be measured. Excitation at 532 nm leads to phycoerythrin (575 nm), phycocyanin (at 650–660 nm), and Chl *a* emission (at 680 nm). However, the fluorescence signal at 680 nm is influenced by the overlap of allophycocyanin (present in all cyanobacteria) and Chl *a* fluorescence. In *S. rubescens*, a strong band is evident at 575 nm (phycoerythrin), but this band is not observed in any other presented species. In the case of the cyanobacteria, the fluorescence emission of the phycobilins is higher than the emission band at 680 nm for Chl *a*, because the energy transfer is less efficient in this case in comparison to the other accessory pigments (Hiller, 1999). In the eukaryotic phycobilin-containing *Cryptomonas* species, the emission band at 620 nm is equal to the Chl *a* emission. Moreover, beside the existence of different phycobilins (e.g., phycoerythrin, phycocyanin, and allophycocyanin) with different fluorescence spectra, differences in the binding to the protein carrier can modify the spectral properties (Becker *et al.*, 2002; Goodwin, 1988; Tevini and Häder, 1985). Thus, there are many spectral differences between phycobilin-containing phytoplankton species, which may be the basis for their identification (and possibly separation by particle sorting) via FCM. Chl *c*-containing and Chl *b*-containing cells normally exhibit only one peak at 680 nm for Chl *a* in the emission spectrum because the energy transfer from *c*-type and *b*-type chlorophylls to Chl *a* is close to 100% in intact cells (Hiller, 1999). In solution, Chl *b* fluorescence can be detected separately from Chl *a* (Meister, 1992),

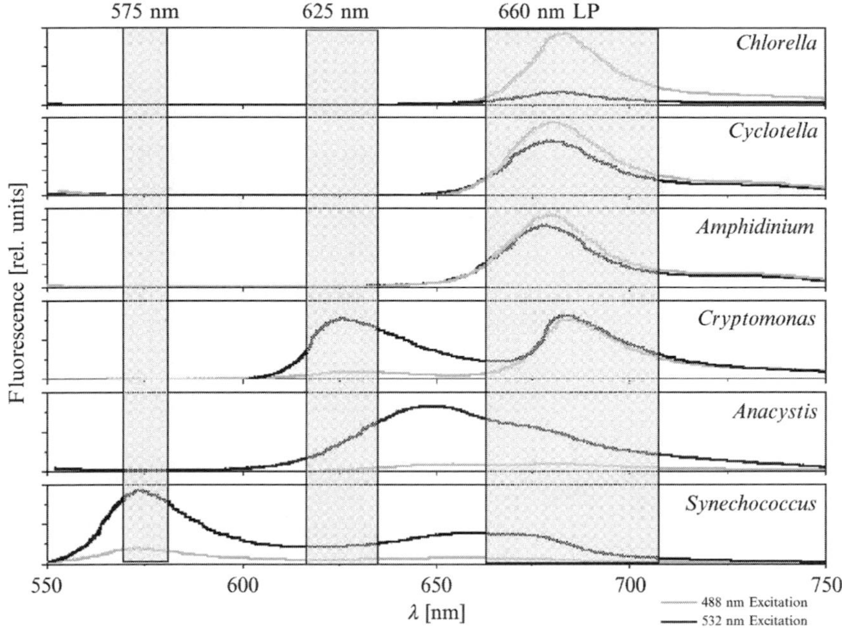

Fig. 2 The emission spectra of six phytoplankton algae excited with 488 or 532 nm, measured via fluorescence emission spectroscopy (Cyclo, *Cyclotella meneghiana*; Amphi, *Amphidinium klebsii*; Crypto, *Cryptomonas ovata*; Chlo, *Chlorella vulgaris*; Ana, *Anacystis nidulans*; Syn, *Synechococcus rubescens*).

but in the living cell, no specific Chl *b* emission can be observed. According to Fig. 1, phycobilin-free algae have taxon-specific absorption differences in the carotenoid region between 500 and 530 nm. Typical carotenoids for this absorption region are fucoxanthin, peridinin, prasinoxanthin, and siphonoxanthin. The absorption efficiency in this region and the excitation transfer efficiency to Chl *a* determine the fluorescence emission intensity at 680 nm. Therefore, the ratio between the 680-nm emission excited at 530 ($E680_{530ex}$) and the 680-nm emission excited at 488 ($E680_{488ex}$) is an effective parameter to differentiate spectrally between various eukaryotic phytoplankton taxa (Steinberg *et al.*, 1998).

The relative fluorescence contribution of six typical freshwater phytoplankton representatives is plotted in Fig. 3. The emission intensities are summarized to 100% to normalize the specific cell size differences in fluorescence intensity. The multiwavelength excitation provides an opportunity to distinguish phytoplankton taxa with just one fluorescence-emitting fluorochrome (Chl *a*), because the absorption by the accessory pigments and the energy transfer to Chl *a* is taxon specific. The resulting fingerprint-like pattern of the relative fluorescence signals for the different species becomes obvious (Fig. 3). In addition, in this presentation, the fluorescence pattern, measured with a double-laser flow cytometer, is

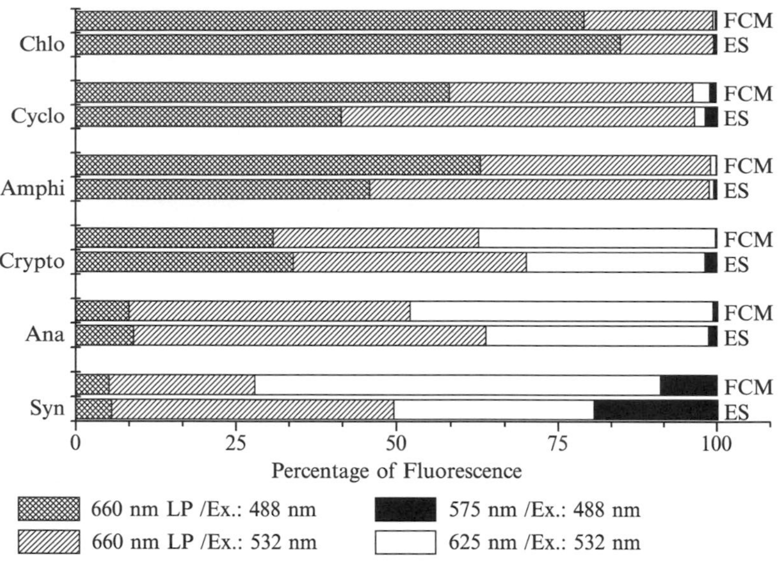

Fig. 3 The fluorescence pattern measured via flow cytometry (FCM) and fluorescence emission spectroscopy (ES) of six phytoplankton algae excited with 488 and 532 nm. The measured fluorescence intensities of the wavelength of interest were summed. (Cyclo, *Cyclotella meneghiana*; Amphi, *Amphidinium klebsii*; Crypto, *Cryptomonas ovata*; Chlo, *Chlorella vulgaris*; Ana, *Anacystis nidulans*; Syn, *Synechococcus rubescens*.)

compared with the fluorescence pattern obtained from a fluorescence emission spectrophotometer (ES). The similarity of the results obtained from both instruments shows that fluorescence emission spectra measured using an ES can be used to define the optical setup in a flow cytometer to optimize the excitation and emission wavelengths for analysis.

B. Problems in Quantifying Chl *a* by Fluorescence Analysis

Although approximately 1% of the dry weight of phytoplankton is composed of photosynthetic pigments with very high absorption coefficients, the fluorescence signal is relatively weak. This is because under *in vivo* conditions the quantum yield of the chlorophyll autofluorescence is reduced to approximately 3% by photochemical and non-photochemical fluorescence quenching (Büchel *et al.*, 1993). Along with the relatively low fluorescence intensity, a second problem for chlorophyll estimations from living cells arises from the variability of the fluorescence quantum yield. Dark-adapted cells illuminated by a short flash exhibit a significantly higher fluorescence yield than light-acclimated cells. In addition, the photochemical quenching induced by light-stimulated photosynthesis can be influenced by nutrient limitation or by inhibitors of photosynthesis, such as the presence of toxic compounds. The calibration between the chlorophyll fluorescence and the amount of chlorophyll determined by spectrophotometric analysis has to be performed under defined conditions. Thus, if phytoplankton is analyzed *in situ*, the fluorescence emission per unit of chlorophyll may depend on the preillumination, the nutrient status, or the exposure to photosynthesis inhibitors. It could be expected that the fluorescence emission of Chl *a* bound to proteins should be uniform, because the pigment-binding protein units are phylogenetically closely related and the binding is highly similar in all cells (Hiller, 1999). However, the fluorescence emission is variable and depends on the algal species. Despite this consideration, the fluorescence emission intensity per chlorophyll molecule may vary over several magnitudes mainly because of changes in the *in vivo* absorption coefficient (Kirk, 1994). The efficiency of light absorption by a chlorophyll molecule in the cell strongly depends on the pigment packing. The cell size limits the density of chlorophyll molecules packed in the chloroplast. An increasing cell size results in a decreasing absorption coefficient per unit chlorophyll in a nonlinear manner. Therefore, in small coccoid cells the *in vivo* absorption coefficient is significantly higher than in large cells. Kirk (1994) calculated that this coefficient decreased from about 0.02 to 0.005 m^2 mg Chl a^{-1} if the cell diameter changed from 2 μm to 80 μm. Taking into account that the size distribution of phytoplankton communities is just in this range, the package effect modifies the absorption efficiency by a factor of about 4. Therefore, accurate estimations of Chl *a* per cell must consider the cell shape and size, parameters that can be characterized by the forward and the side scatter signals (see Fig. 8, later in this chapter).

C. Ecological Reasons for Fast Changes in Phytoplankton Structure

The composition of a phytoplankton community is a balance between growth rate influenced by nutrients, light and temperature, and grazing or sinking rates of dead cells. This balance is determined by the species- and size-specific growth rates. In a first approximation, growth is a function of cell size as long as the water body is nutrient limited (Harris, 1986). Smaller cells have higher relative metabolic activity. However, within a given size fraction, the physiological activity can vary as a result of species-specific temperature optima or species-specific nutrient uptake capacities and efficiencies. These physiological features of nutrient use can be assessed by nutrient uptake rates or affinities characterized by nutrient-specific Michaelis–Menten constants. In addition, light is a nutrient for phytoplankton cells and can, therefore, be treated mathematically as a substrate in a complex enzymatic reaction. Table I shows some key physiological parameters for different species of typical cell sizes between 1 and 20 μm. Obviously, the absolute values of maximal uptake rates in nitrogen and phosphate, as well as the "affinity" to exploit these nutrients, are not simply a matter of cell volume, but a result of genetic differences. Furthermore, photosynthetic efficiency is genetically determined as well. Measuring these key parameters for each predominant species becomes a challenge, not only in the laboratory under artificial conditions but

Table I
Nitrogen and Phosphate Uptake Parameter of Several Phytoplankton Algae

Parameter	Dimension	Value	Species
Urea uptake			
a. V_{max}	nmol mg protein^{-1} * h^{-1}	12	*Chlorella pyrenoidosa*
		88	*Scenedesmus obliquus*
		125	*Cyclotella nana*
b. K_m	μmol	0.62	*Chlorella pyrenoidosa*
		0.77	*Scenedesmus obliquus*
		0.42	*Cyclotella nana*
Nitrate uptake			
a. K_m	μmol	15	*Navicula pelliculosa*
		3	*Scenedesmus obliquus*
		70	*Anabaena cylindrica*
o-Phosphate uptake			
a. V_m	μmol * mm^{-2} * h^{-1}	0.77	*Anabaena flos aquae*
		0.89	*Microcystis* species
		0.02	*Scenedesmus obliquus*
b. K_m	μmol	1.1	*Anabaena flos aquae*
		1.61	*Microcystis* species
		0.60	*Scenedesmus obliquus*
Photosynthesis			
K_m for light saturation	μmol photons * m^{-2} * s^{-1}	65	*Chlorella pyrenoidosa*
		45	*Cryptomonas* species
		90	*Cyclotella* species
		28	*Synechococcus* species

Note: Data taken from different sources summarized in Kohl and Nicklisch, 1988.

also in nature. Sorting via FCM and subsequent analysis of single cells or small fractions of cells with the same properties can yield new information on population development.

The biomass formed by phytoplankton growth has an impact on water chemistry and influences the biological network. The development of phytoplankton indicates a strong dependence on grazing pressure and nutrient load. Again, the nutrient concentration is influenced by the phytoplankton. For example, if dead phytoplankton cells sink to the sediment, the contained nutrients will be eliminated permanently from the euphotic zone if the water body is deep enough and fully oxygenated. However, in many cases the cells become mineralized before reaching the sediment and the release of low molecular organic compounds is a perfect source for planktonic bacteria. This short-circuit cycling of nutrients in the pelagial leads to abrupt changes in the phytoplankton community composition instead of a clear-water stage during the summer and is often observed in eutrophic or hypertrophic lakes (Azam *et al.*, 1995). Therefore, the ratio between sedimentation and mineralization defines the amount of nutrients that cannot be eliminated in a given water body.

The second important factor for phytoplankton growth is the grazing pressure. Phytoplankton cells are the basis of the food chain, although several strategies were developed to prevent grazing (e.g., many nonmobile phytoplanktors, such as *Scenedesmus*, species form spines). It has been shown that phytoplankton can also react to the presence of grazers either by increasing the size of the colonies or by excretion of bioactive molecules that damage the fertility of the grazers (Tosti *et al.*, 2003). Additionally, the cell size and adaptational mechanisms influence the cell-specific retention period in the euphotic zone. The simplest strategy to remain in the euphotic zone is the use of mobility. The relative abundance of flagellates increases under stronger water turbulence when passive motion is not sufficient to reach favorable water layers (Becker *et al.*, 2003). Another strategy is that of gas vacuoles to regulate the motion via density (e.g., *Microcystis* species). Along with gas vacuoles, the accumulation of starch by photosynthesis and respiration of this substrate in low-light layers is an additional method for sinking regulation (Wallace *et al.*, 1999). In addition, *Microcystis* species can form clusters with thousands of cells to avoid grazing and to sink faster in deeper-water layers. From these examples, we can clearly see that phytoplankton composition shows a strong temporal and spatial dynamic. Particularly in lakes, phytoplankton abundance can be patchy in both the horizontal and the vertical dimension. Food web studies require a tremendous number of samples with which to estimate species composition and size distribution.

D. Physiological Investigations via Flow Cytometry

In addition to the phytoplankton dynamic, the determinations of physiological parameters are of interest. In particular, the determination of the photosynthetic capacity and the response to nutrient limitation are main topics in water research. Therefore, several methods were developed to directly determine these parameters in the water environment. A first approach to estimate the primary production of phytoplankton via FCM was established by William (1993). A combination of FCM and fluorometry was shown by Olson *et al.* (1996). They also determined the photosynthetic efficiency of marine phytoplankton samples (Olson *et al.*, 1999, 2000; Sosik *et al.*, 2002), specifically to analyze the influence of iron availability on the photosynthetic quantum yield of photosystem II. Along with the determination of photosynthetic capacity, the metabolic activity (e.g., esterase activity) can be measured via flow cytometric measurements. Jochem (1999) showed a linear relationship between fluorescein diacetate fluorescence and metabolic activity. The nonpolar nonfluorescent molecule (fluorescein diacetate) is split after diffusion into the cells by nonspecific esterases and the originated fluorochrome can be excited with a 488-nm light, leading to a fluorescence detectable at 530 nm. Furthermore, Gregori *et al.* (2002) used FCM to show a linear relationship between oxygen uptake and fluorescence intensity emitted by a fluorescence probe ($DiOC_3$).

To detect the features of specific phytoplankton cells related to nutrient limitation, several authors performed experiments, either via enzyme concentration measurements or via colorimetric methods. Peperzak *et al.* (2000) presented an overview about immunolabeling of phytoplankton cells to demonstrate the possibility to measure physiological parameters on the single-cell level in a mixed population. In addition, Jochem *et al.* (2000) determined nitrate reductase activity in a marine diatom via fluorescence immunolabeling. The ELF 97 dye was used to quantify phosphatase activity and hence to detect phosphate limitation under the microscope (Rengefors *et al.*, 2001, 2003). These studies showed a linear correlation between ELF and the antibody-labeled phosphatase.

E. *In Situ* Hybridization

Fluorescence ISH (FISH) offers the possibility to identify an algal group or a single species in a mixed phytoplankton sample on the basis of genetic information. The FISH method involves application of fluorochrome-labeled oligonucleotide probes (20-bp length approximately) to permeabilized whole cells. The probe inside the cells hybridizes specifically to the complementary target sequence in the small subunit of the ribosomal RNA (18S rRNA). If no target sequence is present in the cell ribosomes, the probe is not able to hybridize and the unbound probe is removed during the subsequent washing steps. Thus, only specifically targeted cells retain the probe under stringent conditions during the hybridization and washing steps (temperature, salt, and formamide concentration). Because of the different cell morphologies and cell wall structures of phytoplankton, it is not possible to give a single protocol and it is necessary to adjust the conditions for each species (Table II). In most cases the oligonucleotide probes are 5′-end–labeled with the fluorochrome fluorescein isothiocyanate (FITC) reaching a natural amplification of the fluorescent signal because of the large numbers of ribosomes in any given target cell. The fluorochrome FITC has the advantage of an emission spectrum that is non-overlapping with the natural pigments, because the emission is about 520 nm after excitation at 488 nm.

The oligonucleotides were designed with the probe calculation function of the ARB software (www.arb-home.de) using a public database of rRNA gene sequence data, like GenBank (www.ncbi.nlm.nih.gov) or EMBL Nucleotide

Table II
Target Site, Temperature Optima, and Sequences of Several 18S rRNA Oligonucleotide Probes

Probe	Target group	Temperature (°C)	Sequence (5′–3′)	Author
EUK120 9R	Eukaryotes	50	GGGCATCACAGACCTG	Giovannoni *et al.*, 1988
HETERO 01	Heterokontophyta	50	ACGACTTCACCTTCCTCT	L. Medlin (2003, personal communication)
CHLO 02	Chlorophyta	54	CTTCGAGCCCCCAACTTT	Simon *et al.*, 2000
MICR 3	Microcystis	62	TCTGCCAGTTTCCACCGCCTTTAGGT	Rudi *et al.*, 2000

Sequence Database (www.ebi.ac.uk/embl/index.html). The Ribosomal Database Project II (RDP II; www.cme.msu.edu/RDP/html/index.html) is another useful resource and gives the possibility to download not only rDNA sequences but also whole alignments and offers various online analysis.

FCM in combination with FISH has been used for species determination in microbiology (Amann *et al.*, 1990; Simon *et al.*, 1995, 2000; Wallner *et al.*, 1993). Several protocols have been developed for ISH (John *et al.*, 2003; Smit *et al.*, 2003). Here we used a modified protocol for FCM (R. Groben, 2003, personal communication).

III. Methods

A. Optical Arrangement

1. Single-Laser Excitation

To avoid excessive differences in the fluorescence intensities of the different pigments, the most strongly fluorescing pigment Chl *a* should be excited relatively weakly by the selected wavelength. This is especially important when only one excitation source is available. The condition is well fulfilled by the 514-nm line of an argon ion laser, which is near the green minimum of Chl *a* absorption/excitation spectrum but excites the phycobilin pigments relatively strongly.

In earlier experiments (Becker *et al.*, 2002), the fluorescence was measured at four wavelengths:

- Chlorophyll *a* at $\lambda = 680$ nm
- Phycocyanin at $\lambda = 650$ nm
- Phycoerythrin of *C. ovata* at $\lambda = 620$ nm
- Phycoerythrin of cyanobacteria at $\lambda = 580$ nm

The practical problem was that the available flow cytometer (FACStar^{PLUS} single-laser system) had only three fluorescence channels. This problem was solved by sacrificing the side scatter channel (SSC) and using it as an additional fluorescence channel.

Furthermore, all photomultipliers were replaced by special red-sensitive ones to increase the detection sensitivity for the mostly red-emitting pigments.

The resulting scheme of single-laser excitation of a suspension of phytoplankton organisms is shown in Fig. 4.

2. Dual-Laser Excitation

Although the single-laser configuration gave quite meaningful results, it is of limited use if pigments emit in the same wavelength range. Two-laser excitation offers the possibility of discriminating algae on the basis of Chl *a* emission, then the excitation spectra are different. In such cases one can use fluorescence ratios

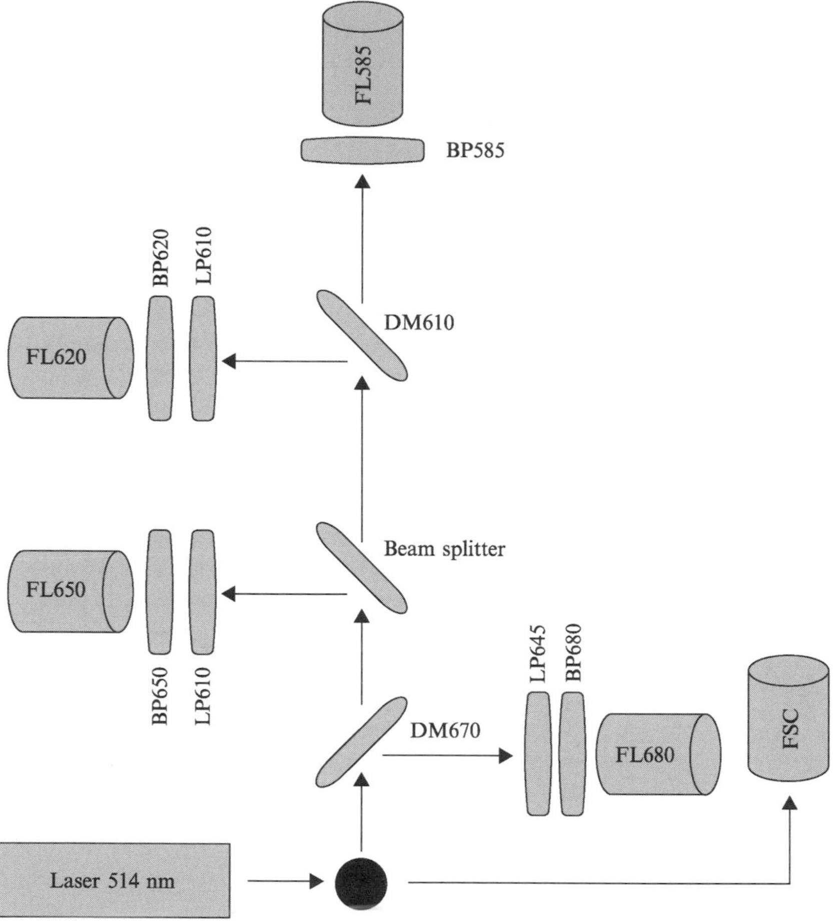

Fig. 4 Single-laser optical configuration of a FACStar^{Plus} flow cytometer and sorter. The forward scatter detector (FSC) is a photodiode. The other detectors of fluorescence at 585 nm (FL585), 620 nm (FL620), 650 nm (FL650), and 680 nm (FL680) are special red-sensitive photomultipliers. The side-scattering channel is used for detection of an additional fluorescence (FL650), so four chromophores can be separately analyzed. BP, bandpass filters with maximum transmission at the indicated wavelength; LP, long-pass filter transmitting light at longer than the indicated wavelength; DM, dichroic mirror transmitting light at shorter and reflecting light at longer than the indicated wavelength.

for identification (see section II.A). An example for dual-laser excitation with two emission channels for each laser is shown in Fig. 5. The two wavelengths used are 488 nm of an argon ion laser (excitation of chlorophyll *a* and phycoerythrin) and 532 nm of a Neodym/YAG laser (excitation of phycocyanin and Chl *a*). The argon ion laser is used for triggering. The corresponding emission wavelengths are more than 660 nm for Chl *a*, 630 nm for phycocyanin, and 575 nm for phycoerythrin.

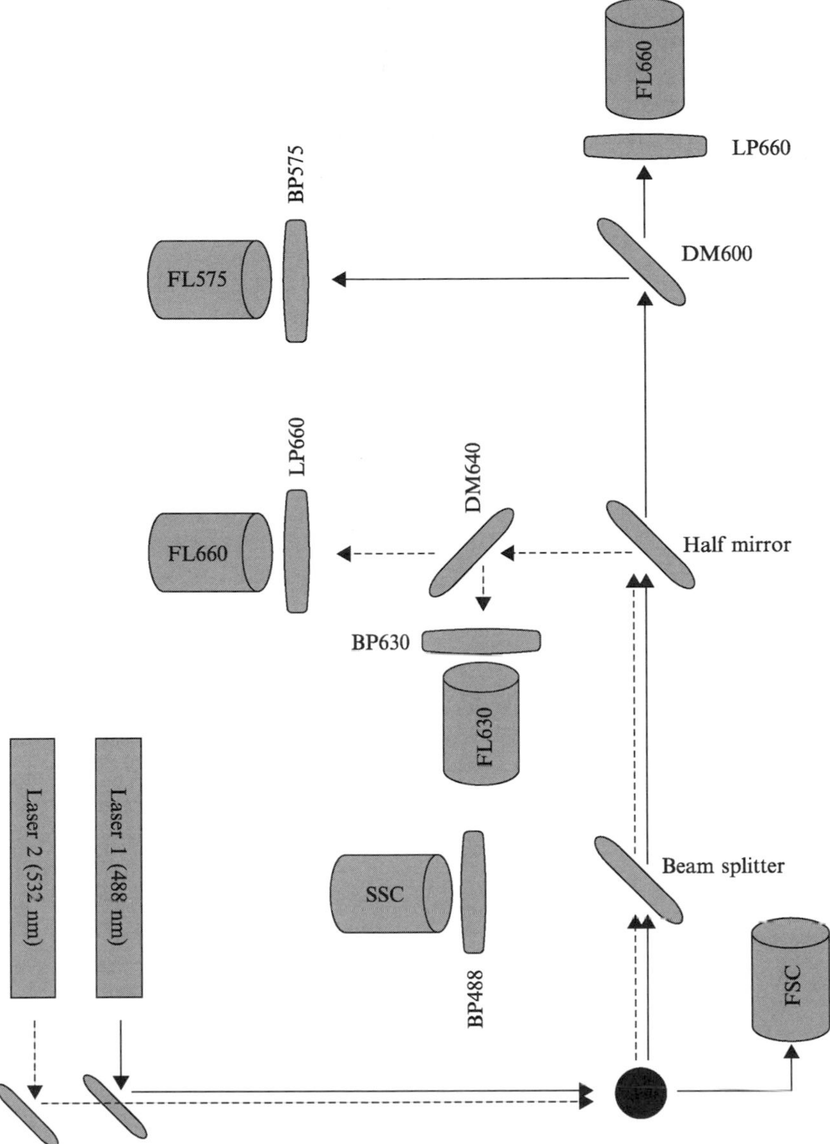

Fig. 5 Dual-laser optical configuration of a FACStar^{Plus} flow cytometer and sorter. The forward scatter detector (FSC) is a photodiode. The other detectors for detection of 90 side scatter (SSC) and fluorescence at 575 nm (FL575), 630 nm (FL630), and 660 nm (FL660) are photomultipliers. BP, bandpass filters with maximum transmission at the indicated wavelength; LP, long-pass filter transmitting light at longer than the indicated wavelength; DM, dichroic mirror transmitting light at longer and reflecting light at shorter than the indicated wavelength; ⟶, optical path of light emitted by laser 1; ----▸, optical path of light emitted by laser 2; For detecting specific labels as fluorescein isothiocyanate, the 575-nm bandpass filter can be replaced by a 535-nm bandpass filter.

By replacing the triggering argon ion laser by a helium neon laser (excitation at 633 nm) or by changing the bandpass filters, we can easily modify this configuration to detect special fluorochromes for DNA (see later discussion) and fluorescence-conjugated antibodies or oligonucleotide probes.

The data acquisition was realized with the Becton Dickinson software Cell Quest. For further data processing, a converting software was developed by Vaulot (Cytowin, www.sb-roscoff.fr/phyto/cyto.html#cytowin). Other programs include the converting program FLOW (A. Becker) or programs at the http://pingu.salk.edu/flow/sitelink.html home page.

3. Special Equipment

Conventional flow cytometers are neither specially designed for phytoplankton analysis nor well suited for work aboard a ship. One technical problem of phytoplankton is the heterogeneity of the objects with regard to cell size and concentrations, which necessitates different configurations for optimal analysis conditions.

Cavender-Bares *et al.* (1998) modified an EPICS V flow cytometer by a dual-sheath flow system, which is able to handle flow rates spanning more than five orders of magnitude. The change between both modes (corresponding to different sample inputs) takes only about 1 minute.

Several instruments were developed specifically for the requirements of phytoplankton analysis. The Optical Plankton Analyser (OPA) (Dubelaar *et al.*, 1989; Peeters, 1989) allows measurement of particles from 0.3 to 500 μm in width and more than 1000 μm in length and has a dynamic range up to 6 decades. Depending on the analyzed organisms, variable excitation and detection optics can be used and a large range of sample throughput rates is possible.

A further development of OPA by a European Community–funded consortium is the European Optical Plankton Analyser (EurOPA) (Jonker *et al.*, 1995). It allows analyses with a total signal range of 7 decades in one run and the processing of large sample volumes. It can be easily transported and operated at sea. Both small single cells and large colonies of cyanobacteria can be analyzed in one run. Absolute cell count is possible. It is equipped with two lasers, 488 nm for chlorophyll and 633 nm for phycocyanin excitation. In addition to the fluorescences, forward and side scattering are also measured. Time-of-flight measurement allows the calculation of particle length.

For continuous monitoring of phytoplankton *in situ*, a special flow cytometer inside a buoy was developed (Cytobuoy) (Dubelaar *et al.*, 1999). A single 20-mW Neodym/YAG laser excites the pigments with 532 nm. Two scatter and two fluorescence signals are measured. A closed recycling system for the sheath fluid is used, and the phytoplankton sample is drawn from just beneath the buoy. The data are radio transmitted to a remote analysis station.

A similar principle is used by the submersible "FlowCytoboot" (Olson *et al.*, 2003; Sosik *et al.*, 2003), which is anchored in 15-m depth on the ground of the ocean and connected with the shore by power and communication cables. The excitation source is also a 532-nm solid-state laser. Forward and side scatter as well as two fluorescences (575 nm = mainly phycoerythrin, 680 nm = mainly Chl *a*) can be measured. Water samples from 5-m depth are pumped continuously through the instrument. It is mainly used for time-series measurements of phytoplankton.

B. Specific Applications

1. DNA and Its Base Composition

Another parameter that can help in the identification of phytoplanktonic microorganisms is their different genomic DNA content. Because DNA has no detectable fluorescence in the visible spectral range, it must be stained with specific fluorescence dyes. Classic dyes are 4′-6-diamidino-2-phenylindole (DAPI) (Bihari *et al.*, 1999) and propidium iodide (PI) (Hsu and Hsu, 1998; Simon *et al.*, 1994). Both have some disadvantages, especially if they are used in combination with pigment autofluorescence. DAPI needs an additional excitation in the UV range and its fluorescence depends not only on DNA content but also on the AT/GC base composition. PI cannot penetrate living cells and needs previous fixation. It is excited in the same range as autofluorescent pigments (at 488 or 514 nm) and emits in the same spectral range as some of them (575–635 nm).

Therefore, other dyes such as SYTO-13, TOTO-1, YOYO-1 (Guindulain *et al.*, 1997), and SYBRGreen I (Zubkov *et al.*, 2000) may be more appropriate because they are excited at 488 nm, like the natural pigments, but emit at about 525 nm, clearly below the autofluorescence range of these pigments. Another possibility for detecting DNA is the use of oxazin 750 (OX750), LD700, and rhodamine 800 (R800) as fluorescent dye (excitation at 633 nm, emission at 665 nm) (Shapiro, 1986). In this case, one of the lasers in the dual-laser system (Fig. 5) is replaced by a helium neon laser with 633-nm emission. The disadvantage is the loss of specificity in the detection of pigment autofluorescence.

Using DNA-specific dyes is also possible in analyzing the cell cycle phases and ploidy levels (Bihari *et al.*, 1999).

Base-specific dyes such as Hoechst 33342 (AT specific) and chromomycin A3 (GC specific) enable an estimation of the base composition of the genomic DNA (Simon *et al.*, 1994). This is an additional parameter to discriminate between species.

2. Species-Specific Labeling of Organisms

By methods such as immunofluorescence or FISH, microorganisms can be specifically stained by attaching a fluorochrome to an antibody or oligonucleotide probe. The most popular fluorochrome for this purpose is FITC, which has the advantage of a non-overlapping emission spectrum with the natural pigments (excitation at 488 nm, emission around 520 nm). If a high level of nonspecific autofluorescence is present, the fluorochromes CY3, phycoerythrin (PE), and tetramethylrhodamine B thioisocyanate (TRITC), all excited at 542 nm and emitting at 575 nm, are preferred (Vesey *et al.*, 1997). The optical scheme of Fig. 5 can be modified accordingly, however, with the omission of one of the pigment autofluorescence components.

3. *In Situ* Hybridization

The ISH with FITC-labeled probes can be performed according to the following protocol.

First, the freshwater samples should be fixed with glutardialdehyde (2%). After this, cells are harvested by centrifugation (5 min, 2000*g*) and fixed with a saline ethanol solution (25 vol EtOH 100%, 2 vol distilled water and 3 vol 25× SET, SET buffer contains 3.75 M NaCl, 25 mM EDTA and 0.5 M Tris/HCl, pH is adjusted to 7.8 and the buffer is filter sterilized through 0.2-μm pore size filters) during 1 hour at room temperature while occasionally shaking the tube. After this incubation, cells are pelleted (5 min, 2000*g*). In this step the centrifugation conditions are very important to avoid destruction of the cells. The pellet is resuspended in 500 μl of hybridization buffer (5× SET, 0.1% (v/v) IGEPAL-CA630 and poly(A) (30 μg/ml), and 0–30% of formamide, depending of the probe) during 5 minutes at room temperature. Next, the cells are pelleted again (5 min, 2000*g*) and the supernatant discarded, and then the hybridization probe is added to a defined determining buffer at saturation concentrations (5–10 ng μl^{-1}). Thus, to 48 μl of the cell suspension in hybridization buffer, 12 μl of probe (50 ng/μl stock) is added. It is necessary to keep the tubes in the dark for the procedure, because the fluorescent-labeled probe is light sensitive. During hybridization, the samples are incubated at temperatures between 50° and 62 °C (depending on the probe) for 2 hours in the dark, using airtight tubes to avoid concentration effects caused by evaporation. The tubes are mixed occasionally during incubation.

To rinse off excess probe molecules to prevent nonspecific binding, a washing step is performed at the same temperature as the hybridization. For this purpose, cells are pelleted (5 min, 4000*g*) and resuspended in 100 μl of SET 1× buffer. Specificity of probe binding to the target site depends on the conditions of hybridization and washing. After washing, hybridization is stopped by the addition of 300–500 μl of cold PBS (0.1 M NaH_2PO_4 and 0.15 M NaCl, pH is adjusted to 7.2, and the buffer is filter sterilized through 0.2-μm pore size filters) and kept on ice until analysis by FCM.

C. Working with Freshwater Samples

If possible, the phytoplankton samples should be analyzed while still fresh. Otherwise, they should be fixed in 2% glutardialdehyde and kept in the dark at 4 °C or in liquid nitrogen for long-term storage (Vaulot *et al.*, 1989). The measurements of the biomass and the Chl *a* contents need a calibration system with unialgal laboratory cultures. Because the fixation with glutardialdehyde leads to an increase of Chl *a* fluorescence up to 25% and a decrease of biomass (up to 29% [Hillebrand *et al.*, 1999]), the cultures used for the model calibration must be prepared the same way as the natural samples, but the analysis of freshwater samples is commonly performed directly to prevent artifacts. Our flow cytometer was equipped with a 70-μm nozzle as sheath fluid and artificial freshwater medium was used with a sheath pressure of 11 psi. Because the nozzle is small (70 μm), it is necessary to filter the water samples collected in the field through a 20-μm filter to remove zooplankton and the phytoplankton fractions with a cell size more than 20 μm. This fraction should be analyzed under the microscope, because the cells in this fraction were destroyed by the stretching forces in the nozzle. However, we can analyze only small cells (<1 μm) with this small nozzle.

IV. Results

A. Population Structure

The analysis of phytoplankton structure was done with two FCM systems, a single-laser and a double-laser module. With the single-laser module we focused on the discrimination of phycobilin-containing algae, and with the double-laser module we tried to discriminate chlorophyceae, diatoms, cyanobacteria, and cryptophytes. Figure 6 shows the possibility of discrimination between 10 laboratory algae on the basis of scatter and fluorescence values excited with the 514-nm laser. The dual-laser system shows different clusters, depending of the pigmentation of the algal group, after plotting the Chl *a* fluorescence values excited with 488 nm (Fl 1) and 532 nm (Fl 3) (Fig. 7). The 488-nm laser strongly excited the Chl *b*, Chl *c*, fucoxanthin, and peridinin (see section II.A), and the 532-nm laser excited more efficiently the phycobilins. Identification of the different clusters was controlled via cell sorting and subsequent microscopic determination of the sorted fractions. In addition, we used a laboratory algal mixture to analyze fluorescence patterns of the different taxa. The single- and the double-laser modules are still limited by the discrimination of algae with similar absorption-like diatoms and dinophyta. Using laboratory-grown algae, we are able to distinguish between these algae, because the pigment ratios and the cell sizes (forward scatter) are different. However, in natural water samples, this method does not show sufficiently different clustering because the pigmentations and the cell sizes are too variable.

The use of internal standards (standard beads) allows the calculation of absolute phytoplankton composition. If the flow cytometer is equipped with a sorter model, the sorting of the different algal groups is an additional control tool. However, the sorting time depends on the abundance of the algae in the population and the desired purity.

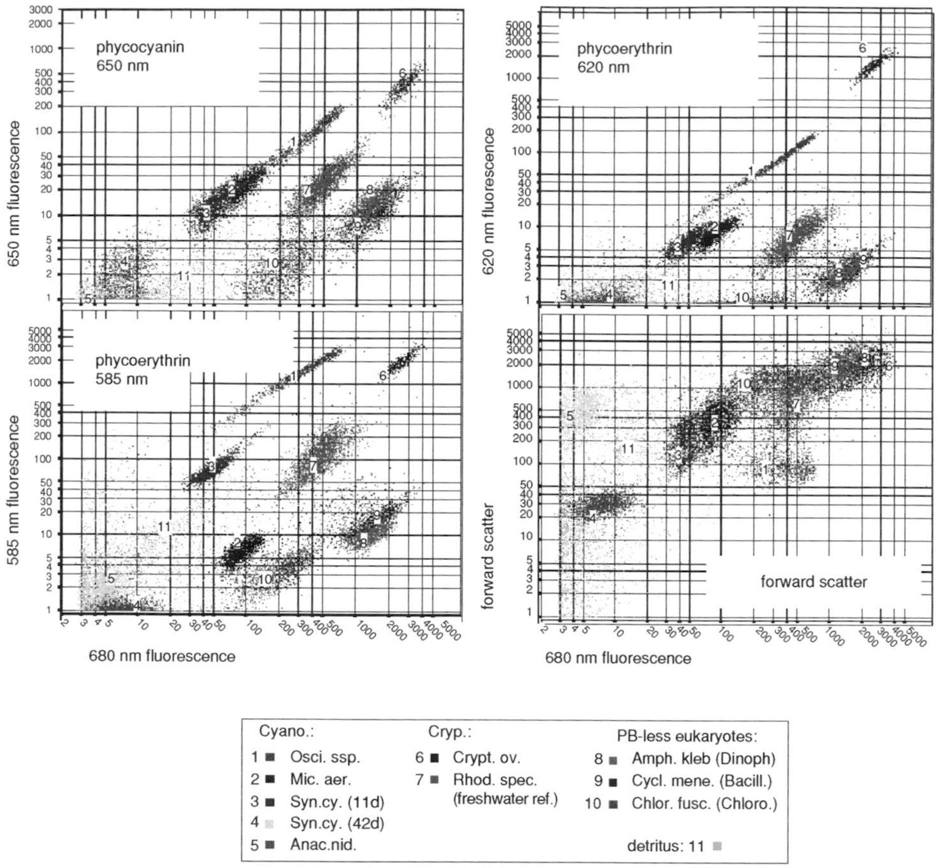

Fig. 6 Discrimination of 10 laboratory phytoplankton algae by fluorescence and scatter signal after excitation with 514 nm. Plotted are the PE (575 nm), PC (620 nm), PC (650 nm) fluorescence, and the forward scatter against the Chl *a* fluorescence at 680 nm (see Becker *et al.*, 2002). (See Color Insert.)

B. Cell Diameter and Biovolume

Cell diameter and biovolume from phytoplankton algae were calculated from the forward scatter signal. However, the cell size (diameter) and the biovolume were calculated via mathematical regression (Fig. 8). The particle shape must be considered when calculating the exact biovolume. Most flow cytometers measure not only the forward scatter channel (FSC) height but also the FSC area. Both parameters were usable for the cell size calculation. Biovolumes of the different taxa were calculated using the mathematical formula described by Hillebrand *et al.* (1999).

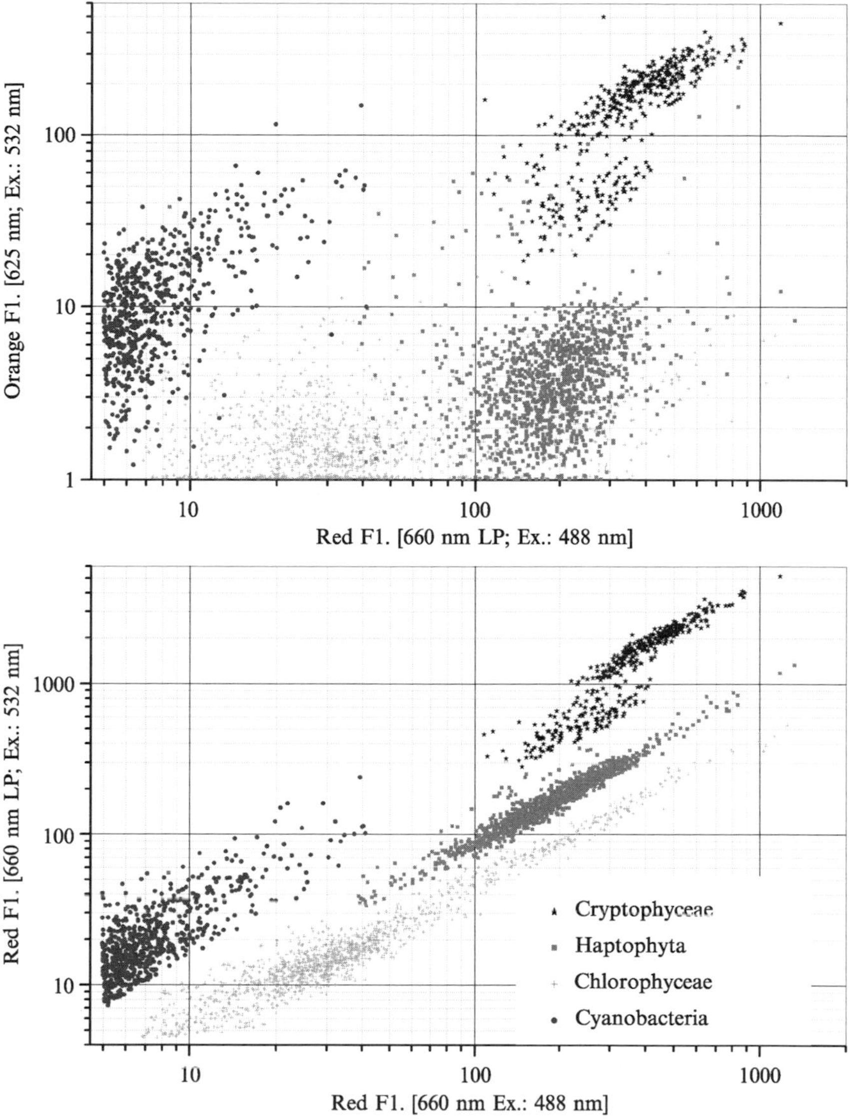

Fig. 7 Discrimination of chlorophyceae, cyanobacteria, haptophyta, and Cryptophyceae on the basis of red fluorescence emission (>660 nm) excited with 488 nm, 532-nm light, and fluorescence emission at 625 nm (excitation wavelength, 532 nm), respectively. The sample was taken from the lake Auensee (Leipzig, Germany); 10,000 cells of the sample were analyzed. Each dot represents a single algal cell and the different clusters represent algal groups with the same fluorescence properties. (See Color Insert.)

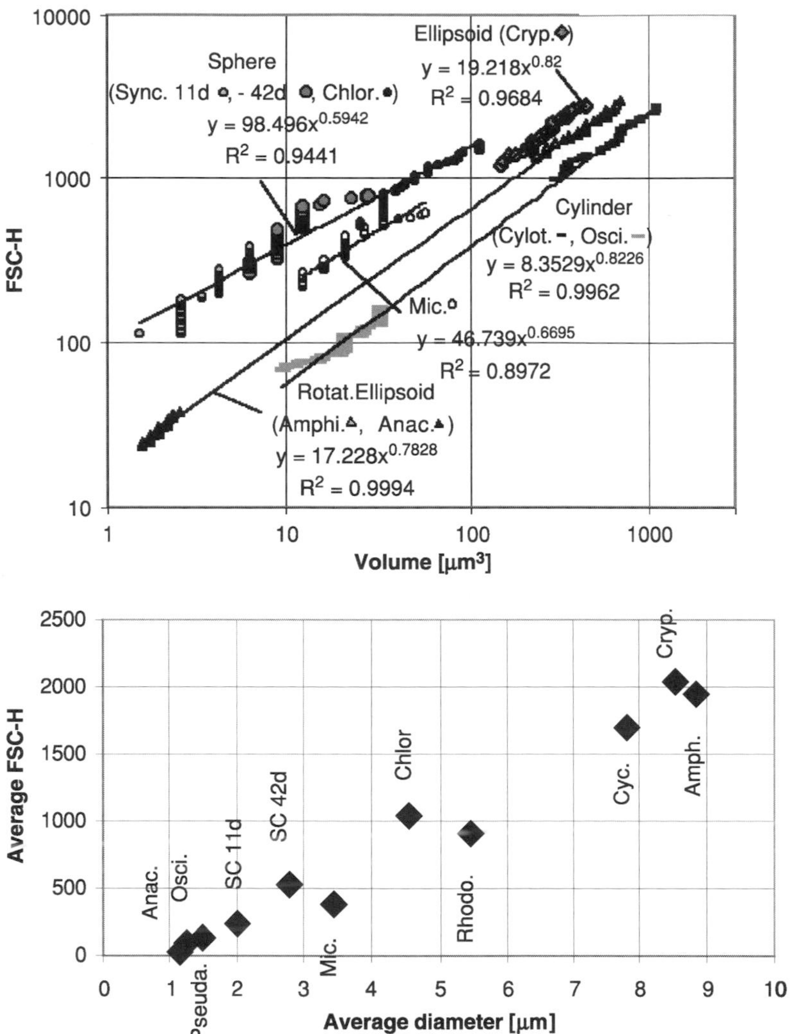

Fig. 8 Relationship between forward scatter (FSC) and biovolume depending on cell shape. Cell diameter of several phytoplankton algae in relation to FSC (excitation wavelength, 514 nm).

C. Chl *a* Content per Cell

This parameter is the crucial link between carbon-related biovolume and photosynthesis-related Chl *a* content. The amount of Chl *a* per cell is also indicative of the light-acclimation state of phytoplankton cells. To measure this parameter, one must use unialgal laboratory samples with a known Chl *a* content per cell for calibration. The fluorescence signals have to be normalized to an internal standard. There are two options: (1) the use of fluorochrome-labeled calibration beads or (2) normalization with glutardialdehyde-fixed algae (2.5%) (*Porphyridium purpureum*), as was done in our study. This alga contains Chl *a*, phycoerythrin, and phycocyanin. Kept cool in the dark, the fluorescence values are stable for months. After normalization, the fluorescence values are used as a basic parameter for the mathematical estimation of the Chl *a* content. In our study, a single fluorescence value model was developed based on Chl *a* fluorescence (excitation, 488 nm). The fluorescence values show a good linear correlation with the Chl *a* content per cell. Because the strong colinearity of all fluorescence parameters prevents a combined multivariate regression analysis for all algal groups, separate linear regressions for all major algal groups are calculated (Fig. 9). With the help of this model, one can determine the Chl *a* content per cell after classification of the different clusters, as shown above (population structure). Cyanobacteria show low fluorescence values with 488 nm excitation, because in these cells the major part of Chl *a* is bound to photosystem I, which has a very low fluorescence yield. The signal can be improved if Chl *a* is excited via the phycobilins at 532 nm.

Note that for all algal groups, a model with an ample sample number must be established. All groups must show a lognormal distribution for cell size and chlorophyll content.

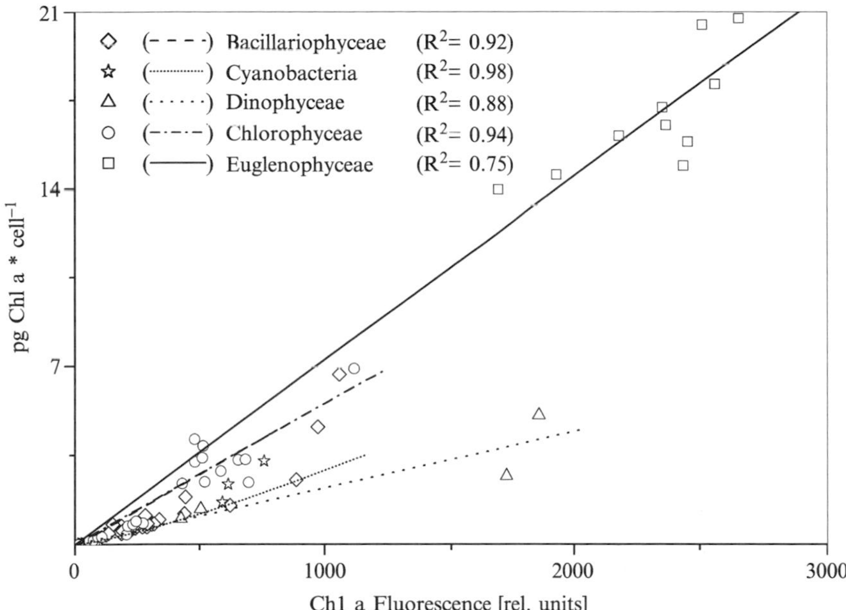

Fig. 9 Relationship between Chl *a* per cell and red fluorescence (>660 nm) for the major algal groups in freshwater (excitation wavelength generally 488 nm, except for cyanobacteria [532 nm]).

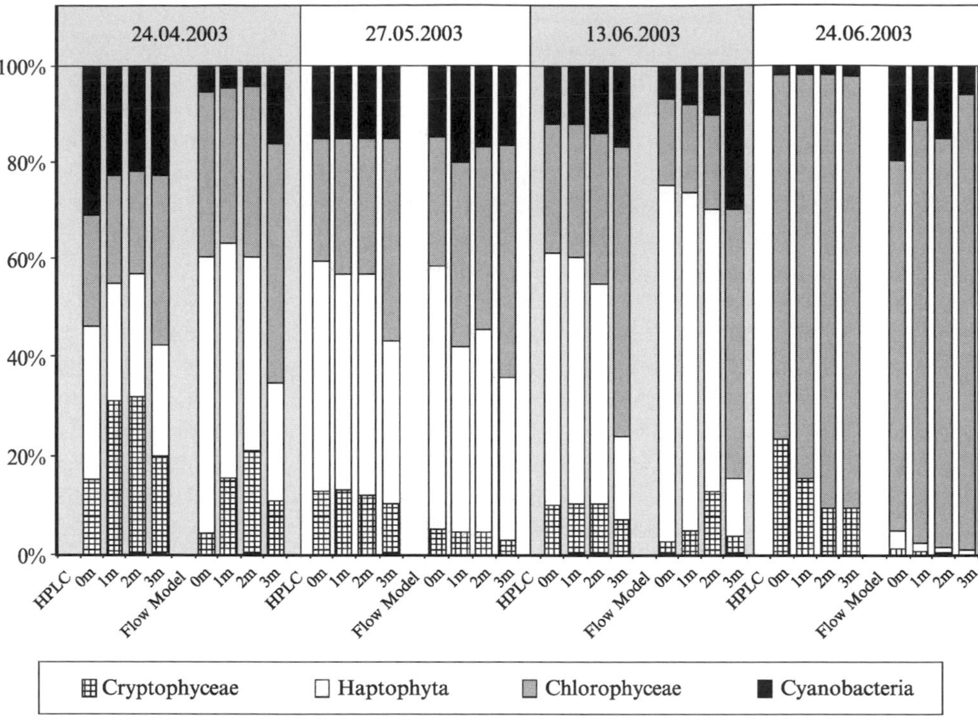

Fig. 10 Phytoplankton composition of a eutrophic lake, Auensee (Leipzig, Germany), in relation to depth during the spring season 2003 calculated via high-pressure liquid chromatography analysis and flow cytometry on the basis of chlorophyll *a* distribution.

Comparative enumeration of a freshwater sample on the basis of Chl *a* via FCM and HPLC analyses yields nearly identical results if the algae with the same pigmentation are added (Fig. 10). Differences in the phytoplankton structure are probably caused by the cell damage of large cryptophytes or underestimation of small cells due to the high detritus load.

D. *In Situ* Hybridization

The efficiency of the ISH is commonly visualized via microscopic analysis. Because the FITC-marked DNA fragments show a green fluorescence that is easy to distinguish from the red autofluorescence emitted mainly by Chl *a*, FCM can be applied for quantification of labeled cells. To detect the green FITC fluorescence, the red fluorescence was minimized by Chl *a* extraction. In Fig. 11, the fluorescence pattern of *Scenedesmus obliquus* hybridized with different probes is demonstrated. The EUK 1209R probe was used in this study as a general eukaryotic probe. The CHLO 02 and Hetero 01 probes were designed to target the division

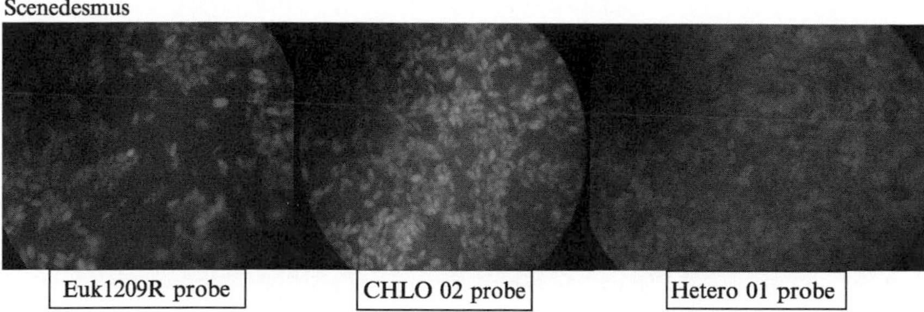

Fig. 11 Epifluorescence microscopy. *Scenedesmus obliquus* cells hybridized with different taxa-specific probes; EUK 1209R probe for eukaryotes, CHLO 02 for chlorophyta, and HETERO 01 probe for heterokontophyta. (See Color Insert.)

chlorophyta and heterokontophyta algae, respectively (Simon *et al.*, 1995; Medlin *et al.*, 2003, personal communication). The fluorescence was strong for *Scenedesmus* hybridized with EUK 1209R and CHLO 02 probes, whereas for the cells hybridized with the EUK 1209R and CHLO 02 probes, whereas for the cells hybridized with the Hetero 01 probe, only autofluorescence resulting from the pigments (mainly Chl *a*) in the cell was observed.

Two reference chlorophyta species (*S. obliquus* and *Chlamydomonas reinhardtii*) and one cyanobacteria species (*Microcystis aeruginosa*) were selected to test the suitability of four probes for whole-cell hybridization with flow cytometric detection (Fig. 12). For *Chlamydomonas* and *Scenedesmus*, an increase in the fluorescence intensity was observed after the hybridization with EUK 1209R and CHLO 02 probes, whereas there were no differences between these cells labeled with MICR 3 probe and the respective controls. *Microcystis aeruginosa* cells showed a strong increase in fluorescence when they were hybridized with MICR 3 probe designed specific for *Microcystis* species.

V. Limitations and Applications

A. Limitations

The biovolume is related to the forward scatter signal, although the determination of the cell size and the biovolume depends on the cell shape (Becker *et al.*, 2002) and the refractive index (Green *et al.*, 2003; Stramski, 1999). For example, because of intracellular gas vacuoles, the refractive index of *Microcystis* is very low, so a preparative cell sorting to determine the cell shape is still necessary. The determination of phytoplankton composition is limited for algae with the same pigmentation. Thus, seven types of phycobilins were found in cryptophytes, and the highly variable phycobilin content makes the identification more difficult. In addition, the cryptophytes are very sensitive to the stretching forces in the nozzle.

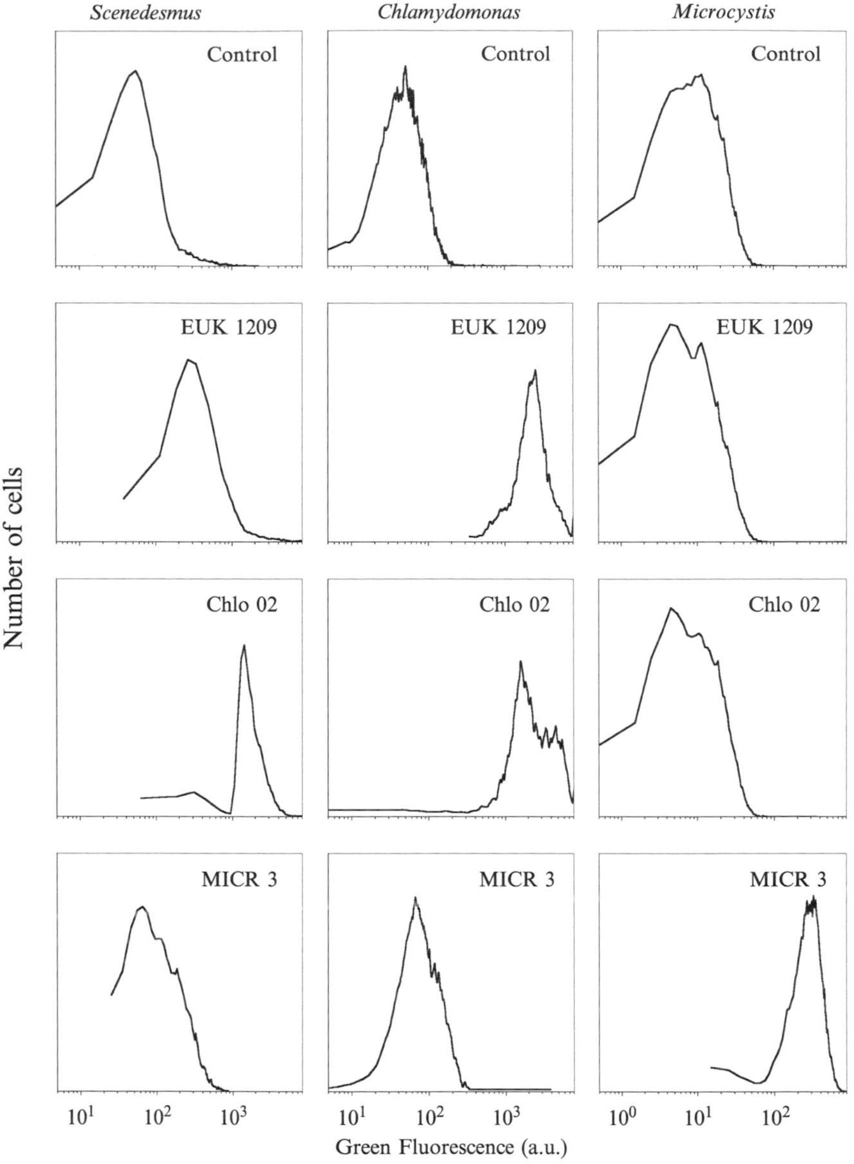

Fig. 12 Flow cytometric analysis of three phytoplankton algae (*Scenedesmus obliquus, Chlamydomonas rheinhardii*, and *Microcystis aeruginosa*) after whole-cell hybridization labeled with eukaryont-, chlorophyta-, and microcystis-specific probes (excitation wavelength, 488 nm/Emission 535 nm). For each species the green fluorescence per cell is plotted in a log scale. For each species the cells were labeled with a eukaryont-specific (EUK 1209), a chlorophyta-specific (CHLO 02), and a microcystis-specific probe.

As mentioned earlier, the Chl *a* fluorescence does not linearly correlate in all cases with the Chl *a* content, because light treatment, nutrient availability, and toxic compounds can influence this signal. Particularly in the case of the cyanobacteria, the highly variable phycobilin excitation transfer is a source of error for the quantitative Chl *a* determination. The values of the phycobilin fluorescence are not stable, because of the different physiological adaptation processes in cyanobacteria (Campell *et al.*, 1998). An example for the variable fluorescence emission is shown in Fig. 13. The two excitation spectra of *S. rubescens* before and after 5 hours of high light treatment vary in the phycoerythrin and in the Chl *a* fluorescence. The uncoupling of the phycobilin antenna leads to an increase in phycobilin fluorescence and a decreasing Chl *a* fluorescence. The allophycocyanin fluorescence at 660–680 nm particularly influences the fluorescence at 680 nm normally

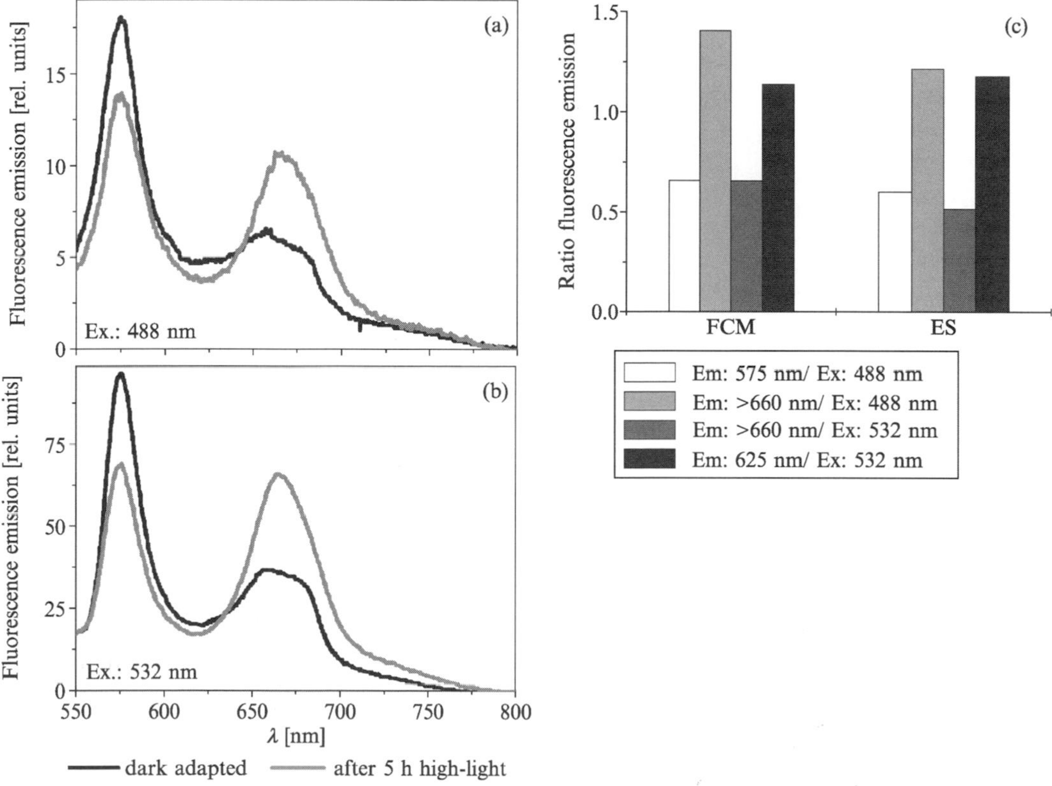

Fig. 13 Fluorescence emission spectra of *Synechococcus rubescens* before (*black line*) and after (*orange line*) 5 hours of high light treatment (1000 μmol photons $\times$ m^{-2} $\times$ s^{-1}) measured via emission spectroscopy (ES) after excitation with 488 nm (a) and 532 nm (b), (c) Ratios of fluorescence emission of *S. rubescens* before and after high light treatment. The fluorescence emission measured via ES (cell suspension) and flow cytometry (mean values of 10,000 single cells) after excitation with 488 and 532 nm. The Chl *a* per cell and the absorption spectra were identical before and after light treatment. (See Color Insert.)

used for Chl *a* determination. In addition, the appearance and the amount of phycoerythrin, phycocyanin, and allophycocyanin as the major accessory pigments in cyanobacteria are very variable intraspecifically and interspecifically. Because the absorption at the excitation wavelength (488 nm or 514 nm) can vary independent of the chlorophyll content, the fluorescence at 680 nm is not simply quantitatively correlated to the amount of Chl *a*.

FCM allows the analysis of thousands of cells per second, so this method offers fast access to statistically reliable information, whereas microscopical analysis is the most powerful for first orientation and qualitative analysis of cells (identity, shape, size). Cell sorting via FCM followed by microscopic analysis can provide a quick control of water samples and combines the advantages of both methods. However, preparative cell sorting of natural water samples (e.g., for HPLC analysis of specific algal groups) can still be time consuming, requiring up to several hours, especially if high purity is necessary.

The main limitations of ISH result from the varying cellular ribosome content and algal-specific cell wall permeability. To avoid the first limitation, the experiments were performed during the exponential phase of growth when the ribosome content is higher. To increase the cell wall permeability, one can use several detergents in the hybridization buffer such as SDS. In addition to ribosome content and cell wall permeability, the FISH signal depends on the accessibility of the rRNA target site to the fluorescently labeled oligonucleotide. Because of the densely packed three-dimensional structure of the ribosome, the probes' access to the target sites can be difficult. Hence, because of the different binding capacities and the target specificity, it must be tested for each probe if a new protocol is necessary.

B. Applications

Because flow cytometers are still expensive and educated technicians are still necessary, at least for the data processing, it is more appropriate for research projects than for routine use in water management laboratories. However, the newly developed wireless cytometer Cytobuoy offers the possibility of continuous water monitoring (see section III.A.3) (Dubelaar *et al.*, 1999), especially for lake sanitation. A similar principle is used by the submersible FlowCytoboot (Olson *et al.*, 2003; Sosik *et al.*, 2003). To use FCM in routine analysis, we need routine protocols, as illustrated by Franqueira *et al.* (2000), who tested the possibility of determining toxic compounds in water samples via FCM. Although flow cytometers are not used in routine analysis, novel findings in marine and basic research demonstrate the capacity of these instruments. One of the most important investigations was the analysis of the deep Chl *a* maxima, especially the discovery of *Prochlorococcus* species (Partensky *et al.*, 1999) with the help of aboard FCM. In addition, the development of FISH techniques was a step forward in understanding the structure and the development of phytoplankton (Not *et al.*, 2002). Presently, the development of automated detection methods of toxic algae species in a water sample is of interest.

VI. Future Directions

As outlined in this chapter, FCM is both an analytical and a preparative method, if sorting is applied. The expected technical progress in the photodiode technology provides a hope that soon expensive laser technology can be replaced by diodes. Because of reductions in costs and handling, this technique will also become attractive for routine water surveillance. In addition, FCM can be the link between laboratory work and investigations of nature water samples. The most interesting perspective of aquatic research lies in the combination of FCM and molecular techniques. Flow identification and sorting using FISH techniques offer great potential (Vives-Rego *et al.*, 2000). Single-cell techniques will cover a broader spectrum of physiological information. Intensifying the fluorescence signal from DNA probes by enzyme-linked fluorescence reactions will allow hybridization of probes, not only with rRNA molecules that carry only taxonomic information, but also with messenger RNA molecules that report the gene activity of metabolic processes. This will open new perspectives for understanding gene expression on the single-cell level. If sorting can be coupled to microscope-based analyzing techniques, the physiological states of a given subpopulation can be analyzed in more detail. Microscope-based techniques are the so-called *microscopy PAM*, which allows the measurement of photosynthetic rates on the single-cell level. Furthermore, Raman spectroscopy, electron spectroscopy, image cytometry, and FT-IR spectroscopy offer options for performing taxonomic and physiological investigations. For example, Fourier-transformation infrared spectroscopy (FT-IR) spectroscopy can deliver information on the composition of the biomass as the ratio of carbohydrates to protein. The more FT-IR data on phytoplankton cells are collected in international data bases, the more this technique can also be used for the identification of cells that cannot be discerned in the normal microscope.

References

Amann, R. I., Binder, B. J., Olson, R. J., Chisholm, S. W., Devereux, R., and Stahl, D. A. (1990). Combination of 16s rRNA-targeted oligonucleotide probes with flow cytometry for analyzing mixed microbial populations. *Appl. Environ. Microbiol.* **56,** 1919–1925.

Azam, F., Smith, D. C., Long, R. A., and Steward, G. F. (1995). Bacteria in carbon cycling as molecular problem. *Mol. Ecol. Aquatic Microbes* **38,** 39–54.

Becker, A., Herschel, A., and Wilhelm, C. (2003). Biological effects of incomplete destratification of hypertrophic freshwater reservoir. *Hydrobiology* (in press).

Becker, A., Meister, A., and Wilhelm, C. (2002). Flow cytometric discrimination of various phycobilin containing phytoplankton groups in a hypertrophic reservoir. *Cytometry* **48,** 45–57.

Beutler, M., Wiltshire, K. H., Meyer, B., Moldaenke, C., Lüring, C., Meyerhöfer, M., Hansen, U.-P., and Dau, H. (2002). A fluorometric method for the differentiation of algal populations *in vivo* and *in situ*. *Photosynthesis Research* **72,** 39–53.

Biegala, I. C., Not, F., Vaulot, D., and Simon, N. (2003). Quantitative assessment of picoeucaryotes in the natural environment using taxon specific oligonucleotide probes in association with TSA-FISH (tyramide signal amplification–fluorescent *in situ* hybridization) and flow cytometry. *Appl. Environ. Microbiol.* **69,** 5519–5529.

Bihari, N., Batel, R., and Zahn, R. K. (1999). Flow cytometry in marine environmental research. *Periodicum. Biologorum.* **101,** 151–155.

Blanchot, J., Andre, J. M., Navarette, C., and Neveux, J. (1997). Picophytoplankton dynamics in the equatorial Pacific: Diel cycling from flow-cytometer observations. *Comptes Rendus de l'Academie des Sciences Series III Sciences de la Vie* **320,** 925–931.

Blanchot, J., Andre, J.-M., Navarette, C., Neveux, J., and Radenac, M.-H. (2001). Picophytoplankton in the equatorial Pacific: Vertical distributions in the warm pool and in the high nutrient low chlorophyll conditions. *Deep Sea Res. I* **48,** 297–314.

Buesseler, K. O., and Boyd, P. W. (2003). Will ocean fertilization work? *Science* **300,** 67–68.

Büchel, C., and Wilhelm, C. (1993). *In Vivo* analysis of slow chlorophyll fluorescence induction kinetics in algae: Progress, problems and perspectives. *Photochem. Photobiol.* **58,** 137–148.

Campell, D., Hurry, V., Clarke, A. K., Gustafsson, P., and Öquist, G. (1998). Chlorophyll fluorescence analysis of cyanobacterial photosynthesis and acclimation. *Microbiol. Mol. Biol. Rev.* **62,** 667–683.

Carr, M. R., Tarran, G. A., and Burkill, P. H. (1996). Discrimination of marine phytoplankton species through the statistical analysis of their flow cytometric signatures. *J. Plankton Res.* **18,** 1225–1238.

Cavender-Bares, K. K., Frankel, S. L., and Chisholm, S. W. (1998). A dual sheath flow cytometer for shipboard analyses of phytoplankton communities from the oligotrophic oceans. *Limnol. Oceanogr.* **43,** 1383–1388.

Cavender-Bares, K. K., Mann, E. L., Chrisholm, S. W., Ondrusek, M. E., and Bidigare, R. R. (1999). Differential responses of equatorial pacific phytoplankton to iron fertilization. *Limnol. Oceanogr.* **44,** 237–246.

Dalton, R. (2002). Oceans tests raise doubts over use of algae as carbon sink. *Nature* **420,** 722.

Dawson, R. M. (1998). The toxicology of microcystins. *Toxicon* **36,** 953–962.

Dyrssen, D. W. (2001). The biochemical cycling of carbon dioxide in the oceans: Perturbations by man. *Sci. Total Environ.* **277,** 1–6.

Dubelaar, G. B. J., Groenewegen, A. C., Stokdijk, W., van den Engh, G. J., and Visser, J. W. M. (1989). Optical plankton analyzer: A flow cytometer for plankton analysis, II. Specifications. *Cytometry* **10,** 529–539.

Dubelaar, G. B. J., Gerritzen, P. L., Beeker, A. E. R., Jonker, R. R., and Tangen, K. (1999). Design and first results of CytoBuoy: A wireless flow cytometer for *in situ* analysis of marine and fresh waters. *Cytometry* **37,** 247–254.

Falkowski, P. G. (2000). Rationalizing elemental ratios in unicellular algae. *J. Phycol.* **36,** 3–6.

Fleming, L. E., Rivero, C., Burns, J., Williams, C., Bean, J. A., Shea, K. A., and Stinn, J. (2002). Blue green algal (cyanobacterial) toxins, surface drinking water, and liver cancer in Florida. *Harmful Algae* **1,** 157–168.

Franqueira, D., Orosa, M., Torres, E., Herrero, C., and Cid, A. (2000). Potential use of flow cytometry in toxicity studies with microalgae. *Sci. Total Environ.* **247,** 119–126.

Garrison, D. L., Gowing, M. M., Hughes, M. P., Campbell, L., Caron, D. A., Dennett, M. R., Shalapyonok, A., Olson, R. J., Landry, M. R., and Brown, S. L. (2000). Microbial food web structure in the Arabian Sea: A US JGOFS study. *Deep Sea Res. II* **47,** 1387–1422.

Geider, R. J., DeLucia, E. H., Falkowski, P. G., Finzi, A. C., Grime, J. P., and Grace, J. (2001). Primary productivity of planet earth: Biological determinants and physical constraints in terrestrial and aquatic habitats. *Global Change Biology* **7,** 849–882.

Giovannoni, S. J., DeLong, E. F., Olsen, G. C., and Pace, N. R. (1988). Phylogenetic group specific oligodeoxynucleodide probes for identification of single microbial cells. *J. Bacteriol.* **170,** 2418.

Green, R. E., Sosik, H. M., Olson, R. J., and DuRand, M. D. (2003). Flow cytometric determination of size and complex refractive index for marine particles: Comparison with independent and bulk estimates. *Appl. Opt.* **42,** 526–541.

Gregori, G., Denis, M., Lefevre, D., and Beker, B. (2002). A flow cytometric approach to assess phytoplankton respiration. *Methods Cell Sci.* **24,** 99–106.

Goodwin, T. W. (1988). The structure and function of phycobilisomes. *In* "Plant pigments." (T. W. Goodwin, ed.). Academic Press, London.

Guindulain, T., Comas, J., and Vives-Rego, J. (1997). Use of nucleic acid dyes SYTO-13, TOTO-1, and YOYO-1 in the study of *Escherichia coli* and marine prokaryotic populations by flow cytometry. *Appl. Exp. Microbiol.* **63,** 4608–4611.

Harris, G. P. (1986) "Phytoplankton ecology" (G. P. Harris, ed.) London.

Hillebrand, H., Dürselen, C.-D., Kirschtel, D., Pollingher, U., and Zohary, T. (1999). Biovolume calculation for pelagic and benthic microalgae. *J. Phycol.* **35,** 403–424.

Hiller, R. G. (1999). The photochemistry of carotenoids. Verlag, Netherlands.

Hipkins, M. F., and Baker, N. R. (1986). Spectroscopy. *In* "Photosynthesis energy transduction. A practical approach" (M. F. Hipkins and N. R. Baker, eds.). IRL Press, Washington, DC.

Hoffmann, P. (1975). Energieleitung. *In* "Photosynthese." (P. Hoffman, ed.). Akademie-Verlag, Berlin.

Hsu, S.-J., and Hsu, B.-D. (1998). Flow cytometry of *Chlorella* after dehydration stress. *Plant Sci.* **134,** 163–169.

Jochem, F. J. (1999). Dark survival strategies in marine phytoplankton assessed by cytometric measurement of metabolic activity with fluorescein diacetate. *Marine Biol.* **135,** 721–728.

Jochem, F. J., Smith, G. J., Zimmermann, R. C., Cabello-Passini, A., Kohrs, D. G., and Alberte, R. S. (2000). Cytometric quantification of nitrate reductase by immunolabeling in the marine diatom *Skeletonema costatum. Cytometry* **39,** 173–178.

John, U., Cembella, A., Hummert, C., Elbrächter, M., Groben, R., and Medlin, L. K. (2003). Discrimination of the toxigenic dinoflagellate species *Alexandrium tamarense* and *Alexandrium ostenfeldii* in co-occurring natural populations from Scottish coastal waters. *Eur. J. Phycol.* **38,** 25–40.

Jonker, R. R., Meulemann, J. T., Dubelaar, G. B. J., Wilkins, M. F., and Ringelberg, J. (1995). Flow cytometry: A powerful tool in the analysis of biomass distribution in phytoplankton. *Water Sci. Technol.* **32,** 177–182.

Kirk, J. T. O. (1994). Light and photosynthesis in aquatic ecosystems. Academic Press, Cambridge.

Kohl, J.-G., and Nicklisch, A. (1988). Ökophysiologic der algen. Fischer Verlag, Stuttgart.

Landsberg, J. H. (2002). The effects of harmful algal blooms on aquatic organisms. *Rev. Fisheries Sci.* **10,** 113–390.

Lebaron, P., Servais, P., Troussellier, M., Courties, C., Muyzer, G., Bernard, L., Schafer, H., Pukall, R., Stackebrandt, E., Guindulain, T., and Vives-Rego, J. (2001). Microbial community dynamics in Mediterranean nutrient-enriched seawater mesocosms: changes in abundances, activity and composition. *FEMS Microbiol. Ecol.* **34,** 255–266.

Leboulanger, C., Dorigo, U., Jacquet, S., Le Berre, B., Paolini, G., and Humpert, J.-F. (2002). Application of a submersible spectrofluorometer for rapid monitoring of freshwater cyanobacterial blooms: A case study. *Aquatic Microbial Ecol.* **30,** 83–89.

Li, W. K. W. (2002). Macroecological patterns of phytoplankton in the northwestern North Atlantic Ocean. *Nature* **419,** 154–157.

Mackey, D. J., Higgins, H. W., Mackey, M. D., and Holdsworth, D. (1998). Algal class abundances in the western equatorial pacific: Estimation from HPLC measurements of chloroplast pigments using CHEMTAX. *Deep Sea Res. I* **45,** 1441–1468.

Mackey, D. J., Blanchot, J., Higgins, H. W., and Neveux, J. (2002). Phytoplankton abundances and community structure in the equatorial pacific. *Deep Sea Res. II* **49,** 2561–2582.

Mackey, M. D., Mackey, D. J., Higgins, H. W., and Wright, S. W. (1996). CHEMTAX: A program for estimating class abundances from chemical markers: Application to HPLC measurements of phytoplankton. *Marine Ecol. Progress Series* **144,** 265–283.

Meister, A. (1992). New fluorimetric method for determination of chlorophyll *a*/*b* ratio. *Photosynthetica* **26,** 533–539.

Nobel, P. S. (1974). Transfers of excitation between photosynthetic pigment. *In* "Introduction to biophysical plant physiology." (P. S. Nobel, ed.). Freeman and Company, San Francisco.

Not, F., Simon, N., Biegala, I. C., and Vaulot, D. (2002). Application of fluorescent *in situ* hybridization coupled with tyramide signal amplification (FISH-TSA) to assess eukaryotic picoplankton composition. *Aquatic Microbial Ecol.* **28,** 157–166.

Olson, R. J., Chekalyuk, A. M., and Sosik, H. M. (1996). Phytoplankton photosynthetic characteristics from fluorescence induction assays of individual cells. *Limnol. Oceanogr.* **41,** 1253–1263.

Olson, R. J., Sosik, H. M., and Chekalyuk, A. M. (1999). Photosynthetic characteristics of marine phytoplankton from pump during probe fluorometry of individual cells at sea. *Cytometry* **35,** 227–234.

Olson, R. J., Sosik, H. M., Chekalyuk, A. M., and Shalapyonok, A. (2000). Effects of iron enrichment on phytoplankton in the Southern Ocean during late summer: Active fluorescence and flow cytometric analyses. *Deep Sea Res. II* **47,** 3181–3200.

Olson, R. J., Shalapyonok, A., and Sosik, H. M. (2003). An automated submersible flow cytometer for analyzing pico- and nanophytoplankton: FlowCytobot. *Deep Sea Res. I* **50,** 301–315.

Partensky, F., Hess, W. R., and Vaulot, D. (1999). *Prochlorococcus*: A marine photosynthetic procaryote of global significance. *Microbiol. Mol. Biol. Rev.* **63,** 106–127.

Peeters, J. C. H., Dubelaar, G. B. J., Ringelberg, J., and Visser, J. W. M. (1989). Optical plankton analyzer: A flow cytometer for plankton analysis, I. Design considerations. *Cytometry* **10,** 522–528.

Peperzak, L., Vrieling, E. G., and Sandee, B. (2000). Immuno flow cytometry in marine phytoplankton research. *Sci. Mar.* **64,** 33–42.

Rengefors, K., Pettersson, K., Blenckner, T., and Anderson, D. M. (2001). Species-specific alkaline phosphatase activity in freshwater spring phytoplankton: Application of a novel method. *J. Plankton Res.* **23,** 435–443.

Rengefors, K., Ruttenberg, K. C., Haupert, C. L., Taylor, C., Howes, B. L., and Anderson, D. M. (2003). Experimental investigation of taxon-specific response of alkaline phosphatase activity in natural freshwater phytoplankton. *Limnol. Oceanogr.* **48,** 1167–1175.

Rudi, K., Skulberg, O. M., Skulberg, R., and Jakobsen, K. (2000). Application of sequence-specific labeled 16S rRNA gene oligonucleotide probes for genetic profiling of cyanobacterial abundance and diversity by array hybridization. *Appl. Environ. Microbiol.* **66/9,** 4004–4011.

Shapiro, H. M., and Stephens, S. (1986). Flow Cytometry of DNA content using oxazine-750 or related laser dyes with 633 nm excitation. *Cytometry* **7,** 107–110.

Simon, N., Barlow, R. G., Marie, D., Partensky, F., and Vaulot, D. (1994). Characterization of oceanic photosynthetic picoeukaryotes by flow cytometry. *J. Phycol.* **30,** 922–935.

Simon, N., LeBot, N., Marie, D., Partensky, F., and Vaulot, D. (1995). Fluorescent *in situ* hybridization with rRNA-targeted oligonucleotide probes to identify small phytoplankton. *Appl. Environ. Microbiol.* **61,** 2506–2513.

Simon, N., Campell, L., Ornolfsdottir, E., Groben, R., Guillou, L., Lange, M., and Medlin, L. K. (2000). Oligonucleotide probes for the identification of three algal groups by dot plot and fluorescent whole-cell hybridization. *J. Eucaryot. Microbiol.* **47,** 76–84.

Smit, E., de Souza, F., Landeweert, R., Jacobsen, C., Ekelund, F., Groben, R., and Medlin, L. K. (2004). Molecular tools and approaches in eukaryotic microbial ecology. *In* "Molecular microbial ecology" (A. M. Osborn, ed.). Bios Scientific, Oxford, UK.

Smith, V. H. (2003). Eutrophication of freshwater and coastal marine ecosystems: A global problem. *Environ. Sci. Poll. Res.* **10,** 1–14.

Sosik, H. M., Olson, R. J., Neubert, M. G., Shalapyonok, A., and Solow, A. R. (2003). Growth rates of coastal phytoplankton from time-series measurements with a submersible flow cytometer. *Limnol. Oceanogr.* **48,** 1756–1765.

Sosik, H. M., and Olson, R. J. (2002). Phytoplankton and iron limitation of photosynthetic efficiency in the Southern Ocean during late summer. *Deep Sea Res. I* **49,** 1195–1216.

Steinberg, C. E. W., Schafer, H., Beisker, W., and Bruggemann, R. (1998). Deriving restoration goals for acidified lakes from ataxonomic phytoplankton studies. *Restor. Ecol.* **6,** 327–335.

Stramski, D. (1999). Refractive index of planktonic cells as a measure of cellular carbon and chlorophyll a content. *Deep Sea Res. I* **46,** 335–351.

Tevini, M., and Häder, D.-P. (1985). Phycobilinproteine. *In* "Allgemeine photobiologie." (M. Tevini and D. Häder, eds.). Thieme, Stuttgart, New York.

Tosti, E., Romano, G., Cuomo, A., Ianora, A., and Miralto, A. (2003). Bioactive aldehydes from diatoms block the fertilization current in ascidian oocytes. *Limnol. Oceanogr.* **66,** 72–80.

Vaulot, D., Courties, C., and Partensky, F. (1989). A simple method to preserve oceanic phytoplankton for flow cytometric analyses. *Cytometry* **10,** 629–635.

Vesey, G., Deere, D., Gauci, M. R., Griffiths, K. R., Williams, K. L., and Veal, D. A. (1997). Evaluation of fluorochromes and excitation sources for immuno-fluorescence in water samples. *Cytometry* **29,** 147–154.

Vives-Rego, J., Lebaron, P., and Nebe-von-Caron, G. (2000). Current and future applications of flow cytometry in aquatic microbiology. *FEMS Microbiol. Rev.* **24,** 429–448.

Wallace, B. B., and Hamilton, D. D. (1999). The effect of variations in irradiance on buoyancy regulation in *Microcystis aeroginosa. Limnol. Oceanogr.* **44,** 273–281.

Wallner, G., Amann, R., and Beisker, W. (1993). Optimizing fluorescent *in situ* hybridization with rRNA-targeted oligonucleotide probes for flow cytometric identification of microorganisms. *Cytometry* **14,** 136–143.

Wilhelm, C., Volkmar, P., Lohmann, C., Becker, A., and Meyer, M. (1995). The HPLC-aided pigment analysis of phytoplankton cells as a powerful tool in water quality control. *J. Water SRT-Aqua* **44,** 132–141.

Wilkins, M. F., Boddy, L., Morris, C. W., and Jonker, R. R. (1999). Identification of phytoplankton from flow cytometry data by using radial basis function neural networks. *Appl. Environ. Microbiol.* **65,** 4404–4410.

William, K. W. L. (1993). Estimation of primary production by flow cytometry. *ICES Mar. Sci. Symp.* **197,** 79–91.

Wiltshire, K. H., Harsdorf, S., Smidt, B., Blöcker, G., Reuter, R., and Schroeder, F. (1998). The determination of algal biomass (as chlorophyll) in suspended matter from Elbe estuary and German bight: A comparison of HPLC, delayed and prompt fluorescence methods. *J. Exp. Mar. Biol. Ecol.* **222,** 113–131.

Yentsch, C. M., Horan, P. K., Muirhead, K., Dortch, Q., Haugen, E., Legendre, L., Murphy, L. S., Perry, M. J., Phinney, D. A., Pomponi, S. A., Spinrad, R. W., Wood, A. M., Yentsch, C. S., and Zahuranec, B. J. (1983). Flow cytometry and cell sorting—a technique for analysis and sorting of aquatic particles. *Limnol. Oceanogr.* **28,** 1275–1280.

Zubkov, M. V., Sleigh, M. A., and Burkill, P. H. (2000). Assaying picoplankton distribution by flow cytometry and underway samples collected along a meridional transect across the Atlantic Ocean. *Aquativ Microbial Ecol.* **21,** 13–20.

CHAPTER 16

Multiplexed Microsphere Assays for Protein and DNA Binding Reactions

Kathryn L. Kellar* and Kerry G. Oliver†

*Biotechnology Core Facility Branch
Scientific Resources Program
National Center for Infectious Diseases
Centers for Disease Control and Prevention
Atlanta, Georgia 30333

†Radix BioSolutions, Ltd.
Georgetown, Texas 78626

0091-679X/04 $35.00

I. Introduction

A flow cytometer has a limit to the size of the molecular or cellular mass that can be detected. Because DNA fragments and immunoglobulin M (IgM) do not fall in that size range for all but the most sophisticated cytometers, these molecules can be detected only when bound or adsorbed onto the surface of a larger particle. This simple concept was demonstrated in the early years of flow cytometry but has seen wide application only in the last few years. Both qualitative and quantitative assays for soluble antigens, antibodies, enzymes, receptors, phosphoproteins, oligonucleotides, viral particles, peptides, and any other ligands can be performed on the surface of microspheres. The multiparametric resolving power of flow cytometry lets us use many combinations of fluorescent dyes in these binding assays. More importantly, we can dye microspheres with different intensities and numbers of fluorochromes or manufacture microspheres in various sizes so we have an array of microspheres to use to capture multiple proteins or DNA sequences simultaneously. This capacity for multiplexing the detection of molecules within one small sample has significantly broadened the range of flow cytometric applications and has fostered the development of new instruments and computer software to automate the procedures.

In a 1977 review of flow cytometry and cell sorting, Horan and Wheeless described a potential application that could identify serum antibodies that were reactive with several antigens coated on the surface of microspheres of different sizes. For flow cytometric analysis, they gated the microspheres individually based on their light-scattering properties and measured the fluorescence of fluorochrome-labeled antihuman immunoglobulin G (IgG) or IgM. In the next 2 decades, McHugh and colleagues developed numerous fluorescent microsphere immunoassays for antigens and antibodies. They used polystyrene microspheres of various sizes. Viral antigens were adsorbed onto the microsphere surfaces for the simultaneous detection of antibodies to cytomegalovirus and herpes simplex virus, hepatitis C core and nonstructural proteins, or four recombinant human immunodeficiency virus (HIV) proteins (McHugh *et al.*, 1988a, 1997; Scillian *et al.*, 1989). McHugh *et al.* (1989) similarly measured antibodies to whole-cell or subcellular extracts of *Candida albicans*. A novel assay employed human C1q-coated microspheres to capture immune complexes sequestering HIV antigens, which were identified by monoclonal antibodies to HIV proteins followed by biotin-conjugated antimouse IgG and streptavidin-phycoerythrin (PE) (McHugh *et al.*, 1986, 1988b). These assays provided improved discrimination between infected and noninfected test subjects compared with other techniques in use at the time (McHugh and Stites, 1991). Additional applications of microspheres to binding assays for single and multiple analytes are reviewed elsewhere (Kellar and Iannone, 2002; McHugh, 1994).

The number of discrete sizes of particles that can be analyzed on a standard flow cytometer is finite, so the capacity of microspheres to adsorb fluorescent dyes was important to the concept of an array of microspheres for measuring multiple analytes. Several variations of this concept have been adopted. Monosized microspheres have been dyed to different fluorescent intensities with a single fluorophore (Camilla *et al.*, 2001; Chen *et al.*, 1999; Cook *et al.*, 2001; Defoort *et al.*, 2000), as have beads of various diameters (Park *et al.*, 2000). Dyeing microspheres with two dyes has been performed to prepare first 64, and subsequently 100, spectrally distinct populations of uniformly sized microspheres for large suspension arrays (Fulton *et al.*, 1997; Kettman *et al.*, 1998). Additionally, quantum dots have been used to tag and differentiate populations of microspheres (Han *et al.*, 2001). The narrow emission bandwidth of quantum dots increases the theoretical number of microsphere populations that can be differentiated.

Commercially, these or similar microspheres are available from numerous sources, including Bangs Laboratories (Fishers, Indiana), Duke Scientific (Palo Alto, California), Luminex Corporation (Austin, Texas), Molecular Probes (Eugene, Oregon), Polysciences (Warrington, Pennsylvania), Serodyn (Indianapolis, Indiana), and Spherotech (Libertyville, Illinois).

II. Technology and Instrumentation

Multiplexed microsphere assays (MMAs) require the following components:

1. An array of microspheres sufficiently large enough to account for the analytes to be multiplexed
2. Capture ligands/molecules that can be bound to the surface of the microspheres, preferably covalently
3. Quantifiable standards for quantitative assays
4. Directly or indirectly fluorochrome-conjugated detection ligands
5. Any intermediate reagents between the captured ligand and final binding/detection step

For most of these assays, the detection fluorochrome can be the same for all analytes in the multiplex and serves as the "reporter," which directly correlates to the amount of analyte bound. The microsphere-embedded and reporter fluorophores must be complementary to the laser's excitation wavelength and the emission filters.

Multiplexed microsphere–based immunoassays have seen wide application. This assay format is an ideal substitute for enzyme-linked immunosorbent assays (ELISAs) because many antigens or antibodies can be detected simultaneously from one small sample, in less time, and with a significant reduction in reagent costs (Camilla *et al.*, 2001; Carson and Vignali, 1999; Cook *et al.*, 2001; De Jager *et al.*, 2003; Kellar *et al.*, 2001; Martins *et al.*, 2002; Oliver *et al.*, 1998; Tárnok *et al.*, 2003). Quantitation of the complex network of human or mouse cytokines, chemokines, and growth factors from culture supernatants, whole blood, sera, plasma, bronchial lavages, and tears has been a logical application for this technology. Several investigators have developed procedures for measuring these important modulatory proteins for studies of infectious and immune diseases (Camilla *et al.*, 2001; Carson and Vignali, 1999; Collins, 2000; Cook *et al.*, 2001; De Jager *et al.*, 2003; Hutchinson *et al.*, 2001; Kellar *et al.*, 2001; Mahanty *et al.*, 2001; Martins *et al.*, 2002; Oliver *et al.*, 1998; Prabhakar *et al.*, 2002; Tárnok *et al.*, 2003; Tripp *et al.*, 2000). Commercially available kits for the measurement of human, mouse, and rat cytokines, chemokines, and growth factors (Bender MedSystems, San Bruno, California; BD Biosciences/Pharmingen, San Diego, California; Bio-Rad Laboratories, Hercules, California; Biosource International, Camarillo, California; Linco Research, St. Charles, Missouri; Qiagen, Valencia, California; R&D Systems, Minneapolis, Minnesota; Upstate, Waltham, Massachusetts) are based on the procedures demonstrated by several of the aforementioned authors and have been employed by Keyes *et al.* (2003) and Weber *et al.* (2003) in their reports of tumor-bearing nude mice and vaccine trials, respectively.

Viral and bacterial antigens (Dunbar *et al.*, 2003; Iannelli *et al.*, 1997), viral and bacterial antibodies (Bellisario *et al.*, 2001; Iannelli *et al.*, 1997; Martins, 2002; Pickering *et al.*, 2002a,b), and heterophilic antibodies in patients with disputed HIV test results (Willman *et al.*, 2001) have been detected by immunoassays on multiplexed microspheres. Competitive immunoassays have proven very specific and rapid for serotyping pneumococcal isolates (Park *et al.*, 2000), quantitating neutralizing antibodies to human papillomavirus genotypes (Opalka *et al.*, 2003), detecting newborn thyroxine concentrations in blood spots (Bellisario *et al.*, 2000), and screening hybridoma supernatants for monoclonal antibodies to interleukin-6 (IL-6) (Seideman and Peritt, 2002).

Another rapidly expanding area for MMAs is the detection and characterization of multiple nucleic acid sequences. Labeled polymerase chain reaction (PCR) or gene-expression products can be directly captured by hybridization to a complementary sequence-coupled microsphere population (Armstrong *et al.*, 2000; Barker *et al.*, 1994; Colinas *et al.*, 2000; Defoort *et al.*, 2000; Dunbar and Jacobson, 2000; Dunbar *et al.*, 2003; Smith *et al.*, 1998; Spiro *et al.*, 2000; Wallace *et al.*, 2003; Yang *et al.*, 2001). Alternatively, oligonucleotide tags that are complementary to sequences at the 5′ end of specific capture probes for PCR amplicons have been attached to microspheres and used to "zip" on biotinylated products for flow cytometric analysis (Cai *et al.*, 2000; Chen *et al.*, 2000; Iannone *et al.*, 2000; Taylor *et al.*, 2001; Ye *et al.*, 2001). These tags, or "zip codes," can be used interchangeably in different multiplexes and allow the initial hybridization to the PCR amplicons to occur in solution.

A. Microspheres

Luminex Corporation provides the largest commercially available array of microspheres. These beads are 5.6 μm polystyrene microspheres cross-linked with divinylbenzene and contain functional carboxylic acid groups. The microspheres are dyed with different concentrations of two spectrally distinct fluorochromes resulting in a 10-by-10 array or 100 fluorescently distinct populations of microspheres.

Fluorescent microspheres produced by other companies resemble Luminex microspheres but differ in the dye excitation and emission characteristics, the number of spectrally distinct dyes, and the surface functionality. For example, Bangs Laboratories markets microspheres of two sizes with five intensities of Starfire Red per size, for a total of 10 distinct sets.

The introduction of quantum dots into microspheres is a realistic approach to significantly increase the number of distinct sets of microspheres available for multiplexed reactions (Han *et al.*, 2001). Practically, 10^4–10^5 microsphere populations should be obtainable using quantum dots; however, the utility of that level of multiplexing remains in question.

B. Instrumentation

Various flow instruments are compatible with MMAs. Any flow cytometer equipped with appropriate excitation sources is applicable for performing multiplexed assays. Luminex Corporation was the first to develop a bench-top cytometer designed solely for running MMAs. The Luminex 100 system uses a 635-nm 10-mW red diode laser and a 532-nm 10-mW yttrium aluminum garnet (YAG) laser for excitation, with four Avalanche Photodiodes (APDs) for discrimination of single microspheres and their classification dyes and a photomultiplier tube (PMT) for quantitation of the reporter fluorochrome. The system also is fitted with a 96-well XY platform that allows uninterrupted analysis of an entire microtiter plate, analogous to an ELISA plate reader. The xMAP system and the various derivations introduced by Luminex strategic partners (list available at www.luminex.com/) include software that classifies each microsphere set, displays the reporter median fluorescent intensity (MFI) values, and either performs curve-fit analyses and extrapolation after acquisition or exports the data to a database or text file for analysis.

One limitation of the Luminex 100 system is its operation as only a microsphere-based assay instrument without providing the capability of performing cell-based flow cytometric assays inherent in other flow cytometers. However, Janossy *et al.* (2002) have demonstrated that the Luminex 100 can be converted to an instrument capable of performing cellular analyses, and Bayer Pharmaceuticals reportedly is working on a project to provide this capability in a production instrument (Shapiro, 2003).

To meet the needs of laboratories that want an all-in-one instrument capable of performing MMAs and cellular analyses, BD Biosciences introduced the FACSArray Bioanalyzer system. As with the Luminex 100 system, the FACSArray is fitted with 532- and 635-nm light sources. The FACSArray has six parameter-detection capabilities (two light scatter and four color measurements). This design provides the ability to detect two microsphere classification fluorochromes and two detection fluorochromes. The instrument was designed to perform its proprietary multiplexed Cytometric Bead Arrays (CBAs), and six-parameter four-color cellular analysis on the same instrument.

To date, there is one source of a liquid handler that was designed specifically for the setup of MMAs on microtiter plates. The BioRobot 3000 (Qiagen) is fully automated, agitates and washes filter-bottom plates, and can be linked to a twister to load the plates onto a Luminex 100.

An alternative to flow-based systems for performing MMAs is laser scanning. A number of vendors (BD Biosciences, Palo Alto, California; Perkin Elmer, Wellesley, Massachusetts; Acumen, Cambridge, United Kingdom) have demonstrated the capability of laser scanning systems to perform MMAs. A scanning-based system provides potential advantages such as the ability to enhance sensitivity through repeated scanning of the same sample. In addition, scanning removes the need for fluidics, which has inherent problems with sample particulates and microsphere aggregates that reduce speed. Currently, none of these systems has reduced MMAs to common practice.

C. Fluorochromes

As with traditional cellular analysis, the fluorochromes best suited for MMAs are dependent on the excitation and emission wavelengths available with the cytometer. Many of the small flow systems, including the Luminex 100, the BD FACSArray, and the Guava system (Guava Technologies, Hayward, California), use a 532-nm YAG laser for excitation. Excitation of R-phycoerythrin (PE) at 532 nm yields a higher extinction coefficient and quantum yield compared with excitation at 488 nm or any other FL2 fluorochrome. There is also low Raman scattering, which occurs at 658 nm and, therefore, interferes minimally with PE emission at 578 nm. Thus, background signals are low and the signal/noise ratios are high with PE as the reporter fluorochrome for MMAs.

An important lesson learned from the original Luminex FlowMetrix system is the importance of the separation of emission spectra between classification dyes and quantitation dyes. The FlowMetrix system was integrated with a BD FACScan or FACSCalibur cytometer and the two classification dyes and the detection dye were excited with a 488-nm laser. Whereas spectra for classification dyes could overlap and still be properly classified, significant spectral overlap between the quantitation dye and classification dyes significantly reduced system usability because of the need for compensation. With multiple microsphere sets that contained variable amounts of classification dyes, compensation to remove the quantitation dye signal overlap into these different populations often became difficult. The current Luminex 100 system has solved this problem by integrating a two-laser system with classification and quantitation fluorochromes that do not overlap spectrally.

III. Methods

Assay development starts with attaching capture molecules to the surface of microspheres. Capture molecules, including nucleic acids, antibodies, peptides, carbohydrates, and other ligands, can be hydrophobically adsorbed or covalently coupled to a surface functionality on the microsphere. Hydrophobic adsorption is performed in much the same manner as adsorption to ELISA plates; however, adsorption is nonspecific and does not covalently bind the capture molecule to the surface. This can be a significant pitfall when performing multiplexed assays because desorption of capture molecules from one microsphere set and resorption to another set can lead to apparent cross-reactivity in assays.

Methods for covalent coupling of capture molecules vary with the microsphere's surface functionality and the available functionality of the capture molecule. Microspheres are available with carboxyl, amine/hydrazide, maleimide, and other groups that can accept amine, carboxyl, carbohydrate, sulfhydryl, and other capture molecule linkages (Bangs Laboratories; Duke Scientific; Luminex Corp.; Molecular Probes; Polysciences; Radix BioSolutions, Georgetown, Texas; Serodyn; Spherotech). There is extensive literature devoted to defining methodologies for coupling molecules of differing functionality (Hermanson, 1996; McHugh, 1994).

An additional approach to covalently attaching specific capture molecules to the surface of microspheres is through high-affinity ligand–ligate reactions such as biotin–avidin, glutathione–GST, poly-histidine–Ni chelate, or anti-immunoglobulin (Bangs Laboratories; Duke Scientific; Luminex Corp.; Molecular Probes; Polysciences; Qiagen; Radix BioSolutions; Serodyn; Spherotech). These ligands can be inherent in the capture molecule, introduced through covalent coupling, or introduced during the molecule's synthesis. The interactions can be high affinity (e.g., biotin–avidin Kd$\sim$10e-15) and can significantly ease the creation of coupled microspheres for the end user.

Once capture molecules are coupled to the surface of the microsphere, development of microsphere-based assays of any format follows the same regimen as assay development for other platforms. The user must decide whether the assay is to be performed as a heterogeneous or homogeneous assay that includes no wash or wash steps, respectively. The format chosen will not drastically alter the development process but will dictate the amount of detection reagent used in the assay and require the user to be aware of the potential for a high-dose "hook effect" in a homogeneous assay. A high-dose "hook effect" is a condition in a homogeneous assay wherein unbound analyte competes with bound analyte for limiting amounts of detection reagent. In this circumstance, a high concentration of unbound analyte reduces the signal generated from the bound analyte, so the signal generated will not correlate with the true analyte concentration.

The following sections describe methods for performing a capture-sandwich immunoassay, a competitive immunoassay, and a direct hybridization nucleic acid detection assay, which are the most common formats being developed for MMAs.

A. Capture-Sandwich Immunoassay

The method for performing a microsphere-based capture-sandwich immunoassay is essentially the same as performing an ELISA. The recommended assay buffer, PBS-0.5% BSA-0.02% sodium azide (PBN), is also used for dilution and storage of reagents. A prewet 96-well filter-bottom microtiter plate (MultiScreen MABVN 1.2 μm, Millipore Corp., Burlington, Massachusetts) is preferred for heterogeneous assays. These assay plates are formatted similarly to an ELISA plate with blanks, standards, and samples. During the entire procedure, the fluorescent microspheres are kept in suspension and in the dark or in subdued light to minimize photobleaching. First, the capture microspheres are mixed with serial dilutions of the standards, blanks, and test samples and incubated at room temperature on a shaker until equilibrium is established, usually 1 hour or less. If the assay is performed as a heterogeneous assay, then the microspheres are washed with PBS-0.05% Tween 20 by vacuum filtration after the capture phase. If tubes or paramagnetic microspheres are used, then centrifugation or magnetic separation can be performed. After the wash, the microspheres are incubated with the detection antibodies. The detection antibodies can be unlabeled, directly labeled with the detection fluorochrome, or biotin-conjugated, depending on the required sensitivity, the assay format, and the sample matrix. Final detection of the assay can occur directly, with a fluorochrome-labeled anti-immunoglobulin or a fluorochrome-labeled streptavidin. If the assay is being performed as a homogeneous assay, the sample matrix will greatly influence the performance of the assay. A serum-derived sample will contain sufficient immunoglobulin to interfere with anti-immunoglobulin detection, and a tissue culture–derived sample will contain sufficient free biotin in the tissue culture media to interfere with streptavidin-based detection. All of these possible interferences must be accounted for by the use of appropriate controls and means of detection.

Initially, assays for each analyte should be developed and optimized separately. Optimization can involve titration of the amount of capture molecule coupled to the microspheres, determination of the range of the standard curves, titration of the amount of the conjugated detection antibody or detection antibody and reporter fluorochrome, as well as the incubation times and the number of wash steps, if used. For most antibody pairs, the first optimization step can be eliminated if the concentrations that have been defined by the manufacturer, and repeatedly tested, are followed (Kellar and Douglass, 2003; Kellar *et al.*, 2001). The range of the standard curves and the amount of the detection antibody will vary with the affinities of the antibody pairs. The most important titration is that of the detection antibody, which may need to be readjusted slightly on a regular basis. Also, fluorescent signals can be very high with the immunoassays, so MFIs should be kept within the linear range of the detectors. Once single assay standard curves are optimized, the assays are multiplexed by combining the reagents for all analytes into the same volumes used for a single analyte. The affinity of the weakest pair of antibodies will define the incubation times and the number of washes required for the multiplexed assay. A detailed protocol for performing a multiplexed assay for eight human cytokines has been reported by Kellar and Douglass (2003).

B. Competitive Immunoassay

The development of a competitive immunoassay requires a single antigen-specific antibody and a competitive antigen. Either the competitive antigen or the antigen-specific antibody can be directly coupled to the microsphere, whereas the other reagent must be labeled and act as a tracer in the presence of the samples (Bellisario *et al.*, 2000; Park *et al.*, 2000). The addition of the sample establishes a competition between the sample antigen and the competitive antigen for binding to the specific antibody. The assay format can be established in a true competitive format in which all reagents are present at the same time. Alternatively, the sample antigen and the specific antibody can be combined, followed by the addition of the competitive antigen to measure the remaining free binding sites.

A competitive assay for thyroxine (T_4) was developed according to the procedure described by Bellisario *et al.* (2000). Microspheres were coupled with T_4-BSA, which was also used as the immunogen for the commercially available antibody. The microspheres (2500 in 20 μl) were added to the wells of a prewet filter plate containing 20 μl of PBN, 40 μl of normal human serum (diluted 1:5), 40 μl of T_4 standards in dilutions ranging from 150 to 0.01 μg/dl, 20 μl of a blocking agent (mouse/rat serum = 1:1), and 50 μl of biotinylated anti-T_4 at 2 mg/ml. The plates were shaken for 30 minutes, washed three times by vacuum filtration, and then 50 μl of PBN was added to each well, followed by 50 μl 8.5 μg/ml streptavidin–PE. After a final wash, 100 μl of PBN was added to each well and the microspheres were analyzed.

C. Nucleic Acid Direct Hybridization Assay

The most common multiplexed microsphere-based nucleic acid detection assay monitors the direct hybridization of a sample nucleic acid to a sequence-specific probe-conjugated microsphere. The nucleic acid sample can be labeled directly during amplification, or a number of other mechanisms, such as primer extension, oligonucleotide ligation, or rolling circle amplification, can be used to introduce a label that is dependent on a specific hybridization event (Cai *et al.*, 2000; Chen *et al.*, 2000; Iannone *et al.*, 2000; Mullinex *et al.*, 2002). To optimize a multiplexed hybridization assay, one can adjust the probe sequence and/or length to equalize each probe's melting temperature in order to maximize the reporter signal. Alternatively, the hybridization can be performed in the presence of one of the many chemicals demonstrated to normalize melting temperatures regardless of G/C ratios (Wood *et al.*, 1985).

To establish a hybridization assay, prepare 10-fold dilutions from 0.1 to 100.0 fmol of a 5′-biotinylated complementary synthetic oligonucleotide target to hybridize to capture oligocoupled microspheres to determine the signal strength and sensitivity of the assay. The synthetic target is diluted in a final volume of 25 μl in 10 mM of Tris–1 mM EDTA buffer, pH of 8.0. Then 50 μl of microspheres (1.5 $\times$ 10^5/ml each specificity), diluted in 1.5 $\times$ hybridization solution (4.5 M tetramethylammonium chloride, 75 mM Tris–HCl-6 mM EDTA, pH of 0.15% Sarkosyl) is added to the biotinylated complementary oligonucleotide dilution. The mixture is heated to 95 °C for 5 minutes and subsequently is transferred to a heat block warmed to an appropriate hybridization temperature (generally 37–55 °C). Reactions are hybridized for 20–30 minutes, pelleted by centrifugation, the supernatants removed, and the reactions are resuspended in 75 μl of streptavidin–PE (20 μg/ml). Reactions are incubated at room temperature for an additional 15 minutes and then are analyzed.

Once the assay is established with synthetic target, assay performance is optimized with PCR-amplified samples. A standard PCR amplification is performed using a biotinylated reverse-strand primer on known positive and known negative samples. Optimization is performed to maximize appropriate and to minimize inappropriate hybridization events. Optimization parameters include increasing or decreasing the volume of amplified material used in the hybridization reaction mixture and determining the optimal hybridization temperature. The optimal hybridization temperature is determined empirically by increasing or decreasing the hybridization temperature until the appropriate signals are maximized and inappropriate cross-hybridization signals are minimized.

IV. Results

Quantitative multiplexed immunoassay data are generated by first gating on the size of the microspheres, selecting only single microspheres, classifying the individual microsphere subsets, and finally measuring the fluorescent intensity of the reporter fluorochrome, usually PE, per microsphere. A statistically significant number of each microsphere population is analyzed to determine the MFI. The standard curves are generated from these values and the concentrations of each analyte in each sample are extrapolated from the curves by applicable curve-fitting software. Standard curves for the chemokine interleukin-8 (IL-8) generated in a multiplexed sandwich immunoassay are shown in Fig. 1. Figure 2 illustrates the standard curves for an MMA that combines a sandwich immunoassay for thyrotropin (thyroid-stimulating hormine) with the competitive immunoassay for T_4, described in section III.

The first immunoassays to be multiplexed on commercially available microsphere arrays were cytokine assays (Camilla *et al.*, 2001; Carson and Vignali, 1999; Cook *et al.*, 2001; Hutchinson *et al.*, 2001; Kellar *et al.*, 2001; Mahanty *et al.*, 2001; Oliver *et al.*, 1998; Tripp *et al.*, 2000). The complex interactions of cytokines, chemokines, and growth factors are ideally suited to simultaneous measurements from the same sample. There is a great potential for MMAs to generate valuable baseline data on the levels of these important analytes in the serum, plasma, and

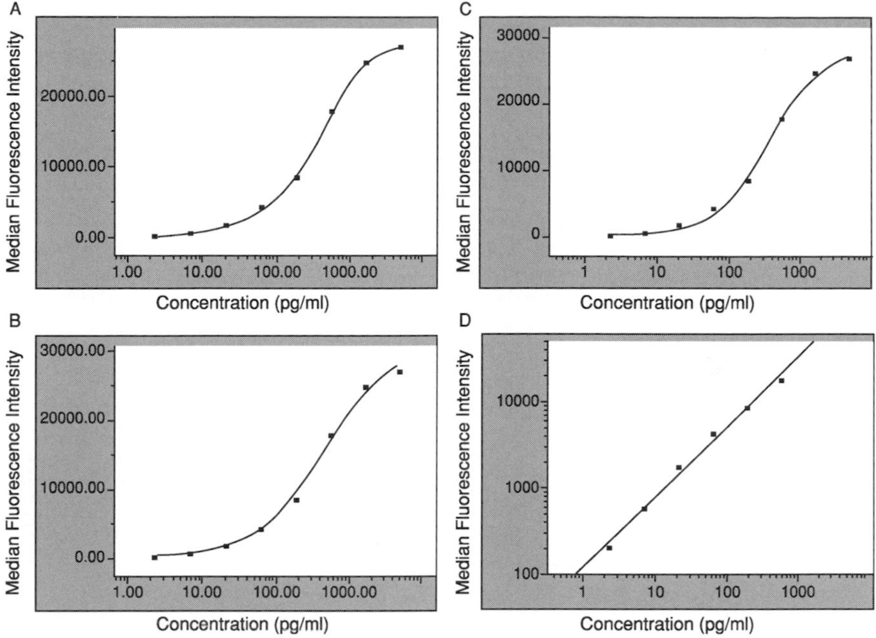

Fig. 1 Standard curves for interleukin-8 (IL-8) generated in a multiplexed microsphere-based assay. Different software packages offer different derivations of the formulas for regression analysis of standard curves. (A) An 8-point curve for IL-8 was transformed by a weighted five-parameter logistic curve fitting model. Data analysis was performed with Bio-Plex Manager 3.0 software (Bio-Rad Laboratories, Hercules, California). (B) The same data presented in (A) was transformed by a four-parameter logistic model with Bio-Plex Manager 3.0. (C) The same data presented in (A) was transformed by a four-parameter logistic formula with Sigma Plot 2000 software (SPSS, Chicago, Illinois). (D) The lower 6 points of the data presented in (A) were subjected to linear regression analysis of a logarithmic transformation of the values with SigmaPlot 2000 software. The lower sensitivity of the assay is visually apparent with this regression model.

other body fluids of healthy subjects and those with a range of infectious and immunological diseases, but thc data are emerging slowly.

Nucleic acid hybridizations are performed by using microspheres coupled with an oligonucleotide specific for the sequence of interest. Labeled complementary oligonucleotides or labeled PCR products are hybridized to these microspheres under experimentally defined conditions. Unless the nucleic acid amplification is performed under linear amplification conditions, the results will be only qualitative. A titration of complementary oligonucleotides defines only the theoretical sensitivity of the assay, because hybridization of an amplicon can be significantly affected by any secondary structure of the amplicon sequence. Figure 3 illustrates hybridization data generated both from titration of a complementary oligonucleotide and from titration of an amplicon.

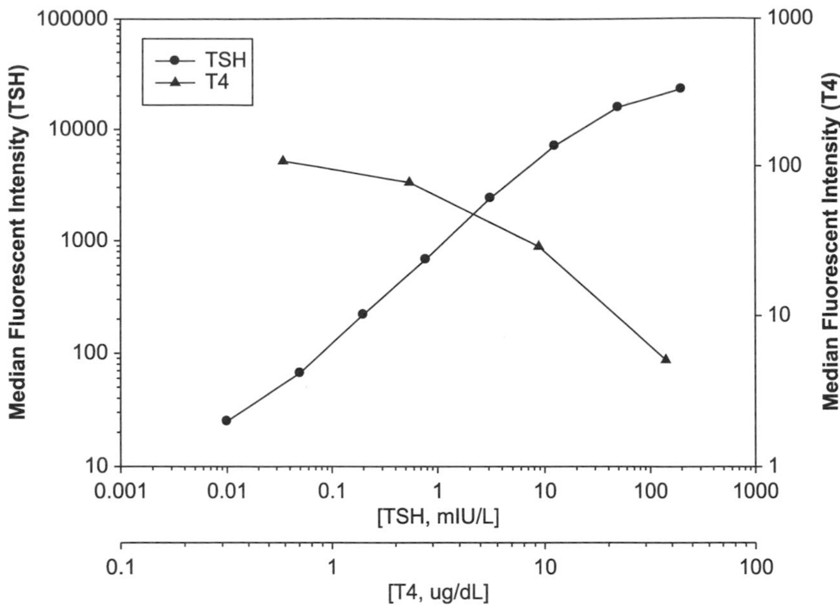

Fig. 2 A competitive immunoassay for thyroxine (T_4) and a sandwich immunoassay for thyrotropin (thyroid-stimulating hormone [TSH]) were combined into one multiplexed immunoassay performed as described by Bellisario *et al.* (2000). The concentration units for each analyte differ. A linear regression analysis was performed on a log transformation of the data with SigmaPlot 2000 software. (Maress Lacuesta, Dr. Carol Worthman, and Dr. Kathryn Kellar, 2003, unpublished data.)

V. Software

For data analysis of quantitative multiplex immunoassays, the choice of software is critical. MMAs can extend over a range of 4 to 5 logs and require a more complex curve-fitting formulation than that used for ELISAs that extend over 2 to 3 logs. A 5-parameter (5-P) logistic curve-fitting model that includes eight weighting algorithms to optimally fit the nonsymmetrical sigmoidal curves has been written for ELISAs and can be interfaced to instruments for multiplexed assay analysis (StatLIA, Brendan Scientific, Grosse Pointe Farms, Michigan). Figure 1 illustrates the transformation of the MFIs for a standard curve for IL-8 with different curve-fitting routines. The weighted 5-P formula provides a better fit than the 4-P formulas based on the fit probability scores (0.2819 vs 0.1047). Other software packages also incorporate 5-P logistical curve-fitting formulations with and without weighting capabilities, as well as the 4-P formulas traditionally used with ELISA data. Table I lists many of the software packages available for postacquisition analysis of multiplexed assay results and those that can be used to import data for analysis.

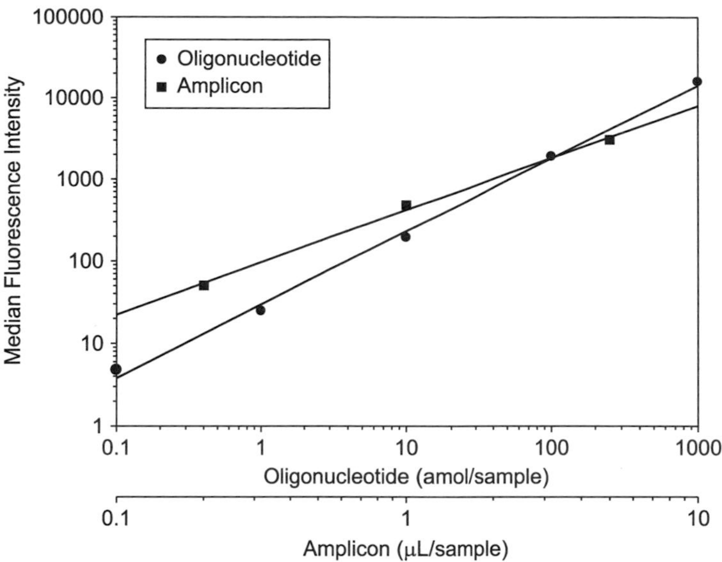

Fig. 3 Titration of oligonucleotide and polymerase chain reaction (PCR) amplicon in a nucleic acid hybridization assay. A sequence-specific oligonucleotide containing a 5′ C12 spacer with an amino terminus was coupled covalently to Luminex 100 carboxylated microspheres. A titration of 5′ biotinylated complementary oligonucleotide or a titration of a 5′ biotinylated PCR amplicon (242 bases) was hybridized and detected as described in the text. Linear regression analysis was applied to the data that defines the sensitivity of the assay. Solid circles correspond to the sequence-specific oligonucleotide hybridization and solid squares represent the PCR amplicon hybridization.

Multiplexed technology not only yields many values per sample but also is applicable to high sample throughput and, thus, can yield considerable data. Undoubtedly, bioinformatics will play an important role in data management as MMAs proliferate.

VI. Critical Aspects of the Methodology

A. Handling the Microspheres

The success of MMAs is often dependent on the handling of the microspheres. Sonicating, vortexing, and shaking the microspheres in tubes or on the filter plate are crucial for maintaining single-bead suspensions that facilitate robust ligand binding, rapid acquisition on the cytometer, and good assay reproducibility. However, the microspheres are susceptible to stripping of molecules from their surface if a filter-bottom microtiter plate is used with a horizontal mixer (Kellar *et al.*, 2001).

Table I
Software for Data Collection and Analysis[a]

Data collection	Source	Data export	Curve fit
Luminex 100 software 1.7[b]	Luminex Corp., Austin, Texas	To Excel	None
Data collection and analysis			
Cytometric Bead Array software[c]	BD Biosciences/Pharmingen, San Diego, California	Excel based	4-P
Bio-Plex Manager 3.0[d]	Bio-Rad, Hercules, California	To Excel p. a.	4-P, 5-Pw
Luminex IS Software[d]	Biosource International, Camarillo, California	To Excel p. a.	4-P, 5-P
	INOVA Diagnostics, San Diego, California		
	LINCO Research, St. Louis, Missouri		
	Mariglen Biosciences, Ijamsville, Maryland		
	QIAGEN, Valencia, California		
	Upstate, Waltham, Massachusetts		
	Zeus Scientific, Raritan, New Jersey		
STarStation[d]	Applied Cytometry Systems, Sacramento, California	From Luminex 100	4-P, 5-P
Data analysis			
MasterPlex QT (Quantitative)[e]	MiraiBio, Alameda, California	From Luminex 100	4-P, 5-Pw
MasterPlex GT (Genotyping)[e]	MiraiBio, Alameda, California	None	
StatLIA[e]	Brendan Scientific, Grosse Pointe Farms, Michigan	From Luminex 100	4-P, 5-Pw
Prism	GraphPad, San Diego, California	From Excel	4-P
SOFTmaxPRO v.3.1.2	Molecular Devices, Sunnyvale, California	From Excel	4-P
SigmaPlot 2000	SPSS, Chicago, Illinois	From Excel	4-P

Notes: Several of the software programs that can be used to collect and analyze data from multiplexed microsphere assays are listed. Most of these programs were written for the Luminex 100 instrument, available from Luminex partners. Curve fitting for most software packages also includes linear, cubic spline, and other logistic regression formulas. 4-P, four-parameter logistic curve-fitting model; 5-P, five-parameter logistic curve-fitting model; 5-Pw, weighted five-parameter logistic curve-fitting model.

[a]Use of trade names and commercial sources is for identification only and does not imply endorsement by the Centers for Disease Control and Prevention, Department of Health and Human Services.

[b]Acquired data are exported to Excel and then to other data analysis software.

[c]Designed for data acquisition and analysis for the Cytometric Bead Array kits available from BD Biosciences.

[d]Data are analyzed postacquisition (p. a.) and then can be exported to Excel for further processing and graphic display.

[e]Modified for use with .csv files from Luminex 100 version 1.7 or IS Software.

B. Optimization and Validation of Immunoassays

The sensitivity and specificity of quantitative multiplexed immunoassays for cytokines or other antigens are directly dependent on the affinity and specificity of the individual antibodies. Monoclonal or polyclonal antibodies have been used effectively, but affinity and specificity vary from lot to lot with polyclonals. For sandwich immunoassays, the antibody pairs should be of high affinity (Kd < 10e-8) and each should react with different epitopes of the antigen. Some antibody pairs that work well with ELISAs may not perform as well in multiplexed immunoassays, so antibodies from different sources may need to be tested as both capture and detection antibodies to determine the best combinations. Most importantly, cross-reactivity between multiple antibodies and antigens should be carefully examined to ensure that the specificity of each antigen for its antibodies is not affected by other components of the assay (Camilla *et al.*, 2001; De Jager *et al.*, 2003; Kellar and Douglass, 2003).

The slopes of the standard curves for an analyte may vary with the sample matrix. For example, if plasma is to be assayed, the standards should be assayed in normal plasma or a complex of plasma proteins that simulates plasma as much as possible. Plasma, serum, and other body fluids contain factors, such as soluble receptors, rheumatoid factor, heterophilic antibodies, and human anti-animal antibodies, which will produce false-positive or false-negative signals and adversely affect quantitation (Hennig *et al.*, 2000; Willman *et al.*, 2001). Spiking of normal matrix proteins with standards and measuring the recovery of each analyte will uncover possible interference from the sample matrix. Blocking proteins may need to be included to significantly reduce or eliminate any known or potential sample matrix effects (Kellar and Iannone, 2002; Kellar *et al.*, 2001).

Multiplexed microsphere-based immunoassays are at least as reproducible as the traditional monoplex ELISA (Camilla *et al.*, 2001; Chen *et al.*, 1999; Kellar *et al.*, 2001). To ensure reproducibility, one must institute proper quality controls. Each preparation of microspheres should be evaluated by preset criteria, and standard curves should be given a trial run to optimize the assay reagents before samples are analyzed (Kellar and Douglass, 2003). Calibration and proper maintenance of the flow cytometer are critical. The MFI of 100 microspheres per analyte per sample is generally reported (Bellisario *et al.*, 2001; Carson and Vignali, 1999; Siedeman and Peritt, 2002). Because each microsphere serves as the solid support for an individual assay, the MFI is highly significant. Replicates can be used to correct for pipetting errors, particularly if small sample volumes are measured.

Competitive immunoassays are relatively simple, relying on one antibody to bind the analyte captured on the microsphere unless competed off by analyte present in the sample (Opalka *et al.*, 2003; Park *et al.*, 2000). A small capture antigen may need to be complexed with a carrier molecule, such as a steroid–BSA conjugate, or linked to the microsphere via a carbon bridge to ensure that it is physically available for binding and not sterically hindered by the microsphere surface.

C. Optimization and Validation of Oligonucleotide Assays

Specificity and hybridization temperature optimums must be determined to ensure proper results. Hybridization temperature optimization is determined empirically based on the amount of signal generated with a positive control sample versus a negative control sample. The temperature is increased to maximize specificity without compromising signal. Specificity is best tested with samples positive for each marker. If not available, then specificity can be tested with individual complementary oligonucleotides. For the analysis of single nucleotide polymorphisms (SNPs), specificity is optimal when the variant base is as close to the middle of the probe sequence as possible. To best validate cross-reactivity, a number of organizations exist that will provide samples of known sequence (Coriell Institute for Medical Research, Camden, New Jersey). In addition, commercial and public software packages can aid in probe design and elucidation of potential cross-reactivity issues (Vector NTI, Informax, Fredrick, Maryland; Visual OMP, DNA Software, Inc., Ann Arbor, Michigan; Primer 3, Whitehead Institute and Howard Hughes Memorial Institute, Massachusetts Institute of Technology, Cambridge, Massachusetts). For hybridization methods that "zip" the probe as opposed to the reaction product onto microspheres for analysis, Kaderali *et al.* (2003) have developed a program to assist in the design and selection of tags and anti-tags.

VII. Comparison with Other Methods

ELISA remains the gold standard for immunoassay measurements of soluble analytes, and it will stay that way until a technology can equal the sensitivity, ease of operation, cost, and availability. MMAs are an attractive alternative to ELISAs because of the multiplexing capability. These assays have been demonstrated to be at least as or more sensitive than, as reliable as, and as easy to perform as ELISAs (Camilla *et al.*, 2001; Carson and Vignali, 1999; Chen *et al.*, 1999; Cook *et al.*, 2001; De Jager *et al.*, 2003; Kellar *et al.*, 2001; Oliver *et al.*, 1998; Prabhakar *et al.*, 2002; Tárnok *et al.*, 2003). In addition to the multiplexing capabilities, the smaller sample requirements and the generally lower cost per data point are desirable features. The major drawbacks are cost of the instrumentation, the limited palettes of commercial kits, and the inflexibility of those kits to provide only the analytes needed by the end user. As more and more kit manufacturers adopt the MMA platform, prices will decrease and the palette of assay kits will increase greatly.

A number of other multiplexed assay platforms are available to challenge the ELISA market. Pierce Biotechnology (Rockford, Illinois), Ray Biotech, Inc. (Norcross, Georgia), and others (Ekins *et al.*, 1990) have developed two-dimensional or flat arrays that involve spotting capture antibodies onto the bottom of microtiter plates, membranes, or glass slides to produce the array. The primary drawbacks of this technology are cost and flexibility. First, the cost involves either the spotting equipment or purchasing prespotted arrays, as well as the cost of a sensitive scanning system capable of discriminating signal from each individual spot. Another drawback is the decreased flexibility compared with MMAs. To modify the components of a microsphere-based array requires the addition or removal of individual microsphere sets, whereas a spotted array requires reformatting and respotting the entire array.

An additional technology that is capable of multiplexing is the multiple high-pressure liquid chromatography affinity column design introduced by Phillips (2001). This system involves generating a separate affinity column for each analyte of interest and performing the analysis in serial. Samples are labeled with Cy5 and then are detected with laser-induced fluorescence following elution from each column. The system demonstrates sensitivity and flexibility comparable to that of MMAs. However, the system is limited because of the complexity of developing and maintaining each column, which can be used only for about 200 analyses. In addition, unlike two-dimensional spotted arrays that compensate for analyte-labeling differences in different samples by comparing the ratio of two samples each labeled with Cy3 and Cy5, this system labels proteins only with Cy5. This format introduces a bias caused by variation in each labeling reaction and variation in labeling due to analyte concentration differences within samples. Many of these issues are minor and can be addressed; however, until a commercial entity manufactures the system, addresses the limitation, and produces a cartridge format that is user friendly, the utility of the system will remain limited.

VIII. Future Directions

Multiplexed assays hold the potential to revolutionize the way biology is studied and understood. The increased knowledge obtained from multiple bits of information about a sample can be synergistic in understanding the biological pathways and events. MMAs hold great promise for becoming the standard format for multiplexed assays because of their flexibility, ease of use, and the relatively small affordable equipment required to perform the assays. Progress toward greater sensitivity leveraged from larger signal/noise ratios and amplification techniques (Mullinex *et al.*, 2002) will provide an enhancement over the current technology. This is especially valuable when comparing diseased individuals with normals who may have extremely low concentrations of specific proteins in their serum. In addition, multiplexing provides a discovery tool for identifying new diagnostic markers of disease and combined with increased sensitivity potentially turns these into prognostic, rather than diagnostic, disease markers.

For drug discovery, smaller, faster, and cheaper is critical and the ability to miniaturize, automate, and multiplex MMAs provides a means for more rapid, cost-efficient drug development. The small size of the microspheres used for MMAs allows for assays to be performed in less than 10 μl, which means less costly reagent is used. Homogeneous formats and the physical characteristics of the microspheres make MMAs amenable to robotic handling, whereas multiplexing generates more information on each compound within a single assay. Advances in sample acquisition, such as that outlined by Ramirez *et al.* (2003), increase throughput to approximately 1.5 seconds per sample, which would provide an enormous enhancement to MMA technology.

Future practical and technological advancements will greatly enhance the adoption of MMA technology. Within clinical diagnostics, a greater understanding of interfering factors in body fluids and how to block them, as well as improvements in sample preparation, will greatly increase the robustness, accuracy, and precision of MMAs. From a practical perspective, as new diagnostic panels for disease are developed, acceptance by the medical community, the Food and Drug Administration, and health care providers will be required for widespread usage. For nucleic acid applications, increases in the size of microsphere arrays and enhancements in signal amplification will broaden the adoption of MMAs into the genomics world. There has been logarithmic growth in the research, development, and commercialization of MMAs since the third edition of *Cytometry*; by the time the fifth edition is written, we may see the need for an entire volume devoted to MMAs.

Acknowledgments

We gratefully acknowledge the contribution of the data in Fig. 2 by collaborators, Maress Lacuesta and Dr. Carol Worthman, Department of Anthropology, Emory University, Atlanta, Georgia. We also appreciate the support of Drs. Robert Wohlhueter and Carolyn Black at the Centers for Disease Control and Prevention.

References

Armstrong, B., Stewart, M., and Mazumder, A. (2000). Suspension arrays for high throughput, multiplexed single nucleotide polymorphism genotyping. *Cytometry* **40,** 102–108.

Barker, R. L., Worth, C. A., and Peiper, S. C. (1994). Cytometric detection of DNA amplified with fluorescent primers: Applications to analysis of clonal *bcl-2* and IgH gene rearrangements in malignant lymphomas. *Blood* **83,** 1079–1085.

Bellisario, R., Colinas, R. J., and Pass, K. A. (2000). Simultaneous measurement of thyroxine and thyrotropin from newborn dried blood-spot specimens using a multiplexed fluorescent microsphere immunoassay. *Clin. Chem.* **46,** 1422–1424.

Bellisario, R., Colinas, R. J., and Pass, K. A. (2001). Simultaneous measurement of antibodies to three HIV-1 antigens in newborn dried blood-spot specimens using a multiplexed microsphere-based immunoassay. *Early Human Develop.* **64,** 21–25.

Cai, H., White, P. S., Torney, D., Deshpande, A., Wang, Z., Marrone, B., and Nolan, P. (2000). Flow cytometry–based minisequencing: A new platform for high-throughput single-nucleotide polymorphism scoring. *Genomics* **66,** 135–143.

Camilla, C., Mély, L., Magnan, A., Casano, B., Prato, S., Debono, S., Montero, F., Defoort, J. P., Martin, M., and Fert, V. (2001). Flow cytometric microsphere–based immunoassay: Analysis of secreted cytokines in whole-blood samples from asthmatics. *Clin. Diagn. Lab. Immunol.* **8,** 776–784.

Carson, R. T., and Vignali, D. A. A. (1999). Simultaneous quantitation of 15 cytokines using a multiplexed flow cytometric assay. *J. Immunol. Methods* **227,** 41–52.

Chen, J., Iannone, M. A., Li, M-S., Taylor, J. D., Rivers, P., Nelson, A. J., Slentz-Kesler, K. A., Roses, A., and Weiner, M. P. (2000). A microsphere-based assay for multiplexed single nucleotide polymorphism analysis using single base chain extension. *Genome Res.* **10,** 549–557.

Chen, R., Lowe, L., Wilson, J. D., Crowther, E., Tzeggai, K., Bishop, J. E., and Varro, R. (1999). Simultaneous quantification of six human cytokines in a single sample using microparticle-based flow cytometric technology. *Clin. Chem.* **45,** 1693–1694.

Colinas, R. J., Bellisario, R., and Pass, K. A. (2000). Multiplexed genotyping of β-globin variants from PCR-amplified newborn blood spot DNA by hybridization with allele-specific oligodeoxynucleotides coupled to an array of fluorescent microspheres. *Clin. Chem.* **46,** 996–998.

Collins, D. P. (2000). Cytokine and cytokine receptor expression as a biological indicator of immune activation: Important considerations in the development of *in vitro* model systems. *J. Immunol. Methods* **243,** 125–145.

Cook, E. B., Stahl, J. L., Lowe, L., Chen, R., Morgan, E., Wilson, J., Varro, R., Chan, A., Graziano, F. M., and Barney, N. P. (2001). Simultaneous measurement of six cytokines in a single sample of human tears using microparticle-based flow cytometry: Allergics vs. non-allergics. *J. Immunol. Methods* **254,** 109–118.

Defoort, J. P., Martin, M., Casano, B., Prato, S., Camilla, C., and Fert, V. (2000). Simultaneous detection of multiplex-amplified human immunodeficiency virus type 1 RNA, hepatitis C virus RNA, and hepatitis B virus DNA using a flow cytometer microsphere-based hybridization assay. *J. Clin. Microbiol.* **38,** 1066–1071.

De Jager, W., te Velthuis, H., Prakken, B. J., Kuis, W., and Rijkers, G. T. (2003). Simultaneous detection of 15 human cytokines in a single sample of stimulated peripheral blood mononuclear cells. *Clin. Diagn. Lab. Immunol.* **10,** 133–139.

Dunbar, S. A., and Jacobson, J. W. (2000). Application of the Luminex LabMAP in rapid screening for mutations in the cystic fibrosis transmembrane conductance regulator gene: A pilot study. *Clin. Chem.* **46,** 1498–1500.

Dunbar, S. A., Vander Zee, C. A., Oliver, K. G., Karem, K. L., and Jacobson, J. W. (2003). Quantitative, multiplexed detection of bacterial pathogens: DNA and protein applications of the Luminex LabMAP system. *J. Microbiol. Methods* **53,** 245–252.

Ekins, R. P., Chu, R., and Biggart, E. (1990). The development of microspot multianalyte radiometric immunoassay using dual fluorescent-labelled antibodies. *Anal. Chin. Acta.* **227,** 73–96.

Fulton, R. J., McDade, R. L., Smith, P. L., Kienker, L. J., and Kettman, J. R., Jr. (1997). Advanced multiplexed analysis with the FlowMetrix system. *Clin. Chem.* **43,** 1749–1756.

Han, M., Gao, X., Su, J. Z., and Nie, S. (2001). Quantum-dot–tagged microbeads for multiplexed optical coding of biomolecules. *Nature Biotech.* **19,** 631–635.

Hennig, D., Rink, L., Fagin, U., Jabs, W. J., and Kirchner, H. (2000). The influence of naturally occurring heterophilic anti-immunoglobulin antibodies on direct measurement of serum proteins using sandwich ELISAs. *J. Immunol. Methods* **235,** 71–80.

Hermanson, G. T. (1996). "Bioconjugate techniques." Academic Press, San Diego.

Horan, P. K., and Wheeless, L. L., Jr. (1977). Quantitative single cell analysis and sorting. *Science* **198,** 149–157.

Hutchinson, K. L., Villinger, F., Miranda, M. E., Ksiazek, T. G., Peters, C. J., and Rollin, P. E. (2001). Multiplex analysis of cytokines in the blood of cynomolgus macaques naturally infected with Ebola virus (Reston serotype). *J. Med. Virol.* **65,** 561–566.

Iannelli, D., D'Apice, L., Cottone, C., Viscardi, M., Scala, F., Zoina, A., Del Sorbo, G., Spigno, P., and Capparelli, R. (1997). Simultaneous detection of cucumber mosaic virus, tomato mosaic virus and potato virus Y by flow cytometry. *J. Virol. Methods* **69,** 137–145.

Iannone, M. A., Taylor, J. D., Chen, J., Li, M-S., Rivers, P., Slentz-Kesler, K. A., and Weiner, M. P. (2000). Multiplexed single nucleotide polymorphism genotyping by oligonucleotide ligation and flow cytometry. *Cytometry* **39,** 131–140.

Janossy, G., Jani, I. V., Kahan, M., Barnett, D., Mandy, F., and Shapiro, H. (2002). Precise CD4 T-cell counting using red diode laser excitation: For richer, for poorer. *Cytometry* **50,** 78–85.

Kaderali, L., Deshpande, A., Nolan, J. P., and White, P. S. (2003). Primer-design for multiplexed genotyping. *Nucleic Acid Res.* **31,** 1796–1802.

Kellar, K. L., and Douglass, J. P. (2003). Protocol. Multiplexed microsphere-based flow cytometric immunoassays for human cytokines. *J. Immunol. Methods* **279,** 277–285.

Kellar, K. L., and Iannone, M. A. (2002). Multiplexed microsphere-based flow cytometric assays. *Exp. Hematol.* **30,** 1227–1237.

Kellar, K. L., Kalwar, R. R., Dubois, K. A., Crouse, D., Chafin, W. D., and Kane, B.-E. (2001). Multiplexed fluorescent bead–based immunoassays for quantitation of human cytokines in serum and culture supernatants. *Cytometry* **45,** 27–36.

Keyes, K. A., Mann, L., Cox, K., Treadway, P., Iversen, P., Chen, Y.-F., and Teicher, B. A. (2003). Circulating angiogenic growth factor levels in mice bearing human tumors using Luminex multiplex technology. *Cancer Chemother. Pharmacol.* **51,** 321–327.

Kettman, J. R., Davies, T., Chandler, D., Oliver, K. G., and Fulton, R. J. (1998). Classification and properties of 64 multiplexed microsphere sets. *Cytometry* **33,** 234–243.

Mahanty, S., Bausch, D. G., Thomas, R. L., Goba, A., Bah, A., Peters, C. J., and Rollin, P. E. (2001). Low levels of interleukin-8 and interferon-inducible protein-10 in serum are associated with fatal infections in acute Lassa fever. *J. Infect. Disease* **183,** 1713–1721.

Martins, T. B. (2002). Development of internal controls for the Luminex instrument as part of a multiplex seven-analyte viral respiratory antibody profile. *Clin. Diagn. Lab. Immunol.* **9,** 41–45.

Martins, T. B., Pasi, B. M., Pickering, J. W., Jaskowski, T. D., Litwin, C. M., and Hill, H. R. (2002). Determination of cytokine responses using a multiplexed fluorescent microsphere immunoassay. *Am. J. Clin. Pathol.* **118,** 346–353.

McHugh, T. M. (1994). Flow microsphere immunoassay for the quantitative and simultaneous detection of multiple soluble analytes. *Methods Cell Biol.* **42,** 575–595.

McHugh, T. M., and Stites, D. P. (1991). Application of bead-based assays for flow cytometry analysis. *Clin. Immunol. Newslett.* **11,** 60–64.

McHugh, T. M., Stites, D. P., Casavant, C. H., and Fulwyler, M. J. (1986). Flow cytometric detection and quantitation of immune complexes using human C1q-coated microspheres. *J. Immunol. Methods* **95,** 57–61.

McHugh, T. M., Miner, R. C., Logan, L. H., and Stites, D. P. (1988a). Simultaneous detection of antibodies to cytomegalovirus and herpes simplex virus by using flow cytometry and a microsphere-based fluorescence immunoassay. *J. Clin. Microbiol.* **26,** 1957–1961.

McHugh, T. M., Stites, D. P., Busch, M. P., Krowka, J. F., Stricker, R. B., and Hollander, H. (1988b). Relation of circulating levels of human immunodeficiency virus (HIV) antigen, antibody to p24, and HIV-containing immune complexes in HIV-infected patients. *J. Infect. Dis.* **158,** 1088–1091.

McHugh, T. M., Wang, Y. J., Chong, H. O., Blackwood, L. L., and Stites, D. P. (1989). Development of a microsphere-based fluorescent immunoassay and its comparison to an enzyme immunoassay for the detection of antibodies to three antigen preparations from *Candida albicans*. *J. Immunol. Methods* **116,** 213–219.

McHugh, T. M., Viele, M. K., Chase, E. S., and Recktenwald, D. J. (1997). The sensitive detection and quantitation of antibody to HCV by using a microsphere-based immunoassay and flow cytometry. *Cytometry* **29,** 106–112.

Mullinex, M. C., Sivakamasundari, R., Feaver, W. J., Krishna, R. M., Sorette, M. P., Datta, H. J., Morosan, D. M., and Piccoli, S. P. (2002). Rolling circle amplification improves sensitivity in multiplex immunoassays on microspheres. *Clin. Chem.* **48,** 1855–1858.

Oliver, K. G., Kettman, J. R., and Fulton, R. J. (1998). Multiplexed analysis of human cytokines by use of the FlowMetrix system. *Clin. Chem.* **44,** 2057–2060.

Opalka, D., Lachman, C. E., MacMullen, S. A., Jansen, K. U., Smith, J. F., Cirmule, N., and Esser, M. T. (2003). Simultaneous quantitation of antibodies to neutralizing epitopes on virus-like particles for human papillomavirus types 6, 11, 16, and 18 by a multiplexed Luminex assay. *Clin. Diagn. Lab. Immunol.* **10,** 108–115.

Park, M. K., Briles, D. E., and Nahm, M. H. (2000). A latex bead–based flow cytometric immunoassay capable of simultaneous typing of multiple pneumococcal serotypes (multibead assay). *Clin. Diagn. Lab. Immunol.* **7,** 486–489.

Phillips, T. M. (2001). Multi-analyte analysis of biological fluids with a recycling immunoaffinity column array. *J. Biochem. Biophys. Methods* **49,** 253–262.

Pickering, J. W., Martins, T. B., Greer, R. W., Schroder, M. C., Astill, M. E., Litwin, C. M., Hildreth, S. W., and Hill, H. R. (2002a). A multiplexed fluorescent microsphere immunoassay for antibodies to pneumococcal capsular polysaccharides. *Microbiol. Infect. Dis.* **117,** 589–596.

Pickering, J. W., Martins, J. B., Schroder, M. C., and Hill, H. R. (2002b). Comparison of a multiplex flow cytometric assay with enzyme-linked immunosorbent assay for quantitation of antibodies to tetanus, diphtheria and *Haemophilus influenzae* type b. *Clin. Diagn. Lab. Immunol.* **9,** 872–876.

Prabhakar, U., Eirikis, E., and Davis, H. M. (2002). Simultaneous quantification of proinflammatory cytokines in human plasma using the LabMAP assay. *J. Immunol. Methods* **260,** 207–218.

Ramirez, S., Aiken, C. T., Andrzejewski, B., Sklar, L. A., and Edwards, B. S. (2003). High-throughput flow cytometry: Validation in microvolume bioassays. *Cytometry* **53A,** 55–65.

Scillian, J. J., McHugh, T. M., Busch, M. P., Tam, M., Fulwyler, M. J., Chien, D. Y., and Vyas, G. N. (1989). Early detection of antibodies against rDNA-produced HIV proteins with a flow cytometric assay. *Blood* **73,** 2041–2048.

Seideman, J., and Peritt, D. (2002). A novel monoclonal antibody screening method using the Luminex-100 microsphere system. *J. Immunol. Methods* **267,** 165–171.

Shapiro, H. M. (2003). "Practical flow cytometry." 4th ed, John Wiley & Sons, Hoboken, NJ.

Smith, P. L., WalkerPeach, C. R., Fulton, R. J., and DuBois, D. B. (1998). A rapid, sensitive, multiplexed assay for detection of viral nucleic acids using the FlowMetrix system. *Clin. Chem.* **44,** 2054–2056.

Spiro, A., Lowe, M., and Brown, D. (2000). A bead-based method for multiplexed identification and quantitation of DNA sequences using flow cytometry. *Appl. Environ. Microbiol.* **66,** 4258–4265.

Tárnok, A., Hambsch, J., Chen, R., and Varro, R. (2003). Cytometric bead array to measure six cytokines in twenty-five microliters of serum. *Clin. Chem.* **49,** 1000–1002.

Taylor, J. D., Briley, D., Nguyen, Q., Long, K., Iannone, M. A., Li, M.-S., Ye, F., Afshari, A., Lai, E., Wagner, M., Chen, J., and Weiner, M. P. (2001). Flow cytometric platform for high-throughput single nucleotide polymorphism analysis. *BioTechniques* **30,** 661–669.

Tripp, R. A., Jones, L., Anderson, L. J., and Brown, M. P. (2000). CD40 ligand (CD154) enhances the Th1 and antibody responses to respiratory syncytial virus in the BALB/c mouse. *J. Immunol.* **164,** 5913–5921.

Wallace, J., Zhou, Y., Usmani, G. N., Reardon, M., Newburger, P., Woda, B., and Pihan, G. (2003). BARCODE-ALL: Accelerated and cost-effective genetic risk stratification in acute leukemia using spectrally addressable liquid bead microarrays. *Leukemia* **17,** 1411–1413.

Weber, J., Sondak, V. K., Scotland, R., Phillip, R., Wang, F., Rubio, V., Stuge, T. B., Groshen, S. G., Gee, C., Jeffrey, G. G., Sian, S., and Lee, P. P. (2003). Granulocyte–macrophage-colony stimulating factor added to a multipeptide vaccine for resected Stage II melanoma. *Cancer* **97,** 186–200.

Willman, J. H., Hill, H. R., Martins, T. B., Jaskowski, T. D., Ashwood, E. R., and Litwin, C. M. (2001). Multiplex analysis of heterophil antibodies in patients with indeterminate HIV immunoassay results. *Am. J. Clin. Pathol.* **115,** 764–769.

Wood, W. I., Gitschier, J., Lasky, L. A., and Lawn, R. M. (1985). Base composition–independent hybridization in tetramethylammonium chloride: A method for oligonucleotide screening of highly complex gene libraries. *Proc. Natl. Acad. Sci. USA* **82,** 1585–1588.

Yang, L., Tran, D. K., and Wang, X. (2001). BADGE, Beads Array for the Detection of Gene Expression, a high-throughput diagnostic bioassay. *Genome Res.* **11,** 1888–1898.

Ye, F., Li, M.-S., Taylor, J. D., Nguyen, Q., Colton, H. M., Casey, W. M., Wagner, M., Weiner, M. P., and Chen, J. (2001). Fluorescent microsphere-based readout technology for multiplexed human single nucleotide polymorphism analysis and bacterial identification. *Hum. Mutation* **17,** 305–316.

PART III

Immunology/T-Cell Responses

CHAPTER 17

Flow Cytometry Applications of MHC Tetramers

John D. Altman

Emory University, School of Medicine
Department of Microbiology and Immunology, and
The Emory Vaccine Center
Atlanta, Georgia 30329

I. Introduction

Quantitative analyses of T-cell responses have proven essential to our understanding of a vast array of immunological phenomena, ranging from the basis of T-cell memory (Hou *et al.*, 1994; Lau *et al.*, 1994; Murali-Krishna *et al.*, 1999) to protection against tumors (Molldrem *et al.*, 2000) to evaluation of vaccine-induced responses (Amara *et al.*, 2001; Barouch *et al.*, 2000; Casimiro *et al.*, 2003; McConkey *et al.*, 2003; Mwau *et al.*, 2004). Until recently, all T-cell responses were measured using bulk assays that often required *in vitro* expansion of low-frequency

METHODS IN CELL BIOLOGY, VOL. 75

0091-679X/04 $35.00

precursor populations. These assays—including ^{3}H-thymidine incorporation for assessment of CD4 proliferative responses and ^{51}Cr-release assays to measure cytotoxicity—suffer from large interassay variation and are somewhat cumbersome and time consuming to perform. Beginning in 1996, two separate and complementary approaches were introduced that have revolutionized analyses of T-cell responses by permitting detection of antigen-specific T cells at the single-cell level. In the first type of assay, cytokine production by individual T cells after short-term *in vitro* stimulation can be detected by either ELISPOT (Anthony and Lehmann, 2003; Czerkinsky *et al.*, 1988; Lalvani *et al.*, 1997) or flow cytometry after intracellular cytokine staining (Waldrop *et al.*, 1997). In the second type of assay, T cells are identified by their ability to bind engineered multimers of their cognate major histocompatibility complex (MHC)–peptide ligand. We first introduced this approach with the introduction of "MHC tetramers" (Altman *et al.*, 1996), and Greten *et al.* (1998) introduced an alternative approach using dimeric MHC–immunoglobulin chimeras in 1998. The vast majority of the published work uses the tetramer alternative, which constitutes the major focus of this chapter.

Why tetramers? Early attempts to stain T cells with monomeric MHC–peptide complexes failed. These failures were explained when the first measurements of the affinity and kinetics of MHC–peptide interactions with specific T-cell receptor (TCR) were made, demonstrating that this interaction has an inherently low affinity (usually in the 1–100 μM range) and that the half-life of this interaction is typically on the order of 10 seconds (Matsui *et al.*, 1991, 1994). We hypothesized that the ability of MHC–peptide reagents to stain T cells could be improved if we could engineer multivalent reagents that would have a higher avidity for T cells compared to simple monomers. We decided to base our reagents around a streptavidin core, which has four biotin-binding sites arranged in a roughly tetrahedral geometry. We then sought methods for biotinylating MHC–peptide complexes in a controlled fashion. When random chemical biotinylation with amine-reactive derivatives of biotin did not lead to reagents with the desired properties, we turned to a technique for enzymatic biotinylation of a short peptide tag (Schatz, 1993), which we attached to the C-terminal end of the soluble domain of a class I MHC heavy chain, located at the opposite end of the molecule from the TCR recognition site (Altman *et al.*, 1996). This technique, coupled with the development of methods for *in vitro* folding of class I MHC molecules produced as inclusion bodies in *Escherichia coli* (Garboczi *et al.*, 1992), proved extremely robust and has been applied to dozens of class I MHC alleles in humans, mice, rhesus macaques, and chimpanzees. The essential features of a class I tetramer are depicted in Fig. 1.

Early attempts to apply the tetramer approach to class II MHC reagents and $CD4^{+}$ T cells focused on the mouse allele I-E^{k} and the cytochrome *c* antigen system (Gutgemann *et al.*, 1998; McHeyzer-Williams *et al.*, 1996). These efforts were successful, prompting optimism that the tetramer approach would be as informative and robust for the analysis of $CD4^{+}$ T-cell responses as it was for CD8 responses. However, successes with class II tetramers have been notable but

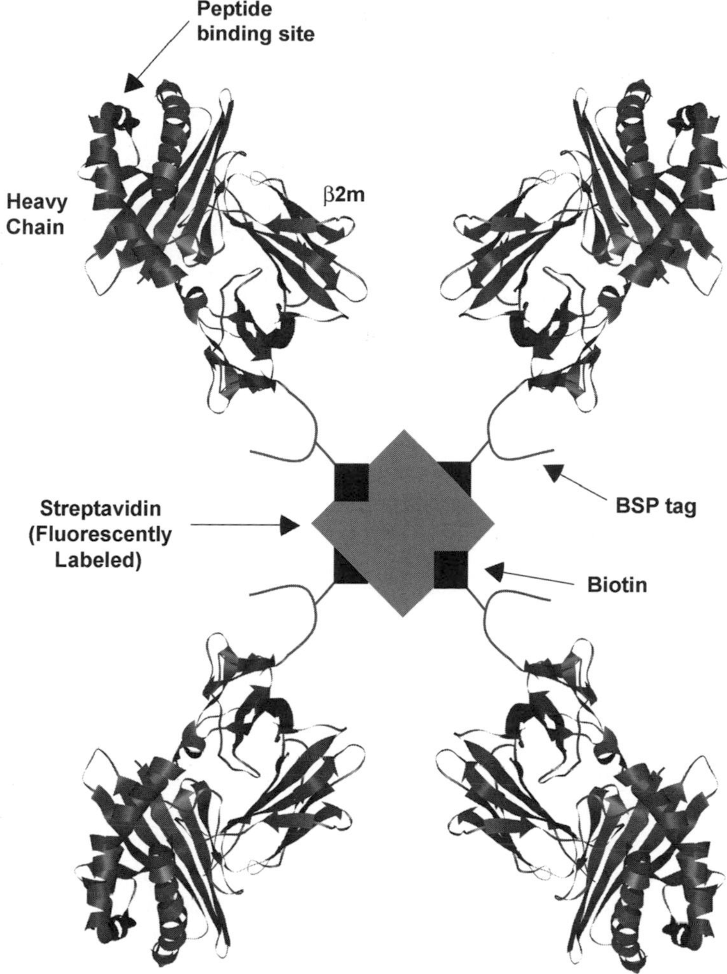

Fig. 1 Cartoon of a typical major histocompatibility complex tetramer. The components of the tetramer are labeled in the figure.

less frequent (Cauley *et al.*, 2002; Day *et al.*, 2003; Meyer *et al.*, 2000; Novak *et al.*, 1999). A number of factors contribute to this lower success rate: (1) Soluble class II MHC–peptide complexes are more difficult to produce than class I complexes; (2) the magnitude of CD4 T-cell responses is generally lower than CD8 responses (Foulds *et al.*, 2002), making them more difficult to analyze; and (3) many biochemically high-quality class II tetramers fail to stain appropriate positive control cells, perhaps indicating that CD4 cells might often have lower effective avidity for MHC–peptide than CD8 cells (Mallet-Designe *et al.*, 2003). As a consequence of failed attempts to stain with conventional class II tetramers, some laboratories have sought to engineer higher-order multimers that should have even greater avidity for appropriate T cells while maintaining low background binding (Mallet-Designe *et al.*, 2003).

As a result of the greater success and collective experience, the rest of this chapter focuses primarily on flow cytometry applications of class I MHC tetramers. Other applications of these reagents, such as their use for *in situ* staining, have been described (Haanen *et al.*, 2000; Skinner *et al.*, 2000) but are beyond the scope of this chapter. Also beyond the scope of this chapter are successful applications of the tetramer approach with nonclassical MHC and MHC-like molecules such as human leukocyte antigen HLA-E, HLA-F, and HLA-G (Allan *et al.*, 2002), Qa-1 (Vance *et al.*, 1998), and CD1d (Benlagha *et al.*, 2000; Matsuda *et al.*, 2000).

II. Materials and Methods

A. MHC Multimer Reagents

1. Production and Availability of MHC Multimers

Detailed descriptions of the preparation of both classical class I MHC tetramers (Altman, 2004) and MHC–immunoglobulin (Schneck *et al.*, 2004) chimeras have been published and are beyond the scope of this chapter. Multimeric MHC reagents are also available from a number of sources, including Beckman-Coulter (www.immunomics.com), Becton-Dickinson (www.bdbiosciences.com/immune_function), and the National Institutes of Health (NIH) MHC Tetramer Core Facility (www.niaid.nih.gov/reposit/tetramer/index.html). Classical class I MHC tetramers are nearly always supplied as homogeneous complexes with a specified peptide ligand, whereas class I MHC–immunoglobulin chimeras are supplied as reagents containing a mixture of peptide ligands that must be replaced by *in vitro* exchange with an appropriate peptide of interest (Schneck *et al.*, 2004).

Various methods have been used to produce class II MHC proteins for tetramers, including (1) *in vitro* folding from subunits produced in *E. coli* (Gutgemann *et al.*, 1998), (2) baculovirus (Crawford *et al.*, 1998), (3) stable transfection of insect cells (Novak *et al.*, 1999), and (4) stable transfection of mammalian cells (Day *et al.*, 2003). In some cases, class II tetramers do not appear to have sufficient avidity to bind to T cells, so methods for the production of higher-order multimers have been sought (Hugues *et al.*, 2002; Mallet-Designe *et al.*, 2003).

2. Naming of Tetramer Reagents

Each classical class I tetramer reagent is composed of three components: the MHC heavy chain, β2-microglobulin (β2m), and a defined peptide. For naming purposes, the β2m is usually ignored (although it should be noted that human β2m is often used with mouse and rhesus macaque heavy chains). Naming of the heavy chain is usually straightforward. However, from the beginning (Altman *et al.*, 1996), naming of the peptide has often been haphazard and done carelessly. The problem is that there is a trade-off between convenience, accuracy, and completeness. For example, consider the reagent that my colleagues and I referred to as "A2/gag" (Altman *et al.*, 1996). In the context of the original publication, it is obvious that this refers to the human immunodeficiency virus (HIV) gag protein, and most experts in the field will correctly guess that it refers to the epitope with the sequence SLYNTVATL, a major immunodominant epitope in many people chronically infected with HIV. However, in hindsight, it is obvious that other HLA-A2–restricted epitopes will be found in the gag protein, leading to potential confusion. As a consequence, I now favor a nomenclature such as A2/HIV. gag.77–85, which indicates not only the pathogen (HIV) and protein (gag) source of the epitope, but also its position in the sequence. Of course, this nomenclature has its own imperfections (what is the reference sequence? what about mutations? who wants to write all of that in their notebook?), but I think it is the best of a bad lot. Alternatives that we and others sometimes use are to indicate the peptide by either its first three amino acids (e.g., the Epstein–Barr virus [EBV]–derived epitope "FLR") or by the first and last amino acids and the peptides length (e.g., the SIV gag–derived peptide "CM9").

3. Choice of Fluorophores for MHC Multimers

The vast majority of published data using MHC tetramers have used tetramers labeled with the phycobiliproteins R-phycoerythrin (PE) or allophycocyanin (APC). Both fluorophores are bright, and most batches of the streptavidin conjugates (available from a wide range of suppliers) produce excellent-quality tetramers. Tandem conjugates to phycobiliproteins can also be used, but they are substantially more expensive than the plain phycobiliproteins. Tetramers with small molecule fluorophores have occasionally been used (Kuroda *et al.*, 1998) and are sometimes required, such as when they are combined with flow–fluorescent *in situ* hybridization (FISH) methods for analysis of telomere lengths (Plunkett *et al.*, 2001). However, there is anecdotal evidence, including from my laboratory, that many commercial preparations of small molecule fluorophore-labeled streptavidin (or other biotin-binding proteins) do not reliably yield high-quality tetramers. The reasons for this are not understood, but it is reasonable to hypothesize that it may be related to differences in aggregation state between phycobiliprotein tetramers and small molecule fluorophore tetramers.

The MHC–immunoglobulin chimeras are usually detected using indirect staining methods, opening up a greater range of options for choice of the fluorophore used to detect the antigen-specific T cells (Schneck *et al.*, 2004). However, the immunoglobulin-derived constant regions of these molecules are usually derived from mouse antibodies, so care must be taken to employ secondary staining protocols that do not also detect primary antibodies specific for other markers, such as CD3 or CD8.

4. Notes on Staining Protocols

In general, the protocols used to stain with MHC tetramers differ little from those used to stain with typical antibody reagents. The quality of the results obtained is not highly sensitive to variation in staining conditions. Typical whole-blood methods work as well as—if not better than—methods that use peripheral blood mononuclear cells (PBMC) or other preparations of lymphocytes. As an example, the steps followed in my laboratory for staining 200 μl of whole blood are as follows:

1. Prepare a 20× stock of the tetramer in ice-cold FACS buffer (2% fetal bovine serum +0.1% sodium azide in PBS). (See the notes below on determination of the optimal titer of a tetramer reagent.)
2. Add 10 μl of the tetramer plus appropriate amounts of antibody reagents (e.g., anti-CD3 and anti-CD8) to the bottom of a 12-by-72 FACS tube.
3. Add 200 μl of whole blood to the FACS tube and vortex for 5 seconds. Note that the anticoagulant used for collection of the blood is not crucial, but we tend to use EDTA or heparin collection tubes.
4. Incubate for 20–30 minutes in the dark at room temperature.
5. Add 2 μl of 1× FACS Lyse solution (Becton Dickinson) and vortex thoroughly.
6. Incubate for 10 minutes at room temperature in the dark.
7. Pellet the cells by centrifugation for 5 minutes at 300×*g*.
8. Decant or aspirate the supernatant. Disrupt the pellet by vortexing for 5 seconds.
9. Wash the cells.
   a. Add 2 ml of ice-cold FACS buffer to the disrupted pellet.
   b. Vortex for 5 seconds.
   c. Pellet the cells by centrifugation for 5 minutes at 300×*g*.
   d. Decant or aspirate the supernatant.
   e. Disrupt the pellet by vortexing for 5 seconds.
10. Repeat step 9 to further reduce the level of background "outlier" events.
11. Add 300 ml of 2% ultrapure formaldehyde (prepared from a 10% stock obtained from Polysciences), and vortex for 5 seconds.
12. Acquire data on the flow cytometer within 24 hours. Keep the samples protected from light before and during acquisition.

Different laboratories use different staining temperatures, especially for staining samples from PBMC or single-cell suspensions of tissues. Some laboratories prefer staining at 37 °C (Whelan *et al.*, 1999) (at which tetramers are internalized), but we generally favor either room temperature or 4 °C for staining. In the early days, we used incubation times as long as 2–3 hours, based on the expectation that it would take longer for a tetramer to gather two to three TCR molecules together than it would take for a high-affinity antibody to bind to its target. However, we now perform room-temperature stainings of both whole-blood and PBMC samples for 20–30 minutes with excellent results. I strongly advise each laboratory to investigate both time and temperature conditions for each new tetramer reagent.

5. Compensation Controls

Because tetramers are almost always used in combination with other reagents, it is necessary to set up appropriate single-stain controls to obtain values for fluorescence compensation. The same antibodies used in combination with the tetramers can usually be used for compensation controls. However, the tetramers will usually not stain a population that is large enough for practical use as a compensation control. This leaves two options. First, one can substitute an antibody such as anti-CD3 or anti-CD8 that is labeled with the same fluorophore as the tetramer for the control. Alternatively, one can stain cells with a biotinylated primary antibody (such as anti-CD3 or anti-CD8), followed by a secondary stain with the same batch of labeled streptavidin that is used to prepare the tetramer. The first option usually works well in practice, especially with PE and APC tetramers. However, for tetramers that are prepared with tandem dyes, the indirect staining method must be used.

6. Controls and Optimizing Staining

Ideally, appropriate positive control cells are available for each tetramer stain. Examples of positive control cells are T-cell clones, lines, hybridomas, TCR transgenic mice, cells transfected with specific TCRs, or lymphocytes from a donor previous shown to have T cells of the indicated specificity. Until one obtains an appropriate sample, each tetramer ought to be regarded with suspicion, and negative results should not be trusted. Once one does obtain an appropriate positive control sample, the reagent can be titered to determine the optimal staining concentration. For these experiments, the best results are obtained if the sample contains abundant internal negative control cells; for example, clones should be diluted into either an irrelevant clone or into PBMCs or another source of lymphocytes of mixed specificities. Dilution of tetramer reagents will lead to a dramatic reduction in the number of nonspecific "outlier" events, in addition to the expected reduction in the mean fluorescence intensity of the bulk negative population. We typically test threefold serial dilutions of tetramers, starting at a concentration of total MHC of 10 μg/ml. It is often necessary to use subsaturating concentrations of tetramer to achieve clean backgrounds. Obviously, it is desirable to maintain a clean "baseline" separation of positive and negative cells while keeping the "outliers" to a minimum, and this can usually be achieved in practice.

There are three types of negative controls that can be used for tetramer staining. First, it is often helpful to have cells from a subject who is not expected to have the

T cells of interest, such as from a donor who is seronegative for HIV if an HIV tetramer is to be used. It is better, but not essential, that the negative control cells come from a donor who shares the HLA allele of interest with the test sample. Perhaps surprisingly, tetramers appear not to stain allogeneic T cells at a significant frequency. Second, some laboratories might want to stain cells with an irrelevant tetramer, which is analogous to an isotype control. For various reasons, we do not believe that this control is informative, and we do not perform it in my laboratory. For gating, it is much better to use "fluorescence minus one" (FMO) controls (Roederer, 2001), in which all of the stains are used except the tetramer.

7. Data Acquisition and Limits of Detection

Tetramers are frequently used to identify low-frequency populations of cells. This means that one must acquire a large number of events on the flow cytometer. The appropriate number of events depends on the frequency of the target population, and how it will be further subdivided by phenotype. It is not unusual for us to acquire data files containing 200,000 or more lymphocytes. As the number of required events increases, one must also stain correspondingly larger volumes of blood or more cells.

The limits of detection of the tetramer staining method are dependent on a number of factors, but the three most important are as follows: (1) the quality of the tetramer reagent and the intensity of its signal, (2) the quality and source of the sample (e.g., human whole-blood and PBMC samples generally give cleaner results than spleen samples from a mouse), and (3) the number of events acquired. In common usage with human PBMC samples, adequate detection of tetramer-positive frequencies as low as 1:5000 $CD8^+$ T cells (0.02%) can be achieved (He *et al.*, 1999). With extraordinary precautions, the limit can be further reduced by a factor of 10.

B. Reagent Combinations

1. Choosing Antibodies to Define "Parental" Subsets

Populations of MHC tetramer-binding cells are almost always "children" of well-defined "parent" populations, such as $CD3^+$ $CD8^+$ cells. When reporting the results of tetramer staining experiments, the parental population defines the denominator in tetramer frequency calculations (e.g., "HLA-A2/EBV.RAK–specific cells are 0.12% of $CD3^+$ $CD8^+$ T cells"), and it is crucial that the reagents used to define this population are reported accurately.

In the case of classical class I MHC tetramers, the most robust approach to defining the parent population is to stain with a combination of antibodies specific for CD3 and CD8 and to gate on $CD3^+$ $CD8^+$ cells (Fig. 2). On a four-color flow cytometer, this approach limits additional phenotyping to one parameter at a time. Fortunately, in many cases one may use only one marker—most often anti-CD8—to define the parent population. Approaches to compensate for the omission of anti-CD3 vary according to the species of the sample. In mice, natural

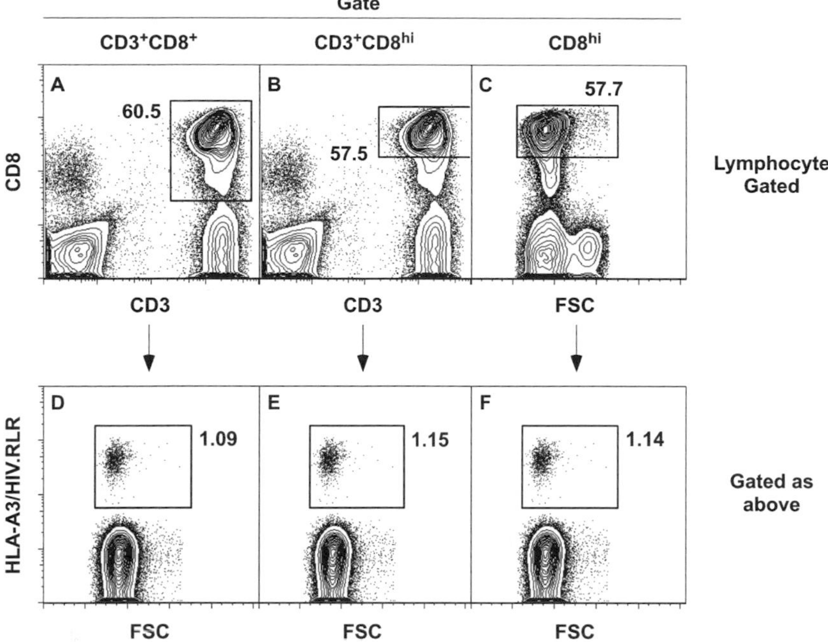

Fig. 2 Approaches to gating on the parent population of a tetramer-positive population. The first row shows three approaches to gating on CD8$^+$ T cells present in a lymphocyte-sized population (gating on forward scatter channel [FSC] and side scatter channel [SSC] not shown). (A) CD3 is plotted versus CD8α and a gate is set to include all CD8$^+$ cells, including those that express intermediate levels of CD8. (B) The CD8^{int} cells are omitted from the gate. The data in (C) are treated as if the cells were not stained with CD3 and shows a gate that is set on CD8^{hi} cells, avoiding both the CD8^{int} T cells and the CD8^{int}CD3$^-$ cells, which are usually natural killer cells. The panels in the second row display cells that fall in the gates of the corresponding plots above. The resulting frequency of the tetramer-positive populations are indicated in the figure, demonstrating that the results, in this case, are only a weak function of the gating approach. In general, we favor the approach of (A).

killer (NK) cells generally do not express CD8, so use of single-parameter gating on CD8$^+$ cells gives an accurate measurement of the most relevant parent population. In humans, NK cells express intermediate levels of CD8$\alpha\alpha$, creating two options. First, it is possible to define the parent population by gating on CD8α^{hi} cells. Alternatively, some laboratories have used anti-CD8β antibodies—which do not stain NK cells—to define the parent population (Schmitz *et al.*, 1998). However, one should use this approach with caution, as our understanding of expression of CD8$\alpha\beta$ versus CD8$\alpha\alpha$ on T cells continues to evolve. Neither of these approaches works as well in rhesus macaques, where NK cells are CD8α^{hi} and many T cells express CD8$\alpha\alpha$. In fact, some investigators use both CD8α and CD8β antibodies (Egan *et al.*, 1999), and it is probably not a bad idea to include a tube containing both in an extended panel in most experiments. The bottom lines are (1) investigators must be aware of the specifics of the animal model and the limitations of their approach when designing appropriate staining combinations and (2) the specifics of the approach used must be accurately reported.

2. Use of Exclusion Markers

The quality of flow cytometric analyses of rare populations is often improved through the use of one or more exclusion markers that are found on one or more cell lineages but never on the subset of interest. Use of an exclusion or "dump" channel will also exclude cells that are nonspecifically "sticky" from the analysis. For example, when using class I tetramers, the investigator might consider using antibodies specific for B-, macrophage-, and even $CD4^+$ T-cell–lineage cells together in the dump channel. However, one must exercise extreme caution when using this approach, because some canonical markers of non–T-cell lineages turn out to be expressed on activated $CD8^+$ T cells. For example, in the mouse, the canonical B-cell marker B220 (Renno *et al.*, 1998) and the macrophage marker Mac-1 (CD11b) (Zimmerman *et al.*, 1996) turn out to be expressed on a high percentage of activated $CD8^+$ T cells and must, therefore, not be used as exclusion markers. For experiments in humans, Beckman-Coulter sells an iMASC gating kit containing a mixture of CD4, CD13, and CD19 antibodies, all conjugated to the PE-Cy5 tandem dye. Others have used the combination of CD4, CD14, and CD19 (Lee *et al.*, 1999) or even unrelated tetramers (Savage *et al.*, 1999) or plain streptavidin–fluorophore conjugates in a dump channel. Live/dead discriminators such as propidium iodide, 7-AAD, or TO-PRO3 can also be used (Savage *et al.*, 1999), although special precautions must be followed if samples are to be fixed before analysis (Schmid *et al.*, 1999); ethidium monoazide bromide is a good alternative when fixation of the samples before analysis is required (Riedy *et al.*, 1991).

3. Interactions between Tetramers and Other Reagents

Although class I tetramers work because they allow discrimination between T cells expressing different $\alpha\beta$ chains of the T-cell antigen receptor, there is strong evidence that they also interact directly with the CD8 coreceptor. In the mouse, different anti-CD8 monoclonal antibodies will either inhibit or—perhaps surprisingly—potentiate binding of K^b tetramers (Daniels and Jameson, 2000). In humans, the level of background binding of HLA tetramers has been shown to be reduced through the use of reagents that have mutations in the CD8-binding site found in the $\alpha 3$ domain of class I HLA molecules (Bodinier *et al.*, 2000), although with routine careful washing procedures, my group continues to use wild-type HLA tetramers with great success. Differential staining with wild-type and mutant HLA tetramers has been used to discriminate between high-avidity, CD8-independent T cells and lower avidity T cells, which require CD8 binding (Choi *et al.*, 2003).

Tetramers are often combined with antibodies that bind to other portions of the TCR, such as CD3 or the Vα or Vβ domains. Although these reagents generally do not compete with tetramers for binding to the same site, the binding sites are close and steric hindrance is a possibility, particularly because all of the reagents are rather large. In fact, some anti-CD3 antibodies have been shown to inhibit HLA tetramer binding (Hoffmann *et al.*, 2000), although this does not appear to be a general problem. In addition, a substantial number of papers have been published, including from my own laboratory (Blattman *et al.*, 2000), demonstrating successful co-staining with tetramers and a large panel of anti-Vβ antibodies. However, although noninterference between tetramers and Vβ antibodies is common, it is not universal, because antibodies specific for mouse Vβ3 have been shown to inhibit binding of I-E^k tetramers containing a moth cytochrome *C* peptide (Baldwin *et al.*, 1999).

Finally, fluorescence resonance energy transfer (FRET) has been reported between a PE-labeled anti-TCR$\alpha\beta$ monoclonal antibody and an APC-labeled tetramer (Batard *et al.*, 2002), reflecting the close proximity of these two reagents when bound to a specific T cell. This interesting observation raises complex practical issues and reminds investigators to be cautious in their choice of reagent combinations. For example, on a FACSCalibur instrument, this particular FRET interaction will result in a positive signal in the FL3 channel that would not occur in the absence of the FRET, complicating use of the FL3 channel. If the FL3 channel were used for exclusion markers, the cells of interest would likely be removed from the analysis. This FRET interaction also precludes the use of the FL3 channel for assessing a variable phenotype, such as CD62L or CD45RA, of the tetramer-positive cells. In fact, about the only thing that could be used in this channel is a solid positive discriminator, such as anti-CD8. We have observed FRET when the fluorophore combinations were reversed (e.g., a PE tetramer and an APC-labeled TCR$\alpha\beta$ antibody). We have also observed FRET between tetramers and other reagents such as anti-CD3 and anti-CD8 (Fig. 3).

4. Analyses of Complex and Variable Phenotypes

Once a specific population has been defined by staining for CD3, CD8, and a tetramer, one can use the remaining channels of the available flow cytometer to determine additional phenotypic properties of the specific T cells. The usual goal of these efforts is to correlate an observed physical phenotype with any of a variety of functions, including cytotoxicity, the ability to produce certain cytokines, the capacity to proliferate when stimulated by antigen or other means, tissue trafficking potential, or the susceptibility to antigen-induced cell death. The list of phenotypic markers that have been analyzed is enormous and the chosen markers often reflect the specific interests of the investigators and/or current trends in the field. The array of markers can be categorized in various ways. For example, phenotypic markers could be classified according to function (e.g., cytokine and chemokine receptors, adhesion molecules, molecules involved

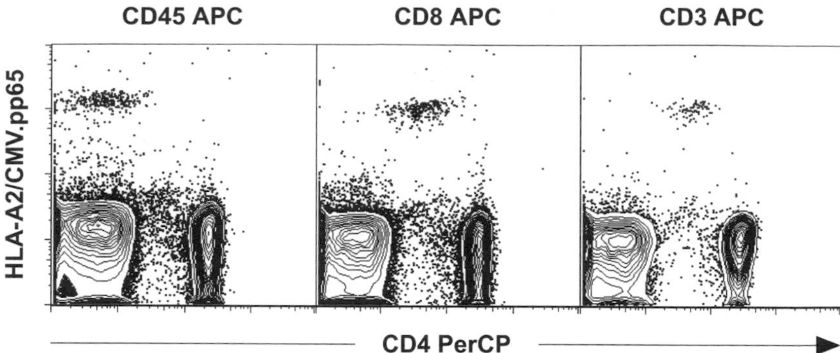

Fig. 3 Tetramers can serve as fluorescent resonance energy transfer (FRET) donors to reagents that are nearby on a T cell. Cells were stained with a phycoerythrin (PE)-labeled human leukocyte antigen-A2 (HLA-A2)/CMV.pp65 tetramer, PerCP-labeled anti-CD4, and APC-labeled antibodies, as indicated in the figure. FRET is not observed when the third stain is CD45-APC, and the tetramer-positive population is located directly above the $CD4^-$ population. In contrast, FRET is observed when both CD8-APC and CD3-APC are used, and the tetramer-positive population appears to express intermediate levels of CD4. This, however, is an artifact caused by the FRET.

in antigen presentation or costimulation, effector molecules involved in cell-mediated cytotoxicity, and cell-cycle status), or they could be classified according to a particular model of differentiation (e.g., markers of naive cells, markers of "central" memory cells, markers of "effector" memory cells, and markers of "maturation"). It is extremely difficult to summarize this literature, which is filled with controversy. It is quite clear that T cells specific for different pathogens have distinct phenotypic propensities (Appay *et al.*, 2002; van Lier *et al.*, 2003). Whether this is a cause of specific patterns of viral persistence (Appay *et al.*, 2000; Champagne *et al.*, 2001) or a consequence of it (Gamadia *et al.*, 2002) is a subject that will remain hotly debated for a long time.

III. Analyzing Tetramer Data

A. Plotting Tetramer Data

In most cases, MHC tetramers will stain a relatively rare population of cells, limiting the types of plots used to analyze tetramer data to those that display dots. Within that subset, the following options are available: (1) simple, conventional dot plots, (2) contour plots with outliers displayed as dots, and (3) pseudo–color dot plots, which combine the visual density information of contour plots with the complete display of conventional dot plots (Fig. 4). In general, the second and third options provide more visual information and are, therefore, preferable. Choosing between these options depends on the context. For paper publications

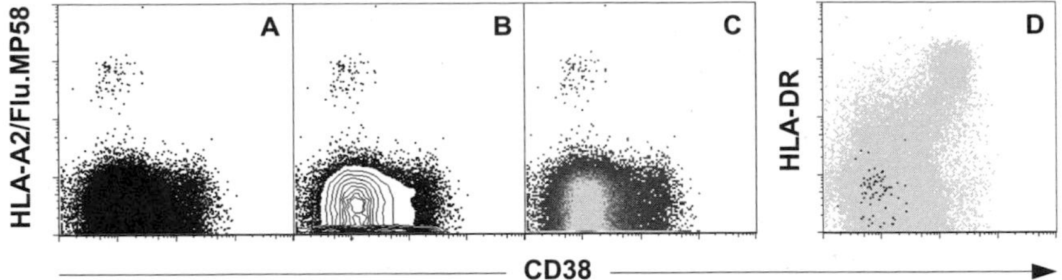

Fig. 4 Four formats for displaying tetramer data. (A) A conventional dot plot. (B) A contour plot with outliers. (C) A pseudo–color dot plot. (D) An overlay of tetramer-positive cells in blue on a background of bulk $CD8^{hi}$ cells, shown in green. All plots were prepared with FlowJo software (Treestar, Eugene, Oregon). (See Color Insert.)

in which there is a significant extra cost for color figures, contour plots with outliers may be preferred. For slide presentations, either format works well.

Finally, some software packages permit overlays of two (or more) bivariate dot plots. We have found these particularly useful for comparing two-parameter phenotypic analyses of tetramer-positive cells with bulk $CD8^+$ T-cell populations, as in Fig. 4D.

B. Gating

There is no one "correct" way to gate on cell populations in a tetramer-staining flow cytometry experiment. Most investigators probably follow a protocol such as this:

1. Plot FSC versus SSC and gate on the lymphocyte population
2. For cells in the lymphocyte gate, plot CD3 versus CD8 and gate on $CD3^+$/$CD8^+$ cells
3. For cells in the lymphocyte and $CD3^+$/$CD8^+$ gate, plot tetramer staining versus another parameter, and gate accordingly

Even with a simple scheme such as this, there are potential pitfalls that the investigator should avoid. First, the appropriate lymphocyte gate is context dependent. If you are attempting to detect acute responders—either *in vivo* or following *in vitro* stimulation—many of the tetramer-binding cells might be blasting, and it is necessary to set generous gates on the FSC versus SSC plot that are sure to include the blasting population. In practice, in these cases it is helpful to begin by plotting tetramer versus both FSC and SSC on an ungated sample to obtain a rough idea of where the tetramer-positive cells fall with respect to the scatter parameters. On the other hand, if you are attempting to detect "resting" populations, setting a tighter gate on small lymphocytes may be appropriate and will certainly be less contaminated with nonspecific binding cells. These same principles apply to setting gates for CD3 and CD8, because genuine tetramer-binding T cells may have downregulated the TCR or co-receptor.

When setting a gate on $CD3^+/CD8^+$ cells—particularly in humans and rhesus macaques—the investigator must decide whether to include the population of $CD3^+$ cells that are clearly $CD8^{int}$, a population that appears to be enriched in CD8$\alpha\alpha$ T cells. Only rarely are tetramer-positive cells found in this subpopulation, but we typically include this population in the $CD3^+/CD8^+$ gate, as in Fig. 2A.

Finally, after setting a gate on CD3 and CD8 (or CD8 only, if anti-CD3 was omitted), it is helpful to plot tetramer versus one of the scatter parameters to set a gate on tetramer-positive cells to determine their frequency. Alternatively, one can plot tetramer versus a fourth reagent (typically used for phenotyping, such as CD62L) and set a tetramer gate that encompasses the entire tetramer-positive population. However, most often one uses these plots to divide the tetramer-positive population into distinct subsets.

IV. Combining Tetramer Staining with Functional Assays

A. Tetramers and Intracellular Cytokine Staining

Tetramer staining is a physical approach to the identification of antigen-specific T cells that does not directly provide functional information. In fact, one of the most important uses of tetramers has been the identification of antigen-specific T cells that are "deficient" in one or more functions (Kostense *et al.*, 2002; Lechner *et al.*, 2000; Zajac *et al.*, 1998). Almost from the beginning, many laboratories sought to combine tetramer staining with one or more functional assays. Because they share the same flow cytometry format, the assay best suited by far for combination with MHC tetramer staining is intracellular cytokine staining. However, the combination is complicated by one simple and important factor. When T cells are stimulated with antigen to induce them to produce cytokines, a significant fraction of the TCR is internalized (Valitutti *et al.*, 1995) and are no longer available for binding to MHC tetramers if they are used after the stimulation. The extent of TCR internalization is usually antigen–dose dependent, and the dose–response curves for internalization are not necessarily the same as for cytokine production. To circumvent these complications, Appay and Rowland-Jones (2002) introduced a method in which cells are stained with tetramers before stimulation. The tetramers are internalized either during the staining period or during the stimulation and remain physically associated with the cells that go on to produce cytokines. There appears to be some loss in tetramer signal intensity, probably as a result of degradation of the tetramer-associated fluorophore after internalization. Although the power of this method is obvious, in our hands the method is quite finicky and we often completely lose the ability to detect tetramer staining. Still, one can often compare the frequency of tetramer-positive cells before stimulation with the frequency of cytokine-positive cells after stimulation and derive useful information (Ravkov *et al.*, 2003).

B. Tetramers and Proliferation by CFSE-Based Methods

Tetramers are also well suited for pairing with the carboxy fluorescein diacetate succinimidyl ester (CFSE) methods that are used to detect cell proliferation (Lyons, 2000). Cells are first labeled with CFSE and then stimulated *in vitro*, usually for 3–7 days, before staining with tetramers and other antibodies. The TCR downregulation that complicates the combination of tetramer staining with intracellular cytokine staining is much less of a factor, because the TCR is usually re-expressed over the extended time course of the proliferation assay. The combination of tetramer staining with the CFSE method has been used to back-calculate precursor frequencies of antigen-specific $CD4^+$ T cells in cases in which the frequency of these cells before expansion was below the limit of detection of tetramer staining (Novak *et al.*, 1999).

T cells from a mouse can be labeled with CFSE and transferred back into a second recipient before future harvest and analysis (Lyons, 2000). This combination was essential in the demonstration that maintenance of memory $CD8^+$ T cells is antigen independent (Murali-Krishna *et al.*, 1999) and revealed that interleukin-15 (IL-15) was required for the process of homeostatic proliferation (Becker *et al.*, 2002).

C. Tetramers and T-Cell Degranulation

Effector and memory $CD8^+$ T cells contain cytoplasmic granules that are released upon activation of the T cell. The granules in effector cells contain high levels of perforin and granzymes, molecules that are responsible for the cytotoxic function of these cells; resting memory cells contain significantly lower concentrations of these molecules (Sallusto *et al.*, 1999). In contrast, the granules in both resting memory and effector cells contain significant levels of CD107a and CD107b, also known as LAMP-1 and LAMP-2 (lysosomal membrane associated protein). Upon degranulation, these molecules remain anchored to the T cell, but they are briefly exposed on the cell surface before they are recycled. Betts *et al.* (2003) have demonstrated that antigen-induced exposure of these molecules can be detected if anti-CD107a and-CD107b are included during the stimulation period, and the antibodies remain associated with the cell upon reinternalization of the CD107. This method has been combined with tetramer staining, using methods similar to those used for the combination of tetramer with intracellular cytokine staining (Appay *et al.*, 2000). Although the degranulation function measured by this assay is closely tied to the cytotoxic effector mechanism, it must still be regarded as a surrogate for true measures of cytotoxic function.

V. Conclusion

MHC tetramers have proven extremely powerful reagents for both simple quantitation and fine-level characterization of T-cell immune responses. Because it is a physical method, tetramers can detect T cells lacking function, cells that are invisible to nearly all other assays. This has provided us with an enhanced understanding of the range of effects of chronic antigen exposure on T cells, an important issue when considering responses against persistent viruses such as HIV or antitumor responses. On the other hand, tetramers are a poor choice for mapping of T-cell epitopes or for characterization of the "complete" response to a pathogen, jobs better suited to cytokine assays that can use pools of extended peptides or viral vector delivery systems.

Many challenges remain in the development of tetramer technology. In addition, whether class II tetramers will prove as generally applicable and useful as the class I reagents remains to be seen. The availability of new flow cytometers capable of measuring many more fluorescent parameters opens the possibility of simultaneous use of tetramers labeled with different fluorophores, but the reagents will have to be developed and characterized. More robust protocols for combining tetramers with functional assays need to be developed. Finally, making sense of the large volume of phenotypic data that has been and will be collected will occupy investigators for a long time.

Acknowledgments

I would like to thank numerous members of my laboratory, past and present, for their contributions to the development and application of the MHC tetramer method. Particularly noteworthy are the contributions of Joe Miller, Lily Wang, Christy Myrick, Eugene Ravkov, Sophia Albott, and Chris Ibegbu. I would also like to thank the long-time members of the MHC Tetramer Core Laboratory, including Donielle Bell, Dale Long, and its two technical directors John Lippolis and Eugene Ravkov for their essential contributions.

References

Allan, D. S., Lepin, E. J., Braud, V. M., O'Callaghan, C. A., and McMichael, A. J. (2002). Tetrameric complexes of HLA-E, HLA-F, and HLA-G. *J. Immunol. Methods.* **268,** 43–50.

Altman, J. D. (2004). MHC-peptide tetramers to visualize antigen-specific T cells. *In* "Current protocols in immunology," (J. E. Coligan, A. M. Kruisbeek, D. H. Margulies, E. M. Shevach, and W. Strober, eds.). Wiley, New York.

Altman, J. D., Moss, P. A., Goulder, P. J., Barouch, D. H., McHeyzer-Williams, M. G., Bell, J. I., McMichael, A. J., and Davis, M. M. (1996). Phenotypic analysis of antigen-specific T lymphocytes. *Science* **274,** 94–96.

Amara, R. R., Villinger, F., Altman, J. D., Lydy, S. L., O'Neil, S. P., Staprans, S. I., Montefiori, D. C., Xu, Y., Herndon, J. G., Wyatt, L. S., Candido, M. A., Kozyr, N. L., Earl, P. L., Smith, J. M., Ma, H. L., Grimm, B. D., Hulsey, M. L., Miller, J., McClure, H. M., McNicholl, J. M., Moss, B., and Robinson, H. L. (2001). Control of a mucosal challenge and prevention of AIDS by a multiprotein DNA/MVA vaccine. *Science* **292,** 69–74.

Anthony, D. D., and Lehmann, P. V. (2003). T-cell epitope mapping using the ELISPOT approach. *Methods* **29,** 260–269.

Appay, V., Dunbar, P. R., Callan, M., Klenerman, P., Gillespie, G. M., Papagno, L., Ogg, G. S., King, A., Lechner, F., Spina, C. A., Little, S., Havlir, D. V., Richman, D. D., Gruener, N., Pape, G., Waters, A., Easterbrook, P., Salio, M., Cerundolo, V., McMichael, A. J., and Rowland-Jones, S. L.

(2002). Memory CD8+ T cells vary in differentiation phenotype in different persistent virus infections. *Nat. Med.* **8,** 379–385.

Appay, V., Nixon, D. F., Donahoe, S. M., Gillespie, G. M., Dong, T., King, A., Ogg, G. S., Spiegel, H. M., Conlon, C., Spina, C. A., Havlir, D. V., Richman, D. D., Waters, A., Easterbrook, P., McMichael, A. J., and Rowland-Jones, S. L. (2000). HIV-specific CD8(+) T cells produce antiviral cytokines but are impaired in cytolytic function. *J. Exp. Med.* **192,** 63–75.

Appay, V., and Rowland-Jones, S. L. (2002). The assessment of antigen-specific $CD8^+$ T cells through the combination of MHC class I tetramer and intracellular staining. *J. Immunol. Methods* **268,** 9–19.

Baldwin, K. K., Trenchak, B. P., Altman, J. D., and Davis, M. M. (1999). Negative selection of T cells occurs throughout thymic development. *J. Immunol.* **163,** 689–698.

Barouch, D. H., Santra, S., Schmitz, J. E., Kuroda, M. J., Fu, T. M., Wagner, W., Bilska, M., Craiu, A., Zheng, X. X., Krivulka, G. R., Beaudry, K., Lifton, M. A., Nickerson, C. E., Trigona, W. L., Punt, K., Freed, D. C., Guan, L., Dubey, S., Casimiro, D., Simon, A., Davies, M. E., Chastain, M., Strom, T. B., Gelman, R. S., Montefiori, D. C., Lewis, M. G., Emini, E. A., Shiver, J. W., and Letvin, N. L. (2000). Control of viremia and prevention of clinical AIDS in rhesus monkeys by cytokine-augmented DNA vaccination. *Science* **290,** 486–492.

Batard, P., Szollosi, J., Luescher, I., Cerottini, J. C., MacDonald, R., and Romero, P. (2002). Use of phycoerythrin and allophycocyanin for fluorescence resonance energy transfer analyzed by flow cytometry: Advantages and limitations. *Cytometry* **48,** 97–105.

Becker, T. C., Wherry, E. J., Boone, D., Murali-Krishna, K., Antia, R., Ma, A., and Ahmed, R. (2002). Interleukin 15 is required for proliferative renewal of virus-specific memory CD8 T cells. *J. Exp. Med.* **195,** 1541–1548.

Benlagha, K., Weiss, A., Beavis, A., Teyton, L., and Bendelac, A. (2000). *In vivo* identification of glycolipid antigen-specific T cells using fluorescent CD1d tetramers. *J. Exp. Med.* **191,** 1895–1903.

Betts, M. R., Brenchley, J. M., Price, D. A., De Rosa, S. C., Douek, D. C., Roederer, M., and Koup, R. A. (2003). Sensitive and viable identification of antigen-specific $CD8^+$ T cells by a flow cytometric assay for degranulation. *J. Immunol. Methods* **281,** 65–78.

Blattman, J. N., Sourdive, D. J., Murali-Krishna, K., Ahmed, R., and Altman, J. D. (2000). Evolution of the T cell repertoire during primary, memory, and recall responses to viral infection. *J. Immunol.* **165,** 6081–6090.

Bodinier, M., Peyrat, M. A., Tournay, C., Davodeau, F., Romagne, F., Bonneville, M., and Lang, F. (2000). Efficient detection and immunomagnetic sorting of specific T cells using multimers of MHC class I and peptide with reduced CD8 binding. *Nat. Med.* **6,** 707–710.

Casimiro, D. R., Chen, L., Fu, T. M., Evans, R. K., Caulfield, M. J., Davies, M. E., Tang, A., Chen, M., Huang, L., Harris, V., Freed, D. C., Wilson, K. A., Dubey, S., Zhu, D. M., Nawrocki, D., Mach, H., Troutman, R., Isopi, L., Williams, D., Hurni, W., Xu, Z., Smith, J. G., Wang, S., Liu, X., Guan, L., Long, R., Trigona, W., Heidecker, G. J., Perry, H. C., Persaud, N., Toner, T. J., Su, Q., Liang, X., Youil, R., Chastain, M., Bett, A. J., Volkin, D. B., Emini, E. A., and Shiver, J. W. (2003). Comparative immunogenicity in rhesus monkeys of DNA plasmid, recombinant vaccinia virus, and replication-defective adenovirus vectors expressing a human immunodeficiency virus type 1 gag gene. *J. Virol.* **77,** 6305–6313.

Cauley, L. S., Cookenham, T., Miller, T. B., Adams, P. S., Vignali, K. M., Vignali, D. A., and Woodland, D. L. (2002). Cutting edge: Virus-specific CD4+ memory T cells in nonlymphoid tissues express a highly activated phenotype. *J. Immunol.* **169,** 6655–6658.

Champagne, P., Ogg, G. S., King, A. S., Knabenhans, C., Ellefsen, K., Nobile, M., Appay, V., Rizzardi, G. P., Fleury, S., Lipp, M., Forster, R., Rowland-Jones, S., Sekaly, R. P., McMichael, A. J., and Pantaleo, G. (2001). Skewed maturation of memory HIV-specific CD8 T lymphocytes. *Nature* **410,** 106–111.

Choi, E. M., Chen, J. L., Wooldridge, L., Salio, M., Lissina, A., Lissin, N., Hermans, I. F., Silk, J. D., Mirza, F., Palmowski, M. J., Dunbar, P. R., Jakobsen, B. K., Sewell, A. K., and Cerundolo, V. (2003). High avidity antigen-specific CTL identified by CD8-independent tetramer staining. *J. Immunol.* **171,** 5116–5123.

Crawford, F., Kozono, H., White, J., Marrack, P., and Kappler, J. (1998). Detection of antigenspecific T cells with multivalent soluble class II MHC covalent peptide complexes. *Immunity* **8,** 675–682.

Czerkinsky, C., Andersson, G., Ekre, H. P., Nilsson, L. A., Klareskog, L., and Ouchterlony, O. (1988). Reverse ELISPOT assay for clonal analysis of cytokine production. I. Enumeration of gamma-interferon–secreting cells. *J. Immunol. Methods* **110,** 29–36.

Daniels, M. A., and Jameson, S. C. (2000). Critical role for CD8 in T cell receptor binding and activation by peptide/major histocompatibility complex multimers. *J. Exp. Med.* **191,** 335–346.

Day, C. L., Seth, N. P., Lucas, M., Appel, H., Gauthier, L., Lauer, G. M., Robbins, G. K., Szczepiorkowski, Z. M., Casson, D. R., Chung, R. T., Bell, S., Harcourt, G., Walker, B. D., Klenerman, P., and Wucherpfennig, K. W. (2003). *Ex vivo* analysis of human memory CD4 T cells specific for hepatitis C virus using MHC class II tetramers. *J. Clin. Invest.* **112,** 831–842.

Egan, M. A., Kuroda, M. J., Voss, G., Schmitz, J. E., Charini, W. A., Lord, C. I., Forman, M. A., and Letvin, N. L. (1999). Use of major histocompatibility complex class I/peptide/beta2M tetramers to quantitate CD8(+) cytotoxic T lymphocytes specific for dominant and nondominant viral epitopes in simian-human immunodeficiency virus–infected rhesus monkeys. *J. Virol.* **73,** 5466–5472.

Foulds, K. E., Zenewicz, L. A., Shedlock, D. J., Jiang, J., Troy, A. E., and Shen, H. (2002). Cutting edge: CD4 and CD8 T cells are intrinsically different in their proliferative responses. *J. Immunol.* **168,** 1528–1532.

Gamadia, L. E., ten Berge, I. J., Picker, L. J., and van Lier, R. A. (2002). Skewed maturation of virus-specific CTLs? *Nat. Immunol.* **3,** 203.

Garboczi, D. N., Hung, D. T., and Wiley, D. C. (1992). HLA-A2-peptide complexes: refolding and crystallization of molecules expressed in Escherichia coli and complexed with single antigenic peptides. *Proc. Natl. Acad. Sci. USA* **89,** 3429–3433.

Greten, T. F., Slansky, J. E., Kubota, R., Soldan, S. S., Jaffee, E. M., Leist, T. P., Pardoll, D. M., Jacobson, S., and Schneck, J. P. (1998). Direct visualization of antigen-specific T cells: HTLV-1 Tax11-19–specific CD8(+) T cells are activated in peripheral blood and accumulate in cerebrospinal fluid from HAM/TSP patients. *Proc. Natl. Acad. Sci. USA* **95,** 7568–7573.

Gutgemann, I., Fahrer, A. M., Altman, J. D., Davis, M. M., and Chien, Y. H. (1998). Induction of rapid T cell activation and tolerance by systemic presentation of an orally administered antigen. *Immunity* **8,** 667–673.

Haanen, J. B., van Oijen, M. G., Tirion, F., Oomen, L. C., Kruisbeek, A. M., Vyth-Dreese, F. A., and Schumacher, T. N. (2000). *In situ* detection of virus- and tumor-specific T-cell immunity. *Nat. Med.* **6,** 1056–1060.

He, X. S., Rehermann, B., Lopez-Labrador, F. X., Boisvert, J., Cheung, R., Mumm, J., Wedemeyer, H., Berenguer, M., Wright, T. L., Davis, M. M., and Greenberg, H. B. (1999). Quantitative analysis of hepatitis C virus-specific CD8(+) T cells in peripheral blood and liver using peptide-MHC tetramers. *Proc. Natl. Acad. Sci. USA* **96,** 5692–5697.

Hoffmann, T. K., Donnenberg, V. S., Friebe-Hoffmann, U., Meyer, E. M., Rinaldo, C. R., DeLeo, A. B., Whiteside, T. L., and Donnenberg, A. D. (2000). Competition of peptide-MHC class I tetrameric complexes with anti-CD3 provides evidence for specificity of peptide binding to the TCR complex. *Cytometry* **41,** 321–328.

Hou, S., Hyland, L., Ryan, K. W., Portner, A., and Doherty, P. C. (1994). Virus-specific CD8+ T-cell memory determined by clonal burst size. *Nature* **369,** 652–654.

Hugues, S., Malherbe, L., Filippi, C., and Glaichenhaus, N. (2002). Generation and use of alternative multimers of peptide/MHC complexes. *J. Immunol. Methods* **268,** 83–92.

Kostense, S., Vandenberghe, K., Joling, J., Van Baarle, D., Nanlohy, N., Manting, E., and Miedema, F. (2002). Persistent numbers of tetramer+ CD8(+) T cells, but loss of interferon-gamma+ HIV-specific T cells during progression to AIDS. *Blood* **99,** 2505–2511.

Kuroda, M. J., Schmitz, J. E., Barouch, D. H., Craiu, A., Allen, T. M., Sette, A., Watkins, D. I., Forman, M. A., and Letvin, N. L. (1998). Analysis of Gag-specific cytotoxic T lymphocytes in simian immunodeficiency virus-infected rhesus monkeys by cell staining with a tetrameric major histocompatibility complex class I-peptide complex. *J. Exp. Med.* **187,** 1373–1381.

Lalvani, A., Brookes, R., Hambleton, S., Britton, W. J., Hill, A. V., and McMichael, A. J. (1997). Rapid effector function in CD8+ memory T cells. *J. Exp. Med.* **186,** 859–865.

Lau, L. L., Jamieson, B. D., Somasundaram, T., and Ahmed, R. (1994). Cytotoxic T-cell memory without antigen. *Nature* **369,** 648–652.

Lechner, F., Wong, D. K., Dunbar, P. R., Chapman, R., Chung, R. T., Dohrenwend, P., Robbins, G., Phillips, R., Klenerman, P., and Walker, B. D. (2000). Analysis of successful immune responses in persons infected with hepatitis C virus. *J. Exp. Med.* **191,** 1499–1512.

Lee, P. P., Yee, C., Savage, P. A., Fong, L., Brockstedt, D., Weber, J. S., Johnson, D., Swetter, S., Thompson, J., Greenberg, P. D., Roederer, M., and Davis, M. M. (1999). Characterization of circulating T cells specific for tumor-associated antigens in melanoma patients. *Nat. Med.* **5,** 677–685.

Lyons, A. B. (2000). Analysing cell division *in vivo* and *in vitro* using flow cytometric measurement of CFSE dye dilution. *J. Immunol. Methods* **243,** 147–154.

Mallet-Designe, V. I., Stratmann, T., Homann, D., Carbone, F., Oldstone, M. B., and Teyton, L. (2003). Detection of low-avidity CD4+ T cells using recombinant artificial APC: following the antiovalbumin immune response. *J. Immunol.* **170,** 123–131.

Matsuda, J. L., Naidenko, O. V., Gapin, L., Nakayama, T., Taniguchi, M., Wang, C. R., Koezuka, Y., and Kronenberg, M. (2000). Tracking the response of natural killer T cells to a glycolipid antigen using CDld tetramers. *J. Exp. Med.* **192,** 741–754.

Matsui, K., Boniface, J. J., Reay, P. A., Schild, H., Fazekas de St Groth, B., and Davis, M. M. (1991). Low affinity interaction of peptide–MHC complexes with T cell receptors. *Science* **254,** 1788–1791.

Matsui, K., Boniface, J. J., Steffner, P., Reay, P. A., and Davis, M. M. (1994). Kinetics of T-cell receptor binding to peptide/I–Ek complexes: correlation of the dissociation rate with T-cell responsiveness. *Proc. Natl. Acad. Sci. USA* **91,** 12862–12866.

McConkey, S. J., Reece, W. H., Moorthy, V. S., Webster, D., Dunachie, S., Butcher, G., Vuola, J. M., Blanchard, T. J., Gothard, P., Watkins, K., Hannan, C. M., Everaere, S., Brown, K., Kester, K. E., Cummings, J., Williams, J., Heppner, D. G., Pathan, A., Flanagan, K., Arulanantham, N., Roberts, M. T., Roy, M., Smith, G. L., Schneider, J., Peto, T., Sinden, R. E., Gilbert, S. C., and Hill, A. V. (2003). Enhanced T-cell immunogenicity of plasmid DNA vaccines boosted by recombinant modified vaccinia virus Ankara in humans. *Nat. Med.* **9,** 729–735.

McHeyzer-Williams, M. G., Altman, J. D., and Davis, M. M. (1996). Enumeration and characterization of memory cells in the TH compartment. *Immunol. Rev.* **150,** 5–21.

Meyer, A. L., Trollmo, C., Crawford, F., Marrack, P., Steere, A. C., Huber, B. T., Kappler, J., and Hafler, D. A. (2000). Direct enumeration of Borrelia-reactive CD4 T cells ex vivo by using MHC class II tetramers. *Proc. Natl. Acad. Sci. USA* **97,** 11433–11438.

Molldrem, J. J., Lee, P. P., Wang, C., Felio, K., Kantarjian, H. M., Champlin, R. E., and Davis, M. M. (2000). Evidence that specific T lymphocytes may participate in the elimination of chronic myelogenous leukemia. *Nat. Med.* **6,** 1018–1023.

Murali-Krishna, K., Lau, L. L., Sambhara, S., Lemonnier, F., Altman, J., and Ahmed, R. (1999). Persistence of memory CD8 T cells in MHC class I-deficient mice. *Science* **286,** 1377–1381.

Mwau, M., Cebere, I., Sutton, J., Chikoti, P., Winstone, N., Wee, E. G., Beattie, T., Chen, Y. H., Dorrell, L., McShane, H., Schmidt, C., Brooks, M., Patel, S., Roberts, J., Conlon, C., Rowland-Jones, S. L., Bwayo, J. J., McMichael, A. J., and Hanke, T. (2004). A human immunodeficiency virus 1 (HIV-1) clade A vaccine in clinical trials: Stimulation of HIV-specific T-cell responses by DNA and recombinant modified vaccinia virus Ankara (MVA) vaccines in humans. *J. Gen. Virol.* **85,** 911–919.

Novak, E. J., Liu, A. W., Nepom, G. T., and Kwok, W. W. (1999). MHC class II tetramers identify peptide-specific human CD4(+) T cells proliferating in response to influenza A antigen. *J. Clin. Invest.* **104,** R63–R67.

Plunkett, F. J., Soares, M. V., Annels, N., Hislop, A., Ivory, K., Lowdell, M., Salmon, M., Rickinson, A., and Akbar, A. N. (2001). The flow cytometric analysis of telomere length in antigen-specific CD8+ T cells during acute Epstein–Barr virus infection. *Blood* **97,** 700–707.

Ravkov, E. V., Myrick, C. M., and Altman, J. D. (2003). Immediate early effector functions of virus-specific CD8+CCR7+ memory cells in humans defined by HLA and CC chemokine ligand 19 tetramers. *J. Immunol.* **170,** 2461–2468.

Renno, T., Attinger, A., Rimoldi, D., Hahne, M., Tschopp, J., and MacDonald, H. R. (1998). Expression of B220 on activated T cell blasts precedes apoptosis. *Eur. J. Immunol.* **28,** 540–547.

Riedy, M. C., Muirhead, K. A., Jensen, C. P., and Stewart, C. C. (1991). Use of a photolabeling technique to identify nonviable cells in fixed homologous or heterologous cell populations. *Cytometry* **12,** 133–139.

Roederer, M. (2001). Spectral compensation for flow cytometry: visualization artifacts, limitations, and caveats. *Cytometry* **45,** 194–205.

Sallusto, F., Lenig, D., Forster, R., Lipp, M., and Lanzavecchia, A. (1999). Two subsets of memory T lymphocytes with distinct homing potentials and effector functions. *Nature* **401,** 708–712.

Savage, P. A., Boniface, J. J., and Davis, M. M. (1999). A kinetic basis for T cell receptor repertoire selection during an immune response. *Immunity* **10,** 485–492.

Schatz, P. J. (1993). Use of peptide libraries to map the substrate specificity of a peptide-modifying enzyme: a 13 residue consensus peptide specifies biotinylation in *Escherichia coli. Biotechnology NY* **11,** 1138–1143.

Schmid, I., Ferbas, J., Uittenbogaart, C. H., and Giorgi, J. V. (1999). Flow cytometric analysis of live cell proliferation and phenotype in populations with low viability. *Cytometry* **35,** 64–74.

Schmitz, J. E., Forman, M. A., Lifton, M. A., Concepcion, O., Reimann, K. A., Jr., Crumpacker, C. S., Daley, J. F., Gelman, R. S., and Letvin, N. L. (1998). Expression of the CD8alpha beta-heterodimer on CD8(+) T lymphocytes in peripheral blood lymphocytes of human immunodeficiency virus$^-$ and human immunodeficiency virus+ individuals. *Blood* **92,** 198–206.

Schneck, J. P., Slansky, J. E., O'Herrin, S. M., and Greten, T. F. (2004). Monitoring Antigen-Specific T Cells Using MHC-Ig Dimers. *In* "Current protocols in immunology," (J. E. Coligan, A. M. Kruisbeek, D. H. Margulies, E. M. Shevach, and W. Strober, eds.). Wiley, New York.

Skinner, P. J., Daniels, M. A., Schmidt, C. S., Jameson, S. C., and Haase, A. T. (2000). Cutting edge: In situ tetramer staining of antigen-specific T cells in tissues. *J. Immunol.* **165,** 613–617.

Valitutti, S., Muller, S., Cella, M., Padovan, E., and Lanzavecchia, A. (1995). Serial triggering of many T-cell receptors by a few peptide–MHC complexes. *Nature* **375,** 148–151.

van Lier, R. A., ten Berge, I. J., and Gamadia, L. E. (2003). Human CD8(+) T-cell differentiation in response to viruses. *Nat. Rev. Immunol.* **3,** 931–939.

Vance, R. E., Kraft, J. R., Altman, J. D., Jensen, P. E., and Raulet, D. H. (1998). Mouse CD94/NKG2A is a natural killer cell receptor for the nonclassical major histocompatibility complex (MHC) class I molecule Qa-1(b). *J. Exp. Med.* **188,** 1841–1848.

Waldrop, S. L., Pitcher, C. J., Peterson, D. M., Maino, V. C., and Picker, L. J. (1997). Determination of antigen-specific memory/effector CD4+ T cell frequencies by flow cytometry: evidence for a novel, antigen-specific homeostatic mechanism in HIV-associated immunodeficiency. *J. Clin. Invest.* **99,** 1739–1750.

Whelan, J. A., Dunbar, P. R., Price, D. A., Purbhoo, M. A., Lechner, F., Ogg, G. S., Griffiths, G., Phillips, R. E., Cerundolo, V., and Sewell, A. K. (1999). Specificity of CTL interactions with peptide–MHC class I tetrameric complexes is temperature dependent. *J. Immunol.* **163,** 4342–4348.

Zajac, A. J., Blattman, J. N., Murali-Krishna, K., Sourdive, D. J., Suresh, M., Altman, J. D., and Ahmed, R. (1998). Viral immune evasion due to persistence of activated T cells without effector function. *J. Exp. Med.* **188,** 2205–2213.

Zimmerman, C., Brduscha-Riem, K., Blaser, C., Zinkernagel, R. M., and Pircher, H. (1996). Visualization, characterization, and turnover of CD8+ memory T cells in virus-infected hosts. *J. Exp. Med.* **183,** 1367–1375.

CHAPTER 18

Use of Peptides and Peptide Libraries as T-Cell Stimulants in Flow Cytometric Studies

Georgy Cherepnev, Hans-Dieter Volk, and Florian Kern

Institute for Medical Immunology, Charite
Universitätsmedizin Berlin
Campus Charité Mitte
Berlin 10117, Germany

METHODS IN CELL BIOLOGY, VOL. 75

0091-679X/04 $35.00

I. Introduction

In this chapter, we focus on the use of peptides and peptide libraries in connection with flow cytometry for analyzing and quantifying T cells that recognize specific antigens. In all applications described here, peptides are used for stimulating T cells. The readout is always flow cytometry. Either T-cell activation or T-cell proliferation is detected. Other methods that also involve peptides and flow cytometry include the direct staining of T cells with multimeric peptide–major histocompatibility complex (MHC) complexes. These are discussed in detail in other chapters.

Until approximately 1995, analysis and quantification of T cells recognizing specific peptides was relatively cumbersome and usually indirect. The methods were based on cell culture and the most common readouts were proliferation and cytotoxicity. Prolonged cell-culture periods introduced a bias toward T-cell subsets with a higher proliferation rate. The introduction of intracellular staining of cytokines (Picker *et al.*, 1995; Waldrop *et al.*, 1998) was a breakthrough, because now T-cell activation could be measured reliably at the single-cell level. Initially this was achieved after short-term stimulation of peripheral blood T cells with protein antigens and pathogen lysates; however, shortly after, Kern *et al.* (1998) reported the use of peptides for T-cell stimulation in the same type of assay. This type of assay clearly had many obvious advantages. Because of the short incubation time, culture artifacts were unlikely, antigen-specific cells were directly visualized, and the phenotype of these cells could be obtained along with their relative or absolute frequency. Collectively, these techniques are referred to as *cytokine flow cytometry* (CFC) (Maino and Picker, 1998). Trapping secreted cytokines in an affinity matrix on the cell surface is an alternative to staining cytokines inside the cell, an approach termed *secretion assay*. Both CFC and the secretion assay can be used to sort antigen-specific T cells. The so-called *secretion assay* even allows for sorting live T cells for subsequent culture (Pittet *et al.*, 2001b; Scheffold *et al.*, 1998).

The advent of multicolor flow cytometry expanded the possibilities of analyzing phenotype and function of antigen-specific T cells in this assay, the limits of which are set only by the technical limits of flow cytometric instrumentation, at least with respect to sensitivity and the number of recorded parameters (De Rosa *et al.*, 2003). With respect to T-cell stimulation by peptides, the limits have to be defined in a different way. There are limits to the number of peptides one can be use in a single assay given a minimum concentration of each peptide, the number of parallel assays one can perform on a given amount of material, and the number of peptides that one can produce on a given budget (Hoffmeister *et al.*, 2003; Picker *et al.*, 1995). Although the latter is not a scientific aspect, its importance is growing in most institutions and it would be shortsighted to neglect the financial aspect.

II. Background

A. T-Cell Stimulation

T-cell stimulation can be achieved and measured in many different ways. Various stimulants are available widely, ranging from completely nonspecific to polyclonal and from polyclonal to highly antigen specific. For example, stimulation with phorbol-myristate-acetate (PMA) and Ionomycin, a combination of a nonspecific phosphokinase activator and a calcium ionophore, leads to the activation of most cells in a totally nonspecific way (Picker *et al.*, 1995). PHA, a plant lectin, by contrast, activates mainly T cells and is thought to nonspecifically cross-link T-cell receptors (TCRs). The peptide NLVPMVATV is derived from the cytomegalovirus (CMV) pp65 protein and is recognized only by human leukocyte antigen (HLA)-A*0201-restricted CD8 T cells in CMV-infected donors (Diamond *et al.*, 1997; Wills *et al.*, 1996). Each of these tests gives certain information on T-cell reactivity that we may use to assess T-cell function. In the clinical setting, one can use PMA–Ionomycin or PHA stimulation to assess general T-cell responsiveness or the cytokine profile of responding T cells (Parslow *et al.*, 2001). By contrast, one can use stimulation with NLVPMVATV to investigate whether an HLA-A*0201-positive person has a memory T-cell response to the CMV pp65 protein (Diamond *et al.*, 1997). If the result is positive, one can conclude that this person has been exposed to CMV, and if it is negative, however, one cannot conclude that this person does not have a T-cell response to CMV. Only if there was previously a response to this peptide may the absence of this response at a later time be interpreted as an impairment of immunity to CMV. The more specific the stimulus, the more specific the information and the less general the conclusions. Currently the exact implications of the presence or absence of a T-cell response to single antigenic determinants are not known. It is not known how much a response to a single antigenic peptide reflects the overall immune response to the antigen or pathogen from which it is derived (Kern *et al.*, 1999a, 2000, 2002). Because of this uncertainty, it may be preferable to use complete antigens or antigen lysates. However, CD8 T cells do not usually respond very well to these, so analyses based on stimulation with complete protein antigens are essentially limited to CD4 T cells. Using pools of peptides that in their entirety span a complete protein may be a way out of this dilemma, at least in part, because such comprehensive peptide pools can reflect the sum of T-cell epitopes within one or several proteins (Kern *et al.*, 2000; Maecker *et al.*, 2001b). To what degree these pools can account for the CD4 and CD8 T-cell responses to the respective protein may partly depend on the particular sequence and peculiarities of certain epitopes but mostly on the design of such peptide pools. This is the topic of the following technical sections of this chapter.

B. Peptides and Peptide Libraries as T-cell Stimulants

T-cell stimulation with peptides and peptide libraries had been used for a long time along with other readout systems before it was introduced to flow cytometry (Hoffmeister *et al.*, 2003). Most peptides used in immunological experimentation represent fragments of (naturally occurring) proteins. Sometimes peptides are modified by an exchange of amino acids. This may be done sporadically or in a systematic fashion (substitution analysis) (Bhargava *et al.*, 2002; Reineke *et al.*, 2002). After stimulation of T cells (e.g., PBMCs, T-cell lines, or T-cell clones) with peptides, typically one measured either T-cell proliferation or the ability of stimulated cytotoxic T cells to kill suitable target cells (Wills *et al.*, 1996), a process usually done with the ^{3}H-thymidine incorporation assay and a chromium release assay, respectively. Nowadays, both assays are often replaced by flow cytometry–based tests. Proliferation can be measured, for example, by bromodeoxyuridine BrdU incorporation (Latt *et al.*, 1977) Fluorochrome-labeled antibodies to BrdU will identify T cells that have incorporated this molecule into their DNA during proliferation. Alternatively, the loss of carboxy-fluorescein signal intensity can be measured (Lyons, 2000; Tesfa *et al.*, 2003) CFDASE is transformed into CFDA after entry into the intracellular space, where it stains proteins nonspecifically. These proteins are divided between daughter cells upon cell division, leaving each daughter cell with half of the initial amount of fluorochrome. Cytotoxicity can be measured by flow cytometry as well. Readouts may be an increase of propidium iodide (PI) or 7-AAD staining (markers of dead or apoptotic cells) (Lecoeur *et al.*, 2001). The purpose of measuring T-cell activation or function (such as proliferation or killing) has not changed much. Proliferation and cytotoxicity are still important readouts in epitope mapping, for example. Thanks to modern flow cytometry, one can characterize cytokine profiles at the single-cell level as well (Picker *et al.*, 1995). Previously cytokines had to be measured in the supernatant of cell cultures, which was more or less specific, depending on whether cell lines or clones were being stimulated or mixes of different cells, such as in PBMCs. Typically cytokines in cell-culture supernatants were determined by plate-based enzyme-linked immunosorbent assay (ELISA) technology; however, nowadays this can also be done by bead-based flow cytometric assays (Carson and Vignali, 1999). It has become difficult to keep an overview of all the flow cytometry–based technologies being used across laboratories only for the purpose of measuring T-cell activation, let alone non–flow cytometry–based techniques. Peptides and peptide libraries may be used with all of them.

C. Requirements for Peptides that Are Recognized by T Cells

The antigenic determinants recognized by the antigen specific receptors on T cells (i.e., TCRs) are referred to as *epitopes*. Although B cells recognize free spatially conformed structures, T cells recognize antigens only in the context of MHC molecules. T-cell epitopes in the true sense are structures that are formed only when peptides—usually between 9 and 18 amino acids in length—are located in the binding groove of MHC molecules (Rammensee *et al.*, 1997). Unfortunately, peptides and epitopes are often confused, so it is important to understand that a peptide alone is never an epitope, because the TCR has to bind to both a section (usually located in the middle) of the peptide and certain structural determinants of the MHC molecule for the T cell to be activated (Rammensee *et al.*, 1997). For example, the peptide NLVPMVATV represents an epitope only if bound in the peptide-binding groove of an HLA-A*0201 molecule and only with regard to HLA-A*0201–restricted T cells. The epitope is formed by the complex of HLA-A*0201 and the peptide. NLVPMVATV alone is just a peptide, not an epitope. CD8 and CD4 T cells do not recognize peptides the same way. CD8 T cells are class I MHC restricted, whereas CD4 T cells are class II MHC restricted. This means that CD8 T cells recognize peptides only in the binding groove of class I MHC molecules, CD4 T cells only in the binding groove of class II MHC molecules. With a view to using peptides as T-cell stimulants, this difference in MHC presentation translates into different requirements regarding peptide length and termini. Class I MHC molecules have a binding groove that is closed at both ends. The N- and C-terminal amino acid of a binding peptide are harbored in special "pockets" formed by the amino acid side-chains of the MHC molecule. Therefore, class I MHC–presented peptides can generally not be longer than the class I MHC-binding groove. In contrast class II MHC molecules have an open binding groove and peptides can stick out at both ends, so the peptide length can exceed the length of the binding groove. Class I MHC-binding peptides are most commonly nine amino acids long, although shorter and longer binders have been reported. Class II MHC-binding peptides, by contrast, are mostly between 15 and 18 amino acids long, but there is considerable variation. Interestingly though, the area of the binding groove that directly interacts with amino acid side-chains of the bound peptide, the binding region, is usually nine amino acids long, both in class I and in class II MHC molecules. The interaction between the amino acid chains forming the floor and the sidewall of the binding groove with the side-chains of the binding peptide will determine whether a peptide is a strong binder, a weak binder, or does not bind at all. The preferred amino acids at the most relevant positions regarding binding to MHC molecules are collectively referred to as *binding motif*. Peptides may be analyzed with respect to the presence of such binding motifs and such analysis can be used for binding peptide MHC-binding predictions (www.syfpeithi.de).

With CD4 T cells the length of peptides is important in a different way, too. CD4 T cells appear to recognize sections of binding peptides that are outside the MHC-binding groove, that is, flanking the core-binding region. These additional amino acids may even be crucial for the recognition of the epitopes. As a result, peptides that are reduced to the MHC-binding portion of 9 amino acids may not be recognized by CD4 T cells. On the other hand, some peptides of 9 amino acids in length were reported to stimulate CD4 T cells very efficiently (Fig. 1). The corresponding 15–amino acid peptide EHPTFTSQYRIQGKL, however, induced slightly higher frequencies of activated CD4 T cells (not shown).

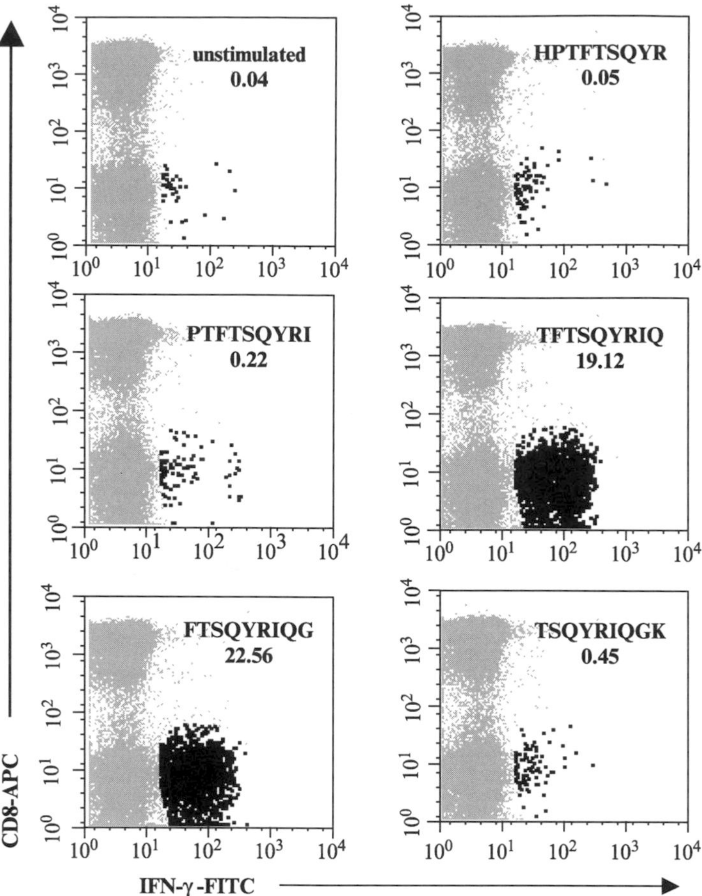

Fig. 1 PBMCs from a cytomegalovirus (CMV)-seropositive donor were stimulated with the 15–amino acid CMV pp65-derived peptide EHPTFTSQYRIQGKL (not shown) and a set of corresponding overlapping 9–amino acid peptides. Of these, only two immediately adjacent 9–amino acid peptides activated CD4 T cells efficiently. CD8 T-cell activation for both 15–amino acid and 9–amino acid peptides was not detectable. We performed the stimulation according to the protocol described in the text.

The relative efficiency of 9- and 15-mer peptides in stimulating specific $CD8^+$ T cells is shown in Fig. 2. Note that although the 9–amino acid peptide generated a maximal response at considerably lower doses than the two 15–amino acid peptides containing the respective 9-mer sequence, the 15-mers could stimulate near-maximal responses at a concentration of 1 μg/ml. Additional experiments (not shown) confirmed that the efficiency of stimulation is reduced as more amino acids are added, making 15–amino acid peptides the longest that can reliably elicit $CD8^+$ T-cell responses. Figure 3 shows an example of dose–response curves

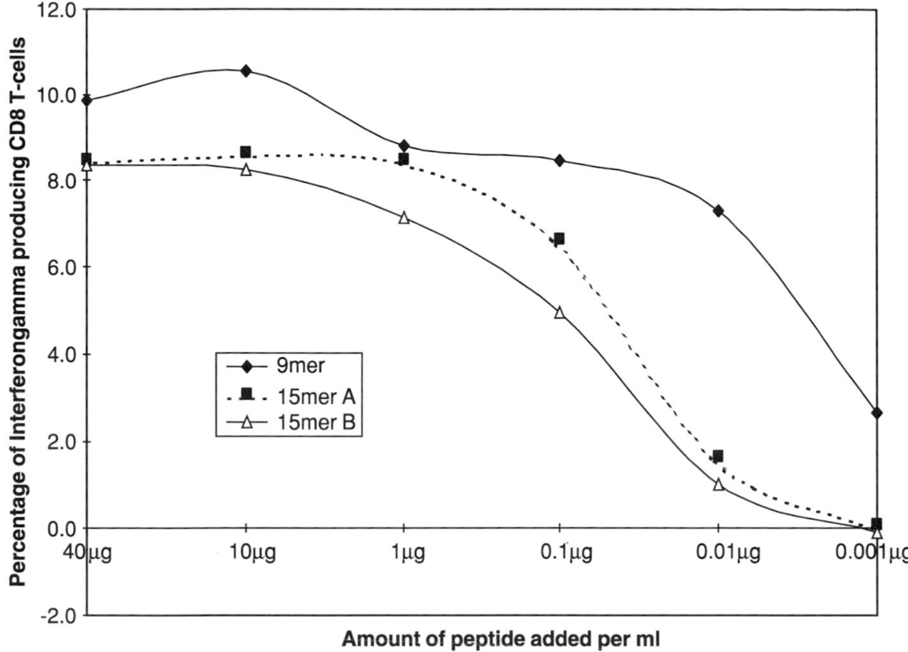

Fig. 2 PBMCs from a cytomegalovirus (CMV)-seropositive donor were stimulated with one 9–amino acid peptide and two 15–amino acid peptides, all including the same known CD8 T-cell epitope. Although the 9–amino acid peptide produces a significant close-to-maximum stimulation at concentrations of 10 and 100 ng/ml, at least 1 μg/ml is required of the 15–amino acid peptides to induce the same amount of interferon-γ (IFN-γ)–positive cells. Values shown are means of duplicate determinations. We stimulated cells according to the protocol described in the text.

obtained with three 15–amino acid peptides that partly overlap and all induce a CD4 T-cell response.

III. Results/Discussion of the Literature

A. Experiments Using Short-Term Stimulation and CFC as Readout

1. Differences between CD4 and CD8 T Cells

To define T-cell epitopes, one can divide protein amino acid sequences into overlapping peptides. These can be used individually or in pools to stimulate T cells. Studies on CD8 T-cell epitopes have frequently used a peptide length of 15 amino acids with variable overlaps. For the aforementioned reasons, peptides of 15 amino acids in length, however, cannot be expected to bind to class I MHC molecules. Nevertheless, many CD8 T-cell epitopes were identified this way. The reason 15–amino acid peptides can be used to stimulate CD8 T cells is not entirely

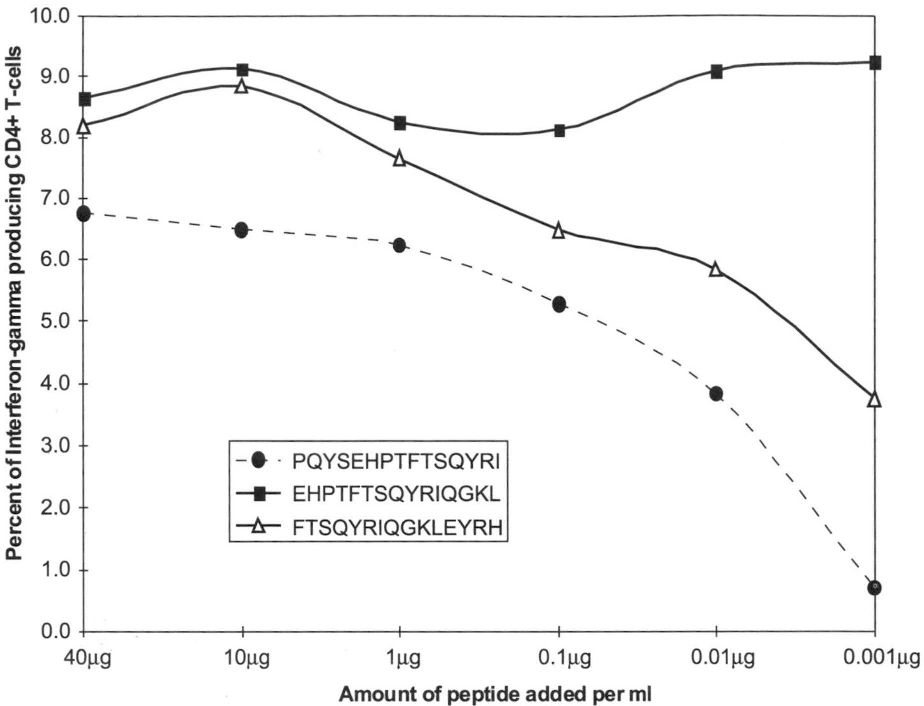

Fig. 3 PBMCs from a human leukocyte antigen (HLA)-DR11–positive and cytomegalovirus (CMV)-seropositive donor were stimulated with three different but consecutive 15–amino acid peptides covering an epitope-containing amino acid stretch of the CMV pp65 protein. Although all three peptides are stimulating, their efficiency is quite variable. The middle peptide (EHPTFTSQYR-IQGKL) best covers two possible theoretical sequences of HLA-DR11–presented epitopes contained in the concerned section of the pp65 sequence, including flanking amino acids at both the N- and the C-terminus (EH**P**TFTSQYR**I**QGKL or EHPT**F**TSQYRIQ**G**KL, [potential anchor amino acids in boldface type]). This peptide is much more efficient than the other two peptides because stimulation is already in a plateau at a concentration of 0.001 μg/ml. Values shown are means of duplicate determinations. We stimulated cells according to the protocol described in the text.

clear. Proteasomal processing of these peptides may play a role in some systems; however, most of the time these peptides are probably loaded externally. Studies show that extracellular antigen processing occurs as well and may contribute to the normal immune response. The specific factors involved in extracellular peptide clipping or trimming are not yet known.

Using intracellular INF-γ as readout, we found that CD8 T-cell stimulation was more efficient, however, with 9–amino acid peptides than with 15–amino acid peptides. The percentage of T cells stimulated within a typical 6-hour incubation period was approximately 20% lower when a 15–amino acid peptide was used instead. Much longer peptides (i.e., 20 or 25 amino acids in length) did not result in effective T-cell stimulation, at least not within the same incubation time.

Because the proteases involved in external peptide clipping are not known, it is impossible to predict how efficiently one or the other 15–amino acid peptides might be reduced to a fitting 9–amino acid peptide. The peptide TRATKMQVI, for example, is contained in two overlapping 15–amino acid peptides in a peptide scan covering the entire pp65 protein (AD 169 strain) with 11 overlaps between peptides. Although the first one of them (VCSMENTRATKMQVI· · ·) stimulated efficiently, the second one (ENTRATQMQVIG· · ·) did not in at least three donors tested. The difference between consecutive 15–amino acid peptides containing the same stimulating shorter sequence is usually not that big in terms of stimulation. In our experience, using peptide scans with 15–amino acid peptides and 11 overlaps has been a successful approach to identifying CD8 T-cell epitopes in two CMV proteins (Fig. 4).

It is noteworthy that within such libraries each stretch of 12 amino acids is contained in at least two consecutive peptides. A given 9 amino acid will not be available only if none of the two 15 amino acids containing it is processed (trimmed or clipped) efficiently.

With respect to CD4 T cells we have sometimes observed the opposite. Both the 15–amino acid peptide and the 9–amino acid peptide are stimulating, but the 15–amino acid peptide stimulates more efficiently. The reason may be that the 15–amino acid peptide contains flanking amino acids recognized by the TCR that are not contained in the 9–amino acid peptide. An alternative explanation for this phenomenon could be that the longer peptide contains several portions that represent epitopes, and the shortened peptide contains only one. In Fig. 1, however, a 15–amino acid peptide was divided into seven overlapping 9–amino acid peptides, and only two immediately adjacent peptides caused stimulation efficiently. Had the long peptide been divided into four peptides of 12 amino acids in length, the result might have been different. Because of these differences between CD4 and CD8 T cells it is far more difficult to predict CD4 T-cell epitopes. Generally speaking, 15–amino acid peptides seem fit for use in peptide sets intended to screen for CD8 and CD4 T-cell epitopes. Studies reporting that longer

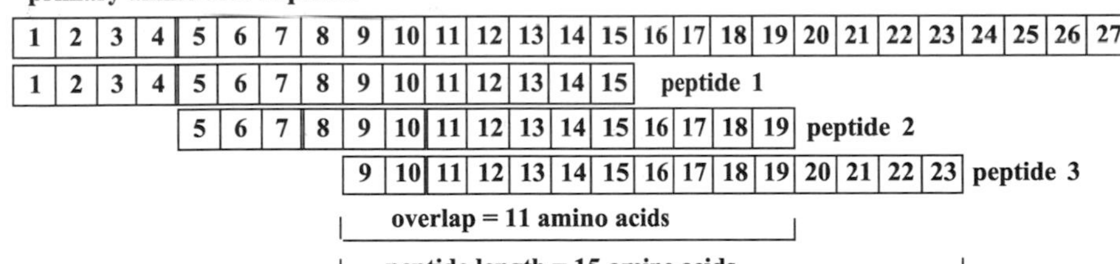

Fig. 4 Schematic representation of a peptide scan covering an amino acid sequence with peptides 15 amino acids in length and with four overlaps. Details about this setup are explained in the text. Boxes represent amino acids.

peptides are optimum for measuring CD4 responses may define *optimum* in a different (not necessarily less justifiable) way. If fewer peptides are needed to cover a certain amino acid sequence, this is, of course, an advantage from the points of view of workload and cost. Thus, using longer peptides may be best for defining CD4 T-cell responses, but this may be at the expense of discovering CD8 T-cell responses. Many studies have addressed this topic and generalizations should be received with caution and always be interpreted within the context of the respective studies.

2. N- and C-Terminal Modifications of Peptides

Because the class I MHC-binding groove is closed at both ends, the C- and N-termini of the peptide have to fit into specific pockets. Changes at these termini may result in the peptide being unable to bind. It is generally recommended and our experience that peptides with free N- and C-termini are best for stimulating CD8 T cells. Regarding CD4 T cells modification of the N- or C-terminus does not seem to affect binding to class II MHC in the same way. CD4 T cells may even recognize peptides bearing fluorochromes or other labels.

3. Measuring T-Cell Stimulation

The approach discussed here is the measurement of intracellular cytokines after secretion inhibition with brefeldin A (BFA), a macrolide antibiotic originating from the fungus, *Eupenicillium brefeldianum*. It specifically and reversibly blocks protein transport from the endoplasmic reticulum to the Golgi apparatus in many cell types and species. Because of its numerous and reversible effects on protein transport and processing, BFA has become an important tool for cell biologists. Another inhibitor of cytokine secretion monensin has not performed as well in our hands, in particular in combination with peptide stimulation (unpublished observations). T-cell stimulation will usually result in changes in surface molecule expression and various intracellular changes. One of these changes is *de novo* synthesis of cytokines such as interferon-γ (IFN-γ), tumor necrosis factor-α (TNF-α), or interleukin-2 (IL-2). These are most frequently used to detect T-cell activation. Nevertheless, other cytokines such as interleukin-4 (IL-4) or interleukin-10 (IL-10) may be used as well. IFN-γ and TNF-α occur more universally in activated T cells than the others and may be of advantage in epitope mapping: nevertheless, the most suitable cytokine should be chosen within the context of the organisms under investigation.

4. Incubation Time

Kinetics studies have shown that T cells produce measurable amounts of cytokines after only 6 hours of incubation with proteins or pathogen lysates. The maximal cytokine response was observed between 6 and 10 hours, depending on the cytokine and with a decline at later times. Because peptides need not be processed intracellularly, the incubation time can be as short as that or even shorter. Most published studies have used an incubation time of 6 hours with peptides or peptide pools for CFC. Incubation times should be chosen sufficiently short so cell division does not occur, especially if frequencies of responding cells are to be determined. No good kinetics studies with peptides have been published. Most researchers seem to rely on the assumption that peptides will not possibly require longer incubation times than proteins or pathogen lysates. It is in fact difficult to answer the question of whether with this type of assay the optimum incubation time in terms of the maximum frequency of responding T cells should be used (which may vary from individual to individual) or in order to compare results between individuals, it is more important that the incubation time is always identical. Ideally, CFC should be able to determine the frequency of T cells responding to a specific peptide, but we know from the combined use of peptide–MHC tetramers and peptide stimulation that a variable major proportion of T cells that recognize a given peptide in this assay responds with IFN-γ synthesis. So, even if the maximum number of IFN-γ–producing T cells is stimulated, this is only part of the response. A detailed and comprehensive study answering these or at least some of these open questions is missing. In the published kinctics studies, the secretion inhibitor BFA was mostly added 4 hours before the end of the total incubation time. However, it was reported that exposure to BFA for up to 12 hours was not detrimental. Some of these studies were done on whole blood rather than PBMCs.

5. Activation of T Cells Not Producing the Cytokine(s) Stained For

Studies with peptide–MHC tetramers (class I MHC) have shown that generally the number of IFN-γ–producing T cells is close to and correlates well with the number of tetramer-binding cells. It is obvious, however, that the measurement of cytokine production alone cannot account for all T cells that have been activated by a given peptide. Under certain circumstances, many T cells bearing the correct TCR may be anergic or at least not produce IFN-γ and, therefore, may not be recognized by this assay. This may be the case in immunosuppressed individuals, for example.

6. Other Sources of "T-Cell Cytokines"

It is important to note that apart from T cells IFN-γ is produced by natural killer (NK) cells and TNF-α is produced by monocytes/macrophages. CD3 or CD4 is required as a marker to discriminate between NK cells and T cells CD8 expression may be markedly downregulated after activation of T cells with peptides so these T cells cannot be discriminated from NK cells with respect to CD8 expression (the CD8-weak T-cell subset can never be discriminated by CD8 expression alone but is sometimes neglected). Additionally, particularly after longer incubation times, scatter characteristics may not allow to distinguish activated T lymphocytes from monocytes. Depending on the incubation time after blastogenesis, activated T lymphocytes are not necessarily found within a typical lymphocyte gate. Monocytes may also stain, at least weakly, with CD4, so it is useful to include additional markers such as CD16/56 or CD14 into the staining panel. Depending on the machine used, this may happen at the crucial expense of one of four fluorescence channels.

B. Alternatives to CFC: Tetramers/Streptamers

Indirectly, tetramers represent another use of peptides in flow cytometry. They are described in detail in chapter 17. For the reasons stated earlier, peptides are essential components of these reagents. Unfortunately, the use cannot exchange peptides in tetramers in a simple way. Similar reagents, referred to as *dimers*, allow the exchange of the peptide, or rather loading, with different peptides, but they do not stain quite as brightly as the tetramers. The use of peptide libraries in connection with tetramers has been described, however, at the stage of refolding of the MHC molecule together with peptide. By providing many peptides at this time, a kind of tetramer library is being produced. Although these reagents represent an important and elegant research tool, they are not discussed here any further.

IV. Technical Considerations

A. Choice of Peptides

Custom peptides can be bought from numerous companies. They are generally produced according to standard Fmoc technology. This usually results in milligram amounts of "raw" peptides in reasonable purity. Peptides can normally be ordered to be delivered with a mass-spectroscopic and high-pressure liquid chromatography (HPLC) analysis. This is an important quality control. After synthesis, these peptides may be purified if HPLC results indicate that purity is low. This is laborious and usually rather costly. In our experience, for most experiments carried out for the purpose of epitope identification or T-cell monitoring, raw peptides of approximately 60% purity are sufficient. It is often inferred that peptides of 15 amino acids in length may be contaminated with shorter peptides. This is clearly possible; however, the following should be noted. Because peptides are synthesized beginning at the C-terminus truncated versions of peptides contaminating the end-product will always include the C-terminus. They are synthesized in steps of one amino acid at a time with the first 4 or 5 steps usually being almost 100% effective. As a result, most contaminating shorter peptides will be at least 4 to 5 amino acids in length and start with the C-terminus. For example, a 9–amino acid peptide located centrally within a synthesized 15–amino acid peptide cannot possibly be a contaminant. The CMV pp65 peptide NLVPMVATV is located within the 15–amino acid peptide, ARNLVPMVATVQGQN. Stimulation with the latter cannot be explained by contamination with the former. With respect to excluding contamination, it is of note that purification of peptide preparations by HPLC can hardly exceed 99% purity because components encompassing less than 1% of a sample may simply not be picked up by HPLC. However, if, for example, 5 mg/ml of a peptide is used for stimulation, a 1% (w/w) contamination with a C-terminal 9–amino acid peptide would result in approximately 50 ng of this peptide being present in the assay. Such short peptides may already be effective stimulants of IFN-γ production in CD8 T cells.

Standard Fmoc technology is best for producing individual peptides in large amounts. However, if large numbers of peptides in small amounts are required, alternative technologies may be of advantage. Protein-spanning peptide pools can be used instead of recombinant proteins for stimulating both CD4 and CD8 T cells. If peptides of 15 amino acids in length with 11 overlaps are used, a protein amino acid sequence encompassing 500 amino acids requires approximately 125 peptides to be made. This would be extremely costly using standard Fmoc technology, and a big financial risk, if it is not known whether the protein in question is a T-cell target or not. If such protein-spanning pools are used to screen a complete proteome for T-cell targets, technologies are needed that can produce large numbers of peptides in small amounts.

It should be mentioned that the term *peptide library* strictly speaking, refers to a set of peptides that is identical with respect to the amino acids in certain positions (e.g., anchor positions) but variable in all other positions. For example, each peptide in a peptide library might have the ideal anchor amino acids for binding to the MHC molecule, and H-2K^d, however, vary in all positions that interact with the TCR. Library, however, is sometimes used for different sets of peptides as well, such as protein-spanning pools of peptides with certain overlaps. These are also sometimes referred to as *peptide scans*. There are two principally different approaches: First, synthesis of peptide libraries in the stricter sense using pooling strategies and second, parallel synthesis. The former yields compound mixes, and the latter results in discrete compounds. A synthesis method addressing a single compound in a spatial fashion bears the advantage that peptides can be identified by position. No tagging or deconvolution procedures are required. Thus, structure-activity data can be obtained directly and follow-up production of larger amounts of materials is facilitated. Keeping multiple single compounds in separate areas (spatially separated) is achieved in different ways. The carriers of these compounds have to be able to divide into segments. Here, we briefly describe two methods that are used commercially: the so-called "spot" technology and the so-called "pin" technology.

1. So-Called "SPOT Synthesis"

The original spot-synthesis concept by Frank involves positionally addressed delivery of small volumes of activated amino acid solutions directly onto a coherent membrane sheet. The areas wet by the resulting droplets can be considered microreactors provided that a nonvolatile solvent system is used. The functional groups fixed onto the membrane surface react with the pipetted reagents and conventional solid-phase syntheses occur. The volumes dispensed and the physical properties of the membrane surface and the solvent system define the size of the resulting spots and together with the minimum distance between the spots the number of peptides that can be synthesized per membrane area. The methodology has been reviewed (Frank, 2002; Reineke *et al.*, 2001; Wenschuh *et al.*, 2000).

Depending on the specifications of the performed method, the peptides resulting from this type of synthesis usually bear a linker group at the C-terminus: frequently this is leucin amide. It has been demonstrated that this type of synthesis can yield reasonable amounts of peptides with free C-termini, provided the membrane can be functionalized directly with one particular amino acid. In this case, all peptides synthesized on one membrane have an identical C-terminus.

2. Multipin Method of Peptide Synthesis

The multipin peptide synthesis approach originated as an immunological tool for epitope mapping (Maeji *et al.*, 1990). This is an effective, low-cost, simultaneous multiple synthesis technology producing 20–25 μmol of a peptide per modular grafted surface (Ede, 2002). Technically, the synthesis is performed on plastic pins arranged in an 8×12 microplate format to allow the simultaneous preparation of sets of 96 peptides. The "pin" consists of a radiation-grafted polypropylene "crown" fitted to an inert polypropylene "stem." The molded pin is the "container" or "synthesizer," and the mobile polymer grafted on the pin is the "resin."

Currently, Multipin technology is widely used for antibody and receptor epitope screening and characterization (Tribbick, 2002) and for multimilligram (commercial) synthesis of peptides (Maeji *et al.*, 1995).

B. Handling Peptides

Peptides may be dissolved in various ways. For highly concentrated stock solutions, it is best to dissolve peptides in dimethyl-sulfoxide (DMSO). Sometimes, particularly for lower concentrations, phosphate-buffered saline (PBS) is sufficient. It is important to keep in mind that on the one hand, peptides must be sufficiently soluble in the solvent, particularly if large pools of 100 or so peptides are used simultaneously for stimulation and that on the other hand, the solvent must be at sufficiently low-end concentration to avoid toxicity. In our experience, most peptides of 15 amino acids in length can be dissolved at 50–100 mg/ml in DMSO. When dissolved in DMSO, peptides should be kept cool. Warming peptides in DMSO solution accelerates oxidization of side-chains, which may alter the peptide's binding characteristics. To prevent oxidization, one can dissolve peptides under nitrogen atmosphere, an advisable practice if peptides are kept for a long time.

C. Standardization

As a prerequisite to integrating HCMV-specific CFC into clinical routine, we performed systematic studies to establish intra-assay, interassay, and diurnal variability.

1. *Intra-assay variability:* The standard assay was performed as described in section VII. The workup of each sample included one tube each for the two peptide libraries (IE-1 and pp65), viral lysate, and the unstimulated CD4 and CD8 controls. Ten replicates of the complete set were run.
2. Examination was concluded when we had five responding donors for each stimulus to be tested. The CV of the CD8-mediated responses to IE-1 and pp65 peptides was observed at around 10%, whereas the response to viral lysate showed slightly greater variability.
3. *Inter-assay variability:* Blood was drawn from six subjects on 5 consecutive days, always at the same time of the day. The standard assay was performed in duplicate for the two peptide libraries (IE-1 and pp65), viral lysate, and unstimulated controls on each day. In some donors, we observed steady patterns, but in others, patterns varied. This is an important observation because it reflects both methodological and biological variation. Purely technical variation may be determined using frozen samples: however, stimulation of thawed PBMCs with virus lysate is not recommended, because APC seems to not recover fully.
4. *Diurnal variation:* Blood was drawn from 13 subjects at 9 a.m. and 4 p.m. on the same day. The CD4-mediated response to viral lysate was subject to substantial diurnal variation in some but not all individuals and more often than not showed a trend to upregulation in the afternoon (up to threefold). Similar results were obtained for CD8 T cells. This observation is in line with the reported circadian variations in antigen-specific proliferation of human T lymphocytes (Hiemke *et al.*, 1995).

The standard protocol per se proved to be a reliable, sensitive, and reproducible diagnostic procedure with a very acceptable intra-assay variability of around 10–13% for all stimuli. The message from the *diurnal variation* section is the necessity for uniform diagnostic blood drawing, that is always at the same time of day. The observed interassay variability over the 5-day period points mainly to biological variation. The purely technical/methodological variation is currently reevaluated on frozen samples.

V. Specific Applications

A. Epitope Mapping

Peptides designed for epitope mapping must take into account the minimum length of epitopes. If the overlap between consecutive peptides is too short, the peptide scan does not cover all peptides that may be presented by MHC molecules.

Peptide scans for identifying CD8 T-cell epitopes should, therefore, have a minimum overlap of eight amino acids between the peptides. This way, every possible stretch of nine amino acids is contained in the library. Peptide length must consider that peptides exceeding 15 amino acids increasingly lose their ability to stimulate (which may be due to insufficient processing in the extracellular space). Some reported CD8 T-cell epitopes are 10 amino acids long. Whether all of these peptides are still stimulating if truncated to a length of nine is unknown. For example, the CMV pp65–derived epitope TPRVTGGGAM remains stimulating even if the C-terminal "T" or the N-terminal "M" is removed. This may be due to several reasons. The 10–amino acid peptide may be recognized by different clones than the two 9–amino acid peptides, and the latter may be recognized by different clones. Or, there may be overlap between the clones recognizing one or the other. Or, some clones may recognize both the 10–amino acid peptide and one or both of the truncated versions. This question could be addressed with T-cell clones but thus far has remained unresolved. This example was chosen to illustrate that using 10–amino acid peptides may conceal the fact that smaller peptides are also stimulating, and vice versa, using 9–amino acid peptides, the stimulating 10–amino acid peptide would not have been found. Ideally, a peptide scan for identifying CD8 T-cell epitopes would contain all possible stimulating peptides potentially contained in a given protein amino acid sequence, that is, all possible 8–amino acid peptides, 9–amino acid peptides, 10–amino acid peptides, and 11–amino acid peptides. Although this is certainly feasible from a technical point of view, the question remains whether it would be so much more efficient than using just 9–amino acid peptides that the benefit would justify the additional effort. Because most ideal stimulating peptides are 9 amino acids long, most of these will be found with the simpler design. Although 9 amino acids may be the one best length for mapping CD8 T-cell epitopes from an efficiency of T-cell stimulation point of view, it is certainly not from a point of view of cost. Published studies suggest that using 15–amino acid peptides is very useful for discovering CD8 T-cell epitopes (Kern *et al.*, 1998, 1999b; Wills *et al.*, 1996). Producing a peptide scan with 9–amino acid peptide length and eight overlaps (length of step = 1) will require about seven times as many peptides as producing a peptide scan with peptides of 15 amino acids in length and eight overlaps (length of step = 7). Because both CD4 and CD8 T-cell epitopes are of interest with a view to immunotherapies, ideally, a peptide scan should cover all potential CD4 T-cell–stimulating peptides as well. Because CD4 T-cell–stimulating peptides are longer, the peptides in the scan will have to be longer as well. The length requirements for MHC binding are apparently less stringent than those for CD8 T cells (15–18 amino acids according to Rammensee, 1997). We recommend the use of 15–amino acid peptides with 11 overlaps for mapping CD4 and CD8 T-cell epitopes. It has been formally shown that any identified CD4 T-cell epitope of 18 amino acids in length could not be replaced by a shorter peptide. Unfortunately in this field, there is a lot more speculation than solid data. It is unlikely, though, that this will change anytime soon because peptides are extremely polymorphic and the ideal length of a peptide may depend on various factors that cannot be accounted for with a simple formula or a few experiments. A peptide scan designed for mapping CD4 T-cell epitopes may well work only if the peptides are 20 amino acids long and the overlap is maybe 15 amino acids. The problem of all studies addressing this question is that they do not know which epitopes they missed by using one or the other approach. Using long peptides for identifying CD4 epitopes may even be misleading. In some experiments trying to confirm known CD4 T-cell epitopes in a protein from *M. tuberculosis*, we discovered that what appeared as just one epitope when using 15–amino acid peptides turned out to be two epitopes when using peptides of 12 amino acids in length (L. Tesfa, 2003, unpublished data). Unfortunately there is no really satisfying answer. For practical purposes, we will mostly have to use a one-size-fits-all approach, simply for lack of being able to account for all possibilities of peptide length in a way that addresses workload, cost, and the requirement of material (blood) in a balanced fashion.

Figure 5 demonstrates an example of a complete mapping procedure accommodating 120 peptides in a three-dimensional pool setup with five horizontal, five frontal, and five sagittal pools ($5 \times 5 \times 5 = 125$). Candidate peptides for confirmation are chosen from the intersections of overlapping positive pools. Dot plots show results with all 15 pools, an IE-1 total pool (containing all peptides at once), and an unstimulated control sample. Percentages indicate IFN-γ-positive events. Tables on the right represent the frontal sections (I–V). The points of intersection with the other pools (VII, VIII, and XI) are indicated by gray shading. Peptides 77 and 78 are identified as stimulating peptides. A square or cross-matrix is a simpler way of setting up pools (Betts *et al.*, 2000; Kern *et al.*, 1999b, 2002).

B. T-Cell Monitoring

Monitoring and epitope mapping are quite different. Usually, epitope mapping is the first step, and monitoring the second. Individual peptides can be used to stimulate CD4 or CD8 T cells in donors or patients of interest and the response in a percentage of IFN-γ–positive T cells, for example, can be measured. An alternative approach to monitoring frequencies is the use of reagents that combine recombinant MHC molecules and peptides (so-called *tetramers, streptamers, or dimers*) (Altman *et al.*, 1996; Knabel *et al.*, 2002; Lebowitz *et al.*, 1999). However, the most important difference is that with these multimeric peptide–MHC complexes, T cells can be stained directly. Thus, a "native" surface marker expression can be analyzed on these cells. However, nothing is learned about the functional status of these cells unless stimulation with a peptide, for example, is performed at the same time (Kostense *et al.*, 2002a,b). The disadvantage of the CFC technology using individual peptides, on the other hand, is that nonfunctional T cells are not accounted for, because cytokine production is a prerequisite for detection. There are no convincing examples that would sustain the assumption that single T-cell specificities reflect the overall immune response to a pathogen. There is no formal proof that this may not be the case either. It is obvious though that single T-cell

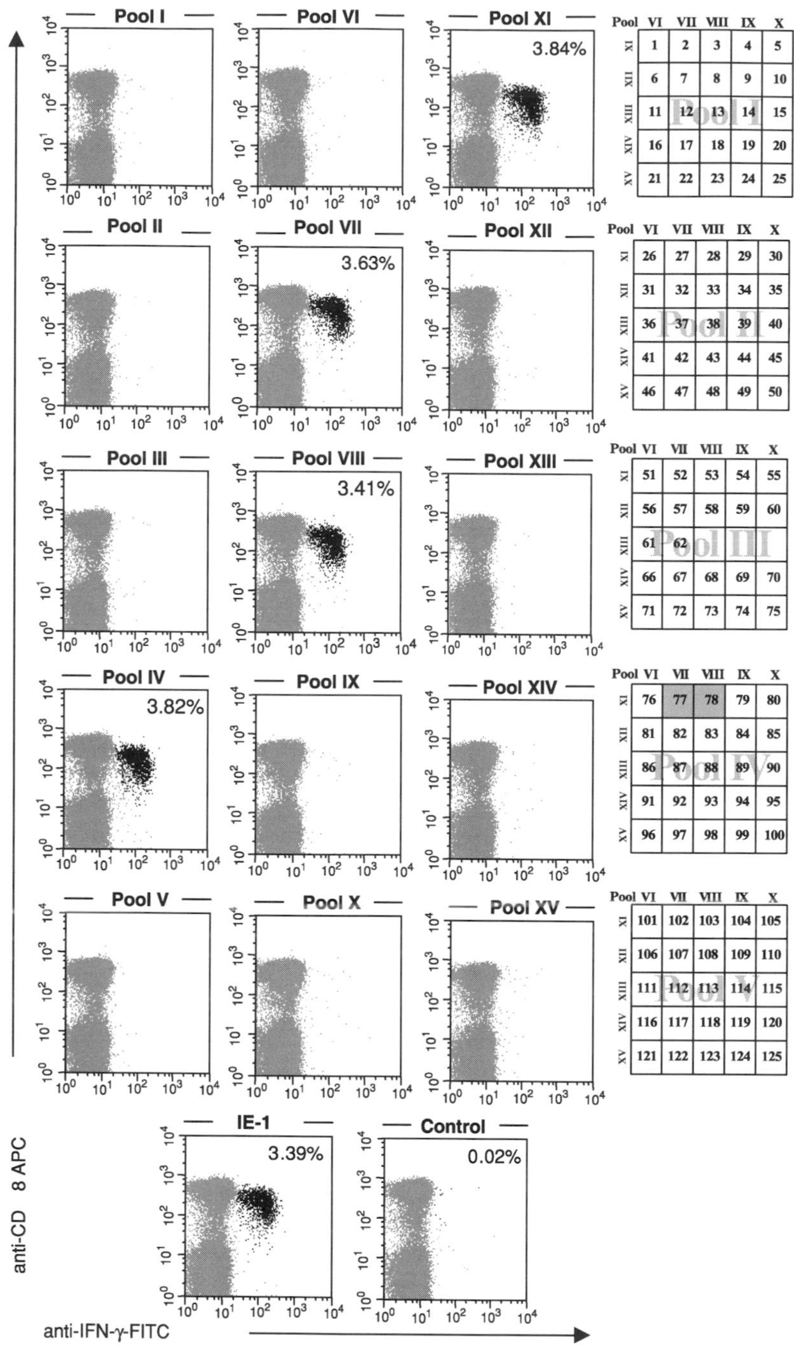
Pool I
Pool VI
Pool XI
3.84%
Pool II
Pool VII
3.63%
Pool XII
Pool III
Pool VIII
3.41%
Pool XIII
Pool IV
3.82%
Pool IX
Pool XIV
Pool V
Pool X
Pool XV
IE-1
3.39%
Control
0.02%
anti-CD 8 APC
anti-IFN-γ-FITC

specificities, such as single peptides, sometimes represent just infinitely small subunits of a pathogen. For example, the HIV-Gag–derived peptide SLYNTVATL, which is represented by HLA-A*0201, is recognized by not even half of HLA-A*0201–positive HIV-positive patients (Betts *et al.*, 2000). It would be totally useless to monitor the response to HIV in these patients using this particular peptide. On the other hand, the peptide NLVPMVATV, which is derived from CMV pp65, was used in several studies to measure the recovery of the CMV-directed immune response in bone marrow recipients (Gratama *et al.*, 2001; Hebart *et al.*, 2002). It appeared from the published studies that there was some correlation between immune system recovery and the number of T cells recognizing this peptide. Several similar studies are ongoing.

No studies using protein-spanning pools for monitoring purposes have been published. However, we have performed several studies in heart transplant recipients who are being evaluated (F. Kern, unpublished data, 2003). Protein-spanning pools have the great advantage that they can be used in any patient without prior knowledge of epitopes or reactivity. Because these pools represent proteins, the inherent limitation is reactivity to these proteins. In regards to CMV-specific responses, however, most donors seem to have T-cell responses to pp65 or IE-1, which is more than could be said about any single peptide. Protein-spanning pools have also been used to evaluate T-cell responses to the ESAT-6 protein of *M. tuberculosis* and various HIV proteins. They are very versatile tools with great potential in clinical diagnostics and research.

VI. Potential Clinical Use

A. Immunosuppressed Patients

Human Cytomegalovirus (HCMV) disease is a relatively rare complication after solid organ transplantation, but it is not uncommon after bone marrow transplantation (Riddell and Greenberg, 1995; Riddell *et al.*, 1994) and a major cause of morbidity and mortality in acquired immunodeficiency syndrome (AIDS) (Barry *et al.*, 1999; Gatell *et al.*, 1996). Unfortunately, clinical signs that would predict

Fig. 5 PBMCs from a cytomegalovirus (CMV)-seropositive donor were stimulated with 15 peptide pools representing a total CMV IE-1 protein spanning peptide scan. The shown diagrams and tables represent one mapping experiment. Tables in the right column show the frontal slices of a "cube" representing the pool setup. Each peptide is contained in and represents the point of intersection of exactly three different pools or "slices," one frontal, one sagittal, one horizontal. Shaded areas in the tables identify stimulating peptides. The positive response with pool XI indicates the stimulating peptides are in the uppermost horizontal pool. Positive responses with pools VII and VIII show the responses occur with peptides localized in the second and third sagittal slice from the left, and finally the positive response with pool IV localizes the stimulating peptides to the second frontal slice from counting from the front. Frequencies of interferon-γ–positive CD8 T cells are shown in percentages. For more details, please refer to the text.

this dangerous complication do not exist. Determinations of virus load are being used by some researchers to predict the risk of HCMV disease in transplant recipients (Boeckh and Boivin, 1998). Whether such determinations are clinically meaningful when used alone is not entirely clear. Viral replication and viral load depend on the efficiency of cellular immune mechanisms, and it may, therefore, be useful to analyze HCMV-specific T-cell responses. There is evidence that HCMV-directed cytotoxic T-lymphocyte (CTL) responses and HCMV antigenemia measured by immune cytology are correlated inversely (Reusser *et al.*, 1999). This underscores the potential advantage of having a measure of cellular immune defense. At the same time, however, the method used for such an assessment ought to be quicker than conventional CTL assays, so therapy can be initiated before HCMV disease develops in patients with an insufficient T-cell response. Monitoring the T-cell reactivity toward HCMV in immunosuppressed individuals may help us define clinically relevant lower limits of protective T-cell reactivity. These will probably be related to virus load.

B. Testing Immunity after Infection or Vaccination

The measurement of T-cell responses after vaccination is not a usual procedure. Because in most cases serology is used to assess whether an individual has established an immune response to a vaccine, knowledge that a helper T-cell response has been induced is, at best, indirect. Once a virus infection is established, antibodies are not helpful in getting rid of the virus. An effective response in most cases will depend on CD8 and CD4 T cells. Such responses can be identified by measuring T-cell responses using either conventional methods or CFC. Because the CD8 T-cell response is an essential part of the defense against CMV, testing this response after vaccination seems more relevant than measuring antibody levels. The presence of a T-cell response to a given pathogen, on the other hand, is clear evidence of prior exposure. A positive tuberculosis skin test, for example, identifies the presence of a T-cell response to mycobacterial antigens. Instead of purified protein derivative (or "tuberculin"), peptides representing proteins from *M. tuberculosis* can be used (Tesfa *et al.*, 2004). Most researchers have used either ELISPOT or cytokine secretion as readout; however, we have shown that CFC may be used as well. Tuberculosis is just one of many infections in which testing for T-cell responses may be useful.

C. Examples of Studies

Viral infections: CFC is most frequently applied to the analysis of antiviral $CD8^+$ T cells. Using CFC, Mollet *et al.* (2000) and Oxenius *et al.* (2002) established, for example, that in patients receiving highly active antiretroviral therapy (HAART) for the first time, the numbers of HIV-specific CD8 T cells rise sharply as viral load decreases but gradually decrease once viral load is efficiently controlled. The use of tetramer staining and CFC in combination demonstrated that many HIV-specific CD8 T cells are not functional, because they were unable to produce IFN-γ in response to stimulation with the respective peptide (Appay *et al.*, 2000). It had previously been shown that HIV-specific CD4 T cells have an important role in this regard, because they were more numerous in nonprogressing patients than in patients progressing to AIDS (Pitcher *et al.*, 1999). The loss of HIV-specific CD4 T cells in particular was accompanied by a loss of IFN-γ–producing HIV-specific CD8 T cells, increasing viral load, and progression to AIDS (Kostense *et al.*, 2002b).

It was also established that the frequencies of CMV-specific CD4 T cells was higher in the blood of HIV-infected individuals compared to healthy carriers of CMV (Waldrop *et al.*, 1997). Even in the latter, considerable changes of the frequencies of CMV-specific T cells were observed over time (Dunn *et al.*, 2002). In stem cell transplant (SCT) recipients, the occurrence of non-IFN-γ–producing CD8 T cells was found to be associated with recurrent CMV antigenemia (Ozdemir *et al.*, 2002). Many more clinical studies were done including T-cell responses to a range of other pathogens (Gratama/Kern, Cytometry, in press). CFC was also used to study antigen-specific T cells in various malignancies including AML (Scheibenbogen *et al.*, 2002) adenocarcinoma (Karanikas *et al.*, 2000) multiple myeloma (Maecker *et al.*, 2001a) and malignant melanoma (Nielsen *et al.*, 2000; Pittet *et al.*, 2001a; Smith *et al.*, 2003). Interestingly, combining CFC with immunophenotyping melanoma-specific T cells were found to display marked phenotypic heterogeneity (Pittet *et al.*, 2001b; Valmori *et al.*, 2002).

VII. Protocols

The following is the standard protocol for T-cell stimulation and intracellular cytokine staining.

Day 1:

- PBMCs are prepared according to standard Ficoll–Paque protocol using heparinized or citrated venous blood, adjusted to a concentration of 5×10^6/ml in RPMI 1640 (low endotoxin) medium supplemented with 10% (v/v) heat-inactivated fetal calf serum (FCS), 2 mmol/l glutamine, and 100 IU/ml penicillin/streptomycin. The cell suspension is transferred into a 50-ml Falcon tube, which is stored in the incubator (37 °C, humidified 5% CO_2 atmosphere) with a loosely tightened cap overnight (16 hours) before the stimulation with peptides or viral lysate.

Day 2:

- Place 100 μl of supplemented medium containing one test volume of peptide or viral lysate solution in sterile Falcon 2054 tubes (we commonly use a test volume of 4 μl [0,25 μl peptide/μl in DMSO] added to 96 μl of supplemented medium). Place 100 μl of supplemented medium containing a corresponding amount of DMSO (4 μl) alone in one tube, which will serve as the unstimulated control.

• Place 400 μl of cell suspension (1.25–2.5 $\times$ 10^6 cells/ml) in each tube (0.5 $\times$ 10^6 to 1 $\times$ 10^6 cells in total).

• Place tubes in a rack and place the rack in the incubator (37 °C, humidified CO_2 atmosphere) at a 5-degree slant (tubes almost horizontal)

• After 2 hours add 500 μl of prewarmed supplemented medium containing 10 μg of BFA to each tube: BFA solution should be freshly prepared.

• Return rack to the incubator and make sure that each tube is in the same slanting position as before (if a tube has been turned, adherent cells may now be dry without medium covering them).

Note: The final concentration of each peptide should be at least 1 μg/ml. The DMSO concentration should not exceed 1% (v/v) at any time. The final concentration of BFA should be 10 μg/ml. BFA stock solution of 5 mg/ml is prepared using DMSO and kept at −80 °C. Two microliters of this stock solution added in 498 μl of supplemented medium will provide an addition of 10 μg of BFA to the assay.

• After an additional 4 hours, add 3 ml of ice-cold PBS to each tube.

• Centrifuge (430 g, 8 minutes, 4 °C) and decant or aspirate supernatant.

• Re-suspend pellets in remaining fluid.

• Add 3 ml of PBS containing 2 mM of EDTA. Make sure the area of the tube wall that may have adherent cells on it is covered with fluid (especially if you use tubes other than the aforementioned, 3 ml may not be sufficient).

• Incubate all tubes for 10 minutes at 37 °C (water bath).

• Vortex at low speed for 30 °C.

• Centrifuge (430 g, 8 minutes, 4 °C) and decant or aspirate supernatant.

• Re-suspend pellets in remaining fluid.

• Add 3 ml of washing buffer (PBS containing 0.5% [w/v] bovine serum albumin [BSA] and 0.1% [w/v] sodium azide [NaN_3]).

• Centrifuge (430 g, 8 minutes, 4 °C) and decant or aspirate supernatant.

Note: Blotting tubes dry ensures that the remaining fluid is minimum and does not dilute the reagents added afterward to variable degrees.

• Re-suspend pellets in remaining fluid.

• Add 1 ml of BD Permeabilizing Solution 2 to each tube and incubate for 10 minutes at room temperature (protect tubes from light). In our experience, this step can be used as a single fixation/permeabilization step.

Alternatively, add 1 ml of BD lysing solution to each tube and incubate for 10 minutes at room temperature (protect tubes from light), or add another fixing solution according to the respective manufacturer's instructions (4% paraformaldehyde in PBS with a 5-minute incubation at 37 °C in a water bath may also be used).

- Add at least 3 ml of washing buffer.
- Centrifuge (430 g, 8 minutes, 4 °C) and decant or aspirate supernatant.
- Re-suspend pellets in remaining fluid.
- Add 1 ml of BD Permeabilizing Solution 2 to each tube and incubate for 10 minutes at room temperature. Or use other permeabilizing solutions according to the respective manufacturer's instructions.
- Add 3 ml of washing buffer.
- Centrifuge (430 g, 8 minutes, 4 °C) and decant or aspirate supernatant.
- Proceed to antibody staining (30 minutes on ice, in the dark). *Tip*: We prefer to stain in a total volume of 100 μl using a premixed "cocktail" of antibodies generally combining anti-IFN-γ–FITC (10 μl), anti-CD69–PE (10 μl), anti-CD3-PerCP (10 μl), anti-CD4- or anti-CD8–APC (2,5 μl), and washing buffer (67,5 μl). Alternatively, one could replace anti-CD69–PE with anti-CD4 or anti-CD8–PE. Applying an antibody cocktail is a great way of saving time compared with adding all antibodies separately.

Note. PerCP conjugates are available from BD only; all other antibodies can be obtained from various companies.

- Add 4 ml of washing buffer.
- Centrifuge (430 g, 8 minutes, 4 °C) and decant or aspirate supernatant. Probes are ready for cytofluorimetric analysis.
- Add 1 ml of 1% paraformaldehyde in PBS solution and incubate tubes for 5 minutes at room temperature protected from light (refixation to enable postponed cytofluorimetric analysis).
- Add 4 ml of washing buffer.
- Centrifuge (430 g, 8 minutes, 4 °C) and decant or aspirate supernatant. Re-suspend pellets in 150 μl of washing buffer. Keep tubes at 4 °C protected from light until reading on a flow cytometer.

References

Altman, J. D., Moss, P. A. H., Goulder, P. J. R., Barouch, D. H., McHeyzer-Williams, M. G., Bell, J. I., McMichael, A. J., and Davis, M. M. (1996). Phenotypic analysis of antigen-specific T lymphocytes [published erratum in *Science* 1998;**274:**94–96]. *Science* **280,** 1821.

Appay, V., Nixon, D. F., Donahoe, S. M., Gillespie, G. M., Dong, T., King, A., Ogg, G. S., Spiegel, H. M., Conlon, C., Spina, C. A., Havlir, D. V., Richman, D. D., Waters, A., Easterbrook, P., McMichael, A. J., and Rowland-Jones, S. L. (2000). HIV-specific CD8(+) T cells produce antiviral cytokines but are impaired in cytolytic function. *J. Exp. Med.* **192,** 63–75.

Barry, S. M., Lipman, M. C., Johnson, M. A., and Prentice, H. G. (1999). Respiratory infections in immunocompromised patients. *Curr. Opin. Pulm. Med.* **5,** 168–173.

Betts, M. R., Casazza, J. P., Patterson, B. A., Waldrop, S., Trigona, W., Fu, T. M., Kern, F., Picker, L. J., and Koup, R. A. (2000). Putative immunodominant human immunodeficiency virus-specific CD8(+) T-cell responses cannot be predicted by major histocompatibility complex class I haplotype. *J. Virol.* **74,** 9144–9151.

Bhargava, S., Licha, K., Knaute, T., Ebert, B., Becker, A., Grotzinger, C., Hessenius, C., Wiedenmann, B., Schneider-Mergener, J., and Volkmer-Engert, R. (2002). A complete substitutional analysis of VIP for better tumor imaging properties. *J. Mol. Recognit.* **15,** 145–153.

Boeckh, M., and Boivin, G. (1998). Quantitation of cytomegalovirus: Methodologic aspects and clinical applications. *Clin. Microbiol. Rev.* **11,** 533–554.

Carson, R. T., and Vignali, D. A. (1999). Simultaneous quantitation of 15 cytokines using a multiplexed flow cytometric assay. *J. Immunol. Methods* **227,** 41–52.

De Rosa, S. C., Brenchley, J. M., and Roederer, M. (2003). Beyond six colors: a new era in flow cytometry. *Nat. Med.* **9,** 112–117.

Diamond, D. J., York, J., Sun, J. Y., Wright, C. L., and Forman, S. J. (1997). Development of a candidate HLA A*0201 restricted peptide-based vaccine against human cytomegalovirus infection. *Blood* **90,** 1751–1767.

Dunn, H. S., Haney, D. J., Ghanekar, S. A., Stepick-Biek, P., Lewis, D. B., and Maecker, H. T. (2002). Dynamics of CD4 and CD8 T cell responses to cytomegalovirus in healthy human donors. *J. Infect. Dis.* **186,** 15–22.

Ede, N. J. (2002). Multiple parallel synthesis of peptides on SynPhase grafted supports. *J. Immunol. Methods* **267,** 3–11.

Frank, R. (2002). The SPOT-synthesis technique. Synthetic peptide arrays on membrane suports—principles and apllications. *J. Immunol. Methods* **267,** 13–26.

Gatell, J. M., Marrades, R., el-Ebiary, M., and Torres, A. (1996). Severe pulmonary infections in AIDS patients. *Semin. Respir. Infect.* **11,** 119–128.

Gratama, J. W., van Esser, J. W., Lamers, C. H., Tournay, C., Lowenberg, B., Bolhuis, R. L., and Cornelissen, J. J. (2001). Tetramer-based quantification of cytomegalovirus (CMV)-specific CD8+ T lymphocytes in T-cell–depleted stem cell grafts and after transplantation may identify patients at risk for progressive CMV infection. *Blood* **98,** 1358–1364.

Hebart, H., Daginik, S., Stevanovic, S., Grigoleit, U., Dobler, A., Baur, M., Rauser, G., Sinzger, C., Jahn, G., Loeffler, J., Kanz, L., Rammensee, H. G., and Einsele, H. (2002). Sensitive detection of human cytomegalovirus peptide-specific cytotoxic T-lymphocyte responses by interferon-gamma-enzyme–linked immunospot assay and flow cytometry in healthy individuals and in patients after allogeneic stem cell transplantation. *Blood* **99,** 3830–3837.

Hiemke, C., Brunner, R., Hammes, E., Muller, H., Meyer zum Buschenfelde, K. H., and Lohse, A. W. (1995). Circadian variations in antigen-specific proliferation of human T lymphocytes and correlation to cortisol production. *Psychoneuroendocrinology* **20,** 335–342.

Hoffmeister, B., Kiecker, F., Tesfa, L., Volk, H. D., Picker, L. J., and Kern, F. (2003). Mapping T cell epitopes by flow cytometry. *Methods* **29,** 270–281.

Karanikas, V., Lodding, J., Maino, V. C., and McKenzie, I. F. (2000). Flow cytometric measurement of intracellular cytokines detects immune responses in MUC1 immunotherapy. *Clin. Cancer Res.* **6,** 829–837.

Kern, F., Bunde, T., Faulhaber, N., Kiecker, F., Khatamzas, E., Khatamzas, E., Rudawski, I. M., Pruss, A., Gratama, J. W., Volkmer-Engert, R., Ewert, R., Reinke, P., Volk, H. D., and Picker, L. J. (2002). Cytomegalovirus (CMV) phosphoprotein 65 makes a large contribution to shaping the T cell repertoire in CMV-exposed individuals. *J. Infect. Dis.* **185,** 1709–1716.

Kern, F., Faulhaber, N., Frömmel, C., Khatamzas, E., Prösch, S., Schönemann, C., Kretzschmar, I., Volkmer-Engert, R., Volk, H.-D., and Reinke, P. (2000). Analysis of CD8 T cell reactivity to cytomegalovirus using protein-spanning pools of overlapping pentadecapeptides. *Eur. J. Immunol.* **30,** 1676–1682.

Kern, F., Faulhaber, N., Khatamzas, E., Frommel, C., Ewert, R., Prosch, S., Volk, H., and Reinke, P. (1999a). Measurement of anti-human cytomegalovirus T cell reactivity in transplant recipients and its potential clinical use: a mini-review. *Intervirology* **42,** 322–324.

Kern, F., Surel, I. P., Brock, C., Freistedt, B., Radtke, H., Scheffold, A., Blasczyk, R., Reinke, P., Schneider-Mergener, J., Radbruch, A., Walden, P., and Volk, H. D. (1998). T-cell epitope mapping by flow cytometry. *Nat. Med.* **4,** 975–978.

Kern, F., Surel, I. P., Faulhaber, N., Frommel, C., Schneider-Mergener, J., Schonemann, C., Reinke, P., and Volk, H. D. (1999b). Target structures of the CD8(+)-T-cell response to human cytomegalovirus: the 72-kilodalton major immediate-early protein revisited. *J. Virol.* **73,** 8179–8184.

Knabel, M., Franz, T. J., Schiemann, M., Wulf, A., Villmow, B., Schmidt, B., Bernhard, H., Wagner, H., and Busch, D. H. (2002). Reversible MHC multimer staining for functional isolation of T-cell populations and effective adoptive transfer. *Nat. Med.* **8,** 631–637.

Kostense, S., Otto, S. A., Knol, G. J., Manting, E. H., Nanlohy, N. M., Jansen, C., Lange, J. M., van Oers, M. H., Miedema, F., and van Baarle, D. (2002a). Functional restoration of human immunodeficiency virus and Epstein-Barr virus-specific CD8(+) T cells during highly active antiretroviral therapy is associated with an increase in CD4(+) T cells. *Eur. J. Immunol.* **32,** 1080–1089.

Kostense, S., Vandenberghe, K., Joling, J., Van Baarle, D., Nanlohy, N., Manting, E., and Miedema, F. (2002b). Persistent numbers of tetramer+ CD8(+) T cells, but loss of interferon-gamma+ HIV-specific T cells during progression to AIDS. *Blood* **99,** 2505–2511.

Latt, S. A., George, Y. S., and Gray, J. W. (1977). Flow cytometric analysis of bromodeoxyuridine-substituted cells stained with 33258 Hoechst. *J. Histochem. Cytochem.* **25,** 927–934.

Lebowitz, M. S., O'Herrin, S. M., Hamad, A. R., Fahmy, T., Marguet, D., Barnes, N. C., Pardoll, D., Bieler, J. G., and Schneck, J. P. (1999). Soluble, high-affinity dimers of T-cell receptors and class II major histocompatibility complexes: biochemical probes for analysis and modulation of immune responses. *Cell Immunol.* **192,** 175–184.

Lecoeur, H., Fevrier, M., Garcia, S., Riviere, Y., and Gougeon, M. L. (2001). A novel flow cytometric assay for quantitation and multiparametric characterization of cell-mediated cytotoxicity. *J. Immunol. Methods* **253,** 177–187.

Lyons, A. B. (2000). Analyzing cell division *in vivo* and *in vitro* using flow cytometric measurement of CFSE dye dilution. *J. Immunol. Methods* **243,** 147–154.

Maecker, H. T., Auffermann-Gretzinger, S., Nomura, L. E., Liso, A., Czerwinski, D. K., and Levy, R. (2001a). Detection of CD4 T-cell responses to a tumor vaccine by cytokine flow cytometry. *Clin. Cancer Res.* **7,** 902s–908s.

Maecker, H. T., Dunn, H. S., Suni, M. A., Khatamzas, E., Pitcher, C. J., Bunde, T., Persaud, N., Trigona, W., Fu, T. M., Sinclair, E., Bredt, B. M., McCune, J. M., Maino, V. C., Kern, F., and Picker, L. J. (2001b). Use of overlapping peptide mixtures as antigens for cytokine flow cytometry. *J. Immunol. Methods* **255,** 27–40.

Maeji, N. J., Bray, A. M., and Geysen, H. M. (1990). Multi-pin peptide synthesis strategy for T cell determinant analysis. *J. Immunol. Methods* **134,** 23–33.

Maeji, N. J., Bray, A. M., Valerio, R. M., and Wang, W. (1995). Larger scale multipin peptide synthesis. *Pept. Res.* **8,** 33–38.

Maino, V. C., and Picker, L. J. (1998). Identification of functional subsets by flow cytometry: Intracellular detection of cytokine expression. *Cytometry* **34,** 207–215.

Mollet, L., Li, T. S., Samri, A., Tournay, C., Tubiana, R., Calvez, V., Debre, P., Katlama, C., and Autran, B. (2000). Dynamics of HIV-specific CD8+ T lymphocytes with changes in viral load. The RESTIM and COMET Study Groups. *J. Immunol.* **165,** 1692–1704.

Nielsen, M. B., Monsurro, V., Migueles, S. A., Wang, E., Perez-Diez, A., Lee, K. H., Kammula, U., Rosenberg, S. A., and Marincola, F. M. (2000). Status of activation of circulating vaccine-elicited CD8+ T cells. *J. Immunol.* **165,** 2287–2296.

Oxenius, A., Sewell, A. K., Dawson, S. J., Gunthard, H. F., Fischer, M., Gillespie, G. M., Rowland-Jones, S. L., Fagard, C., Hirschel, B., Phillips, R. E., and Price, D. A. (2002). Functional discrepancies in HIV-specific CD8+ T-lymphocyte populations are related to plasma virus load. *J. Clin. Immunol.* **22,** 363–374.

Ozdemir, E., St John, L. S., Gillespie, G., Rowland-Jones, S., Champlin, R. E., Molldrem, J. J., and Komanduri, K. V. (2002). Cytomegalovirus reactivation following allogeneic stem cell transplantation is associated with the presence of dysfunctional antigen-specific CD8+ T cells. *Blood* **100,** 3690–3697.

Parslow, T., Stites, D., Terr, A., and Imboden, J. (2001). Medical immunology. *In* McGraw-Hill Companies, East Norwalk, Connecticut, USA.

Picker, L. J., Singh, M. K., Zdraveski, Z., Treer, J. R., Waldrop, S. L., Bergstresser, P. R., and Maino, V. C. (1995). Direct demonstration of cytokine synthesis heterogeneity among human memory/effector T cells by flow cytometry. *Blood* **86,** 1408–1419.

Pitcher, C. J., Quittner, C., Peterson, D. M., Connors, M., Koup, R. A., Maino, V. C., and Picker, L. J. (1999). HIV-1-specific CD4+ T cells are detectable in most individuals with active HIV-1 infection, but decline with prolonged viral suppression. *Nat. Med.* **5,** 518–525.

Pittet, M. J., Speiser, D. E., Lienard, D., Valmori, D., Guillaume, P., Dutoit, V., Rimoldi, D., Lejeune, F., Cerottini, J. C., and Romero, P. (2001a). Expansion and functional maturation of human tumor antigen-specific CD8+ T cells after vaccination with antigenic peptide. *Clin. Cancer Res.* **7,** 796s–803s.

Pittet, M. J., Zippelius, A., Speiser, D. E., Assenmacher, M., Guillaume, P., Valmori, D., Lienard, D., Lejeune, F., Cerottini, J. C., and Romero, P. (2001b). *Ex vivo* IFN-gamma secretion by circulating CD8 T lymphocytes: implications of a novel approach for T cell monitoring in infectious and malignant diseases. *J. Immunol.* **166,** 7634–7640.

Rammensee, H. G., Bachmann, J., and Stefanovic, S. (1997). MHC ligands and binding motifs. Landes Bioscience, Georgetown, Texas.

Reineke, U., Volkmer-Engert, R., and Schneider-Mergener, J. (2001). Applications of peptide arrays prepared by the SPOT-technology. *Curr. Opin. Biotechnol.* **12,** 59–64.

Reineke, U., Ivascu, C., Schlief, M., Landgraf, C., Gericke, S., Zahn, G., Herzel, H., Volkmer-Engert, R., and Schneider-Mergener, J. (2002). Identification of distinct antibody epitopes and mimotopes from a peptide array of 5520 randomly generated sequences. *J. Immunol. Methods.* **267,** 37–51.

Reusser, P., Cathomas, G., Attenhofer, R., Tamm, M., and Thiel, G. (1999). Cytomegalovirus (CMV)-specific T cell immunity after renal transplantation mediates protection from CMV disease by limiting the systemic virus load. *J. Infect. Dis.* **180,** 247–253.

Riddell, S. R., and Greenberg, P. D. (1995). Cellular adoptive immunotherapy after bone marrow transplantation. *Cancer Treat. Res.* **76,** 337–369.

Riddell, S. R., Walter, B. A., Gilbert, M. J., and Greenberg, P. D. (1994). Selective reconstitution of CD8+ cytotoxic T lymphocyte responses in immunodeficient bone marrow transplant recipients by the adoptive transfer of T cell clones. *Bone Marrow Transplant.* **14**(Suppl. 4), S78–S84.

Scheffold, A., Lohning, M., Richter, A., Assenmacher, M., Manz, R., Austrup, F., Hamann, A., and Radbruch, A. (1998). Analysis and sorting of T cells according to cytokine expression. *Eur. Cytokine Netw.* **9,** 5–11.

Scheibenbogen, C., Letsch, A., Thiel, E., Schmittel, A., Mailaender, V., Baerwolf, S., Nagorsen, D., and Keilholz, U. (2002). CD8 T-cell responses to Wilms tumor gene product WT1 and proteinase 3 in patients with acute myeloid leukemia. *Blood* **100,** 2132–2137.

Smith, J. W., 2nd, Walker, E. B., Fox, B. A., Haley, D., Wisner, K. P., Doran, T., Fisher, B., Justice, L., Wood, W., Vetto, J., Maecker, H., Dols, A., Meijer, S., Hu, H. M., Romero, P., Alvord, W. G., and Urba, W. J. (2003). Adjuvant immunization of HLA-A2–positive melanoma patients with a modified gp 100 peptide induces peptide-specific CD8+ T-cell responses. *J. Clin. Oncol.* **21,** 1562–1573.

Tesfa, L., Koch, F., Pankow, W., Volk, H., and Kern, F. (2004). Confirmation of *Mycobacterium tuberculosis* infection by flow-cytometry after *ex-vivo* incubation of peripheral blood T cells with an ESAT6 derived peptide pool. *Cytometry* **60B**(1), 47–53.

Tesfa, L., Volk, H. D., and Kern, F. (2003). Comparison of proliferation and rapid cytokine induction assays for flow cytometric T-cell epitope mapping. *Cytometry* **52A,** 36–45.

Tribbick, G. (2002). Multipin peptide libraries for antibody and receptor epitope screening and characterization. *J. Immunol. Methods* **267,** 27–35.

Valmori, D., Scheibenbogen, C., Dutoit, V., Nagorsen, D., Asemissen, A. M., Rubio-Godoy, V., Rimoldi, D., Guillaume, P., Romero, P., Schadendorf, D., Lipp, M., Dietrich, P. Y., Thiel, E., Cerottini, J. C., Lienard, D., and Keilholz, U. (2002). Circulating tumor-reactive CD8(+) T cells in melanoma patients contain a CD45RA(+)CCR7(−) effector subset exerting *ex vivo* tumor-specific cytolytic activity. *Cancer Res.* **62,** 1743–1750.

Waldrop, S. L., Davis, K. A., Maino, V. C., and Picker, L. J. (1998). Normal human CD4+ memory T cells display broad heterogeneity in their activation threshold for cytokine synthesis. *J. Immunol.* **161,** 5284–5295.

Waldrop, S. L., Pitcher, C. J., Peterson, D. M., Maino, V. C., and Picker, L. J. (1997). Determination of antigen-specific memory/effector CD4+ T cell frequencies by flow cytometry: evidence for a novel, antigen-specific homeostatic mechanism in HIV-associated immunodeficiency. *J. Clin. Invest.* **99,** 1739–1750.

Wenschuh, H., Volkmer-Engert, R., Schmidt, M., Schulz, M., Schneider-Mergener, J., and Reineke, U. (2000). Coherent membrane supports for parallel microsynthesis and screening of bioactive peptides. *Biopolymers* **55,** 188–206.

Wills, M. R., Carmichael, A. J., Mynard, K., Jin, X., Weekes, M. P., Plachter, B., and Sissons, J. G. (1996). The human cytotoxic T-lymphocyte (CTL) response to cytomegalovirus is dominated by structural protein pp65: Frequency, specificity, and T-cell receptor usage of pp65-specific CTL. *J. Virol.* **70,** 7569–7579.

CHAPTER 19

Flow Cytometric Analysis of Human Antigen-Specific T-Cell Proliferation

Jason M. Brenchley and Daniel C. Douek

Human Immunology Section
Vaccine Research Center
National Institute of Allergy and Infectious Disease
National Institutes of Health
Bethesda, Maryland 20874

0091-679X/04 $35.00

I. Introduction and Background

After maturation, naive T cells leave the thymus and reside predominantly in lymphoid tissues, recirculating through the blood where they maintain a naive phenotype (Anderson *et al.*, 1996; Picker, 1994; Picker and Siegelman, 1993). Naive T cells may encounter cognate antigens presented as peptides bound to major histocompatibility complex (MHC) molecules. Antigen recognition via the T-cell receptor (TCR) in conjunction with appropriate costimulation results in activation of T cells leading to proliferation, differential expression of surface proteins, and production of effector cytokines including interferon-γ and tumor necrosis factor-α. As naive T cells divide in response to antigen, they also gain effector function and phenotype. A subset of these activated T cells survive and persist as memory T cells and can respond in an anamnestic manner to future antigenic stimulations (Ahmed and Gray, 1996; Murali-Krishna *et al.*, 1998; Wu *et al.*, 2002).

Examination of stimulated T-cell effector function is one of the most frequently used output readings in immunological research laboratories around the world. Measurement of T-cell effector function can be accomplished via many routes including measurement of secreted cytokines by enzyme-linked immunosorbent assay (ELISA) (Engvall *et al.*, 1971), intracellular analysis of produced cytokines by flow cytometry (Pitcher *et al.*, 1999; Prussin, 1997; Prussin and Metcalfe, 1995), and induction of proliferation (Bloom *et al.*, 1972; Cheers and Sprent, 1973).

Historically, T-cell proliferation has been determined by incorporation of tritiated thymidine into the genome of proliferating T cells (Bloom *et al.*, 1972; Cheers and Sprent, 1973). Here, we review the use of a technique based on serial dilution of a stably binding intracellular fluorochrome, carboxy fluorescein diacetate succinimidyl ester (CFSE), which allows monitoring of cell division by flow cytometry, to monitor antigen-specific T-cell responses. CFSE is capable of freely diffusing across the cell membrane where it reacts with free amine groups of cellular proteins. Carboxyl groups are removed from CFSE by enzymatic action of ubiquitously expressed esterases and CFSE emits strongly at 519 nm after excitation by 488-nm argon laser light.

Monitoring cell division by flow cytometry holds many advantages over traditional thymidine incorporation assays. First, the number of cell divisions can be directly quantified using the flow cytometric assay (Lyons and Parish, 1994; Lyons *et al.*, 2001). Second, the phenotype of cells undergoing proliferation can be directly ascertained (Lyons and Parish, 1994; Lyons *et al.*, 2001). Third, phenotypic changes and other properties that occur with respect to cell division can be monitored (Bird *et al.*, 1998; Brenchley *et al.*, 2002; Doyle *et al.*, 2001; Gett and Hodgkin, 1998; Hasbold *et al.*, 1998; Murali-Krishna and Ahmed, 2000). Finally, flow cytometric analysis of proliferation allows for sorting of viable cells and/or based on rounds of proliferation (Douek *et al.*, 2002b; Lyons, 1997; Michalek *et al.*, 2003).

Initially, CFSE was used to monitor proliferation of T and B cells in response to mitogen and superantigen (Brenchley *et al.*, 2002; Fulcher and Wong, 1999) and T-cell responses of transgenic mice (Bird *et al.*, 1998; Doyle *et al.*, 2001; Hasbold *et al.*, 1998; Murali-Krishna and Ahmed, 2000). This technology has also been used to monitor cytokine-induced proliferation of T cells (Brenchley *et al.*, 2002; Geginat *et al.*, 2001, 2002). Furthermore, this method has been adopted to monitor human antigen-specific T-cell responses (Brenchley *et al.*, 2003; Karandikar *et al.*, 2002; Migueles *et al.*, 2002; van Leeuwen *et al.*, 2002). Taken together, CFSE technology to monitor human antigen-specific proliferation has allowed analysis of drug mechanisms in human disease, determination of fundamental differences between T-cell responses of multiple virally infected individuals, and T-cell senescence in response to viral stimulation. Use of CFSE technology will undoubtedly continue and will be adapted to monitor responses to other antigens or responses induced by vaccines.

II. Methods

A. Solutions

A stock solution of CFSE [5(and 6)-CFSE] (molecular weight 557, Molecular Probes, Eugene, Oregon) is prepared by dissolving 5.57 mg/ml of CFSE in dimethyl sulfoxide (DMSO). This creates a 10-μM solution of CFSE that can be aliquoted into single-use volumes (e.g., 20 μl) and stored at $-20\,^{\circ}$C under desiccating conditions. It is important to keep CFSE stock solutions protected from exposure to light as the dye is readily photobleached by ambient light. Stock solutions can be stored for more than 1 year under these conditions.

B. Staining Procedures

Standard phosphate-buffered saline (PBS) is employed in the staining procedure. Cells to be CFSE labeled are re-suspended at a concentration of 1×10^6/ml in PBS. For *in vitro* studies, CFSE is added so the final concentration is 0.25 μM. For *in vivo* studies, substantially higher concentrations of CFSE are used. Cells are incubated in 0.25 μM of CFSE in a 37 °C water bath for 7 minutes. As staining homogeneity is crucial for monitoring cell divisions, ensuring CFSE and cells are equally distributed throughout the solution is critical. Vortexing cells after addition of CFSE can achieve this. To quench the CFSE staining, filtered, heat-inactivated serum is directly added so the final concentration of serum is 2%. CFSE-labeled cells should be washed twice with Roswell Park Memorial Institute medium (RPMI)/10% serum. CFSE-labeled cells should be kept free from excessive light. Intensity of CFSE labeling differs moderately from time to time; hence, for proper instrument compensation, a small portion of each CFSE-labeled mixture should be saved (unstimulated, undivided) for the duration of incubation periods. It is important to note that CFSE can diffuse from permeabilized cells with time. Therefore, when analyzing CFSE-labeled cells for cytokine production, one should perform data acquisition immediately after staining.

C. Antigenic Stimulations

PBMCs are initially stained with CFSE. PBMCs are then suspended in media supplemented with interleukin-2 (IL-2) (10 U/ml) (Sigma) and 1 μg/ml of anti-CD28 and anti-CD49d (BD) (for costimulation) at 1×10^6/ml. CFSE-labeled cells are rested for at least 5 hours before any stimulation. After rest, CFSE-labeled PBMCs are stimulated with antigen. Unstimulated, CFSE-labeled PBMCs are concurrently treated with IL-2 and costimulatory antibodies as a negative control. Other PBMCs are treated with 1 μg/ml SEB (Sigma) as a positive control. The cells are then incubated for at least 48 hours at 37 °C in a 5% CO_2 incubator. PBMCs can be incubated for an additional 12 hours in the presence of brefeldin A (1 μg/ml) (Sigma) to examine cytokine production. Cells are then stained for flow cytometric analysis with antibodies of interest. After gating through CD3 and CD8 or CD4, the percentage of antigen-specific T cells that had divided are calculated by determining the percentage that had diluted CFSE after subtracting the "background" division (the percentage of T cells that divided in the negative control). Experiments that do not result in proliferation of SEB-stimulated T cells (positive control) are considered not useful information and these experiments should be repeated.

D. Intracellular Staining

For intracellular staining, cells are initially washed and spun after staining for surface molecules. Cells are then permeabilized using one of many permeabilization protocols. However, an important consideration when permeabilizing CFSE-labeled cells is retention of CFSE. Many fixation and permeabilization methods described in the literature are not sufficient for CFSE-labeled cells. Stimulated CFSE-labeled cells are initially treated with a permeabilization/fixation solution (10% BD FACS lyse, 0.05% Tween 20) for exactly 10 minutes in the dark at room temperature, followed by two consecutive washes with wash solution that does not contain detergent. Antibodies directed against intracellular antigens can then be added for 30 minutes at 4 °C. Data collection should be performed as quickly as possible because fidelity of CFSE retention decreases with time.

III. Results

A. Example T-Cell Proliferative Responses to Antigen

CFSE staining and antigenic stimulation have been successfully used to monitor human antigen-specific T-cell proliferation (Brenchley *et al.*, 2002, 2003; Karandikar *et al.*, 2002; Michalek *et al.*, 2003; Migueles *et al.*, 2002). Figure 1

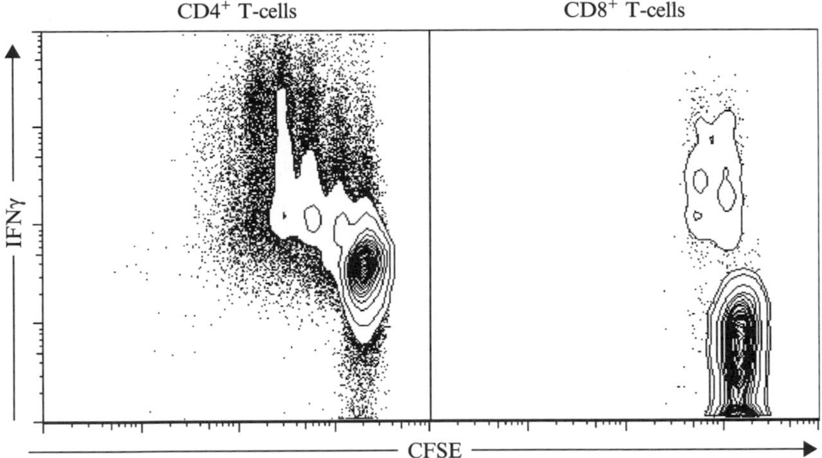

Fig. 1 Antigen-specific proliferation of CD4$^+$ and CD8$^+$ T cells. PBMCs from a cytomegalovirus (CMV)-seropositive individual were labeled with 0.25 μM of carboxy fluorescein diacetate succinimidyl ester (CFSE) and then stimulated with either CMV whole antigen (CD4$^+$ T cells) for 4 days or the CMV human leukocyte antigen (HLA)-A*02–restricted epitope NLVPMVATV CD8$^+$ T cells for 2 days. PBMCs were then stained with CD4– or CD8–phycoerythrin (PE), CD3–PerCP, and interferon-γ (IFN-γ)–allophycocyanin (APC). After lymphocyte, CD3, and CD4 or CD8 gating, CFSE versus IFN-γ is examined to study antigen-specific proliferation.

shows that this approach can be used to monitor CD8$^+$ or CD4$^+$ antigen–specific T-cell proliferative responses. Here, PBMCs from CMV-seropositive individuals are CFSE labeled followed by stimulation with either inactivated whole CMV (to measure the CD4$^+$ T-cell response) or an optimized, HLA-A*02–restricted immunodominant CMV epitope (to measure the CD8$^+$ T-cell response). As shown in Fig. 1, one can ascertain the phenotype of T cells responding with proliferation without initially sorting the cells into individual compartments, making this procedure more similar to *in vivo* stimulations. In addition, as is evident from Fig. 1, one can study the production of effector cytokines within the dividing T-cell populations. Because T-cell responses to antigen are rarely homogeneous, it is often desirable to monitor as many effector functions as possible.

B. Autospecific and Drug-Induced T-Cell Responses

Using CFSE technology, Karandikar *et al.* (2002) monitored T-cell proliferative responses induced by and specific to the drug glatiramer acetate (GA) (Copaxone). GA has been known to inhibit experimental autoimmune encephalomyelitis (EAE) in rodents and primates (Teitelbaum *et al.*, 1971), but the mechanism by which GA exerts its immunomodulatory effect was unclear but was thought to act by suppression of Th1-type CD4$^+$ T cells (Aharoni *et al.*, 1997). Karandikar *et al.* (2002) demonstrated that there are Th1-type GA-specific CD4$^+$ and CD8$^+$ T-cell responses in healthy individuals, but the CD8$^+$ T-cell responses are lacking in patients with multiple sclerosis (MS). GA treatment acts to restore these T-cell responses in patients with MS. These conclusions were achieved by both intracellular cytokine staining and molecular reverse transcriptase polymerase chain reaction (RT-PCR) approaches after viably sorting GA-specific T cells that had proliferated in response to GA.

In subsequent work, Crawford *et al.* (2004) used the same technology to monitor CD8$^+$ and CD4$^+$ T-cell responses to autoantigen. Here, the finding of CD8$^+$ T-cell responses against central nervous system (CNS) antigens implies that CD8$^+$ T cells may play an important role in disease progression.

Importantly, in both aforementioned circumstances, the CFSE technology allowed the researchers to perform phenotypic analyses of proliferating T cells. Prior to these data, we assumed that CNS-specific T-cell responses in MS were restricted to CD4$^+$ T cells based on tritiated thymidine incorporation assays.

C. Allospecific T-Cell Responses

Allogeneic hemopoietic stem cell transplantation (HSCT) is an effective therapy for many hematological malignancies and inherited disorders (Armitage, 1994; Appelbaum, 2001). In the past, prediction of graft-versus-host disease (GVHD) has relied on mixed lymphocyte reactions (MLRs) with incorporation of tritiated thymidine by donor PBMCs as the readout of potential donor-derived alloreactive T cells (Michalek *et al.*, 2003).

Michalek *et al.* (2003) used CFSE technology to identify and monitor graft-versus-host–specific T-cell responses in the HSCT setting. In this situation, the authors used CFSE-labeled donor cells in an MLR with irradiated PBMCs of the recipient. This technology allowed the authors to viably sort alloreactive T cells followed by clonotypic analysis of the allospecific T-cell response (Douek *et al.*, 2002a).

Subsequently, Michalek *et al.* (2003) used the same methodology to identify and monitor donor-generated T cells specific for minor histocompatibility antigens. Notably, identification and manipulation of minor histocompatibility antigen–specific T cells would not have been possible without the CFSE technology.

D. Virus-Specific T-Cell Responses

1. Comparing Responses between Different Groups

CD4 cells are ultimately lost in most HIV-infected individuals, leaving these subjects susceptible to various opportunistic infections (Douek *et al.*, 2003). However, there exist a small percentage of HIV-infected individuals who control viral replication without substantial loss of CD4 cells in the absence of antiretroviral therapy (Vesanen *et al.*, 1996). Mechanisms of protection from HIV within such long-term nonprogressors (LTNPs) has remained controversial (Aiuti and D'Offizi, 1995; Balotta *et al.*, 1997; Boaz *et al.*, 2002; Migueles *et al.*, 2000).

Using CFSE technology, Migueles *et al.* (2000) demonstrated that HIV-specific $CD8^+$ T cells of LTNPs maintain an antigen-specific proliferative response that is coupled to production of the lytic enzyme perforin. The authors infected autologous $CD4^+$ T cells with a laboratory-adapted strain of HIV and used these infected $CD4^+$ T cells to stimulate CFSE-labeled $CD8^+$ T cells in a cohort of HIV-infected individuals. In doing so, the authors found marked differences between the proliferative ability of HIV-specific $CD8^+$ T cells from LTNPs and other HIV-infected individuals, although proliferative responses to mitogens between the two groups were similar.

2. Attributing Lack of Proliferation to Replicative Senescence

T-cell proliferative responses cannot always be used to study antigen-specific T-cell responses reliably. Certain T-cell subsets (especially within $CD8^+$ T cells) can produce effector cytokines but fail to proliferate in response to antigenic stimulation (Evans *et al.*, 1999; Rowbottom *et al.*, 2000; Schmidt *et al.*, 1996; Sze *et al.*, 2001; Wang and Borysiewicz, 1995; Wang *et al.*, 1997; Warrington *et al.*, 2001). Although the provenance and exact phenotype of such $CD8^+$ T cells remain unclear (Coakley *et al.*, 2000; Frassanito *et al.*, 1998; Kern *et al.*, 1999; Rowbottom *et al.*, 2000; Sadat-Sowti *et al.*, 1991, 1994; Vingerhoets *et al.*, 1995), such a failure to proliferate is generally attributed to replicative senescence resulting from continual stimulation by antigen and/or cytokine.

Brenchley *et al.*, 2003 used CFSE technology and phenotypic analysis to study the proliferative ability and phenotypic status of HIV-specific $CD8^+$ T cells from a cohort of HIV-infected individuals. Here, PBMCs from HIV-infected individuals were CFSE labeled followed by stimulation with peptides spanning the HIV proteome (Betts *et al.*, 1999, 2000; Pitcher *et al.*, 1999). After finding specific subsets of $CD8^+$ T cells that failed to proliferate in response to HIV-specific stimulation, the authors were then able to find phenotypic attributes of these $CD8^+$ T cells. After finding phenotypic attributes of these $CD8^+$ T cells, the authors demonstrated that these $CD8^+$ T cells had indeed reached replicative senescence, attributed to repetitive antigenic stimulation *in vivo* evident from their shortened telomeres and very low levels of TCR excision circles.

IV. Critical Aspects of the Methodology

A. Background Proliferation

When using CFSE technology to monitor antigen-specific T-cell responses, we must consider several important points. One of the most important considerations is whether the proliferation being observed is antigen dependent. Indeed, high levels of "background" T-cell proliferation have been observed. This is defined as proliferation that occurs in the absence of antigenic stimulation. Many antigen-specific T cells are present at very low precursor frequencies. In such situations, it is often difficult to demonstrate compelling data that observed proliferation is antigen specific. Clearly, all experiments must include a negative control wherein T cells (or PBMCs) are treated with everything except antigen. Such negative controls often consist of CFSE-labeled PBMCs treated with IL-2 and costimulatory antibodies (e.g., anti-CD28 and anti-CD49d). This combination of reagents along with potential antigens present in fetal calf serum (commonly used to supplement media) creates potential for T-cell stimulation and proliferation. Several suggestions to diminish the amount of background T-cell proliferation have been described. One of the largest sources of background proliferation is believed to be antigenic stimulation induced by fetal calf serum. To diminish serum-induced proliferation, use of human AB serum has been proposed. In addition, the concentration of serum used in media can often be decreased to 5% to aid in diminishing background T-cell proliferation.

Another method one can use to minimize the amount of background T-cell proliferation is to minimize the amount of time in which PBMCs are kept in culture. This approach has been successful for monitoring virus-specific T-cell proliferative responses. In general, during active viral replication, viral-specific T-cell responses are substantial. Therefore, it is relatively unproblematic to monitor a single cell division of virus-specific T cells in a 48-hour period. However, when studying T-cell responses that are quite small, a single cell division is generally not sufficient and longer incubation periods are required.

B. Instrument Compensation

A second key consideration when using CFSE staining is instrumentation compensation. CFSE is routinely used in standard four-color experiments; however, CFSE highly overlaps into the PE, TRPE, Cy5.5–PE, Cy7–PE and PerCP, or Cy5–PE channels (Fig. 2). Importantly, it is impossible to correctly compensate CFSE fluorescence with fluorescein isothiocyanate (FITC) (Fig. 2). CFSE has a moderately different emission spectrum compared to FITC and CFSE is substantially brighter than FITC (Fig. 2). Therefore, proper CFSE compensation requires a CFSE single color control, and FITC single color control will not suffice. Furthermore, intensity of CFSE staining will invariably be inconsistent each time the staining procedure is accomplished. Therefore, the amount of compensation required for each individual experiment can widely vary. It follows that it is crucial to use a subset of the actual CFSE-labeled experimental sample for proper compensation. When calculating the number of PBMCs required for individual experiments, one must include the number of PBMCs required for compensation. The CFSE-labeled PBMCs to be used for compensation should be treated similar to PBMCs used as a negative control, that is, PBMCs for

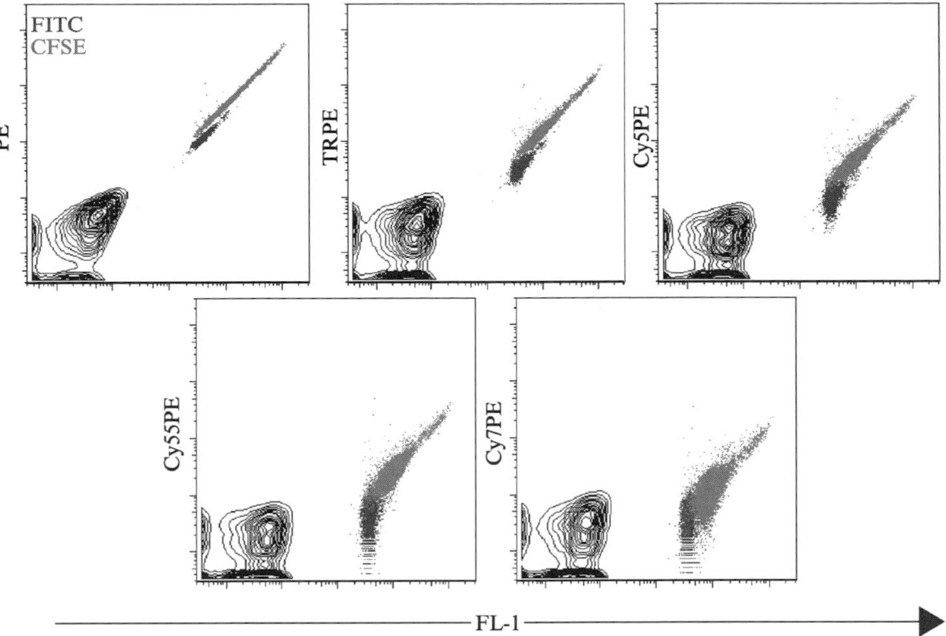

Fig. 2 Carboxy fluorescein diacetate succinimidyl ester (CFSE) and fluorescein isothiocyanate (FITC) require different compensation. PBMCs were CFSE labeled or labeled with anti-CD8–FITC. Both were then collected on a modified FACSVantage and events were superimposed on each other. (See Color Insert.)

compensation are kept in culture for the same time as those under experimental conditions. This ensures that the PBMCs to be used for compensation will have exactly the same CFSE fluorescence as the sample.

Further compensation complications arise when advancing CFSE staining into experiments that incorporate more than four fluorochromes (De Rosa *et al.*, 2003). Instruments equipped with the electronics required to measure more than four colors notoriously lose signal intensity compared to standard four-color instruments (Baumgarth and Roederer, 2000; De Rosa *et al.*, 2001). This signal intensity loss creates unique problems for CFSE staining that can be resolved. The largest obstacle to overcome when adapting CFSE staining to multicolor flow cytometry is the overlap of CFSE fluorescence into the PE channel. Hence, CFSE–PE combinations, which routinely work for standard four-color experiments, may not work well for experiments incorporating more than the standard four colors. Therefore, the PE channel should be reserved to study surface antigens that are brightly expressed by CFSE-labeled cells.

V. Pitfalls and Misinterpretation of the Data

Cell division can clearly be used as a parameter to quantify multiple antigen-specific T-cell responses. However, the mathematics involved in making such calculations is often overlooked, data analysis can be oversimplified, and data can be misinterpreted. For example, when comparing $CD8^+$ or $CD4^+$ T-cell proliferation in response to SEB, simply demonstrating the percentage of $CFSE^{lo}$ T cells would suggest that the $CD4^+$ T-cell response to SEB is 30% greater than the $CD8^+$ T-cell response to SEB (Fig. 3). However, by calculating the percentage of $CD4^+$ and $CD8^+$ T cells that underwent proliferation in response to SEB, we find that the responses between $CD4^+$ and $CD8^+$ T cells differ only by 15% (Fig. 3). This large discrepancy between analysis methods results because the percentage of T cells that divide in response to stimulation is dependent on the number of times the cells divide and the percentage of cells within each division peak. Therefore, when comparing different T-cell responses based on CFSE proliferation, one must use a "proliferation platform." Most flow cytometric data analysis programs offer cell division platforms that can adequately analyze CFSE data.

Proliferation platform programs generally output at least three variables, each of which yields different information regarding the population of cells that have

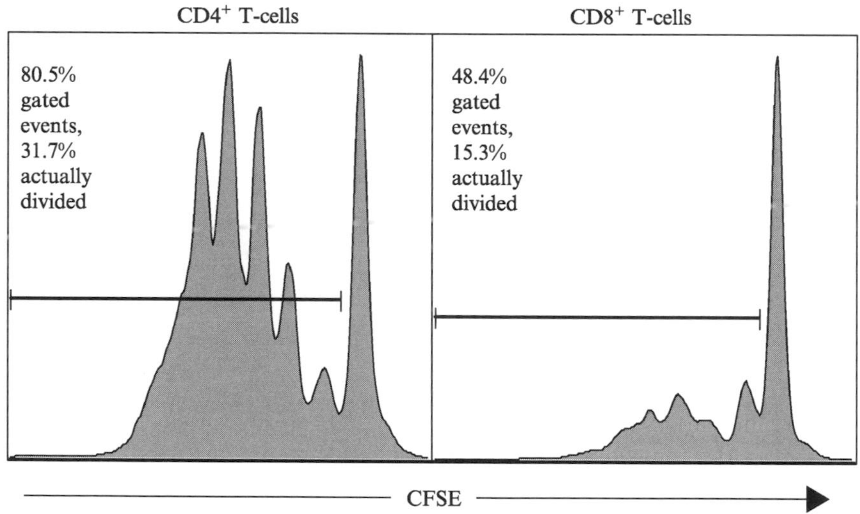

Fig. 3 Analysis of cell division data. PBMCs from a healthy individual were labeled with 0.25 μM of carboxy fluorescein diacetate succinimidyl ester (CFSE) and stimulated with 1 μg/ml of SEB for 5 days. PBMCs were then stained with CD3–PE, CD8-PerCP, and CD3–APC. After lymphocyte and CD3 gating, T cells were then separated based on CD4 or CD8 expression and were then analyzed for cell division history. The percentages of $CD4^+$ and $CD8^+$ T cells that had dilute CFSE were determined as were the percentages of $CD4^+$ and $CD8^+$ T cells that originally underwent cell division using the proliferation platform of Flow Jo (Tree Star, Ashland, Oregon).

proliferated. The percentage divided (mentioned earlier) depicts the percentage of cells that actually underwent proliferation and is arguably the most important parameter to report when using proliferation as a measure of responsiveness. However, proliferative responses might be qualitatively different. In this situation, one group might proliferate more times than another group within the same period. Here, the proliferative index or division index will quantify these proliferative differences. The proliferative index is the average number of divisions that divided cells underwent, whereas the division index is the average number of divisions that a cell (that was present in the starting population) has undergone. Hence, the division index uses information from the undivided peak, and the proliferation index examines only those cells that divided. Arguably, the proliferation index is the more relevant parameter to report when comparing proliferative responses between groups. Importantly, this technology adds a huge advantage over assays using incorporation of tritiated thymidine because these proliferation platform calculations cannot be performed with tritiated thymidine experiments.

VI. Future Directions

A. Monitoring Vaccine Responses

Historically, the success of vaccines relies on their ability to elicit neutralizing antibodies. However, many evolving vaccine designs aim to elicit T-cell responses (Hanke *et al.*, 2002; Letvin, 2002; McMichael and Hanke, 2002). Determining immunogenicity of T-cell responses has depended on tetramer analysis (Barouch *et al.*, 2000; Shiver *et al.*, 2002) or measurement of secreted effector cytokines from vaccine-specific T cells using either ELISPOT (Barouch *et al.*, 2000; Cao *et al.*, 2003; MacGregor *et al.*, 2002) or intracellular cytokine staining (Barouch *et al.*, 2000; Peters *et al.*, 2003; Wee *et al.*, 2002). However, production of effector cytokine is but one effector function elicited by T cells in response to antigenic stimulation. Furthermore, cytokine production can be quite heterogeneous and the choice of which cytokine to monitor may limit the quantity of antigen-specific T cells monitored. Also, antigen-specific T cells may exist but fail to produce effector cytokine. Taken together, using an assay that measures only production of effector cytokines might under-represent the number of antigen-specific T cells elicited by a particular vaccine. To better quantify antigen-specific T cells induced by vaccine antigens, monitoring multiple effector functions will be crucial. Therefore, monitoring cell division in conjunction with production of effector cytokines might enable detection of more vaccine-induced antigen-specific T cells.

B. Multicolor Flow Cytometry: Combining Phenotype and Function

Monitoring T-cell responses to different antigenic stimuli via flow cytometric–based procedures provides direct quantitative information. However, the only qualitative data that usually accompany flow cytometric analyses are those that separate responses into $CD4^+$ or $CD8^+$ T cells. This limitation is due to the fact that most flow cytometric instrumentation is capable of measuring only two physical parameters and four fluorescent parameters simultaneously. However, advances in flow cytometric technology have made it possible to monitor as many as 12 fluorescent parameters simultaneously (De Rosa *et al.*, 2001, 2003). Such advances have made it possible to monitor multiple phenotypic and functional parameters simultaneously. Although flow cytometric technology capable of monitoring more than six fluorescent parameters is not used by the broad scientific community, such advances are slowly becoming more widespread. Such technology will allow dissection of cytokine production and cell division within defined of phenotypically distinct lymphocytes.

References

Aharoni, R., Teitelbaum, D., Sela, M., and Arnon, R. (1997). Copolymer 1 induces T cells of the T helper type 2 that crossreact with myelin basic protein and suppress experimental autoimmune encephalomyelitis. *Proc. Natl. Acad. Sci. USA* **94,** 10821–10826.

Ahmed, R., and Gray, D. (1996). Immunological memory and protective immunity: understanding their relation. *Science* **272,** 54–60.

Aiuti, F., and D'Offizi, G. (1995). Immunologic and virologic studies in long-term non-progressor HIV infected individuals. NOPHROCO Study Group. Non progressors HIV + Roman cohort. *J. Biol. Regul. Homeost. Agents* **9,** 82–87.

Anderson, G., Moore, N. C., Owen, J. J., and Jenkinson, E. J. (1996). Cellular interactions in thymocyte development. *Annu. Rev. Immunol.* **14,** 73–99.

Appelbaum, F. R. (2001). Haematopoietic cell transplantation as immunotherapy. *Nature* **411,** 385–389.

Armitage, J. O. (1994). Bone marrow transplantation. *N. Engl. J. Med.* **330,** 827–838.

Balotta, C., Bagnarelli, P., Riva, C., Valenza, A., Antinori, S., Colombo, M. C., Sampaolesi, R., Violin, M., de Pasquale, M. P., Moroni, M., Clementi, M., and Galli, M. (1997). Comparable biological and molecular determinants in HIV type 1–infected long-term nonprogressors and recently infected individuals. *AIDS Res. Hum. Retroviruses* **13,** 337–341.

Barouch, D. H., Santra, S., Schmitz, J. E., Kuroda, M. J., Fu, T. M., Wagner, W., Bilska, M., Craiu, A., Zheng, X. X., Krivulka, G. R., Beaudry, K., Lifton, M. A., Nickerson, C. E., Trigona, W. L., Punt, K., Freed, D. C., Guan, L., Dubey, S., Casimiro, D., Simon, A., Davies, M. E., Chastain, M., Strom, T. B., Gelman, R. S., Montefiori, D. C., Lewis, M. G., Emini, E. A., Shiver, J. W., and Letvin, N. L. (2000). Control of viremia and prevention of clinical AIDS in rhesus monkeys by cytokine-augmented DNA vaccination. *Science* **290,** 486–492.

Baumgarth, N., and Roederer, M. (2000). A practical approach to multicolor flow cytometry for immunophenotyping. *J. Immunol. Methods* **243,** 77–97.

Betts, M. R., Casazza, J. P., Patterson, B. A., Waldrop, S., Trigona, W., Fu, T. M., Kern, F., Picker, L. J., and Koup, R. A. (2000). Putative immunodominant human immunodeficiency virus–specific CD8(+) T-cell responses cannot be predicted by major histocompatibility complex class I haplotype. *J. Virol.* **74,** 9144–9151.

Betts, M. R., Krowka, J. F., Kepler, T. B., Davidian, M., Christopherson, C., Kwok, S., Louie, L., Eron, J., Sheppard, H., and Frelinger, J. A. (1999). Human immunodeficiency virus type 1–specific cytotoxic T lymphocyte activity is inversely correlated with HIV type 1 viral load in HIV type 1–infected long-term survivors. *AIDS Res. Hum. Retroviruses* **15,** 1219–1228.

Bird, J. J., Brown, D. R., Mullen, A. C., Moskowitz, N. H., Mahowald, M. A., Sider, J. R., Gajewski, T. F., Wang, C. R., and Reiner, S. L. (1998). Helper T cell differentiation is controlled by the cell cycle. *Immunity* **9,** 229–237.

Bloom, B. R., Gaffney, J., and Jimenez, L. (1972). Dissociation of MIF production and cell proliferation. *J. Immunol.* **109,** 1395–1398.

Boaz, M. J., Waters, A., Murad, S., Easterbrook, P. J., and Vyakarnam, A. (2002). Presence of HIV-1 Gag-specific IFN-gamma+IL-2+ and CD28+IL-2+ CD4 T cell responses is associated with nonprogression in HIV-1 infection. *J. Immunol.* **169,** 6376–6385.

Brenchley, J. M., Douek, D. C., Ambrozak, D. R., Chatterji, M., Betts, M. R., Davis, L. S., and Koup, R. A. (2002). Expansion of activated human naive T-cells precedes effector function. *Clin. Exp. Immunol.* **130,** 432–440.

Brenchley, J. M., Karandikar, N. J., Betts, M. R., Ambrozak, D. R., Hill, B. J., Crotty, L. E., Casazza, J. P., Kuruppu, J., Migueles, S. A., Connors, M., Roederer, M., Douek, D. C., and Koup, R. A. (2003). Expression of CD57 defines replicative senescence and antigen-induced apoptotic death of CD8+ T cells. *Blood* **101,** 2711–2720.

Cao, H., Kaleebu, P., Hom, D., Flores, J., Agrawal, D., Jones, N., Serwanga, J., Okello, M., Walker, C., Sheppard, H., El-Habib, R., Klein, M., Mbidde, E., Mugyenyi, P., Walker, B., Ellner, J., and Mugerwa, R. (2003). Immunogenicity of a recombinant human immunodeficiency virus (HIV)-canarypox vaccine in HIV-seronegative Ugandan volunteers: results of the HIV Network for Prevention Trials 007 Vaccine Study. *J. Infect. Dis.* **187,** 887–895.

Cheers, C., and Sprent, J. (1973). Proliferation of activated thymocytes in mixed lymphocyte reactions. *Adv. Exp. Med. Biol.* **29,** 149–156.

Coakley, G., Iqbal, M., Brooks, D., Panayi, G. S., and Lanchbury, J. S. (2000). CD8+, CD57+ T cells from healthy elderly subjects suppress neutrophil development *in vitro:* Implications for the neutropenia of Felty's and large granular lymphocyte syndromes. *Arthritis Rheum.* **43,** 834–843.

Crawford, M. P., Yan, S. X., Ortega, S., Mehta, R. S., Hewitt, R. E., Price, D. A., Stastny, P., Douek, D. C., Koup, R. A., Racke, M. K., and Karandikar, N. J. (2004). High prevalence of autoreactive neuroantigen-specific CD8+ T cells in multiple sclerosis revealed by novel flow cytometric assay. *Blood* **103**(11), 4222–4231.

De Rosa, S. C., Brenchley, J. M., and Roederer, M. (2003). Beyond six colors: a new era in flow cytometry. *Nat. Med.* **9,** 112–117.

De Rosa, S. C., Herzenberg, L. A., and Roederer, M. (2001). 11-color, 13-parameter flow cytometry: Identification of human naive T cells by phenotype, function, and T-cell receptor diversity. *Nat. Med.* **7,** 245–248.

Douek, D. C., Betts, M. R., Brenchley, J. M., Hill, B. J., Ambrozak, D. R., Ngai, K. L., Karandikar, N. J., Casazza, J. P., and Koup, R. A. (2002a). A novel approach to the analysis of specificity, clonality, and frequency of HIV-specific T cell responses reveals a potential mechanism for control of viral escape. *J. Immunol.* **168,** 3099–3104.

Douek, D. C., Brenchley, J. M., Betts, M. R., Ambrozak, D. R., Hill, B. J., Okamoto, Y., Casazza, J. P., Kuruppu, J., Kunstman, K., Wolinsky, S., Grossman, Z., Dybul, M., Oxenius, A., Price, D. A., Connors, M., and Koup, R. A. (2002b). HIV preferentially infects HIV-specific CD4+ T-cells. *Nature* **417,** 95–98.

Douek, D. C., Picker, L. J., and Koup, R. A. (2003). T cell dynamics in HIV-1 infection. *Annu. Rev. Immunol.* **21,** 265–304.

Doyle, A. M., Mullen, A. C., Villarino, A. V., Hutchins, A. S., High, F. A., Lee, H. W., Thompson, C. B., and Reiner, S. L. (2001). Induction of cytotoxic T lymphocyte antigen 4 (CTLA-4) restricts clonal expansion of helper T cells. *J. Exp. Med.* **194,** 893–902.

Engvall, E., Jonsson, K., and Perlmann, P. (1971). Enzyme-linked immunosorbent assay. II. Quantitative assay of protein antigen, immunoglobulin G, by means of enzyme-labelled antigen and antibody-coated tubes. *Biochim. Biophys. Acta* **251,** 427–434.

Evans, T. G., Kallas, E. G., Luque, A. E., Menegus, M., McNair, C., and Looney, R. J. (1999). Expansion of the CD57 subset of CD8 T cells in HIV-1 infection is related to CMV serostatus. *AIDS* **13,** 1139–1141.

Frassanito, M. A., Silvestris, F., Cafforio, P., and Dammacco, F. (1998). CD8+/CD57 cells and apoptosis suppress T-cell functions in multiple myeloma. *Br. J. Haematol.* **100,** 469–477.

Fulcher, D., and Wong, S. (1999). Carboxyfluorescein succinimidyl ester-based proliferative assays for assessment of T cell function in the diagnostic laboratory. *Immunol. Cell Biol.* **77,** 559–564.

Geginat, J., Campagnaro, S., Sallusto, F., and Lanzavecchia, A. (2002). TCR-independent proliferation and differentiation of human CD4+ T cell subsets induced by cytokines. *Adv. Exp. Med. Biol.* **512,** 107–112.

Geginat, J., Sallusto, F., and Lanzavecchia, A. (2001). Cytokine-driven proliferation and differentiation of human naive, central memory, and effector memory CD4(+) T cells. *J. Exp. Med.* **194,** 1711–1719.

Gett, A. V., and Hodgkin, P. D. (1998). Cell division regulates the T cell cytokine repertoire, revealing a mechanism underlying immune class regulation. *Proc. Natl. Acad. Sci. USA* **95,** 9488–9493.

Hanke, T., McMichael, A. J., Mwau, M., Wee, E. G., Ceberej, I., Patel, S., Sutton, J., Tomlinson, M., and Samuel, R. V. (2002). Development of a DNA-MVA/HIVA vaccine for Kenya. *Vaccine* **20,** 1995–1998.

Hasbold, J., Lyons, A. B., Kehry, M. R., and Hodgkin, P. D. (1998). Cell division number regulates IgG1 and IgE switching of B cells following stimulation by CD40 ligand and IL-4. *Eur. J. Immunol.* **28,** 1040–1051.

Karandikar, N. J., Crawford, M. P., Yan, X., Ratts, R. B., Brenchley, J. M., Ambrozak, D. R., Lovett-Racke, A. E., Frohman, E. M., Stastny, P., Douek, D. C., Koup, R. A., and Racke, M. K. (2002). Glatiramer acetate (Copaxone) therapy induces CD8+ T cell responses in patients with multiple sclerosis. *J. Clin. Invest.* **109,** 641–649.

Kern, F., Khatamzas, E., Surel, I., Frommel, C., Reinke, P., Waldrop, S. L., Picker, L. J., and Volk, H. D. (1999). Distribution of human CMV-specific memory T cells among the CD8pos. subsets defined by CD57, CD27, and CD45 isoforms. *Eur. J. Immunol.* **29,** 2908–2915.

Letvin, N. L. (2002). Strategies for an HIV vaccine. *J. Clin. Invest.* **110,** 15–20.

Lyons, A. B. (1997). Pertussis toxin pretreatment alters the *in vivo* cell division behavior and survival of B lymphocytes after intravenous transfer. *Immunol. Cell Biol.* **75,** 7–12.

Lyons, A. B., Hasbold, J., and Hodgkin, P. D. (2001). Flow cytometric analysis of cell division history using dilution of carboxyfluorescein diacetate succinimidyl ester, a stably integrated fluorescent probe. *Methods Cell Biol.* **63,** 375–398.

Lyons, A. B., and Parish, C. R. (1994). Determination of lymphocyte division by flow cytometry. *J. Immunol. Methods* **171,** 131–137.

MacGregor, R. R., Ginsberg, R., Ugen, K. E., Baine, Y., Kang, C. U., Tu, X. M., Higgins, T., Weiner, D. B., and Boyer, J. D. (2002). T-cell responses induced in normal volunteers immunized with a DNA-based vaccine containing HIV-1 env and rev. *AIDS* **16,** 2137–2143.

McMichael, A., and Hanke, T. (2002). The quest for an AIDS vaccine: is the CD8+ T-cell approach feasible? *Nat. Rev. Immunol.* **2,** 283–291.

Michalek, J., Collins, R. H., Hill, B. J., Brenchley, J. M., and Douek, D. C. (2003). Identification and monitoring of graft-versus-host specific T-cell clone in stem cell transplantation. *Lancet* **361,** 1183–1185.

Migueles, S. A., Laborico, A. C., Shupert, W. L., Sabbaghian, M. S., Rabin, R., Hallahan, C. W., Van Baarle, D., Kostense, S., Miedema, F., McLaughlin, M., Ehler, L., Metcalf, J., Liu, S., and Connors, M. (2002). HIV-specific CD8+ T cell proliferation is coupled to perforin expression and is maintained in nonprogressors. *Nat. Immunol.* **3,** 1061–1068.

Migueles, S. A., Sabbaghian, M. S., Shupert, W. L., Bettinotti, M. P., Marincola, F. M., Martino, L., Hallahan, C. W., Selig, S. M., Schwartz, D., Sullivan, J., and Connors, M. (2000). HLA B*5701 is highly associated with restriction of virus replication in a subgroup of HIV-infected long term nonprogressors. *Proc. Natl. Acad. Sci. USA* **97,** 2709–2714.

Murali-Krishna, K., and Ahmed, R. (2000). Cutting edge: naive T cells masquerading as memory cells. *J. Immunol.* **165,** 1733–1737.

Murali-Krishna, K., Altman, J. D., Suresh, M., Sourdive, D. J., Zajac, A. J., Miller, J. D., Slansky, J., and Ahmed, R. (1998). Counting antigen-specific CD8 T cells: a reevaluation of bystander activation during viral infection. *Immunity* **8,** 177–187.

Peters, C., Peng, X., Douven, D., Pan, Z. K., and Paterson, Y. (2003). The induction of HIV Gag-specific CD8(+) T cells in the spleen and gut-associated lymphoid tissue by parenteral or mucosal immunization with recombinant *Listeria monocytogenes* HIV Gag. *J. Immunol.* **170,** 5176–5187.

Picker, L. J. (1994). Control of lymphocyte homing. *Curr Opin Immunol.* **6,** 394–406.

Picker, L. J., and Siegelman, M. H. (1993). Lymphoid tissues and organs. *In* "Fundamental immunology," (W. E. Paul, ed.). Raven Press, New York.

Pitcher, C. J., Quittner, C., Peterson, D. M., Connors, M., Koup, R. A., Maino, V. C., and Picker, L. J. (1999). HIV-1–specific CD4+ T cells are detectable in most individuals with active HIV-1 infection, but decline with prolonged viral suppression. *Nat. Med.* **5,** 518–525.

Prussin, C. (1997). Cytokine flow cytometry: Understanding cytokine biology at the single-cell level. *J. Clin. Immunol.* **17,** 195–204.

Prussin, C., and Metcalfe, D. D. (1995). Detection of intracytoplasmic cytokine using flow cytometry and directly conjugated anti-cytokine antibodies. *J. Immunol. Methods* **188,** 117–128.

Rowbottom, A. W., Garland, R. J., Lepper, M. W., Kaneria, S. S., Goulden, N. J., Oakhill, A., and Steward, C. G. (2000). Functional analysis of the CD8+CD57+ cell population in normal healthy individuals and matched unrelated T-cell–depleted bone marrow transplant recipients. *Br. J. Haematol.* **110,** 315–321.

Sadat-Sowti, B., Debre, P., Idziorek, T., Guillon, J. M., Hadida, F., Okzenhendler, E., Katlama, C., Mayaud, C., and Autran, B. (1991). A lectin-binding soluble factor released by CD8+CD57+ lymphocytes from AIDS patients inhibits T cell cytotoxicity. *Eur. J. Immunol.* **21,** 737–741.

Sadat-Sowti, B., Debre, P., Mollet, L., Quint, L., Hadida, F., Leblond, V., Bismuth, G., and Autran, B. (1994). An inhibitor of cytotoxic functions produced by CD8+CD57+ T lymphocytes from patients suffering from AIDS and immunosuppressed bone marrow recipients. *Eur. J. Immunol.* **24,** 2882–2888.

Schmidt, D., Goronzy, J. J., and Weyand, C. M. (1996). CD4+ CD7-CD28-T cells are expanded in rheumatoid arthritis and are characterized by autoreactivity. *J. Clin. Invest.* **97,** 2027–2037.

Shiver, J. W., Fu, T. M., Chen, L., Casimiro, D. R., Davies, M. E., Evans, R. K., Zhang, Z. Q., Simon, A. J., Trigona, W. L., Dubey, S. A., Huang, L., Harris, V. A., Long, R. S., Liang, X., Handt, L., Schleif, W. A., Zhu, L., Freed, D. C., Persaud, N. V., Guan, L., Punt, K. S., Tang, A., Chen, M., Wilson, K. A., Collins, K. B., Heidecker, G. J., Fernandez, V. R., Perry, H. C., Joyce, J. G., Grimm, K. M., Cook, J. C., Keller, P. M., Kresock, D. S., Mach, H., Troutman, R. D., Isopi, L. A., Williams, D. M., Xu, Z., Bohannon, K. E., Volkin, D. B., Montefiori, D. C., Miura, A., Krivulka, G. R., Lifton, M. A., Kuroda, M. J., Schmitz, J. E., Letvin, N. L., Caulfield, M. J., Bett, A. J., Youil, R., Kaslow, D. C., and Emini, E. A. (2002). Replication-incompetent adenoviral vaccine vector elicits effective anti-immunodeficiency-virus immunity. *Nature* **415,** 331–335.

Sze, D. M., Giesajtis, G., Brown, R. D., Raitakari, M., Gibson, J., Ho, J., Baxter, A. G., Fazekas de St Groth, B., Basten, A., and Joshua, D. E. (2001). Clonal cytotoxic T cells are expanded in myeloma and reside in the CD8(+)CD57(+)CD28(−) compartment. *Blood* **98,** 2817–2827.

Teitelbaum, D., Meshorer, A., Hirshfeld, T., Arnon, R., and Sela, M. (1971). Suppression of experimental allergic encephalomyelitis by a synthetic polypeptide. *Eur. J. Immunol.* **1,** 242–248.

van Leeuwen, E. M., Gamadia, L. E., Baars, P. A., Remmerswaal, E. B., ten Berge, I. J., and van Lier, R. A. (2002). Proliferation requirements of cytomegalovirus-specific, effector-type human CD8+ T cells. *J. Immunol.* **169,** 5838–5843.

Vesanen, M., Stevens, C. E., Taylor, P. E., Rubinstein, P., and Saksela, K. (1996). Stability in controlling viral replication identifies long-term nonprogressors as a distinct subgroup among human immunodeficiency virus type 1–infected persons. *J. Virol.* **70,** 9035–9040.

Vingerhoets, J. H., Vanham, G. L., Kestens, L. L., Penne, G. G., Colebunders, R. L., Vandenbruaene, M. J., Goeman, J., Gigase, P. L., De Boer, M., and Ceuppens, J. L. (1995). Increased cytolytic T lymphocyte activity and decreased B7 responsiveness are associated with CD28 down-regulation on CD8+ T cells from HIV-infected subjects. *Clin. Exp. Immunol.* **100,** 425–433.

Wang, E. C., and Borysiewicz, L. K. (1995). The role of CD8+, CD57+ cells in human cytomegalovirus and other viral infections. *Scand. J. Infect. Dis. Suppl.* **99,** 69–77.

Wang, E. C., Lawson, T. M., Vedhara, K., Moss, P. A., Lehner, P. J., and Borysiewicz, L. K. (1997). CD8high+ (CD57+) T cells in patients with rheumatoid arthritis. *Arthritis Rheum.* **40,** 237–248.

Warrington, K. J., Takemura, S., Goronzy, J. J., and Weyand, C. M. (2001). CD4+,CD28− T cells in rheumatoid arthritis patients combine features of the innate and adaptive immune systems. *Arthritis Rheum.* **44,** 13–20.

Wee, E. G., Patel, S., McMichael, A. J., and Hanke, T. (2002). A DNA/MVA-based candidate human immunodeficiency virus vaccine for Kenya induces multi-specific T cell responses in rhesus macaques. *J. Gen. Virol.* **83,** 75–80.

Wu, C. Y., Kirman, J. R., Rotte, M. J., Davey, D. F., Perfetto, S. P., Rhee, E. G., Freidag, B. L., Hill, B. J., Douek, D. C., and Seder, R. A. (2002). Distinct lineages of T(H)1 cells have differential capacities for memory cell generation *in vivo*. *Nat. Immunol.* **3,** 852–858.

CHAPTER 20

Detection of T-Cell Degranulation: CD107a and b

Michael R. Betts and Richard A. Koup

Immunology Laboratory
Vaccine Research Center
National Institute of Allergy and Infectious Diseases
National Institutes of Health
Bethesda, Maryland 20892

I. Introduction: Methods to Assess CD8$^+$ T-Cell Function

T lymphocyte–mediated cytotoxicity has historically been examined using the ^{51}Cr release assay, wherein antigen-expressing radioactively labeled target cells are recognized and killed by specific T cells (Brunner *et al.*, 1968). Though cumbersome, these assays directly assessed perhaps the most critical function of cytolytic T cells: the ability to kill target cells. Unfortunately, these assays tend to be nonquantitative and insensitive; and no information is gained about the number or more importantly the identity of the effector T cells.

Flow cytometric analysis of $CD8^+$ T lymphocyte–mediated cytotoxicity has similar limitations, most importantly that the identity of cytolytic $CD8^+$ T cells (the cells responsible for eliciting cytotoxic activity) is not directly examined. Several flow-based killing assays have been developed, which we and others have used to detect $CD8^+$ T-cell–mediated cytotoxicity (Betts *et al.*, 2003; Fischer *et al.*, 2002; Lecoeur *et al.*, 2001; Liu *et al.*, 2002; Ritchie *et al.*, 2000; Sheehy *et al.*, 2001). One such assay system uses fluorescent dyes, such as carboxy fluorescein diacetate succinimidyl ester (CFSE), PKH-26, or chloromethyl-benzoyl-amino-tetramethyl-rhodamine (CMTMR) to label peptide-loaded target cells, which are then combined with $CD8^+$ T cells (Betts *et al.*, 2003; Fischer *et al.*, 2002; Ritchie *et al.*, 2000; Sheehy *et al.*, 2001). As the peptide-specific $CD8^+$ T cells kill the target cells, cytotoxicity is measured as the loss of fluorescently labeled target cells within the assay sample. This type of system can be combined with standard measures of apoptosis, such as 7-AAD or annexin V (Lecoeur *et al.*, 1997, 2001) staining to provide additional information about the target cells. Another method to assess cytotoxicity is the caspase assay (Chahroudi *et al.*, 2003; Liu *et al.*, 2002). This method detects the activation of caspase enzymes within target cells, an early process in the apoptotic pathways that is induced upon delivery of the "lethal hit" by cytolytic T cells. Unfortunately, however, none of these assays alone provides sufficient information regarding the identity, phenotype, or frequency of the $CD8^+$ T cells that mediate the killing. Several methods have been developed to identify antigen-specific $CD8^+$ T cells. These include, most notably, major histocompatibility complex (MHC) class I tetrameric complexes (Altman *et al.*, 1996) and intracellular cytokine staining (Betts *et al.*, 2000; Jung *et al.*, 1993; Kern *et al.*, 1998; Waldrop *et al.*, 1997). Although both of these techniques can provide valuable information regarding the frequency, phenotype, and/or the functionality of antigen-specific T cells, they do not examine the ability of the $CD8^+$ T cells to elicit cytotoxic activity.

In this chapter, we describe an assay that serves to help bridge the gap between measurement of $CD8^+$ T-cell cytolytic activity and the identity of the effector $CD8^+$ T cells. To understand how this assay is performed, it is important to first review the fundamental mechanism by which $CD8^+$ T cells mediate cytolytic activity—degranulation. Cytotoxic $CD8^+$ T lymphocytes mediate the killing of target cells via two major pathways: perforin-granzyme–mediated activation of apoptosis and fas–fas ligand–mediated induction of apoptosis (Henkart, 1994; Nagata and Golstein, 1995; Sarin *et al.*, 1997; Trapani *et al.*, 1998). Induction of these pathways depends on the release of cytolytic granules from the responding $CD8^+$ T cells (Bossi and Griffiths, 1999; Yannelli *et al.*, 1986). Degranulation is a necessary prerequisite to perforin-granzyme–mediated killing and is required for immediate lytic function mediated by responding antigen-specific $CD8^+$ T cells. Cytotoxicity does not require *de novo* synthesis of proteins by the effector $CD8^+$ T cell; instead, pre-formed lytic granules located within the cytoplasm are released in a polarized fashion toward the target cell (Barry and Bleackley, 2002; Griffiths, 1997). The lytic granules are membrane-bound secretory lysosomes that contain a

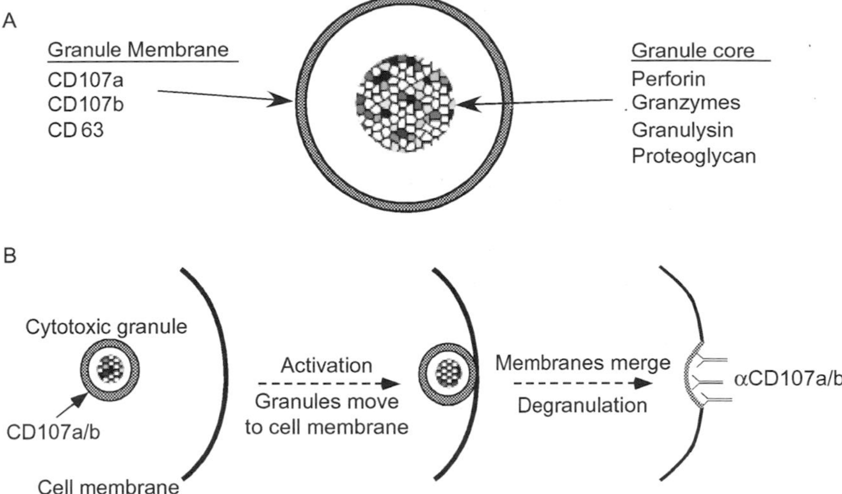

Fig. 1 Schematic representation of a CD8$^+$ T-cell cytolytic granule (A) and the principle mechanism of degranulation as it relates to measurement by CD107a/b expression (B). (A) Cytolytic granules are membrane-bound organelles related to lysosomes. A delimiting outer membrane contains various lysosomal membrane proteins, including CD107a, CD107b, and CD63. This surrounds a densely packed core, containing perforin, various granzymes, and granulysin, complexed with proteoglycan. (B) Upon activation of a CD8$^+$ T-cell, pre-formed cytotoxic granules move toward the outer membrane of the cell in a polarized fashion toward the site of antigen presentation. Upon reaching the plasma membrane, the membrane of the granule and the plasma membrane merge, releasing the contents of the granule into the extracellular space. At this time, the CD107a and CD107b proteins from the granule membrane are accessible at the cell surface for specific antibody binding.

dense core composed of various proteins, including perforin and granzymes (Peters *et al.*, 1991) (Fig. 1A). The granule core is surrounded by a lipid bilayer containing numerous lysosomal-associated membrane glycoproteins (LAMPs), including CD107a (LAMP-1), CD107b (LAMP-2), and CD63 (LAMP-3).

During the process of degranulation, the lytic granule membrane merges with the plasma membrane of the activated CD8$^+$ T cell (Peters *et al.*, 1991) and the contents of the granule are then released into the immunological synapse between the CD8$^+$ T cell and the target cell (Fig. 1B). As a result of this process, the granular membrane, including CD107a, CD107b, and CD63 glycoproteins therein, is incorporated into the plasma membrane of the responding CD8$^+$ T cell (Fukuda, 1991; Peters *et al.*, 1991). High-level expression of CD107a and b on the cell surface of activated T cells requires degranulation, because degranulation inhibitors, such as colchicine, dramatically reduce cell-surface expression of CD107a and b (Betts *et al.*, 2003). Protein synthesis inhibitors, such as cycloheximide, do not prevent expression, indicating that cell-surface CD107a/b expression on activated T cells occurs as a result of degranulation and not *de novo* synthesis of the CD107 proteins (A. Oxenieus, 2003, unpublished observations).

Importantly, these proteins are rarely found on the surface of resting T lymphocytes (Betts *et al.*, 2003; Fukuda, 1991; Peters *et al.*, 1991). Thus, labeling responding cells with antibodies to CD107a and b and measuring their expression by flow cytometry can directly identify degranulating $CD8^+$ T cells.

II. Materials and Methods

The measurement of CD107 expression on the cell surface of activated T cells can be used with numerous other measures of T lymphocyte function. We first present a general protocol for this technique and then present specific applications and pitfalls. The methods presented here all use human cells; however, these methods also apply to studies using murine cells, provided that the appropriate species-specific antibodies are used.

A. Materials

1. *Cells to be analyzed:* Fresh or frozen peripheral blood lymphocytes or T-cell clones can be used; viability is very important for background issues (see pitfalls), so samples with viability less than 85–90% will lead to spurious results.

2. *Antibody to CD107a and/or CD107b:* These antibodies are available from BD-Pharmingen (San Diego, California) in directly conjugated forms; the only commercially available direct conjugates include the following: for CD107a (clone H4A3), biotin, fluorescein isothiocyanate (FITC), phycoerythrin (PE), and Cy5–PE; and for CD107b (cloneH4B4), FITC. The same antibodies are also available from Southern Biotech (Birmingham, Alabama) in purified form; these reagents must be titered (see later discussion) for use in the assay. Note that CD107a is the higher expressed of the two molecules (Betts *et al.*, 2003) and, therefore, may itself be sufficient for degranulation analysis (Rubio *et al.*, 2003). Additional formats of CD107a and b, including allophycocyanin (APC) and Alexa 680 conjugates, were made using standard conjugation techniques (Roederer, 2003).

3. *Antibodies to CD28 and CD49d:* Purified azide-free antibodies to these molecules are added to provide costimulation, as is commonly used in intracellular cytokine assays; these antibodies are available from BD Biosciences (San Diego, California) and are used at a concentration of 1 μg/ml of each antibody.

4. Fluorescently labeled antibodies to various cell-surface markers (CD3, CD4, CD8, CD14, CD19) and cytokines (interferon-γ [IFN-γ], tumor necrosis factor-α [TNF-α], etc.).

5. *Appropriate antigen:* When examining peptide-specific $CD8^+$ T-cell responses, use cytotoxic T lymphocyte (CTL) epitopes or peptide libraries at the desired peptide concentrations; *Staphylococcus* enterotoxin B (1 μg/ml final) serves as a positive control in all assays. Include unstimulated (or costimulation only) samples to control for background issues.

6. *Culture media:* Typically, RPMI-1640 supplemented with 10% fetal calf serum, antibiotics, and L-glutamine.

7. *Monensin:* Used to neutralize pH of endosomal and lysosomal compartments, and to prevent cytokine secretion; available from BD Biosciences (monensin [Golgistop], 0.7 μl/ml final concentration); it is absolutely necessary to include this reagent in all stimulations (see section IV).

8. *Brefeldin A* (BFA): Used as a secretion inhibitor; only necessary if TNF-α production is to be examined (O'Neil-Andersen and Lawrence, 2002); 10 mg/ml in dimethyl sulfoxide (DMSO) stock concentration, used at a final concentration of 10 μg/ml.

9. *Wash buffer:* 1% bovine serum albumin and 0.1% sodium azide in phosphate-buffered saline (PBS), store at 4 °C.

10. *Fixative:* 1% paraformaldehyde in PBS, store at 4 °C.

11. *Permeabilization solution:* Available directly from BD Pharmingen, Cytofix/Cytoperm kit.

B. Initial Assay Setup: Titration of CD107 Antibodies

It is particularly important to initially determine the amount of CD107a/b antibodies to be included in the culture stimulation. Unlike typical antibody-staining procedures, the CD107a and b antibodies must be included throughout the duration of the stimulation period (see section IV).

1. Adjust cells to a concentration of 1×10^6 cells/ml in culture media. Transfer cells to a 5-ml polystyrene tube (Falcon no. 2058, Becton Dickinson), 1 ml/tube.

2. Add 1 μg of SEB to each of six tubes containing 1×10^6 cells.

3. In parallel, prepare six tubes containing 1×10^6 cells with no antigen added.

4. Add 0.7 μl of monensin to each tube.

5. Add in serial dilutions of CD107a or b antibody into each SEB-stimulated and unstimulated tube. If using CD107a and b antibody direct conjugates from BD Pharmingen, start at an initial concentration of 20 μl/ml, followed by twofold serial dilutions.

6. Incubate for 5–6 hours at 37 °C in a 5% CO_2 incubator.

7. Wash cells with wash buffer, 5 minutes at 1500 rpm; re-suspend pellet.

8. *(Optional):* Permeabilize the cells using Cytofix/Cytoperm kit, per manufacturer's instructions. It is especially useful, though not necessary, to co-stain the activated cells with a cytokine marker, to ensure that measurement of CD107a/b expression is found on most cytokine-producing CD8$^+$ T cells. Determination of the optimum amount of CD107a/b antibody to use is also simplified by having another positive response marker.

9. Stain cells with appropriate T-cell markers. It is particularly important to include anti-CD3, to help differentiate background CD107a/b staining. When examining SEB-specific responses, it is helpful to include an antibody to CD8, because the vast majority of T-cell–expressed CD107a/b will be found on this subset. If cells were permeabilized, stain also with anti-IFN-γ, to provide an internal positive control for the SEB response.

10. Wash cells with wash buffer, 5 minutes at 1800 rpm; re-suspend pellet in 1% PFA.

11. Analyze by flow cytometry. Data analysis and gating strategy is shown in Fig. 2. Determine the minimum amount of CD107a/b antibody necessary to provide the best separation and population size of the responding cells while keeping background staining to a minimum.

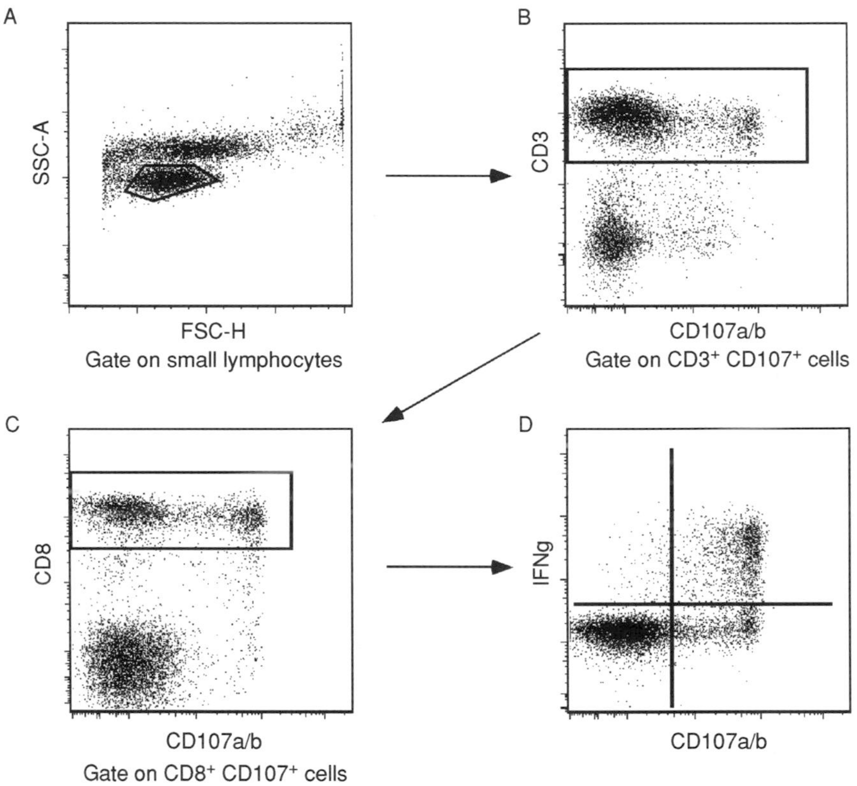

Fig. 2 Gating schematic for analysis of CD107a and b expression on SEB-stimulated cells. The initial gate is placed around small resting lymphocytes on a forward scatter channel versus a side scatter channel plot (A). Next, the gated cells are subgated based on CD3 expression to analyze on T cells (B). The $CD3^+$ T cells are then further subdivided into $CD8^+$ and $CD8^-$ cells, and $CD8^+$ events are gated upon (C). Finally, CD107a/b expression is plotted versus IFN-γ secretion using a quadrant analysis to differentiate various subpopulations of cells based on response type (D).

C. Antigen-Specific CD8$^+$ T-Cell Degranulation

Once the optimum amount of CD107a/b antibody to include in the assay is determined, antigen-specific CD8$^+$ T-cell responses can be examined. It becomes especially important, and more interesting, to combine CD107a/b measurement and intracellular cytokine staining together, thus allowing the examination of separate CD8$^+$ T-cell functions simultaneously. The general assay setup is nearly identical to the methods used when titrating the CD107a and b antibodies.

1. Prepare a master cell mix, containing the following:
   Cells diluted to 1×10^6 cells/ml in culture media
   0.7 μl/ml monensin
   1 μl/ml each of anti-CD28 and anti-CD49d antibodies
   1 μl/ml BFA, if TNF-α production is to be examined
   Appropriate amount of CD107a/b antibodies, as determined by titration
2. Aliquot master cell mix into a 5-ml polystyrene tube, 1 ml/tube.
3. Add the following antigens and controls to each tube containing 1×10^6 cells:
   No antigen (negative control)
   SEB, as previously (positive control)
   Peptide(s)/antigen of interest: The concentration of peptide should be determined previously or a peptide titration curve can be used to determine optimal concentrations. Typically, 1–2 μM peptide concentration should give maximal responses, although this can vary. Note that if peptides are diluted in DMSO, the final concentration of DMSO in the culture volume cannot exceed 1% without a substantial negative impact on cell response and viability.
4. Incubate for 5–6 hours at 37 °C in a 5% CO_2 incubator.
5. Wash cells with wash buffer, 5 minutes at 1500 rpm; re-suspend pellet in minimal volume.
6. Stain cells with any antibodies requiring surface staining.
7. Permeabilize the cells using Cytofix/Cytoperm kit, per manufacturer's instructions.
8. Wash cells with Cytofix wash buffer twice, 5 minutes at 1800 rpm; re-suspend pellet in minimal volume.
9. Stain cells with cytokine antibodies. Note that it is preferable to stain cells with anti-CD3 and anti-CD8 antibodies after they are permeabilized. This serves to partially, if not fully, eliminate the reduced staining of these molecules on the cell surface of activated T cells due to surface downmodulation of these markers after activation.

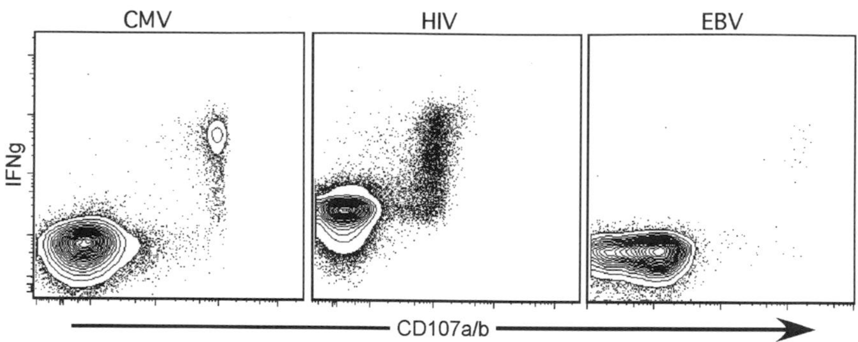

Fig. 3 Representative examples of antigen-specific expression of CD107a/b on $CD8^+$ T cells after activation with viral peptides. PBMCs from three patients were stimulated with peptides derived from cytomegalovirus (left), human immunodeficiency virus (middle), or Epstein–Barr virus (right). After incubation, cells were analyzed as described in Fig. 2. Note the presence of CD107a/b-positive IFN-positive cells in each panel, as well as the different types and level of the responses observed.

10. Wash cells with wash buffer, 5 minutes at 1800 rpm; re-suspend pellet in 1% PFA.

11. Analyze by flow cytometry. Three examples of antigen-specific responses wherein the responding $CD8^+$ T cells both degranulate and produce cytokine upon activation with viral peptides derived from human cytomegalovirus (CMV), human immunodeficiency virus (HIV), or Epstein–Barr virus (EBV) are shown in Fig. 3.

III. Results

An example of the gating strategy to analyze CD107a/b expression in activated $CD8^+$ T cells is shown in Fig. 2. Cells were stimulated with SEB, as described in section II, and collected on a FACSCalibur flow cytometer, using FITC, PE, PerCP, and APC as the fluorescent parameters. The initial gate is on small resting lymphocytes (Fig. 2A), with care taken to avoid events with low side scatter and low forward scatter, as well as events with high side scatter (granulocytes/macrophages). This is particularly important for reducing background staining. Next, $CD3^+$ events are gated upon (Fig. 2B). At this stage, one must account for any downregulation of CD3 that may have occurred in the activated T cells, if the CD3 stain was performed on the cell surface rather than intracellularly. If downregulation has occurred and is not accounted for, a significant proportion of the responding cells may be inadvertently gated out. In the example shown, the CD3 stain was performed on permeabilized cells, thus little downregulation of CD3 from the surface of the activated cells is observed. Also notice in this panel that the T cells expressing CD107a/b are very bright, as compared to the low-level expression observed in the CD3-negative cells. After the $CD3^+$ T-cell gate, the cells are now separated into $CD8^+$ and $CD8^-$ events (Fig. 2C). As was the case for the previous

CD3 gate, care must again be taken to account for any cells that have down-regulated CD8 in the process of becoming activated. In this example, the CD8 stain was again performed on permeabilized cells. It can also be seen here that a small but detectable portion of $CD3^+/CD8^-$ cells also becomes CD107a/b positive in response to SEB; these likely represent degranulating $CD4^+$ T cells, and a similar type of analysis can be performed on this T-cell subset. Finally, if the cells were co-stained with a cytokine marker, such as IFN-γ, the relationship between cytokine production and degranulation can be directly observed by plotting IFN-γ versus CD107a/b expression in the $CD8^+$ (Fig. 2D) or $CD4^+$ T cells. As can be seen, although most SEB-responding $CD8^+$ T cells become CD107a/b positive, the expression of IFN-γ can be quite variable and a substantial proportion of responding cells do not produce IFN-γ. This serves in part to illustrate the diversity of the $CD8^+$ T-cell response, because not all the responding cells exhibit the same functional profile.

Figure 3 shows three examples of responses obtained using specific peptides derived either from CMV, HIV, or EBV and illustrates some of the different response profiles that may be observed and the level of sensitivity that can be achieved when analyzing very small responses. In all three examples, human PBMCs were stimulated with antigenic peptides using the method described in section II. Figure 3 (left panel) shows the response to a CMV peptide in a CMV-positive individual. As can be seen, all the cells producing IFN-γ also become CD107a/b positive. The center panel shows the response to a mixture of overlapping 15-mer peptides derived from HIV Pol in an HIV-positive individual. Unlike the CMV response in the left panel, a significant proportion of responding cells to the HIV peptides do not produce IFN-γ, and a varying degree of IFN-γ is observed in those cells that do produce this cytokine. Both the CMV- and the HIV-specific responses shown here are relatively large responses, on the order of 2–7% of $CD8^+$ T cells. In the right panel, however, a much smaller yet detectable response to an EBV-derived peptide is shown. Although this response is very low, approximately 0.2% of $CD8^+$ T cells, it is still easily distinguishable as a predominantly IFN-γ^+ CD107a/b^+ response.

IV. Critical Aspects

As with other techniques to measure T-cell function by flow cytometry, proper data analysis is a crucial aspect of this technique. Unlike standard intracellular cytokine assays, there are specific issues with examination of CD107a/b expression that must be considered during both the assay setup and the data analysis.

A. Trafficking of CD107a and b

CD107a and b in $CD8^+$ T cells are localized to lysosomal, endosomal, and cytotoxic granular membranes (Peters *et al.*, 1991). This localization occurs as a result of a sorting signal within the cytoplasmic tail of these proteins (Chang *et al.*,

2002; Gough and Fambrough, 1997; Guarnieri *et al.*, 1993; Honing and Hunziker, 1995). Furthermore, cell-surface expression of these proteins on $CD8^+$ T cells is typically very low and is retrieved from the cell surface via an endocytic pathway into an endosomal or lysosomal compartment (Chang *et al.*, 2002; Fukuda, 1991). Because of this active internalization of the CD107a and b molecules, it is necessary to label $CD8^+$ T cells with antibodies to these molecules during the activation period (Betts *et al.*, 2003). Failure to include antibodies during the stimulation can have a dramatic effect on the ability to stain cell-surface CD107a and b. As shown in Fig. 4, cells labeled throughout the stimulation period have a significantly enhanced staining pattern compared to cells stained at the end of the stimulation period.

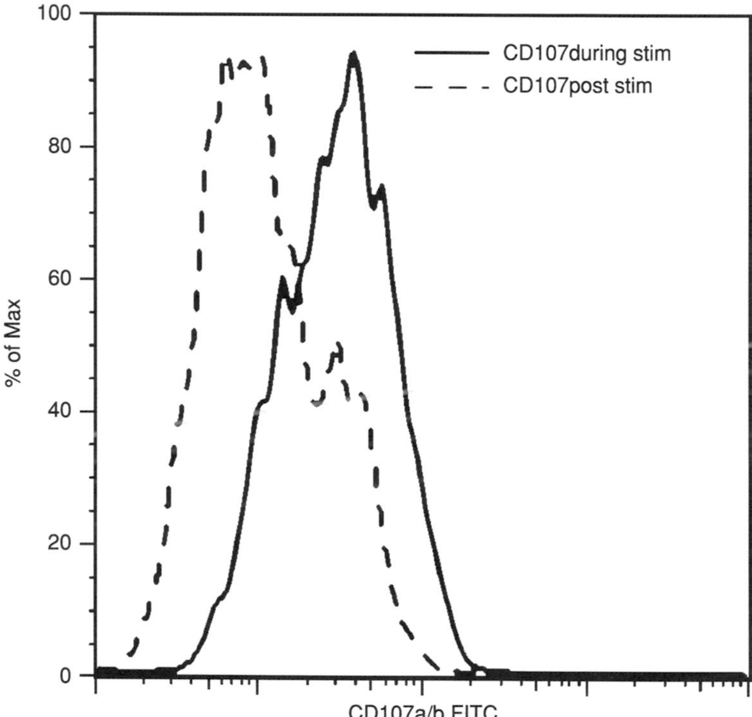

Fig. 4 Histogram analysis of CD107a/b expression on SEB-stimulated cells stained either during the stimulation period (*solid line*) or after the stimulation period (*dashed line*). Analysis of the cells was performed as described in Fig. 2. Gates were placed on $CD3^+/CD8^+$ cells shown to respond to SEB stimulation by producing IFN-γ. Histograms of CD107a/b levels on the IFN-γ–producing cells were then compared, demonstrating that cells stained with CD107a/b after activation have substantially lower levels of surface CD107a/b.

B. Monensin Requirement

The active trafficking of CD107a and b from the cell surface introduces yet another wrinkle into the staining procedure. When the antibody-fluorochrome–bound CD107a or b molecules are reinternalized, they will traffic into endosomal or lysosomal compartments, where they may be exposed to a highly acidic environment and various proteases. This environment has a dramatic effect on the fluorescence of various fluorochromes, because of either acid sensitivity or protease sensitivity. To reduce this effect, one should include monensin throughout the stimulation period. Monensin serves to neutralize the pH of the endosomal and lysosomal compartments (Mollenhauer *et al.*, 1990), reducing the loss of fluorescence of various fluorochromes exposed to this environment. As shown in Fig. 5, this effect is not limited only to small organic fluorochromes sensitive to pH, such as FITC and Alexa 680, but also on proteinaceous fluorochromes such as PE and APC (Roederer *et al.*, 1987). The latter effect likely occurs because of altered trafficking of proteases within the endosomal and lysosomal compartments as a result of monensin-induced deacidification of these compartments (Chapman and Munro, 1994). As can be seen in Fig. 5, stimulation in the absence of monensin leads to the substantial loss of the fluorescent signal obtained from FITC, PE, APC, and Alexa 680–conjugated anti CD107a/b in activated cells. The inclusion of monensin also serves to prevent the secretion of various cytokines, such as IFN-γ, MIP1β, and IL-2 (Jung *et al.*, 1993; Lee *et al.*, 1990). It must be noted, however, that monensin does not adequately prevent the secretion of TNF-α from activated cells (O'Neil-Andersen and Lawrence, 2002). Thus, if CD107a/b expression and TNF-α production are to be monitored in CD8$^+$ T cells, both monensin and BFA must be used in combination.

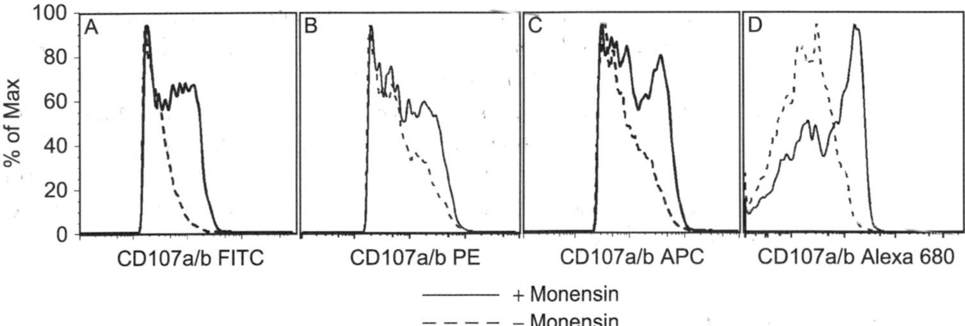

Fig. 5 Monensin is necessary to obtain optimal staining of CD107a/b conjugated with fluorescein isothiocyanate (A), phycoerythrin (B), allophycocyanin (C), or Alexa 680 (D). Cells were stimulated with SEB in the presence of CD107a/b antibodies, with (*solid line*) or without (*dashed line*) monensin. Histograms represent the level of CD107a/b expression on CD3$^+$/CD8$^+$ cells that responded to SEB by IFN-γ production.

C. Background Expression of CD107a and b

Background expression of CD107a and b on the surface of T cells is typically very low in peripheral blood, although they can be detected on the cell surface of mitogen-activated T cells (Kannan *et al.*, 1996). It should be noted, however, that these proteins are not expressed only in T lymphocytes. These proteins are normal residents of endosomal and lysosomal membranes and are detectable at a relatively high copy number in all cell types in humans, mice, rats, and chickens (Fukuda, 1991). Some peripheral blood cells, including activated platelets and monocytes, express these molecules on the cell surface to a higher degree than other cell types (Eskelinen *et al.*, 2003; Silverstein and Febbraio, 1992). Additionally, because all cell types express these proteins, cells with a permeable membrane such as dead or dying cells will stain positive for CD107a and/or b. Thus, it is extremely important for the cells to be tested to have high viability to help reduce background concerns. An additional concern regarding background arises as a result of the CD107a and b staining procedures. By including the antibodies to these proteins during the stimulation period, antigen-presenting cells present within the cells being examined, specifically B cells and monocytes, readily uptake the antibody for processing. Because monensin is present, these cells are unable to proteolytically process the antibody/fluorochromes and, therefore, also stain positive for CD107a and b. This effect can be readily seen in the example shown in Fig. 6 (upper left panel). Monocytes and B cells, labeled here with CD14 or CD19, are all CD107a/b^{dull}. By gating on small lymphocytes (Fig. 6, upper right panel), the majority of the CD107a/b^{dull} cells are removed. By removing the remaining CD14/19-positive events within the small lymphocyte gate (Fig. 6, bottom panels), a substantial proportion of the remaining CD107a/b background staining is eliminated, leaving the antigen-specific T-cell–expressed CD107a and b intact for further analysis. Thus, whenever possible, inclusion of a dump channel, specifically to label monocytes, B cells, and dead cells (if possible), can significantly reduce the CD107a and b background. Despite these measures, occasional patient samples will have higher CD107 background than others, with expression levels varying from 0.01% to more than 0.5% of $CD8^+$ T cells. Once the background has been reduced by the means presented here, the remaining background likely represents either nonspecific or an unknown antigen-specific degranulation by the $CD8^+$ T cells. Typically, however, the expression levels tend to be quite dull. In these cases, it may be pertinent to use a stricter gating method only on CD107a/b^{bright} events.

V. Applications

The general applications for this technique revolve primarily around the examination of antigen-specific T lymphocyte function in humans and mice. This is the first flow cytometry–based assay that directly assesses the ability of T lymphocytes to degranulate. At the most basic level, the functionality of the *ex vivo* $CD8^+$ T-cell response to specific individual peptides of viral origin can be examined in a

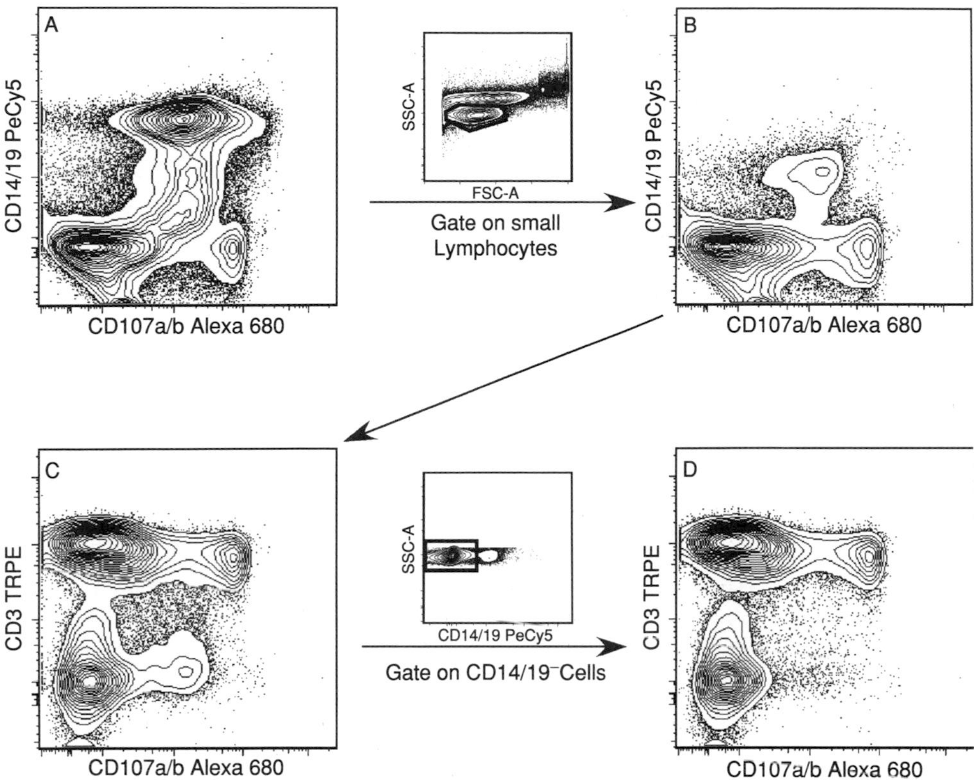

Fig. 6 Removal of background CD107a/b expression through gating and CD14/CD19 staining. (A) Ungated SEB-stimulated cells are shown to demonstrate the level of CD107a/b labeling found on CD14/CD19$^+$ cells. Note that the level of CD107a/b found on the CD14/CD19$^+$ events is intermediate, as compared to the CD107a/b bright cells that do not co-stain with CD14/CD19. (B) By gating on small resting lymphocytes, a large proportion of CD107a/b dull events are removed. Note that this gating did not affect the CD107a/b bright events. Also, a small proportion of CD14/CD19$^+$ cells remain within the small lymphocyte gate and are CD107a/b dull. (C) Changing the parameters compared in (B), we can see that a large proportion of CD107a/b dull events do not co-stain with the T-cell marker CD3. (D) Removing the remaining CD14/CD19$^+$ events by negative gating dramatically reduces the remaining CD107a/b background expression on the non–T cells.

more complete manner by measuring cytokine production and degranulation simultaneously. With polychromatic flow cytometry, multiple cytokines and degranulation can be measured simultaneously in activated CD4$^+$ and CD8$^+$ T lymphocytes.

Aside from simply providing another quantitative measure of functionality for T cells, measurement of degranulation allows a direct link between cytolytic potential and cytokine production. Measurement of degranulation provides a positive marker for cytolytic activity of effector CD8$^+$ T cells, unlike measurement of perforin in CD8$^+$ T cells, which is lost if the cells degranulate as a result of activation (Appay *et al.*, 2000). There is a direct correlation between the ability of human CD8$^+$ T-cell clones to kill, as measured by ^{51}Cr release assay, and the level of cell-surface CD107a expression after stimulation (Rubio *et al.*, 2003).

CD107a/b expression is also highly amenable for use in live cell sorting of activated antigen-specific T cells, because no fixation or permeabilization of the cells is necessary. Current methods to sort live antigen-specific CD8$^+$ T cells relies on either MHC class I tetramer, which is highly restricted by reagent availability, or cytokine-secretion capture systems, which tend to be of low sensitivity. Cell sorting using CD107a/b expression is highly sensitive and is not limited by reagent availability, because peptides or whole antigens are used, obviating the need for patient- or MHC-specific reagents.

References

Altman, J. D., Moss, P. A. H., Goulder, P. J. R., Barouch, D. H., McHeyzer-Williams, M. G., Bell, J. I., McMichael, A. J., and Davis, M. M. (1996). Phenotypic analysis of antigen-specific T lymphocytes. *Science* **274,** 94–96.

Appay, V., Nixon, D. F., Donahoe, S. M., Gillespie, G. M., Dong, T., King, A., Ogg, G. S., Spiegel, H. M., Conlon, C., Spina, C. A., Havlir, D. V., Richman, D. D., Waters, A., Easterbrook, P., McMichael, A. J., and Rowland-Jones, S. L. (2000). HIV-specific CD8(+) T cells produce antiviral cytokines but are impaired in cytolytic function. *J. Exp. Med.* **192,** 63–75.

Barry, M., and Bleackley, R. C. (2002). Cytotoxic T lymphocytes: All roads lead to death. *Nat. Rev. Immunol.* **2,** 401–409.

Betts, M. R., Brenchley, J. M., Price, D. A., DeRosa, S. C., Douek, D. C., Roederer, M., and Koup, R. A. (2003). Sensitive and viable identification of antigen-specific CD8$^+$ T cells by a flow cytometric assay for degranulation. *J. Immunol. Methods* **281,** 65–78.

Betts, M. R., Casazza, J. P., Patterson, B. A., Waldrop, S., Trigona, W., Fu, T. M., Kern, F., Picker, L. J., and Koup, R. A. (2000). Putative immunodominant human immunodeficiency virus–specific CD8(+) T-cell responses cannot be predicted by major histocompatibility complex class I haplotype. *J. Virol.* **74,** 9144–9151.

Bossi, G., and Griffiths, G. M. (1999). Degranulation plays an essential part in regulating cell surface expression of Fas ligand in T cells and natural killer cells. *Nat. Med.* **5,** 90–96.

Brunner, K. T., Mauel, J., Cerottini, J. C., and Chapuis, B. (1968). Quantitative assay of the lytic action of immune lymphoid cells on 51-Cr–labelled allogeneic target cells *in vitro*; inhibition by isoantibody and by drugs. *Immunology* **14,** 181–196.

Chahroudi, A., Silvestri, G., and Feinberg, M. B. (2003). Measuring T cell–mediated cytotoxicity using fluorogenic caspase substrates. *Methods* **31,** 120–126.

Chang, M. H., Karageorgos, L. E., and Meikle, P. J. (2002). CD107a (LAMP-1) and CD107b (LAMP-2). *J. Biol. Regul. Homeost. Agents* **16,** 147–151.

Chapman, R. E., and Munro, S. (1994). Retrieval of TGN proteins from the cell surface requires endosomal acidification. *Embo. J.* **13,** 2305–2312.

Eskelinen, E. L., Tanaka, Y., and Saftig, P. (2003). At the acidic edge: emerging functions for lysosomal membrane proteins. *Trends Cell Biol.* **13,** 137–145.

Fischer, K., Andreesen, R., and Mackensen, A. (2002). An improved flow cytometric assay for the determination of cytotoxic T lymphocyte activity. *J. Immunol. Methods* **259,** 159–169.

Fukuda, M. (1991). Lysosomal membrane glycoproteins. Structure, biosynthesis, and intracellular trafficking. *J. Biol. Chem.* **266,** 21327–21330.

Gough, N. R., and Fambrough, D. M. (1997). Different steady state subcellular distributions of the three splice variants of lysosome-associated membrane protein LAMP-2 are determined largely by the COOH-terminal amino acid residue. *J. Cell Biol.* **137,** 1161–1169.

Griffiths, G. M. (1997). Protein sorting and secretion during CTL killing. *Semin. Immunol.* **9,** 109–115.

Guarnieri, F. G., Arterburn, L. M., Penno, M. B., Cha, Y., and August, J. T. (1993). The motif Tyr-X-X-hydrophobic residue mediates lysosomal membrane targeting of lysosome-associated membrane protein 1. *J. Biol. Chem.* **268,** 1941–1946.

Henkart, P. A. (1994). Lymphocyte-mediated cytotoxicity: Two pathways and multiple effector molecules. *Immunity* **1,** 343–346.

Honing, S., and Hunziker, W. (1995). Cytoplasmic determinants involved in direct lysosomal sorting, endocytosis, and basolateral targeting of rat lgp120 (lamp-I) in MDCK cells. *J. Cell Biol.* **128,** 321–332.

Jung, T., Schauer, U., Heusser, C., Neumann, C., and Rieger, C. (1993). Detection of intracellular cytokines by flow cytometry. *J. Immunol. Methods* **159,** 197–207.

Kannan, K., Stewart, R. M., Bounds, W., Carlsson, S. R., Fukuda, M., Betzing, K. W., and Holcombe, R. F. (1996). Lysosome-associated membrane proteins h-LAMP1 (CD107a) and h-LAMP2 (CD107b) are activation-dependent cell surface glycoproteins in human peripheral blood mononuclear cells which mediate cell adhesion to vascular endothelium. *Cell Immunol.* **171,** 10–19.

Kern, F., Surel, I. P., Brock, C., Freistedt, B., Radtke, H., Scheffold, A., Blasczyk, R., Reinke, P., Schneider-Mergener, J., Radbruch, A., Walden, P., and Volk, H. D. (1998). T-cell epitope mapping by flow cytometry. *Nat. Med.* **4,** 975–978.

Lecoeur, H., Fevrier, M., Garcia, S., Riviere, Y., and Gougeon, M. L. (2001). A novel flow cytometric assay for quantitation and multiparametric characterization of cell-mediated cytotoxicity. *J. Immunol. Methods* **253,** 177–187.

Lecoeur, H., Ledru, E., Prevost, M. C., and Gougeon, M. L. (1997). Strategies for phenotyping apoptotic peripheral human lymphocytes comparing ISNT, annexin-V and 7-AAD cytofluorometric staining methods. *J. Immunol. Methods* **209,** 111–123.

Lee, C. L., Lee, S. H., Jay, F. T., and Rozee, K. R. (1990). Immunobiological study of interferon-gamma–producing cells after staphylococcal enterotoxin B stimulation. *Immunology* **70,** 94–99.

Liu, L., Chahroudi, A., Silvestri, G., Wernett, M. E., Kaiser, W. J., Safrit, J. T., Komoriya, A., Altman, J. D., Packard, B. Z., and Feinberg, M. B. (2002). Visualization and quantification of T cell-mediated cytotoxicity using cell-permeable fluorogenic caspase substrates. *Nat. Med.* **8,** 185–189.

Mollenhauer, H. H., Morre, D. J., and Rowe, L. D. (1990). Alteration of intracellular traffic by monensin; mechanism, specificity and relationship to toxicity. *Biochim. Biophys. Acta.* **1031,** 225–246.

Nagata, S., and Golstein, P. (1995). The Fas death factor. *Science* **267,** 1449–1456.

O'Neil-Andersen, N. J., and Lawrence, D. A. (2002). Differential modulation of surface and intracellular protein expression by T cells after stimulation in the presence of monensin or brefeldin A. *Clin. Diagn. Lab. Immunol.* **9,** 243–250.

Peters, P. J., Borst, J., Oorschot, V., Fukuda, M., Krahenbuhl, O., Tschopp, J., Slot, J. W., and Geuze, H. J. (1991). Cytotoxic T lymphocyte granules are secretory lysosomes, containing both perforin and granzymes. *J. Exp. Med.* **173,** 1099–1109.

Ritchie, D. S., Hermans, I. F., Lumsden, J. M., Scanga, C. B., Roberts, J. M., Yang, J., Kemp, R. A., and Ronchese, F. (2000). Dendritic cell elimination as an assay of cytotoxic T lymphocyte activity *in vivo*. *J. Immunol. Methods* **246,** 109–117.

Roederer, M. (2003). Methods for fluorescent conjugation of monoclonal antibodies. Available at: www.drmr.com/abcon.

Roederer, M., Bowser, R., and Murphy, R. F. (1987). Kinetics and temperature dependence of exposure of endocytosed material to proteolytic enzymes and low pH: Evidence for a maturation model for the formation of lysosomes. *J. Cell Physiol.* **131,** 200–209.

Rubio, V., Stuge, T. B., Singh, N., Betts, M. R., Weber, J. S., Roederer, M., and Lee, P. P. (2003). *Ex vivo* identification, isolation and analysis of tumor-cytolytic T cells. *Nat. Med.* **9,** 1377–1382.

Sarin, A., Williams, M. S., Alexander-Miller, M. A., Berzofsky, J. A., Zacharchuk, C. M., and Henkart, P. A. (1997). Target cell lysis by CTL granule exocytosis is independent of ICE/Ced-3 family proteases. *Immunity* **6,** 209–215.

Sheehy, M. E., McDermott, A. B., Furlan, S. N., Klenerman, P., and Nixon, D. F. (2001). A novel technique for the fluorometric assessment of T lymphocyte antigen specific lysis. *J. Immunol. Methods* **249,** 99–110.

Silverstein, R. L., and Febbraio, M. (1992). Identification of lysosome-associated membrane protein-2 as an activation-dependent platelet surface glycoprotein. *Blood* **80,** 1470–1475.

Trapani, J. A., Jans, D. A., Jans, P. J., Smyth, M. J., Browne, K. A., and Sutton, V. R. (1998). Efficient nuclear targeting of granzyme B and the nuclear consequences of apoptosis induced by granzyme B and perforin are caspase-dependent, but cell death is caspase-independent. *J. Biol. Chem.* **273,** 27934–27938.

Waldrop, S. L., Pitcher, C. J., Peterson, D. M., Maino, V. C., and Picker, L. J. (1997). Determination of antigen-specific memory/effector CD4+ T cell frequencies by flow cytometry: Evidence for a novel, antigen-specific homeostatic mechanism in HIV-associated immunodeficiency. *J. Clin. Invest.* **99,** 1739–1750.

Yannelli, J. R., Sullivan, J. A., Mandell, G. L., and Engelhard, V. H. (1986). Reorientation and fusion of cytotoxic T lymphocyte granules after interaction with target cells as determined by high resolution cinemicrography. *J. Immunol.* **136,** 377–382.

CHAPTER 21

T-Cell Responses to Cancer

Peter P. Lee

Department of Medicine
Division of Hematology
Stanford University School of Medicine
Stanford, California 94305

METHODS IN CELL BIOLOGY, VOL. 75

0091-679X/04 $35.00

I. Introduction

The immunotherapy of cancer holds significant promise, but current efforts have met with limited clinical success. Robust methods to identify, enumerate, isolate, and analyze antitumor T cells elicited by different strategies are vital to better understand why certain immunotherapies may work on some patients but not others, and which immune parameters are critical for clinical responses. Furthermore, evidence is now mounting that endogenous antitumor T-cell responses develop in a significant proportion of patients with various cancers (Lee *et al.*, 1999b; Nagorsen *et al.*, 2003), yet these responses are ultimately nonprotective for the host. A clear understanding of why these endogenous antitumor T-cell responses are nonprotective and what mechanisms tumor cells use to evade immune destruction is key to designing clinically effective immunotherapy strategies.

Over the past 5–10 years, a number of powerful methods have been developed, which are greatly contributing to our understanding of T-cell responses to cancer (Keilholz *et al.*, 2002). Much of the knowledge and available methods are for $CD8^+$ T cells. Although most researchers agree that $CD4^+$ T cells play an important role in eliciting and maintaining a $CD8^+$ T-cell response (Zajac *et al.*, 1998), fewer human leukocyte antigen (HLA) class II–restricted tumor antigens or methods have been developed to identify and study $CD4^+$ T-cell responses. In this chapter, we provide an overall framework to approach T-cell responses to cancer and briefly discuss different available methods to study these responses. This chapter is not meant to be a definitive guide to any individual method, because many methods discussed here are described in greater detail in other chapters of this text. To identify T cells that may be potentially tumor reactive, we should first ask how T cells may recognize cancer.

A. How Do T Cells Recognize Cancer?

It is generally believed that the immune system responds to foreign antigens while ignoring "self." Because cancer cells are of "self" origin, how immune cells recognize a cancer cell from a normal cell remains unclear. In most (if not all) cancer cells, mutations or translocations (e.g., Bcr-Abl in chronic myelogenous leukemia [Buzyn *et al.*, 1997]) arise, which potentially lead to novel antigens that may be recognized by T cells. Indeed, T-cell responses to such unique cancer "neoantigens" have been identified. However, most identified antitumor T cells have been directed against self, nonmutated antigens. Such antigens are commonly referred to as *tumor-associated antigens* (TAAs).

Many of the TAAs identified have been in the setting of melanoma (Kawakami and Rosenberg, 1997; Rosenberg, 2001). The most common ones include melanoma antigen recognized by T cells (MART), gp100, and tyrosinase; others include MAGE, BAGE, GAGE, NY-ESO, and many others. TAAs have also been identified for breast cancer (e.g., HER-2/neu [Sotiropoulou *et al.*, 2003], MUC [Böhm *et al.*, 1998]), leukemia (e.g., proteinase 3 [Molldrem *et al.*, 1999], WT1 [Oka *et al.*, 2000]), and colon cancer (e.g., carcinoembryonic antigen [CEA] [Fong *et al.*, 2001]). A number of proteins have also been suggested to serve as general TAAs: These include p53 (Theobald *et al.*, 1995), telomerase (Minev *et al.*, 2000), cyclin (Dahl *et al.*, 1996), and survivin (Schmitz *et al.*, 2000). Table I summarizes some of the known TAAs in several types of cancer.

How TAAs—which are self nonmutated proteins—break tolerance in the setting of cancer remains poorly understood. Possible explanations include overexpression of TAAs by tumor cells or the expression of these antigens in the context of inflammatory signals in cancer. Once elicited, whether TAA-specific T cells can also lyse non-transformed counterparts (e.g., melanocytes) is unclear. The development of vitiligo (T-cell–mediated destruction of melanocytes) in some patients undergoing immunotherapy for melanoma (Rosenberg and White, 1996) suggests that T cells capable of killing melanoma can also kill normal melanocytes. Hence, tumor immunity is a form of autoimmunity (Pardoll, 1999). Nonetheless, T cells capable of killing tumor cells do not always kill their

Table I
Endogenous T-Cell Responses to Cancer

Cancer	Tumor-associated antigens	Detected in % patients	Method(s)	Study
Melanoma	MART, gp100, tyrosinase	~50%	pMHC tetramers	Lee *et al.* (1999)
	Melanoma targets	~60%	γ ELISPOT	Letsch *et al.* (2000)
Breast cancer	MUC-1, Her-2/neu, carcinoembryonic antigen (CEA), NY-ESO-1, SSX-2	~50%	γ ELISPOT + pMHC tetramers	Rentzsch *et al.* (2003)
Leukemia	WT-1, PR1, Bcr-Abl	~65%	Q-PCR	Rezvani *et al.* (2003)
	WT-1, PR1	~30%	γ ELISPOT + CFC	Scheibenbogen *et al.* (2002)
Head and neck cancer	p53	~75%	γ ELISPOT + pMHC tetramers	Hoffmann *et al.* (2002)
Colorectal cancer	CEA, Her-2/neu, Ep-CAm	~35%	γ ELISPOT	Nagorsen *et al.* (2000)

Note: This table lists the common tumor-associated antigens for different cancers in humans, as well as studies showing endogenous T-cell responses to these antigens in patients with cancer.

normal counterparts, suggesting that there may be mechanisms of selectivity in certain circumstances. Understanding such mechanisms would be important to develop immunotherapeutic strategies that minimize damage to healthy tissues.

B. Immune Dysfunction in Cancer

Cancer affects the immune response of the host in a number of ways. Dysfunction or anergy of tumor-infiltrating lymphocytes (TILs) and circulating tumor-specific T cells may be early events in tumor progression (Lee *et al.*, 1999b; Staveley-O'Carroll *et al.*, 1998), whereas global immune dysfunction develops in most patients with advanced tumor burden (Ochoa and Longo, 1995). Defects include alterations in TCR signaling events, reduced TCR zeta-chain levels, reduced proliferation, and increased apoptosis. In patients with melanoma, breast, oral, and head and neck cancer, spontaneous apoptosis occurs in a high percentage of peripheral $CD8^+$ T cells and natural killer (NK) cells (Bauernhofer *et al.*, 2003; Kuss *et al.*, 2003). What drives the dysfunction or apoptosis of lymphocytes in the cancer setting and what are the underlying molecular defects within T cells in cancer remains unclear. Evidence also exists that cancer may induce a skewing of $CD4^+$ T-cell responses to a Th2 rather than Th1 response (Lee *et al.*, 1997). Because antitumor T cells may be dysfunctional *in vivo,* methods to analyze these responses should ideally be capable of detecting these cells regardless of their functional state.

C. Nomenclature

It is worthwhile to be precise in how best to refer to T cells against cancer. The term *tumor reactive* implies T cells capable of a functional response to tumor. As discussed earlier, antitumor T cells may be dysfunctional in patients with cancer and, therefore, strictly cannot be considered tumor "reactive" until their functional status *in vivo* has been determined. Referring to T cells as *tumor specific* may also be problematic. Because recognition of tumors by T cells is predominantly based on recognition of TAAs, which are also expressed by non–cancer cells, few T cells are truly tumor specific. Exceptions are T cells specific for antigens uniquely expressed by tumor cells, cancer neoantigens, such as mutations or translocation products. Hence, it may be most accurate to refer to T cells as either *antitumor* or *TAA specific,* neither of which suggests a functional status or specificity only to tumor cells but not their normal counterparts.

II. Methods to Identify and Enumerate TAA-Specific T Cells

Methods to detect antitumor T cells mainly rely on peptide specificity for TAAs. These include the enzyme-linked immunospot (ELISPOT) assay, cytokine flow cytometry (CFC), quantitative (real-time) polymerase chain reaction (Q-PCR), and staining with peptide–MHC (pMHC) tetramers (Keilholz *et al.*, 2002). Based on these methods, TAA-specific T cells have indeed been successfully identified and studied, which develop endogenously (Lee *et al.*, 1999b) and after vaccination (Lee *et al.*, 2001). The first three methods (ELISPOT, CFC, and Q-PCR) rely on a functional readout in response to peptide (TAA) or protein stimulation. As such, by definition cells scoring positive with these methods are functional. In contrast, pMHC tetramer staining detects cells based on expression of TCRs with sufficient affinity for a particular pMHC regardless of the cell's functional status. Thus, pMHC tetramer staining detects T cells based on antigen specificity regardless of functional status and, thus, is the only method that can detect anergic T cells. This may be viewed as either a positive or a negative. In the setting of immune monitoring for a cancer vaccine trial, one may be interested only in measuring functional antigen-specific T cells elicited. However, it has been through pMHC tetramer analysis that certain TAA-specific T responses were shown to be anergic *in vivo* in some patients with cancer (Dunn *et al.*, 2002; Lee *et al.*, 1999b). Proliferation and limiting dilution assays are older methods that are being superseded by these newer methods. Two new methods to identify antigen-specific T cells were reported in 2003: Lysispot (Snyder *et al.*, 2003) and an assay based on acquisition of target pMHC complexes by an effector in a productive recognition event (Tomaru *et al.*, 2003). Although these methods hold promise, there is little practical experience with them, so we do not discuss them in detail in this chapter.

A. ELISPOT

ELISPOT assay is the gold standard technique in measuring peptide-specific T-cell responses in immunotherapy trials (Whiteside *et al.*, 2003). This method relies on detection of cytokine secretion by antigen-specific T cells upon stimulation with cognate peptide. The most commonly assayed cytokine is interferon-γ (IFN-γ), although release of other cytokines such as interleukin-5 (IL-5) or cytolytic mediators such as granzyme B may be used. This assay is generally done in 96-well plates with a nitrocellulose membrane bottom, although plates with plastic or nylon bottoms can also be used. Wells are coated with capture antibodies specific for the cytokine of interest. Patient PBMCs are added, generally about 10^5 cells/well, along with the cognate peptide at a high concentration, generally 1 μg/ml final. Because monocytes and B cells are present in PBMCs, these can serve as antigen-presenting cells (APCs) for the added peptides. However, if purified $CD8^+$ or $CD4^+$ T cells are used, then peptides should be pulsed onto APCs (such as T2 cells) and added at a 1:1 or greater ratio to achieve optimal stimulation of T cells. Tumor cells, autologous or partially HLA matched, may also be used as stimuli. However, there is evidence that tumor cells may secrete immunosuppressive factors, which may reduce the data from this assay (Pass *et al.*, 1998). The cells are allowed to incubate undisturbed on the plates for 6–48 hours depending on the antigen being assayed. During this time, cytokines released in response to cognate peptide stimulation are captured by the coated antibodies. The cells are then washed away, bound cytokines are detected via detection antibodies either directly conjugated to an enzyme for detection, or a secondary antibody-enzyme conjugate is added for detection. The bound enzymes catalyze the conversion of a chromogenic reagent, which is deposited onto the nitrocellulose membrane, thus appearing as spots as a semiquantitative measure of each responding T cell. The ability to visualize spots significantly increases the sensitivity of the ELISPOT assay over standard enzyme-linked immunosorbent assay (ELISA), which simply measures the total amount of cytokines released by all cells within a well. A number of computer-assisted image analysis systems allow automated and accurate evaluation of these plates. Data are reported as numbers of spots per well or per number of PBMCs. Background, as defined by responses in unstimulated T cells or T cells stimulated with irrelevant peptides, can be quite variable depending on the cytokine being detected (IFN-γ generally lowest) and operator technique. Published ELISPOT data are generally reported with an average background subtracted. The limit of detection of this assay is generally 1 in 100,000–300,000 PBMCs or approximately 1 in 10,000–30,000 $CD8^+$ T cells. Whole proteins (rather than peptides) may be added as stimuli, which would be taken up into the MHC class II pathway by APCs for presentation and detection of antigen-specific $CD4^+$ T cells. Disadvantages of this method are that it is time and labor intensive, detection cannot be coupled with further analysis, and identified cells cannot be isolated.

B. Cytokine Flow Cytometry

Like ELISPOT, CFC is another method of detecting cytokine production in response to cognate stimulation. In this method, cytokine production is detected intracellularly via flow cytometry and can be applied to detect $CD4^+$ or $CD8^+$ T cells. As with other intracellular cytokine (ICC) staining methods, cells are stimulated with cognate peptides, peptide mixtures, whole proteins, or cell lysates. Brefeldin A (BFA) or monensin is added to block secretion via the endoplasmic reticulum (ER) pathway, such that cytokines produced remain within the cytoplasm. At the end of the stimulation period (generally 6 hours), cells are fixed and permeabilized, and then antibodies specific for the cytokine of interest are added for intracellular staining and detection by flow cytometry. Although IFN-γ is generally used, because of its sensitivity, CFC allows detection of a wider range of cytokines than ELISPOT, including IL-2, IL-4, and tumor necrosis factor-α (TNF-α). To further distinguish T cells that newly respond to the stimulus with cytokine production versus those that produce the cytokine constitutively, one may add an acute activation marker such as CD69 to gate on double-positive cells. Unlike ELISPOT, CFC may be coupled with analysis of other markers by flow cytometry to gain additional phenotypic information on the population of interest. However, identified cells are killed during the fixation/permeabilization process and cannot be isolated for expansion or further analysis.

C. Q-PCR

A newer approach to detect cytokine responses to cognate stimulation uses real-time PCR (i.e., Q-PCR) technology (Panelli *et al.*, 2002a; Rezvani *et al.*, 2003). Rather than detecting the cytokine as a protein via ELISPOT or CFC, messenger RNA (mRNA) encoding for the cytokine of interest may be detected by Q-PCR. This has the advantage of potentially being more sensitive and rapid (requires as little as 2 hours of cognate stimulation) because Q-PCR can detect at a lower level of response than ELISPOT or CFC. However, background issues are more prominent, and the data generated do not directly give information about the numbers of cells responding. This method could be useful for detecting very subtle responses that would otherwise be missed by other assays. Because cells are lysed at the beginning of this assay, cells of interest cannot be further analyzed or isolated.

D. Which Cytokine Is Most Predictive?

An important issue with all three of the aforementioned methods is the choice of cytokine assayed. Different T cells—Th1, Th2, Tc1, Tc2, and so on—respond by producing different cytokines—IFN-γ, TNF-α, IL-2, IL-4, IL-5, IL-10, and so on. There is no single cytokine that is produced by all T cells in response to cognate stimulation. Although IFN-γ is a commonly produced cytokine by many $CD8^+$ and Th1 $CD4^+$ T cells, it nonetheless is not produced by all antigen-specific T cells. Thus, measuring T-cell responses via any single cytokine would underestimate the total response. Furthermore, the correlation between IFN-γ production and cytolytic potential is fairly good but not absolute (Betts *et al.*, 2003; Rubio *et al.*, 2003). The production of IFN-γ (or any other cytokine) is not a direct measure of cytolytic potential. Indeed, IFN-γ secretion and cytotoxicity have been shown to be regulated independently in $CD8^+$ T cells (Bachmann *et al.*, 1999; Betts *et al.*, 2003; Lim *et al.*, 2000; Snyder *et al.*, 2003).

E. MHC Tetramer Staining

The pMHC tetramers were invented and first reported by Altman *et al.* in 1996. It has revolutionized the study of T-cell responses by allowing for accurate and rapid enumeration of antigen-specific T cells, the phenotypic characterization of such cells, and their viable isolation for further analyses. Tetramers have been successfully used to gain new insights into T-cell responses to viruses and cancer (for a complete discussion of the use of tetramers, see chapter 17). Tetramers bind T cells based solely on their expression of TCRs with sufficient affinity for a specific pMHC complex. Hence, tetramers detect antigen-specific T cells regardless of functional status. Although this could be viewed as a potential disadvantage, this could also be a significant advantage when coupled with a functional assay to provide valuable information on the proportion of cells that are functional or nonfunctional within a given response. Important factors in the use of tetramers include reagent validation and titration, staining temperature and time, tetramer concentration, and use of anti-CD8 and other antibodies. Important advantages of tetramers include the ability to combine with antibodies for detailed phenotypic analysis in multicolor flow cytometry and to isolate tetramer-positive cells for further analysis or *ex vivo* expansion.

III. Important Considerations

A. Which Method to Use When, and Which Is Right?

These assays produced discordant data when applied to the same samples (Whiteside *et al.*, 2003), calling into question which assay, if any, provides an accurate reflection of the biologically relevant T-cell response. This very much relates to the underlying biology of these cells and the method by which they are detected. As expected, tetramers generally produce the highest frequencies because they bind all antigen-specific T cells regardless of function. CFC generally produces the lowest frequencies because it detects only those cells that respond by producing a specific cytokine. ELISPOT generally produces intermediate frequencies because it also requires antigen-specific T cells to produce a specific cytokine, but it has a higher sensitivity than CFC. However, ELISPOT requires long incubation periods, during which some antigen-specific T cells could undergo apoptosis. Thus, the data from these assays are subjected to additional factors that are difficult to predict. Table II provides a comparison of these methods.

B. *Ex Vivo* Restimulation and Expansion

Ex vivo restimulation and *in vitro* expansion of T cells had been routinely used to detect small T-cell responses in the past. Although this is useful in providing a yes/no result, these data cannot be regarded as truly quantitative even though some publications claimed them to be as such via back-calculations. Moreover, certain cells may be selectively expanded *in vitro* and alter their biology, so it would be impossible to assess the *in vivo* biology of such cells. Current assays have sufficiently high sensitivity to allow analysis of antigen-specific T cells directly *ex vivo* without restimulation and *in vitro* expansion in most cases. Nonetheless, early or very small responses constituting fewer than 1 in 10,000 T cells may still fall below the limit of detection for most methods.

C. Optimal Time Points

The optimal time points and sites (see later discussion) to sample for antitumor T cells after immunotherapy are also important factors. Careful kinetic studies have been rarely performed in cancer immunotherapy trials. Rather, patients are

Table II
Comparison of Major Methods for Detecting Antitumor T Cells

Method	T cells detectable	Stimulus	Readout	Detection limit	Time required
Enzyme-linked immunospot (ELISPOT) assay	CD8 or CD4	Peptides, proteins, cells	Functional: release of cytokines, granzyme	1/10,000–1/30,000 T cells	2–3 days
Cytokine flow cytometry (CFC)	CD8 or CD4	Peptides, proteins, cells	Functional: intracellular cytokines	1/1000–1/10,000 T cells	6–8 hr
Real-time polymerase chain reaction (Q-PCR)	CD8 or CD4	Peptides, proteins, cells	Functional: messenger RNA for cytokines	Unclear at this time	1–2 days
Tetramers	Mainly CD8 in humans; CD4 less reliable	None needed	T-cell receptor binding by flow cytometry	1/10,000 T cells	2–4 hr
CD107 mobilization	CD8 only	Peptides, cells	Functional: degranulation	1/1000–1/10,000 T cells	6–8 hr

arbitrarily studied at time points before and several weeks after immunotherapy. Kinetic studies examining the dynamics of T-cell responses to tetanus toxoid showed that the $CD4^+$ T-cell response may peak around 7–10 days after injection, whereas the $CD8^+$ T-cell response may peak as much as 3 weeks later (Mosca *et al.*, 2001). Response kinetics to cancer vaccines, either peptides or proteins, may be quite different. Hence, careful kinetic analysis should be done in several patients at the beginning of each clinical trial to determine the optimal time points for analysis. Furthermore, it is important to study the T-cell responses at multiple times to better understand the dynamics of these responses over time.

D. Blood Versus Tumor Versus Lymph Node T Cells

For antitumor T cells to be effective against solid tumors, they need to leave the blood and home to tumor sites. Thus, examining only the blood for antitumor T cells provides an incomplete view of the overall immune response to cancer in a patient. It is important to also assess the tumor sites and draining lymph nodes, which are obviously much more difficult to access. Data suggest that the frequencies of antitumor T cells may be much higher at these sites (Romero *et al.*, 1998). More importantly, the biology of antitumor T cells within the tumor and draining lymph nodes may be very different from that in blood.

E. Correlation with Clinical Outcome

The most important question to ask is whether the immunological responses detected by any of these methods correlate with clinical outcome. Unfortunately, an unequivocal answer to this question is not available and is complicated by the numerous immunotherapy approaches being undertaken and different immune-monitoring methods used in each trial. Although several studies showed some correlation between T-cell responses detected by any of the aforementioned methods, particularly ELISPOT, with clinical outcome (Reynolds *et al.*, 1997; Wang *et al.*, 1999), many other studies showed a lack of correlation (Lee *et al.*, 1999a, 2001). It is important to recall that TAA-specific T cells may be rendered anergic *in vivo*. Furthermore, such cells need to home to tumor sites to mediate a clinically protective effect. Hence, it is important to combine methods of identification and enumeration of TAA-specific T cells with methods to address their *in vivo* biology and homing patterns. A novel important finding is that not all TAA-specific T cells, even those that are functional, can recognize or kill tumor (Rubio *et al.*, 2003). Peptide specificity represents a range of T cells of differing abilities to lyse targets based on peptide requirements.

IV. Methods to Further Analyze *In Vivo* Biology of Antitumor T Cells

In addition to the identification and enumeration of antigen-specific T cells within a heterogeneous cell sample, tetramers may also be used to isolate such cells by sorting, thereby facilitating T-cell cloning, TCR spectratyping, and microarray analyses (Dietrich *et al.*, 2001; Dunbar *et al.*, 1998; Lee *et al.*, 1999b). Tetramers may also be used to discern the functional status of antigen-specific T cells using multiparameter FACS to assess surface (activation, memory, cytotoxic) markers and intracellular or secreted cytokines/chemokines (He *et al.*, 1999; Lee *et al.*, 1999b; Pittet *et al.*, 2001).

A. Isolation by Sorting and Cloning

The viable isolation of antigen-specific T cells for cloning or further analyses is a powerful use of pMHC tetramers. This may be done routinely using a standard FACSort instrument. For example, a tetramer-positive T-cell population may be isolated and directly tested *ex vivo* for cytolytic activity against target cells in a standard chromium release assay. It has been well documented that the tetramer staining and sorting process does not impair the cytolytic activity of T cells. Care should be taken to keep the cells cold and in a serum-rich media to maximize their function. Furthermore, tetramer-positive cells could be sorted singly into wells of tissue culture plates and grown in the presence of feeder cells and cytokines to create T-cell clones. Such cells could then be analyzed for their TCR specificities and structural avidity (see later discussion) and expanded for adoptive cellular immunotherapy (Yee *et al.*, 2002). Sorted T cells have also been analyzed for gene expression profiles by DNA microarrays (Xu *et al.*, 2004).

B. Phenotypic Analysis by Multicolor Flow Cytometry

The advent of multicolor flow cytometry allows the extensive phenotypic characterization of antigen-specific T cells on an individual cell basis. Even with three- or four-color FACS, the expression of several markers by an antigen-specific T-cell population of interest may be determined by performing several stains. With the increasing availability of 6- and even 12-color FACS, large numbers of markers could be determined from only a few stains (Lee *et al.*, 1999b). This is important especially for rare populations and precious patient samples. In general, one channel is taken up by tetramers (usually PE) and another channel for anti-CD8. If cells are limited, two channels could be devoted to two different tetramers (such as PE and APC) so that two separate antigen-specific T-cell populations could be simultaneously assessed per stain. Not all antibodies work well in combination, so considerable effort must be made to optimize phenotypic panels that work well together. For example, because anti-CD3 antibodies may compete with tetramers for binding to the TCR–CD3 complex (Hoffmann *et al.*, 2002), it is advisable to not include this antibody in panels except for validation purposes. Panels could also be optimized for intracellular antigens (see section IV.C).

C. Intracellular Staining

Tetramers can be combined successfully with intracellular staining to analyze expression of cytokines (e.g., IFN-γ, IL-2, IL-4, and TNF-α) or cytolytic granules (perforin, granzyme A, granzyme B) by antigen-specific T cells. After routine surface staining with tetramers and other markers, cells are fixed and permeabilized, then stained with antibodies against intracellular antigens (Lee *et al.*, 1999b). Because this process involves numerous washing steps, it is important to start with large numbers of cells to ensure sufficient cells remaining at the end to detect rare populations. It is also important to note that T cells generally downregulate the TCR complex upon activation. Because many intracellular cytokine-staining protocols involve nonspecific cellular activation with mitogens, the level of tetramer staining will decrease substantially. To circumvent this issue, one could perform tetramer staining before activation so the tetramers are internalized (along with the appropriate fluorophore) during the activation process with its intended TCR complex internalization.

D. Analysis of Gene Expression Profile of Antitumor T Cells Using Microarray

To derive maximum biological information about an antigen-specific T-cell population, one may combine tetramer staining with DNA microarray technology. DNA microarrays allow the simultaneous assessment of 10,000 genes or more by a given cell population. This could provide tremendous information about different antitumor T-cell populations under different conditions, such as within the lymph node versus within the tumor microenvironment. A major technical hurdle is the amount of material needed for microarray analysis. Most microarray technologies require total or mRNA from approximately 10 million cells for each hybridization, which is not feasible for purified T-cell populations. However, robust linear amplification methods make it possible to perform microarray analysis on as few as 10,000 sorted cells after one round of amplification (Wang *et al.*, 2000). Using these methodologies, a number of authors have detected gene expression differences in T cells from patients with cancer under different clinical conditions (Panelli *et al.*, 2002b; Wang *et al.*, 2002).

V. Novel Issues and Methods

A. Not All TAA-Specific T Cells Can Recognize Tumor: Importance of "Recognition Efficiency"

The ability of different peptide-specific T cells to respond to different amounts of peptides is a reflection of their "functional avidity" or "recognition efficiency" (RE). RE of a T cell reflects the cumulative effects of the intrinsic affinity of its TCRs for their cognate peptide displayed on the target cell, expression levels of TCR, CD4/8, and adhesion molecules, as well as redistribution of these molecules on the cell membrane and their recruitment efficacy of the signaling cascade (Cawthon and Alexander-Miller, 2002; Margulies, 2001; Slifka and Whitton, 2001). Identification of peptide-specific T cells using pMHC tetramers, ELISPOT, or CFC does not distinguish between cells of high- or low-recognition efficiency—at least not without careful reagent titration on each sample (Fig. 1). Although previous investigations have attempted to use the intensity of tetramer staining as a measure for binding strength (Crawford *et al.*, 1998; Yee *et al.*, 1999), evidence suggests that tetramer staining intensity alone does not directly correlate with recognition efficiency (Derby *et al.*, 2001; Dutoit *et al.*, 2002; Echchakir *et al.*, 2002). Nonetheless, the rate of dissociation of bound pMHC tetramers from antigen-specific T cells via the addition of a competing antibody (such as anti-TCR) may be used as a relative measure of the difference in TCR affinities between clones; as such, this may be a useful assay to assess the "structural avidity" of a T-cell population (Savage *et al.*, 1999).

B. A Novel Method to Identify Tumor-Reactive T Cells: CD107 Mobilization

We developed a novel method to isolate pure viable populations of tumor-cytolytic T cells directly *ex vivo* from patient blood samples using flow cytometric quantification of the surface mobilization of CD107—an integral membrane protein within cytolytic granules—as a marker for degranulation upon tumor stimulation (Rubio *et al.*, 2003). We showed that tumor-cytolytic T cells are

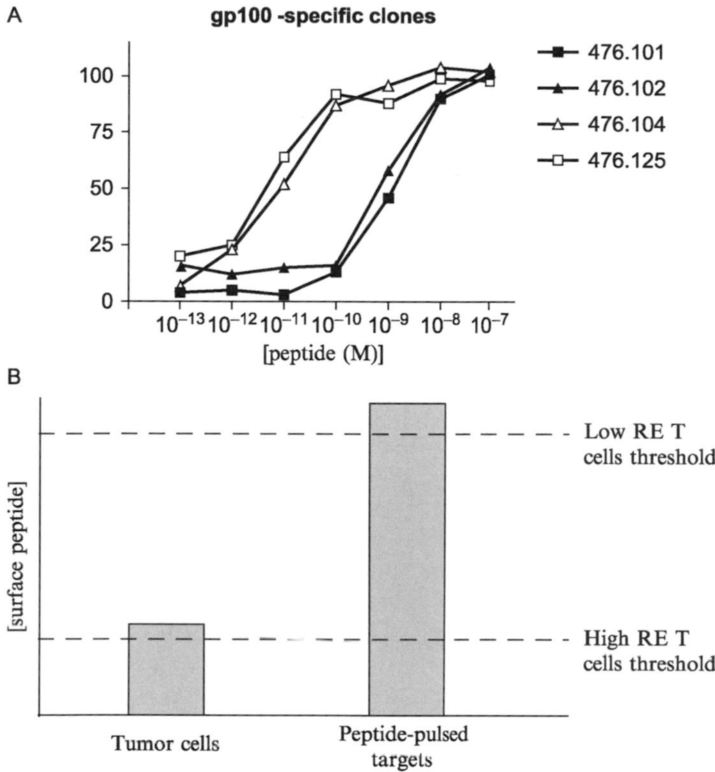

Fig. 1 Significance of recognition efficiency (RE) of T-cell responses. (A) Four gp100-specific T-cell clones (generated based on equivalent tetramer staining) were assayed against a wide titration of peptide concentrations. Two clones (476.104 and 476.125, high RE) required much lower (~100-fold) amounts of peptides for efficient lysis of T2 targets than the other two clones (476.101 and 476.102, low RE). (B) Because high-RE T cells require much lower amounts of cognate peptide on a target cell for efficient lysis, they can lyse both peptide-pulsed and tumor targets. In contrast, low-RE T cells require much higher amounts of cognate peptide on a target cell for efficient lysis and, therefore, could lyse only peptide-pulsed but not tumor targets.

indeed elicited in patients post–cancer vaccination, and that tumor reactivity is strongly correlated with recognition efficiency of the T cells for peptide-bearing targets. Combining CD107 mobilization with pMHC tetramer staining, we directly correlated antigen specificity and cytolytic ability on a single-cell level to show that high-recognition efficiency, tumor-cytolytic T cells represent only a minority of peptide-specific T cells elicited in patients after heteroclytic peptide vaccination (Fig. 2). We have also shown that even high-recognition efficiency, tumor-reactive T cells could be anergized *in vivo,* an important immune evasion mechanism for tumor cells.

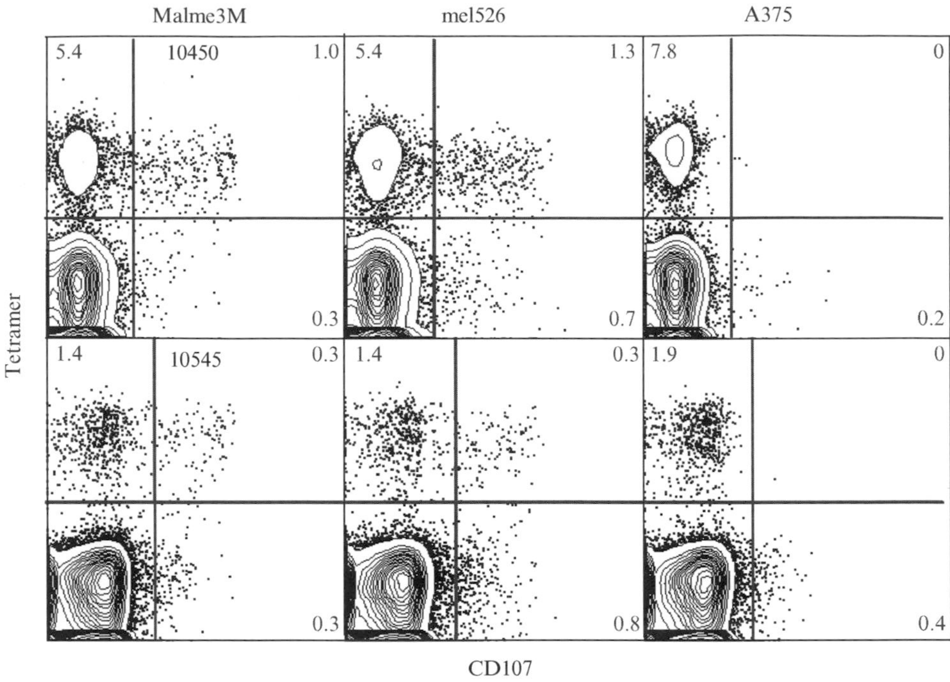

Fig. 2 Tetramer staining and CD107 mobilization. Tetramer staining may be combined with CD107 mobilization to identify peptide-specific T cells that do or do not degranulate in response to tumor stimulation. (Reprinted with permission from Rubio *et al.*, 2003.)

C. Detection of Tumor-Reactive T Cells of Unknown Target Specificity

A key advantage of the CD107 technique is the ability to detect tumor-reactive $CD8^+$ T cells without knowing the pMHC target. Because the assay measures T cells that degranulate in response to tumor cells, there is no *a priori* need to know the actual peptide target, which is required for most current assays. This is an important advantage because only a few TAAs have been identified, mostly in the setting of melanoma. This technique may also be useful for immune monitoring of clinical trials involving vaccination with whole tumor cells, tumor–APC fusions, APCs pulsed with tumor lysates or transfected with tumor RNA, or other novel immunotherapeutic strategies in which the exact peptide targets are undefined. In such instances, the same cells used for vaccination could be used as stimulators in the immune-monitoring assay to reveal tumor-reactive cytolytic T cells.

D. Peptide–MHC Microarrays

A novel advance in detection of antigen-specific T cells is the introduction of pMHC microarrays (Soen *et al.*, 2003). As with DNA microarrays, pMHC complexes are spotted onto a matrix, which then captures T cells in an antigen-specific

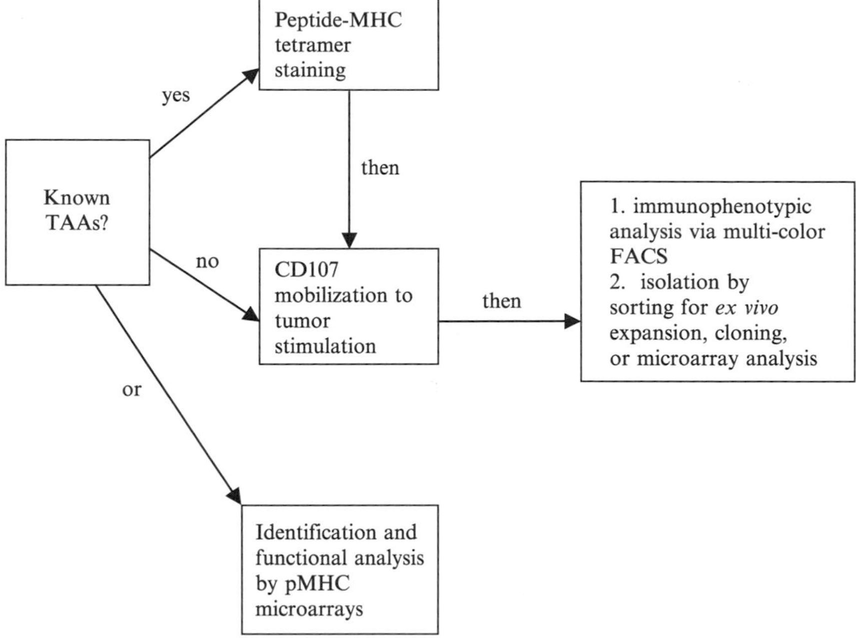

Fig. 3 Schema to characterize antitumor T-cell responses. T-cell responses to known tumor-associated antigens (TAAs) may be detected and enumerated with peptide–major histocompatibility (pMHC) tetramer staining. These cells may then be further characterized via CD107 mobilization to tumor stimulation, immunophenotypic analysis in multicolor flow cytometry, and isolation by sorting. T-cell responses to unknown TAAs may be detected and enumerated with CD107 mobilization to tumor stimulation, then further characterized as stated above. Alternatively, the new pMHC microarrays may be directly used to identify, quantitate, and characterize (limited) TAA-specific T-cell responses.

manner. Such arrays have high sensitivity (detection limit to 1/10,000 [Daniel Chen, 2004, personal communication]) and specificity. The power of this approach is that T cells for numerous pMHC combinations may be assayed simultaneously. Furthermore, antigen-specific T cells bound to individual spots could be further assessed biologically, including the ability to lyse tumor. Thus, this is a promising platform that combines the ability to identify and enumerate T cells specific for many antigens concurrently and to further assess the biology of such cells directly on the array.

VI. Conclusion

Over the past 5–10 years, many powerful new methods have been developed to identify and study antitumor T cells in much greater detail than previously possible. These methods have allowed us to uncover important biology about the complex interactions between cancer cells and the host immune system. Important lessons include the significance of recognition efficiency for T-cell responses, and that tumor cells may actively modulate the immune response *in vivo*. The ability of cancer cells to render responding T cells dysfunctional or anergic *in vivo* warrants the use of methods that can detect antitumor T cells regardless of functional status. Although no one method can completely address these biological issues, a number of these methods could be combined to provide a detailed picture of the T-cell responses to cancer (Fig. 3). Looking beyond T cells, one needs to also consider the important roles of dendritic cells, natural killer cells, and B cells to bring together an orchestrated immune response against cancer. New methods will need to incorporate these key cell populations and provide a dynamic view of these responses over time. Only then will we have a sufficient understanding of a successful immune response against cancer so we can develop the next generations of effective cancer immunotherapies.

References

Altman, J. D., Moss, P. A. H., Goulder, P. J. R., Barouch, D. H., McHeyzer-Williams, M. G., Bell, J. I., McMichael, A. J., and Davis, M. M. (1996). Phenotypic analysis of antigen-specific T lymphocytes. *Science* **274,** 94–96.

Bachmann, M. F., Barner, M., Viola, A., and Kopf, M. (1999). Distinct kinetics of cytokine production and cytolysis in effector and memory T cells after viral infection. *Eur. J. Immunol.* **29,** 291–299.

Bauernhofer, T., Kuss, I., Henderson, B., Baum, A. S., and Whiteside, T. L. (2003). Preferential apoptosis of CD56dim natural killer cell subset in patients with cancer. *Eur. J. Immunol.* **33,** 119–124.

Betts, M. R., Brenchley, J. M., Price, D. A., De Rosa, S. C., Douek, D. C., Roederer, M., and Koup, R. A. (2003). Sensitive and viable identification of antigen-specific CD8+ T cells by a flow cytometric assay for degranulation. *J. Immunol. Methods* **281,** 65–78.

Böhm, C. M., Hanski, M. L., Stefanovi*c, S., Rammensee, H. G., Stein, H., Taylor-Papadimitriou, J., Riecken, E. O., and Hanski, C. (1998). Identification of HLA-A2–restricted epitopes of the tumor-associated antigen MUC2 recognized by human cytotoxic T cells. *Int. J. Cancer* **75,** 688–693.

Buzyn, A., Ostankovitch, M., Zerbib, A., Kemula, M., Connan, F., Varet, B., Guillet, J. G., and Choppin, J. (1997). Peptides derived from the whole sequence of BCR-ABL bind to several class I molecules allowing specific induction of human cytotoxic T lymphocytes. *Eur. J. Immunol.* **27,** 2066–2072.

Cawthon, A. G., and Alexander-Miller, M. A. (2002). Optimal colocalization of TCR and CD8 as a novel mechanism for the control of functional avidity. *J. Immunol.* **169,** 3492–3498.

Crawford, F., Kozono, H., White, J., Marrack, P., and Kappler, J. (1998). Detection of antigen-specific T cells with multivalent soluble class II MHC covalent peptide complexes. *Immunity* **8,** 675–682.

Dahl, A. M., Beverley, P. C., and Stauss, H. J. (1996). A synthetic peptide derived from the tumor-associated protein mdm2 can stimulate autoreactive, high avidity cytotoxic T lymphocytes that recognize naturally processed protein. *J. Immunol.* **157,** 239–246.

Derby, M. A., Wang, J., Margulies, D. H., and Berzofsky, J. A. (2001). Two intermediate-avidity cytotoxic T lymphocyte clones with a disparity between functional avidity and MHC tetramer staining. *Int. Immunol.* **13,** 817–824.

Dietrich, P. Y., Walker, P. R., Quiquerez, A. L., Perrin, G., Dutoit, V., Lienard, D., Guillaume, P., Cerottini, J. C., Romero, P., and Valmori, D. (2001). Melanoma patients respond to a cytotoxic

T lymphocyte–defined self-peptide with diverse and nonoverlapping T-cell receptor repertoires. *Cancer Res.* **61,** 2047–2054.

Dunbar, P. R., Ogg, G. S., Chen, J., Rust, N., van der Bruggen, P., and Cerundolo, V. (1998). Direct isolation, phenotyping and cloning of low-frequency antigen-specific cytotoxic T lymphocytes from peripheral blood. *Curr. Biol.* **8,** 413–416.

Dunn, G. P., Bruce, A. T., Ikeda, H., Old, L. J., and Schreiber, R. D. (2002). Cancer immunoediting: from immunosurveillance to tumor escape. *Nat. Immunol.* **3,** 991–998.

Dutoit, V., Guillaume, P., Cerottini, J. C., Romero, P., and Valmori, D. (2002). Dissecting TCR-MHC/peptide complex interactions with A2/peptide multimers incorporating tumor antigen peptide variants: Crucial role of interaction kinetics on functional outcomes. *Eur. J. Immunol.* **32,** 3285–3293.

Echchakir, H., Dorothee, G., Vergnon, I., Menez, J., Chouaib, S., and Mami-Chouaib, F. (2002). Cytotoxic T lymphocytes directed against a tumor-specific mutated antigen display similar HLA tetramer binding but distinct functional avidity and tissue distribution. *Proc. Natl. Acad. Sci. USA* **99,** 9358–9363.

Fong, L., Hou, Y., Rivas, A., Benike, C., Yuen, A., Fisher, G. A., Davis, M. M., and Engleman, E. G. (2001). Altered peptide ligand vaccination with Flt3 ligand expanded dendritic cells for tumor immunotherapy. *Proc. Natl. Acad. Sci. USA* **98,** 8809–8814.

He, X. S., Rehermann, B., Lopez-Labrador, F. X., Boisvert, J., Cheung, R., Mumm, J., Wedemeyer, H., Berenguer, M., Wright, T. L., Davis, M. M., and Greenberg, H. B. (1999). Quantitative analysis of hepatitis C virus–specific CD8+ T cells in peripheral blood and liver using peptide–MHC tetramers. *Proc. Natl. Acad. Sci. USA* **96,** 5692–5697.

Hoffmann, T. K., Donnenberg, A. D., Finkelstein, S. D., Donnenberg, V. S., Friebe-Hoffmann, U., Myers, E. N., Appella, E., DeLeo, A. B., and Whiteside, T. L. (2002). Frequencies of tetramer+ T cells specific for the wild-type sequence p53(264-272) peptide in the circulation of patients with head and neck cancer. *Cancer Res.* **62,** 3521–3529.

Kawakami, Y., and Rosenberg, S. A. (1997). Human tumor antigens recognized by T-cells. *Immunol. Res.* **16,** 313–339.

Keilholz, U., Weber, J., Finke, J. H., Gabrilovich, D. I., Kast, W. M., Disis, M. L., Kirkwood, J. M., Scheibenbogen, C., Schlom, J., Maino, V. C., Lyerly, H. K., Lee, P. P., Storkus, W., Marincola, F., Worobec, A., and Atkins, M. B. (2002). Immunologic monitoring of cancer vaccine therapy: Results of a workshop sponsored by the Society for Biological Therapy. *J. Immunother.* **25,** 97–138.

Kuss, I., Donnenberg, A. D., Gooding, W., and Whiteside, T. L. (2003). Effector CD8+CD45RO-CD27-T cells have signalling defects in patients with squamous cell carcinoma of the head and neck. *Br. J. Cancer* **88,** 223–230.

Lee, K. H., Wang, E., Nielsen, M. B., Wunderlich, J., Migueles, S., Connors, M., Steinberg, S. M., Rosenberg, S. A., and Marincola, F. M. (1999a). Increased vaccine-specific T cell frequency after peptide-based vaccination correlates with increased susceptibility to *in vitro* stimulation but does not lead to tumor regression. *J. Immunol.* **163,** 6292–6300.

Lee, P. P., Yee, C., Savage, P. A., Fong, L., Brockstedt, D., Weber, J. S., Johnson, D., Swetter, S., Thompson, J., Greenberg, P. D., Roederer, M., and Davis, M. M. (1999b). Characterization of circulating T cells specific for tumor-associated antigens in melanoma patients. *Nat. Med.* **5,** 677–685.

Lee, P., Wang, F., Kuniyoshi, J., Rubio, V., Stuges, T., Groshen, S., Gee, C., Lau, R., Jeffery, G., Margolin, K., Marty, V., and Weber, J. (2001). Effects of interleukin-12 on the immune response to a multipeptide vaccine for resected metastatic melanoma. *J. Clin. Oncol.* **19,** 3836–3847.

Lee, P. P., Zeng, D., McCaulay, A. E., Chen, Y. F., Geiler, C., Umetsu, D. T., and Chao, N. J. (1997). T helper 2–dominant antilymphoma immune response is associated with fatal outcome. *Blood* **90,** 1611–1617.

Letsch, A., Keilholz, U., Schadendorf, D., Nagorsen, D., Schmittel, A., Thiel, E., and Scheibenbogen, C. (2000). High frequencies of circulating melanoma-reactive CD8+ T-cells in patients with advanced melanoma. *Int. J. Cancer* **87,** 659–664.

Lim, D. G., Bieganowska Bourcier, K., Freeman, G. J., and Hafler, D. A. (2000). Examination of CD8+ T cell function in humans using MHC class I tetramers: Similar cytotoxicity but variable proliferation and cytokine production among different clonal CD8+ T cells specific to a single viral epitope. *J. Immunol.* **165,** 6214–6220.

Margulies, D. H. (2001). TCR avidity: It's not how strong you make it, it's how you make it strong. *Nat. Immunol.* **2,** 669–670.

Minev, B., Hipp, J., Firat, H., Schmidt, J. D., Langlade-Demoyen, P., and Zanetti, M. (2000). Cytotoxic T cell immunity against telomerase reverse transcriptase in humans. *Proc. Natl. Acad. Sci. USA* **97,** 4796–4801.

Molldrem, J. J., Lee, P. P., Wang, C., Champlin, R. E., and Davis, M. M. (1999). A PR1-human leukocyte antigen-A2 tetramer can be used to isolate low-frequency cytotoxic T lymphocytes from healthy donors that selectively lyse chronic myelogenous leukemia. *Cancer Res.* **59,** 2675–2681.

Mosca, P. J., Hobeika, A. C., Clay, T. M., Morse, M. A., and Lyerly, H. K. (2001). Direct detection of cellular immune responses to cancer vaccines. *Surgery* **129,** 248–254.

Nagorsen, D., Scheibenbogen, C., Marincola, F. M., Letsch, A., and Keilholz, U. (2003). Natural T cell immunity against cancer. *Clin. Cancer Res.* **9,** 4296–4303.

Nagorsen, D., Keilholz, U., Rivoltini, L., Schmittel, A., Letsch, A., Asemissen, A. M., Berger, G., Buhr, H. J., Thiel, E., and Scheibenbogen, C. (2000). Natural T-cell response against MHC class I epitopes of epithelial cell adhesion molecule, her-2/neu, and carcinoembryonic antigen in patients with colorectal cancer. *Cancer Res.* **60,** 4850–4854.

Ochoa, A. C., and Longo, D. L. (1995). Alteration of signal transduction in T cells from cancer patients. *Imp. Adv. Oncol.* **55,** 43–54.

Oka, Y., Elisseeva, O. A., Tsuboi, A., Ogawa, H., Tamaki, H., Li, H., Oji, Y., Kim, E. H., Soma, T., Asada, M., Ueda, K., Maruya, E., Saji, H., Kishimoto, T., Udaka, K., and Sugiyama, H. (2000). Human cytotoxic T-lymphocyte responses specific for peptides of the wild-type Wilms' tumor gene (WT1) product. *Immunogenetics* **51,** 99–107.

Panelli, M. C., Wang, E., Monsurro, V., and Marincola, F. M. (2002a). The role of quantitative PCR for the immune monitoring of cancer patients. *Expert Opin. Biol. Ther.* **2,** 557–564.

Panelli, M. C., Wang, E., Phan, G., Puhlmann, M., Miller, L., Ohnmacht, G. A., Klein, H. G., and Marincola, F. M. (2002b). Gene-expression profiling of the response of peripheral blood mononuclear cells and melanoma metastases to systemic IL-2 administration. *Genome Biol.* **3,** RESEARCH0035.

Pardoll, D. M. (1999). Inducing autoimmune disease to treat cancer. *Proc. Natl. Acad. Sci. USA* **96,** 5340–5342.

Pass, H. A., Schwarz, S. L., Wunderlich, J. R., and Rosenberg, S. A. (1998). Immunization of patients with melanoma peptide vaccines: Immunologic assessment using the ELISPOT assay. *Cancer J. Sci. Am.* **4,** 316–323.

Pittet, M. J., Speiser, D. E., Valmori, D., Rimoldi, D., Lienard, D., Lejeune, F., Cerottini, J. C., and Romero, P. (2001). *Ex Vivo* analysis of tumor antigen specific CD8+ T cell responses using MHC/peptide tetramers in cancer patients. *Int. Immunopharmacol.* **1,** 1235–1247.

Rentzsch, C., Kayser, S., Stumm, S., Watermann, I., Walter, S., Stevanovic, S., Wallwiener, D., and Guckel, B. (2003). Evaluation of pre-existent immunity in patients with primary breast cancer: Molecular and cellular assays to quantify antigen-specific T-lymphocytes in peripheral blood mononuclear cells. *Clin. Cancer Res.* **9,** 4376–4386.

Reynolds, S. R., Oratz, R., Shapiro, R. L., Hao, P., Yun, Z., Fotino, M., Vukmanovic, S., and Bystryn, J. C. (1997). Stimulation of CD8+ T cell responses to MAGE-3 and Melan A/MART-1 by immunization to a polyvalent melanoma vaccine. *Int. J. Cancer* **72,** 972–976.

Rezvani, K., Grube, M., Brenchley, J. M., Sconocchia, G., Fujiwara, H., Price, D. A., Gostick, E., Yamada, K., Melenhorst, J., Childs, R., Hensel, N., Douek, D. C., and Barrett, A. J. (2003). Functional leukemia-associated antigen-specific memory CD8+ T cells exist in healthy individuals and in patients with chronic myelogenous leukemia before and after stem cell transplantation. *Blood* **102,** 2892–2900.

Romero, P., Dunbar, P. R., Valmori, D., Pittet, M., Ogg, G. S., Rimoldi, D., Chen, J. L., Liénard, D., Cerottini, J. C., and Cerundolo, V. (1998). *Ex vivo* staining of metastatic lymph nodes by class I major histocompatibility complex tetramers reveals high numbers of antigen-experienced tumor-specific cytolytic T lymphocytes. *J. Exp. Med.* **188,** 1641–1650.

Rosenberg, S. A. (2001). Progress in human tumour immunology and immunotherapy. *Nature* **411,** 380–384.

Rosenberg, S. A., and White, D. E. (1996). Vitiligo in patients with melanoma: Normal tissue antigens can be targets for cancer immunotherapy. *J. Immunother. Emphasis Tumor Immunol.* **19,** 81–84.

Rubio, V., Stuge, T. B., Singh, N., Betts, M. R., Weber, J. S., Roederer, M., and Lee, P. P. (2003). *Ex Vivo* identification, isolation and analysis of tumor-cytolytic T cells. *Nat. Med.* **9,** 1377–1382.

Savage, P. A., Boniface, J. J., and Davis, M. M. (1999). A kinetic basis for T cell receptor repertoire selection during an immune response. *Immunity* **10,** 485–492.

Scheibenbogen, C., Letsch, A., Thiel, E., Schmittel, A., Mailaender, V., Baerwolf, S., Nagorsen, D., and Keilholz, U. (2002). CD8 T-cell responses to Wilms tumor gene product WT1 and proteinase 3 in patients with acute myeloid Leukemia. *Blood* **100,** 2132–2137.

Schmitz, M., Diestelkoetter, P., Weigle, B., Schmachtenberg, F., Stevanovic, S., Ockert, D., Rammensee, H. G., and Rieber, E. P. (2000). Generation of survivin-specific CD8+ T effector cells by dendritic cells pulsed with protein or selected peptides. *Cancer Res.* **60,** 4845–4849.

Slifka, M. K., and Whitton, J. L. (2001). Functional avidity maturation of CD8(+) T cells without selection of higher affinity TCR. *Nat. Immunol.* **2,** 711–717.

Snyder, J. E., Bowers, W. J., Livingstone, A. M., Lee, F. E., Federoff, H. J., and Mosmann, T. R. (2003). Measuring the frequency of mouse and human cytotoxic T cells by the Lysispot assay: Independent regulation of cytokine secretion and short-term killing. *Nat. Med.* **9,** 231–236.

Soen, Y., Chen, D. S., Kraft, D. L., Davis, M. M., and Brown, P. O. (2003). Detection and characterization of cellular immune responses using peptide–MHC microarrays. *PLoS Biol.* **1,** E65.

Sotiropoulou, P. A., Perez, S. A., Voelter, V., Echner, H., Missitzis, I., Tsavaris, N. B., Papamichail, M., and Baxevanis, C. N. (2003). Natural CD8+ T-cell responses against MHC class I epitopes of the HER-2/ neu oncoprotein in patients with epithelial tumors. *Cancer Immunol. Immunother.* **52,** 771–779.

Staveley-O'Carroll, K., Sotomayor, E., Montgomery, J., Borrello, I., Hwang, L., Fein, S., Pardoll, D., and Levitsky, H. (1998). Induction of antigen-specific T cell anergy: An early event in the course of tumor progression. *Proc. Natl. Acad. Sci. USA* **95,** 1178–1183.

Theobald, M., Biggs, J., Dittmer, D., Levine, A. J., and Sherman, L. A. (1995). Targeting p53 as a general tumor antigen. *Proc. Natl. Acad. Sci. USA* **92,** 11993–11997.

Tomaru, U., Yamano, Y., Nagai, M., Maric, D., Kaumaya, P. T., Biddison, W., and Jacobson, S. (2003). Detection of virus-specific T cells and CD8(+) T-cell epitopes by acquisition of peptide–HLA–GFP complexes: Analysis of T-cell phenotype and function in chronic viral infections. *Nat. Med.* **9,** 469–476.

Wang, E., Miller, L. D., Ohnmacht, G. A., Liu, E. T., and Marincola, F. M. (2000). High-fidelity mRNA amplification for gene profiling. *Nat. Biotechnol.* **18,** 457–459.

Wang, E., Miller, L. D., Ohnmacht, G. A., Mocellin, S., Perez-Diez, A., Petersen, D., Zhao, Y., Simon, R., Powell, J. I., Asaki, E., Alexander, H. R., Duray, P. H., Herlyn, M., Restifo, N. P., Liu, E. T., Rosenberg, S. A., and Marincola, F. M. (2002). Prospective molecular profiling of melanoma metastases suggests classifiers of immune responsiveness. *Cancer Res.* **62,** 3581–3586.

Wang, F., Bade, E., Kuniyoshi, C., Spears, L., Jeffery, G., Marty, V., Groshen, S., and Weber, J. (1999). Phase I trial of a MART-1 peptide vaccine with incomplete Freund's adjuvant for resected high-risk melanoma. *Clin. Cancer Res.* **5,** 2756–2765.

Whiteside, T. L., Zhao, Y., Tsukishiro, T., Elder, E. M., Gooding, W., and Baar, J. (2003). Enzyme-linked immunospot, cytokine flow cytometry, and tetramers in the detection of T-cell responses to a dendritic cell–based multipeptide vaccine in patients with melanoma. *Clin. Cancer Res.* **9,** 641–649.

Xu, T., Shen, C., Dang, D., Ilsley, D., Holmes, S., and Lee, P. P. (2004). Subtle but consistent gene expression differences between T cells from melanoma patients and healthy subjects. *Cancer Res.* **64,** 3661–3667.

Yee, C., Savage, P. A., Lee, P. P., Davis, M. M., and Greenberg, P. D. (1999). Isolation of high avidity melanoma-reactive CTL from heterogeneous populations using peptide–MHC tetramers. *J. Immunol.* **162,** 2227–2234.

Yee, C., Thompson, J. A., Byrd, D., Riddell, S. R., Roche, P., Celis, E., and Greenberg, P. D. (2002). Adoptive T cell therapy using antigen-specific CD8+ T cell clones for the treatment of patients with metastatic melanoma: *In vivo* persistence, migration, and antitumor effect of transferred T cells. *Proc. Natl. Acad. Sci. USA* **99,** 16168–16173.

Zajac, A. J., Blattman, J. N., Murali-Krishna, K., Sourdive, D. J., Suresh, M., Altman, J. D., and Ahmed, R. (1998). Viral immune evasion due to persistence of activated T cells without effector function. *J. Exp. Med.* **188,** 2205–2213.

PART IV

Mutli-Color Immunophenotyping

CHAPTER 22

Multicolor Flow Cytometric Analysis in SIV-Infected Rhesus Macaque

Joshua M. Walker,* Holden T. Maecker,† Vernon C. Maino,† and Louis J. Picker*

*Vaccine and Gene Therapy Institute and the Oregon National Primate Research Center at the Oregon Health & Science University, West Campus
Beaverton, Oregon 97006

†Becton Dickinson Biosciences
San Jose, California 95131

METHODS IN CELL BIOLOGY, VOL. 75

0091-679X/04 $35.00

I. Introduction

Flow cytometry has been a cornerstone of analysis of acquired immunodeficiency syndrome (AIDS) pathogenesis and immunity since recognition of the disease. Indeed, monitoring of peripheral blood $CD4^+$ T-cell counts remains an important diagnostic and prognostic criterion in the clinical management of human immunodeficiency virus (HIV)–infected patients (Cohen and Fauci, 2001; Masur *et al.*, 1989). Flow cytometric analysis has proven to be equally important to investigators using the SIV/nonhuman primate (NHP) model of AIDS, which recapitulates the key features of HIV infection and disease, and offers the best available animal model for fundamental studies of the pathogenesis and immunity of AIDS-causing viruses (Johnson, 1996; Letvin, 1998; Nathanson *et al.*, 1999; Whetter *et al.*, 1999). Studies in the SIV-NHP model have revealed new intricacies in the interactions between pathogenic lentiviruses and host immune systems. These intricacies suggest that new flow cytometric paradigms are needed, both to follow disease course in experimentally infected animals (e.g., monitoring infection after experimental vaccination) and to reveal the basic mechanisms underlying progressive immunodeficiency and resistance of these infections to adaptive immunity. For example, coreceptor–mediated tropism has been shown to directly determine $CD4^+$ depletion patterns in SIV-infected NHPs, with disease progression being most closely linked not to total $CD4^+$ T-cell counts but to the dynamics of direct viral targets (Douek *et al.*, 2003; Harouse *et al.*, 1999; Karlsson *et al.*, 1998; Picker *et al.*, 2004; Veazey *et al.*, 1998, 2000, 2003). In addition, the relatively poor correlation between conventional measures of lentiviral-specific cellular immunity and viral dynamics (Betts *et al.*, 2001; Draenert *et al.*, 2004) has suggested that more sophisticated assessments need to be developed; assessments that take into account not only the frequency of virus-specific T cells, but also their level of differentiation and specific functional attributes (Champagne *et al.*, 2001; Migueles *et al.*, 2002; van Baarle *et al.*, 2002; Zhang *et al.*, 2003).

Thus, analytical strategies capable of more precise definition and quantification of NHP lymphocyte, particularly T-cell, differentiation and function would allow investigation of whether differences in such precisely defined populations correlate with SIV infection course. Although the NHP immune system closely mirrors that of the human, differences in both population staining patterns and reagent applicability/availability indicate a need for independent assay development and validation in each primate species (Pitcher *et al.*, 2002). However, the greater flexibility of the NHP model facilitates novel observations that can often be "back-validated" to humans (Guadalupe *et al.*, 2003; Veazey *et al.*, 1998). The advent of polychromatic (six or more fluorescence parameter) flow cytometers offers new opportunities for high-resolution immunological analysis, potentially allowing unprecedented phenotypic and functional dissection of complex immunocyte populations (De Rosa *et al.*, 2001, 2003). However, the enormous promise of this technology does not come without a price, and that price is the added complexity of analysis and more numerous pitfalls that could trap the unwary. Here, we review these pitfalls and offer practical protocols for their avoidance, as well as describe our current approach for evaluating disease pathogenesis and antigen-specific immunity in SIV-infected rhesus macaques (RM).

II. Methods

A. Animals

The material presented in this chapter is based on the study of colony-bred RMs (*Macaca mulatta*) of Indian origin, including uninfected animals and animals experimentally infected with either SIVmac239 (5 ng intravenously) (Regier and Desrosiers, 1990) or rhesus cytomegalovirus (RhCMV) strain 68.1 (5×10^6 plaque-forming units, subQ) (Lockridge *et al.*, 1999). Although the approaches described herein are generally applicable to all NHP species (and even humans), specific staining patterns and interpretation criteria may vary from species to species and should be independently validated for each species. All animals were maintained and used in accordance with guidelines of the Animal Care and Use Committee at the Oregon National Primate Research Center and the *National Institutes of Health Guide for the Care and Use of Laboratory Animals*.

B. Antigen Stimulation and Immunofluorescent Staining

Immunophenotyping was performed on 100 μl of whole acid citrate dextrose anticoagulated blood in a 12-by-100 polystyrene tissue culture tube (BD Biosciences, Franklin Lakes, New Jersey). Cells were stained in a two-step process: prefixation staining and postfixation staining. Prefixation staining was used to visualize cell-surface expression of selected markers using monoclonal antibody (mAb) conjugates that are insensitive to the fixation process (see later discussion for explanation). Postfixation staining was used to visualize intracellular markers or cell-surface markers using mAb conjugates that are fixation sensitive. Step 1 was performed by incubating blood with appropriately titered mAb conjugates at room temperature for 30 minutes. After washing in 4 ml cold (4 °C) dPBS with 0.1% of BSA and 5 mm of sodium azide (wash buffer) once, cells were fixed and permeabilized by incubation at room temperature for 10 minutes in 1 ml of FACS Lysing solution (BD Biosciences, San Jose, California), washed, incubated in 0.5 ml of FACS Permeabilizing Solution (BD Biosciences, San Jose, California) for 10 minutes at room temperature, washed, and finally incubated a second time in 0.5 ml of FACS Permeabilizing Solution for 10 min at 4 °C. After these permeabilization steps, cells were washed twice at 4 °C and then incubated for 30 minutes at room temperature with appropriately titered directly conjugated antibodies. Cells were then washed at 4 °C once, resuspended in 200 μl of wash buffer, and analyzed immediately on the LSRII flow cytometer described later.

Cytokine flow cytometry (CFC) was performed as previously described (Pitcher *et al.*, 2002). Briefly, 1×10^6 PBMCs were suspended in 1 ml of complete medium (RPMI 1640 medium; HyClone Laboratories, Logan, Utah) supplemented with 10% heat-inactivated fetal calf serum (HyClone Laboratories), 2 mM L-glutamine (Sigma-Aldrich, St. Louis, Missouri), 1 mM of sodium pyruvate (Sigma-Aldrich), and 50 μM of 2-mercaptoethanol (Sigma-Aldrich) in a 12-by-75 mm polypropylene tissue culture tube (BD Biosciences). These cell preparations were incubated with media alone (negative control) or overlapping peptide mixes (consecutive 15 mers with 11 amino acid overlaps) based on the SIVmac239 gag sequence (NCBI accession no. M33262) or the RhCMV IE1 sequence (NCBI accession no. M93360) for 1 hour at 37 °C (tubes were incubated leaned over at a 5-degree slant from horizontal, as described [Pitcher *et al.*, 2002]). The secretion inhibitor brefeldin A was then added at a concentration of 10 μg/ml, and cells were incubated at 37 °C for an additional 5 hours. After stimulation, cells were washed and then stained as described earlier for the phenotyping samples. Overlapping peptide mixes were formulated and used as previously described (Maecker *et al.*, 2001).

III. Flow Cytometry Instrumentation and Software

A modified "10-color" (12-parameter) LSRII flow cytometer (BD Biosciences, San Jose, California) was used for the development of the analytical paradigms described here. The optical configuration of the LSRII is such that a photomultiplier tube (PMT) array, linked to the flow cell by a fiberoptic filament, collects the fluorescent signals from each laser line. The arrays are designed to detect either three or eight parameters ("trigon" and "octogon" detector arrays, respectively) in series using long-pass dichroic mirrors (Fig. 1 and Table II). Our instrument is configured with three laser lines. The first laser is a 20-mW 488-nm argon-ion laser (Coherent, Santa Clara, California), linked to a PMT octagon, collecting five fluorescent parameters (fluorescein isothiocyanate [FITC], phycoerythrin [PE], PE–Texas Red, PerCP–Cy5.5, PE–Cy7; Table II) and orthogonal light scatter. The second laser is a 25-mW 405-nm solid-state laser (Coherent) linked to a PMT trigon, collecting two fluorescence parameters (Pacific blue and AMCyan). The third light source is a 20-mW 635-nm HeNe (JDS Uniphase, San Jose, California), linked to a PMT trigon collecting three fluorescence parameters (allophycocyanin [APC], Alexa 700, and APC–Cy7). Note that the first and second lasers have two and one currently unused PMTs, respectively, that could potentially be used for development of 13-color (15-parameter) analysis. The dichroic filter configuration of each of these arrays is described in Table II. The instrument is also capable of collecting both area and width measurements in any parameter, which allows us to perform doublet discrimination and cell cycle analysis on the 405-nm laser line, analyzing Hoechst 33342 pulse area versus pulse width (Telford and Frolova, 2004; Telford *et al.*, 2003).

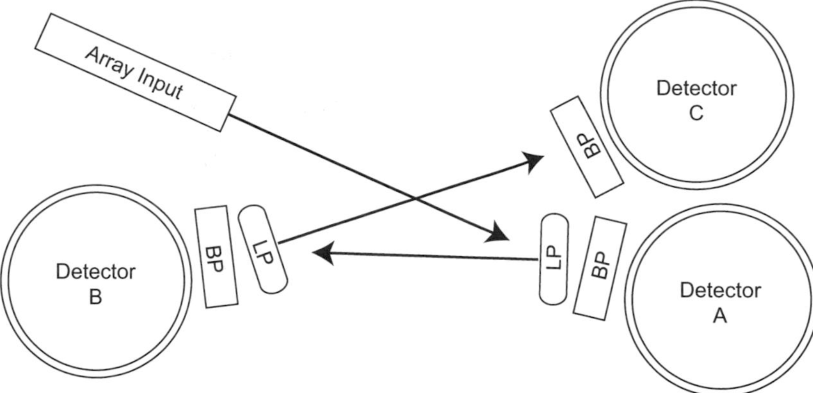

Fig. 1 Detector array configuration within a photomultiplier tube (PMT) trigon. LP, dichroic long-pass mirrors; BP, narrow bandpass filters (see Table II for specifics). The angle of incidence of the fluorescence signals upon each LP filter is 11.25 degrees, and all custom long-pass filters must be manufactured to meet these specifications. The array input indicates where the fiberoptic cable linked to the flow cell introduces the signal. The octagon, which on our instrument collects five fluorescence parameters as well as orthogonal scatter from the 488-nm laser, is identical to the trigon, except that eight PMTs are arranged in a circular pattern.

The digital acquisition system mated with the instrument enables us to collect FACS data with 18-bit precision (generating 262,144 possible fluorescence channels). The data are collected uncompensated in linear format. Compensation and log conversion are performed during analysis. This, as well as several other factors inherent in analog data that have been obviated in digital data, leads to a number of differences between analog and digital data that will be addressed later in the chapter. Flow cytometric data were acquired and spectral overlap compensation values calculated using FACSDiva software (BD Biosciences, San Jose, California). All flow cytometric data were analyzed using FlowJo version 4.5.3 (Tree Star, Ashland, Oregon).

IV. Instrument Setup

In analog flow cytometry systems, instrument set up consists of preacquisition adjustment of both PMT voltages and compensation values. With the use of digital electronics, compensation is performed post hoc at the time of analysis and preacquisition considerations center on the adjustment of the PMT voltages. Although collection of appropriate data on the spectral overlap of the desired fluorochromes remains a critical aspect of experiment planning, PMT voltages, unlike compensation data, cannot be corrected after sample collection. Incorrect PMT voltages undermine the ability to perform compensation and to visualize/separate cell populations of interest.

A. PMT Voltage Calibration

The major considerations of PMT voltage adjustment are to achieve voltages that (1) raise the negative population of a fully stained sample above the electronic noise in each fluorescence channel and (2) keep the populations of interest on scale (no closer than one-quarter log to the maximum channel number). A convenient method of approximating these PMT voltages consists of maximizing the difference in median fluorescence intensity (ΔMFI) between a high-intensity (bright) and a low-intensity (dim) population in each fluorescence channel. This is accomplished by analyzing an equal mixture of bright and dim particles, which fluoresce in the channel to be optimized, at intervals over the full voltage range of the PMT. Once the voltage has been reached where a negative population is discernible from background noise, the ΔMFI will no longer increase, irrespective of how much additional voltage is applied to the PMT. This point is clearly visible if the ΔMFI at each voltage is graphed (Fig. 2). The goal of this process is to find the lowest voltage at which the maximum ΔMFI between the two populations is observed. Several options exist for the generation of high- and low-intensity signals in each of the fluorescence channels; among them are the use of two intensities of rainbow particles or the use of two intensities of anti-immunoglobulin G (IgG) particles that have been incubated with a directly conjugated antibody appropriate for the channel to be optimized. Cells that have been singly stained with a bright conjugate appropriate for the fluorescence channel to be optimized can be used; however, they generate populations with a significantly wider distribution than most particles. However, basing voltage calibration on artificial particles alone can sometimes produce voltages that are inappropriate for certain analyses (e.g., very bright markers can be off scale and certain intracellular markers can have excessively high baseline fluorescence). Therefore, PMT voltage calibration should be verified using cells stained with the mAbs to be used in any given analysis. In some situations, it might be necessary to reduce PMT voltages below theoretically optimal levels to achieve the best "real-world" resolution of stained populations.

Formal PMT voltage adjustment needs to be performed after initial instrument configuration and after the introduction of any changes in this configuration (e.g., filter changes or PMT replacement). To quality control instrument performance, reference values need to be established by analyzing a high-quality medium-intensity rainbow bead (one that fluoresces in all channels simultaneously; e.g., Spherotech UltraRainbow beads) and determining a reference mean fluorescence intensity and coefficient of variation (CV) for each channel (Schwartz *et al.*, 1998). At a given flow rate and (optimized) PMT voltage setting, these reference values should remain stable (MFI values ± 5% of the original value; CVs increasing by no more than 20% of the original values) from day to day. Deviations from this normal variation indicate instrument servicing may be required. These target values are also portable, allowing a user to generate essentially identical data in different instruments with the same configuration by adjusting the voltage of each channel to reach these targets.

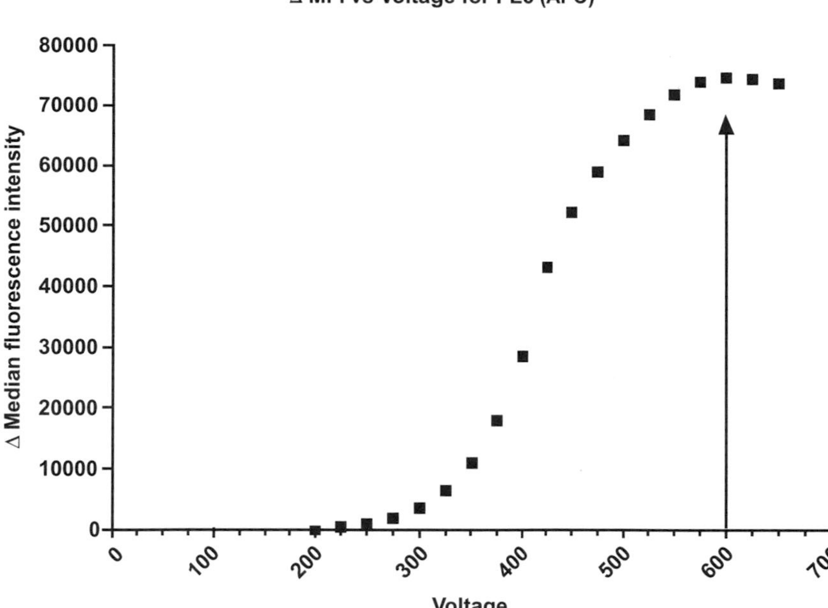

Fig. 2 PMT voltage versus detection sensitivity. The plot shows the difference in median fluorescence intensity between high-intensity and low-intensity bead populations versus voltage. The arrow at 600 V indicates the point at which the maximum difference is seen. Comptrol (Spherotech) low-density and high-density antimouse immunoglobulin-coated particles were used to generate the populations. One drop each of high- and low-density antimouse immunoglobulin particles were each incubated for 30 minutes with appropriately titered CD8α allophycocyanin (APC) monoclonal antibody (mAb) (BD Immunocytometry Systems), washed, and then analyzed. Each data point on the graph represents 5000 events each of the high- and low-intensity particles that were analyzed at that voltage.

B. Compensation

Compensation is the process by which the spectral overlap of the individual fluorochromes used in a polychromatic analysis panel is, in so far as possible, electronically eliminated, allowing signal in any given channel to be interpreted as largely, if not exclusively, reflecting the intensity of the appropriate mAb–fluorochrome conjugate (Bagwell and Adams, 1993). Although this process is associated with certain caveats (see later discussion), correct compensation still remains a critical aspect of polychromatic flow cytometric analysis. Indeed, compensation is even more critical for polychromatic flow cytometry because more fluorochromes must fit within the same light spectrum, dictating more spectral overlap.

The process of compensation involves individual assessment of a positive population for each fluorochrome in all the other fluorescent channels. With these data, compensation programs determine the observed spectral overlap of each fluorochrome in all other channels and electronically adjust experimental data to eliminate this overlap. To achieve accurate and reproducible results, the user prepares compensation specimens for each fluorescence channel by binding each directly conjugated mAb used in an experiment to anti-immunoglobulin particles (Pharmingen, San Diego, California) and individually analyzing each conjugate-coated particle preparation (vs uncoated negative control particles). It should be noted that for a single 10-color analysis, this amounts to 11 (10 + 1) different samples; however, if multiple staining combinations are used in a single experiment, it is advisable to determine compensation values for each mAb conjugate because the spectral characteristics of given fluorochrome (particularly the tandem dyes) may vary among difference conjugates (Baumgarth and Roederer, 2000; Stewart and Stewart, 1999). For example, current-generation tandem fluorochromes vary in their spectral characteristics to the point that compensation values, calculated with different lots of an identical conjugate at the same signal intensity, can vary by as much as 20% (data not shown), and compensation values required for the same tandem fluorochrome from different manufacturers can vary by at least this much. After acquisition of each sample (at the optimized voltages determined during PMT voltage setup), the user calculates compensation values using FACSDiva software.

V. Critical Aspects of the Technology

A. Tandem Fluorochrome Degradation

One serious potential pitfall to be considered when performing polychromatic flow cytometry using present fluorochrome technology is the potential degradation of cyanine dye tandems such as PE–Cy7 and APC–Cy7. For reasons that remain unclear, exposure to paraformaldehyde can cause degradation of these fluorochromes resulting in independent fluorescence of the parent fluorophore (PE, APC). Thus, a PE–Cy7 or APC–Cy7 conjugate will manifest (compensation refractory) fluorescence in the PE or APC channels, respectively (Fig. 3). This tandem fluorochrome degradation can severely affect the ability to accurately determine percentages of cells expressing low-abundance antigens that do not exhibit clearly separated populations (e.g., CCR5, CD25, and Ki-67) as tandem degradation and legitimate staining of the PE or APC conjugate can be impossible to distinguish. The degree to which degradation occurs also tends to vary from sample to sample within the same experiment, making simple subtraction of the tandem degradation observed in control stains problematic. Unfortunately, all current commercially available fixation/permeabilization solutions contain paraformaldehyde and, thus, cause this degradation to some extent. Because the use of intracellular antigens such as Ki-67, tumor necrosis factor-α (TNF-α), and interferon-γ (IFN-γ) necessitate the use of fixation/permeabilization reagents, it is not feasible in many cases to avoid paraformaldehyde treatment entirely. The most effective solution we have found for this problem is to limit the use of cyanine dye tandems to conjugates that must be or can be used after fixation and permeabilization (Table I). However, given inherent

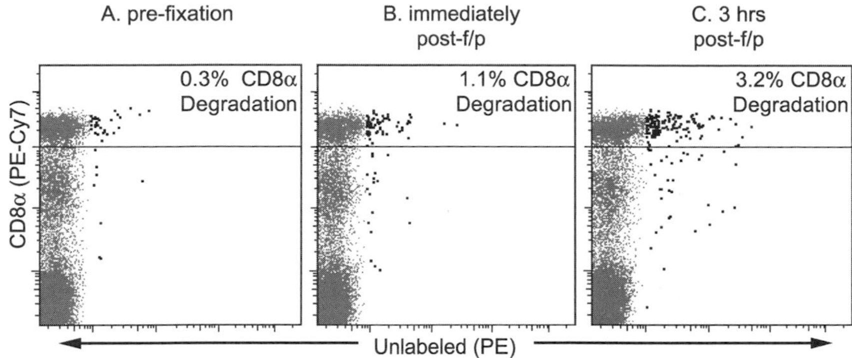

Fig. 3 Cyanine–tandem fluorochrome degradation. 10^6 RM PBMCs were stained with an appropriately titered CD8α phycoerythrin (PE)–Cy7 monoclonal antibody (mAb) alone. The fluorescence of the stained cells is visualized in FL5 (the PE–Cy7 channel) versus FL2 (the PE channel) prefixation (A), immediately after fixation/permeabilization with FACS Lysing Solution and FACS Permeabilizing Solution (B), and 3 hours. after fixation/permeabilization (C) (sample stored at 4 °C in the dark). The three plots clearly show the progression of the cyanine dye tandem fluorochrome degradation after exposure to fixation/permeabilization reagents. The events shown represent gated small lymphocytes, and each dot plot including 10,000 events. The percentages in the figures represent the fraction of the CD8α (PE–Cy7) "bright" population (designated by the solid horizontal line), which also exhibits FL2 (PE) fluorescence.

limitations in conjugate availability and compatibility, this approach may not always be possible. In such situations, interpretable results can be achieved by limiting the parent fluorochrome conjugates to mAbs that delineate either uniformly bright or distinct, widely separated cell populations. In either of these situations, the relatively low-intensity, degradation-associated fluorescence does not significantly alter the results.

B. "Data Spread" Due to Population Variance

One phenomenon, previously obscured by analog instrumentation, which has come to the fore with the advent of polychromatic digital flow cytometry, is the issue of "data spread." *Data spread* refers to the difference in fluorescence variance between bright and negative populations, after optimal digital compensation, in a channel for which compensation has been applied. This difficult concept is most easily conveyed visually, using a two-dimensional dot plot with bi-exponential scaling (Fig. 4A). This example shows the same flow cytometric analysis of a mixture of APC-positive and APC-negative particles either with or without compensation versus APC–Cy7. In the uncompensated sample, the APC-positive particle population appears as a tight cluster in the APC–Cy7 channel, whereas in the compensated sample an apparent large standard deviation is observed. For the APC-negative particle population, such differences are much less marked. In fact, the variance in either the APC-positive or the APC-negative particle

Table I
Monoclonal Antibody Used (RM Validated)

Target	Clone	Works after fixation[a]	Manufacturer
CD3	FN18	No	Biosource
CD3ε	SP34-2	No	BD/Pharmingen
CD4	L200	No	BD/Pharmingen
CD8α	SK1	Yes	BD/Pharmingen
CD8$\alpha\beta$	2ST8.5H7	Yes	Beckman/Coulter
CD14	MoP9	Yes	BD/Pharmingen
CD20	L27	Yes	BD/Pharmingen
CD11a	G-25.2	No	BD/Pharmingen
CD25	M-A251	No	BD/Pharmingen
CD28	CD28.2	No	Multiple Sources
CD49d	9F10	No	BD/Pharmingen
CD62L	SK11	No	BD/Pharmingen
CD95	DX2	Yes	Multiple Sources
CD69	CH/4	Yes	Caltag
CD69	FN50	Yes	BD/Pharmingen
CD69	TP1.55.3	Yes	Beckman/Coulter
β7 Integrin	FIB504	No	BD/Pharmingen
BRDU	3D4	Yes (required)	BD/Pharmingen
CCR5	3A9	No	BD/Pharmingen
CCR7	150503	No	R&D Laboratories
CXCR4	12G5	No	BD/Pharmingen
HLA-DR	L243	Yes	BD/Pharmingen
IFN-γ	B27	Yes (required)	BD/Pharmingen
IL-2	MQ1-17H12	Yes (required)	BD/Pharmingen
IL-4	8D4-8	Yes (required)	BD/Pharmingen
Ki-67	B56	Yes (required)	BD/Pharmingen
TNF-α	MAB11	Yes (required)	BD/Pharmingen

[a]Application of reagent postfixation results in a similar staining pattern as prefixation staining with sufficient resolution to be useful and interpretable.

population in the two profiles is identical (the data are the same), but in the uncompensated dot plot, the variance in the APC-positive population is visually "masked" because of log scaling. Compensation moves the positive population down the y-axis log scale, revealing the much larger standard deviation of the APC-positive population in the APC–Cy7 channel than that of the negative population. Data spread is the result of photon counting error, which is an unavoidable aspect of flow cytometry measurements (Roederer, 2001). Discussion of the statistical underpinnings of this phenomenon is beyond the scope of this chapter but has been the subject of several excellent publications (Shapiro *et al.*, 1998; Wood, 1998). In short, compensation can correct for spectral overlap, but it is unable to correct for the inherent statistical defects that originate with photon counting. The practical implication of data spread is that traditional (rectilinear)

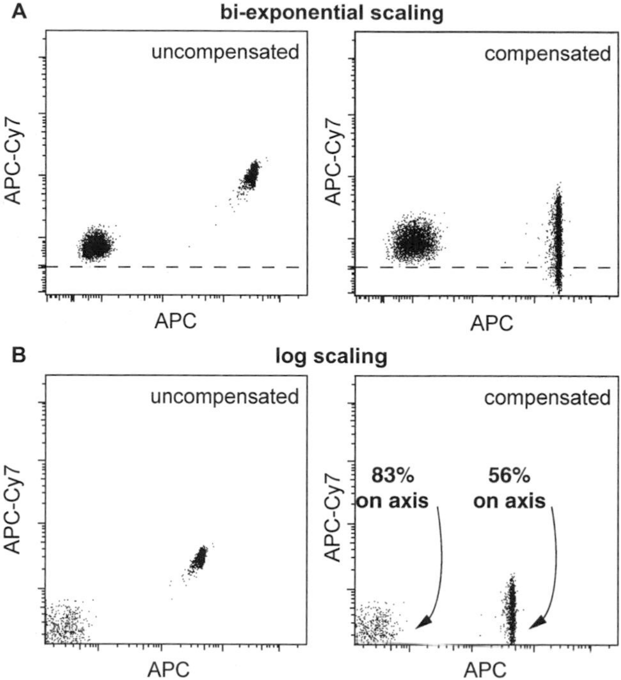

Fig. 4 Postcompensation "data spread." The figure shows the same allophycocyanin (APC) (FL6) versus APC–Cy7 (FL8) fluorescence data of a mixture of APC-positive and APC-negative beads (5000 events of each) before compensation (left panels) and after optimal compensation (right panels), using bi-exponential scaling (A) or traditional log scaling (B). Optimal compensation is defined as equivalence of the median fluorescence intensities of the APC-positive and APC-negative bead populations in the APC–Cy7 channel (FL8). Although the large standard deviation of the APC-positive bead population is not evident before compensation, it becomes so once compensation has been applied. This is most evident with bi-exponential scaling, because most events are not visualized using the traditional log scale. Note that because the sample was collected using digital electronics that retain low and negative values, both the positive and the negative "data spread" of the APC-positive population can be seen using bi-exponential scaling. The positive and negative bead populations in the figure were generated by mixing equal portions of antimouse CompBeads (BD Pharmingen, San Diego, California) that have been incubated with CD8α APC with negative control beads.

quadrant statistics are no longer applicable to many polychromatic analyses. Positive populations are not defined by straight lines, but by curves.

Fortunately, parallel to the development of instrumentation and fluorochrome technology, software and methodologies have been developed to effectively analyze polychromatic flow cytometry data, approaches that take the data spread phenomenon into account. One of the most useful tools now available for use in the analysis of digital polychromatic flow cytometry data is the bi-exponential scaling system developed by David Parks and Wayne Moore at Stanford University (Tung *et al.*, 2004). This scale includes positive and negative values in a combination linear and log scale, becoming progressively more linear as values approach zero and approaching true log scaling as values move away from zero. Bi-exponential scaling allows low and negative values generated in digital flow cytometry to be retained and aids in the visualization of compensated data. The use of this scale is shown in Fig. 4A, where "zero" on the APC–Cy7 axis is shown by the dashed line. With traditional positive log scaling, most events in the compensated plot would stack on "zero" of the APC–Cy7 channel (Fig. 4B), rather than being more correctly displayed as part of a continuous negative population. A second development in analysis and interpretation is the use of fluorescence-minus-one (FMO) controls (Roederer, 2001) to determine the proper analysis of nondiscrete populations (e.g., CCR5, CD25, and Ki-67). These FMO controls are samples that are identical to experimental samples with the exception of the mAb of interest, which has been eliminated from the stain. FMO controls have become an essential aspect of polychromatic flow cytometry data analysis, because they allow accurate determination of the boundary between a positive and negative population by duplicating autofluorescence levels and data spread present in a fully stained sample.

C. Staining Panel Optimization Strategies

The promise of polychromatic flow cytometry is the simultaneous visualization and phenotypic definition of complex cell populations on an unprecedented scale. However, in this type of analysis, the number of experimental variables increases proportionally with analytic potential, making the achievement of this analytic promise much more dependent on strategic planning, particularly in regard to panel development, than traditional three- or four-color flow cytometry (Roederer *et al.*, 2004). Both technical and practical considerations contribute to analytic complexity. We have already discussed the data spread and tandem fluorochrome degradation issues. Even more basic is that the inherent characteristics of the fluorochromes, combined with the extent of compensation required for their use in polychromatic formats, result in each channel having a different dynamic response. Thus, an mAb conjugated to a high dynamic response fluorochrome (e.g., PE) will show much brighter staining on a given population of cells than a low dynamic response fluorochrome (e.g., APC–Cy7). Low-abundance antigens would, therefore, require a high dynamic range fluorochrome channel for their effective visualization, and in panel planning, the high dynamic range channels need to be reserved for their use. The relative dynamic response of each fluorochrome–channel combination used in our LSRII instrument is listed in Table II. In addition, the different fluorochrome conjugates vary in regard to their applicability to intracellular staining, a consideration for any cytokine flow cytometry panel or any phenotyping panel examining cytoplasmic or nuclear antigen expression. Finally, practical considerations enter into strategic planning. With three- or four-color flow cytometry, it is often the case that commonly used mAbs are available in all possible fluorochrome conjugates, and most mAbs are available

Table II
Modified LSRII Filter Configuration

Laser	Photo-multiplier tube	Dichroic	Bandpass filter	Fluorochromes used	Relative dynamic range
488-nm octagon (blue)	SSC	None	488/10BP		
	FL1	505DCLP	530/30DF	Fluorescein isothiocyanate (FITC)	Medium
	FL2	550DCLP	575/26DF	Phycoerythrin (PE)	High
	FL3	600DCLP	610/20DF	PE–Texas red	Medium
	FL4	685DCLP	695/40DF	PerCP–Cy5.5	Low
				PE–Cy5.5	Medium
	FL5	735DCLP	780/60DF	PE–Cy7	High
635-nm trigon (red)	FL6	None	660/20DF	Allophycocyanin (APC)	High
	FL7	685DCLP	730/40DF	Alexa 700	Medium
	FL8	750DCLP	780/60DF	APC–Cy7	Low
405-nm trigon (violet)	FL9	None	440/40DF	Pacific blue, Hoechst	Medium
	FL10	505DCLP	525/50DF	AMCyan	Low

in at least two conjugates. With 10-color flow cytometry, however, it is highly unlikely that all (i.e., all 10) or even most of the possible fluorochrome conjugates will be available for any mAbs. In addition, for the immediate future, it will certainly be true that reagent diversity will be greatest for mAbs conjugated to traditional fluorochromes (FITC, PE, APC). Taken together, these considerations dictate certain approaches to effective polychromatic panel planning. First, commonly used lineage markers—CD3, CD4, CD8, CD20—should be targeted to the newer channels, leaving the traditional channels open for more variable lineage-subset defining markers. In addition, because lineage-defining markers are often abundantly expressed, low dynamic response channels (e.g., APC–Cy7 or PerCP–Cy5.5) are often adequate for their effective visualization. However, in panels including intracellular staining, the tandem fluorochrome degradation issue comes into play. When possible, tandem fluorochromes should be targeted toward the intracellular antigens themselves (e.g., when the dynamic response and intracellular staining characteristics are appropriate) or toward markers that have been validated for use after fixation/permeabilization (CD8α, CD20). Data spread issues raise another complexity. For example, when assessing a dim, partially expressed antigen (e.g., CCR5) in a high dynamic response channel such as APC, it is advisable *not* to have the cell population of interest stained with a bright APC–Cy7 conjugated mAb (Fig. 4). Finally, there will always be idiosyncrasies in the ability of different conjugate combinations to visualize particular populations, necessitating flexibility in conjugate selection and careful validation of staining panels. Several examples of validated staining panels are described in the following sections.

VI. Application of 10-Color Analysis to SIV Infection

The enormous information content of polychromatic (10-color) flow cytometric analysis offers an extraordinary opportunity to address fundamental questions in SIV pathobiology and immunity and to more accurately evaluate new therapeutic and vaccine strategies. Although myriad applications for this technology can be imagined, we have focused on two primary areas: SIV immunopathogenesis analysis and quantitative "dissection" of SIV-specific T-cell responses.

A. SIV Immunopathogenesis Analysis

The objective of SIV immunopathogenesis analysis is to determine the pathogenic pathway(s) by which SIV replication damages the immune system and leads to symptomatic immune dysfunction—essentially, defining the sequence of events that "connect" viral replication to disease. Viral replication is clearly the "driver" of lentiviral pathogenesis, and complete suppression of viral replication mitigates against the development of disease (Lifson *et al.*, 1997; Richman, 2001; Staprans *et al.*, 1999). However, at any given level of viral replication, host factors may also play a crucial role in determining the tempo of pathogenesis. Thus, given that complete viral suppression may not always be obtainable, it is clear that the ability to monitor downstream pathogenic effects of viral replication, before overt disease, would allow a more accurate evaluation of virtually any intervention strategy. In addition, understanding of the immunopathogenic pathways leading to disease provides the opportunity for development of novel therapeutic strategies aimed at "disconnecting" viral replication from disease development—strategies that can only be monitored by immunopathological (as opposed to virological) criteria. Peripheral blood $CD4^+$ T-cell counting comprises a simple example of such an immunopathogenic assessment, but it is well documented that these data are a relatively poor predictor of SIV disease progression in monkeys, particularly in the setting of CCR5-tropic viruses (Dykhuizen *et al.*, 1998; Garber *et al.*, 2004; Hirsch *et al.*, 2004; Picker *et al.*, 2004). Data suggest that more sophisticated monitoring of $CD4^+$ T-cell subset dynamics provides a much clearer view of the effect of SIV on the NHP immune system (Garber *et al.*, 2004; Picker *et al.*, 2004). In this regard, it has been definitively shown that $CD4^+$ T-cell lineage is not a uniform target of SIV infection, but these viruses selectively infect those CD4-expressing cells with appropriate coreceptor expression (Doms, 2001). CCR5 is the dominant coreceptor of most SIV strains (Chen *et al.*, 1997), and thus, T cells expressing this molecule—which include a minority subset of memory T cell in blood and lymphoid tissues, but the vast majority of memory T cells in mucosal tissues (intestinal lamina propria, lung, vaginal mucosa, liver) (Kunkel *et al.*, 2002;

Picker *et al.*, 2004; Veazey *et al.*, 2000, 2003)—constitute the primary targets of CCR5-tropic SIVs. Indeed, infection of RMs with the CCR5-tropic SIVmac251 or SIVmac239 results in the dramatic depletion of this subset by day 21 of infection (Douek *et al.*, 2003; Picker *et al.*, 2004; Veazey *et al.*, 1998, 2000, 2003). In peripheral blood, lymph node, and spleen, this target cell depletion has little effect on overall $CD4^+$ T-cell counts; however, in the mucosal tissues listed earlier, virtually the entire $CD4^+$ T-cell population is destroyed. Regenerative mechanisms (e.g., production of new $CCR5^+/CD4^+$ memory T cells) can often partially and/or transiently restore this population in these sites and appear to protect against overt immunodeficiency (Picker *et al.*, 2004). However, in those animals without regeneration, or in those in which regeneration ultimately fails, tissue depletion of $CD4^+$ memory T cells appears to be a requisite prelude to progressive disease (Picker, 2004, unpublished observations).

Taken together, these data indicate that the extent of SIV infection/ disease progression can be effectively monitored, at least in part, by quantitative evaluation of tissue-homing $CD4^+$ "effector-memory" T cells (Esser *et al.*, 2003; Lanzavecchia and Sallusto, 2000; Masopust *et al.*, 2001; Pitcher *et al.*, 2002; von Andrian and Mackay, 2000), and their regenerative capabilities. Although continuous assessment of effector site T-cell populations via intestinal biopsy or lung lavage is certainly useful, such invasive monitoring is not always practical. Instead, we have developed a 10-color panel for peripheral blood analysis that allows both routine phenotyping and precise quantitative assessment of "target cell" dynamics. This panel evaluates the correlated expression of the lineage-defining makers, CD3, CD20, CD4, CD8α (Fig. 5A); the naive and memory subset markers CD95 (FAS), CD28 (costimulatory counter receptor for CD80/ 86), β7-integrin (intestinal homing receptor), and CCR7 (CD197; lymph node homing chemokine receptor) the primary SIV coreceptor CCR5 (CD195), and the proliferation-associated antigen Ki-67 (moderate to bright expression indicates the occurrence of S phase in the preceding 4 days) (Pitcher *et al.*, 2002; von Andrian and Mackay, 2000; Fig. 5B–E). Figure 5 illustrates the application of this analytic paradigm by comparison of the phenotypic profiles of a normal adult RM (Fig. 5A and B), a juvenile macaque (from an expanded specific pathogen free colony) 14 days into primary infection with rhesus cytomegalovirus (RhCMV; Fig. 5C), and 2 SIVmac239-infected animals, one each with nonprogressive (Fig. 5D) and progressive (Fig. 5E) disease. Significantly, the fraction of $CD4^+$ T cells of total peripheral blood T cells in these four animals were comparable (47%, 64%, 40%, and 54% for uninfected, primary RhCMV, and nonprogressive SIV, and progressive SIV infection, respectively), but as shown in the figure, the composition of these $CD4^+$ T-cell populations demonstrated significant differences. In conventionally reared RMs older than 3 years (Fig. 5B), more than 40–50% of the circulating $CD4^+$ T cells have a memory phenotype (black events) (Pitcher *et al.*, 2002), including (1) a predominant (except in aged monkeys) $CCR7^+/CD28^+$ "central memory" population, usually equally divided between β7-integrin–positive and β7-integrin–subsets (corresponding to their efficiency in

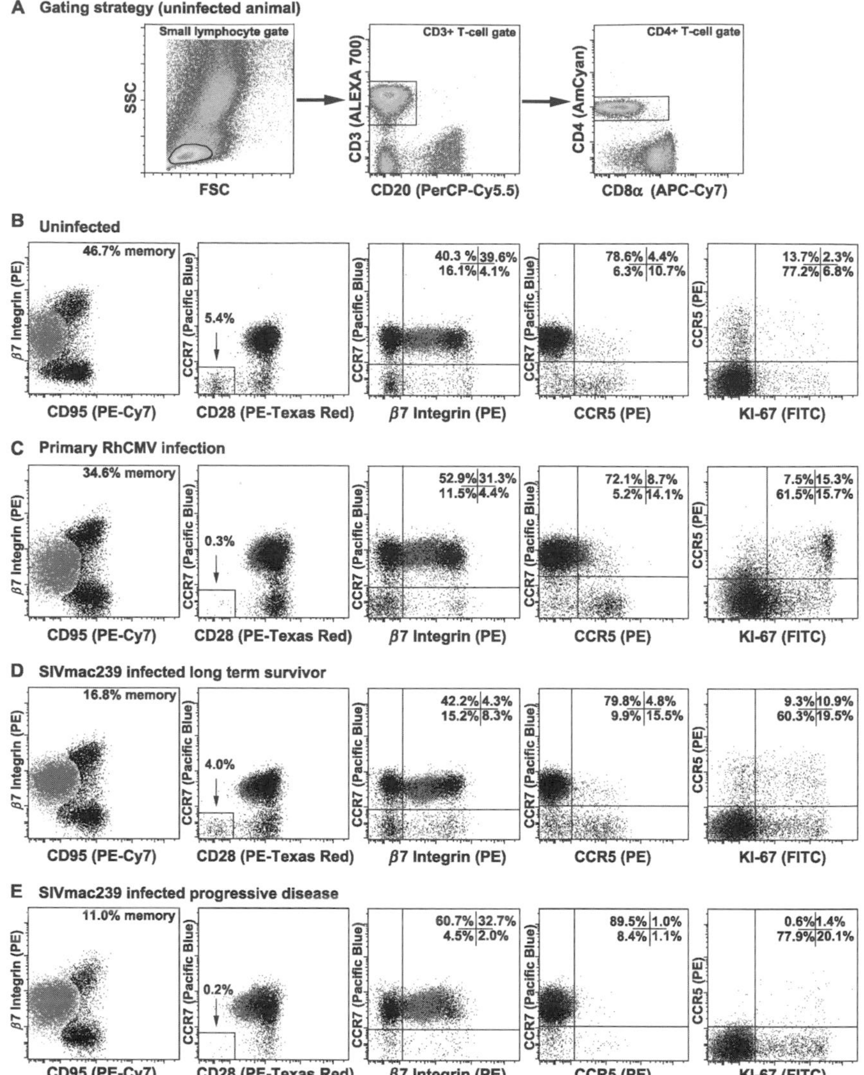

Fig. 5 SIV immunopathogenesis analysis. The staining panel used in this experiment is as follows: CCR7 (Pacific blue)/CD4 (AMCyan)/*Ki-67 (FITC)*/β7-integrin (phycoerythrin [PE])/CD28 (PE–Texas red)/*CD20 (PerCP–Cy5.5)*/*CD95 (PE–Cy7)*/CCR5 (APC)/CD3 (Alexa 700)/*CD8α (APC–Cy7)*. Underlining reflects staining after fixation/permeabilization. The figure demonstrates representative peripheral blood immunophenotyping from the following clinical situations: (A, B) Healthy, uninfected adult RM; (A) delineates the gating strategy used on each sample represented in the figure. (C) Juvenile, expanded specific pathogen-free (previously RhCMV-) RM experimentally infected with RhCMV 14 days before analysis. (D) SIVmac239-infected, (asymptomatic) slow progressor RM. (E) SIVmac239-infected RM with progressive disease (clinically manifested in this animal by chronic diarrhea and wasting). The *%memory* shown in the far left dot plot of (B) through

homing to gastrointestinal lymphoid tissues), (2) a CCR7$^-$ "effector memory" population, representing the CD4$^+$ T cells targeted toward extralymphoid "effector" sites (with β7 expression again delineating homing efficiency to intestinal vs peripheral sites, and which, when fully differentiated, also lack expression of the costimulatory CD28 molecules), (3) a prominent (mean $\pm$ SD = 22 $\pm$ 8%) population of CCR5$^+$, largely "effector memory" phenotype (SIV target) cells, and (4) a variable number (5–15% in most animals) of proliferating Ki-67$^+$ memory cells (including both central and effector memory subsets) (Pitcher *et al.*, 2002). In uninfected animals (Fig. 5B), the CCR5$^+$ population in peripheral blood is largely non-proliferative (Ki-67$^-$), but in primary RhCMV infection a prominent population of CCR5$^+$, Ki-67$^{bright+}$ "new" memory T cells is induced (Fig. 5C). Given that this population is a prime target for productive SIV infection, it is not surprising that SIV coinfection in this situation usually results in rapidly progressive disease (Sequar *et al.*, 2002). It should be noted that the lack of a fully differentiated CCR7$^-$/CD28$^-$ effector memory CD4$^+$ T-cell population in this animal (Fig. 5C) is developmental, reflecting this animal's youth and its relative lack of pathogen exposure secondary to special rearing conditions. Non-progressive SIV infection (Fig. 5D) is associated with a modest reduction in the overall memory CD4$^+$ population, but most memory populations are maintained, including the CCR5$^+$ and β7-integrin$^+$ effector memory (CCR7$^-$) subsets. Of note, the proliferating (Ki-67$^+$) CD4$^+$ memory fraction is increased about threefold, and the majority of the CCR5$^+$/CD4$^+$ memory cells are also Ki-67$^+$, indicating they are newly produced cells. Thus, in stable, non-progressive SIV infection, viral-mediated destruction appears to be chronically compensated by new cell production. In contrast, progressive SIV infection is characterized by profound depletion of circulating effector memory CD4$^+$ T cells, particularly the β7-integrin$^+$ and CCR5$^+$ subsets (Fig. 5E). CD4$^+$ memory T-cell proliferation is initially increased in these animals (note that most of the residual CCR5$^+$/CD4$^+$ cells are Ki-67$^+$) but is clearly insufficient to counter target cell destruction. In some animals, particularly rapid progressors, CD4$^+$ memory T-cell proliferative failure ensues, accelerating the decline of CD4$^+$ effector memory populations. Loss of the circulating effector memory CD4$^+$ T-cell population parallels tissue depletion of CD4$^+$ cells and as indicated earlier appears to be a prerequisite for the stochastic occurrence of symptomatic opportunistic infection. Thus, in CCR5-tropic SIV infection, the tempo of disease progression follows the balance between destruction of CCR5$^+$/CD4$^+$ T-cell targets, particularly the tissue-homing effector memory component and their regeneration. The combination of a decline in CCR5$^+$ cells within the CD4$^+$ total memory subset below 5% and a decline in Ki-67$^+$ memory CD4$^+$ T cells below 30% portends a poor prognosis.

(E) indicates percentage of black-colored events in each plot, previously demonstrated to represent the memory population (Pitcher *et al.*, 2002). The percentages shown in each of the subsequent dot plots of (B) through (E) indicate the percentage of memory cells within the designated gate or quadrant. (See Color Insert.)

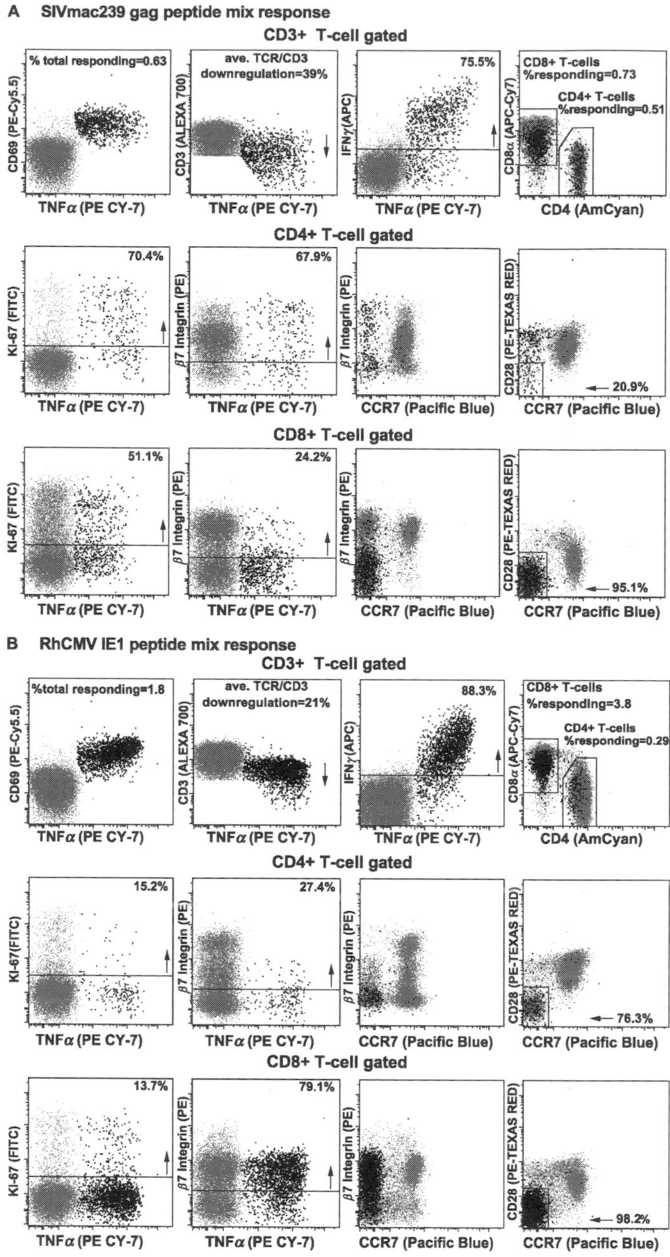

Fig. 6 Viral-specific T-cell responses analysis. PBMCs from an SIVmac239-infected, slow progressor RM and a normal adult (RhCMV-seropositive) RM were stimulated with overlapping peptides comprising the SIVmac239 gag (A) and RhCMV (strain 68.1) IE1 (B) open reading frames, respectively, as described in the methods, and stained with the following panel: CCR7 (Pacific blue)/

B. Dissection of Viral-Specific T-Cell Responses

The second application of 10-color flow cytometry to be discussed here is the unparalleled potential of this technique to dissect SIV-specific T-cell responses. It is increasingly clear that the current techniques (ELISPOT or standard cytokine flow cytometry) used to quantify total antigen-specific $CD4^+/CD8^+$ T cells offer little insight into immunological correlates of protection in HIV infection (Betts *et al.*, 2001; Draenert *et al.*, 2004). One possibility is that the overall frequency of pathogen-specific T cells is less important than their differentiation state, particularly as related to their specific homing and effector characteristics. Thus, such qualitative dissection of antigen-specific populations would require complex assays using large numbers of cells, effectively preventing the routine evaluation of potential criteria that would be necessary to identify the parameters most closely associated with protection. However, 10-color cytokine flow cytometry changes this equation, allowing precise quantification of overall SIV-specific T-cell populations and simultaneously their segregation into functional/phenotypic subsets defined by multiple criteria. One approach to this type of analysis is illustrated in Fig. 6. Two responses are illustrated, one to SIV gag and the other to RhCMV IE-1, both elicited using overlapping mixes of 15mer peptides (11 amino acid overlaps) that allow detection of either or both $CD4^+$ and $CD8^+$ T-cell responses to the protein in question (Betts *et al.*, 2001). The panel evaluates correlated expression of the lineage-defining makers CD3, CD4, CD8α; the response-defining markers CD69, TNF-α, and IFN-γ (all stained intracellularly following brefeldin A–mediated secretion inhibition), the memory subset markers CD28, CCR7, and β7-integrin, and the proliferation marker Ki-67. Responding cells are defined by upregulation of CD69 (a pan activation antigen) and either or both of the cytokines TNF-α and IFN-γ. Assessment of CD3 expression not only helps define the T-cell lineage but also—by the degree of T-cell receptor/CD3 downregulation—delineates the extent of T-cell receptor–mediated signaling experienced by each responding cell (Itoh and Germain, 1997). As described earlier, CD28 and CCR7 define effector memory differentiation; β7-integrin, intestinal homing; and Ki-67, recent proliferative activity. In the examples shown, note that despite evidence of stronger signaling, SIV gag-specific T cells producing TNF-α were less likely to co-produce IFN-γ than the RhCMV IE-1–specific T cells. Not surprisingly, SIV gag-specific T cells, both $CD4^+$ and $CD8^+$, were more proliferative than their RhCMV IE-1–specific counterparts, likely reflecting differences in antigen availability *in vivo*. The vast majority of RhCMV IE-1–specific T cells, both $CD4^+$ and $CD8^+$, demonstrated "full" effector memory differentiation, lacking both CD28 and CCR7

CD4 (AMCyan)/*Ki-67 (FITC)*/β7-integrin (PE)/CD28 (PE-Texas red)/*CD69 (PE–Cy5.5)*/*TNF-α (PE–Cy7)*/*IFN-γ (APC)*/CD3 (Alexa 700)/*CD8α (APC–Cy7)*. Underlining reflects staining after fixation/permeabilization. All plots shown have been gated to include small lymphocytes and exclude CD20-positive B cells. Further gating is indicated above each row of dot plots. Responding cells ($CD69^+$/TNF-α–positive) are black; non-responding cells, red. The percentages shown in the $CD4^+$ and $CD8^+$ T-cell–gated dot plots indicate the relative percentages of responding cells in the indicated areas. (See Color Insert.)

expression. $CD8^+$ SIV gag-specific T cells shared this phenotype, but $CD4^+$ SIV gag-specific T cells were $CCR7^-$, but largely $CD28^+$. Even more intriguingly, whereas $CD4^+$ SIV gag-specific T cells were predominantly β7-integrin$^+$, $CD4^+$ RhCMV IE-1–specific T cells were not, and the reverse was true for the $CD8^+$ SIV gag- and RhCMV IE-1–specific T cells. The significance of these various differences with regard to the protective efficiency of the responses in question is unclear, but 10-color cytokine flow cytometry allows these detailed measurements to be made in concert with standard analyses—in essence, providing exponentially more information using the same samples and similar processing procedures currently applied to experimental NHP cohorts. With routine application of these assays, correlation with clinical outcome will eventually provide insight into the nature of protective cellular immune responses.

VII. Conclusion

The ability of flow cytometry to cross-correlate increasing numbers of parameters on individual lymphocytes—two color over one color, three over two, four over three, and so on—has been a key element in the enormous progress made in the understanding of immune physiology in the last 25 years. At each step of the way, some investigators, citing the added analytic complexity, have questioned the utility of these advances, but history has shown that any technical hurdles are eventually overcome, and the scientific payoff has been enormous. The past 2–3 years have seen a quantum jump in flow cytometer sophistication, with instruments capable of reliably resolving 10 or more parameters now routinely available. As pointed out here, this engineering advance is not without added analytic complexity, but as also shown here, practical application of this technology is possible now and will be facilitated with further experience. In no field of investigation are these advances more important than in SIV/HIV vaccine and immunotherapeutic development. The complexity of the immunobiology of these infections—with its intricate interplay between immune protection and immunopathogenesis—has resisted our attempts at understanding. The advent of 10-color flow cytometry will allow a renewed attack on deciphering the operative primary mechanisms, enabling the field to accumulate vastly more information from the same animals and samples available for study. Although advances in understanding will still require innovative experimentation and insightful analysis, 10-color flow cytometry has moved our efforts to a higher plane, one in which success is much more likely.

Acknowledgments

This publication would not have been possible without the hard work and generous assistance of Larry Duckett, Joseph Trotter, Jonathan Nugen, Ken Davis, and Tim Dubrovsky at BD Biosciences; Brent Wood at the University of Washington; and the efforts of our coworkers Michael Axthelm, Alfred Legasse, Shannon Planer, and Andrew Sylwester.

References

Bagwell, C. B., and Adams, E. G. (1993). Fluorescence spectral overlap compensation for any number of flow cytometry parameters. *Ann. N Y Acad. Sci.* **677,** 167–184.

Baumgarth, N., and Roederer, M. (2000). A practical approach to multicolor flow cytometry for immunophenotyping. *J. Immunol. Methods* **243,** 77–97.

Betts, M. R., Ambrozak, D. R., Douek, D. C., Bonhoeffer, S., Brenchley, J. M., Casazza, J. P., Koup, R. A., and Picker, L. J. (2001). Analysis of total human immunodeficiency virus (HIV)–specific CD4(+) and CD8(+) T-cell responses: Relationship to viral load in untreated HIV infection. *J. Virol.* **75,** 11983–11991.

Champagne, P., Ogg, G. S., King, A. S., Knabenhans, C., Ellefsen, K., Nobile, M., Appay, V., Rizzardi, G. P., Fleury, S., Lipp, M., Forster, R., Rowland-Jones, S., Sekaly, R. P., McMichael, A. J., and Pantaleo, G. (2001). Skewed maturation of memory HIV-specific CD8 T lymphocytes. *Nature* **410,** 106–111.

Chen, Z., Zhou, P., Ho, D. D., Landau, N. R., and Marx, P. A. (1997). Genetically divergent strains of simian immunodeficiency virus use CCR5 as a coreceptor for entry. *J. Virol.* **71,** 2705–2714.

Cohen, O. J., and Fauci, A. S. (2001). Pathogenesis and medical aspects of HIV-1 infection. *In* "Fields virology" (D. M. Knipe and P. M. Howley, eds.), vol. 2, Lippincott Williams & Wilkins, Philiadelphia.

De Rosa, S. C., Brenchley, J. M., and Roederer, M. (2003). Beyond six colors: A new era in flow cytometry. *Nat. Med.* **9,** 112–117.

De Rosa, S. C., Herzenberg, L. A., and Roederer, M. (2001). 11-color, 13-parameter flow cytometry: Identification of human naive T cells by phenotype, function, and T-cell receptor diversity. *Nat. Med.* **7,** 245–248.

Doms, R. W. (2001). Chemokine receptors and HIV entry. *AIDS* **15**(suppl. 1), S34–S35.

Douek, D. C., Picker, L. J., and Koup, R. A. (2003). T cell dynamics in HIV-1 infection. *Annu. Rev. Immunol.* **21,** 265–304.

Draenert, R., Verrill, C. L., Tang, Y., Allen, T. M., Wurcel, A. G., Boczanowski, M., Lechner, A., Kim, A. Y., Suscovich, T., Brown, N. V., Addo, M. M., and Walker, B. D. (2004). Persistent recognition of autologous virus by high-avidity CD8 T cells in chronic, progressive human immunodeficiency virus type 1 infection. *J. Virol.* **78,** 630–641.

Dykhuizen, M., Mitchen, J. L., Montefiori, D. C., Thomson, J., Acker, L., Lardy, H., and Pauza, C. D. (1998). Determinants of disease in the simian immunodeficiency virus–infected rhesus macaque: Characterizing animals with low antibody responses and rapid progression. *J. Gen. Virol.* **79,** 2461–2467.

Esser, M. T., Marchese, R. D., Kierstead, L. S., Tussey, L. G., Wang, F., Chirmule, N., and Washabaugh, M. W. (2003). Memory T cells and vaccines. *Vaccine* **21,** 419–430.

Garber, D. A., Silvestri, G., Barry, A. P., Fedanov, A., Kozyr, N., McClure, H., Montefiori, D., Larsen, C. P., Altman, J. D., Staprans, S. I., and Feinberg, M. B. (2004). Blockade of T cell costimulation reveals interrelated actions of CD4+ and CD8+ T cells in control of SIV replication. *J. Clin. Invest.* **113,** 836–845.

Guadalupe, M., Reay, E., Sankaran, S., Prindiville, T., Flamm, J., McNeil, A., and Dandekar, S. (2003). Severe CD4+ T-cell depletion in gut lymphoid tissue during primary human immunodeficiency virus type 1 infection and substantial delay in restoration following highly active antiretroviral therapy. *J. Virol.* **77,** 11708–11717.

Harouse, J. M., Gettie, A., Tan, R. C., Blanchard, J., and Cheng-Mayer, C. (1999). Distinct pathogenic sequela in rhesus macaques infected with CCR5 or CXCR4 utilizing SHIVs. *Science* **284,** 816–819.

Hirsch, V. M., Santra, S., Goldstein, S., Plishka, R., Buckler-White, A., Seth, A., Ourmanov, I., Brown, C. R., Engle, R., Montefiori, D., Glowczwskie, J., Kunstman, K., Wolinsky, S., and Letvin, N. L. (2004). Immune failure in the absence of profound CD4+ T-lymphocyte depletion in simian immunodeficiency virus–infected rapid progressor macaques. *J. Virol.* **78,** 275–284.

Itoh, Y., and Germain, R. N. (1997). Single cell analysis reveals regulated hierarchical T cell antigen receptor signaling thresholds and intraclonal heterogeneity for individual cytokine responses of CD4+ T cells. *J. Exp. Med.* **186,** 757–766.

Johnson, R. P. (1996). Macaque models for AIDS vaccine development. *Curr. Opin. Immunol.* **8,** 554–560.

Karlsson, G. B., Halloran, M., Schenten, D., Lee, J., Racz, P., Tenner-Racz, K., Manola, J., Gelman, R., Etemad-Moghadam, B., Desjardins, E., Wyatt, R., Gerard, N. P., Marcon, L., Margolin, D., Fanton, J., Axthelm, M. K., Letvin, N. L., and Sodroski, J. (1998). The envelope glycoprotein ectodomains determine the efficiency of CD4+ T lymphocyte depletion in simian-human immunodeficiency virus–infected macaques. *J. Exp. Med.* **188,** 1159–1171.

Kunkel, E. J., Boisvert, J., Murphy, K., Vierra, M. A., Genovese, M. C., Wardlaw, A. J., Greenberg, H. B., Hodge, M. R., Wu, L., Butcher, E. C., and Campbell, J. J. (2002). Expression of the chemokine receptors CCR4, CCR5, and CXCR3 by human tissue–infiltrating lymphocytes. *Am. J. Pathol.* **160,** 347–355.

Lanzavecchia, A., and Sallusto, F. (2000). Dynamics of T lymphocyte responses: Intermediates, effectors, and memory cells. *Science* **290,** 92–97.

Letvin, N. L. (1998). Progress in the development of an HIV-1 vaccine. *Science* **280,** 1875–1880.

Lifson, J. D., Nowak, M. A., Goldstein, S., Rossio, J. L., Kinter, A., Vasquez, G., Wiltrout, T. A., Brown, C., Schneider, D., Wahl, L., Lloyd, A. L., Williams, J., Elkins, W. R., Fauci, A. S., and Hirsch, V. M. (1997). The extent of early viral replication is a critical determinant of the natural history of simian immunodeficiency virus infection. *J. Virol.* **71,** 9508–9514.

Lockridge, K. M., Sequar, G., Zhou, S. S., Yue, Y., Mandell, C. P., and Barry, P. A. (1999). Pathogenesis of experimental rhesus cytomegalovirus infection. *J. Virol.* **73,** 9576–9583.

Maecker, H. T., Dunn, H. S., Suni, M. A., Khatamzas, E., Pitcher, C. J., Bunde, T., Persaud, N., Trigona, W., Fu, T. M., Sinclair, E., Bredt, B. M., McCune, J. M., Maino, V. C., Kern, F., and Picker, L. J. (2001). Use of overlapping peptide mixtures as antigens for cytokine flow cytometry. *J. Immunol. Methods* **255,** 27–40.

Masopust, D., Vezys, V., Marzo, A. L., and Lefrancois, L. (2001). Preferential localization of effector memory cells in nonlymphoid tissue. *Science* **291,** 2413–2417.

Masur, H., Ognibene, F. P., Yarchoan, R., Shelhamer, J. H., Baird, B. F., Travis, W., Suffredini, A. F., Deyton, L., Kovacs, J. A., and Falloon, J. (1989). CD4 counts as predictors of opportunistic pneumonias in human immunodeficiency virus (HIV) infection. *Ann. Intern. Med.* **111,** 223–231.

Migueles, S. A., Laborico, A. C., Shupert, W. L., Sabbaghian, M. S., Rabin, R., Hallahan, C. W., Van Baarle, D., Kostense, S., Miedema, F., McLaughlin, M., Ehler, L., Metcalf, J., Liu, S., and Connors, M. (2002). HIV-specific CD8+ T cell proliferation is coupled to perforin expression and is maintained in nonprogressors. *Nat. Immunol.* **3,** 1061–1068.

Nathanson, N., Hirsch, V. M., and Mathieson, B. J. (1999). The role of nonhuman primates in the development of an AIDS vaccine. *AIDS* **13,** S113–S120.

Picker, L. J., Hagen, S. I., Lum, R., Reed-Inderbitzin, E. F., Daly, L. M., Sylwester, A. W., Walker, J. M., Maino, V. C., Lifson, J. D., Kodama, T., and Axthelm, M. K. (2004). Insufficient production and tissue delivery of CD4+ memory T cells in rapidly progressive SIV infection. (submitted).

Pitcher, C. J., Hagen, S. I., Walker, J. M., Lum, R., Mitchell, B. L., Maino, V. C., Axthelm, M. K., and Picker, L. J. (2002). Development and homeostasis of T cell memory in rhesus macaque. *J. Immunol.* **168,** 29–43.

Regier, D. A., and Desrosiers, R. C. (1990). The complete nucleotide sequence of a pathogenic molecular clone of simian immunodeficiency virus. *AIDS Res. Human Retroviruses* **6,** 1221–1231.

Richman, D. D. (2001). HIV chemotherapy. *Nature* **410,** 995–1001.

Roederer, M. (2001). Spectral compensation for flow cytometry: Visualization artifacts, limitations, and caveats. *Cytometry* **45,** 194–205.

Roederer, M., Brenchley, J. M., Betts, M. R., and De Rosa, S. C. (2004). Flow cytometric analysis of vaccine responses: How many colors are enough? *Clin. Immunol.* **110,** 199–205.

Schwartz, A., Marti, G. E., Poon, R., Gratama, J. W., and Fernandez-Repollet, E. (1998). Standardizing flow cytometry: A classification system of fluorescence standards used for flow cytometry. *Cytometry* **33,** 106–114.

Sequar, G., Britt, W. J., Lakeman, F. D., Lockridge, K. M., Tarara, R. P., Canfield, D. R., Zhou, S. S., Gardner, M. B., and Barry, P. A. (2002). Experimental coinfection of rhesus macaques with rhesus cytomegalovirus and simian immunodeficiency virus: Pathogenesis. *J. Virol.* **76,** 7661–7671.

Shapiro, H. M., Perlmutter, N. G., and Stein, P. G. (1998). A flow cytometer designed for fluorescence calibration. *Cytometry* **33,** 280–287.

Staprans, S. I., Dailey, P. J., Rosenthal, A., Horton, C., Grant, R. M., Lerche, N., and Feinberg, M. B. (1999). Simian immunodeficiency virus disease course is predicted by the extent of virus replication during primary infection. *J. Virol.* **73,** 4829–4839.

Stewart, C. C., and Stewart, S. J. (1999). Four color compensation. *Cytometry* **38,** 161–175.

Telford, W. G., and Frolova, E. G. (2004). Discrimination of the Hoechst side population in mouse bone marrow with violet and near-ultraviolet laser diodes. *Cytometry* **57A,** 45–52.

Telford, W. G., Hawley, T. S., and Hawley, R. G. (2003). Analysis of violet-excited fluorochromes by flow cytometry using a violet laser diode. *Cytometry* **54A,** 48–55.

Tung, J., Parks, D., Moore, W., Herzenberg, L., and Herzenberg, L. A. (2004). New approaches to fluorescence compensation and visualization of FACS data. *Clin. Immunol.* **110,** 277–283.

van Baarle, D., Kostense, S., van Oers, M. H., Hamann, D., and Miedema, F. (2002). Failing immune control as a result of impaired CD8+ T-cell maturation: CD27 might provide a clue. *Trends Immunol.* **23,** 586–591.

Veazey, R. S., DeMaria, M., Chalifoux, L. V., Shvetz, D. E., Pauley, D. R., Knight, H. L., Rosenzweig, M., Johnson, R. P., Desrosiers, R. C., and Lackner, A. A. (1998). Gastrointestinal tract as a major site of CD4+ T cell depletion and viral replication in SIV infection. *Science* **280,** 427–432.

Veazey, R. S., Mansfield, K. G., Tham, I. C., Carville, A. C., Shvetz, D. E., Forand, A. E., and Lackner, A. A. (2000). Dynamics of CCR5 expression by CD4(+) T cells in lymphoid tissues during simian immunodeficiency virus infection. *J. Virol.* **74,** 11001–11007.

Veazey, R. S., Marx, P. A., and Lackner, A. A. (2003). Vaginal CD4+ T cells express high levels of CCR5 and are rapidly depleted in simian immunodeficiency virus infection. *J. Infect. Dis.* **187,** 769–776.

von Andrian, U. H., and Mackay, C. R. (2000). T-cell function and migration. Two sides of the same coin. *N. Engl. J. Med.* **343,** 1020–1034.

Whetter, L. E., Ojukwu, I. C., Novembre, F. J., and Dewhurst, S. (1999). Pathogenesis of simian immunodeficiency virus infection. *J. Gen. Virol.* **80,** 1557–1568.

Wood, J. C. (1998). Fundamental flow cytometer properties governing sensitivity and resolution. *Cytometry* **33,** 260–266.

Zhang, D., Shankar, P., Xu, Z., Harnisch, B., Chen, G., Lange, C., Lee, S. J., Valdez, H., Lederman, M. M., and Lieberman, J. (2003). Most antiviral CD8 T cells during chronic viral infection do not express high levels of perforin and are not directly cytotoxic. *Blood* **101,** 226–235.

CHAPTER 23

Multicolor Immunophenotyping: Human Immune System Hematopoiesis

Brent Wood
Department of Laboratory Medicine
University of Washington
Seattle, Washington 98195

I. Introduction

Hematopoiesis begins with a quiescent stem cell that proliferates under environmental influences giving rise to progeny capable of differentiation along multiple lineages. In humans, this process occurs largely in the bone marrow, the one exception being T-cell differentiation, which occurs principally in the thymus. Differentiation along each lineage occurs in a relatively linear fashion through a sequential series of stages, resulting in fully functional mature forms that may undergo further activation and differentiation as a consequence of their interaction with different microenvironments and stimuli. For instance, a B cell differentiates from stem cells in the bone marrow of humans and matures through a

0091-679X/04 $35.00

series of stages producing the functional mature B cell (McKenna *et al.*, 2001), whereupon it exits the bone marrow and populates lymphoid tissue throughout the body. Additional maturation and processing of these naive B cells occurs in the germinal center of lymph nodes and other lymphoid tissue after antigen exposure, resulting in subsequent maturation to plasma cells or memory B cells (MacLennan, 1994).

The maturation of hematopoietic cells is a result of the tightly regulated and sequential expression of genes and gene products (Payne and Crooks, 2002). As a consequence, the derived protein products exhibit predictable and reproducible patterns of expression with maturation that correlate with morphological or functional stages. Cell-surface proteins represent a subset of these proteins with functions as diverse as extracellular enzymes, growth factor receptors, signal transduction molecules, and adhesion molecules. The functions for many of these molecules are relatively poorly understood, many being identified simply by the presence of antigenic reactivity. Some cell-surface antigens exhibit expression restricted to a particular cell lineage, and others are more widely expressed but with differing levels depending on maturational stage or activation state. Many of these antigens can be used to aid in the identification of maturational stages and lineage assignment once their patterns of expression are known. In addition to defining normal maturational states, disease states often exhibit alterations in antigenic expression reflecting abnormalities in maturation or function, and these alterations can be used in the diagnosis and monitoring of disease states.

Flow cytometry is an excellent technique for the examination of patterns of protein expression, particularly on the cell surface. Using fluorescent-labeled antibodies that bind with high avidity and specificity to selected antigens, a wealth of information regarding hematopoietic maturation has been generated over the past 2 decades. This chapter describes the identification of normal maturational stages of hematopoietic cells in human bone marrow by multiparametric flow cytometry using seven, eight, and nine simultaneous antibodies directed against antigens commonly evaluated in a clinical setting.

II. Methodology

Anticoagulated human bone marrow was obtained from the hematopathology laboratory at the University of Washington Medical Center as residual material sent for clinical evaluation. This material was used in accordance with a Human Subjects Protocol approved by our Institutional Review Board.

First, 100 μl of normal bone marrow containing roughly 10^6 cells was incubated at room temperature for 30 minutes, with 100 μl of a pretitered cocktail of monoclonal antibodies appropriate for the cell lineage being evaluated. The sample was then treated with 1.5 ml of buffered NH_4Cl containing 0.25% formaldehyde for 10 minutes at room temperature followed by a single wash in phosphate-buffered saline (PBS) using centrifugation at 550 g for 5 minutes. The resulting pellet was resuspended in 100 μl of PBS and roughly 1,000,000 events acquired on a modified LSRII (Becton-Dickinson Immunocytometry Systems, San Jose, California) capable of up to 10-color analysis. The instrument contains four spatially separated lasers (22 mW 407 nm diode, 100 mW variable 488 nm diode, 8 mW 594 nm HeNe, and 25 mW 638 nm diode) with independent fiberoptic-coupled detector arrays for each laser. The data were compensated and analyzed after acquisition using software developed in our laboratory. For demonstration of B-cell maturation, serum light chain was removed by prelysing 100 μl of bone marrow with 1.5 ml of buffered NH_4Cl for 10 minutes at room temperature followed by a single wash with PBS; the sample was then prepared as described earlier, with the lyse step omitted. To demonstrate erythroid maturation, 1 ml of bone marrow was subjected to density-gradient centrifugation (Ficoll) and the resulting mononuclear layer was washed twice with PBS before being used in the assay; the subsequent lysis step was omitted.

Fluorochromes used for these experiments included Pacific blue, fluorescein isothiocyanate (FITC), phycoerythrin (PE), PE–Cy5.5, PerCp–Cy5.5, PE–Cy7, Alexa 594, allophycocyanin (APC), Alexa 700, and APC–Cy7. Zenon reagents (Molecular Probes, Eugene, Oregon) were used for CD11b–Alexa 594, and all other antibodies were directly conjugated and obtained from either Beckman-Coulter or Becton-Dickenson.

III. Normal Immunophenotypic Patterns of Maturation

A. CD45 and Side Scatter Gating

CD45 is a transmembrane tyrosine phosphatase ubiquitously expressed by white blood cells (WBCs) and is important in modulating signals derived from integrin and cytokine receptors (Hermiston *et al.*, 2003). On WBCs, the level of CD45 is differentially expressed throughout maturation, being lower on blasts and immature forms and highest on mature myelomonocytic and lymphoid cells. In contrast, with erythroid differentiation, the low level of CD45 seen on blasts rapidly decreases and mature erythrocytes consistently show a lack of CD45 expression.

Orthogonal light scatter in large part reflects the presence of subcellular components capable of scattering light, in particular the presence of granules or vacuoles. It also is correlated with overall cell size or volume and thus shows a mild increase as cells enlarge with maturation. Consequently, the early acquisition of large cytoplasmic primary granules in neutrophilic lineage cells is accompanied by a marked increase in side scatter that allows their ready separation from progenitor cells. Further maturation along the neutrophilic lineage is marked by acquisition of smaller secondary granules and loss of primary granules with a concomitant decrease in side scatter. Monocytic differentiation is marked by a moderate increase in cell size, including increased amounts of cytoplasm containing few vacuoles, resulting in a mild increase in side scatter from that seen in blasts, though not to the degree seen in neutrophilic differentiation. In contrast, lymphoid differentiation is accompanied by a mild decrease in cell size with the small amount of associated cytoplasm containing essentially no granules or vacuoles. Consequently, lymphoid cells show a moderate decrease in side scatter from that seen in blasts. Erythroid differentiation is accompanied by a marked increase in cell size in early forms with a gradual decrease in size to the erythrocyte stage, with all stages having only modest amounts of cytoplasm without significant cytoplasmic structures. This results in an early increase in side scatter with a gradual decline to a final level similar to that seen in lymphocytes.

When combined, CD45 expression and orthogonal light scatter provide a powerful, simple, and reproducible method for distinguishing major cell lineages in bone marrow (Borowitz *et al.*, 1993). As a result, CD45–side scatter gating has become one of the most common techniques used for immunophenotyping in the clinical laboratory. The characteristic patterns seen using this method are illustrated in Fig. 1.

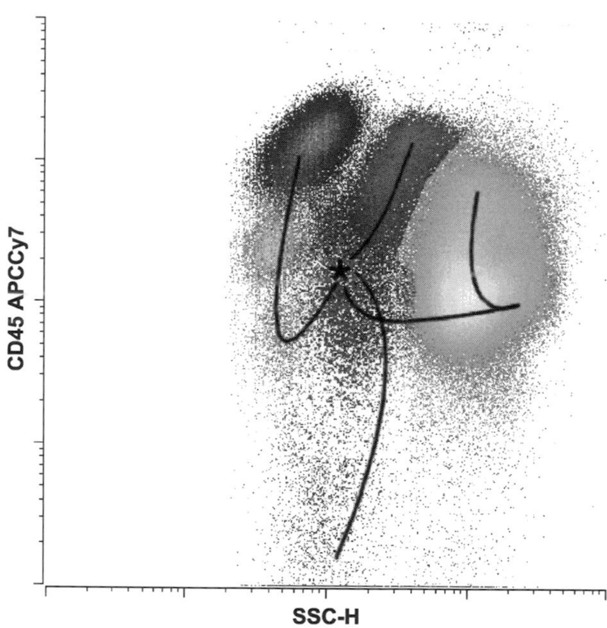

Fig. 1 CD45 versus side scatter representation of normal human bone marrow. Each of the major cell lineages differentiates from the early blast/stem cell population (yellow) (*asterisk*). The major more mature blast population (red) is easily identified by this display. Subsequent maturation toward the neutrophil (green), monocyte (purple), B lymphocyte (blue), and erythrocyte (gray) populations are indicated by the appropriate black line. (See Color Insert.)

B. Stem Cells and Blasts

The earliest easily identifiable hematopoietic cell in the marrow is a quiescent cell having the ability to differentiate along multiple lineages depending on the environmental influences to which it is exposed (i.e., a hematopoietic stem cell). The frequency of this population in normal bone marrow is low but can be routinely identified if a sufficiently large number of cells are evaluated (Bender *et al.*, 1991; D'Arena *et al.*, 1998; Gaipa *et al.*, 2002; Macedo *et al.*, 1995; McGuckin, 2003; Terstappen *et al.*, 1992; Tjonnfjord *et al.*, 1996). This population shows the expression of low forward and side scatter, intermediate CD45, intermediate human leukocyte antigen-DR (HLA-DR), bright CD34, low to absent CD38, intermediate CD133, low to intermediate CD13, low CD117, variable CD90, low CD123 and CD135, and very low CD33 without the expression of antigens characteristic of a particular lineage or later stages of differentiation (Fig. 2). As the cells mature, they begin to proliferate with a gain in the expression of CD38 and CD117, mildly decreased expression of CD34 and CD133, and decreased CD90. Further maturation results in each cell demonstrating a commitment to differentiation along a single lineage with resulting fragmentation of this relatively uniform-appearing population into different components, such as myelomonocytic, erythroid, lymphoid, and others. This is accompanied by the acquisition of lineage-associated antigens and patterns of antigen expression characteristic of that lineage, with all lineages ultimately showing a loss of CD34, CD133, CD117, and CD38. The earliest and most lineage-specific antigen expressed during myelomonocytic differentiation is CD64, with an associated mild increase in CD13, increased CD33, HLA-DR, and CD15, and increased side scatter. Early B cells show a loss of CD33 and CD13, a decrease in CD45 and side scatter, and the acquisition of CD10 and CD19 with retention of HLA-DR. Differentiation along the erythroid lineage is characterized by the acquisition of bright CD71 intermediate CD36 and a decrease in CD45, HLA-DR, CD13, and CD33 while showing a mild initial increase in side scatter.

C. Neutrophilic Maturation

Neutrophils mature from blasts through a linear process, historically divided into morphological stages termed *promyelocytes, myelocytes, metamyelocytes, bands*, and *neutrophils*. It is not until the band/neutrophilic stage that the cells are fully functional and capable of bactericidal action. Although these stages have associated antigenic changes, from an immunophenotypic perspective they are not so clearly defined and largely appear as a single continuum (Elghetany, 2002; Terstappen and Loken, 1990; Terstappen *et al.*, 1990, 1992). A selection of informative and useful antigens and their expression patterns with maturation are presented in Table I. Of particular interest are antigens that exhibit dramatic increases in expression at defined stages of maturation, such as acquisition of CD15 and CD66b by promyelocytes, acquisition of CD11b, CD11c, CD24, and CD66a by myelocytes, gain of CD55 by metamyelocytes, gain of CD35 and CD87

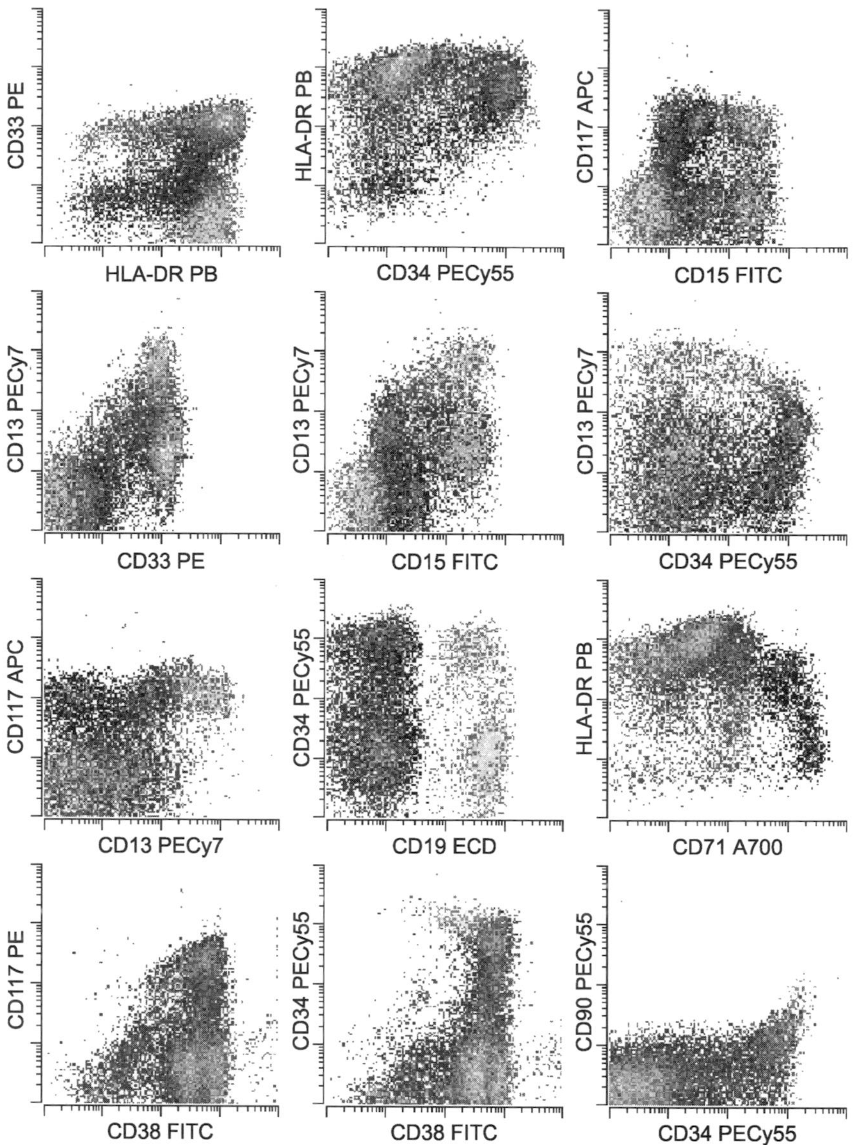

Fig. 2 Maturation and lineage commitment of blasts in human bone marrow. Normal human bone marrow was evaluated with two combinations of antibodies, one containing nine simultaneous fluorochromes (top three rows) and one containing eight simultaneous fluorochromes (bottom row) as labeled. The blast population is displayed following CD45/side scatter gating with stem cell (yellow), blast (red), neutrophil (green), monocyte (lavender), B cell (light blue), and erythroid (blue) populations colored. Note the complex patterns of maturation present in this subset of views that allow the unique identification of each of the major cell lineages. These views represent less than 20% of the total information contained in these antibody combinations. (See Color Insert.)

Table I
Surface Antigen Expression During Neutrophil Maturation

	Blast	Promyelocyte	Myelocyte	Metamyelocyte	Band	Neutrophil
CD10	−	−	−	−	−	++
CD11b	−	−	++	++	++	++
CD11c	−	−	++	++	++	++
CD13	++/+++	+++/++	+	+/++	++	+++
CD14	−	−	−	−	−	+
CD15	−	++	+++	+++	+++	+++
CD16	−	−	−	+	++	+++
CD18	++	+	+++	++	++	++
CD24	−	−	+	++	++	++
CD33	++	+++	++	+	+	+
CD35	−	−	−	−	++	++
CD44	+++	++	+	+	++	+++
CD45	++	+	+	+	++	++
CD55	+++	+	+	+++	+++	+++
CD64	−	+	++	++	−	−
CD65	−	++	++	++	+++	+++
CD66a	−	−	++	++	++	++
CD66b	−	+++	+++	++	++	++
CD87	−	−	−	−	++	++
CD117	++	+	−	−	−	−

by bands, and gain of CD10 by granulocytes. Another useful group of antigens including CD13, CD44, and CD55 is expressed early in maturation and exhibits a decrease at an intermediate stage of maturation only to increase with terminal differentiation. A less common pattern of antigen expression, exemplified by CD64, is characterized by an absence of expression on blasts, an increase at intermediate stages of maturation, and a loss with terminal differentiation. A final group of antigens (e.g., CD33) is expressed at relatively high levels on early myeloid forms and declines with maturation. The combined use of antigens from each of these groups in multiparametric combinations serves a particularly useful function in more precisely defining maturational stages and makes the appreciation of maturational abnormalities in disease states more readily apparent. Examples of these expression patterns are displayed in Fig. 3.

D. Monocytic Maturation

Monocytes mature from blasts as a continuum through one intermediate stage termed the *promonocyte* (Terstappen and Loken, 1990; Terstappen *et al.*, 1992). The expression levels of individual antigens are listed in Table II and illustrated in Fig. 4. With maturation toward promonocytes, blasts gain the expression of CD64, CD33, HLA-DR, CD36, and CD15, with an initial mild decrease in CD13 and an increase in CD45. Further maturation toward mature monocytes

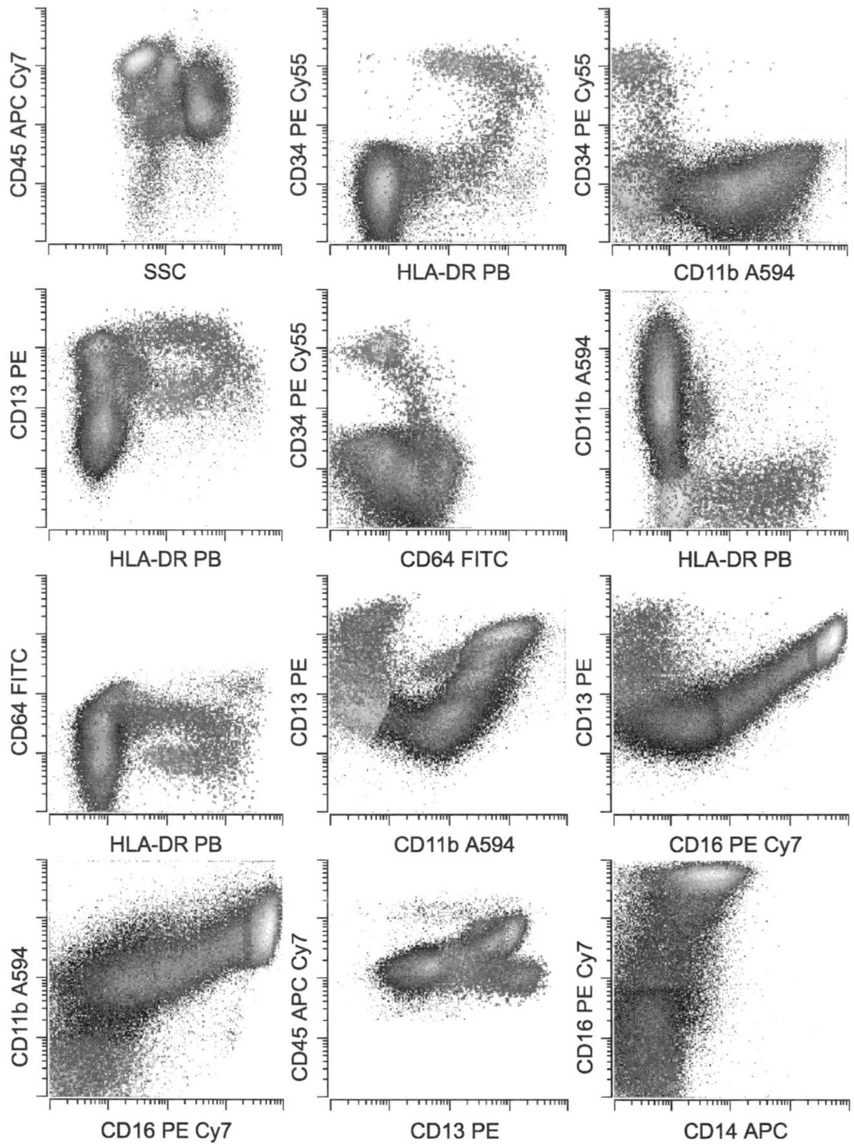

Fig. 3 Maturation of cells during neutrophilic differentiation. Normal human bone marrow was evaluated with a single combination of nine simultaneous antibodies. Maturation from early (yellow) and late (green) blasts proceeds through promyelocyte (light blue), myelocyte (blue), metamyelocyte/band (lavender) and neutrophil (pink) stages. Eosinophils can also be noted by their increased autofluorescence (orange). Though recognizable, these stages are not discretely defined and the subdivisions are somewhat arbitrary. The blasts have been emphasized and other populations largely excluded to improve visibility of the maturational relationships. These views represent less than 20% of the total information contained in these antibody combinations. (See Color Insert.)

Table II
Surface Antigen Expression During Monocyte Maturation

	Blast	Promonocyte	Monocyte
CD4	−/+	+	+
CD11b	−	++	+++
CD13	++	+/++	++/+++
CD14	−	+/++	+++
CD15	−	++	+
CD16	−	−	−/+
CD33	+++	+++	+++
CD36	−	++	+++
CD45	+	++	+++
CD64	−	++	+++
HLA-DR	++	+++	++/+++

shows a progressive increase in CD14, CD11b, CD13, CD36, and CD45, with a mild decrease in HLA-DR and CD15. Unlike with maturing myeloid forms, the expression of CD64, HLA-DR, and CD33 are retained on maturing monocytes at relatively high levels and serve as useful monocytic markers. Mature monocytes show expression of bright CD14, relatively bright CD33, variably bright CD13, bright CD36 and CD64, and low CD15.

E. Erythroid Maturation

Examination of erythroid maturation requires a different method of specimen preparation than the red blood cell lysis methods commonly used to process bone marrow. Commonly used lysing reagents such as NH_4Cl with subsequent washing steps generally leave only very early erythroid precursors (proerythroblasts) to evaluate or at a minimum noticeably compromise more mature erythroid forms as detected by a prominent loss in forward scatter. Additionally, mature red blood cells must also be significantly reduced in number without undue loss of immature forms if one wants to examine antigens highly expressed on mature red blood cells (e.g., CD235a). Density-gradient centrifugation with either Percoll or Ficoll is an acceptable alternative for this purpose.

Erythroid forms have historically been divided into maturational stages based on their morphological appearance with the nucleated forms termed *proerythroblasts, basophilic erythroblasts, polychromatophilic erythroblasts*, and *orthochromatophilic erythroblasts*, and the anucleate forms termed *polychromatophilic erythrocytes* and *erythrocytes*. These morphological stages correlate reasonably well with observed immunophenotypic changes (De Jong *et al.*, 1995; Loken *et al.*, 1987; Rogers *et al.*, 1996; Scicchitano *et al.*, 2003), although the differences between polychromatophilic and orthochromatophilic stages are ill-defined. The earliest erythroid forms to arise from blasts are identified by their acquisition of

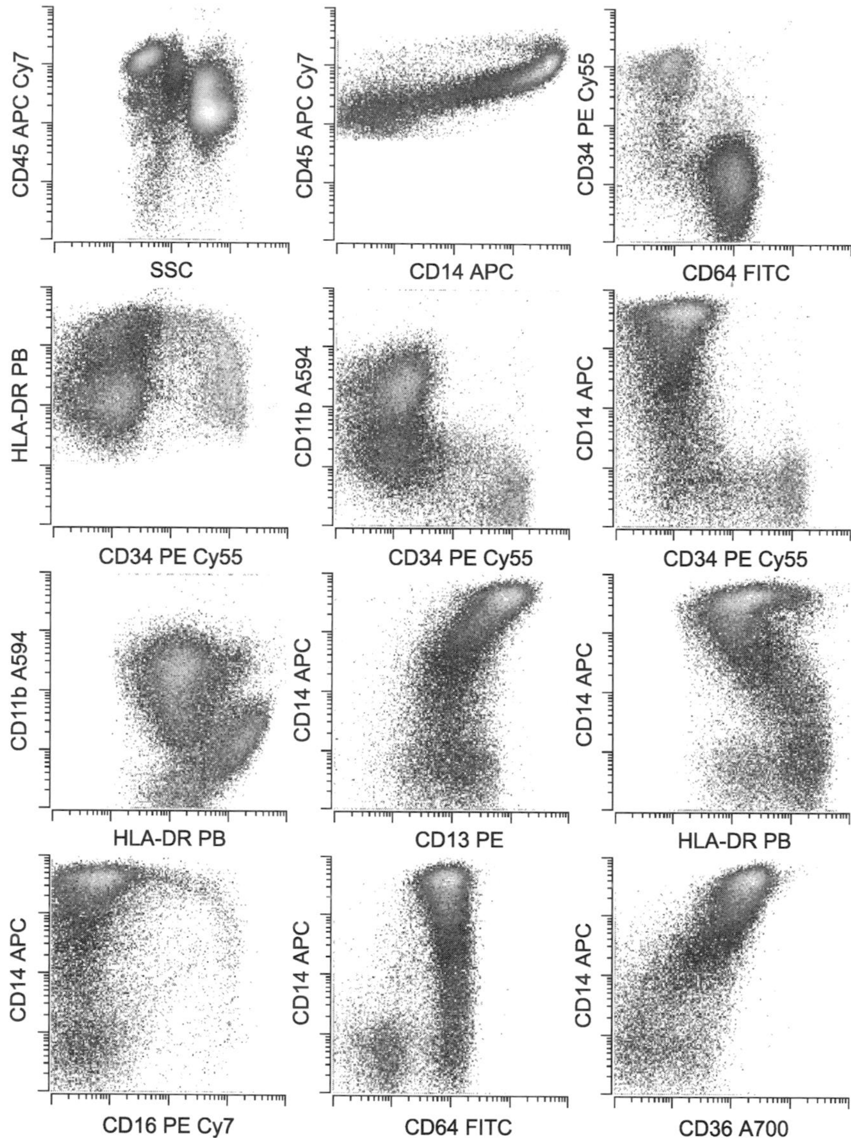

Fig. 4 Maturation of cells during monocytic differentiation. Normal human bone marrow was evaluated with a single combination of nine simultaneous antibodies. Maturation from early (yellow) and late (green) blasts proceeds through promonocyte (light blue) and monocyte (lavender) stages. Though recognizable, these stages are not discretely defined and the subdivisions are somewhat arbitrary. The blasts have been emphasized and other populations largely excluded to improve visibility of the maturational relationships. These views represent less than 20% of the total information contained in these antibody combinations. (See Color Insert.)

Table III
Surface Antigen Expression during Erythrocyte Maturation

	Blast	Proerythroblast	Basophilic	Poly/Ortho	Retic	Mature
CD34	++	−/+	−	−	−	−
CD36	−	++	+++	++	−/+	−
CD38	++	+	−/+	−	−	−
CD45	++	+	−/+	−	−	−
CD71	−	+++	+++	+++	+/++	−
CD117	++	++	−	−	−	−
CD235a	−	+/++	+++	+++	+++	+++
HLA-DR	++	++	+	−	−	−

bright CD71, intermediate CD36 and expression of CD117 with intermediate CD45, and slightly higher forward and side scatter. Glycophorin A (CD235a) is expressed at a low level at this stage. Maturation to the basophilic erythroblast is accompanied by a decrease in CD45 and acquisition of bright CD235a expression to the level seen throughout the remainder of erythroid maturation, including mature erythrocytes. Transition to polychromatophilic/orthochromatophilic erythroblasts shows a further loss of CD45 and HLA-DR and a mild decrease in CD36. After these stages, the cells rapidly lose CD71 and CD36 after becoming anucleate. These changes are described in Table III and illustrated in Fig. 5.

F. B-Cell Maturation

In humans, B-cell maturation occurs from the hematopoietic stem cell in the bone marrow and proceeds through relatively discrete well-defined stages unlike the continuum seen in myelomonocytic and erythroid maturation (Davis *et al.*, 1994; Dworzak *et al.*, 1997, 1998; Loken *et al.*, 1987; Longacre *et al.*, 1989; Lucio *et al.*, 1999; McKenna *et al.*, 2001). Upon reaching a functional stage characterized by expression of immunoglobulin on the cell surface, B cells exit the bone marrow and undergo further maturation in the presence of antigen in peripheral tissues such as lymph node. A subset of these B cells will undergo terminal differentiation to plasma cells and few will return to the bone marrow. Consequently, bone marrow B cells typically consist of a mixture of maturing B cells, a few mature but naive B cells, and a small number of plasma cells.

The immunophenotypic stages of B-cell maturation present in normal bone marrow are detailed in Table IV and examples of the observed patterns of antigen expression are shown in Fig. 6. The antigens expressed by the earliest identifiable B cells include TdT, CD79a, and low CD10 without CD19, but the size of this population is very small and is generally not observed. The next stage (early) is the first that can be readily identified and shows the acquisition of CD19, an antigen that serves as a useful marker of B cells through terminal plasma cell differentiation. At this stage, the B cells express the immature antigens CD34 and nuclear

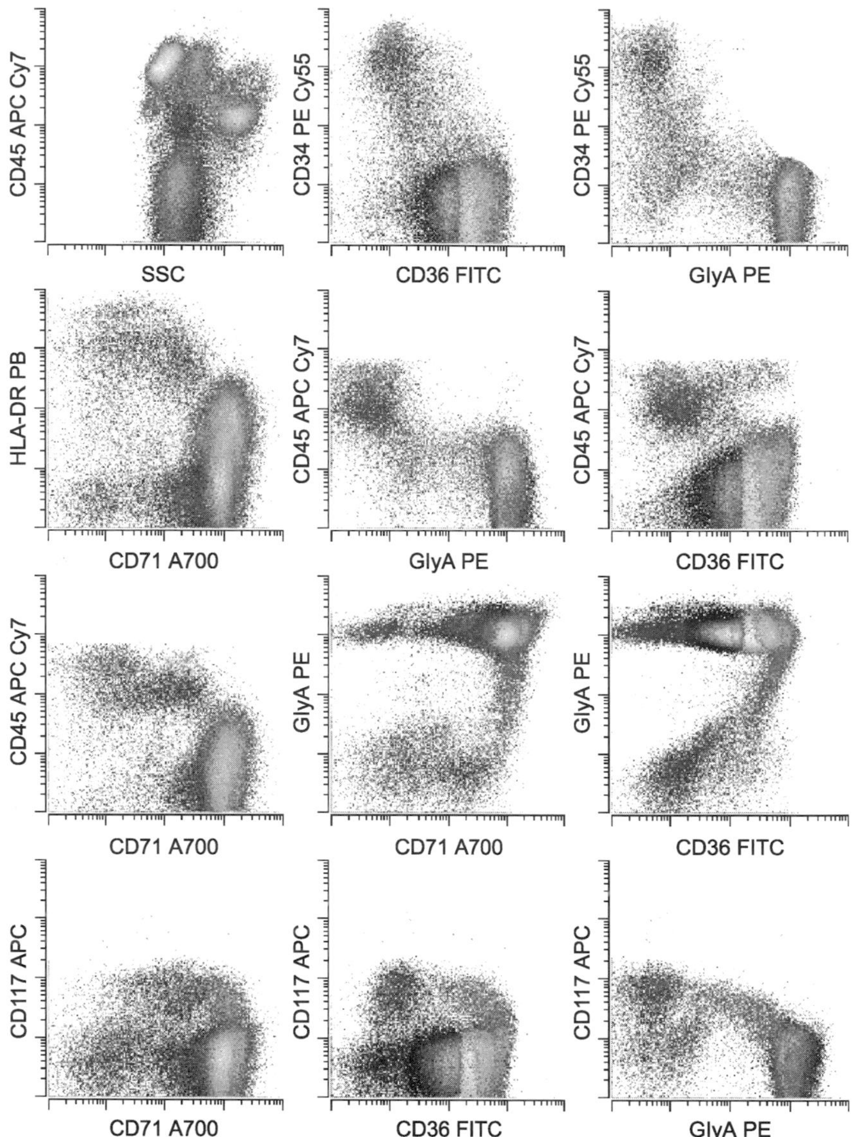

Fig. 5 Maturation of cells during erythroid differentiation. Normal human bone marrow was evaluated with a single combination of seven simultaneous antibodies. Maturation from early (yellow) and late (red) blasts proceeds through proerythroblast (green), basophilic erythroblast (yellow), polychromatophilic and orthochromatophilic (light blue and blue), and erythrocyte (lavender) stages. Though recognizable, these stages are not discretely defined and the subdivisions are somewhat arbitrary. The blasts have been emphasized and other populations largely excluded to improve visibility of the maturational relationships. These views represent less than 20% of the total information contained in these antibody combinations. (See Color Insert.)

Table IV
Surface Antigen Expression during B-Cell Maturation

	Early	Intermediate	Late	Mature	Plasma cell
CD10	+++	++	+	–	–
CD19	++	+++	++	++	++
CD20	–	+/++	+++	++	–
CD21	–	–	–	++	+
CD22	+	+	++	++	–
CD24	+++	+++	+++	++	–
CD34	++	–	–	–	–
CD38	++	+++	++	+/++	++++
CD45	+	++	+++	+++	++
HLA-DR	++	+++	++	++	+/–
cIgM	–	++	++	++	–
Kappa	–	–	+	++	+
Lambda	–	–	+	++	+
TdT	+++	–	–	–	–

TdT with bright CD10 and low CD45 without CD20 or surface immunoglobulin. Transition to the intermediate stage of maturation results in an abrupt decrease in CD10, a gain in CD45, gradual acquisition of CD20, and a loss of CD34 and nuclear TdT. The last immature stage (late) is generally small in number but is characterized by the brightest level of CD20, a further decrease in CD10, increased CD45, and acquisition of surface immunoglobulin. Mature B cells show a loss of CD10, bright CD20, and the brightest CD45 with expression of surface immunoglobulin. Plasma cells are readily identified by their extremely bright expression of CD38 or CD138 without CD20. The relative proportions of each of these maturational stages are dependent on the overall regenerative state of the bone marrow and the presence of ongoing inflammatory reactions.

G. Miscellaneous

A number of other cell populations mature or are present in normal bone marrow and include eosinophils, basophils, dendritic cells, natural killer cells, and megakaryocytes. For most of these populations, the patterns of maturation are either not well defined or difficult to visualize using routine methods of analysis.

Of these populations, eosinophil maturation is the easiest to demonstrate and is characterized by one immature stage (eosinophilic myelocyte), showing expression of high side scatter, intermediate CD45 at a level slightly higher than neutrophilic myelocytes, low to intermediate CD11b, intermediate CD13, and low CD33 with bright CD66b without CD16. Maturation to the mature eosinophil is accompanied by an increased level of CD45, a mild decrease in side scatter, and an increase in CD11b with a decrease in CD33. Eosinophilic maturation is illustrated in Fig. 7.

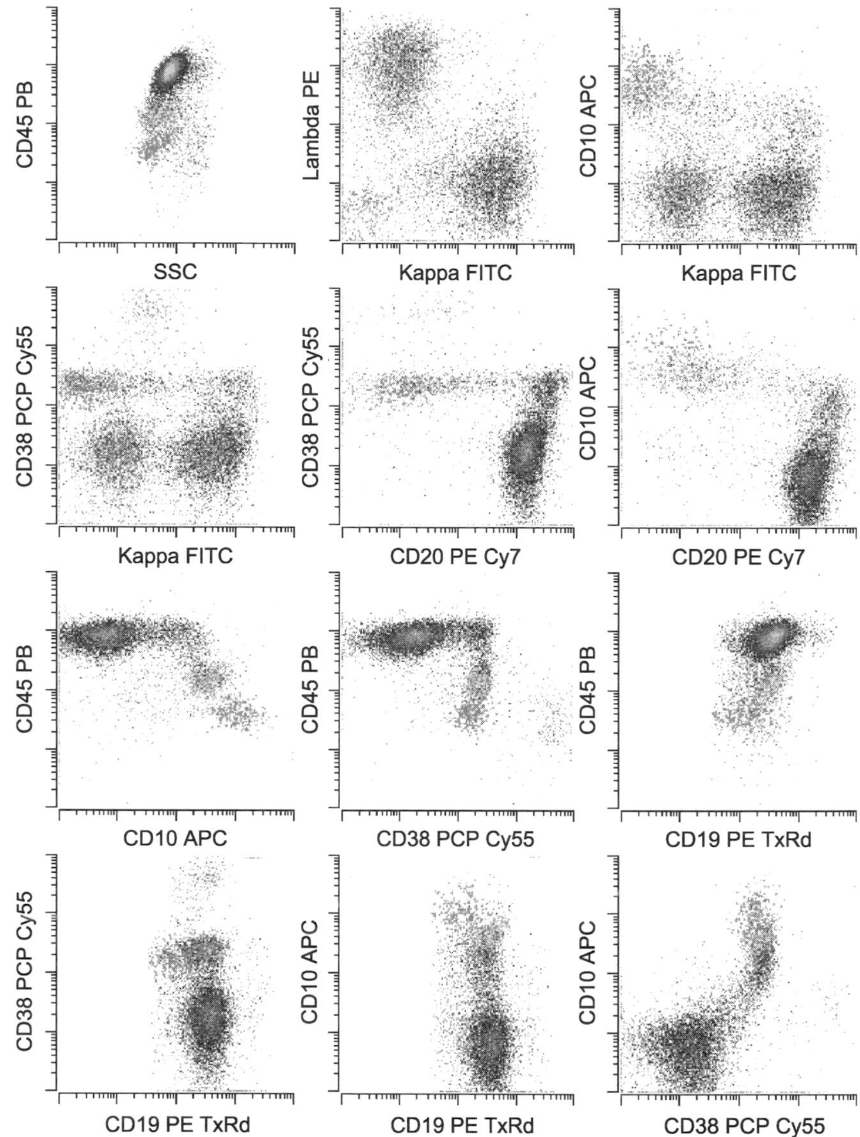

Fig. 6 Maturation of cells during B-cell differentiation. Normal human bone marrow was evaluated with a single combination of eight simultaneous antibodies. Maturation from early (yellow) B cells proceeds through an intermediate stage (light blue) before acquiring surface light-chain expression (blue and red: kappa and lambda, respectively) as late immature B cells (bright CD20) before reaching maturity (bright CD45 without CD10). Rare plasma cells are also present (green). Note the relatively discrete nature of the maturational stages. These views represent less than 20% of the total information contained in these antibody combinations. (See Color Insert.)

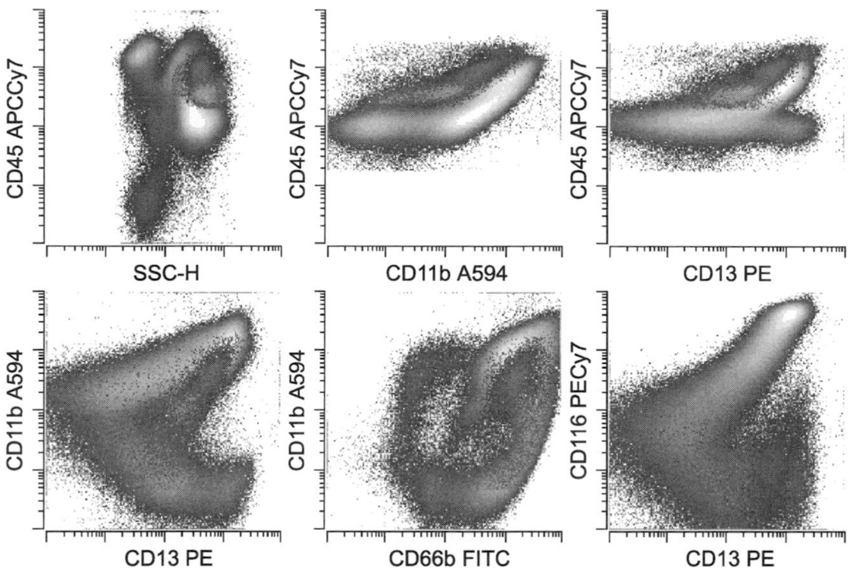

Fig. 7 Maturation of cells during eosinophilic differentiation. Normal human bone marrow was evaluated with a single combination of nine simultaneous antibodies. Maturation from blasts proceeds through the eosinophilic myelocyte stage (yellow) to the mature eosinophil (red). Note the high side scatter present throughout maturation, as well as the increased CD45 and lower CD11b relative to neutrophilic precursors. These views represent less than 20% of the total information contained in these antibody combinations. (See Color Insert.)

IV. Abnormal Immunophenotypic Patterns of Maturation

The normal patterns of antigenic expression detailed earlier reflect the well-orchestrated expression of genes characteristic of normal differentiation. In hematopoietic neoplasms, an increasing variety of specific genetic abnormalities have been described that are either directly or indirectly capable of perturbing these normal patterns of protein expression. Stem cell disorders such as myelodysplastic syndromes serve as an informative model of maturational dysregulation and illustrate the association between genetic and immunophenotypic abnormalities (Kussick and Wood, 2003a; Wells *et al.*, 2003). Figure 8 shows an example of the type of abnormalities seen in myelodysplasia. Immunophenotypic abnormalities can be identified in most hematopoietic neoplasms including acute myeloid and lymphoid leukemia (Weir and Borowitz, 2001), myeloproliferative disorders (Kussick and Wood, 2003b), and B- and T-cell lymphoma (Stetler-Stevenson and Braylan, 2001). These abnormalities can be used to diagnose, classify, and monitor disease following therapy and techniques capable of detecting these abnormalities, such as flow cytometry, play an increasingly important role in clinical medicine.

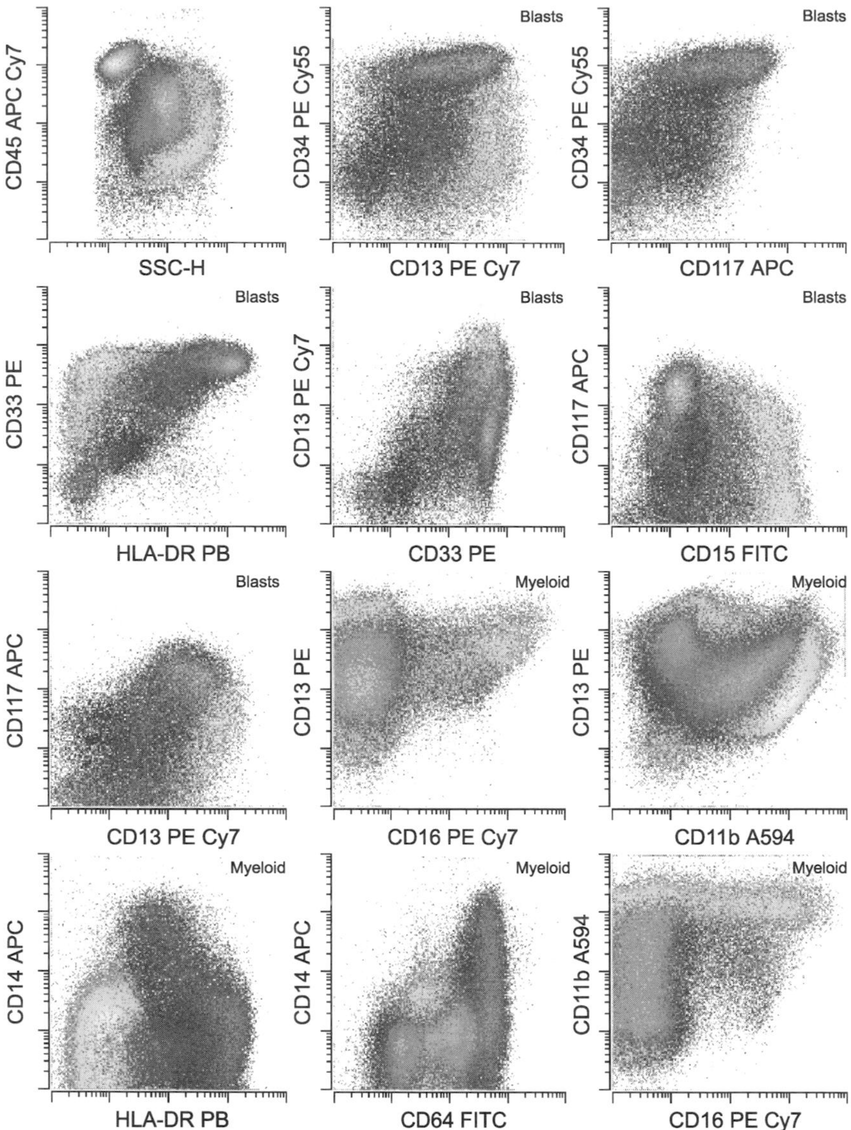

Fig. 8 Myelodysplasia. Human bone marrow was evaluated with a single combination of nine simultaneous antibodies. Blasts are increased in number (red) and show abnormal expression of CD117, CD34, CD13, and CD45. In addition, the neutrophilic lineage shows hypogranularity (decreased side scatter) with abnormal maturational expression of CD13, CD11b, and CD16 with a dyssynchronous pattern of CD13 and CD16 expression. Monocytes also show a relative increase in immature forms with abnormal CD45 and CD14 expression. These findings are characteristic of myelodysplasia. (See Color Insert.)

References

Bender, J. G., Unverzagt, K. L., Walker, D. E., Lee, W., van Epps, D. E., Smith, D. H., Stewart, C. C., and To, L. B. (1991). Identification and comparison of CD34+ cells and their subpopulations from normal peripheral blood and bone marrow using multicolor flow cytometry. *Blood* **77,** 2591–2596.

Borowitz, M. J., Guenther, K. L., Shults, K. E., and Stelzer, G. T. (1993). Immunophenotyping of acute leukemia by flow cytometric analysis. Use of CD45 and right-angle light scatter to gate on leukemic blasts in three-color analysis. *Am. J. Clin. Pathol.* **100,** 534–540.

D'Arena, G., Musto, P., Cascavilla, N., and Carotenuto, M. (1998). Thy-1 (CDw90) and c-Kit receptor (CD117) expression on CD34+ hematopoietic progenitor cells: A five dimensional flow cytometric study. *Haematologica* **83,** 587–592.

Davis, R. E., Longacre, T. A., and Cornbleet, P. J. (1994). Hematogones in the bone marrow of adults. Immunophenotypic features, clinical settings, and differential diagnosis. *Am. J. Clin. Pathol.* **102,** 202–211.

De Jong, M. O., Wagemaker, G., and Wognum, A. W. (1995). Separation of myeloid and erythroid progenitors based on expression of CD34 and c-kit. *Blood* **86,** 4076–4085.

Dworzak, M. N., Fritsch, G., Fleischer, C., Printz, D., Froschl, G., Buchinger, P., Mann, G., and Gadner, H. (1997). Multiparameter phenotype mapping of normal and post-chemotherapy B lymphopoiesis in pediatric bone marrow. *Leukemia* **11,** 1266–1273.

Dworzak, M. N., Fritsch, G., Froschl, G., Printz, D., and Gadner, H. (1998). Four-color flow cytometric investigation of terminal deoxynucleotidyl transferase-positive lymphoid precursors in pediatric bone marrow: CD79a expression precedes CD19 in early B-cell ontogeny. *Blood* **92,** 3203–3209.

Elghetany, M. T. (2002). Surface antigen changes during normal neutrophilic development: A critical review. *Blood Cells Molec. Dis.* **28,** 260–274.

Gaipa, G., Coustan-Smith, E., Todisco, E., Maglia, O., Biondi, A., and Campana, D. (2002). Characterization of CD34+, CD13+, CD33-cells, a rare subset of immature human hematopoietic cells. *Haematologica* **87,** 347–356.

Hermiston, M. L., Xu, Z., and Weiss, A. (2003). CD45: A critical regulator of signaling thresholds in immune cells. *Annu. Rev. Immunol.* **21,** 107–137.

Kussick, S. J., and Wood, B. L. (2003a). Using 4-color flow cytometry to identify abnormal myeloid populations. *Arch. Pathol. Lab. Med.* **127,** 1140–1147.

Kussick, S. J., and Wood, B. L. (2003b). Four-color flow cytometry identifies virtually all cytogenetically abnormal bone marrow samples in the workup of non-CML myeloproliferative disorders. *Am. J. Clin. Pathol.* **120,** 854–865.

Loken, M. R., Shah, V. O., Dattilio, K. L., and Civin, C. I. (1987). Flow cytometric analysis of human bone marrow: I. Normal erythroid development. *Blood* **69,** 255–263.

Longacre, T. A., Foucar, K., Crago, S., Chen, I. M., Griffith, B., Dressler, L., McConnell, T. S., Duncan, M., and Gribble, J. (1989). Hematogones: a multiparameter analysis of bone marrow precursor cells. *Blood* **73,** 543–552.

Lucio, P., Parreira, A., van den Beemd, M. W., van Lochem, E. G., van Wering, E. R., Baars, E., Porwit-MacDonald, A., Bjorklund, E., Gaipa, G., Biondi, A., Orfao, A., Janossy, G., van Dongen, J. J., and San Miguel, J. F. (1999). Flow cytometric analysis of normal B cell differentiation: A frame of reference for the detection of minimal residual disease in precursor-B-ALL. *Leukemia* **13,** 419–427.

Macedo, A., Orfao, A., Ciudad, J., Gonzalez, M., Vidriales, B., Lopez-Berges, M. C., Martinez, A., Landolfi, C., Canizo, C., and San Miguel, J. F. (1995). Phenotypic analysis of CD34 subpopulations in normal human bone marrow and its application for the detection of minimal residual disease. *Leukemia* **9,** 1896–1901.

MacLennan, I. C. (1994). Germinal centers. *Annu. Rev. Immunol.* **12,** 117–139.

McGuckin, C. P., Pearce, D., Forraz, N., Tooze, J. A., Watt, S. M., and Pettengell, R. (2003). Multiparametric analysis of immature cell populations in umbilical cord blood and bone marrow. *Eur. J. Haematol.* **71,** 341–350.

McKenna, R. W., Washington, L. T., Aquino, D. B., Picker, L. J., and Kroft, S. H. (2001). Immunophenotypic analysis of hematogones (B-lymphocyte precursors) in 662 consecutive bone marrow specimens by 4-color flow cytometry. *Blood* **98,** 2498–2507.

Payne, K. J., and Crooks, G. M. (2002). Human hematopoietic lineage commitment. *Immunol. Rev.* **187,** 48–64.

Rogers, C. E., Bradley, M. S., Palsson, B. O., and Koller, M. R. (1996). Flow cytometric analysis of human bone marrow perfusion cultures: Erythroid development and relationship with burst-forming units-erythroid. *Exp. Hematol.* **24,** 597–604.

Scicchitano, M. S., McFarland, D. C., Tierney, L. A., and Narayanan, P. K, and Schwartz, L. W. (2003). *In vitro* expansion of human cord blood CD36+ erythroid progenitors: Temporal changes in gene and protein expression. *Exp. Hematol.* **31,** 760–769.

Stetler-Stevenson, M., and Braylan, R. C. (2001). Flow cytometric analysis of lymphomas and lymphoproliferative disorders. *Semin. Hematol.* **38,** 111–123.

Terstappen, L. W., Buescher, S., Nguyen, M., and Reading, C. (1992). Differentiation and maturation of growth factor expanded human hematopoietic progenitors assessed by multidimensional flow cytometry. *Leukemia* **6,** 1001–1010.

Terstappen, L. W., and Loken, M. R. (1990). Myeloid cell differentiation in normal bone marrow and acute myeloid leukemia assessed by multi-dimensional flow cytometry. *Anal. Cell Pathol.* **2,** 229–240.

Terstappen, L. W., Safford, M., and Loken, M. R. (1990). Flow cytometric analysis of human bone marrow. III. Neutrophil maturation. *Leukemia* **4,** 657–663.

Tjonnfjord, G. E., Steen, R., Veiby, O. P., and Egeland, T. (1996). Lineage commitment of CD34+ human hematopoietic progenitor cells. *Exp. Hematol.* **24,** 875–882.

Weir, E. G., and Borowitz, M. J. (2001). Flow cytometry in the diagnosis of acute leukemia. *Semin. Hematol.* **38,** 124–138.

Wells, D. A., Benesch., M., Loken, M. R., Vallejo, C., Myerson, D., Leisenring, W. M., and Deeg, H. J. (2003). Myeloid and monocytic dyspoiesis as determined by flow cytometric scoring in myelodysplastic syndrome correlates with the IPSS and with outcome after hematopoietic stem cell transplantation. *Blood* **102,** 394–403.

CHAPTER 24

Multicolor Immunophenotyping: Human Mature Immune System

Stephen C. De Rosa

University of Washington School of Medicine
Department of Laboratory Medicine
Seattle, Washington 98109

I. Introduction

Multicolor flow cytometry has advanced remarkably in the last few years. Until a few years ago, flow cytometry using more than four or five colors was available in relatively few laboratories, largely because of the limited availability of appropriate instrumentation and appropriate reagents. Commercial manufacturers are now marketing benchtop instruments and sorters capable of measuring 12 or more colors. In addition, new staining reagents that use a wider variety of fluorochromes are now being introduced into the marketplace. Therefore, although simultaneous measurement of 12 colors may still be limited to a few laboratories, flow cytometric analysis and sorting using more than six colors is now being used more widely.

METHODS IN CELL BIOLOGY, VOL. 75

0091-679X/04 $35.00

The ability to measure many markers simultaneously is resulting in the identification of multiple new cellular subsets and is revealing new levels of complexity of the immune system. In addition to revealing this complexity, it also provides new tools to examine the immune system in greater detail. One major advantage of multicolor flow cytometry is the ability to combine markers that precisely identify cellular subsets with additional markers that access function. This allows functional information to be obtained at the single-cell level for cells that are carefully characterized by surface marker expression. It is expected that increasing amounts of this type of data will appear in the literature as more research groups use this technology.

This chapter focuses on the examination of blood, as opposed to other tissues of the immune system. Blood is conveniently sampled and is the primary source of immune cells in our laboratory. The methods discussed here apply to the study of peripheral blood mononuclear cells (PBMCs) and specifically to lymphocytes. T cells are the primary focus of this chapter, and there is only limited discussion of natural killer (NK) and B cells. Other cell types present in PBMCs, such as granulocytes, monocytes, and dendritic cells, are not discussed.

Several issues need to be considered when studying the immune system with multicolor flow cytometry. Any analysis beyond simply looking at lineage-defining markers requires at least three colors, and more detailed analysis requires six or more colors, so in this chapter, we focus on flow cytometric studies using six or more colors. However, the concepts in general also apply to flow cytometric studies using fewer colors.

The first part of the chapter discusses some of the markers used to identify and classify T-cell subsets. This is not intended to list all markers of interest, but we discuss a large number of the markers commonly used to identify naive and memory T-cell subsets. Most of these are cell-surface markers, but we also discuss some intracellular markers as additional phenotyping markers and markers that provide information about function.

After this is a discussion of some of the issues involved with designing staining panels. This begins with a brief description of the fluorochromes or dyes commonly used and some characteristics of these dyes when used for flow cytometry. Some concepts such as spectral overlap and compensation between dyes are also described briefly. Some practical guidelines are presented for matching markers with dyes and a few special issues related to panel design are discussed.

II. Choice of Markers

A. Overview

There are several approaches to categorizing T cells. T cells can be characterized and subdivided in many different ways. A functional classification is the most useful. Indeed, some functions can be examined on a cell-by-cell basis by flow cytometry (discussed later in this chapter). However, a complete functional profiling of individual cells is not yet possible. Therefore, much of the classification of T cells is based on the differential expression of surface antigens.

For some of the subsets identified by surface marker phenotyping, a few functional attributes have been determined by examining subsets that have been isolated or sorted. For example, there are major functional differences between the broadly defined CD4- and CD8-expressing T-cell subsets. As flow cytometric capabilities expand and more markers are examined simultaneously, many more T-cell subsets are continuing to be described. It is now clear that the functions described for "bulk" T-cell subsets, such as CD4 and CD8 T cells, need to be reevaluated because these bulk subsets can be further subdivided into subsets of cells expressing distinct sets of surface markers. These subpopulations likely express unique functional profiles.

Because of the large number of these T-cell subsets that have been identified, classification of these subsets is a difficult task. One approach to the classification of these subsets has been the attempt to identify cells at different stages of differentiation or maturation. Some of these approaches are discussed briefly in the sections that follow. In the next section, there is an extended discussion concerning the identification of naive T cells. This issue is discussed in more detail because there is considerable confusion over the individual markers or the sets of markers that best identify naive cells.

B. Accurately Identifying Naive T Cells

Naive T cells are those cells that have not yet encountered their cognate antigen. It is often necessary to identify naive cells and to distinguish naive cells from activated or memory cells. Several surface markers are expressed at different levels on naive and non-naive cells. These are used individually or in combination to identify naive cells. Often a single marker, such as CD45RA or CD45RO, is used. However, we and others have shown that the use of one marker is not sufficient to accurately identify naive cells without including a substantial number of non-naive cells (De Rosa *et al.*, 2001, 2004; Rabin *et al.*, 1995; Roederer *et al.*, 1995). A second naive-defining marker, such as CD62L, CD11a, or CD27, needs to be examined. This is especially true for CD8 T cells. Figure 1 shows that among the cells classified as naive based on one marker ($CD45RO^-$), there is a relatively large proportion of cells that are non-naive based on the second marker ($CD62L^-$, $CD11a^{bright}$, or $CD27^{dull}$).

For CD4 T cells, the proportion of non-naive cells included among the cells that are $CD45RO^-$ is generally very small. In some circumstances, this single marker may be sufficiently accurate to identify naive CD4 T cells when high purity of naive cells is not required. However, there is variation among individuals in the proportion of the non-naive CD4 T cells included among the $CD45RO^-$ cells. In addition, in disease states where naive CD4 or CD8 cells are few in number (such as in human immunodeficiency virus [HIV] disease), the use of a single

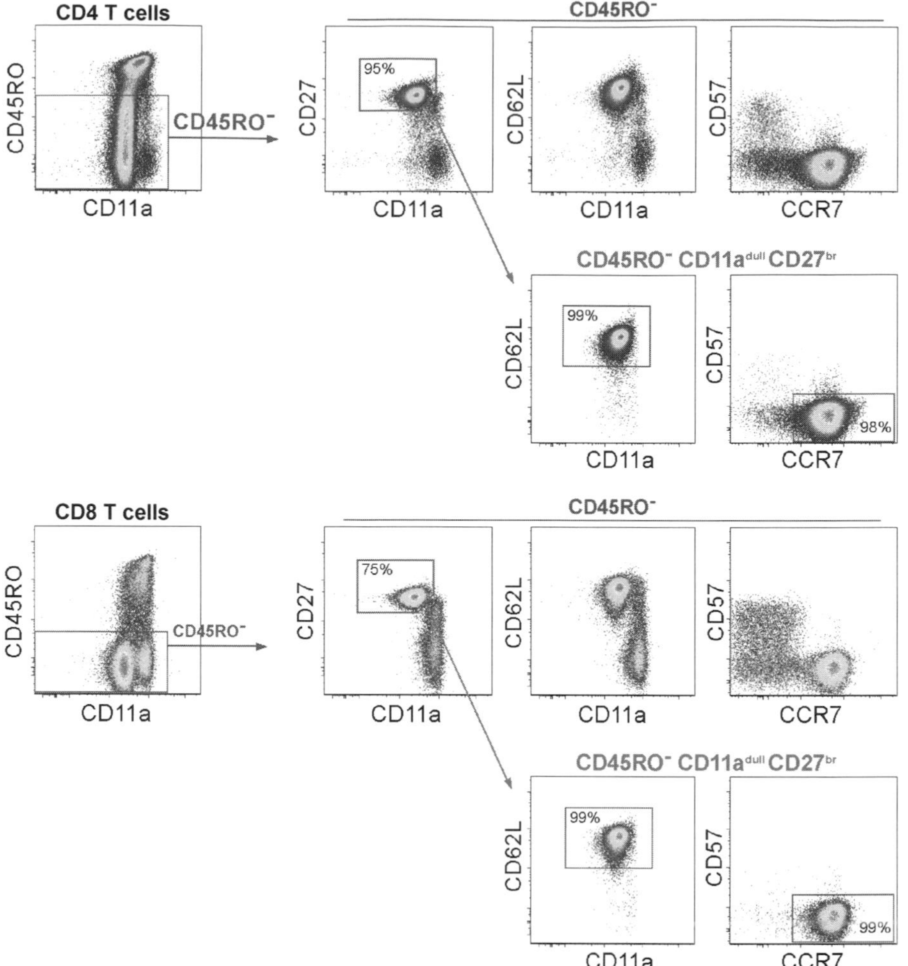

Fig. 1 Naive T cells are accurately identified using three markers. Naive cells do not express CD45RO, so CD4 and CD8 were gated only on CD45RA−cells (left panels). The expression of several other naive-defining markers are shown for the CD45RO−cells (upper three panels for CD4 and CD8 T cells). Naive cells are then gated as CD11a^{dull} and CD27^{br} as shown by the red gate. The expression of the remaining naive markers is shown for the cells gated on CD45RO, CD11a, and CD27 (lower right panels for CD4 and CD8 T cells). The percentages are shown as a percentage of the parent gate. All data are from one 10-color staining panel used to analyze freshly isolated peripheral blood mononuclear cells from one adult donor. (See Color Insert.)

naive-defining marker is clearly inadequate because the relative proportion of non-naive cells among the CD45RO^{-} cells increases, often dramatically.

This is why we never rely on just one naive-defining marker. In fact, we have shown that use of more than two naive-defining markers improves the accuracy of naive cell identification. In addition, the choice of which combinations of markers are used greatly affects the accuracy. We have compared different sets of naive-defining markers used in combination and can make suggestions based on our experience for the combinations of markers that work well. We have found that either expression of CD45RA or nonexpression of CD45RO provides similar information, so there is little need to use these markers together. In other words, we find few, if any, cells in blood that are double negative or double positive for these markers. Note that for both CD45RA and CD45RO, there are cells that express intermediate levels of both these markers (these are non-naive). However, cells expressing bright levels of both markers are rarely observed. Note also that this may not be the case when examining cells in locations other than blood, although we have not examined this.

We always include either CD45RA or CD45RO, along with at least one other marker and preferably a third marker. CD62L works well as the second marker. However, there are situations when CD62L cannot be used. CD62L is proteolytically cleaved off cells upon stimulation. In addition, surface expression is not stable after freeze/thaw. In these situations, an alternative marker should be used. We then use CD11a, which is also useful when used as a third marker along with CD62L. All non-naive cells express higher levels of CD11a as compared with naive cells. CD27 is useful as a third marker when CD11a is used as the second marker. It is important to note that CD27 is expressed at three levels. Naive cells express high levels of CD27. Because it can be difficult to distinguish high from intermediate levels of expression, we feel that CD27 is not the best choice for a second marker. Figure 1 shows an example of the staining profile for these markers. The figure also shows that there are very few non-naive cells that contaminate the naive subset identified using three markers (CD45RO, CD11a, and CD27). Table I summarizes some combinations of two or three markers to use for identifying naive cells. Note that these markers must be used in combination with CD3 and CD4 or CD8.

CCR7 can be used to identify naive cells; however, because it is expressed at varying levels on different subsets, it can be difficult to distinguish the bright levels

Table I
Combinations of Markers Used to Identify Naive T Cells

First marker[a]	Second marker	Third marker
CD45RA/RO	CD62L	
CD45RA/RO	CD11a[b]	
CD45RA/RO	CD62L	CD11a
CD45RA/RO	CD11a	CD27

[a]Either CD45RA or CD45RO can be used.

[b]CD62L cannot be used after cells are stimulated or after cells have been frozen. In these situations, CD11a is useful as the second marker.

on naive cells from the intermediate levels on non-naive cells. CD28 and CD57 are differentially expressed on naive and non-naive cells but do not add information after the aforementioned markers are used. We have not examined in detail the use of CD95 for identifying naive cells. The study of naive T cells may be more complex than described here because studies now indicate that the naive subset of cells as identified with these markers is not homogeneous. For example, expression or nonexpression of CD31 identifies subsets of naive cells with different TREC levels, suggesting different degrees of maturation (Kimmig *et al.*, 2002). Therefore, additional markers may be necessary if these distinctions are needed.

C. Effector/Memory Cell Subsets

In addition to identifying naive cells, the aforementioned markers are used to identify multiple subsets of non-naive or memory/effector cells. Depending on which combinations of markers are used for this purpose, many subsets have been described. For a relatively small number of these subsets, functional information has been determined and correlates with disease states have been described. However, much work remains to be done to fully characterize these subsets.

Because different research groups are using different combinations of markers to identify and classify T-cell subsets, it is difficult to compare results presented by different groups. To determine the extent to which the subsets identified by various combinations of markers may overlap with subsets identified by other combinations, one must examine the markers simultaneously. We and others are beginning these studies. Results of this work are not available, but our preliminary studies indicate that there is complex heterogeneity in the subsets identified using up to eight markers (not including CD3, CD4, and CD8) to identify memory subsets (Fig. 2).

Beginning the discussion of memory cell markers is a listing of several of the combinations of markers used by different groups (Aandahl *et al.*, 2003; Appay *et al.*, 2000, 2002; de Jong *et al.*, 1991; Hamann *et al.*, 1997, 1999; Picker *et al.*, 1993; Roederer *et al.*, 1995; Sallusto *et al.*, 1999). Note that this is not intended to be a complete review of this topic. We have used CD45RA and CD62L to describe several subsets of CD4 and CD8 T cells. CD45RA has been used in combination with CCR7 and CD62L to describe subsets of cells with differing levels of maturation. These have been termed *central* and *effector* memory subsets. CD28 has been used in combination with CD27. CD57 has been used with other markers, and the cells expressing CD57 have been described as terminally differentiated end-stage effector cells. Subsets identified using CD7 have been described. In addition, a large number of other surface markers are not listed (e.g., chemokine receptors). Because of the numerous memory subsets that have been identified using various combinations of these markers, specific discussion of memory subsets is not presented. Instead, in the following sections, guidelines for creating staining panels to look at multiple surface and intracellular markers simultaneously are discussed.

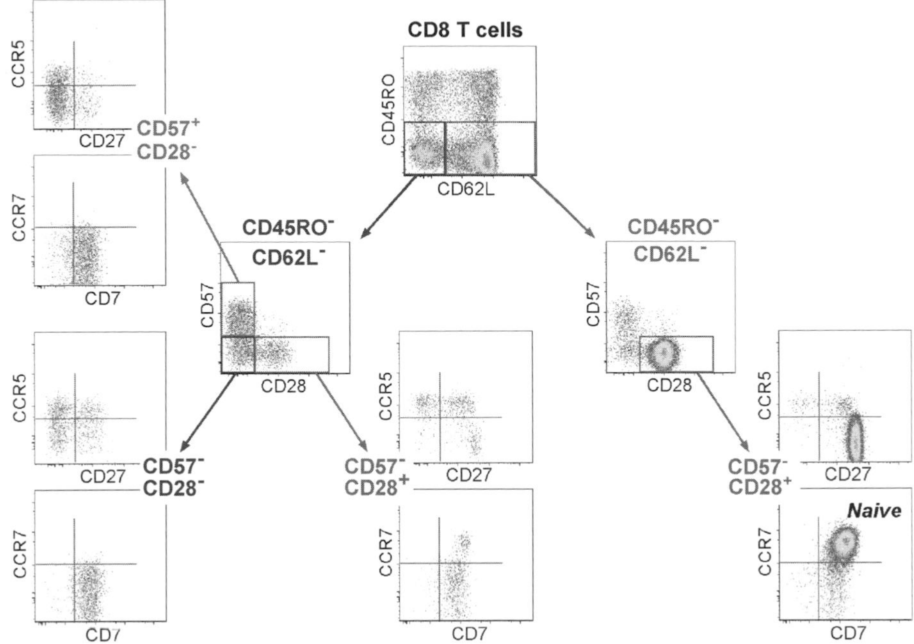

Fig. 2 Simultaneous analysis of multiple memory T-cell markers reveals complex heterogeneity. Peripheral blood mononuclear cells were stained for eight memory T-cell markers, in addition to CD3, CD4, and CD8. (Center) The expression of CD45RO and CD62L on CD8 T cells is shown. As examples, those cells not expressing CD45RO and either expressing CD62L (right) or not expressing CD62L (left) are shown as indicated by the arrows. Expression of CD57 and CD28 reveals further heterogeneity within these subsets. Further gating on the cells expressing or not expressing CD57 and CD28 indicates that there is variable expression of CCR5 and CD27. The expression of CCR5 and CD7 is less variable for the memory subsets shown. (Right) The $CD45RO^-$ cells expressing CD62L are largely naive; however, there is some contamination of memory cells that can be gated out using the other markers. (See Color Insert.)

D. Other Lymphocyte Subsets

This discussion has focused on the CD4 and CD8 T-cell lineages, which comprise most T cells in blood. In addition to these major lineages, co-staining with CD4 and CD8 identifies other T cells with frequencies that vary in different individuals. These are cells that do not express either (double negative [DN]), cells that express both (double positive [DP]), cells that express low levels of CD8 but not CD4, and cells that express low levels of CD8 and CD4. Cells expressing low levels of CD8 often express CD8$\alpha\alpha$ homodimers, rather than the CD8$\alpha\beta$ heterodimer. Because antibody reagents have not been described that specifically recognize only CD8$\alpha\alpha$ cells, CD8$\alpha\alpha$ cells can be identified only by co-staining with anti-CD8$\alpha\beta$ (which recognizes the $\alpha\beta$ complex) and anti-CD8β antibodies. CD8$\alpha\beta$ cells stain for both, and CD8$\alpha\alpha$ cells stain only with CD8$\alpha\beta$. Note that successful co-staining for CD8$\alpha\beta$ and CD8β requires use of both antibodies at the saturating titers. Because both reagents bind to the same molecular complex, there can be competition for binding. Sometimes the correct titer to use needs to be determined empirically.

Because of the presence of these CD4- and CD8-defined subsets that can be identified only by co-staining with CD4 and CD8, it is optimal to include both CD4 and CD8, in addition to CD3, in all T-cell panels. Otherwise, use of only one marker (CD4 or CD8) with positive gating on the marker will include DP cells, and negative gating on the marker will include DN cells. Specific gating on the CD4- and CD8-defined markers may be especially necessary in certain disease states in which the frequencies or relative proportions of these subsets may change substantially.

The previous discussion describes subsets of T cells expressing the $\alpha\beta$ T-cell receptor. These same markers can be used to examine subsets of T cells expressing the $\gamma\delta$ T-cell receptor. The $\gamma\delta$ T cells can be identified by using pan-$\gamma\delta$–specific antibody reagents. In addition, there are antibody reagents that identify $\gamma\delta$ T cells expressing specific T-cell receptor genes. For example, there are antibodies specific for Vδ1 and for Vδ2. Because most $\gamma\delta$ T cells in blood use either of these genes, the use of these two reagents in a staining panel will identify most $\gamma\delta$ T cells. The subsets of $\gamma\delta$ T cells expressing either Vδ1 or Vδ2 have functional differences and change in frequency in certain disease states, so it is often useful to examine these subsets separately rather than looking at $\gamma\delta$ T cells in bulk. An antibody specific for $\gamma\delta$ T cells expressing Vγ9 is also useful. In adult blood, most Vδ2 cells coexpress Vγ9. Co-staining for Vδ2 and Vγ9 allows differentiation of Vδ2 cells coexpressing Vγ9 from Vδ2 cells expressing other Vγ receptors. Note that, as for the CD8$\alpha\alpha$ staining, there can be competition between Vδ2 and Vγ9, and both antibodies must be used at the saturating titers.

Although $\gamma\delta$ T cells are enriched at some peripheral sites such as epithelial surfaces, they are found at low frequencies in blood. Because fewer numbers of these cells are found in PBMC samples, it is important to collect a large number of events on the flow cytometer so that when the $\gamma\delta$ cells are gated, there will be a large enough number to represent the population adequately. Alternatively, it may be helpful to pre-enrich the sample for $\gamma\delta$ cells by sorting out $\alpha\beta$ T cells (e.g., by negative sorting using magnetic beads). Some experimental procedures enrich for $\gamma\delta$ cells by sorting out cells expressing CD4 and CD8, but this procedure is inappropriate. Although most $\gamma\delta$ cells do not express CD4 and CD8, this procedure arbitrarily eliminates all the $\gamma\delta$ cells expressing CD4 and/or CD8. The CD4- and CD8-expressing $\gamma\delta$ T cells, which may have unique functional and phenotypic characteristics, are then not included in the sample to be analyzed.

NK cells are another cell population found within lymphocytes. These have been examined in detail, and the reader is referred to the literature for details concerning the markers used to identify NK cells. A few of the more common NK-associated markers are briefly described here. NK cells are distinguished from T cells because they do not express CD3. Some NK cells express low levels of CD8

(CD8$\alpha\alpha$ homodimers), so it is absolutely necessary to include CD3 in panels designed to examine either T cells or NK cells. Otherwise, it is not possible to identify the dull CD8 cells as T cells or NK cells. This is true even when the panel includes NK-associated markers, because some T cells also express these NK markers. CD16 and CD56 are used as markers to identify NK cells, and when used in combination, they identify several subsets of NK cells. CD57 and CD161 also identify subsets of NK cells. Because many NK cells are cytotoxic, intracellular staining for perforin and the granzymes (see later discussion) also identifies subsets of NK cells. There are now also many reagents specific for NK cell receptors that are not discussed here. Many of the markers listed here are also expressed on some T cells and may be useful for identifying T-cell subsets.

Immunotyping of B cells is discussed in detail by Nicole Baumgarth (see chapter 27). Therefore, only a few brief comments are made here in reference to T-cell staining panels. Because B cells may have high levels of antibody Fc receptors on the cell surface, monoclonal antibody reagents may nonspecifically bind to B cells. If this is a problem, the samples can be treated with an Fc-blocking reagent before staining. Alternatively, a marker that identifies B cells can be included in the T-cell panel, and this marker can be used to gate out the B cells before using any of the other markers to identify the T-cell subsets. CD19 and CD20 are markers that work well for this purpose (as the "dump" channel).

E. Intracellular Markers Associated with Cell Function

During the past several years, several methods have become commonly available that allow for functional information to be determined by flow cytometry. Many of these methods stain for markers mainly expressed inside the cell, but not on the cell surface. The protocols for intracellular staining are now well established and thus are not discussed here. Several manufacturers also offer kits with the reagents needed to prepare cells for intracellular staining. This chapter covers some of the intracellular markers we commonly examine. These markers provide information about cytokine production and cytolytic potential. Therefore, simultaneous examination of these intracellular markers along with the surface markers discussed earlier provides functional information concerning the numerous naive and memory subsets that are identified.

1. Cytokine Production

Cytokine production by individual cells is determined by flow cytometry after intracellular staining with antibody reagents specific for individual cytokines. Before the analysis by flow cytometry, cells are placed in culture with various stimuli for a period of time (typically 6 hours to overnight). A reagent that blocks cellular export of the cytokine (monensin or brefeldin A) is included in the culture so that the cytokine accumulates in the cell and will produce a brighter signal after the cells are intracellularly stained for the cytokine(s).

We have used these methods to detect multiple cytokines after polyclonal T-cell stimulation and antigen-specific stimulation. Some of the antibody reagents that work well for flow studies that we routinely use are interferon-γ (IFN-γ), tumor necrosis factor-a (TNF-α), interleukin-2 (IL-2), interleukin-4 (IL-4), and MIP1β. Our data indicate that T-cell cytokine responses are complex, with cells producing only one of these cytokines or many combinations of these cytokines. The functional significance of these various cytokine profiles and the correlations with surface marker phenotype have yet to be determined.

2. Cytolytic Potential

One method of identifying cells with cytolytic potential is by surface staining cells for CD107a during a period of stimulation in culture. This method was developed by Mike Betts and is described in chapter 20. Another method is to stain cells intracellularly for cytotoxic molecules. We have stained simultaneously for perforin and granzymes A and B. Cells are identified that contain one or various combinations of these cytotoxic compounds. It is important to note that these compounds are stored in cytotoxic granules inside the cell, and that stimulation results in degranulation. Therefore, it is not possible to stain for these compounds after stimulation, which is unfortunate because it would be useful to determine the cytokine production of the subsets identified after staining for perforin and the granzymes. In addition to providing information about cytolytic potential, these intracellular markers are also useful for identifying T-cell subsets (Fig. 3).

III. Staining Panels: Matching Fluorochrome with Cell Marker

A. Overview

The choice of which fluorochrome (or dye) to use with which marker and the choice of which combinations of markers to use for each staining panel are very important for all flow cytometric experiments. The choice of which combinations of markers to include in a panel depends on the subsets that need to be identified. This has been discussed earlier in this chapter for naive cells. For memory cells, many potential markers can be used in combination, and at this time, there are insufficient data to recommend specific combinations. The choice of which dye to use with which marker and the choice of which combinations of dyes to use is dependent on the characteristics of each dye, the characteristics of the fluorescent channel used to collect data from the dye, and the instrument configuration. These topics are briefly discussed as a general guide for designing staining panels.

The sets of fluorochromes that can be used in combination are dependent on the flow instrumentation, and the reader will need to determine this in consultation with the director of the local flow facility or with a representative of the instrument manufacturer. In this chapter, we discuss a configuration we use that allows for

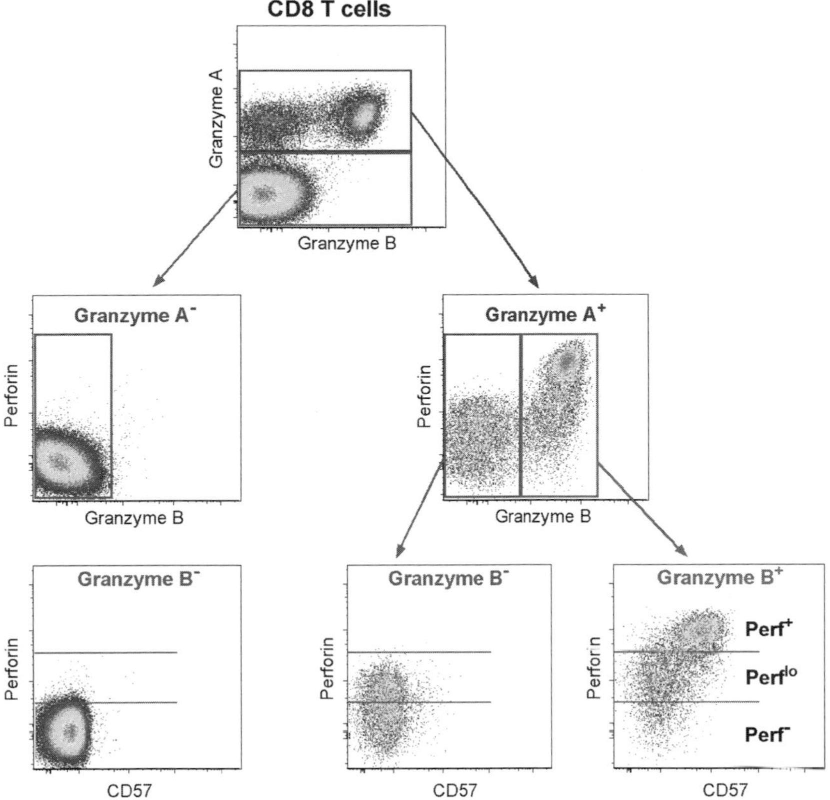

Fig. 3 Simultaneous analysis of perforin and granzymes A and B identifies subsets of CD8 T cells. Freshly isolated peripheral blood mononuclear cells were surface stained for CD3, CD4, CD8, and CD57 and then stained intracellularly for the cytotoxic markers. This representative example demonstrates that these markers are expressed only in certain combinations. Perforin is expressed at three levels, and only the bright perforin cells express CD57. (See Color Insert.)

the collection of 12 fluorescent parameters (and two scatter parameters) simultaneously using excitation with three lasers. Although it may be unlikely that most readers will routinely use this many colors in combination, the concepts involved are relevant when using fewer parameters.

This section serves more as a practical guide rather than a detailed theory-based description of multicolor flow cytometry. However, one concept in particular deserves particular attention—compensation. This is the correction for crossover in the signals from different channels. The theory and consequences of compensation are quite complex and cannot be considered in detail in this chapter. This very important flow concept is often misunderstood, so refer to the literature for a detailed description of compensation (Roederer, 2001).

Table II
Fluorescent Dyes Used for Multicolor Flow Cytometry

Channel[a]	Dyes[b]	Laser excitation[c]
1	Cascade blue, Pacific blue	
2	Cascade yellow, Alexa 430	Violet (605 nM)
3	Quantum Dots[d]	
4	FITC, Alexa 488, CFSE	Blue (488 nM)
5	PE, Alexa 546	
6	TRPE	
7	Cy5–PE, PerCP, PI, EMA	
8	Cy5.5–PE, PerCP–Cy5.5	
9	Cy7–PE	
10	APC, Alexa 647	
11	Cy5.5–APC, Alexa 680	Red (600–635 nM)
12	Cy7–APC	

[a]Channel numbers are labeled arbitrarily. Channels are grouped by laser excitation, and within each laser, they are listed in order of increasing emission wavelength.

[b]Multiple dyes listed in each channel indicate that any of the dyes can be used in that channel, but with the exception of PI, more than one dye cannot be used in each channel in any given staining panel.

[c]Listed is the type of laser used in our flow instruments. There may be other lasers with different wavelengths that can excite these dyes. Some dyes are excited by more than one dye, but only the optimal excitation is indicated here.

[d]Quantum Dots (Quantum Dot Corp., Hayward, California) are fluorescent particles (nanocrystals) available as streptavidin reagents to use in combination with biotin-conjugated staining reagents. The quantum dots are available with several emission spectra so several can be used in combination. All are excited by ultraviolet and violet excitation.

Table II lists the fluorescent dyes that we commonly use, along with the laser excitation used for each dye. These are listed according to the fluorescent channel used to collect the data for each dye. For some of these channels, multiple dyes can be used. The table is organized in this manner to indicate which fluorochromes can be used in combination, and which fluorochromes can be used as substitutes for others. For example, fluorescein isothiocyanate (FITC) and Alexa 488 cannot be used together, but either one of these fluorochromes can be used in this channel. This is one reason that fluorescent channels are often given random names such as FL1 or FL2. We often label channels with the name of a common dye, such as FITC, with the understanding that dyes other than FITC can be collected in that channel.

The channels (or dyes) are grouped in Table II based on the laser excitation, in order of increasing wavelength of excitation. In addition, within these groupings, the dyes are listed in the order of increasing wavelength of emission. This organization is useful for the discussion that follows. For example, adjacent channels are more likely to require greater compensation. Also, dyes that emit at longer wavelengths tend to have greater measurement error, and these are the dyes listed lower in the table (within each laser excitation section). Some general comments concerning the characteristics of these dyes are briefly presented in the following section.

B. Fluorescent Dye Characteristics Relevant to Flow Cytometry

1. *Brightness:* When planning any multicolor experiment, one should begin by considering the characteristics of each of the channels or dyes. With reference to flow cytometry, brightness of the dye refers not only to the absolute intensity of the signal, but also to the separation between positive signal and background. Referring to Table I and the fluorochromes used in our laboratory, phycoerythrin (PE), and allophycocyanin (APC) are among the brightest reagents. TRPE, Cy5–PE, Cy7–PE, and Cy7–APC are relatively bright, depending on the conjugate. FITC, Alexa 680, and cascade blue provide good separation but are generally less bright. Cy5.5–PE and Cy5.5–APC are generally bright, but because of higher background, may provide less separation between positive and negative. Cascade yellow and Alexa 480 are very dull and can be used only with markers expressed at high levels on cells. Although these are general comments referring to these dyes, the "brightness" of each staining reagent conjugated to a dye must be evaluated individually.

2. *Measurement error and compensation requirement:* Another characteristic to be considered is the measurement error associated with each channel. Error is due to a number of factors, but a large component is due to counting statistics and the number of photons collected in each channel. For a detailed description, see the excellent discussion of this topic and compensation by Roederer (2001). Because photomultiplier tubes are less sensitive in the far red, those dyes emitting in the far red will be more likely to have larger measurement errors. This may not be a concern when the dyes are used individually. However, in multicolor combinations, where any individual channel may have compensation corrections resulting from several other channels, the errors accumulate and become more evident. This is not a concern for bright signals, because adequate light is collected. It is a concern for dull signals. The practical consequence is that for some channels, such as APC and the cyanine tandem dyes, the background will be increased and it may not be possible to detect dull-staining events. It is the error associated with measurement, rather than the absolute amount of compensation required, that has the greater effect on the quality of individual channels in multicolor experiments. Because these variables are difficult to predict, the consequences are often evaluated empirically after testing different staining combinations.

3. *Dyes used for intracellular staining:* In our experience, some dye conjugates work well as intracellular staining reagents and others do not. This is likely a consequence of the dye conjugate rather than the dye itself. For example, markers that we have conjugated in our laboratory to phycobiliproteins (PE, APC, or tandem dyes with PE or APC) rarely work well intracellularly, although

commercial reagents work well. Part of the reason for this may be that for the reagents we conjugate, the unconjugated PE or APC is not separated from the labeled reagent. This may result in high backgrounds after intracellular staining. Commercial manufacturers purify the conjugate and this may be the reason many of these reagents work well intracellularly. Our cascade blue conjugates also do not work as intracellular markers for reasons we do not understand. Therefore, markers to be used intracellularly must be conjugated to dyes known to work intracellularly or must be obtained commercially. In any event, the reagents need to be titrated separately for use as intracellular reagents, as noted in a subsequent section in this chapter.

4. *Characteristic of the cell-associated marker to be examined:* The characteristics of the marker of interest must be considered when deciding on the dye to use with the marker. Markers that distinguish separate populations at very different levels of expression (such as CD4 or CD8) can be used on most channels. Markers that are expressed only at dull levels must be conjugated to dyes that are bright and that do not have large measurement errors. This is also true for markers that are expressed at varying levels and for which the distinction between dull expression and no expression is important.

C. Practical Advice for Choosing Reagent Combinations in Staining Panels

First, if a live/dead marker or dump channel is to be used, this should be assigned a channel. In our experience, freshly isolated PBMCs have few dead cells, and a marker to exclude dead cells may not be necessary. In most other situations, we feel a viability marker is important, because many reagents can nonspecifically bind to dead cells. This is especially true when examining cells that were previously frozen or cells that have been stimulated in culture.

Viability markers are available only for a few channels, and one of these channels must be assigned for this marker. When propidium iodide (PI) is used, it may be possible to detect a cell marker and PI on the same channel. In this case, the marker must not be so bright that it overlaps with PI. If cells are to be permeabilized, there are fewer choices for viability markers. We use ethidium monoazide (EMA) to identify dead cells when cells are to be permeabilized. EMA covalently links to DNA after exposure to light, so the label will not leak out of dead cells when the cells are permeabilized. EMA has a spectrum similar to PI and can be detected in one of several channels: TRPE, Cy5–PE, and Cy5.5–PE. EMA is much less bright than PI, so one channel must be devoted to it.

Second, reagents that need to be used intracellularly should be assigned to dyes that work for intracellular staining. In our laboratory, these are restricted to FITC, PE, Cy7–PE, APC, Alexa 680, and Cy7–APC, based on reagents we make or obtain from commercial sources. Intracellular staining is not restricted to markers expressed mainly inside the cell. Other situations when it may be beneficial to use intracellular staining for markers normally expressed on the cell surface are discussed later in this chapter.

Third, surface markers which require the most sensitive discrimination, should be considered. These should be assigned to the dyes that work best in the context of the particular multicolor combination. Sometimes, the decision is to use very bright dyes such as APC. However, note that although APC generally always works well in three- or four-color combinations, it may not work as well for measuring dull-staining markers when many colors are used, likely because of contributions to the error in this channel from compensation due to other channels. This often needs to be evaluated empirically.

In many multicolor combinations, dyes that are less bright such as FITC or cascade blue may be a good option for the markers that require sensitive discrimination. Despite being less bright, they have little measurement error and few other dyes require compensation into these channels. Therefore, FITC and cascade blue have less accumulated error because of the compensation from other channels so that after compensation, the signal to noise separation will not be affected greatly.

Fourth, the remaining channels are used for bright markers with distinct positive and negative populations. For example, we often leave CD8 as the last marker to assign, because this works well in most channels.

Fifth, because much of the panel design is empirical, it is best to test a few combinations. When evaluating which multicolor panel to use, it is useful to know what the correct staining profile is for each marker. If this is not known based on prior experience, then it may be beneficial to set up several one- to three-color panels to observe the staining profile without the potential errors or artifacts introduced by the multicolor combination. There are also a number of controls that may be useful, and these are described in detail elsewhere. Only one control will be mentioned here, the fluorescence-minus-one (FMO) control (Baumgarth and Roederer, 2000), which helps to determine where to set the gate to distinguish between background and dull positive staining. For the markers where this distinction is important, a control including all the markers except the marker of interest should be included. This controls not only for the level of background staining due to the staining reagent (revealed by an isotype control), but also for the contributions to the background due to all the other markers after compensation. When multicolor flow using many colors is used, the background as determined by an isotype control is not relevant. To correctly determine background, an FMO control must be used.

Sixth, there are special notes concerning intracellular staining. Because many staining panels now include at least one intracellular marker, it is often necessary to design panels that work well for identifying surface markers and one or more intracellular markers. In addition to the aforementioned suggestions, several other issues must be considered when designing these panels.

The various intracellular staining procedures share some common features. They require permeabilization of the cells to allow the staining reagents to enter the cells, and they require fixation to maintain cell structure during the permeabilization. Often the staining procedure is separated into two steps. The cells are first stained with the reagents designed to identify cell-surface markers. Then the cells are fixed and permeabilized and stained with the reagents designed to identify intracellular markers.

Because cell-surface markers are accessible during the intracellular staining step, it may be possible to include the reagents intended for the cell-surface markers during the intracellular staining step. There are several reasons this may be more appropriate, but there are also some situations in which this may not work. Including all staining reagents in one step is more convenient because it eliminates one staining step and simplifies the protocol. Another potential advantage of staining cell-surface markers intracellularly is that some cell-surface reagents are downregulated after stimulation and may not be adequately detected on the cell surface. Because this downregulation is often due to internalization of the cell-surface marker, it may be more optimal to detect this marker inside the cell by intracellular staining. Note that any remaining cell surface expression will also be detected during the intracellular staining step. Intracellular staining for CD3 and CD4 works very well for detecting these cell-surface markers that can be markedly downregulated after stimulation. This may also work for CD8. For cell-surface reagents that are lost from the cell surface by means other than internalization, intracellular staining for the marker may not be of additional benefit. For example, CD62L is proteolytically cleaved from the surface after stimulation and is not detected intracellularly.

A few issues must be considered before using a cell-surface reagent as an intracellular reagent. Intracellular staining protocols require fixation, so the labeling reagents used for the surface antigens must work after fixation. Some markers are affected by fixation, so the labeling reagent may no longer recognize the marker. Another factor to consider is the titer of the reagent to use. The titer of the reagent used for intracellular staining may be very different than the titer used for surface staining. The reagents must be titrated under intracellular conditions. For some of our reagents, the intracellular titer is 10-fold lower than the surface titer. In general, all reagents used for intracellular reagents should be titrated and used only at the correct titer. For surface staining, using a large excess of the staining reagent will often not substantially affect the staining pattern. On the other hand, using intracellular reagents well above the titer often results in a large increase in background, with a resulting decrease in the signal/background separation. As one final note concerning titers, intracellular reagents, unlike surface reagents, are not always used at saturating titers. The titer should be chosen that gives low background and good separation between background and positive signal.

IV. Conclusion

Multicolor flow studies of the adult immune system have revealed extraordinary complexity. Because studies using more than six markers simultaneously have been relatively rare, many of the subsets identified have not yet been functionally characterized and the importance of these subsets in different disease states has not been determined. Because of this, staining combinations to identify specific memory cell subsets have not been described here. Instead, some guidelines useful for setting up multicolor panels have been presented. As with all multicolor flow cytometric experiments, successful staining panels are determined in part by using flow-based knowledge and in part by empirical testing. Hopefully, this chapter will increase the likelihood of establishing successful staining experiments while decreasing the amount of empirical testing required.

References

Aandahl, E. M., Sandberg, J. K., Beckerman, K. P., Tasken, K., Moretto, W. J., and Nixon, D. F. (2003). CD7 is a differentiation marker that identifies multiple CD8 T cell effector subsets. *J. Immunol.* **170,** 2349–2355.

Appay, V., Dunbar, P. R., Callan, M., Klenerman, P., Gillespie, G. M., Papagno, L., Ogg, G. S., King, A., Lechner, F., Spina, C. A., Little, S., Havlir, D. V., Richman, D. D., Gruener, N., Pape, G., Waters, A., Easterbrook, P., Salio, M., Cerundolo, V., McMichael, A. J., and Rowland-Jones, S. L. (2002). Memory CD8+ T cells vary in differentiation phenotype in different persistent virus infections. *Nat. Med.* **8,** 379–385.

Appay, V., Nixon, D. F., Donahoe, S. M., Gillespie, G. M., Dong, T., King, A., Ogg, G. S., Spiegel, H. M., Conlon, C., Spina, C. A., Havlir, D. V., Richman, D. D., Waters, A., Easterbrook, P., McMichael, A. J., and Rowland-Jones, S. L. (2000). HIV-specific CD8(+) T cells produce antiviral cytokines but are impaired in cytolytic function. *J. Exp. Med.* **192,** 63–75.

Baumgarth, N., and Roederer, M. (2000). A practical approach to multicolor flow cytometry for immunophenotyping. *J. Immunol. Methods* **243,** 77–97.

de Jong, R., Brouwer, M., Miedema, F., and van Lier, R. A. (1991). Human CD8+ T lymphocytes can be divided into CD45RA+ and CD45RO+ cells with different requirements for activation and differentiation. *J. Immunol.* **146,** 2088–2094.

De Rosa, S. C., Andrus, J. P., Perfetto, S. P., Mantovani, J. J., Herzenberg, L. A., and Roederer, M. (2004). Ontogeny of gammadelta T cells in humans. *J. Immunol.* **172,** 1637–1645.

De Rosa, S. C., Herzenberg, L. A., and Roederer, M. (2001). 11-color, 13-parameter flow cytometry: identification of human naive T cells by phenotype, function, and T-cell receptor diversity. *Nat. Med.* **7,** 245–248.

Hamann, D., Baars, P. A., Rep, M. H., Hooibrink, B., Kerkhof-Garde, S. R., Klein, M. R., and van Lier, R. A. (1997). Phenotypic and functional separation of memory and effector human CD8+ T cells. *J. Exp. Med.* **186,** 1407–1418.

Hamann, D., Kostense, S., Wolthers, K. C., Otto, S. A., Baars, P. A., Miedema, F., and van Lier, R. A. (1999). Evidence that human CD8+CD45RA+CD27− cells are induced by antigen and evolve through extensive rounds of division. *Int. Immunol.* **11,** 1027–1033.

Kimmig, S., Przybylski, G. K., Schmidt, C. A., Laurisch, K., Mowes, B., Radbruch, A., and Thiel, A. (2002). Two subsets of naive T helper cells with distinct T cell receptor excision circle content in human adult peripheral blood. *J. Exp. Med.* **195,** 789–794.

Picker, L. J., Treer, J. R., Ferguson-Darnell, B., Collins, P. A., Buck, D., and Terstappen, L. W. (1993). Control of lymphocyte recirculation in man. I. Differential regulation of the peripheral lymph node homing receptor L-selection on T cells during the virgin to memory cell transition. *J. Immunol.* **150,** 1105–1121.

Rabin, R. L., Roederer, M., Maldonado, Y., Petru, A., and Herzenberg, L. A. (1995). Altered representation of naive and memory CD8 T cell subsets in HIV-infected children. *J. Clin. Invest.* **95,** 2054–2060.

Roederer, M. (2001). Spectral compensation for flow cytometry: visualization artifacts, limitations and caveats. *Cytometry* **45,** 194–205.

Roederer, M., Dubs, J. G., Anderson, M. T., Raju, P. A., and Herzenberg, L. A. (1995). CD8 naive T cell counts decrease progressively in HIV-infected adults. *J. Clin. Invest.* **95,** 2061–2066.

Sallusto, F., Lenig, D., Forster, R., Lipp, M., and Lanzavecchia, A. (1999). Two subsets of memory T lymphocytes with distinct homing potentials and effector functions. *Nature* **401,** 708–712.

CHAPTER 25

Differential Diagnosis of T-Cell Lymphoproliferative Disorders by Flow Cytometry Multicolor Immunophenotyping. Correlation with Morphology

Wojciech Gorczyca
Director of Hematopathology/Oncology Services
Genzyme Genetics/IMPATH
New York, New York 10019

METHODS IN CELL BIOLOGY, VOL. 75

0091-679X/04 $35.00

I. Introduction

T-cell lymphoproliferative disorders are a heterogeneous and relatively rare group of lymphoid tumors that can present as adenopathy, hepatosplenomegaly, skin lesions, a mass involving various other organs, or leukemia (Arrowsmith *et al.*, 2003; Ascani *et al.*, 1997; Chan, 1999; Chan *et al.*, 1999; de Bruin *et al.*, 1997; Farcet *et al.*, 1990; Ferry, 2002; Greenland *et al.*, 2001; Harris *et al.*, 1999, 2000a,b, 2001; Jaffe, 1995, 1996; Jaffe *et al.*, 1999, 2003; Kinney and Kadin, 1999; Lopez-Guillermo *et al.*, 1998; Loughran, 1998; Loughran *et al.*, 1997; Matutes, 2002; Matutes and Catovsky, 2003; Nakamura *et al.*, 1999; Pileri *et al.*, 1995). Morphologically and clinically, they can mimic both benign conditions and nonhematopoietic neoplasms. The diagnosis often requires a multitechnology approach including morphology, flow cytometry (FC), immunohistochemistry, cytogenetics/fluorescence *in situ* hybridization (FISH), and molecular studies (e.g., polymerase chain reaction [PCR] and Southern blot analysis) (Ascani *et al.*, 1997; Ashton-Key *et al.*, 1997; Bakels *et al.*, 1997; Borowitz *et al.*, 1997; Chan *et al.*, 1997, 1999; Ginaldi *et al.*, 1996; Gorczyca *et al.*, 2002, 2003; Hastrup *et al.*, 1991; Jaffe, 1996; Jennings and Foon, 1997; Orfao *et al.*, 1999; Pileri *et al.*, 1998; Weisberger *et al.*, 2000). Although there is no single phenotypic marker specific for T-cell lymphoma/leukemia, FC plays an important role in the diagnosis and subclassification of both mature (peripheral/post-thymic) and immature T-cell disorders. We present the flow cytometric phenotypic data from 220 peripheral T-cell lymphoproliferative disorders and 59 precursor tumors (13 cases of blastic natural killer [NK] cell lymphoma and 46 cases of T-lymphoblastic lymphoma/leukemia [T-ALL/LBL]).

II. Materials

Flow cytometric samples from IMPATH, Inc. (New York division), containing abnormal T-cell populations were submitted for this study. FC data were reanalyzed and correlated with cytomorphology and/or molecular studies. All cases without firm morphological and/or molecular confirmation were excluded. The lesions fulfilling the criteria for bi-lineage lymphoma/leukemia were excluded. Precursor T-ALL/LBL was defined by blastic cytomorphology and expression of TdT and/or CD34 with cytoplasmic CD3, CD1a, or CD10. The neoplasms were classified according to the World Health Organization (WHO) classification of hematopoietic neoplasms. There were 220 cases of peripheral (mature/post-thymic) T-cell lymphoproliferations, 46 cases of precursor T-lymphoblastic lymphoma, and 13 cases of blastic NK cell lymphoma/leukemia (DC-2 acute leukemia). The control population consisted of peripheral blood from patients with acute viral syndromes (5 cases), relative or absolute lymphocytosis (10 cases) or anemia (10 cases), lymph node with reactive changes (10 cases), Hodgkin's lymphoma (10 cases), diffuse large B-cell lymphoma (10 cases), and bone marrow from patients during or after chemotherapy for acute myeloid leukemia (10 cases), for staging of B-cell lymphomas (5 cases), and from patients without specific hematologic diagnosis (10 cases).

III. Methods

A. Flow Cytometry Analysis

Heparinized bone marrow aspirate, peripheral blood, and fresh tissue specimens were used for FC analysis. The specimens were processed within 24 hours of collection. A leukocyte cell suspension was obtained from peripheral blood and bone marrow specimens after red blood cell (RBC) lysis with ammonium chloride lysing solution, followed by 5 minutes of centrifugation. The cell pellet was suspended with an appropriate amount of RPMI 1640 (GIBCO, New York). Fresh tissue samples were disaggregated with a sterile blade, followed by passage through a mesh filter ($<100\,\mu$m). The cells were washed in RPMI media and centrifuged at 1500 rpm for 5 minutes. To minimize nonspecific binding of antibodies, the cells were incubated in RPMI media supplemented with 10% heat-inactivated fetal bovine serum (FBS) in a 37 °C water bath for 30 minutes. The samples were then washed with 0.1% sodium azide/10% FBS (phosphate-buffer saline [PBS] buffer). Viability was assessed using both trypan blue and 7-aminoactinomycin D (Sigma Chemical Co., St. Louis, Missouri) exclusion assays.

Immunophenotypic analysis was performed on FACSCalibur System Instruments equipped with a 15-mW 488-nm air-cooled argon-ion laser supplemented with a 635-nm red-diode laser (Becton Dickinson Immunocytometry System, San Jose, California). Three- and four-color directly labeled antibody combinations consisting of fluorescein isothiocyanate (FITC), phycoerythrin (PE), peridinin-chlorophyl (PerCP), and allophycocyanin (APC) were used for surface staining of fresh tissue, bone marrow, and peripheral blood cell suspensions. Internal negative controls within each tube and isotype controls for immunoglobulin G1 (IgG1), immunoglobulin G2a (IgG2a), and immunoglobulin G2b (IgG2b) were used as negative controls.

The instrument fluorescence detectors' settings and calibration were monitored according to manufacturer's recommendations using Calibrite Beads (Becton Dickinson) and the system linearity was evaluated using Sphero Rainbow Beads (Pharmingen). FC data were collected in list mode and analyzed using CellQuest

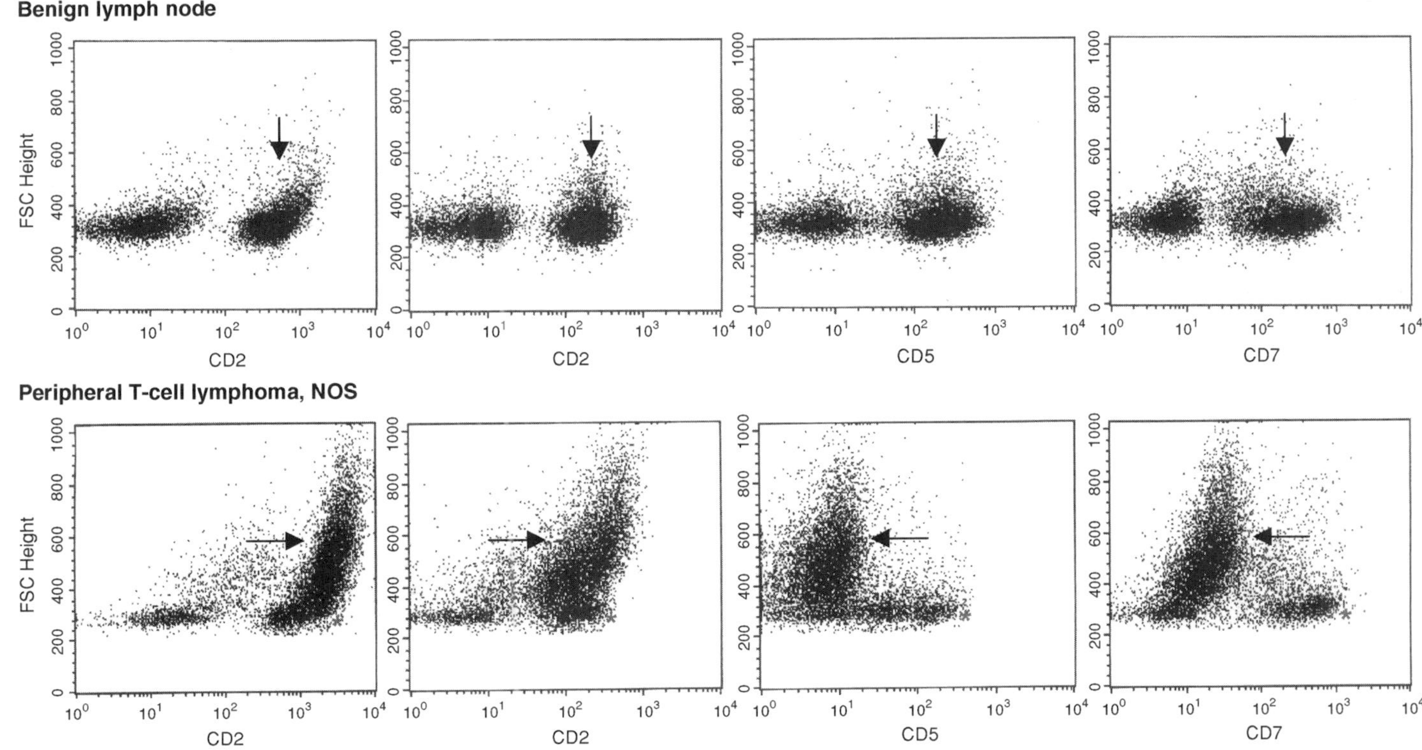

Fig. 1 Comparison of normal expression of pan-T antigens in benign lymph node (upper panels) with aberrant expression from peripheral T-cell lymphoma, unspecified (lower panels). Lymphomatous cells display increased forward scatter (*arrow*) and aberrant expression of CD5 and CD7. Residual normal T cells (*) are positive for all pan-T antigens.

B cells of primary effusion lymphoma. CD5 is aberrantly expressed in B-chronic lymphocytic leukemia/small lymphocytic lymphoma, mantle cell lymphoma, *de novo* diffuse large B-cell lymphoma, and thymoma/thymic carcinoma. CD7 is often positive in acute myeloid leukemia, blastic NK cell lymphoma/leukemia (DC-acute leukemia), and acute monocytic leukemia.

No control cases showed aberrant expression of CD45, whereas 16 (7.3%) of 220 cases of peripheral T-cell disorders, 14 (30%) of 46 cases of precursor T-ALL/LBL, and 3 (23%) of 13 cases of blastic NK cell lymphomas showed aberrant expression of CD45 (dim or negative). Increased forward scatter, though observed more frequently in neoplastic processes (55% of cases), is not entirely specific and was observed in the case of phenytoin-induced adenopathy. The cases in the control group showed only a very small subset of negative cells with an absence of the CD7 antigen (average 1.7% cells). The true absence of one or more pan-T antigens in a significant or major lymphoid population is compatible with a neoplastic process. This is not the case with diminished expression of any of the pan-T antigens, because such an aberrant (dim) expression was observed in benign conditions (viral infections, cases after chemotherapy, or in reactive T cells accompanying large B-cell lymphoma).

V. Precursor T Lymphoblastic Lymphoma/Leukemia

Precursor T-ALL/LBL often shows diminished expression of CD45 as compared to normal T cells. Among 46 cases analyzed, bright (normal) expression was seen in 11 cases (23.9%), moderate (slightly dimmer than normal T cells) expression was seen in 21 cases (45.7%), and dim expression in 14 cases (30.4%). None of the cases showed lack of CD45. All cases showed lack or diminished expression of at least one of the pan-T antigens. Among pan-T antigens, CD7 was most often positive (97.8%) and surface CD3 was least often positive (41.3%). Most of the cases were dual CD4/CD8 negative (41.3%) or dual CD4/CD8 positive (39.4%). CD4-positive cases made up 10.9% and CD8-positive cases 8.7%. The distribution of TCR expression was as follows: TCR negative, 51.4%; TCR$\alpha\beta^+$, 31.4%; and TCR$\gamma\delta^+$, 17.1%. CD1a, CD10, CD13, CD33, CD34, CD56, CD57, CD117, HLA-DR, and TdT were positively expressed in 32%, 23.9%, 17.4%, 21.7%, 36.9%, 23.9%, 4.3%, 10.9%, 8.7%, and 87%, respectively (see Table I for details). Figure 2 presents an example of flow cytometric analysis of T-ALL/LBL.

VI. Peripheral (Mature/Post-thymic) Lymphoma Versus Precursor T-Lymphoblastic Lymphoma/Leukemia

Table I presents a comparison between the mature and precursor T-cell neoplasms. T-ALL/LBL more often showed dimmer expression of CD45, lack of CD2 and CD3 (surface), dual-positive or dual-negative CD4/CD8 expression, lack of

Table I
Peripheral (Mature/Post-Thymic) T-Cell Lymphoma/Leukemia Versus Precursor T-Lymphoblastic Lymphoma/Leukemia

Antigen	Peripheral T-cell lymphomas (%)	Precursor T-lymphoblastic lymphoma/leukemia (%)
CD45		
Moderate/bright+	93.3	69.6
Dim+	4.1	30.4
Negative	2.6	0
Pan T-cell antigens		
CD2+	89.1	65.2
CD3+	73.6	41.3
CD5+	77.7	87.0
CD7+	58.0	97.8
One pan-T antigen negative	39.9	47.8
Two pan-T antigens negative	16.1	21.7
Three pan-T antigens negative	6.7	6.5
All pan-T antigens negative	2.0	0
All pan-T antigens positive	32.1	6.5
CD4/CD8		
CD4+	48.7	10.9
CD8+	29.0	8.7
CD4/CD8+	6.2	39.4
CD4/CD8−	16.1	41.3
TCR		
TCR$\alpha\beta$+	75.6	31.4
TCR$\gamma\delta$+	14.4	17.1
TCR−	10.0	51.4
CD1a+	0	32.0
CD10+	6.2	23.9
CD13+	0	17.4
CD33+	0	21.7
CD34+	0	36.9
CD56+	20.7	23.9
CD57+	24.3	4.3
CD117+	2.6	10.9
HLA-DR+	4.1	8.7
TdT+	0	87.0

TCR, positive expression of CD10, and positive expression of CD117 when compared to mature tumors. Peripheral (mature/post-thymic) T-cell proliferations more often showed positive expression of all four pan-T antigens (32% vs 6.5%) and lack of CD7 (58% vs 97.8%) when compared to T-ALL/LBL. Presence of CD1a, CD13, CD33, CD34, and TdT indicates T-ALL/LBL.

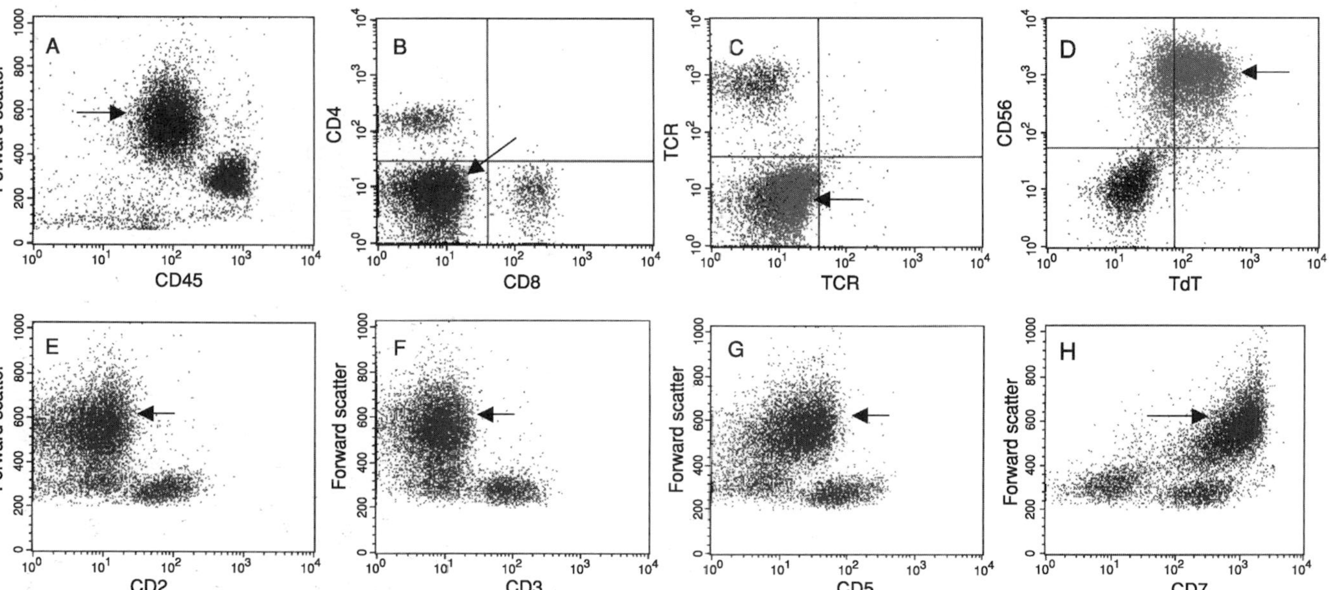

Fig. 2 Precursor T-lymphoblastic lymphoma/leukemia (T-ALL/LBL). Neoplastic cells (*arrow*) show increased forward scatter (A), moderate expression of CD45 (A, residual benign lymphocytes are brightly positive for CD45), negative expression of CD4/CD8 (B), negative expression of T-cell receptor (C), positive expression of CD56 and TdT (D), negative CD2 (E), surface CD3 (F), dim CD5 (G), and moderate to bright CD7 (H).

VII. Thymocytes from Thymic Hyperplasia/Thymoma Versus Precursor T-Lymphoblastic Lymphoma/Leukemia

Dual-positive expression of CD4/CD8 is rarely observed in peripheral (mature/post-thymic) T-cell lymphoproliferative disorders (6.2%) and, therefore, when present is suggestive of an immature T-cell population (precursor T-ALL/LBL shows coexpression of CD4 and CD8 in 39.4% of cases). Thymocytes from either hyperplastic thymus or thymoma (Fig. 3A, B) are always CD4/CD8 positive (Fig. 3C), so the diagnosis of mediastinal T-ALL/LBL cannot be based on the presence of CD4/CD8 coexpression. Although in the case of thymocytes, there is gradual transition from $CD4^{+}/CD8^{-}$ to $CD4^{-}/CD8^{+}$ cells with major population being $CD4^{+}/CD8^{+}$, whereas T-ALL/LBL shows distinct $CD4^{+}/CD8^{+}$ population (Fig. 3F, black arrow) and with well-separated residual (benign) small T lymphocytes (Fig. 3F, blue arrows), the differences are too subtle to constitute a base for definite diagnosis. Whether from thymoma or benign thymic hyperplasia, immature T cells (thymocytes) display characteristic expression of surface CD3 antigen (Fig. 3D). Most cells (small T cells) show positive but variable (smeared-like) CD3 expression (black arrows), whereas larger T cells (red arrow) are CD3 negative. This smeared pattern is never observed in surface $CD3^{+}$ T-ALL/LBL. Small T cells from thymoma/thymic hyperplasia are $CD10^{-}$ and larger T cells (surface $CD3^{-}$) show positive expression of CD10 (Fig. 4).

VIII. Mature T-Cell Lymphoproliferative Disorders

A. T-Cell Prolymphocytic Leukemia

T-cell prolymphocytic leukemia (T-PLL) was diagnosed in 28 cases. The ages ranged from 43 to 92 years. There were 10 female patients and 18 male patients. Most cases showed moderate/bright expression of CD45. Only 7.2% of cases were CD45 negative. Normal expression of pan-T antigens was present in 57% of cases. Abnormal expression of one, two, and three pan-T antigens was noted in 21%, 11%, and 11%, respectively. Rare cases of T-PLL showed CD117 expression. Those cases were always CD8 positive. In the cases with no aberrant expression of pan-T antigens, abnormal CD4/CD8 expression, lack of TCR or CD45, and presence of CD117 expression helped identify abnormal T-cell population. Careful evaluation of peripheral blood smear and correlation with laboratory data (marked lymphocytosis) are helpful in establishing correct diagnosis. Molecular studies (PCR and/or Southern blot analysis) should follow flow cytometric diagnosis of T-PLL. Table II presents detailed flow cytometric findings in T-PLL. Figure 5 shows T-PLL with aberrant loss of surface CD3 and lack of TCR expression.

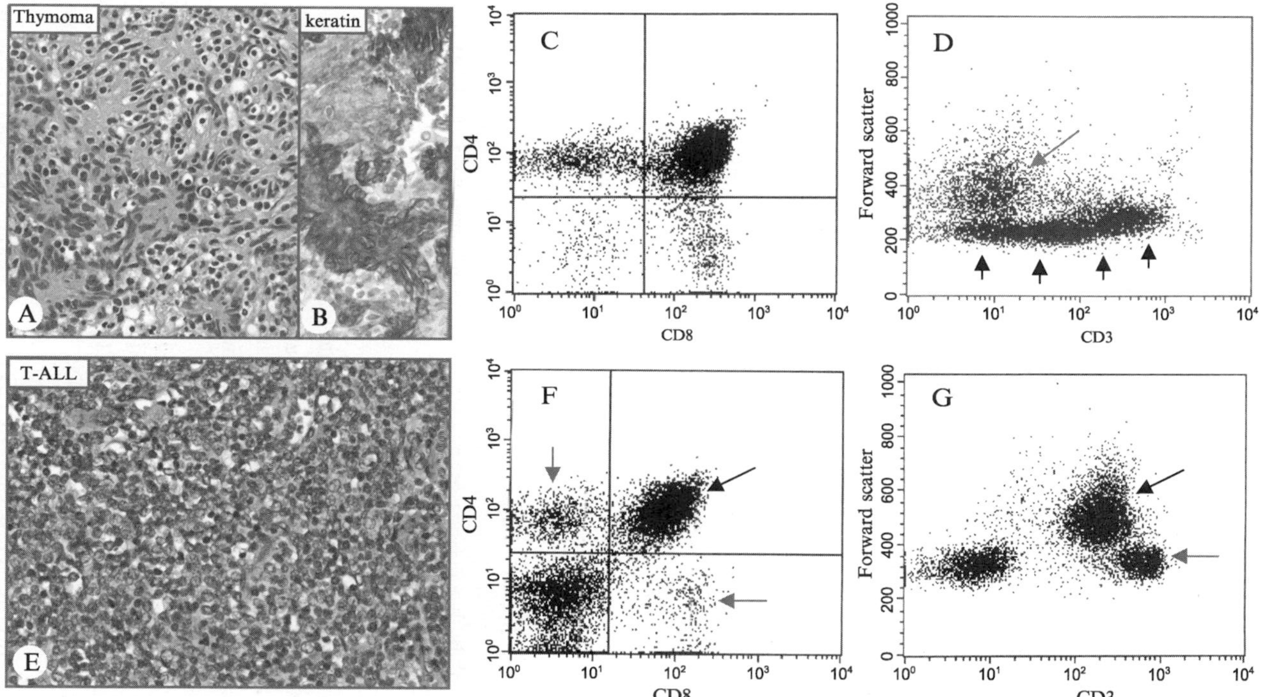

Fig. 3 Comparison of flow cytometric features of immature T cells from thymoma (A) and precursor T-lymphoblastic lymphoma (E). Thymocytes display dual expression of CD4/CD8 with gradual transition of $CD4^{+}/CD8^{-}$ cells to $CD4^{-}/CD8^{+}$ cells and most cells showing $CD4^{+}/CD8^{+}$ phenotype (C). T lymphoblasts are often dual $CD4/CD8^{+}$ or dual $CD4/CD8^{-}$. When coexpression is present (F), there is distinct population of neoplastic cells ($CD4^{+}/CD8^{+}$, upper right quadrant). Benign (residual) small T cells are either $CD4^{+}$ (upper left quadrant) or $CD8^{+}$ (lower right quadrant). The most useful flow cytometric criterion for differentiating thymocytes (thymoma or thymic hyperplasia) from T lymphoblasts is the evaluation of surface CD3 (D, G). Thymocytes from thymoma or thymic hyperplasia (D) show two cell populations: small cells (low forward scatter, *black arrows*) and larger cells (increased forward scatter, *red arrow*). Small thymocytes have a characteristic smeared pattern of CD3. Larger thymocytes are CD3 negative (D). T lymphoblasts (G) show compact cluster of $CD3^{+}$ cells (*black arrow*). Residual (benign) T cells have brighter expression of CD3 and lower forward scatter (*blue arrow*). (See Color Insert.)

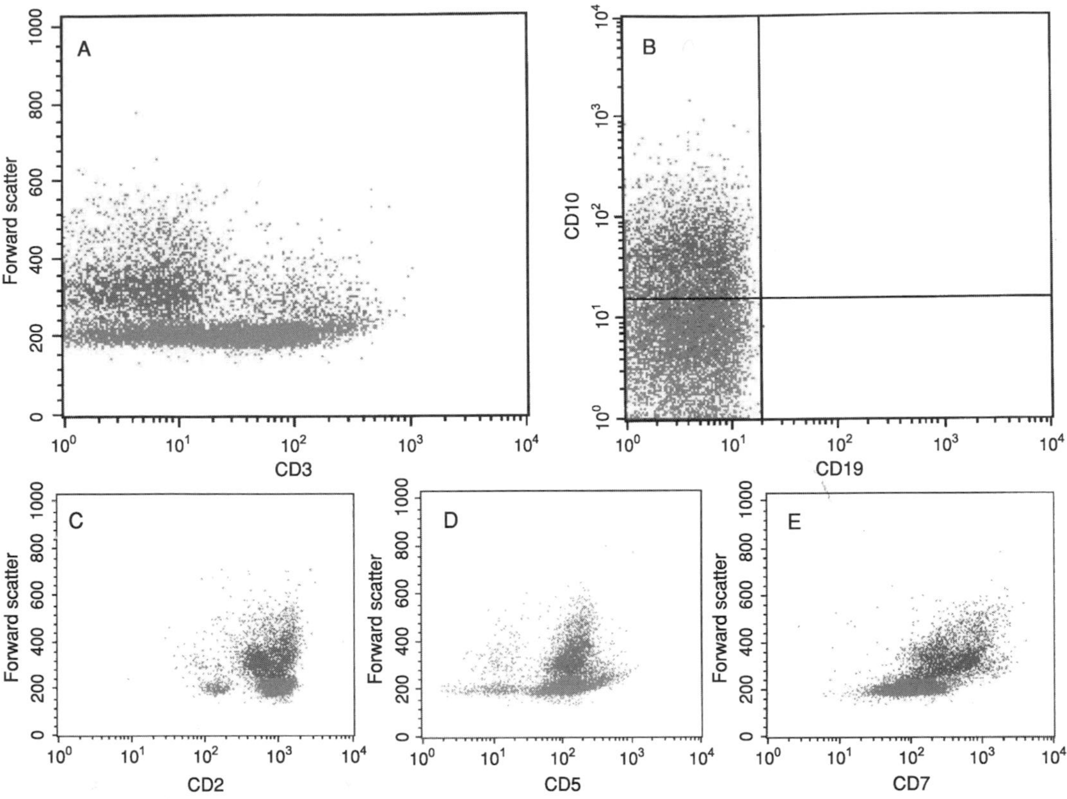

Fig. 4 Thymocytes show two distinct populations (A): Small cells (red) are CD3 positive with characteristic variable (smeared) expression and larger cells (green) are CD3 negative. Larger cells are positive or CD10 (B). Both populations are positive for other pan-T antigens (C–E). (See Color Insert.)

B. Large Granular Lymphocyte Leukemia

LGL leukemia is characterized by lymphoid cells with abundant cytoplasm with prominent azurophilic granules. Two variants of LGL proliferations can be recognized: T-LGL leukemia and NK-LGL leukemia. The former has an indolent clinical course and may be associated with neutropenia and rheumatoid arthritis. We analyzed 38 cases of T-LGL leukemia. Patient ages ranged from 33 to 91 years, with 17 cases occurring in women and 19 in men. On FC analysis, T-LGL leukemia has a mature T-cell phenotype (Table III) with CD8 expression and TCR$\alpha\beta$ expression. Rare cases may be CD4$^+$/CD8$^+$, CD4$^+$/CD8$^-$ or CD4$^-$/CD8$^-$. Figure 6 shows an example of CD4$^+$ T-LGL. Of the NK cell–associated markers, CD57 was most frequently expressed in our series (95%). None of T-LGL cases showed coexpression of CD16 and CD56. NK-LGL leukemias are less common. The phenotypic hallmark of NK-LGL leukemia is lack of CD3 and CD5 and positive expression of CD2 and CD7. This phenotype was observed in all

Table II
T-Cell Prolymphocytic Leukemia (T-PLL)

Antigen	T-PLL (%)
CD45	
Moderate+	92.8
Dim+	0
Negative	7.2
Pan T-cell antigens	
CD2+	96.4
CD3+	89.3
CD5+	92.8
CD7+	89.3
One pan-T antigen negative	17.9
Two pan-T antigens negative	7.1
Three pan-T antigens negative	0
All pan-T antigens negative	0
One pan-T antigen dim (3 others positive)	10.7
Two pan-T antigens dim (2 others positive)	3.6
Three pan-T antigens dim (1 remaining 1 positive)	3.6
Four pan-T antigens dim	0
All pan-T antigens normal (no loss or dim expression)	57.1
CD4/CD8	
CD4+	50.0
CD8+	25.0
CD4/CD8+	14.3
CD4/CD8−	10.7
TCR	
TCR$\alpha\beta$+	92.6
TCR$\gamma\delta$+	0
TCR−	7.4
CD10+	0
CD56+	7.1
CD57+	0
CD117+	14.3
HLA-DR+	7.1

13 analyzed cases (Table III). In contrast to T-LGL leukemias, LGL with NK cell phenotype more often show expression of CD16 and CD56, with 35% of cases coexpressing both antigens. Figure 7 shows NK-LGL leukemia.

C. Adult T-Cell Lymphoma/Leukemia

Adult T-cell leukemia/lymphoma (HTLV-1 positive; six cases), with confirmed seropositivity for HTLV-1 were seen in the blood (three cases) or bone marrow (three cases). All six cases were positive for CD4, CD5, and CD25. None of the cases expressed CD7, NK cell antigens (CD16, CD56, CD57), CD11b, or CD11c.

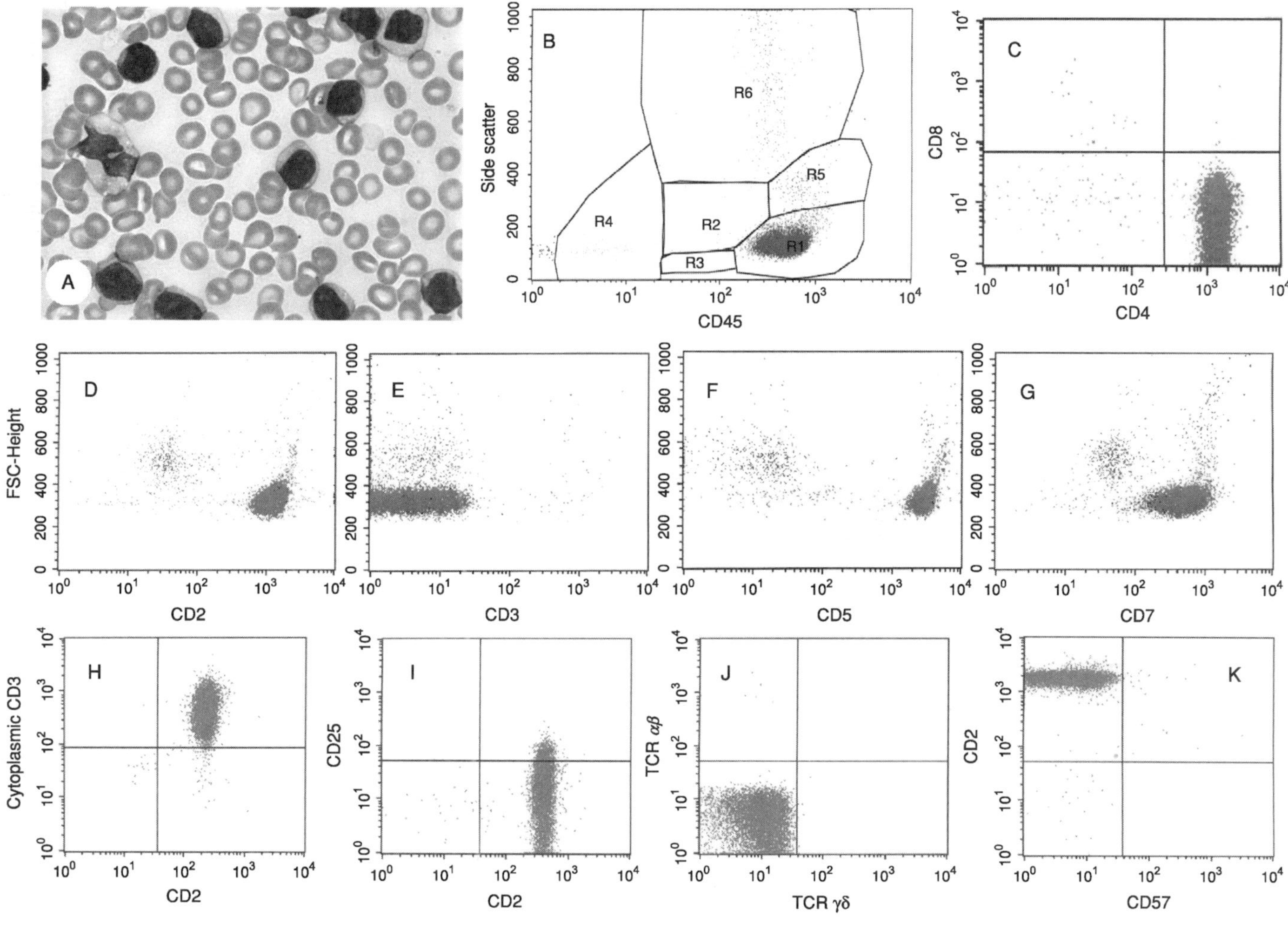

Fig. 5 T-prolymphocytic leukemia (T-PLL). (See Color Insert.)

Table III
Large Granular Lymphocyte Leukemia (T-LGL and NK-LGL)

Antigen	T-LGL (%)	NK-LGL (%)
CD45		
Moderate+	100	100
Dim+	0	0
Negative	0	0
Pan T-cell antigens		
CD2+	100	100
CD3+	100	0
CD5+	78.9	0
CD7+	86.8	100
One pan-T antigen negative	31.6	0
Two pan-T antigens negative	0	100
Three pan-T antigens negative	0	0
All pan-T antigens negative	0	0
All pan-T antigens positive	68.4	0
CD4/CD8		
CD4+	2.6	0
CD8+	86.8	23
CD4/CD8+	7.9	0
CD4/CD8−	2.6	77
TCR		
TCR$\alpha\beta$+	78.9	8
TCR$\gamma\delta$+	21.1	0
TCR−	0	92
CD11b+	26.3	–
CD11c+	47.4	61
CD16+	34.2	70
CD56+	31.6	70
CD57+	94.7	61
Coexpression of CD16 and CD56	0	35
CD117+	0	0
HLA-DR+	7.9	0

Of the six cases, five expressed CD2 and four expressed CD3. TCR$\alpha\beta$ was positive in four of the six cases and TCR$\gamma\delta$ in the remaining two cases. Peripheral blood film from patients with adult T-cell leukemia/lymphoma showed medium-sized lymphocytes with markedly irregular nuclei with occasional "flower-like" appearance (Fig. 8).

D. Anaplastic Large Cell Lymphoma

We analyzed anaplastic large cell lymphoma (ALCL) by FC and/or immunohistochemistry in 26 cases (including 2 cases in which FC data were not diagnostic). The age of patients ranged from 33 to 77 years (average 55.8 years).

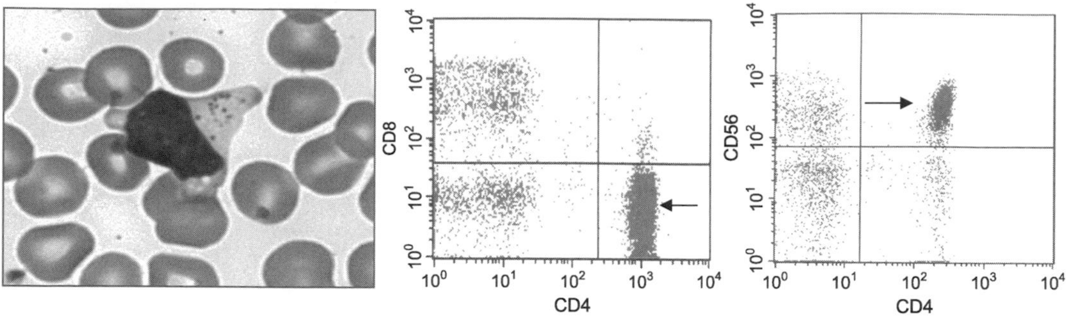

Fig. 6 T-cell large granular lymphocyte leukemia (T-LGL leukemia). (See Color Insert.)

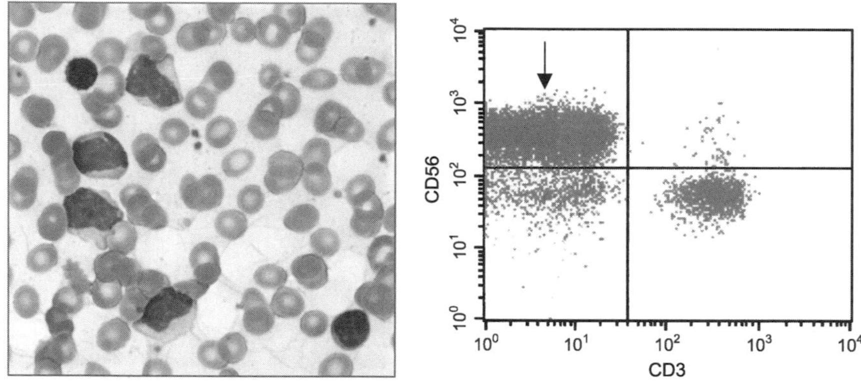

Fig. 7 Large granular lymphocyte leukemia with natural killer cell phenotype (NK-LGL leukemia). (See Color Insert.)

The tumors were most often located in the lymph nodes (20/26 cases), followed by soft tissue, liver, and other extranodal sites. Among cases involving lymph nodes, inguinal location was most common, followed by neck and axillary nodes. Only 11% of cases showed normal expression of pan-T antigens. Remaining cases displayed either dim expression or complete lack of one or more of the pan-T antigens. In 22% all four pan-T antigens were negative. Among pan-T antigens, CD7 was most frequently absent (78%) and CD2 was most frequently positive (67%). CD3 was positive in 33% and CD5 in 50%. Most cases were $CD4^+$ (72%). $CD8^+$, $CD4/CD8^+$ and $CD4/CD8^-$ were present in 17%, 6%, and 5% of cases, respectively. None of the analyzed cases was TCR$\gamma\delta$ positive. Most cases were TCR$\alpha\beta$ positive (67%), and the remaining 33% of cases did not express TCR. The expression of CD30 was present in all cases, and ALK-1 was present in 66% cases

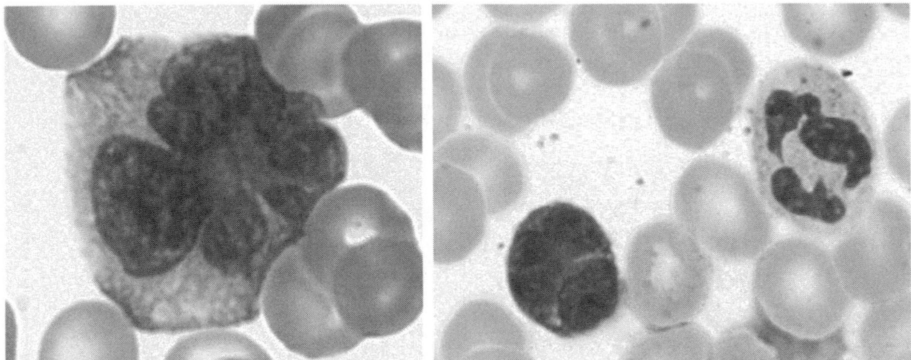

Fig. 8 Adult T-cell leukemia. (See Color Insert.)

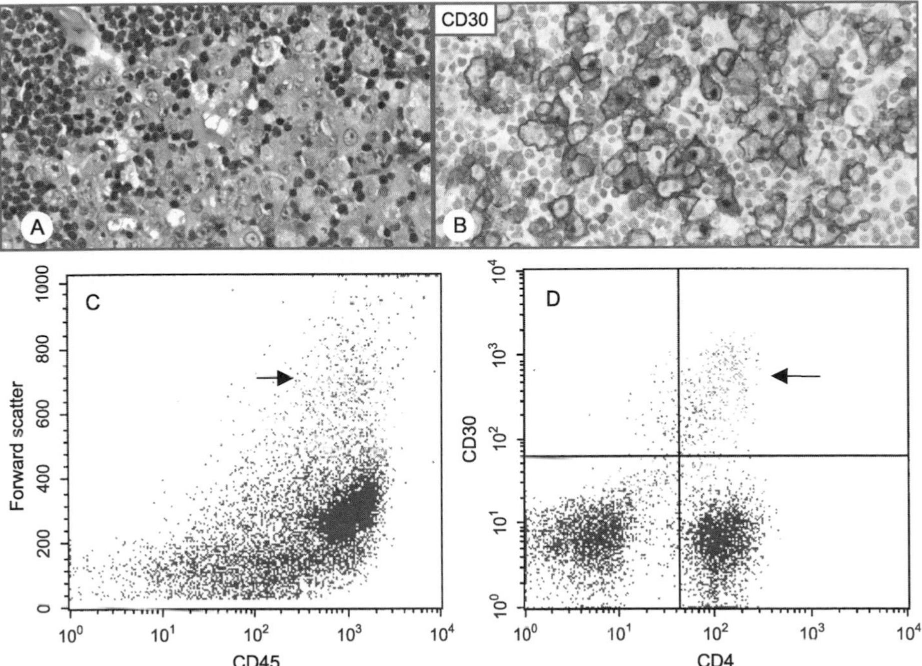

Fig. 9 Anaplastic large cell lymphoma. (See Color Insert.)

(the diagnosis of ALCL without ALK expression is controversial, because it does not share good prognosis with ALK-positive tumors). It is likely that it represents $CD30^+$ peripheral T-cell lymphoma with anaplastic features. EMA expression was found in 78% and CD10 in 12%. Figure 9 presents ALCL.

E. Peripheral T-Cell Lymphoma, Unspecified

Peripheral T-cell lymphoma, unspecified (68 cases), was from lymph nodes (39), bone marrow (16), blood (8), and one each spleen, lung, cerebrospinal fluid (CSF), omentum, and skin. The ages ranged from 19 to 92 years (average age 66.2 years). Table IV presents flow cytometric findings. The expression of CD45 was moderate/bright in 88.2%, dim in 10.3%, and negative in 1.5% of cases. Among pan-T antigens, CD7 was most frequently lost (55.9%) and CD2 was least frequently lost (13.3%). Dim expression of CD2, CD3, CD5, and CD7 was noted in 4.4%, 11.8%, 17.6%, and 8.8%, respectively; 7.5% of cases did not reveal any loss

Table IV
Peripheral T-Cell Lymphoma, not Otherwise Specified

Antigen	Peripheral T-cell lymphoma, not otherwise specified (%)
CD45	
Moderate+	88.2
Dim+	10.3
Negative	1.5
Pan T-cell antigens	
CD2+	86.7
CD3+	58.8
CD5+	82.3
CD7+	44.1
One pan-T antigen negative	48.5
Two pan-T antigens negative	25.0
Three pan-T antigens negative	8.8
All pan-T antigens negative	0
All pan-T antigens positive	17.6
CD4/CD8	
CD4+	60.3
CD8+	11.8
CD4/CD8+	5.8
CD4/CD8−	22.1
TCR	
TCR$\alpha\beta$+	76.0
TCR$\gamma\delta$+	0
TCR−	24.0
CD10+	7
CD25	14.7
CD30	8.0
CD56+	13.2
CD57+	5.9
CD117+	0
HLA-DR+	2.9
EMA	7

of diminution of pan-T antigens. Abnormal expression (dim or complete lack of expression) of one pan-T markers was seen in 35.8%, two markers in 41.8%, three markers in 13.4% and all four markers in 1.5%. Although $CD4^+$ lymphomas predominated (60%), dual-negative expression of CD4/CD8 was observed on significant proportions of tumors (22%). Of additional markers analyzed, CD10, CD11c, CD25, CD30, CD56, CD57, CD117, EMA, and HLA-DR were occasionally expressed (see Table IV for details). Figure 10 presents peripheral T-cell lymphoma, unspecified.

F. Angioimmunoblastic T-Cell Lymphoma

AILD lymphoma (12 cases analyzed) occurred in patients from 49 to 88 years old (average 71.2 years). All tumors (100%) were positive for CD2, CD4, CD5, CD45, and TCR$\alpha\beta$. CD10 was expressed in 50% and Bcl-6 (immunohistochemistry) in 30%. None of the tumors expressed CD8, CD56, CD57, CD11b, CD11c, and HLA-DR. Twenty-five percent of cases showed normal (moderate or bright) expression of all four pan-T antigens. The remaining cases showed loss of either one marker (58.3%) or two markers (16.7%). Lack of aberrant loss of pan-T antigens or mixture of benign (residual) $CD8^+$ cells accompanying most of the AILD lymphomas made flow cytometric evaluation difficult in some cases. Careful analysis of forward scatter, side scatter, dual staining of CD4 with pan-T antigens, and CD10, as well as correlation with histological section, was crucial in establishing correct diagnosis. On histological examination, clusters of atypical cells with clear cytoplasm were found in most cases. CD21 and CD23 staining for follicular dendritic cells (FDCs) was useful in identifying expanded FDC meshwork in most cases. Scattered lager B cells with EBER expression were often present. AILD lymphoma is presented in Fig. 11.

G. Hepatosplenic T-Cell Lymphoma

Hepatosplenic T-cell lymphoma (11 cases) occurred in patients from 35 to 63 years old (average 46.2 years). Eight cases occurred in men and three in women. None of the cases showed normal expression of all pan-T antigens. One marker was negative in six cases (54.5%), two markers were negative in three cases (27.3%), and the remaining two cases (18.2%) had dim expression of at least one marker. All 11 cases showed positive expression of CD3. Most cases were CD5 negative (only two cases had dimly positive expression of CD5). CD2, CD3, CD5, and CD7 were positively expressed in 9 cases (81.8%), 11 cases (100%), 2 cases (18.2%), and 10 cases (90.9%), respectively. All but one case had dual-negative expression of CD4/CD8 (90.9%) and were TCR$\gamma\delta$ positive. The remaining one case was $CD8^+$ and expressed TCR$\alpha\beta$. NK cell–associated antigens were positive as follows: CD16 in 3 cases (27.3%), CD56 in 10 cases (90.9%), and CD57 was negative in all analyzed cases. Figure 12 presents hepatosplenic lymphoma (liver and bone marrow) involvement.

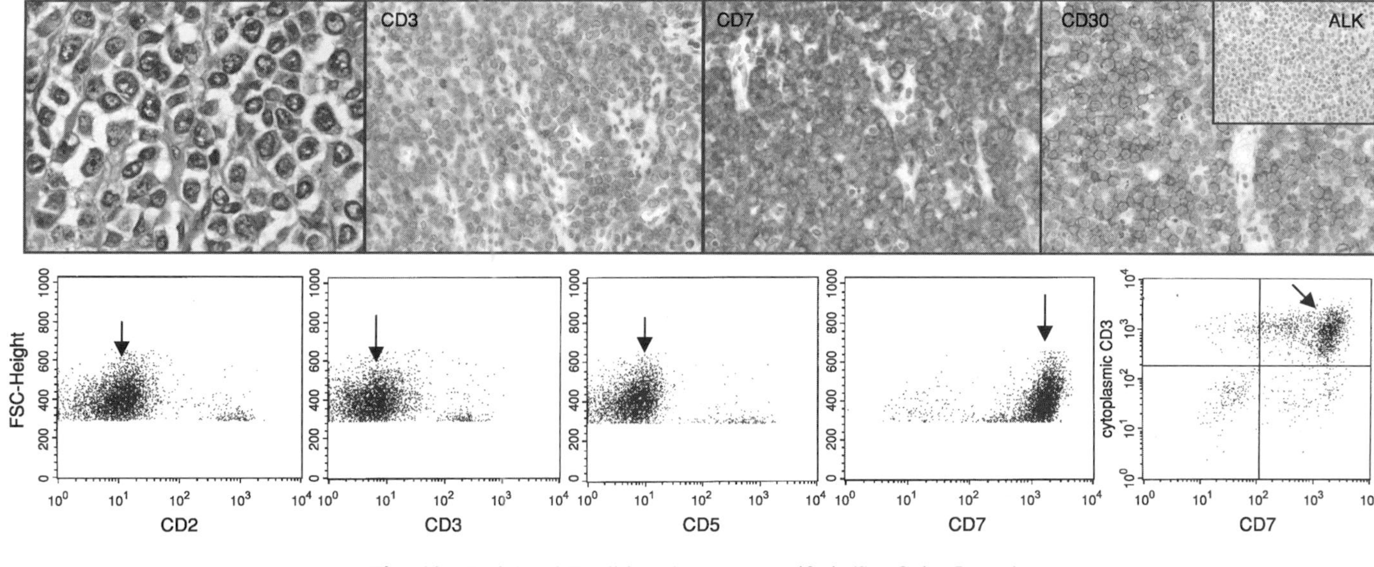

Fig. 10 Peripheral T-cell lymphoma, unspecified. (See Color Insert.)

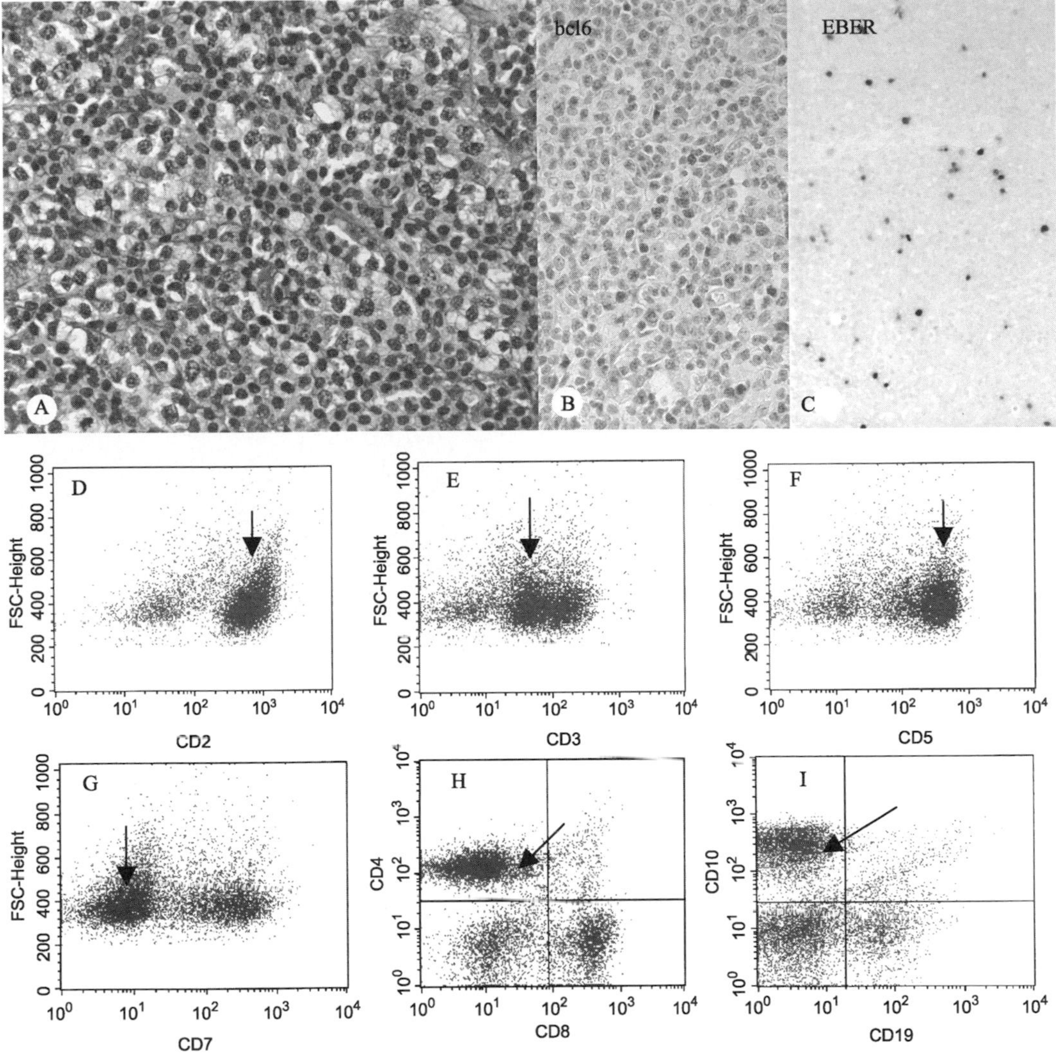

Fig. 11 Angioimmunoblastic T-cell lymphoma. (See Color Insert.)

H. Sézary's Syndrome

Sézary's syndrome (eight cases) were seen in blood (four cases) and bone marrow (four cases) in patients 52–87 years of age (five men and three women). All cases were $CD4^+$ and $TCR\alpha\beta^+$. Only one case showed no aberrant expression of all four pan-T antigens (12.5%). The remaining seven cases showed lack of one pan-T antigen (five cases), two pan-T antigens (one case), and dim expression of one antigen (one case). CD7 was most frequently absent (five of eight cases).

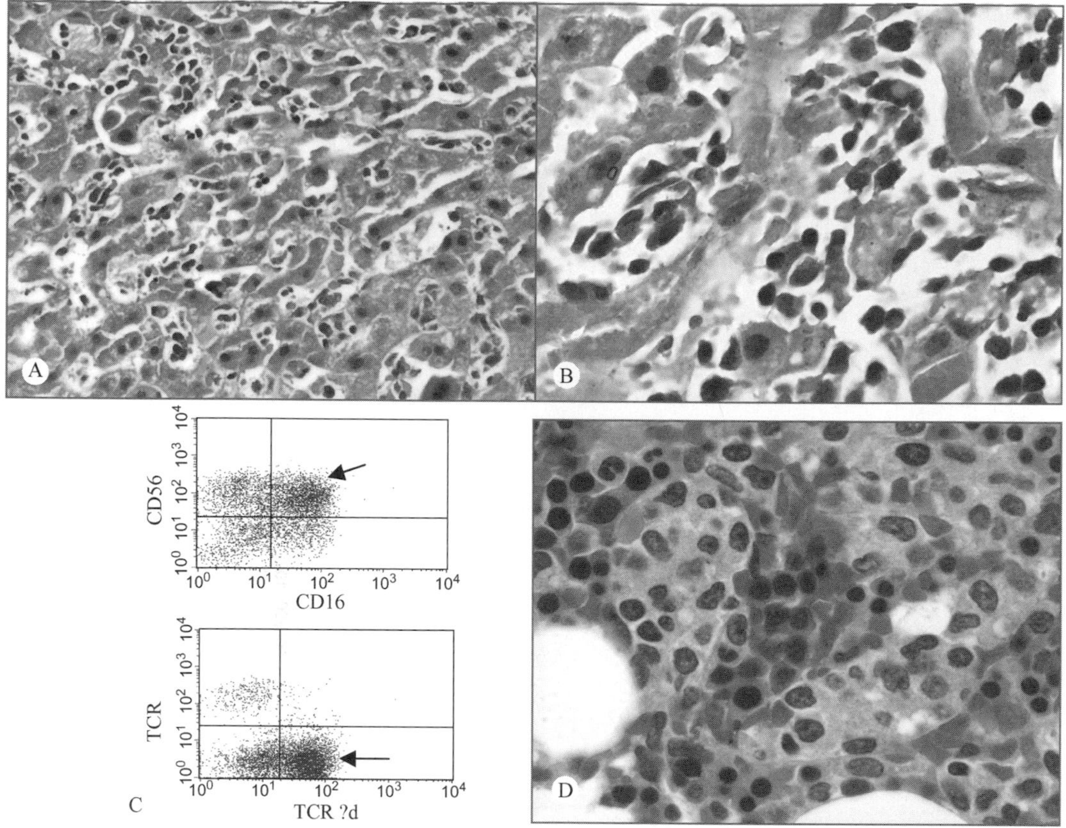

Fig. 12 Hepatosplenic T-cell lymphoma. (See Color Insert.)

I. Enteropathy Type T-Cell Lymphoma

Enteropathy-type T-cell lymphoma (three cases) occurred in the jejunum in two women and one man (ages, 40–82 years). All cases were positive for CD45, CD2, CD3, CD7, CD11c, and CD103. Two cases were $CD8^+$ and one case of CD4/$CD8^-$. CD5 was negative in two cases. One case was positive for CD56 and one case for CD25.

IX. CD117 Expression in T-Cell Lymphoproliferations

CD117 was positive in two cases of peripheral T-cell lymphoma, unspecified (2.9%), four cases of T-PLL (14%), five cases of precursor T-ALL/LBL (11%), and four cases of blastic NK cell lymphoma/leukemia (30%). In precursor neoplasms, the expression of CD117 was dim. Blastic NK cell lymphoma/leukemias were

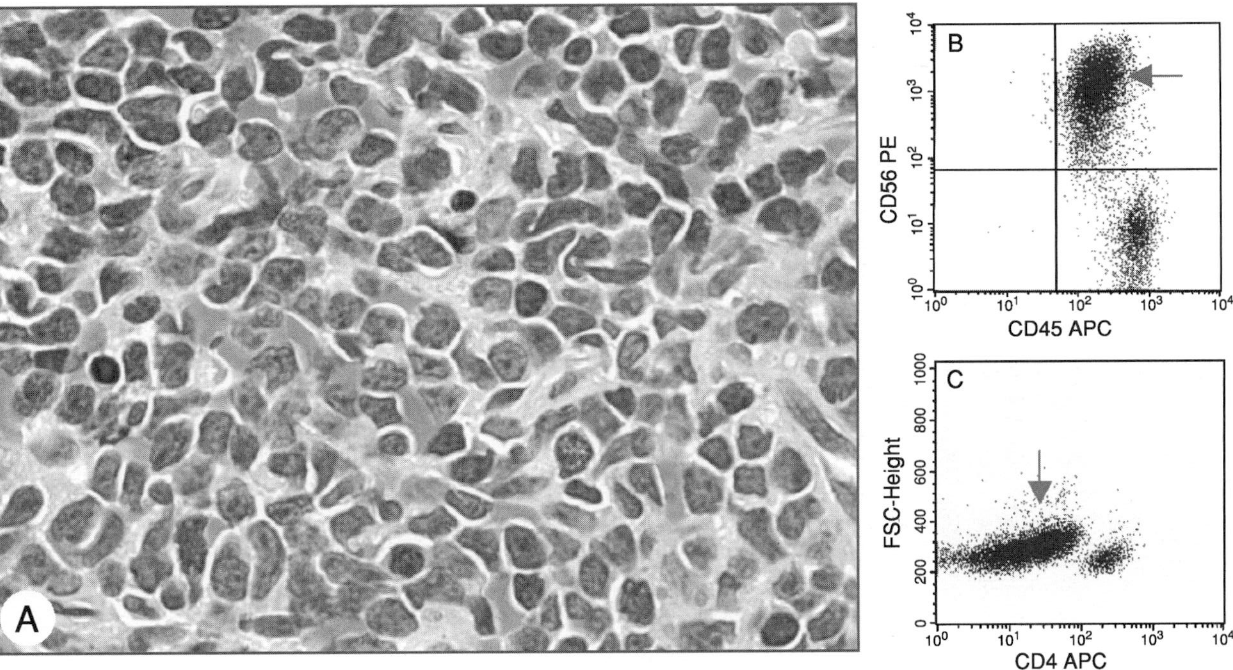

Fig. 13 Blastic natural killer cell lymphoma/leukemia. (See Color Insert.)

MPO negative and all cases expressing CD117 did not express pan-myeloid antigens (CD13, CD33). All cases of blastic NK cell lymphoma/leukemia were $CD4^+$. $CD117^+$ precursor T-ALL/LBLs (T-ALL/LBL) were $CD8^+$ in two cases, $CD4^-/CD8^-$ in two cases and $CD4^+$ in one case. All six cases of $CD117^+$ peripheral T-cell lymphomas were $CD8^+$.

X. Blastic NK Cell Lymphoma (DC-2 Acute Leukemia)

Blastic NK cell lymphoma is composed of medium-sized cells resembling lymphoblasts with coexpression of CD4 and CD56 (Fig. 13). Among 13 analyzed cases, the ages ranged from 67 to 88 years (average 79.8 years). Nine cases occurred in men and four in women in the following locations: bone marrow (five), skin (three), lymph node (two), and peripheral blood (3). Table V presents detailed flow cytometric findings in blastic NK cell lymphoma/leukemia.

XI. Conclusion

FC plays an important role in diagnosis, classification, and monitoring of the T-cell lymphoproliferations. Aberrant expression of pan-T antigens, TCR, and CD45, abnormal CD4/CD8 ratio (including dual expression or dual negativity),

Table V
Blastic Natural Killer Cell Lymphoma

Age (yr)	68	88	81	80	85	78	79	67	79	85	88	76	83
Sex	M	M	M	F	F	M	M	F	M	F	M	M	M
CD2	−	++	++	−	−	+	−	++	−	++	−	+	−
CD3	−	−	−	−	−	−	−	−	−	−	−	−	
CD4	+	+	++	++	+	+++	+	++	++	++	+	+++	++
CD5	−	−	−	−	−	−	−	−	−	−	−	−	−
CD7	+	+	+	++	++	++	+	−	++	++	+	+	+
CD8	−	−	−	−	−	−	−	−	−	−	−	−	−
CD10	−	−	−	−	−	−	−	+	−	−	+	−	−
CD11b or CD11c	−	−	−	−	−	−	−	−	−	−	−	−	−
CD13	−	−	−	−	−	−	−	−	−	−	−	−	−
CD16	−	−	−	−	−	−	−	−	−	−	−	−	−
CD23	−	−	−	−	−	−	−	−	−	−	−	−	−
CD33	−	−	+	+	−	+	−	−	−	−	+	+	−
CD45	+	++	++	++	++	++	+	++	+	++	++	++	++
CD56	+++	+++	+++	++	+++	+++	++	++	++	++	+++	++	+++
CD57	−	−	−	−	−	−	−	−	−	−	−	−	−
CD64	−	−	−	−	−	−	−	−	−	−	−	−	−
HLA-DR	++	+	+	+	+	+	+	+	+	+	+	+++	+
TdT	−	++	−	−	−	−	+	−	−	−	−	−	−
TCRαβ or TCRγδ	−	−	−	−	−	−	−	−	−	−	−	−	−

and presence of additional phenotypic markers help identify abnormal T-cell populations and their further subclassification. Proper classification is important because of the different prognoses and treatment options for patients with different types of T-cell malignancies. Once an aberrant immunophenotype has been detected by FC, cytomorphological correlation is essential for diagnosis. In difficult cases, additional methodology, such as cytogenetics/FISH, PCR, or Southern blot, may be required to confirm the clonal nature of the process.

References

Arrowsmith, E. R., Macon, W. R., Kinney, M. C., Stein, R. S., Goodman, S. A., Morgan, D. S., Flexner, J. M., Cousar, J. B., Jagasia, M. H., McCurley, T. L., and Greer, J. P. (2003). Peripheral T-cell lymphomas: Clinical features and prognostic factors of 92 cases defined by the revised European American lymphoma classification. *Leuk. Lymphoma* **44,** 241–249.

Ascani, S., Zinzani, P. L., Gherlinzoni, F., Sabattini, E., Briskomatis, A., de Vivo, A., Piccioli, M., Fraternali Orcioni, G., Pieri, F., Goldoni, A., Piccaluga, P. P., Zallocco, D., Burnelli, R., Leoncini, L., Falini, B., Tura, S., and Pileri, S. A. (1997). Peripheral T-cell lymphomas. Clinico-pathologic study of 168 cases diagnosed according to the R.E.A.L. Classification. *Ann. Oncol.* **8,** 583–592.

Ashton-Key, M., Diss, T. C., Pan, L., Du, M. Q., and Isaacson, P. G. (1997). Molecular analysis of T-cell clonality in ulcerative jejunitis and enteropathy-associated T-cell lymphoma. *Am. J. Pathol.* **151,** 493–498.

Bakels, V., van Oostveen, J. W., van der Putte, S. C., Meijer, C. J., and Willemze, R. (1997). Immunophenotyping and gene rearrangement analysis provide additional criteria to differentiate between cutaneous T-cell lymphomas and pseudo-T-cell lymphomas. *Am. J. Pathol.* **150,** 1941–1949.

Borowitz, M. J., Bray, R., Gascoyne, R., Melnick, S., Parker, J. W., Picker, L., and Stetler-Stevenson, M. (1997). U.S.-Canadian Consensus recommendations on the immunophenotypic analysis of hematologic neoplasia by flow cytometry: Data analysis and interpretation. *Cytometry* **30,** 236–244.

Chan, A. C., Ho, J. W., Chiang, A. K., and Srivastava, G. (1999). Phenotypic and cytotoxic characteristics of peripheral T-cell and NK-cell lymphomas in relation to Epstein-Barr virus association. *Histopathology* **34,** 16–24.

Chan, D. W., Liang, R., Chan, V., Kwong, Y. L., and Chan, T. K. (1997). Detection of T-cell receptor delta gene rearrangement by clonal specific polymerase chain reaction. *Leukemia* **11**(Suppl. 3), 281–284.

Chan, J. K. (1999). Peripheral T-cell and NK-cell neoplasms: An integrated approach to diagnosis. *Mod. Pathol.* **12,** 177–199.

de Bruin, P. C., Connolly, C. E., Oudejans, J. J., Kummer, J. A., Jansen, W., McCarthy, C. F., and Meijer, C. J. (1997). Enteropathy-associated T-cell lymphomas have a cytotoxic T-cell phenotype. *Histopathology* **31,** 313–317.

Farcet, J. P., Gaulard, P., Marolleau, J. P., Le Couedic, J. P., Henni, T., Gourdin, M. F., Divine, M., Haioun, C., Zafrani, S., Goossens, M., *et al.* (1990). Hepatosplenic T-cell lymphoma: Sinusal/sinusoidal localization of malignant cells expressing the T-cell receptor gamma delta. *Blood* **75,** 2213–2219.

Ferry, J. A. (2002). Angioimmunoblastic T-cell lymphoma. *Adv. Anat. Pathol.* **9,** 273–279.

Ginaldi, L., Matutes, E., Farahat, N., De Martinis, M., Morilla, R., and Catovsky, D. (1996). Differential expression of CD3 and CD7 in T-cell malignancies: A quantitative study by flow cytometry. *Br. J. Haematol.* **93,** 921–927.

Gorczyca, W., Tsang, P., Liu, Z., Wu, C. D., Dong, H. Y., Goldstein, M., Cohen, P., Gangi, M., and Weisberger, J. (2003). CD30-positive T-cell lymphomas co-expressing CD15: An immunohistochemical analysis. *Int. J. Oncol.* **22,** 319–324.

Gorczyca, W., Weisberger, J., Liu, Z., Tsang, P., Hossein, M., Wu, C. D., Dong, H., Wong, J. Y., Tugulea, S., Dee, S., Melamed, M. R., and Darzynkiewicz, Z. (2002). An approach to diagnosis of T-cell lymphoproliferative disorders by flow cytometry. *Cytometry* **50,** 177–190.

Greenland, C., Dastugue, N., Touriol, C., Lamant, L., Delsol, G., and Brousset, P. (2001). Anaplastic large cell lymphoma with the t(2;5)(p23;q35) NPM/ALK chromosomal translocation and duplication of the short arm of the non-translocated chromosome 2 involving the full length of the ALK gene. *J. Clin. Pathol.* **54,** 152–154.

Harris, N. L., Jaffe, E. S., Diebold, J., Flandrin, G., Muller-Hermelink, H. K., Vardiman, J., Lister, T. A., and Bloomfield, C. D. (1999). World Health Organization classification of neoplastic diseases of the hematopoietic and lymphoid tissues: Report of the Clinical Advisory Committee meeting-Airlie House, Virginia, November 1997. *J. Clin. Oncol.* **17,** 3835–3849.

Harris, N. L., Jaffe, E. S., Diebold, J., Flandrin, G., Muller-Hermelink, H. K., Vardiman, J., Lister, T. A., and Bloomfield, C. D. (2000a). The World Health Organization classification of neoplasms of the hematopoietic and lymphoid tissues: Report of the Clinical Advisory Committee meeting–Airlie House, Virginia, November, 1997. *Hematol. J.* **1,** 53–66.

Harris, N. L., Jaffe, E. S., Diebold, J., Flandrin, G., Muller-Hermelink, H. K., Vardiman, J., Lister, T. A., and Bloomfield, C. D. (2000b). The World Health Organization classification of neoplastic diseases of the haematopoietic and lymphoid tissues: Report of the Clinical Advisory Committee Meeting, Airlie House, Virginia, November 1997. *Histopathology* **36,** 69–86.

Harris, N. L., Stein, H., Coupland, S. E., Hummel, M., Favera, R. D., Pasqualucci, L., and Chan, W. C. (2001). New approaches to lymphoma diagnosis. *Hematol. Am. Soc. Hematol. Educ. Progr.*194–220.

Hastrup, N., Hamilton-Dutoit, S., Ralfkiaer, E., and Pallesen, G. (1991). Peripheral T-cell lymphomas: an evaluation of reproducibility of the updated Kiel classification. *Histopathology* **18,** 99–105.

Jaffe, E. S. (1995). Angioimmunoblastic T-cell lymphoma: New insights, but the clinical challenge remains. *Ann. Oncol.* **6,** 631–632.

Jaffe, E. S. (1996). Classification of natural killer (NK) cell and NK-like T-cell malignancies. *Blood* **87,** 1207–1210.

Jaffe, E. S., Krenacs, L., Kumar, S., Kingma, D. W., and Raffeld, M. (1999). Extranodal peripheral T-cell and NK-cell neoplasms. *Am. J. Clin. Pathol.* **111,** S46–S55.

Jaffe, E. S., Krenacs, L., and Raffeld, M. (2003). Classification of cytotoxic T-cell and natural killer cell lymphomas. *Semin. Hematol.* **40,** 175–184.

Jennings, C. D., and Foon, K. A. (1997). Recent advances in flow cytometry: Application to the diagnosis of hematologic malignancy. *Blood* **90,** 2863–2892.

Kinney, M. C., and Kadin, M. E. (1999). The pathologic and clinical spectrum of anaplastic large cell lymphoma and correlation with ALK gene dysregulation. *Am. J. Clin. Pathol.* **111,** S56–S67.

Lopez-Guillermo, A., Cid, J., Salar, A., Lopez, A., Montalban, C., Castrillo, J. M., Gonzalez, M., Ribera, J. M., Brunet, S., Garcia-Conde, J., Fernandez de Sevilla, A., Bosch, F., and Montserrat, E. (1998). Peripheral T-cell lymphomas: Initial features, natural history, and prognostic factors in a series of 174 patients diagnosed according to the R.E.A.L. Classification. *Ann. Oncol.* **9,** 849–855.

Loughran, T. P., Jr. (1998). Chronic T-cell leukemia/lymphoma. *Cancer Control* **5,** 8–9.

Loughran, T. P., Jr., Abbott, L., Gentile, T. C., Love, J., Cunningham, C., Friedman-Kien, A., Huang, Y. Q., and Poiesz, B. J. (1997). Absence of human herpes virus 8 DNA sequences in large granular lymphocyte (LGL) leukemia. *Leuk. Lymphoma* **26,** 177–180.

Matutes, E. (2002). Chronic T-cell lymphoproliferative disorders. *Rev. Clin. Exp. Hematol.* **6,** 401–420. 449–450.

Matutes, E., and Catovsky, D. (2003). Classification of mature T-cell leukemias. *Leukemia* **17,** 1682–1683.

Nakamura, N., Suzuki, S., Tasaki, K., Asano, S., Sai, T., Wakasa, H., and Abe, M. (1999). Peripheral T-cell lymphoma other than angioimmunoblastic T-cell lymphoma (AILD), with follicular dendritic cells proliferation and infection of B immunoblasts with Epstein Barr virus. *Fukushima J. Med. Sci.* **45,** 45–51.

Orfao, A., Schmitz, G., Brando, B., Ruiz-Arguelles, A., Basso, G., Braylan, R., Rothe, G., Lacombe, F., Lanza, F., Papa, S., Lucio, P., and San Miguel, J. F. (1999). Clinically useful information provided by the flow cytometric immunophenotyping of hematological malignancies: current status and future directions. *Clin. Chem.* **45,** 1708–1717.

Pileri, S. A., Ascani, S., Sabattini, E., and Falini, B. (1998). Peripheral T-cell lymphoma: A developing concept. *Ann. Oncol.* **9,** 797–801.

Pileri, S. A., Piccaluga, A., Poggi, S., Sabattini, E., Piccaluga, P. P., De Vivo, A., Falini, B., and Stein, H. (1995). Anaplastic large cell lymphoma: update of findings. *Leuk. Lymphoma* **18,** 17–25.

Weisberger, J., Wu, C. D., Liu, Z., Wong, J. Y., Melamed, M. R., Darzynkiewicz, Z., and Gorczyca, W. (2000). Differential diagnosis of malignant lymphomas and related disorders by specific pattern of expression of immunophenotypic markers revealed by multiparameter flow cytometry. *Int. J. Oncol.* **17,** 1165–1177.

CHAPTER 26

Isolation and Immunophenotyping of Human and Rhesus Macaque Dendritic Cells

Karin Loré

Vaccine Research Center
National Institutes of Health
Bethesda, Maryland 20892-3022

I. Introduction

Dendritic cells (DCs) represent a highly specialized bone marrow–derived subpopulation of leukocytes (Ardavin *et al.*, 2001). They perform an essential role as antigen-presenting cells (APCs) in the induction and regulation of adaptive immune responses (Banchereau and Steinman, 1998). DCs migrate from the blood to peripheral tissues, where they reside in an immature state, awaiting antigen encounter. Immature DCs respond rapidly to foreign antigens and employ

METHODS IN CELL BIOLOGY, VOL. 75

0091-679X/04 $35.00

several antigen internalization methods such as phagocytosis, receptor-mediated endocytosis, and macropinocytosis (Larsson *et al.*, 1997; Sallusto *et al.*, 1995). Upon capture of antigens, DCs process them into peptides, which are then loaded onto major histocompatibility complex (MHC) molecules for presentation to T cells. After tissue damage, inflammation, or pathogen invasion, DCs also receive additional activating signals, many of which are mediated through interaction with toll-like receptors (TLRs) expressed on DCs (Akira *et al.*, 2001; Jarrossay *et al.*, 2001; Kadowaki *et al.*, 2001). This activation induces maturation of the DCs, leading to profound changes in their phenotype and function. Maturation is characterized by upregulation and cell-surface expression of peptide-loaded MHC class I/II molecules, costimulatory molecules such as CD80, CD86, CD40, and adhesion molecules, as well as downregulation of antigen-capturing molecules. Mature DCs express the chemokine receptor CCR7, which directs them from peripheral tissues to regional draining lymph nodes (Sallusto and Lanzavecchia, 1999; Sallusto *et al.*, 1999). Thus, mature DCs efficiently present antigens and stimulate both $CD4^+$ and $CD8^+$ T cells located in the T-cell–rich areas of lymph nodes.

The presence of distinct DCs at different stages of maturation and localization, together with the possibility that DCs may be generated from blood monocytes and bone marrow precursors, makes the determination of clear functional and phenotypic distinctions between subsets of DCs difficult. A more complete detailed description of the differences between these subpopulations is, therefore, needed and is being continuously investigated. Two major subsets of DCs have been identified in the human immune system. $CD11c^+$ myeloid DCs (MDCs) originate from myeloid precursors and, therefore, are related to monocytes/macrophages. A distinctly different subpopulation was identified as CD11c-$CD123^+$ plasmacytoid DCs (PDCs) (Cella *et al.*, 1999; Kohrgruber *et al.*, 1999; Liu, 2001; O'Doherty *et al.*, 1994; Robinson *et al.*, 1999; Siegal *et al.*, 1999). PDCs differ from the $CD11c^+$ MDCs as they lack expression of myeloid markers, including CD33, CD13, and CD11b and they express CD45RA and the interleukin-3 (IL-3) receptor α-chain (CD123). MDCs and PDCs are mutually exclusive professional APCs that differ in some important functional aspects. MDCs exhibit a high ability to activate naive T cells, whereas freshly isolated PDCs are less potent APCs and require *in vitro* maturation to acquire strong T-cell stimulatory function (O'Doherty *et al.*, 1994). MDCs are also the more prevalent DC subset overall and can secrete high levels of interleukin-12 (IL-12) during antigen presentation. MDCs are widely distributed throughout the body, being most prevalent in the skin (*viz* Langerhans cells) and mucosal tissues. PDCs are more sparsely distributed and are found only in blood, lymph nodes, thymus (Cella *et al.*, 1999; Res *et al.*, 1999; Siegal *et al.*, 1999; Summers *et al.*, 2001) and in small amounts in cerebrospinal fluid (CSF) (Pashenkov *et al.*, 2001) under normal conditions but are recruited to sites of inflammation or tumors in pathological conditions (Farkas *et al.*, 2001; Hartmann *et al.*, 2003; Jahnsen *et al.*, 2000, 2001; Zou *et al.*, 2001). PDCs have a strong ability to produce high levels of interferon-α (IFN-α) in response to foreign antigen structures, particularly viruses, and may, therefore, play an important role in linking innate and adaptive immunity (Akira *et al.*, 2001; Jarrossay *et al.*, 2001; Kadowaki *et al.*, 2001).

II. *In Vitro* Differentiation of Monocyte-Derived Dendritic Cells

The critical role of DCs in the initiation and maintenance of immune responses is now widely appreciated. An expanded knowledge about the phenotypic and functional properties of DCs has been achieved during the 1990s due to the development of methods for *in vitro* generation and use of DCs. To date, much of human DC biology has been uncovered by studying *in vitro*-propagated, blood monocyte-derived DCs, generated using a combination of the recombinant cytokines for granulocyte–macrophage colony-stimulating factor (GM-CSF) and interleukin-4 (IL-4). During a period of 5–7 days, purified adherent $CD14^+$ $CD1a^-$ monocytes differentiate into nonadherent $CD14^-$ $CD1a^+$ DCs with characteristic morphology and phenotype of immature DCs. Detailed protocols for their generation can be found elsewhere (Bender *et al.*, 1996; Lore *et al.*, 1998, 2001; Romani *et al.*, 1994; Sallusto and Lanzavecchia, 1994). Provided with adequate cytokines such as tumor necrosis factor-α (TNF-α) or a wide range of other activation signals, such as several microbial products, DCs undergo phenotypic maturation (Bender *et al.*, 1996; Lore *et al.*, 1998, 2001; Romani *et al.*, 1994; Sallusto and Lanzavecchia, 1994). Using similar protocols, one can generate monocyte-derived DCs from nonhuman primates (Mehlhop *et al.*, 2002; Soderlund *et al.*, 2000). Although these *in vitro*–derived DCs resemble *in vivo*–derived DCs, whether such GM-CSF/IL-4–dependent pathways of DC generation exist has not been established.

III. Detection and Sorting Procedures of Subsets of Dendritic Cells

A. Identification and Quantification of Dendritic Cells

Characterization of DC subsets, as they exist *in vivo,* and the ability to manipulate their functions will be important if DCs are to be applied in biologically and clinically relevant models. Flow cytometric methods for enumerating DC subsets *in vivo* are available. Table I shows the frequencies of DCs in blood in normal adult individuals as reported in various studies. In addition, several reports have quantified the frequencies of DCs during the course of human immunodeficiency virus (HIV) infection (Barron *et al.*, 2003; Donaghy *et al.*, 2001; Grassi *et al.*, 1999; Pacanowski *et al.*, 2001), different cancer conditions (Mohty *et al.*, 2001; Savary *et al.*, 1998; Vuckovic *et al.*, 1999), and transplantation strategies (Arpinati *et al.*, 2002; Fearnley *et al.*, 1999; Galy *et al.*, 2000). Still, little is known about the

Table I
Frequencies of Dendritic Cell Subsets in Blood of Normal Donors Reported in Various Studies

Study	MDCs % out of total PBMCs	MDCs (antibodies number) (× 10^3/ml) mean ± SD	PDCs (% out of total PBMCs)	PDCs (antibodies number) (× 10^3/ml) mean ± SD
Galy *et al.*, 2000	0.14 ± 0.19		0.53 ± 0.12	
Upham *et al.*, 2000		17.4		
Pacanowski *et al.*, 2001		16.9 ± 14.3		12.3 ± 6.5
Mohty *et al.*, 2001	0.26 ± 0.23		0.24 ± 0.18	
Arpinati *et al.*, 2002			0.12	
Szabolcs *et al.*, 2003		15.5 ± 9.3		8.2 ± 4.3
Ueda *et al.*, 2003	0.32 ± 0.13		0.11 ± 0.07	
Barron *et al.*, 2003	0.41	8.1	0.30	6.0
Coates *et al.*, 2003 (Rhesus macaques)		18.0 ± 2.6		1.3 ± 0.2

mechanisms that regulate homeostasis of DCs *in vivo.* The turnover of DCs in tissues is substantially increased by various inflammatory stimuli (Macatonia *et al.*, 1987; McWilliam *et al.*, 1994, 1996) and by surgical and physical stress (Ho *et al.*, 2001). Disruptions in the homeostasis of DCs can have significant consequences on immunity. For accurate *in vivo* quantification of DCs, specimens should be manipulated as little as possible before being assessed. Inaccurate results can be obtained if cells are placed in culture overnight. Also, different processing steps may reduce certain DC populations (Borras *et al.*, 2001; Donnenberg and Donnenberg, 2003; Donnenberg *et al.*, 2001; Sorg *et al.*, 1999). Variations between laboratories in cell preparation methods (gradient centrifugation, FACS-lysis solution, etc.) and antibody combination used for staining of DC subsets have led to differences in reported composition and number of DCs (Table I summarizes various studies). Here, we use peripheral blood mononuclear cell (PBMC) separation on Ficoll Hypo Paque (Amersham Biosciences, Uppsala, Sweden) and lymphoid tissue disruption as described in section III.C followed by a staining protocol using antibodies for the lineage markers anti-CD3, CD14 (CD15), CD19, CD56 labeled with the same fluorochromes in addition to anti-CD11c, CD123, and HLA-DR antibodies each labeled with separate fluorochromes (see Table II for suggested clones) for identification of both the DC subsets. The stained cells can thereafter be analyzed by flow cytometry by gating on the $CD123^+$ PDCs and $CD11c^+$ MDCs within the $HLA\text{-}DR^+$ lineage $marker^-$ population (Figs. 1–4). More details about the staining procedure and flow cytometric gating strategies for identification of DCs are described below in sections III D–E.

Table II
Antibodies for Identification and Phenotyping of Human Dendritic Cells

Cell marker	Antibody clone	Used for identification of	Manufacturer
BDCA-1 (CD1c)	AD5-8E7[a]	Myeloid DCs	Miltenyi Biotec
BDCA-2	AC144	Plasmacytoid DCs	Miltenyi Biotec
BDCA-3	AD5-14H12	Myeloid DCs	Miltenyi Biotec
BDCA-4 (Neuropilin-1)	AD5-17F6	Plasmacytoid DCs	Miltenyi Biotec
CD3	SK7	T cells	BD Immunocytometry Systems
CD3	BW264/56	T cells	Miltenyi Biotec
CD3	SP34[a]	T cells	BD Pharmingen
CD11c	S-HCL-3[a]	Myeloid DCs	BD Pharmingen
CD11c	3.9[a]	Myeloid DCs	Serotec, Caltag, Biosource International
CD11c	B-Ly6	Myeloid DCs	BD Immunocytometry Systems
CD11c	BU15	Myeloid DCs	Beckman Coulter, Caltag
CD14	M5E2[a]	Monocytes	BD Pharmingen
CD14	TüK4[a]	Monocytes	Miltenyi Biotec, Caltag
CD14	332A-1[a]	Monocytes	Beckman Coulter
CD14	MoP9	Monocytes	BD Immunocytometry Systems
CD15	MMA	Granulocytes	BD Pharmingen
CD15	VIMC6	Granulocytes	Miltenyi Biotec
CD19	H1B19	B cells	BD Pharmingen
CD19	LT19	B cells	Miltenyi Biotec
CD19	4G7[a]	B cells	BD Immunocytometry Systems
CD19	J4.119[a]	B cells	Beckman Coulter
CD20	2H7[a]	B cells	BD Pharmingen
CD20	B1-RD1[a]	B cells	Beckman Coulter
CD20	LT20[a]	B cells	Miltenyi Biotec
CD20	B-Ly 1[a]	B cells	DakoCytomation
CD34	563[a]	Hematopoietic stem cells	BD Pharmingen
CD34	AC136	Hematopoietic stem cells	Miltenyi Biotec
CD40	5C3[a]	Costimulatory molecule	BD Pharmingen
CD40	14G7	Costimulatory molecule	Caltag
CD40	MAB89[a]	Co-stimulatory molecule	Beckman Coulter
CD56	B159[a]	NK cells	BD Pharmingen
CD56	NCAM16.2[a]	NK cells	BD Immunocytometry Systems
CD56	MY31[a]	NK cells	BD Immunocytometry Systems
CD56	NKH1[a]	NK cells (rhesus macaque monocytes)	Beckman Coulter
CD56	AF12-7H3[a]	NK cells (rhesus macaque monocytes)	Miltenyi Biotec
CD80	L307.4[a]	Costimulatory molecule	BD Immunocytometry Systems
CD80	BB-1	Costimulatory molecule	BD Pharmingen
CD80	DAL-1	Costimulatory molecule	Caltag
CD83	HB15a[a]	DC maturation molecule	Beckman Coulter
CD83	HB15e[a]	DC maturation molecule	BD Pharmingen, Caltag
CD86	FUN-1[a]	Costimulatory molecule	BD Pharmingen
CD86	IT2.2[a]	Costimulatory molecule	BD Pharmingen

(continues)

Table II *(continued)*

Cell marker	Antibody clone	Used for identification of	Manufacturer
CD86	BU63	Costimulatory molecule	Caltag
CD123	7G3[a]	Plasmacytoid DCs	BD Pharmingen
CD123	AC145	Plasmacytoid DCs	Miltenyi Biotec
CD123	9F5	Plasmacytoid DCs	BD Pharmingen
CD208 (DC-LAMP)	104.G4	MIIC protein	Beckman Coulter
CD209 (DC-SIGN)	DCN46	ICAM receptor	BD Pharmingen
CD209 (DC-SIGN)	120507	ICAM receptor	R&D Systems
CCR7	150503	Chemokine receptor	R&D Systems
CCR7	3C12	Chemokine receptor	BD Pharmingen
Fascin, p55	55K-2[a]	Actin-bundling protein	DakoCytomation
HLA-DR	L243[a]	MHC class II	BD Pharmingen
HLA-DR	G46-6[a]	MHC class II	BD Pharmingen
HLA-DR	CR3/43[a]	MHC class II	DakoCytomation

[a]Known cross-reactive antibodies to rhesus macaque.

B. Isolation of Dendritic Cells from Blood

The low frequency of DCs in PBMCs or other tissues makes direct isolation procedures laborious and often results in very low yields. Therefore, using large volumes of cells increases the odds of obtaining a workable number of DCs at the end (Alejandro Lopez *et al.*, 2002). Collection of PBMCs by automated leukapheresis generates the highest number of cells with minimal risk to the donor (Furuta *et al.*, 1999; Sato *et al.*, 2001). Leukocytes are maintained in this process, and erythrocytes and plasma are returned to the donor. Enriched populations of monocytes can efficiently be obtained from the leukapheresis product by counterflow centrifugal elutriation. This is a separation technique based on cellular size and density in which monocyte-like–sized cells and lymphocytes can be divided into different fractions. Relatively high yields of MDCs and PDCs can thereafter be isolated from the elutriated monocyte fraction using a combination of magnetic bead isolation and flow cytometric sorting as described in sections III.D and III.E.

C. Isolation of Dendritic Cells from Lymph Nodes and Tonsil Samples

Mononuclear cells can be isolated from tonsil or lymph node samples by mechanical and enzymatic disruption and DCs can thereafter be isolated using magnetic bead and flow cytometric sorting. The tissue must first be disrupted into small fragments, preferably using a scalpel, scissors, and/or forceps. If further enzymatic digestion is required, it can be performed in serum-free Iscove media (Invitrogen, Carlsbad, California) supplemented with DNAse 25 μg/ml (Worthington Biomedical, Freehold, New Jersey) and 2.4 mg/ml of collagenase type I (Invitrogen) for 30 minutes at 37 °C with gentle agitation. Additional

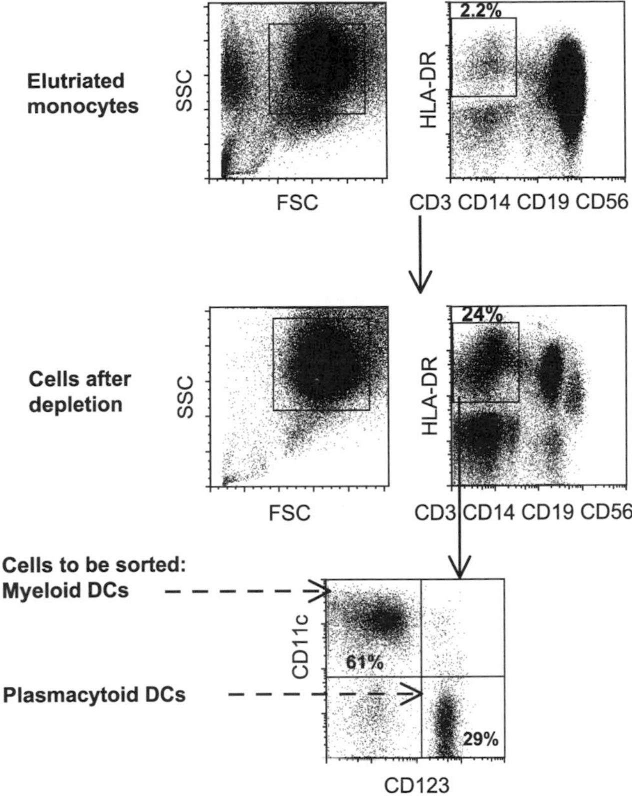

Fig. 1 Flowchart of sorting procedure of blood-derived subsets of dendritic cells from elutriated monocytes. The gate in the upper right flow plot shows the proportion of cells expressing human leukocyte antigen-DR (HLA-DR) but not any of the lineage markers for T cells, monocytes, B cells, or natural killer (NK) cells (CD3, CD14, CD19, CD56) in elutriated monocytes. This population can be enriched by depletion of unwanted cells using magnetic beads specific for the linage markers (i.e., microbeads, see section III.D). The CD11c$^+$ myeloid dendritic cells and CD123$^+$ plasmacytoid dendritic cells can thereafter be sorted by flow cytometric sorting by gating on cells expressing CD11c and CD123, respectively, in the HLA-DR$^+$ and lineage marker$^-$ population. Numbers indicate proportions (percentages) of cells included in the respective gates.

digestion can be performed for 5 minutes at 37 °C of digestion in 5 to 10 mM of EDTA in Hank's balanced salt solution without Ca^+ and Mg^+ with 10% serum. The cells, regardless of having been digested, should then be passed through a 100-μm mesh filter, using a sterile glass rod or a syringe plunger to further dissociate any remaining tissue fragments. Further enrichment can be obtained by density-gradient separation over a metrizamide cushion (1.065 g/ml) at 1600 × g for 10 minutes at 4 °C.

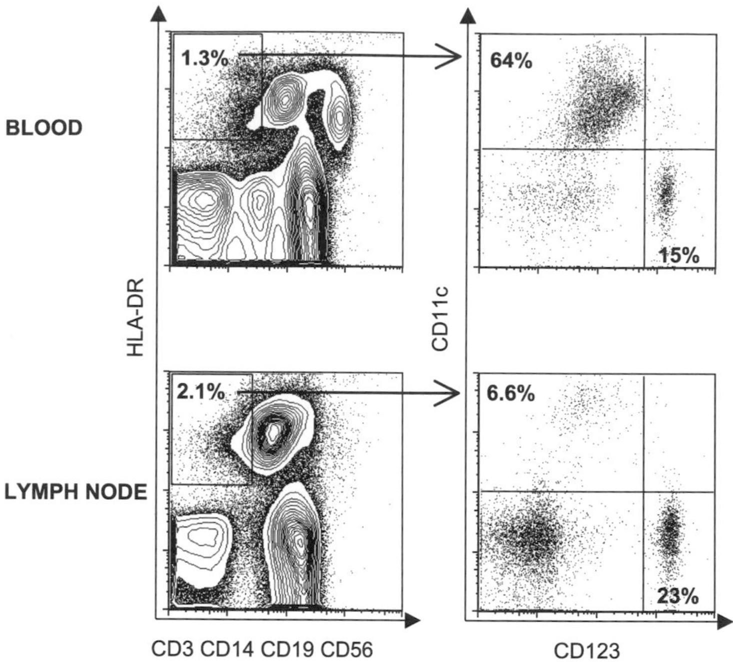

Fig. 2 Characteristic flow cytometric staining for subsets of dendritic cells (DCs) in peripheral blood and lymph node (axillar) from a healthy individual. Samples were stained with lineage marker antibodies (CD3, CD14, CD19, CD56) conjugated with the same fluorochrome, in addition to human leukocyte antigen-DR (HLA-DR), CD11c, and CD123, each labeled with separate fluorochromes. The subsets of DCs fall in the gate for HLA-DR$^+$ lineage marker$^-$ cells and are separated based on CD11c (myeloid DCs [MDCs]) and CD123 (plasmacytoid DCs [PDCs]) expression. MDCs are the most frequent DC subset in blood, but the ratio is switched in lymph nodes. Numbers indicate proportions (percentages) of cells included in the respective gates.

D. Enrichment of Dendritic Cells by Magnetic Bead Depletion before Flow Cytometric Sort

The single-cell suspension, originating either from bulk PBMCs, from elutriated monocytes, or from a tissue sample, should be depleted of cells expressing lineage markers (CD3$^+$ T cells, CD14$^+$ monocytes, CD19$^+$ B cells, CD56$^+$ natural killer [NK] cells). If the sample contains high numbers of polynuclear cells, CD15$^+$ granulocytes may also be depleted. Depletion of CD16$^+$ cells should not be included in this isolation process because it is expressed on a subpopulation of MDCs (Grage-Griebenow *et al.*, 2001a,b). A subpopulation of PDCs has shown CD56 expression in Flt3L-mobilized normal human donors, which is why CD56 may also be excluded from the lineage markers in some cases (Comeau *et al.*, 2002). Hematopoietic stem and progenitor cells comprise extremely small number of cells and usually do not cause any problems in the analysis of DCs. However, including antibodies against CD34 will eliminate the stem cells along with the lineage marker$^+$ cells. The depletion step of lineage marker$^+$ cells is primarily performed to decrease the number of cells that are subsequently sorted by flow cytometric sorting and thereby quicken the isolation process. If the initial cell numbers are low, this step may be excluded. The cells should be incubated with either a cocktail of fluorescein isothiocyanate (FITC)– or phycoerythrin (PE)-labeled antibodies against the lineage markers (i.e., anti-CD3, CD14 (CD15), CD19, CD56 [see Table II for suggested clones]) at more than 4 °C, washed in phosphate-buffer solution (PBS) supplemented with 0.5% bovine serum albumin and 2 mM of EDTA and subsequently incubated with anti-FITC or anti-PE antibody coupled to colloidal paramagnetic microbeads (anti-FITC/PE microbeads, Miltenyi Biotec, Auburn, California). Antibodies to the lineage markers coupled directly to microbeads (Miltenyi Biotec) can also be used. However, one has to consider that if any subsequent staining using antibodies of the same clones are to be used, the prior staining with microbeads can block their binding. Use of antibodies with other epitope specificity is, therefore, recommended (see Table II for suggested clones). The cells can be separated using magnetic separation columns manually or using AutoMacs according to the manufacturer's instructions (Miltenyi Biotec). The unlabeled cell fraction coming through the columns should be depleted of a significant amount of the lineage marker$^+$ cells (Fig. 1). These cells may be stained again with the same lineage markers anti-CD3, CD14 (CD15), CD19, and CD56 antibodies, but most importantly by anti-CD11c, CD123, and HLA-DR antibodies each labeled with separate fluorochromes. The stained cells can thereafter be sorted by flow cytometry after gating on the CD123$^+$ PDCs and CD11c$^+$ MDCs within the HLA-DR$^+$ lineage marker$^-$ population.

E. Flow Cytometric Gating Strategies

Using antibodies against CD11c and CD123, the MDC and PDC fractions can be identified within the lineage marker$^-$ (CD3$^-$ CD14$^-$CD19$^-$ CD56$^-$) HLA-DR$^+$ fraction (Figs. 1–4). This approach offers the advantage of simultaneous identification of both subsets on the same flow cytometric dot plot, thereby facilitating direct comparison of the relative percentage of each subset within the lineage marker$^-$ HLA-DR$^+$ fraction, which usually represents less than 1–2% of total PBMCs or cells in lymphoid tissue (Figs. 2–4). Gating on lineage marker$^-$ cells is essential because CD11c and CD123 are expressed not only on DCs but also on subpopulations of monocytes. Basophilic granulocytes also express both CD11c and CD123. However, basophils are not HLA-DR$^+$ and should not fall into the DC gate even if a marker for granulocytes is excluded. The gating strategy of a typical flow cytometric sort of MDCs and PDCs from elutriated human monocytes is shown in Fig. 1. High yields ($1–15 \times 10^6$) of MDCs and PDCs can be isolated from $0.5–2.5 \times 10^9$ elutriated monocytes. This isolation procedure usually results in more than 98% pure populations of DCs (Lore *et al.*, 2003). Normally, blood MDCs are more common than PDCs (Lore *et al.*, 2003; Mohty *et al.*, 2001; Szabolcs *et al.*, 2003; Ueda *et al.*, 2003) (Table I), but this proportion varies by donor. A ratio of 2–10 MDCs per PDCs is commonly found. In contrast, more PDCs than MDCs are usually found and recovered from lymph node and tonsillar tissue (Hartmann *et al.*, 2003; Stent *et al.*, 2002). Figure 2 shows the frequencies of DC subsets in PBMCs and lymph node from a healthy individual.

F. Direct Isolation of Dendritic Cells Using Dendritic Cell Isolation Kits

A number of alternative DC-specific antibodies were developed for the identification of human DC subsets (Dzionek *et al.*, 2000, 2001; MacDonald *et al.*, 2002). The cell-surface receptors targeted by these antibodies are C-type lectins that are named *blood DC antigens* (BDCAs) types 1–4. BDCA-1 (a myeloid marker that identifies CD1c) and BDCA-3 bind freshly isolated MDCs. In contrast, freshly isolated PDCs bind BDCA-2 and BDCA-4 (neuropilin-1). DC isolation kits based on positive selection of subsets of DCs using these markers can be purchased from Miltenyi Biotec. These kits offer a convenient and rapid isolation technique of DCs. In this regard, sequential magnetic separation for purification of DCs from elutriated monocytes is highly recommended to obtain good purity. When elutriated monocytes are used as the source for blood-derived DC subsets, the BDCA-2 and BDCA-4 isolation kits provide more than 80% pure population of PDCs, and the BDCA-1 and BDCA-3 isolation kits result in more than 90% purity of MDCs (Lore *et al.*, 2003). The contaminating cells are mostly monocytes. Using BDCA-1 microbeads for isolation for CD1a$^+$ Langerhans cells from epidermal skin has also been a reasonably good sorting method (Peiser *et al.*, 2003). Although the kits are developed for isolation of DCs directly from unfractioned PBMCs, using PBMCs rather than elutriated monocytes decreases the purity significantly. Moreover, it has to be taken into consideration that these markers (BDCA-1 through BDCA-4) are also expressed on subpopulations of other cells. BDCA-1 is expressed on a subpopulation of B cells, which is why B cells have to be depleted before positive selection of BDCA-1$^+$ MDCs. BDCA-4 is expressed on a subpopulation of monocytes. BDCA-4 is also upregulated on MDCs upon culture, which is why isolation of PDCs based on BDCA-4 expression should be performed only on freshly isolated cells. BDCA-2 and BDCA-3 bind a subpopulation and not all PDCs and MDCs, respectively. Nevertheless, because using CD11c and CD123 as markers for isolation or enumeration of MDCs and PDCs, respectively, reveals similar frequencies of DCs as using BDCA-1 and BDCA-4, the latter combination may be as useful. Figure 3 shows a comparison of using the respective combinations, CD11c/CD123 versus BDCA-1/BDCA-4, for identification of DC subsets in blood and lymph node of a healthy donor.

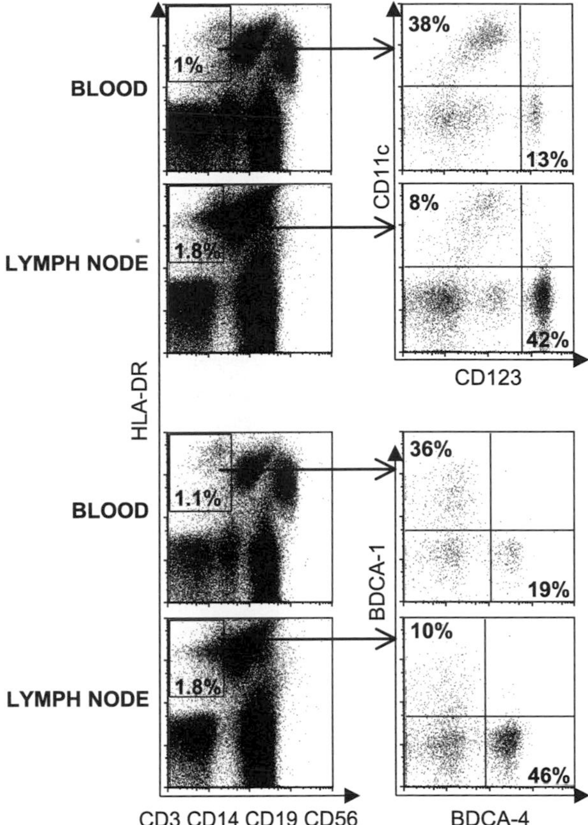

Fig. 3 Identification of subsets of dendritic cells (DCs) in peripheral blood and lymph node (axilla) from a healthy individual using either CD11c/CD123 or BDCA-1/BDCA-4 stainings to separate the myeloid DCs (MDCs) and plasmacytoid DCs (PDCs). Samples were stained with lineage marker antibodies (CD3, CD14, CD19, CD56) and human leukocyte antigen-DR (HLA-DR) in addition to either CD11c and CD123 or BDCA-1 and BDCA-4. The subsets of DCs that fall in the gate for HLA-DR$^+$ lineage marker$^-$ cells give comparable frequencies using either of the different antibody combinations. Higher frequencies of CD11c$^+$ BDCA-1$^+$ MDCs are found in blood, and CD123$^+$ BDCA-4$^+$ PDCs are the predominant DC subset in lymph nodes. Numbers indicate proportions (percentages) of cells included in the respective gates.

G. Identification of Dendritic Cells in Nonhuman Primates

There are differences in the presence of DC subsets and in the phenotype and function of DCs between mice and humans. This is a reason why direct extrapolation from rodent data to humans can be difficult. Alternatively, DCs from nonhuman primates have been shown to closely resemble those of humans (Coates *et al.*, 2002, 2003). Two distinct DC subsets, CD11c$^+$ CD123$^{-/+}$ MDCs and CD11c$^-$ CD123^{+++} PDCs, have been identified in rhesus monkeys (*Macacca*

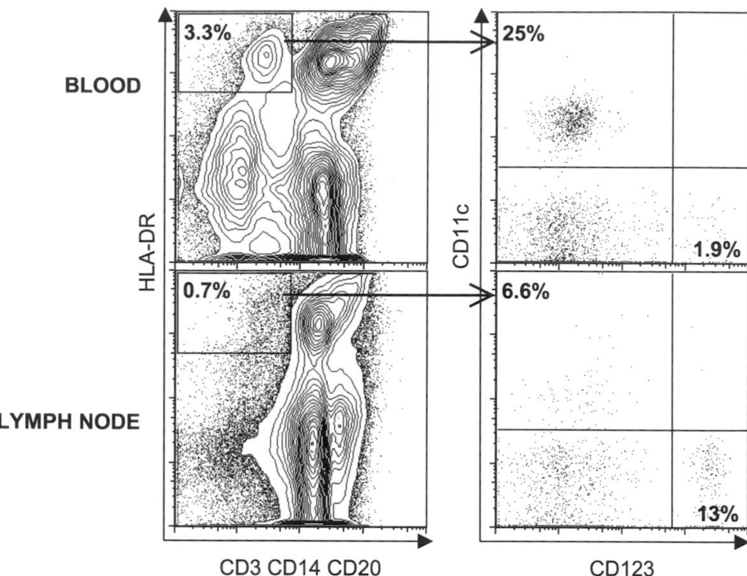

Fig. 4 Characteristic flow cytometric staining for subsets of dendritic cells (DCs) in peripheral blood and lymph node (axillar) from a healthy rhesus macaque monkey. Samples were stained with lineage marker antibodies (CD3, CD14, CD20) conjugated with the same fluorochrome, in addition to human leukocyte antigen-DR (HLA-DR), CD11c, and CD123, each labeled with separate fluorochromes. $CD123^+$ PDCs and $CD11c^+$ myeloid DCs (MDCs) fall in $HLA\text{-}DR^+$ lineage $marker^-$ gate. As in humans, MDCs are the prevalent DC subset in blood, whereas plasmacytoid DCs (PDCs) are most common in lymph nodes in rhesus macaque monkeys. Numbers indicate proportions (percentages) of cells included in the respective gates.

mulatta) whose phenotypes and functions are similar to those in humans. Non-human primates, therefore, serve as important models for assessing the role of human DCs. The staining and gating procedures for identification of rhesus macaque DCs are quite similar to those of human. However, a careful selection of antibody clones must be done because only a proportion of antibody clones designed for human cells are cross reactive for rhesus macaque cells (see Table II for suggested clones). The lineage markers used in monkeys are different from those used in humans. Staining for $CD3^+$ T cells and $CD14^+$ monocytes should be included. CD20 is generally used instead of CD19 for staining B cells. CD56 and CD16 are expressed on rhesus NK cells but also on a significant number of monocytes and on some MDCs (Carter *et al.*, 1999; Coates *et al.*, 2003). CD56 and CD16 should, therefore, not be included as lineage markers. In addition, CD34 is expressed on a subset of rhesus macaque DCs and should also not be included (Coates *et al.*, 2003). Figure 4 shows a gating strategy for identification of DC subsets ($CD11c^+$ MDCs and $CD123^+$ PDCs) in PBMCs and a lymph node from a rhesus macaque monkey. Among the alternative markers (BDCA-1 through BDCA-4) for identification of the subsets of DCs, only BDCA-1 shows cross-reactivity with rhesus macaque cells (Coates *et al.*, 2003). Therefore, BDCA-2 through BDCA-4 should be used only for isolation and analysis of human DC subsets, not for rhesus macaque DCs.

IV. Culture and Activation of Dendritic Cells

A. Phenotyping of Differentiated Dendritic Cell Subsets

Sorted blood or lymph node DCs may not display a characteristic dendritic morphology directly after the isolation procedure. They usually are rounded with reniform nuclei and few or no dendrites. However, short-term culture of the cells leads to the development of dendrites and increases in cellular size (especially among MDCs). MDCs tend to spread out during cell culture and to be semi-adherent to plastic surfaces, whereas most PDCs do not adhere and form cell clusters. Both DC subsets may be cultured in RPMI media supplemented with 10% fetal bovine serum and antibiotics. PDCs are dependent on supplementation with IL-3 (1 ng/ml, R&D Systems, Minneapolis, Minnesota) in the culture media for cell survival. The viability of MDCs is improved if the cells are cultured in GM-CSF (2 ng/ml, PeproTech, Inc., Rocky Hill, New Jersey). Freshly isolated DCs from blood exhibit an immature phenotype and are modest T-cell stimulators in mixed leukocyte reaction assays. However, short-term culture leads to remodeling of the DCs. Culturing of PDCs and MDCs in IL-3 or GM-CSF, respectively, will slowly differentiate the cells. However, to induce a significant maturation of DCs and acquisition of exclusive ability to induce a T-cell response, addition of a maturation stimulus is needed. Because the natural response of DCs to infection and inflammation is largely determined by signals transduced through toll-like receptors (TLRs) expressed in the DC subsets, using compounds binding to TLRs is an efficient way to mature and activate DCs (Lore *et al.*, 2003). The differential expression of the TLRs on PDCs and MDCs may allow plasticity of the cellular immune response to various environmental pathogens but also means that a relevant stimulus must be selected for *in vitro* stimulation of either DC subset (Jarrossay *et al.*, 2001; Kadowaki *et al.*, 2001). MDCs express TLR4, which is part of a complex to which lipopolysaccharide (LPS) binds, and TLR3, which is required for responsiveness to double-stranded RNA (Hornung *et al.*, 2002; Jarrossay *et al.*, 2001; Kadowaki *et al.*, 2001; Krug *et al.*, 2001; Muzio *et al.*, 2000; Visintin *et al.*, 2001; Zarember and Godowski, 2002). In contrast to MDCs, human PDCs do not express TLR3 or TLR4 but express TLR9, which recognizes unmethylated bacterial DNA with immunostimulatory CG motifs, known as CpG oligodeoxynucleotides (ODNs) (Bauer *et al.*, 2001; Ito *et al.*, 2002; Krug *et al.*, 2001). TLR7 may be expressed on both subpopulations of DCs, whereas TLR8 may be more profoundly expressed on MDCs (Hemmi *et al.*, 2002; Hornung *et al.*, 2002; Ito *et al.*, 2002; Jurk *et al.*, 2002). It was reported that the low-molecular-weight imidazoquinoline compounds, imiquimod and R-848 (resiquimod), activate human DCs via TLR7 and TLR8 (Hemmi *et al.*, 2002; Ito *et al.*, 2002; Jurk *et al.*, 2002; Lore *et al.*, 2003). Concentration and activation kinetics have to be monitored for optimal stimulation by each TLR ligand (or other stimulus). There are many types of LPSs (Sigma, St. Louis, Missouri) and they are usually used at 0.01–1.00 μg/ml for maturation of DCs. Double-stranded RNA complex polyinosinic:polycytidylic acid poly(I:C) (Sigma) may be used at 50 ng/ml. There are various CpG ODNs present, usually used at 2–10 μg/ml, and may be purchased from Coley Pharmaceutical Group (Ottawa, Canada). The imidazoquinoline imiquimod (1-[2-methylpropyl]-1 H-imidazo[4,5-c] quinolin-4-amine) or its derivative R-848 (4-amino-2-ethoxymethyl-α, α-dimethyl-1H-imidazoquinoline-1-ethanol) may be obtained from 3M (Minneapolis, Minnesota) and used at 0.25 μg/ml ($1\text{–}10 \times 10^{-8}$ M) and at 1 μg/ml ($1\text{–}3 \times 10^{-7}$ M), respectively. Specific activation induces maturation of the DCs, which includes a substantial transformation of their phenotype and function. Maturation is characterized by upregulation of stable cell-surface MHC class I/II–peptide complexes, costimulatory molecules such as CD80, CD86, CD40, and adhesion molecules, as well as downregulation of antigen-capturing molecules. This can be assessed by flow cytometry using antibodies against these receptors (see Table II for suggested clones). Phenotypic maturation usually occurs within 24 hours if the DCs are stimulated optimally (Lore *et al.*, 2003). Figure 5 shows a characteristic phenotypic remodeling of the DC subsets depicted as upregulation of costimulatory molecules and

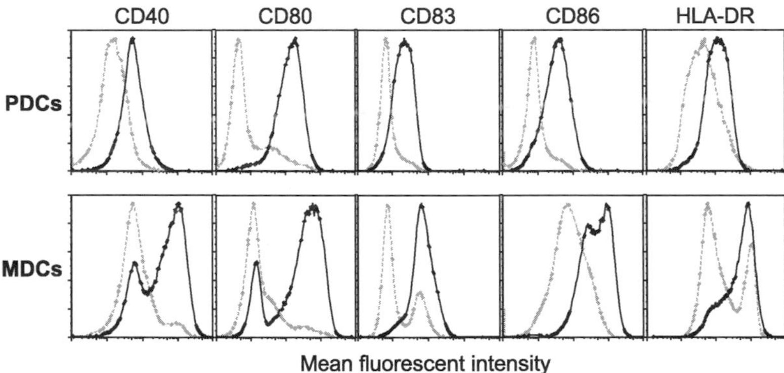

Fig. 5 Plasmacytoid dendritic cells (PDCs) and myeloid dendritic cells (MDCs) differentiate in response to toll-like receptor (TLR) ligation. Flow histograms show cell-surface expression of various maturation markers (CD40, CD80, CD86, CD83, and major histocompatibility complex [MHC] class II [human leukocyte antigen-DR (HLA-DR)]) after exposure of freshly isolated human dendritic cell (DC) subsets to TLR ligands. Phenotypic changes on PDCs (A) and MDCs (B) after exposure to CpG ODN and poly(I:C), respectively, for 24 hours were measured by flow cytometry. Data presented are overlaid histograms representing fluorescence intensity of cell-surface molecules expressed on TLR ligand–stimulated DCs (*solid black lines*) and unstimulated DCs (*dashed gray lines*).

MHC class II (HLA-DR) in response to relevant TLR ligation. Other markers may be useful for assessment of DC maturation state or identification of subpopulations of DCs (Lore *et al.*, 2002). DC-LAMP (CD208) appears during DC maturation in intracellular MHC class II compartments (MIICs) where peptides associate with MHC class II molecules (de Saint-Vis *et al.*, 1998). The chemokine receptor CCR7 is upregulated on mature DCs to direct the DCs to draining lymph nodes (Sallusto and Lanzavecchia, 1999; Sallusto *et al.*, 1999). The 55-kd actin-bundling protein p55 fascin has also been used as a characteristic marker for more mature DCs (Mosialos *et al.*, 1996). CMRF-44 and CMRF-56 antigens are early markers for activation and differentiation of DCs but have also been used for enumeration of DCs (Hock *et al.*, 1994, 1999). DC-SIGN (CD209) is a DC-specific C-type lectin-like cell-surface receptor that binds intercellular adhesion molecule-3 (ICAM-3) and intercellular adhesion molecule-2 (ICAM-2) on T cells, promoting the adhesion of DCs to naive T cells, and may be useful for identification of subpopulations of DCs (Geijtenbeek *et al.*, 2000). DEC-205 is a multilectin receptor that targets late endosomal/lysosomal vacuoles containing MHC class II products (Mahnke *et al.*, 2000). It has to be taken into consideration that phenotypic differences have been found between directly sorted DCs and monocyte-derived DCs (described in section II) in their expression of surface markers such as DC-SIGN and DEC-205, which is why some markers may not be suitable to use for all types of DCs (Kato *et al.*, 2000).

References

Akira, S., Takeda, K., and Kaisho, T. (2001). Toll-like receptors: Critical proteins linking innate and acquired immunity. *Nat. Immunol.* **2,** 675–680.

Alejandro Lopez, J., Crosbie, G., Kelly, C., McGee, A. M., Williams, K., Vuckovic, S., Schuyler, R., Rodwell, R., Wright, S. J., Taylor, K., and Hart, D. N. (2002). Monitoring and isolation of blood dendritic cells from apheresis products in healthy individuals: a platform for cancer immunotherapy. *J. Immunol. Methods* **267,** 199–212.

Ardavin, C., Martinez del Hoyo, G., Martin, P., Anjuere, F., Arias, C. F., Marin, A. R., Ruiz, S., Parrillas, V., and Hernandez, H. (2001). Origin and differentiation of dendritic cells. *Trends Immunol.* **22,** 691–700.

Arpinati, M., Chirumbolo, G., Urbini, B., Martelli, V., Stanzani, M., Falcioni, S., Bonifazi, F., Bandini, G., Tura, S., Baccarani, M., and Rondelli, D. (2002). Use of anti-BDCA-2 antibody for detection of dendritic cells type-2 (DC2) in allogeneic hematopoietic stem cell transplantation. *Bone Marrow Transplant.* **29,** 887–891.

Banchereau, J., and Steinman, R. M. (1998). Dendritic cells and the control of immunity. *Nature* **392,** 245–252.

Barron, M. A., Blyveis, N., Palmer, B. E., MaWhinney, S., and Wilson, C. C. (2003). Influence of plasma viremia on defects in number and immunophenotype of blood dendritic cell subsets in human immunodeficiency virus 1–infected individuals. *J. Infect. Dis.* **187,** 26–37.

Bauer, M., Redecke, V., Ellwart, J. W., Scherer, B., Kremer, J. P., Wagner, H., and Lipford, G. B. (2001). Bacterial CpG-DNA triggers activation and maturation of human CD11c−, CD123+ dendritic cells. *J. Immunol.* **166,** 5000–5007.

Bender, A., Sapp, M., Schuler, G., Steinman, R. M., and Bhardwaj, N. (1996). Improved methods for the generation of dendritic cells from nonproliferating progenitors in human blood. *J. Immunol. Methods* **196,** 121–135.

Borras, F. E., Matthews, N. C., Lowdell, M. W., and Navarrete, C. V. (2001). Identification of both myeloid CD11c+ and lymphoid CD11c-dendritic cell subsets in cord blood. *Br. J. Haematol.* **113,** 925–931.

Carter, D. L., Shieh, T. M., Blosser, R. L., Chadwick, K. R., Margolick, J. B., Hildreth, J. E., Clements, J. E., and Zink, M. C. (1999). CD56 identifies monocytes and not natural killer cells in rhesus macaques. *Cytometry* **37,** 41–50.

Cella, M., Jarrossay, D., Facchetti, F., Alebardi, O., Nakajima, H., Lanzavecchia, A., and Colonna, M. (1999). Plasmacytoid monocytes migrate to inflamed lymph nodes and produce large amounts of type I interferon. *Nat. Med.* **5,** 919–923.

Coates, P. T., Barratt-Boyes, S. M., Donnenberg, A. D., Morelli, A. E., Murphey-Corb, M., and Thomson, A. W. (2002). Strategies for preclinical evaluation of dendritic cell subsets for promotion of transplant tolerance in the nonhuman primate. *Hum. Immunol.* **63,** 955–965.

Coates, P. T., Barratt-Boyes, S. M., Zhang, L., Donnenberg, V. S., O'Connell, P. J., Logar, A. J., Duncan, F. J., Murphey-Corb, M., Donnenberg, A. D., Morelli, A. E., Maliszewski, C. R., and Thomson, A. W. (2003). Dendritic cell subsets in blood and lymphoid tissue of rhesus monkeys and their mobilization with Flt3 ligand. *Blood* **102,** 2513–2521.

Comeau, M. R., Van der Vuurst de Vries, A. R., Maliszewski, C. R., and Galibert, L. (2002). CD123bright plasmacytoid predendritic cells: Progenitors undergoing cell fate conversion? *J. Immunol.* **169,** 75–83.

de Saint-Vis, B., Vincent, J., Vandenabeele, S., Vanbervliet, B., Pin, J. J., Ait-Yahia, S., Patel, S., Mattei, M. G., Banchereau, J., Zurawski, S., Davoust, J., Caux, C., and Lebecque, S. (1998). A novel lysosome-associated membrane glycoprotein, DC-LAMP, induced upon DC maturation, is transiently expressed in MHC class II compartment. *Immunity* **9,** 325–336.

Donaghy, H., Pozniak, A., Gazzard, B., Qazi, N., Gilmour, J., Gotch, F., and Patterson, S. (2001). Loss of blood CD11c(+) myeloid and CD11c(−) plasmacytoid dendritic cells in patients with HIV-1 infection correlates with HIV-1 RNA virus load. *Blood* **98,** 2574–2576.

Donnenberg, V. S., and Donnenberg, A. D. (2003). Identification, rare-event detection and analysis of dendritic cell subsets in broncho-alveolar lavage fluid and peripheral blood by flow cytometry. *Front. Biosci.* **8,** s1175–s1180.

Donnenberg, V. S., O'Connell, P. J., Logar, A. J., Zeevi, A., Thomson, A. W., and Donnenberg, A. D. (2001). Rare-event analysis of circulating human dendritic cell subsets and their presumptive mouse counterparts. *Transplantation* **72,** 1946–1951.

Dzionek, A., Fuchs, A., Schmidt, P., Cremer, S., Zysk, M., Miltenyi, S., Buck, D. W., and Schmitz, J. (2000). BDCA-2, BDCA-3, and BDCA-4: Three markers for distinct subsets of dendritic cells in human peripheral blood. *J. Immunol.* **165,** 6037–6046.

Dzionek, A., Sohma, Y., Nagafune, J., Cella, M., Colonna, M., Facchetti, F., Gunther, G., Johnston, I., Lanzavecchia, A., Nagasaka, T., Okada, T., Vermi, W., Winkels, G., Yamamoto, T., Zysk, M., Yamaguchi, Y., and Schmitz, J. (2001). BDCA-2, a novel plasmacytoid dendritic cell–specific type II C-type lectin, mediates antigen capture and is a potent inhibitor of interferon alpha/beta induction. *J. Exp. Med.* **194,** 1823–1834.

Farkas, L., Beiske, K., Lund-Johansen, F., Brandtzaeg, P., and Jahnsen, F. L. (2001). Plasmacytoid dendritic cells (natural interferon-alpha/beta–producing cells) accumulate in cutaneous lupus erythematosus lesions. *Am. J. Pathol.* **159,** 237–243.

Fearnley, D. B., Whyte, L. F., Carnoutsos, S. A., Cook, A. H., and Hart, D. N. (1999). Monitoring human blood dendritic cell numbers in normal individuals and in stem cell transplantation. *Blood* **93,** 728–736.

Furuta, M., Shimizu, T., Mizuno, S., Kamiya, T., Ozawa, K., Nakase, T., Tadokoro, K., Takenaka, M., Ohkawa, T., Yokoyama, S., Ogawa, Y., Kiyokawa, H., Shimizu, M., Sekine, N., and Yoshimura, I. (1999). Clinical evaluation of repeat apheresis donors in Japan. *Vox Sang.* **77,** 17–23.

Galy, A., Rudraraju, S., Baynes, R., and Klein, J. (2000). Recovery of lymphocyte and dendritic cell subsets after autologous CD34+ cell transplantation. *Bone Marrow Transplant.* **25,** 1249–1255.

Geijtenbeek, T. B., Torensma, R., van Vliet, S. J., van Duijnhoven, G. C., Adema, G. J., van Kooyk, Y., and Figdor, C. G. (2000). Identification of DC-SIGN, a novel dendritic cell–specific ICAM-3 receptor that supports primary immune responses. *Cell* **100,** 575–585.

Grage-Griebenow, E., Flad, H. D., and Ernst, M. (2001a). Heterogeneity of human peripheral blood monocyte subsets. *J. Leukoc. Biol.* **69,** 11–20.

Grage-Griebenow, E., Zawatzky, R., Kahlert, H., Brade, L., Flad, H., and Ernst, M. (2001b). Identification of a novel dendritic cell–like subset of CD64(+)/CD16(+) blood monocytes. *Eur. J. Immunol.* **31,** 48–56.

Grassi, F., Hosmalin, A., McIlroy, D., Calvez, V., Debre, P., and Autran, B. (1999). Depletion in blood CD11c-positive dendritic cells from HIV-infected patients. *AIDS* **13,** 759–766.

Hartmann, E., Wollenberg, B., Rothenfusser, S., Wagner, M., Wellisch, D., Mack, B., Giese, T., Gires, O., Endres, S., and Hartmann, G. (2003). Identification and functional analysis of tumor-infiltrating plasmacytoid dendritic cells in head and neck cancer. *Cancer Res.* **63,** 6478–6487.

Hemmi, H., Kaisho, T., Takeuchi, O., Sato, S., Sanjo, H., Hoshino, K., Horiuchi, T., Tomizawa, H., Takeda, K., and Akira, S. (2002). Small anti-viral compounds activate immune cells via the TLR7 MyD88-dependent signaling pathway. *Nat. Immunol.* **3,** 196–200.

Ho, C. S., Lopez, J. A., Vuckovic, S., Pyke, C. M., Hockey, R. L., and Hart, D. N. (2001). Surgical and physical stress increases circulating blood dendritic cell counts independently of monocyte counts. *Blood* **98,** 140–145.

Hock, B. D., Fearnley, D. B., Boyce, A., McLellan, A. D., Sorg, R. V., Summers, K. L., and Hart, D. N. (1999). Human dendritic cells express a 95 kDa activation/differentiation antigen defined by CMRF-56. *Tissue Antigens* **53,** 320–334.

Hock, B. D., Starling, G. C., Daniel, P. B., and Hart, D. N. (1994). Characterization of CMRF-44, a novel monoclonal antibody to an activation antigen expressed by the allostimulatory cells within peripheral blood, including dendritic cells. *Immunology* **83,** 573–581.

Hornung, V., Rothenfusser, S., Britsch, S., Krug, A., Jahrsdorfer, B., Giese, T., Endres, S., and Hartmann, G. (2002). Quantitative expression of toll-like receptor 1–10 mRNA in cellular subsets of human peripheral blood mononuclear cells and sensitivity to CpG oligodeoxynucleotides. *J. Immunol.* **168,** 4531–4537.

Ito, T., Amakawa, R., Kaisho, T., Hemmi, H., Tajima, K., Uehira, K., Ozaki, Y., Tomizawa, H., Akira, S., and Fukuhara, S. (2002). Interferon-alpha and interleukin-12 are induced differentially by Toll-like receptor 7 ligands in human blood dendritic cell subsets. *J. Exp. Med.* **195,** 1507–1512.

Jahnsen, F. L., Lund-Johansen, F., Dunne, J. F., Farkas, L., Haye, R., and Brandtzaeg, P. (2000). Experimentally induced recruitment of plasmacytoid (CD123high) dendritic cells in human nasal allergy. *J. Immunol.* **165,** 4062–4068.

Jahnsen, F. L., Moloney, E. D., Hogan, T., Upham, J. W., Burke, C. M., and Holt, P. G. (2001). Rapid dendritic cell recruitment to the bronchial mucosa of patients with atopic asthma in response to local allergen challenge. *Thorax* **56,** 823–826.

Jarrossay, D., Napolitani, G., Colonna, M., Sallusto, F., and Lanzavecchia, A. (2001). Specialization and complementarity in microbial molecule recognition by human myeloid and plasmacytoid dendritic cells. *Eur. J. Immunol.* **31,** 3388–3393.

Jurk, M., Heil, F., Vollmer, J., Schetter, C., Krieg, A. M., Wagner, H., Lipford, G., and Bauer, S. (2002). Human TLR7 or TLR8 independently confer responsiveness to the antiviral compound R-848. *Nat. Immunol.* **3,** 499.

Kadowaki, N., Ho, S., Antonenko, S., Malefyt, R. W., Kastelein, R. A., Bazan, F., and Liu, Y. J. (2001). Subsets of human dendritic cell precursors express different toll-like receptors and respond to different microbial antigens. *J. Exp. Med.* **194,** 863–869.

Kato, M., Neil, T. K., Fearnley, D. B., McLellan, A. D., Vuckovic, S., and Hart, D. N. (2000). Expression of multilectin receptors and comparative FITC-dextran uptake by human dendritic cells. *Int. Immunol.* **12,** 1511–1519.

Kohrgruber, N., Halanek, N., Groger, M., Winter, D., Rappersberger, K., Schmitt-Egenolf, M., Stingl, G., and Maurer, D. (1999). Survival, maturation, and function of CD11c− and CD11c+

peripheral blood dendritic cells are differentially regulated by cytokines. *J. Immunol.* **163,** 3250–3259.

Krug, A., Towarowski, A., Britsch, S., Rothenfusser, S., Hornung, V., Bals, R., Giese, T., Engelmann, H., Endres, S., Krieg, A. M., and Hartmann, G. (2001). Toll-like receptor expression reveals CpG DNA as a unique microbial stimulus for plasmacytoid dendritic cells which synergizes with CD40 ligand to induce high amounts of IL-12. *Eur. J. Immunol.* **31,** 3026–3037.

Larsson, M., Majeed, M., Ernst, J. D., Magnusson, K. E., Stendahl, O., and Forsum, U. (1997). Role of annexins in endocytosis of antigens in immature human dendritic cells. *Immunology* **92,** 501–511.

Liu, Y. J. (2001). Dendritic cell subsets and lineages, and their functions in innate and adaptive immunity. *Cell* **106,** 259–262.

Lore, K., Betts, M. R., Brenchley, J. M., Kuruppu, J., Khojasteh, S., Perfetto, S., Roederer, M., Seder, R. A., and Koup, R. A. (2003). Toll-like receptor ligands modulate dendritic cells to augment cytomegalovirus- and HIV-1–specific T cell responses. *J. Immunol.* **171,** 4320–4328.

Lore, K., Sonnerborg, A., Brostrom, C., Goh, L. E., Perrin, L., McDade, H., Stellbrink, H. J., Gazzard, B., Weber, R., Napolitano, L. A., van Kooyk, Y., and Andersson, J. (2002). Accumulation of DC-SIGN+CD40+ dendritic cells with reduced CD80 and CD86 expression in lymphoid tissue during acute HIV-1 infection. *AIDS* **16,** 683–692.

Lore, K., Sonnerborg, A., Spetz, A. L., Andersson, U., and Andersson, J. (1998). Immunocytochemical detection of cytokines and chemokines in Langerhans cells and *in vitro* derived dendritic cells. *J. Immunol. Methods* **218,** 173–187.

Lore, K., Spetz, A. L., Fehniger, T. E., Sonnerborg, A., Landay, A. L., and Andersson, J. (2001). Quantitative single cell methods that identify cytokine and chemokine expression in dendritic cells. *J. Immunol. Methods* **249,** 207–222.

Macatonia, S. E., Knight, S. C., Edwards, A. J., Griffiths, S., and Fryer, P. (1987). Localization of antigen on lymph node dendritic cells after exposure to the contact sensitizer fluorescein isothiocyanate. Functional and morphological studies. *J. Exp. Med.* **166,** 1654–1667.

MacDonald, K. P., Munster, D. J., Clark, G. J., Dzionek, A., Schmitz, J., and Hart, D. N. (2002). Characterization of human blood dendritic cell subsets. *Blood* **100,** 4512–4520.

Mahnke, K., Guo, M., Lee, S., Sepulveda, H., Swain, S. L., Nussenzweig, M., and Steinman, R. M. (2000). The dendritic cell receptor for endocytosis, DEC-205, can recycle and enhance antigen presentation via major histocompatibility complex class II–positive lysosomal compartments. *J. Cell Biol.* **151,** 673–684.

McWilliam, A. S., Napoli, S., Marsh, A. M., Pemper, F. L., Nelson, D. J., Pimm, C. L., Stumbles, P. A., Wells, T. N., and Holt, P. G. (1996). Dendritic cells are recruited into the airway epithelium during the inflammatory response to a broad spectrum of stimuli. *J. Exp. Med.* **184,** 2429–2432.

McWilliam, A. S., Nelson, D., Thomas, J. A., and Holt, P. G. (1994). Rapid dendritic cell recruitment is a hallmark of the acute inflammatory response at mucosal surfaces. *J. Exp. Med.* **179,** 1331–1336.

Mehlhop, E., Villamide, L. A., Frank, I., Gettie, A., Santisteban, C., Messmer, D., Ignatius, R., Lifson, J. D., and Pope, M. (2002). Enhanced *in vitro* stimulation of rhesus macaque dendritic cells for activation of SIV-specific T cell responses. *J. Immunol. Methods* **260,** 219–234.

Mohty, M., Jarrossay, D., Lafage-Pochitaloff, M., Zandotti, C., Briere, F., de Lamballeri, X. N., Isnardon, D., Sainty, D., Olive, D., and Gaugler, B. (2001). Circulating blood dendritic cells from myeloid leukemia patients display quantitative and cytogenetic abnormalities as well as functional impairment. *Blood* **98,** 3750–3756.

Mosialos, G., Birkenbach, M., Ayehunie, S., Matsumura, F., Pinkus, G. S., Kieff, E., and Langhoff, E. (1996). Circulating human dendritic cells differentially express high levels of a 55-kd actin-bundling protein. *Am. J. Pathol.* **148,** 593–600.

Muzio, M., Bosisio, D., Polentarutti, N., D'Amico, G., Stoppacciaro, A., Mancinelli, R., van't Veer, C., Penton-Rol, G., Ruco, L. P., Allavena, P., and Mantovani, A. (2000). Differential expression and regulation of toll-like receptors (TLR) in human leukocytes: Selective expression of TLR3 in dendritic cells. *J. Immunol.* **164,** 5998–6004.

O'Doherty, U., Peng, M., Gezelter, S., Swiggard, W. J., Betjes, M., Bhardwaj, N., and Steinman, R. M. (1994). Human blood contains two subsets of dendritic cells, one immunologically mature and the other immature. *Immunology* **82,** 487–493.

Pacanowski, J., Kahi, S., Baillet, M., Lebon, P., Deveau, C., Goujard, C., Meyer, L., Oksenhendler, E., Sinet, M., and Hosmalin, A. (2001). Reduced blood CD123+ (lymphoid) and CD11c+ (myeloid) dendritic cell numbers in primary HIV-1 infection. *Blood* **98,** 3016–3021.

Pashenkov, M., Huang, Y. M., Kostulas, V., Haglund, M., Soderstrom, M., and Link, H. (2001). Two subsets of dendritic cells are present in human cerebrospinal fluid. *Brain* **124,** 480–492.

Peiser, M., Grutzkau, A., Wanner, R., and Kolde, G. (2003). CD1a and CD1c cell sorting yields a homogeneous population of immature human Langerhans cells. *J. Immunol. Methods* **279,** 41–53.

Res, P. C., Couwenberg, F., Vyth-Dreese, F. A., and Spits, H. (1999). Expression of pTalpha mRNA in a committed dendritic cell precursor in the human thymus. *Blood* **94,** 2647–2657.

Robinson, S. P., Patterson, S., English, N., Davies, D., Knight, S. C., and Reid, C. D. (1999). Human peripheral blood contains two distinct lineages of dendritic cells. *Eur. J. Immunol.* **29,** 2769–2778.

Romani, N., Gruner, S., Brang, D., Kampgen, E., Lenz, A., Trockenbacher, B., Konwalinka, G., Fritsch, P. O., Steinman, R. M., and Schuler, G. (1994). Proliferating dendritic cell progenitors in human blood. *J. Exp. Med.* **180,** 83–93.

Sallusto, F., Cella, M., Danieli, C., and Lanzavecchia, A. (1995). Dendritic cells use macropinocytosis and the mannose receptor to concentrate macromolecules in the major histocompatibility complex class II compartment: Downregulation by cytokines and bacterial products. *J. Exp. Med.* **182,** 389–400.

Sallusto, F., and Lanzavecchia, A. (1994). Efficient presentation of soluble antigen by cultured human dendritic cells is maintained by granulocyte/macrophage colony-stimulating factor plus interleukin 4 and downregulated by tumor necrosis factor alpha. *J. Exp. Med.* **179,** 1109–1118.

Sallusto, F., and Lanzavecchia, A. (1999). Mobilizing dendritic cells for tolerance, priming, and chronic inflammation. *J. Exp. Med.* **189,** 611–614.

Sallusto, F., Palermo, B., Lenig, D., Miettinen, M., Matikainen, S., Julkunen, I., Forster, R., Burgstahler, R., Lipp, M., and Lanzavecchia, A. (1999). Distinct patterns and kinetics of chemokine production regulate dendritic cell function. *Eur. J. Immunol.* **29,** 1617–1625.

Sato, H., Shiobara, S., Yasue, S., Chuhjo, T., and Nakao, S. (2001). Lymphocyte collection for donor leucocyte infusion from normal donors: Estimation of the minimum processed blood volume and safety of the procedure. *Vox Sang.* **81,** 124–127.

Savary, C. A., Grazziutti, M. L., Melichar, B., Przepiorka, D., Freedman, R. S., Cowart, R. E., Cohen, D. M., Anaissie, E. J., Woodside, D. G., McIntyre, B. W., Pierson, D. L., Pellis, N. R., and Rex, J. H. (1998). Multidimensional flow-cytometric analysis of dendritic cells in peripheral blood of normal donors and cancer patients. *Cancer Immunol. Immunother.* **45,** 234–240.

Siegal, F. P., Kadowaki, N., Shodell, M., Fitzgerald-Bocarsly, P. A., Shah, K., Ho, S., Antonenko, S., and Liu, Y. J. (1999). The nature of the principal type 1 interferon-producing cells in human blood. *Science* **284,** 1835–1837.

Soderlund, J., Nilsson, C., Ekman, M., Walther, L., Gaines, H., Biberfeld, G., and Biberfeld, P. (2000). Recruitment of monocyte derived dendritic cells *ex vivo* from SIV infected and non-infected cynomolgus monkeys. *Scand. J. Immunol.* **51,** 186–194.

Sorg, R. V., Kogler, G., and Wernet, P. (1999). Identification of cord blood dendritic cells as an immature CD11c− population. *Blood* **93,** 2302–2307.

Stent, G., Reece, J. C., Baylis, D. C., Ivinson, K., Paukovics, G., Thomson, M., and Cameron, P. U. (2002). Heterogeneity of freshly isolated human tonsil dendritic cells demonstrated by intracellular markers, phagocytosis, and membrane dye transfer. *Cytometry* **48,** 167–176.

Summers, K. L., Hock, B. D., McKenzie, J. L., and Hart, D. N. (2001). Phenotypic characterization of five dendritic cell subsets in human tonsils. *Am. J. Pathol.* **159,** 285–295.

Szabolcs, P., Park, K. D., Reese, M., Marti, L., Broadwater, G., and Kurtzberg, J. (2003). Absolute values of dendritic cell subsets in bone marrow, cord blood, and peripheral blood enumerated by a novel method. *Stem Cells* **21,** 296–303.

Ueda, Y., Hagihara, M., Okamoto, A., Higuchi, A., Tanabe, A., Hirabayashi, K., Izumi, S., Makino, T., Kato, S., and Hotta, T. (2003). Frequencies of dendritic cells (myeloid DC and plasmacytoid DC) and their ratio reduced in pregnant women: Comparison with umbilical cord blood and normal healthy adults. *Hum. Immunol.* **64,** 1144–1151.

Upham, J. W., Lundahl, J., Liang, H., Denburg, J. A., O'Byrne, P. M., and Snider, D. P. (2000). Simplified quantitation of myeloid dendritic cells in peripheral blood using flow cytometry. *Cytometry* **40,** 50–59.

Visintin, A., Mazzoni, A., Spitzer, J. H., Wyllie, D. H., Dower, S. K., and Segal, D. M. (2001). Regulation of Toll-like receptors in human monocytes and dendritic cells. *J. Immunol.* **166,** 249–255.

Vuckovic, S., Fearnley, D. B., Gunningham, S., Spearing, R. L., Patton, W. N., and Hart, D. N. (1999). Dendritic cells in chronic myelomonocytic leukaemia. *Br. J. Haematol.* **105,** 974–985.

Zarember, K. A., and Godowski, P. J. (2002). Tissue expression of human Toll-like receptors and differential regulation of Toll-like receptor mRNAs in leukocytes in response to microbes, their products, and cytokines. *J. Immunol.* **168,** 554–561.

Zou, W., Machelon, V., Coulomb-L'Hermin, A., Borvak, J., Nome, F., Isaeva, T., Wei, S., Krzysiek, R., Durand-Gasselin, I., Gordon, A., Pustilnik, T., Curiel, D. T., Galanaud, P., Capron, F., Emilie, D., and Curiel, T. J. (2001). Stromal-derived factor-1 in human tumors recruits and alters the function of plasmacytoid precursor dendritic cells. *Nat. Med.* **7,** 1339–1346.

CHAPTER 27

B-Cell Immunophenotyping

Nicole Baumgarth
Center for Comparative Medicine
University of California, Davis
Davis, California 95616

I. Introduction

B lymphocytes provide the humoral (i.e., antibody-mediated) arm of the adaptive immune system. The term *B cell* was coined in the late 1950s from the organ of their first identification in chickens, the bursa of Fabricius. In most mammals, including humans and mice, B-cell development after birth occurs in the bone marrow. In the mouse, hematopoietic stem cells can also be found throughout life in the spleen. Whether these cells contribute significantly to adult murine B lymphopoiesis has not been studied in detail. In addition to the steady-state B-cell populations that are generated by the normal B-cell developmental processes, functionally distinct B-cell

METHODS IN CELL BIOLOGY, VOL. 75

0091-679X/04 $35.00

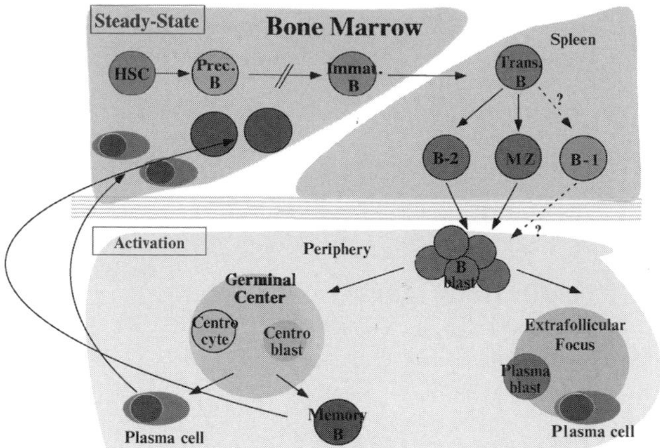

Fig. 1 Steady-state and activation-induced immunophenotypically distinct B-cell subsets of mice. Outline of the different developmental and differentiation stages resulting in functionally and phenotypically distinct B-cell subsets and their location. Bone marrow: HSCs, hematopoietic stem cells; Prec. B, committed B-cell precursor; Immat. B, immature B cell; Spleen: Trans. B, transitional B cell; B-2, follicular recirculating mature B cell; MZ, marginal zone B cell; B-1, B-1 ($CD5^+$, $Ly\text{-}1^+$) cell; Periphery, all secondary lymphoid tissues including spleen, lymph node, and tissue-associated lymphoid aggregates. Note that marginal zone B cells are found only in the spleen.

subsets appear after infection and/or immunization (Fig. 1). As discussed later in this chapter, their phenotypes are at least in part delineated and multicolor flow cytometry can be used for their identification. The aim of this chapter is to provide information on the phenotypes of the different B-cell developmental stages in the bone marrow and on mature B-cell subsets in the periphery (Fig. 1). Information provided here is given mainly for mice, although it appears that mouse and human B-cell development are remarkably similar (Ghia *et al.*, 1998). Fetal B-cell development differs in a number of important aspects from adult development in both mice and humans. Phenotypic characteristics of fetal B cells and B-cell precursors, as well as B cells from other species are less well studied and will not be reviewed in this chapter. Because of the spatial separation of B-cell development on the one hand and the generation of mature B-cell subsets and B-cell activation on the other, I will distinguish between B-cell immunophenotyping of cells from bone marrow and peripheral lymphoid tissues.

II. Immunophenotyping of B-Cell Developmental Stages in the Bone Marrow

Like all leukocytes, B cells develop from hematopoietic stem cells in the bone marrow. A common lymphoid progenitor has been identified in mouse bone marrow that gives rise to T and B cells but not to cells of the myeloid or erythroid lineage (Kondo *et al.*, 1997). The phenotype of the committed B-cell precursor is still a subject of debate (see later discussion). B-cell development from such committed B-cell precursors is characterized by an ordered process of immunoglobulin (Ig) gene rearrangements, resulting in the expression of the B-cell receptor (BCR). Although the genetic rearrangement profile and the developmental potential of a B cell is the ultimate proof of its developmental stage, each gene rearrangement step is characterized by a defined set of alterations in cell-surface immunophenotype. Flow cytometry in conjunction with *in vitro* culture systems that recapitulate B-cell development from the earliest committed precursor to the immature B-cell stage were instrumental in identifying these phenotypic alterations. B-cell development in the bone marrow concludes when a fully rearranged, functional BCR that does not strongly bind to bone marrow–expressed self-antigens is expressed on the cell surface.

A number of different marker combinations have been used to classify the developmental stages of B cells in mice and humans. The most frequently used are those developed by a number of groups (Allman *et al.*, 1999; Hardy *et al.*, 1991; Li *et al.*, 1996; Lu *et al.*, 1998; Melchers *et al.*, 1995; Osmond, 1990; Rolink *et al.*, 1994, 1996). Some confusion and discrepancies exist because each group used different markers and different nomenclatures to identify the various stages of development. It is hoped that with the availability of multicolor flow cytometry and the simultaneous use of larger numbers of markers, many of these discrepancies can be resolved. Table I summarizes the different nomenclatures developed by these groups, their phenotypes, and how they relate to the genotype (Ig gene rearrangement profile) of the developing B cells. Figure 2 shows the analysis of a C57BL/6 wild-type mouse bone marrow using multicolor staining combinations to differentiate the Hardy fractions A–F.

A. Earliest Committed B-Cell Precursor

Some controversy regarding the phenotype of the earliest committed B-cell progenitor exists. According to the Basel Group, expression of the markers CD19 and c-*kit* unequivocally identifies these cells in the bone marrow of adult mice (Rolink *et al.*, 1996) (Table I). In contrast, Allman and Hardy (1999) found B-cell lineage–restricted cells among $CD19^-$ c-kit^- $B220^{dull}$ cells. Both groups have provided evidence that CD19 is expressed on the cell surface after B220 (CD45R) is expressed (Allman *et al.*, 1999; Nikolic *et al.*, 2002; Rolink *et al.*, 1996). The $CD19^+$ c-kit^+ cells, identified as earliest committed B-cell precursor by Melchers *et al.* (1995), appear to constitute a later developmental stage than the fraction A_1 cells ($AA4.1^+$, $B220^{dull}$ $CD4^{+/-}$ $CD11b^{lo}$ $CD19^-$ $CD24^-$ $CD43^-$ $BP1^-$ c-kit^- $Sca1^{lo}$ $IL7R^-$ (Allman *et al.*, 1999; Li *et al.*, 1996; Rolink *et al.*, 1996) identified by Allman *et al.* in 1999 (Table I). Thus, the main discrepancies between these two groups seem to regard the definition of commitment to the B-cell lineage, rather than differences in the phenotypic identification of the cells.

Table I

Nomenclature and Phenotype of B-Cell Developmental Stages in Mouse Bone Marrow

Hardy *et al.*	**Pre-Pro B**	**Pro-B**	**Late Pro-B**	**Early Pre-B**	**Late Pre-B**	**Immature B**	**Mature B**
Fractions:	A_1 A_2	B	C	C′	D	E	F
	$B220^{lo}$ $AA4.1^+$ $CD24^-$ $CD4^{+/-}$	$B220^{lo}$ $CD43^+$ $CD24^+$	$B220^{lo}$ $CD43^+$ $CD24^+$ $BP\text{-}1^+$	$B220^+$ $CD43^+$ $CD24^{hi}$ $BP\text{-}1^+$	$B220^+$ $CD43^-$ IgM^- IgD^-	$B220^+$ $CD43^-$ IgM^+ $IgD^{-/lo}$	$B220^{hi}$ $CD43^-$ IgM^+ IgD^+
Melchers *et al.*	**Pro-B**	**Pre B-I**		**Large Pre-B-II**	**Small Pre-B-II**	**Immature B**	**Mature B**
	$B220^+$ $CD19^+$ c-*kit*$^+$ $CD25^-$	$B220^+$ $CD19^+$ c-*kit*$^+$ $CD25^-$		$B220^+$ $CD19^+$ c-*kit*$^-$ $CD25^+$ $c\mu^+$ large cycling	$B220^+$ $CD19^+$ c-*kit*$^-$ $CD25^+$ small noncycling	$B220^+$ $CD19^+$ IgM^+	$CD19^+$ IgM^+ IgD^+
Osmond *et al.*	**Early pro-B**	**Intermediate Pro-B**	**Late Pro-B**	**Large Pre-B**	**Small Pre-B**	**Immature B**	**Mature B**
	TdT^+ $B220^-$	TdT^+ $B220^+$	TdT^- $B220^+$	TdT^- $B220^+$ $c\mu^+$ large, cycling	TdT^- small $B220^+$ $c\mu^+$	TdT^- $B220^+$ $sIgM^+$	TdT^- $B220^+$ $sIgM^+$ IgD^+
Ig Heavy Chain Rearrangement	Germline (μ sterile transcripts)	$D_H \rightarrow J_HD_HJ_H$	$D_HJ_H \rightarrow V_H$	$V_HD_HJ_H$	$V_HD_HJ_H$	$V_HD_HJ_H$	$V_HD_HJ_H$
Ig Light Chain Rearrangement	Germline	Germline	Germline	Germline	$V_L \rightarrow J_L$	V_LJ_L (kappa or kappa + lambda)	V_LJ_L

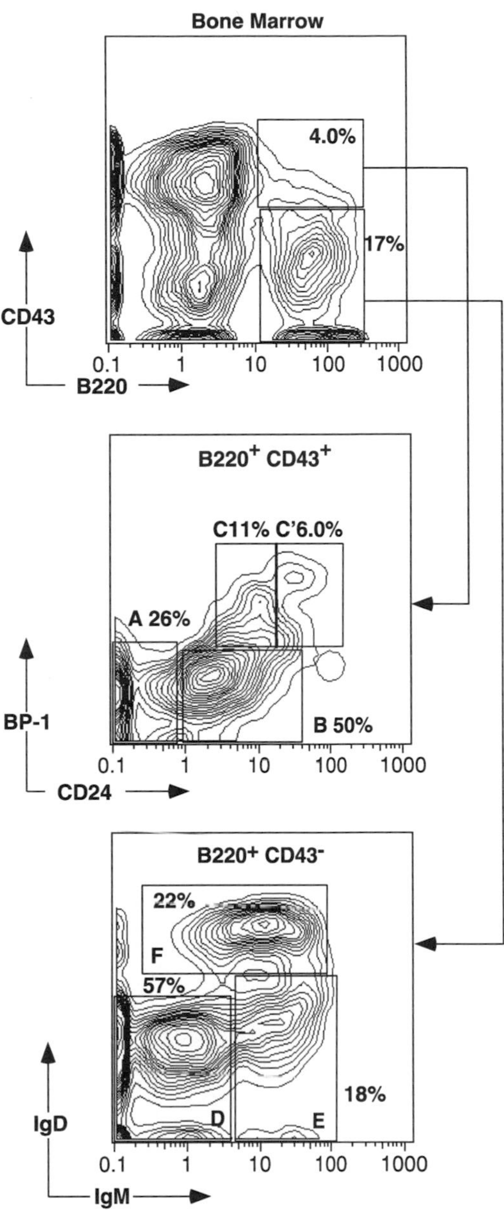

Fig. 2 B-cell development is accompanied by distinct changes in cell-surface immunophenotype. Staining was done according to Hardy *et al.* (1991). Shown are 5% contour plots of bone marrow cells from a 3-month-old C57BL/6 mouse gated for exclusion of propidium iodide (live) and a lymphocyte forward scatter channel/side scatter channel. Early B-cell developmental stages are distinguished from later stages of development by the expression of CD43 on B220$^+$ cells (top). Note that CD43 expression is very low and requires a bright conjugate (best with either phycoerythrin or

In contrast, Montecino-Rodriguez *et al.* (2001) described a AA4.1$^+$ B220$^-$ CD24$^+$ CD43$^+$ CD19$^+$ population in the bone marrow with bipotential B-cell/macrophage progenitor capability, suggesting that a small subset of not yet fully committed B-cell precursors might express CD19 before acquisition of B220. Alternatively, B220 expression might under certain conditions be downmodulated after CD19 acquisition. This population seems to show many of the characteristics described for fraction B cells of Hardy *et al.* (1991) (with the exception of B220), including the findings of D_HJ_H rearrangement (Montecino-Rodriguez *et al.*, 2001). Thus, it is also possible that low expression of B220 that characterizes early B-cell developmental stages according to Hardy *et al.* (1991) was characterized by Montecino-Rodriguez *et al.* as B220 negative. Nonetheless, a small fraction (around 5%) of this population seems to have retained its ability to differentiate into macrophages, but not dendritic cells (Montecino-Rodriguez *et al.*, 2001); a finding not easily reconciled with the findings of Hardy *et al.* (1991) and Melchers *et al.* (1995), but consistent with data from neonatal and adult B-cell lines (Davidson *et al.*, 1992; Martin *et al.*, 1993) and *in vitro* co-cultures of B cells with splenic fibroblasts (Borrello and Phipps, 1995, 1999).

B. Immunoglobulin Gene Rearrangement and B-Cell Immunophenotype

Terminal commitment to the B-cell lineage appears to already have occurred when Ig gene rearrangement is initiated (Allman *et al.*, 1999). Although it has been argued that the presence of macrophages with at least partial Ig heavy-chain rearrangement indicate that a "lineage switch" from B cell to myeloid cell might be possible (Davidson *et al.*, 1992; Martin *et al.*, 1993; Montecino-Rodriguez *et al.*, 2001).

The signals that initiate this complex Ig rearrangement event are not fully understood. Variable region gene recombination (VDJ recombination) occurs first on the Ig heavy chain in the order of $D \rightarrow J$, $V \rightarrow DJ$ (Table I). Expression of RAG1 and RAG2 genes is essential for rearrangement to occur (Schatz *et al.*, 1989; Shinkai *et al.*, 1992). A successfully rearranged heavy-chain variable region is then paired with the Igμ heavy-chain constant region and expressed on the cell surface in conjunction with the surrogate light chains lambda and V-pre-B and the signaling chains Igα and Igβ (CD79a/CD79b) as "pre-B-cell receptor" (Karasuyama *et al.*, 1994). Unfortunately, no good antibodies exist that recognize this unique surface receptor. In general, heavy-chain rearrangement precedes light-chain rearrangement. Light chains are encoded by V and J genes only and use one of two constant regions: kappa and lambda. Lambda rearrangement occurs only when kappa

allophycocyanin). Those CD43$^+$/B220$^+$ B cells (middle) are further divided into fractions A–C′ as shown. (Bottom) The separation of pre-B from immature and mature B cells (fractions D-F) according to their levels of expression of surface immunoglobulin M and immunoglobulin D. See Table I for further explanation. Note that fraction A was found to be heterogeneous and only some of the cells within this fraction are committed to become B cells (see text).

rearrangement results in nonproductive rearrangement, or when the BCR recognizes self-antigens (re-rearrangement). Successful gene rearrangement on one of the two heavy- and four light-chain alleles shuts down further rearrangement on the other allele(s) (allelic exclusion). This ensures that every B cell expresses mature BCR with exactly one specificity. IgM-positive IgD-negative B cells leave the bone marrow at the immature stages and migrate toward the spleen.

Phenotypic changes that characterize each stage in B-cell development have been described (Table I). The appearance of cytoplasmic Igμ signals successful rearrangement of the Ig heavy chain. On the cell surface, this seems to coincide with expression of CD25 and loss of expression of c-*kit* (Rolink *et al.*, 1994). At this stage, pre-B cells undergo strong interleukin-7 (IL-7)–dependent clonal expansion and the cells are relatively large and blastlike (Melchers *et al.*, 1995). Re-expression of RAG occurs after this expansion phase for the induction of light-chain rearrangement. These cells become smaller, lose CD43 (as analyzed with the monoclonal antibody [mAb] S7), express increased levels of CD24, and are BP-1^{+} (Hardy *et al.*, 1991; Li *et al.*, 1996; Osmond *et al.*, 1998). The expression of surface IgM without IgD characterizes the final step of development in the bone marrow. These cells are termed *immature B cells*. Depending on the specificity of the receptor, further rearrangement of the light chain might occur at this developmental stage in the bone marrow (Melchers *et al.*, 1995). The phenotype of B cells that alter their specificity by further rearrangement of their light chains, a process termed *re-rearrangement,* has not been defined. However, eventually many of those cells will express lambda light chains rather than kappa light chains (Meffre *et al.*, 2000).

Apart from the various B-cell precursors outline earlier, mature, B220^{+} CD23^{+} IgM^{lo} IgD^{hi} B cells are found in the bone marrow of mice (Fig. 2). These cells represent mature recirculating follicular B cells that develop from immature B-cell emigrants in the spleen (see later discussion). In addition, both plasma cells and memory B cells, induced after antigen exposure, can home to the bone marrow. Their phenotypes are discussed later in this chapter. In conclusion, the bone marrow is a tissue in which B cells of all developmental stages can be isolated. Although controversy surrounds the phenotype of the earliest committed B cell, good marker combinations for the identification of most of the later developmental stages of B cells are available. As shown in Fig. 2, a single six- to eight-color staining combination allows the near-complete delineation of B-cell development from the pro-B-cell stage to the fully mature B cells.

III. Peripheral B-Cell Populations

Immature B cells travel from the bone marrow to the spleen. In the spleen, these cells are referred to as *transitional B cells*. In addition to these bone marrow emigrants, three functionally and phenotypically distinct mature B-cell subsets can be separated in the spleen: follicular B cells, the main recirculating mature B-cell population; marginal-zone B cells; and B-1 cells (Fig. 3).

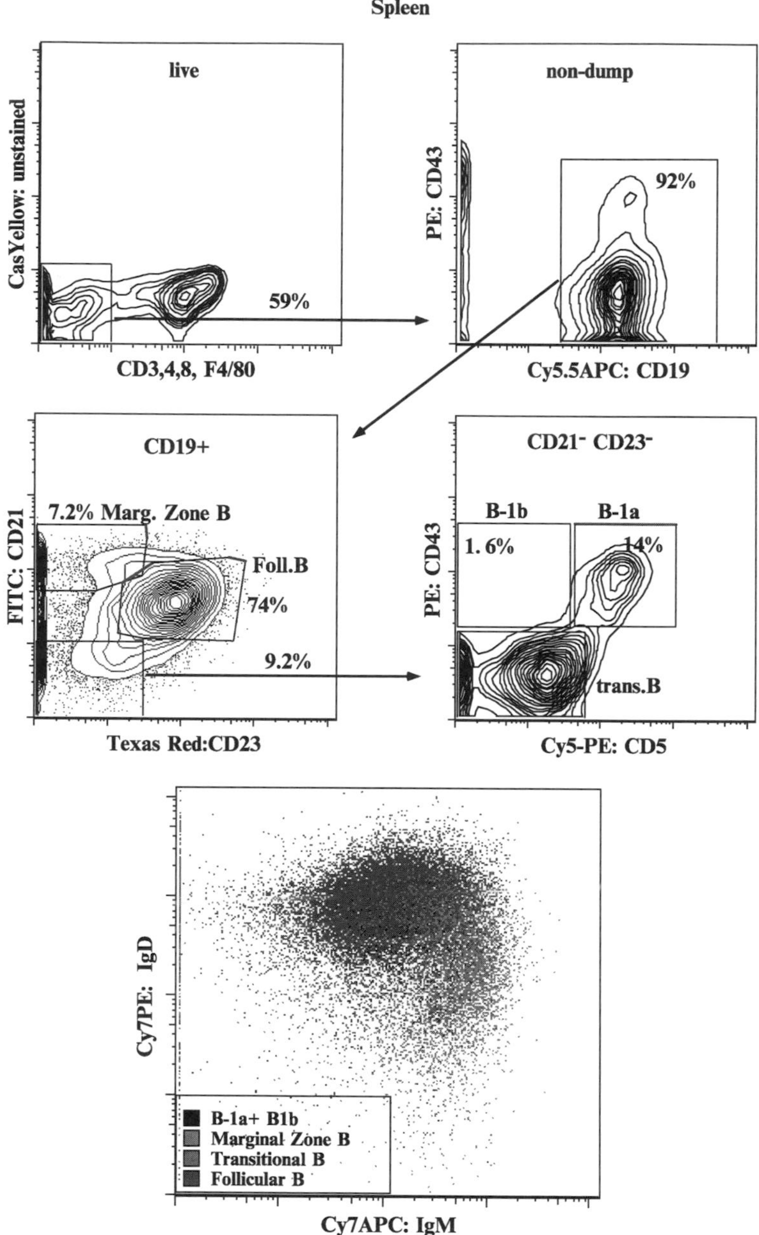

Fig. 3 Identification of major B-cell subsets in murine spleen. Shown are 5% contour plots from the spleen of a 3-month-old female C57BL/6 mouse gated on a lymphocyte forward scatter channel/side scatter channel profile and exclusion of propidium iodide. B cells are identified by their lack of expression of CD3, CD4, CD8, and F4/80 (top left) and their expression of CD19 (top right). Those

Infection- or immunization-induced B-cell activation results in the early accumulation of mature B cells that form antibody-forming foci at the borders to T-cell areas. Somewhat later, anatomical structures termed *germinal centers* (GCs) develop within the primary follicles of spleen and lymph nodes. They contain many rapidly cycling B cells that undergo Ig isotype switching and affinity maturation. GC reactions lead to the development of high-affinity, isotype-switched memory B cells. Both GC reaction and the extrafollicular antibody-forming foci generate antibody-secreting plasma blasts and terminally differentiated plasma cells. There is currently no marker that can distinguish the origin of these end-stage B-cell effector populations.

A. Transitional B Cells

B220^{+} IgM^{hi} IgD$^{-/lo}$ CD21^{-}/^{lo} B cells from the bone marrow undergo a final selection process in the spleen. It has been calculated that only about 20% of the bone marrow emigrants will be selected into the mature B-cell pool. The mechanisms that govern this selection process are unknown. Explained by their high levels of expression of CD95 (Fas), transitional B cells are exquisitely sensitive to antigen-induced apoptosis, presumably as a safeguard mechanism to ensure that strongly self-reactive B cells are eliminated from the B-cell pool. Identification of transitional B cells in the spleen can be made based on the surface markers B220 or CD19, CD21, CD23, CD5, IgM, and IgD (Fig. 3). Loder *et al.* (1999) showed that two discrete subsets of immature B cells can be identified: transitional 1 and transitional 2 B cells. T1 cells lack expression of CD5 (see later discussion), CD21, and CD23 and express either no or very little IgD. T2 cells have gained CD23 and CD21 and express higher levels of IgD but in contrast to mature B cells still express high levels of IgM. Neither of these subtypes is found in peripheral tissues other than blood and spleen.

B. Follicular B Cells

Immature B cells develop to become "mature" B cells. They are also called *follicular B cells*. Once selected into the mature B-cell pool (i.e., recruited into the B-cell follicles), B cells reduce their expression of IgM and further gain expression of IgD. The functional consequences of this change in the ratio of these IgM and IgD on the cell surface have not been fully elucidated. These cells are CD23^{+} CD21^{int} and CD5^{-}. Follicular B cells make up about 70–80% of the splenic B-cell population (Fig. 3) and are the main B-cell population in most peripheral tissues, with the exception of the peritoneal and pleural cavities. Because of their relative

---

cells are then subdivided into marginal zone B cells (*Marg. Zone B*), follicular B cells (*Foll. B*), transitional B cells (*trans. B*), B-1a, and B-1b cells. The bottom overlay dot plot shows the various levels of immunoglobulin M and immunoglobulin D expression levels of the above identified cell subsets. (See Color Insert.)

abundance, their expression of CD23, and high levels of expression for IgD, this population is fairly easy to identify by flow cytometry (Fig. 3). Follicular B cells are remarkably homogeneous for most of the markers studied. One exception is expression of the nonclassic major histocompatibility complex (MHC) molecule CD1. This molecule is expressed at high levels on marginal zone B cells (see later discussion), but also on a small not yet functionally characterized subset of B cells that otherwise resembles follicular B cells (Amano *et al.*, 1998).

C. Marginal Zone B Cells

Marginal zone B cells seem to develop as an alternative pathway from transitional B cells in the spleen. They make up about 5–10% of splenic B cells (Fig. 3). Cells with marginal zone phenotype (see later discussion) are not found in the lymph nodes. The term *marginal zone B cell* was developed from the fact that these cells reside in the marginal zone of the spleen. It is here that bloodborne antigens are filtered through the marginal zone synoids of the spleen. Consistent with this observation is that marginal zone B cells are responding rapidly to antigen exposure (Martin and Kearney, 2002). They are also particularly sensitive to lipopolysaccharide (LPS) stimulation because they express high levels of toll-like receptor molecule CD180 (Baumgarth, 2000, unpublished observations) that facilitates responsiveness to LPS (Nagai *et al.*, 2002) . Furthermore, they express a number of other receptors on their cell surface that facilitate interaction with the innate immune system, including the nonclassic MHC molecule CD1 (Amano *et al.*, 1998; Roark *et al.*, 1998) and the complement receptor CD21 (Takahashi *et al.*, 1997). The latter is known as a potent costimulatory molecule for B cells. Thus, phenotypically marginal zone B cells are characterized by their high levels of expression of IgM, CD1, and CD21, their lack of expression of CD23, and their low expression of IgD. Although it appears that at least some marginal zone B cells (as can be seen from the overlay profile in Fig. 3) (as defined as $CD19^{+}$ $CD23^{-}$ $CD21^{hi}$) express considerable levels of IgD, whether these cells constitute a different developmental or activation stage requires further study.

It is important to note that not all B cells that reside in the marginal zone have the phenotypic characteristics of marginal zone B cells. Other B-cell populations that reside in the marginal zone are some plasma blasts and plasma cells, as well as B-1 cells (see later discussion). Hence, the pathologist's definition of marginal zone B cells is broader than the term defined by their phenotypic characteristics by flow cytometry.

D. B-1 Cells

B-1 cells constitute a very small proportion of B cells in the spleen (roughly 1%), but they produce most of the circulating natural antibodies, particularly IgM (Baumgarth *et al.*, 1999; Hayakawa and Hardy, 2000). They are very rare in lymph nodes (<0.2%) and they make up about 1% of B cells in the spleen

(Fig. 3), but they are the majority of the B-cell population in the peritoneal and pleural cavities of mice (Fig. 4). Depending on the mouse strain and the age and sex of the animal, up to 70% of B cells in the peritoneal cavity are B-1 cells. Peritoneal and pleural cavity B-1 cells are IgM^{hi}, IgD^{lo} CD23^{negative}, CD11b^{+}, and CD43^{+} (Wells *et al.*, 1994) (Fig. 4). A further distinction is made about their expression of CD5. Roughly two thirds to three fourths of these cells express the otherwise T-cell–restricted marker CD5 (Fig. 4). CD5^{+} B-1 cells are termed *B-1a cells* and CD5^{-} B-1 cells *B-1b*. It is important to point out that expression of CD5 can only be seen when bright CD5 conjugates are used, such as phycoerythrin (PE) and allophycocyanin (APC) reagents. Because staining for CD5 on B-1 cells is substantially lower than that observed on T cells (data not shown), CD5-conjugate evaluation for the identification of B-1 cells must be done on those cells. A good source for B-1 cells is the peritoneal cavity (Fig. 4).

Importantly, in contrast to the peritoneal cavity, in the spleen B-1 cells do not express CD11b. This makes identification of splenic B-1b cells particularly challenging. In addition, it has been shown that CD5 expression can be induced in follicular B cells *in vitro* under certain stimulation conditions (Cong *et al.*, 1991). Hence, whether a distinction has to be made between follicular B cells expressing CD5 and B cells that develop into CD5^{+} B-1 cells is unclear. No marker has been described that could distinguish these cells.

The origin of B-1 cells is subject to debate (Berland and Wortis, 2002; Herzenberg, 2000). A particularly good source for B-1 cells is the fetal liver. It has been suggested that developmental processes for B-1 cells are distinct from those that lead to the development of follicular B cells and marginal zone B cells, favoring B-1 cell development during fetal development (Wasserman *et al.*, 1998). B-1 cells seem to be maintained throughout life through "self-replenishment" (i.e., slow proliferation of B-1 cells in the periphery), rather than *de novo* development. Whether in addition the bone marrow is a source for B-1 cells in the adult under physiological conditions is less clear. Expression of a transgenic Ig receptor with B-1–cell specificity does allow B-1–cell development from the bone marrow. However, the number of B-1 cells that accumulate in the periphery is much lower than that observed when a follicular B-cell transgenic Ig is induced (Lam and Rajewsky, 1999). This suggests that development of B-1 cells in the bone marrow is not occurring at the same frequency as that of follicular B cells.

Taken together, at least six distinct B-cell subsets can be identified in the periphery that develop in the absence of antigen challenge: T1 and T2 transitional B cells, follicular B cells, marginal zone B cells, and B-1a and B-1b cells. Combinations of markers can be designed that differentiate all these B-cell populations. Figure 3 shows flow cytometric analysis of mouse splenic B-cell populations using nine-color staining combinations in just one antibody cocktail. Such an approach provides the greatest accuracy, because all populations are quantified in one stain. The challenge in detecting non–antigen-induced splenic/peripheral B-cell subsets is the distinction between the non–follicular B cells, all of which are IgM^{hi} IgD$^{lo/neg}$ and CD23^{-}. Flow cytometric evaluations of these relatively small populations

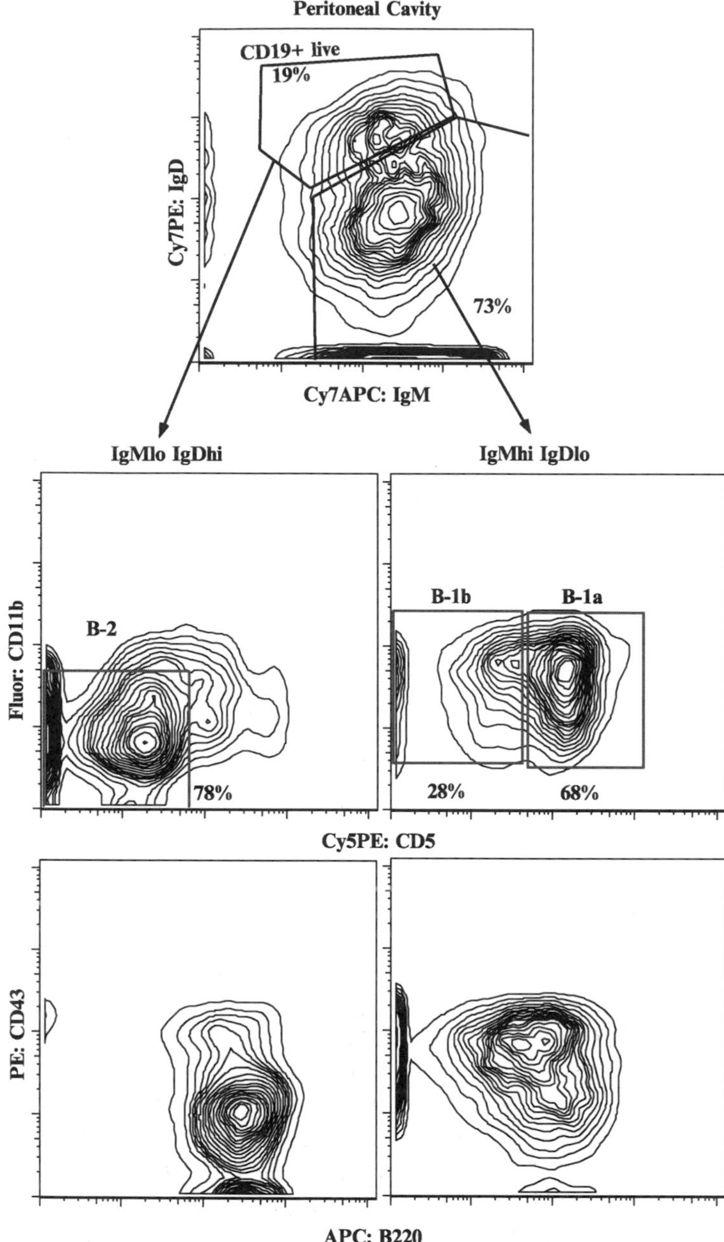

Fig. 4 The peritoneal cavity contains large numbers of B-1 cells. Shown are cells harvested from the peritoneal cavity of a 3-month-old female C57BL/6 mouse. Shown are 5% contour plots of cells gated on exclusion of propidium iodide and a lymphocytic forward scatter channel/side scatter channel profile. B-1 and B-2 cells are separated according to their differential expression of immunoglobulin

should be done by acquiring at least 200,000 events. This ensures high enough event counts for each of the cell populations to be identified. However, if only one particular B-cell population of interest is to be examined, marker combinations with a smaller number of antibodies can be designed.

IV. Antigen-Induced B-Cell Subsets

As shown in Fig. 1, exposure to antigen induces a number of B-cell activation events. Mature B cells respond to antigen encounter and appropriate costimulatory signals with proliferation (clonal expansion) and differentiation to antibody-secreting B cells or plasma cells. Immunohistologically, the cells might form "antibody-producing foci" (APF) at the edges of the T- and B-cell zones in secondary lymphoid tissues at early times after infection or immunization, or they form "GCs" within B-cell follicles. The exact molecular events that determine whether a B cell will form an APF or GC are not clear. It was demonstrated, however, that SAP gene-expressing CD4 T cells are required for GC B-cell differentiation into long-lived plasma cells and memory B cells, but not for the generation of short-lived plasma blasts that develop from APF (Crotty *et al.*, 2003).

A. Germinal Center B Cells

By immunohistochemistry, B cells in GCs can be divided into centrocytes and centroblasts. As the name suggests, centroblasts are relatively large cells that are rapidly proliferating. These cells acquire hypermutations and might undergo isotype switching. Centroblasts give rise to centrocytes, which express a mutated Ig gene and express surface IgG, IgA, or IgE instead of IgM. These cells undergo rapid Fas-mediated apoptosis, unless rescued by signaling through antigen binding via BCR and costimulation via CD40 (MacLennan, 1994). It is believed that this selection process provides the basis for the increased affinity of antibody responses during the course of an infection or antigen challenge.

Identification of centrocytes and centroblasts by immunophenotyping has not been achieved in the mouse. All GC B cells are B220^{+} CD19^{+}, CD21/35^{+} CD22 and CD40^{+}, like most other B-cell subsets. They differ immunophenotypically from other subsets by their increased ability to bind the lectin peanut agglutinin (PNA) (Butcher *et al.*, 1982) and expression of an antigen identified by the mAb GL7 (Han *et al.*, 1996; Shinall *et al.*, 2000). They also express higher levels of CD24 (HSA) compared to follicular B cells. GC B cells do not express CD43 and

---

M and immunoglobulin D (top). The large differences between B-1 and B-2 cells in their expression of CD11b, CD5, CD43, and B220 are indicated (middle and bottom). Note the separation of B-1 cells into B-1a and B-1b according to their expression of CD5. In contrast to B-2 cells, both B-1a and B-1b cells express CD11b, CD43, and low levels of B220.

CD138 (syndecan), consistent with their inability to spontaneously secrete antibodies. Also, GC B cells downregulate CD23 and in the mouse CD38. The latter is in contrast to human GC B cells, which express CD38 at high levels. Thus, high expression of CD38 cannot be used in the mouse to identify GC B cells (Oliver *et al.*, 1997; Ridderstad and Tarlinton, 1998). GC B cells appear heterogeneous with regard to their expression of surface Ig. Although IgD is rapidly downregulated during the GC reaction, some IgD^+ IgM^+ and IgD^- IgM^+ and isotype-switched surface IgG-, IgA- and IgE-positive cells can be found at varying degrees throughout the GC reaction (Shinall *et al.*, 2000). Heterogeneity of PNA^+ $B220^+$ GC B cells has been reported also for a number of other surface markers. How these differences relate to the immunohistochemically identified cell populations of the GC has not been elucidated (Han *et al.*, 1997; Shinall *et al.*, 2000).

GC cells can give rise either to memory B cells or to long- and short-lived (affinity-matured) antibody-secreting cells and/or plasma cells (MacLennan, 1994).

B. Plasma Cells

After antigen encounter, a subset of B cells return to the bone marrow as terminal-differentiated long-lived plasma cells (Benner *et al.*, 1981). Cells of similar phenotype are also located in various mucosal tissues and in the splenic red pulp. The plasma cell migration event to bone marrow is dependent on the chemokine CXCL12 (SDF-1) and on expression of CXCR4 (Hargreaves *et al.*, 2001).

Identification of plasma cells can be tricky. Though cytoplasmic Ig positive, these cells may be surface Ig negative and many have either lost or strongly downregulated the pan-B–cell markers B220 and CD19 and MHC class II molecules (Underhill *et al.*, 2003). Syndecan (CD138) has been widely used as a marker to identify plasma cells. However, although anti-CD138 specifically stains plasma cells in the periphery, not all plasma cells are syndecan positive (Underhill *et al.*, 2003, unpublished observation). CD43 (staining with either antibody clone S7 or S11) seems to be expressed on most plasma cells, as is CD27 (Jung *et al.*, 2000). Because CD43 is also expressed on all T cells, some APC and early B-cell precursors, a staining combination that includes a pan-T–cell marker and markers to exclude macrophages, granulocytes, and monocytes in conjunction with CD43, syndecan, staining for cytoplasmic Ig and CD24 (to exclude precursor B-cell populations) seems to be the best combination to enable enumeration of plasma cells in the bone marrow. CD9 might also help to further differentiate plasma cells from other cell populations, because its expression otherwise seems limited to splenic marginal zone B cells and to B-1 cells (Won and Kearney, 2002).

C. Memory B Cells

The bone marrow has also been identified as a source for long-lived memory B cells (Gray, 1993). Memory B cells are defined as cells that can respond to repeated antigen exposure with faster and stronger antibody responses. Moreover, these cells are able to survive long term after adoptive transfer. The phenotype of the memory B-cell pool has not been fully elucidated. Most but not all cells have lost expression of IgM and IgD and express a downstream Ig isotype, mostly IgG (Gray, 1993; McHeyzer-Williams *et al.*, 1991; Schittek and Rajewsky, 1990). However, the presence of IgM-expressing memory B cells has been reported (Klein *et al.*, 1999), consistent with the fact that antigen-specific IgM can be found even at late points after antigen exposure. In the human, CD27 has been identified as a marker for memory B cells (Klein *et al.*, 1999), and CD27 upregulation was noted on post-GC B cells in the mouse (Jung *et al.*, 2000). Furthermore, ligation of CD27 has been shown to inhibit differentiation of B cells to the plasma cell stage in mice, suggesting that expression of this marker might be necessary for the maintenance or generation of the memory B-cell pool (Raman *et al.*, 2000).

Recent controversy has arisen over whether a subset of memory B cells exist in spleen and bone marrow that has lost surface expression of B220 and expresses CD11b, a marker most commonly observed on myeloid cells (Bell and Gray, 2003; Driver *et al.*, 2001; McHeyzer-Williams *et al.*, 2000), but also on B-1 cells (see later discussion). Identification of this presumed $B220^{-}$ $CD11b^{+}$ memory B-cell subset has been done by staining for antigen-binding and surface Ig expression (Driver *et al.*, 2001; McHeyzer-Williams *et al.*, 2000). However, others suggested that this population consists of "antigen-capture cells" of myeloid origin that carry antigen-binding IgG on their cell surface via Fcγ receptors (Bell and Gray, 2003), consistent with early adoptive transfer experiment results that had not shown any evidence for memory cells within the B220-negative fraction of bone marrow and spleen (Manz *et al.*, 1997). This controversy demonstrates the need for the identification of additional markers that help to unequivocally identify memory B-cell subsets. Much of our understanding on the regulation and development of the immune system has arisen from the ability to isolate and characterize discrete subsets. Multicolor flow cytometry allows us to isolate increasingly smaller subpopulations that previously have resisted analysis. However, without the availability of discrete cell markers, these efforts would be futile.

D. B Cells within Antibody-Forming Foci

APF give rise mainly if not only to short-lived antibody-secreting cells. Whether the phenotype of the B cells seeding the APF reaction is different than those from the mature B cells that initiate the GC reaction is unclear. Marginal zone B cells are reported to quickly respond to antigen encounter with antibody secretion, thus suggesting that there main differentiation pathway is that of the extrafollicular response (Martin *et al.*, 2001). It was shown, however, that they can give rise to both APF as well as GC (Song and Cerny, 2003). B cells in the APF rapidly expand and thus have a blast-like forward side scatter (FSC) profile. Isotype switching occurs in these APF, although initially IgM antibodies are being secreted by plasma blasts. Using an antigen-specific detection system (see later discussion), IgG1-positive NP-specific plasma blasts from extrafollicular foci were identified by the following phenotype: syndecan-1^{+} B220^{-} PNA^{lo} (Smith *et al.*, 1996). This differentiates them from the IgG1-positive NP-specific syndecan^{-} B220^{+} PNA^{hi} GC B cells (Smith *et al.*, 1996). Whether the phenotype of IgM-positive plasma blasts is similar to those having undergone isotype switching has not been reported. Also, it is not clear whether B cells that leave the GC and develop into antibody-secreting plasma blasts and plasma cells differ in phenotype from those that have arisen via the extrafollicular route. It appears that expression of syndecan-1 and CD43 is a shared property of antibody-secreting B cells independent of their differentiation site. However, as discussed earlier, not all antibody-secreting plasma cells are syndecan-1 positive, thus leaving open the possibility that GC-derived plasma cells might not necessarily express this surface receptor.

V. Antigen-Specific B Cells

B cells recognize native unprocessed antigen via their antigen-specific receptor. This can be exploited for the identification of antigen-specific B cells by flow cytometry. Hayakawa *et al.* (1987) were first to demonstrate this using the fluorochrome PE as an immunogen. Using cell suspensions from mice immunized with PE, they were able to identify PE-specific cells by their ability to bind to PE. PE-positive B200^{+} B cells could be identified by flow cytometry. Lalor *et al.* (1992) used a similar strategy; they conjugated the hapten NP to biotin or to a fluorochrome. NP-specific B cells in NP-KLH–immunized mice were identified by binding to NP and expression of surface IgG1.

This strategy is not restricted to haptens or fluorochromes. In principle, any molecule that can be conjugated to a fluorochrome or biotin should be usable to stain antigen-specific B cells. Our experiments (Fig. 5) demonstrated that the hemagglutinin-molecule of influenza virus can be biotinylated and used to identify influenza virus-specific B cells (Doucett, Gerhard, Brown, and Baumgarth, *in preparation*). Thus, identification of antigen-specific B cells is considerably more straightforward than identification of antigen-specific T cells, which requires a haplotype-matched MHC-tetramer reagent that contains a certain peptide of interest. Also, because the native intact antigen is used, B cells with multiple epitope specificities can be identified with one reagent. The only restriction is the necessity for BCR expression by the cell. Thus, it is possible that fully differentiated plasma cells do not express sufficient surface Ig to be labeled. Whether intracytoplasmic staining with antigen-specific reagents could overcome this problem is being tested.

In conclusion, many available markers can be used in conjunction with multicolor flow cytometry to evaluate B-cell developmental stages in the bone marrow and mature B-cell subset in peripheral tissues of mice and similarly of humans. Though not fully exploited, the ability to easily develop staining reagents

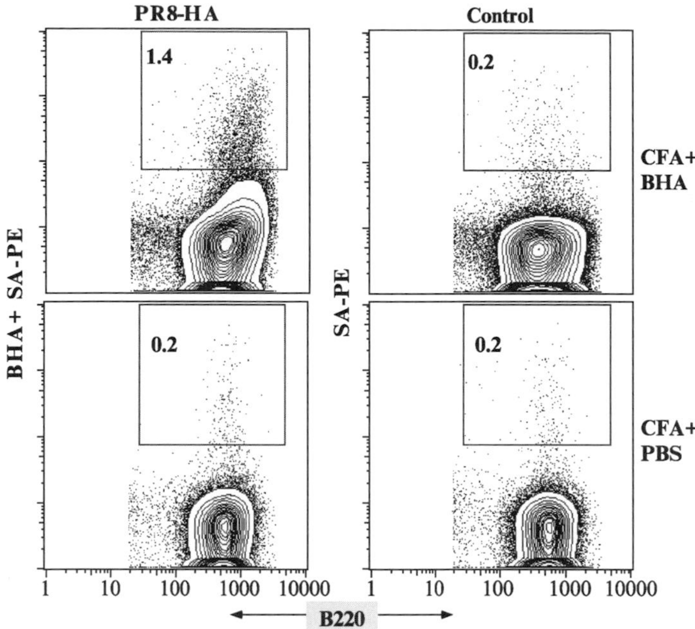

Fig. 5 Identification of influenzavirus-specific cells with biotinylated hemagglutinin. Hemagglutinin from influenzavirus A/PR8 was purified and biotinylated as outlined elsewhere (Doucett, Gerhard, Brown, and Baumgarth, in preparation). Shown are 5% contour plots with outliers of draining inguinal lymph node cells from mice immunized at the base of tail for 15 days with complete Freund's adjuvant (CFA) plus hemagglutinin (BHA) (top), or phosphate-buffered solution (bottom). Cells were gated for exclusion of propidium iodide and expression of B220. (Left) Staining with biotinylated hemagglutinin plus streptavidin–phycoerythrin. (Right) Staining with streptavidin–phycoerythrin only.

for antigen-specific B cells is a very powerful tool for the evaluation of B-cell responses to infection or immunization alike. The challenge lies in developing good reagent combinations that work on the particular instruments of the investigator to provide clear results. The marker combinations used in the examples provided here might be a good starting point for those interested in comprehensively analyzing B-cell subsets in mice. For a discussion on the choice of fluorochromes and fluorochrome combinations in multicolor flow cytometry, see Baumgarth and Roederer (2000).

Acknowledgments

I thank Virginia Doucett for help with the antigen-specific staining of influenzavirus-specific B cells. This work was funded in part by the National Institutes of Health/NIAID grant no. AI051354.

References

Allman, D., Li, J., and Hardy, R. R. (1999). Commitment to the B lymphoid lineage occurs before DH-JH recombination. *J. Exp. Med.* **189,** 735–740.

Amano, M., Baumgarth, N., Dick, M. D., Brossay, L., Kronenberg, M., Herzenberg, L. A., and Strober, S. (1998). CD1 expression defines subsets of follicular and marginal zone B cells in the spleen: Beta 2-microglobulin–dependent and independent forms. *J. Immunol.* **161,** 1710–1717.

Baumgarth, N., Herman, O. C., Jager, G. C., Brown, L., and Herzenberg, L. A. (1999). Innate and acquired humoral immunities to influenza virus are mediated by distinct arms of the immune system. *Proc. Natl. Acad. Sci. USA* **96,** 2250–2255.

Baumgarth, N., and Roederer, M. (2000). A practical approach to multicolor flow cytometry for immunophenotyping. *J. Immunol. Methods* **243,** 77–97.

Bell, J., and Gray, D. (2003). Antigen-capturing cells can masquerade as memory B cells. *J. Exp. Med.* **197,** 1233–1244.

Benner, R., Hijmans, W., and Haaijman, J. J. (1981). The bone marrow: The major source of serum immunoglobulins, but still a neglected site of antibody formation. *Clin. Exp. Immunol.* **46,** 1–8.

Berland, R., and Wortis, H. H. (2002). Origins and functions of B-1 cells with notes on the role of CD5. *Annu. Rev. Immunol.* **20,** 253–300.

Borrello, M. A., and Phipps, R. P. (1995). Fibroblasts support outgrowth of splenocytes simultaneously expressing B lymphocyte and macrophage characteristics. *J. Immunol.* **155,** 4155–4161.

Borrello, M. A., and Phipps, R. P. (1999). Fibroblast-secreted macrophage colony-stimulating factor is responsible for generation of biphenotypic B/macrophage cells from a subset of mouse B lymphocytes. *J. Immunol.* **163,** 3605–3611.

Butcher, E. C., Rouse, R. V., Coffman, R. L., Nottenburg, C. N., Hardy, R. R., and Weissman, I. L. (1982). Surface phenotype of Peyer's patch germinal center cells: Implications for the role of germinal centers in B cell differentiation. *J. Immunol.* **129,** 2698–2707.

Cong, Y. Z., Rabin, E., and Wortis, H. H. (1991). Treatment of murine CD5–B cells with anti-Ig, but not LPS, induces surface CD5: Two B-cell activation pathways. *Int. Immunol.* **3,** 467–476.

Crotty, S., Kersh, E. N., Cannons, J., Schwartzberg, P. L., and Ahmed, R. (2003). SAP is required for generating long-term humoral immunity. *Nature* **421,** 282–287.

Davidson, W. F., Pierce, J. H., and Holmes, K. L. (1992). Evidence for a developmental relationship between CD5+ B-lineage cells and macrophages. *Ann. NY Acad. Sci.* **651,** 112–129.

Driver, D. J., McHeyzer-Williams, L. J., Cool, M., Stetson, D. B., and McHeyzer-Williams, M. G. (2001). Development and maintenance of a B220–memory B cell compartment. *J. Immunol.* **167,** 1393–1405.

Ghia, P., Boekel, E. T., Rolink, A. G., and Melchers, F. (1998). B-cell development: A comparison between mouse and man. *Immunol. Today* **19,** 480–485.

Gray, D. (1993). Immunological memory. *Annu. Rev. Immunol.* **11,** 49–77.

Han, S., Zheng, B., Schatz, D. G., Spanopoulou, E., and Kelsoe, G. (1996). Neoteny in lymphocytes: Rag1 and Rag2 expression in germinal center B cells. *Science* **274,** 2094–2097.

Han, S., Zheng, B., Takahashi, Y., and Kelsoe, G. (1997). Distinctive characteristics of germinal center B cells. *Semin. Immunol.* **9,** 255–260.

Hardy, R. R., Carmack, C. E., Shinton, S. A., Kemp, J. D., and Hayakawa, K. (1991). Resolution and characterization of pro-B and pre-pro-B cell stages in normal mouse bone marrow. *J. Exp. Med.* **173,** 1213–1225.

Hargreaves, D. C., Hyman, P. L., Lu, T. T., Ngo, V. N., Bidgol, A., Suzuki, G., Zou, Y. R., Littman, D. R., and Cyster, J. G. (2001). A coordinated change in chemokine responsiveness guides plasma cell movements. *J. Exp. Med.* **194,** 45–56.

Hayakawa, K., and Hardy, R. R. (2000). Development and function of B-1 cells. *Curr. Opin. Immunol.* **12,** 346–353.

Hayakawa, K., Ishii, R., Yamasaki, K., Kishimoto, T., and Hardy, R. R. (1987). Isolation of high-affinity memory B cells: Phycoerythrin as a probe for antigen-binding cells. *Proc. Natl. Acad. Sci. USA* **84,** 1379–1383.

Herzenberg, L. A. (2000). B-1 cells: The lineage question revisited. *Immunol. Rev.* **175,** 9–22.

Jung, J., Choe, J., Li, L., and Choi, Y. S. (2000). Regulation of CD27 expression in the course of germinal center B cell differentiation: The pivotal role of IL-10. *Eur. J. Immunol.* **30,** 2437–2443.

Karasuyama, H., Rolink, A., Shinkai, Y., Young, F., Alt, F. W., and Melchers, F. (1994). The expression of Vpre-B/lambda 5 surrogate light chain in early bone marrow precursor B cells of normal and B cell–deficient mutant mice. *Cell* **77,** 133–143.

Klein, U., Rajewsky, K., and Kuppers, R. (1999). Phenotypic and molecular characterization of human peripheral blood B-cell subsets with special reference to N-region addition and J kappa-usage in V kappa J kappa-joints and kappa/lambda-ratios in naive versus memory B-cell subsets to identify traces of receptor editing processes. *Curr. Topics Microbiol. Immunol.* **246,** 141–146; discussion 147.

Kondo, M., Weissman, I. L., and Akashi, K. (1997). Identification of clonogenic common lymphoid progenitors in mouse bone marrow. *Cell* **91,** 661–672.

Lalor, P. A., Nossal, G. J., Sanderson, R. D., and McHeyzer-Williams, M. G. (1992). Functional and molecular characterization of single, (4-hydroxy-3-nitrophenyl)acetyl (NP)–specific, IgG1+ B cells from antibody–secreting and memory B cell pathways in the C57BL/6 immune response to NP. *Eur. J. Immunol.* **22,** 3001–3011.

Lam, K. P., and Rajewsky, K. (1999). B cell antigen receptor specificity and surface density together determine B-1 versus B-2 cell development. *J. Exp. Med.* **190,** 471–477.

Li, Y. S., Wasserman, R., Hayakawa, K., and Hardy, R. R. (1996). Identification of the earliest B lineage stage in mouse bone marrow. *Immunity* **5,** 527–535.

Loder, F., Mutschler, B., Ray, R. J., Paige, C. J., Sideras, P., Torres, R., Lamers, M. C., and Carsetti, R. (1999). B cell development in the spleen takes place in discrete steps and is determined by the quality of B cell receptor–derived signals. *J. Exp. Med.* **190,** 75–89.

Lu, L., Smithson, G., Kincade, P. W., and Osmond, D. G. (1998). Two models of murine B lymphopoiesis: A correlation. *Eur. J. Immunol.* **28,** 1755–1761.

MacLennan, I. C. (1994). Germinal centers. *Annu. Rev. Immunol.* **12,** 117–139.

Manz, R. A., Thiel, A., and Radbruch, A. (1997). Lifetime of plasma cells in the bone marrow. *Nature* **388,** 133–134.

Martin, F., and Kearney, J. F. (2002). Marginal-zone B cells. *Nat. Rev. Immunol.* **2,** 323–335.

Martin, F., Oliver, A. M., and Kearney, J. F. (2001). Marginal zone and B1 B cells unite in the early response against T-independent blood-borne particulate antigens. *Immunity* **14,** 617–629.

Martin, M., Strasser, A., Baumgarth, N., Cicuttini, F. M., Welch, K., Salvaris, E., and Boyd, A. W. (1993). A novel cellular model (SPGM 1) of switching between the pre-B cell and myelomonocytic lineages. *J. Immunol.* **150,** 4395–4406.

McHeyzer-Williams, L. J., Cool, M., and McHeyzer-Williams, M. G. (2000). Antigen-specific B cell memory: Expression and replenishment of a novel b220(−) memory b cell compartment. *J. Exp. Med.* **191,** 1149–1166.

McHeyzer-Williams, M. G., Nossal, G. J., and Lalor, P. A. (1991). Molecular characterization of single memory B cells. *Nature* **350,** 502–505.

Meffre, E., Casellas, R., and Nussenzweig, M. C. (2000). Antibody regulation of B cell development. *Nat. Immunol.* **1,** 379–385.

Melchers, F., Rolink, A., Grawunder, U., Winkler, T. H., Karasuyama, H., Ghia, P., and Andersson, J. (1995). Positive and negative selection events during B lymphopoiesis. *Curr. Opin. Immunol.* **7,** 214–227.

Montecino-Rodriguez, E., Leathers, H., and Dorshkind, K. (2001). Bipotential B-macrophage progenitors are present in adult bone marrow. *Nat. Immunol.* **2,** 83–88.

Nagai, Y., Shimazu, R., Ogata, H., Akashi, S., Sudo, K., Yamasaki, H., Hayashi, S., Iwakura, Y., Kimoto, M., and Miyake, K. (2002). Requirement for MD-1 in cell surface expression of RP105/CD180 and B-cell responsiveness to lipopolysaccharide. *Blood* **99,** 1699–1705.

Nikolic, T., Dingjan, G. M., Leenen, P. J., and Hendriks, R. W. (2002). A subfraction of B220(+) cells in murine bone marrow and spleen does not belong to the B cell lineage but has dendritic cell characteristics. *Eur. J. Immunol.* **32,** 686–692.

Oliver, A. M., Martin, F., and Kearney, J. F. (1997). Mouse CD38 is down-regulated on germinal center B cells and mature plasma cells. *J. Immunol.* **158,** 1108–1115.

Osmond, D. G. (1990). B cell development in the bone marrow. *Semin. Immunol.* **2,** 173–180.

Osmond, D. G., Rolink, A., and Melchers, F. (1998). Murine B lymphopoiesis: Towards a unified model. *Immunol. Today* **19,** 65–68.

Raman, V. S., Bal, V., Rath, S., and George, A. (2000). Ligation of CD27 on murine B cells responding to T-dependent and T-independent stimuli inhibits the generation of plasma cells. *J. Immunol.* **165,** 6809–6815.

Ridderstad, A., and Tarlinton, D. M. (1998). Kinetics of establishing the memory B cell population as revealed by CD38 expression. *J. Immunol.* **160,** 4688–4695.

Roark, J. H., Park, S. H., Jayawardena, J., Kavita, U., Shannon, M., and Bendelac, A. (1998). CD1.1 expression by mouse antigen-presenting cells and marginal zone B cells. *J. Immunol.* **160,** 3121–3127.

Rolink, A., Grawunder, U., Winkler, T. H., Karasuyama, H., and Melchers, F. (1994). IL-2 receptor alpha chain (CD25, TAC) expression defines a crucial stage in pre-B cell development. *Int. Immunol.* **6,** 1257–1264.

Rolink, A., ten Boekel, E., Melchers, F., Fearon, D. T., Krop, I., and Andersson, J. (1996). A subpopulation of B220+ cells in murine bone marrow does not express CD19 and contains natural killer cell progenitors. *J. Exp. Med.* **183,** 187–194.

Schatz, D. G., Oettinger, M. A., and Baltimore, D. (1989). The V(D)J recombination activating gene, RAG-1. *Cell* **59,** 1035–1048.

Schittek, B., and Rajewsky, K. (1990). Maintenance of B-cell memory by long-lived cells generated from proliferating precursors. *Nature* **346,** 749–751.

Shinall, S. M., Gonzalez-Fernandez, M., Noelle, R. J., and Waldschmidt, T. J. (2000). Identification of murine germinal center B cell subsets defined by the expression of surface isotypes and differentiation antigens. *J. Immunol.* **164,** 5729–5738.

Shinkai, Y., Rathbun, G., Lam, K. P., Oltz, E. M., Stewart, V., Mendelsohn, M., Charron, J., Datta, M., Young, F., Stall, A. M., and Alt, F. W. (1992). RAG-2–deficient mice lack mature lymphocytes owing to inability to initiate V(D)J rearrangement. *Cell* **68,** 855–867.

Smith, K. G., Hewitson, T. D., Nossal, G. J., and Tarlinton, D. M. (1996). The phenotype and fate of the antibody-forming cells of the splenic foci. *Eur. J. Immunol.* **26,** 444–448.

Song, H., and Cerny, J. (2003). Functional heterogeneity of marginal zone B cells revealed by their ability to generate both early antibody-forming cells and germinal centers with hypermutation and memory in response to a T-dependent antigen. *J. Exp. Med.* **198,** 1923–1935.

Takahashi, K., Kozono, Y., Waldschmidt, T. J., Berthiaume, D., Quigg, R. J., Baron, A., and Holers, V. M. (1997). Mouse complement receptors type 1 (CR1;CD35) and type 2 (CR2;CD21): Expression on normal B cell subpopulations and decreased levels during the development of autoimmunity in MRL/lpr mice. *J. Immunol.* **159,** 1557–1569.

Underhill, G. H., Kolli, K. P., and Kansas, G. S. (2003). Complexity within the plasma cell compartment of mice deficient in both E- and P-selectin: Implications for plasma cell differentiation. *Blood* **102,** 4076–4083.

Wasserman, R., Li, Y. S., Shinton, S. A., Carmack, C. E., Manser, T., Wiest, D. L., Hayakawa, K., and Hardy, R. R. (1998). A novel mechanism for B cell repertoire maturation based on response by B cell precursors to pre-B receptor assembly. *J. Exp. Med.* **187,** 259–264.

Wells, S. M., Kantor, A. B., and Stall, A. M. (1994). CD43 (S7) expression identifies peripheral B cell subsets. *J. Immunol.* **153,** 5503–5515.

Won, W. J., and Kearney, J. F. (2002). CD9 is a unique marker for marginal zone B cells, B1 cells, and plasma cells in mice. *J. Immunol.* **168,** 5605–5611.

PART V

Other Topics in Immunology

CHAPTER 28

Flow Cytometry Immunophenotypic Characteristics of Monocytic Population in Acute Monocytic Leukemia (AML-M5), Acute Myelomonocytic Leukemia (AML-M4), and Chronic Myelomonocytic Leukemia (CMML)

Wojciech Gorczyca
Director of Hematopathology/Oncology Services
Genzyme Genetics/IMPATH
New York, New York 10019

0091-679X/04 $35.00

I. Introduction

Monocytic proliferations comprise a heterogeneous group of disorders ranging from reactive monocytosis to acute monocytic leukemia. Based on the cytomorphological and phenotypic features, differential diagnosis includes acute promyelocytic leukemia (especially microgranular variant), acute myeloid leukemia (AML) without maturation, minimally differentiated AML, chronic myelomonocytic leukemia (CMML), acute megakaryocytic leukemia, and acute myelomonocytic leukemia (AML-M4). Extramedullary myeloid tumors with monocytic differentiation (monoblastic sarcoma) may be mistaken for large cell lymphoma, carcinoma, or sarcoma. Well-prepared fresh bone marrow aspirate with myeloperoxidase staining is helpful in differential diagnosis. Myeloblasts and abnormal promyelocytes are strongly MPO positive, whereas the monocytes are either weakly positive or negative. The monoblasts and promonocytes usually are positive with nonspecific esterase (NSE) staining, but a significant subset of acute monocytic leukemias is NSE negative. Therefore, the definite diagnosis often requires correlation of complete blood cell (CBC) count data, cytological features, and cytochemistry with additional techniques such as immunophenotyping by flow cytometry (FC), cytogenetics/fluorescence *in situ* hybridization (FISH), and molecular tests (e.g., polymerase chain reaction [PCR]). FC immunophenotyping is an accurate method for quantitative and qualitative evaluation of hematopoietic cells. It plays important role in diagnosis, classification, and monitoring of hematopoietic neoplasms, including acute leukemias (Baumgarth and Roederer, 2000; Borowitz *et al.*, 1997; Gorczyca *et al.*, 2002; Jennings and Foon, 1997a,b; Knapp *et al.*, 1994; Kotylo *et al.*, 2000; Kussick and Wood, 2003; Manaloor *et al.*, 2000; Orfao *et al.*, 1999a, 1999b; Weir and Borowitz, 2001; Weisberger *et al.*, 2000). This chapter presents the phenotypic characteristic of monocytic populations from acute monocytic leukemia (AML-M5), CMML and AML-M4.

II. Materials

Flow cytometric samples from IMPATH, Incorporated (New York division), containing abnormal monocytic populations were submitted for this study. FC data were reanalyzed and correlated with cytomorphology and/or bone marrow studies. All cases without firm morphological confirmation were excluded. The neoplasms were classified according to the World Health Organization (WHO) classification of hematopoietic neoplasms (Harris *et al.*, 2000a,b). There were 28 cases of AML-M5, 20 cases of CMML, and 15 cases of AML-M4.

III. Methods

A. Flow Cytometry Analysis

We used heparinized bone marrow aspirate, peripheral blood, and fresh tissue specimens for FC analysis and processed the specimens within 24 hours of collection. We obtained a leukocyte cell suspension from peripheral blood and bone marrow specimens after red blood cell (RBC) lysis with ammonium chloride lysing solution, followed by 5 minutes of centrifugation. The cell pellet was suspended with an appropriate amount of RPMI 1640 (GIBCO, New York). Fresh tissue samples were disaggregated with a sterile blade, followed by passage through a mesh filter ($<100\,\mu$m). The cells were washed in RPMI media and centrifuged at 1500 rpm for 5 minutes. To minimize nonspecific binding of antibodies, we incubated the cells in RPMI media supplemented with 10% heat-inactivated fetal bovine serum (FBS) in a 37 °C water bath for 30 minutes. We then washed the samples with 0.1% sodium azide/10% FBS (phosphate-buffered saline [PBS] buffer) and assessed viability using both trypan blue and 7-aminoactinomycin D (Sigma Chemical Co., St. Louis, Missouri) exclusion assays.

Immunophenotypic analysis was performed on FACSCalibur System Instruments equipped with a 15-mW 488-nm air-cooled argon-ion laser supplemented with a 635-nm red-diode laser (Becton Dickinson Immunocytometry System, San Jose, California). Three- and four-color directly labeled antibody combinations consisting of fluorescein isothiocyanate (FITC), phycoerythrin (PE), peridinin-chlorophyl (PerCP), and allophycocyanin (APC) were used for surface staining of fresh tissue, bone marrow, and peripheral blood cell suspensions. Internal negative controls within each tube and isotype controls for immunoglobulin G1 (IgG1), IgG2a, and IgG2b were used as negative controls.

We monitored the instrument fluorescence detectors' settings and calibration according to manufacturer's recommendations, using Calibrite Beads (Becton Dickinson) and evaluated the system linearity using Sphero Rainbow Beads (Pharmingen). FC data was collected in list mode and analyzed using CellQuest and Cell Quest Pro software (Becton Dickinson). A six-gate strategy was employed, using CD45 PerCP versus side scatter to characterize the lymphocyte, monocyte, granulocyte, blast, hematogone, and nucleated red cell precursor (erythroid) populations. Five- to six-parameter analysis (forward-scatter channel [FSC], side-scatter channel [SSC], FL1, FL2, FL3, and FL4) or multiparameter data analysis of antibody staining patterns was used to assess specific antigen expression.

IV. Acute Monocytic Leukemia

Acute monocytic (monoblastic) *leukemia* (AML-M5) is defined as myeloid leukemia in which 80% or more of the leukemic cells are of monocytic lineage (monoblasts, promonocytes, and monocytes). In this series (28 cases), patients ranged in age from 17 to 85 years. Figure 1 presents typical cytologic and phenotypic features of AML-M5. Leukemic cells have abundant cytoplasm that may show irregular borders with pseudopod and cytoplasmic vacuoles. NSE is positive in most cases, although it may be weak. All AML-M5 are positive for pan-myeloid markers, CD33 (bright expression), CD11c (moderate or bright), and CD45 (moderate or bright). Most cases are positive for human leukocyte antigen-DR (HLA-DR) (96%), CD4 (93%), CD11b (75%), CD13 (78%, usually dim expression), CD56 (86%), and CD64 (89%). A subset of cases show positive expression of CD2, CD7, CD10, CD23, and CD117. Table I presents details of FC immunophenotypic findings.

V. Chronic Myelomonocytic Leukemia

Chronic myelomonocytic leukemia (CMML) is mixed myelodysplastic/myeloproliferative disorder defined by persistent monocytosis (>1 by 10^9/L) in peripheral blood, fewer than 20% blasts, and dysplastic features in one or more myeloid lineages. Molecular/cytogenetic study results are negative for *bcr-abl* fusion gene (Philadelphia chromosome). The monocytes are usually mature with focal nuclear or cytoplasmic atypia. Based on the number of blasts, CMML is divided into two categories: CMML-1 (<5% blasts in blood, <10% blasts in bone marrow) and CMML-2 (5–19 blasts in blood and 10–19% blasts in bone marrow). Figure 2 presents FC analysis of CMML-2. The neoplastic monocytes have the phenotype resembling normal monocytes; they are always positive for CD11b (bright expression), CD11c (bright expression), CD14 (bright expression), CD33 (bright expression), CD45 (bright expression), and CD64 (bright expression). Most cases express CD13 (95%), HLA-DR (71%), and CD4 (76%). CD56 is present in 53% of cases. Lack of HLA-DR and CD13 and presence of aberrant expression of CD10, CD16, CD23, CD56, and CD117 distinguished CMML from reactive monocytosis. Table I presents phenotypic data of all analyzed cases.

VI. Acute Myelomonocytic Leukemia

Acute myelomonocytic leukemia (AML-M5) is an acute leukemia characterized by the proliferation of both neutrophil and monocyte precursors with 20% or more myeloblasts in the bone marrow. Both monocytic and granulocytic lineages must comprise at least 20% of marrow cells. FC analysis from AML-M4 cases shows distinct populations of blasts, monocytes, and residual (maturing) myeloid

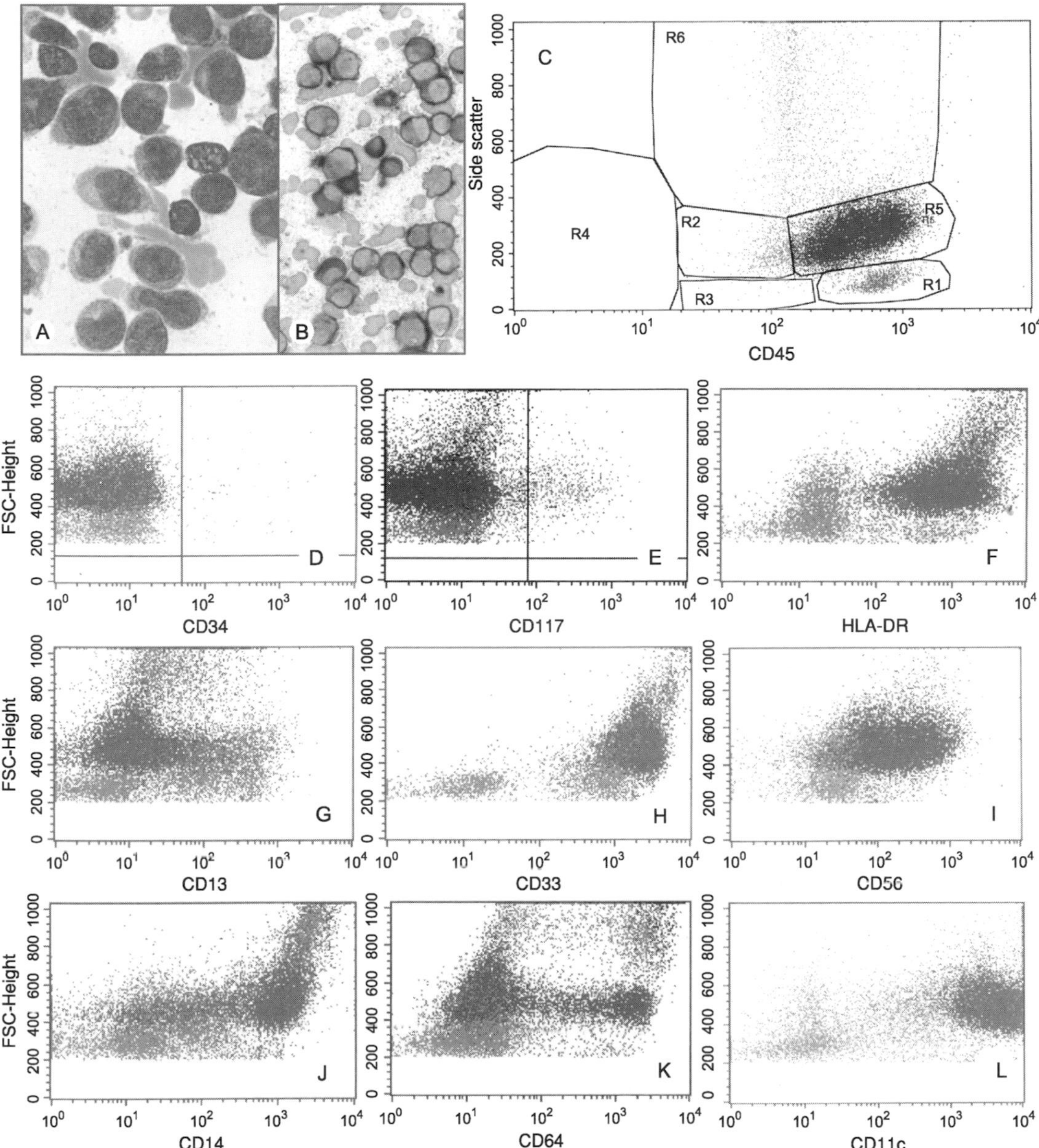

Fig. 1 Acute monocytic leukemia (AML-M5). (A) Bone marrow aspirate shows monoblasts with irregular nuclei. (B) Nonspecific esterase is strongly positive. (C–L) Flow cytometry immunophenotyping. (C) Monoblasts are brightly positive for CD45 and have increased side scatter (*blue dots*). They are negative for CD34 (D) and CD117 (E) and positive for HLA-DR (F), CD33 (H), CD56 (I), CD14 (J), and CD11c (L). CD13 (G) and CD64 (K) are partially expressed. (See Color Insert.)

Table I
Comparison of Flow Cytometric Features of Monocytic Cells in Acute Monocytic Leukemia, Chronic Myelomonocytic Leukemia, and Acute Myelomonocytic Leukemia

Marker	Acute monocytic leukemia (AML-M5) (% positive)	Chronic myelomonocytic leukemia (% positive)	Monocytic population in acute myelomonocytic leukemia (AML-M4) (% positive)
CD2	14	34	40
CD4	93	76	100
CD7	21	9	40
CD10	7	28	0
CD11b	75	100	100
CD11c	100	100	100
CD13	78	95	100
CD14	18 (additional 36% shows positive expression in small subset)	100	100
CD16	4	29	0
CD23	32	9	0
CD33	100	100	100
CD34	14	0	26
CD45	100 (moderate/bright)	100 (bright)	100 (bright)
CD56	86	53	26
CD64	89	100	100
CD117	25	5	0
HLA-DR	96	71	100

cells (Fig. 3). Monocytic cells in AML-M4 are always positive for CD4, CD11b, CD11c, CD13, CD14, CD33, CD45, CD64, and HLA-DR. A subset of leukemias show expression of CD2, CD7, CD34, and CD56 (Table I).

VII. Differential Diagnosis

Figure 4 presents the expression of CD45 versus side scatter in different types of leukemia. Myeloblasts in AMLs (M0–M2) display moderate expression of CD45 and low side scatter (Fig. 4B). Blasts usually predominate and there are no or few other elements like granulocytes, monocytes, and lymphocytes (compare with normal marrow, Fig. 4A). Monoblasts in AML-M5 are characterized by bright CD45 and increased side scatter (blue dots, Fig. 4E). Note rare myeloblasts (green dots) and paucity of other bone marrow elements (especially granulocytes). CMML shows predominance of monocytes (Fig. 4F), with granulocytes displaying decreased side scatter (dysgranulopoiesis). Blasts may be present but

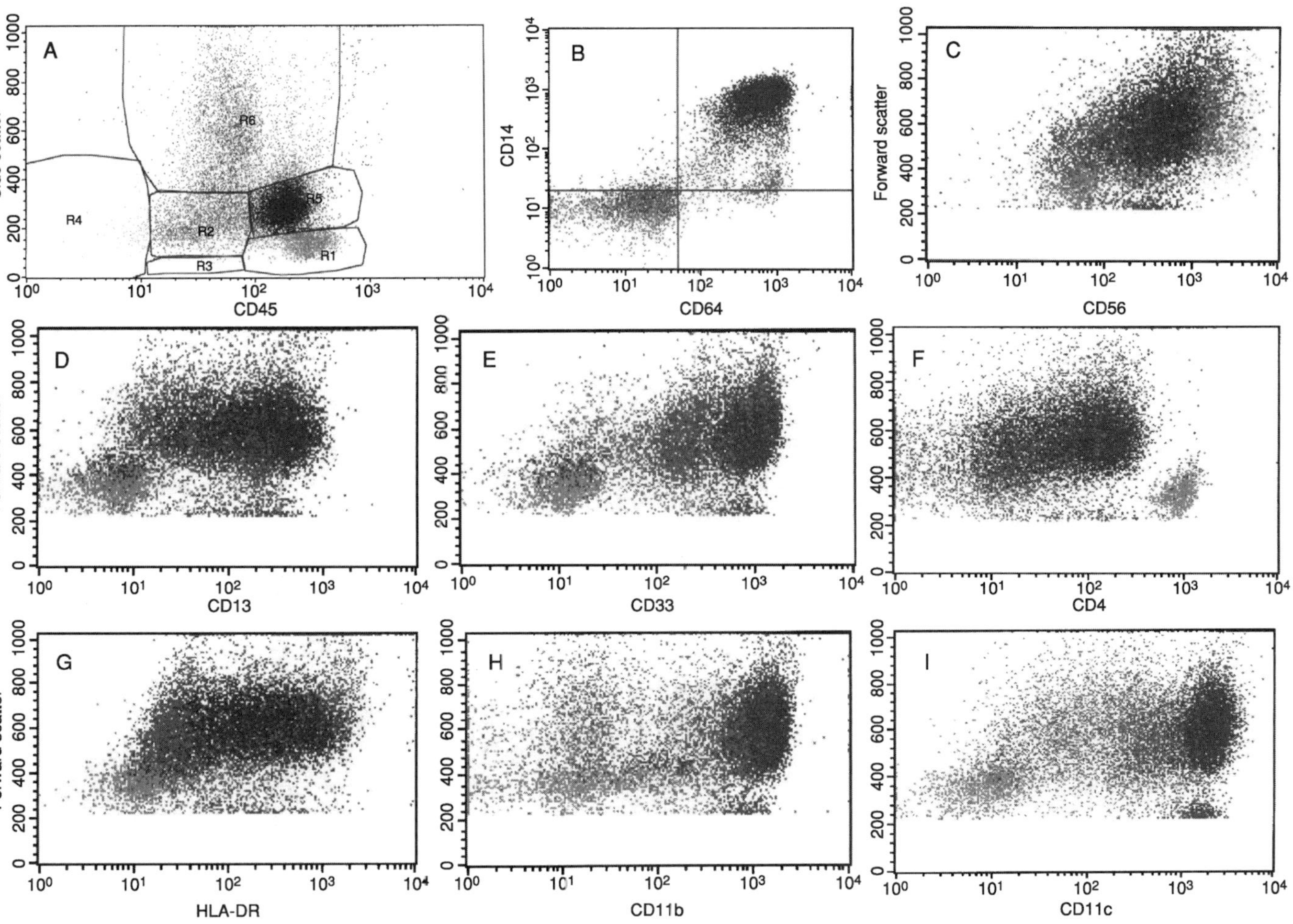

Fig. 2 Chronic myelomonocytic leukemia (CMML) flow cytometry. Atypical monocytes (*blue dots*) are positive for CD45 (A), CD14 and CD64 (B), CD56 (C), CD13 (D), CD33 (E), CD4 (F), HLA-DR (G), CD11b (H), and CD11c (I). Granulocytes (*gray dots*) and lymphocytes (*red dots*) are also present. Blasts (*green dots*) do not exceed 20% by flow cytometry and/or enumeration on fresh bone marrow aspirate. (See Color Insert.)

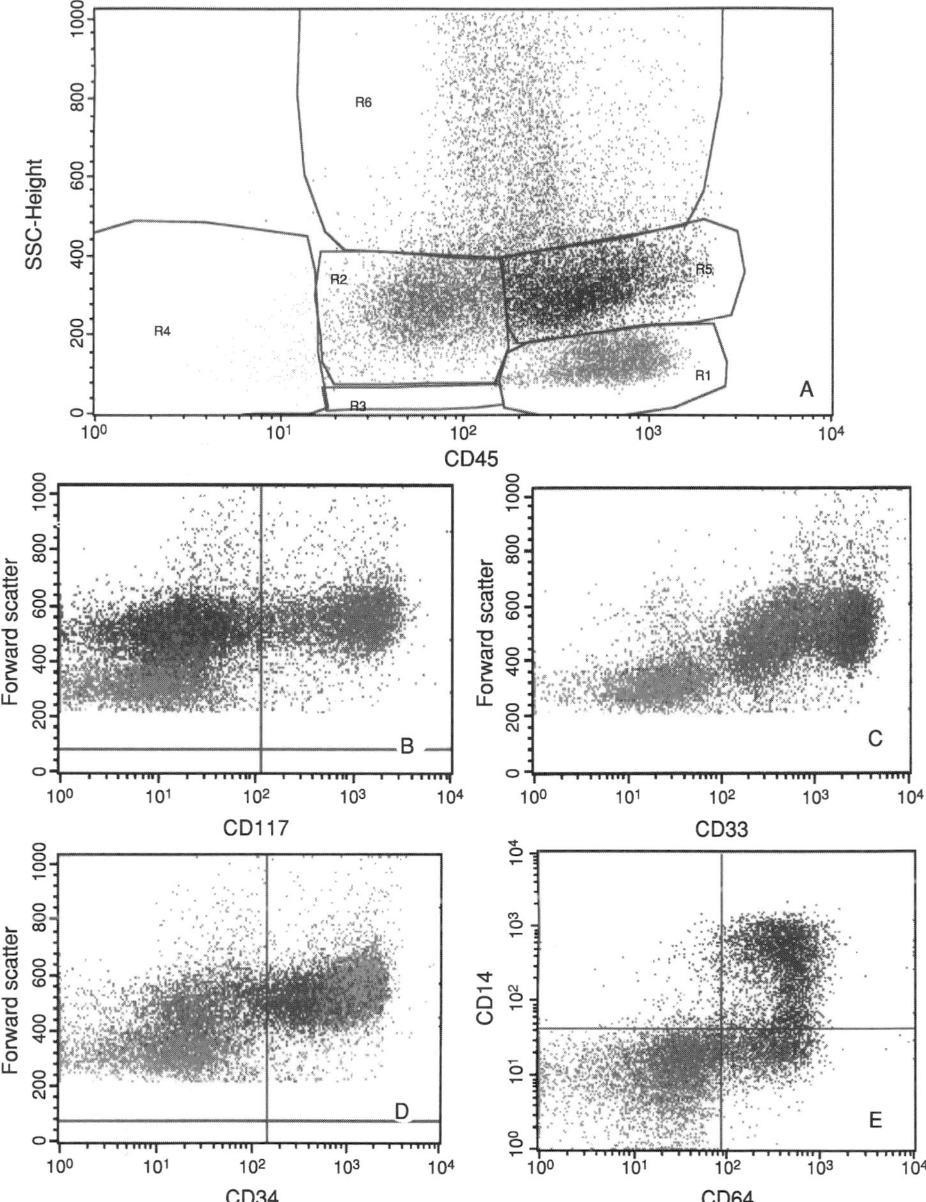

Fig. 3 Acute myelomonocytic leukemia (AML-M4) flow cytometry. Two distinct populations are noted: blasts (A, *green dots*) and monocytic cells (A, *blue dots*). Granulocytes (*gray dots*) and lymphocytes (*red dots*) are also present. Myeloblasts are positive for CD117 (B), CD33 (C), and CD34 (D), whereas monocytic cells are positive for CD33 (C), CD34 (D), and CD64 and CD14 (E). (See Color Insert.)

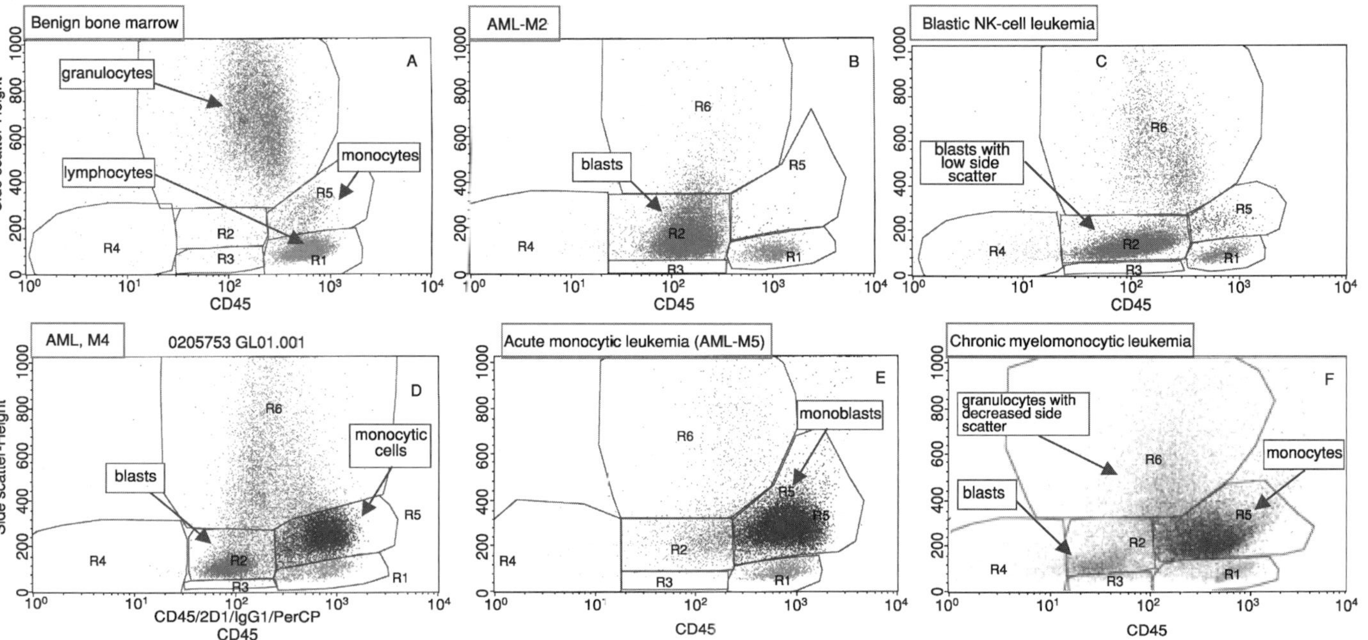

Fig. 4 Comparison of CD45 versus side scatter in acute leukemias. (A) Normal (benign) bone marrow. Note predominance of granulocytes (*gray dots*) and lymphocytes (*red dots*). Rare monocytes (*blue dots*) are also present. (B) Acute myeloid leukemia with differentiation (AML-M2). Myeloblasts (*green dots*) predominate. They have low side scatter and moderate CD45 expression. (C) Blastic natural killer cell leukemia/lymphoma. Blasts have low side scatter and moderate CD45. (D) Acute myelomonocytic leukemia (AML-M4). Two populations are present: myeloblasts (*green dots*) and monocytic cells (*blue dots*). Granulocytes are present (*gray dots*). (E) Acute monocytic leukemia (AML-M5). Monoblasts with increased side scatter and bright CD45 expression predominate (*blue dots*). Only rare myeloblasts (*green dots*) and very few granulocytes (*gray dots*) are present. (F) Chronic myelomonocytic leukemia (CMML). There is predominance of monocytic cells (*blue dots*), but lymphocytes, myeloblasts, and granulocytes are also present. Note decreased side scatter of granulocytes. (See Color Insert.)

are less than 20%. AML-M4 (Fig. 4D) shows two distinct populations of blasts (green dots) and monocytes (blue dots). The proportion of myeloblasts, monoblasts/monocytes, granulocytes, and lymphocytes in AML-M4, CMML, and AML-M4 is presented in Fig. 5. Neoplastic monocytes differ from benign monocytes by expression of CD16 (most often in CMML), CD23 (most often in AML-M5), CD34 (AML-M4.AML-M5), CD56 (AML-M5>CMML>AML-M4), and CD117 (AML-M5) and lack of HLA-DR (most often seen in CMML), CD64 (AML-M5), CD11b (AML-M5), CD14 (AML-M5), and CD64 (AML-M5) (Table I).

Table II summarizes flow cytometric phenotypic characteristics of different types of acute leukemias, which fall into differential diagnosis with AML-M5. Positive expression of HLA-DR differentiates AML-M5 from acute promyelocytic leukemia (APL, AML-M3). Positive expression of CD56 is seen most commonly in AML-M5 and blastic natural killer cell lymphoma/leukemia. Expression of CD11b and CD11c is most typical for AML-M5 (CD11c is often seen also in AML-M0, AML-M1, and AML-M2). CD64 is slightly more brightly expressed in AML-M5 than in megakaryocytic or promyelocytic leukemias. CD23 can be expressed on a subset of AML-M5 and microgranular variants of APL. CD34 expression is negative in hypergranular APL and megakaryocytic leukemia. CD13 is never expressed in blastic natural killer cell lymphoma/leukemia, shows moderate expression in AML-M0 through AML-M2, and dim expression in AML-M5, APL, AML-M6, and AML-M7.

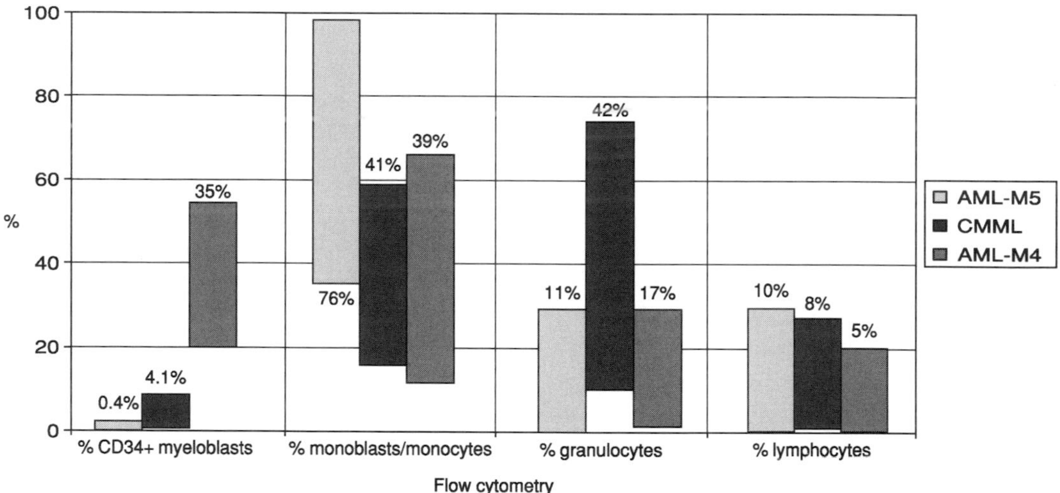

Fig. 5 Comparison of proportion of myeloblasts, monocytic cells, granulocytes, and lymphocytes in acute monocytic leukemia (AML-M5), chronic myelomonocytic leukemia, and acute myelomonocytic leukemia (AML-M4).

Table II
Differential Diagnosis of Acute Leukemias Based on Flow Cytometric Phenotypic Features

Marker	Acute monocytic leukemia (AML-M5)	Acute myeloid leukemia (AML-M2)	Acute promyelocytic leukemia (hypergranular)	Acute megakaryocytic leukemia	Blastic natural killer cell lymphoma/leukemia	Acute promyelocytic leukemia (microgranular)
CD2	−/+	−	−/+	−	−	+
CD4	+ (rare −)	+/−	−/+	−	+	+/−
CD7	−/+	−/rare +	−/+	−	+/very rare −	−
CD10	−/+	−	−	−	−	−
CD11b	+/−	−	−	−	−	−
CD11c	+ (bright)	+ (dim/moderate)	−	−	−	−
CD13	dim	+ (moderate)	+ (dim/moderate)	+ (dim)	−	+ (dim)
CD14	subset cells +	−	−	−	−	−
CD16	rare +	−	−	−	−	−
CD23	~1/3 cases +	−	−	−	−	−/+
CD33	100% + (bright)	+ (variable)	+ (moderate/bright)	+ (bright)	−/very rare +	+ (bright)
CD34	−	+	−	−	+/−	+ (occasionally only in subset)
CD41/61	−	−	−	+	−	−
CD45	100% + (moderate/bright)	+ (dim/moderate)	+ (moderate)	+	+ (dim/moderate)	+ (moderate)
CD56	+ (86%)	−/rare +	−/+	−/rare +	100% +	−/rare +
CD64	+ (dim/moderate)	+/−	+ (dim)	−	−	
CD117	−/rare + (25%)	+	+	+ (dim)	−/rare +	
TdT	−	+/−	−	−	−/+	−
HLA-DR	+ (96% cases)	+/rare −	−	+ (dim)	+ (dim)	−

VIII. Conclusion

FC, particularly CD45 versus side scatter and expression of CD2, CD7, CD10, CD11b, CD11c, CD14, CD16, CD23, CD33, CD34, CD45, CD56, CD64, CD117, and HLA-DR, can identify and differentiate abnormal monocytes and complements cytomorphology and cytochemical staining in diagnosis of myeloid proliferations with monocytic differentiation.

References

Baumgarth, N., and Roederer, M. (2000). A practical approach to multicolor flow cytometry for immunophenotyping. *J. Immunol. Methods* **243,** 77–97.

Borowitz, M. J., Bray, R., Gascoyne, R., Melnick, S., Parker, J. W., Picker, L., and Stetler-Stevenson, M. (1997). U.S.-Canadian Consensus recommendations on the immunophenotypic analysis of hematologic neoplasia by flow cytometry: Data analysis and interpretation. *Cytometry* **30,** 236–244.

Gorczyca, W., Weisberger, J., Liu, Z., Tsang, P., Hossein, M., Wu, C. D., Dong, H., Wong, J. Y., Tugulea, S., Dee, S., Melamed, M. R., and Darzynkiewicz, Z. (2002). An approach to diagnosis of T-cell lymphoproliferative disorders by flow cytometry. *Cytometry* **50,** 177–190.

Harris, N. L., Jaffe, E. S., Diebold, J., Flandrin, G., Muller-Hermelink, H. K., Vardiman, J., Lister, T. A., and Bloomfield, C. D. (2000a). The World Health Organization classification of neoplasms of the hematopoietic and lymphoid tissues: Report of the Clinical Advisory Committee meeting–Airlie House, Virginia, November, 1997. *Hematol. J.* **1,** 53–66.

Harris, N. L., Jaffe, E. S., Diebold, J., Flandrin, G., Muller-Hermelink, H. K., Vardiman, J., Lister, T. A., and Bloomfield, C. D. (2000b). The World Health Organization classification of neoplastic diseases of the haematopoietic and lymphoid tissues: Report of the Clinical Advisory Committee Meeting, Airlie House, Virginia, November 1997. *Histopathology* **36,** 69–86.

Jennings, C. D., and Foon, K. A. (1997a). Flow cytometry: Recent advances in diagnosis and monitoring of leukemia. *Cancer Invest.* **15,** 384–399.

Jennings, C. D., and Foon, K. A. (1997b). Recent advances in flow cytometry: Application to the diagnosis of hematologic malignancy. *Blood* **90,** 2863–2892.

Knapp, W., Strobl, H., and Majdic, O. (1994). Flow cytometric analysis of cell-surface and intracellular antigens in leukemia diagnosis. *Cytometry* **18,** 187–198.

Kotylo, P. K., Seo, I. S., Smith, F. O., Heerema, N. A., Fineberg, N. S., Miller, K., Greene, M. E., Chou, P., and Orazi, A. (2000). Flow cytometric immunophenotypic characterization of pediatric and adult minimally differentiated acute myeloid leukemia (AML-M0). *Am. J. Clin. Pathol.* **113,** 193–200.

Kussick, S. J., and Wood, B. L. (2003). Using 4-color flow cytometry to identify abnormal myeloid populations. *Arch. Pathol. Lab. Med.* **127,** 1140–1147.

Manaloor, E. J., Neiman, R. S., Heilman, D. K., Albitar, M., Casey, T., Vattuone, T., Kotylo, P., and Orazi, A. (2000). Immunohistochemistry can be used to subtype acute myeloid leukemia in routinely processed bone marrow biopsy specimens. Comparison with flow cytometry. *Am. J. Clin. Pathol.* **113,** 814–822.

Orfao, A., Chillon, M. C., Bortoluci, A. M., Lopez-Berges, M. C., Garcia-Sanz, R., Gonzalez, M., Tabernero, M. D., Garcia-Marcos, M. A., Rasillo, A. I., Hernandez-Rivas, J., and San Miguel, J. F. (1999a). The flow cytometric pattern of CD34, CD15 and CD13 expression in acute myeloblastic leukemia is highly characteristic of the presence of PML-RARalpha gene rearrangements. *Haematologica* **84,** 405–412.

Orfao, A., Schmitz, G., Brando, B., Ruiz-Arguelles, A., Basso, G., Braylan, R., Rothe, G., Lacombe, F., Lanza, F., Papa, S., Lucio, P., and San Miguel, J. F. (1999b). Clinically useful information provided by the flow cytometric immunophenotyping of hematological malignancies: Current status and future directions. *Clin. Chem.* **45,** 1708–1717.

Weir, E. G., and Borowitz, M. J. (2001). Flow cytometry in the diagnosis of acute leukemia. *Semin. Hematol.* **38,** 124–138.

Weisberger, J., Wu, C. D., Liu, Z., Wong, J. Y., Melamed, M. R., Darzynkiewicz, Z., and Gorczyca, W. (2000). Differential diagnosis of malignant lymphomas and related disorders by specific pattern of expression of immunophenotypic markers revealed by multiparameter flow cytometry [Review]. *Int. J. Oncol.* **17,** 1165–1177.

CHAPTER 29

Phagocyte Function

Gregor Rothe and Mariam Klouche
Bremer Zentrum für Laboratoriumsmedizin GmbH
D-28205 Bremen, Germany

0091-679X/04 $35.00

I. Introduction

A. Biological Functions and Mechanisms of Activation

Professional phagocytes, including neutrophil, eosinophil, and basophil granulocytes, and monocytes/macrophages play a major role in innate host defense and function as scavengers of debris, removing damaged, senescent, and apoptotic cells, as well as foreign substances, such as latex particles (Djaldetti *et al.*, 2002). Moreover, phagocytes participate in inflammatory reactions and exert tumoricidal activity. Besides their principal function in the first-line defense, monocytes have a critical function regulating hemostasis and wound repair via expression of tissue factor and through their functional capacity to interact with platelets and endothelial cells (Osterud, 1998). Phagocytes may directly engulf microorganisms, particles, or altered endogenous substances; however, their activity is greatly enhanced by opsonizing factors, such as complement, particularly C3b and C3bi, immunoglobulins, or the acute-phase reactants C-reactive protein and pentraxin-3 (Hart *et al.*, 2004; Mantovani, 2003). Although granulocytes primarily degrade phagocytosed materials until completion, ultimately leading to cell death, the principal role of monocytes/macrophages includes the specific degradation to antigenic peptides for presentation via the major histocompatibility complex (MHC), thus linking the innate to the specific immune response.

The functional reaction cascade of the phagocytes to external microorganisms, foreign substances, or endogenous decayed material, such as cellular debris or apoptotic cells, encompasses a complex series of fine-tuned responses from chemotaxis, actin assembly, and migration to cell-surface receptor recognition, receptor assembly, adhesion, aggregation, phagocytosis, degranulation, reactive oxygen species production, and intracellular pH changes (Aderem and Underhill, 1999; Kwiatkowska and Sobota, 1999) (Fig. 1).

Chemotaxis and migration are the initial steps of phagocyte function, which are largely governed by adhesion events, chemokine receptors, and ultimate reorganizations of the actin cytoskeleton. The binding of opsonized and free particles to specific cell-surface receptors limited to phagocytes initiates phagocytosis. Phagocytes are equipped with at least two sets of receptors for recognition of these different kinds of presentation of particles; the first recognizes primarily host opsonin molecules, and the second consists of receptors recognizing

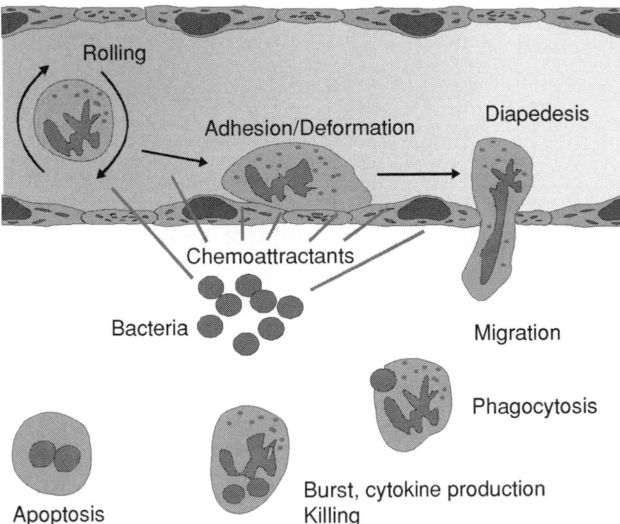

Fig. 1 Stepwise process of neutrophil activation during the extravasation to sites of infection. An upregulation of adhesion is induced by chemoattractants followed by actin polymerization, diapedesis through the endothelium, and extravasation. After migration along a chemotactic gradient, the phagocytosis of bacteria is accompanied by the release of reactive oxidants, secretion of granule-associated proteases, and microbicidal proteins as well as the autocrine synthesis of proinflammatory cytokines or chemokines such as interleukin-8. Apoptosis finally represents an important regulatory step under the control of interleukin-10, which limits the inflammatory reaction.

integral surface or membrane components of particles. The best characterized opsonin-dependent receptors include the immunoglobulin superfamily receptors FcγRI, FcγRII, FcγRIII, the complement receptor-3, CR3, and the integrin receptors $\alpha_5\beta_1$ (VLA-5), and $\alpha_v\beta_3$ (Kwiatkowska and Sobota, 1999). The opsonin-independent receptors encompass receptors for *N*-formylated peptides such as *N*-formyl-Met-Leu-Phe (fMLP), the lectin-class mannose receptor, the β-glucan receptor, the class A scavenger receptors I and II, CD36, and MARCO, as well as CD68 (macrosialin), several toll-like receptors (TLRs), and CD14 (Ozinsky *et al.*, 2000; Palecanda and Kobzik, 2001; Savill, 1998).

The first event triggered by the binding of ligands, irrespective of the nature of the phagocyte receptors, is the clustering of the cell-surface membrane receptors (Allen and Aderem, 1996; Jongstra-Bilen, 2003; Pfeiffer *et al.*, 2001). In fact, receptor clustering as a salient feature of phagocyte activation transmits several signal transduction pathways, including tyrosine kinases, small guanosine triphosphatases (GTPases) of the Rho and ARF families, and phosphatidylinositol-3-kinase (Kwiatkowska and Sobota, 1999). Particularly for the phosphatidylinositol-3-kinase, a crucial and non-redundant role determining the responsiveness of phagocytes to chemoattractants and the process of chemotactic leukocyte recruitment has been demonstrated (Wymann *et al.*, 2000). This action is probably based on the capacity of phosphatidylinositol-3-kinase to modulate the assembly of the submembranous actin filament system locally, leading not only to directed migration but also to particle internalization.

These stimulus-dependent activation steps can be greatly enhanced by phagocyte "priming," a process involving several mediators, which at higher concentrations may also act as direct stimuli, such as interleukin-8 (IL-8), and others, which act primarily as facilitators of subsequent activators, such as tumor necrosis factor-α (TNF-α). Currently, several classes of priming substances have been identified (Fig. 2), including proinflammatory cytokines (TNF-α/IL-1), chemokines such as IL-8, granulocyte–macrophage colony-stimulating factor (GM-CSF), and granulocyte colony-stimulating factor (G-CSF), lipid mediators, such as lipopolysaccharide (LPS), leukotriene B_4 (LTB4), and platelet-activating factor (PAF),

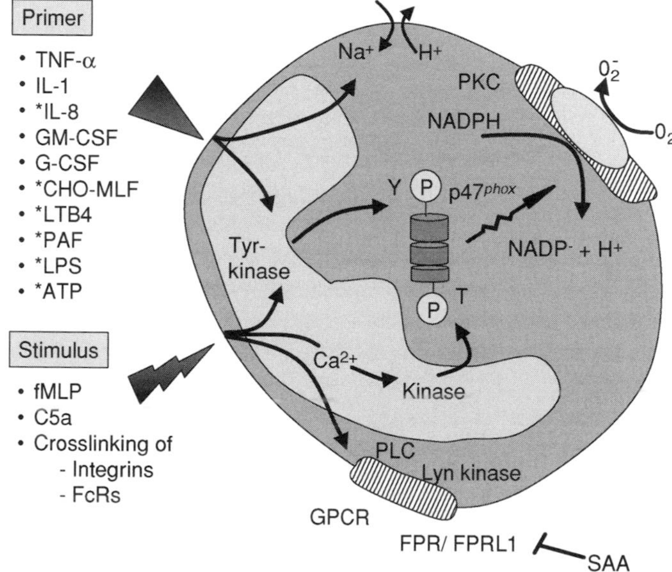

Fig. 2 Cellular signals triggered by priming of phagocytes. Ligand-specific activation may be substantially amplified by priming of phagocytes with multiple activators, including proinflammatory cytokines, colony-stimulating factors, lipid mediators, and adenosine triphosphate. Several interactive priming mechanisms have been discovered, including limited phosphorylation of p47phox, enhanced expression, and microdomain redistribution of the NADPH–oxidase complex at the plasma membrane, increased expression of triggering receptors, including G-protein–coupled receptors redistribution of these receptors and second messengers, as well as adaptor molecules and kinases in rafts. Flow cytometric quantification of responses revealed a homozygous alkalinization as crucial events during the oxidative burst cascade. Ligand involved (*) in priming at substimulatory concentrations are shown, which may act as direct stimulants of the oxidative response at higher concentrations. CHO-MLF, *N*-formyl-methionyl-leucyl-phenylalanine; ATP, adenosine triphosphate; SAA, serum amyloid A.

as well as formylated peptides (Dewas *et al.*, 2003; Hallett *et al.*, 1995, 2003; Wittmann *et al.*, 2004). The mechanisms of neutrophil priming encompass a complex and increasing array of cross-talking cellular events, which ultimately amplify oxidant production and cytoskeletal rearrangements (Fig. 2). Priming may result from increases of formylated receptor (FPR) expression and enhancement of NADPH oxidase assembly and expression at the plasma membrane. This assumption has been supported by the finding of priming-induced redistribution of FPRs with G-protein–coupled receptors (GPCRs), Lyn kinase, and phospholipase C (PLC), as well as recruitment of the NADPH oxidase system and its effector protein kinase C (PKC) in specialized membrane microdomains or rafts (Keil *et al.*, 2003; Shao *et al.*, 2003). In fact kinetic analyses show that depletion of rafts delays the onset of NADPH oxidase activation, supporting the role of rafts as bundling sites for functional receptor clustering and co-localization of signal transduction components (Shao *et al.*, 2003). Moreover, quantitative flow cytometry has revealed a central role of alkalinization for promoting neutrophil activation.

The final step in the phagocyte activation cascade, intracellular killing of microorganisms and degradation of foreign substances, also involves a rearrangement of plasma membrane composition. Upon ligand binding, several resting dispersed cytosolic and membrane-bound enzymatic components of the NADPH oxidase become rapidly assembled in the plasma membrane. NADPH oxidase and the granule-associated enzyme myeloperoxidase generate reactive oxygen species including superoxide anion, hydroxyl radical, singlet oxygen, and hydrogen peroxide in the oxidative burst reaction (Babior *et al.*, 2003; Dahlgren and Karlsson, 1999; Forman and Torres, 2002).

B. Phagocyte Functions and Disease

Primary or acquired phagocyte dysfunctions or quantitative defects are associated with increased susceptibility to infections, leading to recurrent and often life-threatening infections, particularly with encapsulated bacteria, fungi, or parasites. Clinically, cutaneous infections (such as abscesses, furuncles, and carbuncles), respiratory infections (such as sinusitis and pneumonia), and systemic purulent affection (e.g., liver abscesses) are prevalent (Malech and Nauseef, 1997). The World Health Organization (WHO) classification of primary immunodeficiencies is expanded every 5 years, and defects have been described for all stages of phagocyte function, including (1) deficient chemotaxis, mobility, and adhesion, such as actin dysfunction, Chédiak–Higashi syndrome, Shwachman disease, and gp110 deficiency, as well as (2) defective endocytosis and killing, such as X-linked or autosomal recessive chronic granulomatous disease (CGD), myeloperoxidase deficiency, GP150, and neutrophil G6PD deficiency (WHO, 1999). One of the best characterized groups of rare inherited disorders of phagocyte function encompasses CGD, which is characterized by defects in the phagocyte-specific NADPH–oxidase complex resulting in absent or reduced respiratory burst (Dinauer and Orkin, 1992; Segal and Holland, 2000).

Acquired dysfunctions of phagocytes are by far more common and constitute important causes of morbidity and mortality particularly from infections in chronically ill patients, in patients with diabetes mellitus, autoimmune disease, renal or hepatic failure, alcoholism, as well as in immunosuppression in the context of trauma, surgery, burns, and viral infections (Engelich *et al.*, 2001). The mechanisms leading to phagocyte suppression are complex and often less well characterized compared to primary defects of phagocyte function and include chemical toxins (e.g., ethanol), metabolic disturbances (e.g., uremia), or pathological activation in the circulation (e.g., burns and bypass). An impairment of monocytes/macrophage phagocytosis following human immunodeficiency virus (HIV) infection has been well documented and forms the basis of increased susceptibility to opportunistic pathogens (Kedzierska *et al.*, 2003).

Although reactive oxygen derivatives and proteases normally serve a microbicidal function, excessive or inappropriate release of these products contributes to inflammatory reactions and tissue injury (Babior, 2000). This increased activation of phagocytes with concomitant tissue destruction is believed to contribute to the pathogenesis of the adult respiratory distress syndrome (ARDS), as well as the systemic inflammatory response syndrome (SIRS), which may occur in absence of infection (Djaldetti *et al.*, 2002; Hasleton and Roberts, 1999). The expression of tissue factor by monocytes/macrophages contributes to their pathophysiological roles in disseminated intravascular coagulation linked to sepsis, postoperative thrombosis, unstable angina, atherosclerosis, chronic inflammation, and cancer (Osterud, 1998). Tissue factor is now considered the primary physiological activator of the blood coagulation system, and altered expression of tissue factor by monocytes has been recognized as a trigger for pathological intravascular coagulation and has been observed in general inflammatory disorders, such as sepsis and hypoxic shock syndromes.

A further mechanism interfering with phagocyte function includes iatrogenic suppression by several drugs, such as general and regional anesthesia, particularly after using inhalant anesthetics and, paradoxically, many antibiotics (Djaldetti *et al.*, 2002). Antimicrobial drugs may adversely influence the interaction between neutrophils and microorganisms, and conversely, neutrophils may interfere with the action of antimicrobial drugs (Pallister and Lewis, 2000). Moreover, a number of drugs have been reported not only to suppress functional activation but also to trigger the reduction of phagocyte numbers up to agranulocytosis; these include antiviral substances such as zidovudine and ganciclovir, antibiotics such as sulfonamides, the dihydrofolate reductase inhibitors including trimethoprim, pyrimethamine, as well as antiparasitic drugs such as pentamidine (Pisciotta, 1990).

C. Flow Cytometry as a Tool to Study Phagocyte Function

Numerous methods allow the analysis of phagocyte function by flow cytometry at the single-cell level. Phagocytosis of bacteria or latex particles can be quantitated per cell by flow cytometry (Steinkamp *et al.*, 1982). Furthermore, cell-bound and internalized pathogens can be discriminated (Bjerknes and Bassoe, 1983). Quantitative studies of bacterial uptake in this context allow the characterization of mechanisms of opsonization (Bassoe and Bjerknes, 1984), and the decay of microbe-associated fluorescence represents a correlate of bacterial degradation (Bassoe and Bjerknes, 1985).

The specific assessment of the functional activation of phagocytic cells became possible when florigenic substrates were induced, which allow the quantitation of the phagocyte-specific generation of reactive oxidants in the oxidative burst response (Bass *et al.*, 1983). Subsequently, probes were induced that can be used to detect the oxidative response to soluble receptor ligands such as *N*-formylated bacterial peptides (Rothe *et al.*, 1988) and that allow the differential analysis of membrane-associated superoxide anion generation and lysosomal peroxidase activity (Rothe *et al.*, 1990).

These tools allow the characterization of receptor-specific cellular activation processes. Thus, cross-linking of receptors by antibodies combined with analysis of calcium fluxes and the oxidative burst response was applied for the characterization of cellular activation processes by glycosyl-phosphatidylinositol (GPI)-anchored surface receptors (Lund-Johansen *et al.*, 1990). Flow cytometry similarly has been used to characterize the antagonism of the acute-phase protein serum amyloid A and fMLP in stimulating the oxidative burst (Linke *et al.*, 1991), which later was identified to be due to binding to the same receptor FPRL1 (He *et al.*, 2003). The interaction of different activation processes such as the priming of the cellular response to fMLP by proinflammatory cytokines (Fig. 3) can also be analyzed through the flow cytometric oxidative burst analysis at the single-cell level (Wittmann *et al.*, 2004). Finally, regulatory processes in the conformational activation of receptor clusters such as the clustering of the LPS receptor CD14 with integrins or TLRs and other receptors mediating signal transduction have been addressed using fluorescence resonance energy transfer (FRET) (Pfeiffer *et al.*, 2001; Zarewych *et al.*, 1996). The anti-inflammatory downregulation of activation processes by IL-10, finally, has been addressed by the analysis of apoptosis (Keel *et al.*, 1997) and has been linked to intracellular recruitment of IL-10 receptors (Elbim *et al.*, 2001).

The methods now allow direct assessment of basal and activated cell-surface receptor expression, measurement of intracellular pH changes by the use of pH-sensitive fluorescent colors, detection of phagocytic activity based on incubation of phagocytic cells with fluorescent-conjugated particles, and FRET, using suitably labeled antibody pairs, to visualize functional receptor clustering.

This chapter focuses on several of the assays that can be used to characterize the graded response of neutrophils and monocytes to activation. These assays can be grouped in four categories as shown in Table I.

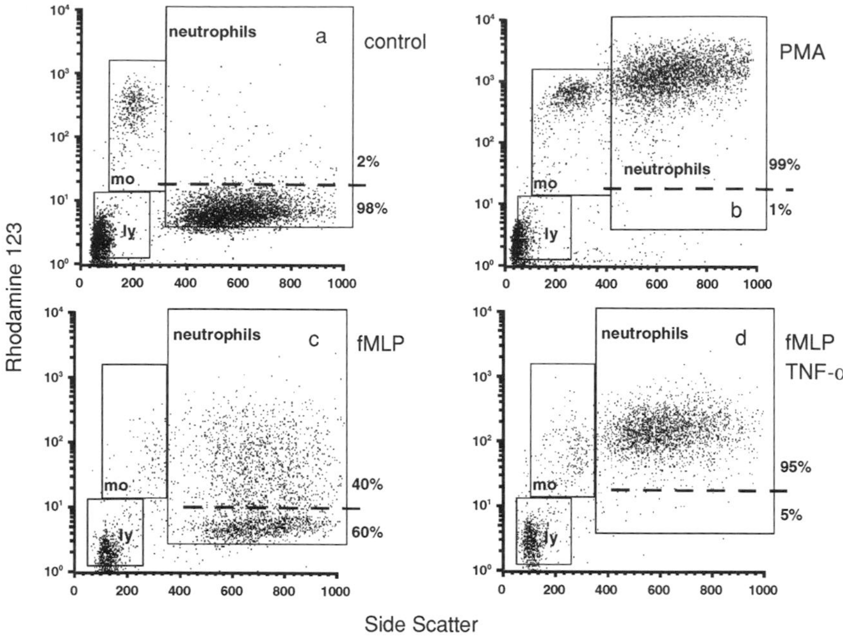

Fig. 3 Oxidative burst response of neutrophils and monocytes (mo) as detected by intracellular oxidation of dihydrorhodamine to rhodamine 123. (A) In the absence of stimulation, only monocytes show rhodamine-123 fluorescence resulting from spontaneous intracellular oxidation of the substrate; (B) PMA induces a homogeneous and maximal response of neutrophils and monocytes, and lymphocytes (ly), which do not express NADPH–oxidase, remain negative; (C) fMLP as a physiological stimulus induces an oxidative burst response in a subpopulation of neutrophils only; (D) preincubation with proinflammatory cytokines such as tumor necrosis factor-α (TNF-α) leads to an enhanced response to stimulation with fMLP.

1. Cytosolic Ca^{2+} Transients

Cytosolic free Ca^{2+} is a universal second messenger, which is increased within seconds of cellular activation (Berridge *et al.*, 2000; da Silva *et al.*, 2000; Lewis, 2001). In resting cells, a low cytosolic free Ca^{2+} concentration is actively maintained at 100–150 nM by a Ca^{2+} adenosine triphosphatase (ATPase), compared to the 10^4-fold higher free Ca^{2+} concentration (1.3 mM) in the extracellular space. Upon surface receptor stimulation, an increase in the cytosolic free Ca^{2+} concentration in nonmuscle cells typically occurs in three phases. First, receptor-coupled PLC activity leads to an initial liberation of calcium from calciosomes, the intracellular calcium stores of nonmuscle cells. In the second phase, an influx of calcium from the extracellular space is induced. This is followed within a few minutes by reuptake of Ca^{2+} by the calciosomes and export through the plasma membrane–bound Ca^{2+} ATPase, leading to the termination of the generalized increase in cytosolic Ca^{2+}. A sustained activation of calcium-dependent cellular responses is then maintained by locally increased calcium fluxes in the submembrane space only.

Table I
Applications of Flow Cytometry in the Study of Neutrophil and Monocyte Activation

Method	Labeling
Analysis of signaling and priming and phagocytic defense cascade	
Ca^{2+} transients	Indo-1, Fluo-3
F-actin generation	phalloidin, phallicidin
Cytosolic alkalinization	SNARF-1
Oxidative burst response	dihydrorhodamine 123, hydroethidine
Lysosomal proteinases	$(Z\text{-Ala-Ala})_2$-R110, $(Z\text{-Arg-Arg})_2$-R110
Analysis of phagocytosis and adhesive phenotype	
Phagocytosis of bacteria	FITC-conjugated bacteria
Expression of adhesion antigen and granule proteins	CD11b, CD62L, CD66b
Characterization of ligands and receptors	
Recruitment of receptors involved in signal transduction	FLPEP, interleukin-10 receptor
Co-assembly of heteromeric receptor complexes	CD14/CD11b, CD14/TLR-4
Autocrine secretion of cytokines	IL-8
Analysis of anti-inflammatory termination of activation	
Induction of apoptosis	annexin V

Ca^{2+} signaling, in general, is regulated by a number of tools (Berridge *et al.*, 2000): (1) Ca^{2+}-mobilizing signals including the intracellular messengers inositol-1,4,5-triphosphate, cyclic adenosine diphosphate (cADP) ribose, nicotinic acid dinucleotide phosphate, and sphingosine-1-phosphate, (2) channels in the plasma membrane or the endoplasmic reticulum membrane, (3) Ca^{2+} sensors, including the ubiquitous Ca^{2+}-binding protein calmodulin and a wide variety of Ca^{2+}-calmodulin-activated proteins, and (4) mechanisms that reduce the free cytoplasmic Ca^{2+} concentration by sequestering/buffering or pumping it out of the cytoplasm.

2. Cytosolic Alkalinization

Intracellular pH is closely regulated in the range of 7.0–7.4 in eukaryotic cells. The regulation of the pH level is tightly related to the function of leukocytes (Hackam *et al.*, 1996) and especially phagocytic cells, which maintain a distinct pH level within their phagocytic organelles dependent on cellular activation (Bernardo *et al.*, 2002; Jankowski *et al.*, 2002). The pH regulation also is closely related to cell survival and apoptosis (Ishaque and Al-Ruebeai, 1998). The most common pH-regulatory mechanism is an amiloride-sensitive, electroneutral Na^+/H^+-antiport, which uses the inward-directed Na^+ gradient for H^+ extrusion. Alkalinization of intracellular pH is induced by upregulation of the antiport after stimulation by hormones or growth factors. In addition, pH-regulatory anion antiport mechanisms such as an $Cl^-/HCO3^-$ antiport seem to be present in most cells. The regulation of the intracellular pH is of major importance for cellular activation as increases in intracellular pH are associated with an increased metabolic activity of cells. In the case of neutrophils, activation of the Na^+/H^+ antiport (e.g., during spreading) is tightly linked to NADPH oxidase activity and the oxidative burst response as the intracellular alkalinization induced by the antiport directly compensates for the acidification induced by the transmembrane transport of electrons during superoxide anion generation (Demaurex *et al.*, 1996).

3. Oxidative Burst Response

Professional phagocytes are well equipped with specific enzyme complexes that raise the production of reactive oxygen species and ozone after ligand-induced assembly in cell membranes (Babior, 2004; Babior *et al.*, 2003; Clark, 1999; Dahlgren and Karlsson, 1999; Forman and Torres, 2002). Reactive oxidants are released into phagosomes and into the extracellular space during the oxidative burst reaction, which contributes to their microbicidal activity. The NADPH oxidase constitutes one of the best characterized phagocyte-specific enzyme complexes, which in resting cells is disassembled in cytosolic and membrane-bound complexes. Upon activation, the enzyme incorporates into the plasma or lysosomal membrane and catalyzes the transmembrane transport of electrons from NADPH to molecular oxygen. Superoxide anion, the product of the one-electron reduction of molecular oxygen, dismutates to hydrogen peroxide, which finally is converted by myeloperoxidase into long-lasting oxidants such as hypochlorous acid or chlorinated tauramines.

$$\mathrm{NADPH} + 2\mathrm{O}_2 \rightarrow \mathrm{NADP}^+ + \mathrm{H}^+ + 2\mathrm{O}_2^-$$

NADPH Oxidase

$$2\mathrm{O}_2^- + 2\mathrm{H}^+ \rightarrow \mathrm{H}_2\mathrm{O}_2 + \mathrm{O}_2$$

Superoxide Dismutase

$$\mathrm{H}_2\mathrm{O}_2 + \mathrm{X}^- + \mathrm{H}^+ \rightarrow \mathrm{HOX} + \mathrm{H}_2\mathrm{O}$$

$$(\mathrm{X}^- = \mathrm{Cl}^-, \mathrm{Br}^-, \mathrm{I}^-, \mathrm{SCN}^-)$$

Myeloperoxidase

Two other oxidants, hydroxyl radical originating either from an iron-catalyzed reaction of O_2^- with H_2O_2, a reaction of O_2^- with HOCl, or a reaction of O_2^- with nitric oxide, and singlet oxygen, generated from a water-induced dismutation of O_2^-, or in a myeloperoxidase-dependent reaction, may also play a role in the respiratory burst reaction. Moreover, activated granulocytes may produce an oxidant with the chemical signature of ozone, a particularly powerful oxidant (Babior *et al.*, 2003). Whether production of hydroxyl radical or singlet oxygen by neutrophils occurs under physiological reactions is unclear.

4. Lysosomal Proteinases

Cellular endopeptidases are composed of four classes of enzymes defined by the action of class-specific inhibitors: (1) serine proteinases inhibitable by diisopropylphosphofluoridate (DFP), (2) cysteine proteinases inhibitable by E-64, (3) aspartic proteinases inhibitable by pepstatin, and (4) metalloproteinases inhibitable by phenanthrolene (Hooper, 2002; Kirschke and Barrett, 1987). Because of a wide substrate specificity, these enzymes are involved in intracellular protein turnover and extracellular protein degradation during inflammation. Cell-lineage–dependent differences in the cellular expression of lysosomal proteases, for example, expression of the serine proteinases elastase and cathepsin G in neutrophils in contrast to expression of the cysteine proteinases cathepsin B and L in monocytes and macrophages, furthermore make them interesting as markers of cellular differentiation.

As a specific function in phagocytic cells, elastase activity is linked to the lysosomal degradation of bacteria in neutrophils, because cells deficient in elastase or cathepsin G are unable to kill bacteria despite an intact oxidative burst response (Reeves *et al.*, 2002). At the same time, extracellular release of elastase is linked to tissue destruction by phagocytes such as in ARDS (Moraes *et al.*, 2003). Cysteine proteases, which are expressed by monocytes and macrophages, play an important role in regulating antigen presentation by both MHC class II and CD1d molecules (Honey and Rudensky, 2003).

5. Analysis of fMLP Binding

N-formylated peptides such as the model tripeptide fMLP can originate from both prokaryotic cells and the mitochondrial proteins of disrupted eukaryotic cells. Their recognition by phagocytes through the formyl peptide receptor and its variant FPRL1, thus, is coupled to the host defense against bacterial infections and the clearance of damaged cells (Le *et al.*, 2002).

Both seven transmembrane domain Gi-protein–coupled receptors are expressed at high levels on neutrophils and monocytes. They activate multiple transduction pathways responsible for various functions such as adhesion, chemotaxis, exocytosis of secretory granules, and superoxide anion production (Dalpiaz *et al.*, 2003). An unresolved question is whether signaling requirements are identical or specific for each physiological function.

6. Co-Assembly of Heteromeric Receptor Complexes in Membrane Microdomains

Cholesterol- and sphingolipid-rich microdomains, also called *rafts*, have been identified as an important regulatory principle in the organization of cell membranes (Simons and Ehehalt, 2002). Such rafts are considered to cluster ligands, receptors, and kinases into small signaling units, and changes in the lateral organization of these platforms are tightly coupled to the process of cell activation. In neutrophils, a translocalization of G-protein subunits, *N*-formyl peptide receptors, Lyn kinase, and phospholipases has been linked to the process of priming (Keil *et al.*, 2003). Also NADPH oxidase has been shown to localize to rafts upon cell activation, and rafts seem to be determinants of the onset of oxidative burst activity (Shao *et al.*, 2003). In macrophages, spatial raft coalescence has been shown to be an early event in Fcγ-receptor signaling (Kono *et al.*, 2002). Glycosyl-phosphatidyl receptors, which lack a transmembrane domain, typically depend in their signaling on receptor co-localization within rafts, and an interaction of the LPS receptor CD14 with integrins has been shown as an early step in the activation of the receptor both in neutrophils (Zarewych *et al.*, 1996) and in monocytes (Pfeiffer *et al.*, 2001). Interestingly, specific patterns of co-association for different ligands binding to the pattern-recognition receptor in monocytes could be identified.

II. Background

The assays are in general based on the analysis of the kinetic response of neutrophils and/or monocytes to stimulation by defined ligands in a physiological environment. A number of criteria have to be met by the methods to obtain physiologically meaningful results:

- Compatibility of staining with whole blood or at least the lack of cell purification
- Potency of ligands for activation of neutrophils and/or monocytes in suspension rather than under adherence only
- Lack of interference of the assay with a physiological activation of the cells

A. Preparation of Cells

Analysis of cells in their physiological environment (i.e., unseparated whole blood) is preferable. Anticoagulants that do not chelate calcium such as heparin do not interfere with calcium-dependent cellular activation processes. Antiproteases, or more specifically thrombin inhibitors, are alternative anticoagulants (McCarthy and Macey, 1996) but are not routinely available for blood drawing.

Cooling and warming of cells leads to upregulation of surface antigens, indicating cellular activation (Forsyth and Levinsky, 1990), and should be avoided. The stability of cell function is limited over time and the maximum transport time that does not interfere with assay results should be determined for each test.

Leukocyte isolation procedures, which depend on a lysis of erythrocytes or on the centrifugation of erythrocytes through density gradients, lead to activation of leukocytes, resulting in a high spontaneous oxidative burst activity and increased responses to cellular stimulation. Whole blood may be depleted of erythrocytes by 1-g sedimentation for 30–40 minutes on Ficoll or dextran (Boyum, 1968). This is followed by withdrawal of a leukocyte-rich supernatant in autologous plasma for further analysis. Alternatively, assays may be performed in whole blood followed by fixation and lysis of erythrocytes before analysis in cases in which fluorescent probes are retained in their cellular localization during the lysis process, for example, by fixation such as for rhodamine 123 or in assays dependent on antibody staining.

Buffers used during incubation of cells should contain physiological amounts of divalent cations similar to cell culture media. Buffering, however, should not depend on carbon dioxide in case of incubation in ambient air. Phagocytic cells tend to adhere to surfaces when incubated at 37 °C at low protein concentrations. Polypropylene tubes are preferable to polystyrene or glass tubes because of a less adhesive surface.

B. Ligands

The bacterial peptide fMLP represents a ligand capable of inducing calcium transients, activation of the Na^+/H^+ antiport leading to cytosolic alkalinization and the oxidative burst response in a dose-dependent manner at concentrations ranging from 10^{-9} to 10^{-5} M. Cytokines such as TNF-α, GM-CSF, and IL-8 can be employed for *in vitro* induction of neutrophil priming similar to the reactivity of neutrophils in inflamed tissues. Phorbol 12-myristate 13-acetate can be used to induce maximum activation of neutrophils or monocytes through direct activation of PKC.

C. Analysis of Ca^{2+} Transients

The cytosolic free calcium concentration can be measured intracellularly by the spectral shift of indo-1 from green fluorescence in the absence of Ca^{2+} to blue fluorescence in the presence of Ca^{2+}. This allows precise determination of the calcium concentration independently of the cell size and cellular dye loading. First, indo-1 is accumulated intracellularly through intracellular cleavage of the membrane-permeable derivative indo-1/acetoxymethyl ester (AM) by esterases. The cellular blue/green fluorescence ratio is then analyzed as a measure of the intracellular free calcium concentration. A molar calibration curve can be obtained from the analysis of cells incubated in buffers of defined Ca^{2+} concentration in the presence of the Ca^{2+} ionophore ionomycin. Intracellular buffering of Ca^{2+} may lead to difficulties in calibrating concentrations of Ca^{2+} lower than the 100–150 nM typically found in resting cells. In addition, protein interaction can modify the dissociation constant (Kd) of Ca^{2+} chelating fluorophores. The spectral response of the flow cytometer when flushed with indo-1 and buffers of known Ca^{2+} concentrations, therefore, also may be recorded to compare the spectral Ca^{2+} calibration curve of the flow cytometer with the Ca^{2+} calibration curve of the cells.

The analysis of the kinetic response of intracellular Ca^{2+} to stimulation is achieved in three steps (do Céu Monteiro *et al.*, 1999; Réthi *et al.*, 2002). First, cells loaded with indo-1 are analyzed at 37 °C without stimulation. Then a stimulus is added, and the increase and decline in intracellular Ca^{2+} concentration are measured kinetically until a stable level is reached again. This typically occurs in 2–5 minutes. Finally, ionomycin is added to record the maximum indo-1 ratio in the presence of Ca^{2+} as a positive control.

If fluorescence excitation at 365 nm is not possible, for example, using argon laser–based instruments, which allow excitation at 488 nm only, the Ca^{2+}-sensitive fluorescein derivative fluo-3 may be used alternatively. Binding of Ca^{2+} does not change the fluorescence emission spectrum of fluo-3 but increases the green fluorescence intensity by 40-fold (Minta *et al.*, 1989) or more depending on the source of the reagent. This allows sensitive detection of the kinetically fast relative increases in cytosolic free Ca^{2+} after stimulation. Absolute Ca^{2+} concentrations, however, are difficult to measure with fluo-3 because the cellular fluorescence intensity is substantially affected by differences in the extent of enzymatic dye loading, cell size, compartmentalization of the dye, and effects of protein binding on the Kd of the probe.

D. Cytosolic Alkalinization

The cytoplasmic pH value can be measured intracellularly through the spectral shift of the 488-nm excitable SNARF-1 to a longer wavelength fluorescence emission with increasing pH value. In the first step, SNARF-1 is accumulated intracellularly by intracellular cleavage of the membrane-permeable derivative SNARF-1/AM by esterases. The cellular red/orange (SNARF-1) fluorescence ratio is then analyzed as a measure of the intracellular pH value. A pH calibration curve can be obtained from the analysis of cells incubated in buffers of defined pH value with equilibration of the intracellular pH by addition of the ionophore nigericin in the presence of a high K^+ concentration (Thomas *et al.*, 1979).

In addition to differences in resting intracellular pH value between cells, the response of the intracellular pH to stimulation may be analyzed. The stimulation-dependent activation of the Na^+/H^+ antiport typically leads to prolonged cytoplasmic alkalinization. This response can be analyzed by addition of the stimulus during the SNARF-1 loading procedure at 37 °C.

E. Oxidative Burst Response

Formation of the reactive oxidants superoxide anion and hydrogen peroxide during the oxidative burst is measured intracellularly by the oxidation of membrane-permeable florigenic substrates such as DHR or hydroethidine (HE) to fluorescent products. The intracellular oxidation of nonfluorescent DHR to green fluorescent rhodamine 123 by hydrogen peroxide and peroxidases is the most sensitive method for the analysis of the oxidative burst response (Lund-Johansen *et al.*, 1990; Rothe *et al.*, 1988, 1991; Vowells *et al.*, 1995). This high sensitivity is reached by the intracellular retention of the positively charged fluorescent product rhodamine 123 at mitochondrial binding sites. The blue fluorescent HE, in contrast to DHR, can be oxidized to red fluorescent ethidium bromide (EB) already by superoxide anion, the primary product of the oxidative burst response (Rothe *et al.*, 1990; Walrand *et al.*, 2003). Cells are first preincubated with the florigenic substrates at 37 °C. An oxidative burst response is then induced by addition of stimulatory ligands such as fMLP with or without response modifiers such as TNF-α. PMA may be used as a positive control or for the analysis of the cellular expression of the oxidative burst enzymes. The incubation is stopped by transferring cells onto ice. The cellular accumulation of the fluorescent oxidation products is then analyzed on the flow cytometer. Because the cellular fluorescence is stable for about 90 minutes on ice, cells may be subsequently labeled with monoclonal antibodies (mAbs). Fixation with paraformaldehyde retains the cellular localization of the fluorescent product.

F. Lysosomal Proteinases

Lysosomal proteinase activity can be analyzed intracellularly by the conversion of nonfluorescent and membrane-permeable specifically bis-substituted R110-peptide derivatives to green fluorescent monosubstituted R110 and free R110 in a two-step reaction (Leytus *et al.*, 1983a,b; Rothe *et al.*, 1992). Cells are incubated kinetically at 37 °C with appropriate substrates such as (Z-Arg-Arg)$_2$-R110 for the cysteine proteinases cathepsin B and L of monocytes or (Z-Ala-Ala)$_2$-R110 for the serine proteinase elastase of neutrophils. The specificity of the reaction is shown by the inhibition of the reaction through preincubation with specific inhibitors such as Z-Phe-Ala-CHN$_2$ for cathepsin B and L (Green and Shaw, 1981) or DFP for neutrophil elastase (Powers, 1986). Because the cellular fluorescence is stable for about 90 minutes on ice, cells may be subsequently labeled with mAbs.

G. Analysis of fMLP Binding

Quantification of binding sites for fMLP can be performed by staining with the fluoresceinated peptide formyl-Nle-Leu-Phe-Nle-Tyr-Lys (FLPEP) (Sklar *et al.*, 1981). The binding of FLPEP is quantitated by flow cytometry after incubation at 4 °C to avoid cellular stimulation (Allen, 1992). Nonspecific binding is assessed by incubation with an excess of fMLP in parallel.

H. Co-Assembly of Heteromeric Receptor Complexes

Receptor co-localization can be assessed by flow cytometry using fluorescence resonance energy transfer (FRET) for the determination of the proximity of antibodies directed at different components of a heteromeric receptor clusters (see chapter 5). In short, in case of R-phycoerythrin (PE) as the donor and Cy5 as the acceptor, spatially resolved dual-laser excitation at 488 and 635 nm is used for the independent quantification of both fluorochromes. Energy transfer is calculated as described in chapter 5 based on the determination of the increased Cy5-specific fluorescence excited at 488 nm after simultaneous staining with donor and acceptor in comparison to the fluorescence of donor-only or acceptor-only stained samples.

A change in the spatial distance to CD11b can be used for the determination of the activation of the pattern-recognition receptor CD14 by LPS or one of its alternative bacterial or inflammatory ligands (Pfeiffer *et al.*, 2001; Zarewych *et al.*, 1996). Pairs of antibodies to the same pairs of proteins, such as CD14 and CD11b, differ in their efficiency for showing FRET because of the relative proximity of epitopes. Indirect staining can improve transfer efficiency because of an increased radius of fluorochromes around epitopes (Koksch *et al.*, 1995).

III. Methods

A. Preparation of Cells

1. Isolation of Leukocytes through Sedimentation of Erythrocytes on Ficoll

Blood is depleted of erythrocytes by a method that avoids contact of leukocytes to any separation media. Undiluted heparinized (10 U/ml) whole blood (3 ml) is layered on top of 3 ml of lymphocyte separation medium (Ficoll, density 1.077). Erythrocytes will aggregate at the interface and sediment at room temperature without centrifugation. After 40 minutes, the upper 800 μl of supernatant plasma is withdrawn, avoiding contact to the plasma fraction near the interface to the separation medium and is stored on ice. This will contain platelets and approximately 2×10^7/ml of unseparated leukocytes in autologous plasma.

B. Analysis of Ca^{2+} Transients

1. Measurement of Cytosolic Free Calcium with Indo-1

1. Incubate 5×10^6 cells/ml in HEPES-buffered medium with 0.5–5 μM indo-1/AM at 37 °C. An equilibrium of dye loading is reached within 20–30 minutes of incubation. Depending on the cell type, select the lowest concentration of substrate that results in a homogeneous fluorescence ratio before stimulation.
2. Add propidium iodide (PI) at 30 μM final concentration and analyze the blue fluorescence of indo-1/Ca^{2+} complexes using a 390- to 440-nm bandpass filter, and the blue-green fluorescence of Ca^{2+}-free indo-1 and the red PI fluorescence of dead cells using a 490-nm long-pass filter with excitation at 351/363 nm by an argon laser or a high-pressure mercury arc lamp with a 360- to 370-nm bandpass filter.
3. For the analysis of stimulation-dependent changes in intracellular Ca^{2+}, add an appropriate stimulus, for example, fMLP (10^{-8} M) in the case of neutrophils,

after an initial measurement for 30 seconds. Continue kinetic data acquisition for approximately 3 minutes.

4. Add ionomycin (2 μM) as a positive control for the maximum Ca^{2+} response of the cells.

2. Measurement of Cytosolic Free Calcium with Fluo-3

1. Incubate 5 $\times$ 10^6 cells/ml in HEPES-buffered medium with 0.5–2.0 μM fluo-3/AM at 37 °C. An equilibrium of dye loading is reached within 20–30 minutes of incubation. Depending on the cell type, select the lowest concentration of substrate that results in a homogeneous fluorescence before stimulation.

2. Add PI, and at 30 μM final concentration, analyze the green fluorescence of fluo-3 using a 515- to 535-nm bandpass filter, and the red PI fluorescence of dead cells using a 620-nm long-pass filter with excitation by the 488-nm line of an argon laser.

3. For the analysis of stimulation-dependent changes in intracellular Ca^{2+}, add an appropriate stimulus, for example, fMLP (10^{-8} M) in the case of neutrophils, after an initial measurement for 30 seconds. Continue kinetic data acquisition for approximately 3 minutes.

4. Add ionomycin (2 μM) as a positive control for the maximum Ca^{2+} response of the cells.

C. Cytosolic Alkalinization

1. Incubate 5 $\times$ 10^6 cells/ml in HEPES-buffered medium with 0.2–1.0 μM SNARF-1/AM at 37 °C. An equilibrium of dye loading is reached within 20–30 minutes of incubation. Depending on the cell type, select the lowest concentration of substrate that results in a homogeneous fluorescence ratio in resting cells.

2. For the analysis of stimulation-dependent changes in intracellular pH, incubate for the final 15 minutes in the presence of appropriate stimuli, for example, fMLP (10^{-8} M) or TNF-α (1 ng/ml) in the case of neutrophils.

3. Add PI at 30 μM final concentration and analyze the orange fluorescence of acidic SNARF-1 using a 575- to 595-nm bandpass filter, and the red fluorescence of basic SNARF-1 and the PI fluorescence of dead cells using a 620-nm long-pass filter with excitation by the 488-nm line of an argon laser or a high-pressure mercury arc lamp with a 470- to 500-nm bandpass filter.

4. For the generation of a calibration curve, incubate cells with SNARF-1/AM as described earlier. Divide the sample into six aliquots and spin down at 60 g and 4 °C for 5 minutes. Resuspend for 5 minutes at room temperature in the high K^+ calibration buffers (pH 6.40, 6.80, 7.20, 7.60, 8.00, and 8.40) supplemented with 10 μM of nigericin followed by flow cytometric analysis.

D. Oxidative Burst Response

1. Incubate 5 × 10^6 cells/ml in HEPES-buffered medium with 1 μM DHR (stock 1 mM in *N*,*N*-dimethylformamide [DMF]) or 10 μM HE (stock 10 mM in DMF) at 37 °C for 5 minutes.

2. Incubate for 15–30 minutes with appropriate stimuli, for example, fMLP (10^{-6} M) or PMA (10^{-7} M) in the case of neutrophils. The cellular "priming" by incomplete stimuli such as TNF-α (10 ng/ml) may be analyzed by 5 minutes preincubation with the cytokine, followed by incubation for 15 minutes in a low concentration of fMLP (10^{-7} M).

3. Add PI at 30 μM final concentration and analyze the green fluorescence of rhodamine 123 using a 515- to 535-nm bandpass filter, and the PI red fluorescence of dead cells or the EB red fluorescence of HE-stained cells using a 620-nm long-pass filter using excitation with the 488-nm line of an argon laser.

E. Lysosomal Proteinases

1. Preincubate 5 × 10^6 cells/ml in HEPES-buffered medium with or without specific proteinase inhibitor such as 100 μM Z-Phe-Ala-CHN_2 (stock 100 mM in DMSO) or 1 mM DFP (stock 1 M in DMSO) at 37 °C for 10 minutes.

2. Incubate for 20 minutes with appropriate proteinase substrate such as 4 μM (Z-Arg-Arg)2-R110 for the cathepsin B/L of monocytes or 4 μM (Z-Ala-Ala) 2-R110 for the elastase of neutrophils.

3. Add PI at 30 μM final concentration and analyze the green fluorescence of R110 using a 515- to 535-nm bandpass filter, and the red PI fluorescence of dead cells using a 620-nm long-pass filter using excitation with 488-nm line of an argon laser.

F. Analysis of fMLP Binding

1. Preincubate 5 × 10^6 cells/ml in HEPES-buffered medium with or without specific agonists for 30 minutes at 37 °C for the study of the upregulation of binding sites. Stop the incubation by cooling to 4 °C.

2. Incubate for 20 minutes with 20 nM FLPEP and wash twice. Co-incubate with fMLP (10^{-4} M) in a parallel set of tubes for determination of nonspecific binding of FLPEP.

3. Analyze the green fluorescence of FLPEP using a 515- to 535-nm bandpass filter and excitation with 488-nm line of an argon laser.

G. Co-Assembly of Heteromeric Receptor Complexes

1. Preincubate whole blood with or without specific agonists, such as LPS from *Salmonellae minnesota*, for 15 minutes at 37 °C. Wash cells with cold PBS containing 0.1% NaN_3.

2. Incubate the washed whole-blood samples for 15 minutes at 4 °C with saturating concentrations of the fluorochrome-conjugated or biotinylated antibodies, such as CD11b (clone D12) as a R-PE conjugate and CD14 (clones ×8) in biotinylated form. A separate sample is incubated with the biotinylated antibody only. Lyse the cells with a one-step fix and lyse solution and wash twice.

3. Divide the samples that have been stained with both antibodies into two parts and stain one part of each sample with a saturating amount of streptavidin–Cy5 for 15 minutes on ice.

4. Wash the cells and analyze the cells using spatially resolved dual-laser excitation at 488 and 635 nm. For all samples, independent of staining, the orange fluorescence of R-PE is analyzed with a bandpass filter centered at 585 nm with 488-nm excitation and the red fluorescence of Cy5 is analyzed at more than 650 nm with separate excitation by both lasers.

5. Energy transfer is calculated separately both as the decrease in donor fluorescence and as the increase in donor-dependent and 488-nm excited acceptor fluorescence as described in detail elsewhere (see chapter 5) (Pfeiffer *et al.*, 2001).

IV. Results

The analysis of the oxidative burst response can serve as a typical result for the activation-dependent characterization of neutrophil function. As shown in Fig. 3A, neutrophils do not show intracellular oxidation of dihydrorhodamine 123 to rhodamine 123 in the absence of stimulation, as indicated by a similar fluorescence to lymphocytes that do not express NADPH oxidase. Monocytes, however, typically show a significant oxidative burst activity representing spontaneous generation of oxidants. Stimulation with PMA induces a more than 100-fold and homogenous increase of rhodamine 123 fluorescence in neutrophils and a more than threefold increase of fluorescence in monocytes, indicating a maximal induction of the respiratory burst response.

fMLP as a physiological agonist of neutrophil activation when applied at a low concentration (e.g., 10^{-7} M) typically induces an oxidative burst response in a subpopulation of neutrophils, only (Fig. 3C). Increasing the amount of fMLP (e.g., incubation with 10^{-6} M fMLP) results in an increase in the size of the population of reactive neutrophils, indicating that the cellular heterogeneity corresponds to the graded induction of an all-or-none response in all cells rather than to a constitutive heterogeneity of cells. Preincubation with an "incomplete" stimulus (e.g., 1 ng/ml of TNF-α, which does not directly induce an oxidative burst response in cells in suspension) results in a significantly enhanced response to fMLP stimulation (Fig. 3D), indicating "priming" of neutrophils by the cytokine (Fig. 2).

A dose-dependent priming of the oxidative burst response of neutrophils is obtained for priming with TNF-α, GM-CSF, or IL-8 (Wittmann *et al.*, 2004).

FMLP receptor expression and an intracellular alkalinization are induced with a similar potency by these cytokines and can serve as a direct indicator of the "priming" of cells without a secondary stimulation, for example, by fMLP.

Transient increases of cytosolic Ca^{2+} concentrations are typically induced by lower concentrations of fMLP (e.g., 10^{-9} M), compared to those necessary for the induction of an oxidative burst response. The assay is particularly useful for the determination of the ability of cells to react to a given ligand or the cross-linking of receptors (Lund-Johansen *et al.*, 1990, 1993) but can hardly be used for quantitative comparison of the potency of ligands.

The determination of intracellular elastase activity in neutrophils and of cathepsin B activity in monocytes or macrophages allows the identification of an altered maturation and differentiation of cells (e.g., in inflammatory disorders or neoplastic diseases of the hematopoietic systems). Increased cathepsin B activity is closely related to the *in vivo* activation of monocytes. Degranulation of cells, which is observed *in vivo* at sites of inflammation, however, is hardly observed in cells circulating in peripheral blood, suggesting that this process is linked to cellular adhesion.

A dose-dependent clustering of CD14 and CD11b *in vitro* occurs in monocytes after stimulation with LPS or alternative ligands of CD14 such as lipoteichoic acid (Pfeiffer *et al.*, 2001). Stimulation with agonists such as fMLP or PMA, which differ in their mechanism of cell activation, does not result in spatial proximity of the receptors, as detected by FRET analysis. Spontaneous clustering of CD14 and CD11b can be observed in the *ex vivo* analysis of blood from patients with inflammatory conditions.

V. Pitfalls and Misinterpretation of the Data

Artificial activation of cells as indicated by a spontaneous oxidative burst response and nonspecific staining, which occurs through redistribution of fluorescent products after incubation, are the most critical problems in the functional analysis of phagocytic cells. Both problems can be addressed by undelayed processing of fresh samples and timely analysis of cells after staining.

A. Analysis of Ca^{2+} Transients

- If only low cellular accumulation of fluorescent indo-1 is reached, dye loading may be improved by addition of the nonionic dispersing agent Pluronic F-127 (Poenie *et al.*, 1986).
- Low changes in the cellular indo-1 fluorescence ratio even after addition of ionomycin may be due to incomplete hydrolysis of indo-1/AM. This may occur in some cell types resulting in Ca^{2+}-insensitive fluorescence. A more complete hydrolysis may be reached by washing the cells and incubating for a prolonged time in a substrate-free medium before analysis.

• A lack of a cellular response to stimulation despite a normal response after addition of ionomycin may be due to a low Ca^{2+} concentration extracellularly, for example, when media such as RPMI 1640 are used, or due to intracellular buffering of Ca^{2+} by the fluorescent indicator. This can occur as a result of high intracellular concentrations of the dyes in the micromolar range compared the 100-fold lower Ca^{2+} concentration and the low kilodalton values of the dyes for Ca^{2+}, for example, 250 nM in the case of indo-1. Therefore, the lowest substrate concentration that results in a homogeneous fluorescence ratio of resting cells should be used. No interaction of the dye loading procedure with the cellular sensitivity for stimulation may be shown by independent cellular responses such as chemotaxis, depolarization of the membrane potential, or an oxidative burst response in phagocytic cells.

B. Cytosolic Alkalinization

Cytosol has a high buffering capacity for intracellular pH. Therefore, the additional buffering of the intracellular pH value by the intracellular indicator usually does not significantly affect the pH measurement.

A lack of a cellular response to stimulation despite pH-sensitive cellular fluorescence as observed by calibration with nigericin may be due to the simultaneous activation of H^+ extruding antiports and metabolic activity, which leads to the intracellular generation of H^+. Na^+/H^+ antiport activity can be shown in this case by pH changes after preincubation with the specific Na^+/H^+ antiport inhibitor amiloride.

C. Oxidative Burst Response

• Only low cellular responses to stimulation may be due to the presence of high (>0.1% [v/v]) concentrations of organic solvents such as dimethylsulfoxide (DMSO) or DMF used in the reagent stock solutions, which both act as scavengers for reactive oxidants and interfere with cellular stimulation.

• High fluorescence in nonphagocytic cells such as lymphocytes indicates spontaneous oxidation of the florigenic substrates, which are unstable when exposed to excessive light or stored at higher temperature.

• Specificity of the intracellular accumulation of the fluorescent products for the oxidative burst response may be shown by preincubating the cells with 100 μM of diphenyl iodonium as a specific inhibitor of the NADPH oxidase of phagocytic cells (Cross, 1987). Hydrogen peroxide (1 mM) may be used as a positive control to show sensitivity of the substrates for intracellular oxidation.

• Cellular staining is usually stable for at least 2 hours at 4 °C because of the intracellular trapping of rhodamine 123 at mitochondrial binding sites. If prolonged stability of the assay samples is required, fixation with 1% paraformaldehyde (w/v) in PBS may be performed to maintain cellular staining.

D. Lysosomal Proteinases

• Selective cell death occurring in samples incubated with the proteinase substrates but not in samples preincubated with the inhibitors before incubation with the proteinase substrates may be a result of the local accumulation of high amounts of the product R110 inside cellular lysosomes. This can be avoided by reducing the substrate concentration or time of incubation.

• The susceptibility of florigenic peptide substrates for cleavage by different enzymes typically strongly depends on the precise assay conditions such as pH or buffer, as well as the fluorophore (Assfalg-Machleidt *et al.*, 1992). It is, therefore, critical to prove the specificity of the intracellular cleavage of substrates using, for example, specific cell-permeant inhibitors or cellular models that selectively do or do not express a specific enzymes such as transfected cells.

E. Co-Assembly of Heteromeric Receptor Complexes

A significant energy transfer between antibodies directed against CD14 and CD11b when analyzing blood samples obtained from healthy individuals in the absence of stimulation indicates artefactual activation of cells. The time after sample drawing and contamination of buffers are the most critical components of the assay.

VI. Comparison with Other Methods

The intracellular oxidation of 2′,7′-dichlorofluorescin (DCFH) to green fluorescent 2′,7′-dichlorofluorescein was used in an earlier flow cytometric assay for the oxidative burst response of phagocytic cells (Bass *et al.*, 1983). DCFH oxidation in comparison to the intracellular oxidation of DHR results in a far lower cellular fluorescence of stimulated phagocytic cells, making the analysis of cellular responses to low amounts of physiological stimuli impossible (Rothe *et al.*, 1991). The assay is further complicated by a high background fluorescence in cells that are not capable of oxidative burst activity, such as lymphocytes, and the need to accumulate the florigenic substrate in a first step through a hydrolytic intracellular cleavage of the membrane-permeable derivative 2′,7′-dichlorofluorescein diacetate. In addition to H_2O_2 and peroxidases, further mechanisms such as reactive nitrogen species are involved in DCFH oxidation (Walrand *et al.*, 2003).

In the analysis of phagocytosis by neutrophils or monocytes, flow cytometry can be used to assess the attachment and internalization of ligands depending on the result of their opsonization as well as secondary reactions such as the phagosomal pH and the oxidative burst response induced. These techniques are reviewed by Lehmann *et al.* (2000). As a major drawback, phagocytosis is a rather strong stimulus for cells that can hardly be applied in a graded manner. Defects in intracellular killing of microbes are not specifically addressed by the available methodologies. These techniques are, therefore, not very sensitive to detect changes of phagocyte function in disease.

VII. Applications and Biomedical Information

A. Applications in Identification of Carriers in Primary Immunodeficiency Screening

Although the role of flow cytometry in the diagnosis of primary immunodeficiencies is undisputed, only few reports address the potential role in detecting heterozygous carriers in affected families. One important example is CGD, which is caused by different mutations in at least five components of the phagocytic NADPH oxidase. The most common mutations, such as the 25% affecting the NCF1 or $p47^{phox}$ gene, causing complete deletions may be addressed by sequencing-based methods (Dekker *et al.*, 2001). However, the majority of the CGD-causing mutations cannot be addressed by gene-scan methods. By contrast, flow cytometric methods determining the phagocyte oxidative burst in response to phagocytosis of bacteria or activation by PMA are suitable as a very sensitive diagnostic screening. (Emmendörffer *et al.*, 1994). Flow cytometry, using dihydrorhodamine 123 oxidation, clearly allows to distinguish heterozygous individuals with one NCF1 gene (carriers) from controls and from NCF1-deficient patients (Atkinson *et al.*, 1997; Crockard *et al.*, 1997; Emmendörffer *et al.*, 1994). Moreover, flow cytometry permits the detection of novel defects of phagocyte function before molecular identification. Thus, flow cytometry may serve as an adjunct for genetic counseling, particularly in sisters of affected patients.

B. Application in Inflammation and Sepsis

Professional phagocytes in the circulation respond weakly to many external ligands. Under conditions of stress, particularly in inflammation, cytokines may prime phagocytes to dramatically increase their responsiveness to bacterial ligands (Figs. 2 and 3). Flow cytometry allowed the characterization of cytokine priming of phagocytes to fMLP-induced oxidative burst and actin polymerization in the presence of rapid changes of fMLP receptor cycling (Elbim *et al.*, 1993; Wittmann *et al.*, 2004). In sepsis and SIRS, flow cytometry may be a tool for the identification of individuals with a hyperresponsive inflammation that may profit from immunomodulatory therapy. Because a dysregulation of fMLP-induced L-selectin shedding and oxidative burst in cytokine-primed granulocytes of asymptomatic patients with HIV could be demonstrated, flow cytometric follow-up may contribute to the prediction of the susceptibility to infections (Elbim *et al.*, 1994).

A central feature of flow cytometry in the characterization of the inflammatory transition from resting over an intermediate primed state with restrained neutrophil activity to full-blown activation of neutrophils is the capability to identify regulatory Ca^{2+} fluxes (Hallett and Lloyds, 1995; Hallett, 2003), as well as actin polymerization, intracellular alkalinization, upregulation of fMLP receptors, and the heterogeneity of the cytokine primed oxidative burst response in a closely correlated way in the same cell (Rothe and Valet, 1994; Wittmann *et al.*, 2004; Yamashiro *et al.*, 2001). This has already led to the identification of a number of mechanisms involved in the altered function of phagocytic cells in inflammation (Fig. 2) and should be of value to identify potential mechanisms for the modulation of the altered function of these cells.

C. Immunopharmacological Applications

Several drugs compromise phagocyte function, including such different chemical compounds as general and inhalational anesthetics, antibiotics, chemotherapeutics, or antiparasitic drugs, increase the risk of infections, or counteract the beneficial effect of antimicrobials. By using flow cytometry, a direct dose-dependent reduction of complement or fMLP-induced oxidative response of neutrophils in patients during general anesthesia nitrous oxide could be demonstrated (Fröhlich *et al.*, 1997). The mechanism leading to functional depression of granulocytes differed with the chemical compounds and involved inhibition of cellular signaling upstream from protein kinase C (Fröhlich *et al.*, 2002). In fact, distinctive inhibition of the neutrophil oxidative function was shown for different anesthetic substances (Fröhlich *et al.*, 2002). Because suppression of granulocyte function after general anesthesia may contribute to the postoperative risk of infection, rapid determination of the extent of the reduction of the oxidative response by flow cytometry may add to optimized and individualized patient care.

D. Application for Characterization of Ligands

Ligand-specific interaction with professional phagocytes and with platelets may affect distinct functional activation profiles, including increased calcium fluxes in the absence of an oxidative burst or conformational changes of receptors without general platelet degranulation. Discriminatory flow cytometric analyses allowed the identification of selective induction of cytoplasmic calcium fluxes and oxidative burst by cross-linking of GPI-anchored receptors using anti-CD14 or anti-CD55 in unprimed myeloid cells, whereas cross-linking of a number of non-GPI–anchored antigens (e.g., CD11a, CD18, and CD45) had no effect (Lund-Johansen *et al.*, 1993). Similarly, in platelets, a selective alteration of the resting and the activated αIIbβ3 (GPIIb/IIIa) receptor conformation in the absence of general platelet activation by the GPIIb/IIIa antagonist tirofiban could be visualized by flow cytometry (Barlage *et al.*, 2002). Moreover, ligand-specific activation mechanisms based on distinctive recruitment of co-receptors (e.g., TLR-2 and TLR-4) were identified using FRET, which allowed the characterization of ligand-specific clusters of pattern-recognition receptors (Pfeiffer *et al.*, 2001). This may be coupled to the analysis of alterations in the spatial organization of membranes into rafts and the lipid composition and microfluidity of membranes in general as determinants of receptor mobility (Triantafilou *et al.*,

2002). Thus, flow cytometry using receptor cross-linking, clustering, or conformational changes may serve to characterize mechanisms of receptor- and ligand-specific signal transduction in conditions of altered phagocyte reactivity in disease and during treatment with drugs and specific inhibitors.

VIII. Future Directions

Flow cytometry is increasingly used to assess the functional status of leukocytes and leukocyte subpopulations, including phagocytes. Going further from multiparameter immunophenotyping, cell function–based flow cytometry has provided many new insights into the ligand-specific functional receptor clustering, intracellular processes, such as actin polymerization, cytokine production, and protein phosphorylation. Direct visualization and quantification of distinct phagocytic cells allows an estimation of the quality of the innate immune response. Because all steps of the phagocytosis cascade are accessible to flow cytometric study, the method is particularly suited to assess the functional status of phagocytes in various human diseases. Flow cytometry may serve not only as a simple and distinguished method to control therapeutic regimens addressed at stimulating phagocytosis, but also to monitor therapies with potential adverse effects on phagocytes.

Besides the ultimate diagnosis of primary phagocyte defects, the monitoring of secondary deficiencies of phagocyte function during chronic and acute diseases, including diabetes or SIRS/sepsis, as well as the control in the course of phagocyte-suppressive therapies provides an important field of application of diagnostic flow cytometry. Particularly for the prediction of the risk of infections associated with chronic diseases, such as diabetes, flow cytometry may gain additional impact. Similarly, depressed phagocytosis in immunosuppressive infectious diseases, particularly with AIDS, predisposes to opportunistic infections. Here, flow cytometry could not only contribute to the assessment of the remaining phagocyte function but also serve to monitor the efficiency of therapeutic measures to reactivate phagocyte activity with interferon-γ (IFN-γ) or GM-CSF. Similarly, the functional capacity of transfused granulocyte concentrates in neutropenic patients may gain importance in predicting the therapeutic outcome. Further potentially very useful applications include testing of adverse effects of antibiotics on phagocyte function, which may counteract the effectiveness of antimicrobial treatment, thus predicting *in vivo* treatment failure.

References

Aderem, A., and Underhill, D.M. (1999). Mechanisms of phagocytosis in macrophages. *Annu. Rev. Immunol.* **17,** 593–623.

Allen, C. A., Broom, M. F., and Chadwick, V. S. (1992). Flow cytometry analysis of the expression of neutrophil fMLP receptors. *J. Immunol. Methods* **149,** 159–164.

Allen, L. A., and Aderem, A. (1996). Molecular definition of distinct cytoskeletal structures involved in complement- and Fc receptor–mediated phagocytosis in macrophages. *J. Exp. Med.* **184,** 627–637.

Assfalg-Machleidt, I., Rothe, G., Klingel, S., Banati, R., Mangel, W. F., Valet, G., and Machleidt, W. (1992). Sensitive determination of cathepsin L activity in the presence of cathepsin B using rhodamine-based fluorogenic substrates. *Biol. Chem. Hoppe-Seyler* **373,** 433–440.

Atkinson, T. P., Bonitatibus, G. M., and Berkow, R. L. (1997). Chronic granulomatous disease in two children with recurrent infections: Family studies using dihydrorhodamine-based flow cytometry. *J. Pediatr.* **130,** 488–491.

Babior, B. M. (2000). Phagocytes and oxidative stress. *Am. J. Med.* **109,** 33–44.

Babior, B. M. (2004). NADPH oxidase. *Curr. Opin. Immunol.* **16,** 42–47.

Babior, B. M., Takeuchi, C., Ruedi, J., Gutierrez, A., and Wentworth, P., Jr. (2003). Investigating antibody-catalyzed ozone generation by human neutrophils. *Proc. Natl. Acad. Sci. USA* **100,** 3031–3034.

Barlage, S., Wimmer, A., Pfeiffer, A., Rothe, G., and Schmitz, G. (2002). MK-383 (tirofiban) induces a GPIIb/IIIa receptor conformation which differs from the resting and activated receptor. *Platelets* **13,** 133–140.

Bass, D. A., Parce, J. W., Dechatelet, L. R., Szejda, P., Seeds, M. C., and Thomas, M. (1983). Flow cytometric studies of oxidative product formation by neutrophils: A graded response to membrane stimulation. *J. Immunol.* **130,** 1910–1917.

Bassoe, C. F., and Bjerknes, R. (1985). Phagocytosis by human leukocytes, phagosomal pH and degradation of seven species of bacteria measured by flow cytometry. *J. Med. Microbiol.* **19,** 115–125.

Bassoe, C. F., and Bjerknes, R. (1984). The effect of serum opsonins on the phagocytosis of *Staphylococcus aureus* and zymosan particles, measured by flow cytometry. *Acta Pathol. Microbiol. Immunol. Scand. C* **92,** 51–58.

Bernardo, J., Hartlaub, H., Yu, X., Long, H., and Simons, E. R. (2002). Immune complex stimulation of human neutrophils involves a novel $Ca^{++}/H+$ exchanger that participates in the regulation of cytoplasmic pH: Flow cytometric analysis of Ca^{++}/pH responses by subpopulations. *J. Leukocyte Biol.* **72,** 1172–1179.

Berridge, M. J., Lipp, P., and Bootman, M. D. (2000). The versatility and universality of calcium signaling. *Nat. Rev. Mol. Cell. Biol.* **1,** 11–21.

Bjerknes, R., and Bassoe, C. F. (1983). Human leukocyte phagocytosis of zymosan particles measured by flow cytometry. *Acta Pathol. Microbiol. Immunol. Scand. C* **91,** 341–348.

Boyum, A. (1968). Isolation of leucocytes from human blood. Further observations. Methylcellulose, dextran, and ficoll as erythrocyte aggregating agents. *Scand. J. Clin. Lab. Invest. Suppl.* **97,** 31–50.

Clark, R. A. (1999). Activation of the neutrophil respiratory burst oxidase. *J Infect. Dis.* **179**(Suppl. 2), S309–S317.

Crockard, A. D., Thompson, J. M., Boyd, N. A., Haughton, D. J., McCluskey, D. R., and Turner, C. P. (1997). Diagnosis and carrier detection of chronic granulomatous disease in five families by flow cytometry. *Int. Arch. Allergy Immunol.* **114,** 144–152.

Cross, A. R. (1987). The inhibitory effects of some iodonium compounds on the superoxide generating system of neutrophils and their failure to inhibit diaphorase activity. *Biochem. Pharmacol.* **36,** 489–493.

da Silva, C. P., and Guse, A. H. (2000). Intracellular Ca^{++} release mechanisms: Multiple pathways having multiple functions within the same cell type? *Biochim. Biophys. Acta* **1498,** 122–133.

Dahlgren, C., and Karlsson, A. (1999). Respiratory burst in human neutrophils. *J. Immunol. Methods* **232,** 3–14.

Dalpiaz, A., Spisani, S., Biondi, C., Fabbri, E., Nalli, M., and Ferretti, M. E. (2003). Studies on human neutrophil biological functions by means of formyl-peptide receptor agonists and antagonists. *Curr. Drug Targets Immune Endocr. Metabol. Disord.* **3,** 33–42.

Dewas, C., Dang, P. M., Gougerot-Pocidalo, M. A., and El-Benna, J. (2003). TNF-alpha induces phosphorylation of p47(phox) in human neutrophils: Partial phosphorylation of p47phox is a common event of priming of human neutrophils by TNF-alpha and granulocyte–macrophage colony-stimulating factor. *J. Immunol.* **171,** 4392–4398.

Dekker, J., de Boer, M., and Roos, D. (2001)). Gene-scan method for the recognition of carriers and patients with p47(phox)-deficient autosomal recessive chronic granulomatous disease. *Exp. Hematol.* **29,** 1319–1325.

Demaurex, N., Downey, G. P., Waddell, T. K., and Grinstein, S. (1996). Intracellular pH regulation during spreading of human neutrophils. *J. Cell Biol.* **133,** 1391–1402.

Dinauer, M. C., and Orkin, S. H. (1992). Chronic granulomatous disease. *Annu. Rev. Med.* **43,** 117–124.

Djaldetti, M., Salman, H., Bergman, M., Djaldetti, R., and Bessler, H. (2002). Phagocytosis—the mighty weapon of the silent warriors. *Microsc. Res. Tech.* **57,** 421–431.

do Céu Monteiro, M., Sansonetty, F., Gonçalves, M. J., and O'Connor, J. E. (1999). Flow cytometric kinetic assay of calcium mobilization in whole blood platelets using Fluo-3 and CD41. *Cytometry* **35,** 302–310.

Elbim, C., Chollet-Martin, S., Bailly, S., Hakim, J., and Gougerot-Pocidalo, M. A. (1993). Priming of polymorphonuclear neutrophils by tumor necrosis factor alpha in whole blood: Identification of two polymorphonuclear neutrophil subpopulations in response to formyl-peptides. *Blood* **82,** 633–640.

Elbim, C., Prevot, M. H., Bouscarat, F., Franzini, E., Chollet-Martin, S., Hakim, J., and Gougerot-Pocidalo, M. A. (1994). Polymorphonuclear neutrophils from human immunodeficiency virus–infected patients show enhanced activation, diminished fMLP-induced L-selectin shedding, and an impaired oxidative burst after cytokine priming. *Blood* **84,** 2759–2766.

Elbim, C., Reglier, H., Fay, M., Delarche, C., Andrieu, V., El Benna, J., and Gougerot-Pocidalo, M. A. (2001). Intracellular pool of IL-10 receptors in specific granules of human neutrophils: Differential mobilization by proinflammatory mediators. *J. Immunol.* **166,** 5201–5207.

Emmendörffer, A., Nakamura, M., Rothe, G., Spiekermann, K., Lohmann-Matthes, M. L., and Roesler, J. (1994). Evaluation of flow cytometrical methods for diagnosis of chronic granulomatous disease (CGD)–variants under routine laboratory conditions. *Cytometry* **18,** 147–156.

Engelich, G., Wright, D. G., and Hartshorn, K. L. (2001). Acquired disorders of phagocyte function complicating medical and surgical illnesses. *Clin. Infect. Dis.* **33,** 2040–2048.

Forman, H. J., and Torres, M. (2002). Reactive oxygen species and cell signaling: Respiratory burst in macrophage signaling. *Am. J. Respir. Crit. Care Med.* **166,** S4–S8.

Forsyth, K. D., and Levinsky, R. J. (1990). Preparative procedures of cooling and re-warming increase leukocyte integrin expression and function on neutrophils. *J. Immunol. Methods* **128,** 159–163.

Fröhlich, D., Rothe, G., Schwall, B., Schmid, P., Schmitz, G., Taeger, K., and Hobbhahn, J. (1997). Effects of volatile anaesthetics on human neutrophil oxidative burst response to the bacterial peptide fMLP. *Brit. J. Anaesth.* **78,** 718–723.

Fröhlich, D., Wittmann, S., Rothe, G., Schmitz, G., and Taeger, K. (2002). Thiopental impairs neutrophil oxidative response by inhibition of intracellular signalling. *Eur. J. Anaesthesiol.* **19,** 474–482.

Green, D. J., and Shaw, E. (1981). Peptidyl diazomethyl ketones are specific inactivators of thiol proteinases. *J. Biol. Chem.* **256,** 1923–1928.

Hackam, D. J., Grinstein, S., and Rotstein, O. D. (1996). Intracellular pH regulation in leukocytes: Mechanisms and functional significance. *Shock* **5,** 17–21.

Hallett, M. B. (2003). Holding back neutrophil aggression; the oxidase has potential. *Clin. Exp. Immunol.* **132,** 181–184.

Hallett, M. B., and Lloyds, D. (1995). Neutrophil priming: The cellular signals that say "amber" but not 'green.' *Immunol. Today* **16,** 264–268.

Hart, S.P., Smith, J.R., and Dransfield, I. (2004). Phagocytosis of opsonized apoptotic cells: roles for "old-fashioned" receptors for antibody and complement *Clin. Exp. Immunol.* **135,** 181–185.

Hasleton, P. S., and Roberts, T. E. (1999). Adult respiratory distress syndrome—an update. *Histopathology* **34,** 285–294.

He, R., Sang, H., and Ye, R. D. (2003). Serum amyloid A induces IL-8 secretion through a G protein–coupled receptor, FPRL1/LXA4R. *Blood* **101,** 1572–1581.

Honey, K., and Rudensky, A. Y. (2003). Lysosomal cysteine proteases regulate antigen presentation. *Nat. Rev. Immunol.* **3,** 472–482.

Hooper, N. M. (2002). Proteases: A primer. *Essays Biochem.* **38,** 1–8.

Ishaque, A., and Al-Rubeai, M. (1998). Use of intracellular pH and annexin-V flow cytometric assays to monitor apoptosis and its suppression by bcl-2 over-expression in hybridoma cell culture. *J. Immunol. Methods* **221,** 43–57.

Jankowski, A., Scott, C. C., and Grinstein, S. (2002). Determinants of the phagosomal pH in neutrophils. *J. Biol. Chem.* **277,** 6059–6066.

Jongstra-Bilen, J., Harrison, R., and Grinstein, S. (2003). Fcγ-receptors induce Mac-1 (CD11b/CD18) mobilization and accumulation in the phagocytic cup for optimal phagocytosis. *J. Biol. Chem.* **278,** 45720–45729.

Kedzierska, K., Azzam, R., Ellery, P., Mak, J., Jaworowski, A., and Crowe, S. M. (2003). Defective phagocytosis by human monocyte/macrophages following HIV-1 infection: Underlying mechanisms and modulation by adjunctive cytokine therapy. *J. Clin. Virol.* **26,** 247–263.

Keel, M., Ungethum, U., Steckholzer, U., Niederer, E., Hartung, T., Trentz, O., and Ertel, W. (1997). Interleukin-10 counterregulates proinflammatory cytokine-induced inhibition of neutrophil apoptosis during severe sepsis. *Blood* **90,** 3356–3363.

Keil, M. L., Solomon, N. L., Lodhi, I. J., Stone, K. C., Jesaitis, A. J., Chang, P. S., Linderman, J. J., and Omann, G. M. (2003). Priming-induced localization of G(ialpha2) in high density membrane microdomains. *Biochem. Biophys. Res. Commun.* **301,** 862–872.

Kirschke, H., and Barrett, A. J. (1987). Chemistry of lysosomal proteases. *In* "Lysosomes: Their role in protein breakdown" (H. Glaumann and F. J. Ballard, eds.). Academic Press, London.

Koksch, M., Rothe, G., Kiefel, V., and Schmitz, G. (1995). Fluorescence resonance energy transfer as a new method for the epitope-specific characterization of anti-platelet antibodies. *J. Immunol. Methods* **187,** 53–67.

Kono, H., Suzuki, T., Yamamoto, K., Okada, M., Yamamoto, T., and Honda, Z. (2002). Spatial raft coalescence represents an initial step in Fc gamma R signaling. *J. Immunol.* **169,** 193–203.

Kwiatkowska, K., and Sobota, A. (1999). Signaling pathways in phagocytosis. *Bioessays* **21,** 422–431.

Le, Y., Murphy, P. M., and Wang, J. M. (2002). Formyl-peptide receptors revisited. *Trends Immunol.* **23,** 541–548.

Lehmann, A. K., Sørnes, S., and Halstensen, A. (2000). Phagocytosis: Measurement by flow cytometry. *J. Immunol. Methods* **243,** 229–242.

Lewis, R. S. (2001). Calcium signaling mechanisms in T lymphocytes. *Annu. Rev. Immunol.* **19,** 497–521.

Leytus, S. P., Melhado, L. L., and Mangel, W. F. (1983a). Rhodamine-based compounds as fluorogenic substrates for serine proteinases. *Biochem. J.* **209,** 299–307.

Leytus, S. P., Patterson, W. L., and Mangel, W. F. (1983b). New class of sensitive and selective fluorogenic substrates for serine proteinases. Amino acid and dipeptide derivatives of rhodamine. *Biochem. J.* **215,** 253–260.

Linke, R. P., Bock, V., Valet, G., and Rothe, G. (1991). Inhibition of the oxidative burst response of N-formyl peptide-stimulated neutrophils by serum amyloid-A protein. *Biochem. Biophys. Res. Commun.* **176,** 1100–1105.

Lund-Johansen, F., Olweus, J., Aarli, A., and Bjerknes, R. (1990). Signal transduction in human monocytes and granulocytes through the PI-linked antigen CD14. *FEBS Lett.* **273,** 55–58.

Lund-Johansen, F., Olweus, J., Symington, F. W., Arli, A., Thompson, J. S., Vilella, R., Skubitz, K., and Horejsi, V. (1993). Activation of human monocytes and granulocytes by monoclonal antibodies to glycosylphosphatidylinositol-anchored antigens. *Eur. J. Immunol.* **23,** 2782–2791.

Malech, H. L., and Nauseef, W. M. (1997). Primary inherited defects in neutrophil function: Etiology and treatment. *Semin. Hematol.* **34,** 279–290.

Mantovani, A., Garlanda, C., and Bottazzi, B. (2003). Pentraxin 3, a non-redundant soluble pattern recognition receptor involved in innate immunity. *Vaccine* **21,** S43–S47.

McCarthy, D. A., and Macey, M. G. (1996). Novel anticoagulants for flow cytometric analysis of live leucocytes in whole blood. *Cytometry* **23,** 196–204.

Minta, A., Kao, J. P. Y., and Tsien, R. Y. (1989). Fluorescent indicators for cytosolic calcium based on rhodamine and fluorescein chromophores. *J. Biol. Chem.* **264,** 8171–8178.

Moraes, T. J., Chow, C. W., and Downey, G. P. (2003). Proteases and lung injury. *Crit. Care Med.* **31,** S189–S194.

Osterud, B. (1998). Tissue factor expression by monocytes: Regulation and pathophysiological roles. *Blood Coagul. Fibrinolysis* **9,** S9–S14.

Ozinsky, A., Underhill, D.M., Fontenot, J.D., Hajjar, A.M., Smith, K.D., Wilson, C.B., Schroeder, L., and Aderem, A. (2000). The repertoire for pattern recognition of pathogens by the innate immune system is defined by cooperation between toll-like receptors. *Proc. Natl. Acad. Sci. USA* **97,** 13766–13771.

Palecanda, A., and Kobzik, L. (2001). Receptors for unopsonized particles: The role of alveolar macrophage scavenger receptors. *Curr. Mol. Med.* **1,** 589–595.

Pallister, C. J., and Lewis, R. J. (2000). Effects of antimicrobial drugs on human neutrophil–microbe interactions. *Br. J. Biomed. Sci.* **57,** 19–27.

Pfeiffer, A., Böttcher, A., Orsó, E., Kapinsky, M., Nagy, P., Bodnár, A., Spreitzer, I., Liebisch, G., Drobnik, W., Gempel, K., Horn, M., Holmer, S., Hartung, T., Multhoff, G., Schütz, G., Schindler, H., Ulmer, A. J., Heine, H., Stelter, F., Schütt, C., Rothe, G., Szöllösi, J., Damjanovich, S., and Schmitz, G. (2001). Lipopolysaccharide and ceramide docking to CD14 provokes ligand specific receptor clustering in rafts. *Eur. J. Immunol.* **31,** 3153–3164.

Pisciotta, A. V. (1990). Drug-induced agranulocytosis. Peripheral destruction of polymorphonuclear leukocytes and their marrow precursors. *Blood Rev.* **4,** 226–237.

Poenie, M., Alderton, J., Steinhardt, R., and Tsien, R. (1986). Calcium rises briefly and throughout the cell at the onset of anaphase. *Science* **233,** 886–889.

Powers, J. C. (1986). Serine proteases of leukocyte and mast cell origin: substrate specificity and inhibition of elastase, chymases, and tryptases. *Adv. Inflammation Res.* **11,** 145–157.

Reeves, E. P., Lu, H., Jacobs, H. L., Messina, C. G., Bolsover, S., Gabella, G., Potma, E. O., Warley, A., Roes, J., and Segal, A. W. (2002). Killing activity of neutrophils is mediated through activation of proteases by K^+ flux. *Nature* **416,** 291–297.

Réthi, B., Detre, C., Gogolák, P., Kolonics, A., Magócsi, M., and Rajnavölgyi, E. (2002). Flow cytometry used for the analysis of calcium signaling induced by antigen-specific T-cell activation. *Cytometry* **47,** 207–216.

Rothe, G., Emmendörffer, A., Oser, A., Roesler, J., and Valet, G. (1991). Flow cytometric measurement of the respiratory burst activity of phagocytes using dihydrorhodamine 123. *J. Immunol. Methods* **138,** 133–135.

Rothe, G., and Valet, G. (1990). Flow cytometric analysis of respiratory burst activity in phagocytes with hydroethidine and 2′,7′-dichlorofluorescin. *J. Leukoc. Biol.* **47,** 440–448.

Rothe, G., and Valet, G. (1994). Flow cytometric assays of oxidative burst activity in phagocytes. *In* "Methods in enzymology: Oxygen radicals in biological systems" (L. Packer, ed.), part C, Vol. 233, Academic Press, Orlando.

Rothe, G., Assfalg-Machleidt, I., Machleidt, W., Klingel, S., Zirkelbach, Ch., Banati, R., Mangel, W. F., and Valet, G. (1992). Flow cytometric analysis of protease activities in vital cells. *Biol. Chem. Hoppe-Seyler* **373,** 547–554.

Rothe, G., Oser, A., and Valet, G. (1988). Dihydrorhodamine 123: A new flow cytometric indicator for respiratory burst activity in neutrophil granulocytes. *Naturwissenschaften* **75,** 354–355.

Savill, J. (1998). Apoptosis. Phagocytic docking without shocking. *Nature* **392,** 442–443.

Segal, B. H., and Holland, S. M. (2000). Primary phagocytic disorders of childhood. *Pediatr. Clin. North Am.* **47,** 1311–1338.

Shao, D., Segal, A. W., and Dekker, L. V. (2003). Lipid rafts determine efficiency of NADPH oxidase activation in neutrophils. *FEBS Lett.* **550,** 101–106.

Simons, K., and Ehehalt, R. (2002). Cholesterol, lipid rafts, and disease. *J. Clin. Invest.* **110,** 597–603.

Sklar, L. A., Oades, Z. G., Jesaitis, A. J., Painter, R. G., and Cochrane, C. G. (1981). Fluoresceinated chemotactic peptide and high-affinity antifluorescein antibody as a probe of the temporal characteristics of neutrophil stimulation. *Proc. Natl. Acad. Sci. USA* **78,** 7540–7544.

Steinkamp, J. A., Wilson, J. S., Saunders, G. C., and Stewart, C. C. (1982). Phagocytosis: Flow cytometric quantitation with fluorescent microspheres. *Science* **215,** 64–66.

Thomas, J. A., Buchsbaum, R. N., Zimniak, A., and Racker, E. (1979). Intracellular pH measurements in Ehrlich ascites tumor cells utilizing spectroscopic probes generated *in situ*. *Biochemistry* **18,** 2210–2218.

Triantafilou, M., Miyake, K., Golenbock, D. T., and Triantafilou, K. (2002). Mediators of innate immune recognition of bacteria concentrate in lipid rafts and facilitate lipopolysaccharide-induced cell activation. *J. Cell Sci.* **115,** 2603–2611.

Vowells, S. J., Sekhsaria, S., Malech, H. L., Shalit, M., and Fleisher, T. A. (1995). Flow cytometric analysis of the granulocyte respiratory burst: A comparison study of fluorescent probes. *J. Immunol. Methods* **178,** 89–97.

Walrand, S., Valeix, S., Rodriguez, C., Ligot, P., Chassagne, J., and Vasson, M. P. (2003). Flow cytometry study of polymorphonuclear neutrophil oxidative burst: A comparison of three fluorescent probes. *Clin. Chim. Acta* **331,** 103–110.

WHO Consensus Group. (1999). Primary immunodeficiency diseases. *Clin. Exp. Immunol.* **118,** 1–28.

Wittmann, S., Rothe, G., Schmitz, G., and Fröhlich, D. (2004). Cytokine upregulation of surface antigens correlates to the priming of the neutrophil oxidative burst response. *Cytometry* **57A,** 53–62.

Wymann, M. P., Sozzani, S., Altruda, F., Mantovani, A., and Hirsch, E. (2000). Lipids on the move: Phosphoinositide 3-kinases in leukocyte function. *Immunol. Today* **21,** 260–264.

Yamashiro, S., Kamohara, H., Wang, J.M., Yang, D., Gong, W. H., and Yoshimura, T. (2001). Phenotypic and functional change of cytokine-activated neutrophils: Inflammatory neutrophils are heterogeneous and enhance adaptive immune responses. *J. Leukoc. Biol.* **69,** 698–704.

Zarewych, D. M., Kindzelskii, A. L., Todd, R. F., and Petty, H. R. (1996). LPS induces CD14 association with complement receptor type 3, which is reversed by neutrophil adhesion. *J. Immunol.* **156,** 430–433.

CHAPTER 30

Neutralizing Antibody Quantification by Flow Cytometry

John R. Mascola

Vaccine Research Center
National Institute of Allergy and Infectious Diseases
National Institutes of Health
Bethesda, Maryland 20892

I. Introduction

Antibody-based immunity is a vital component of protection against many infectious diseases, and most existing vaccines work, at least in part, by inducing antibodies that inactivate or neutralize the invading microbe or its toxin. Thus, there is a long history of *in vitro* measurements of the immunoglobulin-mediated effect on microbes. Antibodies may act directly on cell-free bacteria or viruses to inactivate the infectious inoculum or to prevent the early spread of infection. This is best illustrated by the action of immunoglobulin G (IgG), immunoglobulin A (IgA), or immunoglobulin M (IgM) antibodies that prevent cell-free viruses from infecting host cells. This chapter focuses on the application and advantages of flow cytometry to measure virus neutralization. The methods detailed are for a human immunodeficiency virus type 1 (HIV-1) neutralization assay, but similar methods can be applied to other viruses.

Virus neutralization assays are designed to measure a reduction in virus infectious titer mediated by exposure to antibody. Antibodies bound to viral determinants can interrupt viral infection of a cell by several mechanisms, including the inhibition of cellular attachment, virus-cell fusion, and uncoating after viral entry. Regardless of the mechanism, the result is that the host cell is not productively infected and does not express viral antigens. Early virus neutralization assays relied on the observation that upon virus infection, a plaque would appear in an adherent cell monolayer. The plaque was the result of cell death surrounding an infected cell. Because each plaque is the result of the initial infection of one cell, these assays could calculate the amount of infectious virus by counting the number of plaques produced by a given volume of virus supernatant. These plaque-reduction neutralization assays were time consuming and cumbersome to perform, but they were highly accurate and quantitative. A key feature of these assays is that the fraction of virus neutralized could be directly inferred by counting the reduction in the number of infected cells. The application of flow cytometric methods to virus neutralization assays is based on this same principle. By enumerating cells that are productively infected, one can directly infer the amount of infectious virus in culture and measure the fraction of virus neutralized resulting from interaction with antibody.

II. Background

To develop a rapid and highly quantitative HIV-1 neutralization assay, we took advantage of observations that HIV-1–infected cells could be detected by intracellular staining for the HIV-1 internal p24 Gag antigen (p24 antigen) (Cameron *et al.*, 1998; Cory *et al.*, 1987; Costigliola *et al.*, 1992; Folghera *et al.*, 1994; Jason and Inge, 1999; Kux *et al.*, 1996; Landay *et al.*, 1990; McSharry *et al.*, 1990; Ohlsson-Wilhelm *et al.*, 1990; Steele-Mortimer *et al.*, 1990; Vanham *et al.*, 2000). Most directly, Darden *et al.* (2000) had reported the use of a primary isolate HIV-1 neutralization assay that enumerated the number of HIV-1–infected CD4 T cells after 4–7 days in culture. By including a protease inhibitor in culture with target CD4 T-cells, we developed a neutralization assay that measures single-round infection of individual cells. This enumeration of first-round infection provides quantitative and reproducible data on the number of infectious virus particles, as measured by the number of target cells infected. Importantly, there is a linear relationship between the amount of infectious input virus and the number of CD4 T-cells infected. Thus, this *in vitro* neutralization assay directly measures and quantifies the inactivation of infectious virus mediated by exposure to antibody (Mascola *et al.*, 2002).

Of note, single rounds of infection reporter viruses can also be used in this assay format. For example, we use a recombinant green fluorescent protein

(GFP) reporter virus that is generated by co-transfection of 293T cells with the pNL4-3env$^-$ plasmid (full-length NL4-3 HIV-1 proviral DNA with a frameshift in env and encoding GFP in place of nef) and a DNA plasmid encoding the full length gp160 Env protein. Upon infection of target cells, this HIV-1 pseudovirus encodes for GFP and, thus, allows the target cells to be readily detected by flow cytometry. The advantage of this method is that no protease inhibitor is needed in culture and cells can be directly interrogated without the subsequent steps of permeabilization and intracellular staining.

A. Basic Methods for an HIV-1 Flow Cytometric Neutralization Assay

- *Cells:* This assay uses CD4 T-cells obtained by CD8 T-cell depletion of PBMCs isolated by Ficoll–Hypaque gradient centrifugation. However, any target cell that can be interrogated by flow cytometry cell is suitable.
- *Neutralization assay:* The intracellular p24 antigen neutralization assay is performed in 96-well culture plates by incubating 40 μl of virus stock with 10 μl of antibody. Antibody concentration is defined at this step (i.e., final concentration of antibody with virus). After preincubation for 30 minutes at 37 °C, 20 μl of mitogen-stimulated CD4 T-cells (1.5×10^5 cells) is added to each well. Cells are maintained in interleukin-2 (IL-2) culture medium containing 1 μM of indinavir, and the cells are fed on day 1 with 150 μl of IL-2 culture media. The cells are harvested the next day. They need not be washed before harvest, but are washed before intracellular staining for p24 antigen. For each antibody dilution, two wells are set up and combined to produce enough cells for precise quantitation by flow cytometry. To enumerate infected target cells, cells are washed, fixed and permeabilized, and stained with the KC57 anti-p24 antibody as described later in this chapter. Live cells initially gated by forward and side scatter are analyzed for intracellular expression of p24 antigen. The number of p24 antigen–positive cells is determined using a bivariate plot of fluorescence versus forward scatter; the gate is set on mock-infected cells. We routinely count 50,000 events. Thus, if the infection rate in control wells is 4%, there are 2000 positive events. Final quantitation of p24 antigen–positive cells is done by subtraction of background events in mock-infected cells (usually $<$10 positives per 50,000 events). The percent neutralization is defined as the reduction of p24 antigen–positive cells compared to control wells with no antibody. Of note, antibody dose–response curves can be fit with a nonlinear function, and the inhibitory concentration that neutralizes 50%, 80%, and 90% (IC_{50}, IC_{80}, and IC_{90}) of virus can be calculated by a least-squares regression analysis. Additional methodological details are provided elsewhere (Mascola *et al.*, 2002).
- *Intracellular staining:* For intracellular p24 antigen staining, cells are transferred to V-bottom plates and washed once in phosphate-buffered saline (PBS) containing 1% fetal calf serum (FCS). Cells are fixed and permeabilized using the Cytofix/Cytoperm Kit (BD-Pharmingen, San Diego, California). Permeabilized

cells are washed twice in V-bottom plates using wash buffer provided by the manufacturer and resuspended for 20 minutes at 4 °C with 50 μl of a 1:160 dilution of a phycoerythrin (PE)-conjugated mouse anti-24 monoclonal antibody (KC57-RD1) (Beckman Coulter, Inc.). After two additional washes, HIV-1 or mock-infected PBMCs are analyzed using a FACSCalibur flow cytometer (Becton Dickinson) and data analysis was performed with FlowJo software (Tree Star, Inc., San Carlos, California).

III. Results

A. Intracellular Staining

To evaluate whether the KC57 antibody was detecting most of the HIV-1–infected cells, we infected CD4 T-cells with a single-round HIV-1 envelope pseudo typed GFP reporter virus and monitored cells for expression of GFP and p24 antigen. The contour plot in Fig. 1 shows that more than 90% of GFP-positive cells also stained for intracellular p24 antigen. These data confirm the robust nature of the intracellular staining for HIV-1 p24 antigen and show that pseudo typed viruses, such as this GFP reporter construct, can be used in flow-based neutralization assays.

B. Linear Dose–Response between Virus Input and the Number of p24-Ag Positive Cells

The accurate measurement of antibody-mediated virus neutralization is best determined in an assay with a linear relationship between the amount of infectious virus and the number of target cells infected. To test this, we exposed

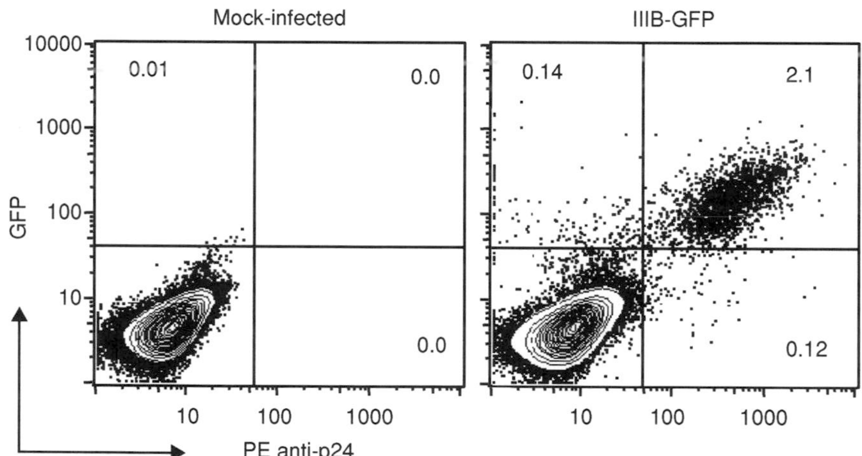

Fig. 1 CD4 T-cells infected with a human immunodeficiency virus type 1 (HIV-1) envelope pseudo-typed green fluorescent protein (GFP) reporter virus express both GFP and HIV p24 antigen. More than 90% of the infected cells are shown in the upper right quadrant.

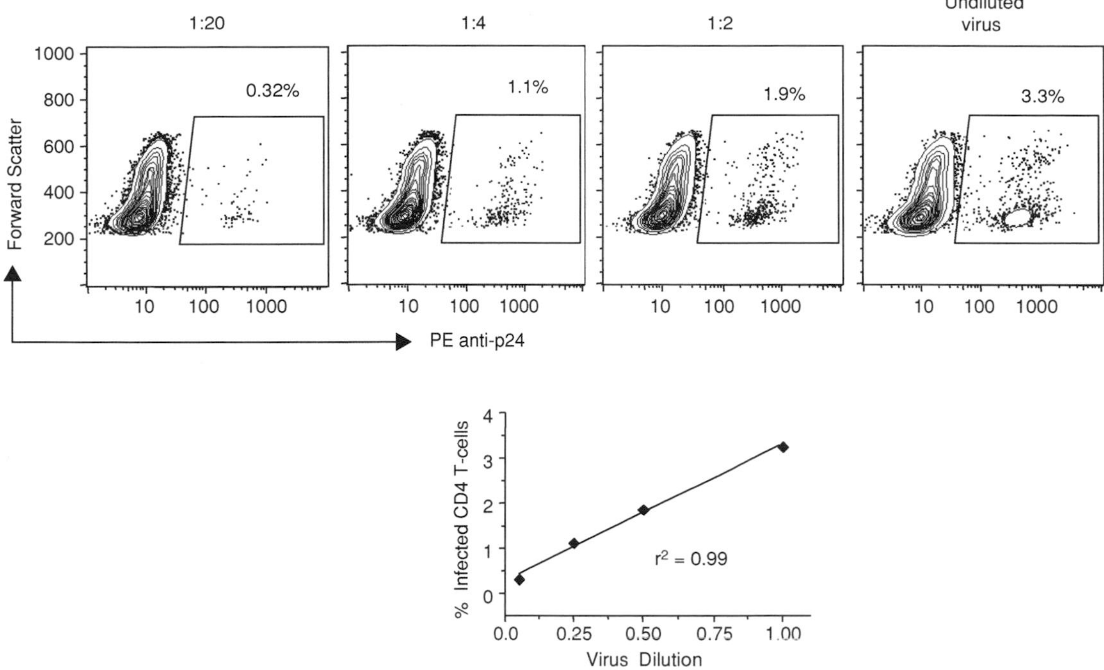

Fig. 2 Linear dose response between virus input and the number of p24 antigen–positive cells. The percentage of target cells infected at each virus dilution (Top panel). The linear regression line for the same data (Bottom panel).

indinavir-treated CD4 T-cells to serial dilutions of HIV-1. As shown in Fig. 2, we observed a consistent linear dose response. Because some viruses may have a plateau-shaped curve at a high multiplicity of infection, it is important to confirm this linear dose relationship before performing neutralization assays.

C. Target Cell Viability

Because some animal sera can produce toxic effects on target cells, forward side-scatter gating can be used as a general indicator that the target cells have not been adversely affected. For example, in our HIV-1 neutralization assay, impaired cell growth or viability could lead to less p24 antigen–positive cells and the possibility of falsely inferring antibody-mediated neutralization. Figure 3 shows an example of a substantial decrement in cell viability caused by a 1:5 dilution of rabbit serum. Thus, an advantage of this flow cytometric assay is that the physical characteristics of the cells can be assessed before gating for virus-infected cells. In addition, the lack of exclusion of ethidium monoazide bromide (EMA) can be used to identify dead cells that should be excluded from further analysis.

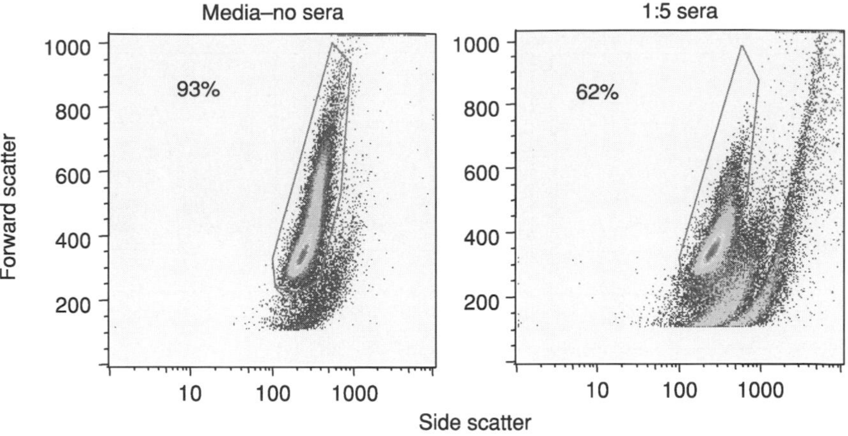

Fig. 3 Forward and side scatter characteristics of CD4 T-cells before gating for p24 antigen expression. Note that a 1:5 dilution of rabbit sera has adversely affected the physical characteristics of the cells. Although not all animal sera has this effect, the ability to observe target cells for possible adverse effects is a distinct advantage of flow cytometric assay. (See Color Insert.)

IV. Discussion and Key Points

The ability to make accurate and precise measurements of antibody-mediated virus neutralization is important for the evaluation of mechanisms of neutralization and for the assessment of antibody responses elicited by immunization. Similar to the plaque-reduction neutralization assays described for many viruses, flow cytometry can measure the amount of infectious virus in culture and the reduction in infectious titer due to reaction with antibody. To achieve this, we developed a single round of replication HIV-1–neutralization assay that enumerates infected cells by the flow cytometric detection of cells expressing p24 antigen. Compared to our prior methodologies that required enzyme-linked immunosorbent assay (ELISA) measurement of p24 antigen in culture after several rounds of replication, the intracellular p24 antigen assay directly identifies infected target cells and provides a more precise and reproducible measurement of antibody-mediated inactivation of infectious virus. Several characteristics of this assay allow for a high level of precision and reproducibility. The number of target cells is directly enumerated by a highly accurate technology, and even infection levels of 1% can be readily detected because of the high number of events counted. Additionally, single-round infection of target cells results in a direct relationship between the amount of infectious virus and the number of target cells infected.

This neutralization uses primary strains of HIV-1 and primary human $CD4^+$ T-cells as target cells. However, a single round of infection recombinant envelope pseudo typed viruses can also be used. Many such HIV-1 reporter viruses have been described, such as those expressing chloramphenicol acetyltransferase, alkaline phosphatase, B-galactosidase, luciferase, or GFP (Chen *et al.*, 1994; Connor *et al.*, 1995; He and Landau, 1995; Helseth *et al.*, 1990). Among these reporter systems, alkaline phosphatase, B-galactosidase, and GFP have the advantage of directly visualizing and enumerating the number of infected target cells (He *et al.*, 1997; Page *et al.*, 1997). In contrast, quantitating chloramphenicol acetyltransferase and luciferase involves lysis of cells, and the data generated reflect relative amounts of reporter gene produced in the entire culture. The use of GFP reporter viruses is optimal for flow cytometric assays; the GFP-expressing cells can be readily detected with minimal manipulation.

Overall, this flow cytometric assay for measuring HIV-1 neutralization is less labor intensive and substantially less expensive than previously described ELISA antigen-capture assays. Although a flow cytometer is required, the cost of the anti-p24 KC57 antibody and related staining reagents is about fivefold less than the per-well cost of a commercial HIV-1 antigen-capture ELISA. We also used a Multiwell autosampler attachment to the FACSCalibur (Multiwell autosampler system, Becton Dickinson). This allows the neutralization assay to be set up in a 96-well plate and analyzed by automated flow cytometric determinations in a 96-well format.

References

Cameron, P. U., Hunter, S. D., Jolley, D., Sonza, S., Mijch, A., and Crowe, S. M. (1998). Specificity of binding of HIV-1 anti-p24 antibodies to CD4+ lymphocytes from HIV-infected subjects. *Cytometry* **33,** 83–88.

Chen, B. K., Saksela, K., Andino, R., and Baltimore, D. (1994). Distinct modes of human immunodeficiency virus type 1 proviral latency revealed by superinfection of nonproductively infected cell lines with recombinant luciferase-encoding viruses. *J. Virol.* **68,** 654–660.

Connor, R. I., Chen, B. K., Choe, S., and Landau, N. R. (1995). Vpr is required for efficient replication of human immunodeficiency virus type-1 in mononuclear phagocytes. *Virology* **206,** 935–944.

Cory, J. M., Ohlsson-Wilhelm, B. M., Brock, E. J., Sheaffer, N. A., Steck, M. E., Eyster, M. E., and Rapp, F. (1987). Detection of human immunodeficiency virus–infected lymphoid cells at low frequency by flow cytometry. *J. Immunol. Methods* **105,** 71–78.

Costigliola, P., Tumietto, F., Ricchi, E., and Chiodo, F. (1992). Detection of circulating p24 antigen-positive CD4+ cells during HIV infection by flow cytometry. *AIDS* **6,** 1121–1125.

Darden, J. M., Polonis, V. R., deSouza, M. S., Chantakulkij, S., Brown, A. E., Birx, D. L., and Pattanapanyasat, K. (2000). A flow cytometric method for measuring neutralization of HIV-1 subtype B and E primary isolates. *Cytometry* **40,** 141–150.

Folghera, S., Fiorentini, S., Martinelli, F., Ravizzola, G., Gargiulo, F., Terlenghi, L., De Francesco, M., Caruso, A., and Turano, A. (1994). Development of a flow cytometric assay for the detection and measurement of neutralizing antibodies against human immunodeficiency virus. *New Microbiol.* **17,** 21–28.

He, J., Chen, Y., Farzan, M., Choe, H., Ohagen, A., Gartner, S., Busciglio, J., Yang, X., Hofmann, W., Newman, W., Mackay, C. R., Sodroski, J., and Gabuzda, D. (1997). CCR3 and CCR5 are co-receptors for HIV-1 infection of microglia. *Nature* **385,** 645–649.

He, J., and Landau, N. R. (1995). Use of a novel human immunodeficiency virus type 1 reporter virus expressing human placental alkaline phosphatase to detect an alternative viral receptor. *J. Virol.* **69,** 4587–4592.

Helseth, E., Kowalski, M., Gabuzda, D., Olshevsky, U., Haseltine, W., and Sodroski, J. (1990). Rapid complementation assays measuring replicative potential of human immunodeficiency virus type 1 envelope glycoprotein mutants. *J. Virol.* **64,** 2416–2420.

Jason, J., and Inge, K. L. (1999). Increased expression of CD80 and CD86 in *in vitro*–infected CD3+ cells producing cytoplasmic HIV type 1 p24. *AIDS Res. Hum. Retroviruses* **15,** 173–181.

Kux, A., Bertram, S., Hufert, F. T., Schmitz, H., and von Laer, D. (1996). Antibodies to p24 antigen do not specifically detect HIV-infected lymphocytes in AIDS patients. *J. Immunol. Methods* **191,** 179–186.

Landay, A., Ohlsson-Wilhelm, B., and Giorgi, J. V. (1990). Application of flow cytometry to the study of HIV infection. *AIDS* **4,** 479–497.

Mascola, J. R., Louder, M. K., Winter, C., Prabhakara, R., De Rosa, S. C., Douek, D. C., Hill, B. J., Gabuzda, D., and Roederer, M. (2002). Human immunodeficiency virus type 1 neutralization measured by flow cytometric quantitation of single-round infection of primary human T cells. *J. Virol.* **76,** 4810–4821.

McSharry, J. J., Costantino, R., Robbiano, E., Echols, R., Stevens, R., and Lehman, J. M. (1990). Detection and quantitation of human immunodeficiency virus-infected peripheral blood mononuclear cells by flow cytometry. *J. Clin. Microbiol.* **28,** 724–733.

Ohlsson-Wilhelm, B. M., Cory, J. M., Kessler, H. A., Eyster, M. E., Rapp, F., and Landay, A. (1990). Circulating human immunodeficiency virus (HIV) p24 antigen-positive lymphocytes: A flow cytometric measure of HIV infection. *J. Infect. Dis.* **162,** 1018–1024.

Page, K. A., Liegler, T., and Feinberg, M. B. (1997). Use of a green fluorescent protein as a marker for human immunodeficiency virus type 1 infection. *AIDS Res. Hum. Retroviruses* **13,** 1077–1081.

Steele-Mortimer, O. A., Meier-Ewert, H., Loser, R., and Hasmann, M. J. (1990). Flow cytometric analysis of virus-infected cells and its potential use for screening antiviral agents. *J. Virol. Methods* **27,** 241–252.

Vanham, G., Penne, L., Allemeersch, H., Kestens, L., Willems, B., van der Groen, G., Jeang, K. T., Toossi, Z., and Rich, E. (2000). Modeling HIV transfer between dendritic cells and T cells: Importance of HIV phenotype, dendritic cell-T cell contact and T-cell activation. *AIDS* **14,** 2299–2311.

PART VI

Cytogenetics

CHAPTER 31

Telomere Length Measurements Using Fluorescence *In Situ* Hybridization and Flow Cytometry

Gabriela M. Baerlocher*,† and Peter M. Lansdorp*,‡

*Terry Fox Laboratory
British Columbia Cancer Agency
Vancouver, British Columbia, V5Z 1L3, Canada

†Department of Hematology and Department of Clinical Research
University Hospital Bern
3010 Bern, Switzerland

‡Department of Medicine
University of British Columbia
Vancouver, British Columbia, V5Z 4E3, Canada

METHODS IN CELL BIOLOGY, VOL. 75

0091-679X/04 $35.00

I. Introduction

A. Telomere Length Dynamics

Linear chromosome ends are composed of TTAGGG repeats and associated proteins, which are assembled into a dynamic three-dimensional structure—the telomere. Most of the telomeric tract consists of double-stranded DNA, but the very end of the chromosome contains a short (50–300 nucleotide) single-stranded G-rich 3′ overhang. This overhang appears essential for telomere function and the formation of a typical fold-back structure called a *telomere loop* or *t-loop* (Griffith *et al.*, 1999). Current data support the concept that telomeres can exist in at least two states: an uncapped or open form and a capped or closed t-loop form (Blackburn, 2001). It is possible that telomeres reversibly switch between these two states. Such an equilibrium would provide entry for regulation of the cell cycle, programmed cell death, DNA damage responses, and possibly other cellular functions.

In the open or uncapped state, the single-stranded overhang at telomeres can serve as a substrate for telomerase, a cellular reverse transcriptase that can add new telomeric hexanucleotides onto chromosome ends. Telomerase action is required to compensate for the loss of telomere repeats with each cell division because of incomplete DNA replication (Olovnikov, 1971; Watson, 1972), oxidative damage (von Zglinicki, 2002), and possibly other causes. In cells that do not express sufficient compensatory telomerase, telomeres shorten with each cell division *in vitro* and *in vivo*. Critically short telomeres trigger apoptosis or irreversible cell cycle arrest but can also result in chromosome fusions and genomic instability. Important evidence for a link between telomere shortening and cellular senescence is derived from experiments with diverse human cell types showing that ectopic expression of telomerase results in telomere elongation and extension of their proliferative potential and replicative life span (Bodnar *et al.*, 1998; Vaziri and Benchimol, 1998). Haploinsufficiency for the human telomerase template gene (hTERC), resulting in a modest reduction in telomerase activity, triggers the autosomal dominant disorder dyskeratosis congenita (DKC). Patients with DKC show clinical manifestations in rapidly proliferating tissues such as nail, skin, and bone marrow and typically die before the age of 50 years from marrow failure, immune deficiencies, or cancer (Vulliamy *et al.*, 2001). In patients with DKC, telomeres are critically short and telomere dysfunction during tissue development and regeneration is the most likely explanation for the proliferative defect and related pathology. Patients with DKC highlight the critical importance of adequate telomerase levels in human cells and point to carefully controlled telomere length regulation to sustain the proliferation in stem cells of various organs.

Current data support the idea that telomere shortening with age evolved as a tumor suppressor mechanism in long-lived species. By suppressing immortal growth, younger and healthier (stem) cells are recruited into action as the propagation of older cells (which are more likely to have acquired detrimental mutations) is suppressed. Human cancer cells have to overcome the restrictions in proliferative potential imposed by progressive telomere shortening. Typically, this is achieved by upregulation of telomerase activity, a notion that is supported by the observation that more than 90% of human tumor cells show readily detectable telomerase activity (Kim *et al.*, 1994; Shay and Bacchetti, 1997).

In summary, telomere length dynamics are very important for the regulation of the replicative life span of cells, especially in long-lived species. Telomere shortening and telomerase activity are believed to be important factors in aging and tumorigenesis. To study the telomere length in different cell types, we developed flow–fluorescence *in situ* hybridization (flow-FISH), a sensitive and reproducible technique to measure the telomere length in individual cells.

B. Methods to Measure Telomere Length

We focus on the three most frequently used methods to measure the length of telomere repeats in cells (Table I): Southern blot, Q-FISH (telomere length

Table I
Methods to Measure Telomere Length

	Southern blot	Quantitative Q-FISH	Quantitative flow-FISH
Cell requirement	10^5–10^6	10^5–10^6 for 10–15 metaphases	10^4–10^5
Cell cycle status	Cell cycle independent	Proliferating cells	Cell cycle independent
Processing time needed	3–5 days	3–5 days	1–2 days
Hybridization targets	Telomeric and subtelomeric repeats within a cell population	Telomeric repeats at specific chromosome ends	Telomeric repeats and intrachromosomal telomeric repeats in single cells
Additional parameter	None	None	Forward light scatter and side scatter, DNA fluorescence, limited cell-surface markers

measurements using digital images of metaphase chromosomes after *in situ* hybridization with fluorescently labeled telomere probe), and flow-FISH (measurements of the average telomere length in cells using flow cytometry after *in situ* hybridization with fluorescently labeled telomere probe). Another promising polymerase chain reaction (PCR)–based method was described to measure telomere length (Baird *et al.*, 2003). This sensitive but laborious method is not discussed in this chapter because of space limitations.

1. Southern Blot Analysis

Telomere length measurement by Southern blot is based on the following steps:

1. Isolation and restriction enzyme digestion of genomic DNA
2. Separation of the DNA fragments containing telomeric DNA based on their size using gel electrophoresis
3. Hybridization and visualization of telomere fragments
4. Estimation of the average length of the fragments containing telomere repeats from the smear of heterogeneously sized fragments

Each step has significant drawbacks:

1. There are no restriction enzyme sites directly at the end of the telomeric repeats, so a variable amount of subtelomeric chromosomal DNA will be included in the telomere fragments. As a consequence, telomere length will be invariably overestimated.

2. Very small and very long telomere fragments cannot be efficiently separated on the same gel. Because of the nonlinear relation between fragment size and migration distance in the gel, very short telomeres will run faster (distributed over a wider range) and their size will be underestimated. Long telomeres (>30 kb) may not migrate into the gel very well and often cannot be resolved. These problems are exacerbated by the fact that the number of labeled telomere repeat oligonucleotide molecules that can hybridize to the fragments is related to the size of the fragment. This is yet another factor that makes it difficult, if not impossible, to avoid overestimations of telomere length using Southern blot analysis.

3. To be able to visualize telomere fragments in a gel, high-quality genomic DNA from relatively large numbers of cells (typically corresponding to 10^5–10^6 cells) is required. The resulting telomere length estimate represents the average telomere length in a population of cells and does not allow estimates of the telomere length in minor subpopulations of cells within the sample.

4. Southern blot is a time-consuming method that typically requires radiolabeled probes.

2. Q-FISH

For telomere length measurements by Q-FISH, the following steps are performed:

1. Stimulation of cells in culture and harvest of cells arrested in metaphase
2. Preparation of slides with fixed metaphase cells
3. *In situ* hybridization with fluorescently labeled telomere peptide nucleic acid (PNA) probe and DNA counterstaining
4. Acquisition and analysis of the telomere fluorescence on individual chromosome ends using digital images acquired with a sensitive camera mounted on a fluorescence microscope

There are two major advantages of this method: Telomere length measurements are highly accurate and sensitive (with appropriate controls) (Martens *et al.*, 1998; Poon *et al.*, 1999) and information on the average telomere length at specific chromosome ends can be obtained. However, each step of the Q-FISH protocol also has its limitations:

1. The requirement for metaphase chromosomes excludes Q-FISH for telomere length analysis in nondividing cells. Presenescent cells or chronic lymphocytic leukemic B cells, which hardly divide, are good examples of cells in which measurement of telomere length by Q-FISH is problematic. Furthermore, to prepare slides with suitable metaphase spreads requires specific technical expertise.
2. Although the nature of telomere-PNA probes allows specific and sensitive hybridization under certain conditions (see section II), the accessibility of the telomere-PNA probe to the condensed target sequences in metaphase chromosomes remains incompletely known. Thus far, no internal standard to control for possible variation in accessibility during *in situ* hybridization has been developed.
3. Expensive fluorescence microscope and digital camera/computer equipment are required and personnel needs to be properly trained to guarantee the quality and standardization of telomere length measurements from captured fluorescence images.
4. Finally, the processes of fluorescence image acquisition and analysis, despite the use of a freely available dedicated computer program (TFL-Telo, available at www.flintbox.com) (Poon *et al.*, 1999), remain cumbersome and time consuming.

3. Flow-FISH

Telomere length measurements by the basic flow-FISH technique (Rufer *et al.*, 1998) rely on the following steps:

1. Isolation of the nucleated cells from whole blood or tissue.

2. *In situ* hybridization with fluorescently labeled telomere-PNA probe and DNA counterstaining of the cells.

3. Acquisition and analysis of the telomere fluorescence on the flow cytometer. Significant advantages of this method are that telomere length measurements can be performed with individual cells in suspension, that telomere length measurements are not compromised by the presence of cells with very long or short telomeres, that cells can be nondividing, that many cells can be analyzed within a short period, and that in samples of nucleated blood cells, two subpopulations of cells (granulocytes and lymphocytes) can be analyzed separately based on differences in light scatter properties.

Nevertheless, each step also has its problems:

1. Hemoglobin in the cell suspension can interfere with telomere fluorescence in cells and the cell isolation procedures that are required can lead to cell loss and fragmentation.
2. Heat and formamide treatment of cells, though resulting to some extent in the fixation of cells, results in loss of cells and loss of most cell-surface antigens.
3. Binding of the telomere probe to cellular organelles and membranes and incomplete removal of excess probe can result in overestimates of telomere length.
4. Accessibility of the telomere-PNA probe to target sequences may vary with the degree of DNA condensation in the interphase nucleus.
5. The autofluorescence of cells (especially granulocytes) and the DNA fluorescence of the DNA counterstain (used for the gating of cells and for distinction between cells with a diploid [2N] chromosome content and cells in S or G_2/M) can interfere with the detection of specific telomere fluorescence.

C. New Method: Automated Multicolor Flow-FISH

Flow-FISH offers many advantages compared to Southern blot and Q-FISH (Table I). However, the first flow-FISH protocols (Rufer *et al.*, 1998, 1999) were still quite time consuming, especially when more than 10 samples were processed at a time, and further problems were encountered when differences in telomere length were small or when cell counts were (very) low. In addition, no information on the telomere length in subsets of cells (except for granulocytes and lymphocytes) within one sample could be obtained unless such subpopulations were sorted before the procedure.

We optimized and automated the basic flow-FISH technique (Baerlocher *et al.*, 2002) and combined the procedure with limited immunophenotyping (Baerlocher and Lansdorp, 2003). Many variables influencing the *in situ* hybridization process and the subsequent results were tested in extensive protocol development steps to find the most optimal protocol for telomere length measurements with high sensitivity and good reproducibility. One of the crucial improvements was the automation of most cell-processing steps by a robotic washing station—the Hydra (Robbins Scientific or Apogent Discoveries). The Hydra device allows to aspirate and dispense very small volumes of liquids simultaneously and gently in and from 96 microtubes arranged in a standard 96 well plate format. Again, each step of our original protocol was adjusted to optimally integrate the Hydra device. Furthermore, we included control cells (bovine thymocytes) in each tube as an internal standard to further control for the inevitable method-related variation between tubes. Finally, flow-FISH was combined with limited immunophenotyping.

D. Chapter Outline

In this chapter, we first provide a brief overview of the development of quantitative *in situ* hybridization techniques and we describe specific properties of PNA probes. We then present our latest improvements in the flow-FISH technique for highly sensitive and reproducible measurements of telomere length in interphase cells. We aim to point out the critical parameters of the technique and important aspects of data acquisition on the flow cytometer. Furthermore, we elaborate on *in situ* hybridization combined with immunophenotyping. Finally, we discuss some current applications of the flow-FISH method.

II. Background

A. Quantitative Fluorescence *In Situ* Hybridization

At the end of the 1970s, recombinant DNA technology, which allowed production of large quantities of known DNA or RNA sequences in bacteria, paved the way for *in situ* hybridization methods to map mammalian genes and study gene expression in cells and tissue sections. The first FISH experiments used messenger RNA (mRNA) molecules, which were linked to fluorescent latex microspheres to visualize globin genes in human metaphase chromosomes (Cheung *et al.*, 1977). Since then, direct labeling of probes with suitable fluorescent dyes was introduced and many other improvements in FISH methods, and in general, fluorescence detection by fluorescence microscopy, were made (Trask, 1991). As a result, FISH has become a standard *qualitative* tool in cytogenetics and gene mapping.

In the late 1980s, the first reports can be found describing *in situ* hybridization of probes to interphase nuclei in suspension followed by the quantification of bound probe by flow cytometry (Trask *et al.*, 1988). The aim in those studies was to combine FISH with the speed and quantitative analysis provided by flow cytometry. Improvements in FISH probes, antibody labeling, fluorescence labels, and flow cytometers contributed to further advances in FISH techniques using either fluorescence microscopy or flow cytometry. However, one of the fundamental drawbacks of conventional FISH techniques for DNA targets remained that the DNA or RNA probes that were used always had to compete for hybridization to the denatured target DNA sequences with the original complementary DNA strand. Although high probe concentrations are used to increase the likelihood of probe hybridization, this was not a satisfactory solution for the detection of repeat sequences (as in telomeric DNA) in which a high concentration of short oligonucleotide probes cannot effectively or reproducibly compete with long complementary strands. This shortcoming of conventional FISH was overcome in 1993 when a new sort of molecule was invented: PNAs (Nielsen and Egholm, 1999). Soon, it was found that directly fluorescently labeled PNA probes could be hybridized to target sequences under conditions (low ionic strength) that *did not allow* reannealing of complementary strands (Lansdorp, 1996). These observations allowed the development of the first *quantitative* FISH (Q-FISH) techniques.

B. Peptide Nucleic Acids

PNA is a nucleic acid analog in which the sugar phosphate backbone of natural nucleic acid has been replaced by a synthetic peptide backbone formed from *N*-(2-amino-ethyl)-glycine units, resulting in an achiral and uncharged moiety that mimics RNA or DNA oligonucleotides (Fig. 1). PNA is chemically stable, resistant to hydrolytic (enzymatic) cleavage, and cannot be degraded inside living cells. PNA is capable of sequence-specific recognition of DNA and RNA, obeying the Watson–Crick hydrogen bonding scheme, and the hybrid complexes exhibit extraordinary thermal stability and unique ionic strength properties (Ray and Norden, 2000). PNA resulted from efforts during the 1980s and early 1990s by Peter Nielsen and Michael Egholm in Ole Buchardt's laboratory in Denmark. The unique properties of PNA–DNA hybrid complexes were exploited in FISH techniques for the first time in 1995 by Peter Lansdorp during a sabbatical visit in the laboratory of Hans Tanke and Ton Raap at Leiden University in the Netherlands. Although directly labeled PNA probes were already commercially available, their use in FISH techniques required the development of conditions that allowed maximal hybridization at low ionic strength (preventing competition between probe hybridization and DNA renaturation) and low background staining of the labeled probe. The latter was particularly problematic because PNA probes are very hydrophobic and strongly bind, for example, to glass and various hydrophobic sites in cells and chromosomes. Solutions to these problems were found by including blocking protein in the hybridization solution and performing hybridization and wash steps in the presence of high concentrations (70%) of formamide (Lansdorp, 1996). As in all protocol development efforts, countless experiments

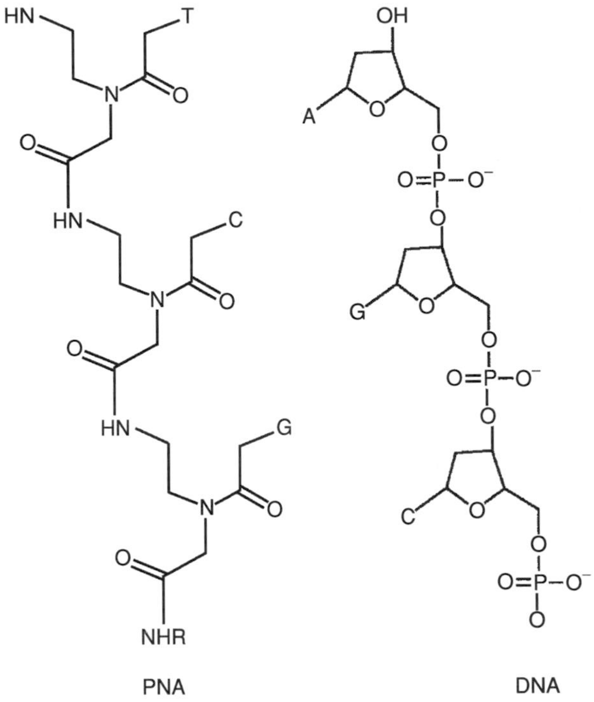

PNA DNA

Fig. 1 The general structure of deoxyribose nucleic acid (DNA) oligonucleotides compared to peptide nucleic acid (PNA) is shown. The PNA monomer consists of *N*-(2-amino-ethyl)-glycine units linked by a methylene carbonyl to one of the four bases (adenine, guanine, thymine, or cytosine) found in DNA. Like amino acids, PNA monomers have amino and carboxy termini. Unlike DNA or RNA, PNAs lack pentose sugar phosphate groups.

had to be performed to delineate the importance of each parameter in the technique, distill the most important variables, and establish a procedure that works reproducibly.

III. Methods

This section is based on many experiments performed over the last several years to optimize flow-FISH for *quantitative* measurements of the telomere repeat content in interphase cells. We try to give an overview of the most important conclusions drawn from the findings in many experiments. For additional details, we refer to our two method papers on this topic (Baerlocher and Lansdorp, 2003; Baerlocher *et al.*, 2002). Table II lists what type of equipment, material, and solutions are needed to set up automated multicolor flow-FISH. In Fig. 2, the

Table II
Equipment, Materials, and Supplies Needed for Automated Multicolor Flow-FISH

Equipment	Materials	Solutions
Rotator	Sodium heparin Vacutainer tubes	Ammonium chloride
Shaker	50-ml tubes	Phosphate buffer solution, pH 7.4
Tube racks	14-ml tubes	Dextrose 5%, 252 mOsm, pH 4.0
Centrifuge for 10/14-ml tubes	1.5-ml tubes	Bovine serum albumin 10%
Microcentrifuge for 1.5-ml tubes	Polymerase chain reaction tubes	Hepes 100 mM
Circulating water bath	96 deep-well plates	Trizma base, pH 7.1
Thermometer	Pipette tips	Deionized formamide
Fume hood	Foil	Formamide
Biohazard hood	Parafilm	Sodium chloride 1 M
Pipettes (1000 μl, 100 μl, 20 μl)	Resin beads	Telomere-PNA-fluorescein at 30 μg/ml
Repeater pipette	0.22 μm polyethersulfone	Tween 20 10%
Washing station (e.g., Hydra)	Vacuum filter system	LDS 751 at 0.2 mg/ml
Flow cytometer	Whatman paper	RNase T1 at 100,000 U/ml
	Sieve/strainer	Sodium acid 20%
		FACSFlow
		Dulbecco modified Eagle's Medium
		DNAse
		Heparin
		Dimethylsulfoxide
		Paraformaldehyde

main steps of the protocol, the most critical parameters, and our recommended processing steps are presented in a diagram.

A. Optimization of Quantitative *In Situ* Hybridization

1. Preparation/Isolation of Nucleated Cells for Flow-FISH

Ideally, one would like to start the flow-FISH procedure with a homogenous single-cell suspension containing sufficient (2–10 $\times$ 10^5), intact, and stable (not necessarily viable) nucleated cells. In reality, the cells of interest for telomere length measurements typically come in a cell mixture with red blood cells (RBCs) (e.g., whole blood) and/or they are embedded in tissue (e.g., thymus, spleen). Both situations make isolation or separation procedures for the cells of interest necessary. Hemoglobin has fluorescent properties and can act as quencher (absorb the fluorescence from other fluorescent molecules). Therefore, a high concentration of RBCs or hemoglobin (corresponding to a RBC suspension with a hematocrit $\geq$2%) can lead to interference with fluorescence detection from the fluorescein-labeled telomere probe. Second, any cell isolation or separation step can lead to changes in the properties (size, shape, autofluorescence), viability, and recovery of cells. Therefore, the aim is to use gentle techniques for cell preparation and keep the preparation time as short as possible.

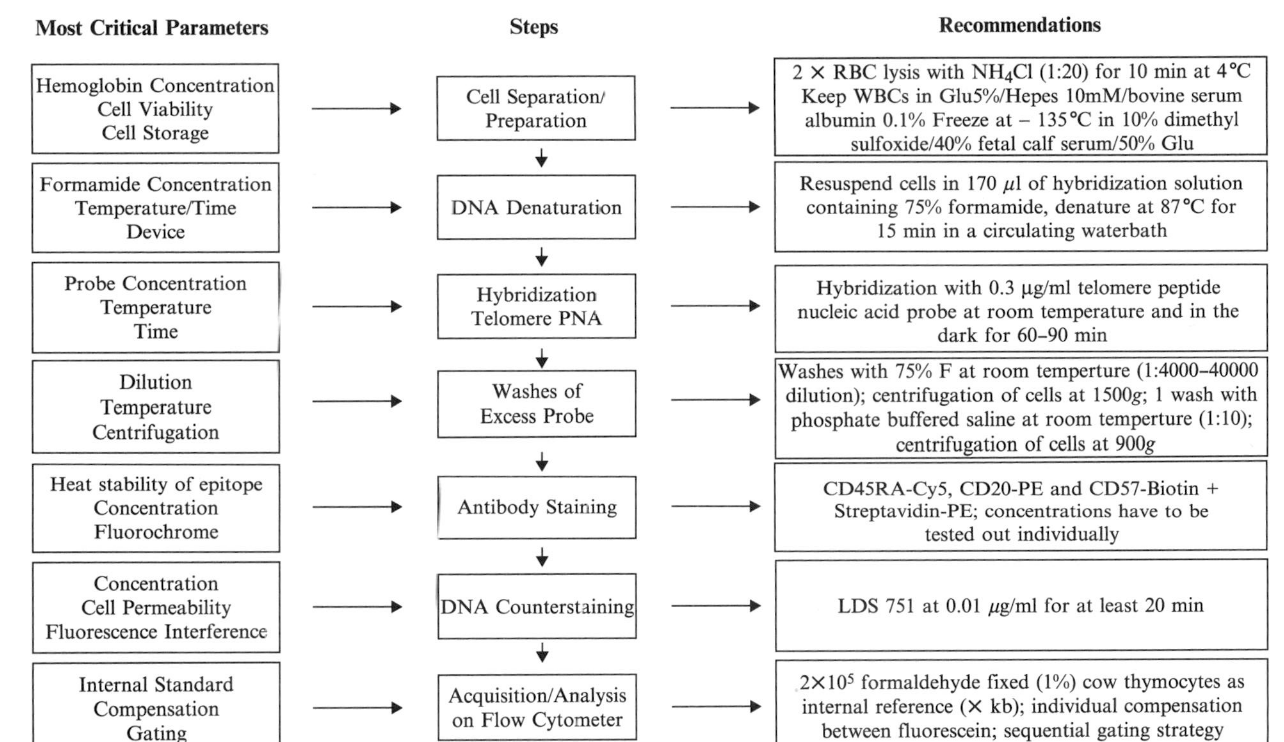

Fig. 2 Diagram of the seven steps in the automated multicolor flow–fluorescent *in situ* hybridization (FISH) protocol. The most critical parameters in each step are indicated on the left and our recommendations on how to perform each step are shown on the right.

Unfortunately, there is no good, easy, and fast method to completely separate nucleated cells from RBCs without any loss or destruction of nucleated blood cells. Two approaches to isolate or separate nucleated cells from RBCs have been explored: (1) direct RBC destruction by solutions that lead to swelling and lysis of the RBC via changes in the RBC osmolality (examples are Zap-oglobulin, water, 3% acetic acid in water or ammonium chloride) and (2) separation/isolation of RBCs based on cell-surface antigens that are expressed specifically on RBCs (immunomagnetic selection).

Many RBC lysing reagents are commercially available to prepare leukocyte cell suspensions for use on a cell counter or flow cytometer. Such lysing solutions have been optimized for those specific applications and appear very tempting for the isolation of nucleated cells for flow-FISH. Unfortunately, many lysing reagents contain variable amounts of various cross-linking or non–cross-linking fixatives to prevent destruction of the nucleated cells by the lysing reagent (e.g., Zap-oglobulin). Such fixatives may affect the access of the fluorescent probe to target sequences and decrease the efficiency of DNA denaturation by subsequent treatment with heat and formamide (see section III.A.2). We tested several lysing reagents without fixatives for their efficiency of RBC lysis and recovery of nucleated blood cells from moderate volumes (3–7 ml) of whole blood sufficient for flow-FISH (Baerlocher *et al.*, 2002). Hypotonic lysis with 3% acetic acid in water or water itself was found to be very effective in lysing RBCs but also resulted in poor recovery of white blood cells (WBCs) (typically only 30–50% of the WBCs). The results were better for RBC lysis with ammonium chloride. Interestingly, the first approaches to lyse RBCs with ammonium chloride date back to the 1940s and 1950s. Then, it was found that ammonium chloride can rapidly enter the RBC and change its osmolality via ion exchanges (OH− or HCO3− with Cl−) over the cell membrane. These processes lead to cell swelling and hemolysis of the RBC within 6–10 minutes at 4 °C without significant loss of WBCs. Because of differences in membrane ion transport between RBCs and WBCs, the lysis with ammonium chloride occurs in the first minutes primarily in the RBCs. Therefore, leukocytes can be preserved and RBCs highly efficiently eliminated using ammonium chloride. A few essential points, however, have to be taken into account: The plasma should be removed from the cell suspension before lysis; the dilution factor of whole blood to ammonium chloride should be at least 20 times; the mechanism of RBC lysis does exhaust after two rounds of ammonium chloride exposure; and resuspending WBCs after ammonium chloride lysis in a solution containing dextrose greatly increases the recovery of WBCs.

Isolation of cells from tissues (spleen, thymus, tonsils) for flow-FISH typically requires either physical (e.g., dissecting the tissue with a knife or scissors or straining pieces of tissue through a sieve or mesh) or enzymatic digestion (e.g., collagenase) procedures. Such processing steps typically result in a certain amount of cell destruction with release of DNA even with the most gentle cell handling and viable cells can be trapped in the resulting clumps. Addition of DNAse at 1 μg/ml to the cell suspension effectively digests DNA in solution and prevents or reverses such cell clumping. Cell losses occurring during the cell isolation procedures also have to be considered. Isolation of mononuclear cells by Ficoll–Hypaque density centrifugation, for example, results in about a 50% loss of mononuclear cells (as is mentioned in the product sheet by the company). If specific subtypes of cells are of interest for telomere length measurements by flow-FISH, isolation of such cells by immunomagnetic isolation procedures or fluorescence-activated cell sorting (FACS) is an option. However, such cell preparation procedures are time consuming and labor intensive and typically require a lot of cells to start with and inevitable cell losses have to be taken into account. Besides attempts to minimize such cell losses, it is crucial to keep cells in as good shape as is possible. For this reason, the time for cell isolation and preparation of cells should be as short as possible and cells should be kept at 4 °C in a balanced salt solution with an optimal pH and some protein (e.g., 0.1% BSA or 2% FCS). Cell losses can be further reduced by minimizing centrifugation steps and avoiding transfer of cells between tubes.

After the isolation or separation of cells and performing cell counts, it is often convenient to store cells before subsequent flow-FISH because the preparation already took some time and because samples often arrive in small batches. Fortunately, it is possible to store cell suspensions after processing in 10% v/v DMSO, 40% v/v fetal calf serum, and 50% v/v glucose 5%/Hepes 20 mM at −70 °C for a few weeks and at −135 °C for years. Thawing cells for 1–2 minutes at 37 °C in a water bath and resuspension in glucose 5%/Hepes 20 mM with 0.1% BSA without DMSO allows quick and efficient recovery of the previously frozen cells for the next steps in the flow-FISH protocol.

2. DNA Denaturation

Many protocols are available for the denaturation of isolated genomic DNA. Optimization of such molecular biology protocols was aimed at complete denaturation of DNA without loss or destruction of the resulting single-stranded DNA. For *in situ* hybridization using whole cells, it is important that complete breakdown of cells and membranes is avoided to allow for subsequent analysis of cells on a flow cytometer and to acquire additional information on the cell type using remaining cellular characteristics. The first experiments with *in situ* hybridization using whole cells were performed with fixatives to preserve nuclear and membrane structures during the harsh conditions required for denaturation of DNA in FISH. Later, it was found that heat and formamide already result in partial fixation of cells, most likely by dehydration and denaturation of proteins. Such fixation can be compared with what is observed when an egg is boiled: Protein denaturation by heat solidifies the liquid egg white. Fixation by heat and formamide does not prevent the PNA-telomere probe to pass through the remains of the cell and nuclear membrane and bind efficiently to complementary target sequences in the denatured DNA. In contrast, prior fixation almost invariably decreases the hybridization efficiency. In our hands, denaturation at

87 °C in a solution containing 75% formamide for 15 minutes followed by hybridization in the same solution at room temperature for at least 90 minutes results in the most optimal hybridization with the least loss or disintegration of DNA. Interestingly, the denaturation parameters are somewhat different for lymphocytes and granulocytes, most likely because of their different nuclear structure. Granulocytes need somewhat higher temperatures for optimal denaturation than lymphocytes (Baerlocher *et al.*, 2002). In addition, there are several ways to apply the heat for the denaturation step. A circulating water bath offers the most efficient and reliable method to denature many samples uniformly at the same time. In contrast, denaturation in a heat block does not allow the same degree of temperature control because wells at the periphery typically have lower temperatures than wells in the middle of the block.

3. *In Situ* Hybridization

Quantitative *in situ* hybridization became feasible only with the introduction of directly labeled PNA probes. PNA probes and other uncharged oligomimetics have a higher specificity and sensitivity than DNA or RNA oligos and can hybridize under conditions (low salt concentration) that do not favor the renaturation of target sequences with complementary DNA strands. The definition of such conditions is a problem when *in situ* hybridization is performed with whole cells in suspension. In our experience, whole cells can maximally withstand 10–20 mM Tris in the hybridization solution containing 75% formamide, 1% BSA, 0.3 μg/ml PNA probe and water. 20 mM Tris in the hybridization solution is the optimal salt concentration to still preserve the cell structure on the one hand and to achieve favorable conditions for denaturation/hybridization of the telomere-PNA probe to telomeric DNA on the other hand. Generally, synthetic PNA probes do not easily pass cell membranes, which is one of the hurdles for antisense therapy with PNA. However, with heat and formamide treatment, PNA probes enter cells and nuclei without prior permeabilization. We calculated that in theory about 70% of the telomere repeats in cells should be hybridized with PNA when telomere-PNA 18-mers (C-probe) are used (Fig. 3). This may be an underestimate because the hydrophobic nature of PNA may force further alignment of PNA probes along target sequences. Optimal hybridization with the use of the telomere-PNA fluorescein-CCC TAA CCC TAA CCC TAA C-terminus probe is reached at a concentration of 0.3 μg/ml. This is an estimated more than 10^4-fold excess of probe relative to the calculated number of target sites available for hybridization in typical flow-FISH samples.

4. Washes of Excess Probe

The concentration of telomere-PNA probe used for hybridization is lower than that which would typically be used for the hybridization with DNA probes. Nevertheless, an excess amount of telomere-PNA probe is needed to guarantee

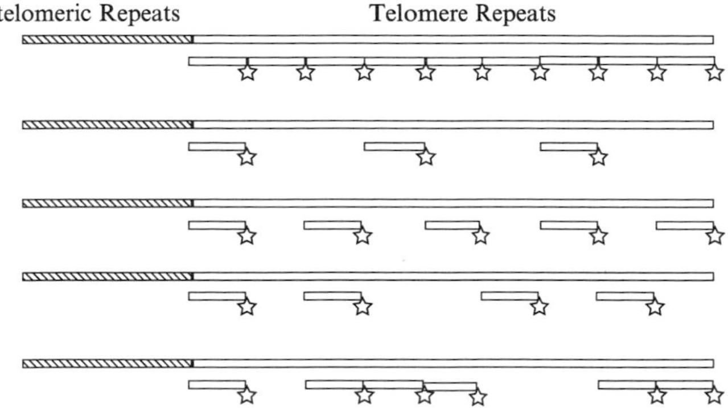

Fig. 3 Possible hybridization patterns for the telomere-peptide nucleic acid (PNA) 18-mer probe. With the assumption that those telomere-PNA probes hybridize with a high specificity and sensitivity and have most optimal accessibility to telomere ends, no bigger unhybridized gaps than 12 nucleotides should exist after hybridization. In theory, at least 70% or more of the telomere repeats will be complexed with PNA-telomere probe.

complete hybridization. Therefore, steps to remove unbound and nonspecifically bound excess fluorescent PNA probes are crucial to only acquire and quantitate the specifically bound probe. When *in situ* hybridization is performed on metaphase chromosomes, one to two wash steps are generally sufficient to dilute the excess telomere-PNA probe to undetectable background levels. However, to eliminate the excess telomere-PNA probe in interphase cells in suspension, more extensive washes are required because the dilution factor with each wash step is much smaller. With our protocol, wash steps need to dilute the excess PNA probe more than 10^4-fold to effectively eliminate most of the background fluorescence resulting from unbound or nonspecifically bound telomere-PNA probe and allow specific detection of hybridization probe. We perform four wash steps with a wash solution containing 75% formamide at room temperature and one wash step with a phosphate-buffered saline (PBS) solution to achieve the required dilution of unbound probe. Wash steps can also be performed with wash solutions containing PBS at 40 to 50 °C. Note that cells in suspension have to be spun down to recover them for the next steps in the protocol. Importantly, washes in formamide solution require a higher gravity force (1500 g) to adequately sediment cells in 5-minute centrifugation steps than washes in aqueous solution (900 g).

5. Antibody Staining

The combination of *quantitative in situ* hybridization for the telomere length measurement with immunostaining has great potential for studies of the replicative histories of cells within one sample and for telomere length measurements in rare cells, which so far have been impossible because of low cell numbers. Unfortunately, the heat and formamide treatment used for the denaturation of the DNA also denatures most membrane and cytoplasmic proteins and epitopes. Most antibodies used for the immunophenotyping of viable or paraformaldehyde-fixed cells fail to react with cells after the flow-FISH procedure. However, a few antibodies (anti-human CD45RA, anti-human CD20, anti-human CD57) are able to specifically bind to cells despite the changes that occur during the heat and formamide treatment. These antibodies are typically used in immunohistochemistry, in which antibody staining is performed on paraffin-embedded tissue sections. We assume that such antibodies recognize peptide or carbohydrate epitopes that are resistant to heat and formamide denaturation. In our hands, any linker or fixative reagent, even in a low concentration, to attach antibodies to the cell membrane before *in situ* hybridization or to preserve epitopes during exposure to heat and formamide, results in a variable decrease in hybridization efficiency that is difficult, if not impossible, to control.

6. DNA Counterstaining

The idea behind counterstaining DNA for flow-FISH is to distinguish diploid (2N) cells from polyploid cells or cells in the S phase or the G_2M (4N) phase of the cell cycle. An optimal DNA dye for flow-FISH should easily enter the flow-FISH–processed cells, bind specifically and tightly to (denatured) DNA, allow discrimination between 2N cells and 4N cells, be optimally excited by a 488-nm laser (available on most flow cytometers), and display a narrow emission spectrum, which is far from the fluorescein isothiocyanate absorbance and emission spectrum (fluorescein-labeled in our protocol to the telomere-PNA). We tested many DNA dyes (propidium iodide, Syto orange dyes, Syto red dyes, 7-AAD, Hoechst H33342, 4,6-diamidino-2-phenylindole dihydrochloride [DAPI], DRAQ5, and LDS 751) and none fulfilled all of the above criteria. LDS 751 at a low concentration (0.01 μg/ml) and DAPI (0.01–1.00 μg/ml) were the only DNA dyes that did not notably interfere with the detection of fluorescein from the telomere-PNA probe and the phycoerythrin (PE)- and Cy5-labeled antibodies that are used in our protocol. However, LDS 751 and Cy5 have largely overlapping fluorescence emission spectra and these dyes require significant cross-compensation.

7. Acquisition on Flow Cytometer

To acquire the telomere fluorescence quantitatively in flow-FISH–processed cells, it is important that the flow cytometer is perfectly set up (warmup, alignment of laser, adjustment of time delay, etc.) according to the manufacturer's recommendations. Before test cells are acquired on the flow cytometer, we usually run a mixture of fluorescein-labeled calibration beads. This mixture of calibration beads contains four subpopulations of beads that are tagged with four defined amounts of molecular equivalents of soluble fluorochrome (MESF). The bead fluorescence is acquired to determine the linear range of the fluorescence detectable in FL1 and to convert telomere fluorescence values into MESF units for comparison of measurements between instruments and from experiment to experiment. Typical dot plots and histograms used for the acquisition of the telomere fluorescence in subtypes of human leukocytes are shown in Fig. 4. For each detector (FL1, FL2, FL3, FL4), optimal settings (values for the voltage of preamplifiers and photo multiplier tubes) have to be determined. Most optimally, the cell populations (granulocytes, lymphocytes, bovine thymocytes) should fall into the center part of the forward and side scatter dot plot and the three cell populations should be easily separable on the side scatter versus DNA dye (FL3) dot plot. In addition, subsets of cells should be easily detectable in forward scatter versus the specific antibody (e.g., anti-human CD45RA–Cy5 in FL4 for naive T cells or anti-human CD20–PE in FL2 for B cells) and the peaks for the telomere fluorescence should be in the lowest hundred units (linear scale) for unstained cells and in the highest 200 units (linear scale) for the telomere-fluorescent bovine thymocytes. To optimize the detection of antibody-labeled cells and to guarantee accurate telomere fluorescence measurements, compensations have to be performed between the diverse detectors. It is important that the fluorescence from PE detected in FL1 is subtracted by compensation to guarantee accurate telomere fluorescence measurements on cells labeled with PE. Each sample pair is, therefore, individually compensated.

8. Calculation and Presentation of Results

For each sample, subtypes of cells are gated according to the example shown in Fig. 4, and the telomere fluorescence histogram of each subtype of cell is generated. In general, we typically use the median value for the telomere fluorescence. However, when histograms are very asymmetrical, the mean values or percentiles can be more useful. For each sample, duplicate measurements are performed and the specific telomere fluorescence is calculated by subtraction of the (auto-) fluorescence of unstained controls from the telomere fluorescence measured in cells hybridized with the fluorescein-labeled telomere-PNA probe (Table III). The specific telomere fluorescence (in MESF) of the cells of interest is then set in relation to the specific telomere fluorescence (in MESF) of the bovine thymocytes analyzed in the same tube according to the following calculation:

$$\frac{(\text{MESF value test cells} - \text{MESF value unstained test cells}) : 92^{a} \times 15.0\ \text{kb}^{d,e}}{(\text{MESF value control cells} - \text{MESF value unstained control cells}) \times 1.22^{b} : 120^{c}},$$

where

[a]Number of chromosome ends in a normal diploid human cell

[b]correction factor to compensate for the measured difference in median telomere fluorescence between fixed and unfixed control cells (fluorescence fixed cells was 82% of that in unfixed control cells in two experiments)

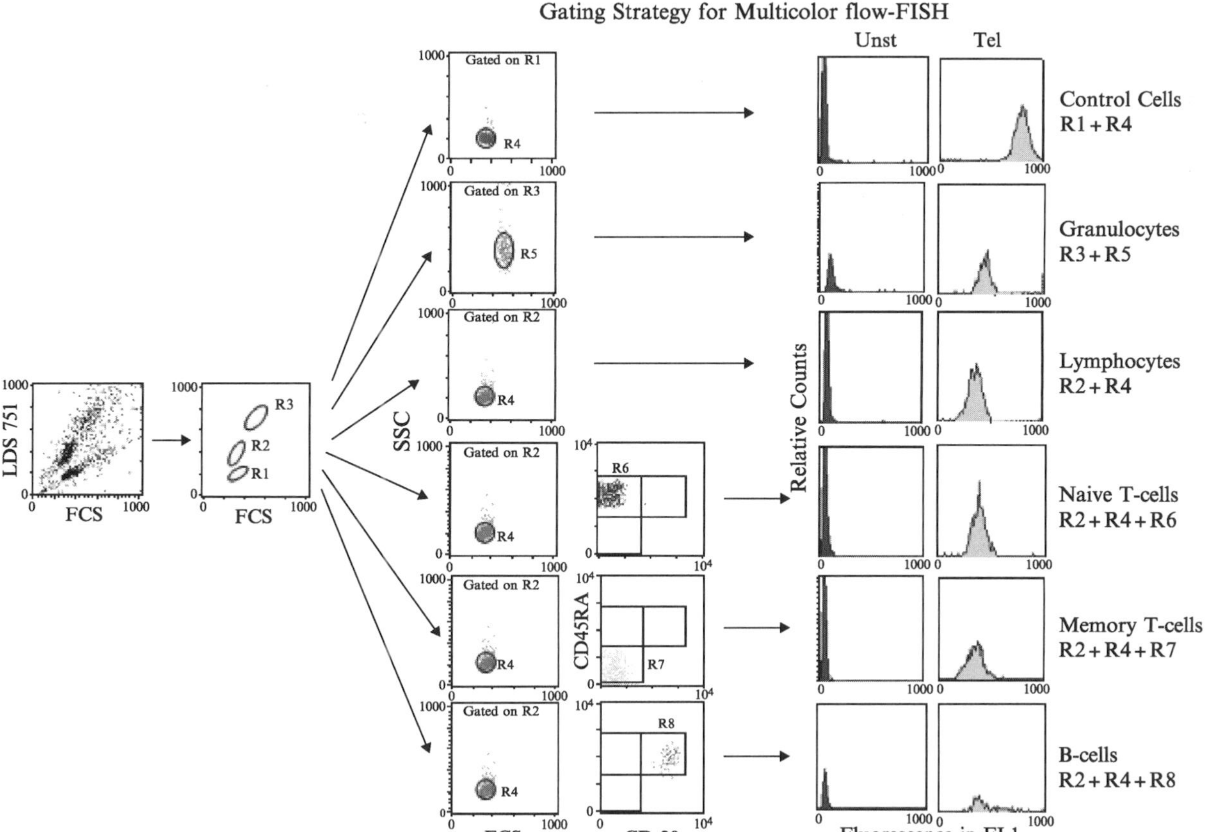

Fig. 4 Flow cytometric dot plots and histograms used for the gating of the internal standard (bovine thymocytes) and the subpopulations of leukocytes and for the analysis of the telomere fluorescence. First, a dot plot with forward scatter channel (FSC) versus LDS 751 is created to gate control cells (R1), lymphocytes (R2), and granulocytes (R3). In a second step, those same populations are gated on FSC versus side scatter channel (SSC). Note that bovine thymocytes and lymphocytes fall into the same gate (R4), whereas granulocytes are somewhat bigger and more granular (R5). Based on CD20 versus CD45RA, three subpopulations of lymphocytes can be gated. CD45RA-positive and CD20-negative naive T cells (R6), CD45RA-negative and CD20-negative memory T cells (R7), and CD45RA-positive and CD20-positive B cells (R8). For each of the subpopulations of leukocytes (bovine thymocytes in gates R1 + R4, granulocytes in gates R3 + R5, lymphocytes in gate R2 + R4, naive T cells in gates R2 + R4 + R6, memory T cells in gates R2 + R4 + R7, and B cells in gates R2 + R4 + R8), the telomere fluorescence shown as a histogram will then be acquired on a unstained sample (Unst = no telomere-PNA probe in the hybridization solution) and on a stained sample (Tel = telomere-PNA probe in the hybridization solution). (See Color Insert.)

Table III
Example of Calculation of Telomere Length in kb from Flow-FISH Telomere Fluorescence Values

	Unst	Unst	Mean Unst	Tel	Tel	Mean Tel	Specific Tel	Corr Factor (×1.22)	Chromosome no. :120 or :92	Telomere length (kb)
Bovine thymocytes	43	45	44	810	803	806.5	762.5	930.3	7.8	15.0
Leukocytes	67	73	70	392	393	392.5	322.5		3.5	6.8
Granulocytes	112	125	118.5	448	438	443	324.5		3.5	6.8
Lymphocytes	60	65	62.5	372	368	370	307.5		3.3	6.5
Naive T lymphocytes	58	63	60.5	381	382	381.5	321		3.5	6.8
Memory T lymphocytes	62	70	66	351	344	347.5	281.5		3.1	5.9
B lymphocytes	61	65	63	383	375	379	316		3.4	6.6

Note: Channel numbers of the F1 1 channel (linear scale) of a FACSCalibur flow cytometer following calibration with molecular equivalents of soluble fluorochrome beads are shown.
Unst, sample without telomere-PNA probe; Tel, sample with telomere-PNA probe; Specific Tel, Tel – Unst; Corr Factor, to compensate for the lower fluorescence due to fixation; Chromosome #, 120 chromosome ends for bovine thymocytes and 92 chromosome ends for human cells; Telomere length in kb, 15 for this specific aliquot of bovine thymocytes.

[c]number of chromosome ends in each control cell (bovine thymocyte)

[d]measured telomere restriction fragment size in DNA from control bovine thymocytes

[e]assuming the length of subtelomeric DNA in terminal restriction fragments is similar to human and bovine DNA and that internal T_2AG_3 repeats in human and bovine chromosomes do not contribute significantly to the telomere fluorescence.

Because we measured the telomere restriction fragment size in DNA from the bovine thymocytes, we are able to convert the fluorescence values of the test cells into kilobases (Table III). With duplicate measurements, the average (mean), standard deviation (SD), and coefficient of variation (CV) can be calculated. Duplicates with CVs of more than 10% are excluded from analysis.

There are many ways to present telomere length data. Depending on the data set and the experimental questions, dot plots, histograms, box plots, or simple data tables can be used.

B. Preparation of Internal Standard

Even with the standardization of all steps in the protocol, a certain variation between measurements remains. To reduce the variation in duplicate or triplicate samples from the same individual, and to increase the reproducibility of measurements at different times, we started to add control cells to each tube as an internal standard (Hultdin *et al.*, 1998). We selected bovine thymocytes as internal controls based on their small size, abundant availability, ease of preparation and storage, and low autofluorescence after heat and formamide treatment. In addition, they are mostly diploid cells and have telomeres that are twice as long as humans and up to 2–10 times shorter telomeres than those in mice. Only one characteristic was not optimal: after heat and formamide treatment, bovine thymocytes have very similar forward and side scatter properties as human, baboon, or mice lymphocytes when analyzed on a flow cytometer. Therefore, separation of those cell types was not feasible using light scatter parameters alone. However, we found that fixation of bovine thymocytes with 1% paraformaldehyde after the isolation procedure (details of the protocol are described in Baerlocher and Lansdorp, 2003) results in weaker staining with the DNA dye LDS 751. Most likely, the modest cross-linking by the paraformaldehyde fixation changes the accessibility of LDS 751 to the chromosomal DNA. This small difference in the DNA counterstaining makes it possible to separate between bovine thymocytes and lymphocytes when they are mixed in the same tube for *in situ* hybridization (Fig. 4, left). As expected, telomere hybridization is compromised by the fixation step, and the resultant telomere fluorescence is lower in fixed cells compared to unfixed cells. The difference in telomere fluorescence was measured (the telomere fluorescence of 1% paraformaldehyde fixed cells is 82% of that in nonfixed cells) and a correction factor ($\times 1.22$) was incorporated into the formula for the calculation of the telomere length in cells of interest (for further details, see Baerlocher and Lansdorp, 2003).

C. Automation

Nowadays, automation systems are widely used in laboratories to standardize and speed up processes. Many repetitive steps such as aspiration and dispensing of cells and fluids are necessary during the flow-FISH procedure, and we were interested in finding a device that can very accurately aspirate and dispense diverse volumes of solutions to and from many tubes at the same time. The Hydra microdispensers by Robbins Scientific are bench-top instruments that feature 96 or 384 syringes arranged in a standard array format (Baerlocher and Lansdorp, 2003). The Hydra can be programmed to very accurately pipette or mix solutions into, from, and in 96- or 384-well plates. Most often, this type of device is used in genome analysis and drug discovery programs, and to our knowledge, it has not been used for processing of cell suspensions. We tested the standard 96-well Hydra instrument with 1-ml syringes and somewhat wider needles (inner diameter 0.0155 inches) for use in flow-FISH. Because of the even and gentle aspiration and dispensing of the machine, cells in suspension can easily be transferred or mixed with various solutions. As a result, the cell recovery is better and cells are more evenly distributed than can be achieved manually. We adapted most of the steps in our flow-FISH protocol specifically for integration with the Hydra. For this purpose, the adequate volume, position (height) of the needles, and the number of cycles were determined for each aspiration, dispensing, or mixing step following the instructions in the manual of the manufacturer for the plates and tubes used in flow-FISH. Using the Hydra, we greatly decreased the variation in telomere fluorescence measurements, and we saved up to about 50% of the solutions and reagents needed for most of the steps in the protocol. Furthermore, we processed about five times as many samples as was possible by hand.

IV. Results

A. Variation/Sensitivity/Accuracy

To analyze which improvements could be achieved by optimizing and automating most steps of the basic flow-FISH protocol, we compared the following parameters for samples processed with the basic flow-FISH protocol and with the automated flow-FISH protocol:

1. Variation in telomere fluorescence values within one experiment and from experiment to experiment
2. Cell recovery
3. Time to process samples
4. Amounts of solutions and reagents used

First, we took aliquots of the paraformaldehyde-fixed bovine thymocytes (internal standard cells) and analyzed the variation in telomere fluorescence acquired from tube to tube and from experiment to experiment with the optimized/automated protocol. In Fig. 5, the variation in telomere fluorescence in MESF units from tube to tube (12 tubes for each experiment) is shown. Within the same experiment, the SD ranged from 0.6 MESF (experiment 2) to 1.5 MESF (experiment 3) (CV ranged from 1.5% to 3.7%). From experiment to experiment ($n = 26$), the MESF SD was 2.8 (7.1% coefficient of variation). Note, that the variation in telomere fluorescence of the control cells (2×10^5) was measured in the presence of test cells (typically around 0.5×10^6 human, baboon, or murine hematopoietic cells) to most accurately mimic a relevant situation. It is very likely that the variation in telomere fluorescence of control cells is influenced by the presence of other cells in the same test tube (dilution of telomere PNA-probe, binding of DNA to cells, etc.).

Two conclusions can be drawn from this experiment: Even with the optimized and most careful execution of each step in the protocol, there still is a certain variation in telomere fluorescence for the same biological sample between tubes within an experiment and from experiment to experiment. The variation within an experiment ranges from 0.3 to 0.7 kb and the variation from experiment to experiment can go up to 1.4 kb. When we looked at unfixed human, baboon, bovine, or murine hematopoietic cells, the variation in telomere fluorescence was in the range of 5–10% within an experiment and 20–30% from experiment to

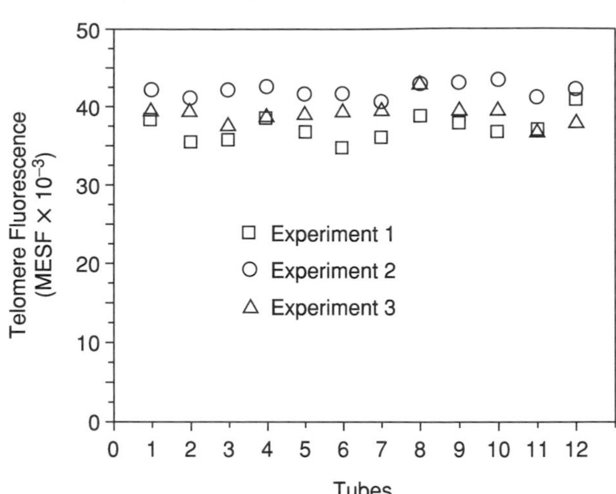

Fig. 5 Variation in telomere fluorescence of control cells within an experiment and from experiment to experiment. Shown are telomere fluorescence values (in molecular equivalents of soluble fluorochrome [MESF] $\times$ 10^{-3}) for 1% paraformaldehyde-fixed bovine thymocytes. For each experiment (experiment 1–3 indicated as different symbols), 12 tubes with the same amount of bovine thymocytes were hybridized and analyzed. Note that the average telomere fluorescence value ($\sim$40 $\times$ 10^3 MESF per cell) corresponds to about 40% of the calculated maximum fluorescence using the following assumptions: one hybridized PNA-fluorescein molecule yields one MESF unit; bovine cells have 60 chromosomes (120 telomeres per cell); with on average 15 kb of telomere repeats per telomere, each cell has 1.8 Mb of telomere "target" DNA allowing a maximum of 10^5 PNA molecules (18-mers) to hybridize. The measured value ($\sim$0.4 $\times$ 10^5 MESF) is even closer to the expected efficiency of about 70% or 0.7 $\times$ 10^5 MESF (Fig. 3) if the estimated 3 kb of subtelomeric DNA included in the TRF measurements used for calibration is subtracted from the estimated 15 kb of telomere repeats.

experiment. By using the bovine thymocytes as internal standards, we reduced this variation in telomere fluorescence to 2–5% within one experiment and to 5–10% from experiment to experiment. As a result, the minimum detectable difference in telomere length for human, baboon, bovine, or murine hematopoietic cells was calculated to be in the order of 0.5 kb.

Figure 6 illustrates the improvement in the reproducibility of telomere length measurements using the basic and current flow-FISH methods. In this experiment, the telomere length in hematopoietic cells from various bovine blood samples was calculated from two flow-FISH experiments (two different runs with the same samples at different times). With the basic method, the CV was 5.2% (n = 8) and with the new protocol 1.9% (n = 14). In addition, the reproducibility of telomere length measurements came into the range of ±0.5 kb between experiments. Many similar experiments with human, baboon, and murine hematopoietic cells have confirmed these findings.

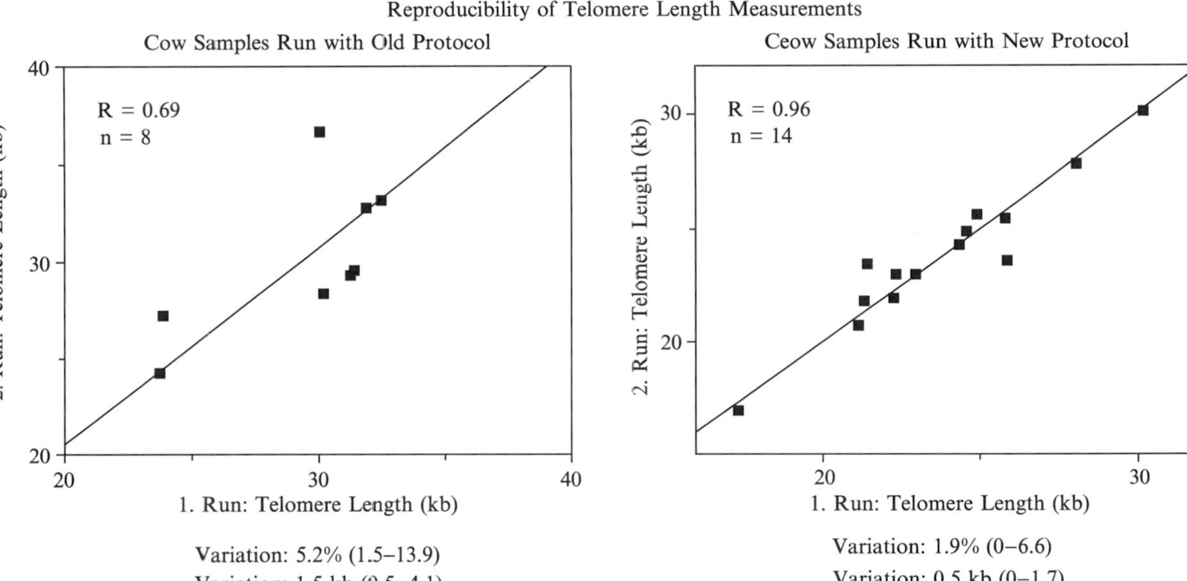

Fig. 6 Comparison of reproducibility in telomere length measurements in various bovine nucleated blood samples analyzed by the old flow–fluorescent *in situ* hybridization (FISH) protocol and the new version. For each sample, two telomere length measurements were performed at different times using each method. The data points on the graphs, where the results from the first run represent measured telomere length values, are plotted against the results from the second run. Correlation analysis was performed for all the data points obtained by the old protocol (n = 8, $R = 0.69$) and by the new protocol (n = 14, $R = 0.89$).

Because of improvements in the solutions used during the flow-FISH protocol (e.g., inclusion of glucose), the use of an internal standard (to guide the position and identify test cells on the flow cytometry plot more readily), and the more gentle mixing of the cell suspensions by the Hydra, we improved the cell recovery of cells for analysis on the flow cytometer from 1–10% to 25–50% of the initial cell count. As a result, telomere length measurements can now be performed with much lower cell numbers (in the range of 10^4) (Baerlocher and Lansdorp, 2003; Van Ziffle *et al.*, 2003).

With the Hydra robotic device, 96 tubes or 24 samples (duplicates of unstained and stained samples) can easily be processed at the same time. This is not feasible with the manual basic flow-FISH protocol. For example, the time to perform a wash step with the processes of aspiration, mixing, and addition of wash solution takes maximally 5 minutes for 96 tubes with the Hydra, whereas aspirating, mixing, and addition of wash solution to 96 tubes by hand (where you have to take each tube out of the rack, open it, and hold it slanted to aspirate fluid, etc.) takes at least 30–60 minutes. With the five wash steps in the flow-FISH protocol, manual processing of 96 tubes clearly becomes impractical. With automated flow-FISH, 24 duplicate samples can easily be processed within 1 day and, if necessary, still be acquired on the flow cytometer in the later afternoon/evening or the next morning.

Because of the smaller volume and diameter of the tubes in a standard 96-well plate, we tested out and adjusted all the volumes of solutions used during the various steps in the flow-FISH protocol. Most importantly, we reduced the volume for the hybridization solution (from 300 μl to 170 μl) and for the solutions with the antibodies (100 μl to 30 μl). The reduced volumes in the automated flow-FISH protocol have allowed significant cost reductions.

V. Critical Aspects of Methodology

A. PNA Probes

In collaboration with scientists at Applied Biosystems, an extensive survey was made of possible PNA probes that can hybridize with mammalian telomeric $(TTAGGG)_n$ repeats. It was found that targeting the G-rich strand was advantageous over targeting the C-rich strand because PNA probes containing TTAGGG sequences are poorly soluble and give rise to a high nonspecific staining/hybridization. It is possible that PNA probes containing the TTAGGG sequence, similar to DNA oligonucleotides containing the same sequence, form guanine-quadruplex structures, a stable four-stranded structure in which the repeating unit is a plane of four guanine bases (Sen and Gilbert, 1988). Using Q-FISH, it was found that 12-mers $(CCCTAA)_2$ PNA can hybridize but, as expected, at a lower efficiency than 18-mers containing the same sequence. The latter were selected for further studies based on their good solubility and predicted high specificity (the stability of PNA/DNA hybrids containing nucleotide mismatches is very poor (Ratilainen *et al.*, 2000)). Various fluorochromes were explored for labeling of $(CCCTAA)_3$ PNA in Q-FISH and flow-FISH techniques. For Q-FISH, the Cy3 dye appears ideal (good excitation at the 546-nm peak emission of the mercury light source in most fluorescence microscopes, very photostable, and an excellent quantum yield), whereas for flow cytometry the 488-nm argon laser available on most flow cytometers favored the use of fluorescein. This dye has some disadvantages because it is rather hydrophobic (as is PNA) and bleaches relatively fast. For flow cytometry, the exposure to the laser excitation is very short and relatively constant, so bleaching is not so much of a concern, but the hydrophobic nature of the fluorescein-labeled $(CCCTAA)_3$ PNA does impose specific requirements for prevention of nonspecific binding, which we found were best addressed by initial wash steps in 75% formamide. Although labeling of $(CCCTAA)_3$ PNA with two fluorescein molecules increased the fluorescence yield (not by a factor of two, presumably as a result of some quenching), the further increase in hydrophobicity and decreased solubility in aqueous solutions did not favor the use of such probes. The preferred way to label fluorescein to PNA is using succinimidyl esters of fluorescein rather than fluorescein because the former yields a more stable bond. Cy5-labeled telomere PNA probes can also be used, but because of the necessity of a helium-neon laser to excite the Cy5 dye, and because of a somewhat lower sensitivity of Cy5 detection on our flow cytometer, we prefer telomere-PNA probes labeled with fluorescein for the most sensitive telomere length measurements. It has not been possible to hybridize to both C and G strands quantitatively with the use of a C-strand and G-strand telomere PNA probes because the combination of both probes results in the formation of PNA hybrids. Phosphoramidite probes for centromeric repeats were described (O'Sullivan *et al.*, 2002). It is possible that synthetic oligonucleotide probes other than PNA can also be used in Q-FISH for telomere repeats if hybridization conditions for such probes can be developed that prevent annealing of complementary DNA sequences.

B. Fixation

In our hands, any fixative or cross-linking solution (paraformaldehyde, ethanol, methanol/acetic acid, BS3 or bis-[sulfosuccinimyl]-substrate) to preserve nuclear and membrane structures led to a decrease in the hybridization efficiency. Furthermore, such reagents did induce agglutination and clumping of cells, which resulted in a higher loss of cells and a greater variability in telomere fluorescence. Difficulties to control for the increased variability in hybridization efficiency led us to abandon such prefixation steps; however, it remains tempting to continue to explore this approach because the possibility to measure the telomere length in rare cell subpopulations, defined by (combinations of) labile cell surface markers, clearly would be advantageous (Batliwalla *et al.*, 2001; Plunkett *et al.*, 2001; Schmid *et al.*, 2002).

C. Choice of Fluorochromes

For the telomere-PNA probe, it is necessary to use heat-stable fluorochromes (excluding the use of highly fluorescent PE and other fluorescent proteins). In addition, it is optimal to choose a fluorochrome for the telomere-PNA probe that can be used on most flow cytometers (availability of lasers to excite the selected fluorochrome). As mentioned earlier, we decided to use fluorescein-labeled PNA as the telomere probe in our flow-FISH protocol. Initially, we used propidium iodide in a subsaturating concentration (0.06 μg/ml) as a DNA counterstain. However, because of its broad emission spectrum, which can lead to interference with the detection of the weak fluorescence from the fluorescein-labeled telomere-PNA probe (spectral overlap, energy transfer), we switched to LDS 751 at 0.01 μg/ml (Baerlocher *et al.*, 2002). LDS 751 can be excited by the 488-nm argon-ion laser available on most flow cytometers, and its emission spectrum is very distinct from that of fluorescein. Next to telomere-PNA-fluorescein and LDS 751, PE and Cy5 or allophycocyanin (APC) can be used as fluorochromes for antibodies. Because some of the bright fluorescence from PE-labeled antibodies will spill into the detector (FL1) used for the detection of telomere fluorescence, each sample needs to be individually compensated for cross fluorescence of FL2 into FL1. We decided to use the PE-labeled antibody for the detection of epitopes with a low abundance per cell (in our protocol, CD20–PE and CD57–PE). Whereas the signals from Cy5- or APC-labeled antibody (in our protocol, CD45RA–Cy5) are generally less bright than those from PE-labeled antibodies, good discrimination above background is typically possible into the FL4 channel of the FACSCalibur, although significant cross compensation with LDS 751 detected in FL3 is required.

VI. Pitfalls and Misinterpretation of Data

A. DNA Content

The telomere fluorescence in a cell has to be correlated to its DNA content to measure the average length of telomeres. To distinguish cells with 2N or 4N DNA, or to identify cells in S phase, cells are usually stained with a DNA dye at saturating concentrations. Unfortunately, most DNA dyes used at saturating concentration (except for DAPI, which is excited by an ultraviolet [UV] laser and, therefore, not useful on most flow cytometers) interfere notably with the detection of the relatively weak telomere fluorescence from PNA fluorescein (especially in human cells with short telomeres). Therefore, we had to drop the DNA dye concentration to subsaturating levels with the disadvantage of losing much of the resolution in the DNA profile (already compromised by the flow-FISH procedure: denaturation and loss of DNA). As a result, it is at this point not possible to directly correlate cell cycle status with telomere length and telomere length measurements by flow-FISH in rapidly dividing cells such as tumor cells or cultured cells remain problematic. No doubt, future flow-FISH protocols will aim to improve the resolution of both DNA and telomere fluorescence while preserving a linear relationship between telomere length and telomere fluorescence.

B. Background Fluorescence

Another major hurdle for accurate measurements of telomere length by flow-FISH is related to the autofluorescence of cells. Various cellular structures influence the autofluorescence properties of a cell. Such factors include various cell organelles and fluorescent molecules in the cytoplasm (such as lipofuscin, hemoglobin, and cytochromes). Some of these factors only marginally increase the autofluorescence in viable cells. However, with heat and formamide treatment, the autofluorescence may increase dramatically and this may interfere with the detection of specific telomere fluorescence. Cultured fibroblasts with a large volume of cytoplasm are an example of highly autofluorescent cells where it is difficult, if not impossible, to measure telomere length by flow-FISH accurately. Another problem is that the PNA telomere probe may bind nonspecifically to cells in ways that may vary between cell types (e.g., as a function of their size). Unfortunately, there are no good ways to control for this source of variation (different PNA "control" probes yield highly variable probe-specific "background" fluorescence). Extreme caution should, therefore, be used to apply the techniques described in this chapter to cell types other than the ones that have been investigated and validated in extensive experiments. In developing protocols for such novel cell types, inspection of cells under a fluorescence microscope to confirm that fluorescence is derived only from distinct (telomere) spots in the nucleus is of critical importance.

VII. Comparison with Other Methods

There is no "gold standard" or optimal method for comparisons of telomere length measurements by flow-FISH. Several times, we have compared the telomere length measured by flow-FISH to telomere length values obtained by Southern blot. Telomere length measurements with either the basic or the optimized/automated flow-FISH protocol typically showed a high correlation (with R values ≥ 0.80) with measurements obtained by Southern blot. However, improvements made in the sensitivity of telomere length measurements by the optimized/automated flow-FISH protocol do not really show up in this type of comparison because the variation in measurements by Southern blot is typically much higher (with an estimated SD of 1–3 kb for measurements of telomeres in the range of 5–15 kb). In addition, only short telomeres (within a range of 5–30 kb) can be optimally assessed by Southern blot. Therefore, no comparisons between methods can be made using cells with very long telomeres, such as murine cells. Telomere length measurements by flow-FISH were also compared to telomere length measurements by Q-FISH. Such results also correlate very well ($R > 0.8$). However, with Q-FISH, cells have to be able to divide to obtain metaphase spreads. The resulting selection of cells combined with errors in measurements of telomere length using limited numbers of metaphase cells can easily explain the differences in telomere length measurements between the two methods.

VIII. Applications

A. Bovine Samples

The regulation of telomere length in the germline and during early embryogenesis is of great interest. At one point, it was believed that organisms created by nuclear transfer, such as the cloned sheep Dolly, might undergo premature senescence and aging resulting from insufficient repair of preexisting short telomeres by the cloning procedure (Shiels *et al.*, 1999). In collaborative studies, we looked at the telomere length in hematopoietic cells of cloned calves derived by nuclear transfer of either senescent or "young donor" nuclei from somatic cells into enucleated oocytes (Lanza *et al.*, 2000). Almost all cloned animals showed longer telomere length values in hematopoietic cells than those measured in age-matched control samples or in the cells used for nuclear transfer. These flow-FISH telomere length data clearly demonstrate that telomeres can be extended upon nuclear transfer and cloning. This is just one example in which we were able to assess the telomere length in cells within a very short period and with a higher sensitivity than is possible by Southern blot. Many other examples, most notably in studies of mutant mice, can be found in the literature.

B. Baboons

Another area of interest is the age-related regulation of telomere length in various subpopulations of cells *in vivo*. Telomere length measurements in lymphocytes and granulocytes of humans at different ages suggest that the decline in telomere length with age is not linear. In humans, a rapid decline of telomere length is observed early in childhood, with a more gradual but steady telomere loss thereafter (Rufer *et al.*, 1999). Because telomere length shows a large variation among individuals at any given age, a highly sensitive and reproducible method to measure telomere length is necessary to exclude method-related variation, and ideally, longitudinal studies of telomere length with age should be performed. Baboons are in many aspects (hematopoietic stem cell turnover, development of immune system, etc.) very similar to humans, and blood samples from baboons are somewhat more easily available for longitudinal studies. For these reasons, we examined the telomere length in lymphocytes and granulocytes of 22 baboons (Papio hamadryas cynocephalus) ranging in age from 2 months to 26 years (Baerlocher *et al.*, 2003a). In addition, we developed multicolor flow-FISH for baboon hematopoietic cells to look at telomere length in diverse subsets of leukocytes (granulocytes, naive T cells, memory T cells, and B cells). Telomere length in granulocytes, B cells, and subpopulations of T cells decreased with age. However, the telomere length kinetics were lineage and cell subset specific. T cells showed the most pronounced overall decline in telomere length, and memory T cells with very short telomeres accumulated in old animals. In contrast, the average telomere length values in B cells remained relatively constant from middle age onward. Individual B cells showed highly variable telomere length, and B cells with very long telomeres were observed after the ages of 1–2 years. To look more specifically at the decline in telomere length after birth inferred from population studies, we also studied the telomere length dynamics in subsets of hematopoietic cells longitudinally every 2 weeks in the first 1 and 1/2, years of four baboons. The telomere length in granulocytes, T lymphocytes, and B lymphocytes was maintained (variation <0.5 kb between serial samples) or even slightly increased in the first 2–3 months for three animals and up to 9 months for one animal. After this period, the telomere length values started to decrease slowly but significantly. In addition, we found that certain B lymphocytes with a much higher fluorescence appeared in the circulation between 9 and 16 weeks of life, and heterogeneity within B cells was clearly detectable after the age of 1 year (Baerlocher *et al.*, 2003b). Again, these telomere length data clearly demonstrate the value of highly sensitive and reproducible measurements of telomere length in combination with immunophenotyping.

C. Humans

A clinically highly interesting and relevant question is whether the replicative demand imposed on transplanted donor stem and progenitor cells may result in substantial telomere shortening, thereby potentially compromising the long-term function of the marrow graft. We looked at a situation in which a 10-year-old male recipient was transplanted with donor cells from his 64-year-old grandmother (Awaya *et al.*, 2002). Histograms for the telomere fluorescence in subtypes of peripheral blood leukocytes from the donor and the recipient are shown in Fig. 7. The telomere length values in subsets of cells obtained from the donor cells are comparable to those of age-matched control individuals. However, values in post-transplantation donor cells harvested from the patient were 0.8–2.0 kb shorter than those in the donor, and these values were much lower than those in normal individuals age-matched with the recipient. Furthermore, the patient presented similar telomere length values for CD45RA-positive naive T cells and CD45RA-negative memory T cells and showed practically no B cells with longer telomeres as can typically be found in normal individuals. These findings suggest that the telomere erosion with cell division in this patient is not compensated by telomerase activity. The very low telomere length values in various lymphocyte subsets could limit cell replication and compromise immune function as a result. In general, studies of the telomere length in subsets of cells in various (autoimmune) immune disorders are in their infancy and the results of such studies are eagerly awaited.

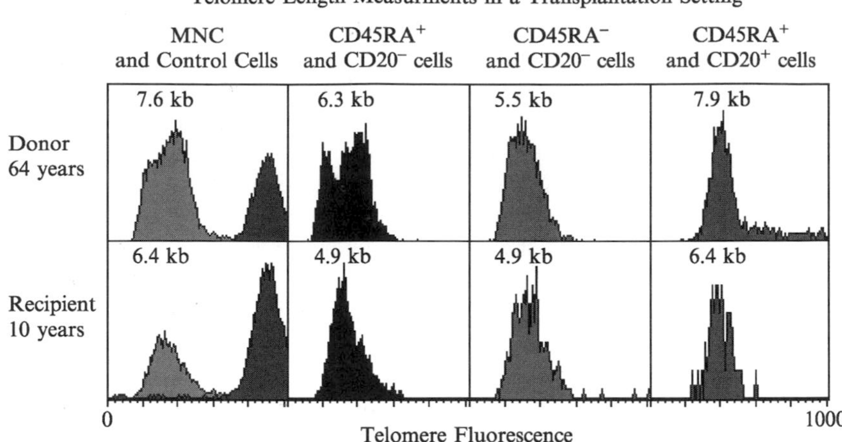

Fig. 7 Telomere length measurements in a transplantation setting. Histograms of telomere fluorescence in control cells and subsets of leukocytes (MNCs, mononuclear cells; CD45RA positive, CD20-negative naive T cells; CD45RA negative, CD20-negative memory T cells; CD45RA positive, CD20-positive B cells) from the donor (64 years old) and the recipient (10 years old) after transplantation. The telomere length in kilobases is calculated from the median telomere fluorescence.

IX. Future Directions

A. DNA Dyes

Improvements in DNA profiles without affecting the acquisition of quantitative telomere fluorescence is one of the next goals for telomere length measurements. Such an improvement will enable studies of the timing of telomere replication during the cell cycle in various cell types and telomere length measurements in dividing cell populations. One way to optimize the DNA profile without affecting the telomere fluorescence might be to use DAPI. However, this DNA dye will require access to a flow cytometer with a UV laser. Furthermore, there is hope that new DNA dyes that fulfill all the requirements of a DNA counterstain in the flow-FISH protocol will become available.

B. Antibodies

To measure telomere length in diverse subsets of hematopoietic cells (CD4-, CD8-, CD19-positive cells, etc.), additional antibodies that are able to recognize and bind specifically to epitopes following the harsh flow-FISH denaturation step would be of great advantage. Another option is to use already available antibodies with better fixatives or linker molecules, which do not affect the hybridization efficiency. As was discussed, such fixation steps require extremely careful control to exclude variable effects on hybridization efficiency.

Acknowledgments

Work in the laboratory of P. M. L. is supported by grants from the National Institutes of Health (AI29524), the Canadian Institutes of Health Research (MOP38075), the National Cancer Institute of Canada (with support from the Terry Fox Run), and the Stem Cell Network. G. M. B. was supported by the Swiss National Foundation and the Bernese Cancer League. Jennifer Mak, Teri Tien, and Irma Vulto are thanked for excellent technical assistance. We would like to thank Jim Coull, Krishan Taneja, Jens Hyldig-Nielsen, and Tish Creasey at Applied Biosystems (Boston) for their generous help in providing various PNA probes. Michael Egholm (Molecular Staging Inc., New Haven, Connecticut) is thanked for initial encouragement and Andre van Agthoven (Beckman Coulter Immunotech, Marseille, France) for generous gifts of CD20-PE.

References

Awaya, N., Baerlocher, G. M., Manley, T. J., Sanders, J. E., Mielcarek, M., Torok-Storb, B., and Lansdorp, P. M. (2002). Telomere shortening in hematopoietic stem cell transplantation: A potential mechanism for late graft failure? *Biol. Blood Marrow Transplant.* **8,** 597–600.

Baerlocher, G. M., and Lansdorp, P. M. (2003). Telomere length measurements in leukocyte subsets by automated multicolor flow-FISH. *Cytometry* **55A,** 1–6.

Baerlocher, G. M., Mak, J., Roth, A., Rice, K. S., and Lansdorp, P. M. (2003). Telomere shortening in leukocyte subpopulations from baboons. *J. Leukoc. Biol.* **73,** 289–296.

Baerlocher, G. M., Mak, J., Tien, T., and Lansdorp, P. M. (2002). Telomere length measurements by fluorescence *in situ* hybridization and flow cytometry: Tips and pitfalls. *Cytometry* **47,** 89–99.

Baerlocher, G. M., Vulto, I. M., Rice, K. S., and Lansdorp, P. M. (2003). Longitudinal measurements of telomere length in leukocyte subsets from baboons. *Blood* **102,** 567a(Abstract).

Baird, D. M., Rowson, J., Wynford-Thomas, D., and Kipling, D. (2003). Extensive allelic variation and ultrashort telomeres in senescent human cells. *Nat. Genet.* **33,** 203–207.

Batliwalla, F. M., Damle, R. N., Metz, C., Chiorazzi, N., and Gregersen, P. K. (2001). Simultaneous flow cytometric analysis of cell surface markers and telomere length: Analysis of human tonsillar B cells. *J. Immunol. Methods* **247,** 103–109.

Blackburn, E. H. (2001). Switching and signaling at the telomere. *Cell* **106,** 661–673.

Bodnar, A. G., Ouellette, M., Frolkis, M., Holt, S. E., Chiu, C.-P., Morin, G. B., Harley, C. B., Shay, J. W., Lichtsteiner, S., and Wright, W. E. (1998). Extension of life-span by introduction of telomerase into normal human cells. *Science* **279,** 349–353.

Cheung, S. W., Tishler, P. V., Atkins, L., Sengupta, S. K., Modest, E. J., and Forget, B. G. (1977). Gene mapping by fluorescent *in situ* hybridization. *Cell Biol. Int. Rep.* **1,** 255–262.

Ding, H., Schertzer, M., Wu, X., Gertsenstein, M., Selig, S., Kammori, M., Pourvali, R., Poon, S., Vulto, I., Chavez, E., Tam, P. P., Nagy, A., and Lansdrop, P. M. (2004). Regulation of murine telomere length by Rtel: An essential gene encoding a helicase-like protein. *Cell* **117**(7), 873–886.

Griffith, J. D., Comeau, L., Rosenfield, S., Stansel, R. M., Bianchi, A., Moss, H., and de Lange, T. (1999). Mammalian telomeres end in a large duplex loop. *Cell* **97,** 503–514.

Hultdin, M., Gronlund, E., Norrback, K., Eriksson-Lindstrom, E., Just, T., and Roos, G. (1998). Telomere analysis by fluorescence *in situ* hybridization and flow cytometry. *Nucl. Acids Res.* **26,** 3651–3656.

Kim, N. W., Piatyszek, M. A., Prowse, K. R., Harley, C. B., West, M. D., Ho, P. L. C., Coviello, G. M., Wright, W. E., Weinrich, S. L., and Shay, J. W. (1994). Specific association of human telomerase activity with immortal cells and cancer. *Science* **266,** 2011–2015.

Lansdorp, P. M. (1996). Close encounters of the PNA kind. *Nat. Biotechnol.* **14,** 1653.

Lanza, R. P., Cibelli, J. B., Blackwell, C., Cristofalo, V. J., Francis, M. K., Baerlocher, G. M., Mak, J., Schertzer, M., Chavez, E. A., Sawyer, N., Lansdorp, P. M., and West, M. D. (2000). Extension of cell life-span and telomere length in animals cloned from senescent somatic cells. *Science* **288,** 665–669.

Martens, U. M., Zijlmans, J. M., Poon, S. S. S., Dragowska, W., Yui, J., Chavez, E. A., Ward, R. K., and Lansdorp, P. M. (1998). Short telomeres on human chromosome 17p. *Nat. Genet.* **18,** 76–80.

Nielsen, P. E., and Egholm, M. (1999). An introduction to peptide nucleic acid. *Curr. Issues Mol. Biol.* **1,** 89–104.

O'Sullivan, J. N., Bronner, M. P., Brentnall, T. A., Finley, J. C., Shen, W. T., Emerson, S., Emond, M. J., Gollahon, K. A., Moskovitz, A. H., Crispin, D. A., Potter, J. D., and Rabinovitch, P. S. (2002). Chromosomal instability in ulcerative colitis is related to telomere shortening. *Nat. Genet.* **32,** 280–284.

Olovnikov, A. M. (1971). Principles of marginotomy in template synthesis of polynucleotides. *Dokl. Akad. Nauk. SSSR.* **201,** 1496–1499.

Plunkett, F. J., Soares, M. V., Annels, N., Hislop, A., Ivory, K., Lowdell, M., Salmon, M., Rickinson, A., and Akbar, A. N. (2001). The flow cytometric analysis of telomere length in antigen-specific CD8+ T cells during acute Epstein–Barr virus infection. *Blood* **97,** 700–707.

Poon, S. S. S., Martens, U. M., Ward, R. K., and Lansdorp, P. M. (1999). Telomere length measurements using digital fluorescence microscopy. *Cytometry* **36,** 267–278.

Ratilainen, T., Holmen, A., Tuite, E., Nielsen, P. E., and Norden, B. (2000). Thermodynamics of sequence-specific binding of PNA to DNA. *Biochemistry* **39,** 7781–7791.

Ray, A., and Norden, B. (2000). Peptide nucleic acid (PNA): Its medical and biotechnical applications and promise for the future. *FASEB J.* **14,** 1041–1060.

Rufer, N., Brummendorf, T. H., Kolvraa, S., Bischoff, C., Christensen, K., Wadsworth, L., Schultzer, M., and Lansdorp, P. M. (1999). Telomere fluorescence measurements in granulocytes and T lymphocyte subsets point to a high turnover of hematopoietic stem cells and memory T cells in early childhood. *J. Exp. Med.* **190,** 157–167.

Rufer, N., Dragowska, W., Thornbury, G., Roosnek, E., and Lansdorp, P. M. (1998). Telomere length dynamics in human lymphocyte subpopulations measured by flow cytometry. *Nat. Biotechnol.* **16,** 743–747.

Schmid, I., Dagarag, M. D., Hausner, M. A., Matud, J. L., Just, T., Effros, R. B., and Jamieson, B. D. (2002). Simultaneous flow cytometric analysis of two cell surface markers, telomere length, and DNA content. *Cytometry* **49,** 96–105.

Sen, D., and Gilbert, W. (1988). Formation of parallel four-stranded complexes by guanine-rich motifs in DNA and its implications for meiosis. *Nature* **334,** 364–366.

Shay, J. W., and Bacchetti, S. (1997). A survey of telomerase activity in human cancer. *Eur. J. Cancer* **33,** 787–791.

Shiels, P. G., Kind, A. J., Campbell, K. H., Waddington, D., Wilmut, I., Colman, A., and Schnieke, A. E. (1999). Analysis of telomere lengths in cloned sheep. *Nature* **399,** 316–317.

Trask, B. (1991). Fluorescence *in situ* hybridization. Application in cytogenetic and gene mapping. *Trends Genet.* **7,** 149–154.

Trask, B., Van Den Engh, G., Pinkel, D., Mullikin, J., Waldman, F., van Dekken, H., and Gray, J. (1988). Fluorescence *in situ* hybridization to interphase cell nuclei in suspension allows flow cytometric analysis of chromosome content and microscopic analysis of nuclear organization. *Hum. Genet.* **78,** 251–259.

Van Ziffle, J. A. G., Baerlocher, G. M., and Lansdorp, P. M. (2003). Telomere length in subpopulations of human hematopoietic cells. *Stem Cells* **21,** 654–660.

Vaziri, H., and Benchimol, S. (1998). Reconstitution of telomerase activity in normal human cells leads to elongation of telomeres and extended replicative life span. *Curr. Biol.* **8,** 279–282.

von Zglinicki, T. (2002). Oxidative stress shortens telomeres. *Trends Biochem. Sci.* **27,** 339–344.

Vulliamy, T., Marrone, A., Goldman, F., Dearlove, A., Bessler, M., Mason, P. J., and Dokal, I. (2001). The RNA component of telomerase is mutated in autosomal dominant dyskeratosis congenita. *Nature* **413,** 432–435.

Watson, J. D. (1972). Origin of concatameric T7 DNA. *Nat. New Biol.* **239,** 197–201.

CHAPTER 32

Detecting Copy Number Changes in Genomic DNA: MAPH and MLPA

Stefan J. White, Martijn H. Breuning, and Johan T. den Dunnen
Center for Human and Clinical Genetics
Leiden University Medical Center
2333 AL Leiden, Nederland

I. Introduction

Copy number changes (i.e., deletions, duplications, and amplifications in genomic DNA) are involved in many genetic diseases. Technically, detection of these rearrangements is rather difficult, mainly because the assay applied should be quantitative. This latter aspect makes many methods used for the detection of rearrangements in genomic DNA unsuitable. In essence, most methods currently applied are polymerase chain reaction (PCR) based, meaning that an endpoint analysis is performed on a reaction that is near saturation and therefore does not provide quantitative information.

0091-679X/04 $35.00

Although many methods for detecting copy number changes have been described (Armour *et al.*, 2002), the most versatile techniques applied have been fluorescent *in situ* hybridization (FISH), Southern blotting, and quantitative PCR. FISH combines hybridization with a fluorescently labeled probe with microscopic analysis of genomic DNA (Ligon *et al.*, 2000; Petrij *et al.*, 2000). Because of its relative reliability and ease of analysis, the technique has been widely used in diagnostic laboratories. Its main disadvantages are the relatively low resolution obtained and the workload involved (i.e., culturing cells, making chromosome preparations, and performing hybridizations).

As an alternative, Southern blotting has also been broadly applied (Koenig *et al.*, 1987; Petrij-Bosch *et al.*, 1997). Good-quality blots allow dosage analysis to detect deletions and duplications. In addition, one can scan for the unique junction fragments of the rearrangements and use these for diagnosis (Den Dunnen *et al.*, 1987). Although this latter possibility requires some effort to identify a junction fragment, when found it does provide a powerful diagnostic tool. In this respect, though technically demanding, the combination of pulsed-field gel electrophoresis (PFGE) and Southern blotting can be very informative, facilitating the use of a nearby probe to detect all rearrangements from a distance (Den Dunnen *et al.*, 1987; Van Der Maarel *et al.*, 1999).

Quantitative PCR has been tried by many but turned out to be technically demanding, difficult to implement, and requiring specialized equipment. The development of real-time PCR technology has revived the diagnostic application of quantitative PCR (Anhuf *et al.*, 2003; Wilke *et al.*, 2000), but it failed to take away another drawback, the difficulty of multiplexing (i.e., measuring copy number changes of several DNA sequences in parallel).

In this chapter, we describe multiplex amplifiable probe hybridization (MAPH) (Armour *et al.*, 2000) and multiplex ligation-dependent probe amplification (MLPA) (Schouten *et al.*, 2002), two new methodologies developed specifically to detect copy number changes in many target sequences simultaneously.

II. Reactions

The MAPH and MLPA protocols are outlined in Fig. 1 and discussed in more detail in the following sections.

A. MAPH

The MAPH assay is based on the quantitative recovery of probes after hybridization to immobilized genomic DNA. Probes for MAPH analysis were initially created by cloning small DNA fragments into a vector (Armour *et al.*, 2000). These fragments were generated either by PCR or by restriction digestion of larger DNA constructs (e.g., PACs or BACS). The specific probes were then prepared by amplification from the vector using primers specific for the vector sequence. The

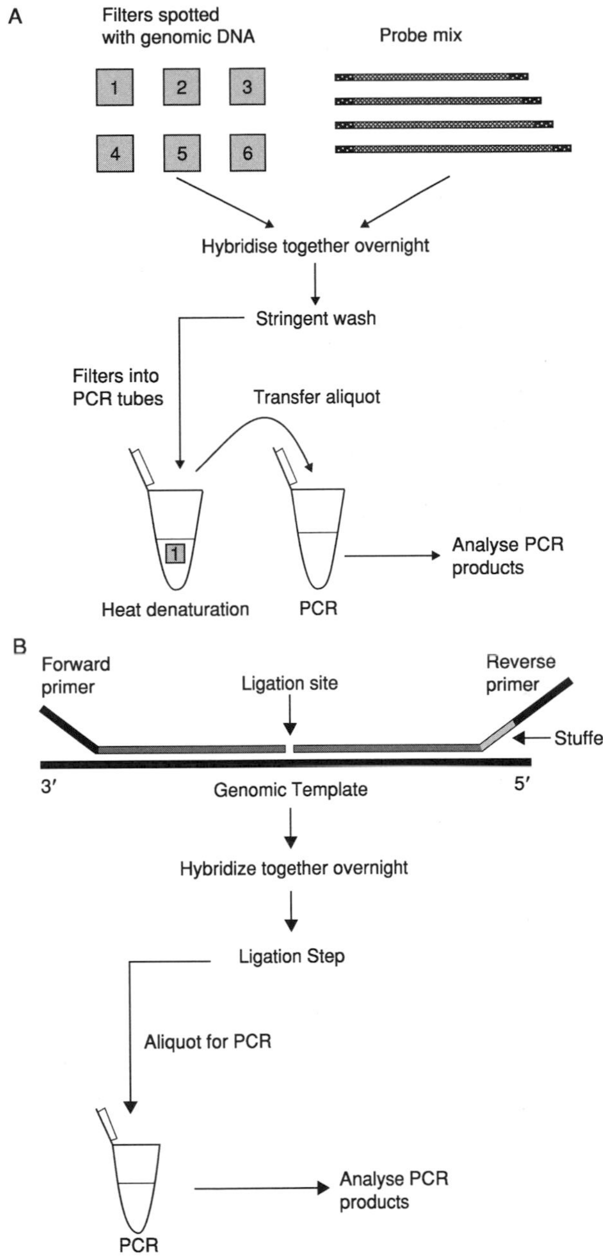

Fig. 1 An outline of multiplex amplifiable probe hybridization (MAPH) and multiplex ligation-dependent probe amplification (MLPA). (A) In the MAPH technique, a series of probes with identical ends are hybridized to genomic DNA immobilized on a filter. After stringent washing, the specifically bound probes are released into solution and an aliquot of this is used to seed a polymerase chain reaction (PCR). (B) In the MLPA technique, only if the two half-probes hybridize adjacently on a target sequence can they be ligated and amplified using the universal primers in a subsequent PCR. Probes can be distinguished by using stuffer sequences of different lengths.

drawback of the cloning approach is that the preparation of a large number of probes is arduous and time consuming.

An alternative method of probe preparation is to design primers to consist of two parts (Reid *et al.*, 2003): a unique section, which is used for the amplification from genomic DNA, and an identical 5′ priming sequence, which is the same for each probe. Once each probe is amplified from genomic DNA, the products can then be combined into a probe mix. Because they all have the same ends, the subsequent amplification can take place in exactly the same manner as for cloned probes.

A potential disadvantage of this approach is that during the original PCR amplification from genomic DNA, more than one PCR product could be amplified. These extra products would then be present in the probe mix, and the extra peaks would appear on any subsequent trace. In practice, however, we have not seen this as a common problem, and any problematic probes could be cloned and/ or gel purified if necessary.

There are several criteria to which the probes need to conform. First, each sequence must be unique. This can be tested using the BLAST (Altschul *et al.*, 1990) or BLAT (Kent, 2002) program to check the sequence against the genomic database. Second, the hybridizing segment of each probe should ideally have a GC content of between 40% and 60%. Some regions of the genome do not fall within these limits, and the probe length should be adjusted accordingly. Probes for regions with a GC content as low as 30% have been successfully made, but these were all longer than 500 base pairs. For difficult regions of the genome, it may be necessary to try several sequences before a good probe is found. Third, each probe must be a different length so the probes can be distinguished after electrophoretic separation.

Mixes are prepared by combining all probes so that the final concentration of each is about 0.5–1.0 ng/μl. We routinely store all probe mixes at −20 °C, and these have been successfully used up to 18 months after preparation with repeated freeze/thaw cycles.

Protocols have been outlined in previous articles (Armour *et al.*, 2000; White *et al.*, 2002), and complete descriptions are available at the following web sites (www.nottingham.ac.uk/~pdzjala/maph/ and www.dmd.nl/DMD_MAPH.html). Because it is necessary to remove the unbound probes after the hybridization step, the genomic DNA has to be immobilized. This is done by denaturing the DNA in NaOH and applying it to a small nylon filter. Because the filters are combined in a single tube during the hybridization step, they must be marked so that they can be easily distinguished from each other. It is possible to cut the filters into different shapes, but the most convenient method is to number each filter with a sharp pencil. This mark will remain visible after all the washing steps.

We have found that the easiest way of applying the DNA to the filter is to pipette each DNA sample into a well of micotiter plate, already containing the NaOH. Each filter can then be put into the appropriate well, and the DNA allowed to adsorb onto the filter (Fig. 2).

After the filters are dry, the DNA is fixed to the filter, either by ultraviolet (UV) cross-linking or by heating the filters at 80 °C for 1 hour. The filters are then combined in a prehybridization mixture from 2 hours to overnight, followed by addition of the hybridization mix. This mixture contains the probes of interest along with competitive DNA (e.g., C_0t1, fragmented *Escherichia coli* or herring sperm DNA, as well as sequences that block any interactions between the common ends of the probes). After overnight incubation, the hybridization mixture is removed, followed by stringent washing. To avoid accidental loss of the filters during this step, we first pour off washing solution into a beaker, because it is easier to recover filters from this than the U-bend of a sink. Alternatively, a tea strainer or something similar can be used.

After the washing steps, each filter is placed individually into a PCR tube containing 1 × PCR buffer. After heating to 95 °C for 5 minutes, the previously bound probes are released into solution, and an aliquot of this is added to a PCR mix. Optimization of the PCR regarding such factors as cycle number and denaturation time may influence the final results.

Fig. 2 The use of a microtiter plate for applying denatured genomic DNA to the filters. The number on the filter held in the forceps will be visible after all the washing steps.

B. MLPA

MLPA (Schouten *et al.*, 2002) is based on the ligation of two adjacently annealing oligonucleotides, followed by a quantitative PCR amplification of the ligated products. The left-hand half-probe is usually the shorter of the two probes (typically 45–70 nt) and is chemically synthesized. This half-probe is composed of two sections: a unique annealing sequence and terminal priming sequences common to all probes. The right-hand half-probe can be up to 440 nt, which cannot be routinely synthesized. Instead, a series of M13 vectors has been created, each with a "spacer" sequence of a different length (Schouten *et al.*, 2002). Two complementary oligonucleotides are annealed to each other, followed by ligation into the modified M13 vector. It is the spacer that determines the final length of the half probe. The resulting M13 construct containing the annealed oligonucleotides is isolated in single-stranded form, and the cloned sequence, including the spacer and priming sequence, can be isolated from the vector by restriction digestion.

It is also possible to use chemically synthesized oligos for the right-hand half-probe as well, but to facilitate ligation it is necessary that this sequence is 5′ phosphorylated. Although chemically synthesizing both probes greatly accelerates the probe production process, because of size constraints, the number of probes that can be subsequently combined within a single set is limited.

The MLPA protocol is described in full in Schouten *et al.* (2002) and is available at the following web site (www.mrc-holland.com/). As outlined in the notes of the protocol, several modifications are possible. To save on reagents, we can reduce the PCR volume from 50 μl to 25 μl without influencing the results. Combining the two pre-PCR mixes and adding 5 μl of the ligated products on ice makes the overall procedure easier, particularly when working with many samples. As with MAPH, we may need to adjust the number of PCR cycles to obtain sufficient product for analysis.

III. Analysis

For both MAPH and MLPA, the methods of data analysis are effectively identical. The PCR products are separated by electrophoresis and each probe is quantified, with the relative amount of each product being proportional to the copy number of the locus being tested.

A. Product Separation

In the original MAPH protocol (Armour *et al.*, 2000), the probes were labeled with a ^{33}P-5′-end–labeled primer and separated on a polyacrylamide gel, with the bands being quantified using a PhosphorImager screen. A faster and easier method is to use a fluorescently labeled primer during the PCR amplification (White *et al.*, 2002) and separate the products on a sequencing system, either a polyacrylamide gel or through polymer-filled capillaries (Fig. 3). This approach

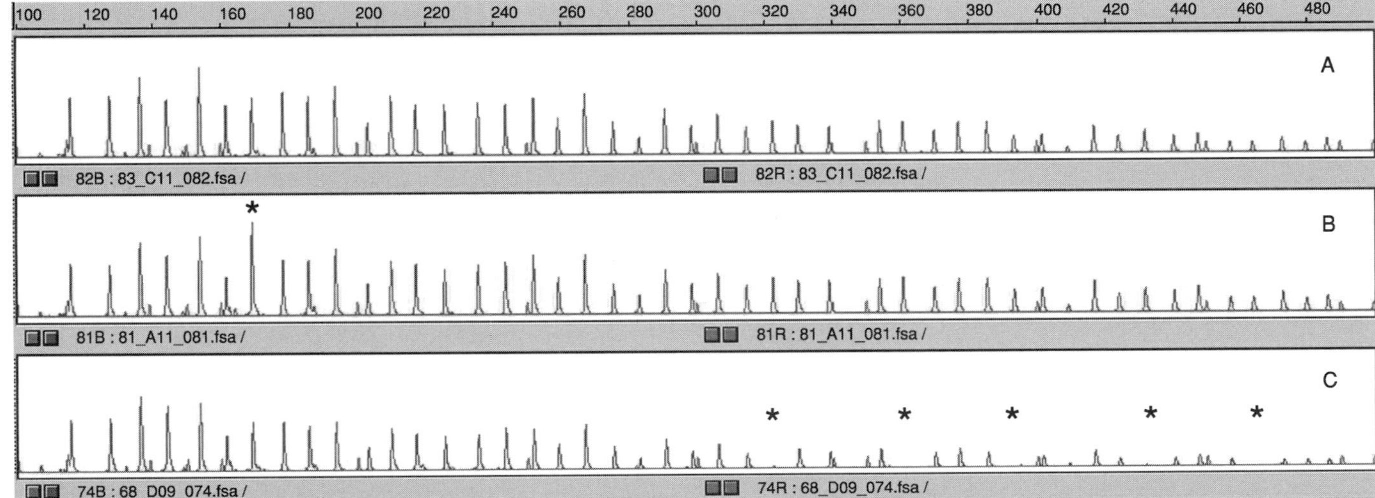

Fig. 3 Peaks obtained following separation of one of the multiplex ligation-dependent probe amplification probe sets for the DMD gene on the ABI 3700. Shown are typical patterns from (A) an unaffected individual, (B) a male patient with a single exon duplication (*) and (C) a male deleted for 5 exons, with the missing peaks being indicated (*). (Probe set kindly provided by Jan Schouten, MRC-Holland, Nederland.) (See Color Insert.)

also simplifies the analysis of the products, as software (e.g., GeneScan and Genotyper from Applied Biosystems) has been integrated. To obtain sufficient signal, it may be necessary to concentrate the PCR, because the amount of product to be added depends on the sensitivity of the system.

Another possibility is to use the Lab-on-a-Chip from Agilent (Fig. 4). This DNA chip can separate 12 samples in 45 minutes. The resolution on the DNA500 chip is about 20 bp, meaning that up to 15 probes can be analyzed, and it is not

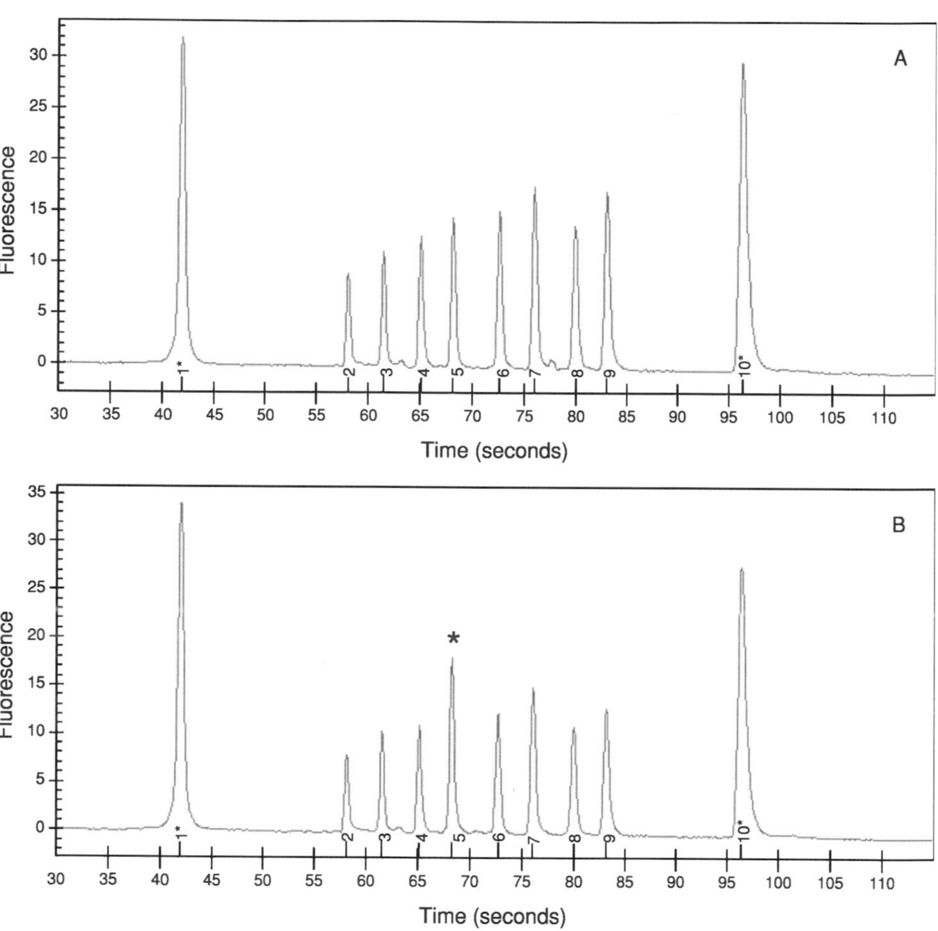

Fig. 4 An example of traces obtained by analyzing multiplex amplifiable probe hybridization (MAPH) products on the Lab-on-a-Chip from Agilent. In this example, the probes have a spacing of 30–40 base pairs. As can be seen, this spacing could be reduced if necessary to allow more probes to be used. A typical pattern from an unaffected individual is shown (A), and a single locus duplication is indicated (*) (B).

necessary to use a fluorescently labeled primer. This approach might be attractive when relatively few loci and/or samples need to be tested (White *et al.*, 2003).

B. Data Analysis

Several methods have been described for data analysis. At the most basic level, peaks can be visually compared (Schouten *et al.*, 2002), for example, by overlaying control and test traces. In theory, any deletion or duplication should be sufficiently obvious without any calculations being performed. Although this is usually true for deletions, duplications involving a single probe will not always be apparent. For this reason and to automate the analysis, it is recommended to perform a computer-based analysis.

Software such as GeneScan provides data about the peak height and peak area, both of which have been used in different reports (Hollox *et al.*, 2002; Montagna *et al.*, 2003; Rooms *et al.*, 2004; White *et al.*, 2002) with equivalent results. Most methods of calculation described have been based on dividing the value of a given probe by the sum of 2 or more other probes to obtain a ratio. These are preferably control probes that are unlinked to the loci in question. The "nearest neighbor" method can be used, in which the values of the four nearest peaks are added. This means that each probe will be normalized against a different group of probes. It is also possible to add the values of all probes in a trace (a global approach) and divide each probe by that value. Which method to use depends on several factors, primarily how many probes and samples you expect to be affected. For example, subtelomeric rearrangements are found in about 5% of mentally retarded patients, and it is unlikely that more than two probes would be affected. It should, therefore, be no problem to use the nearest-neighbor or global approach, because the vast majority of probes and samples will be unaffected. In contrast, sets composed of probes for all exons within one gene should be analyzed by another method. It is not impossible for most if not all of the exons within a gene to be deleted, which means that none of these probes should be used for normalization purposes. It is, therefore, essential to add control probes for loci elsewhere in the genome.

The ratio obtained for each probe is averaged across a series of samples to obtain a normalized value, usually corresponding to a copy number of 2 (one for X-linked probes in a male). The original ratio for each sample is then divided by the normalized value and in an unaffected situation should give a number distributed around 1. Under ideal circumstances, a deletion (1:2) would give a value 50% lower (i.e., 0.5), and a duplication (3:2) a value 50% higher (i.e., 1.5) (Fig. 5). Of course each probe will show a certain level of variation, and it is this variation that determines the degree of certainty of each measurement. The variation is dependent on several factors. Amplification by PCR will introduce a certain level of variability; this can be determined by performing several amplifications from a single pre–MAPH/MLPA. Each probe itself will have a certain amount of variance, depending on several factors such as GC content, degree of homology with

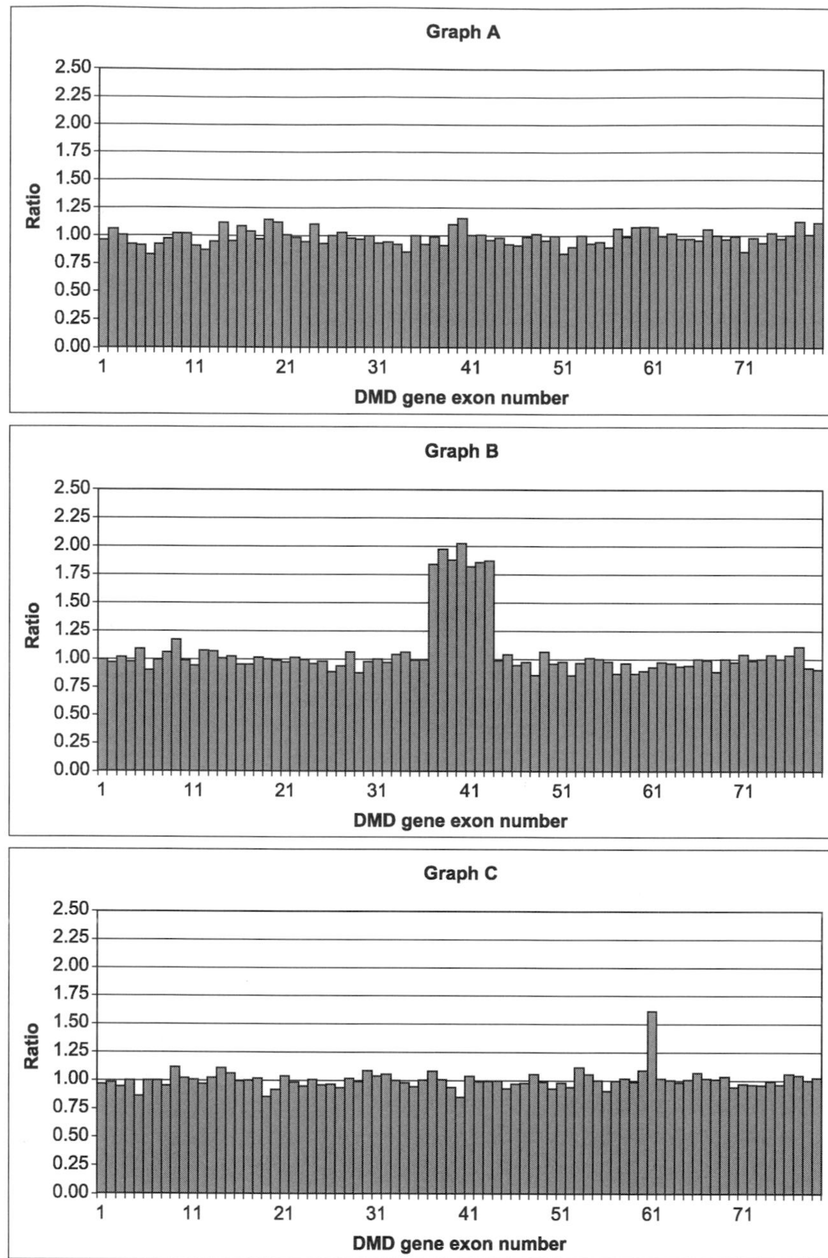

Fig. 5 Analysis of multiplex ligation-dependent probe amplification (MLPA) reactions with two DMD probe sets after normalization. Graph A shows the ratios of all probes to be clustered around 1.0, indicating an unaffected individual. Graph B has the ratios of seven probes around 2.0, corresponding to a duplication of exons 37–43 in a male patient. A single probe has a ratio about 1.5 in graph C; this is from a female carrier of an exon 61 duplication.

other regions in the genome, possible interactions with other probes, and amplification efficiency. These influences can be estimated by calculating the standard deviation for each probe over a series of control samples. The accuracy of each probe in detecting a true copy number is also important. For this reason, it is desirable to be able to test each probe on samples with known mutations.

Different criteria can be applied for deciding whether a given result is significant.

At the most basic level, thresholds can be set, usually at 0.75 and 1.25. A more statistical approach can be used by calculating the standard deviation either for each probe across all samples or for all probes within a sample. Because the variation of each probe also depends on the quality of the DNA being tested, it does not always follow that the standard deviation seen on control samples will be the same as that seen on test samples. This usually, however, provides a good estimate as to the overall reliability of any given probe.

The use of bivariate analysis has also been described (Hollox *et al.*, 2002). This relies on duplicate testing of each sample and allows the user to decide beforehand what the false-negative rate will be. A full explanation is provided at the following web site (www.nottingham.ac.uk/~pdzjala/maph/ststats.pdf).

The exact criteria used for determining a significant result, be it a fixed figure or a certain number of standard deviations, will determine the false-positive and false-negative rate. Setting the thresholds at a relatively low level will lead to a correspondingly low false-negative rate. However, this will lead to a relatively higher false-positive rate. One of the strengths of both MAPH and MLPA is that processing many samples does not take much more time than processing a few samples. Thus, it should be easy to test all samples twice, and only probes that deviate in a significant manner in both duplicates should then be retested with another technique. If most of the samples do not show any changes, however, then it can be argued that routine duplicate testing, at least on the same DNA isolate, is not necessary. If both tests need to show the same deviation and the first is normal, then the second test is not going to change the conclusion. Therefore, only samples that show significant changes in the first round need to be retested.

Although detecting deletions (1:2) and duplications (3:2) is relatively straightforward, distinguishing relatively smaller changes (e.g., in mixed or mosaic samples) requires testing each sample several times to determine confidence limits. This is one of the strengths of both MAPH and MLPA, because multiple testing requires little extra time or expense. It is also advantageous to be able to use data from more than one probe, because this strengthens the statistical analysis.

In addition to false positives and false negatives because of variations affecting the calculations, incorrect conclusions can also be drawn due to inherent limitations of each technique. If only part of the genomic region that binds an MAPH probe is deleted, then the signal obtained from that probe will be proportionally lower, but perhaps not outside the predefined thresholds. We have seen an example of this in a patient with Duchenne muscular dystrophy (DMD), in which a single exonic probe was consistently at a height of only 70% of that seen in healthy controls. Although this suggested a mosaic deletion, testing with an MLPA probe for the same exon gave no signal, indicating that the region around the ligation site of the two half-probes was deleted. It is most likely that the ligation site for the MLPA probe fell within the deleted region (hence, no signal), but that enough genomic sequence remained that could bind the MAPH probe, thereby giving the unclear result.

Because of the sensitivity of the MLPA reaction to mismatches at the ligation site of the two half-probes, even a single nucleotide change may be enough to prevent successful ligation. In such a case, the result will be scored as a deletion, a conclusion that will not be changed by repeated analyses. For this reason, it is critical to confirm all single-probe deletions found with MLPA by using another technique (e.g., sequence analysis of the affected region) to reveal potential variations that affect ligation or probe hybridization.

IV. Applications

There have been several reports on the use of MAPH and MLPA for detecting copy number changes. These have focused primarily on screening either single genes for exonic deletions and duplications or chromosomal regions for rearrangements. Some of these are discussed in the following sections.

A. MAPH

Because of the ability to screen up to 50 loci simultaneously, an obvious target was to develop probes for each of the subtelomeric regions of the human chromosomes (43 in total). This probe set can be used to detect trisomies and unbalanced translocations. In addition, these regions have been shown to be involved in mental retardation (MR), with rearrangements found in about 5% of cases (Flint and Knight, 2003). Subtelomeric screening using MAPH has been described in two reports. The first used a combination of FISH and MAPH (Sismani *et al.*, 2001). A total of 70 samples from mentally retarded patients had been screened with FISH, and one deletion was detected. Analysis with MAPH gave the same result. The second report was of the MAPH screening of 37 DNA samples from patients referred for fragile X screening (Hollox *et al.*, 2002). In this study, six rearrangements were found (16%), one of the highest percentages found in subtelomeric screening of mentally retarded patients. No FISH screening was performed, and the rearrangements were confirmed by semiquantitative PCR. A problem with FISH confirmation of MAPH results is that a negative result with FISH does not mean that the MAPH finding is not correct. The difference in resolution afforded by the techniques (a minimum of 40 kb vs ~100 bp) means that MAPH might have detected a genuine but small alteration that simply cannot be resolved with FISH. For this reason, confirmation with a high-resolution technique such as MAPH, MLPA, quantitative PCR, or Southern blotting may be preferable.

MAPH probe sets have been developed covering each of the 79 exons of the DMD gene (White *et al.*, 2002). Mutations in this gene cause DMD/Becker's muscular dystrophy (DMD/BMD), and in about 70% of cases the causative mutation is a deletion or duplication of one or more exons. Analysis of more than 100 samples showed a wide range of mutations, including several that had not been previously detected using other techniques. Noteworthy here is that a duplication of exon 2 was the most common duplication found but had never been reported using alternative methods such as Southern blotting. This emphasizes the high degree of sensitivity of MAPH.

MAPH has also been applied to the analysis of deletions and amplifications in chronic myeloid leukemia (Reid *et al.*, 2003). This clonal malignancy is characterized by the generation of a BCR-ABL1 fusion gene after a translocation involving chromosomes 9 and 22 (the Philadelphia chromosome). The marked difference in disease progression seen in these patients was thought to be at least partly due to deletions at the breakpoints of the chromosome 9 derivative. Reid *et al.* (2003) made probes at the breakpoints (9q34 and 22q12) and used these to test DNA samples with known deletions and duplications. Using titration experiments, the authors were able to show that they could still see deletions and amplifications when only 60–70% of the cells were affected. This demonstrates the ability of MAPH to detect mutations in mosaic cases.

Another application of MAPH, testing its limits of sensitivity, was the analysis of a previously identified polymorphic region on 8p23.1 (Hollox *et al.*, 2003). This region is flanked by olfactory repeats, and it had been shown that up to 25% of the normal population carried an inversion polymorphism between these repeats (Giglio *et al.*, 2002). Additionally, an apparently benign duplication of the region had been described, with no obvious effect on the carriers. By designing several MAPH probes within this region, along with the use of semiquantitative FISH, we could estimate the copy number of this region. It could be shown that most individuals had between 2 and 7 copies of the allele, whereas carriers of the apparent duplication had 9–12 copies.

B. MLPA

MLPA probe sets are commercially available, and there have been several reports detailing their application. These have been primarily focused on screening for exonic deletions and duplications in specific genes. The first genes screened were those known to be commonly deleted in different cancers. These include BRCA1 (Hogervorst *et al.*, 2003; Montagna *et al.*, 2003), which is involved in breast cancer, and MSH2 and MLH1, which are involved in colon cancer (Gille *et al.*, 2002; Taylor *et al.*, 2003). Two sets together covering all exons of the DMD gene have also become available.

Subtelomeric screening has also been performed using MLPA (Rooms *et al.*, 2004). A total of four rearrangements were found in a group of 75 DNA samples from mentally retarded patients, which is similar to the average percentage found in other studies using alternative methods. An additional rearrangement was suspected, but further examination showed that there was a 3-bp deletion at the probe site, illustrating the importance of confirming apparent deletions found with a single probe, because even a single nucleotide change may disturb the ligation of the two half-probes.

An interesting application lies in the detection of aneuploidy directly from amniotic fluid lysates (Slater *et al.*, 2003). In a study of 492 samples, there were 18 aneuploidies identified, with no false positives or false negatives. The probe set used contained eight probes for each of chromosomes 13, 18, and 21, four probes for chromosomes X and Y, and eight probes for other chromosomes. To determine whether an aneuploidy was present, the mean normalized ratio of all probes for a given chromosome was used, because it was stated that not all relevant probes showed the aneuploidy. By combining the results of multiple probes, the sensitivity of the assay is enhanced, and in this case, the vast majority of samples were tested only once.

To circumvent the time-consuming and expensive cloning required in MLPA probe preparation, we have looked at the possibility of using synthetic oligonucleotides for both half-probes (White *et al.*, 2004). Because of length limitations during the chemical synthesis, the maximum size of ligated probes is about 130 bp, considerably less than the 490 bp obtained from cloned products. To partially circumvent these restrictions, we combined the single base-pair resolution of capillary electrophoresis with the ability to analyze multiple fluorescent labels simultaneously. We made two sets of probes: each being amplified with a defined primer pair and each set being labeled with a different fluorophore. Because the primers function under the same PCR amplification conditions, we could combine all probes in the same reaction. In this way, we have been able to analyze more than 30 loci in a single reaction.

V. Future Possibilities

It is easy to foresee that new MLPA/MAPH assays will be developed that will focus not only on specific genes but also on high-resolution analysis of chromosomal regions associated with a range of diseases. Several large deletions and duplications are known to be associated with a specific spectrum of disorders (e.g., microdeletion syndromes and contiguous gene disorders). Such rearrangements can be on the order of several megabases, making them cytogenetically visible. Often many genes are found within these areas and the extent of the rearrangement is directly correlated with disease severity (i.e., the sum of the genes affected determines the overall phenotype). Using MAPH/MLPA assays, it should be possible to determine the extent of the deletion/duplication. In addition, a higher resolution can be achieved by testing individual genes, thereby resolving smaller changes, which are beneath the detection limits of current methods. Indeed, a detailed analysis of patients with no cytogenetically visible deletion showed that haploinsufficiency of RAI1 is sufficient to cause the phenotypic characteristics of Smith–Magenis syndrome (Slager *et al.*, 2003).

Current methodologies facilitate the versatile detection of copy number changes in a selected set of up to 50 target sequences. What is currently lacking is a technology that can be used in a scanning mode and genome wide. In many cases, it is not yet known where genes involved in specific disorders reside (e.g., MR). As a first step to identify the genes responsible, one would want to scan the entire genome for cases in which a (large) deletion/duplication is involved. These rare cases might then point to the genes involved, which could then be analyzed in more detail and in a much larger set of patients. Similarly, it would be desirable to have such a scanning tool to analyze genomic DNA of newborns presenting with unclear health problems. In a significant fraction of these newborns, *de novo* genomic copy number changes might be involved, which when detected could provide a much earlier diagnosis and thereby valuable information for health care.

To fill in this gap, arrayCGH (array-based CGH) has been developed. It has been successfully applied by spotting BACs/PACs (or PCR products thereof) onto a glass slide and hybridizing labeled genomic DNA (Albertson and Pinkel, 2003; Snijders *et al.*, 2001; Vissers *et al.*, 2003) to pick up genomic copy number changes. The method could be successfully applied not only to detect large rearrangements, especially in cancer-related samples, but also to reveal rearrangements in patients with MR and dysmorphisms. However, these arrays have a limited resolution. First, on average only one probe per megabase of DNA is used, and second, only large (>50–100 kb) deletions/duplications will be detected. Furthermore, because of the presence of repetitive sequences, some regions of the genome are not amenable to analysis in this way.

An obvious approach to resolve the issue of resolution is to design an array-based MLPA/MAPH assay. In such an assay, each probe no longer would need to be of a different length for subsequent electrophoretic separation. This feature might help resolve the main technical hurdle of this approach, namely the multiplicity of the assay (i.e., the number of loci that can be simultaneously amplified and resolved when combined in a single reaction). Effectively, this aspect of the MLPA/MAPH assay has not yet been questioned in practice, but preliminary results from our laboratory indicate that a 250-plex, array-based MLPA assay is possible (Kalf, White, den Dunnen, 2004, unpublished data).

An approach very similar to MLPA, also based on the ligation and quantitative amplification of oligonucleotides, has been developed (Yeakley *et al.*, 2002). The primary difference lies in the analysis method. Instead of separating products of a different length by electrophoresis, the PCR products are captured by complementary sequences attached to microspheres. Because up to 1500 microspheres can be distinguished using a specific color-coding system, it should be possible to analyze 1500 loci in genomic DNA simultaneously. As the system has been set up for microtiter plate analysis, up to 96 samples can be processed in parallel. This then tackles another important issue of these technologies, which is the number of different patient samples that can be processed in a specific time. In that respect, array-based technologies have significant limitations.

For genome-wide analysis of copy number changes, the most promising developments lie in the use of whole-genome SNP-typing assays. The development of a 10,000 human SNP-typing array was reported (Kennedy *et al.*, 2003), with a 120,000 SNP-typing array recently released and a 300,000 array announced. Deletions can be detected with such tools by first using the SNP itself, that is, inheritance of "null alleles" or indirectly derived from the presence of exceptionally large homozygous regions, which might point to a hidden deletion of the second allele. Rearrangements should also be detectable using the amount of signal generated per SNP with lower signal (and homozygosity) pointing to deleted regions and increased signals to duplicated regions. Bearing the financial investments in SNP-typing technology in mind, it is highly likely that genome-wide copy number detection using such high-throughput SNP-typing tools should soon become feasible. Even when the quantitative aspects of these technologies might be relatively weak, the sheer number of DNA sequences tested in these assays should already enable the detection of copy number changes, that is, a whole range of physically linked probes, each individually giving a weak signal but when taken together shouting where to look.

Acknowledgments

We acknowledge the help of Margot Kalf, Marjolein Kriek, and Gert-Jan van Ommen in developing these techniques in our laboratory, Jan Schouten for providing different probe sets and several interesting discussions regarding the MLPA technique, and ZorgOnderzoek Nederland (ZON) and PamGene BV (Den Bosch, the Netherlands) for financial support.

References

Albertson, D. G., and Pinkel, D. (2003). Genomic microarrays in human genetic disease and cancer. *Hum. Mol. Genet.* **12**(Spec. no. 2), R145–R152.

Altschul, S. F., Gish, W., Miller, W., Myers, E. W., and Lipman, D. J. (1990). Basic local alignment search tool. *J. Mol. Biol.* **215,** 403–410.

Anhuf, D., Eggermann, T., Rudnik-Schoneborn, S., and Zerres, K. (2003). Determination of SMN1 and SMN2 copy number using TaqMan technology. *Hum. Mutat.* **22,** 74–78.

Armour, J. A., Barton, D. E., Cockburn, D. J., and Taylor, G. R. (2002). The detection of large deletions or duplications in genomic DNA. *Hum. Mutat.* **20,** 325–337.

Armour, J. A., Sismani, C., Patsalis, P. C., and Cross, G. (2000). Measurement of locus copy number by hybridisation with amplifiable probes. *Nucl. Acids Res.* **28,** 605–609.

Den Dunnen, J. T., Bakker, E., Klein-Breteler, E.G., Pearson, P. L., and Van Ommen, G. J. B. (1987). Direct detection of more than 50% Duchenne muscular dystrophy mutations by field-inversion gels. *Nature* **329,** 640–642.

Flint, J., and Knight, S. (2003). The use of telomere probes to investigate submicroscopic rearrangements associated with mental retardation. *Curr. Opin. Genet. Dev.* **13,** 310–316.

Giglio, S., Calvari, V., Gregato, G., Gimelli, G., Camanini, S., Giorda, R., Ragusa, A., Guerneri, S., Selicorni, A., Stumm, M., Tonnies, H., Ventura, M., Zollino, M., Neri, G., Barber, J., Wieczorek, D., Rocchi, M., and Zuffardi, O. (2002). Heterozygous submicroscopic inversions involving olfactory receptor-gene clusters mediate the recurrent t(4;8)(p16;p23) translocation. *Am. J. Hum. Genet.* **71,** 276–285.

Gille, J. J., Hogervorst, F. B., Pals, G., Wijnen, J. T., van Schooten, R. J., Dommering, C. J., Meijer, G. A., Craanen, M. E., Nederlof, P. M., de Jong, D., McElgunn, C. J., Schouten, J. P., and Menko, F. H. (2002). Genomic deletions of MSH2 and MLH1 in colorectal cancer families detected by a novel mutation detection approach. *Br. J. Cancer* **87,** 892–897.

Hogervorst, F. B., Nederlof, P. M., Gille, J. J., McElgunn, C. J., Grippeling, M., Pruntel, R., Regnerus, R., van Welsem, T., van Spaendonk, R., Menko, F. H., Kluijt, I., Dommering, C., Verhoef, S., Schouten, J. P., van't Veer, L. J., and Pals, G. (2003). Large genomic deletions and duplications in the BRCA1 gene identified by a novel quantitative method. *Cancer Res.* **63,** 1449–1453.

Hollox, E. J., Armour, J. A., and Barber, J. C. (2003). Extensive normal copy number variation of a beta-defensin antimicrobial-gene cluster. *Am. J. Hum. Genet.* **73,** 591–600.

Hollox, E. J., Atia, T., Cross, G., Parkin, T., and Armour, J. A. (2002). High throughput screening of human subtelomeric DNA for copy number changes using multiplex amplifiable probe hybridisation (MAPH). *J. Med. Genet.* **39,** 790–795.

Kennedy, G. C., Matsuzaki, H., Dong, S., Liu, W. M., Huang, J., Liu, G., Su, X., Cao, M., Chen, W., Zhang, J., Liu, W., Yang, G., Di, X., Ryder, T., He, Z., Surti, U., Phillips, M. S., Boyce-Jacino, M. T., Fodor, S. P., and Jones, K. W. (2003). Large-scale genotyping of complex DNA. *Nat. Biotechnol.* **21,** 1233–1237.

Kent, W. J. (2002). BLAT–the BLAST-like alignment tool. *Genome Res.* **12,** 656–664.

Koenig, M., Hoffman, E. P., Bertelson, C. J., Monaco, A. P., Feener, C. A., and Kunkel, L. M. (1987). Complete cloning of the Duchenne muscular dystrophy (DMD) cDNA and preliminary genomic organization of the DMD gene in normal and affected individuals. *Cell* **50,** 509–517.

Ligon, A. H., Kashork, C. D., Richards, C. S., and Shaffer, L. G. (2000). Identification of female carriers for Duchenne and Becker muscular dystrophies using a FISH-based approach. *Eur. J. Hum. Genet.* **8,** 293–298.

Montagna, M., Dalla, P. M., Menin, C., Agata, S., De Nicolo, A., Chieco-Bianchi, L., and D'Andrea, E. (2003). Genomic rearrangements account for more than one-third of the BRCA1 mutations in northern Italian breast/ovarian cancer families. *Hum. Mol. Genet.* **12,** 1055–1061.

Petrij, F., Dauwerse, H. G., Blough, R. I., Giles, R. H., van der Smagt, J. J., Wallerstein, R., Maaswinkel-Mooy, P. D., van Karnebeek, C. D., van Ommen, G. J., Van Haeringen, A., Rubinstein, J. H., Saal, H. M., Hennekam, R. C., Peters, D. J., and Breuning, M. H. (2000). Diagnostic analysis of the Rubinstein-Taybi syndrome: five cosmids should be used for microdeletion detection and low number of protein truncating mutations. *J. Med. Genet.* **37,** 168–176.

Petrij-Bosch, A., Peelen, T., Van Vliet, M., Van Eijk, R., Olmer, R., Drusedau, M., Hogervorst, F. B. L., Hageman, S., Arts, P. J., Ligtenberg, M. J., Meijers-Heijboer, H., Klijn, J. G., Vasen, H. F., Cornelisse, C. J., Van 't Veer, L. J., Bakker, E., Van Ommen, G. J. B., and Devilee, P. (1997). BRCA1 genomic deletions are major founder mutations in Dutch breast cancer patients. *Nat. Genet.* **17,** 341–345.

Reid, A. G., Tarpey, P. S., and Nacheva, E. P. (2003). High-resolution analysis of acquired genomic imbalances in bone marrow samples from chronic myeloid leukemia patients by use of multiple short DNA probes. *Genes Chromosomes Cancer* **37,** 282–290.

Rooms, L., Reyniers, E., van Luijk, R., Scheers, S., Wauters, J., Ceulemans, B., Van Den, E. J., Van Bever, Y., and Kooy, R. F. (2004). Subtelomeric deletions detected in patients with idiopathic mental retardation using multiplex ligation-dependent probe amplification (MLPA). *Hum. Mutat.* **23,** 17–21.

bb>Schouten, J. P., McElgunn, C. J., Waaijer, R., Zwijnenburg, D., Diepvens, F., and Pals, G. (2002). Relative quantification of 40 nucleic acid sequences by multiplex ligation-dependent probe amplification. *Nucl. Acids Res.* **30,** e57

Sismani, C., Armour, J. A., Flint, J., Girgalli, C., Regan, R., and Patsalis, P. C. (2001). Screening for subtelomeric chromosome abnormalities in children with idiopathic mental retardation using multiprobe telomeric FISH and the new MAPH telomeric assay. *Eur. J. Hum. Genet.* **9,** 527–532.

Slager, R. E., Newton, T. L., Vlangos, C. N., Finucane, B., and Elsea, S. H. (2003). Mutations in RAI1 associated with Smith-Magenis syndrome. *Nat. Genet.* **33,** 466–468.

Slater, H. R., Bruno, D. L., Ren, H., Pertile, M., Schouten, J. P., and Choo, K. H. (2003). Rapid, high throughput prenatal detection of aneuploidy using a novel quantitative method (MLPA). *J. Med. Genet.* **40,** 907–912.

Snijders, A. M., Nowak, N., Segraves, R., Blackwood, S., Brown, N., Conroy, J., Hamilton, G., Hindle, A. K., Huey, B., Kimura, K., Law, S., Myambo, K., Palmer, J., Ylstra, B., Yue, J. P., Gray, J. W., Jain, A. N., Pinkel, D., and Albertson, D. G. (2001). Assembly of microarrays for genome-wide measurement of DNA copy number. *Nat. Genet.* **29,** 263–264.

Taylor, C. F., Charlton, R. S., Burn, J., Sheridan, E., and Taylor, G. R. (2003). Genomic deletions in MSH2 or MLH1 are a frequent cause of hereditary non-polyposis colorectal cancer: identification of novel and recurrent deletions by MLPA. *Hum. Mutat.* **22,** 428–433.

Van Der Maarel, S. M., Deidda, G., Lemmers, R. J., Bakker, E., van der Wielen, M. J., Sandkuijl, L., Hewitt, J. E., Padberg, G. W., and Frants, R. R. (1999). A new dosage test for subtelomeric 4;10 translocations improves conventional diagnosis of facioscapulohumeral muscular dystrophy (FSHD). *J. Med. Genet.* **36,** 823–828.

Vissers, L. E., De Vries, B. B., Osoegawa, K., Janssen, I. M., Feuth, T., Choy, C. O., Straatman, H., Van, D., V, Huys E. H., Van Rijk, A., Smeets, D., Ravenswaaij-Arts, C. M., Knoers, N. V., Van, D. B., I, De Jong, P. J., Brunner, H. G., van Kessel, A. G., Schoenmakers, E. F., and Veltman, J. A. (2003). Array-based comparative genomic hybridization for the genomewide detection of submicroscopic chromosomal abnormalities. *Am. J. Hum. Genet.* **73,** 1261–1270.

White, S., Kalf, M., Liu, Q., Villerius, M., Engelsma, D., Kriek, M., Vollebregt, E., Bakker, B., van Ommen, G. J., Breuning, M. H., and Den Dunnen, J. T. (2002). Comprehensive detection of genomic duplications and deletions in the DMD gene, by use of multiplex amplifiable probe hybridization. *Am. J. Hum. Genet.* **71,** 365–374.

White, S. J., Sterrenburg, E., van Ommen, G. J., Den Dunnen, J. T., and Breuning, M. H. (2003). An alternative to FISH: detecting deletion and duplication carriers within 24 hours. *J. Med. Genet.* **40,** e113.

White, S. J., Vink, G. R., Kriek, M., Wuyts, W., Schouten, J., Bakker, B., Breuning, M. H., and den Dunnen, J. T. (2004). Two-colour MLPA; Detecting genomic rearrangements in hereditary multiple exostoses. *Hum. Mut.* **24,** 86–92.

Wilke, K., Duman, B., and Horst, J. (2000). Diagnosis of haploidy and triploidy based on measurement of gene copy number by real-time PCR. *Hum. Mutat.* **16,** 431–436.

Yeakley, J. M., Fan, J. B., Doucet, D., Luo, L., Wickham, E., Ye, Z., Chee, M. S., and Fu, X. D. (2002). Profiling alternative splicing on fiber-optic arrays. *Nat. Biotechnol.* **20,** 353–358.

CHAPTER 33

Genomic Array Technology

Heike Fiegler and Nigel P. Carter

The Wellcome Trust Sanger Institute
Wellcome Trust Genome Campus
Hinxton, Cambridge CB10 1SA, United Kingdom

I. Introduction

The sequencing of the human genome in the public domain has generated an extensive resource of mapped and sequenced clones that are revolutionizing molecular cytogenetics. This clone resource finds particular use in the construction of large insert clone DNA microarrays spotted onto glass microscope slides. Originally, these arrays were produced to replace metaphase chromosomes in comparative genomic hybridization (CGH) experiments to detect copy number changes along the genome at an increased resolution (Albertson and Pinkel, 2003). In this way, the resolution of CGH becomes limited only by the insert size and the density of the mapped sequences used. However, because of the increased sensitivity and resolution compared to conventional methodologies, genomic arrays are also being applied for cytogenetic studies of microdeletion and microduplication, as well as for rapid mapping of breakpoints (Fiegler *et al.*, 2003b).

METHODS IN CELL BIOLOGY, VOL. 75

0091-679X/04 $35.00

II. Construction of Genomic Clone Arrays Using DOP-PCR Amplification

Cosmid, P1, P1 artificial chromosome (PAC), and bacterial artificial chromosome (BAC) clones are being widely used for the construction of genomic DNA microarrays. DNA was originally extracted from large-scale cultures (Pinkel *et al.*, 1998; Solinas-Toldo *et al.*, 1997), which becomes a costly and time-consuming process when expanded to the number of clones required to construct an array with a resolution of 1 Mb (~3500 clones).

To overcome this problem, several approaches have been applied to amplify the clone DNA enzymatically for spotting, thereby removing the requirement for large-scale DNA preparations. These have included methods such as linker adapter polymerase chain reaction (PCR) (Snijders *et al.*, 2001), rolling circle PCR using Phi29 (Buckley *et al.*, 2002), or degenerate oligonucleotide primed PCR (DOP-PCR) (Hodgson *et al.*, 2001) using an amine-modified version of the standard DOP-PCR primer 6MW (Telenius *et al.*, 1992a,b). *Escherichia coli* genomic DNA, however, is a common contaminant of DNA preparations of large insert clones. The degree of contamination has been estimated by real-time PCR to be between 6% and 26%, depending on the method used for purification (Foreman and Davis, 2000). This contaminating *E. coli* DNA will reduce the capacity of each probe spotted on the array to hybridize with the DNA of interest and may contribute to an increased nonspecific background signal. To overcome this problem, we have designed three new DOP-PCR primers (DOP1, DOP2, and DOP3) that were chosen to be efficient in amplifying human genomic DNA but inefficient in amplifying the contaminating *E. coli* DNA (Fiegler *et al.*, 2003a). The sequence and the frequency of the six bases at the 3′ end of the new primers in human and *E. coli* DNA compared to the standard DOP-PCR primer 6MW is shown in Table I. The new primers retain the ability to amplify human sequence but demonstrate poor amplification of the *E. coli* sequence compared to the standard DOP-PCR primer 6MW. Furthermore, the 3′ ends of the primers DOP1 and DOP3 are the reverse complement of each other, which allows the amplification from the complementary strand in the opposite direction, thereby

Table I
Degenerate Oligonucleotide Primed Polymerase Chain Reaction Primer for the Amplification of Clones Containing Large Human Inserts

Primer	3′ sequence	*Escherichia coli*	Human
DOP1	CTAGAA	0.02	0.63
DOP2	TAGGAG	0.05	0.49
DOP3	TTCTAG	0.02	0.63
6MW	ATGTGG	0.40	0.65

providing different but complementary representations of the template sequence. The use of these three new DOP-PCR primers, particularly in combination, revealed a significant increase in sensitivity and reproducibility in genomic hybridizations compared to arrays constructed with the standard DOP-PCR primer 6MW (Fiegler *et al.*, 2003a).

Following this strategy, we have constructed a large insert clone DNA microarray composed of 3523 sequencing clones selected from the Golden Path of the human genome (Lander *et al.*, 2001) located at approximately 1-Mb intervals across the human genome. The clones were picked from libraries held at the Wellcome Trust Sanger Institute to cover each chromosome omitting the short arms of acrocentric chromosomes. Essential for the correct interpretation of the array CGH experiments is the exact mapped position of each clone along the chromosomes; although this information is available through various genome browsers on a clone-by-clone basis, it is not easily assembled for large clone collections. We have, therefore, generated a database (Cytoview) within the Ensembl genome browser (Cytoview, www.ensembl.org/homo_sapiens/cytoview) that displays the 1-Mb clone set in relation to the Golden Path sequencing clones. Cytoview also offers the possibility to download not only lists of clones but also their mapping information, including fluorescence *in situ* hybridization (FISH) and BAC end data when available. In addition, Ensembl provides automatic updating of all this information with every new assembly of the human genome (Fiegler *et al.*, 2003a).

III. Validation of Array Performance and Data Analysis

It is of great importance to prove the reproducibility of the genomic hybridization and to establish characteristics of the clones used to construct the array before applying this method in large sample studies. We, therefore, performed a series of validation experiments including self–self and male–female hybridizations. For a self–self hybridization, the linear ratio of all clones on the array (autosomal clones and chromosome X/Y clones) is expected to be 1:1, which corresponds to a ratio of 0 on a log2 scale (test vs reference, see Fig. 1).

In contrast, for a female–male hybridization, the expected linear ratio for chromosome X clones is 2:1, although there should be no copy number changes detectable on autosomal clones. However, although most of the autosomal clones reported the expected 1:1 ratio, we have found several clones showing copy number changes between the two individuals (Fig. 2). Further investigation of these clones by FISH analysis revealed that some clones had been mismapped and could be subsequently placed correctly onto the genome. Others, however, mapped to their correct position, thereby suggesting copy number polymorphism between the two individuals used in the study.

Although we were able to detect an increase of ratios for the clones representing chromosome X, most of those clones did not report the expected ratio of 2 (ratio

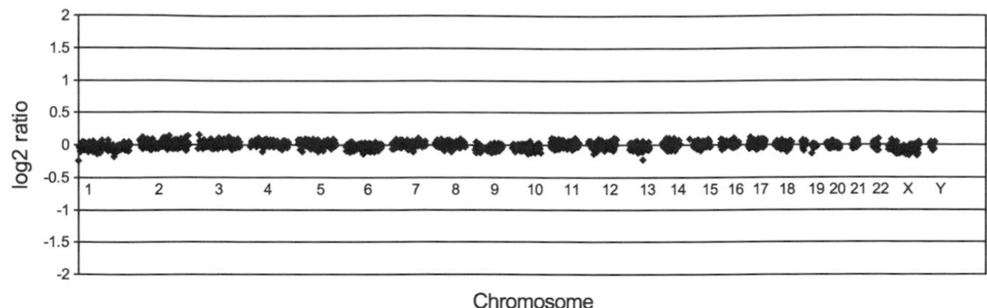

Fig. 1 Genome-wide plot of ratios for a self–self hybridization.

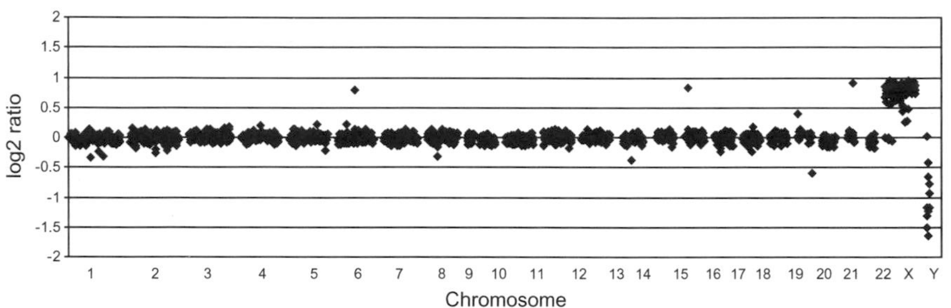

Fig. 2 Genome-wide plot of ratios for a female–male hybridization.

value of 1 on a log2 scale, see Fig. 3). It is, however, interesting to note that all published methods for human array CGH underestimate female/male ratios for chromosome X clones. The mean linear ratios for chromosome X clones in female–male hybridizations reported by different groups were found to be between 1.65 and 1.73 (Fiegler *et al.*, 2003a; Snijders *et al.*, 2001). It has been suggested that this underestimation of the true ratio on chromosome X clones might be due to incomplete suppression of shared repeat sequences and homology between chromosomes X and Y. It has also been suggested that the inactivation of one chromosome X in female DNA might impair the labeling reaction (Fiegler *et al.*, 2003a).

DNA derived from tumor samples often consists of a mixed population of cells (tumor cells and infiltrating normal cells), which can interfere with the correct detection of copy number changes. To test whether our arrays would be sensitive enough to detect single copy number changes even in a mixed cell population, we performed experiments hybridizing female DNA against a mix of 50% male and 50% female DNA onto a small custom array consisting of fifteen chromosome 1 and nine chromosome X clones, which previously reported ratios close to 2:1 in female–male hybridizations (Fig. 4). Although the expected linear ratio for chromosome 1 clones should again be 1:1, the ratios for the chromosome X clones

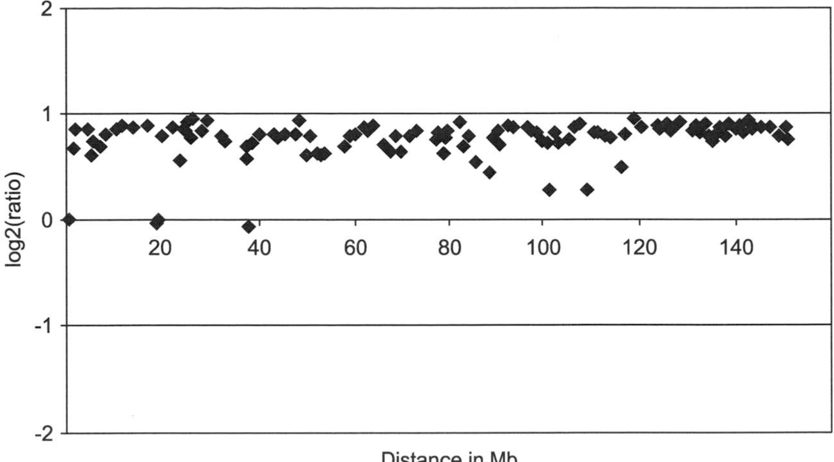

Fig. 3 Chromosome X plot of ratios for a female–male hybridization.

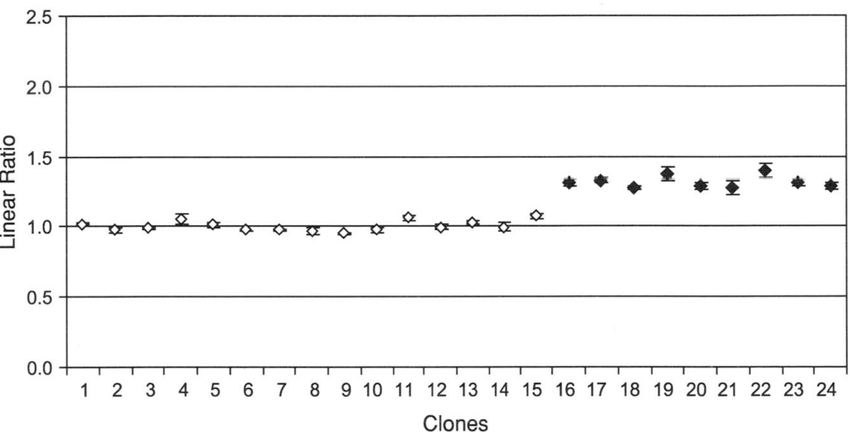

Fig. 4 Hybridization of female DNA versus a mix of 50% male and 50% female DNA. Chromosome 1 clones (*open diamonds*) and chromosome X clones (*filled diamonds*) are shown. The error bars are the standard deviation of triplicates.

should now only be 1.33:1 (instead of 2:1 female vs male). As shown in Fig. 4, all chromosome X clones report an approximate ratio of 1.3, demonstrating that the arrays are indeed sensitive enough to detect single copy number changes even when DNA isolated from mixed populations consists of 50% tumor cells and 50% normal infiltrating cells.

For clones to accurately report copy number changes in genomic hybridizations, the nonrepetitive sequence of each clone has to be unique to its location.

Regions of homology or duplication (gene families, paralogs, pseudogenes, segmental duplications, etc.) will contribute to the hybridization signal and could potentially mask copy number changes at the specific locus. Although the presence of duplications can be detected by, for example, FISH analysis onto metaphase chromosomes, only a small proportion of the Golden Path clones have been FISH mapped so far, and because this method is time consuming, it is not suited for the analysis of the whole 1-Mb clone set (~3500 clones). As an alternative, we have developed a method that allows the analysis of the response characteristics for all clones directly by hybridization on the array. The basis of this method is to use PCR to amplify flow-sorted chromosomes and add copies of chromosomal material to a series of normal versus normal genomic hybridizations, thereby increasing the copy number of a specific chromosome in the test sample. The hybridization profile for each clone can then be generated by plotting a standard curve of ratio versus additional chromosomal material. By repeating these hybridizations for each chromosome in turn, clones that do not respond appropriately to the addition of the chromosome can be identified easily (Fig. 5). In addition, reduction in the slope of the response curve identifies clones that share homologous sequences with other sites of the genome. Unfortunately, intrachromosomal duplications will not be identified using this strategy (Fiegler *et al.*, 2003a).

Another important factor for the reliable detection of single copy number changes in genome-wide array CGH experiments is the successful suppression of repeat sequences present in both target and probe DNA. This is achieved by use of high-quality human Cot1 DNA. The Cot1 fraction of human genomic DNA

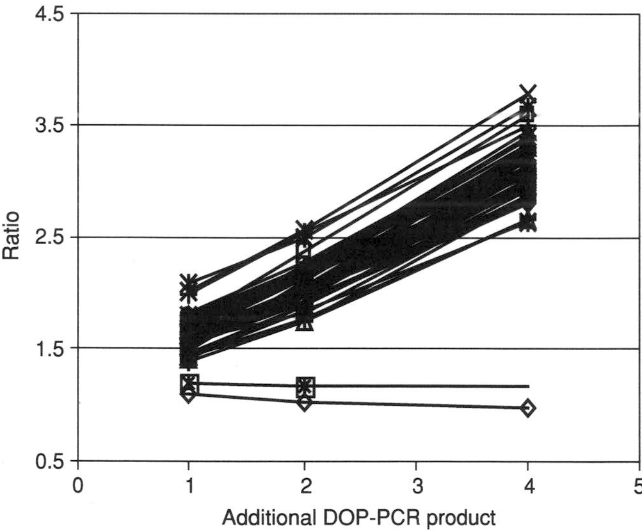

Fig. 5 Characteristics for chromosome 1 clones in response to additional copies of chromosome 1 in a normal versus normal hybridization.

consists largely of rapidly annealing sequences such as small interspersed nuclear elements (SINEs) and large interspersed nuclear elements (LINEs) repeats. These highly repetitive elements contain either repeat sequences consisting of less than 500-bp elements (SINEs, e.g., Alu repeats) or sequences consisting of more than 5-kb elements (LINEs). When differentially labeled male and female DNA was co-hybridized in two independent experiments with Cot1 DNA derived from two batches, we observed different hybridization profiles after analyzing the arrays (Fig. 6A). Although co-hybridization with Cot1 DNA derived from batch 1 showed a clear copy number gain on chromosome X clones (expected linear ratio, 2:1; observed mean ratio of all chromosome X clones, 1.72:1), only a slight change in copy number could be seen when batch 2 was used to suppress repeat sequences (observed mean ratio of all chromosome X clones, 1.43:1). Although the expected ratio of 2:1 could not be obtained with any of the Cot1 samples, the observed ratio of 1.72:1, however, compares well with previously reported values of 1.73:1 and 1.65:1, as mentioned previously. Agarose gel electrophoresis of the two batches revealed that the average size of the DNA fragments varied between 100 bp (batch 2) and 400 bp (batch 1) (Fig. 6B). This suggests that Cot1 DNA containing larger DNA fragments (up to 400 bp) might be more efficient at annealing to the repeat sequences present in both probe and target sequences and in generating specific

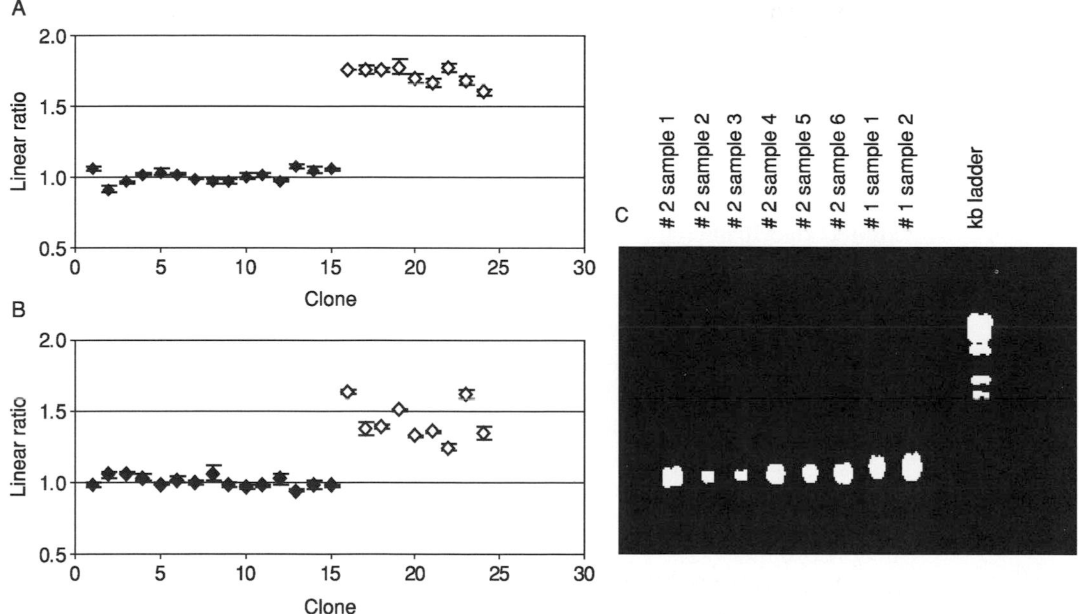

Fig. 6 Performance of different batches of human Cot1 DNA in a female versus male hybridization. (A) Ratio plot for Cot1 DNA batch 1; (B) ratio plot for Cot1 DNA batch 2; (C) agarose gel electrophoresis of different samples of Cot1 DNA batches 1 and 2.

hybridization signals. However, the fragment size alone is not the only indicator of Cot1 DNA quality because we have also seen similar variation in the ability of Cot1 DNA batches containing DNA fragments of the same size (data not shown) to suppress repeat sequences.

IV. Array CGH

Tumor development and progression is often associated with dramatic copy number changes. Although regions of DNA amplification commonly harbor oncogenes, regions of deletion can potentially harbor tumor suppressor genes. Although considerable information about genome-wide copy number changes in tumors has been provided by conventional CGH, its resolution is limited to typically 10–20 Mb and at best to approximately 3–5 Mb (Lichter *et al.*, 2000). However, as mentioned earlier, by replacing the metaphase chromosomes as the hybridization targets with spatially mapped sequences arrayed onto glass slides, the resolution becomes limited only by the size and spacing of the sequences used (Pinkel *et al.*, 1998; Solinas-Toldo *et al.*, 1997). To compare conventional CGH and array CGH, we have hybridized DNA derived from a female renal cell carcinoma cell line (769P) onto the 1-Mb array described earlier. This cell line has previously been analyzed by conventional CGH and M-FISH. Microarray analysis reliably detected previously identified copy number changes, for example, a single copy deletion on the p-arm on chromosome 1 and a single copy gain on the q-arm on chromosome 1 (Fig. 7). In addition, we detected small copy number changes that had not been detected by any of the conventional methods, for example, a single copy loss on chromosome 9p between 18 and 25 Mb that failed to reach significance in conventional CGH (Fiegler *et al.*, 2003a) (Table II).

In addition to detecting single copy number changes, by hybridizing a female colorectal cell line, Colo320 HSR, we could also reliably detect a previously described highly amplified region on chromosome 8q24 containing *CMYC* (Fig. 8). Our results show an approximately 50-fold copy number increase of this region, which compares well with a previously described study that reported an amplification of approximately 40-fold using the same cell line (Wessendorf *et al.*, 2002). Differences in the observed results could be explained using the cell line at a different passage stage and different arrays and hybridization conditions.

We have now employed the 1-Mb array in several large-scale studies, one of which involved the screening of 22 bladder tumor-derived cell lines and one normal urothelium-derived cell line (Hurst *et al.*, 2004). The array results were in concordance with numerous genetic changes previously identified by conventional CGH, M-FISH, or LOH analyses. In addition, we confirmed previously identified homozygous deletions harboring tumor suppressor genes such as *CDKN2A* on 9p21.3, *DBCCR1* on 9q33.1, and *PTEN* on 10q in some of the cell lines used in this study. We also identified several potentially new homozygous deletions and high-level amplifications in this study. Further analysis is

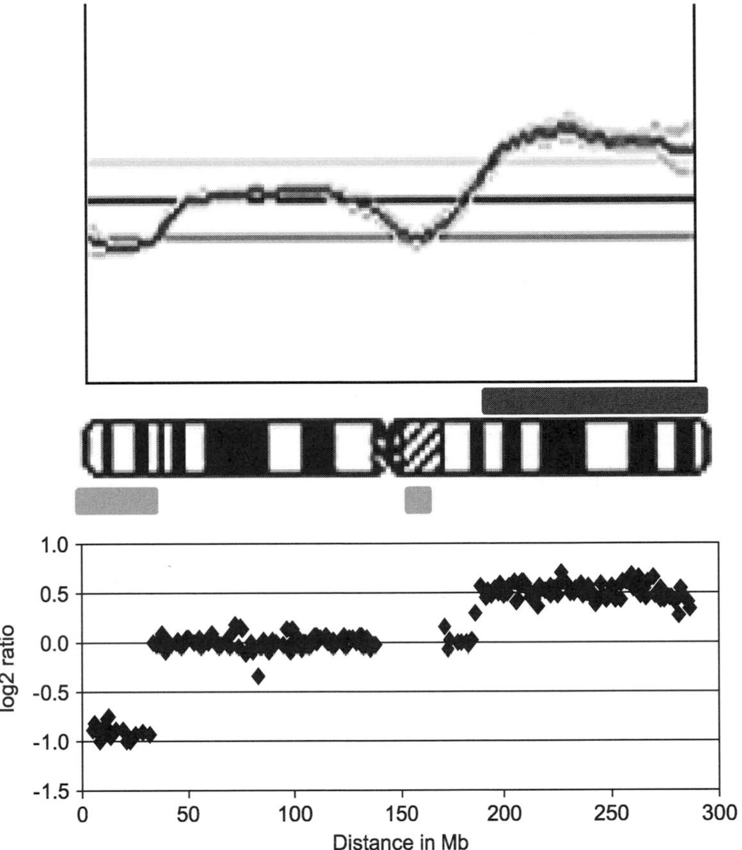

Fig. 7 Chromosome 1 profile of ratios for DNA from the cell line 769P versus female DNA. (A) Conventional comparative genomic hybridization (CGH); (B) array CGH.

Table II
Comparison of Conventional Comparative Genomic Hybridization (CGH) and Array CGH Analysis of Copy Number Changes in the Cell Line 769P

Chromosome	Conventional CGH	Array CGH
1	Deletion 1p35->1pter	Single copy deletion 0.2–28.0 Mb
	Gain 1q22->1qter	Single copy gain 160–256 Mb
3	Deletion 3p14->3pter	Single copy deletion 0.5–72.0 Mb
5	Gain 5q31->5qter	Single copy gain 137–184 Mb
8	Gain 8q11.2->8qter	Single copy gain 52–145 Mb
9		Single copy deletion 18–25 Mb
11	Deletion 11q14->11qter	Single copy deletion 90–141 Mb
14	Deletion of chromosome 14	Single copy deletion 18–104 Mb

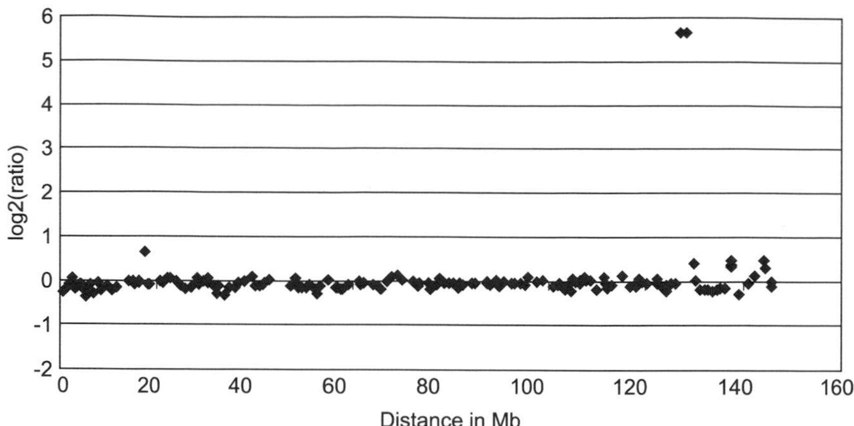

Fig. 8 Chromosome 8 plot of ratios for DNA from the cell line Colo320 HSR versus male DNA.

necessary to study the role of candidate genes located in areas of deletion or amplification.

V. Array CGH for Cytogenetic Analyses

Large insert clone arrays have proved to be of great value in the analysis of copy number changes in tumor-derived cell lines and tumor samples. In addition, they are being more widely used in cytogenetic analyses to define regions of deletion or amplification in the genome that might harbor genes contributing to cytogenetically defined syndromes. The most common of these is Down syndrome. Down syndrome is caused by trisomy of chromosome 21 and is characterized by, for example, cognitive impairment, hypotonia, specific phenotypic features such as flat facies and variations in digits, and ridge formation on hands and feet (Epstein, 2001). Patients with Down syndrome are also at risk of congenital heart disease, Hirschsprung's disease, and other developmental abnormalities. The Down syndrome critical region responsible for the physical phenotype of the disease is thought to be located between 21q22.1 and 21q22.3; however, imbalances of other regions on chromosome 21 might also contribute to the phenotype (Epstein, 2001). The analysis of patients harboring copy number changes of parts of chromosome 21 is, therefore, extremely important because correlation of these partial imbalances with the expressed phenotypes will help identify candidate genes that play a key role in the aspects of Down syndrome (Korenberg *et al.*, 1994). We have analyzed the DNA of a patient with an additional marker chromosome derived from chromosome 21 using the 1-Mb array. The patient did not show any of the dysmorphic features characteristic of Down syndrome but had learning disability with cognitive defects comparable to those seen in patients with Down syndrome. The analysis revealed a partial tetrasomy of chromosome 21 with the location of the breakpoint within chromosome band 21q21.2. The partial tetrasomy did not involve the Down syndrome critical region (Rost *et al.*, 2004). Thus, genes located in the proximal region of chromosome 21 might contribute to aspects of the learning disability but not the dysmorphic features associated with Down syndrome.

VI. Array Painting

Although array CGH is becoming a more widespread technique in cytogenetic and tumor analyses to identify genome copy number changes, balanced translocations will not be detected by this methodology. To overcome this problem, we have developed a technique termed *array painting,* which uses the array technology with flow-sorted chromosome material to rapidly map the constitution and the breakpoints is aberrant chromosomes (Fiegler *et al.*, 2003b). As in array CGH, the resolution is dependent only on the size and the density of the clones spotted onto the array. In array painting, each of the derivative chromosomes in the translocation is flow sorted, amplified, differentially labeled with two fluorochromes, and then hybridized to the array. Only clones containing sequences present in the flow-sorted chromosomes will show fluorescence above background and the ratio determines from which derivative chromosome the hybridizing DNA has been derived. Intermediate values will be generated if a breakpoint spanning clone is present on the array because sequences present on both derivatives will hybridize to the same clone. For example, the DNA of a patient with a *de novo* 46,XY, t(17;22)(q21.1;q12.2) translocation was analyzed by array painting (Fiegler *et al.*, 2003b). The two derivative chromosomes were flow sorted, amplified, and hybridized to the array. Hybridization signals were obtained only from clones representing chromosomes 17 and 22 with the exception of weak signals on chromosome 19. This can be explained by the fact that chromosome 19 is positioned very close to the derivative 17 in the flow karyotype and, therefore, will contaminate the derivative 17 sort. The profiles for both, chromosomes 17 and 22, obtained by plotting the fluorescence ratios against the position of the clones along the chromosomes, clearly show a transition from low to high ratio values on chromosome 17 and vice versa on chromosome 22, thereby identifying the breakpoints on both chromosomes within 1-Mb intervals (Fig. 9). Intermediate ratios that would identify breakpoint spanning clones could not be observed in this case (Fiegler *et al.*, 2003b).

To identify the breakpoint spanning region, the same derivatives were hybridized to a custom array containing overlapping clones within the previously identified 1-Mb breakpoint intervals. This hybridization identified one clone on chromosome 22 with an intermediate value, and subsequent FISH analysis onto chromosomes from the patient confirmed this clone as spanning the breakpoint. For the chromosome 17 breakpoint, three clones with consistent but different

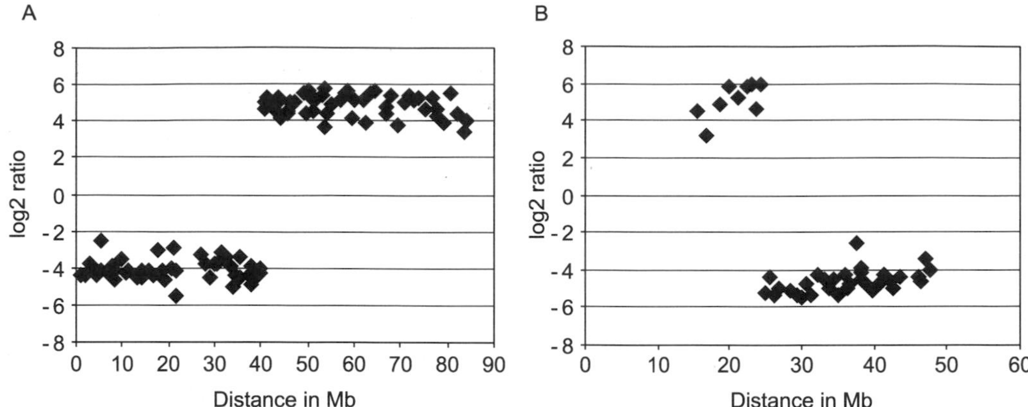

Fig. 9 Array painting results for chromosomes 17 and 22 in the analysis of a t(17;22) patient. (A) Chromosome 17 profile; (B) chromosome 22 profile.

intermediate ratio values were identified, all three being confirmed as spanning the breakpoint. Fiber FISH analysis revealed that the different absolute values of the intermediate ratios of the breakpoint spanning clones on the array were due to the position of the breakpoint within each clone (Fiegler *et al.*, 2003b).

The use of array painting to identify breakpoints in balanced translocations, particularly in more complex cases, is far less time consuming compared to conventional methods, which involve individual hybridizations of large insert clones onto metaphase chromosomes.

VII. Application of Array CGH and Array Painting for Complete Cytogenetic Analyses

After analyzing several balanced translocations by array painting, we found that apparently simple translocations are often more complicated than expected. Array painting in combination with array CGH has proven to be a powerful combination of methods to investigate these balanced translocations at a molecular level. We have previously described a case in which a patient had been referred for cytogenetic investigation for triangular facies and the failure to thrive. Conventional cytogenetic analysis revealed a simple balanced translocation with a 46,XX, t(11;12)(q21;p13.33) *de novo* chromosome constitution. Array painting using the 1-Mb array easily identified the breakpoint on chromosome 12. The most proximal clone mapping 1.02 Mb from the p-terminus of chromosome 12 showed an intermediate ratio (Fig. 10A) and subsequent FISH analysis confirmed that it spanned the breakpoint (Fiegler *et al.*, 2003b).

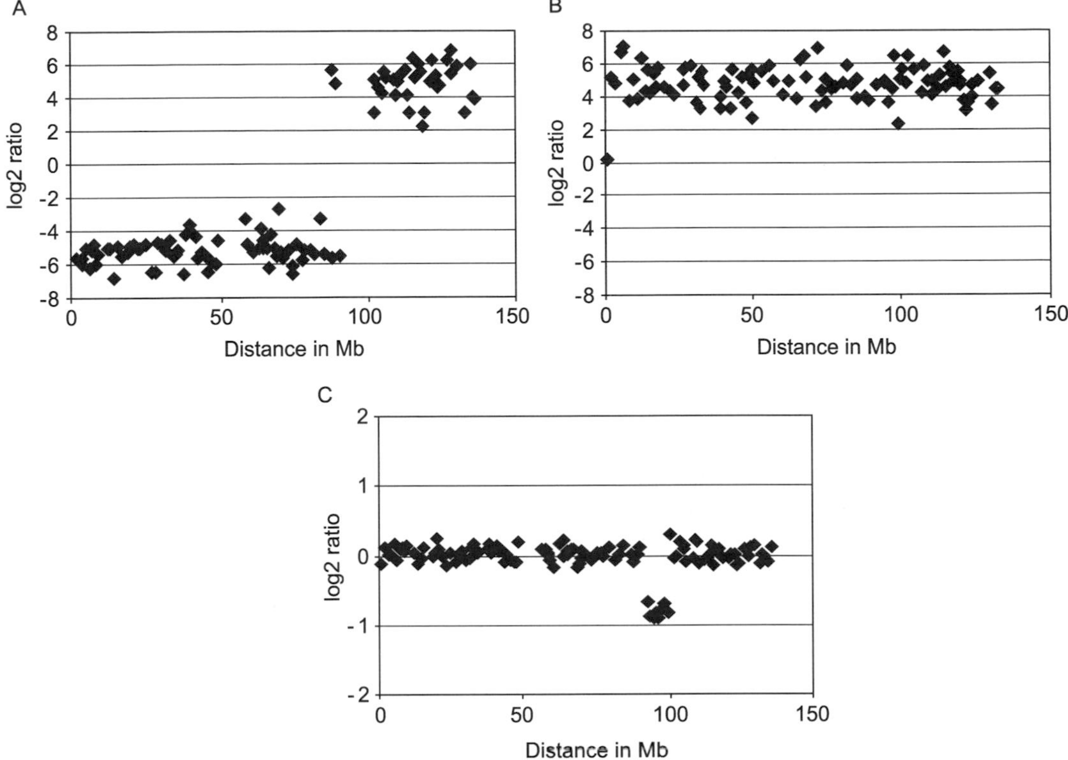

Fig. 10 Array painting and array comparative genomic hybridization (CGH) results in the analysis of a t(11;12) patient. (A) Chromosome 12 array painting profile; (B) chromosome 11 array painting profile; (C) chromosome 11 array CGH profile.

However, the pattern for chromosome 11 was much more complicated. Array painting suggested the breakpoint on chromosome 11 to be between 87 and 101 Mb along the chromosome. For a balanced translocation, we would expect to see a transition of ratios around the breakpoint, but in this case, a complex pattern of ratios within a 14-Mb region, including a potential deletion and inversion, was observed (Fig. 10B). Array CGH confirmed a deletion of about 6 Mb within the breakpoint spanning region (Fig. 10C). Further analysis using FISH to confirm the position of the clones on the derivatives led to the hypothesis that initially one chromosome 11 experienced an inversion/deletion event. This was then followed by a translocation with chromosome 12. It is at this stage impossible to tell whether the deletion followed the inversion or vice versa or if both occurred as a result of the same event (Fiegler *et al.*, 2003b).

Cases such as the translocation described earlier demonstrate that the combination of array CGH and array painting for the analysis of balanced translocations not only identifies breakpoints but also helps reveal deletions and insertions associated with the translocation event. In addition, array CGH will identify intrachromosomal duplications within a single derivative chromosome that will not be detected by array painting, as well as screening the entire genome for unexpected copy number changes that might be associated with the phenotype of a patient.

VIII. ChIP on Genomic Clone Arrays

An increasingly popular application for DNA microarrays is the study of DNA–protein interactions. DNA bound to DNA-binding proteins is cross-linked and then immunoprecipitated using an antibody against the protein of interest. DNA is then isolated and subjected to fluorescent labeling before hybridizing onto the arrays. We have used this application to study DNA damage checkpoint-mediated response in telomere-initiated senescence (Fagagna Fd *et al.*, 2003). *Senescence* is defined as the exhaustion of proliferative potential and can be triggered by telomere erosion whereby proteins such as phosphorylated H2AX, which are usually involved in DNA double-stranded break repair, directly associate with uncapped telomeres. To prove the direct association between dysfunctional telomeres and DNA damage response in senescent cells, we performed chromatin immunoprecipitation using an antibody against phosphorylated H2AX and hybridized the immunoprecipitated DNA against the original input DNA onto the 1-Mb array (Fig. 11). By calculating the difference between the ratios (immunoprecipitated DNA vs input DNA) obtained in senescent and quiescent cells, we could show that phosphorylated H2AX accumulates at a subset of subtelomeric regions in senescent cells, with a preference toward chromosome ends known to harbor short telomeres. Performing the same hybridization on a chromosome 22q tiling path array consisting of overlapping large insert clones that cover the whole of chromosome 22q not only confirmed the association of phosphorylated H2AX to the subtelomeric region of chromosome 22q, but also revealed that phosphorylated H2AX spreads more than 270 kb inward from the chromosome end (Fig. 12).

IX. Conclusion

We have described some of the methods and applications of genomic clone microarrays for the study of chromosome rearrangements. The analysis of genomic changes in this way is clearly becoming increasingly more widespread and the resolution provided by this approach is increasing. Large insert clone arrays are being produced with complete chromosome or even whole genome tiling path coverage. We predict that even higher resolution analysis will become possible with the development of quantitative genomic hybridization to arrays

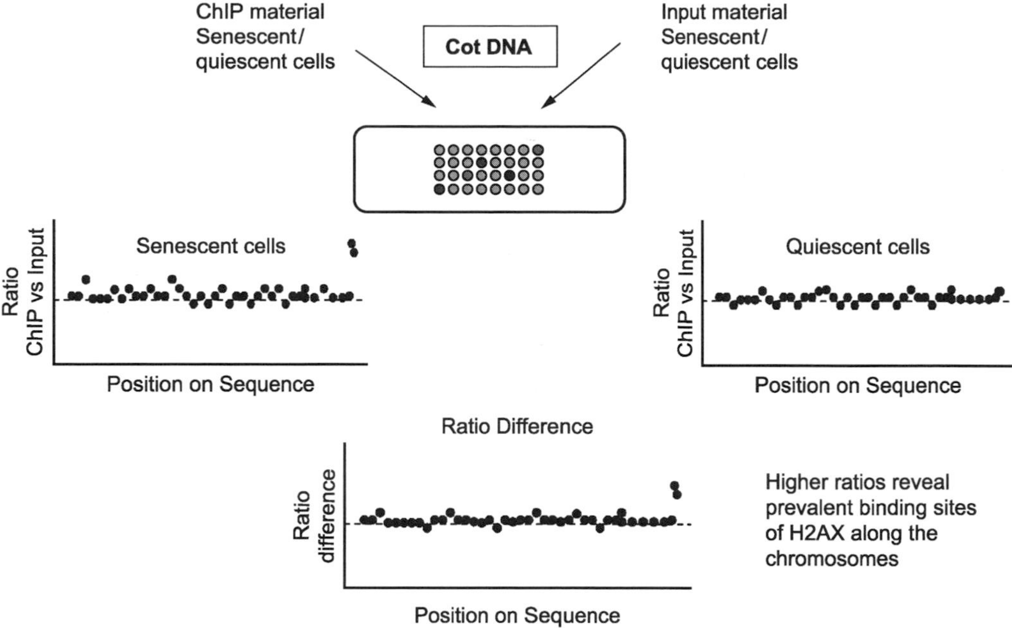

Fig. 11 Principle of ChIP on ChIP analysis using genomic microarrays.

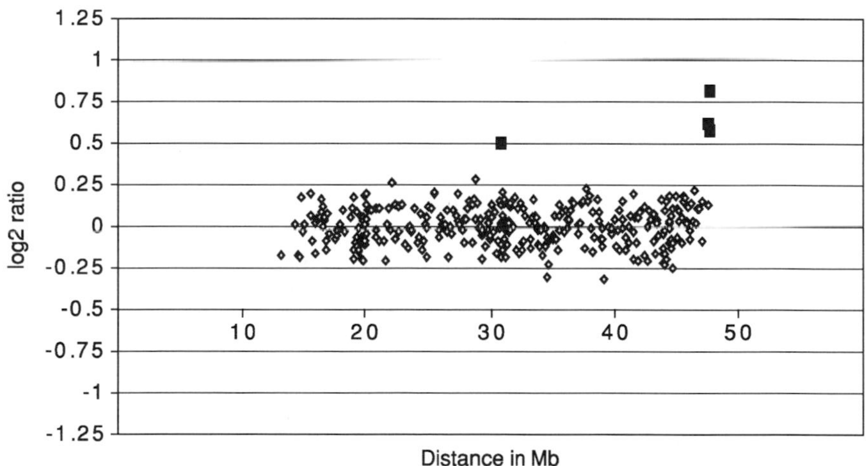

Fig. 12 ChIP analysis using an antibody against phosphorylated histone H2AX in senescent cells. Chromosome 22 profile using a 22q tiling path resolution array.

consisting of short genomic fragments. These developments will allow increasingly subtle genomic changes to be identified and correlated with disease phenotypes.

References

Albertson, D. G., and Pinkel, D. (2003). Genomic microarrays in human genetic disease and cancer. *Hum. Mol. Genet.* **12**(Spec. no. 2), R145–R152.

Buckley, P. G., Mantripragada, K. K., Benetkiewicz, M., Tapia-Paez, I., Diaz De Stahl, T., Rosenquist, M., Ali, H., Jarbo, C., De Bustos, C., Hirvela, C., Sinder Wilen, B., Fransson, I., Thyr, C., Johnsson, B. I., Bruder, C. E., Menzel, U., Hergersberg, M., Mandahl, N., Blennow, E., Wedell, A., Beare, D. M., Collins, J. E., Dunham, I., Albertson, D., Pinkel, D., Bastian, B. C., Faruqi, A. F., Lasken, R. S., Ichimura, K., Collins, V. P., and Dumanski, J. P. (2002). A full-coverage, high-resolution human chromosome 22 genomic microarray for clinical and research applications. *Hum. Mol. Genet.* **11,** 3221–3229.

Epstein, C. J. (2001). The metabolic and molecular bases of inherited disease. McGraw-Hill, New York.

Fagagna Fd, F., Reaper, P. M., Clay-Farrace, L., Fiegler, H., Carr, P., Von Zglinicki, T., Saretzki, G., Carter, N. P., and Jackson, S. P. (2003a). A DNA damage checkpoint response in telomere-initiated senescence. *Nature* **426,** 194–198.

Fiegler, H., Carr, P., Douglas, E. J., Burford, D. C., Hunt, S., Scott, C. E., Smith, J., Vetrie, D., Gorman, P., Tomlinson, I. P., and Carter, N. P. (2003a). DNA microarrays for comparative genomic hybridization based on DOP-PCR amplification of BAC and PAC clones. *Genes Chromosomes Cancer* **36,** 361–374.

Fiegler, H., Gribble, S. M., Burford, D. C., Carr, P., Prigmore, E., Porter, K. M., Clegg, S., Crolla, J. A., Dennis, N. R., Jacobs, P., and Carter, N. P. (2003b). Array painting: A method for the rapid analysis of aberrant chromosomes using DNA microarrays. *J. Med. Genet.* **40,** 664–670.

Foreman, P. K., and Davis, R. W. (2000). Real-time PCR-based method for assaying the purity of bacterial artificial chromosome preparations. *Biotechniques* **29,** 410–412.

Hodgson, G., Hager, J. H., Volik, S., Hariono, S., Wernick, M., Moore, D., Nowak, N., Albertson, D. G., Pinkel, D., Collins, C., Hanahan, D., and Gray, J. W. (2001). Genome scanning with array CGH delineates regional alterations in mouse islet carcinomas. *Nat. Genet.* **29,** 459–464.

Hurst, C. D., Fiegler, H., Carr, P., Williams, S., Carter, N. P., and Knowles, M. A. (2004). High-resolution analysis of genomic copy number alterations in bladder cancer by microarray-based comparative genomic hybridization. *Oncogene.* **23,** 2250–2263.

Korenberg, J. R., Chen, X., Schipper, R., Sun, Z., Gonsky, R., Gerwehr, S., Carpenter, N., Daumer, C., Dignan, P., Disteche, C., Graham Jr, J. M., Hudgins, L., McGillivray, B., Miyazaki, K., Ogasawara, N., Park, J. P., Pagon, R., Pueschel, S., Sack, G., Say, B., Schuffenhauer, S., Soukup, S., and Yamanaka, T. (1994). Down syndrome phenotypes: The consequences of chromosomal imbalance. *Proc. Natl. Acad. Sci. USA* **91,** 4997–5001.

Lander, E. S., Linton, L. M., Birren, B., Nusbaum, C., Zody, M. C., Baldwin, J., Devon, K., Dewar, K., Doyle, M., FitzHugh, W., Funke, R., Gage, D., Harris, K., Heaford, A., Howland, J., Kann, L., Lehoczky, J., LeVine, R., McEwan, P., McKernan, K., Meldrim, J., Mesirov, J. P., Miranda, C., Morris, W., Naylor, J., Raymond, C., Rosetti, M., Santos, R., Sheridan, A., Sougnez, C., Stange-Thomann, N., Stojanovic, N., Subramanian, A., Wyman, D., Rogers, J., Sulston, J., Ainscough, R., Beck, S., Bentley, D., Burton, J., Clee, C., Carter, N., Coulson, A., Deadman, R., Deloukas, P., Dunham, A., Dunham, I., Durbin, R., French, L., Grafham, D., Gregory, S., Hubbard, T., Humphray, S., Hunt, A., Jones, M., Lloyd, C., McMurray, A., Matthews, L., Mercer, S., Milne, S., Mullikin, J. C., Mungall, A., Plumb, R., Ross, M., Shownkeen, R., Sims, S., Waterston, R. H., Wilson, R. K., Hillier, L. W., McPherson, J. D., Marra, M. A., Mardis, E. R., Fulton, L. A., Chinwalla, A. T., Pepin, K. H., Gish, W. R., Chissoe, S. L., Wendl, M. C., Delehaunty, K. D., Miner, T. L., Delehaunty, A., Kramer, J. B., Cook, L. L., Fulton, R. S.,

Johnson, D. L., Minx, P. J., Clifton, S. W., Hawkins, T., Branscomb, E., Predki, P., Richardson, P., Wenning, S., Slezak, T., Doggett, N., Cheng, J. F., Olsen, A., Lucas, S., Elkin, C., Uberbacher, E., Frazier, M., Gibbs, R. A., Muzny, D. M., Scherer, S. E., Bouck, J. B., Sodergren, E. J., Worley, K. C., Rives, C. M., Gorrell, J. H., Metzker, M. L., Naylor, S. L., Kucherlapati, R. S., Nelson, D. L., Weinstock, G. M., Sakaki, Y., Fujiyama, A., Hattori, M., Yada, T., Toyoda, A., Itoh, T., Kawagoe, C., Watanabe, H., Totoki, Y., Taylor, T., Weissenbach, J., Heilig, R., Saurin, W., Artiguenave, F., Brottier, P., Bruls, T., Pelletier, E., Robert, C., Wincker, P., Smith, D. R., Doucette-Stamm, L., Rubenfield, M., Weinstock, K., Lee, H. M., Dubois, J., Rosenthal, A., Platzer, M., Nyakatura, G., Taudien, S., Rump, A., Yang, H., Yu, J., Wang, J., Huang, G., Gu, J., Hood, L., Rowen, L., Madan, A., Qin, S., Davis, R. W., Federspiel, N. A., Abola, A. P., Proctor, M. J., Myers, R. M., Schmutz, J., Dickson, M., Grimwood, J., Cox, D. R., Olson, M. V., Kaul, R., Raymond, C., Shimizu, N., Kawasaki, K., Minoshima, S., Evans, G. A., Athanasiou, M., Schultz, R., Roe, B. A., Chen, F., Pan, H., Ramser, J., Lehrach, H., Reinhardt, R., McCombie, W. R., de la Bastide, M., Dedhia, N., Blocker, H., Hornischer, K., Nordsiek, G., Agarwala, R., Aravind, L., Bailey, J. A., Bateman, A., Batzoglou, S., Birney, E., Bork, P., Brown, D. G., Burge, C. B., Cerutti, L., Chen, H. C., Church, D., Clamp, M., Copley, R. R., Doerks, T., Eddy, S. R., Eichler, E. E., Furey, T. S., Galagan, J., Gilbert, J. G., Harmon, C., Hayashizaki, Y., Haussler, D., Hermjakob, H., Hokamp, K., Jang, W., Johnson, L. S., Jones, T. A., Kasif, S., Kaspryzk, A., Kennedy, S., Kent, W. J., Kitts, P., Koonin, E. V., Korf, I., Kulp, D., Lancet, D., Lowe, T. M., McLysaght, A., Mikkelsen, T., Moran, J. V., Mulder, N., Pollara, V. J., Ponting, C. P., Schuler, G., Schultz, J., Slater, G., Smit, A. F., Stupka, E., Szustakowski, J., Thierry-Mieg, D., Thierry-Mieg, J., Wagner, L., Wallis, J., Wheeler, R., Williams, A., Wolf, Y. I., Wolfe, K. H., Yang, S. P., Yeh, R. F., Collins, F., Guyer, M. S., Peterson, J., Felsenfeld, A., Wetterstrand, K. A., Patrinos, A., Morgan, M. J., Szustakowki, J., de Jong, P., Catanese, J. J., Osoegawa, K., Shizuya, H., Choi, S., and Chen, Y. J. (2001). International Human Genome Sequencing Consortium. Initial sequencing and analysis of the human genome. *Nature* **409,** 860–921.

Lichter, P., Joos, S., Bentz, M., and Lampel, S. (2000). Comparative genomic hybridization: Uses and limitations. *Semin. Hematol.* **37,** 348–357.

Pinkel, D., Segraves, R., Sudar, D., Clark, S., Poole, I., Kowbel, D., Collins, C., Kuo, W. L., Chen, C., Zhai, Y., Dairkee, S. H., Ljung, B. M., Gray, J. W., and Albertson, D. G. (1998). High resolution analysis of DNA copy number variation using comparative genomic hybridization to microarrays. *Nat. Genet.* **20,** 207–211.

Rost, I., Fiegler, H., Fauth, C., Carr, P., Bettecken, T., Kraus, J., Meyer, C., Enders, A., Wirtz, A., Meitinger, T., Carter, N. P., and Speicher, M. R. (2004). Tetrasomy 21pter–>q21.2 in a male infant without typical Down's syndrome dysmorphic features but moderate mental retardation. *J Med Genet.* **41,** e26.

Snijders, A. M., Nowak, N., Segraves, R., Blackwood, S., Brown, N., Conroy, J., Hamilton, G., Hindle, A. K., Huey, B., Kimura, K., Law, S., Myambo, K., Palmer, J., Ylstra, B., Yue, J. P., Gray, J. W., Jain, A. N., Pinkel, D., and Albertson, D. G. (2001). Assembly of microarrays for genome-wide measurement of DNA copy number. *Nat. Genet.* **29,** 263–264.

Solinas-Toldo, S., Lampel, S., Stilgenbauer, S., Nickolenko, J., Benner, A., Dohner, H., Cremer, T., and Lichter, P. (1997). Matrix-based comparative genomic hybridization: Biochips to screen for genomic imbalances. *Genes Chromosomes Cancer* **20,** 399–407.

Telenius, H., Carter, N. P., Bebb, C. E., Nordenskjold, M., Ponder, B. A., and Tunnacliffe, A. (1992a). Degenerate oligonucleotide-primed PCR: General amplification of target DNA by a single degenerate primer. *Genomics* **13,** 718–725.

Telenius, H., Pelmear, A. H., Tunnacliffe, A., Carter, N. P., Behmel, A., Ferguson-Smith, M. A., Nordenskjold, M., Pfragner, R., and Ponder, B. A. (1992b). Cytogenetic analysis by chromosome painting using DOP-PCR amplified flow-sorted chromosomes. *Genes Chromosomes Cancer* **4,** 257–263.

Wessendorf, S., Fritz, B., Wrobel, G., Nessling, M., Lampel, S., Goettel, D., Kuepper, M., Joos, S., Hopman, T., Kokocinski, F., Dohner, H., Bentz, M., Schwaenen, C., and Lichter, P. (2002). Automated screening for genomic imbalances using matrix-based comparative genomic hybridization. *Lab. Invest.* **82,** 47–60.

CHAPTER 34

Prospects for *In Situ* Analyses of Individual and Complexes of DNA, RNA, and Protein Molecules with Padlock and Proximity Probes

Ulf Landegren, Mats Nilsson, Mats Gullberg, Ola Söderberg, Malin Jarvius, Chatarina Larsson, and Jonas Jarvius

Department of Genetics and Pathology
University of Uppsala, S-751-85 Uppsala, Sweden

I. Need for Single-Molecule *In Situ* Analyses

With the completion of the Human Genome Project, all macromolecules encoded in our genomes are now in principle accessible to analysis. To capitalize on this new situation, we need novel molecular procedures and sets of specific

0091-679X/04 $35.00

reagents. In particular, methods are required that permit molecules to be investigated directly in cells and tissues in which they exert their functions, in the context of all other molecules present. Rather than averaging across many cells or whole tissues and all phases of the cell cycle, it is important to develop methods that allow molecular processes to be monitored in individual cells and subcellular compartments and with sufficient precision to do this even at the single-molecule level. Cellular responses can be reflected in the appearance of secondary modifications of proteins and in rapid redistribution of molecules to form novel complexes of interacting proteins. Such processes are investigated using methods like mass spectrometry and immune coprecipitation and by studying the relations between fusion proteins modified with interaction or fluorescent domains. Although these methods provide valuable information about potential properties of and relations between subcellular components, actual effects and interactions should be studied in healthy tissues and preferably at the single-cell or even single-molecule level, allowing the nature and distribution of the participating molecules to be visualized.

II. Problem of Specificity and Sensitivity

In 1942, Coons *et al.* published a seminal paper demonstrating detection of pneumococci-infected human tissue by fluorescence microscopy, using a fluorophore-labeled antibody. Despite continuous improvements of detection reagents and signal amplification (Mayer and Bendayan, 2001), *in situ* detection methods remain limited to fairly abundant or extensive target macromolecules, and distinction between closely similar molecules tends to be difficult or impossible. Moreover, demonstration of spatial relations between macromolecules is typically limited by the light microscopic resolution of several hundred nanometers, which is insufficient to reveal direct molecular interactions.

Accordingly, the extension of *in situ* analyses to the detection of any macromolecules, specific DNA or RNA sequences or proteins occurring at widely different levels, represents a considerable challenge, but one that could provide invaluable insights in physiological and pathological molecular processes. To consider the scale of the difficulty, it is pertinent to point out that a human nucleus includes a total DNA sequence of about 13 billion nucleotides and that methods are needed that can detect changes of even single base pairs. Almost half of our genomes can be expressed as RNA molecules (Scherer *et al.*, 2003), but copy numbers of RNA sequences vary over wide ranges, complicating analysis. The billion or so protein molecules in a human cell, finally, represent abundance classes that vary even more widely from single molecules to abundant molecular species. It is also necessary to distinguish a variety of differently processed or modified variants of specific proteins, and because the functions of proteins are critically determined by their interaction with other proteins or with specific DNA or RNA sequences, relative positional information is required.

A fundamental challenge of *in situ* analyses concerns the dual requirement for extremely high specificity and sensitivity to detect most biomolecules. Detection typically represents a precarious balance between the requirement to reveal a sufficient proportion of the desired target molecules, and avoiding cross reactions for nontarget sequences. Signal to noise is further limited by nonspecific binding of detection reagents and endogenous factors like autofluorescence or haptens used for detection that may contribute to noise levels. Although a number of methods exist to augment detection signals *in situ* (Kerstens *et al.*, 1995), these regularly result in a proportionate increase also of background signals, limiting detection sensitivity.

A. *In Situ* Detection of DNA or RNA Sequences

Nucleic acid probes of tens of kilobases have a high probability of detecting target DNA sequences *in situ* and provide strong detection signals. However, such probes are insensitive to sequence variation, which could spell the difference between health and disease. Shorter hybridization probes have been used to distinguish single nucleotide variation *in situ*, but the signal seen over a background of cross-reactive or completely nonspecific binding by probes is insufficient to interrogate other than multicopy gene sequences (O'Keefe and Matera, 2000). Similarly, although allele-specific RNA expression has been demonstrated *in situ* (Adam *et al.*, 1996), more easily implemented methods are required. The combination of short hybridization probes and enzymatic DNA synthesis by polymerases in the primed *in situ* reaction can improve specificity and sensitivity (Koch *et al.*, 1989), but the sensitivity still falls short of that required for single-copy gene detection.

B. Detection of Proteins *In Situ*

In a similar manner, the specificity of *in situ* protein detection *via* binding by antibodies is constrained by the tendency of antibodies also to bind related or distinct determinants on irrelevant proteins and by their finite affinity. Although monoclonal reagents offer opportunities to select optimal binders and represent renewable resources suitable to be shared among laboratories, the limited target sequence recognition and finite affinity can result in cross reactions. The binding by polyclonal antisera from immunized animals, on the other hand, typically has the virtue of summing binding across several antigenic determinants on target proteins enhancing specific detection, but this can also create increased opportunities for cross reactions.

III. Single- or Dual-Recognition Assays

It is striking that *in situ* detection of nucleic acids or protein still today almost exclusively rests on the binding by single affinity probes or groups of affinity probes, like polyclonal antisera or fragmented long hybridization probes. By contrast, in solution-phase analyses of nucleic acids or proteins, assays in which each target molecule must be detected by two affinity probes dominate. This is true for sandwich immunoassays for soluble proteins in which two different epitopes are simultaneously recognized (Wide *et al.*, 1967), for sandwich DNA hybridization assays (Virtanen *et al.*, 1984), and for DNA detection using pairs of primers in the polymerase chain reaction (PCR) (Saiki *et al.*, 1985) or in oligonucleotide ligation assays (Landegren *et al.*, 1988; Nilsson *et al.*, 1994). Dual-recognition assays reduce the risk of cross reactions because the probability that both recognized molecular motifs are present in a molecule different from that of interest is far lower than for each single probe binding reaction, and they, therefore, offer a substantially improved level of specificity and signal over background. Dual-recognition reactions could confer great advantages also for *in situ* analyses.

IV. Some Possible Assay Formats

Some molecular procedures can bring the required dual-recognition to localized detection of either target nucleic acid or protein molecules. Two such strategies are discussed here, along with a method to amplify detection signals.

A. Padlock Probes as Dual-Recognition *In Situ* Probes

Padlock probes are linear oligonucleotide probes with target-specific sequences at each end, separated by a nontarget complementary segment. The end-sequences are designed to hybridize to two immediately adjacent target DNA sequences so a DNA ligase can join the two ends of each probe molecule upon their binding to the correct sequence (Nilsson *et al.*, 1994) (Fig. 1). Recognition of a specific target sequence results in the formation of a circular probe molecule in a highly specific reaction, capable of revealing single-nucleotide differences in total human genomic DNA samples (Antson *et al.*, 2000; Baner *et al.*, 2003; Hardenbol *et al.*, 2003; Lizardi *et al.*, 1998). The high specificity is the consequence of the requirement for dual-target recognition. Furthermore, the exquisite substrate requirement of DNA ligases ensures that probes correctly base paired at the ligation junction are joined at a rate 1000-fold greater than ones with a single mismatch (Landegren *et al.*, 1988; Luo *et al.*, 1996; Nilsson *et al.*, 2002). Ligation of oligonucleotides can also be used to interrogate RNA molecules, either directly (Nilsson *et al.*, 2000, 2001) or after the RNA sequences have been reverse transcribed to complementary DNA.

Padlock probes become topologically linked to the target strand upon circularization of the probe molecules because they are threaded around the target sequence as many times as the turns in the probe-target DNA duplex (Nilsson *et al.*, 1994). As a consequence of this topological link, circularized probes do not dissociate from the target strands under denaturing washing conditions that

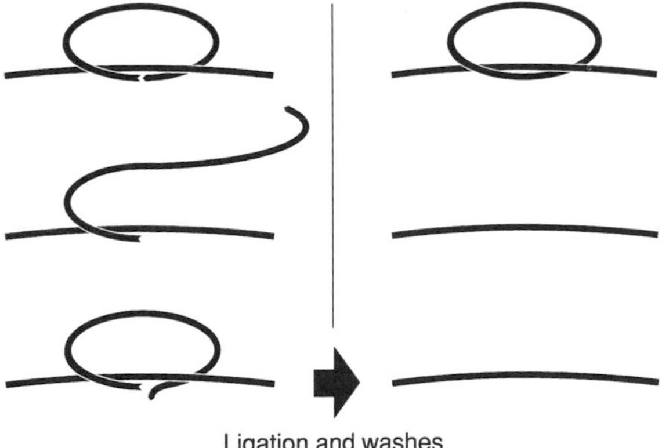

Fig. 1 Padlock probe ligation. The figure illustrates the recognition of a target DNA sequence by a correctly hybridized probe, by one hybridizing via one arm only, and one mismatched at the ultimate 3′ position. After attempted ligation and washes, only the successfully circularized probe remains bound to the target.

remove hybridizing linear probes. Allele-specific differentially labeled padlock probes have proven useful to genotype *in situ* single-nucleotide variation in centromeric-repeated sequences, enabling analysis of the parental origin of homologous human metaphase chromosomes (Antson *et al.*, 2000, 2003; Nilsson *et al.*, 1997). However, because of remaining problems with limited detection signals and nonspecifically bound padlock probes, single-copy gene detection has regrettably not proven possible using this strategy.

B. Proximity Probes for Detection of Proteins

Proximity probes are sets of reagents, each consisting of one binder moiety (e.g., an antibody or an aptamer) and an attached oligonucleotide. Binding to the same protein molecule by a pair of proximity probes brings the attached oligonucleotides in close proximity (Fig. 2). Next, these oligonucleotides are joined by ligation, guided by a connector oligonucleotide added in molar excess that hybridizes to the free ends of nearby DNA strands. The proximity ligation translates detected proteins into amplifiable information-carrying DNA molecules, and the dual-recognition format results in excellent specificity. The dependence on high relative concentration of ligatable ends through coincident target binding ensures high signal/noise ratios, whereas amplified DNA detection enables sensitive detection of small numbers of ligation products (Fredriksson *et al.*, 2002; Gullberg *et al.*, 2004). So far the assay has been demonstrated for homogenous or solid-phase–based detection of soluble proteins, but as discussed later in this chapter, *in situ* detection formats are also possible.

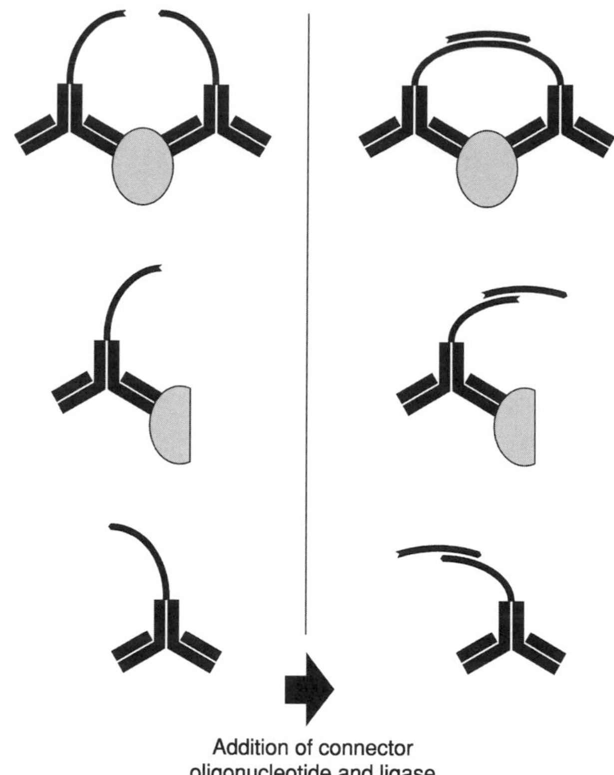

Fig. 2 Proximity probe ligation. Oligonucleotides attached to antibodies are brought in proximity when pairs of antibodies bind the same target protein. Proteins that have bound single antibody reagents or reagents free in solution fail to bring oligonucleotide ends in proximity. Upon addition of a connector oligonucleotide and a ligase, only oligonucleotide ends that have been brought close to one another can be joined by ligation.

C. Signal Amplification through Rolling-Circle Replication

Padlock probe ligation results in the formation of circular DNA strands, and by a modification of the protocol, the proximity ligation process can also result in the formation of circular DNA strands. The mechanism of rolling-circle replication (RCR) affords a convenient means to amplify detection signals from circularized probes in a manner suitable for localized detection. In a 1-hour incubation, the enzyme Phi29 polymerase can replicate a 100-mer DNA circle to generate a single-stranded product representing a concatamer of around 1000 complements of the DNA circle (Fig. 3) (Fire and Xu, 1995; Liu *et al.*, 1996; Zhang *et al.*, 1998). The sequence of the single-stranded DNA reporter molecules that form can include sequence elements that may be decoded by hybridizing labeled oligonucleotides to reveal the identity of the circles that have formed in target-specific padlock or

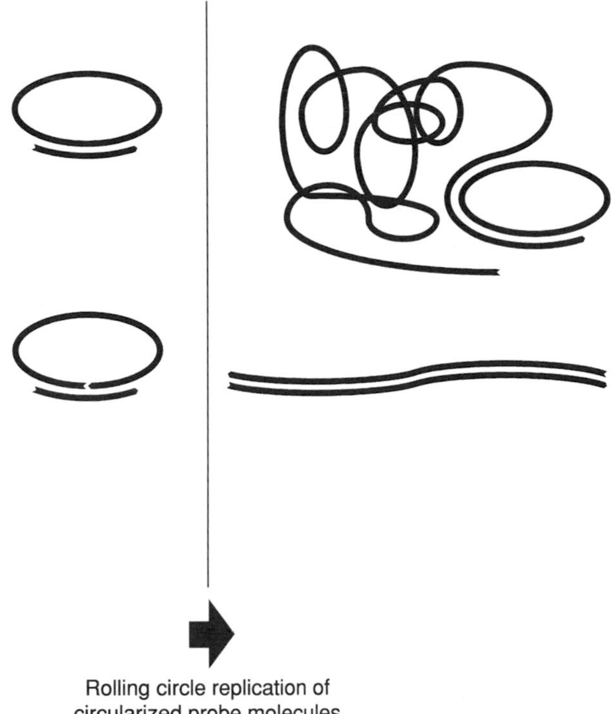

Fig. 3 Rolling-circle replication. Circular DNA strands can be replicated by extension from a hybridized primer, resulting in a long single-stranded concatamer of the complement of the circular DNA strand, and can be detected by hybridization of probes specific for the repeated sequence. By contrast, replication of probes that have failed to circularize terminates at the 5′ end and results in a duplex form of the probe molecule, unavailable for hybridization.

proximity ligation reactions. The RCR products represent individual spatially confined repeats of the replicated probe sequence and can be detected by hundreds of hybridizing fluorescence-labeled oligonucleotides. Such amplification products of reaction products are conveniently observed and enumerated by fluorescence microscopy over any background of nonspecifically bound fluorescent oligonucleotides (Blab *et al.*, 2004; Lizardi *et al.*, 1998; Zhong *et al.*, 2001).

The RCR mechanism has been used for enhanced detection of either nucleic acids or proteins by attaching to hybridization probes or antibodies, respectively, oligonucleotides having free 3′ ends and capable of priming RCR of preformed DNA circles (Gusev *et al.*, 2001; Lizardi *et al.*, 1998; Schweitzer *et al.*, 2000, 2002; Wiltshire *et al.*, 2000; Zhong *et al.*, 2001). Any probes bound through cross reactions or via nonspecific interactions or even non–probe DNA strands present in the sample could prime RCR of the preformed DNA circles, however, and the method, therefore, risks amplifying both background and specific signals. By contrast, when circular DNA strands are formed only as a consequence of specific recognition reactions, signals can be amplified by RCR without increasing nonspecific background. In the following section, we discuss the combination of padlock and proximity probes with RCR for enhanced *in situ* detection.

V. Specific Detection of Macromolecules followed by Selective *In Situ* Amplification

The combination of highly specific DNA-, RNA-, or protein-dependent probe ligation by padlock and proximity probes with amplification of ligation products by RCR constitutes an integral series of reaction steps that preserve specificity and discourage the appearance of nonspecific signals.

A. *In Situ* RCR of Padlock Probes

Both Lizardi *et al.* (1998) and Christian *et al.* (2001) have demonstrated the amplification of specifically reacted padlock probes *in situ* using RCR. The approach is almost ideal because the locally amplified replication product is dependent on a target-specific circularization event. There is one possible experimental hurdle, however, and that is that replication of probes that remain linked to their target strands can be subject to topological inhibition preventing replication (Baner *et al.*, 1998), although there is some controversy in the literature about the level of this inhibition (Kuhn *et al.*, 2002). Once probes are released from the targets, efficient RCR does commence, but then the replication products may not remain at the site of the target molecules and may disappear entirely. Through a judicious choice of template preparation steps, these problems can be overcome by allowing the target strand to serve as the primer for RCR, firmly anchoring the amplified reaction product (Larsson *et al.*, in preparation). In this manner, *in situ* detection of gene sequences has proven possible with sufficient resolution to distinguish even single nucleotide variants of the target sequences.

B. Localized Detection of Proximity Ligation Products

In a manner similar to *in situ* detection of target DNA sequences via RCR of reacted padlock probes, the proximity ligation reaction can be geared to specifically generate circular DNA strands that can then be amplified in a localized RCR reaction (Gullberg and Söderberg *et al.*, in preparation). By making the creation of circular DNA strands conditional on proximity ligation reactions, an element of dual and proximate recognition is introduced. This yields an increased specificity, coupled with an essentially background-free amplification step via the RCR, thereby permitting even single reaction products to be detected. In contrast to standard protein detection by antibody binding *in situ*, the use of proximity ligation ensures dual recognition for enhanced specificity. The reaction mechanism also enables dual recognition of single proteins *in situ* or demonstration of interacting pairs of proteins at a micrometer scale, revealing colocalization of endogenous proteins with a resolution of tens of nanometers. The method can also reveal the presence of modifications of specific protein (Gullberg, 2003; Söderberg *et al.*, in preparation).

VI. Future Directions

A new generation of dual-specificity *in situ* probing mechanisms, resulting in the formation of amplifiable reaction products, promises to permit highly specific detection of even single target DNA, RNA, or protein molecules *in situ*, in individual samples, or in arrays of samples (Kononen *et al.*, 1998). With time, more target molecules should be accessible to analysis in parallel in single experiments by using multiple fluorophores. The methods could be used to investigate protein complexes, and macromolecules of different classes may be investigated in parallel, revealing interactions between specific proteins and target DNA sequences *in situ* as a function of differentiation, cell signaling, or disease processes. For diagnostic purposes, the ability to monitor all macromolecules and their interactions will expand the scope from mere measurement of markers of disease to analyses of their functions and their correlation to disease. This information will provide improved criteria for identification and prognostication of disease, and it may prove instrumental for the selection of new therapeutic targets and pharmaceutical candidates.

In conclusion, new *in situ* analysis methods will offer radical means of investigating the many ways the genome orchestrates cellular functions. The methods will extend biological investigations to microenvironments within tissues and cells and, ultimately, at the level of single target molecules; and they will yield powerful new diagnostic means to characterize and monitor disease processes.

References

Adam, G. I., Cui, H., Miller, S. J., Flam, F., and Ohlsson, R. (1996). Allele-specific *in situ* hybridization (ASISH) analysis: A novel technique which resolves differential allelic usage of H19 within the same cell lineage during human placental development. *Development* **122**(3), 839–847.

Antson, D. O., Isaksson, A., Landegren, U., and Nilsson, M. (2000). PCR-generated padlock probes detect single nucleotide variation in genomic DNA. *Nucl. Acids Res.* **28**(12), E58.

Antson, D. O., Mendel-Hartvig, M., Landegren, U., and Nilsson, M. (2003). PCR-generated padlock probes distinguish homologous chromosomes through quantitative fluorescence analysis. *Eur. J. Hum. Genet.* **11,** 357–363.

Baner, J., Nilsson, M., Mendel-Hartvig, M., and Landegren, U. (1998). Signal amplification of padlock probes by rolling circle replication. *Nucl. Acids Res.* **26**(22), 5073–5078.

Baner, J., Isaksson, A., Waldenstrom, E., Jarvius, J., Landegren, U., and Nilsson, M. (2003). Parallel gene analysis with allele-specific padlock probes and tag microarrays. *Nucl. Acids Res.* **31**(17), e103.

Blab, G. A., Schmidt, T., and Nilsson, M. (2004). Homogeneous detection of single rolling circle replication products. *Anal. Chem.* **76**(2), 495–498.

Christian, A. T., Pattee, M. S., Attix, C. M., Reed, B. E., Sorensen, K. J., and Tucker, J. D. (2001). Detection of DNA point mutations and mRNA expression levels by rolling circle amplification in individual cells. *Proc. Natl. Acad. Sci. USA* **98**(25), 14238–14243.

Coons, A. H., Creech, H. J., Jones, R. N., and Berliner, E. (1942). Demonstration of pneumococcal antigen in tissues by use of fluorescent antibody. *J. Immunol.* **45,** 159–170.

Fire, A., and Xu, S. Q. (1995). Rolling replication of short DNA circles. *Proc. Natl. Acad. Sci. USA* **92**(10), 4641–4645.

Fredriksson, S., Gullberg, M., Jarvius, J., Olsson, C., Pietras, K., Gustafsdottir, S. M., Ostman, A., and Landegren, U. (2002). Protein detection using proximity-dependent DNA ligation assays. *Nat. Biotechnol.* **20**(5), 473–477.

Gullberg, M. (2003). Proximity ligation as a universal protein detection technique. *Genetics and Pathology*. Uppsala, Uppsala University: 51.

Gullberg, M., Fredriksson, S., Taussig, M., Jarvius, J., Gustafsdottir, S., and Landegren, U. (2003). A sense of closeness: Protein detection by proximity ligation. *Curr. Opin. Biotechnol.* **14**(1), 82–86.

Gullberg, M., Gustafsdottir, S. M., Schallmeiner, E., Jarvius, J., Bjarnegard, M., Betsholtz, C., Landegren, U., and Fredriksson, S. (2004). Cytokine detection by antibody-based proximity ligation. *Proc. Natl. Acad. Sci. USA* **101**(22), 8420–8424.

Gusev, Y., Sparkowski, J., Raghunathan, A., Ferguson, H., Jr., Montano, J., Bogdan, N., Schweitzer, B., Wiltshire, S., Kingsmore, S. F., Maltzman, W., and Wheeler, V. (2001). Rolling circle amplification: A new approach to increase sensitivity for immunohistochemistry and flow cytometry. *Am. J. Pathol.* **159**(1), 63–69.

Hardenbol, P., Baner, J., Jain, M., Nilsson, M., Namsaraev, E. A., Karlin-Neumann, G. A., Fakhrai-Rad, H., Ronaghi, M., Willis, T. D., Landegren, U., and Davis, R. W. (2003). Multiplexed genotyping with sequence-tagged molecular inversion probes. *Nat. Biotechnol.* **21**(6), 673–678.

Kerstens, H. M., Poddighe, P. J., and Hanselaar, A. G. (1995). A novel *in situ* hybridization signal amplification method based on the deposition of biotinylated tyramine. *J. Histochem. Cytochem.* **43**(4), 347–352.

Koch, J. E., Kolvraa, S., Petersen, K. B., Gregersen, N., and Bolund, L. (1989). Oligonucleotide-priming methods for the chromosome-specific labelling of alpha satellite DNA *in situ*. *Chromosoma* **98**(4), 259–265.

Kononen, J., Bubendorf, L., Kallioniemi, A., Barlund, M., Schraml, P., Leighton, S., Torhorst, J., Mihatsch, M. J., Sauter, G.,., and Kallioniemi, O. P. (1998). Tissue microarrays for high-throughput molecular profiling of tumor specimens. *Nat. Med.* **4**(7), 844–847.

Kuhn, H., Demidov, V. V., and Frank-Kamenetskii, M. D. (2002). Rolling-circle amplification under topological constraints. *Nucl. Acids Res.* **30**(2), 574–580.

Landegren, U., Kaiser, R., Sanders, J., and Hood, L. (1988). A ligase-mediated gene detection technique. *Science* **241,** 1077–1080.

Liu, D., Daubendiek, S. L., Zillman, M. A., Ryan, K., and Kool, E. T. (1996). Rolling circle DNA synthesis: Small oligonucleotides as efficient templates for DNA polymerases. *J. Am. Chem. Soc.* **118,** 1587–1594.

Lizardi, P. M., Huang, X., Zhu, Z., Bray-Ward, P., Thomas, D. C., and Ward, D. C. (1998). Mutation detection and single-molecule counting using isothermal rolling-circle amplification. *Nat. Genet.* **19**(3), 225–232.

Luo, J., Bergstrom, D. E., and Barany, F. (1996). Improving the fidelity of Thermus thermophilus DNA ligase. *Nucl. Acids Res.* **24**(15), 3071–3078.

Mayer, G., and Bendayan, M. (2001). Amplification methods for the immunolocalization of rare molecules in cells and tissues. *Prog. Histochem. Cytochem.* **36**(1), 3–85.

Nilsson, M., Malmgren, H., Samiotaki, M., Kwiatkowski, M., Chowdhary, B. P., and Landegren, U. (1994). Padlock probes: Circularizing oligonucleotides for localized DNA detection. *Science* **265**(5181), 2085–2088.

Nilsson, M., Krejci, K., Koch, J., Kwiatkowski, M., Gustavsson, P., and Landegren, U. (1997). Padlock probes reveal single-nucleotide differences, parent of origin and *in situ* distribution of centromeric sequences in human chromosomes 13 and 21. *Nat. Genet.* **16,** 252–255.

Nilsson, M., Barbany, G., Antson, D. O., Gertow, K., and Landegren, U. (2000). Enhanced detection and distinction of RNA by enzymatic probe ligation. *Nat. Biotechnol.* **18**(7), 791–793.

Nilsson, M., Antson, D. O., Barbany, G., and Landegren, U. (2001). RNA-templated DNA ligation for transcript analysis. *Nucl. Acids Res.* **29**(2), 578–581.

Nilsson, M., Baner, J., Mendel-Hartvig, M., Dahl, F., Antson, D. O., Gullberg, M., and Landegren, U. (2002). Making ends meet in genetic analysis using padlock probes. *Hum. Mutat.* **19**(4), 410–415.

O'Keefe, C. L., and Matera, A. G. (2000). Alpha satellite DNA variant-specific oligoprobes differing by a single base can distinguish chromosome 15 homologs. *Genome Res.* **10**(9), 1342–1350.

Saiki, R. K., Scharf, S., Faloona, F., Mullis, K. B., Horn, G. T., Erlich, H. A., and Arnheim, N. (1985). Enzymatic amplification of beta-globin genomic sequences and restriction site analysis for diagnosis of sickle cell anemia. *Science* **230**(4732), 1350–1354.

Scherer, S. W., Cheung, J., MacDonald, J. R., Osborne, L. R., Nakabayashi, K., Herbrick, J. A., Carson, A. R., Parker-Katiraee, L., Skaug, J., Khaja, R., Zhang, J., Hudek, A. K., Li, M., Haddad, M., Duggan, G. E., Fernandez, B. A., Kanematsu, E., Gentles, S., Christopoulos, C. C., Choufani, S., Kwasnicka, D., Zheng, X. H., Lai, Z., Nusskern, D., Zhang, Q., Gu, Z., Lu, F., Zeesman, S., Nowaczyk, M. J., Teshima, I., Chitayat, D., Shuman, C., Weksberg, R., Zackai, E. H., Grebe, T. A., Cox, S. R., Kirkpatrick, S. J., Rahman, N., Friedman, J. M., Heng, H. H., Pelicci, P. G., Lo-Coco, F., Belloni, E., Shaffer, L. G., Pober, B., Morton, C. C., Gusella, J. F., Bruns, G. A., Korf, B. R., Quade, B. J., Ligon, A. H., Ferguson, H., Higgins, A. W., Leach, N. T., Herrick, S. R., Lemyre, E., Farra, C. G., Kim, H. G., Summers, A. M., Gripp, K. W., Roberts, W., Szatmari, P., Winsor, E. J., Grzeschik, K. H., Teebi, A., Minassian, B. A., Kere, J., Armengol, L., Pujana, M. A., Estivill, X., Wilson, M. D., Koop, B. F., Tosi, S., Moore, G. E., Boright, A. P., Zlotorynski, E., Kerem, B., Kroisel, P. M., Petek, E., Oscier, D. G., Mould, S. J., Dohner, H., Dohner, K., Rommens, J. M., Vincent, J. B., Venter, J. C., Li, P. W., Mural, R. J., Adams, M. D., and Tsui, L. C. (2003). Human chromosome 7: DNA sequence and biology. *Science* **300**(5620), e767–772(epub 2003 Apr 10)..

Schweitzer, B., Roberts, S., Grimwade, B., Shao, W., Wang, M., Fu, Q., Shu, Q., Laroche, I., Zhou, Z., Tchernev, V. T., Christiansen, J., Velleca, M., and Kingsmore, S. F. (2002). Multiplexed protein profiling on microarrays by rolling-circle amplification. *Nat. Biotechnol.* **20**(4), 359–365.

Schweitzer, B., Wiltshire, S., Lambert, J., O'Malley, S., Kukanskis, K., Zhu, Z., Kingsmore, S. F., Lizardi, P. M., and Ward, D. C. (2000). Inaugural article: Immunoassays with rolling circle DNA amplification: A versatile platform for ultrasensitive antigen detection. *Proc. Natl. Acad. Sci. USA* **97**(18), 10113–10119.

Wide, L., Bennich, H., and Johansson, S. G. (1967). Diagnosis of allergy by an *in-vitro* test for allergen antibodies. *Lancet* **2**(7526), 1105–1107.

Wiltshire, S., O'Malley, S., Lambert, J., Kukanskis, K., Edgar, D., Kingsmore, S. F., and Schweitzer, B. (2000). Detection of multiple allergen-specific IgEs on microarrays by immunoassay with rolling circle amplification. *Clin. Chem.* **46**(12), 1990–1993.

Virtanen, M., Syvanen, A. C., Oram, J., Soderlund, H., and Ranki, M. (1984). Cytomegalovirus in urine: Detection of viral DNA by sandwich hybridization. *J. Clin. Microbiol.* **20**(6), 1083–1088.

Zhang, D. Y., Brandwein, M., Hsuih, T. C., and Li, H. (1998). Amplification of target-specific, ligation-dependent circular probe. *Gene* **211**(2), 277–285.

Zhong, X. B., Lizardi, P. M., Huang, X. H., Bray-Ward, P. L., and Ward, D. C. (2001). Visualization of oligonucleotide probes and point mutations in interphase nuclei and DNA fibers using rolling circle DNA amplification. *Proc. Natl. Acad. Sci. USA* **98**(7), 3940–3945.

CHAPTER 35

The Use of Subtelomeric Probes to Study Mental Retardation

Samantha J. L. Knight and Jonathan Flint
The Wellcome Trust Centre for Human Genetics
Churchill Hospital
Headington, Oxford
Oxfordshire OX3 7BN, United Kingdom

In this chapter, we focus on the genetic basis of mental retardation (MR), specifically the use of subtelomeric probes to provide new diagnoses in idiopathic MR. We discuss both the background to the clinical demand for diagnoses and the

METHODS IN CELL BIOLOGY, VOL. 75

0091-679X/04 $35.00

technological advances that culminated in the development of subtelomeric testing strategies. We explain the theory behind these strategies and briefly outline the protocols involved, giving the advantages, limitations, and pitfalls of the analyses. Finally, we give an overview of the MR subtelomeric studies to date and how subtelomeric testing has become a widely used tool in clinical diagnostic laboratories, particularly in the diagnosis of unexplained MR, but also in other fields of clinical medicine. The conclusion addresses the overall impact that subtelomeric testing has had on the diagnosis of MR, the implications for patients and their families, and future research avenues for exploring the genetic causes of MR and improving our overall understanding of neurocognitive development.

I. Introduction

A. Unexplained Mental Retardation and Genetic Testing

MR is an extremely common condition, affecting about 3% of the population. Virtually all of those individuals with moderate to severe MR (intelligence quotient [IQ] under 50) need lifelong support and about half of those with mild MR (IQ 50–70) are significantly impaired throughout life. Although MR carries with it immense clinical, social, and psychological burdens, the origins remain poorly understood.

One avenue of research that has been increasingly explored is the investigation of possible genetic origins of MR, particularly the identification of small chromosomal rearrangements. The standard test for suspected chromosomal rearrangements is cytogenetic analysis of patients' chromosomes at a 400- to 550-band resolution and this has been available for many years. However, it cannot routinely detect rearrangements smaller than 5 Megabases (Mb), and much larger abnormalities escape notice if they occur in regions where the chromosomal banding pattern is indistinct. This situation would still be the case had it not been for growing evidence that a substantial proportion of MR may well be due to smaller cytogenetically invisible chromosomal anomalies, undetectable using available tests. The latter gave the impetus to develop molecular techniques that would generate new diagnostic assays with a significantly higher level of resolution and reliability.

B. Historical Perspective

The diagnostic potential of using subtelomeric probes to study MR evolved after the convergence of important advances in molecular and clinical genetics. The critical molecular advance was the discovery that human telomeres and adjacent subtelomeric regions can be cloned by functional complementation in yeast (Bates *et al.*, 1990; Brown, 1989; Cross *et al.*, 1989; Riethman *et al.*, 1989). The characterization of these so-called half-yeast artificial chromosome (YAC) clones resulted in the development of a valuable set of molecular resources for analyzing most, if not all, chromosome ends. They paved the way for the identification of telomere-specific microsatellite markers and for the subcloning of telomere-specific cosmids, bacteriophage P1 artificial chromosomes (PACs) and bacterial artificial chromosomes (BACs) (Knight *et al.*, 2000; National Institutes of Health and Institute of Molecular Medicine collaboration, 1996; Rosenberg *et al.*, 1997). The clinical impetus came following the suggestion that cloned telomere resources, such as telomere-specific hypervariable DNA probes (HVPs), might be used to examine the integrity of subtelomeric regions as a possible cause of MR (Wilkie, 1993). Testing telomeres for rearrangements held a number of attractions. First, most translocations involve chromosome ends, so an assay that targeted telomeres would detect these with 100% sensitivity regardless of size. Second, the regions adjacent to telomeres were known to be gene rich (Saccone *et al.*, 1992), so rearrangements involving such regions would be more likely to have phenotypic consequences than rearrangements in many other parts of the genome. Finally, rearrangements involving telomeres were beginning to emerge as an important cause of human genetic diseases. For example, Wolf-Hirschhorn syndrome (chromosome 4p), Cri du Chat syndrome (chromosome 5p), Miller-Dieker syndrome (chromosome 17p), and α-thalassemia with MR (ATR-16, chromosome 16p) were all known to be due to the unbalanced products of subtelomeric translocations (Altherr *et al.*, 1991; Kuwano *et al.*, 1991; Lamb *et al.*, 1993; Overhauser *et al.*, 1989; Wilkie *et al.*, 1990; Wong *et al.*, 1997).

In 1995, Flint *et al.* reported the results of a pilot study in which HVPs were selected to study the subtelomeric regions of 28 chromosome ends in 99 mentally retarded individuals. The results of the study suggested that at least 6% of cases of idiopathic MR (IMR) might be explained by submicroscopic rearrangements involving telomeres. This was a significant result indicating that subtelomeric rearrangements might be the second most common genetic cause of MR after Down syndrome. But for confirmation, it was important to extend the studies to include all possible telomeres and a larger sample set. Collecting more samples presented no difficulty, but as outlined in section II.A, the prospect of analyzing all possible telomeres using the HVP approach was problematic, requiring more than 200 HVPs and samples not only from the patients but also from their parents. These problems were first overcome by the development of a fluorescence *in situ* hybridization (FISH)–based methodology known as the Chromoprobe Multiprobe-T System (Cytocell Limited), which allowed multiple simultaneous hybridizations of telomere-specific cosmid, P1, and PAC probes using only a single microscope slide per patient. The technique required no novel or expensive equipment or materials, making it a simple and immediately feasible proposition available to all cytogenetic diagnostic laboratories already equipped for FISH studies (Knight *et al.*, 1997).

Knight *et al.* (1999) used this method to extend the studies of Flint *et al.* (1995) and showed that subtelomeric rearrangements contribute to the MR in 0.5% cases with mild MR and in 7.4% cases with moderate to severe MR and dysmorphic features and in half of those with a positive family history. These findings have since been corroborated through a vast number of independent patient studies, such that the importance of subtelomeric anomalies in unexplained MR is now well recognized in clinical genetics (Knight, 2004).

C. Alternative Methodologies

For any assay to prove useful in the clinical diagnostic environment, it must be robust and cost effective with a high degree of sensitivity and specificity. The complex structure of subtelomeric regions and the practical difficulties of testing 41 telomeres in a single procedure have proved particularly challenging when developing screening strategies for these regions (Knight and Flint, 2000). The Chromoprobe Multiprobe-T System is still one of the most popular methodologies for assaying telomere integrity and has been modified little since its inception. However, a number of effective alternative methods are available and subtelomeric rearrangements have been detected using at least nine other approaches, including (1) multiplex or multicolor FISH (M-TEL and TM-FISH) (Brown *et al.*, 2001; Henegariu *et al.*, 2001), (2) telomere spectral karyotyping (telomere SKY) (Anderlid *et al.*, 2002; Clarkson *et al.*, 2002), (3) subtelomeric combined binary ratio labeling FISH (S-COBRA FISH) (Engels *et al.*, 2003), (4) primed *in situ* labeling (PRINS) (Bonifacio *et al.*, 2001), (5) high-resolution chromosome analysis (Joyce *et al.*, 2001), (6) comparative genome hybridization (CGH) to chromosomes (Ghaffari *et al.*, 1998; Joly *et al.*, 2001), (7) scanning short tandem repeat polymorphisms (Biorgione *et al.*, 2000; Colleaux *et al.*, 2001; Rosenberg *et al.*, 2000a,b), (8) locus copy number measurement by multiplex amplifiable probe hybridization (MAPH) (Armour *et al.*, 2000; Hollox *et al.*, 2002; Sismani *et al.*, 2001) or multiplex ligation-dependent probe amplification (MLPA) (Rooms *et al.*, 2004; Schouten *et al.*, 2002), and (9) CGH to telomere-specific microarrays (Veltman *et al.*, 2002). Of these, only the high-resolution chromosome analysis and CGH to chromosomes negate the need for subtelomeric probes. The Chromoprobe Multiprobe-T System of Cytocell Limited, the ToTel Vysion Multicolor FISH Probe Panel of Vysis/Abbott, Inc., the MAPH probes of Genetix, and the MLPA SALSA P036 Human Telomere Test Kit of MRC-Holland are all commercially available systems. Telomere-specific clone sets to be prepared in house are available to buy from the American Type Cell and Tissue Collection.

II. Background

Methods that test for telomere integrity rely on the ability to detect submicroscopic unbalanced translocations (monosomies and trisomies) and preferably balanced translocations at as high a resolution as possible. Because FISH-based methods allow the direct visualization/imaging of telomeric regions (through fluorescence microscopy or imaging), they are regarded as the gold standard when testing for such rearrangements. Methodologies that are not FISH based cannot detect balanced translocations and often require confirmation by FISH but nevertheless have proven effective and valuable as screening tools for unbalanced rearrangements. The basic theory behind each of the methodologies is discussed in this section.

A. Basic Theory behind FISH-Based Methods

All subtelomeric FISH methods exploit the use of fluorescent molecules, referred to as *fluorophores*, which are linked to nucleotides or antibodies and allow direct visualization or computer imaging of the hybridization results. Telomere-specific DNA probes are directly or indirectly labeled by nick translation or by the polymerase chain reaction (PCR) to incorporate fluorescently tagged di-deoxy nucleotide triphosphates (dNTPs) or nonfluorescently hapten tagged dNTPs, respectively (Morrison *et al.*, 2002). The probes are then hybridized to their cognate telomeric sequences present in fixed metaphase chromosome preparations immobilized on glass slides. Using the direct labeling methods, a number of wash steps are performed after hybridization and then the directly labeled hybridized probes can be viewed immediately by fluorescence microscopy. The telomeres are assessed for copy number changes or balanced rearrangements by visual inspection and/or computerized imaging of each of the telomere-specific fluorescent signals. Using indirect labeling methods, nonfluorescent hapten tagged dNTPs such as biotin-16-dUTP (BIO) or digoxigenin-11-dUTP (DIG) are incorporated into the probe DNA. After hybridization, the bound probe is detected by the addition of one to three layers of the appropriate fluorescently tagged antibodies before direct visual assessment and/or computerized imaging as described above.

These labeling and detection approaches provide the underlying principle for all FISH-based assays of telomere integrity. Traditionally, the use of two differentially labeled probes (e.g., one visualized as red fluorescence and the other as green), co-hybridized to a single spot on a single microscope slide, is satisfactory for many FISH applications. However, this two-fluorophore-per-slide approach is impractical for the analysis of all subtelomeric regions of an individual because it requires at least 21 microscope slides and 21 hybridizations for the complete analysis of 41 telomeric regions (the remaining five human telomeres consist of satellited ribosomal DNA). Hence, alternative FISH-based telomere assays were developed. The first of these, the commercially available Chromoprobe Multiprobe-T System (Cytocell Limited), overcomes the difficulties of conventional FISH by retaining a two-fluorophore detection system and allowing 23 dual hybridizations to be performed against 23 separate metaphase spots (two telomere hybridizations per spot) on a single customized microscope slide. The ToTelVysion Multicolor FISH probe system marketed by Vysis/Abbott, Inc., works in a similar way, except that three fluorophores are used for labeling and the hybridizations are split over three slides, each slide having five target spots that must be hybridized. Most other systems, developed subsequently, use a strategy known as multiplex or multicolor *in situ* hybridization (M-FISH), spectral karyotyping

(SKY), or modifications thereof (Speicher *et al.*, 1996). In M-FISH, more fluorophores are used in different combinations and/or ratios so that multiple probes can be hybridized to only one or two target metaphase spots on one or two microscope slides, respectively. The probes are later detected by virtue of their individually unique color scheme and banding. Because direct visual interpretation of so many colors is extremely difficult, the detection is achieved by computer imaging and then, if necessary, manual identification of the chromosomes can be performed. As described in section III.A.6, telomeric SKY is similar to M-FISH except that the imaging process is different.

The subtelomeric testing methods that are modifications of M-FISH are M-TEL, TM-FISH, and S-COBRA FISH. The differences between these methodologies lie in how many fluorophores or combinations of fluorophores are used and in what ratios for labeling or post-hybridization detection of the telomere-specific probes. This directly dictates how many probes can be detected in a single hybridization and, therefore, how many might be hybridized simultaneously to a single target spot of metaphase chromosomes on a microscope slide.

Whatever the FISH-based method, once hybridized and detected, the probe signals are analyzed either by direct visual assessment or by computerized imaging. Unlike other approaches that will need additional assays for confirmation, subtelomeric FISH assays in most cases should give unambiguous results that are easily interpreted: hybridization signal from only one chromosome homolog diagnoses monosomy; and from three homologs, diagnoses trisomy.

B. Basic Theory behind Non–FISH-Based Methods

A number of non–FISH-based methods have been developed for the study of telomere integrity. Although all of these can use fluorophores, there is no *in situ* hybridization to immobilized chromosomes, and with one exception, they fundamentally differ from FISH in that it is patient DNA samples (rather than chromosome preparations) that provide the templates for the analyses. The actual methodologies involve either PCR or hybridization and overviews of the theory behind each of these methods are given in this section.

1. Short Tandem Repeat Polymorphisms

Short tandem repeat polymorphisms (STRPs) are small tracts of tandemly repeated DNA sequences that are polymorphic and generally of high heterozygosity in the human population. By designing PCR primers to locus-specific telomeric STRPs, the sequence across the repeats can be amplified and the allele sizes measured. PCR can be performed using radioactive or fluorescence-based procedures (Colleaux *et al.*, 2001; Rosenberg *et al.*, 2000b). The aim is to genotype the test DNA samples and look for non-Mendelian inheritance of the alleles in an affected individual. Because it relies on comparing those alleles inherited by the affected individual with those of his or her parents, it is necessary to genotype both parental and patient samples to gain maximum information. When both parents are heterozygous for different alleles, the method is sensitive and usually highly specific. If a patient has not inherited one allele from each parent, then there is an anomaly. A single allele in the child diagnoses monosomy, three alleles diagnose trisomy, and two alleles inherited from only one parent diagnose uniparental disomy.

2. Primed *In Situ* Labeling

The PRINS approach involves the use of an oligonucleotide primer $(CCCTAA)_7$, the sequence of which is common to every human telomere. The primer is annealed *in situ* to complementary telomere sequences present in metaphase chromosomes that have been prepared from patient blood samples and immobilized on microscope slides (Bonifacio *et al.*, 2001). Fluorescently labeled nucleotides are then added by *in situ* primer extension, creating a product that can be analyzed by fluorescence microscopy. Those telomeres that lack fluorescent signal represent putative regions of deletion. The method cannot detect balanced translocations, unbalanced trisomies, or uniparental isodisomy.

3. Multiplex Amplifiable Probe Hybridization

In MAPH (Armour *et al.*, 2000; Hollox *et al.*, 2002; Sismani *et al.*, 2001), sets of PCR-amplified telomere-specific probes, collectively representing all telomeres, are hybridized in small volumes to patient DNA immobilized on tiny strips of nylon filter. Probes that have bound specifically during hybridization are then recovered and amplified by dosage-sensitive PCR before electrophoretic gel separation and probe detection by radioactivity, phosphoimaging, or fluorescence imaging (Armour *et al.*, 2000; Hollox *et al.*, 2002; Sismani *et al.*, 2001). By measuring the intensity of the gel bands (one band per probe) and comparing it relative to expected disomic profiles, the copy number of the corresponding target sequence in the patient's DNA can be determined and putative copy number changes identified.

4. Multiplex Ligation-Dependent Probe Amplification

MLPA is a refinement of MAPH that uses two synthetic oligonucleotides per telomere as probes (rather than a single telomere-specific fragment) and involves their hybridization to a patient DNA sample in solution (rather than to DNA immobilized on nylon membrane strips) (Rooms *et al.*, 2004; Schouten *et al.*, 2002). The hybridized oligonucleotides are joined to give products of unique length that are PCR amplified and separated by electrophoresis. As in MAPH, the relative amounts of the probe amplification products reflect the copy number of the target sequences and can be used to assess subtelomeric integrity.

5. Telomere Array CGH

In telomere array CGH, test and control DNA samples are differentially labeled with fluorescently tagged nucleotides (e.g., one red and one green) and co-hybridized to telomere-specific DNA probes spotted in an array format on a specialized microscope slide (a microarray) (Veltman *et al.*, 2002; Yu *et al.*, 2003). During hybridization, the differentially labeled DNA samples contend for the probe sequences in a competitive manner. The raw fluorescence signal data from the hybridized spots are then collected using a specialized scanning machine equipped with the appropriate laser systems. Because of the competitive nature of the hybridization, it is possible to compare the control DNA fluorescence signal data with that of the test fluorescence signal data for each arrayed telomere-specific probe. If there is a significant deviation from the expected 1:1 control versus test hybridization fluorescence signal ratio, then a copy number alteration of the telomere is suspected in the test sample. In theory, a control versus test ratio of 1.0:0.5 denotes monosomy, whereas 1.0:1.5 denotes trisomy, although in practice the ratios are probe dependent, tending towards 1.0:0.7 and 1.0:1.3, respectively.

III. Methods

In this section, the subtelomeric testing methodologies are given as a brief overview because in most cases the technical details regarding the procedures have been described elsewhere.

A. FISH-Based Methods

1. Chromoprobe Multiprobe-T System

The methodology for the Chromoprobe Multiprobe-T System is described in detail in Knight and Flint (2002). Briefly, the approach uses a customized microscope slide that is divided into 24 squares and a multiprobe device that is essentially a coverslip with 24 raised bosses. Each assay involves adding a drop of fixed chromosome suspension from the patient sample onto each square of the slide. To each square of the device, the appropriately labeled telomere probes are added so that each square has the differentially labeled p-arm and q-arm telomere probes of a different chromosome. The slide and the device are then sandwiched together, the probe and chromosomal DNA are denatured simultaneously at 75 °C on a hot plate, and hybridization is allowed to proceed overnight in a light-proof box floating in a water bath at 37 °C. After hybridization, a series of wash steps are performed and the resulting slides are mounted in Vectashield anti-fade containing 4′,6-diamidino-2-phenylindole (DAPI) counterstain and analyzed by fluorescence microscopy. The method requires the used of confocal or CCD microscopes equipped with triple bandpass filter sets. The numbers of signals specific to each telomere probe are counted visually and any anomalies indicating possible rearrangements are investigated further. Further FISH investigation may be achieved either by using individual telomere specific clones that are commercially available ready labeled or alternatively by acquiring a glycerol set of telomere-specific clones and preparing and labeling the DNA "in house". Methodology for the latter together with clone resources are given by Knight and Flint (2002).

2. ToTelVysion Multicolor FISH Probe Panel system

The ToTelVysion Multicolor FISH Probe Panel marketed by Vysis/Abbott, Inc., consists of 15 tubes containing multicolor combinations of telomere probes with other probe markers. The multicolor combinations come from the use of three fluorophores: SpectrumOrange, SpectrumGreen and SpectrumAqua. The full procedure can be obtained from Vysis/Abbott, Inc. (www.vysis.com/ToTelVysion_32961.asp), but in brief, metaphase chromosome preparations are applied to three slides so that five good-quality target areas can be identified per slide. For each slide, the ToTelVysion probe mixes are denatured in a heated waterbath and the immobilized metaphases denatured by immersing the slides in denaturing solution. The individual probe mixes are then applied to each of the target areas in known format, covered with a coverslip, and allowed to hybridize at 42 °C overnight. After hybridization, unbound probe is removed by a rapid 2-minute wash, DAPI counterstain is added, and the hybridized target areas are viewed by fluorescence microscopy and assessed as described earlier. Analysis of the ToTelVysion hybridizations requires single bandpass filter sets for visualization of each of the fluorophores. The bandpass filters needed to independently view the probe panel and DAPI counterstain are DAPI single bandpass, green single bandpass, orange single bandpass, and aqua single bandpass. An epifluorescence microscope equipped with a 100-Watt mercury lamp is recommended.

3. Multiplex or Multicolor Telomere (M-TEL) FISH

In generic M-FISH–based strategies, a number of fluorophores are used in different combinations and/or ratios so that multiple chromosomal regions can be detected individually by computer imaging after simultaneous hybridization to single metaphase spots on one or two microscope slides (Jalal and Law, 2002; Speicher *et al.*, 1996). M-TEL is an adaptation of M-FISH devised for the study of subtelomeric regions (Brown *et al.*, 2001). In M-TEL, the telomere probes are divided into two sets, M-TEL1 and M-TEL2 and the probes in each set are labeled by nick translation using one fluorophore (Cy3-dUTP) and three haptens (estradiol 15-dUTP [EST], BIO, and DIG) according to a predetermined combinatorial labeling scheme (Brown *et al.*, 2001). For each complete assay, a single spot of metaphase chromosomes is immobilized on each of two slides and the spreads treated with RNAse and peptin and then fixated. Immediately before hybridization, the two probe sets (in a formamide hybridization mix) are denatured by high temperature and then repeated sequences competed out at 37 °C for 1 hour before hybridization to the immobilized denatured metaphase target spots in a humid chamber for 1–2 days at 37 °C. The slides are washed, and the appropriate layers of fluorescently tagged antibodies applied: (1) avidin Cy3.5 to detect the BIO, (2) rabbit anti-estradiol to detect the EST followed by donkey anti-rabbit Cy5.5 to detect the rabbit anti-estradiol, and (3) donkey anti-sheep fluorescein isothiocyanate (FITC) to detect the DIG. The slides are finally mounted in Vectashield containing DAPI counterstain. Metaphase images are then captured using the appropriate imaging equipment and software. The results for each fluorophore channel are viewed individually and the signal for each telomere recorded manually. Brown *et al.* (2001) used a CCD camera with a Kodak KAF 1400 chip mounted on an Olympus AX70 epifluorescence microscope equipped with an eight-position filter turret with individual filter blocks specific for DAPI, FITC, Cy3, Cy3.5, Cy5, and Cy5.5 fluorophores. A 75-Watt Xenon lamp was used to capture all of the fluorochrome channels. For analysis, the results for each fluorophore channel are viewed individually and the signal for each telomere recorded manually.

4. Telomeric Multiplex FISH (TM-FISH)

TM-FISH (Henegariu *et al.*, 2001) is similar to M-TEL except that combinatorial labeling is performed in alternative ways to suit differently equipped fluorescence microscopes. By dividing the telomere probes into two or three sets and detecting using three fluorophores, any fluorescence microscope equipped with the three common filters (Texas red, FITC, and DAPI) is suitable. By keeping the probes as a single set and detecting using five fluorophores, a microscope equipped with five fluorescence filters and a CCD camera is necessary.

In the two-probe-set scenario, one probe pool includes probes labeled by nick translation with BIO and DIG haptens and the other includes those labeled with BIO, DIG, and dinitrophenyl (DNP) haptens according to predetermined combinations (Henegariu *et al.*, 2001). In the three-probe-set scenario, each of the sets is labeled with combinations of BIO, DIG, and DNP haptens. In the one-probe-set scenario, the probes are labeled with combinations of the fluorophores FITC and carboxyrhodamine 6G (R6G) and the nonfluorescent haptens BIO, DIG, and DNP. For all three combinatorial labeling protocols, Cot-1 is added to the probes and the mixture denatured before applying to the appropriate denatured metaphase target areas on a single slide (one, two, or three areas depending on the number of probe sets). After hybridization overnight, the slides are washed and the appropriate layers of fluorescently tagged antibodies applied. For the two- and three-probe-set hybridizations, mouse anti-DIG FITC is used to detect the DIG haptens, avidin–Cy5 to detect the BIO haptens, and rat anti-DNP–Cy3.5 to detect the DNP haptens. For the one-probe-set hybridizations, FITC is detected with goat anti-FITC followed by donkey anti-goat–DTAF, rhodamine is detected with rabbit anti-rhodamine–Cy3, BIO is detected with avidin–Cy5, DNP with rat anti-DNP–Cy3.5, and DIG with sheep–anti-DIG–DEAC followed by donkey anti-sheep–DEAC. All antibodies are combined into a maximum of three detections steps. After the detection steps, the slide is finally mounted in anti-fade solution containing DAPI counterstain and viewed using a fluorescence microscope equipped with the appropriate filter sets.

5. Subtelomere COmbined Binary RAtio Labeling FISH (S-COBRA FISH)

The S-COBRA FISH methodology was devised by Engels *et al.* (2003). In brief, it combines ratio and pairwise combinatorial labeling to obtain two telomere probe sets. The probes used are the second-generation PAC, P1, and BAC clones described by Knight *et al.* (2000). The first probe set consists of 11 telomere-specific probes labeled using three fluorophores in different ratios and another 10 probes labeled identically to the first set but combined with a fourth fluorophore (the "binary" label). The second set consists of a further two groups of 10 probes labeled using the same strategy as the first set. The probe DNAs are prepared for labeling by digestion with DNAse I followed by purification using standard kits. The prepared probes are labeled according to the predetermined scheme (Engels *et al.*, 2003) using the Universal Linkage System (ULS, KREATECH Biotechnology BV, Amsterdam, The Netherlands) described by Morrison *et al.* (2002). dGreen–ULS, Cy3–ULS, and Cy5–ULS are used as the ratio fluorophores. The binary label is DNP. Once labeled and purified, the probe cocktails are precipitated with ethanol in the presence of Cot-1 DNA and resuspended in a formamide-based hybridization solution. Before hybridization, prepared slides with test metaphase chromosomes are treated with RNAse A and pepsin and then the chromosomes denatured by incubation at 80 °C in 60% formamide, 2xSSC, pH7. The slides are then dehydrated through an ethanol series and air-dried. The probes are added to the slides (two slides total, one for each probe set), and a coverslip applied and sealed with rubber cement. Hybridization is allowed to proceed for 2 days in a humid chamber at 37 °C. The slides are washed after hybridization and a step included to detect the DNP–ULS using Laserpro IR790–conjugated antibody. Hybridized probes are visualized by digital fluorescence imaging as described by Tanke *et al.* (1999), and the acquired images (usually four metaphases per slide) are analyzed using specialized software. Some visual classification or scatterplot analysis may be necessary for any ambiguous classifications that occur (estimated to be about one to two per case). Putative copy number changes are further investigated by additional FISH analyses with the relevant probes.

6. Spectral Karyotyping

SKY is a FISH-based method that emerged concurrently with M-FISH. Standard SKY, like M-FISH, uses combinatorially labeled chromosome-specific paints to investigate chromosome integrity. However, it differs from M-FISH in the imaging process. In SKY, multiple fluorophores are simultaneously excited and an interferometer is used to determine the spectral profile of each pixel of an image, whereas in M-FISH, the fluorophores are detected separately using narrow bandpass excitation/emission filters, the presence or absence of each fluorophore at a given pixel of an image is evaluated, and computerized superimposition of the five fluor layers is then performed. Both SKY and M-FISH subsequently identify chromosomal material by comparing the spectral information obtained with the labeling scheme of the probe set used. Protocols for SKY are outlined in Bayani and Squire (2002) or web pages such as the Reid Laboratory web site for SKY protocols (www.ncbi.nlm.nih.gov/sky/show_html_frag.cgi?filename=tech.html_frag&header=SKY+or+M-FISH+and+CGH&title=SKY+or+M-FISH+and+CGH+Techniques#sky)

In telomeric SKY, the test metaphases are prepared and spotted onto microscope slides using standard approaches. Before hybridization, the slides are pretreated with RNAse and pepsin, washed, and dehydrated through an ethanol series. The subtelomeric probes used undergo two rounds of DOP PCR amplification and the second-round amplification product is used as the template for DOP PCR labeling using predetermined combinations of the fluorophores SpectrumOrange-dUTP, Texas red–dUTP rhodamine 110–dUTP, and the haptens BIO and DIG. As the scheme caters for 24 probes, the telomere probes are split into two sets for labeling. The labeled probe sets are precipitated and resuspended in hybridization solution containing Cot-1. Immediately before hybridization, the probe sets are denatured at 85 °C for 5 minutes and preannealed at 37 °C for 1 hour before addition to the slide. A coverslip is then applied (sealed with rubber cement) and hybridization proceeds in a humidified chamber at 37 °C for 48 hours. After hybridization, the coverslips are removed and the slides washed before treatment with blocking solution. The slides are then washed again before hapten detection steps with the appropriate antibodies. BIO is detected with avidin–Cy5 and DIG with mouse anti-DIG followed by sheep anti-mouse Cy5.5. The slides are washed, stained with DAPI, washed again, dehydrated through an ethanol series, and mounted in anti-fade solution prior to imaging. Acquired images are analyzed and any putative copy number changes investigated further by FISH using the relevant probes.

B. Non–FISH-Based Methods

1. Short Tandem Repeat Polymorphisms

Subtelomeric STRP analysis can be performed using radioactive or fluorescence-based genotyping procedures. Radioactive methodology is described by Rosenberg *et al.* (2000) and Slavotinek *et al.* (1999), whereas detailed fluorescence genotyping methodology is given by Colleaux *et al.* (2002). The subtelomeric microsatellite marker panels used for STRP analysis have been selected primarily by their position on physical maps (Colleaux *et al.*, 2002; Rosenberg *et al.*, 1997). The first step in the methodology is to amplify by PCR the telomeric markers from the test DNA sample. Each subtelomeric primer pair is included in a separate standard PCR mix. For the fluorescence-based procedure, the forward primers are modified with 6-FAM, HEX, or NED phosphoramidites, whereas for the radioactive procedures the primers are unmodified and α^{32}P-dCTP is included in the reaction mix. Radioactive PCR products are separated by electrophoresis through polyacrylamide and urea gels and then exposed directly to autoradiography without drying. The exposed autoradiographs are interpreted manually to determine whether there have been any allele copy number changes between the patient and his or her parents. In the fluorescence procedure, the fluorescent PCR products are pooled in equal amounts according to predetermined size sets and a loading cocktail (containing the GS-400 HD ROX size standard) added before separation through polyacrylamide and urea gels using an ABI automated sequencer. The data are collected using run module GS 36D-2400 and analysis performed using the GeneScan software as recommended by the manufacturer. The size of the ROX-labeled standard is manually checked in every lane before determining the allele sizes using the Genotyper software. The results are then assessed for putative copy number changes, suspected if a patient has not inherited one allele from each parent. To confirm putative changes, further investigations such as genotyping additional markers, quantitative PCR, and/or FISH are necessary. The advantage of FISH for confirmation is that it will also distinguish *de novo* from inherited anomalies.

2. Primed *In Situ* Labeling

In PRINS (Bonifacio *et al.*, 2001), metaphase chromosome spreads are prepared from the patient samples and immobilized on microscope slides according to standard procedures. The reaction primer for the PRINS procedure is the sequence $CCCTAA_7$, the canonical human telomeric sequence. This primer is mixed with dATP, dCTP, dGTP, and BIO in a standard PCR mixture containing *Taq* polymerase and 0.01% BSA. The mixture is then applied to the slide, using a special assembly tool prewarmed to 70 °C, and the slide sealed using specialized discs and metallic clips. The arrangement is inserted into a specialized thermal cycler that performs *in situ* PCR. Annealing is performed at 60 °C for 10 minutes followed by extension at 72 °C for 20 minutes. The reaction is stopped and the slides washed briefly before detection of the biotinylated PCR extension products using avidin–Cy3. The chromosomes are counterstained with DAPI in anti-fade solution and visually inspected for deleted regions using a fluorescence microscope equipped with the appropriate Cy3 and DAPI filters.

3. Multiplex Amplifiable Probe Hybridization

The methodology for MAPH (Armour *et al.*, 2000; Hollox *et al.*, 2002; Sismani *et al.*, 2001) can be found at www.nott.ac.uk/~pdzjala/maph/maph.html and is covered in detail in chapter 32. In brief, the probes used for MAPH were originally generated by cloning PCR products and restriction fragments from telomere-specific clones and then by PCR amplifying the clone inserts using vector-specific oligonucleotides. The sizes of the inserts were designed specifically so they could be pooled into sets that would migrate to independent positions during electrophoresis. To date there have been a number of generations of MAPH probe sets, the latest being "ST3F" which includes a total of 52 probes: 49 from subtelomeric regions (one probe representative for each subtelomeric region and two probes each for 1p, 10p, 12q, 15q, 19p, 19q, 22q, and Xp/Yp) in addition to 3 control probes. The three control probes are from the X chromosome, the Y chromosome, and a nonhuman zebrafish source (the absence of signal from the zebrafish probe provides a control for the specificity of washing).

The original MAPH procedure involved a radiation detection method (Armour *et al.*, 2000), but this was superseded by the use of fluorophores (Hollox *et al.*, 2002) and it is the latter that is described here. In the first step, 1.0 μg genomic DNA in a volume less than 5 μl is denatured with NaOH and spotted on to a nylon filter 2-by-4 mm and cross-linked by ultraviolet irradiation. Up to 10 filters (i.e., 10 different patient samples) are prehybridized together in 1 ml of prehybridization solution at 65° C and then in 200 μl of hybridization solution (which contains Cot-1 DNA for competing out repetitive elements). Probe mixtures containing 100 pg of each sequence, *Escherichia coli* DNA, and end-blocking primers are then denatured before addition to the hybridization solution containing the filter. After hybridization overnight at 65 °C, the filters are washed and transferred to separate amplification reactions in which the bound DNA is PCR amplified at low level. This product is used to seed further 20-μl amplifications using 5′-FAM–labeled primer PZA (for detection by the ABI 377 Genetic Analyser) or 5′ Cy5–labeled primer PZA (for detection by the Amersham Pharmacia ALFexpress machine); the PZA sequence corresponds to vector sequence adjacent to the cloned telomere-specific inserts. The second-round amplification products are either separated using PAGE LongRanger gels (ABI 377) and quantified using GeneScan or separated using ReproGel high-resolution gels (ALFexpress) and quantified using the package software.

The first stage of the data analysis is to produce normalized ratios for every set of patient results. For each sample, this is achieved by taking the measured band intensity (the area under the peak) and dividing it by the sum of the four nearest autosomal band intensities within that gel lane. This ratio value reflects the band intensity relative to the neighboring bands and is constant for that probe across all the samples (provided there is no copy number change). By dividing each relative band intensity value by the mean for that probe, a normalized ratio is obtained and can be compared with the mean normalized ratio value of 1.0 (equivalent to a dosage of two copies per diploid cell). Probe results giving significant deviations from a ratio value of 1.0 indicate a putative copy number change and can be further investigated. This is a univariate analysis. For observations in duplicate, bivariate analysis can be performed (www.nott.ac.uk/~pdzjala/maph/maph.html).

4. Multiplex Ligation-Dependent Probe Amplification

MLPA is a refinement of MAPH (Rooms *et al.*, 2004; Schouten *et al.*, 2002) and is described in detail in chapter 32. Test kits and further information are available from MRC-holland (http://www.mrc-holland.com/). In MLPA, each probe comprises one short synthetic oligonucleotide and one long M13-derived oligonucleotide (rather than a single telomere-specific fragment) and these are hybridized to immediately adjacent target sequences of the patient DNA sample in solution (rather than to DNA immobilized on nylon membrane strips). A total of 20–500 ng of patient DNA is denatured and fragmented by heating at 98 °C for 5 minutes. The MLPA probes are then added and allowed to hybridize at 65 °C for 16 hours in a PCR machine with a heated lid. After hybridization, dilution buffer containing ligase-65 enzyme is then added and the annealed oligonucleotides ligated together to give products of unique length. The ligase is heat inactivated and the appropriate PCR reagents added (one of the two PCR primers is either fluorescently labeled with D4 or radioactively labeled with LICOR IR2). After PCR analysis, the products are detected and quantified either by capillary electrophoresis if fluorescently labeled (e.g., using a Beckman CEQ2000 capillary electrophoresis system) or by denaturing acrylamide gel electrophoresis (LICOR IR2 system) if radioactively labeled. The data may be analyzed as described earlier for MAPH.

5. Telomere Array CGH

A number of methodologies are available for performing array CGH. Detailed protocols can be found at the Wellcome Trust Sanger Institute web site (www.sanger.ac.uk/Projects/ Microarrays/arraylab/methods.tml) and are presented in chapter 33. The array probe preparation and hybridization protocols summarized here are those of Knight *et al.* (www.well.ox.ac.uk/~sknight/CYTOMETRY/). The probe preparation protocols are adapted from those of the Wellcome Trust Sanger Institute and the labeling and hybridization protocols from those of Dr. Patrick Brown's Laboratory (cmgm.stanford.edu/pbrown/protocols/index.html). The method of analysis is that of Veltman *et al.* (2002).

In brief, arrays are fabricated by extracting telomere clone DNA (from the first- or second-generation clone sets) amplifying the insert DNA by degenerate oligonucleotide primed (DOP) PCR using the primers DOP2 and DOP3 (Fiegler *et al.*, 2003). The DOP2 and DOP3 products for each clone are then combined, and the DNA precipitated and resuspended in dH_2O to a final concentration of 400 ng/μl. This DNA is then used as a template to seed further PCR reactions using custom amino-linking primers (Fiegler *et al.*, 2003). The amino-linked products (the array probes) for each clone are precipitated and resuspended in a spotting solution such as 50% dH_2O/50% DMSO; the optimal spotting solution depends on the arraying machine available. The probes are then spotted in duplicate, triplicate, or quadruplicate on to the appropriate slides (typically CodeLink Activated Slides slides [surModics], which chemically attach the probes via the amino-link sequence, or CMT-GAPs slides [Corning Limited], which are positively charged and attract the negatively charged DNA probes). CodeLink Activated Slides arrays are processed according to manufacturer's instructions and CMT-GAPs arrays as described by Knight *et al.* (www.well.ox.ac.uk/~sknight/CYTOMETRY/). The microarrays are stored in a plastic slide container at RT until ready to hybridize.

For each CGH experiment, a sex-matched control DNA sample and a test DNA sample are either digested with Sau IIIA or sonicated and subsequently purified. A total of 900 ng of each purified DNA is differentially labeled with Cy3–dCTP and Cy5–dCTP using the BioPrime Labeling Kit (Invitrogen) as described by Knight *et al.* (www.well.ox.ac.uk/~sknight/CYTOMETRY/). Before hybridization, each labeled target mixture is denatured at 97 °C for 5 minutes and immediately transferred to a 37 °C hot block for 55 minutes to allow repetitive sequences to be blocked. The hybridization mix is then applied to a coverslip and a prewarmed array slide lowered on to the mix. Hybridization is carried out in a specialized hybridization chamber (Corning Limited) submerged in a 65 °C shaking waterbath for 36–48 hours. After hybridization, the coverslip is removed and the microarray slide is then passed through a series of washes on a shaking platform. The slide is dried by centrifugation and stored in a light-proof container ready for image acquisition.

Cy3 and Cy5 fluorescence intensity data are collected by scanning the hybridized microarrays in a compatible scanning machine. The software settings for data collection depend on the scanning machine used. Using the PackardBell Biochip scanning machine and the application ScanArray, the data are collected at a 10-micron resolution with the laser set at 95% and the scanning rate at 100%. The photomultiplier tube (PMT) gains are typically set at 65 for Cy3 images and 55 for Cy5 images. The images are then stored as TIFF files ready for analysis. For the analysis, the fluorescence intensity of each spot (measured as the median of the pixel intensities), together with the local background (measured as the median of the pixel intensities in the region immediately surrounding the spot), is extracted from the scanned image files using the appropriate software. Spots of poor quality are identified by eye and flagged. The net intensities of each of the Cy3 and Cy5 channels are then calculated as the raw spot intensity minus the local background and are further analyzed.

The intensity data are analyzed on an individual clone basis using the method of Veltman *et al.* (2002). First, the data are normalized so the median ratio value for arrayed probes representing disomic regions of the genome is 1.0 (calculated from nontelomeric autosomal or X clone control array probes). Then, control/test intensity ratios are calculated and the ratio data edited by excluding flagged spots and, if necessary, removing the replicate data from probes yielding $SD > 0.2$ until $SD \leq 0.2$ for the remaining replicates. The minimum number of spots accepted for analysis is two. The mean control/test intensity ratios is calculated for each clone by averaging over all accepted spot ratios.

To measure normal clone-to-clone variation, normalized data from control versus control hybridizations are used to calculate the overall mean of the fluorescence ratios and the SD from this mean for each arrayed clone. To correct for the intrinsic variability between clones, the mean intensity ratio yielded by each clone in the control versus test hybridizations is divided by the mean intensity ratio of that particular clone as calculated from the control versus control hybridizations. This correction is also applied to the mean intensity ratios of each of the clones from the control versus control hybridizations, resulting in a ratio value of 1.0 for all clones. In this way, the corrected ratio obtained for every probe of the control versus patient hybridizations can be easily plotted and compared on the same plot with the corrected control versus control ratio (always a value of 1.0) together with the calculated SD for each clone ratio. Using this method, lower and upper threshold values of 0.8 and 1.2 are used as evidence of deletion and polysomy, respectively.

IV. Results

All subtelomeric testing strategies that use telomere-specific probes for the identification of subtelomeric anomalies also use fluorophores and/or radioactivity for probe detection and quantitation. The results are typically presented in one of two ways. For those methods that involve visualization by fluorescence microscopy and/or computerized imaging of patients' metaphase chromosomes, the results are generally presented as images of hybridized metaphase chromosome spreads or sometimes with pseudo-colored aligned karyogram images of hybridized chromosome pairs. When testing patients' DNA samples, the results are typically presented in the form of autoradiographs (as in the case of radioactive STRP analyses) or graphical outputs or traces showing sets of peaks/data points corresponding to the subtelomeric probe results for each patient (as in the case of fluorescence STRP analysis, MAPH, MPLA, and array CGH). Examples of typical results and their representation for the Chromoprobe Multiprobe-T System, the ToTelVysion Multicolor FISH Probe Panel system, M-TEL, and PRINS are shown in Fig. 1. Those for autoradiographic STRP, fluorescence MAPH, and array CGH are shown in Fig. 2. The results for TM-FISH, SKY, and S-COBRA FISH would be displayed similarly to M-TEL results. The results for MLPA would be displayed similarly to those for MAPH.

V. Critical Aspects of the Methodologies

Each of the protocols described in this chapter have proved successful for the detection of subtelomeric copy number changes. All require specialized training for the investigator and the greater the level of experience, the more consistent, efficient, and reliable the results output will be. Not only do the technical aspects of the protocols need to be mastered, but also a thorough knowledge regarding the collection and analysis of data, and particularly the interpretation of results since a number of apparently anomalous subtelomeric results will be benign variants of no phenotypic consequence (see Section VI). Some of the most

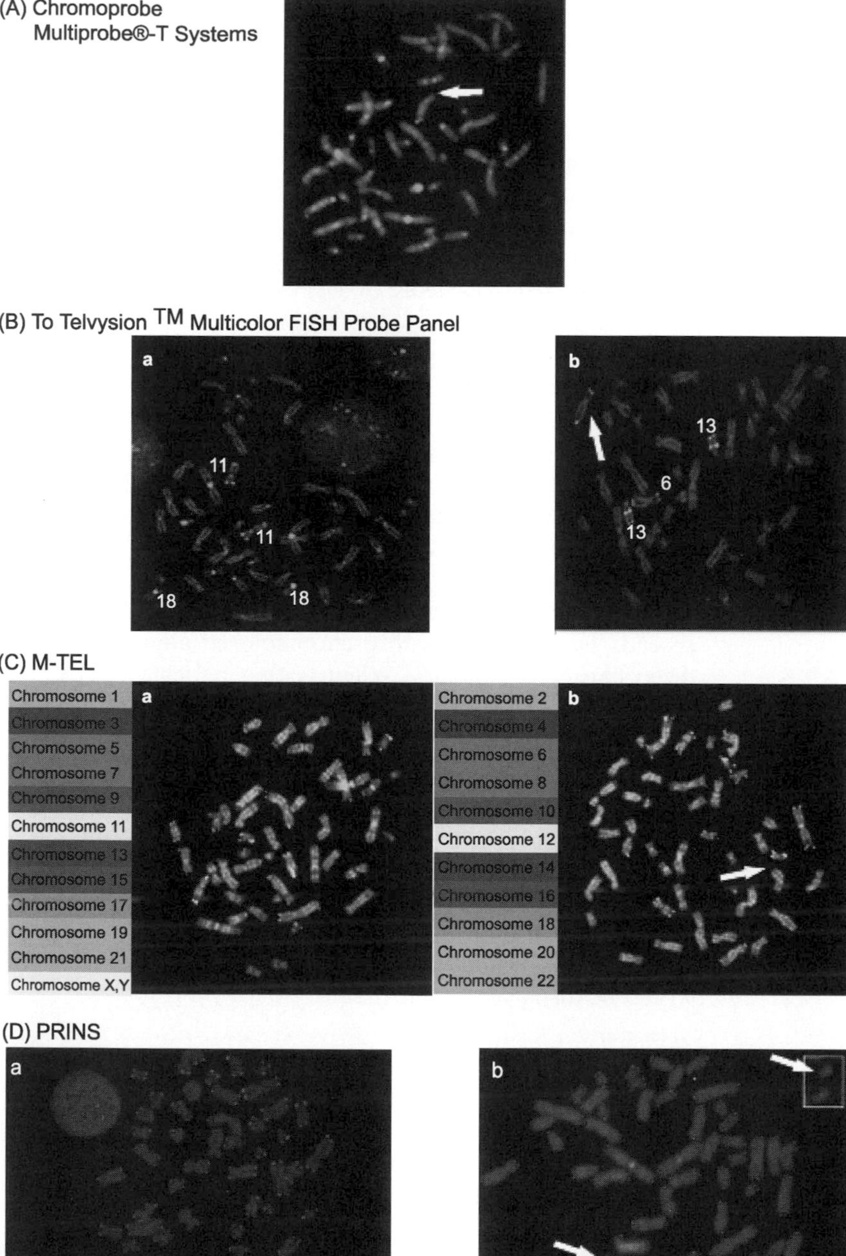

Fig. 1 Representation of results obtained using subtelomeric testing methods that require patient chromosomes for analysis: (A) Chromoprobe Multiprobe-T System (Cytocell Limited). Fluorescence

commonly encountered and technically critical aspects of the methodologies are listed below.

A. Methods that Use Chromosome Preparations

1. Quality of Cell Suspensions

For all of the methods that use patient chromosome preparations, the quality of the chromosome preparations and the slides produced from them is often critical to the experimental outcome. The quality of the preparations will depend on the age of the incoming blood/tissue sample or of previously stored preparations and on the skill and experience of the laboratory worker. Also important here can be the percentage of the prepared cells that are in metaphase, the mitotic index (MI). For the Chromoprobe Multiprobe-T System, the amount of fixed cell suspension required is low (only 2 μl is used per spread), but the optimal MI is more than 5%. The situation is similar for the ToTelVysion Multicolor FISH Probe Panel system except that a slightly lower MI can be tolerated because the size of the spreads is larger. By contrast, M-TEL, TM-FISH, S-COBRA FISH, and SKY can tolerate very low MIs because a minimum of two metaphase spreads are required for

---

in situ hybridization (FISH) result showing a single metaphase result from a patient with a subtelomeric rearrangement involving chromosome 8. The p-arm signals are red and the q-arm signals are green. A red signal is missing from the arrowed chromosome 8 homolog, indicating that this patient is monosomic for the 8p subtelomeric region. (B) Example of ToTelVysion Multicolor FISH Probe Panel (Vysis/Abbott, Inc.) result kindly provided by Dr. Christa Lese Martin. (a) Normal subtelomere FISH results for the probe set containing 11p (labeled in SpectrumGreen), 11q (labeled in SpectrumOrange), and 18p (labeled in SpectrumGreen and SpectrumOrange and shown in yellow). A centromere probe for chromosome 18 (labeled in SpectrumAqua) is included as a control probe for the 18p subtelomere probe. Two hybridization signals can be observed for each probe. (b) Abnormal subtelomere FISH results for the probe set containing 6p (labeled in SpectrumGreen), 6q (labeled in SpectrumOrange), and 13q (labeled in SpectrumGreen and SpectrumOrange and shown in yellow). A locus-specific probe for the long arm of chromosome 13 (labeled in SpectrumAqua) is included as a control probe for the 13q subtelomere probe. A normal hybridization pattern is observed for the 13q subtelomere probe and control probe. However, there is only one normal chromosome 6 homolog. The other chromosome 6 homolog (*arrow*) has a deletion of the 6q subtelomere probe with an extra copy of the 6p subtelomere probe located at the 6q subtelomere region. This unbalanced rearrangement results in trisomy for the 6p subtelomere region (green) and monosomy for the 6q subtelomere region (red). (C) Example of multicolor telomere (M-TEL) color classification result kindly provided by Dr. Jill Brown and with copyright permission from Nature Publishing Group. (a) The M-TEL1 probes identify a deletion of 7q on one chromosome 7 homolog (*arrow*). (b) The M-TEL2 probes identify an additional chromosome 2 signal (orange) on the derivative chromosome 7 (*arrow*). This unbalanced rearrangement results in monosomy for the 7q subtelomeric region and trisomy for the 2q subtelomeric region. (D) Example of PRINS result from Bonifacio *et al.* (2001, Fig. 1 and Fig. 3, p17) with copyright permission from S. Karger AG, Basel. (a) PRINS reaction signals for all telomeres (red) in a control sample metaphase. (b) PRINS reaction results showing a metaphase from a patient monosomic for the 10p subtelomeric region; the p-arm signal from one of the chromosome 10 homolog is missing (*arrow*). The upper right panel shows confirmation of the result by FISH using a 10p subtelomeric probe. (See Color Insert.)

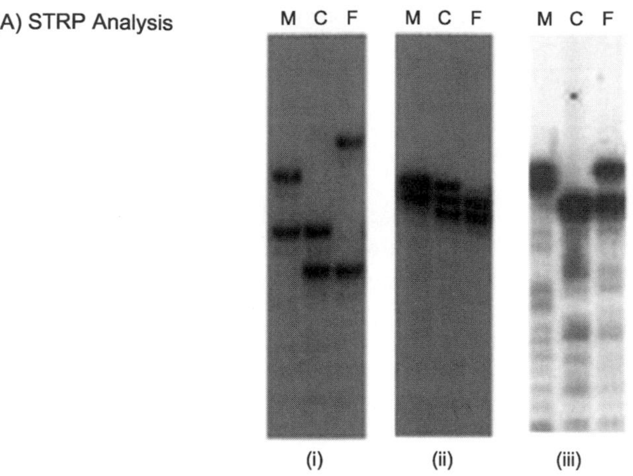
(A) STRP Analysis
M C F
M C F
M C F
(i)
(ii)
(iii)

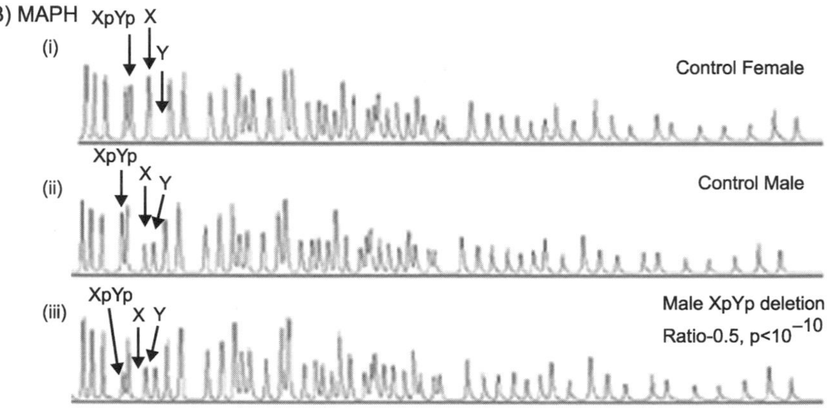
(B) MAPH
(i)
XpYp X Y
Control Female
(ii)
XpYp X Y
Control Male
(iii)
XpYp X Y
Male XpYp deletion
Ratio-0.5, p<10−10

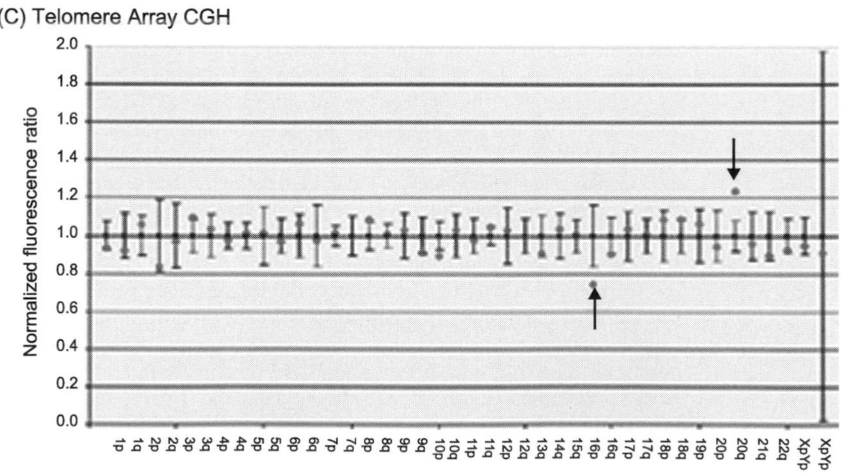
(C) Telomere Array CGH
Normalized fluorescence ratio
2.0
1.8
1.6
1.4
1.2
1.0
0.8
0.6
0.4
0.2
0.0
1p 1q 2p 2q 3p 3q 4p 4q 5p 5q 6p 6q 7p 7q 8p 8q 9p 9q 10p 10q 11p 11q 12p 12q 13q 14q 15q 16p 16q 17p 17q 18p 18q 19p 20p 20q 21q 22q XpYp XpYp

analyses and the multicolor detection strategies mean that multiple telomeres can be analyzed per hybridized metaphase. In the case of PRINS, only one chromosome spread is theoretically necessary. For all of these methods, good-quality slides generally give good signal/noise ratios. For those methods that are tractable to computer imaging and automated analysis, good-quality slides with good-quality chromosomes that give good-quality banding patterns mean that the results can be classified automatically with little manual intervention. However, the poorer the slide quality, the more manual intervention is required and the more labor intensive the analysis becomes.

2. Probe Preparation

It is always advisable to confirm optimal performance of the probes and reagents required by carrying out initial test hybridizations using good-quality normal control chromosome suspensions before proceeding with valuable test samples. This is true of all methodologies regardless of whether probes and reagents have been prepared "in house" or whether they have been obtained commercially.

3. Chromosome Denaturation

Another critical factor for analyses that require immobilized metaphase spreads is achieving optimal denaturation of the chromosomal DNA. When chromosomes are under-denatured, there may be little if any hybridization signal, whereas too

---

Fig. 2 Representation of results obtained using subtelomeric testing methods that require patient DNA for analysis. (A) Autoradiographic short tandem repeat polymorphism (STRP) results kindly provided by Dr. Marjorie Lindhurst. M, Mother; C, Child; F, Father. (i) Normal results obtained using the 12ptel marker D12S391. In this family, each parent has two alleles and the child has inherited one allele from each of the parents. (ii) Results using the 4qtel marker D4S3338. In this family, the child is monosomic for the subtelomeric region of chromosome 4q, inheriting only a single 4q allele (the maternal allele is missing). (iii) Results using the 12ptel marker D12S374. In this family, the child has received both paternal alleles and one maternal allele resulting in trisomy for the 12p subtelomeric region. (B) Multiplex amplifiable probe hybridization (MAPH) results obtained by Jess Tyson and Hema Bulsara and kindly provided by Dr. John Armour. The traces are from (i) a control male, (ii) a control female, and (iii) a male with a heterozygous deletion of the XpYp probe. The reduced size of the XpYp peak in trace (iii) is clearly visible compared with those in traces (i) and (ii). (C) Telomere array comparative genome hybridization (CGH) results. Chart showing the mean normalized hybridization ratios obtained from telomere probes (pink data points) for one patient compared with the mean normalized hybridization ratios and corresponding standard deviations (black error bars) from six control versus control hybridizations (dark blue data points). Each data point represents the mean data for a single arrayed probe and has been corrected for clone-to-clone variation. The standard deviation bars for the XqYq array probe reflects the highly polymorphic nature of the sequences represented by this probe. The patient data points for the p-arm subtelomeric regions of chromosomes 16 and 20 give ratio values below the threshold of 0.8 and above the threshold of 1.2, respectively. The result indicates that the patient has an unbalanced translocation resulting in monosomy for the 16p subtelomeric region and trisomy for the 20p subtelomeric region.

much denaturation results in bloating of the chromosomes and some tracking of the hybridization signal. The latter is particularly true of those procedures that achieve denaturation simply by heat, for example, the Chromoprobe Multiprobe-T System. Here, it can be difficult to get the correct level of denaturation because the age and quality of the sample affect the time needed for optimal denaturation; older preparations generally need more time, whereas fresh samples need less. For the Chromoprobe Multiprobe-T System, some degree of over-denaturation can be tolerated because few telomeres are analyzed per metaphase and the identity of the telomeres is immediately recognizable by the fluorophore identification. However, in those methods that analyze multiple telomeres per metaphase, the assignment of the chromosomes may be affected because over-denaturation affects the overall morphology and the identifying banding patterns can be lost.

4. Use of Antibodies for Detection

A number of the FISH-based methods described use fluorescently tagged antibodies to detect hybridized probes. The important note here is that there is often variation between antibody batches, particularly regarding the concentration. Too little antibody can result in weak signals, whereas too much can give a "speckled" appearance that interferes with the identification of the cognate hybridization signals. Thus, the working dilutions for each new batch should be optimized by hybridizing reliable control probes to good-quality control chromosome suspensions before proceeding with valuable test samples.

B. Methods that Use DNA Samples

All methods that use patient DNA samples are more likely to succeed if the samples are of good quality, that is, high molecular weight and pure. Otherwise, with appropriately trained staff, the STRP, MAPH, and MLPA methodologies should pose few technical difficulties. Telomere array CGH is technically more challenging, but the most notable consideration is the need to minimize dust particles on the array surfaces. This may be achieved by allocating "dust-free" rooms for microarray work or by taking care to eliminate obvious particle sources, for example, by filter sterilizing microarray solutions.

VI. Limitations, Pitfalls, and Misinterpretation of Data

None of the methods available for subtelomeric testing is completely free from limitations, pitfalls, or problems related to interpreting the data. The issues that arise can be divided into (1) those specific to the individual methodologies and (2) the broader issue of data interpretation, in terms of assessing the likely pathogenicity/clinical relevance of confirmed copy number changes.

A. Methodological Issues

A summary of the limitations, pitfalls, and issues concerning misinterpretation of data for each methodology is given below. The advantages and disadvantages are also discussed and are summarized in Table I.

Table I
Comparison of Methodologies that Use Subtelomeric Probes for Testing Telomere Integrity (adapted from Knight *et al.*, 2004 with copyright permission from Marcel Dekker, Inc.)

Method	Advantages	Disadvantages
Chromoprobe Multiprobe-T System	Interpretation straightforward Tolerates poor chromosome morphology	Labor intensive Commercial kits expensive, but probes can be produced "in house" Knowledge of polymorphisms and cross-hybridizations required
M-FISH and SKY: ToTelVysion Multicolor FISH Probe Panel System, Telomere SKY, M-TEL, TM-FISH, S-COBRA FISH	Tolerates chromosome preparations with low mitotic index Useful for samples where chromosomal material is limited TM-FISH designed to suit differently equipped fluorescence microscopes	Often labor intensive Requires specialized, expensive imaging equipment and software Knowledge of polymorphisms and cross-hybridizations required
PRINS	Uses only a single probe for all telomeres	Only detects monosomy
STRP analysis	Uses DNA sample Can detect uniparental isodisomy as well as copy number changes Can be automated	Requires parental samples as well as probands Labor intensive (can be many false positives to follow up) Requires highly informative markers (even with >85% heterozygosity, only 72% monosomies, 75% uniparental isodisomy, and 50% trisomies are detected)
MAPH and MLPA	Uses DNA sample FISH workload significantly reduced	False-positive rate of about 4% False-negative rate of about 2% (depending on the frequency of genuine copy number changes expected)
Telomere array CGH	Uses DNA sample	Experimentally expensive and requires access to specialized arraying and scanning equipment False-positive rate high and some probes give false negatives Some probes detect polymorphisms Rigorous evaluation and optimization pending

1. FISH-Based Approaches

The main concern over commercial FISH approaches is their cost, although there are published protocols for obtaining the telomere clones and working up subtelomeric FISH in house (Knight and Flint, 2002). From the point of view of interpretation, there are two main considerations to be aware of with any form of subtelomeric FISH. The first are homologous sequences that cross-hybridize (e.g., sequences from the 12p probe cross-hybridize with 6p and 20q). The second are length polymorphisms that may result in null alleles (e.g., the 2q polymorphism). These can cause difficulty when interpreting results and the investigator should be aware. A list of known cross-hybridization patterns and polymorphisms have been well documented (Knight and Flint, 2002).

2. Non–FISH-Based Approaches

a. STRP Analysis

Genotyping, whether radioactive or fluorescence based, looks for non-Mendelian inheritance of alleles in an affected individual and, therefore, requires additional family samples, ideally both parents. Thus, for every complete analysis, three individuals must be tested. Although the increased sample testing is a burden, it does bring with it the one advantage that the technique has over all others: it is the only one capable of detecting rare uniparental isodisomy events (the inheritance of two both alleles from the same parent). When both parents are heterozygous for different alleles, the overall method is sensitive and usually highly specific. However, this is not always the case and a number of limitations have been encountered. First, even with four alleles segregating, not all abnormalities can be detected; duplicate copies of the same allele, which occurs in some trisomies, could be missed. Second, the method relies on highly heterozygous markers and even with heterozygosities of greater than 85%, only 72% of those with monosomy, 75% with uniparental disomy, and 50% with trisomy are detected (Wilkie *et al.*, 1993). Third, the method has been found to generate a problematical number of false positives due to the intrinsic instability of microsatellites. Each potentially positive result has to be pursued by genotyping additional markers or using an alternative techniques such as quantitative PCR and/or FISH, and even when a phenotypically significant copy number change has been identified, FISH is required to distinguish *de novo* from inherited anomalies and to identify carriers. Thus, STRP analyses are generally considered prohibitively labor intensive and the advantages of the methodology are rather outweighed by the disadvantages.

b. PRINS

The PRINS procedure (Bonifacio *et al.*, 2001) is fast and easy to perform, but it cannot detect pure trisomies and it is unclear whether it can detect all subtelomeric monosomies (it is possible that some immediately subtelomeric, though interstitial anomalies, may be missed). Although pure subtelomeric trisomies are rare, unbalanced rearrangements account for about half of diagnosed IMR cases and so all cases with monosomies identified by PRINS would require a full set of telomere analyses for the identification of trisomies.

c. *MAPH and MLPA*

For both MAPH and MLPA, sensitivity and specificity issues are key when it comes to quantifying DNA copy number (Hollox *et al.*, 2002). MAPH has been reported to give a false-positive rate of about 4% and a false-negative rate of 2% (the latter depending on the frequency of genuine copy number changes expected). However, despite this relatively high false-positive rate, it has been emphasized that the method is useful in reducing the FISH workload by 96% (Hollox *et al.*, 2002) and there are continuing efforts to improve the probe sets used.

d. *Telomere Array CGH*

Veltman *et al.* (2002) used array CGH for diagnosing telomeric rearrangements and demonstrated that the CGH method not only detected the rearrangements in a panel of known anomalies but also found additional anomalies that were cytogenetically invisible. However, not all of the subtelomeric probes on the arrays were interrogated and other studies have shown that at least some probes will yield false negatives because the sequence of the probe concerned bears similarity with other sequences located elsewhere in the genome and the target DNA can cross-hybridize with these. For example, 1p probes that contain terminal associated repeats are known to cross-hybridize to the subtelomeric region of 8p, thus explaining why other telomere array CGH studies have failed to reveal known cases of 1p monosomy (Knight and Flint, 2000, 2002; Yu *et al.*, 2003). Furthermore, positive results from array CGH do require confirmation using a second approach, usually FISH, and given that each array CGH test is more expensive than, say, the Chromoprobe Multiprobe-T System, it is not yet cost effective for subtelomeric analyses alone. In addition, one false-positive result is likely to occur in each complete analysis, and for some arrayed probes, 15% of the analyses results would be scored as abnormal. Therefore, at this stage, array CGH methods solely directed at telomere testing cannot yet be considered suitable for a clinical diagnostic setting.

B. Data Interpretation: Pathogenicity

An important question is whether subtelomeric rearrangements cause MR or any other phenotype in the patients in whom they are found. In the case of polymorphisms, if one phenotypically normal parent also carries the subtelomeric anomaly, then that anomaly is generally considered a benign variant unrelated to the MR. A list of subtelomeric rearrangements already noted in apparently phenotypically normal individuals is given in Table II. For most of the rearrangements reported to be clinically significant though, either the phenotypes segregate within families or the rearrangements have been noted previously in the literature and are associated with a similar phenotype, or the measured sizes of the rearrangements are so large that they are unlikely to be without phenotypic consequence.

Table II
Phenotypically Neutral Subtelomeric Rearrangements

Pure monosomies	Unbalanced monosomies	Pure trisomies	Unbalanced trisomies	Reference
4q	Yp		17q	Joyce *et al.* (2001)
2q; Xp				van Karnebeck *et al.* (2002)
10q; 17p				Martin *et al.* (2002)
2q; Xp; Yp				Knight *et al.* (2000)
2Q; 7p; Xp; Yp				Jalal *et al.* (2003)
Xp		3q		Riegel *et al.* (2001)
2q; 9p; 7q				Rosenberg *et al.* (2001)

Thus, this information is helpful when considering pathogenicity. However, it still leaves a proportion of cases, those with small previously undocumented *de novo* rearrangements, for whom it is impossible to determine the diagnostic implications because the phenotypic consequences cannot yet be predicted. This situation should improve as the numbers of studies and noted anomalies grows.

VII. Comparison with Other Methods

The methodologies described in this chapter all rely on the use of telomere-specific probes for the detection of subtelomeric anomalies. However, two alternative methodologies negate the use of such probes: (1) high-resolution chromosome banding analysis and (2) high-resolution CGH.

A. High-Resolution Chromosome Banding Analysis

Arguably the most straightforward and cost-effective assay of telomeric integrity is direct visualization by high-resolution chromosome analysis (850-band level). This does not require additional expensive equipment or probe kits and can be viewed as an extension of standard karyotypic analysis at the 450- to 550-band level. High-resolution analysis is strongly advocated by Joyce *et al.* (2001), who suggest that most of the subtelomeric anomalies can be detected in this way. Despite this, the major drawback is that the success of the approach is largely reliant on the quality of the samples received, the laboratory concerned, and the diligence and skill of the cytogenetic analyst. In the United Kingdom, the recommended requirement for the detection of microdeletions is a quality score of 7, as laid out by the National External Quality Assessment Scheme (NEQAS) (for more information, see www.well.ox.ac.uk/~sknight/NEQAS/NEQAS BandingCriteria.htm and www.ukneqas.org.uk/Directory/GENET/clincyto.htm). In practice, few laboratories achieve this standard routinely but would endeavor to do so for selected cases, on request. In 2000, the U.K. NEQAS for Clinical Cytogenetics report showed that sample slides produced by all 40 of the U.K.

Clinical Cytogenetics laboratories gave scores ranging from a quality score of 4 to a quality score of 8 (Dr. Ros Hastings, U.K. NEQAS Organizer for Clinical Cytogenetics, personal communication, 2004).

B. High-Resolution Metaphase CGH

CGH is traditionally used for the detection of copy number differences using a modified FISH technique in which test and reference DNAs are differentially labeled and co-hybridized to normal metaphase chromosomes on a glass microscope slide. On analysis, deviation from the expected 1:1 hybridization ratio may indicate a copy number difference. Until recently, telomeric regions have been excluded from CGH analysis because the fluorescence intensities of the hybridized DNAs decrease toward the telomeres and can give unreliable ratio changes. However, CGH analyses that can detect unbalanced cryptic telomeric translocations of at least some chromosomes have been reported with a resolution as low as 1.8 Mb in one case (Joly *et al.*, 2001; Schoumans *et al.*, 2004). The method has the advantage that the whole genome can be assayed for copy number changes, but the disadvantage is that the average resolution at telomeric regions is still only about 4 Mb, so smaller rearrangements would be missed. In addition, efficacy has yet to been shown for all telomeric regions.

VIII. Applications

Subtelomeric testing was originally developed for the study of IMR, and there are now published results from more than 3000 patients that unequivocally demonstrate the importance of subtelomeric rearrangements in the etiology of IMR. Rearrangements can be detected in about 5% of unselected cases, but the prevalence is higher in patients with moderate to severe MR (about 8%). In about half of those with subtelomeric anomalies, the rearrangements appear to have arisen *de novo*, whereas in half they are familial unbalanced rearrangements arising from balanced translocations in a parent (Knight, 2004).

Although the study of IMR is by far the most popular application for subtelomeric testing, there are also other clinical applications for which testing chromosome ends is becoming increasingly useful. This is particularly the case in diagnostic laboratories where abnormal karyotypes have been noted in patients, but the existence of additional submicroscopic rearrangements have not been ruled out because of the known limitations of the cytogenetic resolution. The use of subtelomeric probes in these cases has resulted in a plethora of published studies demonstrating their utility in resolving such partially characterized or semicryptic karyotypes (Brkanac *et al.*, 1998; Davies *et al.*, 2003; Horsley *et al.*, 1998; Joyce *et al.*, 1999). Another field in which subtelomeric probes have proved a useful tool is in the evaluation of hematological cancers such as the myelodysplastic syndromes, acute myeloid leukemia, chronic myeloid leukemia, and secondary leukemia (Foot *et al.*, 1999; Jaju *et al.*, 1999; Tosi *et al.*, 1999).

Furthermore, subtelomeric probes are also finding potential in preimplantation and prenatal diagnosis. In selected cases, preimplantation screening of cleavage stage embryos with the appropriate subtelomeric probes can detect unbalanced embryos and thus increase the chances of proceeding with a successful pregnancy (Handyside *et al.*, 1998; Mackie-Ogilvie *et al.*, 1999; Scriven *et al.*, 1998). The strategy can also be extended to allow the rapid screening of amniotic fluid in the prenatal period or CVS samples in established pregnancies or where one partner carries a known chromosome rearrangement (Souter *et al.*, 2003).

In all of these applications, subtelomeric FISH probes are proving invaluable for the identification and refined characterization of subtle rearrangements that could not be detected by conventional cytogenetics or whole chromosome painting alone.

IX. Future Directions

The identification of subtelomeric rearrangements represents an important step toward understanding the genetic causes of unexplained MR. The findings thus far have been particularly relevant for providing clinical diagnoses and more accurate genetic counseling but ultimately will be important for understanding cognitive development as a whole and for developing new therapeutic avenues. In this respect, subtelomeric studies have already opened important avenues for future research. For example, one crucial question is whether small rearrangements occurring elsewhere in the genome might also be responsible for a significant proportion of unexplained MR cases. Indeed, there are already indications that whole-genome testing for copy number changes would be worthwhile (Rosenberg *et al.*, 2000; Shaw-Smith *et al.*, 2004; Vissers *et al.*, 2003). One of the most promising new technologies that may prove suitable for this in a diagnostic setting is CGH to microarrays. Here, a single array CGH experiment has the potential to identify copy number changes anywhere in the test genome at a resolution of at least 1 Mb, although even higher resolution is possible using whole tiling path arrays (arrays spotted with probes from overlapping clones that achieve coverage of the entire genome). Using a 1Mb resolution array, Vissers *et al.* (2003) tested 20 mentally retarded, dysmorphic patients with normal karyotypes and a phenotype suggestive of a chromosomal abnormality, whereas Shaw-Smith *et al.* (2004) tested 50 patients with learning disability and dysmorphism. Although both sample sizes were small, the studies suggest that copy number changes throughout the genome (including telomeres) may account for 14–15% of idiopathic learning disability with dysmorphism. However, one issue has yet to be evaluated: that of false-negative and false-positive rates and of benign variants/polymorphisms. The STRP studies of Rosenberg *et al.* (2000) demonstrate the importance of this. STRP analysis, though extremely effective, was also found to be impractical because the number of false positives that had to be followed up made it unacceptably labor intensive to implement. It may be relevant then that the array CGH results of Vissers *et al.* (2003) and Shaw-Smith *et al.* (2004) each yielded 10% anomalies that had to be followed up but were later found to be benign variants. Thus, although whole-genome array CGH may also be extremely effective, the suitability of the approach for use in a clinical diagnostic setting remains to be fully evaluated. In the meantime, it is also important to consider other technologies that might be tractable for whole genome analysis, MAPH or MLPA, for example, or array CGH to a tiling path of probes encompassing the whole human genome.

Whatever the chosen methodology and whether the focus is on subtelomeric regions or the entire genome, one diagnostically important issue will be to resolve how much true monosomy or trisomy (even nullisomy or polysomy) can be tolerated without phenotypic effect. In the case of telomeric regions, this is being addressed by establishing a set of markers and clones to measure the size of monosomic/trisomic regions, thereby providing the foundations for future studies designed to determine whether a given subtelomeric rearrangement would be expected to be benign or pathogenic (Martin *et al.*, 2002). In the case of the entire genome, efforts are underway to provide database resources that can be utilized to find out whether anomalies mapping to certain regions have been reported previously and whether they are thought to be benign or associated with a recognized phenotype.

In conclusion, the wide selection of testing methods described in this chapter attest to the considerable impact that subtelomeric testing has already had in providing diagnosis in mental retardation as well as initiating broader studies into the spectrum of telomeric involvement and possible genotype–phenotype profiles in IMR. In addition, subtelomeric testing is also proving useful for other clinical applications, including the full characterization of partially defined chromosomal abnormalities, the study of spontaneous recurrent miscarriages, infertility, and hematological malignancies, as well as for preimplantation diagnosis (in selected cases). For the future, the trend is already toward the development of further new tests aimed toward testing the whole genome for copy number changes. This will undoubtedly lead to further advanced diagnostic, counseling, and therapeutic capabilities together with a greater scientific understanding of unexplained MR, as well as finding application in a wide range of other human conditions with genetic origins.

Acknowledgments

We are very grateful to all of those who contributed the examples of methodological results included in Fig. 1 of this chapter, namely Dr. Christa Lese Martin, Emory University School of Medicine, Atlanta, Georgia (ToTelVysion Multicolor FISH Probe Panel system), Dr. Marjorie Lindhurst, National Human Genome Research Institute, Bethesda, Maryland (STRPs), Dr. Jill Brown, Weatherall Institute of Molecular Medicine, Oxford, United Kingdom (Telomere Multiplex FISH), and Dr. John Armour, Institute of Genetics, Queen's Medical Centre, Nottingham, United Kingdom (locus copy number measurement by hybridization with amplifiable probes). We also thank Dr Niki Meston for reading and discussing the manuscript. Dr Samantha Knight is funded by the Health Foundation, UK and is a member of the Oxford Genetics Knowledge Park.

References

Altherr, M. R., Bengtsson, U., Elder, F. F. B., Ledbetter, D. H., Wasmuth, J. J., McDonald, M. E., Gusella, J. F., and Greenberg, F. (1991). Molecular confirmation of Wolf-Hirschhorn syndrome with a subtle translocation of chromosome-4. *Am. J. Hum. Genet.* **49,** 1235–1242.

Anderlid, B. M., Schoumans, J., Anneren, G., Sahlen, S., Kyllerman, M., Vujic, M., Hagberg, B., Blennow, E., and Nordenskjold, M. (2002). Subtelomeric rearrangements detected in patients with idiopathic mental retardation. *Am. J. Med. Genet.* **107,** 275–284.

Armour, J. A., Sismani, C., Patsalis, P. C., and Cross, G. (2000). Measurement of locus copy number by hybridisation with amplifiable probes. *Nucl. Acids Res.* **28,** 605–609.

Bates, G. P., MacDonald, M. E., Baxendale, S., Sedlacek, Z., Youngman, S., Romano, D., Whaley, W. L., Allitto, B. A., Poustka, A., Gusella, J. F., and Lehrach, H. (1990). A yeast artificial chromosome telomere clone spanning a possible location of the Huntington's disease gene. *Am. J. Hum. Genet.* **46,** 762–775.

Bayani, J., and Squire, J. A. (2002). Spectral karyotyping. *In* "Methods in molecular biology, molecular cytogenetics: Protocols and applications" (Y.-S. Fan, ed.), Vol. 204, Humana Press, Totowa.

Borgione, E., Giudice, M. L., Galesi, O., Castiglia, L., Failla, P., Romano, C., Ragusa, A., and Fichera, M. (2000). How microsatellite analysis can be exploited for subtelomeric chromosomal rearrangement analysis in mental retardation. *J. Med. Genet.* **38,** E1.

Bonifacio, S., Centrone, C., Da Prato, L., Scordo, M. R., Estienne, M., and Torricelli, F. (2001). Use of primed *in situ* labeling (PRINS) for the detection of telomeric deletions associated with mental retardation. *Cytogenet. Cell Genet.* **93,** 16–18.

Brkanac, Z., Cody, J. D., Leach, R. J., and DuPont, B. R. (1998). Identification of cryptic rearrangements in patients with 18q-deletion syndrome. *Am. J. Hum. Genet.* **62,** 1500–1506.

Brown, W. R. A. (1989). Molecular cloning of human telomeres in yeast. *Nature* **338,** 774–776.

Brown, J., Saracoglu, K., Uhrig, S., Speicher, M. R., Eils, R., and Kearney, L. (2001). Subtelomeric chromosome rearrangements are detected using an innovative 12-color FISH assay (M-TEL). *Nat. Med.* **7,** 497–501.

Clarkson, B., Pavenski, K., Dupuis, L., Kennedy, S., Meyn, S., Nezarati, M. M., Nie, G., Weksberg, R., Withers, S., Quercia, N., Teebi, A. S., and Teshima, I. (2002). Detecting rearrangements in children using subtelomeric FISH and SKY. *Am. J. Med. Genet.* **107,** 267–274.

Colleaux, L., Rio, M., Heuertz, S., Moindrault, S., Turleau, C., Ozilou, C., Gosset, P., Raoult, O., Lyonnet, S., Cormier-Daire, V., Amiel, J., Le Merrer, M., Picq, M., de Blois, M. C., Prieur, M., Romana, S., Cornelis, F., Vekemans, M., and Munnich, A. (2001). A novel automated strategy for screening cryptic telomeric rearrangements in children with idiopathic mental retardation. *Eur. J. Hum. Genet.* **9,** 319–327.

Colleaux, L., Heurertz, S., Molinari, F., and Rio, M. (2002). Fluorescence genotyping for screening cryptic telomeric rearrangements. *In* "Methods in molecular biology, molecular cytogenetics: Protocols and applications" (Y.-S. Fan, ed.), Vol. 204, Humana Press, Totowa.

Cross, S. H., Allshire, R. C., McKay, S. J, McGill, N. I., and Cooke, H. J. (1989). Cloning of human telomeres by complementation in yeast. *Nature* **338,** 771–774.

Davies, A. F., Kirby, T. L., Docherty, Z., and Ogilvie, C. M. (2003). Characterization of terminal chromosome anomalies using multisubtelomere FISH. *Am. J. Med. Genet.* **1,** 483–489.

Engels, H., Ehrbrecht, A., Zahn, S., Bosse, K., Vrolijk, H., White, S., Kalscheuer, V., Hoovers, J. M., Schwanitz, G., Propping, P., Tanke, H. J., Wiegant, J., and Raap, A. K. (2003). Comprehensive analysis of human subtelomeres with combined binary ratio labelling fluorescence *in situ* hybridisation. *Eur. J. Hum. Genet.* **11,** 643–651.

Fiegler, H., Carr, P., Douglas, E. J., Burford, D. C., Hunt, S., Scott, C. E., Smith, J, Vetrie, D., Gorman, P., Tomlinson, I. P., and Carter, N. P. (2003). DNA microarrays for comparative genomic hybridization based on DOP-PCR amplification of BAC and PAC clones. (Published erratum in: *Genes Chromosomes Cancer* **36,** 361–374) *Genes Chromosomes Cancer* **37,** 223.

Flint, J., Wilkie, A. O., Buckle, V. J., Winter, R. M., Holland, A. J., and McDermid, H. E. (1995). The detection of subtelomeric chromosomal rearrangements in idiopathic mental retardation. *Nat. Genet.* **9,** 132–140.

Foot, N., Neat, M. J., Kearney, L., Burridge, M. M., Byrne, L., Consoli, C., Lister, T. A., Young, B. D., and Lillington, D. M. (1999). Multiple FISH technology to clarify abnormal leukaemic karyotypes. *J. Med. Genet.* **36,** S38.

Ghaffari, S. R., Boyd, E., Tolmie, J. L., Crow, Y. J., Trainer, A. H., and Connor, J. M. (1998). A new strategy for cryptic telomeric translocation screening in patients with idiopathic mental retardation. *J. Med. Genet.* **35,** 225–233.

Handyside, A. H., Scriven, P. N., and Ogilvie, C. M. (1998). The future of preimplantation genetic diagnosis. *Hum. Reprod.* **13**(suppl 4), 249–255.

Henegariu, O., Artan, S., Greally, J. M., Chen, X. N., Korenberg, J. R., Vance, G. H., Stubbs, L., Bray-Ward, P., and Ward, D. C. (2001). Cryptic translocation identification in human and mouse using several telomeric multiplex fish (TM-FISH) strategies. *Lab. Invest.* **81,** 483–491.

Hollox, E. J., Atia, T., Cross, G., Parkin, T., and Armour, J. A. (2002). High throughput screening of human subtelomeric DNA for copy number changes using multiplex amplifiable probe hybridisation (MAPH). *J. Med. Genet.* **39,** 790–795.

Horsley, S. W., Knight, S. J. L., Nixon, J., Huson, S., Fitchett, M., Boone, R. A., Hilton-Jones, D., Flint, J., and Kearney, L. (1998). Del(18p) shown to be a cryptic translocation using a multiprobe FISH assay for subtelomeric chromosome rearrangements. *J. Med. Genet.* **35,** 722–726.

Jalal, S. M., and Law, M. E. (2002). Multicolor FISH. *In* Methods in molecular biology, molecular cytogenetics: protocols and applications (Y.-S. Fan, ed.), vol 204, Humana Press, Totowa.

Jaju, R. J., Haas, O. A., Neat, M., Harbott, J., Saha, V., Boultwood, J., Brown, J. M., Pirc-Danoewinata, H., Krings, B. W., Muller, U., Morris, S. W., Wainscoat, J. S., and Kearney, L. (1999). A new recurrent translocation, t(5;11)(q35;p15. 5), associated with del(5q) in childhood acute myeloid leukemia. *Blood* **94,** 773–780.

Joly, G., Lapierre, J. M., Ozilou, C., Gosset, P., Aurias, A., de Blois, M. C., Prieur, M., Raoul, O., Colleaux, L., Munnich, A., Romana, S., Vekemans, M., and Turleau, C. (2001). Comparative genomic hybridisation in mentally retarded patients with dysmorphic features and a normal karyotype. *Clin. Genet.* **60,** 212–219.

Joyce, C. A., Dennis, N. R., Cooper, S., and Browne, C. E. (2001). Subtelomeric rearrangements: results from a study of selected and unselected probands with idiopathic mental retardation and control individuals by using high-resolution G-banding and FISH. *Hum. Genet.* **109,** 440–451.

Joyce, C. A., Hart, H. H., Fisher, A. M., and Browne, C. E. (1996). Use of subtelomeric FISH probes to detect abnormalities in patients with idiopathic mental retardation and characterize rearrangements at the limit of cytogenetic resolution. *J. Med. Genet.* **36**(suppl), S16.

Joyce, C. A., Hart, H. H., Fisher, A. M., and Browne, C. E. (1999). Use of subtelomeric FISH probes to detect abnormalities in patients with idiopathic mental retardation and characterize rearrangements at the limit of cytogenetic resolution. *J. Med. Genet.* **36,** S16.

Knight, S. J. L. (2004). Subtelomeric rearrangements in unexplained mental retardation. *In* "Encyclopedia of medical genomics and proteomics" (J. Fuchs and M. Podda, eds.). Marcel Dekker, New York.

Knight, S. J. L., Horsley, S. W., Regan, R., Lawrie, N. M., Maher, E. J., Cardy, D. L., Flint, J., and Kearney, L. (1997). Development and clinical application of an innovative fluorescence *in situ* hybridization technique which detects submicroscopic rearrangements involving telomeres. *Eur. J. Hum. Genet.* **5,** 1–8.

Knight, S. J., Regan, R., Nicod, A., Horsley, S. W., Kearney, L., Homfray, T., Winter, R. M., Bolton, P., and Flint, J. (1999). Subtle chromosomal rearrangements in children with unexplained mental retardation. *Lancet* **354,** 1676–1681.

Knight, S. J., and Flint, J. (2000). Perfect endings: a review of subtelomeric probes and their use in clinical diagnosis. *J. Med. Genet.* **37,** 401–409.

Knight, S. J., and Flint, J. (2002). Multi-telomere FISH. *In* "Methods in molecular biology, molecular cytogenetics: Protocols and applications" (Y.-S. Fan, ed.), Vol. 204, Humana Press, Totowa.

Knight, S. J., Lese, C. M., Precht, K. S., Kuc, J., Ning, Y., Lucas, S., Regan, R., Brenan, M., Nicod, A., Lawrie, N. M., Cardy, D. L., Nguyen, H., Hudson, T. J., Riethman, H. C., Ledbetter, D. H., and

Flint, J. (2000). An optimized set of human telomere clones for studying telomere integrity and architecture. *Am. J. Hum. Genet.* **67,** 320–332.

Kuwano, A., Ledbetter, S. A., Dobyns, W. B., Emanuel, B. S., and Ledbetter, D. H. (1991). Detection of deletions and cryptic translocations in Miller-Dieker syndrome by *in situ* hybridization. *Am. J. Hum. Genet.* **49,** 707–714.

Lamb, J., Harris, P. C., Wilkie, A. O. M., Wood, W. G., Dauwerse, J. G., and Higgs, D. R. (1993). De novo truncation of chromosome 16p and healing with (TTAGGG)n in the α-thalassemia/mental retardation syndrome (ATR-16). *Am. J. Hum. Genet.* **52,** 668–676.

Mackie-Ogilvie, C., Harrison, R. H., Handyside, A. H., and Scriven, P. N. (1999). The use of subtelomeric probes in preimplantation genetic diagnosis. *J. Med. Genet.* **36**(suppl), S15.

Martin, C. L., Waggoner, D. J., Wong, A., Uhrig, S., Roseberry, J. A., Hedrick, J. F., Pack, S. D., Russell, K., Zackai, E., Dobyns, W. B., and Ledbetter, D. H. (2002). "Molecular rulers" for calibrating phenotypic effects of telomere imbalance *J. Med. Genet.* **39,** 734–740.

Morrison, L. E., Ramakrishnan, R., Ruffalo, T. M., and Wilbur, K. A. (2002). Labelling fluorescence *in situ* hybridisation probes for genomic targets. *In* "Methods in molecular biology, molecular cytogenetics: Protocols and applications" (Y.-S. Fan, ed.), Vol. 204, Humana Press, Totowa.

National Institutes of Health and Institute of Molecular Medicine collaboration (1996). A complete set of human telomeric probes and their clinical application. National Institutes of Health and Institute of Molecular Medicine collaboration. (Published erratum in: *Nat Genet.* **14,** 86–89) *Nat Genet* **14,** 487.

Overhauser, J., Bengtsson, U., McMahon, J., Ulm, J., Butler, M. G., Santiago, L., and Wasmuth, J. J. (1989). Prenatal diagnosis and carrier detection of a cryptic translocation by using DNA markers from the short arm of chromosome 5. *Am. J. Hum. Genet.* **45,** 296–303.

Riegel, M., Baumer, A., Jamar, M., Delbecque, K., Herens, C., Verloes, A., and Schinzel, A. (2001). Submicroscopic terminal deletions and duplications in retarded patients with unclassified malformation syndromes. *Hum Genet.* **109,** 286–294.

Riethman, H. C., Moyzis, R. K., Meyne, J., Burke, D. T., and Olson, M. V. (1989). Cloning human telomeric DNA fragments into *Saccharomyces cerevesiae* using a yeast artificial chromosome vector. *Proc. Natl. Acad. Sci. USA* **86,** 6240–6244.

Rooms, L., Reyniers, E., van Luijk, R., Scheers, S., Wauters, J., Ceulemans, B., Van Den Ende, J., Van Bever, Y., and Kooy, R. F. (2004). Subtelomeric deletions detected in patients with idiopathic mental retardation using multiplex ligation-dependent probe amplification (MLPA). *Hum. Mutat.* **23,** 17–21.

Rosenberg, M., Hui, L., Ma, J., Nusbaum, H. C., Clark, K., Robinson, L., Dziadzio, L., Swain, P. M., Keith, T., Hudson, T. J., Biesecker, L. G., and Flint, J. (1997). Characterization of short tandem repeats from thirty-one human telomeres. *Genome Res.* **7,** 917–923.

Rosenberg, M. J., Vaske, D., Killoran, C. E., Ning, Y., Wargowski, D., Hudgins, L., Tifft, C. J., Meck, J., Blancato, J. K., Rosenbaum, K., Pauli, R. M., Weber, J., and Biesecker, L. G. (2000). Detection of chromosomal aberrations by a whole-genome microsatellite screen. *Am. J. Hum. Genet.* **66,** 419–427.

Rosenberg, M. J., Killoran, C. Dziadzio, L., Chang, S., Stone, D. L., Meck, J., Aughton, Bird, L. M., Bodhurtha, J., Cassidy, S. B., Graham Jnr, J. M., Grix, A., Guttmacher, A. E., Hudgins, L., Kozma, C., Michaelis, R., Pauli, R., and Biesecker, L. G. (2001). Scanning for telomeric deletions and duplications and uniparental disomy using genetic markers in 120 children with malformations. *Hum. Genet.* **109,** 311–318.

Saccone, S., De Sario, A., Della Valle, G., and Bernardi, G. (1992). The highest gene concentrations in the human genome are in telomeric bands of metaphase chromosomes. *Proc. Natl. Acad. Sci. USA* **89,** 4913–4917.

Schoumans, J., Anderlid, B. M., Blennow, E., The, B. T., and Nordenskjold, M. (2004). The performance of CGH array for the detection of cryptic constitutional chromosome imbalances. *J. Med. Genet.* **41,** 198–220.

Schouten, J. P., McElgunn, C. J., Waaijer, R., Zwijnenburg, D., Diepvens, F., and Pals, G. (2002). Relative quantification of 40 nucleic acid sequences by multiplex ligation-dependent probe amplification. *Nucl. Acids Res.* **30,** e57.

Scriven, P. N., Handyside, A. H., and Ogilvie, C. M. (1998). Chromosome translocations: Segregation modes and strategies for preimplantation genetic diagnosis. *Prenat. Diagn.* **18,** 1437–1449.

Shaw-smith, C., Redon, R., Rickman, L., Rio, M., Willatt, L., Fiegler, H., Firth, H., Sanlaville, D., Winter, R., Colleaux, L., Bobrow, M., and Carter, N. P. (2004). Microarray based comparative genomic hybridization (array-CGH) detects submicroscopic chromosomal deletions and duplications in patients with learning disability/mental retardation and dysmorphic features. *J. Med. Genet.* **41,** 241–248.

Sismani, C., Armour, J. A., Flint, J., Girgallia, C., Regan, R., and Patsalis, P. C. (2001). Screening for subtelomeric chromosome abnormalities in children with idiopathic mental retardation using the multiprobe telomeric FISH and the new MAPH telomeric assay. *Eur. J. Hum. Genet.* **9,** 527–532.

Slavotinek, A., Rosenberg, M., Knight, S., Gaunt, L., Fergusson, W., Killoran, C., Clayton-Smith, J., Kingston, H., Campbell, R. H., Flint, J., Donnai, D., and Biesecker, L. (1999). Screening for submicroscopic chromosome rearrangements in children with idiopathic mental retardation using microsatellite markers for the chromosome telomeres. *J. Med. Genet.* **36,** 405–411.

Souter, V. L., Glass, I. A., Chapman, D. B., Raff, M. L., Parisi, M. A., Opheim, K. E., and Disteche, C. M. (2003). Multiple fetal anomalies associated with subtle subtelomeric chromosomal rearrangements. *Ultrasound Obstet. Gynecol.* **21,** 609–615.

Speicher, M. R., Ballard, S. G., and Ward, D. C. (1996). Karyotyping human chromosomes by combinatorial multi-fluor FISH. *Nat. Genet.* **12,** 368–375.

Tanke, H. J., Wiegant, J., van Gijlswijk, R. P., Bezrookove, V., Pattenier, H., Heetebrij, R. J., Talman, E. G., Raap, A. K., and Vrolijk, J. (1999). New strategy for multi-colour fluorescence in situ hybridisation: COBRA: COmbined Binary RAtio labelling. *Eur. J. Hum. Genet.* **7,** 2–11.

Tosi, S., Scherer, S. W., Giudici, G., Czepulkowski, B., Biondi, A., and Kearney, L. (1999). Delineation of multiple deleted regions in 7q in myeloid disorders. *Genes Chromosomes Cancer* **25,** 384–392.

van Karnebeek, C. D., Koevoets, C., Sluijter, S., Bijlsma, E. K., Smeets, D. F., Redeker, E. J., Hennekam, R. C., and Hoovers, J. M. (2002). Prospective screening for subtelomeric rearrangements in children with mental retardation of unknown aetiology: the Amsterdam experience. *J. Med. Genet.* **39,** 546–553.

Veltman, J. A., Schoenmakers, E. F., Eussen, B. H., Janssen, I., Merkx, G., van Cleef, B., van Ravenswaaij, C. M., Brunner, H. G., Smeets, D., and van Kessel, A. G. (2002). High-throughput analysis of subtelomeric chromosome rearrangements by use of array-based comparative genomic hybridization. *Am. J. Hum. Genet.* **70,** 1269–1276.

Vissers, L. E. L. M, de Vries, B. B. A., Osoegawa, K., Janssen, I. M., Feu, T., Choy, C. O., Straatman, H., van der Vliet, W., Huys, E. H. L. P. G., R, A., Smeets, D., van van Ravenswaaij-Arts, C. M. A., Knoers, N. V., van der Burgt, I., de Jong, P. J., Brunner, H. G., van Kessel, A. G., Schoenmakers, E. F. P. M., and Veltman, J. A. (2003). Array-based comparative genomic hybridisation for the genome wide detection of submicroscopic chromosome abnormalities. *Am. J. Hum. Genet.* **73,** 1261–1270.

Wilkie, A. O. M. (1993). Detection of cryptic chromosomal abnormalities in unexplained mental retardation: A general strategy using hypervariable subtelomeric DNA polymorphisms. *Am. J. Hum. Genet.* **53,** 688–701.

Wilkie, A. O. M., Lamb, J., Harris, P. C., Finney, R. D., and Higgs, D. R. (1990). A truncated chromosome 16 associated with α thalassemia is stabilized by the addition of telomeric repeats (TTAGGG)n. *Nature* **346,** 868–871.

Wong, A. C., Ning, Y., Flint, J., Clark, K., Dumanski, J. P., Ledbetter, D. H., and McDermid, H. E. (1997). Molecular characterization of a 130-kb terminal microdeletion in a child with mild mental retardation. *Am. J. Hum. Genet.* **60,** 113–120.

Yu, W., Ballif, B. C., Kashork, C. D., Heilstedt, H. A., Howard, L. A., Cai, W. W., White, L. D., Liu, W., Beaudet, A. L., Bejjani, B. A., Shaw, C. A., and Shaffer, L. G. (2003). Development of a comparative genomic hybridization microarray and demonstration of its utility with 25 well-characterized 1p36 deletions. *Hum. Mol. Genet.* **12,** 2145–2152.

INDEX

C

M

VOLUMES IN SERIES

Founding Series Editor
DAVID M. PRESCOTT

Volume 1 (1964)
Methods in Cell Physiology
Edited by David M. Prescott

Volume 2 (1966)
Methods in Cell Physiology
Edited by David M. Prescott

Volume 3 (1968)
Methods in Cell Physiology
Edited by David M. Prescott

Volume 4 (1970)
Methods in Cell Physiology
Edited by David M. Prescott

Volume 5 (1972)
Methods in Cell Physiology
Edited by David M. Prescott

Volume 6 (1973)
Methods in Cell Physiology
Edited by David M. Prescott

Volume 7 (1973)
Methods in Cell Biology
Edited by David M. Prescott

Volume 8 (1974)
Methods in Cell Biology
Edited by David M. Prescott

Volume 9 (1975)
Methods in Cell Biology
Edited by David M. Prescott

Volume 10 (1975)
Methods in Cell Biology
Edited by David M. Prescott

Volume 11 (1975)
Yeast Cells
Edited by David M. Prescott

Volume 12 (1975)
Yeast Cells
Edited by David M. Prescott

Volume 13 (1976)
Methods in Cell Biology
Edited by David M. Prescott

Volume 14 (1976)
Methods in Cell Biology
Edited by David M. Prescott

Volume 15 (1977)
Methods in Cell Biology
Edited by David M. Prescott

Volume 16 (1977)
Chromatin and Chromosomal Protein Research I
Edited by Gary Stein, Janet Stein, and Lewis J. Kleinsmith

Volume 17 (1978)
Chromatin and Chromosomal Protein Research II
Edited by Gary Stein, Janet Stein, and Lewis J. Kleinsmith

Volume 18 (1978)
Chromatin and Chromosomal Protein Research III
Edited by Gary Stein, Janet Stein, and Lewis J. Kleinsmith

Volume 19 (1978)
Chromatin and Chromosomal Protein Research IV
Edited by Gary Stein, Janet Stein, and Lewis J. Kleinsmith

Volume 20 (1978)
Methods in Cell Biology
Edited by David M. Prescott

Advisory Board Chairman
KEITH R. PORTER

Volume 21A (1980)
Normal Human Tissue and Cell Culture, Part A: Respiratory, Cardiovascular, and Integumentary Systems
Edited by Curtis C. Harris, Benjamin F. Trump, and Gary D. Stoner

Volume 21B (1980)
Normal Human Tissue and Cell Culture, Part B: Endocrine, Urogenital, and Gastrointestinal Systems
Edited by Curtis C. Harris, Benjamin F. Trump, and Gray D. Stoner

Volume 22 (1981)
Three-Dimensional Ultrastructure in Biology
Edited by James N. Turner

Volume 23 (1981)
Basic Mechanisms of Cellular Secretion
Edited by Arthur R. Hand and Constance Oliver

Volume 24 (1982)
The Cytoskeleton, Part A: Cytoskeletal Proteins, Isolation and Characterization
Edited by Leslie Wilson

Volume 25 (1982)
The Cytoskeleton, Part B: Biological Systems and *in Vitro* Models
Edited by Leslie Wilson

Volume 26 (1982)
Prenatal Diagnosis: Cell Biological Approaches
Edited by Samuel A. Latt and Gretchen J. Darlington

Series Editor
LESLIE WILSON

Volume 27 (1986)
Echinoderm Gametes and Embryos
Edited by Thomas E. Schroeder

Volume 28 (1987)
***Dictyostelium discoideum:* Molecular Approaches to Cell Biology**
Edited by James A. Spudich

Volume 29 (1989)
Fluorescence Microscopy of Living Cells in Culture, Part A: Fluorescent Analogs, Labeling Cells, and Basic Microscopy
Edited by Yu-Li Wang and D. Lansing Taylor

Volume 30 (1989)
Fluorescence Microscopy of Living Cells in Culture, Part B: Quantitative Fluorescence Microscopy—Imaging and Spectroscopy
Edited by D. Lansing Taylor and Yu-Li Wang

Volume 31 (1989)
Vesicular Transport, Part A
Edited by Alan M. Tartakoff

Volume 32 (1989)
Vesicular Transport, Part B
Edited by Alan M. Tartakoff

Volume 33 (1990)
Flow Cytometry
Edited by Zbigniew Darzynkiewicz and Harry A. Crissman

Volume 34 (1991)
Vectorial Transport of Proteins into and across Membranes
Edited by Alan M. Tartakoff

Selected from Volumes 31, 32, and 34 (1991)
Laboratory Methods for Vesicular and Vectorial Transport
Edited by Alan M. Tartakoff

Volume 35 (1991)
Functional Organization of the Nucleus: A Laboratory Guide
Edited by Barbara A. Hamkalo and Sarah C. R. Elgin

Volume 36 (1991)
***Xenopus laevis:* Practical Uses in Cell and Molecular Biology**
Edited by Brian K. Kay and H. Benjamin Peng

Series Editors
LESLIE WILSON AND PAUL MATSUDAIRA

Volume 37 (1993)
Antibodies in Cell Biology
Edited by David J. Asai

Volume 38 (1993)
Cell Biological Applications of Confocal Microscopy
Edited by Brian Matsumoto

Volume 39 (1993)
Motility Assays for Motor Proteins
Edited by Jonathan M. Scholey

Volume 40 (1994)
A Practical Guide to the Study of Calcium in Living Cells
Edited by Richard Nuccitelli

Volume 41 (1994)
Flow Cytometry, Second Edition, Part A
Edited by Zbigniew Darzynkiewicz, J. Paul Robinson, and Harry A. Crissman

Volume 42 (1994)
Flow Cytometry, Second Edition, Part B
Edited by Zbigniew Darzynkiewicz, J. Paul Robinson, and Harry A. Crissman

Volume 43 (1994)
Protein Expression in Animal Cells
Edited by Michael G. Roth

Volume 44 (1994)
***Drosophila melanogaster:* Practical Uses in Cell and Molecular Biology**
Edited by Lawrence S. B. Goldstein and Eric A. Fyrberg

Volume 45 (1994)
Microbes as Tools for Cell Biology
Edited by David G. Russell

Volume 46 (1995) (in preparation)
Cell Death
Edited by Lawrence M. Schwartz and Barbara A. Osborne

Volume 47 (1995)
Cilia and Flagella
Edited by William Dentler and George Witman

Volume 48 (1995)
***Caenorhabditis elegans:* Modern Biological Analysis of an Organism**
Edited by Henry F. Epstein and Diane C. Shakes

Volume 49 (1995)
Methods in Plant Cell Biology, Part A
Edited by David W. Galbraith, Hans J. Bohnert, and Don P. Bourque

Volume 50 (1995)
Methods in Plant Cell Biology, Part B
Edited by David W. Galbraith, Don P. Bourque, and Hans J. Bohnert

Volume 51 (1996)
Methods in Avian Embryology
Edited by Marianne Bronner-Fraser

Volume 52 (1997)
Methods in Muscle Biology
Edited by Charles P. Emerson, Jr. and H. Lee Sweeney

Volume 53 (1997)
Nuclear Structure and Function
Edited by Miguel Berrios

Volume 54 (1997)
Cumulative Index

Volume 55 (1997)
Laser Tweezers in Cell Biology
Edited by Michael P. Sheez

Volume 56 (1998)
Video Microscopy
Edited by Greenfield Sluder and David E. Wolf

Volume 57 (1998)
Animal Cell Culture Methods
Edited by Jennie P. Mather and David Barnes

Volume 58 (1998)
Green Fluorescent Protein
Edited by Kevin F. Sullivan and Steve A. Kay

Volume 59 (1998)
The Zebrafish: Biology
Edited by H. William Detrich III, Monte Westerfield, and Leonard I. Zon

Volume 60 (1998)
The Zebrafish: Genetics and Genomics
Edited by H. William Detrich III, Monte Westerfield, and Leonard I. Zon

Volume 61 (1998)
Mitosis and Meiosis
Edited by Conly L. Rieder

Volume 62 (1999)
Tetrahymena thermophila
Edited by David J. Asai and James D. Forney

Volume 63 (2000)
Cytometry, Third Edition, Part A
Edited by Zbigniew Darzynkiewicz, J. Paul Robinson, and Harry Crissman

Volume 64 (2000)
Cytometry, Third Edition, Part B
Edited by Zbigniew Darzynkiewicz, J. Paul Robinson, and Harry Crissman

Volume 65 (2001)
Mitochondria
Edited by Liza A. Pon and Eric A. Schon

Volume 66 (2001)
Apoptosis
Edited by Lawrence M. Schwartz and Jonathan D. Ashwell

Volume 67 (2001)
Centrosomes and Spindle Pole Bodies
Edited by Robert E. Palazzo and Trisha N. Davis

Volume 68 (2002)
Atomic Force Microscopy in Cell Biology
Edited by Bhanu P. Jena and J. K. Heinrich Hrber

Volume 69 (2002)
Methods in Cell–Matrix Adhesion
Edited by Josephine C. Adams

Volume 70 (2002)
Cell Biological Applications of Confocal Microscopy
Edited by Brian Matsumoto

Volume 71 (2003)
Neurons: Methods and Applications for Cell Biologist
Edited by Peter J. Hollenbeck and James R. Bamburg

Volume 72 (2003)
Digital Microscopy: A Second Edition of Video Microscopy
Edited by Greenfield Sluder and David E. Wolf

Volume 73 (2003)
Cumulative Index

Volume 74 (2004)
Development of Sea Urchins, Ascidians, and Other Invertebrate Deuterostomes: Experimental Approaches
Edited by Charles A. Ettensohn

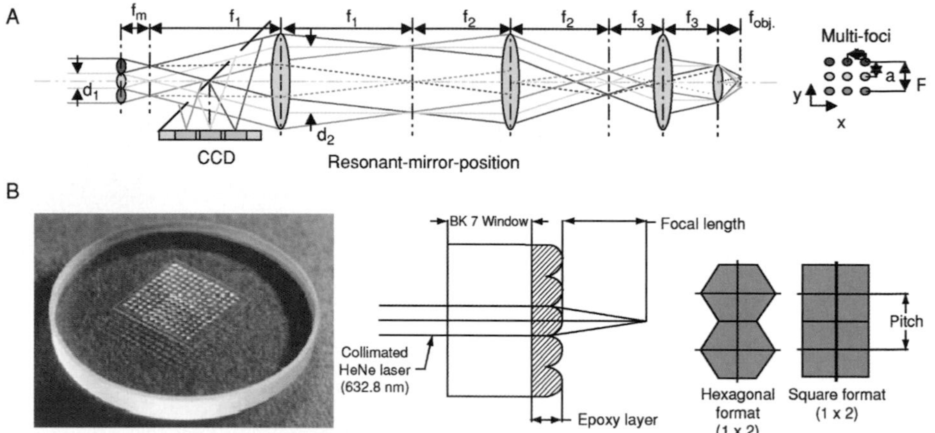

Fig. 2.3. Ragan *et al.* (See Legend in Text).

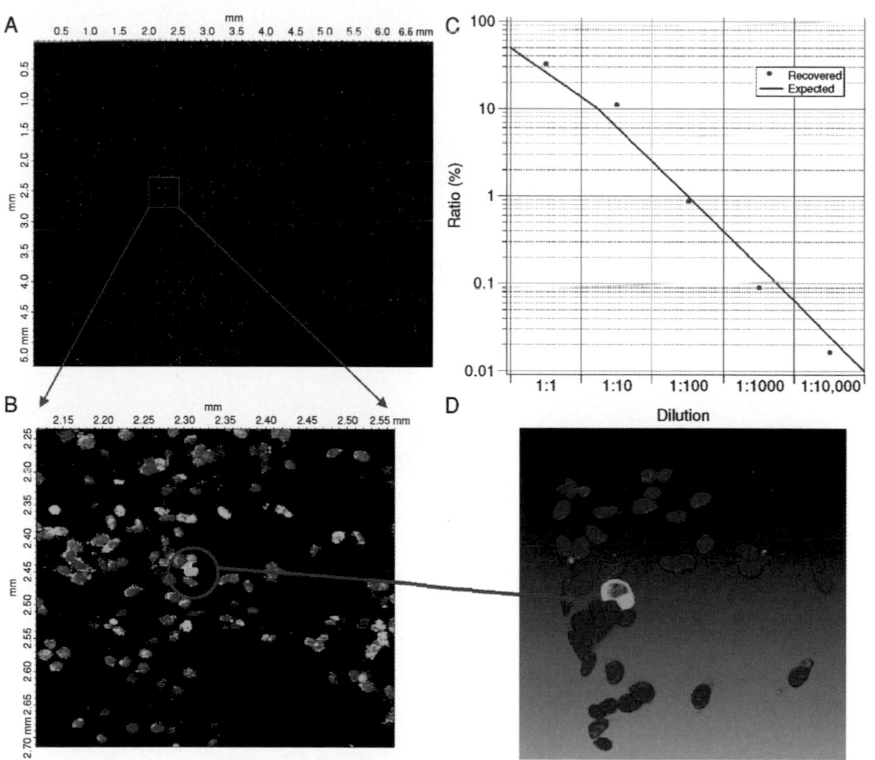

Fig. 2.5. Ragan *et al.* (See Legend in Text).

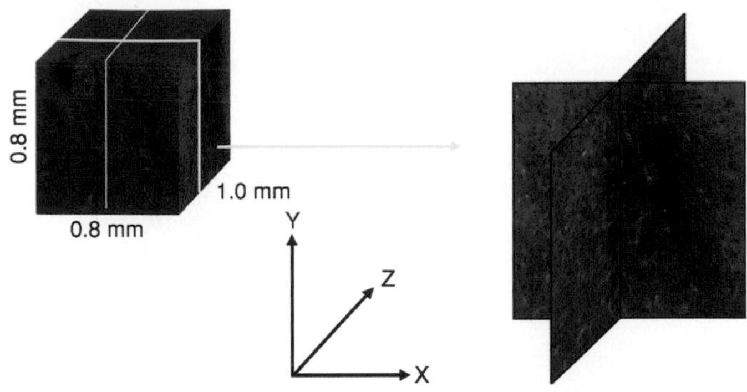

Fig. 2.6. Ragan *et al.* (See Legend in Text).

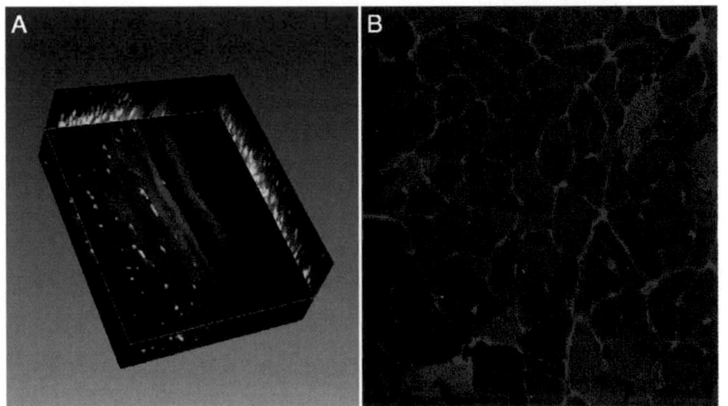

Fig. 2.7. Ragan *et al.* (See Legend in Text).

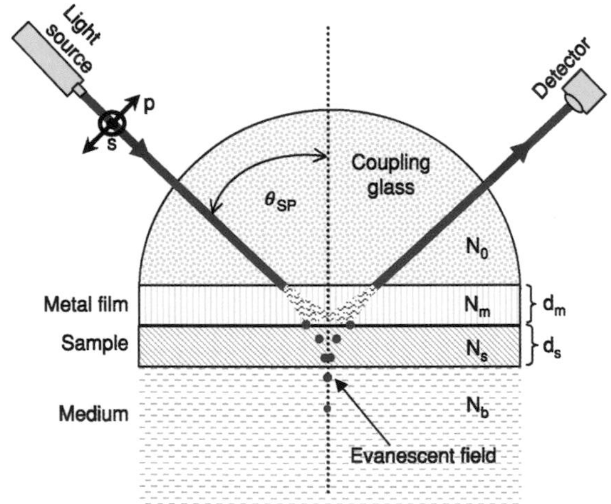

Fig. 4.1. Gryczynski *et al.* (See Legend in Text).

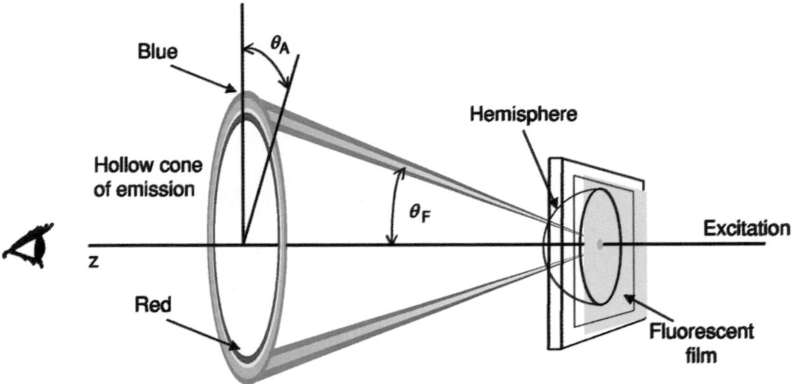

Fig. 4.7. Gryczynski *et al.* (See Legend in Text).

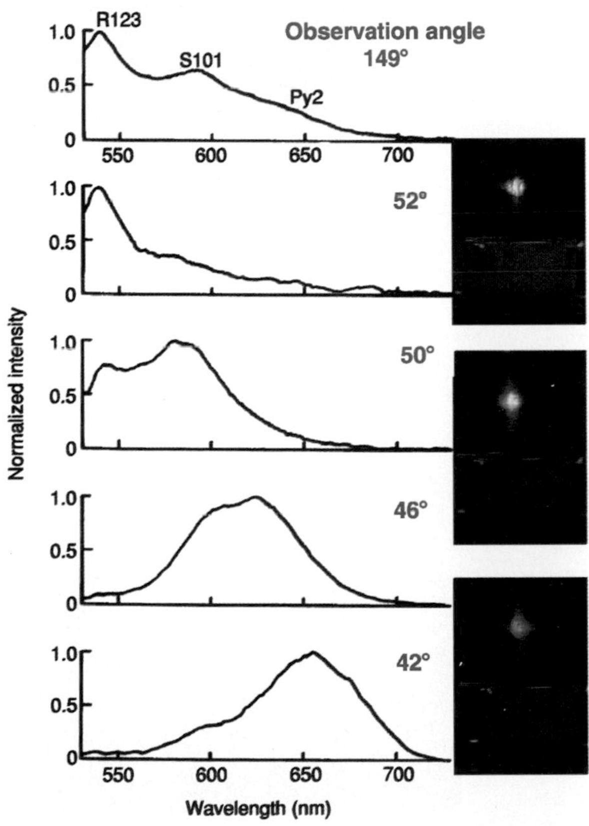

Fig. 4.13. Gryczynski *et al.* (See Legend in Text).

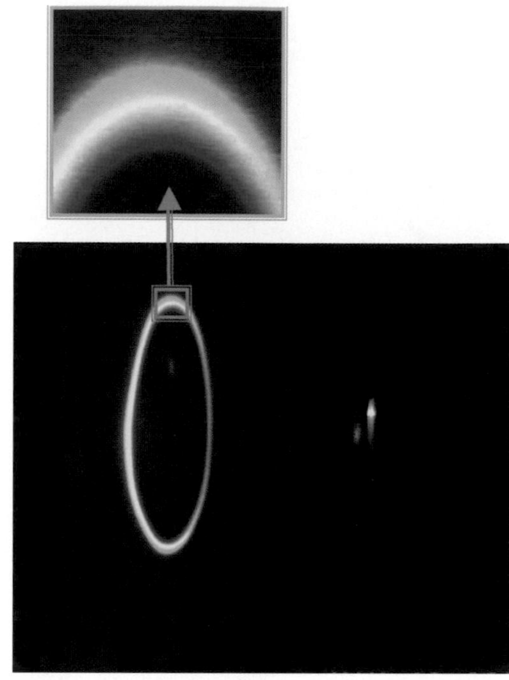

Fig. 4.14. **Gryczynski *et al.*** (See Legend in Text).

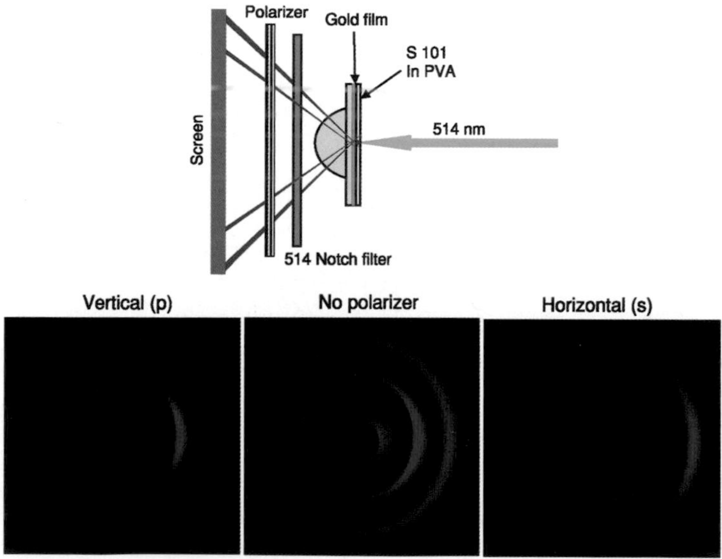

Fig. 4.17. **Gryczynski *et al.*** (See Legend in Text).

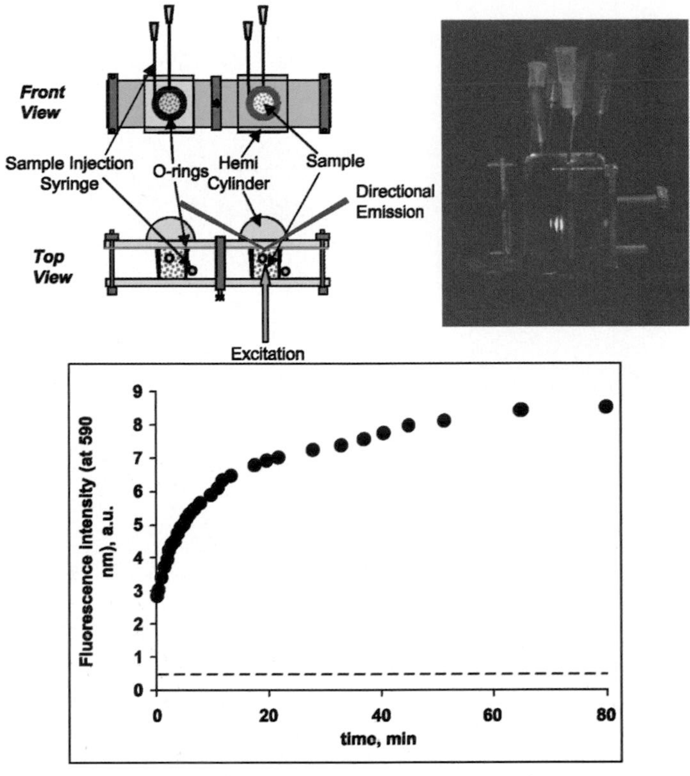

Fig. 4.22. Gryczynski *et al.* (See Legend in Text).

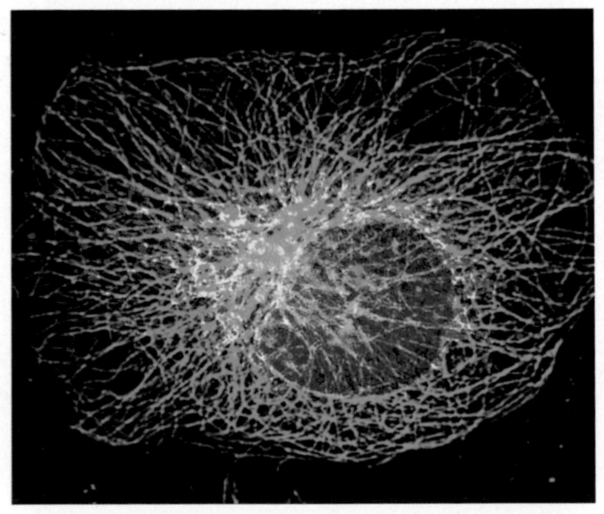

Fig. 7.1. Wu and Bruchez (See Legend in Text).

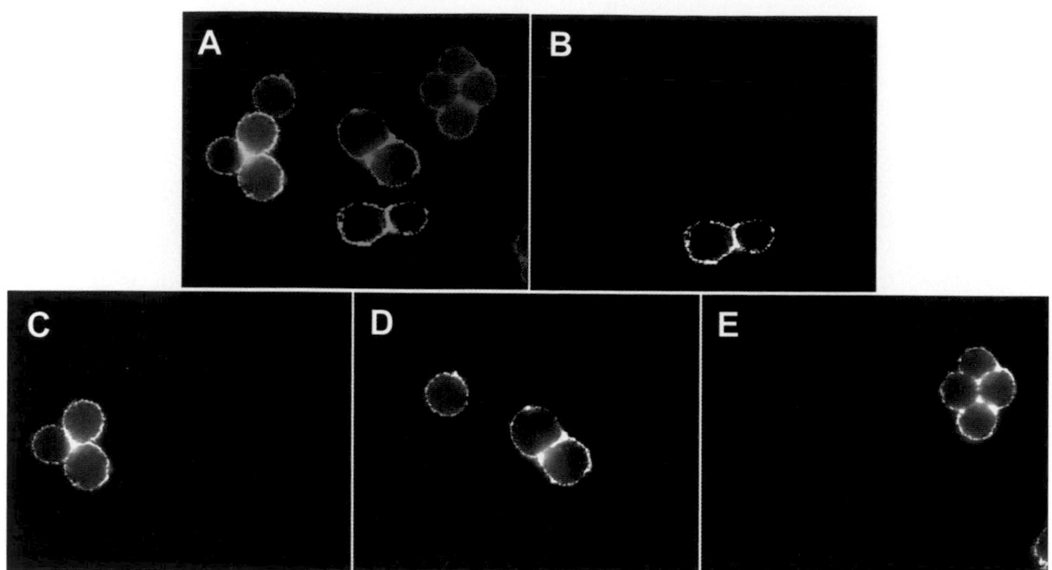

Fig. 7.2. Wu and Bruchez (See Legend in Text).

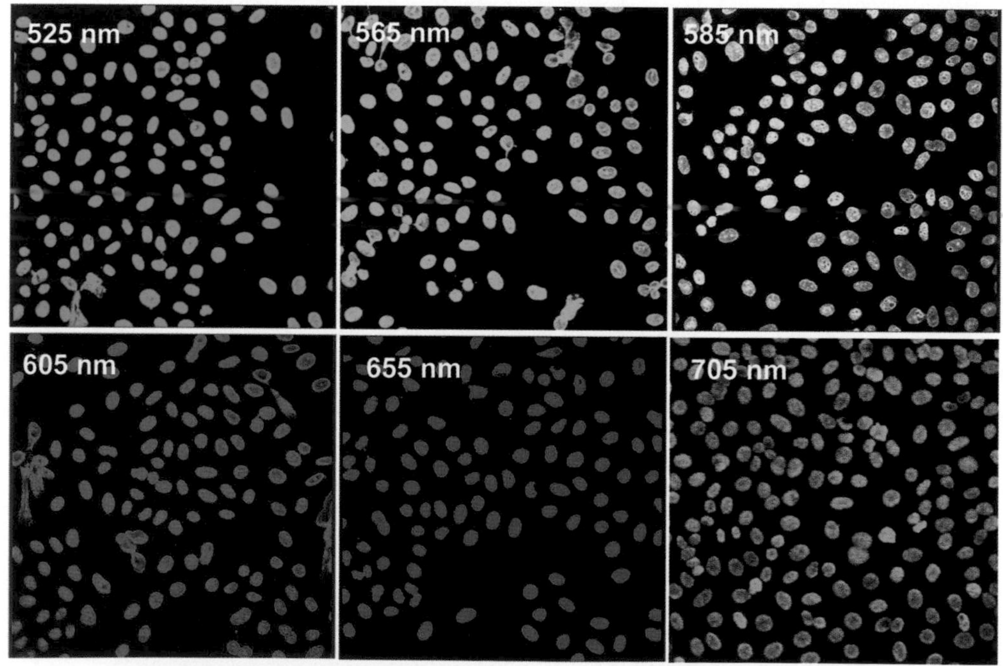

Fig. 7.3. Wu and Bruchez (See Legend in Text).

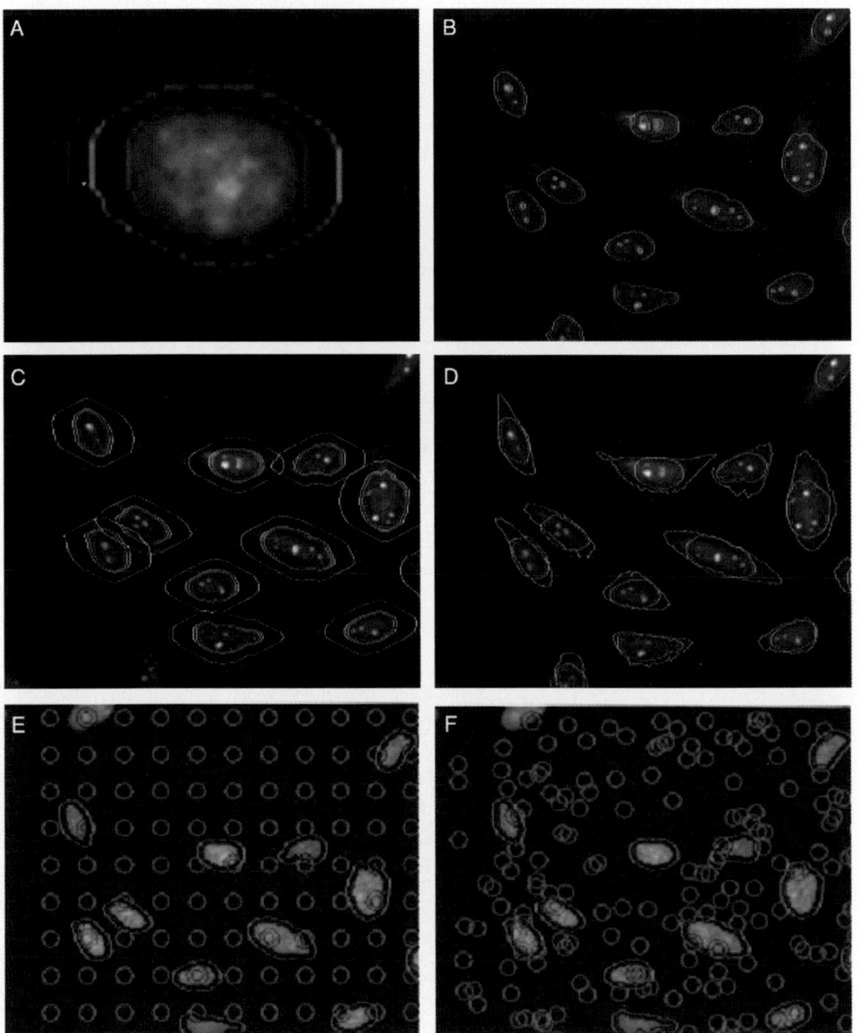

Fig. 8.6. Luther *et al.* (See Legend in Text).

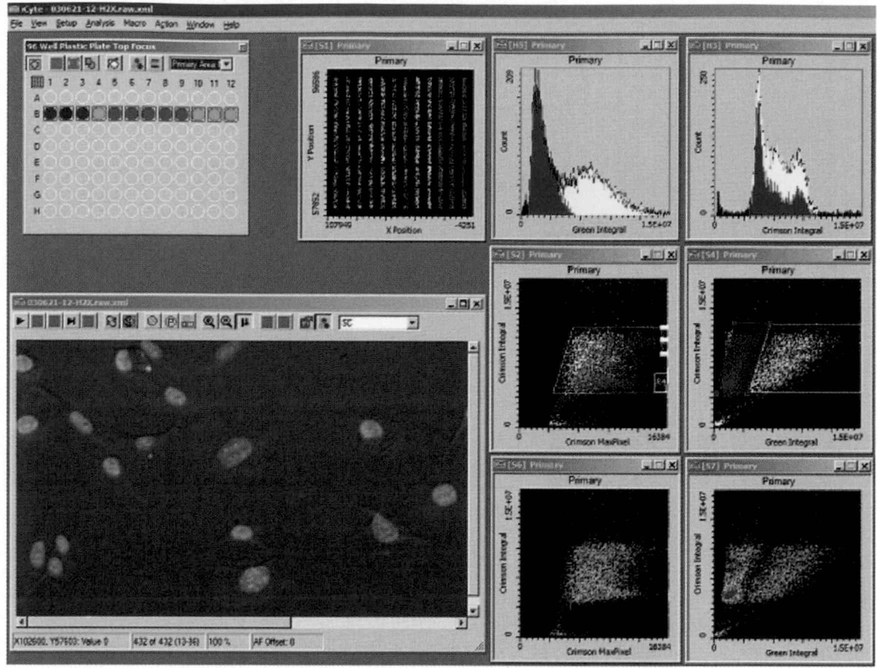

Fig. 8.12. Luther *et al.* (See Legend in Text).

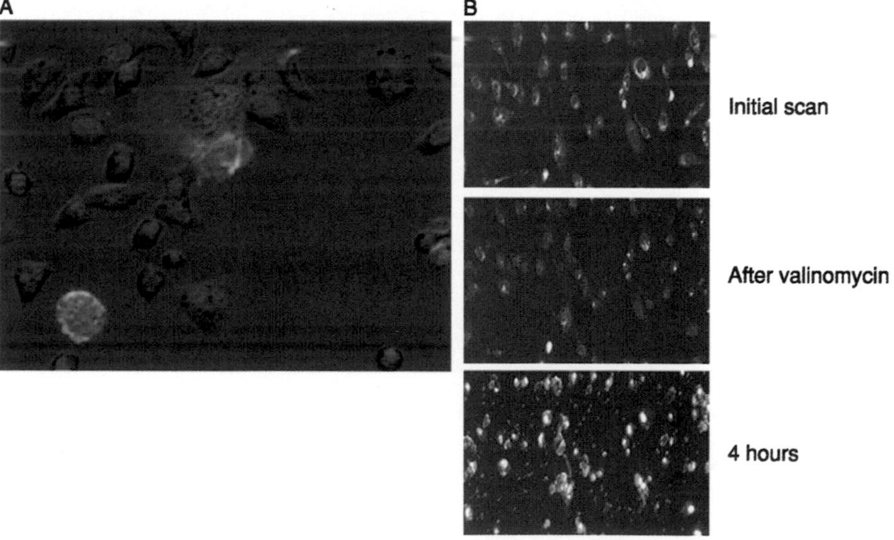

Fig. 8.16. Luther *et al.* (See Legend in Text).

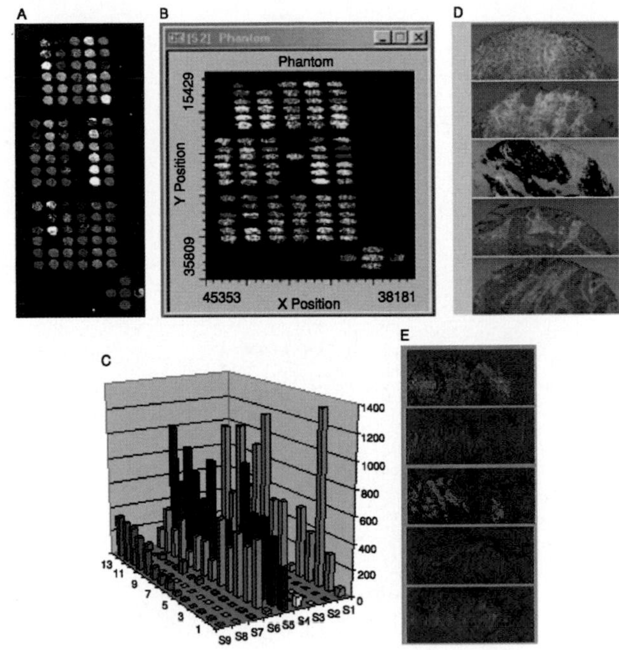

Fig. 8.19. Luther *et al.* (See Legend in Text).

Number of events displayed:	Unsmoothed: Dot Plot	Unsmoothed: Density Plot	Unsmoothed: Contour Plot	Smoothed: Contour Plot	Smoothed: Outlier Contour Plot	Smoothed: Outlier Zebra Plot	Smoothed: Outlier Density Plot
559							
5319							
53220							
532365							

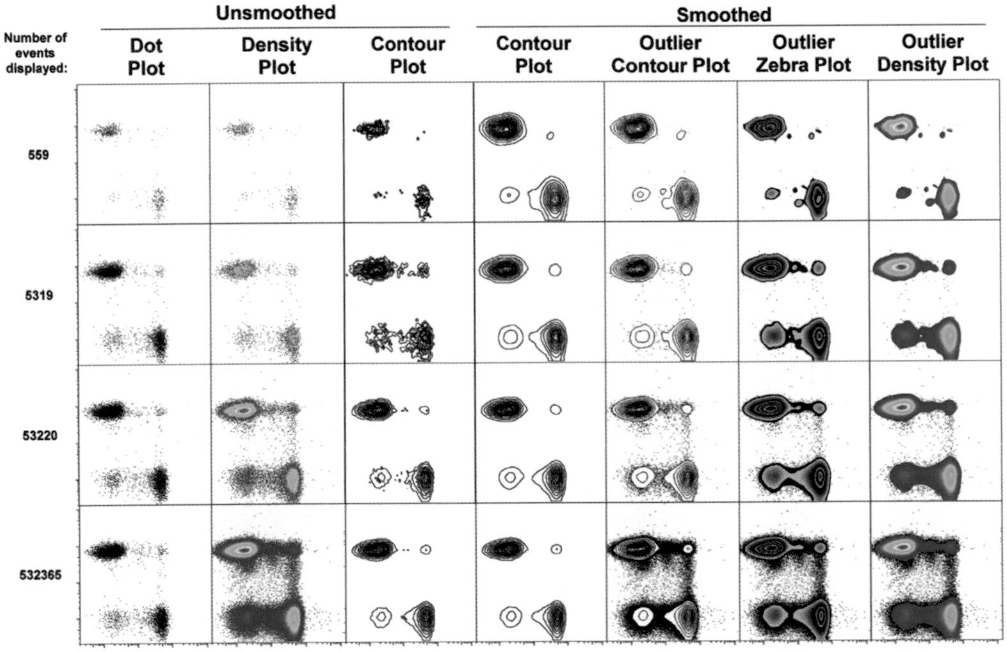

Fig. 10.1. Roederer *et al.* (See Legend in Text).

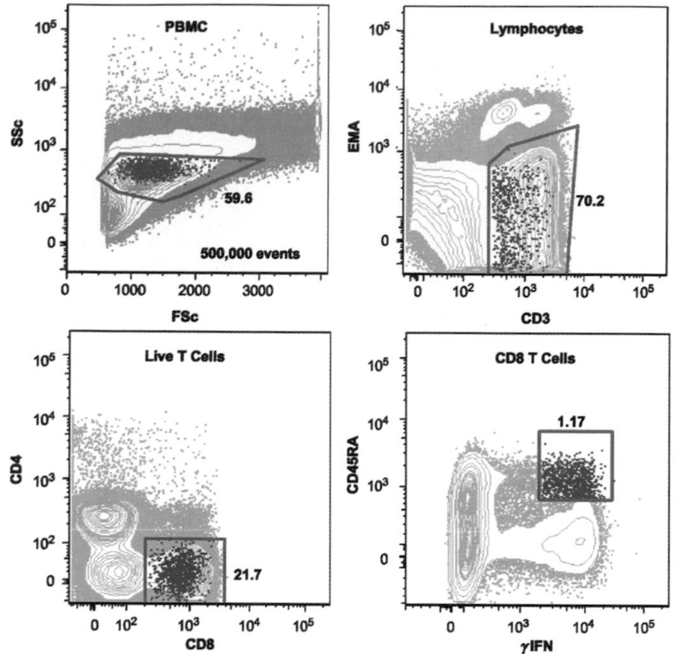

Fig. 10.2. Roederer *et al.* (See Legend in Text).

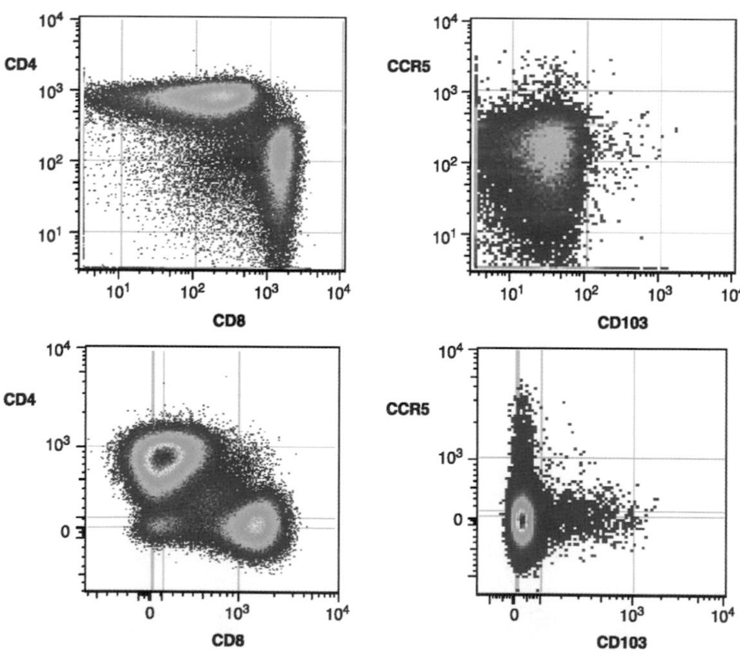

Fig. 10.3. Roederer *et al.* (See Legend in Text).

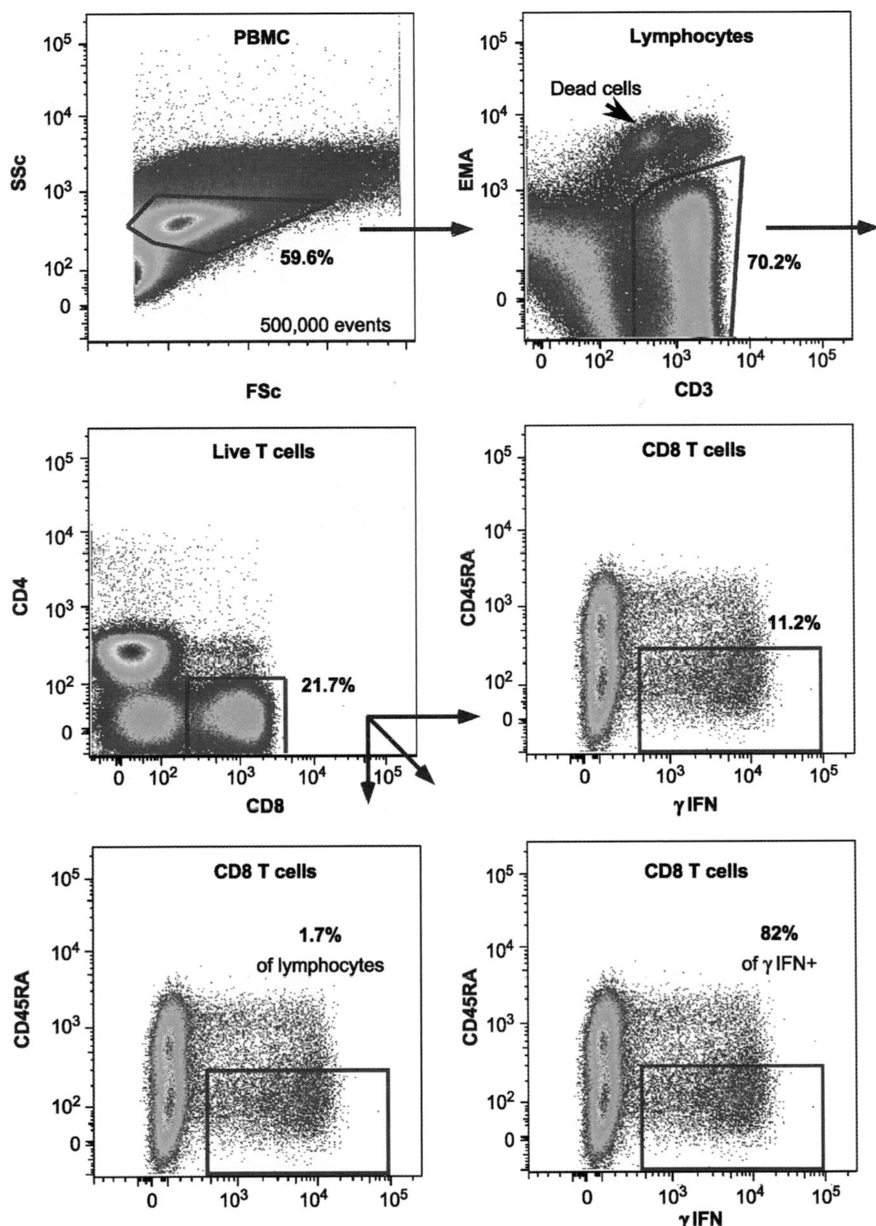

Fig. 10.4. Roederer *et al.* (See Legend in Text).

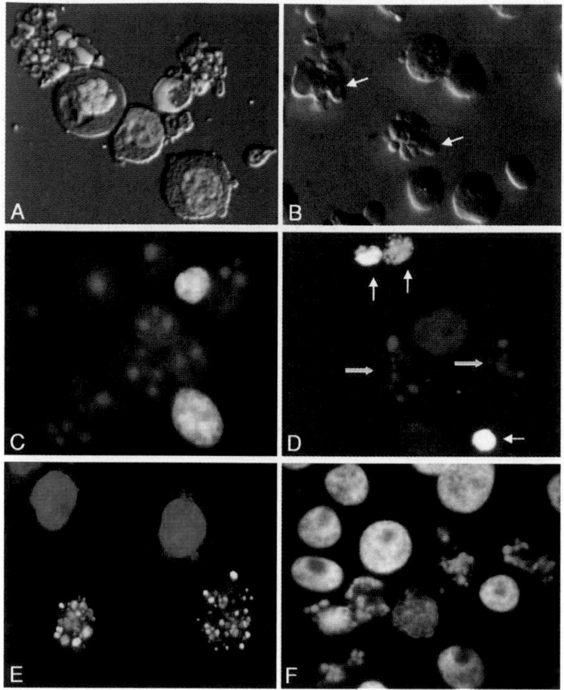

Fig. 12.1. Darzynkiewicz *et al.* (See Legend in Text).

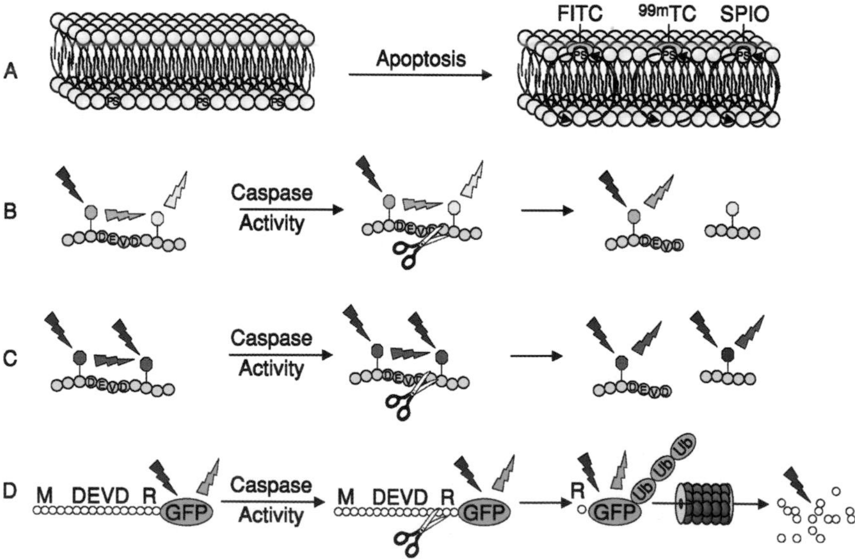

Fig. 13.1. Lee and Segal (See Legend in Text).

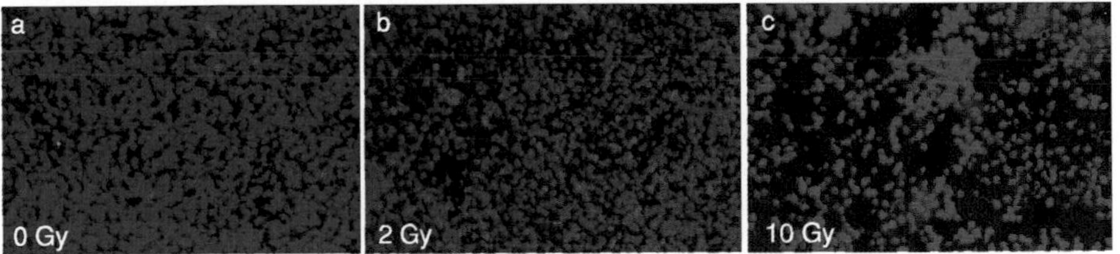

Fig. 14.8. Olive (See Legend in Text).

phycocyanin 650 nm

phycoerythrin 620 nm

phycoerythrin 585 nm

forward scatter

650 nm fluorescence

620 nm fluorescence

585 nm fluorescence

680 nm fluorescence

Cyano.:
1 Osci. ssp.
2 Mic. aer.
3 Syn.cy. (11d)
4 Syn.cy. (42d)
5 Anac.nid.

Cryp.:
6 Crypt. ov.
7 Rhod. spec. (freshwater ref.)

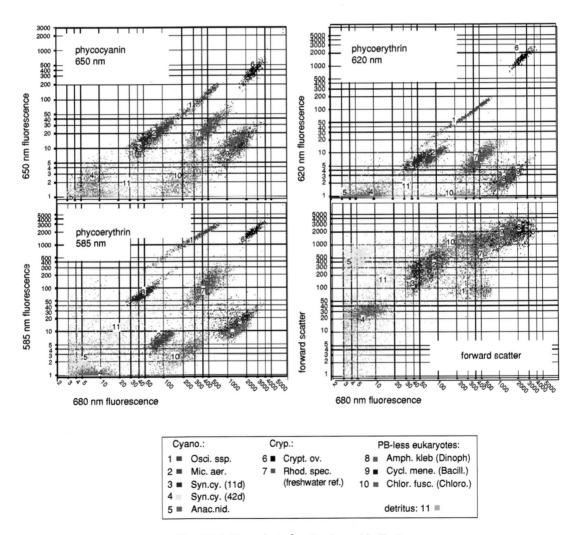

PB-less eukaryotes:
8 Amph. kleb (Dinoph)
9 Cycl. mene. (Bacill.)
10 Chlor. fusc. (Chloro.)

detritus: 11

Fig. 15.6. Toepel *et al.* (See Legend in Text).

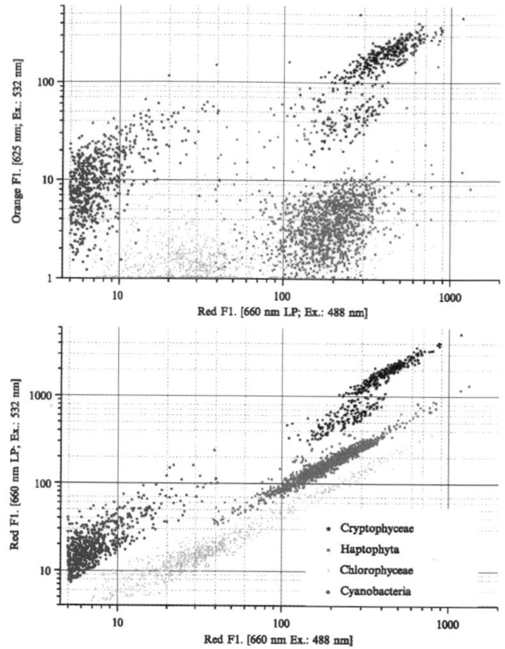

Fig. 15.7. Toepel *et al.* (See Legend in Text).

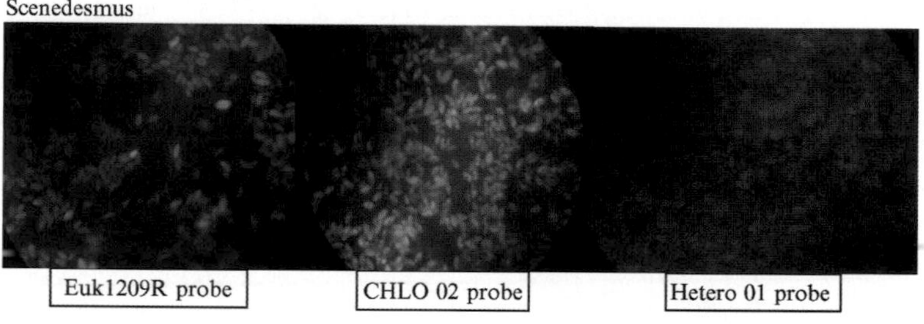

Fig. 15.11. Toepel *et al.* (See Legend in Text).

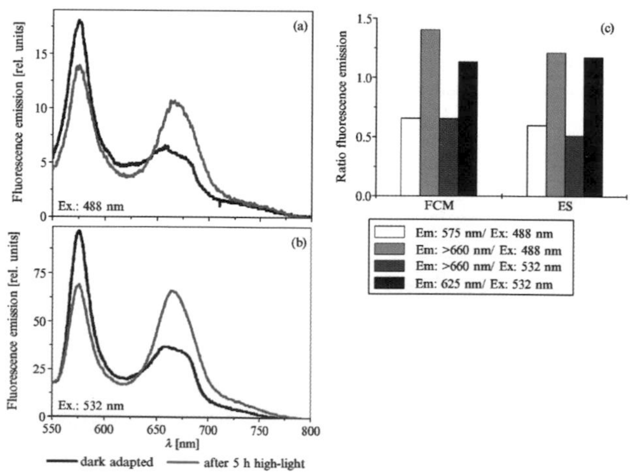

Fig. 15.13. Toepel *et al.* (See Legend in Text).

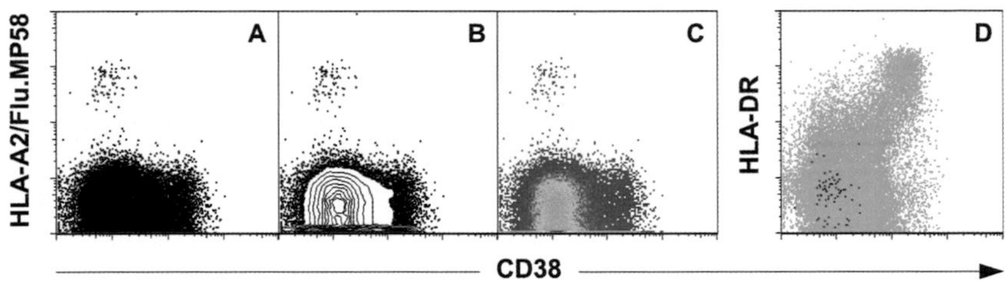

Fig. 17.4. Altman (See Legend in Text).

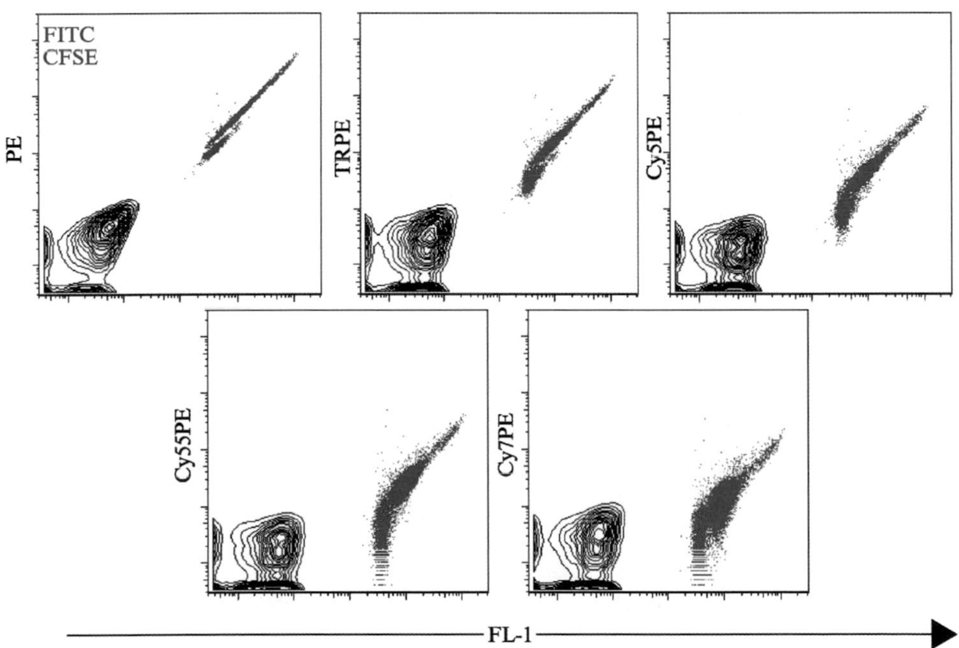

Fig. 19.2. Brenchley and Douek (See Legend in Text).

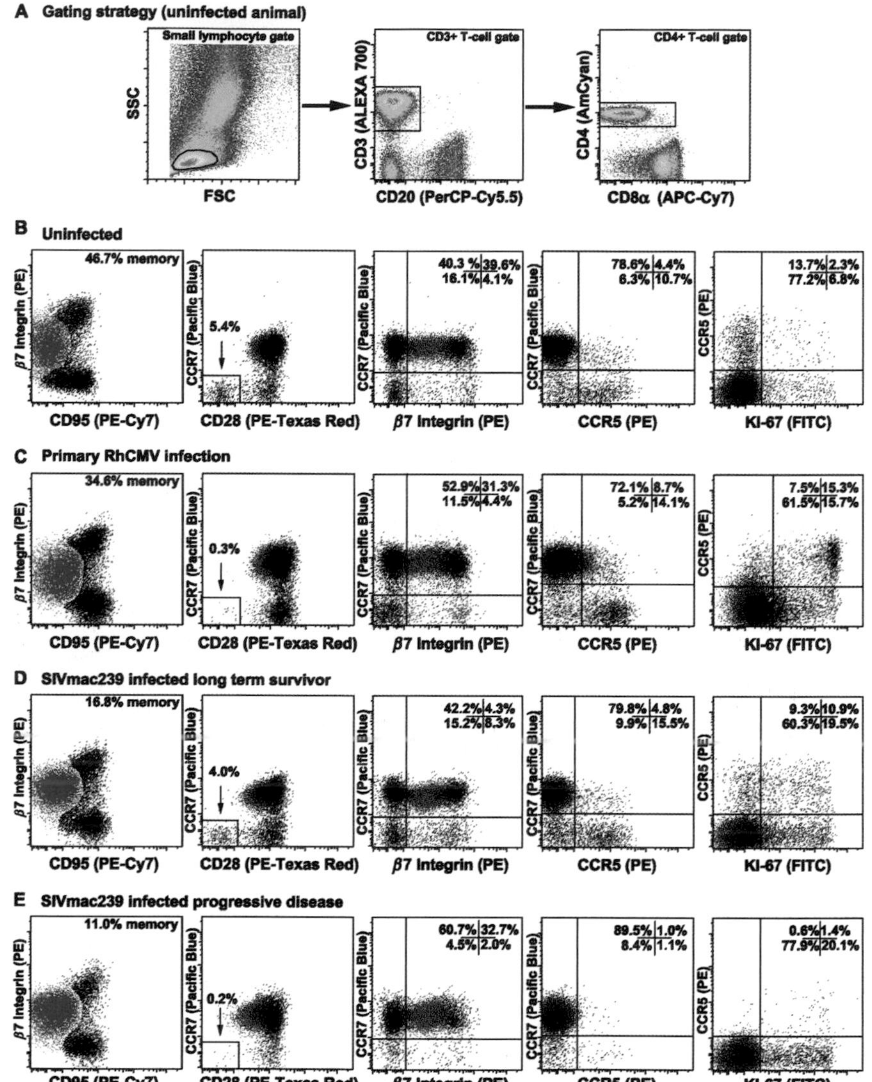

Fig. 22.5. **Walker *et al.*** (See Legend in Text).

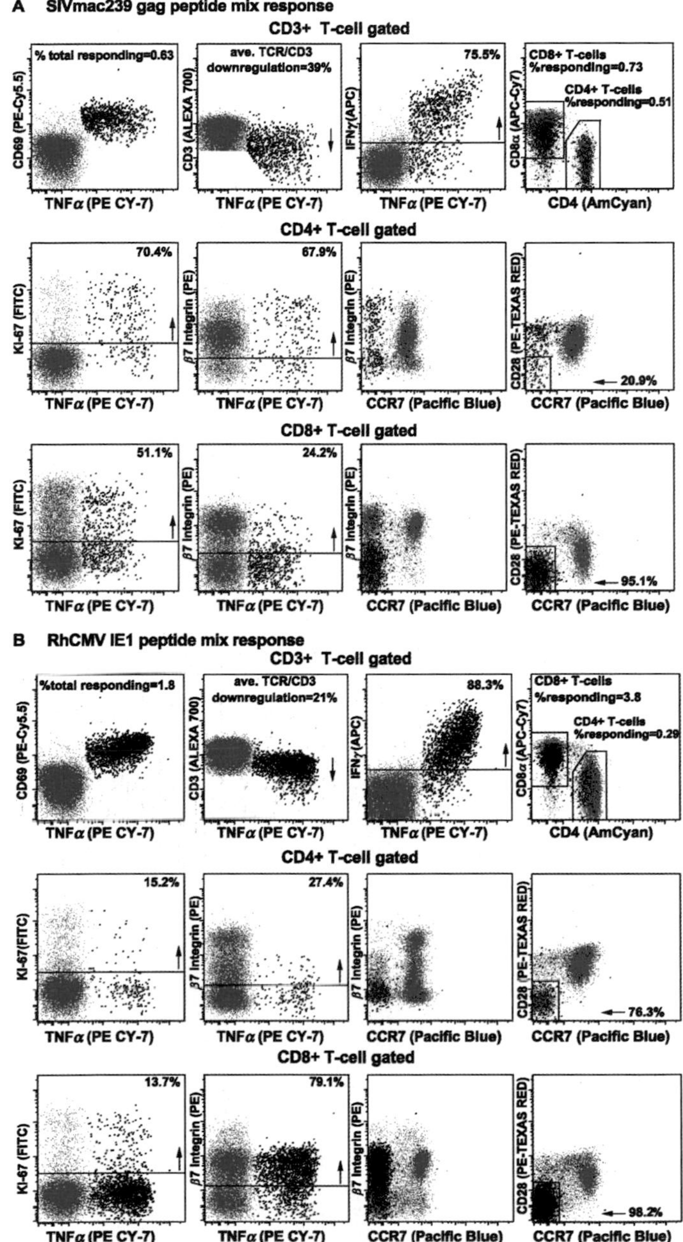

Fig. 22.6. Walker *et al.* (See Legend in Text).

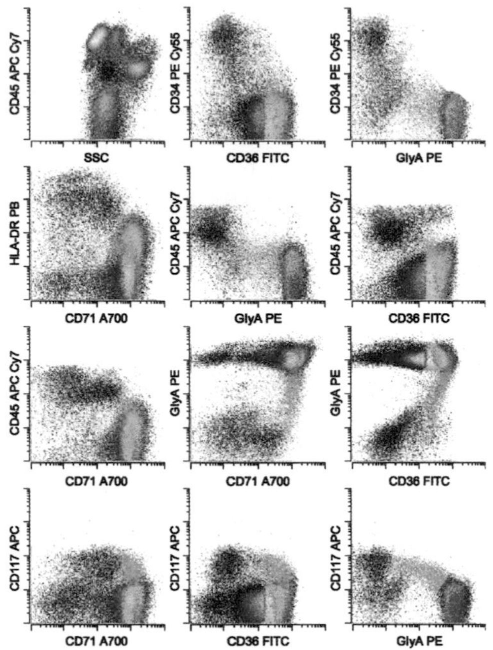

Fig. 23.5. Wood (See Legend in Text).

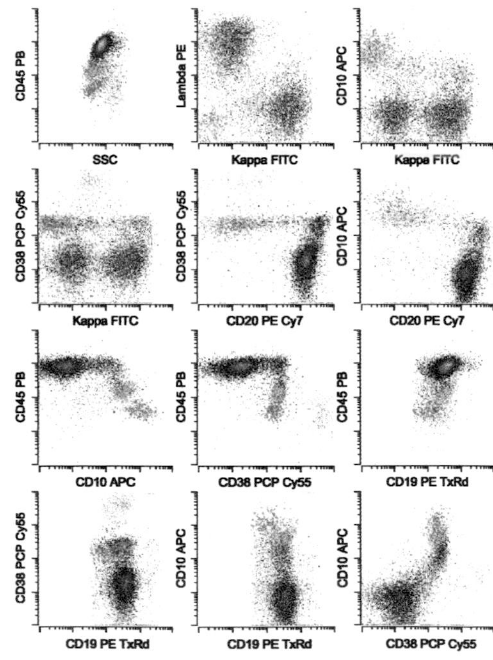

Fig. 23.6. Wood (See Legend in Text).

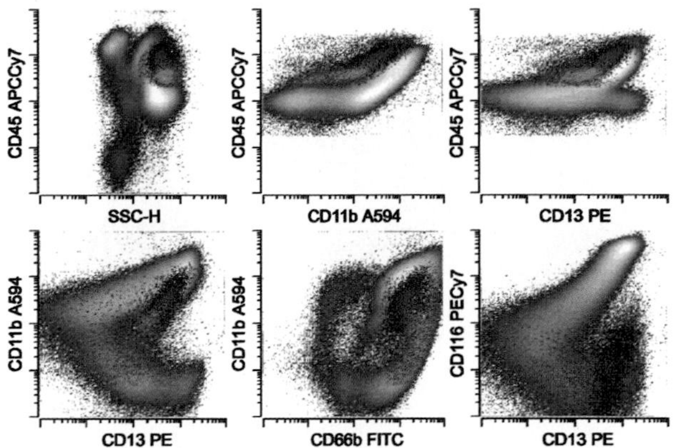

Fig. 23.7. Wood (See Legend in Text).

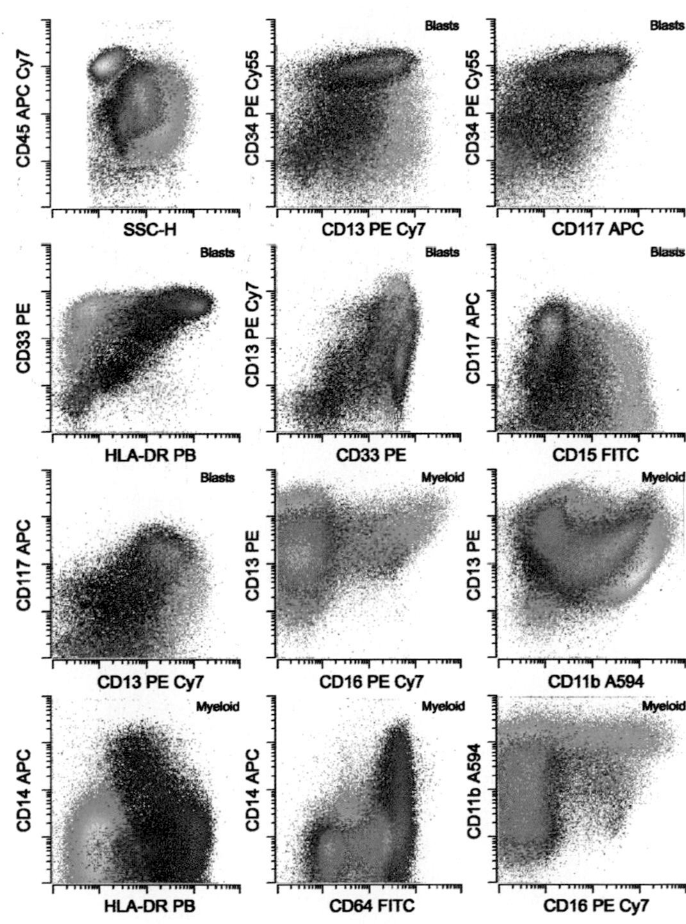

Fig. 23.8. Wood (See Legend in Text).

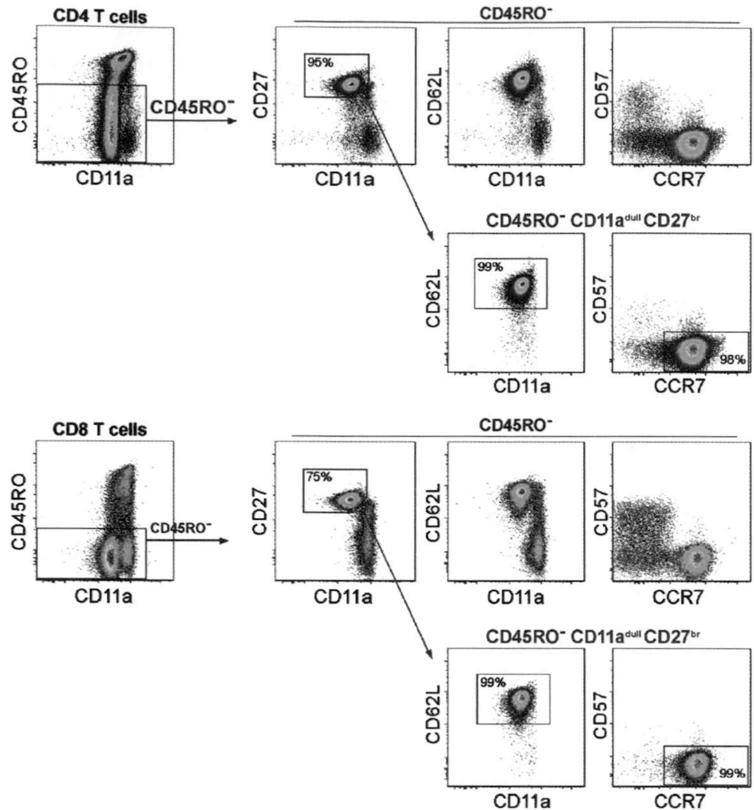

Fig. 24.1. De Rosa (See Legend in Text).

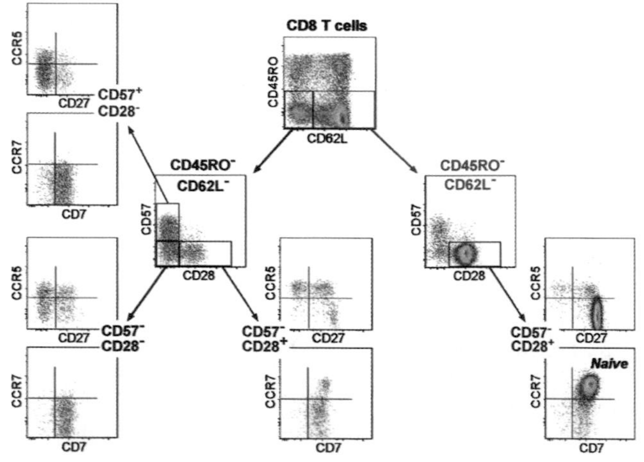

Fig. 24.2. De Rosa (See Legend in Text).

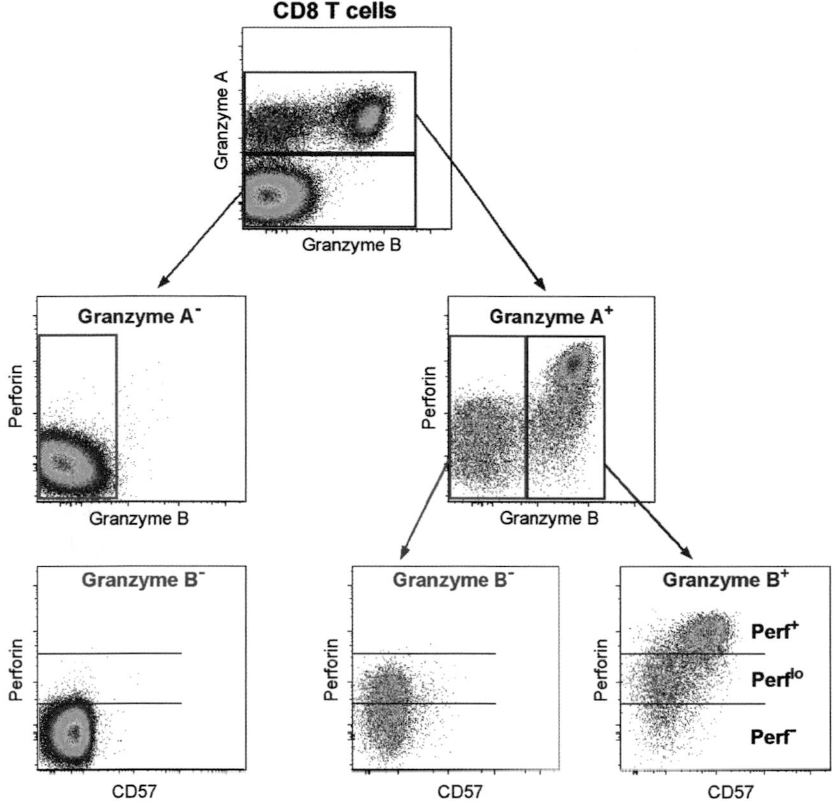

Fig. 24.3. De Rosa (See Legend in Text).

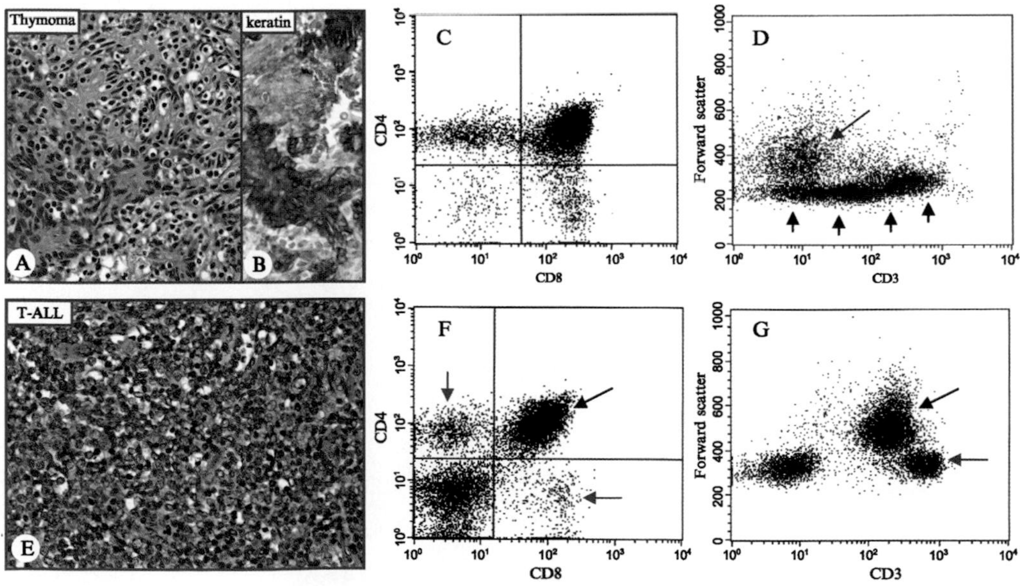

Fig. 25.3. Gorczyca (See Legend in Text).

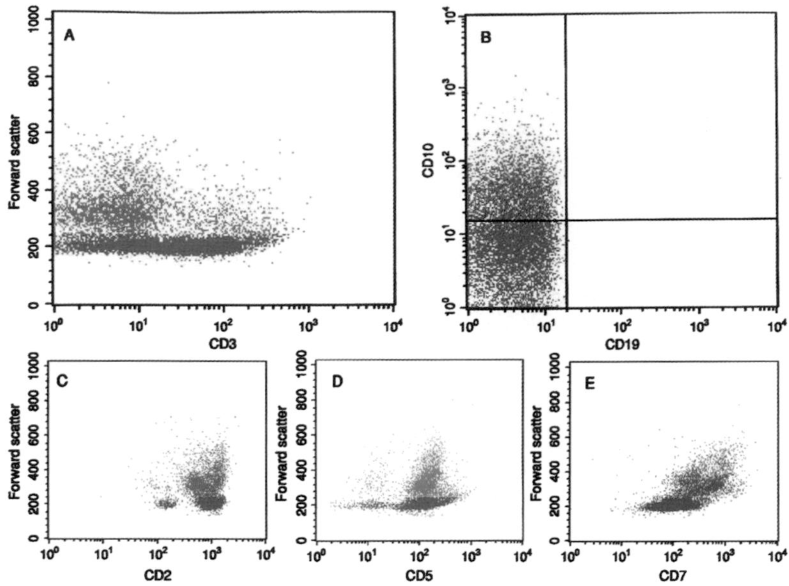

Fig. 25.4. Gorczyca (See Legend in Text).

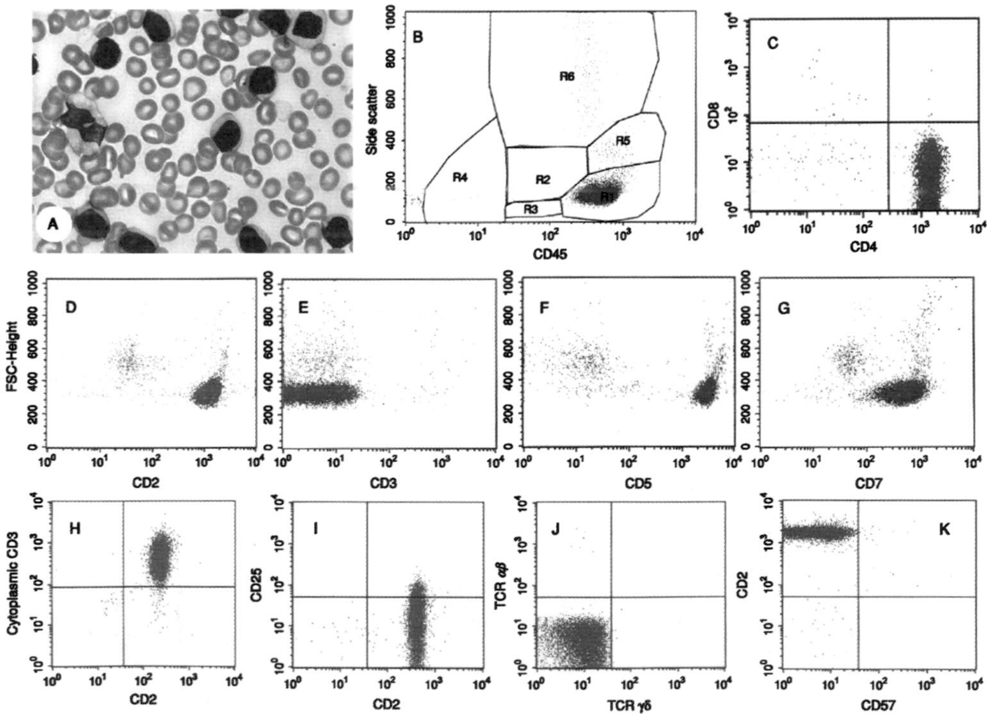

Fig. 25.5. Gorczyca (See Legend in Text).

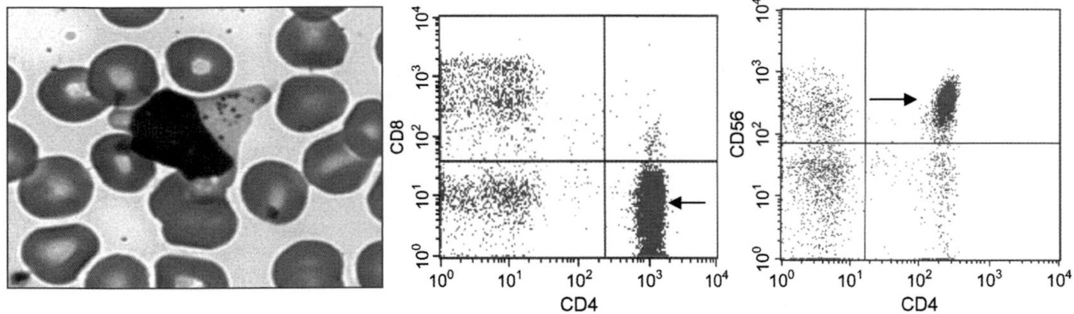

Fig. 25.6. Gorczyca (See Legend in Text).

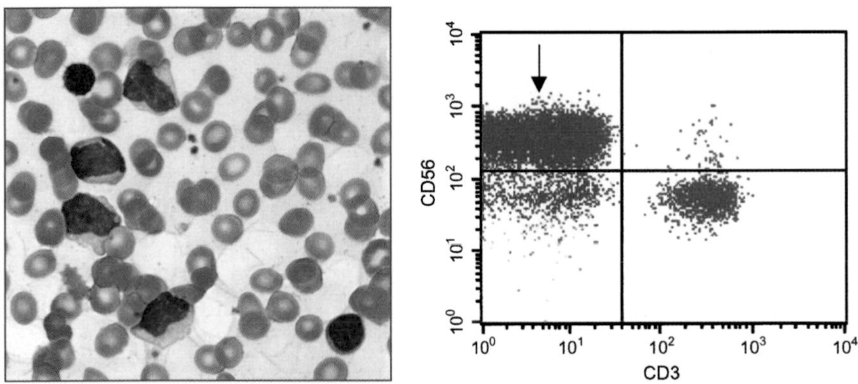

Fig. 25.7. Gorczyca (See Legend in Text).

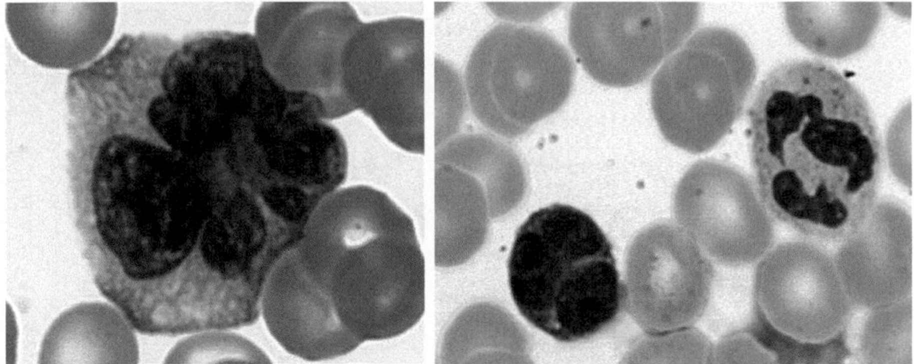

Fig. 25.8. Gorczyca (See Legend in Text).

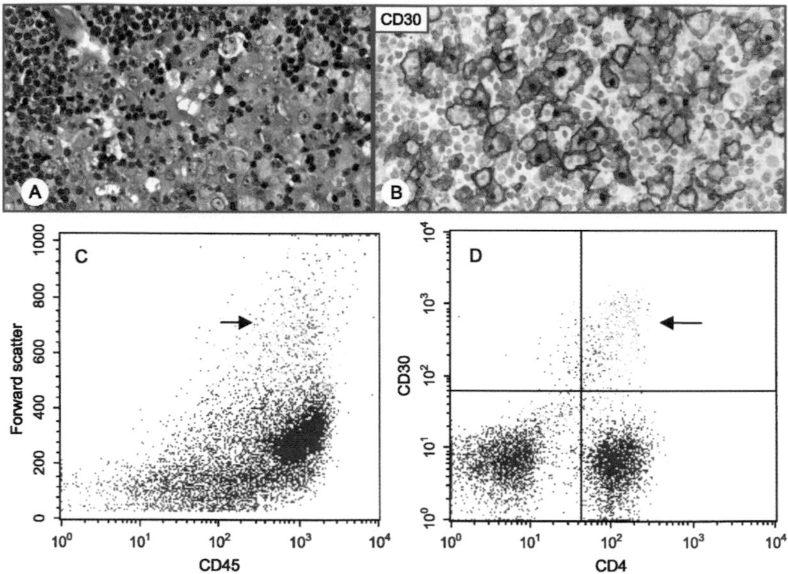

Fig. 25.9. Gorczyca (See Legend in Text).

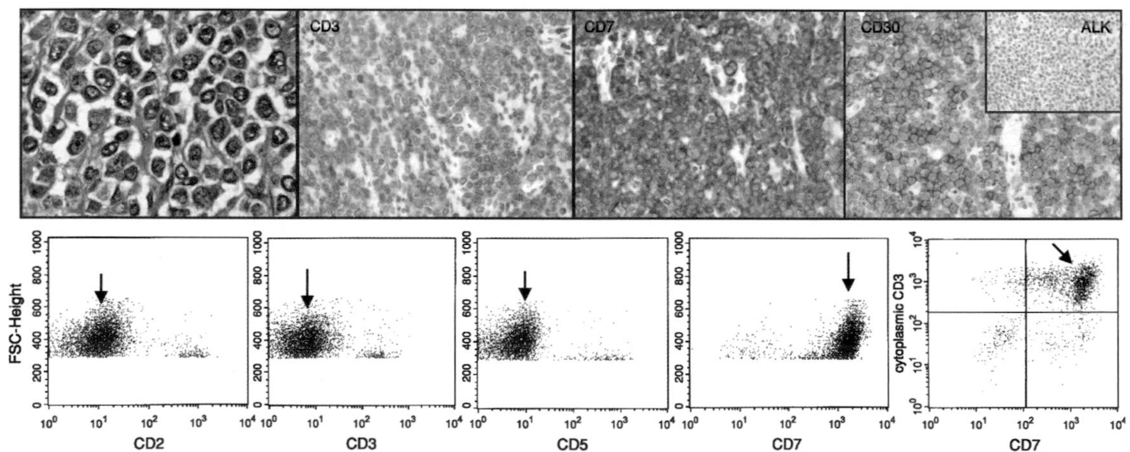

Fig. 25.10. Gorczyca (See Legend in Text).

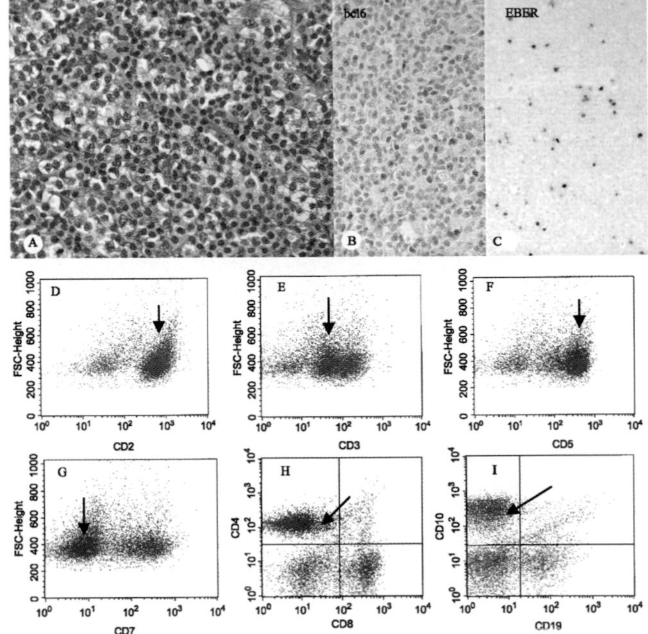

Fig. 25.11. Gorczyca (See Legend in Text).

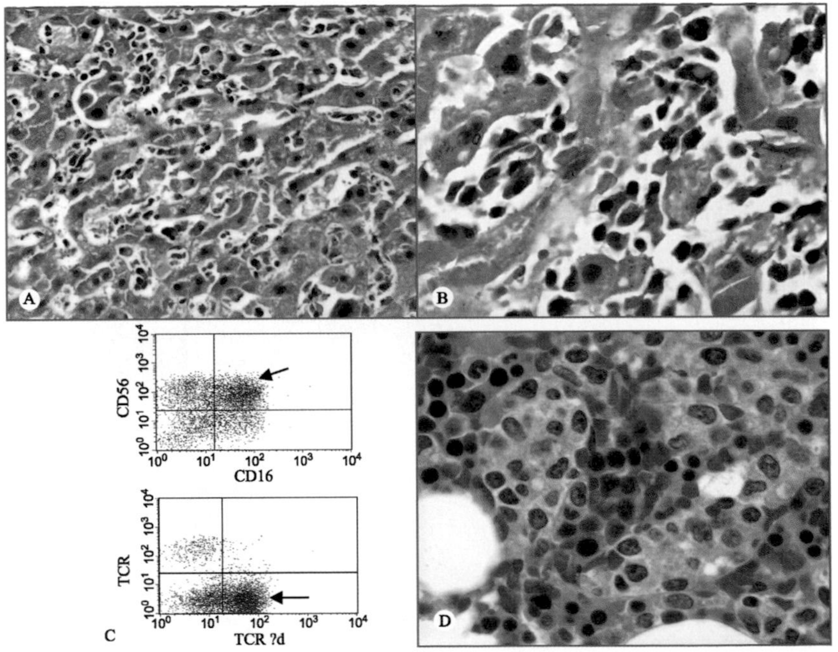

Fig. 25.12. Gorczyca (See Legend in Text).

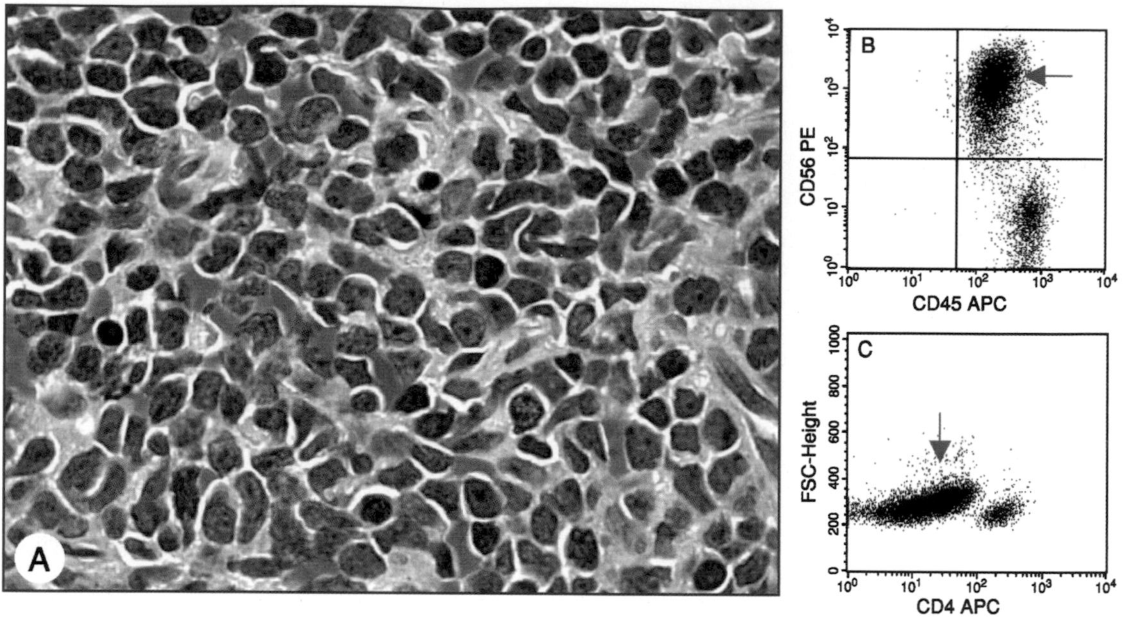

Fig. 25.13. Gorczyca (See Legend in Text).

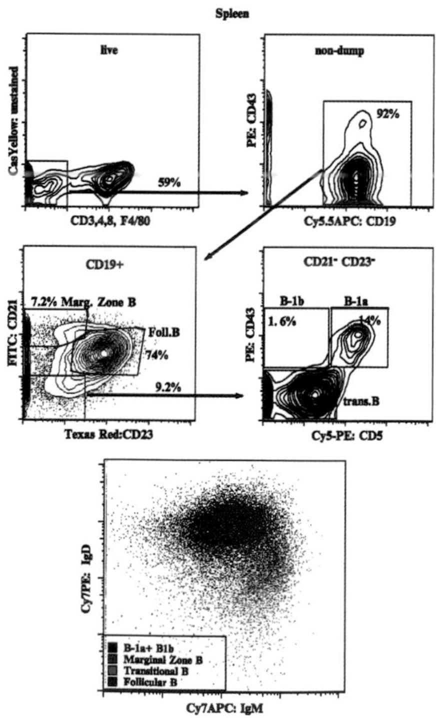

Fig. 27.3. Baumgarth (See Legend in Text).

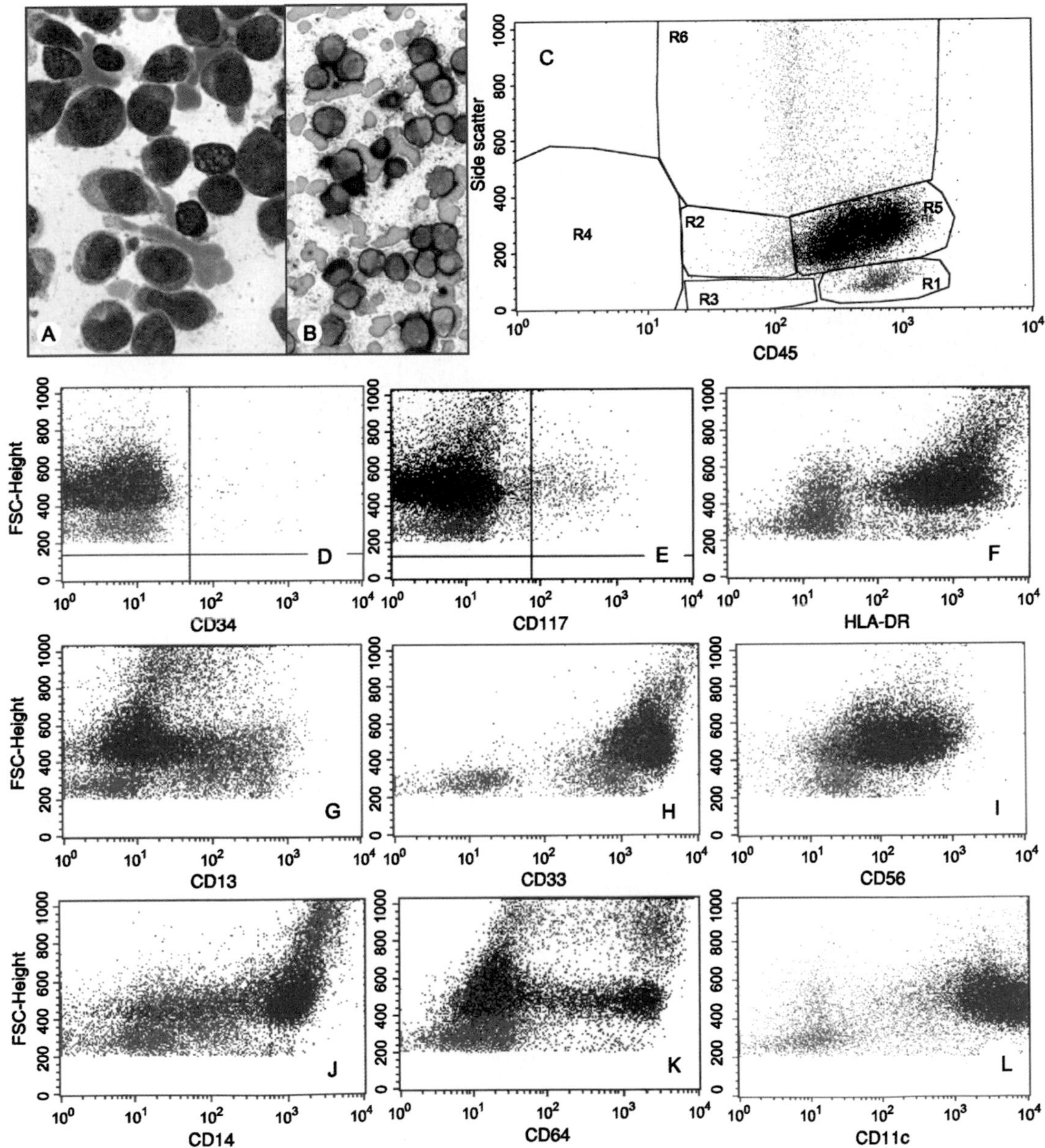

Fig. 28.1. Gorczyca (See Legend in Text).

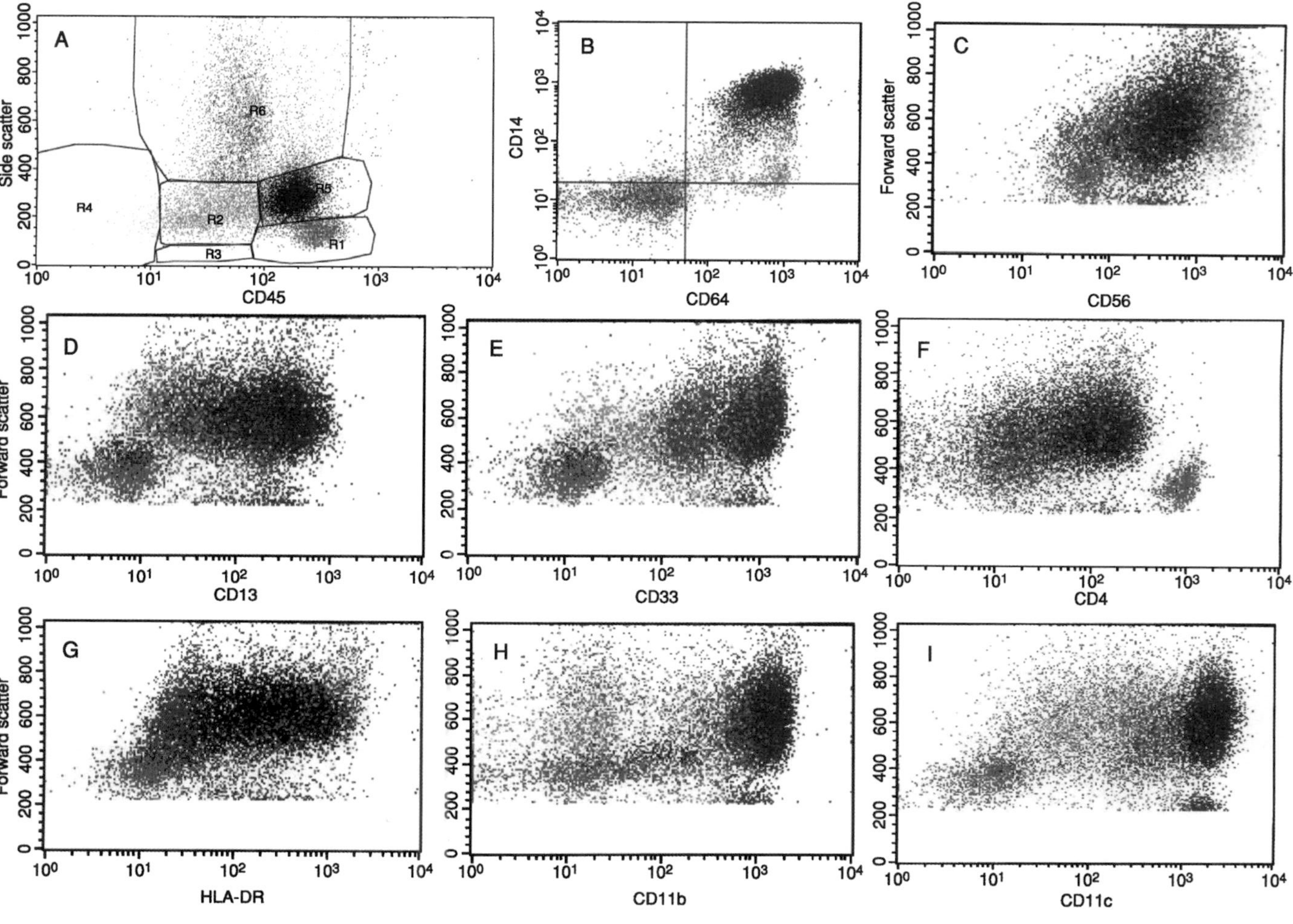

Fig. 28.2. Gorczyca (See Legend in Text).

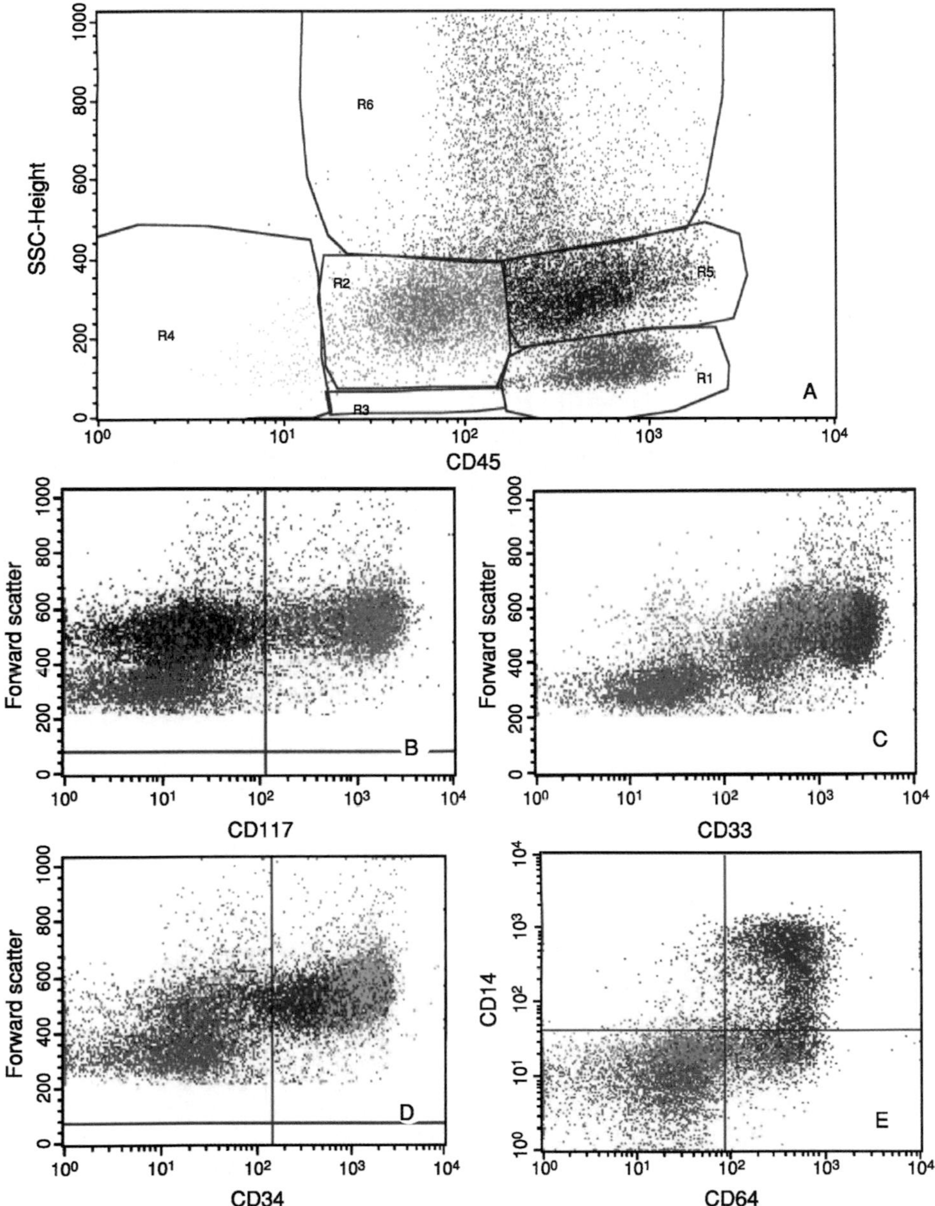

Fig. 28.3. Gorczyca (See Legend in Text).

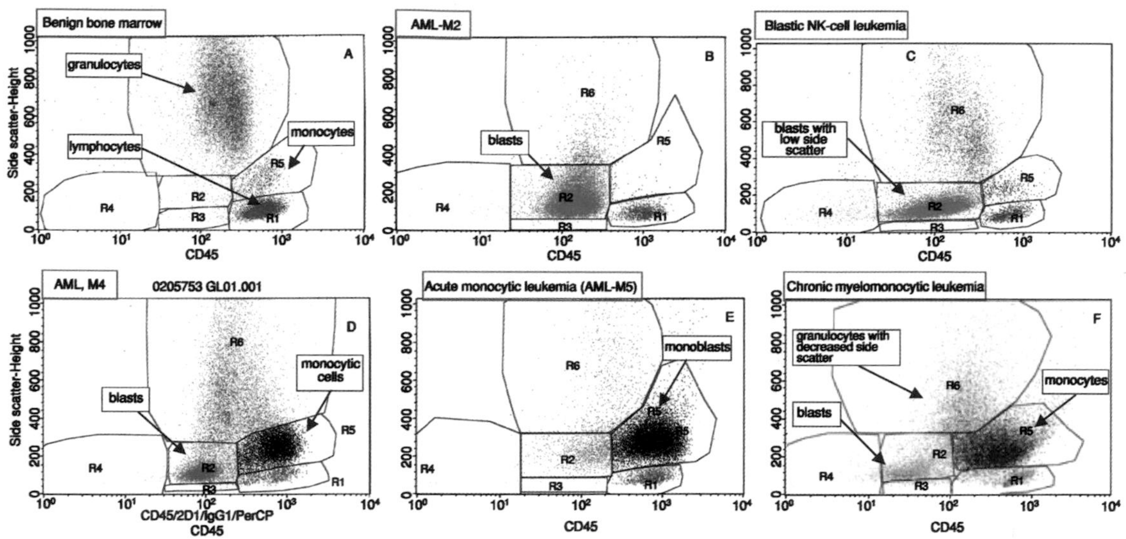

Fig. 28.4. Gorczyca (See Legend in Text).

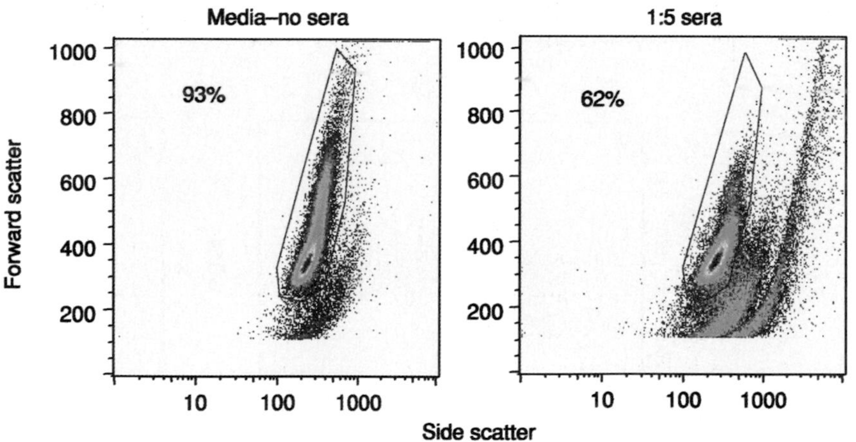

Fig. 30.3. Mascola (See Legend in Text).

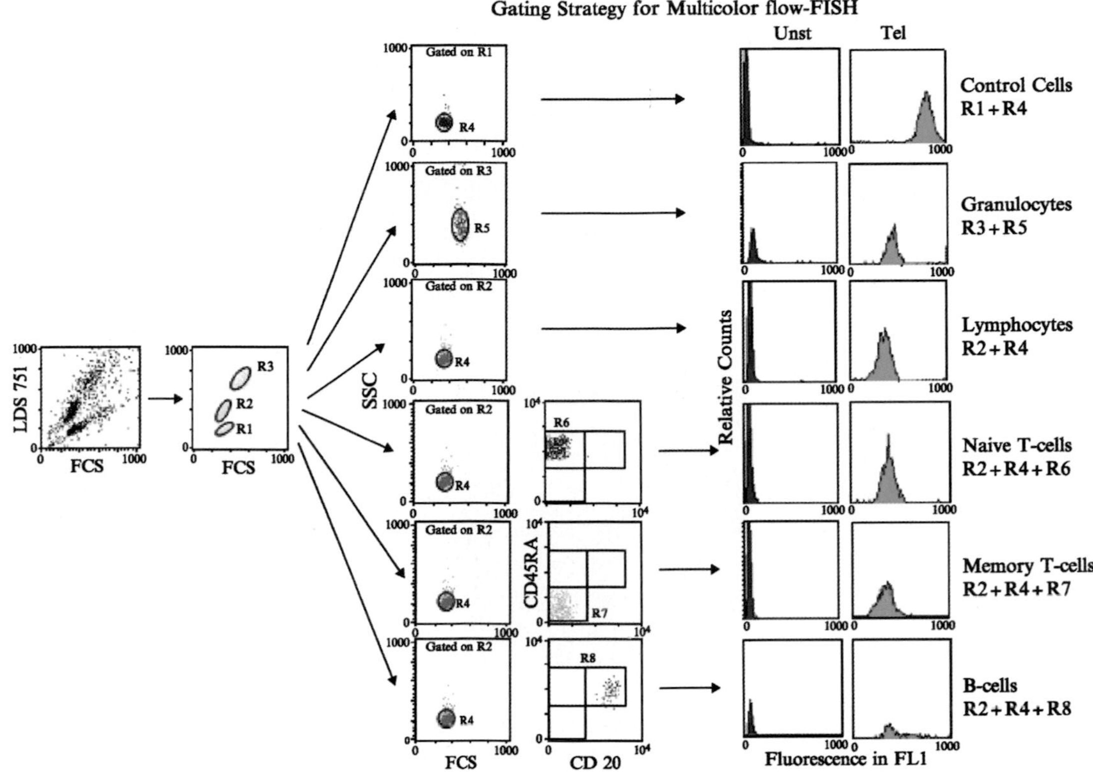

Fig. 31.4. Baerlocher and Lansdorp (See Legend in Text).

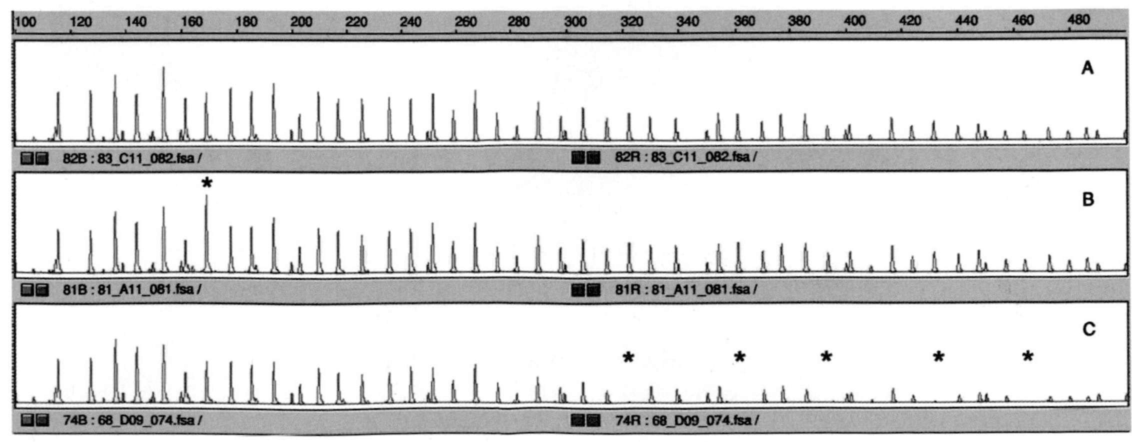

Fig. 32.3. White *et al.* (See Legend in Text).

(A) Chromoprobe Multiprobe®-T Systems

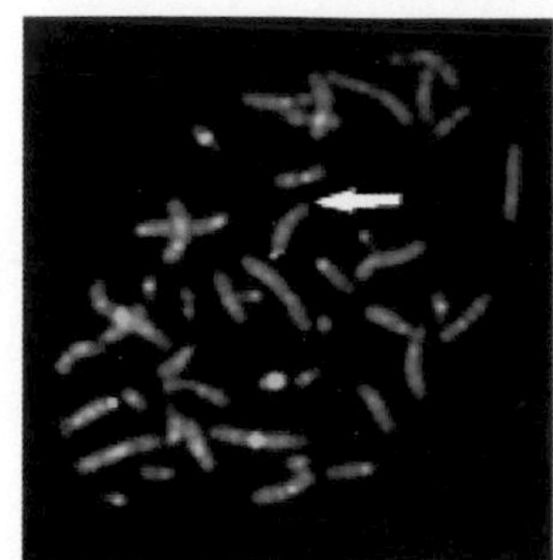

(B) To Telvysion TM Multicolor FISH Probe Panel

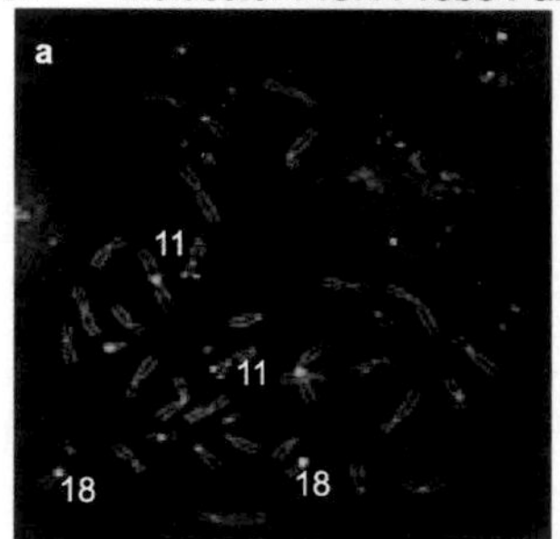

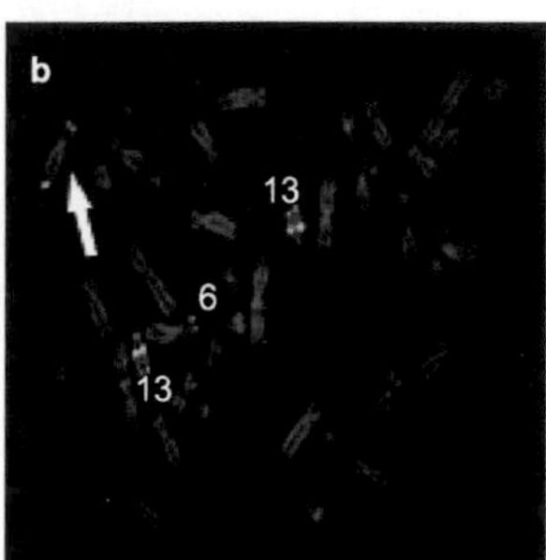

(C) M-TEL

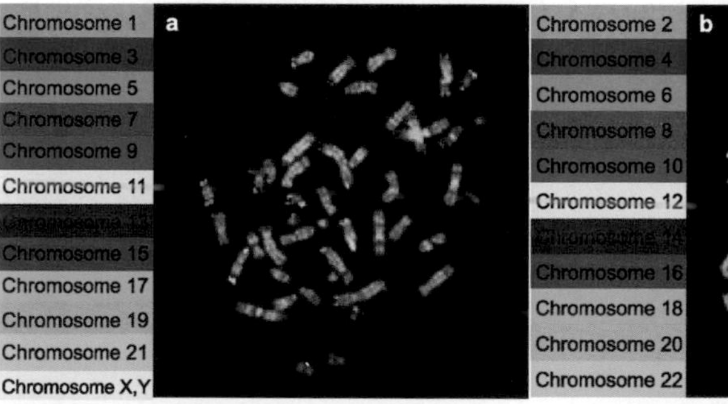

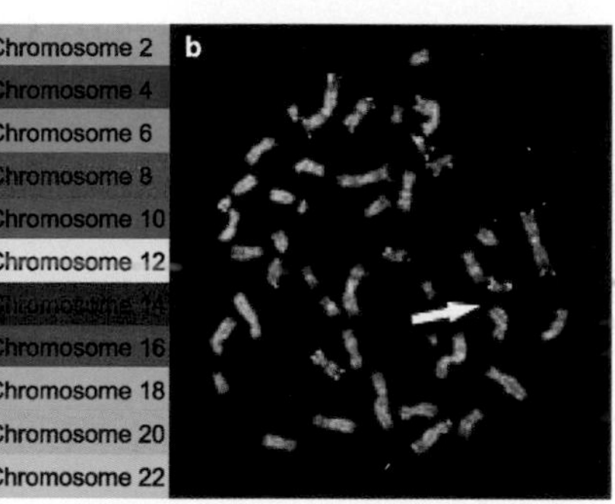

(D) PRINS

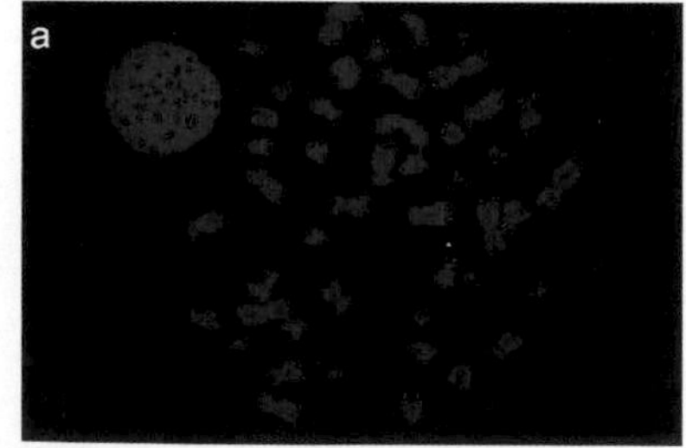

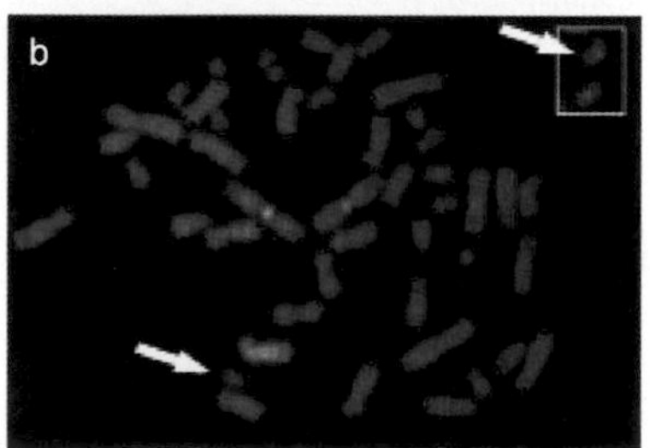

Fig. 35.1. Knight and Flint (See Legend in Text).